ECOLOGY

SIXTH EDITION

ECOLOGY

SIXTH EDITION

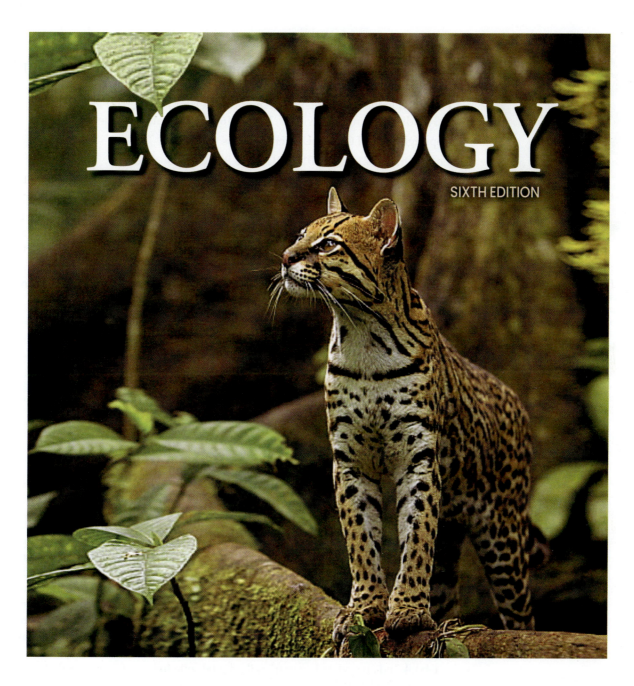

William D. Bowman
University of Colorado

Sally D. Hacker
Oregon State University

OXFORD
UNIVERSITY PRESS

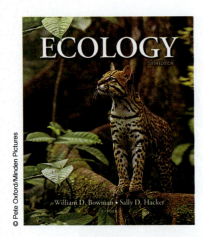

About the Cover

An ocelot (*Leopardus pardalis*) stands on a buttress root on the forest floor in the Amazon rainforest, Ecuador. Ocelots occur in thickly vegetated habitats throughout tropical and sub-tropical areas of the Americas. They are generally nocturnal, hunting small rodents, amphibians, and reptiles using their acute senses of sight and hearing to detect prey. Demand for their beautiful coats have made them a target for hunters, prompting protections in many countries where they occur.

Oxford University Press is a department of the University of Oxford. It furthers the University's objective of excellence in research, scholarship, and education by publishing worldwide. Oxford is a registered trademark of Oxford University Press in the UK and certain other countries.

Published in the United States of America by Oxford University Press, 198 Madison Avenue, New York, NY 10016, United States of America

Address editorial correspondence to:

Oxford University Press
23 Plumtree Road
Sunderland, MA 01375 U.S.A.

Library of Congress Cataloging-in-Publication Data

Names: Bowman, William D., author. | Hacker, Sally D., author.

Title: Ecology / William D. Bowman, University of Colorado, Sally D. Hacker, Oregon State University.

Description: Sixth edition. | New York : Oxford University Press, [2024] | Includes bibliographical references and index. | Summary: "This sixth edition of Ecology, written for undergraduate students taking their first course in ecology, provides comprehensive yet concise coverage of fundamental ecological principles, with attention to relevant issues including climate change, spread of invasive species, and pollution. The text utilizes a variety of learning tools-such as Case Studies, Connections in Nature, Climate Change Connection vignettes, Ecological Toolkit boxes, and Learning Objectives-to engage students, highlight critical information, and make real-world connections to the source material. The text is complemented by an enhanced e-book and an updated, user-friendly digital suite full of interactive activities, quizzes, videos, and layered figures to reinforce key concepts"-- Provided by publisher.

Identifiers: LCCN 2022034923 (print) | LCCN 2022034924 (e-book) | ISBN 9780197614044 (hardcover) | ISBN 9780197614082 (epub)

Subjects: LCSH: Ecology--Textbooks.

Classification: LCC QH541 .E31933 2024 (print) | LCC QH541 (ebook) | DDC 577--dc23/eng/20220728

LC record available at https://lccn.loc.gov/2022034923

LC ebook record available at https://lccn.loc.gov/2022034924

9 8 7 6 5 4 3 2 1

Printed in the United States of America

For Jen, Gordon, and Miles and their unwavering support,
and to my students for pushing me as much as I pushed them.

WDB

For my family and my students, whose gift of time has made all the difference.

SDH

About the Authors

WILLIAM D. BOWMAN is a Professor Emeritus at the University of Colorado at Boulder, affiliated with the Department of Ecology and Evolutionary Biology, Mountain Research Station, and the Institute of Arctic and Alpine Research. He earned his Ph.D. from Duke University. Dr. Bowman has taught courses in introductory ecology, plant ecology, plant–soil interactions, and ecosystems ecology, and for over three decades he directed undergraduate summer field courses and research programs. His research focuses on the intersections of physiological ecology, community dynamics, and ecosystem function, particularly in the context of environmental change.

SALLY D. HACKER is a Professor at Oregon State University, Corvallis, where she has been a faculty member since 2004. She has taught courses in introductory ecology, community ecology, and marine biology. She is particularly interested in promoting active and experiential learning for students interested in ecology and field experiences. Dr. Hacker's research explores the structure, function, and services of natural and managed ecosystems under varying contexts of species interactions and global change. She has conducted research with plants and animals in rocky intertidal, estuarine, and coastal dune ecosystems. Her work has most recently focused on the protective role of coastal ecosystems in mitigating the vulnerability from climate change. In addition to the textbooks *Ecology* (OUP) and *Life: The Science of Biology* (Macmillan), she is an author or coauthor on numerous articles and book chapters exploring themes in community ecology and ecosystem functions and services.

Brief Contents

Preface

Ecology is at the heart of understanding our world—it serves as the glue that brings together information from a multitude of different scientific disciplines, and it integrates this information in a way that informs us about how nature works. As our environment continues to change because of fossil fuel burning, agriculture, resource extraction, and pollution, it becomes increasingly critical that we understand the repercussions of these changes through an ecological perspective. Stewardship of ecosystem goods and services that humanity requires—food, clean air, clean water, and many others—is important to understand through the lens of ecological knowledge. Our goal with this book is to present that ecological knowledge in the best way possible to promote ecological literacy in students with diverse backgrounds and career goals.

Advances in ecology occur regularly, facilitated by technological and computational breakthroughs, as well as creative experimental approaches. This continued advancement, along with the diversity of subjects that form its basis, makes ecology a potentially daunting and complicated subject to learn. We set out to introduce our readers to the beauty of nature and importance of ecology and provide content in a way that engages students without overwhelming them in the process. To achieve these goals, the book's two core principles of "**Learning Comes First**" and "**Less Is More**" guided our every step.

Core Principles Guiding *Ecology*, Sixth Edition

Enabling effective learning is our primary goal and motivation in writing the Sixth Edition of *Ecology*. The structure and content of our chapters are designed primarily to be motivating to read and easy to comprehend. For example, to introduce the content and capture student interest, each chapter begins with an engaging story (a "Case Study," as described more fully later about an important ecological topic or interesting real-life problem. Once one is drawn in by the Case Study, the "storyline" that it initiates is maintained throughout the rest of the chapter. We use a narrative writing style to link the sections of the chapter to one another, thus helping the reader keep the big picture in mind. The sections of the chapter are organized around a small number of Key Concepts (also described more thoroughly later) that are carefully selected to summarize current knowledge and provide students with a clear overview of the subject at hand. Additionally, our Learning Objectives help students focus on the main concepts that they should take away from their reading. When designing the art, pedagogy comes first. Many people are visual learners, so we worked hard to ensure that each figure "tells a story" that can be understood on its own.

In addition, we follow a "less is more" philosophy to help us achieve our primary goal of student learning. We are guided by the principle that if we cover less material but present it in a narrative style, students will increase their reading comprehension. Hence, our chapters are relatively short, and they are built around a small number of Key Concepts and Learning Objectives (typically, three to five each). This approach required some difficult choices, but it has enabled us to focus on teaching students what is currently known about ecology without burdening them with excessive detail.

We also recognize that many instructors are choosing to "flip" their instructional style, with an emphasis on hands-on activities during classroom time and a greater reliance on student learning of core material outside of the classroom. In previous editions, *Ecology* has consistently served this purpose well with its clear, easy-to-read, and well-organized presentation of material, coupled with several of its quantitative active learning features such as the Analyzing Data, Hone Your Problem-Solving Skills, and Hands-On Problem Solving exercises. In this edition, we have expanded our offering with all-new **Oxford Insight** courseware. **Oxford Insight** courseware is an online instructional tool that provides an e-book version of the textbook as well as digital resources that facilitate student engagement and evaluation. **Oxford Insight** hosts a rich tapestry of fascinating videos, interactive exercises, and formative and summative assessments, all while bringing the hallmark quantitative features to life in a dynamic digital environment. **Oxford Insight** gets students reading and interacting with the core content prior to attending lecture, so class time can be spent in small and large group interactions, rather than simply on review of the textbook content. (For more information on **Oxford Insight**, see "Digital Resources for *Ecology*, Sixth Edition.")

New to *Ecology*, Sixth Edition

In addition to the new Oxford Insight courseware, and to keep the book to a manageable length, we have updated, replaced, revised, and, in some cases, cut sections of the book with each new edition. In the Sixth Edition, particular attention was given to updating content on climate change:

the causes, the ecological responses, and the projected impacts. In addition, new examples are included incorporating current issues, such as the impacts of the COVID19 pandemic on the volume of bird songs or the effects of warming on non-native dune grass species interactions.

Examples of more extensive revisions include reorganization of portions of the Life History, Population Distribution and Abundance, and Population Dynamics chapters and updating examples on optimal foraging in the Behavioral Ecology chapter and on the design of nature reserves in the Landscape Ecology and Ecosystem Management chapter. In addition, three new Climate Change Connection vignettes are included such that all chapters have at least one example of this feature.

Hallmark Features

In addition to the changes we just described, we have revised and strengthened the key pedagogical features of *Ecology*, introduced in previous editions:

Pedagogical Excellence Students taking their first course in ecology are exposed to a great deal of material. To help them manage this content, each chapter of *Ecology* is organized around a small number of Key Concepts and Learning Objectives that provide up-to-date summaries of fundamental ecological principles.

Links to Ecological Applications Many students taking introductory ecology are interested in use-inspired aspects of ecology. Thus, ecological applications (including conservation biology) are featured and woven into each chapter, helping to capture and retain student interest.

Links to Evolution Evolution is a central, unifying theme of all biology, and its connections with ecology are strong. Thus, we incorporate evolution throughout the book and make it the focus of some chapters. For example, Chapter 6 (Evolution and Ecology) explores the ecology of evolution both at the population level and as observed in the sweeping history of life on Earth. Other topics in evolutionary ecology are explored in Chapter 7 (Life History) and in Chapter 8 (Behavioral Ecology).

Case Studies Each chapter opens with an interesting vignette—a Case Study. By presenting an engaging story or interesting application, the Case Study captures the reader's attention while introducing the topic of the chapter. Later, the reader is brought full circle with the corresponding "Case Study Revisited" section at chapter's end. Each Case Study relates to multiple levels of ecological hierarchy, thereby providing a nice lead-in to the Connections in Nature feature, described next.

Connections in Nature In most ecology textbooks, connections among levels of the ecological hierarchy are discussed briefly, perhaps only in the opening chapter. As a result, many opportunities are missed to highlight the fact that events in natural systems *really are* interconnected. To facilitate the concept of how events in nature are interconnected, each chapter of *Ecology* closes with a section that discusses how the material covered in that chapter affects and is affected by interactions at other levels of the ecological hierarchy. Where appropriate, these interconnections are also emphasized in the main body of the text.

Climate Change Connections Recognizing the increasing evidence for and effects of climate change on ecological systems, Climate Change Connection examples are included for each chapter topic of the book. These vignettes help students appreciate the many consequences of global climate change on the distributions and functions of organisms as well as the ecosystems in which they depend.

Ecological Inquiry Our understanding of ecology is constantly changing due to new observations, experiments, and models. All chapters of the book emphasize the active, inquiry-based nature of what is known about ecology. This occurs throughout the narrative and is further highlighted by the quantitative and applied Analyzing Data exercises, Hone Your Problem-Solving Skills exercises, and Figure Legend Questions described later.

Analyzing Data Exercises Our understanding of ecology is constantly changing due to new observations, experiments, and models. Thus, all chapters of the book emphasize active, inquiry-based exercises. Toward that end, we have included Analyzing Data exercises in each chapter of the book. These exercises give students extra practice with essential skills, such as performing calculations, making graphs, designing experiments, and interpreting results.

Hone Your Problem-Solving Skills The Hone Your Problem-Solving Skills exercises in each chapter expose students to hypothetical situations or existing data sets and allow them to work through data analysis and interpretation to better understand key ecological concepts and relate these concepts to real-life situations.

Figure Legend Questions Each chapter includes 3–6 Figure Legend Questions that appear in the end of the legend. These questions encourage students to grapple with the concepts of the figure to increase learning comprehension. The questions range from testing whether students understand the axes or other simple aspects of the figure to asking students to develop or evaluate hypotheses.

Ecological Toolkits Nearly half of the chapters include an Ecological Toolkit, a box inset in the chapters that describes ecological "tools" such as experimental design, remote sensing, mark–recapture techniques, stable isotope analysis, DNA fingerprinting, or calculating of species–area curves.

Ecology Is a Work in Progress

This book, like the field of ecology, does not consist of a set of unchanging ideas or a fixed amount of information. Instead, the book has developed and changed over time as we respond to new discoveries and new ways learning. We would love to hear from you—what you like about the book, what you do not like, and any questions or suggestions you may have for how we can improve the learning experience. You can send your comments and suggestions to the senior author at william.bowman@colorado.edu.

Accessible Color Content Every opportunity has been taken to ensure that the content herein is fully accessible to those who have difficulty perceiving color. Exceptions are cases where the colors provided are expressly required because of the purpose of the illustration.

Acknowledgments

We wish to express our gratitude to the staff at Oxford University Press, with whom we worked closely during the writing, revision, and production of the Sixth Edition. Jason Noe expertly oversaw the entire revision from start to finish, with assistance from Sarah D'Arienzo and Jessica McLaughlin. Production Editors Stephanie Nisbet and Johannah Walkowicz organized and managed the project workflow and scheduling and provided editorial assistance. Wendy Walker and Lou Doucette provided expert copyediting and ensured consistency and clarity throughout the book. Elizabeth Morales provided the beautiful illustrations. Production Specialist Donna DiCarlo created the wonderful new cover. Photo Editor Mark Siddall provided excellent choices for instructive and interesting images. Permissions Supervisor Cailen Swain diligently tracked down original sources for permissions, and Media Production Editors Karissa Venne, Sarah D'Arienzo, and Lauren Thompson oversaw the revision of the digital resources.

And finally, we'd like to thank some of the many people who helped us turn our ideas into a book in print and digital formats. We are grateful to our colleagues who generously critiqued one or more chapters over the lifetime of the book; their names follow. We also wish to thank the hundreds of people we contacted while researching this book for their guidance and generous gifts of time and expertise.

William D. Bowman
william.bowman@colorado.edu

Sally D. Hacker
sally.hacker@oregonstate.edu

Reviewers

Reviewers for the Sixth Edition

Eyualem Abebe, Elizabeth City State University
Susan P. Bratton, Baylor University
Michael Collins, Rhodes College
Geoffrey Cook, University of Central Florida
Emily Dangremond, Roosevelt University
Ruth A. Darling, Westfield State University
Leah S. Dudley, East Central University
Nisse Goldberg, Jacksonville University
Shinijini Goswami, Lees-McRae College
Julie Harper, University of Southern California
Karen Henman, Brenau University
William Hintz, The University of Toledo
Andrew Laughlin, University of North Carolina, Asheville
Jason Macrander, Florida Southern College
Jay Mager, Ohio Northern University
Thomas M. McCarthy, Utica College
William R. Norris, Western New Mexico University
Joan Petersen, Queensborough Community College
Alex Pilote, Tennessee Wesleyan University
Pushpa Soti, University of Texas, Rio Grande Valley
Paul Stapp, California State University Fullerton
Arron Stoler, Stockton University
Tim Tibbetts, Monmouth College
Stephen Mitchell Wagener, Western Connecticut State University
Jeffrey Walck, Middle Tennessee State University
Tom Wassmer, Siena Heights University
Stefan Woltmann, Austin Peay State University

Reviewers for the First, Second, Third, Fourth, and Fifth Editions

David Ackerly, University of California, Berkeley; Gregory H. Adler, University of Wisconsin, Oshkosh; Stephano Allesina, University of Chicago; Stuart Allison, Knox College; Kama Almasi, University of Wisconsin, Stevens Point; Peter Alpert, University of Massachusetts, Amherst; Diane Angell, St. Olaf College, David Armstrong, University of Colorado; Anita Baines, University of Wisconsin, La Crosse; Robert Baldwin, Clemson University; Betsy Bancroft, Southern Utah University; Jeb Barrett, Virginia Polytechnic Institute and State University; James Barron, Montana State University; Christopher Beck, Emory University; Beatrix Beisner, University of Quebec at Montreal; Mark C. Belk, Brigham Young University; Michael A. Bell, Stony Brook University; Eric Berlow, University of California, Merced; Kim Bjorgo-Thorne, West Virginia Wesleyan College; Charles Blem, Virginia Commonwealth University; Steve Blumenshine, California State University, Fresno; Carl Bock, University of Colorado; Daniel Bolnick, University of Texas, Austin; Michael Booth, Principia College; April Bouton, Villanova University; Steve Brewer, University of Mississippi; David D. Briske, Texas A&M University; Judie Bronstein, University of Arizona; Linda Brooke Stabler, University of Central Oklahoma; Kenneth Brown, Louisiana State University; Romi Burks, Southwestern University; Stephen Burton, Grand Valley State University; Aram Calhoun, University of Maine; Mary Anne Carletta, Georgetown College; Walter Carson, University of Pittsburgh; Peter Chabora, Queens College, CUNY; David D. Chalcraft, East Carolina University; Gary Chang, Gonzaga University; Colin A. Chapman, University of Florida; Elsa Cleland, University of California, San Diego; Cory Cleveland, University of Montana; Liane Cochran-Stafira, Saint Xavier University; Rob Colwell, University of Connecticut; William Crampton, University of Central Florida; James Cronin, Louisiana State University; Todd Crowl, Utah State University; Sarah Dalrymple, University of California, Davis; Anita Davelos Baines, University of Texas, Pan American; Mark A. Davis, Macalester College; Andrew Derocher, University of Alberta; Megan Dethier, University of Washington; Abby Grace Drake, Skidmore College; Joseph D'Silva, Norfolk State University; John Ebersole, University of Massachusetts, Boston; Bret D. Elderd, Louisiana State University; Erle Ellis, University of Maryland, Baltimore County; Sally Entrekin, University of Central Arkansas; Jonathan Evans, University of the South; Mara Evans, University of California, Davis; John Faaborg, University of Missouri; William F. Fagan, University of Maryland; Jennifer Fox, Georgetown University; Stephanie Foré, Truman State University; Johanna Foster, Wartburg College; Jennifer Fox, Georgetown University; Kamal Gandhi, University of Georgia; Rick Gillis, University of Wisconsin, La Crosse; Thomas J. Givnish, University of Wisconsin; Elise Granek, Portland State University; Martha Groom, University of Washington; Jack Grubaugh, University of Memphis; Vladislav Gulis, Coastal Carolina University; Jessica Gurevitch, Stony Brook University; Bruce Haines, University of Georgia; Nelson Hairston, Cornell University; Jenifer Hall-Bowman, University of Colorado; Jason Hamilton, Ithaca College; Christopher Harley, University of British Columbia; Bradford Hawkins, University of California, Irvine; Christiane Healey, University of Massachusetts, Amherst; Mike Heithaus, Florida International University; Kringen Henein, Carleton University, Ontario; Kevin Higgins, University of South Carolina; Hopi Hoekstra, Harvard University; Nat Holland, Rice University; Stephen Howard, Middle Tennessee State University; Randall Hughes, Florida State University; Vicki Jackson, Central Missouri State University; John Jaenike, University of Rochester; Bob Jefferies, University of Toronto;

Art Johnson, Pennsylvania State University; Jerry Johnson, Brigham Young University; Pieter Johnson, University of Colorado; Melanie Jones, University of British Columbia; Vedham Karpakakunjaram, University of Maryland; Michael Kinnison, University of Maine; Timothy Kittel, University of Colorado; Jeff Klahn, University of Iowa; Gregg Klowden, University of Central Florida; Astrid Kodric-Brown, University of New Mexico; Michelle Koo, University of California, Berkeley; Tom Langen, Clarkson University; Jennifer Lau, Michigan State University; Jack R. Layne, Jr., Slippery Rock University; Jeff Leips, University of Maryland, Baltimore County; Stacey Lettini, Gwynedd-Mercy College; Gary Ling, University of California, Riverside; Scott Ling, University of Tasmania; Dale Lockwood, Colorado State University; Svata Louda, University of Nebraska; Sheila Lyons-Sobaski, Albion College; Karen Mabry, New Mexico State University; Richard Mack, Washington State University; Lynn Mahaffy, University of Delaware; Daniel Markewitz, University of Georgia; Michael Mazurkiewicz, University of Southern Maine; Andrew McCall, Denison University; Shannon McCauley, University of Michigan; A. Scott McNaught, Central Michigan University; Mark McPeek, Dartmouth College; Scott Meiners, Eastern Illinois University; Bruce Menge, Oregon State University; Thomas E. Miller, Florida State University; Sandra Mitchell, Western Wyoming College; Gary Mittelbach, Kellogg Biological Station, Michigan State University; Russell Monson, University of Colorado; Daniel Moon, University of North Florida; David Morgan, University of West Georgia; William F. Morris, Duke University; Kim Mouritsen, University of Aarhus; Shannon Murphy, George Washington University; Courtney Murren, College of Charleston; Shahid Naeem, Columbia University; Jason Neff, University of Colorado; Scott Newbold, Colorado State University; Shawn Nordell, Saint Louis University; Timothy Nuttle, Indiana University of Pennsylvania; Mike Palmer, Oklahoma State University; Kevin Pangle, The Ohio State University; Rick Paradis, University of Vermont; Christopher Paradise, Davidson College; Matthew Parris, University of Memphis; William D. Pearson, University of Louisville; Jan Pechenik, Tufts University; Keith Pecor, The College of New Jersey; Karen Pfennig, University of North Carolina, Chapel Hill; Jeff Podos, University of Massachusetts, Amherst; David M. Post, Yale University; Joe Poston, Catawba College; Andrea Previtalli, Cary Institute of Ecosystem Studies; Seth R. Reice, University of North Carolina; Alysa Remsburg, Unity College; Heather Reynolds, Indiana University, Bloomington; Jason Rohr, University of South Florida; Willem Roosenburg, Ohio University, Athens; Richard B. Root, Cornell University; Scott Ruhren, University of Rhode Island; Natalia Rybczynski, Canadian Museum of Nature; Nathan Sanders, University of Tennessee; Mary Santelmann, Oregon State University; Tom Sarro, Mount Saint Mary College; Dov Sax, Brown University; Maynard H. Schaus, Virginia Wesleyan College; Sam Scheiner; Thomas Schoener, University of California, Davis; Janet Schwengber, SUNY Delhi; Erik P. Scully, Towson University; Catherine Searle, Oregon State University; Dennis K. Shiozawa, Brigham Young University; Jonathan Shurin, University of California, San Diego; Andy Sih, University of California, Davis; Frederick Singer, Radford University; Richard Spellenberg, New Mexico State University; John J. Stachowicz, University of California, Davis; Christopher Steiner, Wayne State University; Cheryl Swift, Whittier College; Ethan Temeles, Amherst College; Michael Toliver, Eureka College; Diana Tomback, University of Colorado, Denver; Bill Tonn, University of Alberta; Kathleen Treseder, University of Pennsylvania; Monica Turner, University of Wisconsin; Thomas Veblen, University of Colorado; Betsy Von Holle, University of Central Florida; Don Waller, University of Wisconsin; Carol Wessman, University of Colorado; Jake F. Weltzin, University of Tennessee; Jon Witman, Brown University; Stuart Wooley, California State University, Stanislaus; Brenda Young, Daemen College; Richard Zimmerman, Old Dominion University; Tobias Züst, Cornell University; Anita Baines, University of Wisconsin, La Crosse; Nate Bickford, University of Nebraska at Kearney; Brian Butterfield, Freed-Hardeman University; Kathleen Curran, Wesley College; John Fauth, University of Central Florida; Natalie Hyslop, University of North Georgia; Taegan McMahon, The University of Tampa; Luis Ruedas, Portland State University; Kathleen Schnaars Uvino, University of Jamestown; Dr. David Allard, Texas A&M University, Texarkana; Stanton G. Belford, PhD, Martin Methodist College; Brian Benscoter, Florida Atlantic University; Scott Bergeson, Purdue University Fort Wayne; Gerick S. Bergsma, California State University, Monterey Bay; Sharon Bewick, Clemson University; Robert F. Bode, Saint Martin's University; Jennifer L. Bouldin, PhD, Arkansas State University; Brian Buma, PhD, University of Colorado, Denver; Eric Butler, Shaw University; Brian Butterfield, Freed-Hardeman University; Patrick W. Cain, Georgia Gwinnett College; Dr. Stephen D. Conrad, Indiana Wesleyan University; Randi (Ruth) Darling, Westfield State University; Johnathan G. Davis, Young Harris College; Walter Dodds, Kansas State University; Lisa M. Ellsworth, Oregon State University; Nancy Emery, University of Colorado, Boulder; Danielle E. Garneau, PhD, SUNY Plattsburgh; Dr. Kevin Geedey, Augustana College; Laura Gonzalez-Guzman, University of Texas, Austin; Daniel Gonzalez-Socoloske, Andrews University; Blaine D. Griffen, Brigham Young University; Emily Grman, Eastern Michigan University; Catherine J.B. Hanna, Robert Morris University; Carmen Harjoe, Oregon State University; Mary A. Heskel, PhD, Macalester College; Carlos Iudica, Susquehanna University; David J. Janetski, Indiana University of Pennsylvania; Christie Klimas, DePaul University; Michael Lares, University of Mary; Susan Lewis, Carroll University; Hangkyo Lim, Notre Dame of Maryland University; Y. Kirk Lin, National Taiwan University; Mark Manteuffel, PhD, Washington University; Dr. Jennie R. McLaren, University of Texas, El Paso; Dr. Taegan A. McMahon, University of Tampa; Joe Milanovich, Loyola University Chicago; Emily Mooney, University of Colorado, Colorado Springs; James E. Moore, Christian Brothers University; Andrea L. Moore, PhD, Savannah State University; Cy L. Mott, Eastern Kentucky University; Dev K. Niyogi, Missouri University of Science and Technology; Robert Northington, Husson University; M. Brady Olson, Western Washington University; Benjamin S. Ramage, Randolph-Macon College; Uli Reinhardt, Eastern Michigan University; Julian Resasco, University of Colorado, Boulder; David L. Stokes, University of Washington, Bothell; Stephen Sumithran, Eastern Kentucky University; Thomas J. Trott, Suffolk University; Neal J. Voelz, St. Cloud State University; Daniel M. Wolcott, University of Central Missouri; Lisa Wood, PhD, University of Northern British Columbia; Lan Xu, South Dakota State University;

Digital Resources for *Ecology*, Sixth Edition

Ecology, Sixth Edition, is available in **Oxford Insight**. **Oxford Insight** delivers the trusted content of *Ecology* within a powerful, data-driven learning experience designed to increase student success. A guided and curated learning environment—delivered either via learning management system/virtual learning environment (LMS/VLE) integration or standalone—**Oxford Insight** provides access to the e-book, multimedia resources, assignable/gradable activities and exercises, and analytics on student achievement and progress. As students work through the course material, **Oxford Insight** automatically sets personalized learning paths for them, based on their specific performance.

Developed with applied social, motivational, and personalized learning research, **Oxford Insight** enables instructors to deliver an immersive experience that empowers students by actively engaging them with assigned reading. This approach, paired with real-time actionable data about student performance, helps instructors ensure that all students are best supported along their unique learning paths.

With Oxford Insight, instructors can:

- Assign auto-scored multiple-choice, fill-in, and other machine-gradable questions
- Score specific items (including open-ended questions) with feedback
- Export grades and change grading points
- Establish a course roster and add/drop students
- Share courses and resources with students and faculty
- Sync real-time assignments with LMS/VLE gradebooks
- Author new content and/or customize the publisher-provided content

Contents include:

For the Student

- **Self-Assessment Quizzes** reinforce understanding of chapter material through end-of-section questions.
- **Case Study Videos** bring fifteen case studies to life with author narration and are accompanied by self-assessment questions that are new to this edition.
- An expanded set of **Hands-On Problems** (10 brand-new) provide practical experience working with experimental data and interpreting results from simulations and models.
- **Interactive Figures** illustrate selected images from the text in a walkthrough format and include new self-assessment questions.
- **Chapter Review, Flashcards,** and **Suggested Readings** highlight chapter takeaways, terminology, and additional reference materials.
- **Chapter Tests** assess comprehension of the chapter's key material in multiple-choice format and are assignable by the instructor.

For the Instructor

- **Figure** and **Lecture Note PPTs** are presentations for course instruction with high-resolution images from the text, enabling instructors and/or lecturers to spend less time preparing class materials and more time with students.
- **Hands-On Problems with Answers** are Instructor versions of all Hands-On Problems in Microsoft Word format and ready-to-use PowerPoint presentations.
- **Online Analyzing Data** resources allow instructors to give students extra practice with quantitative skills and are companions to the in-book exercises. The instructor version in Microsoft Word includes answers to questions.
- **Online Climate Change Connections** explore the links between ecological concepts and climate change.
- **Test Banks** include over 1,500 questions covering key facts and concepts in each chapter with multiple-choice and short-answer questions included. Questions are categorized by Bloom's level and are also aligned with the textbook's Learning Objectives.
- **Web Extensions** expand on the coverage of selected topics introduced in the textbook.
- **Web Stats Review** is a brief statistics primer for ecology.

For more information on how *Ecology*, Sixth Edition, powered by **Oxford Insight**, can enrich the teaching and learning experience in your course, please visit oxfordinsight.oup.com or contact your Oxford University Press representative.

Oxford Learning Link

Oxford Learning Link delivers a wealth of engaging digital learning tools and resources to help both instructors and students get the most of their Oxford University Press title. Instructors can view instructor resources at oup.com/he/bowman6e.

Instructors: This title can be integrated directly into learning management systems. To find out more about integration, or if you have any questions about the course content, please contact your OUP representative at (800) 280–0280 or http://learninglink.oup.com/support.

Contents

UNIT 1   Organisms and Their Environment   21

UNIT 2   Evolutionary Ecology 131

UNIT 3   Populations   203

UNIT 4 Species Interactions 265

UNIT 5 Communities 353

UNIT 6 Ecosystems 449

UNIT 7   Applied and Large-Scale Ecology   519

ECOLOGY
SIXTH EDITION

The Web of Life

Deformity and Decline in Amphibian Populations: A Case Study

On a field trip in the summer of 1995, a group of elementary and middle school students from Henderson, Minnesota, made a gruesome discovery as they caught leopard frogs (*Lithobates pipiens*) for a summer science project: 11 of the 22 frogs they found were severely deformed. Some of the frogs had missing or extra limbs, others had legs that were too short or bent in odd directions, and still others had bony growths coming out of their backs (**FIGURE 1.1**). The students reported their findings to the Minnesota Pollution Control Agency, which investigated and found that 30%–40% of the frogs in the pond the students studied were deformed.

News of the students' discovery traveled fast, capturing public attention and spurring scientists to check for similar deformities in other parts of the country and in other amphibian species. It soon became apparent that the problem was widespread. In the United States, misshapen individuals were found in 46 states and in more than 60 species of frogs, salamanders, and toads. In some localities, more than 90% of the individuals were deformed. Deformed amphibians were also found in Europe, Asia, and Australia. Worldwide, the frequency of amphibian deformities was high and increasing.

Adding to the alarm caused by the gruesome deformities were observations, beginning in the late 1980s, of another disturbing trend: global amphibian populations seemed to be in decline. By 1993, over 500 populations of frogs and salamanders from around the world were reported to be decreasing in size, and some were under threat of extinction. In some cases, entire species were in danger; across the globe, hundreds of species were extinct, missing, or critically endangered (**FIGURE 1.2**). Since 1970, an estimated 200 species of frogs have gone extinct, and the rate of extinctions is increasing (Alroy 2015).

Species in other groups of organisms were also showing signs of decline, but scientists were especially worried about amphibians for three reasons. First, the decline appeared to have started recently across wide regions of the world. Second, some of the populations in decline were located in protected or pristine regions, seemingly far from the effects of human activities. Third, some scientists view amphibians as "biological indicators" of environmental conditions. They hold this view in part because amphibians have permeable skin and eggs that lack shells or other protective coverings, which increases their sensitivity to toxic pollutants. In addition, most amphibians spend part of their lives in water and part on land. As a result, they are exposed to a wide range of potential threats, including water and

KEY CONCEPTS

CONCEPT 1.1 Events in the natural world are interconnected.

CONCEPT 1.2 Ecology is the scientific study of interactions between organisms and their environment.

CONCEPT 1.3 Ecologists evaluate competing hypotheses about natural systems with observations, experiments, and models.

© Craig Line/Associated Press

FIGURE 1.1 Deformed Leopard Frog With its misshapen legs, these frogs show one of the types of limb deformities that have become common in leopard frogs and other amphibian species.

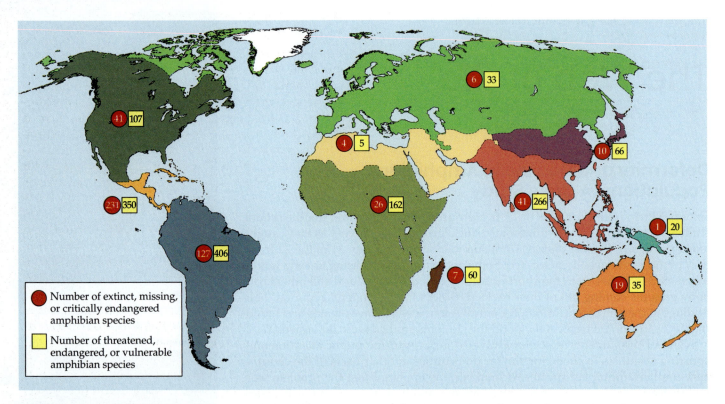

FIGURE 1.2 Amphibians in Decline In many regions of the world, amphibian species face increased risk of extinction. Each pair of numbered circles and squares is associated with one color-coded region on the map. (Map after AmphibiaWeb. 2019. https://amphibiaweb.org/declines/declines.html. University of California, Berkeley, CA, USA. Accessed 25 Sep 2019; B. G. Holt et al. 2013. *Science* 339: 74–78. Data archived at http://macroecology.ku.dk/resources/wallace.)

Legend:
- ● Number of extinct, missing, or critically endangered amphibian species
- ▢ Number of threatened, endangered, or vulnerable amphibian species

air pollution as well as changes in temperature and in the amount of ultraviolet (UV) light in their environment. Moreover, many amphibians remain close to their birthplace throughout their lives, so the decline of a local population is likely to indicate a deterioration of local environmental conditions.

Because amphibians worldwide were showing declining numbers and frequent deformities, scientists initially tried to find one or a few global causes that might explain these problems. However, as we'll see in this chapter, the story turned out to be more complicated than that: a single "smoking gun" has not emerged. What, then, has caused the global decline of amphibian populations?

Introduction

We humans have an enormous impact on our planet. Our activities have transformed over half of Earth's land surface and have altered the composition of the atmosphere, leading to global climate change. We have introduced many species to new regions, an action that can have severe negative effects on both native species and human economies. Even the oceans, seemingly so vast, show many signs of environmental degradation due to human activities, leading to declining fish stocks;

widespread bleaching of once-spectacular coral reefs; the formation of large "dead zones," regions where oxygen concentrations have dropped to levels low enough to kill many species; and acidification.

Global changes like those we've just described can occur when humans take actions without giving much thought to how our actions might affect the environment. In such situations, we have repeatedly been surprised by the unintended and harmful side effects of our actions. Fortunately, we are beginning to realize that a better understanding of how the natural systems of our environment work can help us to anticipate the consequences of our actions and fix the problems we have already caused.

Our growing realization that we must understand how natural systems work brings us to the subject of this book. Natural systems are driven by the ways in which organisms interact with one another and with their physical environment. Thus, to understand how natural systems work, we must understand those interactions. *Ecology* is the scientific study of how organisms affect—and are affected by—other organisms and their environment.

In this chapter, we'll introduce the study of ecology and its relevance for humans. We'll begin by exploring a theme that runs throughout this book: connections in nature.

LEARNING OBJECTIVE

1.1.1 Explain how interactions between organisms and their environment can affect other organisms and potentially lead to unexpected consequences.

Connections in Nature

From what you have read or observed about nature, can you think of examples that might illustrate the phrase *connections in nature*? In this book, we use that phrase to refer to the fact that events in the natural world can be linked or connected to one another. These connections occur as organisms interact with one another and with their physical environment. This does not necessarily mean that there are strong connections among all the organisms that live in a given area. Two species may live in the same area but have little influence on each other. But all organisms are connected to features of their environment. For example, they all require food, space, and other resources, and they all interact with other species and the physical environment as they pursue what they need to live. As a result, two species that do not interact directly with each other can be connected indirectly by shared features of their environment, including other organisms.

Connections in nature are revealed as ecologists ask questions about the natural world and examine what they've learned. To illustrate what this process can teach us about connections in nature, let's return to our discussion of amphibian deformities.

Early observations suggest that parasites cause amphibian deformities

Nine years before the Minnesota students made their startling discovery, Stephen Ruth was exploring ponds in Northern California when he found Pacific tree frogs (*Pseudacris regilla*) and long-toed salamanders (*Ambystoma macrodactylum*) with extra limbs, missing limbs, and other deformities. He asked Stanley Sessions, an expert in amphibian limb development, to examine his specimens. Sessions found that the deformed amphibians all contained a parasite, now known to be *Ribeiroia ondatrae*, a trematode flatworm. Sessions and Ruth hypothesized that the parasite caused the deformities. As an initial test of this hypothesis, they implanted small glass beads near the developing limb buds of tadpoles. These beads were meant to mimic the effects of *Ribeiroia*, which produces cysts near the areas where limbs form in a tadpole as it transitions into an adult frog. In a 1990 paper, Sessions and Ruth reported that the beads caused deformities similar to (but less severe than) those Ruth had found.

A laboratory experiment tests the role of parasites

When Ruth first observed deformed amphibians in the mid-1980s, he assumed that they were an isolated, local phenomenon. By 1996, Pieter Johnson, then an undergraduate at Stanford University, had learned of the Minnesota students' findings and of the paper by Sessions and Ruth. Although Sessions and Ruth provided indirect evidence that *Ribeiroia* may have caused amphibian deformities, they did not infect *P. regilla* or *A. macrodactylum* with *Ribeiroia* and show that deformities resulted. Furthermore, the two amphibian species they used in their experiments (the African clawed frog, *Xenopus laevis*, and the axolotl salamander, *A. mexicanum*) were not known to have limb deformities in nature. Building on the work done by Sessions and Ruth (1990), Johnson and his colleagues set out to provide a more direct test of whether *Ribeiroia* parasites can cause limb deformities in amphibians.

They began by surveying 35 ponds in Santa Clara County, California. They found Pacific tree frogs in 13 of the surveyed ponds, and 4 of these ponds had deformed frogs. Concentrating on 2 of the ponds with deformed frogs, they found that 15%–45% of the tadpoles undergoing metamorphosis had extra limbs or other deformities (Johnson et al. 1999). One source of concern was that the deformities might be caused by pollutants, such as pesticides, polychlorinated biphenyls (PCBs), or heavy metals. However, none of these substances were found in water from the 2 ponds.

Johnson and his colleagues then turned their attention to other factors that might cause the deformities. Aware that Sessions and Ruth had hypothesized that parasites could be the cause, Johnson et al. noted that of the 35 ponds they surveyed, the 4 ponds with deformed frogs were the only ponds that contained both tree frogs and the aquatic snail *Helisoma tenuis*. As shown in **FIGURE 1.3**, this snail is the first of two intermediate hosts required for the *Ribeiroia* parasite to complete its life cycle and produce offspring. The parasite also requires an amphibian or fish as a second intermediate host. In addition, dissections of abnormal frogs collected from the two ponds they studied in detail revealed *Ribeiroia* cysts in all the frogs with deformed limbs.

Like the findings of Sessions and Ruth, Johnson's observations provided only indirect evidence that *Ribeiroia* caused deformities in Pacific tree frogs. Next, Johnson and his colleagues returned to the laboratory to perform a more rigorous test of that idea. They did this by using a standard scientific approach: they performed a **controlled experiment** in which an *experimental group* (that has the factor being tested) was compared with a *control group* (that lacks the factor being tested). Johnson et al. collected *P. regilla* eggs from a region not known to have frog deformities, brought the eggs into the laboratory, and placed the tadpoles that hatched from them in 1-liter containers with one tadpole per container. Each tadpole was then assigned at random to one of four *treatments*, in which 0 (the

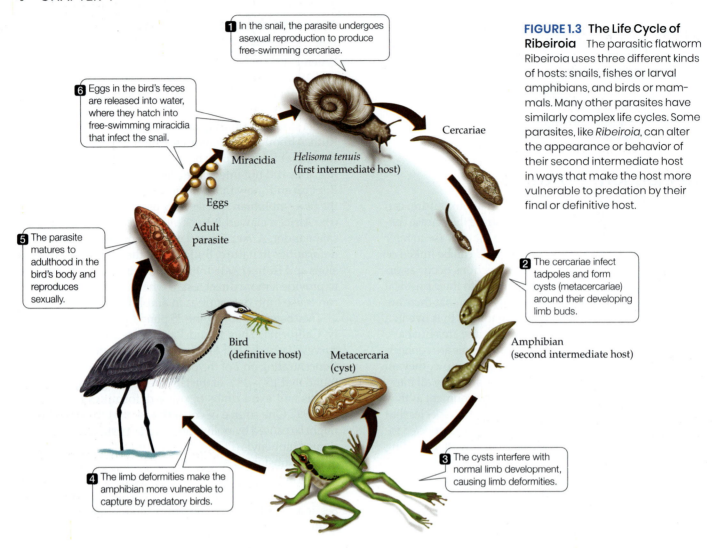

FIGURE 1.3 The Life Cycle of Ribeiroia The parasitic flatworm *Ribeiroia* uses three different kinds of hosts: snails, fishes or larval amphibians, and birds or mammals. Many other parasites have similarly complex life cycles. Some parasites, like *Ribeiroia*, can alter the appearance or behavior of their second intermediate host in ways that make the host more vulnerable to predation by their final or definitive host.

1 In the snail, the parasite undergoes asexual reproduction to produce free-swimming cercariae.

6 Eggs in the bird's feces are released into water, where they hatch into free-swimming miracidia that infect the snail.

5 The parasite matures to adulthood in the bird's body and reproduces sexually.

4 The limb deformities make the amphibian more vulnerable to capture by predatory birds.

2 The cercariae infect tadpoles and form cysts (metacercariae) around their developing limb buds.

3 The cysts interfere with normal limb development, causing limb deformities.

Cercariae

Miracidia

Helisoma tenuis (first intermediate host)

Eggs

Adult parasite

Bird (definitive host)

Metacercaria (cyst)

Amphibian (second intermediate host)

control group), 16, 32, or 48 *Ribeiroia* parasites were placed in its container; these numbers were selected to match parasite levels that had been observed in the ponds.

Johnson and his colleagues found that as the number of parasites increased, fewer of the tadpoles survived to metamorphosis, and more of the survivors had deformities (**FIGURE 1.4**). In the control group (with zero *Ribeiroia*), 88% of the tadpoles survived, and none had deformities (Johnson et al. 1999). The link had been made: *Ribeiroia* could cause frog deformities. Furthermore, since exposure to *Ribeiroia* killed up to 60% of the tadpoles, the results also suggested that the parasites could contribute to amphibian declines.

A field experiment suggests that multiple factors influence frog deformities

A few years after Johnson and his colleagues published their research, other scientists showed that *Ribeiroia* parasites could cause limb deformities in other amphibian species, including western toads (*Anaxyrus boreas*), wood frogs (*Lithobates sylvaticus*), and leopard frogs (*L. pipiens*, the species in which the Minnesota students had discovered

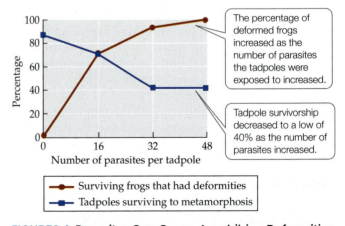

The percentage of deformed frogs increased as the number of parasites the tadpoles were exposed to increased.

Tadpole survivorship decreased to a low of 40% as the number of parasites increased.

→ Surviving frogs that had deformities
→ Tadpoles surviving to metamorphosis

FIGURE 1.4 Parasites Can Cause Amphibian Deformities The graph shows the relationship between the numbers of *Ribeiroia* parasites that tadpoles were exposed to and their rates of survival and deformity. Initial numbers of tadpoles were 35 in the control group (0 parasites) and 45 in each of the other three treatments. (After P. T. J. Johnson et al. 1999. *Science* 284: 802–804.)

? Estimate the number of tadpoles in the control group that survived, as well as the number that had deformities.

deformities). While *Ribeiroia* was clearly important, some researchers suspected that other factors might also play a role. Pesticides, for example, were known to contaminate some of the ponds in which deformed frogs were found. To examine the possible joint effects of parasites and pesticides, Joseph Kiesecker conducted a field experiment in six ponds, all of which contained *Ribeiroia*, but only some of which contained pesticides (Kiesecker 2002).

Three of the ponds in Kiesecker's study were close to farm fields, and water tests indicated that each of these ponds contained detectable levels of pesticides. The other three ponds were not as close to farm fields, and none of them showed detectable levels of pesticides. In each of the six ponds, Kiesecker placed wood frog tadpoles in cages made with a mesh through which water could flow but tadpoles could not escape. Six cages were placed in each pond; three of the cages had a mesh through which *Ribeiroia* parasites could pass, while the other three had a mesh too small for the parasites. Thus, in each pond, the tadpoles in three cages were exposed to the parasites, while the tadpoles in the other three cages were not.

The results showed that *Ribeiroia* caused limb deformities in the field (**FIGURE 1.5**). No deformities were found in frogs raised in cages whose small mesh size, 75 micrometers (μm), prevented the entry of *Ribeiroia*, regardless of which pond the cages were in. Deformities were found in some of the frogs raised in cages whose larger mesh size (500 μm) allowed the entry of *Ribeiroia*. In addition, dissections revealed that every frog with a deformity was infected by *Ribeiroia*. However, a greater percentage of frogs had deformities in the ponds that contained pesticides than in the ponds that did not (29% vs. 4%). Overall, the results of this experiment indicated that (1) exposure to *Ribeiroia* was necessary for deformities to occur, and (2) when frogs were exposed to *Ribeiroia*, deformities were more common in ponds with detectable levels of pesticides than in ponds without detectable levels of pesticides.

Based on these results, Kiesecker hypothesized that pesticides might decrease the ability of frogs to resist infection by parasites. To test whether pesticides had such an effect, Kiesecker (2002) brought wood frog tadpoles into the laboratory, where he reared some in an environment with pesticides and others in an environment without pesticides, then exposed all of them to *Ribeiroia*. The tadpoles exposed to pesticides had fewer white blood cells (indicating a suppressed immune system) and a higher rate of *Ribeiroia* cyst formation (**FIGURE 1.6**). Together,

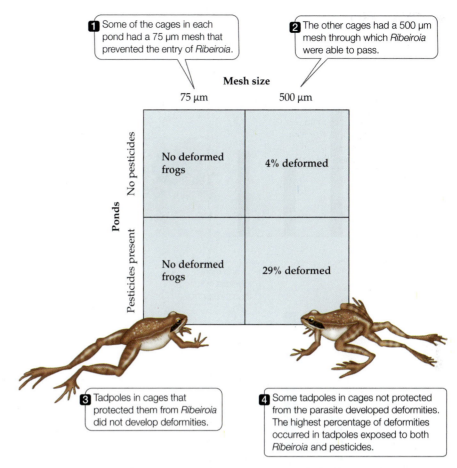

1 Some of the cages in each pond had a 75 μm mesh that prevented the entry of *Ribeiroia*.

2 The other cages had a 500 μm mesh through which *Ribeiroia* were able to pass.

3 Tadpoles in cages that protected them from *Ribeiroia* did not develop deformities.

4 Some tadpoles in cages not protected from the parasite developed deformities. The highest percentage of deformities occurred in tadpoles exposed to both *Ribeiroia* and pesticides.

FIGURE 1.5 Do the Effects of *Ribeiroia* and Pesticides Interact in Nature? To test the effects of *Ribeiroia* and pesticides on frog deformities in the field, screened cages were placed in six ponds. Three of the six ponds contained detectable levels of pesticides; the other three did not. (After J. M. Kiesecker. 2002. *Proc Natl Acad Sci USA* 99: 9900–9904. © 2002 National Academy of Sciences, U.S.A.)

? Based on the results shown here, do pesticides acting alone cause frog deformities? Do the results indicate that pesticides affect frogs? If so, do they indicate how? Explain.

Kiesecker's laboratory and field results suggested that pesticide exposure can affect the frequency with which parasites cause deformities in amphibian populations. This conclusion has since been supported by other studies. Field surveys and laboratory experiments reported in Rohr et al. (2008), for example, indicated that exposure to pesticides can increase the number of trematode infections and decrease survival rates in several frog species. As in Kiesecker's study, one reason for the increased number of parasitic infections appeared to be that the frogs' immune response was suppressed by the pesticide.

Connections in nature can lead to unanticipated impacts

As we have seen, the immediate cause of amphibian deformities is often infection by *Ribeiroia* parasites. But we also noted in the Case Study that amphibian deformities are occurring more often now than in the past. Why has the frequency of amphibian deformities increased?

One possible answer is suggested by the results of Kiesecker (2002) and Rohr et al. (2008): pesticides may decrease the ability of amphibians to ward off parasite

(A) Eosinophils

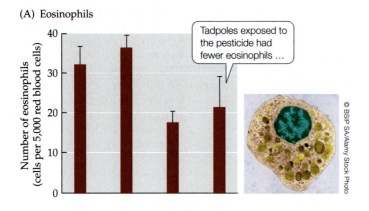

Tadpoles exposed to the pesticide had fewer eosinophils …

© BSIP SA/Alamy Stock Photo

(B) *Ribeiroia*

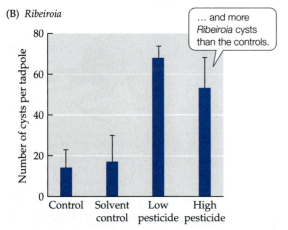

… and more *Ribeiroia* cysts than the controls.

FIGURE 1.6 Pesticides May Weaken Tadpole Immune Systems   In a laboratory experiment, wood frog (*Lithobates sylvaticus*) tadpoles were exposed to low or high concentrations of the pesticide esfenvalerate and then exposed to 50 *Ribeiroia* parasites per tadpole. The tadpoles were then examined for (A) numbers of eosinophils (a type of white blood cell used in the immune response) and (B) numbers of *Ribeiroia* cysts. Two types of controls were used: one in which only parasites were added to the tadpoles' containers ("control"), and another in which both parasites and the solvent used to dissolve the pesticide were added ("solvent control"). Error bars show one standard error (SE) of the mean. (After J. M. Kiesecker. 2002. *Proc Natl Acad Sci USA* 99: 9900–9904. © 2002 National Academy of Sciences, U.S.A.)

?   What was the purpose of using two types of controls in this experiment?

attack, and hence deformities are more likely in environments that contain pesticides. The first synthetic pesticides were developed in the late 1930s, and their use has risen dramatically since that time. Thus, it is likely that amphibian exposure to pesticides has increased considerably through time, which may help to explain the recent rise in the frequency of amphibian deformities.

Other environmental changes may also contribute to the observed increase in amphibian deformities. For example, the addition of nutrients to natural or artificial ponds (used to store water for cattle or crops) can lead to increases in parasite infections and amphibian deformities (Johnson et al. 2007). Nutrients can enter a pond when rain or snowmelt washes fertilizers from an agricultural field into it.

Fertilizer inputs often stimulate increased growth of algae, and the snails that harbor *Ribeiroia* parasites eat algae (to refresh your memory of the parasite's life cycle, see Figure 1.3). Thus, as the amount of algae increases, so do the snail hosts of *Ribeiroia*. An increase in snails tends to increase the number of *Ribeiroia* found in the pond.

Here, a chain of events that begins with increased fertilizer use by people ends with increased numbers of *Ribeiroia*, and hence increased numbers of deformed amphibians. As this example illustrates, events in the natural world are connected. As a result, when people alter one aspect of the environment, we can cause other changes that we do not intend or anticipate. When we increased our use of pesticides and fertilizers, we did not intend to increase the frequency of deformities in frogs. Nevertheless, we seem to have done just that.

The indirect and unanticipated effects of human actions include more than bizarre deformities in frogs. Indeed, some changes we are making to our local and global environment appear to have increased human health risks. The damming of rivers in Africa has created favorable habitat for snails that harbor trematode parasites that cause schistosomiasis, thereby increasing the spread of an infection that can weaken or kill people. Globally, the past few decades have seen an increase in the appearance and spread of new diseases, such as COVID 19, AIDS, Lyme disease, hantavirus pulmonary syndrome, Ebola hemorrhagic fever, and West Nile virus. Many public health experts think that the effects of human actions on the environment have contributed to the emergence of these and other new diseases (Weiss and McMichael 2004).

For example, West Nile virus, which is transmitted by mosquitoes and infects birds and humans, is thought to have been introduced into North America by people in 1999 (**FIGURE 1.7**). Furthermore, the incidence of West Nile virus in humans is influenced by factors such as human population size, the extent of land development, the abundance and identity of mosquito and bird species, and variations in temperature and rainfall (Reisen et al. 2006; Landesman et al. 2007; Allan et al. 2009). Each of these factors can be affected by human actions, either directly (e.g., by urban or agricultural development) or indirectly (e.g., as a result of climate change; see Concepts 25.2 and 25.3).

As we've seen, connections in nature can cause human actions to have unanticipated side effects. Moreover, if you live in a city, it can be easy to forget the extent to which everything you do depends on the natural world. Your house or apartment shelters you from the elements and keeps you warm in winter and cool in summer. Similarly, you obtain food from a grocery store, clothes from a store or online supplier, water from a faucet. Ultimately, however, each of these items—and everything else you use or own—comes from or depends on the natural environment. No matter how far from the natural world our day-to-day activities take us, people, like all other organisms on Earth, are part of an interconnected web of life. Let's turn now to the study of these connections, the scientific discipline of *ecology*.

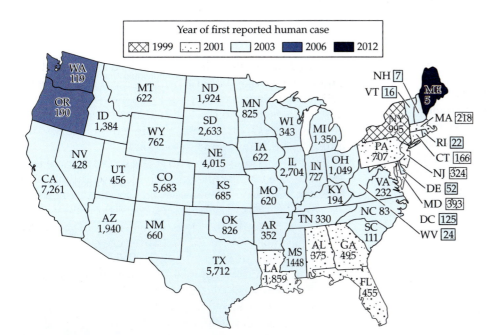

FIGURE 1.7 Rapid Spread of a Deadly Disease Within 13 years, West Nile virus had spread from its North American point of entry (New York City) to all of the lower 48 states. Birds are a primary host for West Nile virus, which may help to explain its rapid spread. Mosquitoes transmit the disease from birds and other animal hosts to people. Numbers show the cumulative number of human cases in each state by December 31, 2020. Not shown: Data for Alaska (2 cases; first reported case in 2018), Hawaii (1 case in 2014), and Puerto Rico (1 case in 2012). (Data from Centers for Disease Control and Prevention.)

CONCEPT 1.2 Ecology is the scientific study of interactions between organisms and their environment.

LEARNING OBJECTIVES

1.2.1 Summarize how the inquiries of ecologists and environmental scientists differ.

1.2.2 Outline how ecologists use spatial and temporal scales when testing their hypotheses.

What Is Ecology?

In this book, **ecology** is defined as the scientific study of interactions between organisms and their environment. This definition is meant to include the interactions of organisms with one another because, as we have seen, organisms are an important part of one another's environment. Ecology can also be defined in a variety of other ways, such as the scientific study of interactions that determine the distribution (geographic location) and abundance of organisms. As will become clear as you read this book, these definitions of ecology can be related to one another, and each emphasizes different aspects of the discipline. A more important point for our purpose here is that the term *ecology*, as used by ecologists, refers to a scientific endeavor.

We emphasize this point because *ecology* has other meanings in its public usage. People who are not scientists may assume that an "ecologist" is an environmental activist. Some ecologists are activists, but some are not. Furthermore, as a scientific discipline, ecology is related to—yet different from—other disciplines such as environmental science. Ecology is a branch of biology, while **environmental science** is an interdisciplinary field that incorporates concepts from the natural sciences (including ecology) and the social sciences (e.g., politics, economics, ethics). Compared with ecology, environmental science is focused more specifically on how people affect the environment and how we can address environmental problems. While an ecologist might examine pollution as one of several factors that influence the reproductive success of wetland plants, an environmental scientist might focus on how economic and political systems could be used to reduce pollution.

Public and professional ideas about ecology often differ

Surveys have shown that many people think that there is a "balance of nature," in which natural systems are stable and tend to return to an original, preferred state after a disturbance, and that each species in nature has a distinct role to play in maintaining that balance. Such ideas about ecological systems can have moral or ethical implications. For example, the view that each species has a distinct function can lead people to think that each species is important and irreplaceable, which in turn can cause people to feel that it is wrong to harm other species. As summarized by one interviewee in a survey on the meaning of ecology (Uddenberg et al. 1995, as quoted in Westoby 1997), "There is a certain balance in nature, and there is a place for all species. There is a reason for their existence and we are not free to exterminate them."

Public views on the balance of nature with stable, orderly systems were once held by many ecologists. However, ecologists now recognize (1) that natural systems do not necessarily return to their original state after a disturbance and (2) that random effects often play important roles in nature. For example, as we will see in Unit 5, current evidence suggests that different communities can form in the same area under similar environmental conditions. Therefore, unless they provide careful qualifications, few ecologists today speak of a balance of nature.

Some ecological concepts have remained unchanged through time. In particular, early ecologists and contemporary ecologists would agree that events in nature are interconnected (via the physical environment and via interactions among species). As a result, a change in one part of an ecological system can alter other parts of that system, including those that govern life-supporting processes such as the provision of food, clean water, and shelter from the environment.

Overall, although the natural world may not be as predictable or as tightly interconnected as early ecologists may have thought, species are linked to one another. For some people, the fact that events in nature are interconnected provides an ethical imperative to protect natural systems. A person who feels an ethical obligation to protect human life, for example, may also feel an ethical obligation to protect the natural systems on which human life depends.

The scale of an ecological study affects what can be learned from it

Whether they study individual organisms or the diversity of life on Earth—or anything in between—ecologists always draw boundaries around what they observe. An ecologist interested in frog deformities might ignore the birds that migrate above the study site, while an ecologist studying bird migrations might ignore the details of what occurs in the ponds below. It is not possible or desirable to study everything at once.

When they seek to answer a particular question, ecologists must select the most appropriate dimension, or **scale**, in both time and space, for collecting observations. Every ecological study addresses events at some scales but ignores events at other scales. A study on the activities of soil microorganisms, for example, might be conducted at a small spatial scale (e.g., measurements might be collected at centimeter to meter scales). For a study addressing how atmospheric pollutants affect the global climate, on the other hand, the scale of observation might include Earth's entire atmosphere. Ecological studies also differ greatly in the time scales they cover. Some studies, such as those that document how leaves respond to momentary increases in the availability of sunlight, concern events on short time scales (seconds to hours). Others, such as studies that use fossil data to show how the species found in a given area have changed over time, address events at much longer time scales (centuries to millennia or longer).

Ecology is broad in scope

Ecologists study interactions in nature across many levels of biological organization. For example, some ecologists are interested in how particular genes or proteins enable organisms to respond to environmental challenges. Other ecologists study how hormones influence social interactions in animals, or how specialized tissues or organ systems allow animals to cope with extreme environments. However, even among ecologists whose research is focused on lower levels of biological organization (e.g., from molecules to organ systems), ecological studies usually emphasize one or more of the following levels: individuals, populations, communities, ecosystems, landscapes, or the entire biosphere (**FIGURE 1.8**).

A **population** is a group of individuals of a single species that live in a particular area and interact with one

An **individual** American alligator (*Alligator mississippiensis*)

… is part of a **population** of this species. The alligator population, in turn …

… is one of the interacting populations of different species found in this **community** or **ecosystem** (if we also consider the physical environment).

The marsh ecosystem is nested within a **landscape** composed of multiple ecosystems (including the near-shore waters and the land in this photograph).

The **biosphere** includes all of Earth's ecosystems.

FIGURE 1.8 An Ecological Hierarchy As suggested by this series of photographs, life in the rocky intertidal ecosystem can be studied at a number of levels, from individuals to the biosphere. These levels are nested within one another, in the sense that each level is composed of groups of the entity found in the level below it.

FIGURE 1.9 A Few of Earth's Many Communities These photographs show (A) a desert community in Peru; (B) a temperate rainforest in Canada; (C) walruses on an arctic beach in Norway; and (D) a coral reef with a variety of corals and sponges in Hawaii.

(A) A desert community in the Salinas y Aguada Blanca National Reserve in Peru

(B) A temperate rainforest near Ucluelet, Canada

(C) Walruses (*Odobenus rosmarus*) on an arctic beach in Spitsbergen, Norway

(D) A tropical coral reef in Hawaii

another. Many of the central questions in ecology concern how and why the locations and abundances of populations change over time. To answer such questions, it is often helpful to understand the roles played by other species. Thus, many ecologists study nature at the level of the **community**, which is an association of interacting populations of different species that live in the same area at the same time. Communities can cover large or small areas, and they can differ greatly in terms of the numbers and types of species found within them (**FIGURE 1.9**).

Ecological studies at the population and community levels often examine not only the effects of the **biotic**, or living, components of a natural system, but also those of the **abiotic**, or physical, environment. For example, a population or community ecologist might ask whether features of the abiotic environment, such as climate and soils, influence the growth of individuals or the relative abundances of the different species found in a community. Other ecologists are particularly interested in how ecosystems work. An **ecosystem** is a collection of communities of organisms (e.g., plants, birds) plus the physical environment in which they live. An ecologist studying ecosystems might want to know the rate at which a chemical (such as nitrogen) moves through a particular community, as well as how the species living in the ecosystem affect what happens to the chemical once it enters the community. For example, ecosystem ecologists studying amphibian deformities might document the rates at which nitrogen from fertilizers enters ponds that do and do not contain deformed amphibians, or they might determine how the presence or absence of algae affects what happens to nitrogen once it has entered the ponds.

Across larger spatial regions, ecologists study **landscapes**, which are areas that vary substantially from one place to another, typically including multiple ecosystems. Finally, global patterns of element, air, and water circulation (see Concept 2.2) link the world's ecosystems into the **biosphere**, which consists of all living organisms on Earth plus the environments in which they live. The biosphere forms the highest level of biological organization. Over recent decades, as we will see in Unit 7, ecologists have acquired new tools that improve their ability to study the big picture: how the biosphere works. As just one example, ecologists can now use satellite data to answer questions such as, How do different ecosystems contribute to ongoing changes in the global concentration of carbon dioxide (CO_2) in the atmosphere?

Some key terms are helpful for studying connections in nature

Whether we are discussing individuals, populations, communities, or ecosystems, all chapters of this book

TABLE 1.1

Key Terms for Studying Connections in Nature

Term	Definition
Adaptation	A feature of an organism that improves its ability to survive or reproduce in its environment
Natural selection	An evolutionary process in which individuals that possess particular characteristics survive or reproduce at a higher rate than other individuals because of those characteristics
Producer	An organism that uses energy from an external source, such as the sun, to produce its own food without having to eat other organisms or their remains
Consumer	An organism that obtains its energy by eating other organisms or their remains
Net primary production (NPP)	The amount of energy (per unit of time) that producers fix by photosynthesis or other means, minus the amount they use in cellular respiration
Nutrient cycle	The cyclic movement of a nutrient between organisms and the physical environment

incorporate the principle that events in the natural world are interconnected. For example, in Unit 3, we will see how an explosion in the population size of an introduced species (the comb jelly *Mnemiopsis leidyi*) altered the entire Black Sea ecosystem. Because we stress connections in nature in every chapter, and hence may discuss ecosystems in a chapter about organisms, or vice versa, we describe here a handful of key terms that you will need to know as you begin your study of ecology. These terms are also summarized in **TABLE 1.1**.

A universal feature of living systems is that they change over time, or *evolve*. Depending on the questions or time scale of interest, evolution can be defined as (1) a change in the genetic characteristics of a population over time or as (2) *descent with modification*, the process by which organisms gradually accumulate differences from their ancestors. We will discuss evolution in the context of ecology more fully in Chapter 6, but here we define two key evolutionary terms: *adaptation* and *natural selection*.

An **adaptation** is a genetically based characteristic of an organism that improves its ability to survive or reproduce within its environment. Adaptations are of critical importance for understanding how organisms function and interact with one another. As we'll see in Concept 6.3, although several mechanisms can cause evolutionary change, only natural selection can produce adaptations consistently. In the process of **natural selection**, individuals with particular characteristics tend to survive and reproduce at a higher rate than other individuals *because of those characteristics*. If the characteristics being selected for are heritable, then the offspring of individuals favored by natural selection will tend to have the same characteristics that gave their parents an advantage. As a result, the frequency of those characteristics in a population may increase over time. If that occurs, the population will have evolved.

Consider what happens within the body of a person taking an antibiotic. Some of the bacteria that live inside that person may possess genes that provide resistance to the antibiotic. The bacteria with the resistance genes will survive and reproduce at a higher rate than will nonresistant bacteria lacking the genes (**FIGURE 1.10**). Because the trait on which natural selection acts (antibiotic resistance) is heritable, the offspring of the resistant bacteria will tend

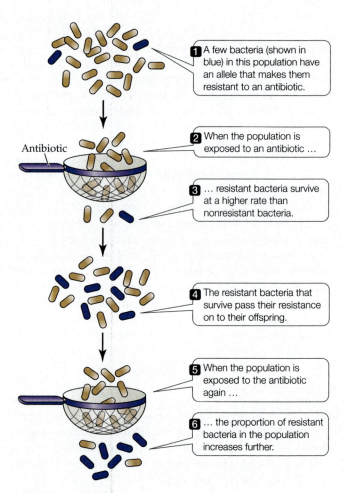

1 A few bacteria (shown in blue) in this population have an allele that makes them resistant to an antibiotic.

Antibiotic

2 When the population is exposed to an antibiotic …

3 … resistant bacteria survive at a higher rate than nonresistant bacteria.

4 The resistant bacteria that survive pass their resistance on to their offspring.

5 When the population is exposed to the antibiotic again …

6 … the proportion of resistant bacteria in the population increases further.

FIGURE 1.10 Natural Selection in Action As shown in this diagram, in which a sieve represents the selective effects of an antibiotic, natural selection can cause the frequency of antibiotic resistance in bacteria to increase over time.

to be resistant. As a result, the proportion of resistant bacteria in the person's body will increase over time, and the bacterial population will have evolved.

The remaining four key terms that we'll introduce here concern ecosystem processes. One way to look at how ecosystems work is to consider the movement of energy and materials through a community. Energy enters the community when an organism such as a plant or bacterium captures energy from an external source, such as the sun, and uses that energy to produce food. An organism that can produce its own food from an external energy source without having to eat other organisms or their remains is called a **producer** (such organisms are also called *primary producers* or *autotrophs*). An organism that obtains its energy by eating other organisms or their remains is called a **consumer** (or a *heterotroph*). Per unit of time, the amount of energy that producers capture by photosynthesis or other means, minus the amount they lose in cellular respiration, is called **net primary production** (**NPP**). Changes in NPP can have large effects on ecosystem function, and NPP varies greatly from one ecosystem to another.

Each unit of energy captured by producers is eventually lost from the ecosystem as respiration (**FIGURE 1.11**). As a result, energy moves through ecosystems in a single direction only—it cannot be recycled. Nutrients, however, are recycled from the physical environment to organisms and back again. The cyclic movement of a nutrient such as phosphorus between organisms and the physical environment is referred to as a **nutrient cycle**. Life as we know it would cease if nutrients were not cycled, because the molecules organisms need for their growth and reproduction would be much less readily available.

Whether they are concerned with adaptations or NPP, populations or ecosystems, the scientists who study ecological systems have not produced a fixed body of knowledge. Instead, what we know about ecology changes constantly as ideas are tested and, if necessary, revised or discarded as new information emerges. As we will see in the next section, ecology, like all branches of science, is about answering questions and seeking to understand the underlying causes of natural phenomena.

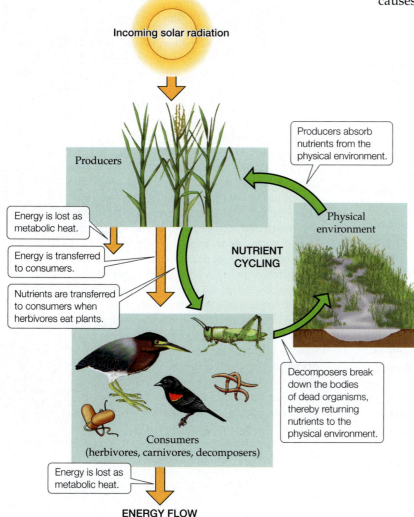

Incoming solar radiation

Producers

Producers absorb nutrients from the physical environment.

Energy is lost as metabolic heat.

Energy is transferred to consumers.

Nutrients are transferred to consumers when herbivores eat plants.

NUTRIENT CYCLING

Physical environment

Decomposers break down the bodies of dead organisms, thereby returning nutrients to the physical environment.

Consumers (herbivores, carnivores, decomposers)

Energy is lost as metabolic heat.

ENERGY FLOW

FIGURE 1.11 How Ecosystems Work Each time one organism eats another, a portion of the energy originally captured by a producer is lost as heat given off during the chemical breakdown of food by cellular respiration. As a result, energy flows through the ecosystem in a single direction and is not recycled. Nutrients such as carbon and nitrogen, on the other hand, cycle between organisms and the physical environment.

? Describe the three main steps by which a nutrient cycles through an ecosystem.

LEARNING OBJECTIVES

1.3.1 Compare the advantages and disadvantages of using field observations, field experiments, and lab experiments to test ecological hypotheses.

1.3.2 Describe the importance of hypotheses, controls, replication, and data analysis to the scientific process.

Answering Ecological Questions

The studies of amphibian deformities that we discussed earlier in this chapter illustrate several ways in which ecologists seek to answer questions about the natural world. The study by Johnson and his colleagues (1999), for example, had two key components: observational studies in the field and a controlled experiment in the laboratory. In the observational part of their work, the researchers surveyed ponds, noted the species present, and observed that tree frog deformities were found only in ponds that contained both tree frogs and a snail that harbored the *Ribeiroia* parasite. These observations suggested that *Ribeiroia* might cause deformities, so Johnson and his colleagues performed a laboratory experiment to test whether that was the case (it was).

Kiesecker (2002) extended these results in two experiments, one performed in the field, the other in the laboratory. To examine the effects of pesticides on frog deformities, Kiesecker compared results from three ponds containing pesticides with results from three ponds without detectable levels of pesticides. While this approach had the advantage of allowing the effects of *Ribeiroia* to be examined under different field conditions (in ponds with and without pesticides), Kiesecker could not control the conditions as precisely as he did in his laboratory experiment. The constraints of working in the field meant, for example, that he could not start out with six identical ponds, then add pesticides to three of them but not to the other three—an experiment that would test more directly whether pesticides were responsible for the results he obtained. As this example suggests, no single approach works best in all situations, so ecologists use a variety of methods when seeking to answer ecological questions.

Ecologists use experiments, observations, and models to answer ecological questions

In an ecological experiment, an investigator alters one or more features of the environment and observes the effect of that change, a procedure that allows scientists to test whether one factor has a cause-and-effect relationship with another. When possible, such experiments include both a control group (which is not subjected to alterations) and one or more experimental groups. When performing experiments, ecologists have a range of types

and scales to choose from, including laboratory studies, small-scale field studies that cover a few square meters, and large-scale field studies in which entire ecosystems, such as lakes or forests, are manipulated (**FIGURE 1.12**).

In some cases, however, it can be difficult or impossible to perform an appropriate experiment. For example, when ecologists are seeking to understand events that cover large geographic regions or occur over long periods, experiments can provide useful information, but they cannot provide convincing answers to the underlying questions of interest. As an example, let's consider global warming.

CLIMATE CHANGE CONNECTION

APPROACHES USED TO STUDY GLOBAL WARMING As we will see in Figure 25.11, Earth's climate is warming as a result of the emission of greenhouse gases, but the future magnitude and effects of global warming remain uncertain. We are not sure, for example, how the geographic ranges of different species will change as a result of the projected temperature increases. There is only one Earth, so of course even if we wanted to, we could not apply different levels of global warming to copies of the planet and then observe how the ranges of species change over time in each of our experimental treatments.

Instead, we must approach such problems using a mixture of observational studies, experiments, and modeling approaches. Field observations reveal that many species have shifted their ranges poleward or up the sides of mountains in a manner that is consistent with the amount of global warming that has already occurred (Lenoir and Svenning 2015). Field observations can also be used to summarize the environmental conditions under which species are currently found, and experiments can be used to examine the performance of species under different environmental conditions. To put all this information together, scientists can use results from observational studies and experiments to develop quantitative models that predict how the geographic ranges of species will change depending on how much the planet actually warms in the future.

The observation that global warming has already altered the geographic ranges of species brings us to a topic addressed in many chapters of this book: **climate change**. This term refers to a directional change in climate (such as warming and/or precipitation) that occurs over three decades or longer. As you'll read in later chapters, climate affects nearly all aspects of ecology, such as the growth and survival of individuals, interactions between members of different species, the relative abundances of species in ecological communities, and the functioning of ecosystems. These observations suggest that changes to climate have far-reaching effects, as shown by the changes that have already occurred in the physiology, survival, reproduction, or geographic ranges of hundreds of species (Parmesan 2006). ☀

(A)

Courtesy of Tim Cooper and Richard Lenski

(B)

Courtesy of Simone DesRoches and Dolph Schluter

(C)

Courtesy of the U.S. Forest Service

FIGURE 1.12 Ecological Experiments The spatial scale of experiments in ecology ranges from (A) laboratory experiments to (B) small-scale field experiments conducted in natural or artificial environments to (C) large-scale experiments that alter major components of an ecosystem, as seen in this clear-cut watershed.

Experiments are designed and analyzed in consistent ways

When ecologists perform experiments, they often take the three additional steps described in **ECOLOGICAL TOOLKIT 1.1**: they replicate each treatment, they assign treatments at random, and they analyze the results using statistical methods.

Replication means that each treatment, including the control, is performed more than once. An advantage of replication is that as the number of replicates increases, it becomes less likely that the results are due to a variable that was not measured or controlled in the study. Imagine that Kiesecker had performed his field experiment with only two ponds, one with detectable levels of pesticides and the other without. Suppose he had found that frog deformities were more common in the single pond that contained pesticides. While pesticides might have been responsible for this result, the two ponds could have differed in many other ways, too, one or more of which might have been the real cause of the deformities. By using three ponds with pesticides and three ponds without pesticides, Kiesecker made it less likely that each of the three ponds with pesticides also contained something else—some variable not controlled in his experiment—that increased the chance of frog deformities. In his experiment, Kiesecker accounted for the possible effects of some uncontrolled variables: he showed, for example, that the number of snails and the frequency of their infection by *Ribeiroia* were similar in all six ponds, thus making it unlikely that the ponds with pesticides had many more *Ribeiroia* than the ponds without pesticides.

Ecologists also seek to limit the effects of unmeasured variables by assigning treatments at random. Suppose an investigator wanted to test whether insects that eat plants decrease the number of seeds the plants produce. One way to test this idea would be to divide an area into a series of plots (see Ecological Toolkit 1.1), some of which would be sprayed regularly with an insecticide (the experimental plots) while others would be left alone (the control plots). The decision about whether a particular plot would be sprayed (or not) would be made at random at the start of the experiment. Assigning treatments at random would make it less likely that the plots that receive a particular treatment share other characteristics that might influence seed production, such as high or low levels of soil nutrients.

Finally, ecologists use statistical analyses to determine whether their results are "significant." To understand why, let's turn again to Kiesecker's experiment. It would have been surprising if Kiesecker had found that rates of frog deformities in ponds with pesticides were exactly equal to those in ponds without pesticides. But how different would those rates have to be to show that the pesticides are having an effect? Since the results of different experimental treatments will rarely be identical, the investigator must ask whether an observed difference is due to the experimental treatments and not to chance. Statistical methods are often used as a standardized way to help make this decision. There are many different types of statistical analyses; books such as those by Zar (2009), Sokal and Rohlf (2011), and Gotelli and Ellison (2013) provide examples of which statistical methods to use under various circumstances.

ECOLOGICAL TOOLKIT 1.1

Designing Ecological Experiments

A key step in any ecological experiment occurs well before it is performed: the experiment must be designed carefully. In a controlled experiment, an experimental group, which has the factor being tested, is compared with a control group, which does not. Different levels of the factor being tested are often referred to as different treatments. For example, in the experiment by Johnson et al. (1999) discussed earlier in this chapter, the control group received a treatment of 0 parasites per container, while members of the experimental group were assigned to one of three other treatments (16, 32, or 48 parasites per container).

The design of many ecological experiments includes three additional steps: replication, random assignment of treatments, and statistical analyses. Replication and random assignment of treatments are used to reduce the chance that variables not under the control of the experimenter will unduly influence the results of the experiment. Once the experiment has been completed, statistical analyses are used to assess the extent to which the results from the different treatments differ from one another.

Several features of experimental design can be illustrated by the layout used in field studies performed by Richard B. Root, Walter Carson, and colleagues at Cornell University. In one such study, Carson and Root (2000) examined how herbivorous (plant-eating) insects affected a plant community dominated by the goldenrod *Solidago altissima*. Their first step was to define their research question: Does plant abundance, growth, or reproduction differ between insecticide-treated and control plots? To find out, they divided a field of goldenrods into the grid of 5 × 5 m plots shown in **FIGURE A**. The experiment ran for 10 years and used two treatments: a control, in which natural processes were left undisturbed, and an insect removal treatment, in which an insecticide was applied annually to reduce the numbers of herbivorous insects. Carson and Root selected 30 plots at random for use in the experiment; half of those plots were then selected at random to receive the insecticide treatment, while the remaining plots served as controls. Thus, there were 15 replicates for each treatment. Statistical analyses of the results indicated that herbivorous insects had major effects on the plant community, as is also suggested by the photograph in **FIGURE B**.

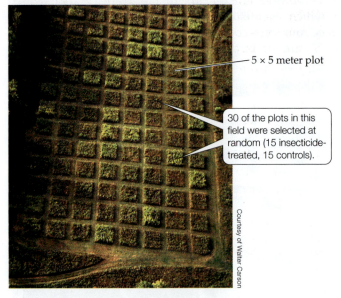

5 × 5 meter plot

30 of the plots in this field were selected at random (15 insecticide-treated, 15 controls).

Courtesy of Walter Carson

FIGURE A Carson and Root's Field Experiment   This aerial photograph shows the field divided (by mowing) into 112 plots, each 5 × 5 m. Thirty of these plots were used in the experiment described here; the rest of the plots were used in other experiments.

Goldenrods (flowering in yellow) are taller and more abundant in this insecticide-treated plot than in the surrounding control plots.

Courtesy of Walter Carson

FIGURE B Carson and Root's Results   A plot sprayed with insecticide (right) is shown surrounded by several control plots.

What we know about ecology is always changing

The information in this book is not a static body of knowledge. Instead, like the natural world itself, our understanding of ecology is constantly changing. Like all scientists, ecologists observe nature and ask questions about how nature works. For example, when the existence of amphibian deformities became widely known in 1995, some scientists set out to answer a series of questions about those deformities. There were many things they wanted to know: How many species were afflicted by deformities? Did amphibian deformities occur in a few or many geographic regions? What caused the deformities, and did these causes differ among species or geographic regions?

The questions stimulated by the discovery of amphibian deformities illustrate the first in a series of four steps by which scientists can learn about the natural world. These four steps constitute the **scientific method**, which can be summarized as follows:

1. Observe nature and ask a well-framed question about those observations.

2. Use previous knowledge or intuition to develop possible answers to that question. In science, such possible explanations of a well-framed question are called **hypotheses**.

3. Evaluate competing hypotheses by performing experiments or gathering carefully selected observations.

4. Use the results of those experiments or observations or of models to modify one or more of the hypotheses, to pose new questions, or to draw conclusions about the natural world.

This four-step process is iterative and self-correcting. New observations lead to new questions, which stimulate ecologists to formulate and test new ideas about how nature works. The results from such tests can lead to new knowledge, still more questions, or the abandonment of ideas that fail to explain the results. Although this four-step process is not followed exactly in all scientific studies, the back-and-forth between observations, questions, and results—potentially leading to a reevaluation of existing ideas—captures the essence of how science is done.

We've already seen some examples of how the process of scientific inquiry works: as answers to some questions about amphibian deformities were found, new questions arose, and new discoveries were made. You can explore one such discovery in **ANALYZING DATA 1.1**, which examines whether introduced species can cause amphibian populations to decline. Indeed, new discoveries occur in all fields of ecology, suggesting that our understanding of ecological processes is, and always will be, a work in progress.

ANALYZING DATA 1.1

Are Introduced Predators a Cause of Amphibian Decline?

Introduced predators are one of many factors thought to have contributed to amphibian population declines, although only a few studies have tested this hypothesis. In one such study, Vance Vredenburg* assessed the effects of two introduced fish species, the rainbow trout (*Oncorhynchus mykiss*) and the brook trout (*Salvelinus fontinalis*), on a frog species in decline, the mountain yellow-legged frog (*Rana muscosa*). Prior to any experimental manipulations, Vredenburg surveyed 39 lakes. For each lake, he noted whether introduced trout were present and then estimated frog abundance; the data from his survey include the following:

Lake status	Average frog density (per 10 m of shoreline)
Trout absent	184.8
Trout present	15.3

Vredenburg then performed experiments in which he compared frog abundances in three categories of lakes: removal lakes (from which he removed introduced trout),

fishless control lakes (that had never contained trout), and fish control lakes (that still contained trout). The data obtained from these experiments appear in the graph. Error bars show one standard error (SE) of the mean.

1. From the survey data in the table, construct a bar graph showing the average density of frogs in lakes with and without trout. What can you conclude from these data? In your answer, distinguish between causation and correlation.

2. Explain why two types of control lakes were used in the experiment.

3. Consider the data for removal lakes 1, 2, and 3. For each of these lakes, calculate (a) the average number of frogs (per 10 m of shoreline) for the 1-year period that ends just before the time frame during which trout were removed and (b) the average number of frogs (per 10 m of shoreline) for the 1-year period that starts a year after the removal of trout began. What can you conclude from these calculations?

4. What do the survey and experimental results suggest about (a) the effect of introduced trout on amphibian populations and (b) prospects for population recovery once trout are removed?

*Vredenburg, V. T. 2004. Reversing introduced species effects: Experimental removal of introduced fish leads to rapid recovery of a declining frog. *Proceedings of the National Academy of Sciences U.S.A.* 101: 7646–7650. © 2004 National Academy of Sciences, U.S.A.

(Continued)

ANALYZING DATA 1.1 (continued)

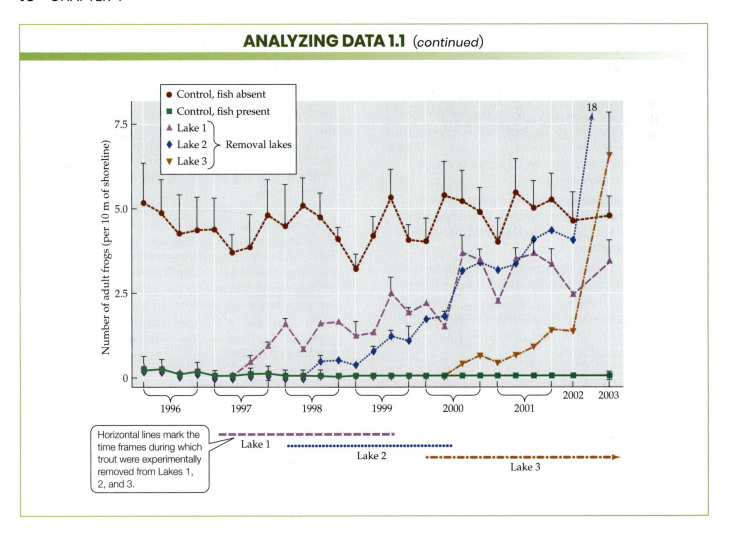

Horizontal lines mark the time frames during which trout were experimentally removed from Lakes 1, 2, and 3.

Deformity and Decline in Amphibian Populations

As we've seen in this chapter, amphibian deformities are often caused by parasites, but they can also be influenced by other factors, such as exposure to pesticides or fertilizers. Studies have also suggested that a range of factors can cause amphibian abundances to drop. Such factors include habitat loss, parasites and diseases, pollution, climate change, overexploitation, and introduced species.

A consensus has yet to be reached on the relative importance of these and other factors that affect amphibian declines. For example, Stuart et al. (2004) analyzed the results of studies on 435 amphibian species that have experienced rapid declines since 1980. Habitat loss was the primary cause of decline for the largest number of species (183 species), followed by overexploitation (50 species). The cause of decline for 207 species was listed as "enigmatic": populations of these species were declining rapidly for reasons that were poorly understood. Skerratt et al. (2007) argued that many such enigmatic declines were caused by pathogens such as the chytrid *Batrachochytrium dendrobatidis*, a fungus that causes a lethal skin disease. This conclusion has now been supported by many other studies (e.g., Voyles et al. 2009; Berger et al. 2016). Although the fungus continues to spread rapidly and has driven hundreds of amphibian populations to extinction, there are signs of hope. For example, McMahon et al. (2014) have shown that some amphibians can acquire resistance to *B. dendrobatidis* when exposed to live or dead fungus, while others have found evidence of resistance in wild populations (Eskew et al. 2015; Voyles et al. 2018).

Other researchers have emphasized the importance of ongoing climate change. Hof et al. (2011a), for example, project that by 2080, climate change will harm more amphibian species than will *B. dendrobatidis*. The impacts of factors such as disease and climate change are not mutually exclusive, however. Indeed, Rohr and Raffel (2010) found that while disease often led to amphibian declines, climate change also played a key role. In particular, the impact of increased temperature variability appears to have decreased the resistance of frogs to *B. dendrobatidis* (Raffel et al. 2012).

Collectively, these and other studies of amphibian population declines suggest that no single factor can explain most of them. Instead, the declines seem to be caused by complex factors that often act together and may vary from place to place. Consider, for example, the effects of pesticides. Although pesticides appear to increase the incidence of frog deformities, many studies have failed to directly link pesticides to decreases in the size of amphibian populations. However, many of these negative findings came from laboratory studies that held other factors constant and examined the effect of pesticides alone on amphibian growth or survival. Rick Relyea, of Rensselaer Polytechnic Institute, repeated such experiments, but with an added twist: predators. In two of six amphibian species studied, pesticides became up

to 46 times more lethal if tadpoles sensed the presence of a predator (Relyea 2003). The predators were kept separate from the tadpoles by netting, but the tadpoles could smell them.

In Relyea's experiments, the ability of some tadpoles to cope with pesticides was reduced by stress caused by the presence of a predator. The mechanism by which these two factors act together is unknown. In general, although we know that a broad set of factors can cause frog deformities and declines (**FIGURE 1.13**), relatively little is known about the extent to which these factors interact or how any such interactions exert their effects. In this and many other areas of ecology, we have learned enough to solve parts of the mystery, yet more remains to be discovered.

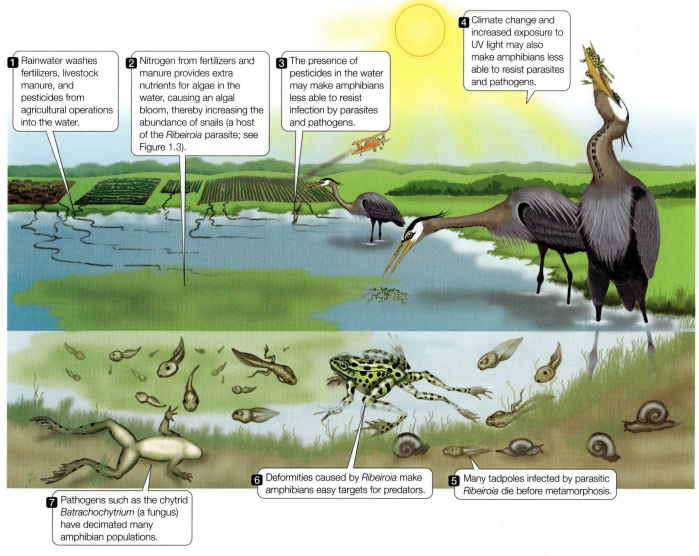

1 Rainwater washes fertilizers, livestock manure, and pesticides from agricultural operations into the water.

2 Nitrogen from fertilizers and manure provides extra nutrients for algae in the water, causing an algal bloom, thereby increasing the abundance of snails (a host of the *Ribeiroia* parasite; see Figure 1.3).

3 The presence of pesticides in the water may make amphibians less able to resist infection by parasites and pathogens.

4 Climate change and increased exposure to UV light may also make amphibians less able to resist parasites and pathogens.

6 Deformities caused by *Ribeiroia* make amphibians easy targets for predators.

5 Many tadpoles infected by parasitic *Ribeiroia* die before metamorphosis.

7 Pathogens such as the chytrid *Batrachochytrium* (a fungus) have decimated many amphibian populations.

FIGURE 1.13 Complex Causation of Amphibian Deformities and Declines As we have seen, amphibian deformities can be caused by parasites such as *Ribeiroia*. However, other factors—many of them a result of human actions—may interact to cause amphibian deformities and declines. (After A. R. Blaustein and P. T. J. Johnson. 2003. *Sci Am* 288: 60–65.)

CONNECTIONS *in* NATURE

MISSION IMPOSSIBLE?   As we emphasized in the opening pages of this chapter, people have begun to realize that it is important for us to understand how nature works, if only to protect ourselves from inadvertently changing our environment in ways that cause us harm. Does the fact that the natural world is vast, complex, and interconnected mean that it is impossible to understand? Most ecologists do not think so. Our understanding of natural systems has improved greatly over the last 100 years. Ongoing efforts to understand how nature works are sure to be challenging, but such efforts are also enormously exciting and important. What we learn, and how we use that knowledge, will have a great impact on the current and future well-being of human societies. Whatever your career path, we hope this book will help you understand the natural world in which you live, as well as how you affect—and are affected by—that world.

SUMMARY

CONCEPT 1.1 Events in the natural world are interconnected.

1.1.1 Explain how interactions between organisms and their environment can affect other organisms and potentially lead to unexpected consequences.

- Laboratory and field experiments on the effects of parasites on amphibian deformities illustrate how events in nature can be connected with one another.

- Because events in the natural world are interconnected, any action can have unanticipated side effects.

- People both depend on and affect the natural environment.

CONCEPT 1.2 Ecology is the scientific study of interactions between organisms and their environment.

1.2.1 Summarize how the inquiries of ecologists and environmental scientists differ.

- Ecology is a scientific discipline that is related to, but differs from, disciplines such as environmental science.

- Public and professional ideas about ecology often differ.

1.2.2 Outline how ecologists use spatial and temporal scales when testing their hypotheses.

- Ecology is broad in scope and encompasses studies at many levels of biological organization.

- All ecological studies address events on some spatial and temporal scales while ignoring events at other scales.

CONCEPT 1.3 Ecologists evaluate competing hypotheses about natural systems with observations, experiments, and models.

1.3.1 Compare the advantages and disadvantages of using field observations, field experiments, and lab experiments to test ecological hypotheses.

- In an ecological experiment, an investigator alters one or more features of the environment and observes the effect of that change on natural processes.

- Some features of the natural world are best investigated with a combination of field observations, experiments, and quantitative models.

1.3.2 Describe the importance of hypotheses, controls, replication, and data analysis to the scientific process.

- Experiments are designed and analyzed in consistent ways: typically, each treatment, including the control, is replicated; treatments are assigned at random; and statistical methods are used to analyze the results.

Review Questions

1. Describe what the phrase *connections in nature* means, and explain how such connections can lead to unanticipated side effects. Illustrate your points with an example discussed in the chapter.

2. What is ecology, and what do ecologists study? If an ecologist studied the effects of a particular gene, how might the emphasis of that researcher's work differ from the emphasis of the work of a geneticist or cell biologist.

3. How does the scientific method work? Include in your answer a description of a controlled experiment.

Hone Your Problem-Solving Skills

Twenty tadpoles of a certain frog species, 15 adult snails (collected from ponds containing deformed frogs), and identical amounts of phytoplankton (small algae that grow suspended in water) and attached algae (that grow on rocks) were added to each of ten freshwater tanks. Atrazine (an herbicide that is often found in ponds) was added to five of these tanks; no atrazine was added to the other five tanks. After 6 months, the mean values shown in the table were determined for phytoplankton abundance, attached algae abundance, water clarity (measured on a scale of 1–5, where 1 is most cloudy and 5 is mostly clear), number of tadpole eosinophils (a type of white blood cell used in the tadpole's immune response), tadpole survival, and number of cysts of the trematode parasite *Ribeiroia*.

	Treatment	
Variable	**Atrazine added**	**No atrazine**
Phytoplankton (abundance index)	2.1	6.4
Attached algae (abundance index)	1.4	0.9
Water clarity (scale 1–5)	4.9	3.8
Eosinophils (number/ml blood)	1.2	2.7
Tadpole survival (%)	45.0	72.0
Ribeiroia cysts (number/tadpole)	28.6	6.9

1. Identify the control used in this experiment. How would investigators use the controls to help them interpret experimental results?

2. Describe the results for phytoplankton abundance, attached algae abundance, and water clarity. Interpret these results.

3. Describe the results for eosinophil number, tadpole survival, and number of *Ribeiroia* cysts. Interpret these results.

4. Based on the results in the table and the *Ribeiroia* life cycle (see Figure 1.3), describe a chain of events that could affect a frog population after atrazine was added to a pond.

ADDITIONAL RESOURCES

This textbook provides resources such as **Web Extensions** and additional **Climate Change Connections** for instructors who can share them with students. Please contact your instructor, if you would like to access that content.

Organisms and Their Environment UNIT 1

2 The Physical Environment

KEY CONCEPTS

CONCEPT 2.1 Climate is the most fundamental component of the physical environment.

CONCEPT 2.2 Winds and ocean currents result from differences in solar radiation across Earth's surface.

CONCEPT 2.3 Large-scale atmospheric and oceanic circulation patterns establish global patterns of temperature and precipitation.

CONCEPT 2.4 Regional climates reflect the influence of oceans and continents, mountains, and vegetation.

CONCEPT 2.5 Seasonal and decadal climate variation are associated with changes in Earth's position relative to the sun and the strength of atmospheric pressure cells.

CONCEPT 2.6 Salinity, acidity, and oxygen concentrations are major determinants of the chemical environment.

Climate Variation and Salmon Abundance: A Case Study

Grizzly bears of the Pacific Northwest feast seasonally on the salmon that arrive in huge numbers to reproduce in the streams of the region (**FIGURE 2.1**). Salmon are *anadromous;* that is, they are born in freshwater streams, spend their adult lives in the ocean, and then return to spawn in the freshwater habitats where they were born. Grizzlies capitalize on the salmon's reproductive habits, gorging themselves on this rich food resource. These normally aggressive bears will forgo their usual territorial behavior and tolerate high densities of other bears while fishing for salmon.

Bears are not the only species that rely on salmon for food. Salmon have been an important part of the human economy of the Pacific Northwest for millennia. The fish were a staple of the diets of Native Americans in this region as well as a central part of their cultural and spiritual lives. Salmon are now fished commercially in the waters of the North Pacific Ocean, providing a $3 billion economic base for coastal communities across the North Pacific. Commercial salmon fishing is a risky venture, however. Successful reproduction for salmon depends on the health of the streams in which they spawn. The construction of dams, increased stream sediments due to forest clear-cutting, water pollution, and overharvesting have all been blamed for declines in salmon populations, primarily from the California coast northward to British Columbia (Pacific Fishery Management Council 2008). Despite efforts to mitigate this environmental degradation, the recovery of salmon stocks has been marginal at best in the southern portion of the region.

Researchers, environmental advocates, and government policy experts have focused primarily on the deterioration of freshwater habitat as a cause for the declines in salmon. In 1994, however, Steven Hare and Robert Francis at the University of Washington suggested that changes in the marine environment, where salmon spend the majority of their adult lives, could be contributing to the declines in salmon abundance. In particular, they noted that records of fish harvests covering more than a century indicated that multi-decadal periods of low or high fish production have occurred repeatedly, separated by abrupt changes in production rather than gradual transitions (**FIGURE 2.2**). In addition, Nathan Mantua and colleagues (1997) noted that periods of high salmon production in Alaska corresponded with periods of low salmon production at the southern end of the salmon range, particularly in Oregon and Washington. They found telling quotes in commercial fishing publications that told the same story: when the fishing was poor in Washington and Oregon, it was good in Alaska, and vice versa.

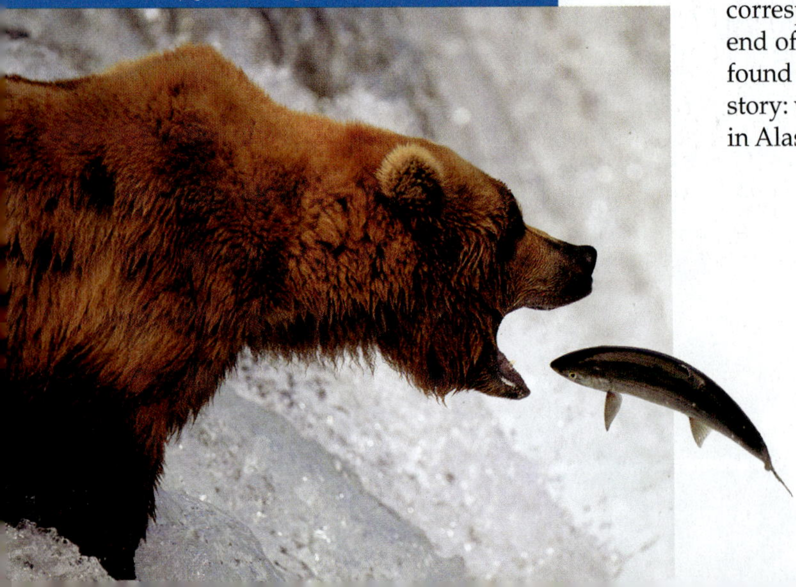

FIGURE 2.1 A Seasonal Opportunity Grizzly bears feed on salmon migrating upstream in streams and rivers in Alaska to reproduce. The size of the salmon run each year depends in part on physical conditions in the Pacific Ocean, many miles away.

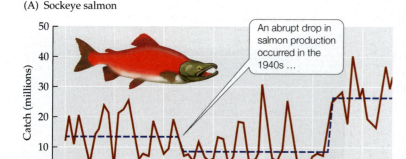

(A) Sockeye salmon

An abrupt drop in salmon production occurred in the 1940s …

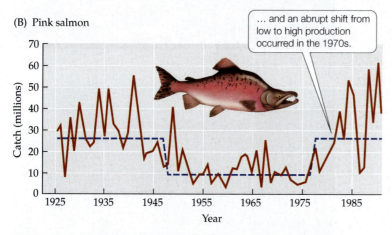

(B) Pink salmon

… and an abrupt shift from low to high production occurred in the 1970s.

FIGURE 2.2 Changes in Salmon Harvests over Time Records of commercial harvests of (A) sockeye salmon and (B) pink salmon in Alaska over 65 years show abrupt drops and increases in production. Solid lines represent annual catch; dashed lines are a statistical fit to the data. (After S. R. Hare and R. C. Francis. 1994. In *Climate Change and Northern Fish Populations. Can Spec Publ Fish Aquat Sci* 121. R. J. Beamish [Ed.], pp. 357–372. National Research Council of Canada: Ottawa. © Canadian Science Publishing or its licensors.)

From *Pacific Fisherman* (September 1915):

> *"Never before have the Bristol Bay* [Alaska] *salmon packers returned to port after the season's operations so early."* [That is, it was a bad year, with few fish to catch.]

> *"The spring* [Chinook salmon] *fishing season on the Columbia River* [Washington and Oregon] *closed at noon on August 25, and proved to be one of the best for some years."*

From *Pacific Fisherman* (1939 Yearbook):

> *"The Bristol Bay Red* [Alaska sockeye salmon] *run was regarded as the greatest in history."*

> *"The* [Chinook] *catch this year is one of the lowest in the history of the Columbia* [Washington]."

Hare and Francis hypothesized that the abrupt shifts in salmon production were associated with long-term climate variation in the North Pacific. The nature and cause(s) of these underlying climate shifts, however, were unclear. Additional work by Mantua and colleagues found good correspondence between the multi-decadal shifts in salmon production and changes in sea surface temperatures in the North Pacific.

How widespread is this variation in climate and its effects on salmon and the associated marine ecosystem? As we will see at the end of this chapter, the research on variation in salmon production led to the discovery of an important long-term cyclic climate pattern that affects a large area.

Introduction

The physical environment is the ultimate determinant of where organisms can live, the resources that are available to them, and the rate at which their populations can grow. Therefore, an understanding of the physical environment is key to understanding all ecological phenomena, from the outcome of interactions between bacteria and fungi in the soil, to the exchange of carbon dioxide between the biosphere and the atmosphere.

The physical environment includes climate, which consists of long-term trends in temperature, wind, and precipitation. Radiation from the sun ultimately drives the climate system, as well as biological energy production. Another aspect of the physical environment is the chemical composition of air and water, which includes salinity (concentrations of dissolved salts), acidity, and concentrations of gases in the atmosphere and dissolved in water. Soil is an important component of the physical environment because it is a medium in which microorganisms, plants, and animals live. Soil also influences the availability of critical resources, particularly water and nutrients. This chapter will focus on climate and the chemical environment; we will cover soil development and nutrient supply in Chapter 22.

This chapter will provide a framework for characterizing the physical environment, including its variability, at a variety of spatial and temporal scales. We will begin by exploring the processes that create the climate patterns we see at global to regional scales.

CONCEPT 2.1 Climate is the most fundamental component of the physical environment.

LEARNING OBJECTIVES

2.1.1 Outline the difference between weather and climate, with specific reference to their temporal scales.

2.1.2 Explain the importance of weather variability for ecological processes.

2.1.3 Summarize how temperature is determined by the gains and losses of energy at Earth's surface.

Climate

Each day we experience the **weather** around us: the current temperature, humidity, precipitation, wind, and cloud cover. Weather is an important determinant of our behavior: what we wear, the activities we engage in, and our mode of transportation. **Climate** is the long-term description of weather at a given location, based on averages and variation measured over decades. Climate variation includes the daily and seasonal cycles associated with changes in solar radiation as Earth rotates on its axis and orbits the sun. Climate variation also includes changes over years or decades, such as large-scale cyclic weather patterns related to changes in the atmosphere and oceans (El Niño Southern Oscillation, discussed later in this chapter, is one example). Longer-term climate change occurs as a result of changes in the intensity and distribution of solar radiation reaching Earth's surface, as well as changes in the overall energy balance. Earth's climate is currently changing because of increases in concentrations of gases such as carbon dioxide, methane, and nitrous oxide that are emitted into the atmosphere as a result of human activities. These gases absorb energy and radiate it back to the surface, creating a **greenhouse effect**.

Climate controls where and how organisms live

Where organisms live, their geographic distribution, and how they function are determined by climate. Temperature regulates the rates of biochemical reactions and physiological activity for all organisms. Water supplied by precipitation is an essential resource for terrestrial organisms. Freshwater organisms are dependent on precipitation for the maintenance and quality of their habitats. Marine organisms depend on ocean currents that influence the temperature and chemistry of the waters they live in.

We usually characterize climate—or any aspect of the physical environment—at a given location by the average conditions. However, the geographic distributions of organisms are influenced by *extreme* conditions more than average conditions because extreme events are important determinants of mortality. Temperature and moisture extremes can affect even long-lived organisms such as forest trees. For example, record high temperatures, along with a severe drought from 2000 to 2003, contributed to widespread mortality in large stands of piñon pines (*Pinus edulis*) in the southwestern United States (Breshears et al. 2005) (**FIGURE 2.3**). These long-lived plants could no longer survive in the region where they had existed for centuries. Thus, the physical environment must also be characterized by its *variability* over time, not just by average conditions, if we are to understand its ecological importance. The frequency and severity of extreme climate events have increased in association with global climate change (Seneviratne et al. 2021; Jentsch et al. 2007). Climate change has increased the probability of large-scale mortality of vegetation such as the die-off in piñon pines, as well as extreme fires and floods.

(A)

(B)

Courtesy of Craig Allen

FIGURE 2.3 **Widespread Mortality in Piñon Pines** Extreme high temperatures and a historic drought from 2000 to 2003 killed large areas of piñon pines (*Pinus edulis*) throughout the southwestern United States. (A) Here, stands in the Jemez Mountains, New Mexico, begin to show substantial needle death due to water and temperature stress, combined with a bark beetle outbreak in October 2002. (B) By May 2004, most of the trees had died.

The *timing* of changes in the physical environment is also ecologically important. The seasonality of rainfall, for example, is important in determining the availability of water for terrestrial organisms. In regions with a "Mediterranean-type" climate, the majority of precipitation falls in winter. Although these regions receive more precipitation than most desert areas, they experience regular dry periods during summer. Lack of water during summer limits the potential growth of plants and promotes fires. In contrast, some grasslands have the same *average annual temperature* (the average temperature measured over an entire year) and precipitation as these Mediterranean-type ecosystems, but precipitation during the summer is higher.

Climate also influences the rates of abiotic processes that affect organisms. The rate at which rocks and soil are broken down to supply nutrients to plants and microorganisms, for example, is determined by climate. Climate can also influence the rates of periodic *disturbances*, such as fires, floods, and avalanches. These events kill organisms and disrupt biological communities, but they subsequently create opportunities for the establishment and growth of new organisms and communities. As noted above, climate change is increasing the frequency and intensity of disturbances.

Global energy balance drives the climate system

The energy that drives the global climate system is ultimately derived from solar radiation. On average, the top of Earth's atmosphere receives 342 watts (W) of solar radiation per square meter each year. About a third of this solar radiation is reflected back out of the atmosphere by clouds, fine atmospheric particles called *aerosols*, and Earth's surface. Another fifth of the incoming solar radiation is absorbed by ozone, clouds, and water vapor in the atmosphere. The remaining half is absorbed by land and water at Earth's surface (**FIGURE 2.4**).

If Earth's temperature is to remain the same, these energy gains from solar radiation must be balanced by energy losses. Much of the solar radiation absorbed by Earth's surface is emitted to the atmosphere as infrared radiation (also known as *longwave* radiation). Earth's surface also

loses energy and is cooled when water evaporates, because the change in phase from liquid water to water vapor absorbs energy. Heat loss due to evaporation is known as **latent heat flux**. Energy is also transferred through the exchange of kinetic energy by molecules in direct contact with one another (**conduction**) and by the movement of currents of air (wind) and water (**convection**). Energy transfer from the warm air immediately above Earth's surface to the cooler atmosphere by convection and conduction is known as **sensible heat flux**.

The atmosphere absorbs much of the infrared radiation emitted from Earth's surface (and from clouds) and reradiates it back to Earth's surface. This reradiation represents a major energy gain. The atmosphere contains gases that absorb and reradiate infrared radiation, known as **greenhouse gases**. These gases include water vapor (H_2O), carbon dioxide (CO_2), methane (CH_4), and nitrous oxide (N_2O). Some of these greenhouse gases are produced through biological activity (e.g., CO_2, CH_4, N_2O), linking the biosphere to the climate system. Without these greenhouse gases, Earth's climate would be considerably cooler than it is (by approximately 33°C, or 59°F). As noted earlier, increases in atmospheric concentrations of greenhouse gases due to human activities are altering Earth's energy balance, changing the climate system, and causing global climate change (**FIGURE 2.5**; see Concept 25.2).

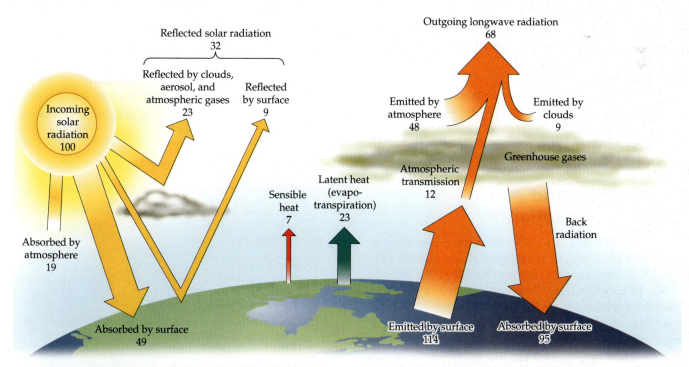

FIGURE 2.4 Earth's Energy Balance Average annual energy balance for Earth's surface and atmosphere, including gains from solar radiation and gains and losses due to emission of infrared radiation, latent heat flux, and sensible heat flux. The numbers are gains and losses of energy, given as percentages of the average annual incoming solar radiation at the top of Earth's atmosphere (342 W/m²). (After J. T. Kiehl and K. E. Trenberth. 1997. *Bull Am Meteorol Soc* 78: 197–208. © American Meteorological Society. Used with permission.)

? What component of Earth's energy balance would be influenced by an increase in greenhouse gases? What would the effect on Earth's energy balance be if there were an increase in atmospheric aerosols?

FIGURE 2.5 **Increasing Atmospheric Carbon Dioxide**   The trend in monthly atmospheric carbon dioxide concentrations measured at Mauna Loa Observatory. Average annual carbon dioxide concentrations have risen by 301% since they were first monitored at the Mauna Loa Observatory in 1958 by Charles Keeling. Similar measurements are now made globally by the U.S. National Oceanic and Atmospheric Administration. (After U.S. NOAA, Earth System Research Laboratory, Global Monitoring Division. https://www.esrl.noaa.gov/gmd/ccgg/trends/full.html; C. D. Keeling et al. 2001. *I. Global Aspects*, SIO Reference Series, No. 01-06. Scripps Institution of Oceanography: San Diego, CA. Data last updated August 2019.)

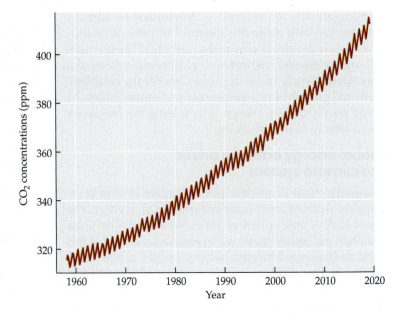

Our discussion of Earth's energy balance has focused on average annual transfers of energy to and from Earth as a whole. But not every location on Earth receives the same amount of energy from the sun. Let's consider how these differences in solar radiation affect the circulation of Earth's atmosphere and ocean waters.

> **CONCEPT 2.2**   Winds and ocean currents result from differences in solar radiation across Earth's surface.

LEARNING OBJECTIVES

2.2.1   Draw connections between differential heat gain across Earth's surface and the development of atmospheric circulation cells.

2.2.2   Explain how surface winds and ocean currents move heat between the tropics and the poles.

Atmospheric and Oceanic Circulation

It's hot near the equator and cold at the poles. Why is this true, and how does it relate to global climate patterns? Near the equator, the sun's rays strike Earth's surface perpendicularly. Toward the North and South Poles, the angle of the sun's rays becomes steeper, so the same amount of energy is spread over a progressively larger area of Earth's surface (**FIGURE 2.6**). In addition, the amount of atmosphere the rays must pass through increases toward the poles, so more radiation is reflected or absorbed before it reaches the surface. As a result, more solar energy is received per unit of area in the tropics (between latitudes 23.5°N and S) than in regions closer to the poles. This differential input of solar radiation not only establishes latitudinal gradients in temperature, but also is the driving force for climate dynamics such as warm and

cold fronts and large storms (e.g., hurricanes). In addition, the movement of Earth around the sun, in combination with the tilt of Earth's axis of rotation, results in changes in the amount of solar radiation received at any location over the course of the year, as we'll see in Concept 2.5. These changes are the cause of seasonal climate variation: winter–spring–summer–fall changes at high latitudes and wet–dry shifts in tropical regions.

Atmospheric circulation cells are established in regular latitudinal patterns

A surface warmed by the sun emits infrared radiation and warms the air above it. As we have just seen, the heating of Earth's surface varies with latitude, and it can also vary with topography. Such differential warming creates

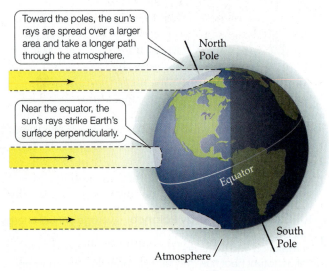

FIGURE 2.6 **Latitudinal Differences in Solar Radiation at Earth's Surface**   The angle of the sun's rays affects the intensity of the solar radiation that strikes Earth's surface.

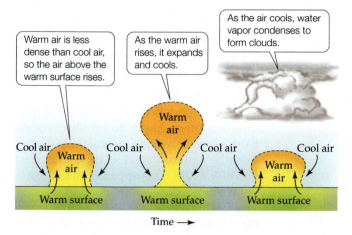

FIGURE 2.7 Surface Heating and Uplift Differential solar heating of Earth's surface leads to the uplift of pockets of air over the warmest surfaces.

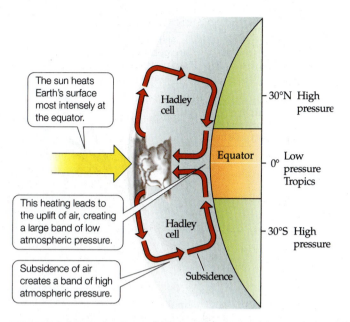

FIGURE 2.8 Tropical Heating and Atmospheric Circulation Cells The heating of Earth's surface in the tropics causes air to rise and release precipitation.

pockets of warm air surrounded by cooler air. Warm air is less dense (has fewer molecules per unit of volume) than cool air, so as long as a pocket of air remains warmer than the surrounding air, it will rise (a process called **uplift**; **FIGURE 2.7**). **Atmospheric pressure** is the force exerted by air's molecules on the air and surface below it. This pressure decreases with increasing altitude, so as a pocket of warm air rises, it expands. This expansion cools the rising air. Cool air cannot hold as much water vapor as warm air, so as the air continues to rise and cool, the water vapor contained within it begins to condense into droplets and form clouds.

The condensation of water into clouds is a warming process (another form of latent heat flux), which may act to keep the pocket of air warmer than the surrounding atmosphere and enhance its uplift, despite its cooling due to expansion. You may have observed this process on a warm summer day when bubble-shaped cumulus clouds formed thunderstorms. When there is substantial heating of Earth's surface and a progressively cooler atmosphere above the surface, the uplifted air will form clouds with wedge-shaped tops. The clouds reach to the boundary between the *troposphere*, the atmospheric layer above Earth's surface, and the *stratosphere*, the next atmospheric layer above the troposphere. This boundary is marked by a transition from progressively cooler temperatures in the troposphere to warmer temperatures in the stratosphere. Thus, the air pocket ceases to rise once it reaches the warmer temperatures at the boundary of the stratosphere.

Differential heating and storm formation explain why the tropics receive the most precipitation of any area on Earth. The tropics receive the most solar radiation and thus experience the greatest amount of surface heating, uplift of air, and cloud formation. The uplift of air in the tropics creates a band of low atmospheric pressure relative to zones to the north and south. When air rising over the tropics reaches the boundary between the troposphere

and stratosphere, it flows toward the poles (**FIGURE 2.8**). Eventually, this poleward-moving air cools as it exchanges heat with the surrounding air and meets cooler air moving from the poles toward the equator. Once the air reaches a temperature similar to that of the surrounding atmosphere, it descends toward Earth's surface, a process known as **subsidence**. Subsidence creates regions of high atmospheric pressure around latitudes 30°N and S, which inhibit the formation of clouds. Thus Earth's major deserts are found at these latitudes.

The tropical uplift of air creates a large-scale pattern of atmospheric circulation in each hemisphere known as a **Hadley cell**, named after George Hadley, the eighteenth-century British meteorologist and physicist who first proposed its existence. Additional atmospheric circulation cells are formed at higher latitudes (**FIGURE 2.9**). The **polar cells**, as the name indicates, occur at the North and South Poles. Cold, dense air subsides at the poles and moves toward the equator when it reaches Earth's surface. The descending air at the poles is replaced by air moving through the upper atmosphere from lower latitudes. Subsidence at the poles creates an area of high pressure, so the polar regions, despite the abundance of ice and snow on the ground, actually receive little precipitation and are known as polar deserts. An intermediate **Ferrell cell** (named after American meteorologist William Ferrell) exists at midlatitudes between the Hadley and polar cells. The Ferrell cell is driven by the movement of the Hadley and polar cells and by exchange of energy between tropical and polar air masses in a region known as the *polar front*.

These three atmospheric circulation cells establish the major climate zones on Earth. Between 30°N and S is

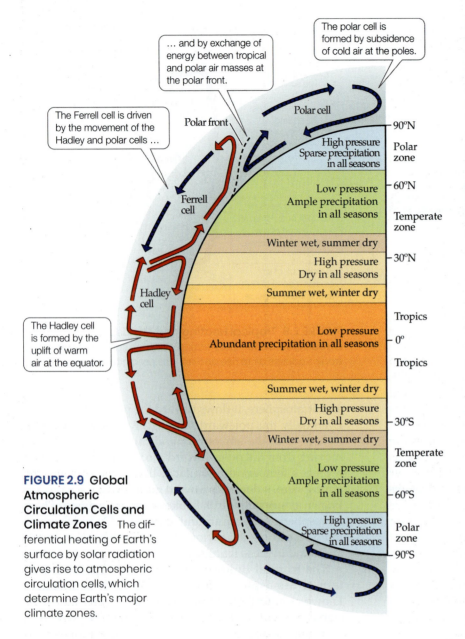

... and by exchange of energy between tropical and polar air masses at the polar front.

The polar cell is formed by subsidence of cold air at the poles.

The Ferrell cell is driven by the movement of the Hadley and polar cells ...

Polar front

Polar cell

90°N

High pressure
Sparse precipitation
in all seasons

Polar zone

Ferrell cell

60°N

Low pressure
Ample precipitation
in all seasons

Temperate zone

The Hadley cell is formed by the uplift of warm air at the equator.

Hadley cell

Winter wet, summer dry

30°N

High pressure
Dry in all seasons

Summer wet, winter dry

Tropics

Low pressure
Abundant precipitation in all seasons

0°

Tropics

Summer wet, winter dry

High pressure
Dry in all seasons

30°S

Winter wet, summer dry

Temperate zone

Low pressure
Ample precipitation
in all seasons

60°S

High pressure
Sparse precipitation
in all seasons

Polar zone

90°S

FIGURE 2.9 Global Atmospheric Circulation Cells and Climate Zones The differential heating of Earth's surface by solar radiation gives rise to atmospheric circulation cells, which determine Earth's major climate zones.

the **tropical zone**, or simply the **tropics**. The **temperate zones** lie between 30° and 60°N and S, and the **polar zones** are above 60°N and S (see Figure 2.9).

Atmospheric circulation cells create surface wind patterns

We've seen how the differential heating of Earth leads to zones of high and low atmospheric pressure. These pressure differences are important in explaining the movement of warm and cold air masses across Earth's surface. Winds flow from areas of high pressure to areas of low pressure. Thus, the areas of high and low pressure formed by atmospheric circulation cells give rise to consistent patterns of air movement at Earth's surface, known as *prevailing winds*. We might expect these winds to blow in straight lines from

high- to low-pressure zones. However, from the standpoint of an observer on Earth, the prevailing winds appear to be deflected to the right (clockwise) in the Northern Hemisphere and to the left (counterclockwise) in the Southern Hemisphere (**FIGURE 2.10A**). The apparent deflection is associated with the rotation of Earth: to an observer on Earth's surface rotating around the planetary axis, the path of the wind appears curved (**FIGURE 2.10B**). This apparent deflection is known as the **Coriolis effect**. To an observer in a fixed position in outer space, however, there is no apparent deflection in the direction of the wind.

As a result of the Coriolis effect, surface winds blowing toward the equator from the high-pressure zones at 30°N and S are deflected to the west from the perspective of Earth's surface. These winds are known as the *trade winds* because of their importance to the global transport of trade goods in sailing ships during the fifteenth through the nineteenth centuries. Winds blowing toward the poles from those zones of high pressure, called *westerlies*, are deflected to the east. The presence of continental land masses interspersed with oceans complicates this idealized depiction of prevailing wind patterns (**FIGURE 2.11**).

Water has a higher **heat capacity** than land, so it absorbs and stores more energy with less temperature change than land. For this reason, the land surface warms up more than ocean water in summer, but in winter the oceans retain more heat, and thus remain warmer, than land at the same latitude. As a result, seasonal air temperature changes are less extreme over the oceans than they are on land. In summer, air over the oceans is cooler and denser than that over land, and semipermanent zones of high pressure (*high-pressure cells*) form over the oceans, particularly around 30°N and S. In winter, the opposite situation exists: the air over the continents is cooler and denser than that over the oceans, so high-pressure cells develop in the temperate zones over large continental areas. Because winds blow from areas of high pressure to areas of low pressure, these seasonal shifts in pressure cells influence the direction of the prevailing winds. The effect of land areas on the development of these semipermanent pressure cells is more pronounced in the Northern Hemisphere than in the Southern Hemisphere because continental land masses make up a larger proportion of Earth's surface in the Northern Hemisphere.

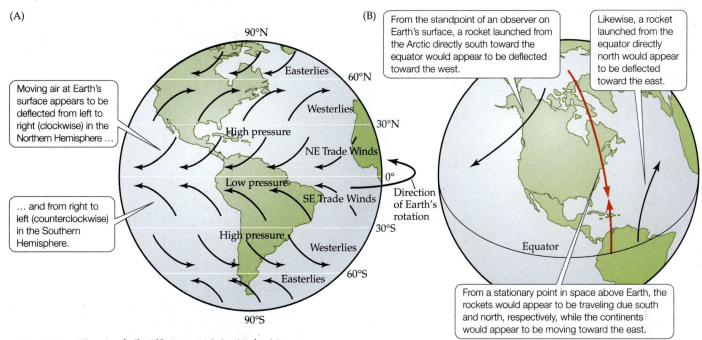

FIGURE 2.10 The Coriolis Effect on Global Wind Patterns
(A) The Coriolis effect results from Earth's rotation. (B) Visualization of the Coriolis effect using rockets.

Ocean currents are driven by surface winds

Wind moving across the ocean surface pushes the surface water. As a result of the Coriolis effect, the water appears to move at an angle to the wind. From the perspective of an observer on Earth, it is deflected to the right in the Northern Hemisphere and to the left in the Southern Hemisphere. For this reason, the pattern of ocean surface currents is similar to, but not identical to, the pattern of prevailing winds. The speed of ocean currents is usually only about 2%–3% of the wind speed. An average wind speed of 10 m/s (22 miles per hour) would therefore produce an ocean current moving at 30 cm/s (0.7 miles per hour). In the North Atlantic Ocean, current velocities may be as high as 200 cm/s (4.5 miles per hour).

Like air in the atmosphere, water in the ocean can move vertically as well as horizontally. Generally, the surface and deep layers of ocean water do not mix, because of differences in their temperature and salinity (concentration of dissolved salts). The surface waters—those above 75–200 m (250–600 feet)—are warmer and less saline, and therefore less dense, than the deeper, cooler ocean waters. When warm tropical surface currents reach polar regions, particularly the coasts of Antarctica and Greenland, their water loses heat to the surrounding environment and becomes cooler and denser. The water eventually cools enough for ice to form, which increases the salinity of the remaining unfrozen water. This combination of cooling and increasing salinity increases the density of the water, which sinks to deeper layers. The dense **downwelling** currents that result move

toward the equator, carrying cold polar water toward the warmer tropical oceans.

These deep ocean currents connect with surface currents again at zones of **upwelling**, where deep ocean water rises to the surface. Upwelling occurs where prevailing winds blow nearly parallel to a coastline, such as off the western coasts of North and South America. The force of the wind, in combination with the Coriolis effect, causes surface waters to flow away from the coast (**FIGURE 2.12**), and deeper, colder waters rise to replace them. Upwelling also occurs in the westward-flowing equatorial Pacific Ocean. As a result of the Coriolis effect, water just to the north and south of the equator is deflected slightly away from the equator, causing divergence of surface water and a zone of upwelling.

Upwelling has important consequences for the local climate, creating a cooler, moister environment. Upwelling also has a strong effect on biological activity in the surface waters. When organisms in the surface waters die, their bodies—and the nutrients they contain—sink. Thus, nutrients tend to accumulate in deep water and in sediments at the ocean bottom. Upwelling brings these nutrients back to the *photic zone*, the layer of surface water where there is enough light to support photosynthesis. Upwelling zones are among the most productive open ocean ecosystems because these nutrients increase the growth of *phytoplankton* (small, free-floating algae and other photosynthetic organisms), which provide food for *zooplankton* (free-floating animals and protists), which in turn support the growth of their consumers such as fish.

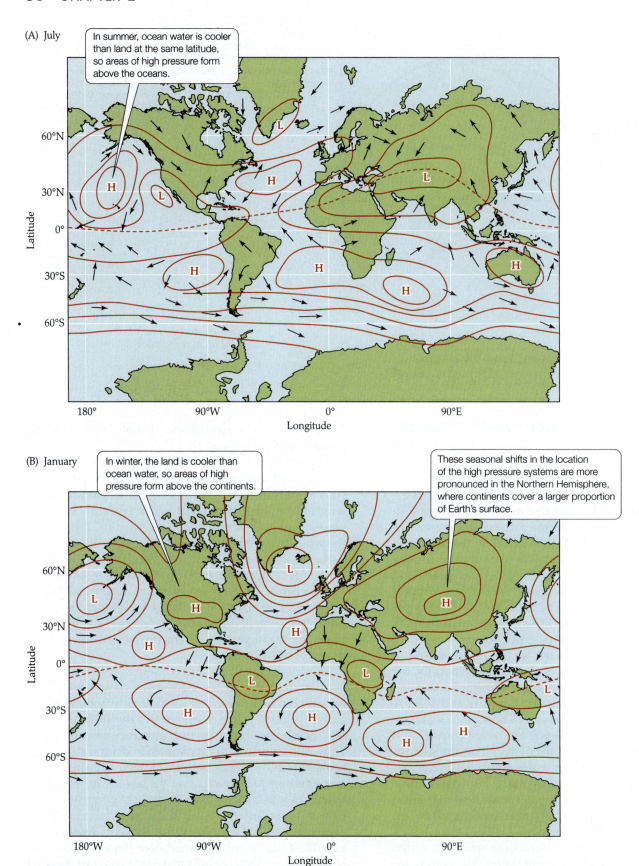

FIGURE 2.11 Prevailing Wind Patterns The difference in heat capacity between the oceans and the continents leads to seasonal changes in atmospheric pressure cells that influence prevailing wind patterns.

(A)

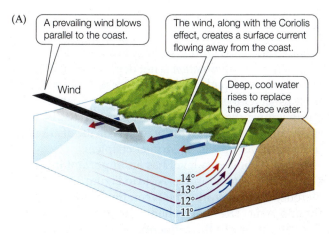

A prevailing wind blows parallel to the coast.

The wind, along with the Coriolis effect, creates a surface current flowing away from the coast.

Wind

Deep, cool water rises to replace the surface water.

14°
13°
12°
11°

(B)

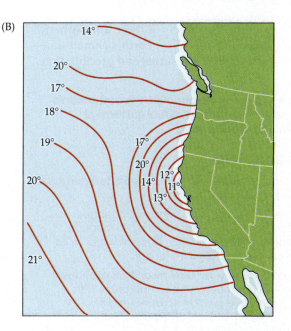

FIGURE 2.12 Upwelling of Coastal Waters (A) Wind blowing parallel to the coast causes surface water to flow away from the coast, pulling deep water upward to replace it. (B) Upwelling influences surface water temperatures off the west coast of North America. Ocean temperatures are shown in °C.

Ocean currents influence the climates of the regions where they flow. For example, the Gulf Stream and North Atlantic Drift, a current system that flows from the tropical Atlantic northward to the North Atlantic contributes to warmer winters in Scandinavia than in locations at the same latitude in North America. In addition, winds blowing eastward across the Atlantic pick up heat from the ocean, which also contributes to a warmer climate in northern Europe. Winter temperatures on the west coast of Scandinavia are approximately 15°C (22°F) warmer than those on the coast of Labrador. This temperature difference is reflected in the vegetation: deciduous forests are common on the Scandinavian coast, while boreal forests of spruce and pine dominate the coast of Labrador. The

Gulf Stream also keeps the North Atlantic ice-free most of the winter, whereas sea ice forms at the same latitude off the North American coast.

Ocean currents are responsible for about 40% of the heat exchanged between the tropics and the polar regions. Thus, ocean currents are sometimes referred to as the "heat pumps" or "thermal conveyers" of the planet. A large system of interconnected surface and deep ocean currents that links the Pacific, Indian, and Atlantic Oceans, sometimes called the great ocean conveyor belt, is an important means of transferring heat to the polar regions (**FIGURE 2.13**).

Now that we have seen how the differential heating of Earth's surface generates prevailing winds and ocean

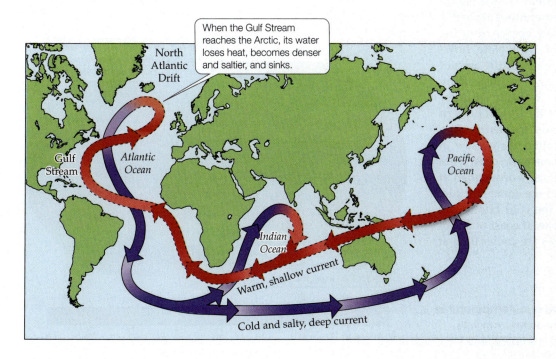

When the Gulf Stream reaches the Arctic, its water loses heat, becomes denser and saltier, and sinks.

North Atlantic Drift

Gulf Stream

Atlantic Ocean

Pacific Ocean

Indian Ocean

Warm, shallow current

Cold and salty, deep current

FIGURE 2.13 The Great Ocean Conveyor Belt An interconnected system of surface and deep ocean currents transfers energy between tropical and polar regions. The red arrows with dashed outlines represent shallow currents, and the blue arrows with solid outlines represent deeper currents. (After Hugo Ahlenius, UNEP/GRID-Arendal. 2007. http://maps.grida.no/go/graphic/world-ocean-thermohaline-circulation1.)

currents, let's examine the effects of these atmospheric and oceanic circulation patterns on Earth's climates, including global patterns of temperature and precipitation.

> **CONCEPT 2.3** Large-scale atmospheric and oceanic circulation patterns establish global patterns of temperature and precipitation.

LEARNING OBJECTIVES

2.3.1 Outline the determinants of global temperature and precipitation patterns.

2.3.2 Explain how a region's seasonal changes in temperature are affected by its location, whether it is near a large body of water or at the center of a large continent.

2.3.3 Summarize how air density and air exchange cause a decrease in air temperature with increases in elevation on a mountain.

Global Climate Patterns

Earth's climates reflect a variety of temperature and precipitation regimes, from the warm, wet climate of the tropics to the cold, dry climate of the Arctic and Antarctic. Ultimately this variation in climate determines the location of the major biological zones (Concept 3.1). In this section, we examine these global patterns of temperature and precipitation and explore how both climate averages and climate variation are influenced by prevailing winds and ocean currents.

Oceanic circulation and the distribution and topography of continents influence global temperatures

The global pattern of solar radiation (see Figure 2.6) largely explains why temperatures at Earth's surface become progressively cooler from the equator to the poles (**FIGURE 2.14**). Note, however, that these changes in temperature are not exactly parallel with changes in latitude. Why do temperatures vary across the same latitude? Three major influences alter the global pattern of temperature: ocean currents, the distribution of land and water, and elevation. As we saw in the previous section, ocean currents contribute to a warmer climate in northern Europe than at North American locations of the same latitude. Similarly, the influence of the cold Humboldt Current is noticeable on the west coast of South America, where temperatures are cooler than at similar latitudes elsewhere.

The difference in heat capacity between the oceans and the continents is not reflected in the average annual temperatures shown in Figure 2.14. Why is this so? Because the annual temperature *variation* is not depicted in that figure. Air temperatures over land show greater seasonal variation, with warmer temperatures in summer and colder temperatures in winter, than those over the oceans (**FIGURE 2.15**). This seasonal change has a major impact on the distribution of organisms, as we will see in later chapters.

Elevation above sea level has an important influence on continental temperatures. Note in Figure 2.14 the large difference in temperature between the Indian subcontinent and Asia. The sharp change in air temperature in this region is due to the influence of the Himalayas and the Tibetan Plateau. The change in elevation is extreme here, from about 150 m (500 feet) on the Ganges Plain in India to over 8,000 m (28,000 feet) in the highest peaks of the Himalayas in only 200 km (120 miles).

Why is it colder in mountains and highlands than in surrounding lowlands? Two factors contribute to the colder climates found at higher elevations. First, at higher elevations there are fewer air molecules to absorb the infrared energy radiating from Earth's surface. Thus, even though highlands may receive as much solar radiation as nearby lowlands, the heating of air by the ground surface is less effective because of the lower air density. Second, highlands exchange air more effectively with cooler air in the surrounding atmosphere. Because the atmosphere is warmed mainly by infrared radiation emitted by Earth's surface, the temperature of the atmosphere decreases with increasing distance from the ground. This decrease in temperature with increasing height above the surface is known as the **lapse rate**. In addition, wind velocity increases with increasing elevation because there is less friction with the

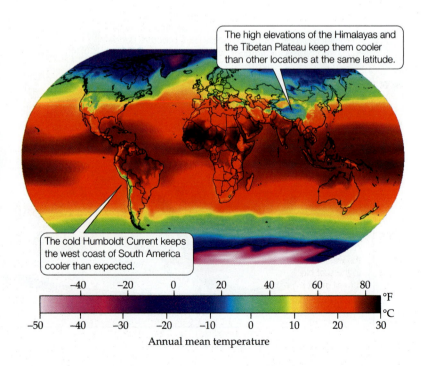

The high elevations of the Himalayas and the Tibetan Plateau keep them cooler than other locations at the same latitude.

The cold Humboldt Current keeps the west coast of South America cooler than expected.

Annual mean temperature

FIGURE 2.14 Global Average Annual Temperatures Average annual air temperatures tend to vary with latitude, but oceanic circulation and topography alter this pattern.

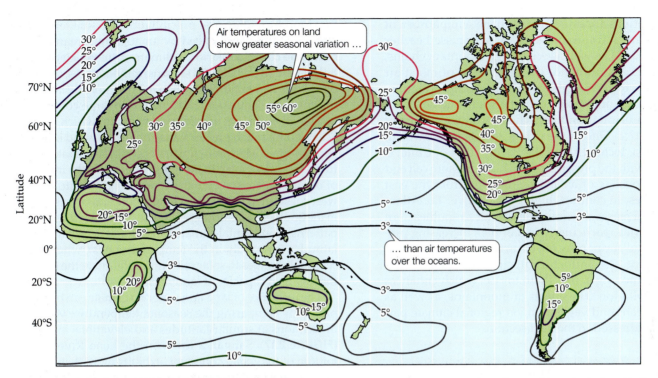

FIGURE 2.15 Annual Seasonal Temperature Variation
Seasonal temperature variation is expressed as the difference in average monthly temperature between the warmest and coldest months (in °C). (After A. H. Strahler and A. N. Strahler. 2005. *Physical*

Geography, 3rd ed. John Wiley and Sons: Hoboken, NJ. Compiled by John E. Oliver.)

? What is the effect of continent size on the magnitude of seasonal temperature variation?

ground surface. As a result, the decrease in air temperature with increasing elevation tends to follow the lapse rate.

Patterns of atmospheric pressure and topography influence precipitation

The locations of the Hadley, Ferrell, and polar circulation cells suggest that precipitation should be highest in the tropical latitudes between 23.5°N and S and in a band at about 60°N and S and should be lowest in zones around 30°N and S (see Figure 2.9). The African continent displays the pattern closest to this idealized precipitation distribution. However, there are substantial deviations from the expected latitudinal precipitation pattern in other areas, particularly in the Americas (**FIGURE 2.16**).

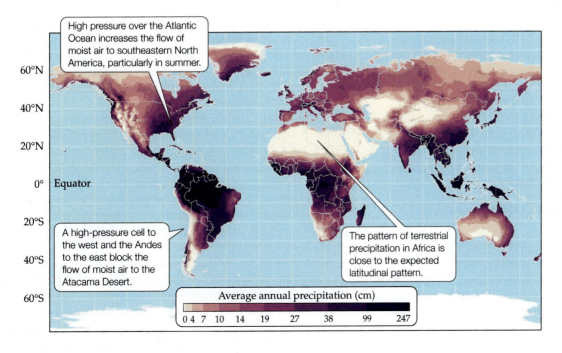

FIGURE 2.16 Average Annual Terrestrial Precipitation The latitudinal pattern of precipitation deviates from what would be expected based on atmospheric circulation patterns alone (see Figure 2.9). (Courtesy of the Center for Sustainability and the Global Environment [SAGE] through their *Atlas of the Biosphere*, https://nelson.wisc.edu/sage/data-and-models/maps.php. Data from CRU 0.5 Degree Dataset [M. G. New et al. 2000. *J Climate* 13: 2217–2238].)

These deviations are associated with the semipermanent high-pressure and low-pressure cells discussed earlier (see Figure 2.11) as well as with large mountain chains.

Pressure cells influence the movement of moist air from oceans to continents, as well as cloud formation. For example, high pressure over the South Pacific Ocean decreases precipitation along the central west coast of South America. One of the driest deserts in the world, the Atacama, located along the Pacific coast of Chile, is associated with the presence of this high-pressure cell and with the blockage of air masses moving from the east by the Andes. In contrast, high pressure over the Atlantic Ocean increases the flow of moist air to southeastern North America, particularly in summer, increasing precipitation and supporting the occurrence of forests there.

Mountains also influence precipitation patterns by forcing air moving across them to rise, which enhances local precipitation. The effects of mountains, as well as those of oceans and vegetation, on regional climate patterns are addressed in the next section.

> **CONCEPT 2.4** Regional climates reflect the influence of oceans and continents, mountains, and vegetation.

LEARNING OBJECTIVES

2.4.1 Describe the changes in an air mass that moves from a maritime zone across mountains, on both the leeward and windward slopes.

2.4.2 Illustrate how energy exchange components are influenced by vegetation and subsequently affect climate.

Regional Climate Influences

You may have noticed the changes in climate as you travel from a coastal area to an inland location. This change in climate can be abrupt, particularly when you travel across a mountain chain. The daily variation in air temperature increases, the humidity decreases, and precipitation decreases. These climate differences result from the effects of oceans and continents on regional energy balance and the influence of mountains on airflow and temperature. The vegetation often reflects these regional climate differences, exemplifying the effects of climate on the distributions of species and biological communities. The vegetation also has important effects on the climate through its influence on energy and water balance.

Proximity to oceans influences regional climates

Earlier we noted that water requires greater energy input to change its temperature (i.e., it has a higher heat capacity) than land. As a result, seasonal temperature changes are smaller over oceans than over continental areas (see Figure 2.15). In addition, oceans provide a source of moisture for cloud formation and precipitation. Coastal terrestrial regions that are influenced by an adjacent ocean have a **maritime climate**. Maritime climates are characterized by little variation in daily and seasonal temperatures, and they often have higher humidity than regions more distant from the coast. In contrast, areas centered in large continental land masses have a **continental climate**, which is characterized by much greater variation in daily and seasonal temperatures. Maritime climates occur in all climate zones, from tropical to polar. In the temperate zones, the influence of oceans on coastal climates tends to be accentuated on west coasts in the Northern Hemisphere and on east coasts in the Southern Hemisphere because of the prevailing wind patterns. Continental climates are limited to mid and high latitudes (primarily in the temperate zones), where large seasonal changes in solar radiation accentuate the effect of the low heat capacity of land masses.

The influence of land and water on climate can be exemplified by comparing the seasonal temperature variation in locations at similar latitudes and elevations in Siberia (**FIGURE 2.17**). Sangar, a town on the Lena River in the middle of the Asian continent, exhibits more than double the seasonal temperature variation of Khatyrka, on the Pacific coast. Note that the maximum and minimum temperatures occur slightly later in the year in the maritime climate (Khatyrka), another reflection of the high heat capacity of the ocean and its effect on local climate.

Mountains influence wind patterns and gradients in temperature and precipitation

The effects of mountains on climate are visually apparent in the elevational patterns of vegetation, particularly in arid regions. As we move up a mountain, grasslands may abruptly change to forests, and at higher elevations, forests may give way to alpine grasslands. These abrupt shifts in vegetation patterns reflect the rapid changes in climate that occur over short distances in mountains as temperatures decrease, precipitation increases, and wind speed increases with elevation. What causes these abrupt changes? The climates of mountains are the product of the effects of topography and elevation on air temperatures, the behavior of air masses, and their own generation of unique local wind patterns.

Air moving across Earth's surface is forced upward when it encounters a mountain range. This uplifted air cools as it rises, and water vapor condenses to form clouds and precipitation. As a result, the amount of precipitation increases with elevation. This enhancement of precipitation in mountains is particularly apparent in north–south-trending mountain ranges on the slopes that face into the prevailing wind (the *windward* slopes). In the temperate zones, where the prevailing winds blow toward the east, moving air encounters the western slopes of mountain ranges (such as the Sierra Nevada and coastal ranges in North America) and loses most of its moisture as precipitation before cresting over the summits.

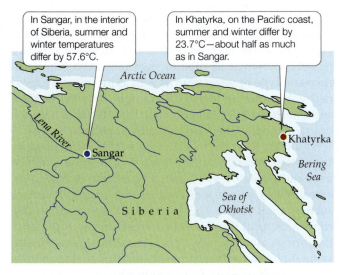

In Sangar, in the interior of Siberia, summer and winter temperatures differ by 57.6°C.

In Khatyrka, on the Pacific coast, summer and winter differ by 23.7°C—about half as much as in Sangar.

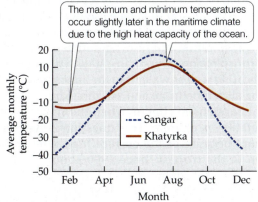

The maximum and minimum temperatures occur slightly later in the maritime climate due to the high heat capacity of the ocean.

FIGURE 2.17 Average Monthly Temperatures in a Continental and a Maritime Climate The difference in seasonal temperature variation between two locations in Siberia at about the same latitude and elevation illustrates the effect of the high heat capacity of ocean water. (Data from NOAA GHCN-Monthly, version 2; T. C. Peterson and R. S. Vose. 1997. *Bull Am Meteorol Soc* 78: 2837–2849.)

The loss of moisture, as well as the warming of the air as it moves down the eastern slopes, dries the air mass (**FIGURE 2.18A**). This **rain shadow effect** results in lower precipitation and soil moisture on the slopes facing away from the prevailing wind (the *leeward* slopes) and higher precipitation and soil moisture on the windward slopes. The rain shadow effect influences the types and amounts of vegetation on mountain ranges: lush, productive plant communities tend to be found on the windward slopes, and sparser, more drought-resistant vegetation on the leeward slopes (**FIGURE 2.18B**).

Mountains can also generate local wind and precipitation patterns. Differences in the direction that mountain slopes face (referred to as the slope exposure or *aspect*) can cause differences in the amounts of solar radiation the slopes and surrounding flatlands receive. As we saw in the case of the large-scale circulation patterns that generate Hadley cells (see Figure 2.8), differences in solar heating

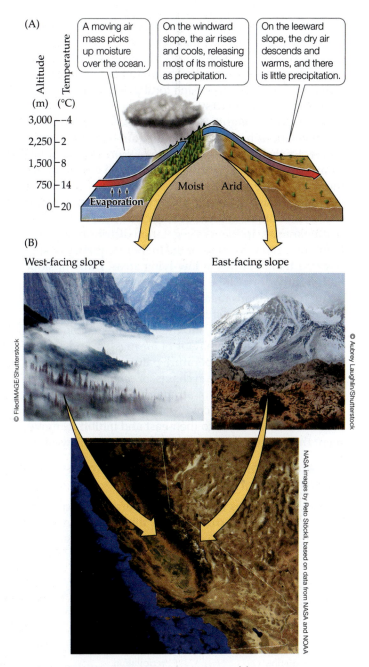

FIGURE 2.18 The Rain Shadow Effect (A) Precipitation tends to be greater on the windward slope of a mountain range than on the leeward slope. (B) Vegetation on west-facing and east-facing slopes in the Sierra Nevada of California reflects the rain shadow effect.

? Which slope aspect (north, south, east, or west) on a north–south-trending mountain range in the tropical zone would have the highest precipitation, and which aspect would be in the rain shadow?

of the ground surface can cause the uplift of air pockets that are warmer than the surrounding air. In the morning, east-facing slopes receive more solar radiation from the rising sun and thus become warmer than the surrounding slopes and lowlands. This differential heating creates localized upslope winds in the mountains. Depending on the moisture content of the air and the prevailing winds at higher elevations, clouds may form on the eastern flanks of the mountains. These clouds can generate local thunderstorms that may move off the mountains and into surrounding lowlands, increasing local precipitation.

At night, the ground surface cools, and the air above it becomes denser. Nighttime cooling is more pronounced at high elevations because the thinner atmosphere absorbs and reradiates less energy and allows more heat to be lost from the ground surface. Air can flow like water, with the cold, dense air moving downslope and pooling in low-lying areas. As a result, valley bottoms are the coldest sites in mountainous areas during clear, calm nights. This *cold air drainage* influences vegetation distributions in the temperate zones because of the higher frequency of subfreezing temperatures in low-lying areas. Daily upslope and nightly downslope winds are a common feature of many mountainous areas, particularly in summer when the input of solar radiation is highest.

At continental scales, mountains influence the movement, position, and behavior of air masses, and as a result they influence temperature patterns in surrounding lowlands. Large mountain chains, or *cordilleras*, can act to channel the movement of air masses. The Rocky Mountains, for example, steer cold Arctic air through the central part of North America to their east and inhibit its movement through the intermountain basins to their west.

Vegetation affects climate via surface energy exchange

Climate determines where and how organisms can live, but organisms, in turn, influence the climate system in several ways. First, the amount and type of vegetation influence how the ground surface interacts with solar radiation and wind and how much water it loses to the atmosphere. The amount of solar radiation that a surface reflects, known as its **albedo**, is influenced by the presence and type of vegetation as well as by soil and topography. A coniferous forest, for example, is darker in color, and thus has a lower albedo, than most types of bare soil or grasslands, so the forest absorbs more solar energy.

The texture of Earth's surface is also influenced by vegetation. A rough surface, such as a savanna of mixed trees and grasses, allows greater transfer of energy to the atmosphere by wind (convection) than a smooth surface such as a grassland. This is because the vegetation disrupts airflow at the ground surface, causing turbulence that brings more surface air into the atmosphere. Finally, vegetation can cool the atmosphere through *transpiration* (evaporation of water from inside a plant via its leaves). The amount of transpiration increases with the amount of leaf area per unit of ground surface area. The sum of water loss by transpiration and by evaporation is referred to as **evapotranspiration**. Evapotranspiration transfers energy (latent heat) as well as water into the atmosphere, thereby reducing air temperature and soil moisture.

What happens to climate when the type or amount of vegetation is altered? This question is particularly important because of the current high rates of deforestation in the tropics: since 1990 about 129 million hectares (500,000 square miles) of tropical forest have been cut (FAO 2020). Loss of the trees increases the albedo of the land surface as bare soil is exposed and the trees are partially replaced with lighter-colored grasses (**FIGURE 2.19**). The higher albedo decreases the absorption of solar radiation, resulting in less heating of the

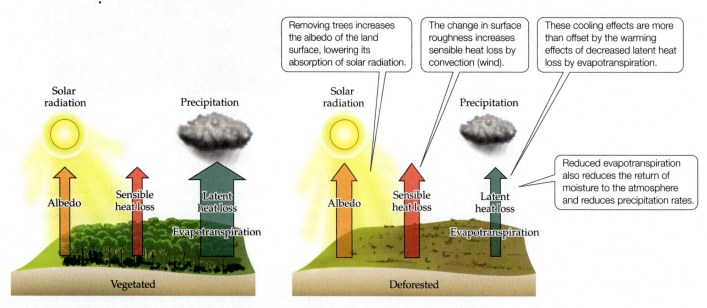

Removing trees increases the albedo of the land surface, lowering its absorption of solar radiation.

The change in surface roughness increases sensible heat loss by convection (wind).

These cooling effects are more than offset by the warming effects of decreased latent heat loss by evapotranspiration.

Reduced evapotranspiration also reduces the return of moisture to the atmosphere and reduces precipitation rates.

Solar radiation

Precipitation

Albedo

Sensible heat loss

Latent heat loss

Evapotranspiration

Vegetated

Solar radiation

Precipitation

Albedo

Sensible heat loss

Latent heat loss

Evapotranspiration

Deforested

FIGURE 2.19 The Effects of Deforestation Illustrate the Influence of Vegetation on Climate
The conversion of forest to pasture in the tropics results in a number of changes in energy exchange with the atmosphere. (After J. A. Foley et al. 2003. *Front Ecol Environ* 1: 38–44.)

land surface. However, the lower heat gain from solar radiation is more than offset by lower evapotranspirative cooling (lower latent heat flux) due to loss of leaf area (Foley et al. 2003). Lower evapotranspiration rates not only reduce surface cooling, but also lead to lower precipitation because less moisture is returned from the ground surface to the atmosphere. Thus, the outcome of tropical deforestation may be a warmer, drier regional climate. Widespread deforestation may lead to climate change that is significant enough to inhibit reforestation and may thus lead to long-term changes in tropical ecosystems. The conversion of natural grasslands to crop production—a widespread human practice—can also affect climate, as you can evaluate in **ANALYZING DATA 2.1**.

In Chapter 25 we will return to the effects of human activities on climate, especially over the past two centuries. Human activities, however, are not the only cause of long-term climate change. We turn next to the natural climate variation that has occurred throughout Earth's history.

ANALYZING DATA 2.1

How Do Changes in Vegetation Cover Influence Climate?

We've learned that the type and amount of vegetation can influence energy exchange at Earth's surface. As a result, human alteration of the land surface, such as tropical deforestation, can lead to changes in regional climate. Determining whether temperatures are likely to get warmer or cooler after such an alteration requires knowledge of the magnitude and direction of the changes in energy balance components.

For example, what happens when humans replace short-grass steppe, a type of grassland characteristic of the western Great Plains of the United States, with croplands? This vegetation change occurred along the South Platte River of northeastern Colorado in the latter part of the twentieth century, and its effects were evaluated by Chase and colleagues (1999).* Some of their data are presented here in the form of questions for your evaluation.

1. First consider changes in albedo. When sparse stands of light-colored grass (albedo = 0.26, meaning that 26% of incoming solar radiation is reflected) are replaced by dark green irrigated crops (albedo = 0.18), how does this influence absorption of solar radiation? If the incoming solar radiation is 470 watts per square meter (W/m^2), what is the difference in energy gain due to solar radiation as a result of the vegetation change? Would this change in albedo alone cause warming or cooling?

2. Next consider heat exchange due to sensible heat flux, including convection, which is related to the roughness of the surface. A dryland (nonirrigated) crop has approximately three times greater surface roughness than short-grass steppe. Which surface would have greater heat loss due to convection, assuming that surface temperatures are warmer than the atmosphere: a cropland or short-grass steppe? The estimated difference in heat exchange due to sensible heat flux associated with

Satellite Image of the South Platte River Drainage Basin, Colorado The Rocky Mountains are to the west. The green circles and rectangles are irrigated cropland found along the South Platte River flowing eastward. The surrounding area is a mix of dryland crops and short-grass steppe.

the land use change to a *dryland* crop is about 40 W/m^2. Would a combination of change in albedo (Question 1) and in surface roughness cause cooling, no net change, or warming?

3. Replacing short-grass steppe with *irrigated* crops, which have a higher leaf area per area of ground surface and higher soil moisture, alters the amount of energy lost via evapotranspiration (latent heat flux). Would this change result in more or less heat loss to the atmosphere relative to the short-grass steppe?

4. Taking both sensible and latent heat flux into account, the combined estimated difference in heat exchange associated with the land use change to irrigated cropland is about 60 W/m^2. Including the change in albedo from Question 1, would an irrigated crop surface have cooler or warmer temperatures relative to short-grass steppe?.

*Chase, T. N., R. Pielke Sr., T. G. F. Kittel, J. S. Baron, and T. J. Stohlgren. 1999. Impacts on Colorado Rocky Mountain weather due to land use changes on the adjacent Great Plains. *Journal of Geophysical Research* 104: 16673–16690.

Google/Landsat/Copernicus

CONCEPT 2.5 Seasonal and decadal climate variation are associated with changes in Earth's position relative to the sun and the strength of atmospheric pressure cells.

LEARNING OBJECTIVES

2.5.1 Explain how the tilt of Earth's axis influences (1) seasonal changes in air temperature in temperate and polar zones and (2) seasonal changes in precipitation in the tropics.

2.5.2 Outline how seasonal changes in surface heating in temperate and polar lakes influence water density and result in the stratification of water.

2.5.3 Describe how cyclic change in the position and strength of high- and low-pressure cells influences weather and climate variability.

Climate Variation over Time

As noted at the beginning of this chapter, understanding climate variation is critical to understanding ecological phenomena such as the distributions of organisms. Climate variation at daily to multi-decadal time scales determines the range of environmental conditions experienced by organisms as well as the availability of the resources and habitats they need to survive. Long-term climate variation over hundreds and thousands of years influences the evolutionary history of organisms and the development of ecosystems. In this section, we will review climate variation, from seasonal to decadal time scales.

Seasonality results from the tilt of Earth's axis

The amount of sunlight striking any point on Earth's surface varies as Earth makes its 365.25-day journey around the sun. Earth's axis is tilted at an angle of 23.5° relative to the sun's direct rays (**FIGURE 2.20**). Thus, the angle and intensity of the rays striking any point on Earth change as Earth orbits the sun. This influence of the tilt of Earth's axis overrides the variation associated with seasonal changes in the distance between Earth and the sun due to Earth's slightly elliptical orbit. Earth is closest to the sun in January (at a point called the *perihelion*: 237 million km or 147 million miles) and farthest away in July (at the *aphelion*: 245 million km or 152 million miles).

The temperate and polar zones experience pronounced changes in temperature associated with variation in solar radiation over the year. Summer occurs in the Northern Hemisphere from June to September, when that hemisphere is tilted toward the sun; at the same time, the Southern Hemisphere is oriented away from the sun and experiences its winter. The difference in solar radiation, and thus the temperature variation, between summer and winter increases from the tropics toward the poles. The seasonal changes in the angle of the sun affect not only the intensity of solar radiation, but also the length of the day.

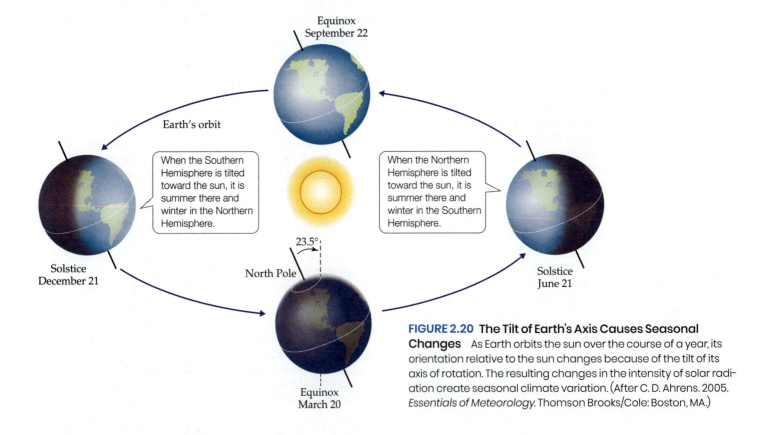

FIGURE 2.20 The Tilt of Earth's Axis Causes Seasonal Changes As Earth orbits the sun over the course of a year, its orientation relative to the sun changes because of the tilt of its axis of rotation. The resulting changes in the intensity of solar radiation create seasonal climate variation. (After C. D. Ahrens. 2005. *Essentials of Meteorology*. Thomson Brooks/Cole: Boston, MA.)

Above 66.5°N and S, the sun does not set for several days, weeks, or even months in summer. During the winter at these same latitudes, the sun does not rise high enough to warm the surface. Because air temperatures regularly drop below freezing during winter in the temperate and polar zones, seasonality in these zones is an important determinant of biological activity and strongly influences the distributions of organisms.

Seasonal changes in solar radiation are relatively small in the tropics compared with those in the temperate and polar zones. As a result, seasonality in the tropics is marked primarily by changes in precipitation rather than by changes in temperature. These seasonal changes are associated with the movement of the zone of maximum air uplift and precipitation in the Hadley cells, known as the **intertropical convergence zone**, or **ITCZ**. This zone of maximum uplift corresponds with the part of the tropics where the sun strikes Earth most directly. Thus, the ITCZ moves from 23.5°N in June to 23.5°S in December, bringing the wet season with it (**FIGURE 2.21**).

Seasonal changes in aquatic environments are associated with changes in water temperature and density

Aquatic environments in the temperate and polar zones also experience seasonal changes in temperature, but as we have seen, they are not as extreme as those on land. Liquid water becomes denser as it gets colder, and it has the unique property of being most dense at 4°C. Ice is less dense than liquid water and therefore forms on the surfaces of water bodies in winter. Because it has a higher albedo than open water, ice on the surface of lakes or polar oceans effectively prevents warming of the water below it.

Differences in water temperature (and thus water density) with depth result in the **stratification**, or layering, of water in oceans and lakes. Stratification has important implications for aquatic organisms because it determines

the movement of nutrients and oxygen. Surface waters in lakes and oceans mix freely, but they are underlain by colder, denser layers of water that do not mix easily with the surface waters. In oceans, the surface waters mix with the subsurface layers only rarely—for example, in upwelling zones.

In temperate-zone lakes, seasonal changes in water temperature and density result in seasonal changes in stratification (**FIGURE 2.22**). In summer, the surface layer, or **epilimnion**, is the warmest and contains active populations of phytoplankton and zooplankton. The epilimnion is underlain by a zone of rapid temperature decline, called the **thermocline**. Below the thermocline is a stable layer of the densest, coldest water in the lake, known as the **hypolimnion**. In summer, dead organisms from the epilimnion will drop to the hypolimnion and bottom (*benthic*) zone, carrying nutrients and energy away from the surface layers.

During the fall, the air above the water surface cools, and the lake loses heat to the atmosphere. As the epilimnion cools, its density increases until it is the same as that of the layers below it. Eventually, the water at all depths of the lake has the same temperature and density, and winds blowing on the surface lead to a mixing of surface and deep layers, known as lake **turnover**. This mixing is important for recycling of the nutrients that are lost from

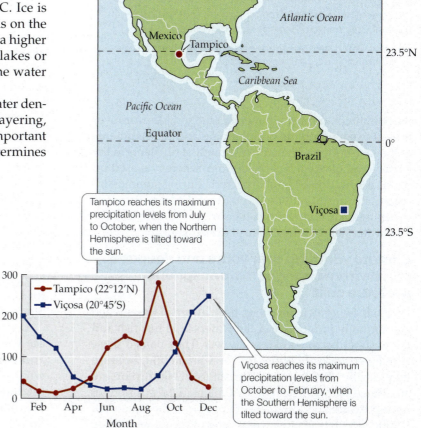

FIGURE 2.21 Wet and Dry Seasons and the ITCZ Seasonality of precipitation in the tropics is associated with movement of the intertropical convergence zone (ITCZ) between the tropics of the Northern and Southern Hemispheres. Thus, Tampico, Mexico, reaches its maximum precipitation levels from July to October and has a dry season from November to April, whereas Viçosa, Brazil, has a wet season from October to February and a dry season from April to August. (Data from NOAA GHCN-Monthly, version 2; T. C. Peterson and R. S. Vose. 1997. *Bull Am Meteorol Soc* 78: 2837–2849.)

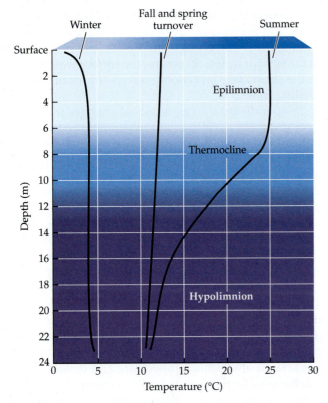

FIGURE 2.22 Lake Stratification Lake stratification, which occurs primarily in summer in temperate and polar regions, results from the effects of temperature on water density. Seasonal changes in water temperature result in the turnover of water that mixes little during summer and winter. (After S. Dodson. 2004. *Introduction to Limnology.* McGraw Hill: New York.)

? Why would seasonal changes in lake stratification be unlikely to occur in tropical lakes?

the epilimnion during summer. In addition, lake turnover moves oxygen into the hypolimnion and the sediments at the lake bottom. The replenishment of nutrients at the surface and of oxygen at the bottom, where it is used up by the respiration of aerobic bacteria during summer, increases biological activity throughout the lake. Turnover occurs again in spring when the surface ice melts and the lake water has a uniform density once again.

Climate variation over years and decades results from changes in atmospheric pressure cells

Peruvian fishermen have long been aware of times when the normally productive ocean waters hold few fish and the weather becomes extremely wet. They named these climate episodes El Niño, for the Christ child, because they usually started around Christmas. El Niño events are associated with a switch (or oscillation) in the positions of high-pressure and low-pressure cells over the equatorial Pacific, which leads to a weakening of the easterly trade

winds that normally push warm water toward Southeast Asia. Climatologists refer to this oscillation and the climate changes associated with it as **El Niño Southern Oscillation**, or **ENSO**. Its underlying causes are still not well understood. The frequency of ENSO is somewhat irregular, but it occurs at intervals of 3–8 years and generally lasts for about 18 months. During El Niño events, the upwelling of deep ocean water off the coast of South America ceases as the easterly winds weaken or, in some events, shift to westerly winds. ENSO also includes La Niña events, which are stronger-than-average phases of the normal pattern, with high pressure off the coast of South America and low pressure in the western Pacific. La Niña events usually follow El Niño events but tend to be less frequent.

ENSO is associated with unusual climate conditions, even at localities distant from the tropical Pacific, through its complex interactions with atmospheric circulation patterns (**FIGURE 2.23**). El Niño events are associated with unusually dry conditions in the Malay Archipelago, other parts of Southeast Asia, and Australia. The likelihood of fires in the grasslands, shrublands, and forests of these areas increases as precipitation decreases and vegetation dries out. In contrast, in the southern United States and northern Mexico, El Niño events may increase precipitation, while the ensuing La Niña events bring drought conditions. The increased plant growth associated with an El Niño event, followed by dry La Niña conditions, intensifies fires in the southwestern United States (Veblen et al. 2000).

Similar atmospheric pressure–ocean current oscillations occur in the North Atlantic Ocean. The **North Atlantic Oscillation** affects climate variation in Europe, in northern Asia, and on the east coast of North America. Another long-term oscillation in sea surface temperature and atmospheric pressure, known as the **Pacific Decadal Oscillation**, or **PDO**, was described for the North Pacific after its influence on salmon numbers was discovered, as described in the Case Study earlier in this chapter. The PDO affects climate in ways similar to ENSO and can moderate or intensify the effects of ENSO. The effects of the PDO are felt primarily in northwestern North America, although southern parts of North America, Central America, Asia, and Australia may also be affected. The PDO and the North Atlantic Oscillation have been linked to long-term droughts in the United States (e.g., the U.S. Dust Bowl in the 1930s; see the Case Study for Chapter 25). We will return to the PDO in the Case Study Revisited.

Long-term changes in climate have occurred throughout Earth's history, including glacial–interglacial cycles and extended periods of much warmer climate than what is occurring now. These long-term climate fluctuations are associated with differences in the amount of solar radiation received and the concentrations of greenhouse gases.

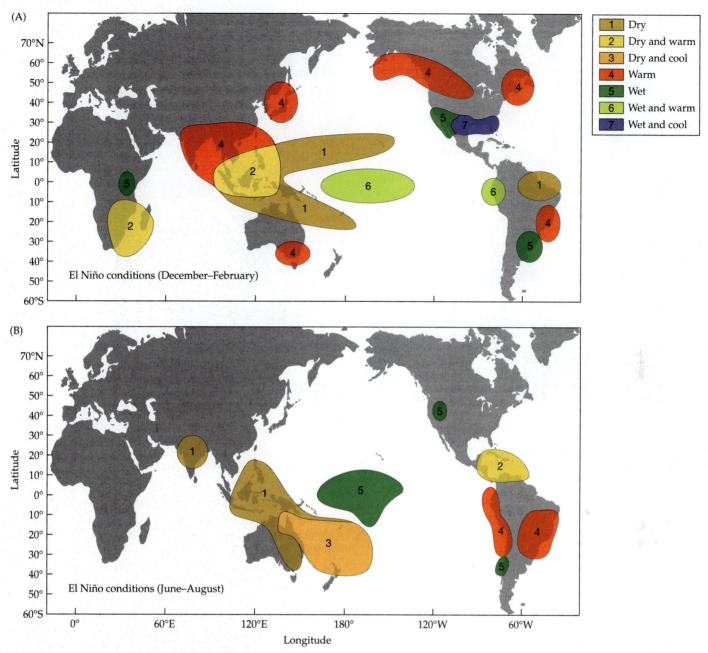

FIGURE 2.23 El Niño Southern Oscillation (ENSO) El Niño events have widespread climate effects that vary seasonally, altering temperature and precipitation patterns at a global scale. (Courtesy of NOAA Tropical Atmosphere Ocean Project.)

Legend:
1 Dry
2 Dry and warm
3 Dry and cool
4 Warm
5 Wet
6 Wet and warm
7 Wet and cool

(A) El Niño conditions (December–February)

(B) El Niño conditions (June–August)

CONCEPT 2.6 Salinity, acidity, and oxygen concentrations are major determinants of the chemical environment.

LEARNING OBJECTIVES

2.6.1 Outline what determines the salinity and acidity of soils and waters.

2.6.2 Explain why oxygen concentrations vary depending on elevation, the influence of water on diffusion, and biological consumption.

The Chemical Environment

All organisms are bathed in a matrix of chemicals. Water is the primary chemical constituent of aquatic environments, along with variable amounts of dissolved salts and gases. Small differences in the concentrations of these dissolved chemicals can have important consequences for the functioning of aquatic organisms, as well as for terrestrial plants and microorganisms that are dependent on water and dissolved chemicals in the soil. Terrestrial organisms are immersed in a gaseous atmosphere that is relatively invariant, consisting primarily of nitrogen (78%), oxygen (20%), water vapor (1%), and argon (0.9%). The atmosphere also

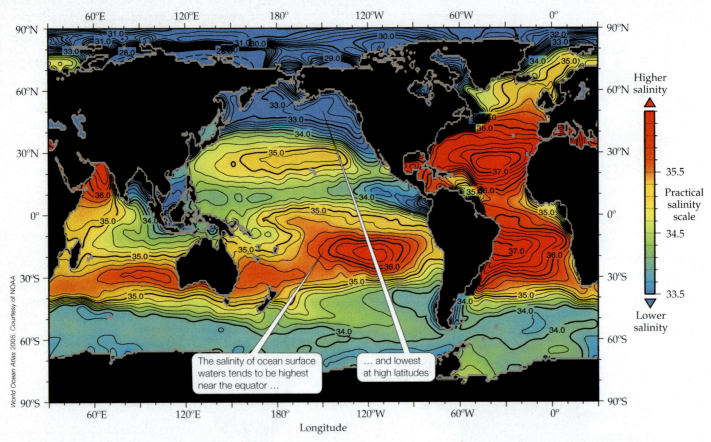

The salinity of ocean surface waters tends to be highest near the equator …

… and lowest at high latitudes

FIGURE 2.24 Global Variation in Salinity at the Ocean Surface Variations in the salinity of ocean surface waters reflect the concentrating effect of evaporation, dilution by melting sea ice, and precipitation.

contains trace gases, including the greenhouse gases, which play a critical role in Earth's energy balance, and pollutants derived from human activities, which can have important effects on atmospheric chemistry. We will discuss the effects of air pollutants and greenhouse gases in Chapter 25. Here we briefly review three chemical variables that influence biological and ecological function: salinity, acidity, and the availability of oxygen.

All waters contain dissolved salts

Salinity refers to the concentration of dissolved salts in water. *Salts* are ionic compounds, composed of cations (positively charged ions) and anions (negatively charged ions) that disassociate when placed in water. Dissolved salts are important from a biological perspective because they influence properties of water that affect the ability of organisms to absorb it, as we will see in Concept 4.3. Salts also have direct influences on organisms as nutrients (as we will see in Concept 22.1) and can inhibit metabolic activity if their concentrations are too high or too low.

Although all waters contain dissolved salts, we often think about salinity in the context of oceans, which account for 97% of the water on Earth; 70% of Earth's surface is under salty ocean waters. The salinity of the oceans varies between 33 and 37 parts per thousand; this variation is a result of evaporation, precipitation, and the freezing and melting of sea ice (**FIGURE 2.24**). The salinity of ocean surface waters is highest near the equator and lowest at high latitudes.

What are the salts that make water saline, and where do they come from? Ocean salts consist mainly of sodium, chloride, magnesium, calcium, sulfate, bicarbonate, and potassium. These salts come from gases that were emitted by volcanic eruptions early in Earth's history, when its crust was cooling, and from the gradual breakdown of minerals in the rocks that make up Earth's crust.

The salinity of water bodies is determined by the balance of inputs and losses of salts and water. Most landlocked bodies of water become more saline over time, reflecting a balance between water inputs from precipitation, water losses due to evaporation, and inputs of salts. When these inland "seas" occur in arid areas (e.g., the Great Salt Lake and the Dead Sea), their salinities usually exceed that of ocean water because of high rates of evaporation and its concentrating effect. The types of salts that contribute to their salinity vary, reflecting the chemistry of the minerals in the rocks that make up their basins. Despite the high salinity levels in these inland lakes, some organisms have managed to thrive in their waters, including algae and cyanobacteria.

High levels of salinity occur naturally in waterlogged soils adjacent to oceans, such as those in salt marshes. Soils may also become more saline in arid regions as water from deeper soil layers is brought to the surface by plant roots or through pumping of groundwater for irrigation. As this transported water evaporates, it leaves its salts behind. If there is little precipitation to leach the salts to deeper soil layers, or if drainage of the water is impeded by impervious layers beneath the soil, high rates of evapotranspiration

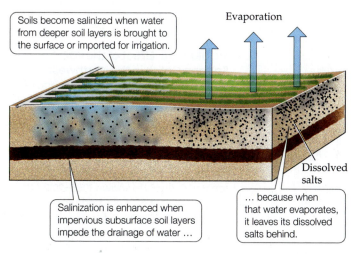

Soils become salinized when water from deeper soil layers is brought to the surface or imported for irrigation.

Evaporation

Dissolved salts

Salinization is enhanced when impervious subsurface soil layers impede the drainage of water …

… because when that water evaporates, it leaves its dissolved salts behind.

FIGURE 2.25 Salinization Salinization of soils is disrupting agricultural production in many areas, especially in arid regions.

will result in a progressive buildup of salts at the soil surface. This process, known as **salinization**, occurs naturally in some desert soils and is a common occurrence in irrigated agricultural soils of arid regions (**FIGURE 2.25**). Salinization contributed to agricultural decline in ancient Mesopotamia (now Iraq) and is a problem today in California's Central Valley, Australia, and other regions.

Organisms are sensitive to the acidity of their environment

Acidity and its converse, **alkalinity**, are measures of the ability of a solution to behave as an acid or a base, respectively. *Acids* are compounds that give up protons (H^+) to the water they are dissolved in. *Bases* take up protons or give up hydroxide ions (OH^-). Examples of common acids include the citric, tannic, and ascorbic acids found in fruits. Examples of common bases include sodium bicarbonate (baking soda) and other carbonate minerals in rock. Acidity and alkalinity are measured as pH, which is equal to the negative of the logarithm ($-\log_{10}$) of the concentration of H^+. Thus, one pH unit represents a tenfold change in the concentration of H^+. Pure water has a neutral pH of 7.0. Solutions with pH values higher than 7.0 are alkaline (basic), and solutions with pH values lower than 7.0 are acidic.

The pH values of water have important effects on organismal function. Changes in pH values can directly affect metabolic activity. The pH values of water also determine the chemistry and availability of nutrients, as we will see in Concept 22.4. Organisms have a limited range of pH values that they can tolerate. Natural levels of alkalinity (when the pH of the environment exceeds 7) tend not to be as important as levels of acidity as a constraint on organismal function and distributions.

Under natural conditions the pH of ocean water does not vary appreciably, because the chemistry of seawater *buffers* changes in pH—that is, the salts in seawater bind free protons and thereby minimize changes in pH. Thus, pH tends to be more variable in terrestrial and freshwater ecosystems than in the ocean. However, increases in atmospheric CO_2 concentrations due to human activities are increasing the acidity of the oceans, with negative effects on marine ecosystems. Marine animals that build shells using calcium carbonate are less able to construct and maintain their shells under more acidic conditions (Orr et al. 2005). We will discuss this phenomenon more thoroughly in Concept 25.1.

On land, the pH of surface waters and soils varies naturally. What causes this variation? Water can become more acidic over time through the input of acidic compounds derived from several sources, most associated with soil development (which is covered in more detail in Concept 22.1). Two of the main components of soil are mineral particles from the breakdown of rocks, and organic matter from the decomposition of dead plants and other organisms. Some rock types, such as granites, generate acidic salts, while other rock types, such as limestones, generate basic salts. Soils become more acidic as they age because the basic salts leach away more easily and because decomposition and leaching of plant matter adds organic acids to the soil. The emission of acidic pollutants into the atmosphere by the burning of fossil fuels, as well as overuse of agricultural fertilizers, can increase the acidity of soil and water. We will cover these sources of acidity in more detail in Concept 25.3.

Oxygen concentrations vary with elevation, diffusion, and consumption

There was no oxygen in the atmosphere when life on Earth first evolved, and oxygen was toxic to the earliest forms of life. Even today, there are organisms that are intolerant of oxygen. However, with the exception of some archaea, bacteria, and fungi, most organisms require oxygen to carry out their metabolic processes and cannot survive in **hypoxic** (low-oxygen) conditions. Hypoxic conditions can also promote the formation of chemicals (e.g., hydrogen sulfide) that are toxic to many organisms. In addition, oxygen levels are important for chemical reactions that determine the availability of nutrients.

Oxygen concentrations in the atmosphere have been stable at about 21% for the past 65 million years, so most terrestrial environments have invariant oxygen concentrations. However, the availability of atmospheric oxygen decreases with elevation above sea level. As we have seen, the overall density of air decreases with elevation, so there are fewer molecules of oxygen in a given volume of air at higher elevations. We will discuss the repercussions of this variation for human health in Concept 4.1.

Oxygen concentrations can vary substantially in aquatic environments and in soils. The rate of diffusion of oxygen into water is slow and may not keep pace with its consumption by organisms. Waves and currents mix oxygen from the atmosphere into ocean surface waters, so its concentration is usually stable there. Oxygen concentrations are low in the deep ocean and in marine sediments, where biological uptake is greater than replenishment from surface waters. The same holds true in deep lakes, lake sediments, and flooded soils (e.g., in wetlands). Oxygen concentrations are highest in freshwater ecosystems with moving water (streams and rivers) because mixing with the atmosphere is greatest there.

(A)

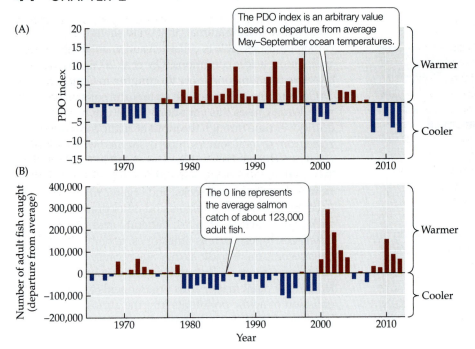

The PDO index is an arbitrary value based on departure from average May–September ocean temperatures.

(B)

The 0 line represents the average salmon catch of about 123,000 adult fish.

FIGURE 2.26 **Effect of the PDO on Salmon Catch in the Northwest United States** (A) Summer average PDO index, 1965–2012. Red and blue bars indicate ocean temperatures that are warmer or cooler than average, respectively. (B) Departures from the average (123,131 fish) in numbers of adult Chinook salmon returning to the Columbia River (Washington and Oregon) to spawn, 1965–2012. (After W. T. Peterson et al. 2013. *Ocean Ecosystem Indicators of Salmon Marine Survival in the Northern California Current*. National Marine Fisheries Service: Newport, OR; Seattle, WA.)

? How frequently does the cool phase of the PDO correspond to a greater-than-average catch of salmon? Conversely, how often does a warm phase of the PDO correspond to a lower-than-average catch of salmon?

A CASE STUDY REVISITED

Climate Variation and Salmon Abundance

The research of Steven Hare and Robert Francis on salmon production in the North Pacific contributed to the discovery of the Pacific Decadal Oscillation (PDO). As noted earlier, the PDO is a multi-decadal shift in sea surface temperature and atmospheric pressure cells. A review of existing records of sea surface temperatures over the past century indicated that the PDO was associated with alternating 20- to 30-year periods of warm and cool temperatures in the North Pacific (**FIGURE 2.26A**). The length of the phases of the PDO differentiates it from other climate oscillations, whose phases tend to be much shorter (e.g., 18 months–2 years for ENSO). The warm and cool phases of the PDO influenced the marine ecosystems that Pacific salmon depended on and thus shifted salmon production north or south, depending on the phase (**FIGURE 2.26B**).

The PDO has been linked to changes in the abundances and distributions of many marine organisms and, through its climate effects, changes in the functioning of terrestrial ecosystems (Mantua and Hare 2002). Its effects have been found primarily in western North America and eastern Asia, but effects have also been reported in Australia. Thus, the influence of the PDO on climate extends throughout the Western Hemisphere. Evidence for the existence of climate changes associated with the PDO dates back to instrumental temperature records from the 1850s, and to the 1600s using climate signals from corals and tree rings. The mechanisms underlying the PDO are unclear, but its effect on climate is significant and widespread (**TABLE 2.1**).

TABLE 2.1

Summary of Climate Effects of the Pacific Decadal Oscillation (PDO)

Climate effect	Warm phase PDO	Cool phase PDO
Ocean surface temperature in the northeastern and tropical Pacific	Above average	Below average
October–March northwestern North American air temperature	Above average	Below average
October–March southeastern U.S. air temperature	Below average	Above average
October–March southern U.S./northern Mexico precipitation	Above average	Below average
October–March northwestern North American and Great Lakes precipitation	Below average	Above average
Northwestern North American spring snowpack and water year (October–September stream flow)	Below average	Above average
Winter and spring flood risk in the Pacific Northwest	Below average	Above average

Source: N. J. Mantua. 2001. In *The Encyclopedia of Global Environmental Change*, Vol. 1, M. C. McCracken and J. S. Perry (Eds.), pp. 592–594. Wiley: New York.

CONNECTIONS *in* NATURE

CLIMATE VARIATION AND ECOLOGY Two aspects of the PDO are particularly important in the context of ecology. First, the realization that the PDO existed was driven initially by an attempt to understand variation in the size of an animal population. This observation underscores the relationship between physical conditions (the topic of this chapter), the functioning of individual organisms and their growth and reproduction (Chapters 4 and 5), and population and community processes (Units 2 and 5, respectively). This relationship is one of the central themes of ecology that will form a common thread throughout this book. Ultimately, the physical environment, including climate and the myriad factors, such as the PDO, that control it, determines whether an organism can exist in a given location (as we'll see in Chapter 3). Extremes in the physical environment, including those that are driven by climate oscillations, play a critical role in our understanding of ecological phenomena.

Second, the time scale of the climate variation associated with the PDO is long relative to the human life span. The abrupt changes in climate, and the associated ecological responses of the marine ecosystem, were therefore perceived by people as unusual events. Indeed, the phases of the PDO may be longer than the life spans of most of the organisms affected by it, limiting their ability to adapt to this climate oscillation. As a result, from the perspective of an ecological community, the PDO represents a disturbance, an event that detrimentally affects the populations of some species and disrupts the community.

Although we don't yet understand what causes it, the PDO has been a part of the climate system for at least the last 400 years. A better understanding of its effects will help us place other climate phenomena, including global climate change, in perspective.

SUMMARY

CONCEPT 2.1 Climate is the most fundamental component of the physical environment.

2.1.1 Outline the difference between weather and climate, with specific reference to their temporal scales.

- Weather refers to the current conditions of temperature, precipitation, humidity, wind, and cloud cover. Climate is the long-term average and variation in weather at a given location.

2.1.2 Explain the importance of weather variability for ecological processes.

- Through their impact on mortality, extreme weather events influence the geographic distributions of organisms.

2.1.3 Summarize how temperature is determined by the gains and losses of energy at Earth's surface.

- The climate system is driven by the balance between energy gains from solar radiation and reradiation by the atmosphere and energy losses due to infrared radiation from Earth's surface, latent heat flux, and sensible heat flux.

CONCEPT 2.2 Winds and ocean currents result from differences in solar radiation across Earth's surface.

2.2.1 Draw connections between differential heat gain across Earth's surface and the development of atmospheric circulation cells.

- Latitudinal differences in the intensity of solar radiation at Earth's surface establish atmospheric circulation cells.
- The Coriolis effect and the difference in heat capacity between the oceans and the continents act on atmospheric circulation cells to determine the pattern of prevailing winds at Earth's surface.

2.2.2 Explain how surface winds and ocean currents move heat between the tropics and the poles.

- Ocean currents are driven by surface winds and by differences in water temperature and salinity.
- Winds and ocean currents transfer energy from the tropics to higher latitudes.

CONCEPT 2.3 Large-scale atmospheric and oceanic circulation patterns establish global patterns of temperature and precipitation.

2.3.1 Outline the determinants of global temperature and precipitation patterns.

- Global temperature patterns are determined by latitudinal variation in solar radiation, but they are also influenced by oceanic circulation patterns and by the distribution of continents.
- Global patterns of terrestrial precipitation are determined by atmospheric circulation cells, but they are also influenced by semipermanent pressure cells.

2.3.2 Explain how a region's seasonal changes in temperature are affected by its location, whether it is near a large body of water or at the center of a large continent.

- Seasonal variation in temperature is greater in the middle of a continent than on the coast because ocean water has a higher heat capacity than land.

2.3.3 Summarize how air density and air exchange cause a decrease in air temperature with increases in elevation on a mountain.

- The heating of air by the ground surface is less effective in the highlands because of the lower air density; therefore, temperature decreases as the elevation of the land surface increases.

CONCEPT 2.4 Regional climates reflect the influence of oceans and continents, mountains, and vegetation.

2.4.1 Describe the changes in an air mass that moves from a maritime zone across mountains, on both the leeward and windward slopes.

- Mountains force air masses passing over them to rise and drop most of their moisture as precipitation, resulting in moister environments on windward slopes and drier environments on leeward slopes.

(Continued)

2.4.2 Illustrate how energy exchange components are influenced by vegetation and subsequently affect climate.

- Vegetation influences regional climates through its effects on energy exchange associated with albedo, evapotranspiration (latent heat transfer), and surface winds (sensible heat transfer).

CONCEPT 2.5 Seasonal and decadal climate variation are associated with changes in Earth's position relative to the sun and the strength of atmospheric pressure cells.

2.5.1 Explain how the tilt of Earth's axis influences (1) seasonal changes in air temperature in temperate and polar zones and (2) seasonal changes in precipitation in the tropics.

- The tilt of Earth's axis as it orbits the sun causes seasonal temperature changes in temperate and polar regions and precipitation changes in tropical regions.

2.5.2 Outline how seasonal changes in surface heating in temperate and polar lakes influence water density and result in the stratification of water.

- Temperature-induced differences in water density result in nonmixing layers of water in oceans and lakes. In temperate-zone lakes, these layers break down in fall and spring, allowing the movement of oxygen and nutrients.

2.5.3 Describe how cyclic change in the position and strength of high- and low-pressure cells influences weather and climate variability.

- Variations in climate over years to decades are caused by cyclic changes in atmospheric pressure cells. The oscillation in the position of high- and low-pressure cells can lead to a weakening of the easterly trade winds or a shift to westerly winds. These changes have widespread effects beyond the regions where the pressure cells are located.

CONCEPT 2.6 Salinity, acidity, and oxygen concentrations are major determinants of the chemical environment.

2.6.1 Outline what determines the salinity and acidity of soils and waters.

- The salinity of Earth's waters, including water in soils, is determined by the balance between inputs of salts and gains (by precipitation) and losses (by evaporation) of water.
- The pH of soils and surface waters is determined by inputs of salts from the breakdown of rock minerals, organic acids from plants, and acidic pollutants.

2.6.2 Explain why oxygen concentrations vary depending on elevation, the influence of water on diffusion, and biological consumption.

- Oxygen concentrations are stable in most terrestrial ecosystems, but oxygen availability decreases as elevation increases. Concentrations of oxygen in aquatic ecosystems are low where its consumption by organisms exceeds its slow rate of diffusion into water.

Review Questions

1. Why is the variability of physical conditions potentially more important than average conditions as a determinant of ecological patterns, such as species distributions?

2. Describe the factors that determine the major latitudinal climate zones (the tropic, temperate, and polar zones).

3. Why are deserts more prone to salinization from irrigation than areas with greater precipitation?

Hone Your Problem-Solving Skills

As we will see in Chapter 3, information presented in this chapter describing variation in climate can be useful in predicting where specific collections of plant types can be found. For the following descriptions, draw a graph that portrays the following important climate features.

1. Draw a graph that shows the seasonal change in temperature for both the center of the Australian continent and the center of Eurasia, with time (12 months) on the *x*-axis and temperature on the *y*-axis. Use Figure 2.15 as a guide, and don't worry about the actual temperatures, but instead focus on the magnitude of the seasonal change.

2. Construct a graph showing the rain shadow effect (see Figure 2.18) along a west coast in the Northern Hemisphere. Use distance for the *x*-axis, spanning a location near the coastline, moving into a mountain range, and ending on the eastern side of the mountain, indicating where each is along the *x*-axis. Use both annual average temperature and precipitation on the *y*-axis.

3. Graph the trend in annual precipitation for northern Mexico in an average year and for an El Niño year (see Figure 2.23); use time on the *x*-axis and precipitation on the *y*-axis (hint: use Figure 2.21 as a guide).

The Biosphere

3

The American Serengeti—Twelve Centuries of Change in the Great Plains: A Case Study

Today, the region covering the central part of North America, known as the Great Plains, bears little resemblance to the Serengeti Plain of Africa. Biological diversity is very low in many parts of the current landscape, which contains large stands of uniform crop plants (which are often even genetically identical) and a few species of domesticated herbivores (mostly cattle). In the Serengeti, on the other hand, some of the largest and most diverse herds of wild animals in the world roam a picturesque savanna (**FIGURE 3.1**). If not for a series of important environmental changes, however, the two ecosystems might look superficially very similar.

Biological communities in the temperate and polar zones have been subjected to natural, long-term climate change, which has led to latitudinal or elevational shifts in their positions and species composition. Eighteen thousand years ago, at the last glacial maximum of the Pleistocene epoch, ice sheets covered the northern portion of North America. Over the next 12,000 years, the climate warmed and the ice receded. Vegetation followed the retreating ice northward and colonized the newly exposed substrate. Grasslands in the center of the continent expanded into former spruce and aspen woodlands. These grasslands contained species of grasses, sedges, and low-growing herbaceous plants similar to those found in the natural grasslands that exist today.

The animal inhabitants of those earlier grasslands were, however, strikingly different from today's. A diverse collection of *megafauna* (animals larger than 45 kg, or 100 pounds) existed in North America, rivaling the diversity found today in the Serengeti (Martin 2005) (**FIGURE 3.2**). Thirteen thousand years ago—a relatively short time in an evolutionary context—North American herbivores included woolly mammoths and mastodons (relatives of elephants), as well as several species of horses, camels, and giant ground sloths. Predators included saber-toothed cats with 18 cm (7-inch) incisors, cheetahs, lions, and giant short-faced bears that were larger and faster than grizzly bears.

About 10,000–13,000 years ago, as the extensive grasslands of the Great Plains were developing, many of the large mammals of North America suddenly went extinct (Barnosky et al. 2004). The rapidity of the disappearance of approximately 28 genera (40–70 species) made this extinction unlike any extinction event during the previous 65 million years. Another unusual aspect of this extinction was that nearly all the animals that went extinct belonged to the same group: large mammals. The causes of this extinction are a mystery.

Several hypotheses have been proposed to account for the disappearance of the North American megafauna. Changes in the climate during the extinction period were rapid and could have led to changes in habitat or food supply

KEY CONCEPTS

CONCEPT 3.1 Terrestrial biomes are characterized by the growth forms of the dominant vegetation.

CONCEPT 3.2 Biological zones in freshwater ecosystems are associated with the velocity, depth, temperature, clarity, and chemistry of the water.

CONCEPT 3.3 Marine biological zones are determined by ocean depth, light availability, and the stability of the bottom substrate.

FIGURE 3.1 The Serengeti Plain of Africa
Large, diverse herds of native animals migrate across the Serengeti in search of food and water.

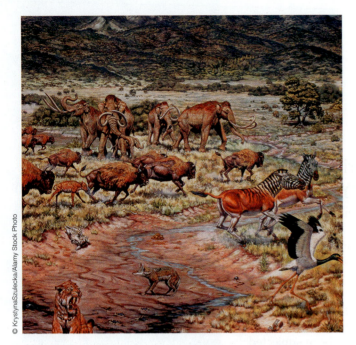

FIGURE 3.2 Pleistocene Animals of the Great Plains
The animals of the grasslands of central North America 13,000 years ago included woolly mammoths, horses, and giant bison. Many of these large mammals went extinct within a short time between 13,000 and 10,000 years ago.

that would have negatively affected the animals. Another hypothesis, which has generated substantial controversy, suggests that the arrival of humans in North America may have hastened the demise of the animals (Martin 1984). When this hypothesis was first proposed, it was met with widespread skepticism, and the initial supporting evidence was considered weak. Although humans first appeared in the central part of North America about 16,000 years ago, it is unclear how hunters bearing stone and wooden tools could have driven so many species of large mammals to extinction. What evidence is there to support the hypothesis that humans were involved in this extinction event?

Introduction

Living things can be found in remarkable places. Birds such as ravens, lammergeyers (Eurasian vultures), and alpine choughs (crows) fly over the highest summits of the Himalayas, over 8,000 m (26,000 feet) above sea level. Fish such as the "fangtooth" (*Abyssobrotula galatheae*) live more than 8,000 m below the ocean surface. Bacteria and archaea can be found almost everywhere on Earth, in hot sulfur springs at the extreme chemical and temperature limits for life, under glaciers, on dust particles many kilometers above Earth's surface, and kilometers deep in ocean sediments. However, most living things occur in a range of habitats that cover a thin veneer of Earth's surface, from the tops of trees to the surface soil layers in terrestrial environments and within 200 m of the surface of the oceans.

The **biosphere**—the zone of life on Earth—is sandwiched between the *lithosphere*, Earth's surface crust and upper mantle, and the *troposphere*, the lowest layer of the atmosphere. Biological communities can be studied at multiple scales of varying complexity, as we saw in Concept 1.2. Here, we will use the biome concept to introduce the amazing diversity of terrestrial life. The diversity of aquatic life is not as easily categorized, but we will describe several freshwater and marine biological zones, which, like terrestrial biomes, reflect the physical conditions where they are found.

> **CONCEPT 3.1** Terrestrial biomes are characterized by the growth forms of the dominant vegetation.

LEARNING OBJECTIVES

3.1.1 Explain why ecologists use plant growth forms to categorize global terrestrial biomes.

3.1.2 Describe how global patterns of precipitation and temperature, and their variability, influence the location of terrestrial biomes.

3.1.3 Evaluate how human activities impact the actual distributions of biomes relative to the potential distributions.

Terrestrial Biomes

Biomes are large-scale biological communities shaped by the physical environment in which they are found. In particular, they reflect the climate variation described in Chapter 2. Biomes are categorized by the most common growth forms of plants distributed across large geographic areas. The categorization of biomes does not take taxonomic relationships among organisms into account; instead, it relies on similarities in the morphological responses of organisms to the physical environment. A biome includes similar biotic assemblages on distant continents, indicating similar responses to similar climate conditions in different locations. In addition to providing a useful introduction to the diversity of life on Earth, the biome concept provides a convenient biological unit for modelers simulating the effects of environmental change on biological communities, as well as for those simulating the effects of vegetation on the climate system (see Concept 2.4). The numbers and categories of biomes used vary from source to source, depending on the preferences and goals of the authors. Here, we use a system of nine biomes: *tropical rainforest, tropical seasonal forest and savanna, desert, temperate grassland, temperate shrubland and woodland, temperate deciduous forest, temperate evergreen forest, boreal forest,* and *tundra*. This system provides a teaching tool for linking biological systems to the environments that shape them.

Terrestrial communities vary considerably—from those in the warm, wet tropics to those in the cold, dry polar regions. Tropical forests have multiple verdant layers, high growth rates, and tremendous species diversity. Lowland tropical forests in Borneo have an estimated 10,000 species of vascular plants, and most other tropical forest communities have about 5,000 species. In contrast, polar regions have a scattered cover of tiny plants clinging to the ground, reflecting a harsh climate of high winds, low temperatures, and dry soils. High-latitude Arctic communities contain about 100 species of vascular plants. Tropical rainforest vegetation may reach over 75 m (250 feet) in height and contain over 400,000 kg (882,000 pounds) of aboveground biomass in a single hectare (about 2.5 acres). Plants of polar regions, on the other hand, rarely exceed 5 cm (2 inches) in height and contain less than 1,000 kg (2,200 pounds) of aboveground biomass per hectare.

Terrestrial biomes are classified by the *growth form* (size and morphology) of the dominant plants (e.g., trees, shrubs, or grasses) (**FIGURE 3.3**). Characteristics of their leaves, such as *deciduousness* (seasonal shedding of leaves), thickness, and *succulence* (development of fleshy water storage tissues), may also be used. Why use plants rather than animals to categorize terrestrial biomes? Plants are immobile, so in order to occupy a site successfully for a long time, they must be able to cope with its environmental extremes as well as its biological pressures, such as competition for water, nutrients, and light. Plant growth forms are therefore good indicators of the physical environment, reflecting the climate zones discussed in Concept 2.2 as well as rates of disturbance (e.g., fire frequency). In addition, animals are a less visible component of most large landscapes, and their mobility allows them to avoid exposure to adverse environmental conditions. Microorganisms (archaea, bacteria, and fungi) are important components of biomes, and the composition of microbial communities reflects physical conditions as plant growth forms do. The tiny size of these organisms, however, as well as rapid temporal and spatial changes in their community composition, makes them impractical for classifying biomes.

Since their emergence from the oceans about 500 million years ago, plants have taken on a multitude of different forms in response to the selection pressures of the terrestrial environment (see Figure 3.3). These selection pressures include aridity, high and subfreezing temperatures, intense solar radiation, nutrient-poor soils, grazing

Growth form **Environment**

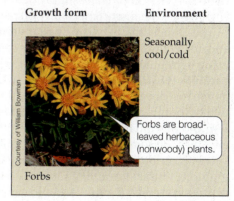

Seasonally dry/moist and warm/cool

Sclerophyllous shrubs have tough, leathery leaves.

Sclerophyllous shrubs

FIGURE 3.3 Plant Growth Forms   The growth form of a plant is an evolutionary response to the environment, particularly climate and soil fertility.

Growth form **Environment**

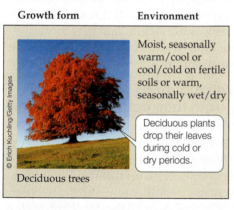

Moist, seasonally warm/cool or cool/cold on fertile soils or warm, seasonally wet/dry

Deciduous plants drop their leaves during cold or dry periods.

Deciduous trees

Growth form **Environment**

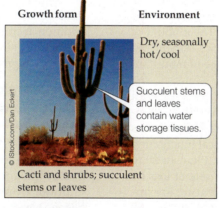

Dry, seasonally hot/cool

Succulent stems and leaves contain water storage tissues.

Cacti and shrubs; succulent stems or leaves

Growth form **Environment**

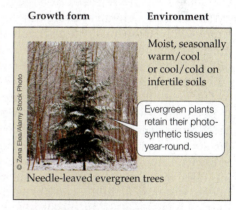

Moist, seasonally warm/cool or cool/cold on infertile soils

Evergreen plants retain their photosynthetic tissues year-round.

Needle-leaved evergreen trees

Growth form **Environment**

Moist, seasonally warm/cool, with fire

Grasses grow from the base of their leaves.

Grasses, sedges

Growth form **Environment**

Wet, warm year-round

Evergreen leaves in tropical regions carry out photosynthesis year-round.

Evergreen broad-leaved trees

Growth form **Environment**

Seasonally cool/cold

Forbs are broad-leaved herbaceous (nonwoody) plants.

Forbs

by animals, and crowding by neighbors. Having deciduous leaves, for example, is one solution to seasonal exposure to subfreezing temperatures or extended dry periods. Trees and shrubs invest energy in woody tissues in order to increase their height and ability to capture sunlight and to protect their tissues from damage by wind or large amounts of snow. Perennial grasses, unlike most other plants, can grow from the bases of their leaves and keep their vegetative and reproductive buds below the soil surface, which facilitates their tolerance of grazing, fire, subfreezing temperatures, and dry soils. Similar plant growth forms appear in similar climate zones on different continents, even though the plants may not be genetically related. The evolution of similar growth forms among distantly related species in response to similar selection pressures is called **convergence**.

Terrestrial biomes reflect global patterns of precipitation and temperature

Chapter 2 described Earth's climate zones and their association with the atmospheric and oceanic circulation patterns that result from the differential heating of Earth's surface by the sun. These climate zones are major determinants of the distribution of terrestrial biomes.

The tropics (between 23.5°N and S) are characterized by high rainfall and warm, invariant temperatures. In the subtropical regions that border the tropics, rainfall becomes more seasonal, with pronounced dry and wet seasons. The major deserts of the world are associated with the zones of high pressure at about 30°N and S and with the rain shadow effects of large mountain ranges. Subfreezing temperatures during winter are an important feature of the temperate and polar zones. The amount of precipitation north and south of 40° varies depending on proximity to the ocean and the influence of mountain ranges (see Figure 2.16).

The locations of terrestrial biomes are correlated with these variations in temperature and precipitation. Temperature influences the distribution of plant growth forms directly through its effect on the physiological functioning of plants. Precipitation and temperature act in concert to influence the availability of water and its rate of loss by plants. Water availability and soil temperature are important in determining the supply of nutrients in the soil, which is also an important control on plant growth form.

The association between spatial variation in climate and terrestrial biome distribution can be visualized using a graph of average annual precipitation and temperature (**FIGURE 3.4**). While these two factors predict biome distributions reasonably well, this approach fails to incorporate seasonal temporal variation in temperature and precipitation. As we saw in Concept 2.1, climate extremes are sometimes more important in determining species distributions than average annual conditions. For example, grasslands and shrublands have wider global distributions than Figure 3.4 would suggest, occurring in

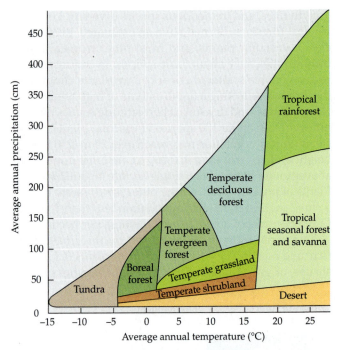

FIGURE 3.4 Biomes Vary with Average Annual Precipitation and Temperature When plotted on a graph of precipitation and temperature, the nine major terrestrial biomes form a triangle. (After R. H. Whittaker. 1975. *Communities and Ecosystems.* Macmillan: London.)

❓ What factor(s) might result in grasslands or shrublands "invading" climate space occupied by forest or savanna?

regions with relatively high average annual precipitation but regular dry periods (e.g., Mediterranean-type shrublands; grasslands at the margins of deciduous forests). In addition, factors such as soil texture and chemistry as well as proximity to mountains and large bodies of water can influence biome distribution.

The potential distributions of terrestrial biomes differ from their actual distributions due to human activities

The effects of land conversion and resource extraction by humans are increasingly apparent on the land surface. These human effects are collectively described as **land use change**. Human modification of terrestrial ecosystems began at least 10,000 years ago with the use of fire as a tool to clear forests and enhance the size of game populations. The greatest changes have occurred over the last 150 years, since the onset of mechanized agriculture and logging and an exponential increase in the human population (see Figure 10.2) (Harrison and Pearce 2001). About 75% of Earth's land surface has been altered by human activities, primarily agriculture, forestry, and livestock grazing; a smaller amount (2%–3%) has been transformed by urban development and transportation corridors (Venter et al. 2016; Sanderson et al. 2002). As a result of these human influences, the potential and the actual distributions of biomes are markedly different (**FIGURE 3.5**). Temperate

(A)

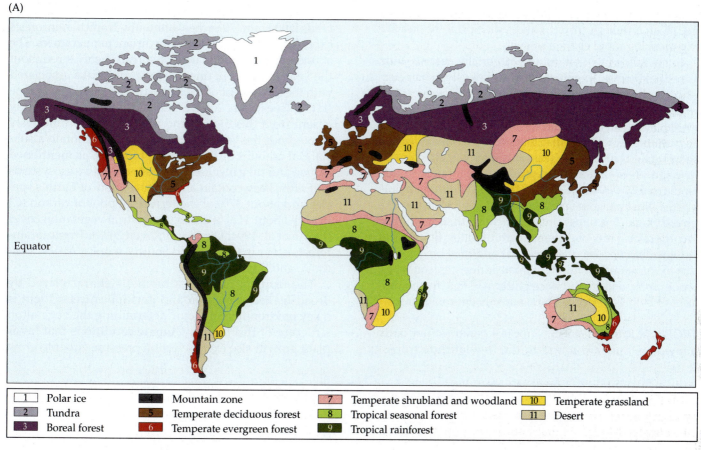

1	Polar ice	4	Mountain zone	7	Temperate shrubland and woodland	10	Temperate grassland
2	Tundra	5	Temperate deciduous forest	8	Tropical seasonal forest	11	Desert
3	Boreal forest	6	Temperate evergreen forest	9	Tropical rainforest		

(B)

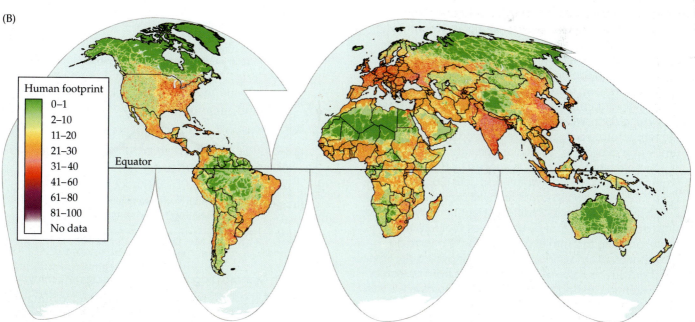

Human footprint

0–1
2–10
11–20
21–30
31–40
41–60
61–80
81–100
No data

FIGURE 3.5 Global Biome Distributions Are Affected by Human Activities The potential distributions of biomes differ from their actual distributions because human activities have altered much of Earth's land surface. (A) The potential global distribution of biomes. (B) Alteration of terrestrial biomes by human activities. The "human footprint" is a quantitative measure (100 = maximum) of the overall human impact on the environment based on geographic data describing human population size, land development, and resource use. (B from E. W. Sanderson et al. 2002. *BioScience* 52: 891–904.)

? Which biomes in North America and Eurasia appear to have been most affected by human activities? In other words, which biomes in (A) overlap most with areas of high human impact in (B)? In South America and on the Indian subcontinent, which biome has been most degraded by human activity?

biomes, particularly grasslands, have been transformed the most, although tropical and subtropical biomes are experiencing rapid change as well.

In the following sections, we will briefly describe nine terrestrial biomes, their biological and physical characteristics, and the human activities that influence the actual amount of natural vegetation cover that remains in each biome. The description of each biome begins with a map of its potential geographic distribution and a *climate diagram* showing the characteristic seasonal patterns of air temperature and precipitation at a representative location in that biome (see **ECOLOGICAL TOOLKIT 3.1**). In addition, sample photos illustrate some of the vegetation types that make up the biome. It is important to remember that each biome incorporates a mix of different communities. Boundaries between biomes are often gradual and may be complex due to variations in regional climate influences, soil types, topography, and disturbance patterns. Thus, the boundaries of biomes portrayed here are only approximations.

TROPICAL RAINFORESTS Tropical rainforests are aptly named, as they are found in the low-latitude tropics (between 10°N and S) where precipitation usually exceeds 2,000 mm (79 inches) annually. Tropical rainforests experience warm, seasonally invariant temperatures. The abundant precipitation may be spread evenly throughout the year or occur in one or two main peaks associated with the movement of the intertropical convergence zone (ITCZ) (see Figure 2.21). Seasonal rhythms are generally absent from this biome, and plants grow continuously throughout the year. Tropical rainforests contain a substantial amount of living plant biomass, as mentioned earlier, and they include the most productive ecosystems on Earth. They contain an estimated 50% of Earth's species and about 37% of the terrestrial pool of carbon (C) in only about 11% of Earth's terrestrial vegetation cover (Dixon et al. 1994; Dirzo and Raven 2003). Tropical rainforests occur in Central and South America, Africa, Australia, and Southeast Asia.

The tropical rainforest biome is characterized by broad-leaved evergreen and deciduous trees. Light is a key environmental factor determining the vegetation structure of this biome. Climate conditions that favor plant growth also exert selection pressure either to grow

---

ECOLOGICAL TOOLKIT 3.1

Climate Diagrams

A climate diagram is a graph of the average temperatures and amounts of precipitation, by month, at a particular location. Climate diagrams are useful for depicting seasonal patterns of climate conditions. In particular, they provide an indication of when temperatures are below freezing for extended periods (blue-shaded areas in the figure) and when precipitation is insufficient for plant growth. When the precipitation curve falls below the temperature curve (orange-shaded area in the figure), water availability limits plant growth.

Climate diagrams were developed by Heinrich Walter and Helmut Lieth (Walter and Lieth 1967), who used them to show the consistency of climate patterns within the same biomes in different locations. Walter and Lieth demonstrated that by using axes scaled with 1°C corresponding to 2 mm of precipitation, a coarse approximation of time periods when water availability limits plant growth could be made. (Water loss from terrestrial ecosystems is related to temperature, a topic we will take up in more detail in Chapter 4.) For example, the tropical seasonal forest and temperate shrubland and woodland biomes show distinct seasonal periods when water is in short supply, and some temperate grasslands also have predictable low-water seasons. Climate diagrams also show when temperatures are conducive to plant growth. It is apparent that there is a latitudinal trend toward longer periods of subfreezing temperatures with more extreme lows (larger areas of blue shading).

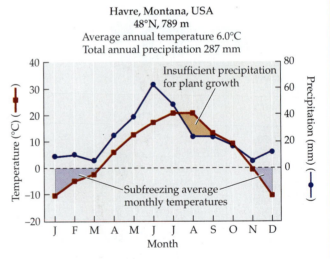

Havre, Montana, USA
48°N, 789 m
Average annual temperature 6.0°C
Total annual precipitation 287 mm

A Sample Climate Diagram A climate diagram contains the name of the climate station where conditions were recorded (Havre, Montana, in this example), its geographic location in latitude, and its elevation. In Havre, there are extended periods of subfreezing temperatures from November to March (blue shaded areas). Frosts do occur outside this time frame, but these isolated events are not reflected in average monthly temperatures. A period of low water availability (orange area) typically occurs from mid-July to October. (Data from NOAA GHCN-Monthly, version 2; T. C. Peterson and R. S. Vose. 1997. *Bull Am Meteorol Soc* 78: 2837–2849.)

Tropical Rainforests

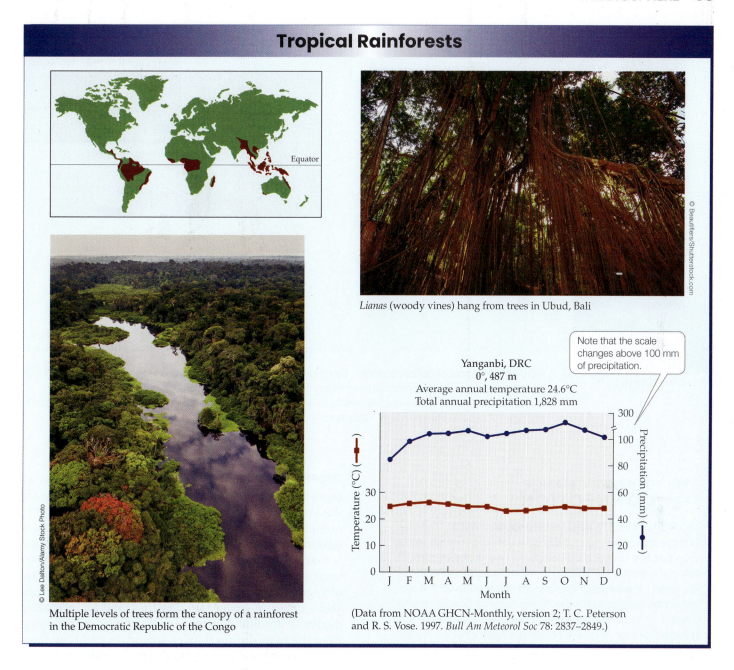

Lianas (woody vines) hang from trees in Ubud, Bali

Note that the scale changes above 100 mm of precipitation.

Yanganbi, DRC
0°, 487 m
Average annual temperature 24.6°C
Total annual precipitation 1,828 mm

Multiple levels of trees form the canopy of a rainforest in the Democratic Republic of the Congo

(Data from NOAA GHCN-Monthly, version 2; T. C. Peterson and R. S. Vose. 1997. *Bull Am Meteorol Soc* 78: 2837–2849.)

tall above neighboring plants or to adjust physiologically to low light levels. As many as five layers of plants occur in tropical rainforests. *Emergent trees* rise above the majority of the other trees that make up the *canopy* of the forest. The canopy consists primarily of the leaves of evergreen trees, which form a continuous layer approximately 30 to 40 m above the ground. Below the canopy, plants that use trees for support and to elevate their leaves above the ground, including *lianas* (woody vines) and *epiphytes* (plants that grow on tree branches), are found draped over or clinging to the canopy and emergent trees. *Understory* plants grow in the shade of the canopy, further reducing the light that finally reaches the forest floor. Shrubs and *forbs* (broad-leaved herbaceous plants) occupy the forest floor, where they must

rely on light flecks that move across the forest floor during the day for photosynthesis.

Globally, tropical rainforests are disappearing rapidly because of logging and conversion of forests to pasture and croplands (**FIGURE 3.6**). Approximately half of the tropical rainforest biome has been altered by deforestation (Asner et al. 2009; Keenan et al. 2015). Rainforests in Africa and Southeast Asia have been altered the most, and rates of deforestation continue to be greatest in those areas (Wright 2005). In some cases, rainforests have been replaced by disturbance-maintained pastures of forage grasses. In other cases, rainforest is regrowing, but the recovery of the previous rainforest structure is uncertain. Rainforest soils are often nutrient poor, and recovery of nutrient supplies may take a very long time, hindering forest regrowth.

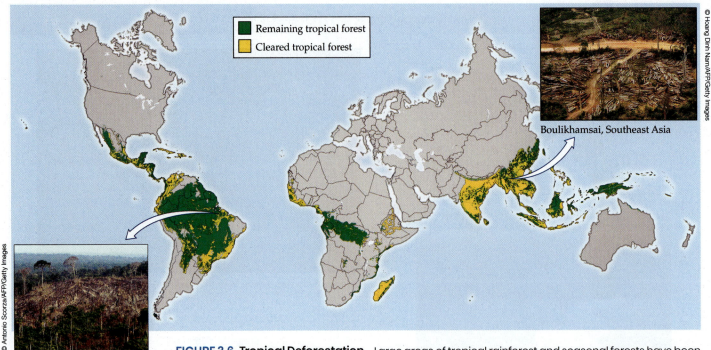

FIGURE 3.6 **Tropical Deforestation** Large areas of tropical rainforest and seasonal forests have been cleared over the past 40 years, primarily for agricultural and pastoral development. The loss of these tropical forests has large consequences for loss of biodiversity, regional climate, and carbon uptake and storage. (Map based on data from 2005. After S. L. Pimm and C. Jenkins. 2005. *Sci Am* 293: 66–73.)

Brazil, South America

Boulikhamsai, Southeast Asia

CLIMATE CHANGE CONNECTION

TROPICAL FORESTS AND GREENHOUSE GASES The loss of tropical forests to cutting and burning means more than just the loss of biodiversity. As noted above, almost 40% of the terrestrial carbon is in the tropical forest biome. The loss of the forests means both lower ability of the terrestrial biosphere to take up C from the atmosphere and greater emissions of greenhouse gases into the atmosphere from soils and decaying vegetation (Guo and Gifford 2002). Restoration projects are ongoing in some countries to help address concerns for the loss of diversity and C sequestration abilities associated with tropical forest loss, although their success has not been as effective as natural regeneration (Crouzeilles et al. 2017). How quickly can tropical forests recover and the pools of C be restored once regrowth is started? A review of more than 600 sites indicates that recovery of the plant biomass above the soil surface occurs within 85 years of regrowth, but a longer time is required for recovery of plant biomass in the soil (Martin et al. 2013). This analysis provides optimism for potential reversal of the contribution of tropical deforestation to atmospheric greenhouse gas concentrations. However, the analysis also found that while tree diversity recovers after 50 years, more than a century is required for full plant species recovery, including lianas and epiphytes.

TROPICAL SEASONAL FORESTS AND SAVANNAS As we move to the north and south of the wet tropics toward the Tropics of Cancer (23.5°N) and Capricorn (23.5°S), rainfall becomes seasonal, with pronounced wet and dry seasons associated with shifts in the ITCZ. This region is marked by a large gradient in climate primarily associated with the seasonality of rainfall. The responses of vegetation to the greater seasonal variability include shorter stature, lower tree densities, and an increasing degree of drought deciduousness, with leaves dropping from the trees during the dry season. In addition, there is a greater abundance of grasses and shrubs and fewer trees than in rainforests.

The tropical seasonal biome includes several different vegetation complexes, including *tropical dry forests*, *thorn woodlands*, and *tropical savannas*. The frequency of fires, which increases with the length of the dry season, influences the vegetation growth forms. Recurrent fires, sometimes set by humans, promote the establishment of **savannas**, communities dominated by grasses with intermixed trees and shrubs. In Africa, large herds of herbivores, such as wildebeests, zebras, elephants, and antelopes, also influence the balance between trees and grasses and act as an important force promoting the establishment of savannas. On the floodplains of the Orinoco River in South America, seasonal flooding contributes to the establishment of savannas, as trees are intolerant of long periods of soil saturation. Thorn woodlands

Tropical Seasonal Rainforests and Savannas

Grandidier's baobab trees on a dry, sunny day in Menabe, Madagascar

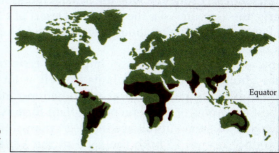

Equator

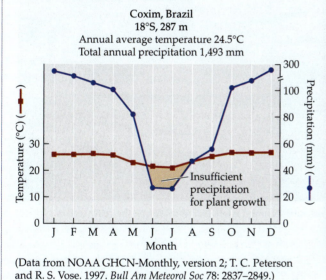

Coxim, Brazil
18°S, 287 m
Annual average temperature 24.5°C
Total annual precipitation 1,493 mm

Insufficient precipitation for plant growth

(Data from NOAA GHCN-Monthly, version 2; T. C. Peterson and R. S. Vose. 1997. *Bull Am Meteorol Soc* 78: 2837–2849.)

Tropical forest in Northern Thailand during the dry season

(communities dominated by widely spaced trees and shrubs) get their name from the protective thorns on the trees, which act as a deterrent to herbivores that would consume the vegetation. Thorn woodlands typically occur in regions with climates intermediate between tropical dry forests and savannas.

Tropical seasonal forests and savannas once covered more area than tropical rainforests, but today less than half of this biome remains intact. Increasing human demand for wood and agricultural land has resulted in loss of tropical seasonal forests and savannas at rates equal to or greater than those for tropical rainforests (Bullock et al. 1995). Large increases in human populations in tropical dry forest regions have had a particularly large effect.

Large tracts of tropical dry forest in Asia and Central and South America have been converted to cropland and pasture to meet the needs of growing human populations for food and earnings from agricultural goods exported to more-developed countries.

DESERTS In contrast to the tropical ecosystems, deserts contain sparse populations of plants and animals, reflecting sustained periods of high temperatures and low water availability. The subtropical positions of hot deserts correspond with the high-pressure zones of the Hadley cells (see Figure 2.8) around 30°N and S, which inhibit the formation of storms and their associated precipitation. Low precipitation levels, coupled with high temperatures and high rates of evapotranspiration, result in a limited supply of water for desert organisms. The major desert zones include the Sahara; the Arabian deserts; the Gobi in Asia; the Atacama Desert of Chile and Peru; and the Chihuahuan, Sonoran, and Mojave Deserts of North America.

Low water availability is an important constraint on the abundance of desert plants as well as an important influence on their form and function. One of the best examples of convergence in plant form is the occurrence of stem succulence in desert plants. Stem succulence occurs in both the cacti of the Western Hemisphere and the euphorb family of the Eastern Hemisphere (**FIGURE 3.7**). Plants with succulent stems can store water in their tissues to help the plants continue to function during dry periods. Other plants of the desert biome include drought-deciduous shrubs and grasses. Some short-lived annual plants are active only after sufficient precipitation has fallen. These annual plants carry out their entire life cycle, from germination through flowering and seed production, in a few short weeks. Although the abundance of organisms may be low, species diversity can be high in some deserts. The Sonoran Desert, for example, has over 4,500 plant species, 1,200 bee species, and 500 bird species (Nabhan and Holdsworth 1998).

Deserts

Equator

Flowers at Sossusvlei, Namib-Desert, Namibia, during the rainy season

Cardon and smaller cacti on Baja California peninsula, Mexico

Ouargla, Algeria
31°N, 150 m
Average annual temperature 22.3°C
Total annual precipitation 39 mm

Insufficient precipitation for plant growth

(Data from NOAA GHCN-Monthly, version 2; T. C. Peterson and R. S. Vose. 1997. *Bull Am Meteorol Soc* 78: 2837–2849.)

(A) Cactus

© Tim Gainey/Alamy Stock Photo

(B) Euphorb

© Organica/Alamy Stock Photo

FIGURE 3.7 Convergence in the Forms of Desert Plants (A) The blue candle cactus (*Myrtillocactus geometrizans*) is native to the Chihuahuan Desert of Mexico. (B) *Euphorbia polyacantha* has cactus-like characteristics. Although only distantly related, both species have succulent stems, water-conserving photosynthetic pathways, upright stems that minimize midday sun exposure, and spines that protect them from herbivores. These traits evolved independently in each species.

Humans have used deserts for livestock grazing and agriculture for centuries. Agricultural development in desert areas is dependent on irrigation, often using water that flows in from distant mountains or is extracted from deep underground. Unfortunately, irrigated agriculture in deserts has repeatedly failed because of salinization (see Concept 2.6). Livestock grazing in deserts is also a risky venture because of the unpredictable nature of the precipitation needed to support plant growth for herbivores. Long-term droughts in association with unsustainable grazing practices can result in loss of plant cover and soil erosion, a process known as **desertification**. Desertification is a concern in populated regions at the margins of deserts, such as the Sahel region in the southern portion of the Sahara in Africa.

TEMPERATE GRASSLANDS Large expanses of grasslands once occurred throughout North America and Eurasia (the Great Plains and the steppes of Central Asia) at latitudes between 30° and 50°N. Southern Hemisphere grasslands are found at similar latitudes on the east coasts of South America, New Zealand, and Africa. These vast, undulating expanses of grass-dominated landscape have often been compared to a terrestrial ocean, with wind-driven "waves" of plants bending to the gusts blowing through them.

Temperate climates have greater seasonal temperature variation than tropical climates, with increasing periods of subfreezing temperatures toward the poles. Within the temperate zone, grasslands are usually associated with warm, moist summers and cold, dry winters. Precipitation in some grasslands is high enough to support forests, as at the eastern edge of the Great Plains. However, frequent fires and grazing by large herbivores such as bison prevent the establishment of trees and thus maintain the dominance of grasses in these environments. The use of fire to manage grasslands near the edges of forests was probably one of the first human activities with a widespread effect on a terrestrial biome.

The world's grasslands have been a major focus for agricultural and pastoral development. In order to acquire adequate supplies of water and nutrients, grasses grow more roots than stems and leaves. The rich organic matter that accumulates in the soils as a result enhances their fertility, so grassland soils are particularly well suited for agricultural development. Most of the fertile grasslands of central North America and Eurasia have been converted to agriculture. The diversity of the crop species grown on these lands is far less than the diversity of the grasslands they replaced. In more arid grasslands, rates of grazing by domesticated animals can exceed the capacity of the plants to produce new tissues, and grassland degradation, including desertification, may occur. As in deserts, irrigation of some grassland soils has resulted in salinization, decreasing their fertility over time. In parts of Europe, cessation of centuries-old grazing practices has resulted in increased forest invasion into grasslands. This long legacy of use for agriculture and grazing has made grasslands the most human-influenced biome on Earth. You can evaluate the possible effects of climate change in **ANALYZING DATA 3.1**.

Temperate Grasslands

Single male plains bison on the prairie in Grasslands National Park, Saskatchewan, Canada

Equator

Denison, Iowa, USA
41°N, 389 m
Average annual temperature 9.1°C
Total annual precipitation 727 mm

Subfreezing average monthly temperatures

(Data from NOAA GHCN-Monthly, version 2; T. C. Peterson and R. S. Vose. 1997. *Bull Am Meteorol Soc* 78: 2837–2849.)

Impala grazing the Mara Triangle, South Africa

TEMPERATE SHRUBLANDS AND WOODLANDS The seasonality of precipitation is an important control on the distribution of temperate biomes. Woodlands (characterized by an open canopy of short trees) and shrublands occur in regions with a winter rainy season (in contrast to grasslands, with a summer rainy season). *Mediterranean-type climates*, which occur on the west coasts of the Americas, Africa, Australia, and Europe between 30° and 40°N and S, are an example of such a climate regime. As we saw in Concept 2.1, these Mediterranean-type climates are characterized by asynchrony between precipitation and the summer *growing season* (the period of time with suitable temperatures to support growth). Precipitation falls primarily in winter, and hot, dry weather occurs throughout the late spring, summer, and fall. The vegetation of

Mediterranean-type climates is characterized by evergreen shrubs and trees. Evergreen leaves allow plants to be active during cooler, wetter periods and also lower their nutrient requirements, since they do not have to grow new leaves every year. Many plants of Mediterranean-type climates have *sclerophyllous* leaves, which are tough, leathery, and stiff. These plants are well adapted to dry soils and may continue to photosynthesize and grow at reduced rates during the hot, dry summer. Sclerophyllous leaves also help to deter consumption by herbivores and prevent wilting as water is lost. Sclerophyllous shrublands are found in each of the zones characterized by a Mediterranean-type climate, including the *mallee* of Australia, the *fynbos* of South Africa, the *matorral* of Chile, the *maquis* around the Mediterranean Sea, and the *chaparral* of North America.

ANALYZING DATA 3.1

How Will Climate Change Affect the Grasslands Biome?

The climate diagrams shown for each of the terrestrial biomes exemplify the climate patterns with which they are associated (see Ecological Toolkit 3.1). In particular, they show periods of potential plant stress due to low water availability and subfreezing temperatures, which are particularly important in shaping the types of plants that grow in a given location. Global climate change is altering both temperature and precipitation patterns worldwide. Eventually, therefore, the species composition of the biome at a given site will change, as happened following the end of the last ice age.

The world's remaining grasslands are particularly threatened by climate change. Much of this biome has already been lost and fragmented by land use change for agricultural and pastoral activities. Current predictions for the tallgrass prairie of the central United States suggest that by 2050, its average annual temperature will increase by 2.3°C and total annual precipitation will not change.

1. Assuming that the changes in temperature occur evenly across the year, draw climate diagrams representing the current and year 2050 conditions for Ellsworth, Kansas, a grassland site in the southern Great Plains. Use the data in the table below for the *current* climate.

2. Redraw the climate diagram assuming that winter (December, January, February) precipitation increases by 20% and that summer (June, July, and August) precipitation decreases by 20%, as predicted by some climate change models.

3. Does the diagram from Question 2 show changes in the periods of possible water and temperature stress? If so, how do you think these changes will influence the vegetation composition of the tallgrass prairie? Use the information in Ecological Toolkit 3.1 to assist your reasoning.

4. What factors other than climate should be considered in a prediction of the future fate of the grassland biome?

Ellsworth, Kansas, 38°43′ N, 98°14′ W, 466 m elevation

	J	F	M	A	M	J	J	A	S	O	N	D
Average monthly temperature (°C)	−2.1	0.9	6.9	13.1	18.3	23.8	26.9	25.7	20.7	14.3	6.1	−0.2
Average monthly precipitation (mm)	15.2	19.8	56.6	61.5	104.1	102.4	81.8	84.1	79.0	56.1	27.7	19.8

Source: http://www.ellsworth.climatemps.com/

Fire is a common feature in Mediterranean-type shrublands and, as it does in some grasslands, may promote their persistence. Some of the shrubs recover after fires by resprouting from woody storage organs protected from the heat below the ground surface. Other shrubs produce seeds that germinate and grow quickly after a fire. Without regular fires at 30- to 40-year intervals, some temperate shrublands may be replaced by forests of oaks, pines, junipers, or eucalypts. Regular disturbance by fire, combined with the unique climate of temperate shrublands, is thought to promote high species diversity.

Shrublands and woodlands are also found in the continental interior of North America and Eurasia, where they are associated with rain shadow effects and seasonally cold climates. The Great Basin, for example, occupies the interior of North America between the Sierra Nevada and the Cascade Range to the west and the Rocky Mountains to the east. Large expanses of sagebrush (*Artemisia tridentata*), saltbush (*Atriplex* spp.), creosote bush (*Larrea*

tridentata), and piñon pine and juniper woodland occur throughout this area.

Humans have converted some regions of temperate shrublands and woodlands to croplands and vineyards. However, their climates and nutrient-poor soils have limited the extent of agricultural and pastoral development. In the Mediterranean basin, agricultural development using irrigation was attempted but failed because of the infertile soils. Urban development has reduced the cover of shrublands in some regions (e.g., Southern California). Increases in local human populations have increased the frequency of fires, which has decreased the ability of shrubs to recover and may lead to their replacement by invasive annual grasses.

TEMPERATE DECIDUOUS FORESTS Deciduous leaves are a solution to the extended periods of freezing weather in the temperate zone. Leaves are more sensitive to freezing than other plant tissues because of the high level

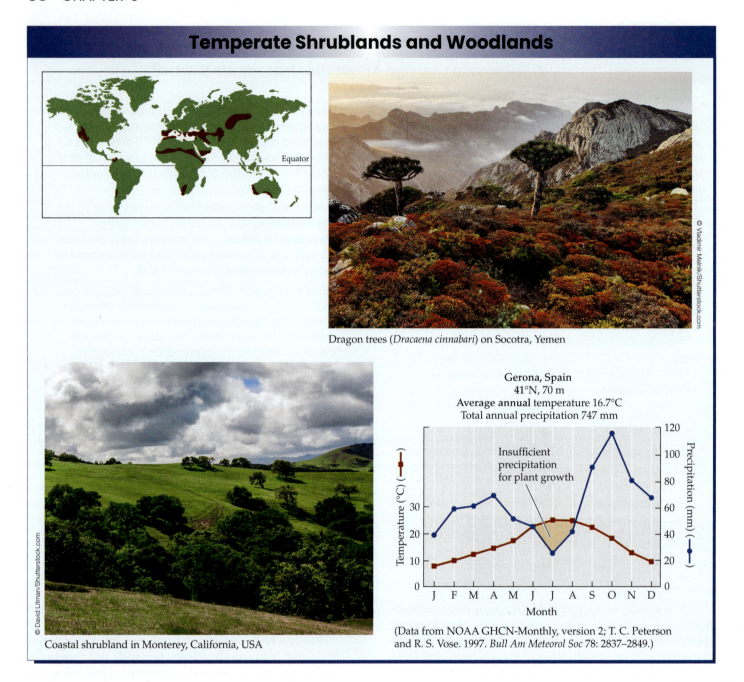

Temperate Shrublands and Woodlands

Equator

Dragon trees (*Dracaena cinnabari*) on Socotra, Yemen

© Vladimir Melnik/Shutterstock.com

Coastal shrubland in Monterey, California, USA

© David Litman/Shutterstock.com

Gerona, Spain
41°N, 70 m
Average annual temperature 16.7°C
Total annual precipitation 747 mm

Insufficient precipitation for plant growth

Temperature (°C)

Precipitation (mm)

Month

(Data from NOAA GHCN-Monthly, version 2; T. C. Peterson and R. S. Vose. 1997. *Bull Am Meteorol Soc* 78: 2837–2849.)

of physiological activity associated with photosynthesis. Temperate deciduous forests occur in areas where there is enough rainfall to support tree growth (500–2,500 mm, or 20–100 inches, per year) and where soils are fertile enough to supply the nutrients lost when leaves are shed in the fall. Temperate deciduous forests are primarily limited to the Northern Hemisphere, as the Southern Hemisphere contains less land area and lacks extensive areas with the continental climates associated with the deciduous forest biome.

Deciduous forests occur at 30° to 50°N on the eastern and western edges of Eurasia and in eastern North America, extending inland to the continental interior before diminishing because of lack of rainfall and, in some cases, increased

fire frequency. Similar species occur on each of these continents, reflecting a common biogeographic history (see Chapter 18). Oak, maple, and beech trees, for example, are components of this forest biome on each continent. The vertical structure of the forest includes canopy trees as well as shorter trees, shrubs, and forbs below the canopy. Species diversity is lower than in tropical forests but can be as high as 3,000 plant species (e.g., in eastern North America). Disturbances such as fire and outbreaks of herbivorous insects do not play a major role in determining the development and persistence of temperate deciduous forest vegetation, although they can influence its boundaries, and periodic outbreaks of herbivores (e.g., the gypsy moth, a non-native insect introduced to North America) do occur.

Temperate Deciduous Forests

Beech trees (*Fagus sylvatica*) in the Sonian Forest, Belgium

Wellsboro, Pennsylvania, USA
41°N, 567 m
Average annual temperature 7.6°C
Total annual precipitation 848 mm

Subfreezing average monthly temperatures

(Data from NOAA GHCN-Monthly, version 2; T. C. Peterson and R. S. Vose. 1997. *Bull Am Meteorol Soc* 78: 2837–2849.)

Deciduous forest in Algonquin Provincial Park, Canada

The temperate deciduous forest biome has been a focus for agricultural development for centuries. The fertile soils and climate are conducive to the growth of crops. Forest clearing for crop and wood production has historically been widespread in this biome. Very little old-growth temperate deciduous forest remains on any continent. Since the early twentieth century, however, agriculture has gradually shifted away from temperate-zone forests toward temperate grasslands and the tropics, particularly in the Americas. Abandonment of agricultural fields has resulted in reforestation in some parts of Europe and North America. However, the species composition of the second-growth forests often differs from what was present prior to agricultural development. Nutrient loss from soils due to long-term agricultural use is one reason for this difference. Another is the loss of some species due to

introductions of invasive species. For example, the chestnut blight fungus, introduced from Asia, nearly wiped out the chestnut trees (*Castanea* spp.) of North America in the early twentieth century (see Figure 13.14). As a result, oak species are more widespread than they were prior to agricultural development.

TEMPERATE EVERGREEN FORESTS Evergreen forests span a wide range of environmental conditions in the temperate zone, from warm coastal regions to cool continental and maritime climates. Precipitation also varies substantially among these forests, from 500 to 4,000 mm (20–150 inches) per year. Some temperate evergreen forests with high levels of precipitation, which are typically located on west coasts at latitudes between 45° and 50°N and S, are referred to as "temperate rainforests" (**FIGURE 3.8**).

FIGURE 3.8 Temperate Rainforest Rainforests occur in temperate zones with high precipitation (over 5,000 mm, or 200 inches) and relatively mild winter temperatures. Here, understory tree ferns grow beneath the canopy trees at Horseshoe Falls in western Tasmania, Australia.

Temperate evergreen forests are commonly found on nutrient-poor soils, whose condition is in part related to the acidic nature of the leaves of the evergreen trees. Some evergreen forests are subject to regular fires at intervals of 30–200 years, which may promote their persistence.

Temperate evergreen forests are found in both the Northern and Southern Hemispheres between 30° and 50° latitude. Their diversity is generally lower than that

Temperate Evergreen Forests

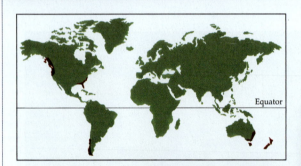

Araucaria (monkey puzzle trees) growing in a forest.

Giant Douglas Fir, red cedartrees, and ferns in Cathedral Grove. MacMillian Provincial Park, Vancouver Island, BC

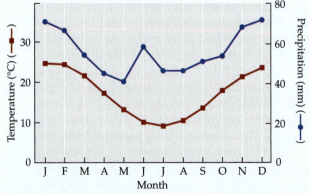

Tamworth, Australia
31°S, 405 m
Average annual temperature 17.5°C
Total annual precipitation 672 mm

(Data from NOAA GHCN-Monthly, version 2; T. C. Peterson and R. S. Vose. 1997. *Bull Am Meteorol Soc* 78: 2837–2849.)

of deciduous and tropical forests. In the Northern Hemisphere, the tree species include needle-leaved conifers such as pines, junipers, and Douglas fir (*Pseudotsuga menziesii*). In the Southern Hemisphere, on the west coasts of Chile and Tasmania, in southeastern and southwestern Australia, and in New Zealand, there is a greater diversity of tree species, including southern beeches (*Nothofagus* spp.), eucalypts, the Chilean cedar (*Austrocedrus*), and podocarps (see Figure 18.4).

Conifers provide a source of high-quality wood and pulp for paper production. The temperate evergreen forest biome has been subjected to extensive clearing, and little old-growth forest remains. Some forestry practices tend to promote sustainable use of these forests, although in some regions the planting of non-native species (such as Monterey pine in New Zealand), the uniform age and density

of the trees, and losses of formerly dominant species have created forests that are ecologically very different from their pre-logging condition. The suppression of naturally occurring fires in western North America has increased the density of some forest stands, which has resulted in more intense fires in association with climate change (Dennison et al. 2014) and has increased the spread of insect pests (e.g., pine beetles) and pathogens. In industrialized countries, the effects of air pollution have damaged some temperate evergreen forests (see Figure 25.18) and made them more susceptible to other stresses.

BOREAL FORESTS   Above 50°N, the severity of winters increases. Minimum temperatures of –50°C (–58°F) are common in continental locations such as Siberia, and continuous subfreezing temperatures may last up to

Boreal Forests

Boreal forest in the Kluane National Park and Reserve, from Alaska Highway, Yukon Territory, Canada

Fort Simpson, Northwest Territories, Canada
61°N, 169 m
Average annual temperature –4.6°C
Total annual precipitation 333 mm

(Data from NOAA GHCN-Monthly, version 2; T. C. Peterson and R. S. Vose. 1997. *Bull Am Meteorol Soc* 78: 2837–2849.)

Boreal forest in the Murmansk region, Russia

6 months. The extreme weather in these subarctic regions is an important determinant of the vegetation structure. Not only must the plants cope with low air temperatures, but soils may regularly freeze, leading to the formation of **permafrost**, defined as a subsurface soil layer that remains frozen year-round for at least 3 years. Although precipitation is low, the permafrost impedes water drainage, so soils are moist to saturated.

The biome that occupies the zone between 50° and 65°N is the boreal (far northern) forest. This biome is also known as *taiga*, the Russian word for this northern forest. It is composed primarily of coniferous species, including spruces, pines, and larches (deciduous needle-leaved trees), but also includes extensive deciduous birch forests in maritime locations, particularly in Scandinavia. Conifers tend to resist damage from winter freezing better than angiosperm trees, despite maintaining green leaves year-round. Although the boreal forest is found only in the Northern Hemisphere, it is the largest biome in area and contains one-third of Earth's forested land.

Boreal forest soils are cold and wet, limiting the decomposition of plant material such as leaves, wood, and roots. Thus, the rate of plant growth exceeds the rate of decomposition, and the soils contain large amounts of organic matter. During extensive summer droughts, forests are more susceptible to fires ignited by lightning. These fires may burn both the trees and the soil (**FIGURE 3.9**). Soil fires may continue to burn slowly for several years, even through the cold winters. In the absence of fire, forest growth enhances permafrost formation by lowering the amount of sunlight absorbed by the soil surface. In low-lying areas, soils become saturated, killing the trees and forming extensive peat bogs.

Boreal forests have been less affected by human activities than other forest biomes. Logging occurs in some regions, as does oil and gas development, including the mining of oil sands. These activities pose an increasing threat to the boreal forest as demands for wood and energy increase. In addition, the large store of organic matter in the soil makes boreal forests an important component of the global carbon cycle. Climate warming may result in more rapid decomposition and thus higher rates of carbon release from boreal forest soils, increasing atmospheric greenhouse gas concentrations that in turn cause additional warming (see Concept 25.2).

TUNDRA Trees cease to be the dominant vegetation beyond approximately 65° latitude. The tree line that marks the transition from boreal forest to tundra is associated with low growing-season temperatures, although the causes of this transition are complex and can also include other climate and soil conditions. The tundra biome occurs primarily in the Arctic but can also be found on the edges of the Antarctic Peninsula and on a few islands in the Southern Ocean. The poleward decrease in temperature and precipitation across the tundra biome is associated with the zones of high pressure generated by the polar atmospheric circulation cells (see Figure 2.8).

The tundra biome is characterized by sedges, forbs, grasses, and low-growing shrubs such as heaths, willows, and birches. Lichens and mosses are also important components of this biome. Although the summer growing season is short, the days are long, with continuous periods of light for 1 to 2 months of the summer. The plants and lichens survive the long winter by going dormant, maintaining living tissues under the snow or soil, where they are insulated from the cold air temperatures.

The tundra and the boreal forest have several similarities: temperatures are cold, precipitation is low, and permafrost is widespread. Despite the low precipitation, many tundra areas are wet, as the permafrost keeps the precipitation that does fall from percolating to deeper soil layers. Repeated freezing and thawing of surface soil layers over several decades results in sorting of soil materials according to their texture. This process forms polygons of soil at the surface with upraised rims and depressed centers (**FIGURE 3.10**). Where soils are coarser or permafrost does not develop, the soils may be dry, and plants must be able to cope with low water availability. These polar deserts are most common at the high latitude limit of the tundra biome.

Herds of caribou and musk oxen, as well as predators such as wolves and brown bears, inhabit the tundra. Many species of migratory birds nest in the tundra during the summer. Human inhabitants are scattered in sparse settlements. As a result, this biome contains some of the largest pristine regions on Earth. The influence of human activities on the tundra is increasing, however. Exploration and development of energy resources has accelerated. A key to limiting the effects of energy development is preventing damage to the permafrost, which can cause long-term

Courtesy of U.S. Forest Service Research

FIGURE 3.9 Fire in the Boreal Forest Despite the cold climate of the boreal forest, fire is an important part of its environment.

Tundra

Polar Bear Pass and tundra landscape on Bathurst Island, Nunavut

Tundra at Eqip Sermia on the west Greenland coast north of Ilulissat

Olenek, Russia
73°N, 11 m
Average annual temperature −14.3°C
Total annual precipitation 184 mm

Subfreezing average monthly temperatures

(Data from NOAA GHCN-Monthly, version 2; T. C. Peterson and R. S. Vose. 1997. *Bull Am Meteorol Soc* 78: 2837–2849.)

erosion. The Arctic has experienced climate warming almost double the global average during the late twentieth and early twenty-first centuries. Increased losses of permafrost, catastrophic lake drainage, and reduced carbon storage in the soil have been linked to climate change.

Now that we've completed our tropics-to-tundra tour of terrestrial biomes, let's consider the influence of mountains on more local-scale patterning of biological communities. In some mountainous locations, elevational changes result in a smaller version of our latitudinal description of biomes.

FIGURE 3.10 Soil Polygons and Pingo Pingos are small hills found in the Arctic, formed by an intrusion of water that freezes in the subsurface permafrost zone, thrusting the soil above it upward. The polygons on the periphery of the pingo result from freezing and thawing of soils, a process that pushes coarse soil materials toward the edges and finer soil to the middle of the polygons.

Biological communities in mountains occur in elevational bands

Approximately one-fourth of Earth's land surface is mountainous. Mountains create climate gradients that change more rapidly over a given distance than those associated with changes in latitude. Temperatures decrease with elevation (for reasons described in Concept 2.3); for example, temperatures in temperate continental mountain ranges decrease approximately 6.4°C for every 1,000 m increase in elevation (or 3.6°F per 1,000 feet), a decrease equivalent to that over approximately a 13° change in latitude, or a distance of 1,400 km (870 miles). As we might expect from our consideration of biomes and their close association with climate, coarse biotic assemblages similar to biomes occur in elevational bands on mountains. Finer-scale biotic distinctions are found in association with slope aspect (e.g., north-facing vs. south-facing), proximity to streams, and the orientation of slopes in relation to prevailing winds (see Concept 2.4).

The biological communities that occur from the base to the summit of a temperate-zone mountain range resemble what we would find along a latitudinal gradient toward higher latitudes. An elevational transect on the eastern slope of the southern Rocky Mountains in Colorado, for example, includes grassland to alpine vegetation across a 2,200 m (7,200-foot) increase in elevation (**FIGURE 3.11**). The changes in climate and vegetation are similar to the transition from grassland to tundra that occurs with a 27° increase in latitude, from Colorado to the Northwest Territories of Canada. Grasslands occur at the base of the mountains, but they give way to pine savannas on the initial slopes (the lower montane zone). Fire plays an important role in determining the vegetation structure of both montane grasslands and savannas. With increasing elevation, the pine savannas are replaced by denser stands of mixed pine-and-aspen forests (the montane zone), which resemble temperate evergreen and deciduous forest biomes. Spruce and fir trees make up the forests of the subalpine zone, which resemble the boreal forest biome. Mountain tree lines are similar to the transition from boreal forest to tundra, although topography can play an important role through its influence on snow distribution and avalanches. The alpine zone above the tree line includes diminutive plants such as sedges, grasses, and forbs, including some of the same species that occur in the Arctic tundra. Although the alpine zone resembles the tundra, its physical environment is different, with higher wind speeds, more intense solar radiation, and lower atmospheric partial pressures of oxygen (O_2) and carbon dioxide (CO_2).[1]

Mountains are found on all continents and at all latitudes. As indicated in the example above, the changes in climate associated with changes in elevation alter the composition of the local vegetation. Not all of the vegetation assemblages that occur in mountains resemble major terrestrial biomes, however. Some mountain-influenced biological communities have no biome analogs. For example, daily temperature changes at high-elevation sites in the tropics (e.g., Kilimanjaro and the tropical Andes) are greater than seasonal temperature changes. Subfreezing

---

[1] The *partial pressure* of a gas is defined as the pressure exerted by a particular component of a mixture of gases. The concentrations of CO_2 and O_2 are the same at high elevations as they are at sea level, but their partial pressures are lower because total atmospheric pressure is lower. The exchange of a gas between an organism and the atmosphere is determined by its partial pressure rather than its concentration.

FIGURE 3.11 Mountain Biological Zones An elevational transect on the eastern slope of the southern Rocky Mountains passes through climate conditions and biome-like assemblages similar to those found along a latitudinal gradient between Colorado and northern Canada. (Data from J. W. Marr. 1967. *Ecosystems of the East Slope of the Front Range of Colorado.* University of Colorado Press: Boulder, CO.)

❓ Would you expect the same biological zonation on east-facing and west-facing slopes in a temperate mountain range near the west coast of a continent?

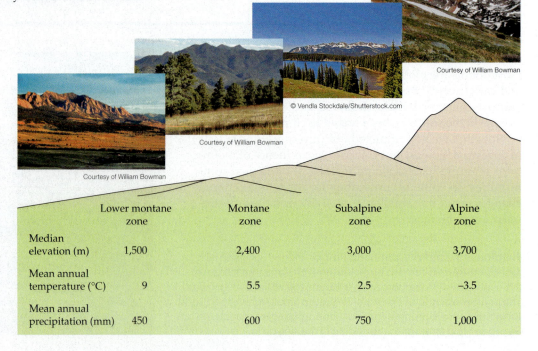

Courtesy of William Bowman

© Vendla Stockdale/Shutterstock.com

Courtesy of William Bowman

Courtesy of William Bowman

	Lower montane zone	Montane zone	Subalpine zone	Alpine zone
Median elevation (m)	1,500	2,400	3,000	3,700
Mean annual temperature (°C)	9	5.5	2.5	−3.5
Mean annual precipitation (mm)	450	600	750	1,000

FIGURE 3.12 Tropical Alpine Plants Frailejón (*Espeletia* spp.) grows in alpine grasslands in the Ecuadorian Andes. Its growth form, characterized by a circle of leaves (rosette), is typical of plants in the tropical alpine zones of South America and Africa. The adult leaves help protect the developing leaves and stems at the apex of the plant from nightly frosts. Such giant rosettes are found exclusively in the tropical alpine zone and do not have analogs in the Arctic or Antarctic.

temperatures occur on most nights in the tropical alpine zone. As a result of these unique climate conditions, tropical alpine vegetation does not resemble that of the temperate alpine zone or the Arctic tundra (**FIGURE 3.12**).

CONCEPT 3.2 Biological zones in freshwater ecosystems are associated with the velocity, depth, temperature, clarity, and chemistry of the water.

LEARNING OBJECTIVES

3.2.1 Summarize how the size of particles on the bottom of streams, as well as water velocity and clarity, change from source streams to large rivers and subsequently influence the organisms that inhabit different zones of moving waters.

3.2.2 Explain how the depth and amount of light penetration in a pond or lake influence the distribution of photosynthetic and nonphotosynthetic organisms.

Freshwater Biological Zones

Although they occupy a small portion of the terrestrial surface, freshwater streams, rivers, and lakes are a key component in the connection between terrestrial and marine ecosystems. Rivers and lakes process inputs of chemical elements from terrestrial ecosystems and transport them to the oceans. The biota of these freshwater ecosystems reflect the physical characteristics of the water, including its velocity (flowing streams and rivers vs. lakes and ponds), its temperature (including seasonal changes),

how far light can penetrate it (clarity), and its chemistry (salinity, oxygen concentrations, nutrient status, and pH).

In this section, we will explore the biota and associated physical conditions found in freshwater ecosystems. In contrast to terrestrial biomes, for which only plants are used as indicators, the biological assemblages of freshwater ecosystems are characterized by both plants and animals, reflecting the greater proportional abundance of animals in aquatic ecosystems.

Biological communities in streams and rivers vary with stream size and location within the stream channel

Water flows downhill over the land surface in response to the force of gravity. The land surface is partly shaped by the erosional power of water, which cuts valleys as it heads toward a lake or ocean. The descending water converges into progressively larger streams and rivers, called **lotic** (flowing-water) ecosystems. The smallest streams at the highest elevations in a landscape are called *first-order streams* (**FIGURE 3.13**). Two first-order streams may converge to form a second-order stream. Large rivers such as the Nile or Mississippi are equal to or greater than sixth-order streams.

Individual streams tend to form repeated patterns of riffles and pools along their paths. *Riffles* are fast-moving portions of the stream flowing over coarse particles on the streambed, which increase oxygen input into the water. *Pools* are deeper portions of the stream where water flows more slowly over a bed of fine sediments. Biological communities in lotic ecosystems are associated with different physical locations within the stream and their related

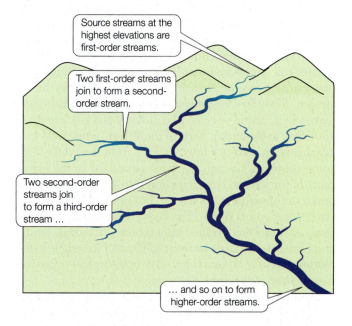

Source streams at the highest elevations are first-order streams.

Two first-order streams join to form a second-order stream.

Two second-order streams join to form a third-order stream …

… and so on to form higher-order streams.

FIGURE 3.13 Stream Orders Stream order affects environmental conditions, community composition, and the energy and nutrient relationships of communities within the stream.

environments (**FIGURE 3.14**). Organisms that live in the flowing water of the main channel are generally swimmers, such as fish. The bottom of the stream, called the **benthic zone**, is home to invertebrates; some of these, such as some mayfly and fly (dipteran) larvae, consume **detritus** (dead organic matter), and some, such as some caddisflies and crustaceans, hunt other organisms. Some organisms, such as rotifers, copepods, and insects, are found in the substrate below and adjacent to the stream, where water, either from the stream or from groundwater moving into the stream, still flows. This area is known as the **hyporheic zone**.

The composition of biological communities in streams and rivers changes with stream order (see Figure 3.13) and channel size. The *river continuum concept* was developed to describe these changes in both the physical and biological characteristics of a stream (Vannote et al. 1980). This conceptual model holds that as a stream flows downslope and increases in size, the input of detritus from the vegetation adjacent to the stream (known as *riparian vegetation*) decreases relative to the volume of water, and the particle size in the streambed decreases, from boulders and coarse rock in the higher portions to fine sand at the lower end, facilitating greater establishment of aquatic plants in the downstream direction. As a result, the importance of terrestrial vegetation as a food source for stream organisms decreases in the downstream direction. Coarse terrestrial detritus is most important near the stream source, while the importance of fine organic matter, algae, and rooted and floating aquatic vascular plants (known as **macrophytes**, from *macro*, "large"; *phyte*, "plant") increases downstream. The general feeding styles of organisms change accordingly as the river flows downstream. *Shredders*, organisms able to tear up and chew leaves (e.g., some species of caddisfly larvae), are most abundant in the higher parts of the stream, while *collectors*, organisms that collect fine particles from the water (e.g., some dipteran larvae), are most abundant in the lower parts of the stream. The river continuum concept applies best to temperate river systems, but not as well in boreal, Arctic, or tropical rivers or in rivers with high concentrations of dissolved organic substances (including tannic and humic acids) derived from wetlands. Nonetheless, the model provides a basis for studying biological organization in stream and river systems.

Human effects on lotic ecosystems have been extensive. Most fourth- and higher-order rivers have been altered by human activities, including pollution, increases in inputs of sediments, and introductions of non-native species. Streams and rivers have been used as conduits

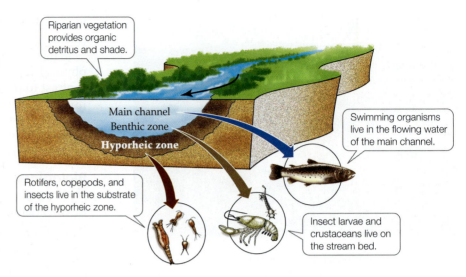

FIGURE 3.14 Spatial Zonation of a Stream Biological communities in a stream vary according to water velocity, inputs of plant material from riparian vegetation, the size of particles on the streambed, and the depth of the stream.

Where in this stream would you expect oxygen concentrations to be highest and lowest?

for the disposal of sewage and industrial wastes in most parts of the world inhabited by humans. These pollutants often reach levels that are toxic to many aquatic organisms. Excessive application of fertilizers to croplands results in runoff into rivers as well as leaching of nutrients into groundwater that eventually reaches rivers. Inputs of nitrogen and phosphorus from fertilizers alter the composition of aquatic communities. Deforestation increases inputs of stream sediment, which can reduce water clarity, alter benthic habitat, and inhibit gill function in many aquatic organisms. Introductions of non-native species, such as sport fishes (e.g., bass and trout), have lowered the diversity of native species in both stream and lake ecosystems. The construction of dams on streams and rivers tremendously alters their physical and biological properties, converting them into still waters—the topic of the next section.

Biological communities in lakes vary with depth and light penetration

Lakes and other still waters, called **lentic** ecosystems, occur where natural depressions have filled with water or where humans have dammed rivers to form reservoirs. Lakes and ponds may be formed when glaciers gouge out depressions and leave behind natural dams of rock debris (moraines), or when large chunks of glacial ice break off, become surrounded by glacial debris, and then melt. Most temperate and polar lakes are formed by glacial processes. Lakes may also form when meandering rivers cease to flow through a former channel, leaving a section stranded, called an *oxbow lake*. Geologic phenomena, such as extinct volcanic calderas and sinkholes, form natural depressions that may fill with water. Lakes and ponds of biological origin, in addition to reservoirs, include beaver dams and animal wallows.

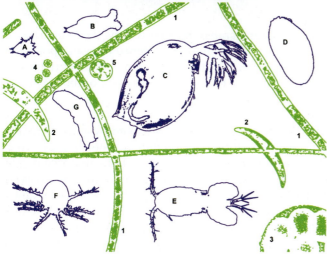

© Laguna Design/Science Source

FIGURE 3.15 Examples of Lake Plankton In this composite image of plankton from a pond, phytoplankton (green in the key) include filamentous algae (1), *Closterium* sp. (2), *Volvox* sp. (3), and other green algae (4, 5). Zooplankton (blue in the key) include a larval copepod (A), rotifer (B), water flea (*Daphnia* sp., C), ciliated protist (D), adult copepod (*Cyclops* sp.) with egg sacs (E), mite (F), and tardigrade (G).

macrophytes join with floating and benthic phytoplankton to produce energy by photosynthesis. Fish and zooplankton also occur in the littoral zone.

In the benthic zone, detritus derived from the littoral and pelagic zones serves as an energy source for animals, fungi, and bacteria. The benthic zone is usually the coldest part of the lake, and its oxygen concentrations are often low.

Let's move from fresh waters to the biological zones of the oceans. You will see that some of those zones have names and characteristics similar to those in freshwater lakes but have much greater spatial cover. As in freshwater communities, physical characteristics are used to differentiate marine biological zones.

> **CONCEPT 3.3** Marine biological zones are determined by ocean depth, light availability, and the stability of the bottom substrate.

LEARNING OBJECTIVES

3.3.1 Describe how substrate stability at the bottom of nearshore and shallow ocean zones determines which types of organisms are present, particularly emergent and nonemergent vascular plants and large algae.

3.3.2 Explain how different sources of energy and food affect the type and number of ocean organisms that exist in the water's surface and in the deepest depths.

Marine Biological Zones

Oceans cover 71% of Earth's surface and contain a rich diversity of life. The vast area and volume of the oceans and their environmental uniformity make them considerably different from terrestrial ecosystems in terms of biological organization. Marine organisms are more widely dispersed, and marine communities are not as easily organized into broad biological units as terrestrial biomes are. Instead, marine biological zones are coarsely categorized by their physical locations relative to shorelines and the ocean bottom (**FIGURE 3.16**). The distributions of the organisms that inhabit these zones reflect differences in temperature, as we saw for terrestrial biomes, as well as other important factors, including light availability, water depth, stability of the bottom substrate, and interactions with other organisms.

Lakes vary tremendously in size, from small, ephemeral ponds to the massive Lake Baikal in Siberia, which is 1,600 m (5,200 feet) deep and covers 31,000 km² (12,000 square miles). The size of a lake has important consequences for its nutrient and energy status and therefore for the composition of its biological communities. Deep lakes with little surface area tend to be nutrient poor compared with shallow lakes with much surface area (see Concept 22.4).

Lake biotic assemblages are associated with depth and degree of light penetration. The open water, or **pelagic zone**, is inhabited by **plankton**: small, often microscopic organisms that are suspended in the water (**FIGURE 3.15**). Photosynthetic plankton (called **phytoplankton**) are limited to the surface layer of water where there is enough light for photosynthesis, called the **photic zone**. **Zooplankton**—tiny animals and nonphotosynthetic protists—occur throughout the pelagic zone, as do other consumers such as bacteria and fungi, feeding on detritus as it falls through the water. Fish patrol the pelagic zone, scouting for food and predators that might eat them.

The nearshore zone where the photic zone reaches to the lake bottom is called the **littoral zone**. Here,

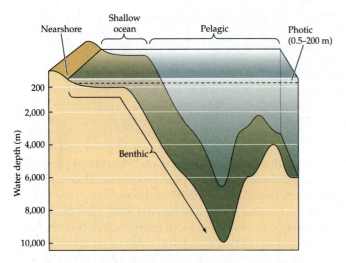

FIGURE 3.16 Marine Biological Zones Biological zones in the ocean are categorized by water depth and by their physical locations relative to shorelines and the ocean bottom.

In this section, we will take a tour of the biological zones of the oceans, from the margins of the land to the deep, dark, cold ocean bottom. We will examine the physical and biological factors that characterize the different zones and the major organisms found in them.

Nearshore zones reflect the influence of tides and substrate stability

Marine biological zones adjacent to the continents are influenced by local climate, by the rise and fall of ocean waters associated with tides, by wave action, and by the influx of fresh water and terrestrial sediments from rivers. **Tides** are generated by the gravitational attraction between Earth and the moon and sun. Ocean water rises and falls in most nearshore zones twice daily. The magnitude of the tidal range varies greatly among different locations because it is related to shoreline morphology and ocean bottom structure. Tides produce unique transition zones between terrestrial and marine environments and influence salinity and nutrient availability in these nearshore habitats.

ESTUARIES The junction of a river with the ocean is called an *estuary* (**FIGURE 3.17**). Estuaries are characterized by variations in salinity associated with the flow of fresh river water into the ocean and the influx of salt water flowing inland from the ocean as tides rise. Rivers bring terrestrial sediments containing nutrients and organic matter to the ocean, and the interaction of tidal and river flows acts to trap these sediments in estuaries, enhancing their productivity. The varying salinity of estuaries is an important determinant of the organisms that occur there. Many commercially valuable fish species spend their juvenile stages in estuaries, away from fish predators that are not as tolerant of the changes in salinity. Other inhabitants of estuaries include shellfish (e.g.,

FIGURE 3.17 Estuaries Are Junctions between Rivers and Oceans The mixing of fresh and salt water gives estuaries a unique environment with varying salinity. Rivers bring in energy and nutrients from terrestrial ecosystems.

clams, oysters, crabs), marine worms, and seagrasses. Estuaries are increasingly threatened by water pollution carried by rivers. Nutrients from upstream agricultural sources can cause local dead zones (see Concept 25.3) and losses of biological diversity.

SALT MARSHES Terrestrial sediments carried to shorelines by rivers form shallow marsh zones (**FIGURE 3.18**) that are dominated by vascular plants that rise out of the water, including grasses, rushes, and broad-leaved herbs. In these salt marshes, as in the estuaries that they often border, the input of nutrients from rivers enhances productivity. Periodic flooding of the marsh at high tide results in a gradient of salinity: the highest portions of the marsh can be the most saline because infrequent flooding and evaporation of water from the soil lead to a progressive buildup of salts. Salt marsh plants grow in distinct zones that reflect this salinity gradient, with the most salt-tolerant species in the highest portions of the

FIGURE 3.18 Salt Marshes Are Characterized by Salt-Tolerant Vascular Plants Emergent vascular plants form salt marshes in shallow nearshore zones.

© Damsea/Shutterstock.com

FIGURE 3.19 Salt-Tolerant Evergreen Trees and Shrubs Form Estuarine Mangrove Forests   The mangrove roots trap mud and sediments and provide habitat for other marine organisms.

© Vara I/Shutterstock.com

FIGURE 3.20   The Rocky Intertidal Zone: Stable Substrate, Changing Conditions   Rocky shorelines provide a stable substrate to which organisms can anchor themselves, but those organisms must cope with the shift from terrestrial to marine conditions that occurs with each tide, as well as wave action. Sessile organisms must be resistant to temperature changes and desiccation. Mobile organisms often take refuge in tide pools to avoid exposure to the terrestrial environment.

marsh. Salt marshes provide food and protection from predators for a wide variety of animals, including fishes, crabs, birds, and mammals. Organic matter trapped in salt marsh sediments may serve as a nutrient and energy source for nearby marine ecosystems.

MANGROVE FORESTS   Shallow coastal estuaries and nearby mudflats in tropical and subtropical regions are inhabited by salt-tolerant evergreen trees and shrubs (**FIGURE 3.19**). These woody plants are collectively referred to as mangroves, but "mangroves" include species from 16 different plant families, not a single taxonomic group. Mangrove roots trap mud and sediments carried by the water, which build up and modify the shoreline. Like salt marshes, mangrove forests provide nutrients to other marine ecosystems and habitat for numerous animals, both marine and terrestrial. Among the unique animals associated with mangroves are manatees, crab-eating monkeys, fishing cats, and monitor lizards. Mangrove forests are threatened by human development of coastal areas—particularly the development of shrimp farms—as well as by water pollution, diversion of inland freshwater sources, and cutting of the forests for wood.

ROCKY INTERTIDAL ZONES   Rocky shorelines provide a stable substrate to which a diverse collection of algae and animals can anchor themselves to keep from being washed away by the pounding waves (**FIGURE 3.20**). The physical environment of the **intertidal** zone—the part of the shoreline affected by the rise and fall of the tides—alternates between marine and terrestrial. Between the high- and low-tide marks, a host of organisms are arranged in zones

associated with their tolerance for temperature changes, salinity, *desiccation* (drying out), wave action, and interactions with other organisms. *Sessile* (attached) organisms such as barnacles, mussels, and seaweeds must cope with these stresses in order to survive. Mobile organisms, such as sea stars and sea urchins, may move to tide pools in order to minimize exposure to these stresses.

SANDY SHORES   Except for a few scurrying crabs and shorebirds and the occasional bit of seaweed washed ashore, sandy beaches appear devoid of life. Unlike the rocky shore, the sandy substrate provides no stable anchoring surface, and the lack of attached seaweeds limits the supply of potential food for herbivorous animals. Tidal fluctuations and wave action further limit the

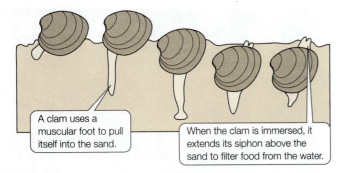

FIGURE 3.21 Burrowing Clams Clams, like most animals of sandy shorelines, live in the sandy substrate.

FIGURE 3.22 A Coral Reef Corals, like this one off North Sulawesi, Indonesia, create habitat for a diverse assemblage of marine organisms.

potential for the development of biological communities. Beneath the sand, however, invertebrates such as clams, sea worms, and mole crabs find suitable habitat (**FIGURE 3.21**). Smaller organisms, such as polychaete worms, hydroids (small animals related to jellyfishes), and copepods (tiny crustaceans), live on or among the grains of sand. These organisms are protected from temperature changes and desiccation at low tide and from the turbulent water at high tide. When the sand is immersed in seawater, some of these organisms emerge to feed on detritus or other organisms, while others remain buried and filter detritus and plankton from the water.

Shallow ocean zones are diverse and productive

Near the coastline, enough light may reach the ocean bottom to permit the establishment of sessile photosynthetic organisms. Like terrestrial plants, these photosynthetic organisms provide energy that supports communities of animals and microorganisms, as well as a physical structure that creates habitat for those organisms, including surfaces to which they can anchor and places where they can find refuge from predators. The diversity and complexity of the habitats provided by the photosynthesizers support considerable biological diversity in these shallow ocean environments.

CORAL REEFS In warm, shallow ocean waters, corals (animals related to jellyfishes), living in a close association with algal partners (a symbiotic mutualism; see Concept 15.1), form large colonies. The corals obtain most of their energy from algae that live within their bodies, while the algae receive protection from grazers and some nutrients from the corals. Many corals build a skeleton-like structure by extracting calcium carbonate from seawater. Over time, these coral skeletons pile up into massive formations called *reefs* (**FIGURE 3.22**). The formation of reefs is aided by other organisms that extract other minerals from seawater, such as sponges that precipitate silica. The unique association of these reef-building organisms gives rise to a structurally complex habitat that supports a rich marine community.

Coral reefs grow at rates of only a few millimeters per year, but they have shaped the face of Earth (Birkeland 1997). Over millions of years, corals have constructed thousands of kilometers of coastline and numerous islands (**FIGURE 3.23**). The rate of production of living biomass in coral reefs is among the highest on Earth. The accretions of coral skeletons are as much as 1,300 m (4,300 feet) thick in some places, and they currently cover a surface area of 600,000 km² (23,000 square miles), approximately 0.2% of the ocean surface.

As many as a million species are found in coral reefs worldwide, including more than 4,000 fishes. Many economically important fish species rely on coral reefs for habitat, and reef fishes provide a source of food for fishes of the open ocean, such as jacks and tuna. The taxonomic and morphological diversity of animals in coral reefs is greater than in any other ecosystem on Earth (Paulay 1997). The full diversity of coral reefs has yet to be explored and described, however. The potential for development of medicines from coral reef organisms is great enough that the U.S. National Institutes of Health established a laboratory in Micronesia to explore it.

Human activities threaten the health of coral reefs in a number of ways. Sediments carried by rivers can cover and kill the corals, and excess nutrients increase the growth of algae on the surfaces of the corals, increasing coral mortality. Changes in ocean temperatures associated with climate change can result in the loss of the corals' algal partners, a condition called *bleaching*. Increased atmospheric CO_2 has increased ocean acidification (discussed in more detail in Concept 25.1), which inhibits the ability of corals to form skeletons (Orr et al. 2005). Another threat is an increased incidence of fungal infections, possibly related to increased environmental stress.

SEAGRASS BEDS Although we typically associate flowering plants with terrestrial environments, some

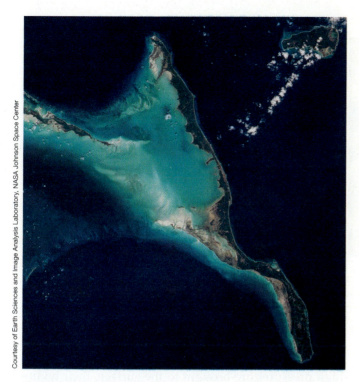

FIGURE 3.23 Coral Reefs Can Be Seen from Outer Space Long Island, in the Bahamas, was formed by coral reefs, which can be seen on the fringes of the island in this satellite photograph.

FIGURE 3.24 A Kelp Bed Giant kelp are brown algae (order Laminariales) that attach themselves to the solid bottom in shallow ocean waters, providing food and habitat for many other marine organisms.

flowering plants are important components of shallow (<5 m) underwater communities. These submerged flowering plants are called seagrasses, although they are not closely related to plants in the grass family. Morphologically, they are similar to their relatives on land, with roots, stems, and leaves as well as flowers, which are pollinated underwater. Seagrass beds are found on subtidal marine sediments composed of mud or fine sand. The plants reproduce primarily by vegetative growth, although they produce seeds as well. Marine algae and animals grow on the surfaces of the plants, and the larval stages of some organisms, such as mussels, are dependent on them for habitat. Inputs of nutrients from upstream agricultural activities can harm seagrass beds by increasing the density of algae in the water and on the surfaces of the seagrasses. Seagrasses are also susceptible to periodic outbreaks of fungal diseases.

KELP BEDS In clear, shallow (<15 m) temperate ocean waters, large stands of seaweed, known as kelp beds or kelp forests (**FIGURE 3.24**), support a rich and dynamic community of marine life. Kelp are large brown algae of several different genera. They have specialized tissues resembling leaves (fronds), stems (stipes), and roots (holdfasts). Kelp are found where a solid substrate is available for anchoring. Residents of kelp beds include sea urchins, lobsters, mussels, abalones, numerous other seaweeds, and sea otters. Interactions among these organisms, both

direct and indirect, influence the abundance of the kelp (see the Case Study in Chapter 9). In the absence of grazing, kelp beds can become so dense that light reaching the bottom of the canopy is not sufficient to support photosynthesis.

Open ocean and deep benthic zones are determined by light availability and proximity to the bottom

Beyond the continental shelves, the vastness and depth of the open ocean, known as the pelagic zone, make it difficult to differentiate distinct biological communities there. Light availability determines where photosynthetic organisms can occur, which in turn determines the availability of food for animals and microorganisms. Thus, the surface waters with enough light to support photosynthesis (the photic zone) contain the highest densities of organisms (see Figure 3.16). The photic zone extends about 200 m downward from the ocean surface, depending on water clarity. Below the photic zone, the supply of energy, mainly in the form of detritus falling from the photic zone, is much lower, and life is far less abundant.

The diversity of life in the pelagic zone varies considerably. Its **nekton** (swimming organisms capable of overcoming ocean currents) include cephalopods such as squids and octopuses, fishes, sea turtles, and mammals such as whales and porpoises. Most of the photosynthesis in the pelagic zone is carried out by phytoplankton, which include green algae, diatoms, dinoflagellates, and cyanobacteria (**FIGURE 3.25A**). Zooplankton include protists such as ciliates, crustaceans such as copepods and krill, and jellyfishes (**FIGURE 3.25B**). Many species of pelagic seabirds, including albatrosses, petrels, fulmars, and boobies, spend the majority of their lives flying over open ocean waters, feeding on marine prey (fish and zooplankton) and detritus found on the ocean surface.

(A) Marine phytoplankton

(B) Marine zooplankton

A,B © D.P. Wilson/FLPA/Science Source

FIGURE 3.25 Plankton of the Pelagic Zone (A) This sample of marine phytoplankton includes several species of diatoms, including *Biddulphia sinensis* (the rectangular cells with the concave ends) and *Thalassiothrix*. (B) These marine zooplankton include adult copepods and the larval stages of various organisms, including the zoea (spherical) larva of a crab.

Organisms that live in the pelagic zone must overcome the effects of gravity and water currents that could force them to progressively greater depths. Photosynthetic organisms, and those directly dependent on them as a food source, must stay in the photic zone where sunlight is sufficient to maintain photosynthesis, growth, and reproduction. Swimming is an obvious solution to this problem, used by organisms such as fishes and squids. Seaweeds such as *Sargassum* and some fish species have gas-filled bladders that keep them buoyant. Large mats of *Sargassum* sometimes form "floating islands" that host rich and diverse biological communities. Some plankton retard their sinking by decreasing their density relative to seawater (e.g., by altering their chemical composition) or through shapes that lower their downward velocity (e.g., having a cell wall with projections).

Beneath the photic zone, the availability of energy decreases, and the physical environment becomes more demanding as temperatures drop and water pressure rises. As a result, organisms are few and far between. Crustaceans such as copepods graze on the rain of falling detritus from the photic zone. Crustaceans, cephalopods, and fishes are the predators of the deep sea. Some fishes take on frightening forms, appearing to be mostly mouth (**FIGURE 3.26**). The meanings of the scientific names given to some of the sea creatures at this depth, such as "vampire squid from hell" (*Vampyroteuthis infernalis*), "stalked toad with many filaments" (*Caulophryne polynema*), and "Prince Axel's wonder fish" (*Thaumatichthys axeli*), testify to the unusual forms found there. Most deep-sea fishes have weak bone structure to reduce their weight and lack the gas bladder found in most fishes, since the high pressures would collapse it.

The ocean bottom (the benthic zone) is also very sparsely populated. Temperatures are near freezing, and pressures are great enough to crush any terrestrial organism. Conversely, if deep-sea creatures adapted to these high pressures are brought to the surface, their bodies may expand and burst. The sediments of the benthic zone, which are rich in organic matter, are inhabited by archaea, bacteria, protists, and sea worms. Sea stars and sea cucumbers graze the ocean floor, consuming organic matter or organisms in the sediments or filtering food from the water. Benthic predators, like those of the deep pelagic zone, use bioluminescence to lure prey. Unique communities of organisms can be found in hydrothermal vents scattered in the benthic zone in association with volcanic

© Norbert Wu/Minden Pictures

FIGURE 3.26 A Denizen of the Deep Pelagic Zone Anglerfish (*Melanocetus* spp.) are named for their unique strategy for capturing prey. In the lightless depths, the bioluminescent organ on the fish's forehead attracts prey to a position where they are easily engulfed by the huge, tooth-filled mouth.

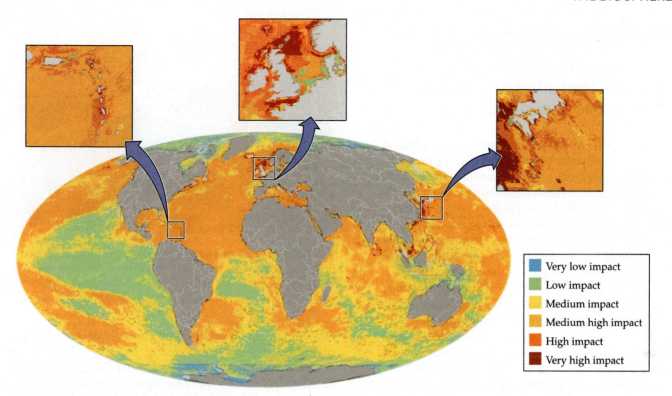

FIGURE 3.27 Human Impacts on the Oceans The impacts of greenhouse gas emissions, pollutant inputs, and overfishing have varied in different regions of the oceans. The colors represent the degree of impact, which was quantified using expert judgments of 17 different environmental impact factors. The enlarged areas from the Caribbean Sea (left), North Atlantic Ocean (center), and western Pacific Ocean (right) show greater detail of more heavily impacted areas. Note the correspondence between the areas of high and very high impact with areas of significant human impact in the adjacent terrestrial regions in Figure 3.5. (From B. S. Halpern et al. 2008. *Science* 319: 948–952.)

activity. Seawater that is chemically altered by magma provides chemical energy that supports these relatively rich and diverse communities (see the Case Study for Chapter 20). The benthic zone has received increasing attention in the past two decades but still remains one of the least explored marine biological zones.

Marine biological zones have been impacted by human activities

Our discussion of marine biological zones has alluded to several services they provide to humans, including food production (e.g., fisheries in the nearshore and open ocean zones), protection of coastal areas from erosion (e.g., mangrove forests), uptake and stabilization of pollutants and nutrients (estuaries and marshes), and recreational benefits (Barbier et al. 2011). These services, along with ocean biodiversity, are increasingly threatened by human activities.

Despite the vastness of the ocean, human activities have affected it to varying degrees over the majority of its area (**FIGURE 3.27**). These include land-based activities that release nutrients, pollutants, and trash into rivers; ocean-based activities such as commercial fishing; and emissions of greenhouse gases. The effects of these activities include changes in water temperature and ocean acidification due to increases in greenhouse gases, increases in UV radiation due to the loss of protective stratospheric ozone, inputs of pollutants, and overharvesting of sea creatures, particularly fishes and whales (Halpern et al. 2008). (See Concepts 25.2 and 25.4 for more discussion of ozone loss and the greenhouse effect.) These impacts have the potential to influence the services on which humans depend, as well as the composition and abundance of the biota that inhabit different marine biological zones. The greatest estimated impacts are in nearshore marine ecosystems (estuaries, rocky intertidal zones, and sandy shores) near terrestrial regions that are sources of pollutants and nutrients, such as the regions adjacent to northern Europe and eastern Asia. Concern is increasing about the role of discarded plastics in the marine environment, with plastic trash found in nearly all marine zones, imparting a high potential to adversely impact marine organisms (Rochman et al. 2016; Law 2017). Despite the widespread nature of human impacts, large areas of the ocean remain only moderately affected, and greater recognition of these impacts could lead to increased conservation and more sustainable use of ocean resources.

The American Serengeti—Twelve Centuries of Change in the Great Plains

Humans have been implicated in several major biological changes in the grasslands of the world. One of the earliest was the disappearance of large mammals from North America during the late Pleistocene. Paul Martin, an early proponent of this hypothesis, noted the strong correspondence between extinction events on several continents and the arrival of humans on those continents, principally Europe, North and South America, and Australia (Martin 1984, 2005). Martin suggested that the rapidity of the extinctions and the greater proportion of large animals that disappeared reflected the hunting efficiency of those early humans. Larger animals have lower reproductive rates than smaller animals, so they cannot recover from increases in predation as quickly. Martin's suggestion therefore took on the unfortunate label of "the overkill hypothesis."

Since it was first proposed, the overkill hypothesis has received increasing support (Surovell et al. 2016). Archaeological research has uncovered numerous butchering sites containing remains of extinct animals. Spearheads have been found among the bones, and some of the bones have scrape marks made by tools found at the sites. Other strong evidence indicates that human arrival on small, isolated oceanic islands led to large numbers of extinctions due to predation by humans and by other animals they introduced (e.g., rats and snakes). While most scientists now accept that hunting of megafauna by humans had a role in some of the continental extinctions in the late Pleistocene, other causes have been proposed as well. These causes include the spread of diseases carried by humans and possibly by the domesticated dogs that accompanied them (MacPhee and Marx 1997). Another hypothesis suggests that the loss of some animals on which other species depended, such as mastodons, led to more widespread extinctions (Owen-Smith 1987). No one hypothesis explains the extinctions of all the megafauna on all the continents, however. A combination of climate change and the arrival of humans probably contributed to their demise (Barnosky et al. 2004).

Although the diversity of large mammals on the Great Plains was greatly diminished following the Pleistocene, large mammals were still abundant. Bison may have numbered 30 million, and numerous elk (wapiti), pronghorn, and deer roamed the plains. These animals continued to be hunted by humans, who also began to use fire on the eastern edge of the Great Plains as a tool for managing the habitat of their prey, as well as for small-scale agriculture (Delcourt et al. 1998). The writings of travelers to the Great Plains in the early 1800s indicate that the eastern deciduous forest began farther

FIGURE 3.28 Buffalo Hunting The arrival of large numbers of Euro-Americans in the Great Plains in the nineteenth century led to a mass slaughter of bison, facilitated by the construction of railroad lines and the use of high-powered rifles.

east than it does today, probably because of the influence of human-set fires.

Between 1700 and 1900, ecological changes occurred in the Great Plains that profoundly transformed both the plants and the animals. The reintroduction of horses into North America by Spanish explorers facilitated the development of a Native American culture centered on the hunting of bison. The arrival of Euro-Americans, and their subsequent conflicts with Native Americans, led to the near extinction of bison and other large Plains animals by the late 1800s (**FIGURE 3.28**). With the arrival of cattle and mechanized agriculture after 1850, the Great Plains became a domesticated landscape. The moister eastern tallgrass prairie was converted into monocultures of corn, wheat, soybeans, and other crops; today, only 4% of tallgrass prairie remains. A larger proportion of the mixed-grass and short-grass prairies to the west remained intact, but overgrazing and unsustainable agricultural practices led to serious degradation of some of these areas during the Dust Bowl of the 1930s, when drought and massive windstorms resulted in substantial losses of fertile topsoil (see the Case Study in Chapter 25).

🌿 CONNECTIONS *in* NATURE

LONG-TERM ECOLOGICAL RESEARCH Most terrestrial biomes and marine biological zones across the globe are experiencing changes due to human activities (see Figures 3.5 and 3.27). Even remote, seemingly pristine areas are subject to the effects of climate change and air pollution. Recognizing the effects of human activities on these systems, as well as our incomplete understanding of those effects, the U.S. National Science Foundation initiated the Long-Term Ecological Research (LTER) Network of study sites in 1980. Initially consisting of 5 sites, the network has grown to 28 sites representing a diversity of terrestrial biomes, from tropical to polar, as well as marine biological zones, croplands,

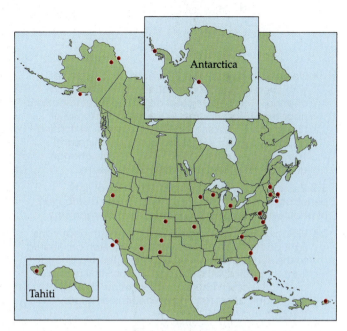

FIGURE 3.29 Long-Term Ecological Research Sites
Twenty-eight research sites constitute the U.S. Long-Term Eco-logical Research (LTER) Network. These sites encompass deserts, grasslands, forests, mountains, lakes, estuaries, agricultural systems, and cities. Researchers measure long-term changes in ecosystems and perform experiments at these sites to better understand ecological dynamics over decades to centuries.

and urban centers (**FIGURE 3.29**). The formation of the U.S. LTER program has spurred the formation of an international network of LTER sites, facilitating international collaborative research to better understand Earth's ecological systems.

Long-term ecological research has advanced our un-derstanding of ecological changes that occur at decadal and longer time scales. For example, research at LTER sites in the western United States has led to an understanding of the influence of El Niño Southern Oscillation and Pa-cific Decadal Oscillation (two climate cycles discussed in Concept 2.5) on the grassland biome. The legacy of climate change since the last glacial maximum, discussed in the Case Study at the opening of this chapter, is also better understood as a result of this research. Finally, research at LTER sites is providing a view of how environmental change, including climate change, may influence terrestrial biomes and marine biological zones in the future.

In this chapter, we've learned that grasslands are the biome most heavily impacted by human activities due to agricultural development. The Konza Prairie LTER site, lo-cated in the Flint Hills of northeastern Kansas, is a remnant tallgrass prairie—a very heavily impacted grassland type with very little of its original cover remaining. Research at the Konza Prairie site has focused on conserving this en-dangered biome in the face of rapid climate and land use change by examining the interactive roles of fire, grazing, and climate in the tallgrass prairie ecosystem. This research has included experiments varying the frequencies of fire

FIGURE 3.30 Research at the Konza Prairie LTER Site
Long-term research and experiments are investigating the effects of the frequencies of (A) grazing, (B) fire, and (C) precipitation on the diversity and function of the tall-grass prairie ecosystem.

and grazing in large landscape units to investigate their importance in maintaining the dominance of the grasses that characterize the grassland biome (**FIGURE 3.30**). Re-searchers have also examined the potential effects of changes in precipitation by varying the amount, intensity, and timing of watering. Results from this research have pro-vided important insights into how climate change may af-fect the grassland biome, indicating that extremes in rain-fall are important controls on its diversity and function (Knapp et al. 2002). Research at this and other LTER sites will enhance our ability to conserve native biodiversity in the face of accelerating environmental change.

SUMMARY

CONCEPT 3.1 Terrestrial biomes are characterized by the growth forms of the dominant vegetation.

3.1.1 Explain why ecologists use plant growth forms to categorize global terrestrial biomes.

- Terrestrial biomes are characterized by plant growth forms, particularly of dominant plants. Plants are immobile, so they must be able to cope with the biome's environmental extremes as well as its biological pressures, such as competition for water, nutrients, and light. Plant growth forms are therefore good indicators of the physical environment, reflecting climate zones as well as rates of disturbance.

3.1.2 Describe how global patterns of precipitation and temperature, and their variability, influence the location of terrestrial biomes.

- Terrestrial biomes reflect global patterns of precipitation and temperature. Temperature influences the distribution of plant growth forms directly through its effect on the physiological functioning of plants. Precipitation influences the availability of water, which is important in determining the supply of nutrients in the soil, which is also an important control on plant growth form.

3.1.3 Evaluate how human activities impact the actual distributions of biomes relative to their potential distributions.

- The potential and actual distributions of terrestrial biomes differ because of human activities, particularly conversion of land for agriculture, forestry, and grazing.

3.1.4 List the nine major terrestrial biomes.

- There are nine major terrestrial biomes: rainforests, seasonal forests and savannas, and hot deserts in tropical and subtropical zones; grasslands, shrublands and woodlands, deciduous forests, and evergreen forests in the temperate zones; and boreal forests and tundra in polar regions.

- Biological communities with close analogs to global biomes occur in mountains along elevational bands associated with climate gradients.

CONCEPT 3.2 Biological zones in freshwater ecosystems are associated with the velocity, depth, temperature, clarity, and chemistry of the water.

3.2.1 Summarize how the size of particles on the bottom of streams, as well as water velocity and clarity, change from source streams to large rivers and subsequently influence the organisms that inhabit different zones of moving waters.

- Biological communities in streams and rivers vary with stream order and location within the stream channel. Coarse terrestrial detritus is most important near the stream source, while the importance of fine organic matter, algae, and rooted and floating aquatic vascular plants increases downstream. The general feeding styles of organisms change accordingly as the river flows downstream.

3.2.2 Explain how the depth and amount of light penetration in a pond or lake influence the distribution of photosynthetic and nonphotosynthetic organisms.

- Biological communities in lakes vary with depth and light penetration. Photosynthetic plankton are limited to the surface layer of water where there is enough light for photosynthesis. Zooplankton—tiny animals and nonphotosynthetic protists—occur throughout the pelagic zone, feeding on detritus as it falls through the water.

CONCEPT 3.3 Marine biological zones are determined by ocean depth, light availability, and the stability of the bottom substrate.

3.3.1 Describe how substrate stability at the bottom of nearshore and shallow ocean zones determines which types of organisms are present, particularly emergent and nonemergent vascular plants and large algae.

- Estuaries, salt marshes, and mangrove forests occur in shallow zones at the margins between terrestrial and marine ecosystems. They are influenced by inputs of fresh water and sediments from nearby rivers.

- Biological communities at the shoreline reflect the influence of tides and the stability of the substrate (sandy vs. rocky).

- Coral reefs and kelp and seagrass beds are productive communities with high diversity associated with the habitat complexity provided by their photosynthesizers.

3.3.2 Explain how different sources of energy and food affect the type and number of ocean organisms that exist in the water's surface and in the deepest depths.

- Biological communities of the open ocean and deep benthic zones contain sparse populations of organisms, whose distributions are determined by light availability and proximity to the bottom.

Review Questions

1. Why are terrestrial biomes characterized using the growth forms of the dominant plants that occupy them?

2. Describe the close association between the distribution of biomes and the major climate zones described in Chapter 2. In particular, consider how seasonality of both temperature and precipitation influences biome distribution.

3. As streams flow from their sources to the oceans, what physical changes occur that affect the distribution of their biological communities?

4. Why do ocean depth and the stability of the substrate play important roles in determining the composition of marine biological communities?

Hone Your Problem-Solving Skills

Chapter 2 introduced the concept of the rain shadow effect on regional climates. Mountains intercepting air masses enhance precipitation on the windward slopes, with drier and warmer climates occurring on the leeward slopes. The change in temperature and precipitation with elevation therefore differs on west- versus east-facing slopes. The Cascade Range of the Pacific Northwest exemplifies this well. The windward west-facing slopes experience a temperature change with elevation (known as the environmental lapse rate) of 4.5°C per 1,000 m, while the leeward east-facing slopes have an environmental lapse rate of 6.5°C per 1,000 m.

1. Describe the vegetation types you would expect to encounter moving along a line from the base of the Cascade Range on the western side to the summit ridges at 3,000 m, and then descending to the bottom of the mountains on the eastern side. Use the information given in Figure 3.4 to determine the potential vegetation types. The annual average temperature for sites at the base of the west slopes (at sea level) is 12°C, and the average annual precipitation is 120 cm. Annual average precipitation on the summit ridges (3,000 m) increases to about 180 cm (assume the precipitation increase with elevation is linear). Annual average precipitation decreases to 50 cm at the base of the mountains on the leeward slope (elevation is 300 m).

2. Using the same approach you used in Question 1, describe the vegetation types you would expect to find along the same transect with the projected warming of 4°C and a 30% decrease in precipitation associated with climate change over the next 50 years.

ADDITIONAL RESOURCES

This textbook provides resources such as **Web Extensions** and additional **Climate Change Connections** for instructors who can share them with students. Please contact your instructor, if you would like to access that content.

4 Coping with Environmental Variation: Temperature and Water

Frozen Frogs: A Case Study

The idea of suspended animation—life being put on hold temporarily—has captured the imagination and hopes of people waiting for medical science to develop ways to cure untreatable diseases or reverse the ravages of aging. *Cryonics* is the preservation of the bodies of deceased people at subfreezing temperatures with the goal of eventually bringing them back to life and restoring them to good health. Proponents of cryonics exist throughout the world, some more visible than others. In Nederland, Colorado, there is a yearly "Frozen Dead Guy Days" festival, considered to be the "Mardi Gras of cryonics." This festival commemorates the efforts of a former resident who had his grandfather frozen immediately after his death from heart failure, hoping that one day his grandfather could be brought back to life and given a heart transplant (as documented in the movie *Grandpa's Still in the Tuff Shed*).

To some, cryonics seems far-fetched, a thing of science fiction and comedy (such as the TV show *Futurama* and the Austin Powers movies). Bringing life to a halt and then restarting it after a long period of quiescence doesn't seem plausible. Yet strange tales from nature provide examples of life apparently springing out of death. While seeking the existence of the Northwest Passage in the boreal and Arctic zones of Canada in 1769–1772, the English explorer Samuel Hearne found frogs under shallow layers of leaves and moss in winter "frozen as hard as ice, in which state the legs are as easily broken off as a pipe-stem" (Hearne 1911) (**FIGURE 4.1**). Hearne wrapped the frogs in animal skins and placed them next to his campfire. Within hours, the rock-hard amphibians came to life and began hopping around. The American naturalist John Burroughs found frozen frogs under a shallow cover of dead leaves in a New York forest in winter. Return visits to the same locations over a period of months indicated that the frogs hadn't moved, yet by spring they had disappeared (J. Burroughs 1914). Could a complex organism like a frog, with a sophisticated circulatory and nervous system, have achieved cryonic preservation as an evolutionary response to a harsh winter climate?

Courtesy of J. M. Storey

FIGURE 4.1 A Frozen Frog Wood frogs (*Rana sylvatica*) spend winters in a partially frozen state, without breathing and with no circulation or heartbeat.

Organisms of the temperate and polar zones face tremendous challenges imposed by a seasonal climate that includes subfreezing temperatures in winter. Amphibians are unlikely candidates to have solved this challenge by allowing their bodies to partially freeze. Aside from their aforementioned complex organ and tissue systems, amphibians are "cold-blooded" (generating little heat internally) and, as a group, first evolved in tropical and subtropical biomes. Yet two frog species, the wood frog (*Rana sylvatica*, also known as *Lithobates sylvaticus*) and the boreal chorus frog (*Pseudacris maculata*), live in the tundra biome (**FIGURE 4.2**) (Pinder et al. 1992). These frogs survive extended periods of subfreezing air temperatures in shallow burrows in a semifrozen state, with no heartbeat, no blood circulation, and no breathing. Among the vertebrates, only a few species of amphibians (four frogs and one salamander) and one turtle species can survive a long winter in a semifrozen state. Freezing in most organisms results in substantial damage to tissues as ice crystals perforate cell membranes and organelles. How do these vertebrates survive being frozen without turning to mush in spring when they thaw out and reinitiate their blood circulation and breathing?

Introduction

Siberian spruce trees (*Picea obovata*) experience the extreme range of seasonal temperatures characteristic of a continental climate. In the Siberian boreal forest, air temperatures regularly drop below –50°C (–58°F) in winter, and in summer they reach 30°C (86°F). Being an immobile tree, the Siberian spruce lacks the option to move to Florida for the winter or head to the coast to cool off in summer. The spruce must *tolerate* these temperature extremes, surviving the 80°C (144°F) seasonal change in its body temperature. Other organisms can *avoid* these extremes through some behavior or physiological change. These two options for coping with environmental change, **tolerance** and **avoidance**, provide a useful framework for thinking about how organisms cope with the environmental extremes they face.

The range of physical environmental conditions described in Chapter 2 establishes the variation in biomes and marine biological zones described in Chapter 3. In this chapter and the next, we will examine the interactions between organisms and the physical environment that

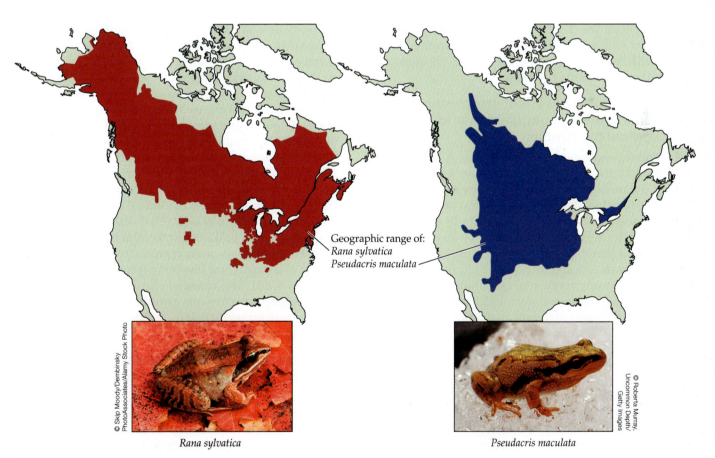

Geographic range of:
Rana sylvatica
Pseudacris maculata

Rana sylvatica

Pseudacris maculata

FIGURE 4.2 **Northern Exposure**   Wood frogs (*Rana sylvatica*) and boreal chorus frogs (*Pseudacris maculata*) have geographic ranges that extend into the boreal forest and tundra biomes. (Range data from IUCN [International Union for Conservation of Nature], Conservation International & NatureServe. 2008.

The IUCN Red List of Threatened Species Version 2019-2. https://www.iucnredlist.org/species/58728/78907321 and https://www.iucnredlist.org/species/136004/78906835. Downloaded on 14 June 2019.)

influence their survival and persistence, and therefore their geographic ranges. The study of these interactions is known as **physiological ecology**.

> **CONCEPT 4.1** Each species has a range of environmental tolerances that determines its potential geographic distribution.

LEARNING OBJECTIVES

4.1.1 Explain why the physical environment is the ultimate determinant of the geographic distribution of a species.

4.1.2 Differentiate between adaptation and acclimatization by explaining how both individual organisms and populations respond, but differently, to changes in the environment.

4.1.3 Illustrate how adaptation and acclimatization may result in trade-offs with other functions.

Responses to Environmental Variation

A fundamental principle in ecology is that the geographic ranges of species are related to constraints imposed by the physical and biological environments. In this section, we will discuss the general principles of organismal responses to the physical environment.

Species distributions reflect environmental influences on energy acquisition and physiological tolerances

The potential geographic range of a species is ultimately determined by the physical environment, which influences an organism's *ecological success* (its survival and reproduction) in two important ways. First, the physical environment affects an organism's ability to obtain the energy and resources required to maintain its metabolic functions, and therefore to grow and reproduce. Rates of photosynthesis and abundances of prey, for example, are controlled by environmental conditions. Therefore, the ability of a species to maintain a viable population is constrained at the limits of its potential geographic range. Second, as we saw in Concept 2.1, an organism's survival can be affected by extreme environmental conditions. If temperature, water supply, chemical concentrations, or other physical conditions exceed what an organism can tolerate, the organism will die. These two influences—the availability of energy and resources and physical tolerance limits—are not mutually exclusive, as energy supply influences an organism's ability to tolerate environmental extremes. Furthermore, it is important to keep in mind that the *actual* geographic distribution of a species differs from its *potential* distribution because of other factors, such as dispersal ability (see Concept 18.1), disturbance (e.g., fire; see Concept 17.1), and interactions with other organisms, such as competition (see Unit 4) (**FIGURE 4.3**).

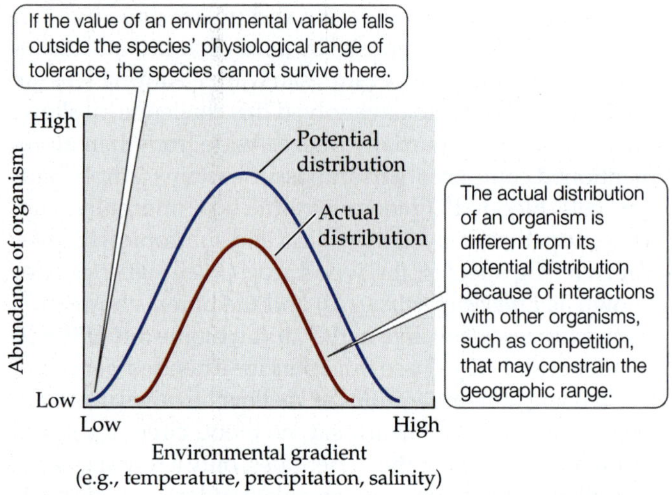

FIGURE 4.3 Abundance Varies across Environmental Gradients The abundance of an organism reaches a theoretical maximum at some optimal value across an environmental gradient and drops off at either end at values that constrain the potential geographic distribution of the organism. The actual abundance curve is likely to differ from the potential abundance curve because of biological interactions.

As we saw in Concept 3.1, the immobility of plants makes them good indicators of the physical environment. Farmers are acutely aware of the effects of extreme events on the survival of crop plants, which are often grown outside the geographic ranges where they evolved. Frosts or extreme droughts can result in catastrophic crop losses. Aspen (*Populus tremuloides*) provides a good example of a native species whose geographic range is related to its climate tolerance. Aspen occurs in boreal forests and mountain zones throughout North America. Its geographic distribution can be predicted fairly accurately from the observed effects of climate on its survival and reproduction (Morin et al. 2007) (**FIGURE 4.4A**). The climate factors that limit its distribution include the effects of low temperatures on its reproductive success and the effects of drought and low temperatures on its survival (**FIGURE 4.4B**). The range of climate conditions under which a species occurs—its **climate envelope**—provides a useful tool for predicting its response to climate change (see Chapter 25).

Individuals respond to environmental variation through acclimatization

Any physiological process, such as growth or photosynthesis, has a set of optimal environmental conditions most conducive to its functioning. Deviations from those optimal conditions cause a decrease in the rate of the process (**FIGURE 4.5**). **Stress** is the condition in which an environmental change results in a decrease in the rate of an important physiological process, thereby lowering the potential for an organism's survival, growth, or reproduction. For example, when you travel to high elevations, typically above 2,400 m (8,000 feet), the lower partial

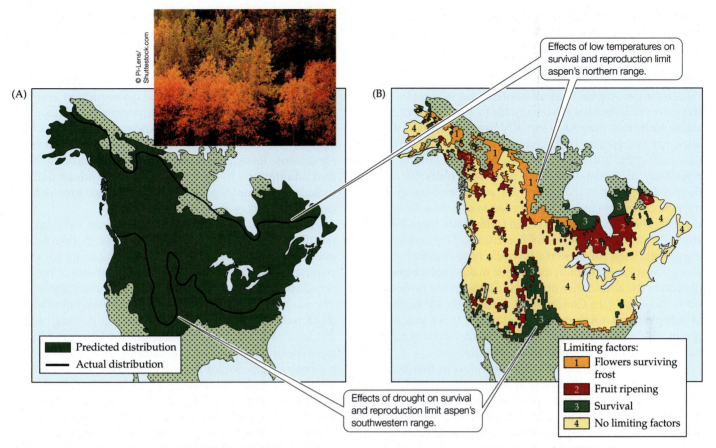

Effects of low temperatures on survival and reproduction limit aspen's northern range.

Effects of drought on survival and reproduction limit aspen's southwestern range.

(A)

Predicted distribution

Actual distribution

(B)

Limiting factors:

1 Flowers surviving frost

2 Fruit ripening

3 Survival

4 No limiting factors

FIGURE 4.4 Climate and Aspen Distribution The geographic distribution of the aspen (*Populus tremuloides*; golden trees in the photo) is associated with climate. (A) Predicted distribution of aspen, based on the effects of climate factors on survival and reproduction observed in natural populations, mapped with the actual distribution. (B) Climate factors limiting the distribution of aspen, based on observations of natural populations. (After X. Morin et al. 2007. *Ecology* 88: 2280–2291.)

? The future climate is predicted to be warmer throughout the interior of western North America and drier in the central portions of the continent. How will these changes influence the geographic distribution of aspen?

pressure of oxygen in the atmosphere (see Concept 3.2) results in the delivery of less oxygen to your tissues by your circulatory system. This condition, known as *hypoxia*, results when the amount of oxygen picked up by hemoglobin molecules in your blood decreases. Hypoxia causes "altitude sickness," a type of physiological stress, decreasing your ability to exercise and think clearly and making you feel nauseated.

Many organisms have the ability to adjust their physiology, morphology, or behavior to lessen the effect of an environmental change and minimize the associated stress. This kind of adjustment, known as **acclimatization**,[1] is usually a short-term, reversible process. Your body acclimatizes to a high elevation if you remain there for several weeks (but only below 5,500 m, or 18,000 feet). Acclimatization to high elevations involves higher breathing rates, greater production of red blood cells and associated

hemoglobin, and higher pressure in the pulmonary arteries to circulate blood into areas of the lung that are not used at lower elevations (Hochochka and Somero 2002). The outcome of these physiological changes is the

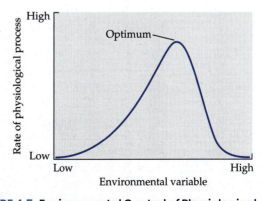

FIGURE 4.5 Environmental Control of Physiological Processes The rates of physiological processes are greatest under a set of optimal environmental conditions (e.g., optimal temperature, optimal water availability). Deviations from the optimum cause a decrease in the rates of physiological processes.

---

[1] Animal physiologists use the term *acclimatization* to refer to the short-term response of an animal to changes in the physical environment under field conditions and use *acclimation* to refer to a short-term response under controlled laboratory conditions.

delivery of more oxygen to your tissues. The acclimatization process reverses when you return to lower elevations.

Populations respond to environmental variation through adaptation

Within the geographic range of a species, particular populations occur in unique environments (e.g., cool climates, saline soils) that may have initially been stressful to the organisms when they first occupied them. Genetic variation among the individuals within such populations, in physiological, morphological, or behavioral traits that influenced their survival, functioning, and reproduction in the new environment, would have led to natural selection favoring those individuals whose traits made them best able to cope with the stressful conditions. The underlying genetic basis for these traits would have resulted in a change over generations in the genetic makeup of the population as the abundance of individuals with the favored traits increased (see Concept 6.3). Such traits are known as **adaptations**. Over many generations, these unique, genetically based solutions to environmental stress would have become more frequent in the population.

Adaptation is similar to acclimatization in that both processes involve a change that lowers stress, and the ability to acclimatize represents a type of adaptation. However, adaptation differs from acclimatization in being a long-term, genetic response of a population to environmental stress that increases its ecological success under the stressful conditions (**FIGURE 4.6**). Populations with adaptations to unique environments are called **ecotypes**. Ecotypes may represent responses to both abiotic (e.g., temperature, water availability, soil type, salinity) and biotic (e.g., competition, predation) environmental factors. Ecotypes can eventually become separate species as the physiology and morphology of individuals in different populations diverge and the populations eventually become reproductively isolated.

Returning to our previous example of stress at high elevations, some human populations have lived continuously in the Andean highlands for at least 10,000 years.

When Spanish explorers first settled in the Andes alongside the native people in the sixteenth and seventeenth centuries, their birth rates were lower than those of the natives for two to three generations, probably because of poor oxygen supply to developing fetuses (Ward et al. 1995). The same held true for the domesticated animals they brought with them. This comparison provides anecdotal evidence that the native Andean populations had become adapted to the low-oxygen conditions at high elevations. Research in the twentieth century showed that adaptations to high elevations by Andean natives include higher red blood cell production and greater lung capacity (Ward et al. 1995).

Adaptations to environmental stress can vary among populations. In other words, the solution to a particular environmental problem may not be the same for each population, as demonstrated by a comparison of human populations native to the Andean and Tibetan highlands. The adaptations to high elevations in Andean populations (high red blood cell concentration and large lung capacity) are not the same as those found in Tibetan populations (Beall 2007). Tibetan populations have red blood cell concentrations similar to, and blood oxygen concentrations lower than, populations at sea level, but they have a higher breathing rate, which enhances the exchange of oxygen with the blood system, and higher blood flow, which improve delivery of oxygen to vital organs such as the brain. Thus, there are at least two different ways in which human populations have adapted to the hypoxic stress imposed by living at high elevations.

Acclimatization and adaptation are not "free"; they require an investment of energy and resources by the organism. They represent possible *trade-offs* with other functions of the organism that may also affect its survival and reproduction. Acclimatization and adaptation must therefore increase the survival and reproductive success of the organism under the specific environmental conditions in order to be favored over other patterns of energy and resource investment. Trade-offs in energy and resource allocation are discussed in Concept 7.3.

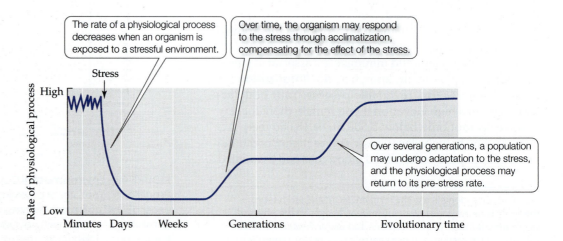

FIGURE 4.6 Organismal Responses to Stress Organisms respond to stress over different time scales. (After H. Lambers et al. 1998. *Plant Physiological Ecology.* Springer: New York.)

The rate of a physiological process decreases when an organism is exposed to a stressful environment.

Over time, the organism may respond to the stress through acclimatization, compensating for the effect of the stress.

Over several generations, a population may undergo adaptation to the stress, and the physiological process may return to its pre-stress rate.

In the remaining two sections, we will examine the factors that determine organisms' temperatures, water content, and water uptake, and we will consider examples of acclimatization and adaptation that allow organisms to function in the face of varying temperatures and water availability.

> **CONCEPT 4.2** The temperature of an organism is determined by exchanges of energy with the external environment.

LEARNING OBJECTIVES

4.2.1 Describe how the body temperature of an organism influences its functioning.

4.2.2 Use information about the gains and losses of energy to determine whether an organism's temperature is rising or dropping.

4.2.3 Identify the heat exchange mechanisms used by plants and animals to regulate their body temperatures.

4.2.4 Contrast ectothermy and endothermy, and explain how each influences the geographic distributions of organisms, along with organisms' sensitivities to changes in body temperature.

Variation in Temperature

Environmental temperatures vary greatly throughout the biosphere, as we saw in Chapter 2. The Siberian boreal forest described earlier in this chapter represents one extreme of seasonal variation, with as much as an 80°C (144°F) swing from summer to winter. Tropical forests, on the other hand, experience far less seasonal variation in temperature, about 15°C (22°F). Soil environments, which are home to many species of microorganisms, plant roots, and animals, are buffered from aboveground environmental temperature extremes, although soil surface temperatures may change as much as or more than air temperatures. Aquatic environments also experience temperature changes over seasonal and daily time scales. Open ocean environments tend to have very little temporal variation in temperature, because of the ocean's massive volume and heat capacity. In contrast, tide pools experience large variations in water temperature as the tides rise and fall, with as much as a 20°C (36°F) change over a 5-hour period.

The survival and functioning of organisms are strongly tied to their internal temperatures. The extreme upper limit for metabolically active multicellular plants and animals is about 50°C (122°F) (**FIGURE 4.7**). Some archaea and bacteria that live in hot springs can function at 90°C (194°F) (Willmer et al. 2005). The extreme lower limit for organismal function is tied to the temperature at which water in cells freezes, typically between −2°C and −5°C (28°F–23°F). Some organisms can survive periods of extreme heat or cold by entering a state of **dormancy**, in which little or no metabolic activity occurs.

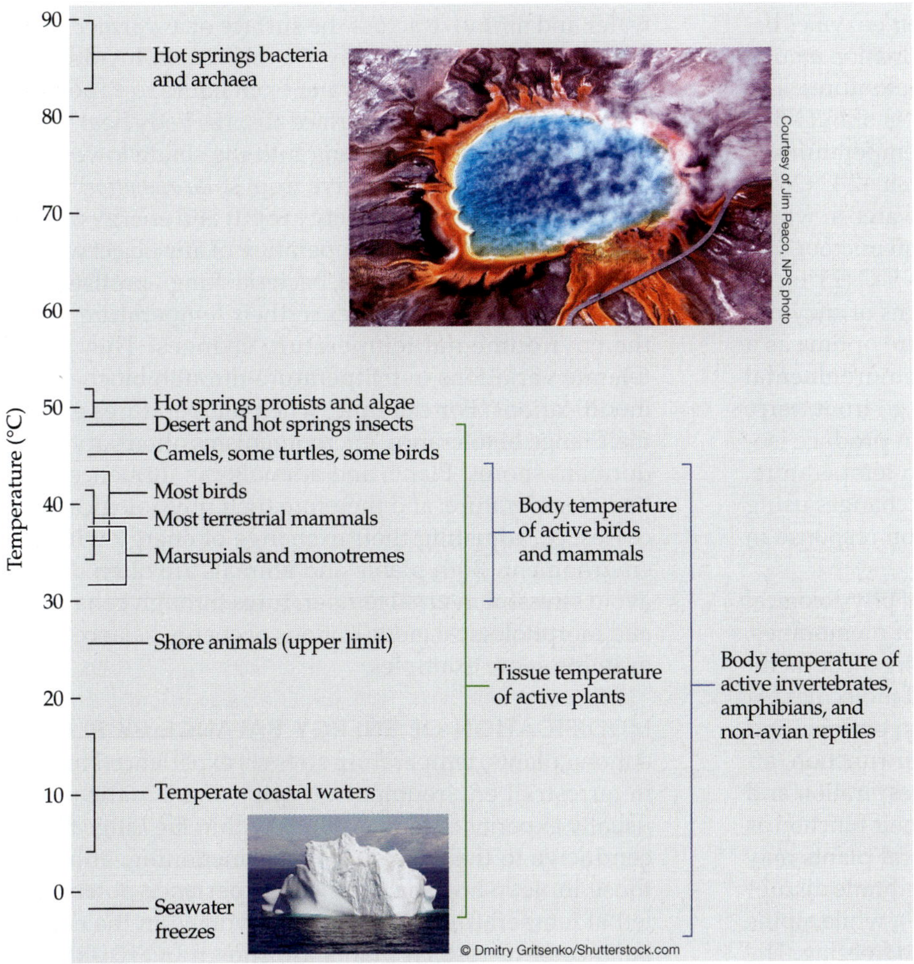

FIGURE 4.7 Temperature Ranges for Life on Earth Living organisms are known to exist in extreme environments, ranging from hot springs to freezing seas. (After P. Willmer et al. 2005. *Environmental Physiology of Animals*. Blackwell Publishing: Malden, MA.)

The internal temperature of an organism is determined by the balance between the energy it gains from and the energy it loses to the external environment. Thus, organisms must either tolerate changes in their internal temperature as the temperature of the external environment changes or modify their internal temperature by using some physiological, morphological, or behavioral means of adjusting these gains and losses. Environmental temperatures—particularly their extremes—are therefore important determinants of the distributions of organisms, as demonstrated by the relationship between biomes and global climate patterns discussed in Chapters 2 and 3.

Temperature controls physiological activity

Key biochemical reactions important to maintenance of life are temperature sensitive. Each reaction has an optimal temperature that is related to the activity of *enzymes*, protein-based molecules that catalyze biochemical reactions. Enzymes are structurally stable under a limited range of temperatures. At high temperatures, the constituent proteins lose their structural integrity, or become *denatured*, as their bonds break. Most enzymes become denatured at temperatures between 40°C and 70°C (104°F–158°F), but enzymes in bacteria inhabiting hot springs can remain stable at temperatures up to 100°C (212°F). The upper lethal temperature for most organisms is lower than the temperature at which their enzymes become denatured, because metabolic coordination among biochemical pathways is lost at these temperatures. The extreme lower limit for enzyme activity is about –5°C (23°F) (Willmer et al. 2005). The internal temperatures of Antarctic fishes and crustaceans may reach –2°C (28°F) because the salt concentration of the seawater in which they live lowers its freezing point. Some soil microorganisms are active at temperatures as low as –5°C (23°F).

Some species can produce different forms of enzymes (called *isozymes*) with different temperature optima as a means of acclimatization to changes in environmental temperature. For example, some fishes (e.g., trout, carp, goldfish) and trees (e.g., loblolly pine) can produce isozymes in response to seasonal changes in temperature. However, acclimatization to temperature changes using isozymes does not appear to be a common response in animals (Willmer et al. 2005).

Temperature also determines the rates of physiological processes by influencing the properties of membranes, particularly at low temperatures. Cell and organelle membranes are composed of two layers of lipid molecules. At low temperatures, these layers can solidify; proteins and enzymes embedded in them can lose their function, affecting processes such as mitochondrial respiration and photosynthesis, and membranes can lose their function as filters, leaking cellular metabolites. Tropical plants may suffer loss of function associated with membrane disruption at temperatures as high as 10°C (50°F), while alpine plants can function at temperatures close to freezing. The sensitivity of membrane function to low temperatures is related to the chemical composition of the membrane lipid molecules. Plants of cooler climates have a higher proportion of unsaturated membrane lipids (with greater numbers of double bonds between carbon molecules) than plants of warmer climates.

Finally, temperature influences physiological processes in terrestrial organisms by affecting water availability. As we saw in Concept 2.2, the warmer the air, the more water vapor it can hold. As a result, the rate at which terrestrial organisms lose water from their bodies increases as temperature becomes warmer. We will return to this point later when we discuss how organisms cope with variations in water availability.

Organisms influence their temperature by modifying energy balance

On a hot day, jumping into a swimming pool and then sitting in the shade in a light breeze brings relief from the oppressive heat. Elephants follow a similar routine, wading into ponds and using their trunks to spray water onto their backs. This kind of behavior facilitates heat loss in several ways. First, the contact of warm skin with cool water causes heat energy to be lost from the body through the process of *conduction*: the direct transfer of energy from warmer, more rapidly moving molecules to cooler, more slowly moving molecules. Also, when cool water and air move across the surface of a warmer body, heat energy is carried away via *convection*. In addition, the change in the state of water from liquid to vapor as it evaporates on the skin's surface absorbs body heat (*latent heat transfer*). Finally, moving into the shade lowers the amount of energy you receive from solar *radiation*.

The balance between energy input and energy output determines whether the temperature of any object will increase or decrease. Archaea, bacteria, fungi, protists, and algae cannot avoid changes in their temperature when the environmental temperature changes. They must tolerate variations in temperature through biochemical modifications. For example, when temperatures exceed their range of tolerance, microorganisms often survive as dormant spores. Plants and animals can influence their body temperature, and therefore their physiological processes, by adjusting their exchange of energy with the environment. Both plants and animals are often able to avoid stressful internal temperatures through behavioral and morphological modifications of energy balance. Let's examine some examples.

MODIFICATION OF ENERGY BALANCE BY PLANTS
Among plants, temperature stress is experienced mainly in terrestrial environments. Marine and aquatic plants usually experience temperatures within the range that is conducive to their physiological functioning, although those in nearshore habitats can experience potentially lethal temperatures. The factors involved in the energy balance of terrestrial plants are shown in **FIGURE 4.8**. Energy inputs that warm the plant include sunlight and

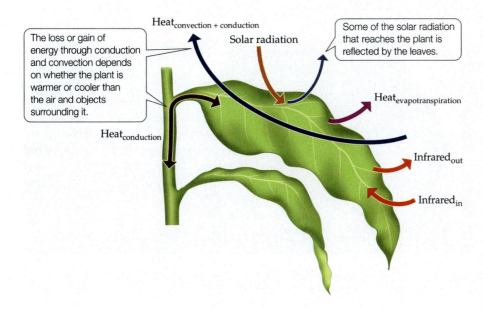

FIGURE 4.8 Energy Exchange in Terrestrial Plants The temperature of a plant is determined by the balance between inputs of energy from and outputs of energy to the environment. (After P. S. Nobel. 1983. *Biophysical Plant Physiology and Ecology.* W. H. Freeman: New York.)

infrared radiation from surrounding objects. If the ground or air is warmer than the plant, energy inputs also include conduction and convection. Losses of energy from the plant include the emission of infrared radiation to the surrounding environment and, if the ground or air is cooler than the plant, conduction and convection. Heat loss also occurs through transpiration (evaporation of water from inside the plant) and surface evaporation, collectively referred to as *evapotranspiration*.

We can put these inputs and outputs together to determine whether the temperature of the plant is changing:

$$\Delta H_{plant} = SR + IR_{in} - IR_{out} \pm H_{conv} \pm H_{cond} - H_{et} \quad (4.1)$$

where ΔH_{plant} is the heat energy change of the plant (the Greek delta usually signifies "change in"), SR is solar radiation, IR_{in} is the input of infrared radiation, IR_{out} is the output of infrared radiation, H_{conv} is convective heat transfer, H_{cond} is conductive heat transfer, and H_{et} is heat transfer through evapotranspiration. A negligible loss of energy occurs as the plant uses solar radiation for photosynthesis. If the plant is warmer than the surrounding air, H_{conv} and H_{cond} are negative. If the sum of the energy inputs exceeds the sum of the outputs, ΔH_{plant} is positive, and the plant's temperature is increasing. Conversely, if more heat is being lost than gained, ΔH_{plant} is negative, and the plant's temperature is decreasing.

Plants can modify their energy balance to control their temperature by adjusting these energy inputs and outputs. Leaves are most often associated with these adjustments because they are the primary photosynthetic organs of the plant and typically are the most temperature-sensitive tissue. The most important and common adjustments include changes in the rate of transpirational water loss. In addition, changes in leaf surface reflective properties (color) or in leaf orientation toward the sun can alter the amount of solar radiation absorbed by the plant. Finally, changes

in convective heat transfer can be accomplished by changing surface roughness.

Transpiration is an important evaporative cooling mechanism for leaves. As we saw in Chapter 2, its effectiveness is especially evident in the canopies of tropical forests, which are subjected to warm air temperatures and high levels of solar radiation. Without transpirational cooling, the leaves of tropical canopy plants could reach temperatures over 45°C (>113°F), which would be lethal. The rate of transpiration is controlled by specialized *guard cells* surrounding pores, called **stomates**, leading to the interior of the leaf. Stomates are the gateway for both transpirational water loss and the uptake of carbon dioxide for photosynthesis; we will return to the latter function in Concept 5.2. Variation in the degree of stomatal opening, as well as in the number of stomates, controls the rate of transpiration and therefore exerts an important control on leaf temperature (**FIGURE 4.9**).

Transpiration requires a steady supply of water. Where the amount of water in the soil is limited—as it is over a substantial part of Earth's land surface—transpiration is not a reliable cooling mechanism. As we saw in Concept 3.1, some plants shed their leaves during dry seasons, thereby avoiding both temperature and water stress. However, the high demand for the resources (e.g., soil nutrients) needed to replace fallen leaves may favor protecting existing leaves rather than shedding them. Plants that maintain their leaves during long dry periods require mechanisms other than transpiration to dissipate heat energy. One option is to alter the reflective properties of leaves via **pubescence**, the presence of light-colored or white hairs on the leaf surface, which lowers the amount of solar radiation absorbed by the leaf surface. Pubescence can also lower the effectiveness of convective heat loss, however, and thus represents a trade-off between two opposing heat exchange mechanisms.

One of the best studies addressing the adaptive significance of leaf pubescence for temperature regulation has focused on shrubs of the genus *Encelia* (members of the daisy family, common name brittlebush). Jim Ehleringer and his colleagues described the role of pubescence in leaf temperature regulation among species of *Encelia* that occupy different geographic ranges. *Encelia farinosa*, a native of the Sonoran and Mojave Deserts, maintains a high amount of leaf pubescence relative to *Encelia* shrubs

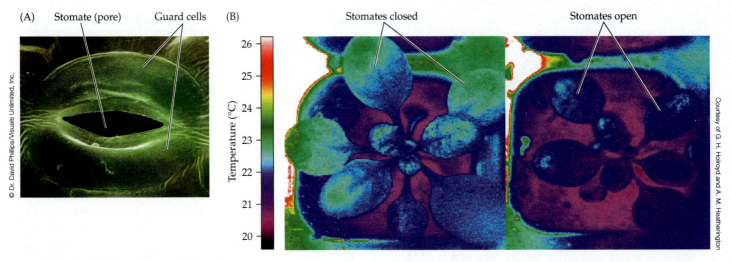

(A) Stomate (pore) Guard cells (B) Stomates closed Stomates open

Temperature (°C)

© Dr. David Phillips/Visuals Unlimited, Inc.

Courtesy of G. H. Holroyd and A. M. Heatherington

FIGURE 4.9 Stomates Control Leaf Temperature by Controlling Transpiration (A) Specialized guard cells control a stomate's degree of opening. Open stomates allow CO_2 to diffuse in for photosynthesis, and they allow water to transpire out, cooling the leaves. (B) Leaf temperatures vary with the degree of stomatal opening. The plant on the right has open stomates and is transpiring freely, while the plant on the left, kept under identical conditions, has closed stomates, a lower transpiration rate, and a temperature 1°C–2°C (2°F–4°F) higher, as indicated by thermal infrared imaging.

? Cooling of leaves using transpiration may be particularly important in what biomes?

from moister, cooler environments. Ehleringer and his colleague Craig Cook (1990) evaluated the relative roles of leaf pubescence and transpiration in the cooling of leaves of *E. farinosa* and two other species whose leaves lack pubescence: *E. frutescens*, which occurs in desert washes (which have more moisture than the rest of the desert), and *E. californica*, native to the cooler, moister coastal sage community of California and Baja California. To control for environmental variation that could influence the morphology and physiology of the plants, they grew plants of each species from seed together in experimental plots in the Sonoran Desert and on the California coast. Half of their experimental plants were watered while the other half were left under natural conditions. They measured the leaf temperatures, the degree of stomatal opening, and the amount of sunlight absorbed.

The three *Encelia* species showed few differences in leaf temperature and stomatal opening when grown in the cooler, moister California coastal garden. In the desert garden, however, *E. californica* and *E. frutescens* shed their leaves during the hot summer months under natural conditions, but *E. farinosa* did not. *Encelia frutescens* did not shed its leaves when the shrubs were watered, and its leaves maintained sublethal temperatures using transpirational cooling. *Encelia farinosa* leaves reflected about twice as much solar radiation as those of the other two species (**FIGURE 4.10A**), which facilitated the shrub's ability to maintain leaf temperatures lower than the air temperature.

Ehleringer and Cook's field experiment provides correlative evidence of the adaptive value of leaf pubescence to *E. farinosa* under hot desert conditions. Additional work by Darren Sandquist and Ehleringer has supported its adaptive value, indicating that natural selection has acted on variation in pubescence among ecotypes of *E. farinosa*. Populations in drier environments have more leaf pubescence, and reflect more solar radiation, than populations from moister environments (Sandquist and Ehleringer 2003).

In addition to varying among species and populations, leaf pubescence can also vary seasonally, exemplifying acclimatization to environmental conditions. *Encelia farinosa* shrubs produce smaller, more pubescent leaves in summer and larger, less pubescent leaves in winter (**FIGURE 4.10B**). There are costs to being pubescent, associated with the construction of the hairs and the loss of solar radiation that could be used for photosynthesis. Thus, when temperatures are cooler or when reliable soil water is present, *E. farinosa* plants construct leaves with fewer hairs.

Heat can be lost from a leaf by convection when the air temperature is lower than the temperature of the leaf. The effectiveness of convective heat loss is related to the speed of the air moving across a surface. As the moving air experiences more friction closer to the surface of an object, the flow becomes more turbulent, forming eddies (**FIGURE 4.11**). This zone of turbulent flow, called the **boundary layer**, lowers convective heat loss. The thickness of the boundary layer around a leaf is related to its size and its surface roughness. Small, smooth leaves have thin boundary layers and lose heat more effectively than large or rough leaves. This relationship between the boundary layer and convective heat loss is one reason for the rarity of large leaves in desert ecosystems.

Excessive heat loss by convection can be a problem for plants (and animals) in cold, windy environments such as the alpine zone in mountains. Convection is the largest source of heat loss from the land surface in temperate alpine environments, and high winds can shred leaves in

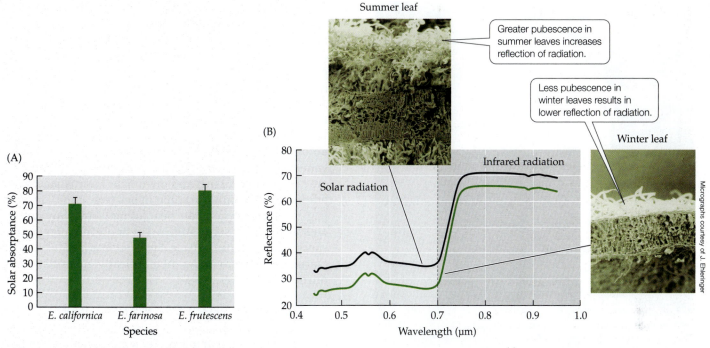

(A)

(B)

FIGURE 4.10 Sunlight, Seasonal Changes, and Leaf Pubescence (A) Solar heating of leaves varies according to the amount of pubescence on those leaves. The pubescent leaves of the desert shrub *Encelia farinosa* absorb a lower percentage of the incoming solar radiation than the leaves of two nonpubescent species: *E. californica*, native to the coastal sage community of California, and *E. frutescens*, an inhabitant of moister desert wash communities. *Encelia farinosa* is therefore less dependent on transpiration for leaf cooling than the other two species. Error bars show 1 standard error of the mean. (B) *Encelia farinosa* produces greater amounts of pubescence on its leaves in summer than in winter, representing acclimatization to hot summer temperatures. The photos are scanning electron micrographs of leaf cross sections. (A after J. R. Ehleringer and C. S. Cook. 1990. *Oecologia* 82: 484–489.)

? Why might temperature regulation associated with greater reflection of solar radiation via pubescence be more important in deserts than in a warm, moist biome such as the tropical rainforest?

exposed sites. Most alpine plants grow close to the ground surface to avoid the high wind velocities. Some alpine plants produce a layer of insulating hair on their surface to lower convective heat loss. The snow lotus of the Himalayas (*Saussurea medusa*) produces a series of very densely pubescent leaves that surround the flowers of the plant

The layer of turbulent air flow close to the leaf surface (boundary layer) loses less heat energy to convection.

Air flow

FIGURE 4.11 A Leaf Boundary Layer Air flowing close to the surface of a leaf is subject to friction, which causes the flow to become turbulent and lowers convective heat loss from the leaf to the surrounding air.

(**FIGURE 4.12**). Although they project above the ground surface and are exposed to more wind than ground-hugging plants, the flowers of *S. medusa* remain as much as 20°C (36°F) warmer than the air by absorbing and retaining solar radiation (Tsukaya et al. 2002). The plant not only keeps its photosynthetic tissues warm, but also provides a warm environment for potential pollinators, which are in short supply in cold, windy alpine environments.

MODIFICATION OF ENERGY BALANCE BY ANIMALS

Animals are subject to the same energy inputs and outputs described for plants in Equation 4.1, with one key difference: some animals—in particular, birds and mammals—have the ability to generate heat internally. As a result, another term is needed in the energy balance equation to represent this metabolic heat generation:

$$\Delta H_{animal} = SR + IR_{in} - IR_{out} \pm H_{conv} \pm H_{cond} - H_{evap} + H_{met}$$ (4.2)

where ΔH_{animal} is the heat energy change of the animal, SR is solar radiation, IR_{in} is the input of infrared radiation, IR_{out} is the output of infrared radiation, H_{conv} is convective heat transfer, H_{cond} is conductive heat transfer, H_{evap} is heat transfer through evaporation, and H_{met} is

Courtesy of Kenzo Okawa

FIGURE 4.12 A Woolly Plant of the Himalayas The snow lotus (*Saussurea medusa*) has dense pubescence surrounding its emergent flowering stems, which provides them with thermal insulation.

metabolic heat generation. In contrast to plants, evaporative heat loss is not widespread among animals. Notable examples of evaporative cooling in animals include sweating in humans, panting by dogs and other animals, and licking of the body by some marsupials under conditions of extreme heat.

The internal generation of heat by some animals represents a major ecological advance. Animals capable of metabolic heat generation can maintain relatively constant internal temperatures near the optimum for physiological functioning under a wide range of external temperatures, and as a result, they can expand their geographic ranges. Varying degrees of reliance on internal heat generation exist throughout the animal kingdom. Animals that regulate their body temperature primarily through energy exchange with the external environment, which includes the majority of animal species, are called **ectotherms**. Animals that rely primarily on internal heat generation, which are called **endotherms**, include, but are not limited to, birds and mammals. Internal heat generation is also found in some fishes (e.g., tuna), insects (e.g., bees, which generate heat for metabolic function as well as for defense; **FIGURE 4.13**), and even a few plant species (e.g., skunk cabbage, *Symplocarpus foetidus*, which warms its flowers using metabolically generated heat during the spring).

TEMPERATURE REGULATION AND TOLERANCE IN ECTOTHERMS   Generally, ectotherms have a greater tolerance for variation in their body temperature than endotherms (see Figure 4.7), possibly because they are less able to adjust their body temperature than endotherms. The exchange of heat between an animal and the environment, whether for cooling or heating, depends on the amount of surface area relative to the volume of the animal. A larger surface area relative to volume allows greater heat exchange but makes it harder to maintain a constant internal temperature in the face of variable external temperatures. A smaller surface area relative to volume decreases the animal's ability to gain or lose heat. This relationship between surface area and volume imposes a constraint on the body size and shape of ectothermic animals. Generally speaking, the surface-area-to-volume ratio decreases as body size increases, and the animal's ability to exchange

(A)

© iStock.com/Queserasera99

(B)

Takahashi/CC BY-SA 3.0

FIGURE 4.13 Internal Heat Generation as a Defense   Bees can generate heat by contracting their flight muscles. Japanese honeybees (*Apis cerana*) use internal heat generation as a defense against Asian giant hornets (*Vespa mandarinia*) that attack bee colonies. (A) When a hornet enters a nest, the honeybees swarm the larger invader. (B) The defensive ball of bees surrounding an invading hornet generates enough heat that temperatures in the center exceed the upper lethal temperature for the hornet (about 47°C, 117°F), thus killing the invader.

heat with the environment decreases as well. As a result, large ectothermic animals are considered improbable.

Small aquatic ectotherms (e.g., most invertebrates and fishes) generally remain at the same temperature as the surrounding water. Some larger aquatic animals, however, can maintain a body temperature warmer than that of the surrounding water (**FIGURE 4.14**). For example,

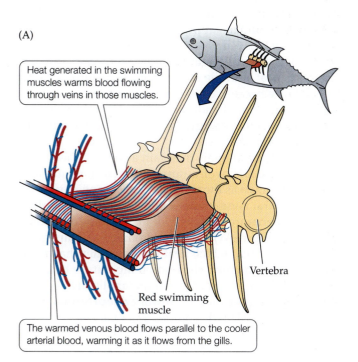

(A)

> Heat generated in the swimming muscles warms blood flowing through veins in those muscles.

Red swimming muscle

Vertebra

> The warmed venous blood flows parallel to the cooler arterial blood, warming it as it flows from the gills.

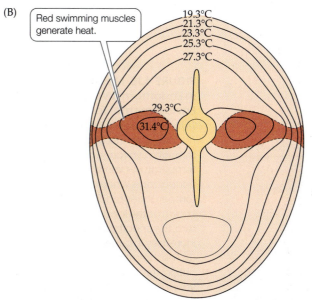

(B)

> Red swimming muscles generate heat.

19.3°C
21.3°C
23.3°C
25.3°C
27.3°C

29.3°C

31.4°C

FIGURE 4.14 Internal Heat Generation by Tuna (A) Heat generated in the red swimming muscles of the skipjack tuna, used for cruising through the water, warms blood flowing through them, which is carried toward the body surface in veins. Those veins run parallel to arteries carrying cool oxygenated blood from the gills, warming that blood before it reaches the swimming muscles. (B) A cross section of the tuna shows that its core remains warmer than the surrounding water.

skipjack tuna (*Katsuwonus pelamis*) use muscle activity, in conjunction with heat exchange between blood vessels, to maintain a body temperature as much as 14°C (25°F) warmer than the surrounding seawater. Other large oceanic fishes use similar circulatory heat exchange mechanisms to keep their muscles warm. Such mechanisms are particularly important for predatory species that depend on rapid acceleration to capture prey, which is aided by having warmer muscles.

The mobility of many terrestrial ectotherms allows them to adjust their body temperature by moving to places that are warmer or cooler than they are. Basking in the sun or moving into the shade allows these animals to adjust their energy gains and losses via solar radiation, conduction, and infrared radiation. For example, reptiles and insects emerging from hiding places after a cool night will bask in the sun to warm their bodies prior to initiating their daily activities (**FIGURE 4.15**). This basking behavior, however, increases their risk of being found by predators. Many of these animals rely on camouflage (also called *crypsis*) to escape detection while basking. In addition to moving between locations with different temperatures, reptiles may regulate their body temperatures by altering their coloration and changing their orientation to the sun.

Because they rely on the external environment for temperature regulation, the activities of ectothermic animals are limited to certain temperature ranges. When temperatures are warm, ectotherms in sunny environments (e.g., deserts) may gain enough energy from the environment to push their body temperatures to lethal levels. Increases in temperature associated with climate change over the past two decades appear to have limited the daily foraging periods of several species of Mexican lizards, whose abundances have decreased significantly during this period (also see Concept 25.2).

© Genevieve Vallee/Alamy Stock Photo

FIGURE 4.15 Mobile Animals Can Use Behavior to Adjust Body Temperature An adult female saltwater crocodile (*Crocodylus porosus*) sunning itself on the riverbank, Daintree River, Daintree National Park, Far North Queensland.

? What components of energy exchange are affected by the behavior of this crocodile?

In temperate and polar regions, temperatures drop below freezing for extended periods. Ectotherms inhabiting these regions must either avoid or tolerate exposure to subfreezing temperatures. Avoidance may take the form of seasonal migration (e.g., moving to a lower latitude) or movement to local microhabitats where temperatures stay at or above freezing (e.g., burrowing into the soil). Tolerance of subfreezing temperatures involves minimizing the damage associated with ice formation in cells and tissues. If ice forms as crystals, it will puncture cell membranes and disrupt metabolic functioning. Some insects inhabiting cold climates contain high concentrations of glycerol, a chemical compound that minimizes the formation of ice crystals and lowers the freezing point of body fluids. These insects spend winter in a semifrozen state, emerging in spring when temperatures are more conducive to physiological activity. Vertebrate ectotherms generally do not tolerate freezing to the degree that invertebrate ectotherms do, because of their larger size and greater physiological complexity. A very few amphibians, however, can survive being partially frozen, as described in the Case Study at the opening of this chapter.

TEMPERATURE REGULATION AND TOLERANCE IN ENDOTHERMS

Endotherms tolerate a narrower range of body temperatures (30°C–45°C, 86°F–113°F) than ectotherms. However, the ability of endotherms to generate heat internally has allowed them to greatly expand their geographic ranges and the times of year they can be active. Endotherms can remain active at subfreezing environmental temperatures, something that most ectotherms cannot do. The cost of being endothermic is a high demand for food to supply energy to support metabolic heat production. The rate of metabolic activity in endotherms is associated with the external temperature and the rate of heat loss. The rate of heat loss, in turn, is related to body size because of its influence on surface-area-to-volume ratio. Small endotherms have higher metabolic rates, require more energy, and have higher feeding rates than large endotherms.

Endothermic animals maintain a constant *basal* (resting) *metabolic rate* over a range of environmental temperatures known as the **thermoneutral zone**. Within the thermoneutral zone, minor behavioral or morphological adjustments are sufficient for maintaining an optimal body temperature. When the environmental temperature drops to a point at which heat loss is greater than metabolic heat production, the body temperature begins to drop, triggering an increase in metabolic heat generation. This point is called the **lower critical temperature** (**FIGURE 4.16A**). The thermoneutral zone and the lower

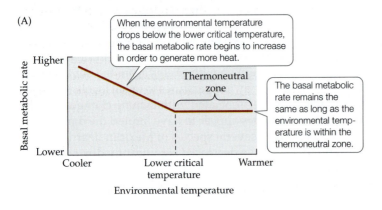

(A)

When the environmental temperature drops below the lower critical temperature, the basal metabolic rate begins to increase in order to generate more heat.

The basal metabolic rate remains the same as long as the environmental temperature is within the thermoneutral zone.

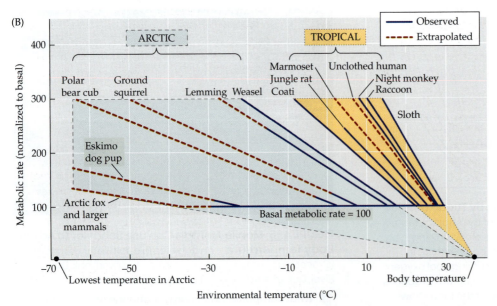

FIGURE 4.16 Metabolic Rates in Endotherms Vary with Environmental Temperatures
(A) An endotherm's resting, or basal, metabolic rate stays constant throughout a range of environmental temperatures known as the thermoneutral zone. When environmental temperatures reach a lower limit, known as the lower critical temperature, the endotherm's metabolic rate increases to generate additional heat. (B) The thermoneutral zones and lower critical temperatures of endotherms vary with their habitats. The lower critical temperatures of Arctic endotherms are lower than those of tropical endotherms, and their metabolic rates increase more slowly below those lower critical temperatures, as shown by the shallower slopes of the curves. (B after P. F. Scholander et al. 1950. *Biol Bull* 99: 237–258.)

critical temperature differ among mammal species (**FIGURE 4.16B**). As one would expect, mammals from the Arctic have lower critical temperatures below those of animals from tropical regions. Note also that the rate of metabolic activity (slope of the line) increases more rapidly below the lower critical temperature in tropical than in Arctic mammals.

What causes these differences in metabolic adjustments between endotherms of different biomes? For endothermy to work efficiently, animals must be able to retain their metabolically generated heat. Thus, the evolution of endothermy in birds and mammals required insulation: feathers, fur, or fat. These insulating layers provide a barrier limiting conductive (and, in some cases, convective) heat loss. Fur and feathers insulate primarily by providing a layer of still air, similar to a plant's boundary layer, adjacent to the skin. Differences in insulation help explain the differences among the endotherms in Figure 4.16B. Arctic mammals generally maintain thick fur. In warmer climates, however, the ability to cool off through conduction and convection is inhibited by insulation, and thick fur can be an impediment to maintaining an optimal body temperature. Some endotherms acclimatize to seasonal temperature changes by growing thicker fur in winter and shedding fur when temperatures get warmer (a fact that most pet owners know well) (**ANALYZING DATA 4.1**). Our human ancestors evolved in the hot tropical regions of Africa and lost much of their hairy insulating layer about 2 million years ago (Jablonski 2006).

Cold climates are tough on small endotherms. Small mammals, by necessity, have thin fur, since thick fur would inhibit their mobility. The high demand for metabolic energy below the lower critical temperature, the low insulation values of their fur, and their low capacity to store energy make small mammals improbable residents of polar, alpine, and temperate habitats. However, the faunas of many of these cold climates contain many species of small endotherms, sometimes in high abundances. What explains this apparent discrepancy? Small endotherms, such as rodents and hummingbirds, are able to alter their lower critical temperature during cold periods by entering a state of dormancy known as **torpor**. The body temperatures of animals in torpor may drop as much as 20°C (36°F) below their normal temperatures. The metabolic rate of an animal in torpor is 50%–90% lower than its basal metabolic rate, providing substantial energy savings (Schmidt-Nielsen 1997). However, energy is still needed to arouse the animal from torpor and bring its body temperature back up to its usual set point. Thus, the length of time an animal can remain in torpor is limited by its reserves of energy. Small endotherms may undergo *daily torpor* to minimize the energy needed during cold nights. Torpor lasting several weeks during the winter, sometimes referred to as **hibernation**, is possible only for animals that have access to enough food and can store enough energy reserves, such as marmots (**FIGURE 4.17**). Hibernation is somewhat rare in polar climates because few animals have access to enough food to provide enough stored energy (in the form of fat) to get

---

ANALYZING DATA 4.1

How Does Fur Thickness Influence Metabolic Activity in Endotherms?

Some endotherms exhibit seasonal changes in fur thickness, which helps to enhance heat loss during summer and retain heat generated by the body during winter. This seasonal change in fur thickness in individual animals is an example of acclimatization to changes in temperature.

The graph* shows the insulation value (how well heat is retained) versus fur thickness for two animals of the boreal forest biome, a red squirrel (*Tamiasciurus hudsonicus*) and a wolf (*Canis lupus*). Both animals are endotherms that exhibit acclimatization to seasonal temperature changes by changes in fur thickness.

1. Each animal is represented by fill or no-fill. Which do you think belongs to which animal (circles and triangles), and why?

2. Which season (summer or winter) is represented by the circles, and which by the triangles? Which animal experiences greater seasonal acclimatory changes in fur length? In what additional way might the animal with the smaller change in fur length cope with extreme winter cold?

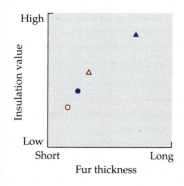

- ● Animal #1, Season one
- ▲ Animal #1, Season two
- ○ Animal #2, Season one
- △ Animal #2, Season two

---

*After P. Willmer et al. 2005. *Environmental Physiology of Animals*, 2nd ed. Blackwell Publishing: Malden, MA.

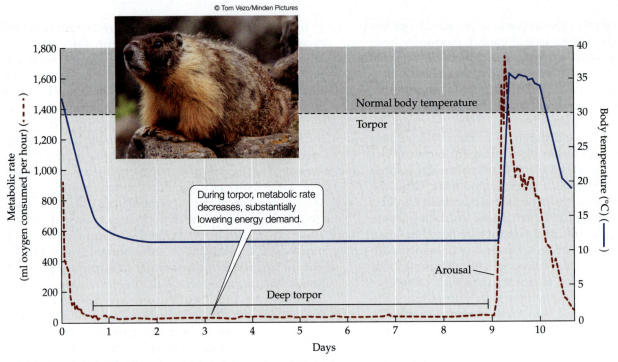

© Tom Vezo/Minden Pictures

During torpor, metabolic rate decreases, substantially lowering energy demand.

Normal body temperature

Torpor

Deep torpor

Arousal

FIGURE 4.17 Long-Term Torpor in Marmots Torpor allows yellow-bellied marmots (*Marmota flaviventris*) to conserve energy during winter, when food is scarce and the demand for metabolic energy to keep warm is high. Regular cycles of arousal and return to torpor occur for unknown reasons. (After K. Armitage et al. 2003. *Comp Biochem Phys* 134A: 101–114.)

through winter without eating. Some large animals, such as bears, enter a long-term winter sleep (sometimes called denning) during which the body temperature decreases only slightly, rather than going into torpor.

Just as organisms must balance energy input and output to maintain an optimal temperature, they must balance the movement of water into and out of their bodies to maintain optimal conditions for physiological functioning.

CONCEPT 4.3 The water balance of an organism is determined by exchanges of water and solutes with the external environment.

LEARNING OBJECTIVES

4.3.1 List the three factors that influence the movement of water from a high-energy state to a low-energy state (i.e., with reference to water potential gradients) in biological systems.

4.3.2 Explain how organisms can control water gains and losses by adjusting resistance to water movement, and describe how high resistance may involve trade-offs with other functions.

4.3.3 Describe how salt and water balances can become challenges for organisms exposed to hyperosmotic and hypoosmotic environments.

Variation in Water Availability

Water is essential for life. Water is the medium in which all biochemical reactions necessary for physiological functioning occur. Water has unique properties that make it a universal solvent for biologically important *solutes* (compounds that are dissolved in water, including salts). The range of organismal water content conducive to physiological functioning is relatively narrow, between 60% and 90% of body mass. Maintaining water content within this range is a challenge primarily to organisms of freshwater and terrestrial environments. Marine organisms seldom gain or lose too much water, because they exist in a medium that is conducive to maintaining water balance: the oceans in which life first evolved.

In addition to maintaining a suitable water balance, organisms must also balance the uptake and loss of solutes, primarily salts. Salt balance is closely tied to water balance because the movements of water and salts influence each other. Aquatic environments may be more saline than an organism's cells or blood (*hyperosmotic; hyper,* "greater"), of similar salinity (*isoosmotic; iso,* "same"), or less saline (*hypoosmotic; hypo,* "less"). Most marine invertebrates rarely face problems with water and solute balance, because they tend to be isoosmotic.

Terrestrial organisms face the problem of losing water to a dry atmosphere, while freshwater organisms may lose solutes to, and gain water from, their environment. The evolution of freshwater and terrestrial organisms is very much a story of dealing with the need to maintain water balance. In this section, we will review some basic principles related to water and solute balance and provide some examples of how freshwater and terrestrial organisms maintain a water balance that is conducive to physiological functioning.

Water flows along energy gradients

Water flows along energy gradients, from high-energy to low-energy conditions. What is an energy gradient in the context of water? Gravity represents one example that is intuitively obvious: liquid water flows downhill, following a gradient of *potential energy*. Another type of energy influencing water movement is *pressure*. When elephants spray water out of their trunks, the water is flowing from a condition of higher energy inside the trunk (where muscles exert pressure on it) to a condition of lower energy outside of the trunk (where muscle pressure is not present).

Other, less obvious factors that influence the flow of water are important to organismal water balance. When solutes are dissolved in water, the solution loses energy. Thus, if the water in a cell contains more solutes than the water surrounding it, water will flow into the cell to equilibrate the energy difference. Alternatively, solutes may flow into the surrounding medium, but most biological membranes selectively block the flow of many solutes. In biological systems, the energy associated with dissolved solutes is called **osmotic potential**. The energy associated with gravity is called **gravitational potential**, but in a biological context it is important in water movement only in very tall trees. The energy associated with the exertion of pressure is called **pressure** (or **turgor**) **potential**. Finally, the energy associated with attractive forces on the surfaces of large molecules inside cells or on the surfaces of soil particles is called **matric potential**.

The sum of these energy components within an aqueous system determines its overall water energy status, or **water potential**. The water potential of a system can be defined mathematically as

$$\Psi = \Psi_o + \Psi_p + \Psi_m \qquad (4.3)$$

where Ψ is the total water potential of the system (in units of pressure; usually megapascals, MPa), Ψ_o is the osmotic potential (a negative value because it lowers the energy status of the water), Ψ_p is the pressure potential (a positive value if pressure is exerted on the system; a negative value if the system is under tension), and Ψ_m is the matric potential (a negative value). Water will always move from a system of higher Ψ to a system of lower Ψ, following the energy gradient. This terminology is most often used in plant, microbial, and soil systems, but it works in animal systems as well.

The atmosphere has a water potential that is related to humidity. From a biological perspective, air with a relative humidity of less than 98% of saturation has a very low water potential, so the gradient in water potential between most terrestrial organisms and the atmosphere is very high. Without some barrier to water movement, terrestrial organisms would lose water rapidly to the atmosphere. Any force that impedes the movement of water (or other substances, such as carbon dioxide) along an energy gradient is called **resistance**.[2] Barriers that increase organisms' resistance to water loss include the waxy cuticle of plants and insects and the skin of amphibians, reptiles, birds, and mammals.

Water losses and solute gains and losses must be compensated

Terrestrial plants and soil microorganisms rely on water uptake from soils to replace the water they lose to the atmosphere. Soils are important reservoirs of water that support a multitude of ecological functions. The amount of water that soils can store is related to the balance between water inputs and outputs, soil texture, and topography (**FIGURE 4.18**). Water inputs include precipitation that infiltrates into the soil and overland flow of water. Water losses include percolation to deeper layers below the plant rooting zone and evapotranspiration.

The water storage capacity of most soils is dominated by their pore space and matric potential, which is related to the attractive forces on the surfaces of the soil particles. Sandy soils store less water than fine-textured soils, but fine soil particles also have a higher matric potential and thus hold on to water more tightly. Soils with mixed coarse and fine particles are generally most effective in storing water and supplying it to plants and soil organisms. When the volume of water in the soil drops below a certain point (25% of total soil mass in fine-textured soils, 5% in sandy soils), the matric forces are strong enough that most of the remaining water is unavailable to organisms. The osmotic potential of some soils also can be important, particularly where dissolved salts are found, as in soils near marine environments or where salinization (see Figure 2.25) has occurred.

WATER BALANCE IN MICROORGANISMS Single-celled microorganisms, which include archaea, bacteria, algae, and protists, are active primarily in aqueous environments. Their water balance is dependent on the water potential of the surrounding environment, which is determined mainly by its osmotic potential. In most marine and freshwater ecosystems, the osmotic potential of the environment changes little over time. Some environments, however, such as estuaries, tide pools, saline lakes, and soils, experience frequent changes in osmotic potential due to evaporation or variable influxes of fresh

---

[2] Many physiologists prefer using conductance rather than resistance to express the influence of a barrier on the movement of water or gases between an organism and its environment. Mathematically, conductance is the reciprocal of resistance.

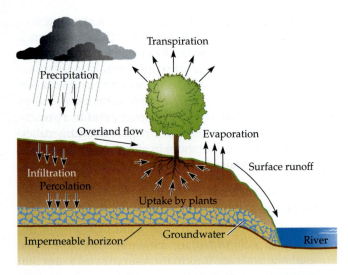

FIGURE 4.18 What Determines the Water Content of Soil? The water content of soil is determined by the balance between water inputs (infiltration of precipitation and overland flow of water) and outputs (percolation to deeper layers, evapotranspiration) and by the capacity of the soil to hold water. Soil water storage capacity and the rate of percolation are dependent on soil texture. (After P. J. Kramer. 1983. *Water Relations of Plants.* Academic Press: Cambridge, MA.)

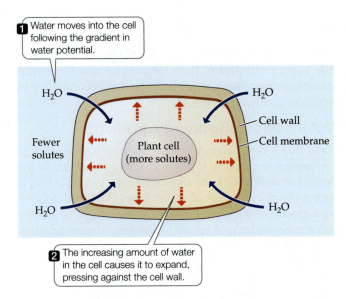

FIGURE 4.19 Turgor Pressure in Plant Cells When a plant cell is surrounded by water with a solute concentration lower than its own, water moves into the cell, while solutes in the cell are prevented from moving out by the cell membrane. The increasing amount of water in the cell causes the cell to expand, pressing against the cell wall.

and salt water. Microorganisms in these environments must respond to these changes by altering their cellular osmotic potential if they are to maintain a water balance suitable for physiological functioning. They accomplish this through **osmotic adjustment**, an acclimatization response that involves changing their solute concentration, and thus their osmotic potential. Some microorganisms synthesize organic solutes to adjust their osmotic potential, which also helps to stabilize enzymes. Others use inorganic salts from the surrounding medium for osmotic adjustment. The ability to adjust osmotic potential in response to changes in external water potential varies substantially among microorganisms: some completely lack this ability, while others (such as *Halobacterium* spp.) can adjust to even the extremely saline conditions in landlocked saline lakes.

As noted above, terrestrial environments are too dry for any organism that is unable to restrict cellular water loss to the atmosphere. Many microorganisms avoid exposure to dry conditions by forming dormant resistant spores, encasing themselves in a protective coating that prevents water loss to the environment. Some microorganisms with filamentous forms, such as fungi and yeasts, are very tolerant of low water potentials and can grow in dry environments. Most terrestrial microorganisms, however, are found in soils, which have a higher water content and humidity than the air above them.

WATER BALANCE IN PLANTS One of the distinguishing characteristics of plants is a rigid cell wall composed of cellulose. Bacteria and fungi also have cell walls, composed of materials such as chitin (in fungi) or peptidoglycans

and lipopolysaccharides (in bacteria). Cell walls are important to water balance because they facilitate the development of positive **turgor pressure**. When water follows a gradient of water potential into a plant cell, it causes the cell to expand and press against the cell wall, which resists the pressure because of its rigidity (**FIGURE 4.19**). Turgor pressure is an important structural component of plants, and it is also an important force for growth, promoting cell division. When nonwoody plants lose turgor pressure due to dehydration, they wilt. Wilting is generally a sign that a plant is experiencing water stress.

Plants take up water from sources with a water potential higher than their own. For aquatic plants, the source is the surrounding aqueous medium. In freshwater environments, the presence of solutes in the plant's cells creates a water potential gradient from the surrounding water to the plant. In marine environments, plants must lower their water potential below that of seawater to take up water. Marine plants, as well as terrestrial plants of salt marshes and saline soils, adjust their osmotic potential in a manner similar to that of microorganisms by synthesizing solutes and taking up inorganic salts from their environment. Inorganic salts must be taken up selectively, however, because some, such as sodium (Na^+) and chloride (Cl^-), can be toxic at high concentrations. The cell membranes of plants act as a solute filter, determining the amounts and types of solutes that move into and out of the plant.

Terrestrial plants acquire water from the soil through their roots, as well as through associations with mutualistic fungi, called *mycorrhizae*, that grow into their roots from the soil (see Concept 15.1). The earliest land plants, which had not yet evolved roots, used mycorrhizal fungi to take

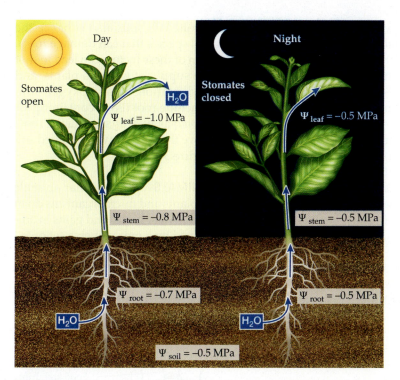

FIGURE 4.20 The Daily Cycle of Dehydration and Rehydration
During the day, when the stomates are open, transpiration results in a gradient of water potential from leaf to stem, stem to roots, and roots to soil. At night, when the stomates are closed, water potential equilibrates as the plant rehydrates.

up water and nutrients from the soil. The majority of modern terrestrial plant species use a combination of roots and mycorrhizae to take up water. Only the finest roots can take up water from the soil, because older, thicker roots develop a water-resistant waxy coating that limits their ability to absorb water as well as to lose water to the soil. Mycorrhizae provide a greater surface area for absorption of water and nutrients for the plant and allow greater exploration of the soil for these resources. In turn, the mycorrhizal fungi obtain energy from the plant.

Plants lose water by transpiration when their stomates open to allow carbon dioxide (CO_2) from the atmosphere to diffuse into their leaves. Water moves out through the stomates, following the water potential gradient from the inside of the leaf (100% relative humidity) to the air. As we saw in the previous section, transpiration is an important cooling mechanism for leaves. The plant must replace the water lost by transpiration, however, if it is to avoid water stress. As a leaf loses water, the water potential of its cells decreases, creating a water potential gradient between the leaf and the xylem in the stem to which it is attached, so water moves through the xylem into the leaf. In this way, when the plant is transpiring, it creates a gradient of decreasing water potential from the soil through the roots and stems to the leaves (**FIGURE 4.20**). Water therefore flows from the soil, which has the highest water potential, into the roots, the xylem, and eventually the leaves, from which it is lost to the atmosphere via transpiration. Because there is greater resistance to the movement of water into the roots and through the xylem than out through the stomates, the water supply from the soil cannot keep up with water loss by transpiration. As a result, the water content of the plant decreases during the day. At night the stomates close, and the water supply from the soil rehydrates the plant until it reaches near equilibrium with the soil water potential. This daily cycle of daytime dehydration and nighttime rehydration can go on indefinitely if the supply of water in the soil is adequate. The availability of water decreases when precipitation is not sufficient to replace the water lost from the soil through transpiration and evaporation. The water content of a plant will then decrease, and its turgor pressure will decrease as its cells become dehydrated (**FIGURE 4.21**).

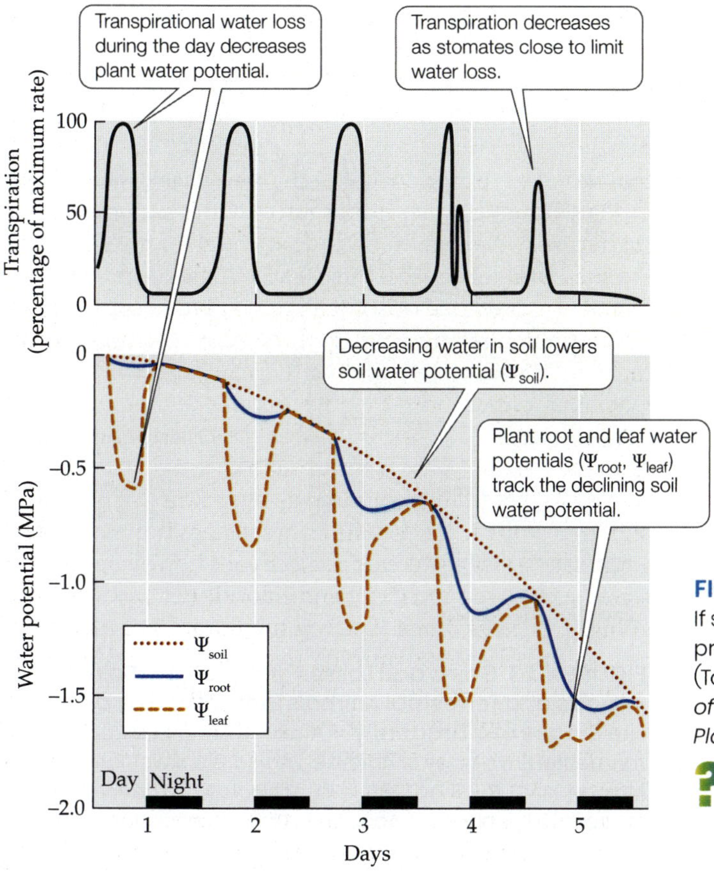

Transpirational water loss during the day decreases plant water potential.

Transpiration decreases as stomates close to limit water loss.

Decreasing water in soil lowers soil water potential (Ψ_{soil}).

Plant root and leaf water potentials (Ψ_{root}, Ψ_{leaf}) track the declining soil water potential.

FIGURE 4.21 How Plants Cope with Depletion of Soil Water
If soil water is not recharged, transpiration will deplete it, leading to progressive drying of the soil and a decrease in soil water potential. (Top, after A. H. Fitter and R. K. Hay. 1987. *Environmental Physiology of Plants*. Academic Press: London; bottom, after R. D. Slatyer. 1967. *Plant-Water Relationships*. Academic Press: Cambridge, MA.)

? As the soil dries, stomates may close at midday and reopen later in the afternoon, as seen on day 4 in the graph. Assuming the air temperature is cooler later in the day, what influence would this have on plant water loss?

To avoid reaching a detrimentally or even lethally low water content, the plant must restrict its transpirational water loss. If leaf cells become so dehydrated that turgor is lost, the stomates close. This level of water stress can harm the plant, causing impairment of physiological functions such as photosynthesis. Extremely dry conditions can cause loss of xylem function.

Some plants of seasonally dry environments shed their leaves during long dry periods to eliminate transpirational water loss. Others have a signaling system that helps prevent the onset of water stress. As the soil dries out, the roots send a hormonal signal (abscisic acid) to the guard cells, which close the stomates, lowering the rate of water loss. Plants of dry environments, such as deserts, grasslands, and Mediterranean-type ecosystems, generally have better control of stomatal opening than plants of wetter climates. Plants of dry environments also have a thick waxy coating (cuticle) on their leaves to prevent water loss through the nonporous regions of the leaves. Additionally, plants of dry environments maintain a higher ratio of root biomass to biomass of stems and leaves than plants of moister environments, enhancing the rate of water supply to transpiring tissues (Mokany et al. 2006) (**FIGURE 4.22**). Some plants are capable of acclimatization by altering the growth of their roots to match the availability of soil moisture and nutrients.

Can plants have too much water? Technically, no, but saturation inhibits the diffusion of oxygen and can cause hypoxia in plant roots. Thus, waterlogged soils inhibit aerobic respiration in roots. Wet soils also enhance the growth of harmful fungal species that can damage roots. Ironically, the combination of these factors can lead to root death, which cuts off the supply of water to plants, and eventually to wilting. Adaptations to low oxygen concentrations in wet soils include root tissue containing air channels (called *aerenchyma*) as well as specialized roots that extend vertically above the water or waterlogged soil (as in mangroves; see Figure 3.19).

WATER BALANCE IN ANIMALS   Multicellular animals face the same challenges plants and microorganisms do in maintaining water balance. Water losses and gains in animals, however, are governed by a more diverse set of exchanges than in plants and microorganisms (**FIGURE 4.23**). Many animals have the added complexity of specialized organs for gas exchange, ingestion and digestion, excretion, and circulation, all of which create areas of localized

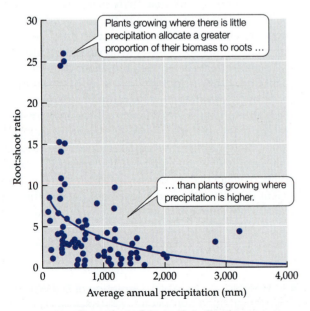

FIGURE 4.22 Allocation of Growth to Roots versus Shoots Is Associated with Precipitation Levels   The ratio of root biomass to leaf and stem (shoot) biomass increases with decreasing precipitation in shrubland and grassland biomes. Allocation of more biomass to roots in dry soils provides more water uptake capacity to support leaf function. (After K. Mokany et al. 2006. *Global Change Biol* 12: 84–96.)

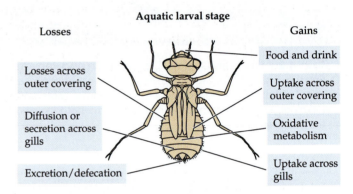

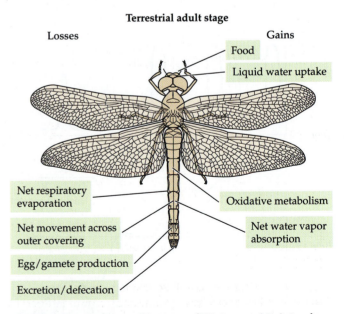

FIGURE 4.23 Gains and Losses of Water and Solutes in Aquatic and Terrestrial Animals Exemplified by Different Life Stages of a Dragonfly   (After P. Willmer et al. 2005. *Environmental Physiology of Animals*, 2nd ed. Blackwell Publishing: Malden, MA; E. B. Edney. 1980. In *Insect Biology in the Future*, M. Locke [Ed.], pp. 39–58. Academic Press: Cambridge, MA.)

water and solute exchange as well as gradients of water and solutes within the animal's body. Most animals are mobile and can seek out environments conducive to maintaining a favorable water and solute balance, an option not available to plants or to most microorganisms.

Many animals must be able to maintain favorable water and solute balances under conditions of varying salinity. A marine animal that lacks this ability will die if transferred to brackish or fresh water. Although most marine invertebrates are isoosmotic to seawater, the specific types of solutes in their bodies can vary. Many invertebrates that are capable of adjusting to changes in the solute concentration of their environment do so by exchanging solutes with the surrounding seawater. Like plants, these animals must selectively control this exchange of specific solutes because some external solutes are toxic at the concentrations at which they are found in seawater and because some internal solutes are needed for biochemical reactions. Jellyfishes, squids, and crabs, for example, have Na^+ and Cl^- concentrations similar to those of seawater, but their sulfate (SO_4^{2-}) concentrations may be one-half to one-fourth of those found in seawater.

Marine vertebrates include animals that are isoosmotic and hypoosmotic to seawater. The cartilaginous fishes, including the sharks and rays, have blood solute concentrations similar to those of seawater, although, as in invertebrates, their concentrations of specific solutes differ from those in seawater. In contrast, marine teleost (bony) fishes and mammals evolved in fresh water and later moved into marine environments. Their blood is hypoosmotic to seawater. Fish exchange water and salts with their environment through drinking and eating, and across the gills, which are also the organs of oxygen (O_2) and CO_2 exchange (**FIGURE 4.24A**). Salts that diffuse into or are ingested by marine teleost fishes must be continuously excreted in urine and through the gills against an osmotic gradient, which requires an expenditure of energy. Water lost across the gills must be replaced by drinking. Marine mammals, such as whales and porpoises, produce urine that is hyperosmotic to seawater and avoid drinking seawater to minimize salt uptake.

Freshwater animals are hyperosmotic relative to their environment; therefore, they tend to gain water and lose salts. Most salt exchange occurs at the gas exchange surfaces, including the skin of some invertebrates (e.g., freshwater worms) and the gills of many vertebrates and invertebrates. These animals must compensate for salt losses by taking up solutes in their food, and some groups, such as

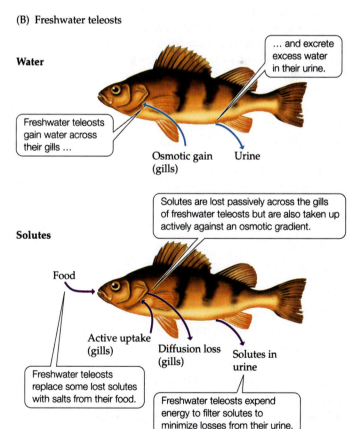

(A) Marine teleosts

Water

Marine teleosts lose water across their gills and in their urine …

Drinking

… which they must replace by drinking seawater.

Osmotic loss (gills)   Urine

Solutes

… and must expend energy to excrete excess solutes across their gills and in their urine.

Salts from seawater

Marine teleosts take up solutes by drinking seawater …

Na^+ and Cl^- (gill secretion) Mg^{2+} and SO_4^{2-} (in urine)

(B) Freshwater teleosts

Water

… and excrete excess water in their urine.

Freshwater teleosts gain water across their gills …

Osmotic gain (gills)   Urine

Solutes

Solutes are lost passively across the gills of freshwater teleosts but are also taken up actively against an osmotic gradient.

Food

Active uptake (gills)   Diffusion loss (gills)   Solutes in urine

Freshwater teleosts replace some lost solutes with salts from their food.

Freshwater teleosts expend energy to filter solutes to minimize losses from their urine.

FIGURE 4.24 Water and Salt Balance in Marine and Freshwater Teleost Fishes Marine and freshwater teleost fishes face opposite challenges in maintaining water and solute balance. (A) Marine teleosts are hypoosmotic to their environment: they tend to lose water and gain solutes. (B) Freshwater teleosts are hyperosmotic to their environment: they tend to gain water and lose solutes. (After K. Schmidt-Nielsen. 1979. *Animal Physiology: Adaptation and Environment.* Cambridge University Press: Cambridge.)

teleost fishes, must take up solutes actively through the gills against an osmotic gradient (**FIGURE 4.24B**). Excess water is excreted as dilute urine, from which the excretory system actively removes solutes to minimize their loss.

Terrestrial animals face the challenge of exchanging gases (O_2 and CO_2) in a dry environment with a very low water potential. These animals lower their evaporative water loss and exposure to water stress by having skin with a high resistance to water loss or by living in environments where they can compensate for high water losses with high water intake. Both approaches involve risks and trade-offs, however. A high resistance to water loss may compromise the animal's ability to exchange gases with the atmosphere. Reliance on a steady water supply puts the animal at risk if the source of water fails (e.g., during a severe drought). Tolerance for water loss varies substantially among groups of terrestrial animals. Generally, invertebrates have a higher tolerance for water loss than vertebrates. Within the vertebrates, amphibians have a higher tolerance for, but lower resistance to, water loss than mammals and birds (**TABLE 4.1**).

Amphibians, including frogs, toads, and salamanders, rely primarily on stable water supplies to maintain their water balance. They can be found in a wide variety of biomes, from tropical rainforests to deserts, as long as there is a reliable source of water, such as regular rains or ponds. Amphibians depend on gas exchange through the skin to a greater degree than other terrestrial vertebrates. Therefore, amphibian skin is often thin, with a low resistance to water loss (**FIGURE 4.25**). However, some adult amphibian species have adapted to dry environments

by developing specialized skin with higher resistance to water loss. For example, the southern foam-nest tree frog (*Chiromantis xerampelina*), which occurs throughout Africa, has skin that resists water loss in a manner similar to that of lizards. To compensate for reduced gas exchange through the skin, it has a higher breathing rate (Stinner and Shoemaker 1987). As a group, tree frogs have higher skin resistance to water loss than ground frogs, reflecting their drier habitat. Some ground frogs of seasonally dry environments, such as the northern snapping frog (*Cyclorana australis*) of Australia, lower their rates of water loss by forming a "cocoon" of mucous secretions consisting of proteins and fats that increases resistance to water loss.

Reptiles have been extremely successful at inhabiting dry environments. The thick skin of desert snakes and lizards provides protection for the internal organs as well as an effective barrier to water loss. The outer skin, made up of multiple layers of dead cells with a fatty coating, is overlain by plates or scales. These layers give reptilian skin a very high resistance to water loss.

TABLE 4.1

Ranges of Tolerances for Water Loss in Selected Animal Groups

Group	Weight loss (%)
Invertebrates	
Mollusks	35–80
Crabs	15–18
Insects	25–75
Vertebrates	
Frogs	28–48
Small birds	4–8
Rodents	12–15
Human	10–12
Camel	30

Source: P. Willmer et al. 2005. *Environmental Physiology of Animals*, 2nd ed. Blackwell Publishing: Malden, MA.

Note: Values are maximum percentages of body weight lost as water that can be tolerated, based on observations of a range of exemplary species in each group.

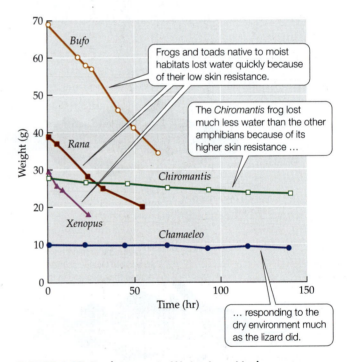

FIGURE 4.25 Resistance to Water Loss Varies among Frogs and Toads Amphibians were kept under uniform dry environmental conditions (25°C, 20%–30% relative humidity) to examine their rates of water loss, measured as loss of body weight. A lizard (*Chamaeleo*) was also tested for comparative purposes. (After K. Schmidt-Nielsen. 1979. *Animal Physiology: Adaptation and Environment*. Cambridge University Press: Cambridge; based on J. P. Loveridge. 1970. *Arnoldia* [*Rhodesia*] 5: 1–6. National Museum of Southern Rhodesia.)

? How could you estimate the resistances of these species to water loss quantitatively using this graph?

TABLE 4.2

Ranges of Resistance of External Coverings (Skin, Cuticle) to Water Loss

Group	Resistance (s/cm)
Crabs (marine)	6–14
Fish	2–35
Frogs	3–100
Earthworms·	9
Birds	50–158
Desert tortoises	120
Desert lizards	1,400
Desert scorpions, spiders	1,300–4,000

Source: P. Willmer et al. 2005. *Environmental Physiology of Animals*, 2nd ed. Blackwell Publishing: Malden, MA.

Mammals and birds have skin anatomy similar to that of reptiles but have hair or feathers covering the skin rather than scales. The presence of sweat glands in mammals represents a trade-off between resistance to water loss and evaporative cooling. The highest resistances to water loss among terrestrial animals are found in the arthropods (e.g., insects and spiders), which are characterized by an outer exoskeleton made of hard chitin and coated with waxy hydrocarbons that prevents water movement (**TABLE 4.2**).

An instructive example of how animals use a variety of integrated adaptations to cope with arid environments involves kangaroo rats (*Dipodomys* spp.), found throughout the deserts of North America. A combination of efficient water use and low rates of water loss greatly diminishes these rodents' water requirements (Schmidt-Nielsen and Schmidt-Nielsen 1951) (**FIGURE 4.26**). Kangaroo rats rarely drink water. A large proportion of their water requirement is met by eating dry seeds and by oxidative metabolism—that is, by metabolically converting carbohydrates and fats into water and carbon dioxide (Schmidt-Nielsen 1964). The animals also consume water-rich foods, such as insects or succulent vegetation, if they are available.

Kangaroo rats minimize water loss through several physiological and behavioral adaptations. During the hottest periods of the year, they are active only at night, when air temperatures are lowest and humidities highest. During the day, they stay in their underground burrows, which are cooler and more humid than the desert surface. In some parts of their range, however, temperatures even in their burrows can rise high enough to expose kangaroo rats to significant evaporative water loss (Tracy and Walsberg 2002). To increase their resistance to this loss, kangaroo rats have thicker, oilier skin, with fewer sweat glands, than related rodents of moister environments.

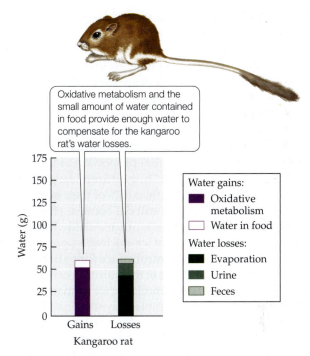

FIGURE 4.26 Water Balance in the Kangaroo Rat
Under dry laboratory conditions (25°C, 25% relative humidity), kangaroo rats, native to deserts of western North America, do not require liquid water to survive. (After K. Schmidt-Nelson. 1997. *Desert Animals*. Clarendon Press: Oxford.)

They minimize water losses in their urine and feces through effective removal of water by their kidneys and intestines. Kangaroo rats produce some of the most concentrated urine of any animal. The combination of these characteristics allows kangaroo rats to inhabit very arid environments without exposure to water stress, even without access to drinking water.

A CASE STUDY REVISITED

Frozen Frogs

The existence of amphibians above the Arctic Circle seems improbable, given their reliance on a steady supply of liquid water to maintain their water balance and the high potential for damage associated with freezing. Several problems must be overcome in order for complex organisms to survive freezing. First, when water freezes, it forms needlelike crystals that can penetrate and damage or destroy cell membranes and organelles. Second, the supply of oxygen to tissues is severely restricted by the lack of circulation and breathing. Finally, as ice forms, pure water is pulled from cells, resulting in shrinkage and an increase in solute concentration. Any one of these factors, or all of them working in combination, will kill tissues and organisms in subfreezing temperatures. Yet the frogs described in the Case Study, as well as many species of invertebrates, can tolerate the freezing of a substantial amount of their body water.

Wood frogs and other freeze-tolerant amphibians spend winter in shallow depressions under leaves, moss, or logs, which do not protect them from subfreezing temperatures. Several adaptations facilitate the survival of these amphibians through the winter and allow them to emerge from their frozen state in spring unharmed. Freezing of water in these animals is limited to the spaces outside the cells. A substantial proportion of their body water, from 35% to 65% in "fully frozen" frogs, freezes (Pinder et al. 1992). If more than 65% of their body water is frozen, most individuals will die because of excessive cell shrinkage. The formation of ice outside the cells is enhanced by the existence of ice-nucleating proteins that serve as the site of slow, controlled ice formation (Storey 1990). Solute concentrations in the unfrozen cells increase as the cells lose water to extracellular ice formation. In addition, freeze-tolerant amphibians synthesize additional solutes, including glucose and glycerol derived from the breakdown of liver glycogen. The resulting increase in solute concentrations lowers the freezing point inside the cells, allowing the intracellular solution to remain liquid at subfreezing temperatures. The concentrated solutes also stabilize the cell volume and the structures of organelles, proteins, and enzymes. As freezing proceeds, the frog's heart stops, and its lungs cease to pump air. Once it reaches this semistable state of partial freezing, the frog can remain frozen for several weeks, as long as the temperature does not drop below about –5°C (23°F). Although their winter "quarters" are not far below the surface of the ground, the insulating cover of leaves and snow keeps the frogs above that temperature.

The freezing process is initiated in wood frogs within minutes of ice formation within the animal, although the full process occurs over several days to weeks (Layne and Lee 1995). Thawing, on the other hand, may be rapid, with normal body functioning returning within 10 hours. This amazing amphibian feat of spending winter in a semifrozen state and emerging unharmed in spring has provided information to medical science that has facilitated the preservation of human tissues and organs at low temperatures (Costanzo et al. 1995), as well as optimism to proponents of whole-body cryonics, who hope that someday Grandpa can finally leave the Tuff Shed.

 ## CONNECTIONS *in* NATURE

DESICCATION TOLERANCE, BODY SIZE, AND RARITY As we saw in Chapter 3, there is a close association between organisms' adaptations to climate conditions and their distribution among terrestrial biomes. While subfreezing temperatures are an important constraint on the distribution

and functioning of organisms in high-latitude and high-elevation biomes, low water availability is a more widespread challenge. Arid conditions can occur in most terrestrial biomes (see the climate diagrams in Concept 3.1), and they regularly occur over more than 60% of the land surface. As we have seen, the majority of terrestrial organisms, particularly animals, avoid exposure to dry conditions and rely on minimizing water losses to the environment. Some organisms, however, can tolerate arid conditions in much the same way that frozen frogs tolerate subfreezing winter conditions: by entering a dormant state while allowing themselves to dry out. This adaptive approach is common in microorganisms, including bacteria, fungi, and protists, but is also found in some multicellular animals and some plants, including mosses, liverworts, and a few flowering plants (Alpert 2006).

Desiccation-tolerant organisms can survive extreme dehydration, losing 80%–90% of their water as they equilibrate with the humidity of the air, then regain metabolic function shortly after they are rehydrated (**FIGURE 4.27**). As its cells dry out, the organism synthesizes sugars, which are the key to protecting its cell and organelle structures (Alpert 2006). Once dehydration proceeds beyond a certain threshold, metabolism ceases, and the sugars and the small amount of remaining water form a glassy coating over the cellular constituents. As with recovery from freezing, recovery from dehydration is rapid, occurring in hours to days.

The prevalence of dry conditions in terrestrial environments suggests that desiccation tolerance should be more common than it is. Why hasn't such tolerance evolved in more plants and animals? A clue to this puzzle may be the small size of the organisms that are desiccation tolerant (Alpert 2006). Small organisms (less than 5 mm in animals) do not require structural reinforcements, such as a skeletal system, that would restrict the necessary shrinking of the organism as it dehydrates. In addition, water loss during dehydration must be slow enough to allow sugar synthesis to occur, but not so slow that the organism spends a long time with a low water content while metabolism is still occurring, which can cause physiological stress. Small organisms have surface-area-to-volume ratios and thicknesses favorable for the water loss rates required.

These arguments explain why desiccation tolerance is more common in small organisms, but not why they are rare (see Chapter 23). The two characteristics—small size and rarity—are intimately linked. As we will see in Chapter 14, small size is often associated with slow growth rates and poor competitive ability under conditions of low resource availability. Thus, natural selection for desiccation tolerance may involve trade-offs with other ecological characteristics, such as competitive ability, that might prevent these organisms from being successful in competitive environments.

(A) *Selaginella lepidophylla* (club moss)

Both photos courtesy of David McIntyre

(B) Tardigrade (water bear)

Both photos © Eye of Science/Science Source

FIGURE 4.27 Desiccation-Tolerant Organisms (A) The leaves of the club moss *Selaginella lepidophylla* reach a very low moisture content during prolonged periods without rain (left); within 6 hours of receiving water, the leaves are functional and carrying out photosynthesis (right). (B) Water bears (tardigrades) are small invertebrates (less than 1 mm in length) found in aqueous environments, including oceans, lakes and ponds, soil water, and the water films on vegetation. Water bears contract and cease metabolism when they and their environment dry up (left) but rehydrate when moisture returns (right).

SUMMARY

CONCEPT 4.1 Each species has a range of environmental tolerances that determines its potential geographic distribution.

4.1.1 Explain why the physical environment is the ultimate determinant of the geographic distribution of a species.

- The physical environment affects an organism's ability to obtain energy and resources, thereby determining its growth and reproduction and, more immediately, its ability to survive the extremes of that environment. The physical environment is therefore the ultimate constraint on the geographic distribution of a species.

4.1.2 Differentiate between adaptation and acclimatization by explaining how both individual organisms and populations respond, but differently, to changes in the environment.

- Individual organisms can respond to environmental change through acclimatization, a short-term adjustment of the organism's physiology, morphology, or behavior that lessens the effect of the change and minimizes the associated stress.

- A population may respond to unique environmental conditions through natural selection for physiological, morphological, and behavioral traits, known as adaptations, that enhance individuals' survival, growth, and reproduction under those conditions.

4.1.3 Illustrate how adaptation and acclimatization may result in trade-offs with other functions.

- Acclimatization and adaptation are not "free"; they require an investment of energy and resources by the organism. They represent possible trade-offs with other functions of the organism that may also affect its survival and reproduction.

CONCEPT 4.2 The temperature of an organism is determined by exchanges of energy with the external environment.

4.2.1 Describe how the body temperature of an organism influences its functioning.

- Temperature controls physiological processes through its effects on enzymes and membranes.

(Continued)

SUMMARY *(continued)*

4.2.2 Use information about the gains and losses of energy to determine whether an organism's temperature is rising or dropping.

- Gains of energy from and losses of energy to the external environment determine an organism's temperature. Modifying this exchange of energy with the environment allows an organism to control its temperature.

4.2.3 Identify the heat exchange mechanisms used by plants and animals to regulate their body temperatures.

- Terrestrial plants may modify their energy balance by controlling transpiration, increasing or decreasing absorption of solar radiation, or adjusting the effectiveness of convective heat loss.

- Animals modify their energy balance mainly through behavior and morphology to adjust heat losses and gains and, in the case of endothermic animals, metabolic heat generation and insulation to lower heat loss.

4.2.4 Contrast ectothermy and endothermy, and explain how each influences the geographic distributions of organisms, along with organisms' sensitivities to changes in body temperature.

- Ectothermy is the regulation of body temperature through energy exchange with the external environment, while endothermy is the regulation of body temperature through internal heat generation.

- Generally, ectotherms have a greater tolerance for variation in their body temperature than endotherms.

CONCEPT 4.3 The water balance of an organism is determined by exchanges of water and solutes with the external environment.

4.3.1 List the three factors that influence the movement of water from a high-energy state to a low-energy state (i.e., with reference to water potential gradients) in biological systems.

- Water flows along energy gradients determined by solute concentration (osmotic potential), pressure or tension (pressure potential), and the attractive force of surfaces (matric potential).

4.3.2 Explain how organisms can control water gains and losses by adjusting resistance to water movement, and describe how high resistance may involve trade-offs with other functions.

- Plants and microorganisms can influence water potential by adjusting the solute concentration in their cells (osmotic adjustment).

- Terrestrial organisms can control their gains or losses of water by adjusting their resistance to water movement, as by the opening or closing of stomates in plants or adaptations of the skin in animals.

4.3.3 Describe how salt and water balances can become challenges for organisms exposed to hyperosmotic and hypoosmotic environments.

- Aquatic animals that are hypoosmotic to the surrounding water must expend energy to excrete salts against an osmotic gradient. On the other hand, aquatic animals that are hyperosmotic to their environment must take up solutes from the environment to compensate for solute losses to the surrounding water.

Review Questions

1. Organisms exhibit different degrees of tolerance for environmental stresses. How does tolerance for variation in body temperature vary among plants, ectothermic animals, and endothermic animals? What factors influence the differences in tolerance among these groups? Can plants exhibit avoidance of temperature extremes?

2. Organismal adaptations to environmental conditions often affect multiple ecological functions, leading to associated trade-offs. The following are two different trade-offs to consider.

   a. Plants transpire water through their stomates. What effects does transpiration have on temperature regulation in leaves? What is the trade-off with transpirational temperature regulation in terms of leaf physiological function?

   b. Animals can more effectively warm their bodies by absorbing solar radiation if they are a dark color. Many animals, however, are not dark but instead have a coloration close to that of their habitat (camouflage, as in the case of the basking crocodile in Figure 4.15). What is the trade-off between animal coloration and heat exchange?

3. List several ways in which plants and animals in terrestrial environments influence their resistance to water loss to the atmosphere.

Hone Your Problem-Solving Skills

Higher albedo leads to lower heat gain from solar radiation, and our earlier discussion of leaf pubescence in desert brittlebush (*Encelia*) species described the link between water availability, air temperature, and the amount of leaf pubescence. Where water is limiting for transpirational cooling and air temperatures are high, the native *Encelia* species have a higher amount of leaf pubescence relative to species from cooler, moister sites. Because there is a cost to building the leaf hairs, and a thick layer of hair potentially lowers photosynthesis rates, we might predict that in *E. farinosa*, which lives in the hot, dry desert, there is variation in the amount of leaf pubescence among populations in environments differing in water availability.

1. Populations of *E. farinosa* occur in Superior and Oatman, Arizona, and at Death Valley, California. Annual precipitation at the three sites is 453 mm, 111 mm, and 52 mm, respectively. What sites would you expect to have the most, least, and intermediate amounts of leaf pubescence? What site should exhibit the most seasonal change in leaf pubescence?

2. Using a graph of time (*x*-axis) versus leaf absorption of solar radiation (*y*-axis, using a relative scale from low to high), plot your predictions of the differences in the three populations as determined by variation in pubescence. Indicate any seasonal changes in pubescence you might expect as soils dry out.

3. The following data are from a controlled experiment run by Sandquist and Ehleringer (2003) to test whether there was ecotypic differentiation among the *E. farinosa* populations. Plants from each population were grown from seed in a common garden under a uniform environment for 2 months under well-watered conditions. The soils were then allowed to dry out for a month under warm temperatures. Leaf absorptance of solar radiation (as a percentage of the incoming light) was measured at three times: under well-watered conditions, halfway into the drydown, and under dry conditions after a month of no added water. Use the data to test the hypotheses you developed for Question 2 by plotting them on a graph with soil conditions on the *x*-axis and absorptance on the *y*-axis. Do the data support or refute your hypotheses about differences in pubescence among the populations and seasonal changes associated with decreased water availability and increased temperature?

Soil condition/ Population	Leaf absorptance (%)		
	Superior	Oatman	Death Valley
Well-watered	74	70	66
Halfway	57	53	49
Dry	46	43	42

Source: Data from D. R. Sandquist and J. R. Ehleringer. 2003. *Am J Bot* 90: 1481–1486.

Coping with Environmental Variation: Energy

Toolmaking Crows: A Case Study

Humans employ a multitude of tools to enhance our ability to gather food to meet our energy needs. We use a highly mechanized system of planting, fertilizing, and harvesting crops to feed ourselves or the livestock that we consume. For thousands of years, we have used specialized tools to increase our efficiency of hunting prey, including spears, bows and arrows, and rifles. We view our toolmaking capacity as something that differentiates us from other animals.

However, we humans are not alone in using tools to enhance our food acquisition ability. In the 1920s, Wolfgang Köhler, a psychologist studying the behavior of chimpanzees, observed that chimps in captivity made tools to retrieve bananas stashed in areas that were difficult to reach (Köhler 1927). Jane Goodall, a prominent primatologist, reported observing chimpanzees in the wild using grass blades and plant stems to "fish" for termites in holes in the ground and in decaying wood (**FIGURE 5.1**). Although these reports challenged the commonly held belief that modern humans were the only makers of tools to enhance food acquisition, it was perhaps comforting to those clinging to this notion that the observations were associated with one of our closest extant relatives. No one would ever have suspected similar behavior in birds, touted as one of the least intelligent vertebrates, as evidenced by the dubious insult "birdbrain" exchanged between humans.

The corvids, a family of birds that includes crows, ravens, magpies, jays, and jackdaws, enter our cultural heritage with a reputation for being clever. Even so, the discovery that crows use food-collecting tools manufactured from plants was unexpected. Gavin Hunt reported in 1996 that the crows (*Corvus moneduloides*) of New Caledonia, an island in the South Pacific, used tools to snag insect larvae, spiders, and other arthropods and pull them from the wood of living and decomposing trees (Hunt 1996) (**FIGURE 5.2A**). Hunt found that individual birds used one of two types of tools, either (1) a hooked twig fashioned from a shoot stripped of its leaves and bark (**FIGURE 5.2B**) or (2) a serrated leaf clipped from a *Pandanus* tree (**FIGURE 5.2C**). Both tools were therefore manufactured, rather than just collected from materials lying on the ground.

Hunt described a unique foraging style used by the New Caledonian crows. The birds probed tree cavities or areas of dense foliage using their tools as extensions of their bills. The birds used the tools repeatedly, carrying them from tree to tree. The presence of hooks on both types of tools suggested an innovative element that might increase the birds' efficiency in extracting prey from their refuges in the trees. The tools also appeared to be uniform in their

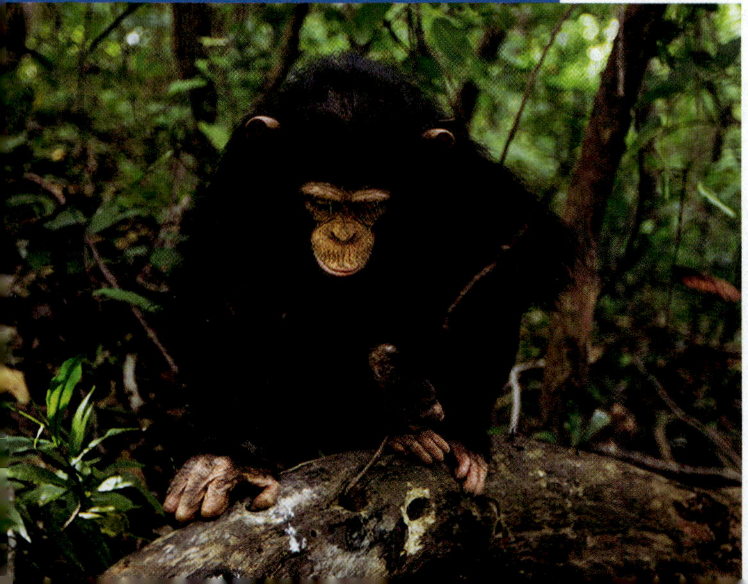

FIGURE 5.1 Nonhuman Tool Use This chimpanzee uses a plant stem as a tool to forage for termites. Chimpanzees were the first nonhuman animals documented using tools to forage for food.

(A)

(B)

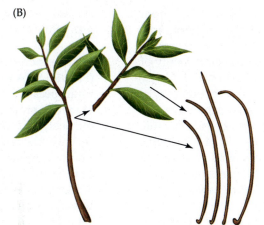

(C)

FIGURE 5.2 Tools Manufactured by New Caledonian Crows (A) Crows use the tools they make to probe for food in the cavities and crevices of trees. (B) Hooked twig tools, made from shoots of trees. The birds use their bills to form the hook while holding the stick with their feet. (C) The crows also can create tools from the serrated leaves of *Pandanus* plants. (B after G. R. Hunt. 1996. *Nature* 379: 249–251.)

construction; Hunt examined 55 tools manufactured by different birds and found that they differed little. When New Caledonian crows were captured and brought into the laboratory, they made hooked tools from wire, and experiments showed that the tools increased their food retrieval efficiency (Weir et al. 2002).

Toolmaking at a skill level equivalent to that shown by the crows appeared in humans only in the late Stone Age, approximately 450,000 years ago (Mellars 1989). How have these birds achieved a similar level of sophistication in their tool construction? The high numbers of New Caledonian crows using tools, and the consistency in the construction of the tools, indicate a cultural phenomenon—a skill learned socially within a population of animals—that had never before been observed in birds. How much of an energetic benefit do the crows gain by using tools rather than just their bills?

Introduction

Energy is one of the most basic requirements for all organisms. Physiological maintenance, growth, and reproduction all depend on energy acquisition. Organisms are complex systems, and if energy input stops, so does biological functioning. Enzyme systems fail if replacement proteins are not made. Cell membranes degrade and organelles cease to operate without energy to maintain and repair them. In this chapter, we will review the different ways in which organisms acquire energy to meet the demands of cellular maintenance, growth, reproduction, and survival. We'll focus on the major mechanisms that allow organisms to obtain energy from their environment, including the capture of sunlight and chemical energy and the acquisition and use of organic compounds synthesized by other organisms.

CONCEPT 5.1 Organisms obtain energy from sunlight, from inorganic chemical compounds, or through the consumption of organic compounds.

LEARNING OBJECTIVE

5.1.1 Differentiate autotrophy from heterotrophy in the context of building energy compounds using external sources of energy versus consuming them from organic matter.

Sources of Energy

We sense energy in our environment in a variety of forms. Light from the sun, a form of *radiant energy*, illuminates our world and warms our bodies. Objects that are cold or warm to our touch have different amounts of *kinetic energy*, which is associated with the motion of the molecules that make up the objects. A grasshopper eating a leaf and a coyote eating a meadow vole both represent the transfer of *chemical energy*, which is stored in the food that is being consumed. Radiant energy and chemical energy are the forms organisms use to meet the demands of growth and maintenance, while kinetic energy, through its influence on the rate of chemical reactions and temperature, is important for controlling the rate of activity and metabolic energy demand of organisms. A cold endotherm needs to warm its body to the optimal temperature for physiological functioning. It does this by "burning" chemical energy from its food during cellular respiration. Ultimately, this food was derived from the radiant energy of sunlight, converted into chemical energy by plants. Most of the energy used to support industrial development, fuel our cars, and heat our homes originated ultimately with photosynthesis, which produced the organisms that became the fossil fuels we pump out of the ground.

Autotrophs are organisms that assimilate energy from sunlight (*photosynthetic* organisms) or from inorganic chemical compounds in their environment (*chemosynthetic* archaea and bacteria).[1] Autotrophs convert the energy of sunlight or inorganic compounds into chemical energy stored in the carbon–carbon bonds of organic compounds, typically carbohydrates. **Heterotrophs** are organisms that obtain their energy by consuming energy-rich organic compounds made by other organisms—all of which ultimately originated with organic compounds synthesized by autotrophs. Heterotrophs include organisms that consume nonliving organic matter (*detritivores*); examples include earthworms and fungi in soil that feed on detritus derived mainly from dead plants, as well as bacteria in lakes that consume dissolved organic compounds. Heterotrophs also include organisms that consume living organisms but do not necessarily kill them (*parasites* and *herbivores*), as well as consumers (*predators*) that capture and kill their food source (*prey*).

On the surface, the distinction between autotrophs and heterotrophs would seem to be clear-cut: all plants are autotrophs, all animals and fungi are heterotrophs, and archaea and bacteria include both autotrophs and heterotrophs. Things are not always so simple, however. Some plants have lost their photosynthetic function and obtain their energy by parasitism. Such plants, known as

---

[1] *Organic* chemical compounds have carbon–hydrogen bonds and are usually biologically synthesized. All other compounds are considered *inorganic* compounds.

holoparasites (*holo*, "entire, whole"), have no photosynthetic pigments and are heterotrophs. Dodder (genus *Cuscuta*, with approximately 150 different species), for example, is a common plant parasite found throughout the world (**FIGURE 5.3A,B**) and is considered a major pest of agricultural species. Dodder attaches to its host plant by growing in spirals around the stem and penetrates the phloem of the host, using modified roots called haustoria, to take up carbohydrates. Other plants, known as *hemiparasites*, are photosynthetic but obtain some of their energy, as well as nutrients and water, from host plants (**FIGURE 5.3C**).

Conversely, animals can act as autotrophs, although this phenomenon is relatively rare. Their photosynthetic capacity is acquired by consuming photosynthetic organisms or by living with them in a close relationship known as a *symbiosis* (see Concept 15.1). Some sea slugs, for example, have fully functional chloroplasts that supply them with carbohydrates through photosynthesis. These animals, in the order Ascoglossa, take intact chloroplasts from the algae they feed on into their digestive cells (**FIGURE 5.4**). The chloroplasts are maintained intact for up to several months, providing energy as well as camouflage to the sea slug.

In the next two sections, we'll take a more detailed look at the mechanisms autotrophs use to capture energy and at some of the adaptations that make that process more efficient. We'll do the same more generally for heterotrophs in the final section of this chapter. Chapters 12 and 13 will provide more detailed considerations of energy capture by heterotrophs, and Chapter 16 will look at the energetic relationships among the species in a community.

(A)

(B)

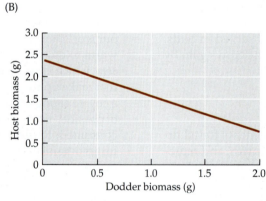

(C)

FIGURE 5.3 Plant Parasites (A) Dodder (*Cuscuta* sp.), a holoparasite that lacks chlorophyll, is shown here wrapped around the stem of a jewelweed plant. (B) Increasing amounts of European dodder (*Cuscuta europaea*) biomass result in decreasing growth of its host plant, stinging nettle (*Urtica dioica*). (C) Mistletoe, like the green mistletoe (*Ileostylus micranthus*) seen here, is a hemiparasite: despite having photosynthetic tissues of its own, mistletoe draws water, nutrients, and some of its energy from its host tree. (B after T. Koskela et al. 2002. *Evolution* 56: 899–908.)

FIGURE 5.4 Green Sea Slug The green color of this lettuce sea slug (*Elysia crispata*) is associated with the chloroplasts it has taken into its digestive system. The chloroplasts can supply enough energy to the sea slug to maintain it for several months without food.

CONCEPT 5.2 Radiant and chemical energy captured by autotrophs is converted into stored energy in carbon–carbon bonds.

LEARNING OBJECTIVES

5.2.1 Summarize chemosynthesis, which results in the synthesis of energy-rich carbon–carbon bonds.

5.2.2 Outline the steps in the light-driven reactions and carbon reactions of photosynthesis, describing their outcomes and how they produce energy-rich compounds in photoautotrophs.

5.2.3 Illustrate how photosynthetic organisms acclimatize and adapt to variations in the intensity of light.

5.2.4 Evaluate the trade-offs that result when a plant controls water loss.

5.2.5 Describe how temperature influences photosynthetic rates through its effect on enzymes and chloroplast membranes.

Autotrophy

The vast majority of the autotrophic production of chemical energy on Earth occurs through **photosynthesis**, a process that uses sunlight to provide the energy needed to take up carbon dioxide and synthesize organic compounds, principally carbohydrates. Although its contribution to the global energy picture is smaller, **chemosynthesis** (also known as *chemolithotrophy*), a process that uses energy from inorganic compounds to produce carbohydrates, is important to some key bacteria involved in nutrient cycling (see Concept 22.2) and in some unique ecosystems, such as hydrothermal vent

communities (see the Case Study in Chapter 20). Because the energy derived from photosynthesis and chemosynthesis is stored in the carbon–carbon bonds of the organic compounds produced by these processes, ecologists often use carbon as a measure of energy.

Chemosynthesis harvests energy from inorganic compounds

The earliest autotrophs on Earth were probably chemosynthetic bacteria or archaea that evolved when the composition of the atmosphere was markedly different than it is today: low in oxygen, but rich in hydrogen, with significant amounts of carbon dioxide (CO_2) and methane (CH_4). A diverse group of archaea and bacteria still use energy from inorganic compounds to take up CO_2 and synthesize carbohydrates. Chemosynthetic bacteria are often named according to the inorganic substrate they use for energy (**TABLE 5.1**).

During chemosynthesis, organisms obtain electrons from the inorganic compound—in other words, they oxidize[2] the inorganic substrate. They use the electrons to synthesize two high-energy compounds: adenosine triphosphate (ATP) and nicotinamide adenine dinucleotide phosphate (NADPH). They then use energy from ATP and NADPH to take up carbon from gaseous CO_2 (a process known as **fixation** of CO_2). The fixed carbon is used to synthesize carbohydrates or other organic molecules, which are stored to meet later demands for energy or biosynthesis (manufacture of chemical compounds, membranes, organelles, and tissues). Alternatively, some bacteria can use electrons from the inorganic substrate directly to fix carbon. The biochemical pathway most commonly

---

[2] Oxidation–reduction reactions involve the exchange of electrons between chemical compounds. The compound that gives up, or donates, electrons is oxidized, while the compound that accepts electrons is reduced.

TABLE 5.1

Inorganic Substrates Used by Chemosynthetic Bacteria as Electron Donors for Carbon Fixation

Substrate (chemical formula)	Type of bacteria
Ammonium NH_4^+	Nitrifying bacteria
Nitrite (NO_2^-)	Nitrifying bacteria
Hydrogen sulfide (H_2S/HS^-)	Sulfur bacteria (purple and green)
Sulfur (S)	Sulfur bacteria (purple and green)
Ferrous iron (Fe^{2+})	Iron bacteria
Hydrogen (H_2)	Hydrogen bacteria
Phosphite (HPO_3^{2-})	Phosphite bacteria

Source: M. T. Madigan and J. M. Martinko. 2005. *Brock Biology of Microorganisms*. Prentice Hall: Upper Saddle River, NJ.

Sulfur bacteria generate energy from hydrogen sulfide, leaving behind a residue of elemental sulfur.

FIGURE 5.5 Sulfur Deposits from Chemosynthetic Bacteria Sulfur bacteria thrive in sulfur hot springs with water temperatures as high as 110°C (230°F).

used to fix carbon is the **Calvin cycle**, named for Melvin Calvin, the biochemist who first described it. The Calvin cycle is catalyzed by several enzymes, and it occurs in both chemosynthetic and photosynthetic organisms.

One of the most widespread and ecologically important groups of chemosynthetic organisms is the nitrifying bacteria (e.g., *Nitrosomonas, Nitrobacter*), which are found in both aquatic and terrestrial ecosystems. In a two-step process, these bacteria convert ammonium (NH_4^+) into nitrite (NO_2^-), then oxidize it to nitrate (NO_3^-). These chemical conversions of nitrogen compounds are an important component of nitrogen cycling and plant nutrition, and we will discuss them in more detail in Concept 22.2. Another important chemosynthetic group is the sulfur bacteria, associated with volcanic deposits, sulfur hot springs, and acidic mine wastes. Sulfur bacteria initially use the higher-energy forms of sulfur, H_2S and HS– (hydrogen sulfide), producing elemental sulfur (S), which is insoluble and highly visible in the environment (**FIGURE 5.5**). Once the H_2S and HS– are exhausted, these bacteria use elemental S as an electron donor, producing SO_4^{2-} (sulfate).

Photosynthesis is the powerhouse for life on Earth

Prior to 1650, most people believed that plants obtained the raw material needed for their growth from the soil. Jan Baptist van Helmont (1579–1644), a Flemish scientist, tested this theory experimentally. He carefully measured the mass of dry soil in a pot (200 pounds or 91 kg) and then planted a willow sapling weighing 5 pounds (2.3 kg). Van Helmont watered the sapling using only rainwater for 5 years as it grew into a small tree. At the end of that time, the tree had gained 164 pounds (74 kg), and the

soil had lost only 2 ounces (0.06 kg). Although he incorrectly concluded that the tree had gained its mass from the water, van Helmont's experiment established the basis for the later discovery that photosynthetic uptake of CO_2 from the air—not material from the soil—was the source of the tree's weight gain.

The vast majority of biologically available energy on Earth is derived from the conversion of sunlight into energy-rich carbon compounds by photosynthesis. Photosynthetic organisms include some archaea, bacteria, and protists and most algae and plants. Leaves are the principal photosynthetic tissue in plants, but photosynthesis may also occur in stem and reproductive tissues. Like chemosynthesis, photosynthesis involves the conversion of CO_2 into carbohydrates that are used for energy storage and biosynthesis. Photosynthesis is also responsible for the largest movements of CO_2 between Earth and the atmosphere, and it is therefore critically important to the global climate system and climate change (as we'll see in Concept 25.1). Here, we will briefly review the major steps of plant photosynthesis and consider some ecologically relevant constraints on photosynthetic rates. In Concept 5.3, we will examine some variations in plant photosynthetic pathways.

LIGHT–DRIVEN AND CARBON REACTIONS Photosynthesis has two major steps. The first is the harvesting of energy from sunlight, which is used to split water to provide electrons for generating ATP and NADPH. This step is often referred to as the *light-driven reactions* of photosynthesis. The second step is the fixation of carbon and the synthesis of sugars and subsequently carbohydrates. This step is often referred to as the *carbon reactions* of photosynthesis.

Sunlight harvesting is accomplished by several pigments, principally chlorophyll. Chlorophyll gives photosynthetic organisms their green appearance because it absorbs red and blue light and reflects green wavelengths (**FIGURE 5.6**). Plants and photosynthetic bacteria have similar chlorophyll pigments, but they absorb light at slightly different wavelengths. Additional pigments associated with photosynthesis, called accessory pigments, include the carotenoids, which are characteristically red, yellow, or orange in appearance. All of these photosynthetic pigments are embedded in a membrane, along with other molecules involved in the light-driven reactions. In plants, this membrane lies within specialized organelles called chloroplasts, while in photosynthetic bacteria the pigments are embedded in the cell membrane. The pigment molecules are arrayed like antennae, with each array containing 50–300 molecules. The pigments absorb energy from discrete units of light, called *photons*. That energy is used to split water and provide electrons. The electrons are passed on to molecular complexes on the membranes, where they are used to synthesize ATP and NADPH.

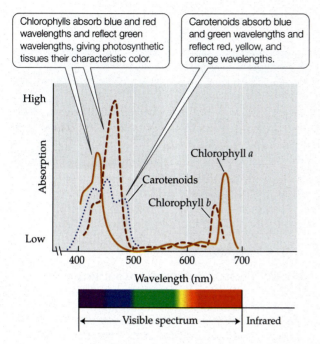

Chlorophylls absorb blue and red wavelengths and reflect green wavelengths, giving photosynthetic tissues their characteristic color.

Carotenoids absorb blue and green wavelengths and reflect red, yellow, and orange wavelengths.

FIGURE 5.6 Absorption Spectra of Plant Photosynthetic Pigments Plants typically contain several light-absorbing pigments, which absorb light of different wavelengths. (After C. J. Avers. 1985. *Molecular Cell Biology*. Addison-Wesley: Boston, MA.)

The splitting of water (H_2O) to provide electrons for the light-driven reactions generates oxygen (O_2). The evolution of photosynthesis, and the accompanying release of O_2 into the atmosphere, was a critical step in the development of the chemistry of the modern atmosphere and lithosphere as well as the evolution of life on Earth. Atmospheric oxygen led to the creation of a layer of ozone (O_3) high in the atmosphere that shields organisms from high-energy ultraviolet radiation (described in Concept 25.4). The evolution of aerobic respiration, in which O_2 is used as an electron acceptor, facilitated great evolutionary changes for life on Earth.

In the carbon reactions of photosynthesis, energy from ATP and NADPH is used in the Calvin cycle to fix carbon. Carbon dioxide is taken up from the atmosphere through the stomates of vascular plants, or it diffuses across the cell membranes in nonvascular plants, algae, and photosynthetic bacteria and archaea. A key enzyme associated with the Calvin cycle is ribulose 1,5 bisphosphate carboxylase/oxygenase, thankfully usually referred to by its abbreviation, rubisco. Rubisco, the most abundant enzyme on Earth, catalyzes the uptake of CO_2 and the synthesis of a three-carbon compound: phosphoglyceraldehyde, or PGA. PGA is eventually converted into a six-carbon sugar [glucose ($C_6H_{12}O_6$) in most plants]. The net reaction of photosynthesis is therefore

$$6CO_2 + 6H_2O \rightarrow C_6H_{12}O_6 + 6O_2 \qquad \text{(5.1)}$$

ENVIRONMENTAL CONSTRAINTS AND SOLUTIONS

The rate of photosynthesis determines the supply of energy and substrates for biosynthesis available in the environment. Because this rate influences the growth and reproduction of photosynthetic organisms—often equated with their ecological success (their abundance and geographic range)—environmental controls on the rate of photosynthesis are a key topic in physiological ecology. It should be noted, however, that net energy (carbon) gain is also influenced by CO_2 losses associated with cellular respiration.

Light is clearly an important influence on rates of photosynthesis in both terrestrial and aquatic habitats. The relationship between the light level and a plant's photosynthetic rate can be portrayed by a *light response curve* (**FIGURE 5.7A**). When there is enough light that the plant's photosynthetic CO_2 uptake is balanced by its CO_2 loss by respiration, the plant is said to have reached the *light compensation point*. As the light level increases above the light compensation point, the photosynthetic rate also increases; in other words, photosynthesis is limited by the availability of light. The photosynthetic rate levels off at a *light saturation point*, which is typically reached at a level below full sunlight.

How do plants cope with light variation? For example, how would an understory forest plant respond to shading by canopy trees? Could that plant acclimatize to more light if the canopy tree fell, allowing full sunlight to reach the ground? In a series of classic studies using controlled growth conditions, Olle Björkman demonstrated that acclimatization to different light levels involves a shift in the light saturation point (Björkman 1981) (**FIGURE 5.7B**). Morphological changes associated with this acclimatization include alterations in the thickness of leaves and variation in the number of chloroplasts available to harvest light (**FIGURE 5.8**). Photosynthetic organisms may also alter the density of their light-harvesting pigments—a strategy analogous to changing the size of the antenna on a radio—and the amounts of photosynthetic enzymes available for the carbon reactions. Typically, the average light level a plant experiences, integrated over the course of the day, is near the transition point between light limitation and light saturation (see **ANALYZING DATA 5.1**).

Some specialized bacteria are especially well adapted to photosynthesis at low light levels, which allows them to thrive in dimly lit environments such as relatively deep ocean water (down to about 20 m). A previously undescribed form of chlorophyll, called chlorophyll *f*, was found in samples of the marine cyanobacteria that form sediments in the shallow waters of Shark Bay, Australia (Chen et al. 2010), and has subsequently been found in cyanobacteria of other low-light habitats, including hot springs, rice paddies, and caves. Chlorophyll *f* absorbs light in the near-infrared region, just beyond

(A)

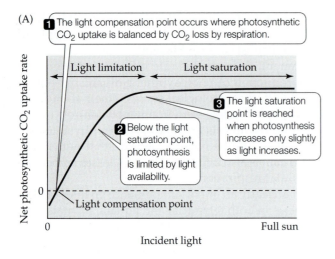

1 The light compensation point occurs where photosynthetic CO_2 uptake is balanced by CO_2 loss by respiration.

Light limitation — Light saturation

2 Below the light saturation point, photosynthesis is limited by light availability.

3 The light saturation point is reached when photosynthesis increases only slightly as light increases.

Net photosynthetic CO_2 uptake rate

0

Light compensation point

0 — Incident light — Full sun

(B)

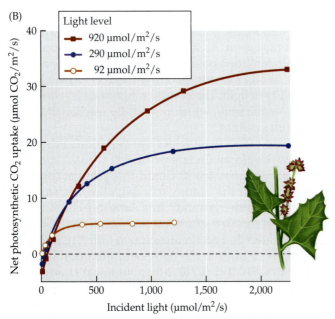

Light level
- 920 μmol/m²/s
- 290 μmol/m²/s
- 92 μmol/m²/s

Net photosynthetic CO_2 uptake (μmol CO_2/m²/s)

Incident light (μmol/m²/s)

FIGURE 5.7 Plant Responses to Variations in Light Levels
(A) Photosynthetic light response curve. (B) Spearscale (*Atriplex triangularis*) plants grown at different light levels in growth chambers acclimatized to those light levels. Their light response curves indicate that adjustments in the light saturation point occurred. Small, but ecologically significant, changes in the light compensation point occur in many other species, facilitating CO_2 uptake at low light levels. (B after O. Björkman. 1981. In *Physiological Plant Ecology I: Encyclopedia of Plant Physiology*, O. L. Lange et al. [Eds.], pp. 57–101. Springer: New York.)

❓ Why might the light saturation point of a plant be below the maximum light level the plant is likely to be exposed to?

the red wavelengths used by other forms of chlorophyll (see Figure 5.6). Chlorophyll *f* is an adaptation that allows cyanobacteria possessing it to grow underneath other photosynthetic organisms that use light in the blue and red wavelengths, as it lets them harvest energy at wavelengths that pass through those other

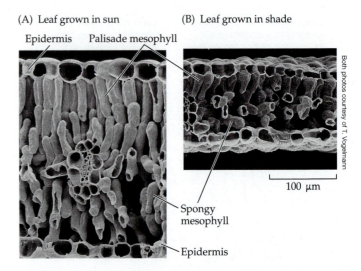

(A) Leaf grown in sun   (B) Leaf grown in shade

Epidermis   Palisade mesophyll

100 μm

Spongy mesophyll

Epidermis

Both photos courtesy of T. Vogelmann

FIGURE 5.8 Effects of Light Level on Leaf Structure
Golden banner (*Thermopsis montana*) leaves adjust morphologically to changes in light levels. Leaves grown at high light levels (A) are thicker, have more photosynthetic cells (palisade and spongy mesophyll), and have greater numbers of chloroplasts than leaves grown at low light levels (B).

photosynthetic organisms (Nürnberg et al. 2018). The discovery of a pigment that can harvest near-infrared energy has implications for increasing the efficiency of photovoltaic panels used to generate electricity, which may help lower emissions of CO_2.

Water availability is an important control on the supply of CO_2 for photosynthesis in terrestrial plants. As we saw in Concept 4.3, low water availability results in closure of the stomates, restricting the entry of CO_2 into leaves. Stomatal control represents an important trade-off for the plant: water conservation versus energy gain through photosynthesis as well as cooling of the leaf through transpiration. Keeping stomates open while tissues lose water can permanently impair physiological processes in the leaf. Closing stomates, however, not only limits photosynthetic CO_2 uptake, but also increases the chances of light damage to the leaf. When the Calvin cycle is not operating, energy continues to accumulate in the light-harvesting arrays, and if enough energy builds up, it can damage the photosynthetic membranes. Plants have evolved a number of ways of dissipating this energy safely, including the use of carotenoids to release it as heat.

Temperature influences photosynthesis in two main ways: through its effects on the rates of chemical reactions and by influencing the structural integrity of membranes and enzymes. Autotrophs acclimatize and adapt to temperature variation by changing properties of the Calvin cycle enzymes and/or the photosynthetic membranes. Different photosynthetic organisms have different forms of the same photosynthetic enzymes that operate best under the environmental temperatures

ANALYZING DATA 5.1

How Does Acclimatization Affect Plant Energy Balance?

Many plants can adjust their morphology and biochemistry to match the light conditions under which they are grown. The curves depicted in the figure are from Olle Björkman's* classic studies and show the net photosynthetic CO_2 uptake for spearscale plants (*Atriplex triangularis*) grown under high-light (920 $\mu mol/m^2/s$ of photosynthetically active radiation) and low-light (92 $\mu mol/m^2/s$) conditions.

1. Assuming no further physiological changes occur, calculate the daily carbon balance for leaves of the high-light and low-light plants grown under the following conditions:

   a. Plants are kept at a light level (irradiance) of 200 $\mu mol/m^2/s$ for 2 hours, then switched to an irradiance of 1,500 $\mu mol/m^2/s$ for 10 hours, then switched back to 200 $\mu mol/m^2/s$ for 2 hours. The lights are then turned off for 10 hours. (This light regime approximates sunny conditions in an open subtropical savanna.)

   b. Plants are kept at an irradiance of 50 $\mu mol/m^2/s$ for 2 hours, then switched to an irradiance of 200 $\mu mol/m^2/s$ for 10 hours, then switched back to 50 $\mu mol/m^2/s$ for 2 hours. The lights are then turned off for 10 hours. (This light regime is similar to that expected in a tropical rainforest understory.)

2. High-light and low-light plants exhibit differences in maximum net photosynthesis rates, light compensation points, and nighttime respiration. Which

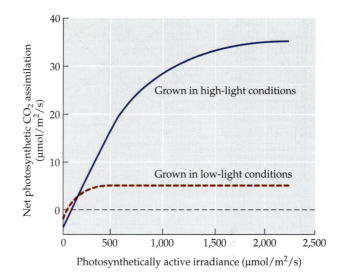

of these three differences contributes the most to the distinction in carbon balance under high-light conditions (calculated in part a of question 1) and low-light conditions (calculated in part b)?

3. What do you think might contribute to the differences in nighttime respiration rates?

*Björkman, O. 1981. Responses to different quantum flux densities. In *Physiological Plant Ecology I: Encyclopedia of Plant Physiology*, O. L. Lange et al. (Eds.), pp. 57–101. Springer-Verlag, Berlin.

---

where the organisms occur. These differences result in markedly different temperature ranges for photosynthesis in organisms from different climates (**FIGURE 5.9A**). Lichens and plants of Arctic and alpine environments can photosynthesize at temperatures close to freezing, while desert plants may have their highest photosynthetic rates at temperatures that are hot enough to denature most other plants' enzymes (40°C–50°C or 104°F–122°F). Plants that acclimatize to changes in temperature synthesize different forms of photosynthetic enzymes with different temperature optima (**FIGURE 5.9B**). Temperature also influences the fluidity of the cell and organelle membranes (see Concept 4.2). Cold sensitivity in plants of tropical and subtropical biomes is associated with loss of membrane fluidity, which inhibits the functioning of the light-harvesting molecules embedded in the chloroplast membranes. And as we have seen, high temperatures, particularly in combination with intense sunlight, can damage photosynthetic membranes.

Nutrient concentrations in leaves reflect their photosynthetic potential because most of the nitrogen in plants is associated with rubisco and other photosynthetic enzymes. Thus, higher amounts of nitrogen in leaves are correlated with higher photosynthetic rates. Why, then, don't all plants allocate more nitrogen to their leaves to increase their photosynthetic capacity? There are two main reasons. First, the supply of nitrogen is low relative to the demand, and nitrogen is needed for growth and other metabolic functions in addition to photosynthesis (see Chapter 22). Second, increasing the nitrogen concentration of a leaf increases the risk that herbivores will consume the leaf, as plant-eating animals are often nitrogen starved (see Concept 22.1). Plants must balance the competing demands of photosynthesis, growth, and protection from herbivores.

Over evolutionary time, some plants have dealt with environmental constraints on photosynthesis with adaptations in their photosynthetic pathways, as we will see next.

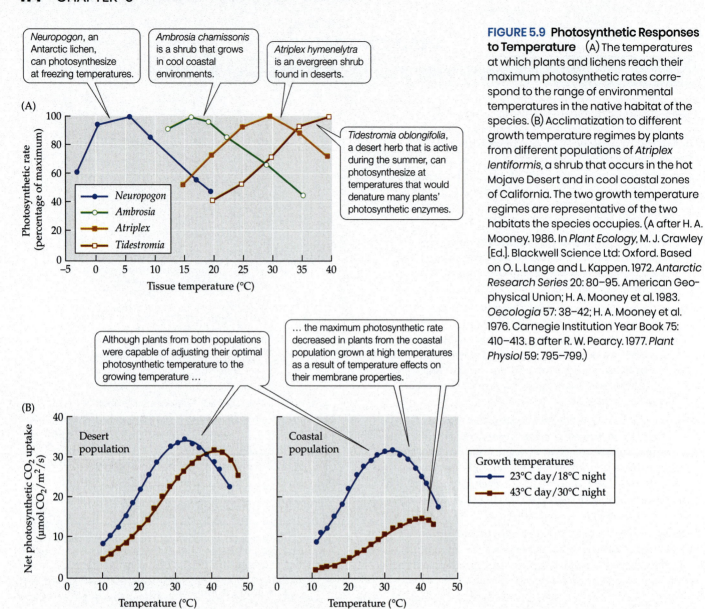

Neuropogon, an Antarctic lichen, can photosynthesize at freezing temperatures.

Ambrosia chamissonis is a shrub that grows in cool coastal environments.

Atriplex hymenelytra is an evergreen shrub found in deserts.

Tidestromia oblongifolia, a desert herb that is active during the summer, can photosynthesize at temperatures that would denature many plants' photosynthetic enzymes.

(A)

Legend:
- Neuropogon
- Ambrosia
- Atriplex
- Tidestromia

y-axis: Photosynthetic rate (percentage of maximum)
x-axis: Tissue temperature (°C)

Although plants from both populations were capable of adjusting their optimal photosynthetic temperature to the growing temperature …

… the maximum photosynthetic rate decreased in plants from the coastal population grown at high temperatures as a result of temperature effects on their membrane properties.

(B)

Desert population

Coastal population

y-axis: Net photosynthetic CO_2 uptake (μmol CO_2/m^2/s)
x-axis: Temperature (°C)

Growth temperatures
- 23°C day / 18°C night
- 43°C day / 30°C night

FIGURE 5.9 Photosynthetic Responses to Temperature (A) The temperatures at which plants and lichens reach their maximum photosynthetic rates correspond to the range of environmental temperatures in the native habitat of the species. (B) Acclimatization to different growth temperature regimes by plants from different populations of *Atriplex lentiformis*, a shrub that occurs in the hot Mojave Desert and in cool coastal zones of California. The two growth temperature regimes are representative of the two habitats the species occupies. (A after H. A. Mooney. 1986. In *Plant Ecology*, M. J. Crawley [Ed.]. Blackwell Science Ltd: Oxford. Based on O. L. Lange and L. Kappen. 1972. *Antarctic Research Series* 20: 80–95. American Geophysical Union; H. A. Mooney et al. 1983. *Oecologia* 57: 38–42; H. A. Mooney et al. 1976. Carnegie Institution Year Book 75: 410–413. B after R. W. Pearcy. 1977. *Plant Physiol* 59: 795–799.)

CONCEPT 5.3 Environmental constraints have resulted in the evolution of biochemical pathways that improve the efficiency of photosynthesis.

LEARNING OBJECTIVES

5.3.1 Explain the difference between photosynthesis and photorespiration and evaluate conditions where photorespiration is detrimental to plant growth.

5.3.2 Summarize how biochemical and anatomical adaptations associated with the C_4 photosynthetic pathway minimize photorespiration, thereby enhancing photosynthesis rates.

5.3.3 Describe how crassulacean acid metabolism reduces water loss relative to the C_3 or C_4 photosynthetic pathways.

Photosynthetic Pathways

Anything that influences energy gain by photosynthesis has the potential to affect the survival, growth, and reproduction of the organism. As we have just seen, rates of photosynthesis are influenced by environmental conditions, particularly temperature and water availability. In addition, an apparent biochemical inefficiency in the initial step of the Calvin cycle limits energy gain by photosynthetic organisms. In this section, we will examine some evolutionary responses to these environmental constraints on photosynthesis. We will describe two specialized photosynthetic pathways, the C_4 pathway and crassulacean acid metabolism (CAM), that make photosynthesis more efficient under particular potentially stressful environmental conditions. Plants that lack these specialized pathways use the **C₃ photosynthetic pathway**. The C_3 and C_4 photosynthetic pathways take their names from the number of carbon atoms in their first stable chemical products. First, we'll examine photorespiration, a process that operates in opposition to the Calvin cycle and lowers its efficiency.

Photorespiration lowers the efficiency of photosynthesis

Earlier, we described a key enzyme in the Calvin cycle, rubisco, and noted that the "o" in the abbreviation stands for "oxygenase." Rubisco can catalyze two competing reactions. One is a carboxylase reaction, in which CO_2 is taken up, leading to the synthesis of sugars and the release of O_2 (i.e., photosynthesis; see Equation 5.1). The other is an oxygenase reaction, in which O_2 is taken up, leading to the breakdown of carbon compounds and the release of CO_2. This oxygenase reaction is part of a process called **photorespiration**, which results in a net loss of energy and is thus potentially detrimental for plants.

The balance between photosynthesis and photorespiration is related to two main factors: (1) the relative amounts of O_2 and CO_2 in the atmosphere and (2) temperature. As the atmospheric concentration of CO_2 decreases relative to that of O_2, the rate of photorespiration increases relative to the rate of photosynthesis (**FIGURE 5.10**). Since the evolution of C_3 photosynthesis over 3 billion years ago, atmospheric CO_2 concentrations have changed repeatedly over periods of hundreds of thousands of years in response to major global geologic and climate events (see Concepts 25.1 and 25.2). These shifts in atmospheric CO_2 concentrations would have influenced the balance between photosynthesis and photorespiration. Furthermore, as temperatures increase, the rate of O_2 uptake catalyzed by rubisco increases relative to the rate of CO_2 uptake. As a result of these two processes, photorespiration increases more rapidly at high temperatures than photosynthesis does. Thus, energy loss due to photorespiration is particularly acute at high temperatures and low atmospheric CO_2 concentrations.

If photorespiration is detrimental to the functioning of photosynthetic organisms, why hasn't a new form of rubisco evolved that minimizes uptake of O_2? Is it possible that photorespiration provides some benefit to the plant? A possible clue comes from experiments with *Arabidopsis thaliana*. *Arabidopsis* plants with a genetic mutation that knocks out photorespiration die under normal light and CO_2 conditions (Ogren 1984). One hypothesis for a potential benefit of photorespiration is that it protects the plant from damage to the photosynthetic machinery at high light levels. This hypothesis is supported by the results of a study by Akiko Kozaki and Go Takeba, who used tobacco plants (*Nicotiana* sp.) that they genetically altered to elevate or lower the plants' rates of photorespiration (Kozaki and Takeba 1996). They subjected

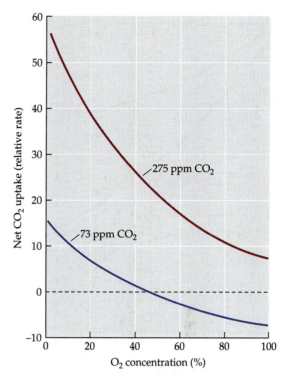

FIGURE 5.10 Influence of Oxygen Concentration on Photosynthesis As the atmospheric oxygen concentration increases, net photosynthetic uptake of CO_2 decreases because of greater photorespiration, as shown here for soybean leaves in light levels equal to about 20% of full sun. (After M. L. Forrester et al. 1966. *Plant Physiol* 41: 428–431.)

? Why does the net rate of CO_2 uptake drop below zero at high oxygen levels for leaves exposed to 73 ppm CO_2?

these experimental plants to high-intensity light and recorded the damage to their photosynthetic machinery. Plants with higher rates of photorespiration showed less damage than control plants with normal rates of photorespiration (**FIGURE 5.11**) or plants with depressed rates of photorespiration.

Despite this possibility that photorespiration plays a role in protecting plants from damage at high light levels, there are conditions in which the decrease in photosynthetic CO_2 uptake it causes could be a serious problem for the plant. If atmospheric CO_2 concentrations are low and temperatures high, photosynthetic energy gain might not keep pace with photorespiratory energy loss. Such conditions existed 7 million years ago, at about the time when plants with a unique biochemical pathway, C_4 photosynthesis, became far more abundant (Cerling et al. 1997).

C₄ photosynthesis lowers photorespiratory energy loss

The **C₄ photosynthetic pathway** reduces photorespiration. C_4 photosynthesis evolved independently several times in different plant species. It is found in 18 plant

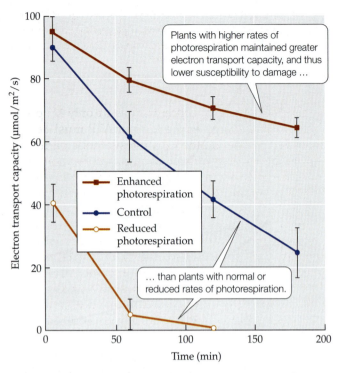

FIGURE 5.11 **Does Photorespiration Protect Plants from Damage by Intense Light?** The ability of plants to process light energy for photosynthesis (electron transport capacity) under conditions that promote damage to photosynthetic membranes (high light levels, low CO_2 concentrations) is greater in genetically altered plants with high rates of photorespiration than in control plants or in genetically altered plants with low rates of photorespiration. Error bars show ± one standard error (SE) of the mean. (After A. Kozaki and G. Takeba. 1996. *Nature* 384: 557–580.)

families (**FIGURE 5.12**) but is most closely associated with the grass family, including several important crop plants such as corn, sugarcane, and sorghum.

C_4 photosynthesis involves both biochemical and morphological specialization. The biochemical specialization can be thought of as a pump that provides high concentrations of CO_2 to the Calvin cycle. This greater supply of CO_2 lowers the rate of O_2 uptake by rubisco, substantially reducing photorespiration. The morphological specialization involves spatial separation of the regions in the leaf where CO_2 is taken up (mesophyll) and where the Calvin cycle operates (bundle sheath), which increases the concentration of CO_2 where rubisco is found.

In C_4 plants, CO_2 is initially taken up by an enzyme called phosphoenolpyruvate carboxylase, or PEPcase, that has a greater capacity to take up CO_2 than rubisco and lacks oxygenase activity. PEPcase fixes CO_2 in the mesophyll tissue of the plant. Once the CO_2 is taken up, a four-carbon compound is synthesized and transported to a group of cells surrounding the vascular tissues (xylem and phloem), known as the bundle sheath, where the Calvin cycle occurs. The four-carbon compound is

(A) Switchgrass (*Panicum virgatum*)

(B) *Cleome gynandra*

FIGURE 5.12 **Plants with the C_4 Photosynthetic Pathway** The C_4 photosynthetic pathway has evolved multiple times. It is found in plants of 18 different families encompassing a variety of growth forms, from switchgrass (*Panicum virgatum*) (A) to eudicots such as *Cleome gynandra*, commonly found in Africa (B).

broken down in the bundle sheath cells, releasing CO_2 to the Calvin cycle, and a three-carbon compound is transported back to the mesophyll to continue the C_4 cycle. The bundle sheath is surrounded by a waxy coating that keeps CO_2 from diffusing out (**FIGURE 5.13**). As a result, CO_2 concentrations inside the bundle sheath may reach a high of 5,000 parts per million (ppm), even

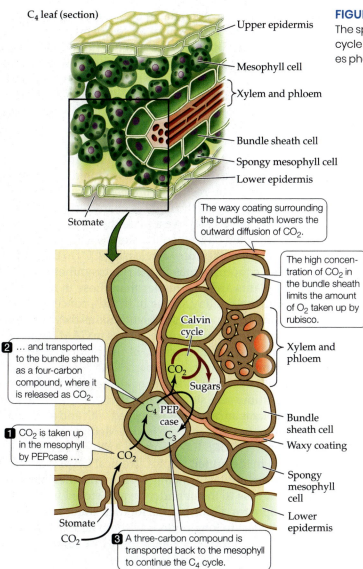

C₄ leaf (section)

- Upper epidermis
- Mesophyll cell
- Xylem and phloem
- Bundle sheath cell
- Spongy mesophyll cell
- Lower epidermis

Stomate

The waxy coating surrounding the bundle sheath lowers the outward diffusion of CO_2.

The high concentration of CO_2 in the bundle sheath limits the amount of O_2 taken up by rubisco.

Calvin cycle

2 … and transported to the bundle sheath as a four-carbon compound, where it is released as CO_2.

CO_2
Sugars
C_4 PEP case
C_3
CO_2

1 CO_2 is taken up in the mesophyll by PEPcase …

- Xylem and phloem
- Bundle sheath cell
- Waxy coating
- Spongy mesophyll cell
- Lower epidermis

Stomate
CO_2

3 A three-carbon compound is transported back to the mesophyll to continue the C_4 cycle.

FIGURE 5.13 Morphological Specialization in the Leaves of C₄ Plants The spatial separation of CO_2 uptake (in the mesophyll cells) and the Calvin cycle (in the bundle sheath cells) minimizes photorespiration and maximizes photosynthetic rates under high temperatures.

though external CO_2 concentrations are only 412 ppm. Additional energy in the form of ATP must be expended to operate the C_4 photosynthetic pathway, but the increased efficiency of carbon fixation usually compensates for the higher energy requirement.

As is apparent from the discussion above, plants with the C_4 photosynthetic pathway can photosynthesize at higher rates than C_3 plants under environmental conditions that elevate rates of photorespiration, such as high temperatures. In addition, most C_4 plants have lower rates of transpiration at a given photosynthetic rate, known as *water use efficiency*, than C_3 plants. This difference is due to the ability of PEPcase to take up CO_2 under the lower CO_2 concentrations that exist when stomates are not fully open.

If we assumed that photosynthetic rates determine ecological success, we could use climate patterns to predict where C_4 plants should predominate over C_3 plants. Such an analysis would be overly simplistic, however, because multiple factors other than temperature influence the biogeography of C_3 and C_4 plants, including abiotic factors such as light levels and biotic factors such as competitive ability and the pool of species available to colonize an area. However, analyses of similar communities across latitudinal and elevational gradients provide support for the benefit of C_4 photosynthesis at high temperatures and for the role this benefit plays in C_4 plant distribution (Ehleringer et al. 1997). In particular, studies of grass- and sedge-dominated communities in Australia suggest a close correlation between growing-season temperature and the proportion of C_3 and C_4 species in the community (**FIGURE 5.14**). As atmospheric CO_2 concentrations continue to increase because of burning of fossil fuels, however, photorespiration rates are likely to decrease, and the advantages of C_4 over C_3 photosynthesis may be diminished in some regions, leading to changes in the proportions of C_3 and C_4 plants.

The increasing proportions of C_4 plants with higher minimum growing-season temperatures suggest that C_4 photosynthesis is especially beneficial at high temperatures.

The decreasing proportions of C_4 plants with lower minimum growing-season temperatures are associated with the cold sensitivity of a key C_4 enzyme.

FIGURE 5.14 C₄ Plant Abundance and Growing-Season Temperatures The proportions of C_4 plants in Australian grass- and sedge-dominated communities correlate with the average minimum growing-season temperatures in the different locations. (After P. W. Hattersley. 1983. *Oecologia* 57: 113–128.)

? Using the data in this graph and the seasonal temperature trends from the climate diagrams in Concept 3.1 (assume that the monthly minimum temperature is 5°C cooler than the monthly average), what biome(s) should lack C_4 species?

CAM photosynthesis enhances water conservation

When plants first colonized the terrestrial environment, they evolved adaptations to restrict water losses to a dry atmosphere. Among these adaptations is a unique photosynthetic pathway called **crassulacean acid metabolism (CAM)**, which occurs in over 10,000 plant species belonging to 33 families. While C_4 photosynthesis separates CO_2 uptake and the Calvin cycle spatially, CAM separates these two steps temporally (**FIGURE 5.15**). CAM plants

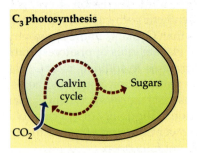

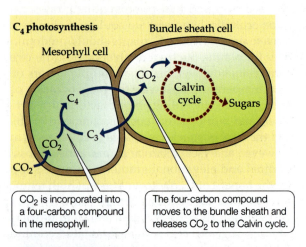

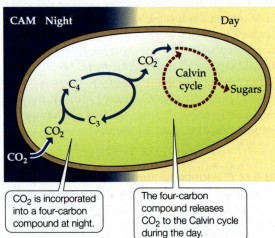

FIGURE 5.15 C_3, C_4, and CAM Photosynthesis Compared
All three photosynthetic pathways fix carbon and produce sugars, but C_4 photosynthesis separates these steps spatially, while CAM separates them temporally.

open their stomates at night, when C_3 and C_4 plants have their stomates closed. Because air temperatures at night are cooler, humidity is higher. Higher humidity results in a lower water potential gradient between the leaf and the air (see Concept 4.3), so the plant loses less water by transpiration than it would during the day. CAM plants close their stomates during the day, when the potential for water loss is highest.

During the night, when the stomates are open, CAM plants take up CO_2 using PEPcase and incorporate it into a four-carbon organic acid, which is stored in vacuoles (**FIGURE 5.16**). The resulting increase in acidity in the plants' tissues during the night is characteristic of CAM plants and can be used to estimate their photosynthetic rates. During the day, when the stomates are closed, the organic acid is broken down, releasing CO_2 to the Calvin cycle. CO_2 concentrations in the photosynthetic tissues of CAM plants are thus higher than those in the atmosphere during the day. These high CO_2 concentrations increase the efficiency of photosynthesis as they suppress photorespiration. Photosynthetic rates in CAM plants are usually related to the capacity of the plant to store the four-carbon organic acid, so many CAM plants are succulent, with thick, fleshy leaves or stems, which enhances their nighttime acid storage capacity.

CAM plants are typically associated with arid and saline environments, such as deserts and Mediterranean-type ecosystems (**FIGURE 5.17**). Some CAM plants, however, are found in the humid tropics. Tropical CAM plants are typically epiphytes growing on the branches of trees, without access to the abundant water stored in the soil. These epiphytes rely on rainfall for their water supply and may be subject to long periods without access to water.

The CAM pathway is also found in some aquatic plants, such as quillworts (*Isoetes*), which are closely related to the club mosses. This observation suggests that water conservation was probably not the only driving force for the evolution of CAM, which evolved independently in at least 35 different families. The rate of CO_2 diffusion into water is low, and CAM has been hypothesized to facilitate the uptake of CO_2 at the low concentrations found in the aquatic environment.

A unique property of some CAM plant species is the ability to switch between C_3 and CAM photosynthesis, known as *facultative* CAM. When conditions are favorable for daytime gas exchange (i.e., abundant water is available), these plants utilize the C_3 photosynthetic pathway, which allows greater carbon gain than CAM. As conditions become more arid or more saline, the plants switch over to CAM. The reversibility of the transition from C_3 to CAM varies among species. For example, the common ice plant (Mesembryanthemum crystallinum), which has been intensively studied as a facultative CAM model system, undergoes an irreversible transition from C_3 to CAM photosynthesis when salinity increases or the soil dries out (Osmond et al. 1982).

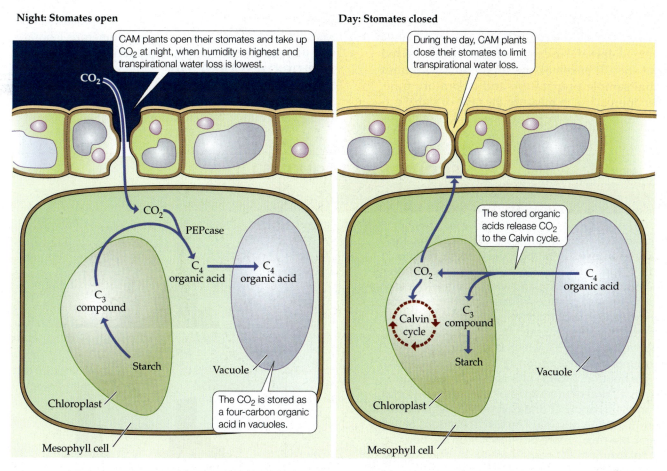

FIGURE 5.16 Crassulacean Acid Metabolism Plants using CAM open their stomates and take up CO_2 at night, then run the Calvin cycle during the day.

In contrast, some species in the genus Clusia can switch relatively rapidly between C_3 and CAM (Borland et al. 1992). These plants start out as epiphytes in canopy trees but grow toward the base of their host tree, eventually strangling it and taking on a tree growth form. The capacity to switch between C_3 and CAM facilitates the change from epiphyte to tree form, and it supports continued photosynthesis during the transition from wet season to dry season characteristic of some tropical locations.

How can we tell what photosynthetic pathway a plant is using? The morphology of the plant gives us a clue: succulent plants suggest CAM photosynthesis, and plants

FIGURE 5.17 Examples of Plants with the CAM Photosynthetic Pathway Most CAM plants are found in arid and saline regions or in other habitats where water availability is periodically low.

with a well-developed bundle sheath suggest C_4 photosynthesis. These clues provide a starting point, but they are far from foolproof. We can measure the presence and activity of specific enzymes, but this approach requires substantial sample preparation and laboratory time. A simpler approach is to measure the ratio of stable carbon isotopes ($^{13}C/^{12}C$) in plant tissues. Although the isotopic technique uses sophisticated equipment, sample preparation is simple, and there are numerous laboratories that can routinely analyze plant tissue samples (see **ECOLOGICAL TOOLKIT 5.1**).

Now that we have reviewed the ways in which autotrophs acquire energy, let's turn our attention to how that energy is acquired by heterotrophs.

ECOLOGICAL TOOLKIT 5.1

Stable Isotopes

Many biologically important elements, including carbon, hydrogen, oxygen, nitrogen, and sulfur, have an abundant "light" isotopic form and one or more "heavy" nonradioactive isotopic forms, which contain additional neutrons. Because isotopes of these elements do not decay over time as radioactive isotopes do, they are referred to as *stable* isotopes. An example of a stable isotope is carbon-13 (^{13}C), which is heavier than the more abundant form, carbon-12 (^{12}C), because it has one more neutron. Groups of stable isotopes include hydrogen (H) and deuterium (D or ^2H); nitrogen-14 and nitrogen-15 (^{14}N and ^{15}N); and oxygen-16, oxygen-17, and oxygen-18 (^{16}O, ^{17}O, and ^{18}O). The lighter isotopes of these elements are much more abundant than the heavier forms. For example, ^{12}C constitutes 98.9%, and ^{13}C only 1.1%, of the C on Earth. Similarly, ^{14}N constitutes 99.6%, and ^{15}N 0.4%, of the N on Earth.

The isotopic composition of a material is usually expressed as delta (δ), the difference between the ratio of the isotopic forms in a sample (R_{sample}) and that in a standard material ($R_{standard}$), divided by the ratio in the standard, multiplied by 1,000 [to give parts per thousand (‰) difference]:

$$\delta = \frac{R_{sample} - R_{standard}}{R_{standard}} \times 1,000$$

Examples of the standard materials chosen for stable isotopes include a limestone rock from South Carolina for C, atmospheric N_2 for N, and ocean water for O and H.

Naturally occurring stable isotopes have become an important tool in ecological research (Fry 2007). Stable isotopes have been used to determine photosynthetic pathways in plants, identify food sources for animals, and track the movements of elements and rates of nutrient cycling in ecosystems. Because of differences in mass, the isotopes are affected differently by biological and physical processes. Generally, the heavier isotope is discriminated against and the lighter isotope enriched. For example, when rubisco catalyzes the uptake of CO_2, it favors $^{12}CO_2$ over $^{13}CO_2$. As a result, plants are enriched in ^{12}C, and depleted in ^{13}C, relative to the C in atmospheric CO_2: atmospheric CO_2 has a δ ^{13}C value

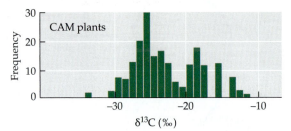

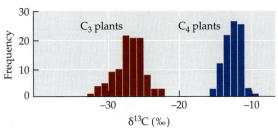

Carbon Isotopic Composition of Plants with Different Photosynthetic Pathways   Plants with the C_3 photosynthetic pathway show the greatest discrimination against ^{13}C (and thus the most negative δ ^{13}C, expressed in parts per thousand), while C_4 and CAM plants are more enriched in ^{13}C (have a less negative δ ^{13}C). (After M. A. Maslin and E. Thomas. 2003. *Quat Sci Rev* 22: 1729–1736.)

❓ Why is the range of δ 13C values for CAM plants larger, bridging the values for C_3 and C_4 plants?

of −7 parts per thousand (in other words, it is 7 parts per thousand more depleted in ^{13}C than the standard), and C_3 plants have a δ ^{13}C value of about −27 parts per thousand. C_4 and CAM plants, however, have less ^{12}C and more ^{13}C than C_3 plants. That is because initial CO_2 uptake in these plants is catalyzed by PEPcase, which discriminates against $^{13}CO_2$ less than rubisco does, and rubisco in C_4 and CAM plants takes up CO_2 in a semi-closed system (in the bundle sheath or with stomates closed), which inhibits enzymatic discrimination. As a result, measurement of the C isotope ratio in plant tissues can be used to determine the photosynthetic pathway used by a plant species, as shown in the figure.

Stable isotopes have also been used to determine food sources for animals. The isotopic ratios of C, N, and S

ECOLOGICAL TOOLKIT 5.1 (continued)

in various potential food sources may differ significantly, and measurement of one or more of these isotopes in potential food sources and in consumer tissues can determine what is being eaten. For example, in this chapter's Case Study Revisited, we will see how isotopic ratios were used to determine the diet of New Caledonian crows. In Concept 20.4, we will describe how N and C isotopes were used to study the diets of both modern North American grizzly bears and extinct cave bears.

Stable isotopes can also be added to the environment to help trace the movements of elements. This

approach is often used to trace the fate of nutrients in ecosystems.

Isotopic analysis of biological samples is relatively straightforward. For C and N, the samples are dried, ground, and burned in a closed furnace. The gases liberated by the combustion are then analyzed for isotopic composition using an instrument called a mass spectrometer. Many commercial laboratories specialize in the isotopic analysis of biological materials, owing in part to the demand for such analyses from ecologists and other environmental scientists.

CONCEPT 5.4 Heterotrophs have adaptations for acquiring and assimilating energy efficiently from a variety of organic sources.

LEARNING OBJECTIVES

5.4.1 Illustrate how the chemical makeup of a food item determines the benefit it provides to the consumer eating it.

5.4.2 Explain how morphological and behavioral adaptations enable heterotrophs to obtain food more efficiently.

5.4.3 Describe how increasing complexity in the digestive systems of heterotrophs makes the assimilation of energy and nutrients more efficient.

Heterotrophy

Heterotrophy is all about eating and being eaten, which are major themes in ecology. The first organisms on Earth were probably heterotrophs that consumed amino acids and sugars, which formed spontaneously in the early atmosphere and rained down on the surface or formed in the oceans near hydrothermal vents. Since that time, the diversity of strategies for obtaining energy by heterotrophs has expanded tremendously. Three general steps are associated with heterotrophic energy acquisition: finding and obtaining food, consuming food, and absorbing its energy and nutrients. The organic matter that provides energy for heterotrophs includes living and freshly killed organisms as well as **detritus**—organic material derived from dead organisms in various stages of decomposition (see Concept 20.4). In this section, we will examine food sources, the ways in which heterotrophs obtain energy, and factors that influence absorption of food. There is a wide range of variation in the complexity of heterotrophic energy acquisition and assimilation processes that is associated with heterotroph body size and physiology. In

Chapters 8, 12, and 13, we will take a more in-depth look at the various types of consumers (predators, herbivores, and parasites), how they forage, and how the food they consume affects their growth and reproduction as well as the distributions and abundances of both the consumers themselves and their food resources (prey and hosts).

Food sources differ in their chemistry and availability

Heterotrophs consume energy-rich organic compounds (food) from their environment and convert them into usable chemical energy—primarily ATP—by processes such as *glycolysis*, which breaks down carbohydrates. The heterotroph's energy gain from food depends on the chemistry of the food, which determines its digestibility and its energy content. The effort invested in finding and obtaining the food also influences how much benefit the heterotroph gets from consuming it. For example, microorganisms that consume detritus in the soil invest little energy in obtaining food. However, the energy content of this decomposing plant matter is low compared with the energy content of live organisms. Living prey are rarer than detritus, and they may have defensive mechanisms that their predators must expend energy to overcome. Thus, a cheetah hunting a gazelle invests substantial energy in finding, chasing, capturing, and killing its prey, but it obtains a substantial, energy-rich meal if the hunt is successful.

The benefit of a food source to a heterotroph is partly related to the chemical compounds that the food contains. The chemical constituents of food can be placed into several categories based on their energy content and ease of assimilation (**FIGURE 5.18**). While water can be an important part of an animal's food, as we saw in Concept 4.3, it does not provide energy. The energy in food is found in the "dry matter" fraction (i.e., what is left when all the water is removed). *Fiber* includes compounds such as cellulose (the primary constituent of plant cell walls) and other structural

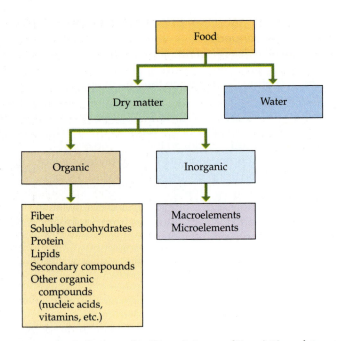

FIGURE 5.18 Categorical Breakdown of Food Chemistry
Food chemistry can be complex, but these simple categories help ecologists understand how groups of chemicals influence the benefits of food for heterotrophs. (After W. H. Karasov and C. Martinez del Rio. 2007. *Physiological Ecology: How Animals Process Energy, Nutrients, and Toxins*. Princeton University Press: Princeton, NJ.)

components of organisms. It is generally a poor energy source because of its chemical structure and the inability of many heterotrophs to break it down chemically. Most of the energy in food is found in carbohydrates, proteins, and fats. Fats are richer in energy than carbohydrates per unit of mass, and carbohydrates provide more energy than the amino acids that make up proteins do. However, amino acids also provide nitrogen, a nutrient that is often in high demand. The ratio of carbon to specific nutrients (usually nitrogen) often provides a good indication of the nutritional quality of the food: a higher amount of nutrients relative to the carbon indicates better-quality food. Secondary compounds (chemicals not used in growth or development) are generally not a good energy source for animals, and some secondary compounds may actually decrease energy intake by binding to digestive enzymes or by being directly toxic to the heterotrophs consuming them.

The differing concentrations of the compounds described in Figure 5.18 among food types are associated with the tissues, cell types, and organisms from which the food is derived. Animal tissues are generally more energy-rich than plant, fungal, or bacterial cells, which tend to have higher concentrations of fiber. As a result, *herbivores* (animals that eat plants) generally have to eat more food to get the same benefit that *carnivores* (animals that eat other animals) do. However, carnivores may expend substantially more energy finding food than herbivores do, as we will see in later chapters.

Heterotrophs obtain food using diverse strategies

Heterotrophs vary in size from archaea and bacteria (as small as 0.5 μm) to blue whales (up to 25 m long). The ratio of body size to food ingested varies widely, but it generally increases as body size increases. Bacteria may be bathed in their food, while food for larger heterotrophs is usually more diffuse and smaller relative to the consumer. Feeding methods and the complexity of food absorption are accordingly very diverse among heterotrophs.

Prokaryotic heterotrophs typically absorb food directly through their cell membranes. Archaea, bacteria, and fungi excrete enzymes into the environment to break down organic matter, acting in effect to digest their food outside their cells. Heterotrophic bacteria have adapted to a wide variety of organic energy sources and produce a large number of enzymes capable of breaking down organic compounds. This capacity of microorganisms as a group to use diverse energy sources has been exploited in environmental waste management as an approach to cleaning up toxic chemical wastes, a process known as *bioremediation*. Spills of fuels, pesticides, sewage, and other toxins have been effectively contained by using microorganisms to break down these harmful compounds. Consumption of oil by marine bacteria is thought to have been an important contributor to cleaning up the oil spill in the Gulf of Mexico that resulted when the *Deepwater Horizon* oil drilling rig exploded in 2010, releasing about 4.9 million barrels (780 × 103 liters) of oil. Much of the oil was released directly to the deeper layers of the ocean from the wellhead, which flowed for 87 days unabated until it was finally capped (**FIGURE 5.19**). The oil spill posed a substantial hazard to marine life, and it was feared that its impact would be long-term, as the impacts of other oil spills had been. Some reports suggest that up to half the oil released in the *Deepwater Horizon* spill was consumed and respired by marine microorganisms (Du and Kessler 2012), although others suggest that the blooms of microorganisms observed after the spill resulted from consumption of natural gas that leaked from the well rather than the oil itself (Valentine et al. 2010). While the magnitude of consumption is still debated, it is clear that the environmental impact of the oil spill was lessened by the action of marine microorganisms that used the spilled oil as an energy source.

Multicellular heterotrophs usually must seek out food, or move it toward themselves in the case of some sessile marine animals. The evolution of mobility was probably associated with the need to seek out food sources, as well as with the need to avoid being eaten by other consumers. Continued morphological and behavioral adaptations for efficiently finding and capturing food in different environments led to additional diversification of form and

FIGURE 5.19 An Environmental Disaster Oil pours from the fractured wellhead of the *Deepwater Horizon* oil drilling rig at the seafloor 1,700 m (5,700 feet) below the surface. About 57,000 barrels (9.1 million liters) were released each day for more than 3 months. The impact of this disaster may have been somewhat lessened by the activities of marine microorganisms that were able to use the oil as an energy source.

function. Animals display tremendous diversity in their specialized feeding adaptations, which reflect the diversity of the foods they consume. Here we present several examples that serve to demonstrate the morphological diversification of heterotrophs; we will take a closer look at behavioral adaptations for feeding in Concept 8.2.

MORPHOLOGICAL DIVERSITY OF INSECT MOUTHPARTS

Insects display tremendous diversity in facial appearance, which reflects the diversity of their food sources, which include detritus, plants, and other animals. They may eat animal prey whole or suck out their body fluids. All insects have the same basic set of mouthparts, consisting of several paired appendages that are used to seize, handle, and consume their food. Morphological variation in these mouthparts reflects the feeding specializations that have evolved within different insect groups (**FIGURE 5.20**). Common houseflies have "sponging" mouthparts that release saliva onto their food, then soak up and ingest the partially digested solution. Female mosquitoes and aphids have piercing and sucking mouthparts for extracting fluids from their food sources—blood from animals and sap from plants. Biting flies have razor-sharp appendages that cut through skin to draw blood for drinking, similar to the cutting mouthparts of insects that consume leaves.

MORPHOLOGICAL ADAPTATION IN BIRD BILLS Like those of insects, the mouthparts of birds—that is, their bills—display morphological adaptations that reflect the multitude of ways they capture, manipulate, and consume their food (**FIGURE 5.21**). The morphology of a bird's bill is closely associated with the taxonomic group to which the bird belongs. In other words, the flat bills of ducks and the hooked bills of raptors vary little within those groups. However, subtle differences in bill morphology among closely related species reflect slight differences in food acquisition and handling. This

The foot-long nectar-drawing proboscis of the hawkmoth *Xanthopan morgani* evolved in concert with the floral anatomy of a specific type of orchid on the island of Madagascar.

Blood-filled abdomen

The mouthparts of hornets, the largest wasps, crush and masticate the larvae of other insects, which worker hornets regurgitate to feed larvae in their own hive.

Mosquitoes use piercing mouthparts to penetrate the skin of animals and extract blood.

FIGURE 5.20 Variations on a Theme: Insect Mouthparts Differences in the morphology of insect mouthparts reflect different strategies for effectively acquiring and consuming the food types the insects prefer.

Finches such as the rose-breasted grosbeak have short bills suitable for cracking seeds.

Skimmers capture fish in their open bills as they fly low across the water's surface.

Hummingbirds extract nectar from flowers through their long bill.

© Kevin Collison/Shutterstock.com

© Steve Byland/Shutterstock.com

© anto1977/Shutterstock.com

FIGURE 5.21 Variations on a Theme: Bird Bills Bird bill morphology is associated with the feeding behavior of a species and enhances the acquisition of its preferred food resources.

variation reflects adaptations that help to optimize food acquisition and minimize competition among species (see Concept 14.2).

Craig Benkman studied the relationship between differences in bill morphology among crossbills as they relate to differences in the conifer seeds they use as food (Benkman 1993, 2003). As their name indicates, crossbills have unique asymmetrical bills with crossing tips (**FIGURE 5.22A**). Crossbills are adept at using their bills to open the cones of coniferous trees and pull out seeds for consumption. Across their geographic range, crossbills have multiple conifer species available as potential food sources; however, the tree species that are most abundant vary across this range. Benkman wondered if there were differences in the bill morphologies of crossbills that were associated with the morphologies of the cones of their preferred conifer species.

Benkman tested this hypothesis experimentally using captive and wild birds from five incipient species (subspecies that are in the process of becoming species) of the red crossbill species complex (*Loxia curvirostra*). He showed in a series of studies that a bird's speed of seed extraction from a given conifer's cone was associated with its bill depth. In addition, Benkman demonstrated that the speed of seed husking (removing the outer cover) was associated with the width of the groove in the bill where the seed is held (Benkman 1993, 2003). Each incipient crossbill species extracted and husked the seeds of one conifer species more efficiently than the seeds of other conifers. The study showed an association between the bill depth of an incipient species and the depth at which the seeds are held in the cones of its preferred conifer species. Furthermore, Benkman found that the annual survival rate for each incipient crossbill species was related to its feeding efficiency, which varied according to the conifer species it was feeding on. When

he put these results together, Benkman found a series of five "adaptive peaks," showing that bill morphology of each incipient species was associated with the conifer species on which it fed most efficiently and survived best (**FIGURE 5.22B**). Benkman (2003) concluded that red crossbills are currently undergoing evolutionary divergence (speciation; see Concept 6.4) as a result of selection associated with differences in available food resources across their range and the effects of those differences on bill morphology.

Heterotrophs vary in the complexity of their digestion and assimilation

As we have seen, food consumed by heterotrophs consists of a mix of complex compounds that must be chemically transformed into simpler compounds before they can be used as energy sources. Digestion breaks down proteins, carbohydrates, and fats into their component amino acids, simple sugars, and fatty acids. The evolution of digestion and assimilation is related to improving the efficiency of energy and nutrient extraction and to meeting the specific needs of physiological functions. Insect flight, for example, has a high energy demand, and some insects must maintain fat storage bodies to supply the energy required for initiation of flight. Humans require carbohydrates to fuel brain activity, which explains why a low blood sugar level can lead to poor cognitive ability. Thus, digestion and absorption of food are important steps in the energy acquisition and functioning of heterotrophs.

The evolution of feeding in heterotrophic protists and animals has led to increasing complexity in the ingestion, digestion, and absorption of food. Small protozoans such as amoebas and ciliates ingest food particles into their cells, where the food is digested in special organelles. With the advent of multicellular animals, specialized tissues for absorption, digestion, transport, and excretion evolved, and the efficiency of energy assimilation increased. Digestive systems evolved from simple chambers with a single

(A)

(B)

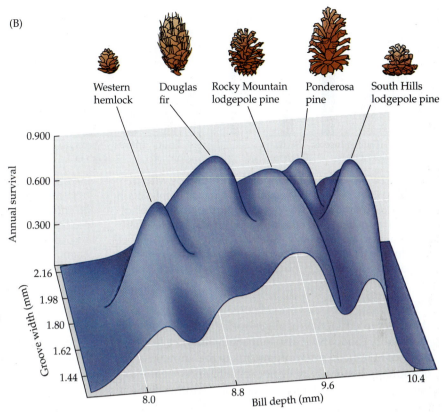

FIGURE 5.22 Crossbill Morphology, Food Preference, and Survival Rates
(A) Red crossbill (*Loxia curvirostra*). (B) A three-dimensional plot of Craig Benkman's data shows the relationship between bill morphology (groove width and bill depth) and annual survival rates in five incipient crossbill species. Each incipient species shows an "adaptive peak" in association with the conifer species it preferentially feeds on; that is, each incipient species has higher survival rates when feeding on the conifer species its bill morphology is best suited to exploit. The cones shown are drawn to relative scale. (B after C. W. Benkman. 2003. *Evolution* 57: 1176–1181.)

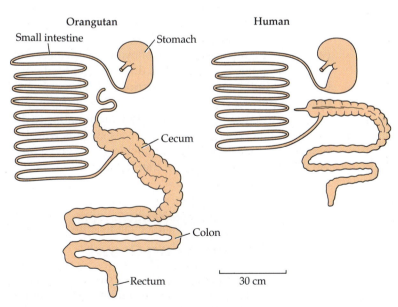

FIGURE 5.23 Herbivores Have Long Digestive Systems   Compared with omnivorous humans, herbivorous primates such as the orangutan have longer digestive systems. The greater volume and absorptive area of herbivore digestive tracts serve to enhance energy absorption from poor-quality food. (After O. M. Wrong et al. 1981. *The Large Intestine: Its Role in Mammalian Nutrition and Homeostasis.* Halsted: New York.)

input and output port, such as those in hydroid animals, to a tube with an input port (mouth) and an output port (anus). Further advancements included chambers specializing in specific digestive steps (e.g., stomachs) and absorption (e.g., intestines). Mechanisms evolved for breaking food down into smaller bits to increase the surface area exposed to digestion, including gizzards (which contain small rocks for grinding food) in earthworms and birds and molar teeth in mammals.

As you might guess from the discussion of food chemistry above, the diet of an animal can influence its digestive adaptations. For example, herbivores consume a food source—plants—that contains a large amount of fiber and small amounts of carbohydrates and proteins. To cope with this poor-quality diet, most herbivores have digestive tracts that are longer than those of carnivores, which increases food processing time and increases the surface area for absorbing energy (**FIGURE 5.23**). In order to further increase the exposure of food to the digestive tract, some herbivores, including many small vertebrate herbivores such as rabbits, reingest their feces (a strategy called *coprophagy*). Young animals may also ingest the feces of older animals. While this feeding strategy might seem disgusting to humans, it

(A)

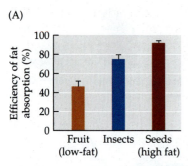

(B)

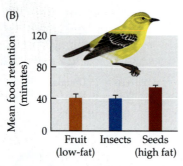

(C)

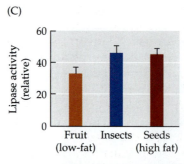

FIGURE 5.24 Adjustment of Digestion Efficiency with a Changing Diet Migrating warblers consume different diets in different parts of their ranges. To investigate the influence of fat content in the diet on their efficiency of fat absorption, researchers fed captive birds diets that were high (seed), medium (insect), or low (fruit) in fat, then measured the efficiency of fat absorption (the proportion of the fat in the diet taken up by the birds). The increase in the efficiency of fat absorption that accompanied a high-fat diet (A) was associated with longer food retention times (B) and greater production of a fat-degrading enzyme (lipase) by the pancreas (C). Error bars show one SE of the mean. (After W. H. Karasov and C. Martínez del Rio. 2007. *Physiological Ecology: How Animals Process Energy, Nutrients, and Toxins.* Princeton University Press: Princeton, NJ.)

enhances the efficiency of digestion and absorption of poor-quality food, and it also helps to maintain a healthy colony of beneficial microorganisms in the animal's gut. Coprophagy generally does not seem to enhance the digestion of fiber in food, but instead is more important for capturing vitamins and nutrients (Karasov and Martinez del Rio 2007).

Some herbivores have bacterial symbionts that greatly enhance the efficiency of digestion. Most animal digestive tracts are inhabited by archaea, bacteria, fungi, and even some protists, although the roles of many of these organisms in helping or hurting their hosts are unknown. For some animals, this relationship between the herbivore and its gut biota is clear: both benefit from the relationship. *Ruminants*, a group of herbivorous mammals that include cattle and giraffes, have a specialized stomach compartment (the rumen) in which large populations of bacteria facilitate the chemical breakdown of cellulose into simple sugars. The rumen acts like a fermentation chamber, providing environmental conditions that favor the growth of these beneficial bacteria. Material from the rumen is eventually passed into another stomach chamber, which absorbs not only the compounds released from digested plant matter, but also the compounds released from the bacteria that accompany the mass of digested food. Ruminants also exhibit *rumination*, or cud chewing, which is the regurgitation of material from a forestomach for additional chewing. Rumination allows these animals to "eat on the run," consuming large amounts of plant material in a short time and thereby minimizing their exposure to predators that might consume them. They can then more thoroughly chew and digest their food at a later time when the threat of being eaten is lower.

We've seen several examples of digestive adaptations to different food types. Can organisms acclimatize to eating different foods? The answer for some animals is yes. Organisms that consume a diverse diet of both plants and animals (*omnivores*) can adjust their digestive morphologies and produce different enzymes as needed to enhance digestion of their food. For example, warblers in the genus *Setophaga* make seasonal migrations that are associated with changes in their diet. The birds spend their breeding season (May–September) in forests of North America, eating mostly insects, and the rest of the year in Central America, consuming fruit and nectar. An experiment with captive warblers, including the pine warbler (*Setophaga pinus*), showed that their diets influenced the efficiency of fat assimilation. Compared with birds raised on diets of insects and fruit (which have a moderate and a low fat content, respectively), birds raised on seeds (which have a high fat content) showed the greatest ability to take up fats from their food due to longer food retention times in the gut and production of higher amounts of fat-degrading enzymes (**FIGURE 5.24**) (Karasov and Martinez del Rio 2007). This ability to acclimatize to different food sources allows omnivores such as warblers to select the best food source available at any given time. We'll discuss other aspects of diet flexibility and specialization in Concept 12.1.

A CASE STUDY REVISITED

Toolmaking Crows

We've seen that foraging animals often display behavioral as well as morphological and biochemical specializations that increase their efficiency in harvesting and digesting food. The specialized bill of crossbills is a morphological adaptation that improves their feeding efficiency. Warblers are able to adjust their digestive efficiencies to match their food source. Does tool use by crows enhance their ability to gain energy by allowing them to obtain food more efficiently or obtain food of higher quality?

New Caledonian crows are omnivores with a wide variety of food sources to select from, including vertebrate and invertebrate prey, plants, and dead animals (*carrion*). As we discussed earlier, the benefit a foraging animal gets from its food is determined by the effort it invests in finding and obtaining the food, the chemistry of the food, and the ability of the animal to digest and absorb it. There is a cost to tool use: collecting materials and fashioning the tools can be time-consuming, and young crows may not initially be adept at using them. Evaluating the benefit of tool use to the crows requires knowledge of their energy requirements, the energetic benefits of their potential food sources, and the crows' actual diet.

The crows' shy nature and their tropical forest habitat make observational studies difficult. To evaluate the energetic benefit of toolmaking and tool use, Christian Rutz and colleagues (2010) used stable isotope measurements (see Ecological Toolkit 5.1) to evaluate what the birds were eating and then used measurements of the lipid content of their potential food sources to estimate the energetic benefits of each. They also estimated the energy demands of the crows. Initial observations suggested that the birds relied on two high-quality food

items, both of which had a lipid content of about 40%: nuts from candlenut trees, which the crows crack open by dropping them onto rocks; and beetle larvae, which these birds obtain by using tools. Stable isotope measurements of N and C in the crows' blood and feathers and in their potential food sources indicated that they used a variety of food resources (**FIGURE 5.25A**) but that over 80% of their lipid intake was coming from the nuts and larvae (**FIGURE 5.25B**). This result indicates that a large proportion of the crows' energetic demand is met using two behaviors: tool use and nut cracking.

To address whether tool-aided beetle larva extraction alone could meet the energetic demand of the crows, Rutz and his colleagues determined the minimum number of beetle larvae needed on a daily basis to sustain a crow of average weight. They found that only three larvae per day were needed, because of their high lipid content. Observations indicated that most adult crows can easily obtain three larvae per day; one competent adult crow was able to extract 15 larvae in 80 minutes. Tool use clearly provides a substantial benefit to the New Caledonian crows, giving them access to a high-quality food item that would otherwise not be available to them, or would at least require a very high investment of energy to obtain.

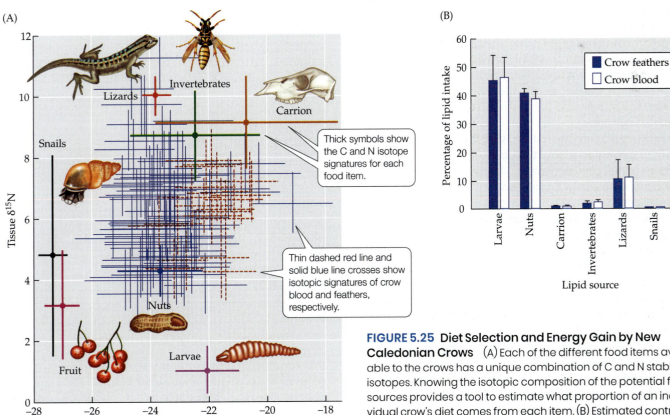

FIGURE 5.25 Diet Selection and Energy Gain by New Caledonian Crows (A) Each of the different food items available to the crows has a unique combination of C and N stable isotopes. Knowing the isotopic composition of the potential food sources provides a tool to estimate what proportion of an individual crow's diet comes from each item. (B) Estimated contributions of the food items to dietary lipid intake based on the isotopic composition of crow blood and feathers. Error bars show one SE of the mean. (After C. Rutz et al. 2010. *Science* 329: 1523–1526.)

🌱 CONNECTIONS *in* NATURE

TOOL USE: ADAPTATION OR LEARNED BEHAVIOR? How widespread is tool use among birds and other nonprimate animals? Many anecdotes of toolmaking and other innovative foraging techniques have been reported, but few have been examined thoroughly. The orange-winged sittella (*Daphoenositta chrysoptera*) of Australia uses sticks to forage for insect larvae, much like the New Caledonian crows. Egyptian vultures (*Neophron percnopterus*) crack open ostrich eggs using rocks. There are additional reports of tool use by insects, mammals, and other bird species (Beck 1980). The multitude of reports involving a wide range of animal species thoroughly dispels the notion of human monopoly on tool use. But how do these tool-using skills develop? Are these behaviors learned from other animals, or are they innate (i.e., determined genetically)? Several studies indicate that both learning and genetic inheritance can influence the development of tool use in animals.

As we learned above, tool use has a clear energetic benefit for New Caledonian crows, but does that benefit exert strong enough selection pressure to have resulted in a behavioral adaptation—are the birds inheriting the ability to use tools? To address this question, Ben Kenward and colleagues reared New Caledonian crows in captivity, without exposure to adult birds. Some of the birds received "tutoring" in toolmaking and tool use by human foster parents, while a control group did not (Kenward et al. 2005). To evaluate the birds' toolmaking abilities, the researchers placed supplemental food in tight crevices in the birds' aviaries, where it was not accessible to the birds without the assistance of tools. Twigs and leaves were also left in their aviaries. The captive crows developed the ability to make and use tools to retrieve the food in the crevices, whether they had been tutored or not (**FIGURE 5.26**). Kenward and colleagues concluded that the ability of New Caledonian crows to manufacture tools is at least partly inherited, rather than an acquired skill learned from adult birds in the wild. Very similar results were reported for experiments with captive woodpecker finches, birds endemic to the Galápagos archipelago that use twigs and cactus spines to forage for arthropods (Tebbich et al. 2001). Additional evidence that toolmaking is part of the genetic makeup of New Caledonian crows comes from an evaluation of their bill morphology, which has unique structural features consistent with tool manufacture and use as a selective force in its design (Matsui et al. 2016).

An additional twist to the crow toolmaking story is the apparent variation in tool styles among different crow populations on New Caledonia. In other words, there appears to be the potential for technological evolution in the styles of tools manufactured by crows. Gavin Hunt and Russell Gray conducted a survey of 21 sites on New Caledonia and examined 5,550 different cutting tools constructed by crows from Pandanus leaves (see Figure 5.2C) (Hunt and

FIGURE 5.26 Untutored Tool Use in Captive Crows A captive New Caledonian crow (*Corvus moneduloides*) uses a stick tool to retrieve food from artificial crevices in a laboratory setting, despite never having been exposed to tool use, either by humans or by other birds.

Gray 2003). They found three distinct widths of tools: wide, narrow, and stepped. Most of the tools found at a given site were very similar, and the geographic ranges of the tool types showed little overlap. There were no apparent correlations between where a tool type was found and local ecological factors such as forest structure or climate. Hunt and Gray suggested that the three tool designs were derived from a single original tool (of the wide type) subjected to additional modifications, including additional stripping of leaf material. Their study suggests ongoing innovation in toolmaking by the New Caledonian crows. This crow engineering challenges our traditional view of technological advancement in nonhuman animals.

Learned behavior is also important for toolmaking in some species. A notable example comes from studies of bottlenose dolphins in Shark Bay, Australia. Researchers observed that some dolphins swim with sponges plucked from the ocean floor on their noses (technically, their rostra) (**FIGURE 5.27**). The sponges appear to protect the sensitive rostra from sharp objects and stinging animals such as stonefish as the dolphins probe the seafloor for fish. The group of dolphins displaying this innovation is part of a larger group under study. The researchers' knowledge of the genetics and family structure of these dolphins allowed them to address the question of whether this unique behavior is learned or inherited. Michael Krützen and colleagues found that the majority of "sponging" dolphins were female. They reasoned that a single sex-linked gene (the kind of genetic basis one might expect for a trait occurring in only one sex) was a highly unlikely cause for a

FIGURE 5.27 Dolphin Nose Gear in Shark Bay, Australia
A bottlenose dolphin wears a sponge to protect its rostrum while foraging on the seafloor.

complex trait such as sponging. A comparison of the genetic fingerprints of individuals that sponged with those of nonsponging dolphins indicated that most of the sponging occurred within a single family line (Krützen et al. 2005). The combination of these results led Krützen and colleagues to conclude that sponging was a learned behavior passed from mother to daughter. This finding supports the idea of a cultural phenomenon in animals that influences the efficiency of their feeding behavior and challenges the notion that cultural learning is unique to humans. ⬭

SUMMARY

CONCEPT 5.1 Organisms obtain energy from sunlight, from inorganic chemical compounds, or through the consumption of organic compounds.

5.1.1 Differentiate autotrophy from heterotrophy in the context of building energy compounds using external sources of energy versus consuming them from organic matter.

- Autotrophs convert energy from sunlight (by photosynthesis) or inorganic chemicals (by chemosynthesis) into energy stored in the carbon–carbon bonds of carbohydrates.

- Heterotrophs acquire energy by consuming organic compounds from other organisms, living or dead.

CONCEPT 5.2 Radiant and chemical energy captured by autotrophs is converted into stored energy in carbon–carbon bonds.

5.2.1 Summarize chemosynthesis, which results in the synthesis of energy-rich carbon–carbon bonds.

- During chemosynthesis, bacteria and archaea oxidize inorganic substrates to obtain energy, which they use to fix carbon and synthesize sugars.

5.2.2 Outline the steps in the light-driven reactions and carbon reactions of photosynthesis, describing their outcomes and how they produce energy-rich compounds in photoautotrophs.

- Photosynthesis has two main steps: the absorption of sunlight by pigments to produce energy in the form of ATP and NADPH (the light-driven reactions) and the use of that energy in the Calvin cycle to fix CO_2 and synthesize carbohydrates (the carbon reactions).

5.2.3 Illustrate how photosynthetic organisms acclimatize and adapt to variations in the intensity of light.

- Photosynthetic responses to variation in light levels, water availability, and nutrient availability include both short-term acclimatization and long-term adaptation.

5.2.4 Evaluate the trade-offs that result when a plant controls water loss.

- Keeping stomates open while tissues lose water can permanently impair physiological processes in the leaf. Closing stomates, however, not only limits photosynthetic CO_2 uptake, but also increases the chances of light damage to the leaf.

5.2.5 Describe how temperature influences photosynthetic rates through its effect on enzymes and chloroplast membranes.

- Autotrophs acclimatize and adapt to temperature variation by changing properties of the Calvin cycle enzymes and/or the photosynthetic membranes. Different photosynthetic organisms have different forms of the same photosynthetic enzymes that operate best under the environmental temperatures where the organisms occur.

CONCEPT 5.3 Environmental constraints have resulted in the evolution of biochemical pathways that improve the efficiency of photosynthesis.

5.3.1 Explain the difference between photosynthesis and photorespiration and evaluate conditions where photorespiration is detrimental to plant growth.

- Photorespiration operates in opposition to photosynthesis, lowering the rate of energy gain, particularly at high temperatures and low atmospheric CO_2 concentrations.

5.3.2 Summarize how biochemical and anatomical adaptations associated with the C_4 photosynthetic pathway minimize photorespiration, thereby enhancing photosynthesis rates.

- The C_4 photosynthetic pathway concentrates CO_2 at the site of the Calvin cycle, minimizing photorespiration.

5.3.3 Describe how crassulacean acid metabolism reduces water loss relative to the C_3 or C_4 photosynthetic pathways.

(Continued)

SUMMARY (continued)

- CAM plants reduce transpirational water loss by opening their stomates at night to take up CO_2 and releasing it to the Calvin cycle during the day, when the stomates are closed.

CONCEPT 5.4 Heterotrophs have adaptations for acquiring and assimilating energy efficiently from a variety of organic sources.

5.4.1 Illustrate how the chemical makeup of a food item determines the benefit it provides to the consumer eating it.

- Variations in the chemistry and availability of food determine how much energy heterotrophs gain from different food sources.

5.4.2 Explain how morphological and behavioral adaptations enable heterotrophs to obtain food more efficiently.

- Heterotrophs display tremendous diversity in behavioral, morphological, and physiological adaptations that enhance their efficiency of energy acquisition and assimilation.

5.4.3 Describe how increasing complexity in the digestive systems of heterotrophs makes the assimilation of energy and nutrients more efficient.

- Complex digestive systems, such as a tube with an input port and an output port, or additional chambers specializing in specific digestive steps (e.g., stomachs) and absorption (e.g., intestines), make the assimilation of energy and nutrients more efficient.

Review Questions

1. Define autotrophy and heterotrophy, and provide a few examples of each that illustrate the diversity of the ways in which organisms obtain energy.

2. How does the CAM photosynthetic pathway influence water loss from plants?

3. What are the trade-offs associated with heterotrophic consumption of live animals versus dead plant material?

Hone Your Problem-Solving Skills

Our earlier comparison of the C_3 and C_4 photosynthetic pathways emphasized the benefit of the C_4 pathway through its capacity to minimize photorespiration by increasing the concentration of CO_2 inside the leaf. With the advent of elevated atmospheric CO_2 concentrations over the past century, ecologists have speculated whether C_3 plants might benefit more than C_4 plants from possible CO_2 fertilization. Such speculation is based on laboratory measurements and modeling of plant responses to variation in CO_2 concentrations. An example is such as shown in the figure. (After H. Lambers et al. 2008. *Plant Physiological Ecology*, 2nd ed. Springer: New York.)

1. Based on the modeled response in the figure, what are the absolute and relative (as a percent) increases in photosynthesis for C_3 and C_4 plants from preindustrial levels of CO_2 (280 ppm) to the current level of CO_2 (412 ppm)? What are the possible consequences of any differences in the responses of C_3 and C_4 plants to elevated CO_2 in plant communities where they co-occur and compete for limited resources?

2. A review of 62 studies examining the responses of C_3 and C_4 grasses to elevated CO_2 found that on average photosynthesis increased by 33% and 25% in these groups, respectively (Wand et al. 1999). How does this compare with your estimate from question 1? What might account for any discrepancies? Possible contributions may be related to the influence of CO_2 on lowering stomatal opening and the demand for nitrogen to make enzymes.

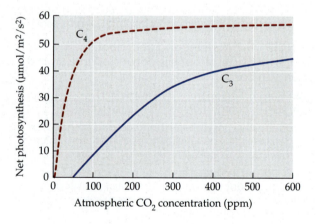

ADDITIONAL RESOURCES

This textbook provides resources such as **Web Extensions** and additional **Climate Change Connections** for instructors who can share them with students. Please contact your instructor, if you would like to access that content.

Evolutionary Ecology

6

Evolution and Ecology

Previous page, © Paul Sawer/Minden Pictures; This page, © Jason Savage

Trophy Hunting and Inadvertent Evolution: A Case Study

Bighorn sheep (*Ovis canadensis*) are magnificent animals, beautifully suited for life in the rugged mountains in which they are found. Despite their substantial size (males can weigh up to 127 kg, or 280 pounds), these sheep can balance on narrow ledges and can leap 6 m (20 feet) from one ledge to another. Bighorn sheep are also noted for the male's large curl of horns, which are used in combat over females (**FIGURE 6.1**). Rams run at speeds of up to 20 miles per hour and crash their heads into each other, battling over the right to mate with a female.

Ram horns have been collected as trophies for many centuries without drastically affecting sheep populations. Over the last 200 years, however, human actions such as habitat encroachment, hunting, and the introduction of domesticated cattle have reduced populations of bighorn sheep by 90%. As a result, the hunting of bighorn sheep has been restricted throughout North America. These restrictions make a world-class trophy ram (one with a large, full curl of horns) extremely valuable: permits to shoot one of these rams, which are sold at auction, can cost over $100,000.

Although funds raised by the auction of hunting permits are used to preserve bighorn sheep habitat, scientists have expressed concern that trophy hunting is having negative effects on today's small populations of bighorn sheep. Trophy hunting removes the largest and strongest males: in a population from which about 10% of the males were removed by hunting each year, both the average size of males and the average size of their horns decreased over a 30-year period (**FIGURE 6.2**). Large and strong males are preferred by females and tend to sire more offspring than other males, so killing the largest and strongest males can make it harder for small populations to recover in abundance.

Hunting, fishing, and other forms of harvest have affected a wide range of other species, including fishes, invertebrates, and plants (Darimont et al. 2009). For example, by targeting older and larger fish, commercial fishing for cod has led to a reduction in the age and size at which these fish become sexually mature. To see why this happens, first note that cod that mature at a younger age and smaller size are more likely to reproduce before they are caught than are fish that mature when they are older and larger. As a result, the genes of fish that mature at a younger age and smaller size are more likely to be passed on to the next generation than are the genes of other fish—hence, we would predict that over time, more and more fish will have genes that encode sexual maturity at a younger age and smaller size. Indeed, in experimental populations of guppies in which small or

FIGURE 6.1 Fighting over the Right to Mate Two bighorn rams butt heads to establish dominance and mating rights. Large horns are beneficial to a ram's success with this dominance ritual.

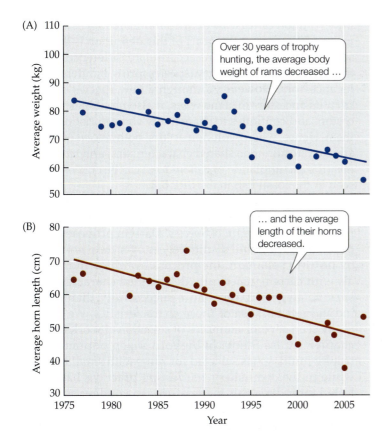

FIGURE 6.2 Trophy Hunting Decreases Ram Body and Horn Size Coltman and colleagues tracked the body weights (A) and horn lengths (B) of rams in a bighorn sheep population on Ram Mountain (Alberta, Canada) that was subjected to trophy hunting over a 30-year period. The changes in horn length occurred across multiple generations of sheep and thus indicate a change in the average characteristics of the sheep born from one generation to the next. (After D. W. Coltman et al. 2003. *Nature* 426: 656–658.)

large individuals were selectively removed for harvest, van Wijk et al. (2013) documented such genetic changes in genes known to affect body size. Similarly, poaching for ivory appears to have resulted in genetic changes that have caused the proportion of female African elephants in a South African park that lack tusks to increase from 62% to 90% over a 20-year period.

The unintended effects of human harvesting on bighorn sheep, cod, and elephants illustrate how populations can change, or *evolve*, over time. What biological mechanisms cause these evolutionary changes? Do human actions other than harvesting produce evolutionary change?

Introduction

As news reports often emphasize, humans have a large effect on the environment. We change the global climate, pollute the water and air, convert large tracts of natural habitat into farmland and urban areas, drain wetlands, and reduce the population sizes of species we hunt for food (e.g., fishes) or use as resources (e.g., trees). Although we have taken steps to limit some of the damage we cause to biological communities, human actions have consequences that we have barely begun to recognize, much less address: we cause evolutionary change.

In this chapter, we'll examine what evolution is and how it affects ecological interactions, as well as how ecological processes influence evolutionary change. At the close of the chapter, we'll focus specifically on how humans cause evolutionary change. Our goal in this chapter is not to provide a comprehensive survey of evolutionary biology—for that, see the textbooks on evolution listed in the Suggested Readings on the book's website. Instead, our aim is to show that ecology and evolution are interconnected, a theme to which we will return in later chapters of this book. We'll begin by considering two ways in which evolution can be defined.

> **CONCEPT 6.1** Evolution can be viewed as genetic change over time or as a process of descent with modification.

LEARNING OBJECTIVES

- **6.1.1** Summarize the genetic basis for the evolution of traits in organisms.
- **6.1.2** Explain how evolution can be considered the accumulation of trait differences between populations.

What Is Evolution?

In the most general sense, biological evolution is change in organisms over time. Evolution includes the relatively small fluctuations that occur continually within populations, as when the genetic makeup of a population changes from one year to the next. But evolution can also refer to the larger changes that occur as species gradually become increasingly different from their ancestors. Let's explore these two ways of looking at evolution in more detail, focusing first on genetic changes (*allele frequency change*) and then on how organisms accumulate differences from their ancestors (*descent with modification*).

Evolution is allele frequency change

Figure 6.2B shows that the average horn size of male bighorn sheep has decreased over time, but it does not reveal the cause of that decline. A clue to the cause comes from an additional observation (Coltman et al. 2003): horn size is an *inherited* trait. This means that rams with large horns tend to have offspring that have large horns and that rams with small horns tend to have offspring that have small horns. Because trophy hunting selectively eliminates rams with large horns, it favors rams whose genetic characteristics lead to the production of small horns. Hence, it seems likely that trophy hunting is causing the genetic characteristics of the bighorn sheep population to change, or evolve, over time—a conclusion supported in a recent analysis of data from a bighorn sheep population subjected to intense hunting for 23 years (Pigeon et al. 2016).

As suggested by the trophy-hunting example, biologists often define evolution in terms of genetic change. To make such a definition more precise (and to introduce terms that will be used throughout this chapter), let's review some basic principles of genetics:

- Genes are composed of DNA, and they specify how to build (encode) proteins.

- A given gene can have two or more forms (known as **alleles**) that result in the production of different versions of the protein that the gene encodes.

- We can designate the **genotype** (genetic makeup) of an individual with letters that represent the individual's two copies of each gene (one inherited from its mother, the other from its father). For example, if a gene has two alleles, designated *A* and *a*, the individual could be of genotype *AA*, *Aa*, or *aa*.

With these principles as background, we can define **evolution** as change over time in the *frequencies* (proportions) of different alleles in a population. To illustrate how this definition is applied, consider a population of 1,000 individuals and a gene with two alleles (*A* and *a*). Suppose there are 360 individuals of genotype *AA*, 480 of genotype *Aa*, and 160 of genotype *aa*. The frequency of the *a* allele in this population is 0.4, or 40%;[1] hence, since there are only two alleles in the population (*A* and *a*), the frequency of the *A* allele must be 1 − 0.4 = 0.6, or 60%. If the frequency of the *a* allele were to change over time, say, from 40% to 71%, then the population would have evolved at that gene. (In scientific studies, researchers often use an approach based on the *Hardy–Weinberg equation* to test whether a population is evolving at one or more genes.)

Evolution is descent with modification

In many parts of this book, when we refer to evolution, we will be referring to allele frequency change over time.

But evolution can also be defined more broadly as *descent with modification*. At the heart of this definition is the observation that populations accumulate differences over time, and hence, when a new species forms, it differs from its ancestors. However, although a new species differs from its ancestors in some ways, it also resembles its ancestors and continues to share many characteristics with them. Hence, when evolution occurs, both *descent* (shared ancestry, resulting in shared characteristics) and *modification* (the accumulation of differences) can be observed, as seen in the fossil fish in **FIGURE 6.3**.

Charles Darwin (1859) used the phrase "descent with modification" to summarize the evolutionary process in his book *The Origin of Species*. Darwin proposed that populations accumulate differences over time primarily by **natural selection**, the process by which individuals with certain genetically determined characteristics survive and reproduce more successfully than other individuals because of those characteristics. We've already seen several examples of selection at work in this chapter's Case Study. In bighorn sheep populations, trophy hunting has selected for rams with small horns, while in the cod fishery, harvesting practices have selected for individuals that mature at a younger age and a smaller size.

How can natural selection explain the accumulation of differences between populations? Darwin argued that if two populations experience different environmental conditions, individuals with one set of characteristics may be favored by natural selection in one population, while individuals with a different set of characteristics may be favored in the other population (**FIGURE 6.4**). By favoring individuals with different heritable characteristics in different populations, natural selection can cause populations to diverge genetically from one another over time; that is, each population will accumulate more and more genetic differences. Thus, natural selection can be responsible for the *modification* part of "descent with modification."

Populations evolve, individuals do not

Natural selection acts as a sorting process, favoring individuals with some heritable traits (e.g., bighorns with small horns) over others (e.g., bighorns with large horns). Individuals with the favored traits tend to leave more offspring than do individuals with other traits. As a result, from one generation to the next, a greater proportion of the individuals in the population will have the traits favored by natural selection. When these traits have a genetic basis, this process can cause the allele frequencies of the population to change over time, thereby causing the population to evolve. But the *individuals* in the population do not evolve—either they have the traits favored by selection or they don't.

---

[1] Each of the 1,000 individuals in the population has two alleles, giving a total of 2,000 alleles in the population. Each of the 360 individuals of genotype AA has zero a alleles, each of the 480 individuals of genotype Aa has one a allele, and each of the 160 individuals of genotype aa has two a alleles. Thus, the frequency of the a allele is $(0 \times 360 + 1 \times 480 + 2 \times 160)/2{,}000 = 0.4$. The frequency of the a allele can also be calculated using genotype frequencies, in which case we have $[(0 \times 0.36) + (1 \times 0.48) + (2 \times 0.16)]/2 = 0.4$, where $360/1{,}000 = 0.36$ is the frequency of genotype AA, 0.48 is the frequency of genotype Aa, and 0.16 is the frequency of genotype aa.

(A)

UNIQUE FOSSILS

These 10 million-year-old fossils of the stickleback fish *Gasterosteus doryssus* were collected from a lake bed in Nevada (USA). The fossils could be dated to the nearest 250 years because the rocks in which they were found showed exceptionally clear annual layers of sediments.

DESCENT

Fossil evidence suggests that *G. doryssus* colonized open waters of the lake about 10 million years ago. Over the next 16,000 years, many of the bones in these fish did not change in size, shape, or position. The resulting similarities in their overall bone structure illustrate common descent—the fish descended from the original colonists and hence shared many characteristics with them.

MODIFICATION

The fossil sticklebacks also show how organisms become modified from their ancestors over time. For example, in less than 5,000 years, the pelvic bone—originally the largest single bone in the body of these fish—became greatly reduced in size. This reduction has also occurred in modern lakes and probably resulted from natural selection.

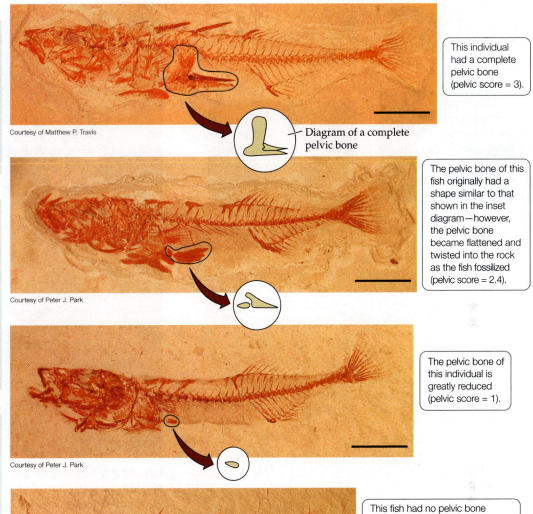

Courtesy of Matthew P. Travis

Diagram of a complete pelvic bone

This individual had a complete pelvic bone (pelvic score = 3).

Courtesy of Peter J. Park

The pelvic bone of this fish originally had a shape similar to that shown in the inset diagram—however, the pelvic bone became flattened and twisted into the rock as the fish fossilized (pelvic score = 2.4).

Courtesy of Peter J. Park

The pelvic bone of this individual is greatly reduced (pelvic score = 1).

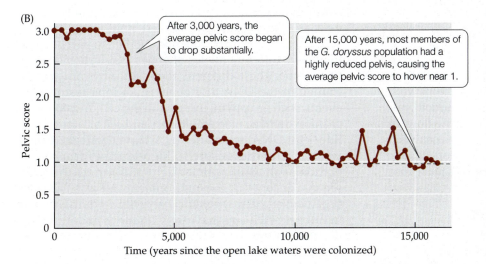

Courtesy of Peter J. Park

This fish had no pelvic bone (pelvic score = 0). Such individuals first appear in the fossil record 11,000 years after open lake waters were colonized.

(B)

After 3,000 years, the average pelvic score began to drop substantially.

After 15,000 years, most members of the *G. doryssus* population had a highly reduced pelvis, causing the average pelvic score to hover near 1.

Time (years since the open lake waters were colonized)

Pelvic score

FIGURE 6.3 Descent with Modification Michael Bell and colleagues have analyzed thousands of 10-million-year-old fossils of the stickleback fish *Gasterosteus doryssus*. Their specimens are unique in that the lake bed in which they were found is so finely layered that the ages of the fossils can be determined to the nearest 250-year interval. (A) Representative *G. doryssus* fossils, showing how the pelvic bone became reduced over time; the scale bar for each fossil is 1 cm. (B) The average pelvic score at different times. Fossil pelvic bones were scored by size according to a scale that ranged from 3 (complete bone) to 0 (no bone). (B after M. A. Bell et al. 2006. *Paleobiology* 32: 562–577.)

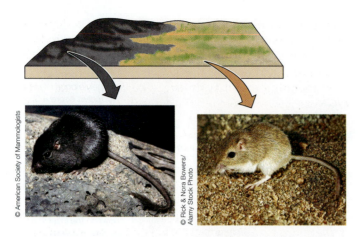

FIGURE 6.4 Natural Selection Can Result in Differences between Populations Populations of rock pocket mice (*Chaetodipus intermedius*) that live on dark lava formations in Arizona and New Mexico have dark coats, while nearby populations that live on light-colored rocks have light coats. In each population, natural selection has favored individuals whose coat colors match their surroundings, making them less visible to predators.

> **CONCEPT 6.2** Natural selection, genetic drift, and gene flow can cause allele frequencies in a population to change over time.

LEARNING OBJECTIVES

6.2.1 Summarize how mutations contribute to the process of evolution.

6.2.2 Compare the effects of stabilizing selection and disruptive selection on the temporal changes in the genetic structure of a population.

6.2.3 Evaluate how random events can affect populations through time via genetic drift.

6.2.4 Describe the role of gene flow among populations in terms of homogenizing genetic structure as well as enhancing evolutionary change.

Mechanisms of Evolution

Although natural selection is often the cause of evolutionary change, it is not the only one. In this section, we'll examine four key processes that influence evolution: mutation, natural selection, genetic drift, and gene flow. In general, mutation is the source of the new alleles on which all of evolution depends, while natural selection, genetic drift, and gene flow are the main mechanisms that cause allele frequencies to change over time.

Mutation generates the raw material for evolution

Individuals in populations may differ from one another in their **phenotype**, the observable characteristics of an organism, such as size or color (**FIGURE 6.5**). Many aspects of an organism's phenotype, including its physical

FIGURE 6.5 Individuals in Populations Differ in Their Phenotypes Poison dart frogs (*Dendrobates tinctorius*) show great variation in color and pattern. Native to northern South America, these frogs live in isolated patches of forest. Their bright colors are thought to serve as a warning to predators of a poison excreted from their skin. Individual frogs likely also differ in other morphological traits as well as in biochemical, behavioral, and physiological traits.

features, metabolism, growth rate, susceptibility to disease, and behavior, are influenced by its genes. As a result, individuals differ from one another, in part because they have different alleles of genes that influence their phenotype. These different alleles arise by **mutation**, a change in the DNA of a gene. Mutations result from events such as copying errors during cell division, mechanical damage when molecules and cell structures collide with DNA, exposure to certain chemicals (called *mutagens*), and exposure to high-energy forms of radiation such as ultraviolet light and X rays. As we'll see in Concept 7.1, the environment can also affect an organism's phenotype. For example, a plant growing in nutrient-rich soil may grow larger than another individual of the same species growing in nutrient-poor soil, even if both have the same alleles of genes that influence size. In this chapter, however, we will focus on phenotypic differences that result from genetic, not environmental, factors.

The formation of new alleles by mutation is critical to evolution. If a species lacked mutations, each gene would have only one allele, and all members of a population would be genetically identical. If this were the case, evolution could not occur: allele frequencies cannot possibly change over time unless the individuals in a population differ genetically. Individuals in a population can differ genetically not only because of mutation, but also because of **recombination**, the production of offspring that have combinations of alleles that differ from those in either of their parents. We can think of mutation as providing the raw material (new alleles) on which evolution is based, and recombination as rearranging that raw material into unique new combinations. Together, these processes provide the genetic variation among individuals that is required for evolution to occur.

Despite its importance to evolution, mutation usually occurs too rarely in most cases to be the direct cause of significant allele frequency change over short periods of time. Mutations typically occur at rates of 10^{-4} to 10^{-6} new mutations per gene per generation (Hartl and Clark 2007). In other words, in each generation, we can expect one mutation to occur in every 10,000 to 1,000,000 copies of a gene. At these rates, in one generation, mutation acting alone causes virtually no change in the allele frequencies of a population. Eventually, mutation can cause appreciable allele frequency change, but typically it takes thousands of generations for it to do so. Overall, in terms of its direct effects, the background mutation rate is a weak agent of allele frequency change. But because it provides new alleles on which natural selection and other mechanisms of evolution can act, mutation is central to the evolutionary process. It should be noted, however, that some environmental factors, such as exposure to high-energy radiation (e.g., radioactivity or X rays) and some mutagenic chemicals, can greatly increase mutation rates.

The evolution of antibiotic resistance is an example where mutation rates are frequent enough to influence allele frequencies in a population. There are around 40 trillion (4×10^{13}) bacterial cells in a human body, consisting of around 500 distinct species. Given the mutation rate described above, we could expect an appreciable number of alleles to appear in a population in each generation. The majority of these mutations are deleterious—that is, they lower growth and reproduction of the bacteria. However, some alleles confer greater resistance to antibiotics applied to kill them. As a result, the efficacy of antibiotics is potentially compromised, particularly with regular application, which enhances the potential for natural selection to favor the allele conferring resistance.

Natural selection increases the frequencies of advantageous alleles and decreases the frequencies of deleterious alleles

Natural selection occurs when individuals with particular heritable traits consistently leave more offspring than do individuals with other heritable traits. But some traits may give organisms an advantage only under certain environmental conditions. Indeed, as we'll see later in this chapter, traits that are advantageous in one environment can be disadvantageous in another.

Depending on what traits are favored, we can categorize natural selection into three types (**FIGURE 6.6**). **Directional selection** occurs when individuals with one extreme of a heritable phenotypic trait (e.g., large size) are favored over other individuals (small and medium-sized individuals). In **stabilizing selection**, individuals with an intermediate phenotype (e.g., medium-sized individuals) are favored, while in **disruptive selection**, individuals with a phenotype at either extreme are favored (e.g., both small and large individuals have an advantage over medium-sized individuals). However, in all three types of natural selection, the fundamental process is the same: some individuals have heritable phenotypes that give them an advantage in survival or reproduction, causing them to leave more offspring than other individuals.

When selection favors a particular phenotype, individuals with alleles that encode that phenotype are likely to leave more offspring than are individuals with other alleles. As a result, alleles that encode a favored phenotype can increase in frequency from one generation to the next. In some cases, the end result of this process is that most or all of the individuals in a population have an allele that encodes a trait favored by selection. A well-studied example is the Andean goose (*Chloephaga melanoptera*), which lives at high elevations. These birds have evolved a version of the oxygen transport protein hemoglobin that has an unusually high affinity for oxygen and hence provides an advantage at high altitudes with low oxygen partial pressure (Weber 2007; McCracken et al. 2009). The allele that encodes this version of hemoglobin occurs at a frequency of 100% in Andean goose populations. An allele such as this that occurs in a population at a frequency of 100% is said to have reached **fixation**.

To recap, natural selection can cause the frequency of an allele that confers an advantage to increase over time, as has occurred in populations of the Andean goose. We'll consider the consequences of such increases in the frequencies of advantageous alleles later in this chapter. But first, we'll look at two other mechanisms that can cause allele frequencies to change: *genetic drift* and *gene flow*.

Genetic drift results from random events

Allele frequencies in populations can be influenced by random events. Imagine a population of ten plants in which three individuals have genotype *AA*, four have genotype *Aa*, and three have genotype *aa*. Thus, the initial frequencies of both the *A* and *a* alleles are 50%. Assume that the *A* and *a* alleles encode two different versions of a protein that function equally well. Although neither allele is more advantageous than the other (and hence natural selection does not affect this gene), random events could alter their frequencies. For example, suppose that a moose walking through the woods happened to step on four of

(A) Directional selection

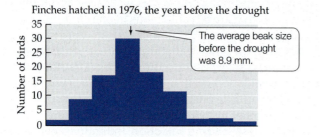

Finches hatched in 1976, the year before the drought

The average beak size before the drought was 8.9 mm.

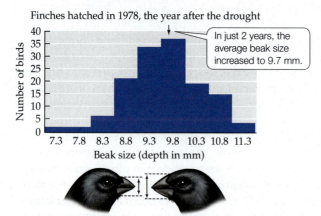

Finches hatched in 1978, the year after the drought

In just 2 years, the average beak size increased to 9.7 mm.

Beak size (depth in mm)

(B) Stabilizing selection

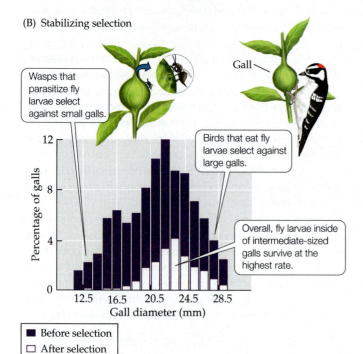

Wasps that parasitize fly larvae select against small galls.

Gall

Birds that eat fly larvae select against large galls.

Overall, fly larvae inside of intermediate-sized galls survive at the highest rate.

Gall diameter (mm)

■ Before selection
□ After selection

(C) Disruptive selection

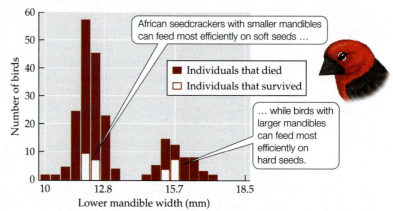

African seedcrackers with smaller mandibles can feed most efficiently on soft seeds …

■ Individuals that died
□ Individuals that survived

… while birds with larger mandibles can feed most efficiently on hard seeds.

Lower mandible width (mm)

FIGURE 6.6 Three Types of Natural Selection (A) Directional selection favors individuals at one phenotypic extreme. A prolonged drought in the Galápagos archipelago resulted in directional selection on the beak size of the seed-eating medium ground finch (*Geospiza fortis*). As a result of the drought, most of the available seeds were large and hard to crack, so birds with large beaks, which could more easily crack those seeds, had an advantage over birds with smaller beaks. (B) Stabilizing selection favors individuals with an intermediate phenotype. *Eurosta* flies parasitize goldenrod plants, causing the plant to produce a gall in which the fly larva matures as it feeds on the plant. The preferences of *Eurosta*'s own predators and parasites result in stabilizing selection on gall size. Field observations showed that wasps that parasitize and kill the fly larvae prefer small galls, while birds that eat the fly larvae prefer large galls. As a result, larvae in galls of intermediate size have an advantage. (C) Disruptive selection favors individuals at both extremes. African seedcrackers (*Pyrenestes ostrinus*) depend on two major food plants in their environment. Birds with smaller mandible sizes can feed on one plant's soft seeds most efficiently, while birds with larger mandibles can feed on the other plant's hard seeds most efficiently. Thus, individuals with mandible sizes that are either relatively small or relatively large have an advantage. (A after B. R. Grant and P. R. Grant. 2003. *BioScience* 53: 965–975; B after A. E. Weis and W. G. Abrahamson. 1986. *Am Nat* 127: 681–695; C after T. B. Smith. 1993. *Nature* 363: 618–620.)

In (B), do birds or wasps appear to provide stronger selection pressure on gall size? Explain.

the wildflowers (two of genotype *AA* and two of genotype *Aa*), killing them, but not harming any of the three plants of genotype *aa*. As a result, the frequency of the *a* allele in the population would increase from 50% to 67% *due to a random event*.

When random events affect which alleles are passed from one generation to the next, **genetic drift** is said to occur. Although random events occur in populations of all sizes, the effects of genetic drift on allele frequency changes is greater in small populations than in large ones. To see why, imagine that our plant population had 10,000 individuals, 3,000 of genotype *AA*, 4,000 of genotype *Aa*, and 3,000 of genotype *aa*. If (as before) a moose stepped on a random sample of 40% of the individuals in this larger population, there is virtually no possibility that all of the 3,000 individuals of genotype *aa* would be spared. Instead, it is likely that many individuals of each genotype would be killed and, hence, that the frequencies of the *A* and *a* alleles would change little, if at all.

Genetic drift has four related effects on evolution, the strength of which is larger in small populations:

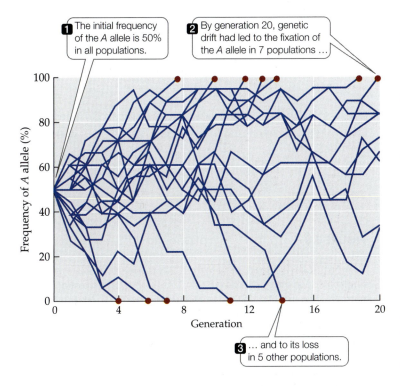

1 The initial frequency of the *A* allele is 50% in all populations.

2 By generation 20, genetic drift had led to the fixation of the *A* allele in 7 populations …

3 … and to its loss in 5 other populations.

FIGURE 6.7 **Genetic Drift Causes Allele Frequencies to Fluctuate at Random** Results of a computer simulation of genetic drift in 20 populations for a gene with two alleles, *A* and *a*. Each population has nine diploid individuals (18 alleles) in each generation. In small populations such as these, genetic drift has rapid effects. (After D. Hartl and A. Clark. 1989. *Principles of Population Genetics*, 2nd ed. Oxford University Press/Sinauer: Sunderland, MA.)

? At the start of the simulation, how many A alleles and how many a alleles did each population have? At generation 20, how many populations still had both alleles? Predict what would eventually happen to the frequency of the A allele in those populations.

1. Because it acts by chance alone, genetic drift can cause allele frequencies to fluctuate randomly over time in small populations (**FIGURE 6.7**). When this occurs, eventually some alleles disappear from small populations experiencing genetic drift, while others reach fixation.

2. By causing alleles to be lost from a population, genetic drift reduces the genetic variation of the population, making the individuals within the population more similar genetically to one another.

3. Genetic drift can increase the frequency of a harmful allele. This may seem counterintuitive because in general, genetic drift acts on alleles that neither harm nor benefit the organism, and we would expect natural selection to reduce the frequency of a harmful allele. However, if the population size is very small and the allele has only slightly deleterious effects, genetic drift can "overrule" the effects of natural selection, causing the harmful allele to increase or decrease in frequency randomly.

4. Genetic drift can increase genetic differences between populations because random events may cause an allele to reach fixation in one population yet be lost from another population (see Figure 6.7).

The second and third of these effects can have dire consequences for small populations. A loss of genetic variation can reduce the capacity of a population to evolve in response to changing environmental conditions, potentially placing it at risk of extinction. Likewise, an increase in the frequency of harmful alleles in a population can hinder the ability of its members to survive or reproduce, again increasing the risk of extinction. This effect presents an ongoing problem for small populations. Although mutation is unlikely to produce harmful alleles of any particular gene from one generation to the next (because mutations are rare), it is highly likely to produce new deleterious alleles in *some* of an organism's many genes—and genetic drift can cause those alleles to increase in frequency.

Such negative effects of genetic drift are thought to have contributed to the near extinction of the Illinois populations of the greater prairie chicken (*Tympanuchus cupido*). In the early 1800s, there were millions of these birds in Illinois. Over time, their numbers plummeted as more than 99% of the prairie habitat on which they depend was converted to farmland and other uses. By 1993, fewer than 50 greater prairie chickens remained in Illinois. By comparing the DNA of birds in the 1993 Illinois population with that of birds that lived in Illinois in the 1930s (obtained from museum specimens), Juan Bouzat and colleagues (1998) showed that the drop in population size had reduced the genetic variation of the population (**FIGURE 6.8**). In addition, more than 50% of the eggs laid by birds in the 1993 Illinois population failed to hatch, suggesting that genetic drift had possibly led to the fixation of harmful alleles. This hypothesis was supported by the results of experiments begun in 1992: when greater prairie chickens from other populations were brought to Illinois, new alleles entered the Illinois population, and egg-hatching rates increased from less than 50% to more than 90% in just 5 years (Westemeier et al. 1998). Unfortunately, since the late 1990s genetic diversity of the greater prairie chickens has declined back to the pre-introduction levels and is a cause of concern for the conservation of those species that fall within the former range (Mussmann et al. 2017). (Concept 11.3 covers the increased risk of extinction borne by small populations in greater detail.)

Gene flow is the transfer of alleles between populations

Gene flow occurs when alleles are transferred from one population to another via the movement of individuals or gametes (e.g., plant pollen). Gene flow has two important effects. First, by transferring alleles between populations, it tends to make populations more similar to one another genetically. This homogenizing effect of gene flow is one

(A)

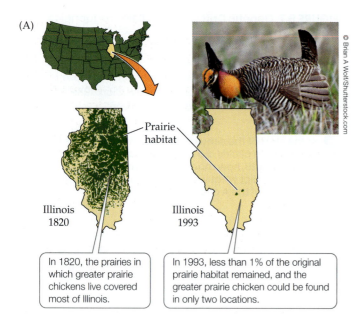

Prairie habitat

Illinois 1820

Illinois 1993

In 1820, the prairies in which greater prairie chickens live covered most of Illinois.

In 1993, less than 1% of the original prairie habitat remained, and the greater prairie chicken could be found in only two locations.

© Brian A Wolf/Shutterstock.com

(B)

By 1993, the greater prairie chicken population in Illinois had dropped to fewer than 50 birds and was experiencing the negative effects of genetic drift on small populations.

	Illinois		Kansas	Nebraska	Minnesota
	1933	1993			
Population size	25,000	<50	750,000	75,000–200,000	4,000
No. of alleles at six genes	31	22	35	35	32
Percentage of eggs that hatch	93	<50	99	96	85

FIGURE 6.8 Harmful Effects of Genetic Drift (A) As a result of habitat loss, the Illinois population of greater prairie chickens dropped from millions of birds in the 1800s to 25,000 in 1933 and, finally, to fewer than 50 birds in 1993. (B) As the Illinois population shrank in size, genetic drift led to a loss of alleles and to a rise in the frequencies of harmful alleles, thereby reducing egg-hatching rates. The table compares the 1993 Illinois populations with historical populations in Illinois and with populations in Kansas, Nebraska, and Minnesota, none of which experienced as severe a drop in population size. (After J. L. Bouzat et al. 1998. *Am Nat* 152: 1–6; R. C. Anderson. 1970. *Trans Illinois State Acad Sci* 63: 214. CC BY-NC-SA 4.0.)

reason why individuals in different populations of the same species resemble one another: alleles are exchanged often enough that relatively few differences accumulate between the populations.

Second, gene flow can introduce new alleles into a population. When this occurs, gene flow acts in a manner similar to mutation (although mutation remains the original source of new alleles). One example of gene flow involves consequences for human health. Before the 1960s, the mosquito *Culex pipiens* was not resistant to organophosphate insecticides. This mosquito transmits West Nile virus, the most common mosquito-borne disease in North America. Insecticides were often used to destroy its populations to prevent disease transmission. In the late 1960s, however, new alleles that provided resistance to organophosphate insecticides were produced by mutation in a few *C. pipiens* populations, probably in Africa or Asia (Raymond et al. 1998). Mosquitoes carrying these alleles were blown by storms or transported accidentally by humans to new locations, where they bred with mosquitoes from the local populations. In populations of mosquitoes exposed to insecticides, the frequency of these introduced alleles then increased rapidly because insecticide resistance was favored by natural selection (**FIGURE 6.9**). The global spread of these alleles by gene flow has allowed

billions of mosquitoes to survive the application of insecticides that otherwise would have killed them.

Evolutionary change that results in a closer match between the traits of organisms and the conditions of their environment, such as the increase in the frequency of insecticide resistance in a mosquito population exposed to insecticides, is an example of adaptive evolution, the topic we'll consider next.

FIGURE 6.9 Gene Flow: Introducing Alleles for Insecticide Resistance In this idealized scenario, an allele that causes resistance to organophosphate insecticides arises by mutation in one population of mosquitoes and then spreads by gene flow to two other populations. If mosquitoes in those two other populations are exposed to the insecticide, natural selection causes the frequency of the resistance allele to increase rapidly.

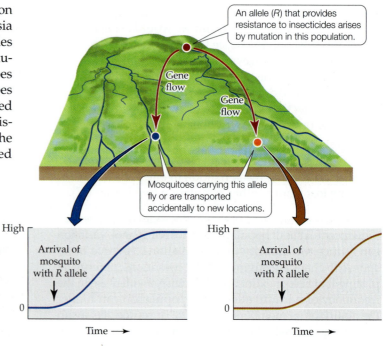

An allele (*R*) that provides resistance to insecticides arises by mutation in this population.

Gene flow

Gene flow

Mosquitoes carrying this allele fly or are transported accidentally to new locations.

Frequency of *R* allele in mosquito population exposed to insecticides

High

Arrival of mosquito with *R* allele

0

Time →

High

Arrival of mosquito with *R* allele

0

Time →

LEARNING OBJECTIVES

6.3.1 Explain how natural selection can lead to adaptations in populations.

6.3.2 Evaluate the conditions in which gene flow can promote or deter adaptations.

6.3.3 Describe factors that limit the development of adaptations in populations.

Adaptive Evolution

The natural world is filled with striking examples of organisms that are well suited for life in their environments. This match between organisms and their environments highlights their *adaptations*, which are features that evolve by natural selection and improve an organism's ability to survive and reproduce in its environment (see Concept 4.1). Examples of adaptations include remarkable features like those shown in **FIGURE 6.10** but also include less visually striking characteristics—such as an enzyme in a desert plant that can function at high temperatures that would denature most enzymes, enabling the plant to thrive in its environment. There are literally millions of other examples of adaptations. How do these adaptations arise?

Adaptations are the result of natural selection

Unlike genetic drift, natural selection is not a random process. Instead, when natural selection operates, individuals with certain alleles predictably have higher survival and produce more offspring than do individuals with other alleles. By consistently favoring individuals with some alleles over individuals with other alleles, natural selection causes **adaptive evolution**, a process of change in which traits that confer survival or reproductive advantages tend to increase in frequency over time. Although gene flow and genetic drift *can* improve the effectiveness of an adaptation (by increasing the frequency of an advantageous allele), they can also do the reverse (by increasing the frequency of a disadvantageous allele). Thus, natural selection is the only evolutionary mechanism that consistently results in adaptive evolution.

An example of adaptive evolution is provided by changes in populations of the soapberry bug (*Jadera haematoloma*) (Carroll and Boyd 1992; Carroll et al. 1997). This insect uses its needle-like beak to feed on seeds located within the fruits of several different plant species. Soapberry bug populations in southern Florida feed on the seeds of the insect's native host, the balloon vine (*Cardiospermum corindum*). Balloon vines, however, are rare in central Florida. Thus, in that region, soapberry bugs do not feed on balloon vines, but instead feed on the seeds of a species introduced from eastern Asia, the goldenrain tree (*Koelreuteria elegans*). A few specimens of the goldenrain tree were brought to Florida in 1926, but it was not commonly planted until the 1950s. The oldest goldenrain trees in the central Florida populations studied by Carroll and colleagues were 35 years old, suggesting that the soapberry bugs there have fed on this species for 35 years or less.

(A)

(B)

(C)

FIGURE 6.10 Some Striking Adaptations (A) The extensive skin extending from the neck to the limbs and to the toes and fingers of the Sundra flying lemur (*Galeopterus variegatus*) allows this animal to glide from tree to tree in the rainforest canopies of Southeast Asia. (B) The thorny devil (*Moloch horridus*) has adapted to withstand the dry scrubland and desert of central Australia. The animal's scales are ridged so that it can absorb water by simply touching it. (C) This archerfish (*Toxotes chatareus*) catches a spider by shooting a jet of water into the air. Field observations show that these fish will squirt repeatedly at potential prey and that they can reliably hit targets at heights of up to eight times their body length.

Soapberry bugs feed most efficiently when the length of a bug's beak matches the depth to which it must pierce a fruit to reach the seeds. Since goldenrain tree fruits are smaller than balloon vine fruits, the introduction of the goldenrain tree 35 years ago can be viewed as a natural experiment on the effect of selection on the insect's beak length. Carroll and Boyd predicted that as a result of natural selection, beak lengths would evolve to be *shorter* in soapberry bug populations that fed on goldenrain tree fruits than in populations that fed on the native host, balloon vines. Carroll and Boyd also studied soapberry bugs in Oklahoma and Louisiana, where the insect had begun to feed on several other new host plants that had been introduced within the past 100 years. However, in Oklahoma and Louisiana, the fruits of the introduced hosts were larger than those of the native hosts, leading to the prediction that in those two states, the beaks of insects that ate the introduced species would be *longer* than those of insects that ate the native species.

In all three locations, Carroll and Boyd found that soapberry bug beak lengths evolved in the direction predicted by fruit size, decreasing in central Florida (**FIGURE 6.11**) and increasing in both Oklahoma and Louisiana. The changes in beak length were substantial:

compared with historical values, average beak lengths dropped by 26% in central Florida and increased by 8% (on one introduced host species) and 17% (on another introduced host species) in Oklahoma and Louisiana. In addition, Carroll et al. (1997) showed that beak length is a heritable characteristic, so the observed changes in beak length must have been due at least in part to changes in the frequencies of alleles that affect beak length. Thus, we can conclude that in a relatively short time (35–100 years, or approximately 35–200 generations), natural selection in soapberry bug populations caused adaptive evolution in which a characteristic of the organism (beak length) evolved to match an aspect of its environment (fruit size) more closely.

Adaptive evolution can occur rapidly

Soapberry bugs are not unique: studies on populations of a wide range of other organisms show that natural selection can lead to rapid increases in the frequency of advantageous traits. Examples include the evolution of increased antibiotic resistance in bacteria (in days to months); increased insecticide resistance in insects (in months to years); drabber coloration in guppies, which makes them harder for visually hunting predators to find (several years); and increased beak size in medium ground finches (several years; see Figure 6.6A). A study of anole lizards in the Turks and Caicos archipelago found that hurricanes can induce strong selection pressure for morphological traits that enhance the ability of the lizards to cling to trees (**FIGURE 6.12**; Donihue et al. 2018). These and many other examples of apparently rapid evolution are described by Endler (1986), Thompson (1998), and Kinnison and Hendry (2001); collectively, these studies suggest that what we think of as "rapid" evolution may in fact be the norm, not the exception.

(A) Key Largo (southern Florida)

In southern Florida, where soapberry bugs continue to feed on balloon vine seeds, beak lengths remain close to the historical average.

Beak length
■ On balloon vine
□ On goldenrain tree

(B) Lake Wales (central Florida)

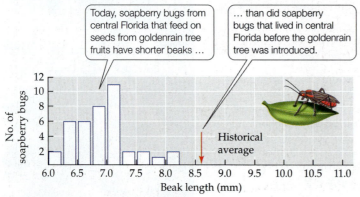

Today, soapberry bugs from central Florida that feed on seeds from goldenrain tree fruits have shorter beaks …

… than did soapberry bugs that lived in central Florida before the goldenrain tree was introduced.

FIGURE 6.11 Adaptive Evolution in Soapberry Bugs Soapberry bug populations in southern Florida feed on the seeds of their native host, the balloon vine (A), while soapberry bug populations in central Florida feed on the seeds of an introduced plant, the goldenrain tree (B). The beak lengths of insects feeding on the goldenrain tree decreased by 26% in 35 years, providing a better match to the smaller fruits of this introduced plant. Red arrows indicate beak length historical averages (obtained from museum specimens collected before the introduction of goldenrain trees). (After S. P. Carroll and C. Boyd. 1992. *Evolution* 46: 1052–1069.)

(A) (B)

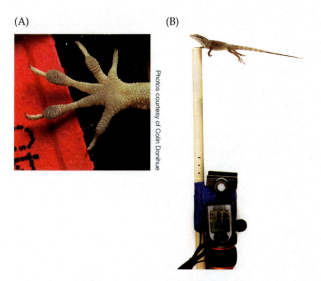

Photos courtesy of Colin Donihue

FIGURE 6.12 Rapid Adaptive Evolution in Anole Lizards
Hurricanes can be a very strong selective force for anole lizards found on small islands in the Caribbean Sea. Following two hurricanes in a 2-week period, researchers found that, compared to the lizards analyzed prior to the hurricane, the surviving lizards had wider footpads and shorter legs (A), which are two genetically based traits. These traits were experimentally shown to enhance the ability of the lizards to cling to dowels resembling branches under high winds (B).

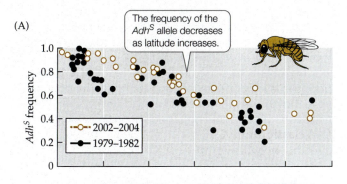

(A)

> The frequency of the Adh^S allele decreases as latitude increases.

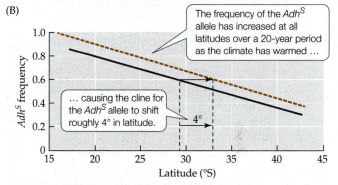

(B)

> The frequency of the Adh^S allele has increased at all latitudes over a 20-year period as the climate has warmed …

> … causing the cline for the Adh^S allele to shift roughly 4° in latitude.

FIGURE 6.13 Rapid Adaptive Evolution on a Continental Scale The *Adh* gene encodes a metabolically important enzyme, alcohol dehydrogenase, used to detoxify alcohol. Previous field and laboratory studies indicate that the Adh^S allele of this gene is selected against in cooler environments, such as those found at high latitudes. (A) Frequencies of the Adh^S allele in coastal Australian *Drosophila melanogaster* populations in 1979–1982 and in 2002–2004. (B) Regression lines calculated from the data in part A show that between 1979–1982 and 2002–2004, the cline of the Adh^S allele shifted 4° toward the South Pole as the region's average temperatures increased by 0.5°C. (After P. A. Umina et al. 2005. *Science* 308: 691–693.)

CLIMATE CHANGE CONNECTION

EVOLUTIONARY RESPONSES TO CLIMATE CHANGE
Rapid, apparently adaptive evolution also has been documented in response to climate change. Several such studies have focused on **clines**, patterns of change in a characteristic of an organism over a geographic region. For example, in the fruit fly *Drosophila melanogaster*, the alcohol dehydrogenase (*Adh*) gene exhibits a cline in which the Adh^S allele decreases in frequency as latitude increases (**FIGURE 6.13A**). This pattern has been found in both the Northern and the Southern Hemisphere. Previous studies indicated that this cline results from natural selection on the Adh^S allele, which codes for a form of the enzyme that is more effective in warmer temperatures at lower latitudes and hence is more common there.

Over a 20-year period in coastal Australia, the *Adh* cline shifted about 4° in latitude toward the South Pole (Umina et al. 2005), a movement of roughly 400 km (**FIGURE 6.13B**). During the same period, mean temperatures in the region increased by 0.5°C. Since the Adh^S allele is favored at higher temperatures, the 4° shift in latitude appears to have been a rapid, adaptive increase in the frequency of this allele in response to climate change. Rapid evolutionary changes that are correlated with global warming have also been observed in worldwide populations of another fruit fly species, *Drosophila subobscura* (Balanyá et al. 2006). Evolutionary responses

to climate change over short periods have also been documented in pitcher-plant mosquitoes (Bradshaw and Holzapfel 2001), red squirrels (Réale et al. 2003), tawny owls (Karell et al. 2011), tufted knotweed (Sultan et al. 2013), and the mustard plant *Brassica rapa* (Franks et al. 2007).

Finally, hundreds of species have altered the timing of key events in their lives in ways that may be a response to global warming, such as delaying the onset of winter dormancy or reproducing earlier in the spring (Parmesan 2006). In most of these cases, it is not yet known whether the observed changes are due to *phenotypic plasticity* (in which the same genotype can produce different phenotypes in different environments; see Concept 7.1), an evolutionary response (in which the genetic constitution of the population changes over time), or both. Research has begun to address this issue. For example, Jill Anderson and colleagues (2012) examined the contributions of phenotypic plasticity and evolution to changes in the flowering time of *Boechera stricta*, a mustard plant native to the U.S. Rocky Mountains. Data from a 38-year field survey of *B. stricta* populations show that the date at which flowers first came into bloom was about 13 days earlier in 2011 than it was in 1973. Both adaptive evolution (flowers opened earlier in populations that experienced warming) and phenotypic plasticity contributed to the earlier flowering times observed for this species. ☀

Gene flow can promote as well as limit local adaptation

Although many populations are strikingly well matched to their environments, others are not. Gene flow is one of the factors that can both promote and limit the extent to which a population is adapted to its local environment. For example, some plant species have tolerant genotypes that can grow on soils at former mine sites containing high concentrations of heavy metals; such soils are toxic to intolerant genotypes. On "normal" soils off of the mine wastes the tolerant genotypes grow poorly compared with the intolerant genotypes. Thus, we would expect the frequencies of tolerant genotypes to approach 100% on mine soils (where they are advantageous) and 0% on normal soils (where they are disadvantageous). Researchers found that a population of the bentgrass *Agrostis tenuis* growing on mine soils was dominated by tolerant genotypes, as expected. However, a population growing on normal soils downwind from the mine site contained more tolerant genotypes than expected (McNeilly 1968). Bentgrass is wind-pollinated, and each year, pollen from the plants growing on mine soils carried alleles for heavy metal tolerance into the population growing on normal soils, preventing that population from becoming fully dominated by intolerant genotype. The population growing on mine soils also received pollen from plants growing on normal soils. In this population, however, gene flow had relatively little effect on allele frequencies, because selection against intolerant genotypes was so strong (they survived poorly on mine soils). In general, whenever alleles are transferred between populations that live in different environments, the extent to which adaptive evolution occurs in each population depends on whether natural selection is strong enough to overcome the effects of ongoing gene flow.

Adaptations are not perfect

As we have just seen, gene flow can limit the extent to which a population adapts to its local environment. But even when gene flow does not have this effect, natural selection does not result in a perfect match between organisms and their environments. In part, this occurs because an organism's environment is not static—it is a moving target because the abiotic and biotic components of the environment change continually. In addition, organisms face a number of constraints on adaptive evolution:

- *Lack of genetic variation.* If none of the individuals in a population has a beneficial allele of a particular gene that influences survival and reproduction, adaptive evolution cannot occur at that gene. For example, the mosquito *C. pipiens* initially lacked alleles that provided resistance to organophosphate insecticides, as described earlier. For decades, this lack of genetic variation prevented adaptive evolution in response to insecticides, allowing humans to destroy mosquito populations at will—at least up until the time when insecticide resistance alleles arose by mutation and spread by gene flow. Note that in this and in all other cases, advantageous alleles arise randomly; they are not produced as needed or "on demand."

- *Evolutionary history.* Natural selection does not craft the adaptations of an organism from scratch. Instead, if the necessary genetic variation is present, it works by modifying the traits already present in an organism. Organisms have certain traits and lack others because of their ancestry. It would be advantageous, for example, for an aquatic mammal such as a dolphin to be able to obtain oxygen using gills. Dolphins lack this capacity, however, in part because of constraints imposed by their evolutionary history: they evolved from terrestrial vertebrates that had lungs and breathed air. Natural selection can bring about great changes, as seen in the mode of life and streamlined body form of the dolphin, but it does so by modifying traits that are already present in the organism, not by creating advantageous traits de novo.

- *Ecological trade-offs.* To survive and reproduce, organisms must perform many essential functions, such as acquiring food, escaping predators, warding off disease, and finding mates. Energy and resources are required for each of these essential functions. Hence, organisms face **trade-offs** in which the ability to perform one function reduces the ability to perform another (**FIGURE 6.14**). Trade-offs occur in all organisms, and as a result adaptations are the product of compromises in the abilities of organisms to perform many different and sometimes conflicting functions.

Despite these constraints, adaptive evolution is a key component of the evolutionary process. What does the importance of adaptive evolution tell us about the link between ecology and evolution? As we saw in the case of soapberry bug populations (see Figure 6.11), natural selection, and the adaptive evolution that results, is driven by the interactions of organisms with one another and with their environment. Any such interaction is an ecological interaction, and hence ecology serves as a basis for understanding natural selection. Next, we'll consider how ecological interactions influence broader evolutionary changes, such as the formation of new species and the great changes that have occurred during the history of life on Earth.

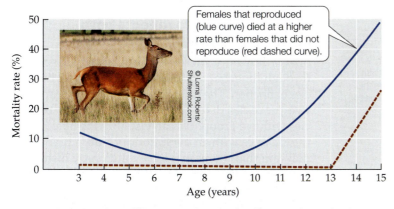

FIGURE 6.14 Trade-Off between Reproduction and Survival
Female red deer that reproduced had a lower probability of surviving to the next year than did females that did not reproduce, as the energy and resources invested into rearing young made reproducing red deer more susceptible to disease and environmental stress. (After T. H. Clutton-Brock et al. 1983. *J Anim Ecol* 52: 367–383.)

? Is the additional risk of mortality that results from reproduction the same for females of all ages? Explain.

CONCEPT 6.4 Long-term patterns of evolution are shaped by large-scale processes such as speciation, mass extinction, and adaptive radiation.

LEARNING OBJECTIVES

6.4.1 Describe the process by which isolation of populations can lead to speciation.

6.4.2 Evaluate the roles of speciation and extinction in determining the diversity of species.

6.4.3 Explain how mass extinctions and rapid adaptations have influenced long-term patterns in diversity.

The Evolutionary History of Life

Earth is home to roughly 1.5 million species[2] that have been named by taxonomists and it is estimated there are another 8.2 million that have yet to be discovered or named. This tremendous diversity serves as a foundation for all of ecology, which, as we saw in Concept 1.2, is the study of how species interact with one another and with their environment. But the causation runs both ways: while it is true that ecological interactions are affected by the diversity of species, it is also true that the diversity of species is shaped by ecological interactions. To see why, let's examine the origin of species and some of the other processes that have affected the history of life on Earth.

The genetic divergence of populations over time can lead to speciation

Each of the millions of species alive today originated by **speciation**, the process by which one species splits into

---

[2] A *species* can be defined as a group of organisms whose members have similar characteristics and can interbreed.

two or more species. Speciation most commonly occurs when a barrier prevents gene flow between two or more populations of a species. The barrier may be geographic, as when a new population becomes established far from the parental population, or when isolation is introduced by continental drift (see Concept 18.2). Barriers may also be ecological, as when some members of an insect population begin to feed on a different host plant. When a barrier to gene flow is established between populations, they diverge genetically over time (**FIGURE 6.15**).

New species can also form in several other ways, such as when members of two different species produce fertile hybrid offspring (see Figure 6.21 for an example in sunflowers). Whether it is produced by genetic divergence, hybridization, or other means, the key step in the creation of a new species is the formation of barriers that prevent its members from breeding freely with members of the parental species. Such reproductive barriers arise when a population accumulates so many genetic differences from the parental species that its members rarely produce viable, fertile offspring if they mate with members of the parental species.

The accumulation of genetic differences that lead to the formation of a new species can be an incidental by-product of selection. For example, an experiment with fruit flies demonstrated the beginnings of reproductive barriers between populations selected for growth on different sources of food, but no such barriers were observed between control populations that had not been subjected to selection (**FIGURE 6.16**). Natural selection has produced similar changes in plant populations growing on soils with differing concentrations of heavy metals (Macnair and Christie 1983), in bird populations living in environments with different food resources (Smith and Benkman 2007), and in fish populations exposed to low or high

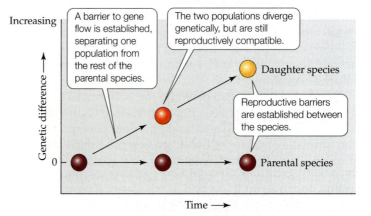

FIGURE 6.15 Speciation by Genetic Divergence Once genetic divergence begins, the time required for speciation varies tremendously, from a single generation (perhaps a single year), to a few thousand years, to millions of years in most cases.

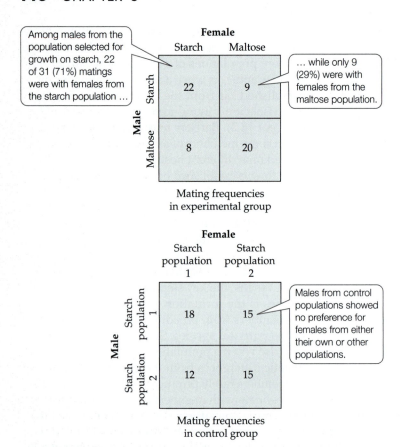

FIGURE 6.16 Reproductive Barriers Can Be a By-Product of Selection After 1 year (about 40 generations) in which experimental populations of *Drosophila pseudoobscura* fruit flies were selected for growth on different sources of food, most matings occurred between flies selected to feed on the same food source. No such mating preference was observed in control populations that were not subjected to selection, regardless of whether the control populations were reared on starch (shown here) or maltose (not shown). To reduce the chance that the food eaten by the larvae would produce a body odor in adults that influenced the results, all flies used in the mating preference tests were reared for one generation on a standard cornmeal medium. (After D. M. B. Dodd. 1989. *Evolution* 43: 1308–1311.)

levels of predation (Langerhans et al. 2007). In each of these cases, reproductive barriers arose as a by-product of selection in response to a feature of the environment, such as food source, heavy metal concentration, temperature, or presence of predators.

Genetic drift can also promote the accumulation of genetic differences between populations (see Figures 6.7 and 6.8). As a result, like natural selection, genetic drift can ultimately lead to the evolution of reproductive barriers and hence to the formation of new species. Gene flow, on the other hand, typically acts to slow down or prevent speciation, because populations that exchange many alleles tend to remain genetically similar to one another, making it less likely that reproductive barriers will evolve.

The diversity of life reflects both speciation and extinction rates

Speciation may increase the number of species through time, but this increase may be offset by species extinction. The balance between the rates of these two processes determines the net diversity of species within a group. We can visualize the outcome of this balance with an **evolutionary tree**, a branching diagram that represents the evolutionary history of a group of organisms. **FIGURE 6.17A** shows an evolutionary tree for the pinnipeds, a group of aquatic mammals consisting of seals, sea lions, and the walrus. The pinniped common ancestor lived about 20 million years ago, and its descendants include the 34 species of living pinnipeds along with a diversity of extinct species. The walrus group, for example, contains only a single species today—the walrus—but it once contained *Gomphotoria pugnax* and as many as 18 other species, all of which are now extinct.

Extinction can also help us to understand the large morphological differences that occur between some closely related groups of organisms. Seals and other pinnipeds, for example, differ greatly from their closest living relatives, members of the weasel family (the mustelids). However, fossils of *Puijila darwini* (Rybczynski et al. 2009), an extinct close relative of the pinnipeds, show that extinct relatives of pinnipeds were similar morphologically to some living mustelids, such as otters (**FIGURE 6.17B**). Over time, repeated speciation events led to the origin of fully aquatic pinnipeds—but because *P. darwini* and other such species have become extinct, there are no living species that "fill the gap" between living pinnipeds and living mustelids.

Speciation and extinction events also have affected the rise and fall of different groups of organisms over long periods, as we'll see in the next section.

Mass extinctions and adaptive radiations have shaped long-term patterns of evolution

Thus far in this chapter, much of our focus has been on the *process* of evolution—the mechanisms by which evolutionary change occurs. But evolution can also be defined as an observed *pattern* of change. Evolutionary patterns are revealed by observations of the natural world, such as data on the changing allele frequencies of a population over time. Patterns of evolutionary change are also documented in the fossil record, which shows that life on Earth has changed greatly over long periods (**FIGURE 6.18**).

The earliest known fossils are those of 3.5-billion-year-old bacteria, while the most ancient fossils of complex multicellular organisms are of red algae that lived 1.2 billion years ago. Animals first appear in the fossil record about 600 million years ago, and complex animals with

(A)

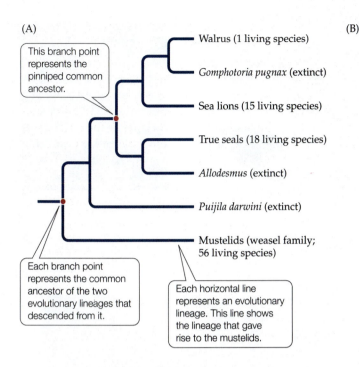

This branch point represents the pinniped common ancestor.

Walrus (1 living species)

Gomphotoria pugnax (extinct)

Sea lions (15 living species)

True seals (18 living species)

Allodesmus (extinct)

Puijila darwini (extinct)

Mustelids (weasel family; 56 living species)

Each branch point represents the common ancestor of the two evolutionary lineages that descended from it.

Each horizontal line represents an evolutionary lineage. This line shows the lineage that gave rise to the mustelids.

(B)

10 cm

Both reconstructions © Canadian Museum of Nature

FIGURE 6.17 An Evolutionary Tree of the Pinnipeds (A) This branching tree is a representation of the evolutionary history of modern seals and their close relatives that is based on recent fossil finds. This research indicates that the marine mammals known as pinnipeds probably share a common ancestor with modern weasels and their relatives. (B) Reconstructions of *Puijila darwini* based on fossils show that extinct close relatives of pinnipeds were similar morphologically to some living mustelids, such as otters. *P. darwini* appears to have foraged both on land (above) and in the water (below). (A after N. Rybczynski et al. 2009. *Nature* 458: 1021–1024.)

bilateral symmetry (in which the body has two equal but opposite halves, as in most living animals) arose roughly 25 million years later (Fedonkin et al. 2007; Chen et al. 2009). These and many other great changes in the history of life resulted from descent with modification as new species arose that differed from their ancestors. Over millions of years, these differences gradually accumulated, leading eventually to the formation of major new groups of organisms, such as terrestrial plants, amphibians, and reptiles.

For example, a rich variety of fossils have been discovered that illustrate steps in the origin of *tetrapods* (vertebrates with four limbs, a group whose living members include amphibians, reptiles, and mammals) from fishes; the fossil of one such species is shown in Figure 6.18E. Similarly, the fossil record contains dozens of fossil species that show how mammals arose over a 120-million-year period (300–180 million years ago) from an earlier group of tetrapods, the synapsids (Allin and Hopson 1992; Sidor 2003). The fossil record also documents cases in which the rise to prominence of one group of organisms was associated with the decline of another group. For example, 265 million years ago, reptiles and dinosaurs

replaced amphibians as the ecologically dominant group of tetrapods, and then, 66 million years ago, the dinosaurs were replaced in turn by the mammals.

The rise and fall of different groups of organisms over time has been heavily influenced by mass extinctions and adaptive radiations. The fossil record documents five **mass extinction** events in which large proportions of Earth's species were driven to extinction in a relatively short time—a few million years or less, sometimes much less (**FIGURE 6.19**). The most recent mass extinction occurred 66 million years ago, likely caused by a large asteroid that struck Earth, setting in motion cataclysmic environmental changes that led to the demise of dinosaurs and many other groups of organisms.

Each of the five mass extinctions was followed by great increases in the diversity of some of the surviving groups of organisms; for example, mammal diversity increased greatly after the extinction of dinosaurs. Mass extinctions can promote such increases in diversity by removing competitor or predator groups, thus allowing the survivors to give rise to new species that expand into new habitats or new ways of life. Great increases in diversity can also occur when a group of organisms evolves major new

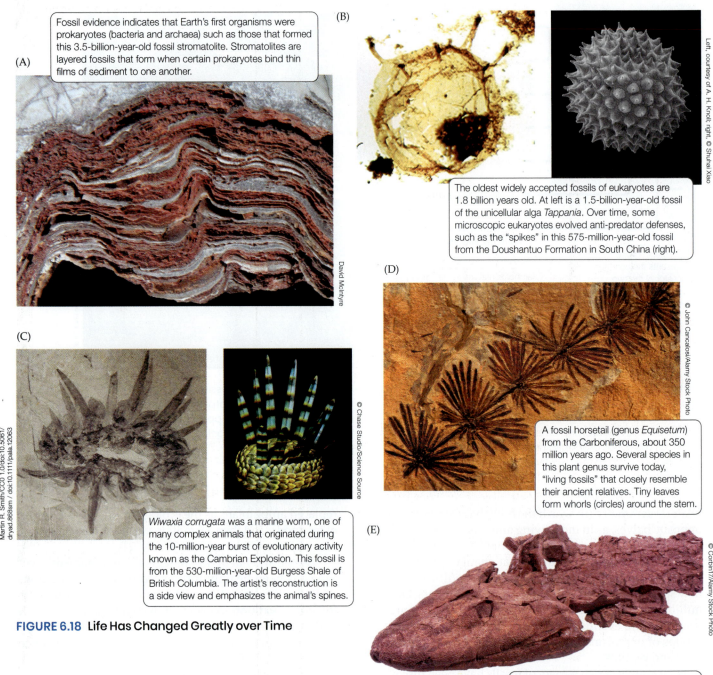

(A)

(B)

Fossil evidence indicates that Earth's first organisms were prokaryotes (bacteria and archaea) such as those that formed this 3.5-billion-year-old fossil stromatolite. Stromatolites are layered fossils that form when certain prokaryotes bind thin films of sediment to one another.

David McIntyre

Left, courtesy of A. H. Knoll; right, © Shuhai Xiao

The oldest widely accepted fossils of eukaryotes are 1.8 billion years old. At left is a 1.5-billion-year-old fossil of the unicellular alga *Tappania*. Over time, some microscopic eukaryotes evolved anti-predator defenses, such as the "spikes" in this 575-million-year-old fossil from the Doushantuo Formation in South China (right).

(C)

(D)

Martin R. Smith/CC0 1.0/doi:10.5061/dryad.868sm / doi:10.1111/pala.12063

© Chase Studio/Science Source

© John Cancalosi/Alamy Stock Photo

Wiwaxia corrugata was a marine worm, one of many complex animals that originated during the 10-million-year burst of evolutionary activity known as the Cambrian Explosion. This fossil is from the 530-million-year-old Burgess Shale of British Columbia. The artist's reconstruction is a side view and emphasizes the animal's spines.

A fossil horsetail (genus *Equisetum*) from the Carboniferous, about 350 million years ago. Several species in this plant genus survive today, "living fossils" that closely resemble their ancient relatives. Tiny leaves form whorls (circles) around the stem.

(E)

© Corbin17/Alamy Stock Photo

FIGURE 6.18 Life Has Changed Greatly over Time

In 2006, researchers reported the discovery of this 380-million-year-old fossil of *Tiktaalik roseae*, one of dozens of fossil species that document the origin of tetrapods from lobe-finned fishes. The tetrapods are four-legged terrestrial vertebrates whose living members include amphibians, reptiles, and mammals.

adaptations, such as the stems, waxy cuticles, and stomates on leaves that provided early land plants with support against gravity and protection from desiccation (see Concept 4.3). Whether stimulated by a mass extinction, new adaptations, or other factors (such as migration to an island that lacks competitors), an event in which a group of organisms gives rise to many new species that expand into new habitats or new ecological roles in a relatively short time is referred to as an **adaptive radiation**.

Fossil evidence also suggests that many of the great changes in the history of life were caused by ecological interactions. For example, the fossil record shows that for over 60 million years, early animals were small or soft-bodied, or both, and that all of the larger species

were herbivores, filter feeders, or scavengers. However, beginning 535 million years ago, this safe, soft-bodied world disappeared forever with the appearance of large, well-armed, mobile predators and large, well-defended prey. This major step in the history of life appears to have resulted from an "arms race" between predators and prey. Early predators equipped with claws and other

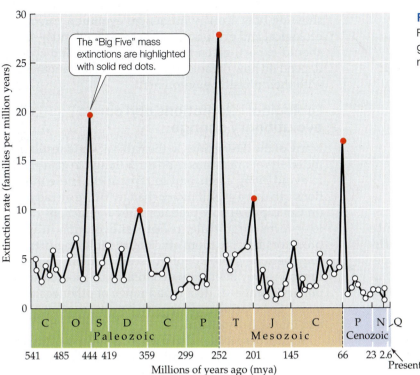

FIGURE 6.19 The "Big Five" Mass Extinctions
Five peaks in extinction rates are revealed by a graph of extinction rates over time in families of marine invertebrates.

adaptations for capturing large prey provided powerful selection pressure that favored heavily armored prey species. That armor, in turn, promoted further increases in the effectiveness of the predators, and so on. Such reciprocal evolutionary change in interacting species, known as coevolution, is discussed in more detail in Concept 13.3.

Ecological interactions have shaped the history of life in many other ways. For example, the origin of new species in one group of organisms can lead to increases in the diversity of other groups, especially those that can escape from, eat, or compete effectively with the new species (Farrell 1998; Benton and Emerson 2007). An example of this process can be seen in parasitic wasps that feed on the apple maggot fly (*Rhagoletis pomonella*), a species that eats fruits (**FIGURE 6.20**). Following the introduction of apple trees to North America 200 years ago, some *Rhagoletis* populations began to eat apples. As these populations adapted to their new food plant, they diverged from the parent species genetically and now appear to be well on the way to forming a new fly species (Feder 1998).

In addition, populations of the wasp have emerged that specialize on the incipient fly species (Forbes et al. 2009). These wasps have become reproductively isolated from the parent wasp species, thereby providing evidence of a sequence of speciation events that is in progress today and appears to be driven by ecological interactions.

We turn next to a more detailed look at an idea that we have already encountered in this chapter: while ecological interactions influence evolution, evolution also influences ecological interactions.

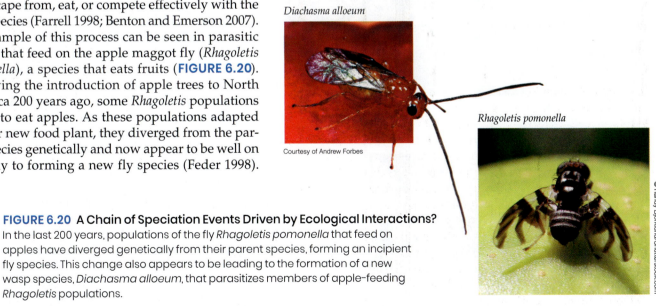

Diachasma alloeum

Courtesy of Andrew Forbes

Rhagoletis pomonella

© Randy Bjorklund/Shutterstock.com

FIGURE 6.20 A Chain of Speciation Events Driven by Ecological Interactions?
In the last 200 years, populations of the fly *Rhagoletis pomonella* that feed on apples have diverged genetically from their parent species, forming an incipient fly species. This change also appears to be leading to the formation of a new wasp species, *Diachasma alloeum*, that parasitizes members of apple-feeding *Rhagoletis* populations.

CONCEPT 6.5 Ecological interactions and evolution exert a profound influence on one another.

LEARNING OBJECTIVES

6.5.1 Evaluate how ecological processes can result in evolutionary changes in populations.

6.5.2 Describe how an evolutionary change in a population has the potential to impact ecological processes.

Joint Effects of Ecology and Evolution

Ecological and evolutionary processes can influence each other greatly. Consider the sunflower species *Helianthus anomalus*. This species originated from a speciation event in which two other sunflowers, *H. annuus* and *H. petiolaris*, produced hybrid offspring. As Loren Rieseberg and colleagues have shown in a series of experiments and genetic analyses (Rieseberg et al. 2003), the new gene combinations generated by hybridization appear to have facilitated a major ecological shift in *H. anomalus*. This hybrid species grows in a much drier environment than does either of its two parental species (**FIGURE 6.21**)—an ecological shift that illustrates how evolution influences ecology. Simultaneously, however, life under different ecological conditions provided the selection pressures that molded the hybrid offspring of *H. annuus* and *H. petiolaris* into a new species, *H. anomalus*,

showing how ecology influences evolution. Such joint ecological and evolutionary effects are common—as we should expect, given that both evolution and ecology depend on how organisms interact with one another and with their physical environment.

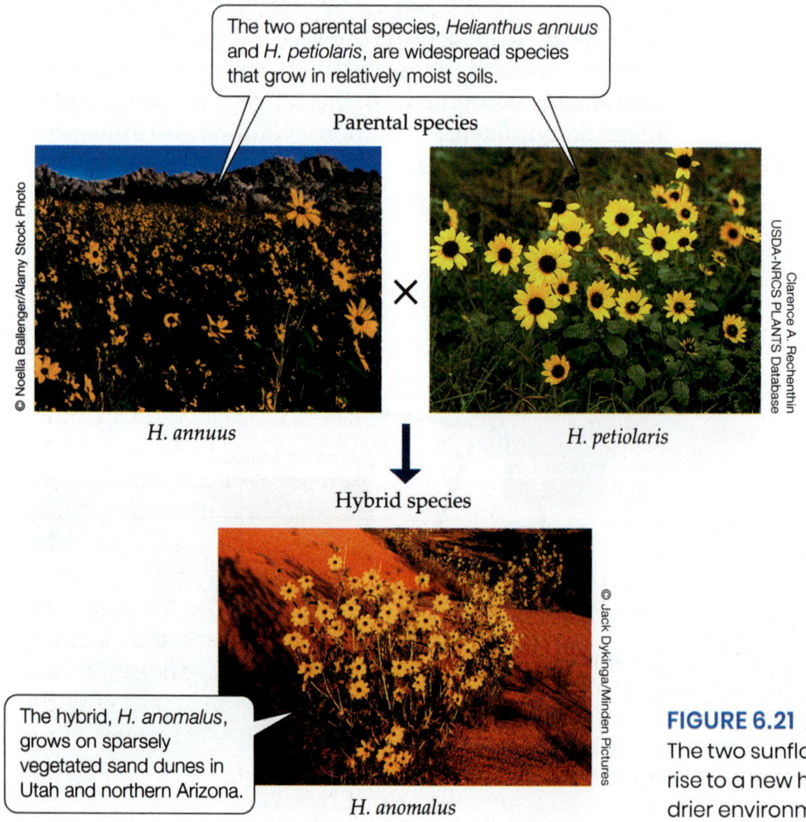

Ecological interactions can cause evolutionary change

Many of the interactions in the natural world result from the efforts of organisms to do three things: to gain energy, to avoid being eaten, and to reproduce. These interactions can drive evolutionary change. We've already seen (in Concept 6.4) how predator–prey interactions caused long-term, large-scale, reciprocal evolution in which predators became more efficient at capturing prey and prey became more adept at escaping their predators. Predator–prey interactions continuously promote evolutionary change, as do a broad range of other ecological interactions, including herbivory, parasitism, competition, and mutualism (see Unit 4).

Studies of speciation have led to a similar conclusion: it may be common for speciation to be caused by ecological factors (Schluter 1998; Funk et al. 2006). The effect of ecology on evolution is also clear from studies of relatively small evolutionary changes in populations. Examples discussed earlier in this chapter include directional selection on soapberry bugs caused by interactions with their food plants (see Figure 6.11) and genetic drift in greater prairie chickens caused by habitat loss (see Figure 6.8).

Evolution can alter ecological interactions

Whenever a group of organisms evolves a new, highly effective adaptation, the outcome of ecological interactions may change, and that change may have a ripple effect that alters the entire community. For example, if a predator evolves a new way of capturing prey, some prey species may be driven to extinction, while others may decrease in abundance, migrate to other areas, or evolve new ways to cope with the more efficient predator. Similar changes can occur among species that compete for resources.

Evolutionary changes that occur over long time scales also affect ecological interactions. For example, the origin and subsequent evolutionary diversification of plants altered the composition and stability of soils, the sources of food available to other organisms, and the cycling of nutrients—each of which had major effects on ecological

The two parental species, *Helianthus annuus* and *H. petiolaris*, are widespread species that grow in relatively moist soils.

Parental species

H. annuus × *H. petiolaris*

Hybrid species

The hybrid, *H. anomalus*, grows on sparsely vegetated sand dunes in Utah and northern Arizona.

H. anomalus

FIGURE 6.21 A Hybrid That Lives in a New Environment
The two sunflower species *Helianthus annuus* and *H. petiolaris* gave rise to a new hybrid species, *H. anomalus*. This species grows in a drier environment than either of the two parental species.

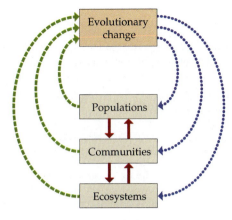

FIGURE 6.22 Rapid Feedback Effects Can Occur between Ecological and Evolutionary Factors Ecological change in a population, community, or ecosystem can drive evolutionary change over short periods of time (green, dashed arrows). Similarly, evolutionary change can alter events at the population, community, or ecosystem level (blue, dotted arrows). A change at one level of ecological organization can cause additional changes at other levels (red, solid arrows), as when an increase in the population size of one species alters nutrient cycling in ecosystems.

interactions. By affecting soils, for example, early plants literally helped to build the habitats in which later communities of microorganisms, plants, and animals would eventually live and interact with one another.

Eco-evolutionary feedbacks can occur over short periods of time

As we discussed earlier in this chapter, evolution can occur over short periods of time (as little as two generations). Because evolution occurs as organisms interact with each other and their physical environment, this suggests that reciprocal feedback effects between ecological and evolutionary factors also can occur over short periods of time. Let's take a closer look at the causes of these rapid feedback effects.

Feedback effects between ecological and evolutionary factors can occur when an ecological change, such as the addition or removal of a predator, alters the selective pressures that organisms face, thereby leading to evolutionary changes (**FIGURE 6.22**). Such evolutionary changes, in turn, can modify key aspects of populations, communities, or ecosystems. For example, in a 3-year field experiment (Agrawal et al. 2013), evolutionary changes in life span and flowering time in populations of the evening primrose (*Oenothera biennis*) led to consistent changes in the abundance of the moth *Mompha brevivittella*, which ate the seeds of this plant (**FIGURE 6.23**)—a demonstration that rapid evolution can cause rapid ecological change in a natural setting. Likewise, in the mountain streams of Trinidad, predator removal (an ecological change) leads to the evolution of larger body size in guppies over short periods of time, an evolutionary change that may increase the rate at which guppy populations add nitrogen to this freshwater ecosystem (El-Sabaawi et al. 2015).

Oenothera biennis

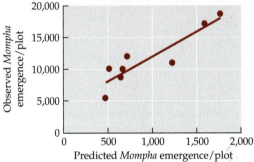

FIGURE 6.23 Feedback of Food Plant Evolution on Insect Abundance Caterpillars of the moth *Mompha brevivittella* eat the seeds of the evening primrose (*Oenothera biennis*). Some plant genotypes are more resistant to moth attack than others, indicating that moth abundance could change depending on plant genotype frequencies. In a 3-year field experiment, evolutionary changes in *O. biennis* genotype frequencies were correlated to moth abundance, indicating a feedback from evolution to ecology. (After A. A. Agrawal et al. 2013. *Am Nat* 181: S35–S45.)

 Suppose that eco-evolutionary feedbacks between changes in plant genotype frequency and moth abundance did not occur. Redraw this figure assuming that was the case.

A CASE STUDY REVISITED

Trophy Hunting and Inadvertent Evolution

Trophy hunters of bighorn sheep prefer to kill large males that carry a full curl of horns. The majority of these males are killed when they are between 4 and 6 years old, often before they have sired many offspring. As a result, hunting decreases the chance that alleles carried by males with a full curl of horns will be passed on to the next generation. Instead, it is males with relatively small horns who father most of the offspring, transmitting their alleles to the next generation. This change has caused the frequency of alleles encoding small horns to increase, thus leading to the observed 30-year decrease in average horn size (see Figure 6.2). Overall, trophy hunting has inadvertently caused directional selection in bighorn sheep, favoring small males with small horns and changing allele frequencies in the population over time.

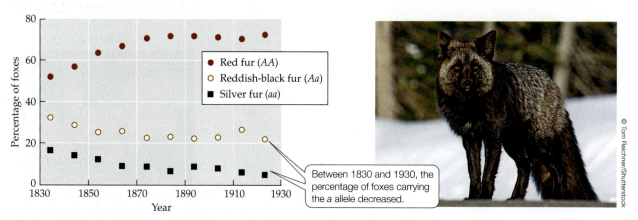

Between 1830 and 1930, the percentage of foxes carrying the *a* allele decreased.

FIGURE 6.24 **Hunting Resulted in the Decline of Silver Foxes** Individual red foxes (*Vulpes fulva*) of genotype *AA* have red fur, and individuals of genotype *Aa* have reddish-black fur. Individuals of genotype *aa* are known as "silver foxes" because the tips of their hairs have a silver tint (photo). Hunters preferentially killed silver foxes because their furs yielded 2.5–4 times the price of other red fox furs. (After C. S. Elton. 1942. *Voles, Mice and Lemmings: Problems in Population Dynamics*. Oxford University Press: Oxford.)

? Based on the graph, estimate the initial (ca. 1832) and final (ca. 1923) frequencies of genotypes AA, Aa, and aa. Next, use the genotype frequencies that you estimated to compute the initial and final frequencies of the a allele. Hint: See footnote in Concept 6.1.

Humans have caused unintended evolutionary changes in a wide variety of other populations. An early example was provided by the decline in the frequency of red foxes (*Vulpes fulva*) with coats that have a silver tint, a color preferred by hunters (**FIGURE 6.24**). In a medical example, shortly after antibiotics were discovered (ca. 1940), their use was highly effective against bacteria that cause diseases and lethal infections. But the use of antibiotics provided a strong source of directional selection, leading to the evolution of antibiotic resistance in bacterial populations (see Figure 1.10). Today, as a result of this directional selection, antibiotic treatments sometimes fail, even when very high doses are administered. Antibiotic resistance has enormous consequences for human health as well as financial costs; in the United States alone, efforts to cure patients infected with antibiotic-resistant strains costs an estimated $2 billion in medical expenses *each year* (Thorpe et al. 2018).

We have seen throughout this chapter that human actions such as trophy hunting and antibiotic use act as selection pressures and hence may cause evolutionary change. But does our influence on evolution extend beyond cases in which we selectively kill other organisms?

 CONNECTIONS *in* NATURE

THE HUMAN IMPACT ON EVOLUTION Many human actions alter the environment and hence have the potential to alter the course of evolution. As we've seen, actions such as trophy hunting, antibiotic use, and commercial fishing are themselves powerful sources of selection. Other human actions, such as emissions of pollutants or introductions of invasive species, change aspects of the abiotic or biotic environment. By changing features of the environment, these and many other human actions can cause evolutionary change. In **ANALYZING DATA 6.1**, you'll analyze data related to a classic example of this process, in which the emission (and subsequent control) of pollutants caused evolution by natural selection in populations of the peppered moth (*Biston betularia*).

Still other human actions, such as *habitat fragmentation* (in which portions of a species' habitat are destroyed, leaving spatially isolated fragments of the original habitat), can also cause large evolutionary changes (**FIGURE 6.25**). In general, human actions that affect the environment can alter each of the three main mechanisms of evolution: natural selection, genetic drift, and gene flow. Because we know with certainty that our actions are causing great changes to environments worldwide, we can infer that they are also causing evolutionary changes in populations worldwide.

As another example of human-caused evolution, consider the effects of adding nutrients such as nitrogen from sewage and fertilizers to lakes. Such nutrient inputs can cause the clarity and oxygen concentration of the water to drop (see Concept 22.4), leading to unintended evolutionary effects. For example, nutrient inputs to European lakes have reduced the effectiveness of reproductive barriers that once isolated species of whitefish (Vonlanthen et al. 2012). Murky (low-clarity) waters can hinder the ability of females to recognize males of their own species, thus making it more likely that a female will select a male from another whitefish species as her mate. When interspecific mating is common, a "speciation reversal" can occur in which two previously isolated species fuse into a single, hybrid species. Vonlanthen et al. concluded that nutrient inputs have caused such speciation reversals, leading to the extinction of eight whitefish species. As we'll see in later chapters of this book, such reductions in the diversity of species can have wide-ranging ecological effects.

Human actions also have the potential to alter patterns of evolution over long time scales. For example, the extinction rate of species today is 100 to 1,000 times higher than the usual, or background, extinction rate seen in the

ANALYZING DATA 6.1

Does Predation by Birds Cause Evolution in Moth Populations?

The peppered moth (*Biston betularia*) has a light-colored and a dark-colored form. The first dark-colored moth was observed in 1848 near Manchester, England; 50 years later, most moths in the area were dark in color. Researchers hypothesized that dark-colored moths increased in frequency because when the moths rested on trees whose bark had been darkened by pollution, it was harder for predators to find dark moths than light moths. In particular, field studies by Kettlewell (1955, 1956) indicated that natural selection by birds favored dark-colored moths in regions where tree bark was blackened by pollution, whereas light-colored moths were favored elsewhere.

After clean air legislation was enacted in England in 1956, tree surfaces lightened over time because of the reduction in soot and the growth of lichens on the trees' bark (lichens are light in color, and they grow poorly in polluted air). During this period, the dark-colored moths decreased in frequency.

Although the rise and fall in the frequency of dark-colored moths were consistent with typical results from natural selection by bird predation, criticisms have been leveled against aspects of this hypothesis. For example, abnormally high densities of moths were released in some experiments, potentially increasing the impact of predation, because some predators preferentially attack abundant prey. Over the course of a 6-year experiment designed to address such criticisms, Michael Majerus released thousands of moths in an area where tree surfaces had been lightened. He determined the number of light- and dark-colored moths that were eaten. His results are reported in the table.

1. The densities (and proportions) of the light- and dark-colored moths that Majerus released were similar to those he observed in the field. Why is this important to the validity of the experiment?

2. Use the proportions of dark moths that Majerus released to determine whether dark-colored moths were increasing or decreasing in frequency in the area where he conducted the experiment (Cambridge, England).

3. Calculate the percentages of released dark- and light-colored moths that were eaten each year, and graph those percentages versus time. Do the results support the hypothesis that evolution by natural selection caused the frequency of dark-colored moths to change over time? Explain.

Year	No. of light moths released	No. of dark moths released	No. of light moths eaten	No. of dark moths eaten
2002	706	101	162	31
2003	731	82	204	24
2004	751	53	128	17
2005	763	58	166	18
2006	774	34	145	6
2007	797	14	158	4

Source: Cook, L. M., et al. 2012. Selective bird predation on the peppered moth: The last experiment of Michael Majerus. *Biology Letters* 8: 609–612.

fossil record for times when no mass extinction was taking place. Human actions such as habitat destruction, overharvesting, and introductions of invasive species are among the main reasons for this rise in the extinction rate (see Concepts 23.3 and 24.2). Climate change is likely to become a leading cause for extinction in the next several decades. Extinction is forever, so when human actions drive a species to extinction, the future course of evolution is altered in a way that cannot be reversed. If human activities cause a sixth mass extinction in the next few centuries or millennia, our actions will greatly and irreversibly change the evolutionary history of life on Earth.

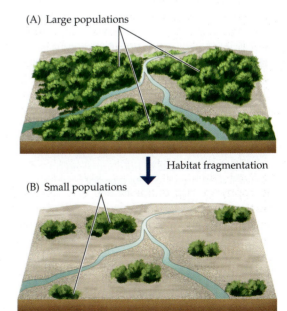

(A) Large populations

Habitat fragmentation

(B) Small populations

FIGURE 6.25 Evolutionary Effects of Habitat Fragmentation on a Hypothetical Species (A) Prior to habitat fragmentation, there are many individuals in each population of the species, and the distances between populations are short. (B) When human activities remove large portions of the habitat, the population sizes shrink, and the distances between populations increase, causing evolutionary changes that decrease the potential for adaptive evolution of the species and increase its risk of extinction.

SUMMARY

CONCEPT 6.1 Evolution can be viewed as genetic change over time or as a process of descent with modification.

6.1.1 Summarize the genetic basis for the evolution of traits in organisms.

- Biologists often define evolution in a relatively narrow sense as change over time in the frequencies of alleles in a population.

6.1.2 Explain how evolution can be considered the accumulation of trait differences between populations.

- Evolution can also be viewed as descent with modification, a process in which populations accumulate differences over time and hence differ from their ancestors.

- Natural selection modifies populations by favoring individuals with some heritable traits over other individuals.

- Although natural selection acts on individuals, an individual does not evolve—either it has a favored trait or it does not. Only populations evolve.

CONCEPT 6.2 Natural selection, genetic drift, and gene flow can cause allele frequencies in a population to change over time.

6.2.1 Summarize how mutations contribute to the process of evolution.

- Mutation and recombination are the sources of new alleles and new combinations of alleles, thereby providing the genetic variation on which evolution depends.

6.2.2 Compare the effects of stabilizing selection and disruptive selection on the temporal changes in the genetic structure of a population.

- Natural selection occurs when individuals with certain heritable phenotypic traits survive and reproduce more successfully than individuals with other traits.

- Natural selection may favor the average traits (stabilizing selection) or traits near the extremes in a population (disruptive selection).

6.2.3 Evaluate how random events can affect populations through time via genetic drift.

- Genetic drift, which occurs when random events determine which alleles are passed from one generation to the next, can have negative effects on small populations.

6.2.4 Describe the role of gene flow among populations in terms of homogenizing genetic structure as well as enhancing evolutionary change.

- Gene flow, the transfer of alleles between populations, makes populations more genetically similar to one another and can introduce new alleles into populations, enhancing evolutionary change.

CONCEPT 6.3 Natural selection is the mechanism for adaptive evolution.

6.3.1 Explain how natural selection can lead to adaptations in populations.

- By favoring individuals that have advantageous alleles over individuals that have other alleles, natural selection can cause adaptive evolution, in which the frequency of an advantageous trait in a population increases over time.

- Natural selection can increase the frequency of advantageous traits rapidly—in days to years, depending on the organism and the selection pressure.

6.3.2 Evaluate the conditions in which gene flow can promote or deter adaptations.

- Gene flow can promote the extent to which a population is adapted to its local environment by introducing new alleles into a population.

- Gene flow can also limit the extent to which a population is adapted to its local environment by preventing fixation of the favored allele.

6.3.3 Describe factors that limit the development of adaptations in populations.

- Constraints on adaptive evolution result from factors such as lack of genetic variation, evolutionary history, and ecological trade-offs.

CONCEPT 6.4 Long-term patterns of evolution are shaped by large-scale processes such as speciation, mass extinction, and adaptive radiation.

6.4.1 Describe the process by which isolation of populations can lead to speciation.

- The genetic divergence of populations over time can lead to speciation, the process by which one species splits into two or more species. Speciation requires the evolution of reproductive barriers between populations.

6.4.2 Evaluate the roles of speciation and extinction in determining the diversity of species.

- The number of species in a group of organisms increases when more species are produced by speciation than are lost to extinction, and decreases when the reverse is true. The outcome of this process can be visualized with an evolutionary tree.

6.4.3 Explain how mass extinctions and rapid adaptations have influenced long-term patterns in diversity.

- Biological communities can lose much of their diversity in mass extinctions, global events in which large proportions of Earth's species are driven to extinction in a relatively short time.

- An adaptive radiation occurs when a group of organisms gives rise to many new species over a short period of time that expand into new habitat or fill new ecological roles.

- Adaptive radiations can be promoted by factors such as the removal of competitor groups by a mass extinction or the evolution of a major new adaptation.

CONCEPT 6.5 Ecological interactions and evolution exert a profound influence on one another.

6.5.1 Evaluate how ecological processes can result in evolutionary changes in populations.

- Ecological interactions among organisms and between organisms and their environment can cause evolutionary changes, ranging from allele frequency changes in populations to the formation of new species.

6.5.2 Describe how an evolutionary change in a population has the potential to impact ecological processes.

- Evolutionary change can alter the outcomes of ecological interactions, thus having a large influence on biological communities.

Review Questions

1. Natural selection acts on individuals, yet one of the points made in this chapter is that *populations evolve, but individuals do not.* Explain how natural selection works and why the italicized statement is true.

2. What causes adaptive evolution? Explain in your answer why each of the three primary mechanisms of allele frequency change in populations causes or does not cause adaptive evolution.

3. What large-scale processes determine patterns of evolution observed over long time scales? Explain how each process that you describe has this effect.

4. Explain why ecological interactions and evolutionary change have joint effects, each affecting the other.

5. More than 100 years ago, Rutter (1902) expressed concern about the effects of fishing on river salmon. He wrote (p. 134), "A large fish is worth more on the market than a small fish; but so are large cattle worth more on the market than small cattle, yet a stock raiser would never think of selling his fine cattle and keeping only the runts to breed." From an evolutionary perspective, summarize the reasons for Rutter's concern, and describe how harvesting-induced evolution is thought to affect fish populations today.

Hone Your Problem-Solving Skills

Stuart and colleagues (2014) studied how the invasion of islands in Florida by the Cuban brown anole lizard (*Anolis sagrei*) affected the native anole, *Anolis carolinensis*. After *A. sagrei* invaded, *A. carolinensis* moved to higher tree perches. Stuart and colleagues tested whether this change in habitat use caused evolution in the *A. carolinensis* toepad area; a larger toepad area improves the ability of lizards to grasp the slender branches found high in trees. They measured the toepad area of *A. carolinensis* caught in the wild on islands that were uninvaded and on islands invaded by *A. sagrei*. They also measured *A. carolinensis* toepad area in a "common garden experiment" in which *A. carolinensis* eggs collected on uninvaded and on invaded islands were reared to adulthood under identical conditions. Average toepad areas are shown in the table.

	Wild caught		Common garden	
	Uninvaded islands	Invaded islands	Uninvaded islands	Invaded islands
Toepad area (size-corrected index)	1.04	2.55	0.96	2.21

(Continued)

1. Make a bar graph of results for wild-caught and common garden *A. carolinensis*. Compare the results for uninvaded versus invaded islands.

2. When traits such as toepad area differ consistently between lizards living on uninvaded versus invaded islands, those differences may be due to evolution, phenotypic plasticity,* or both. Suppose that evolution had been the primary cause of differences in toepad area between lizards from uninvaded versus invaded islands. Under that assumption, predict whether wild-caught results would differ from common garden results. Explain.

3. Suppose that phenotypic plasticity had been the primary cause of differences in toepad area between lizards from uninvaded versus invaded islands. Under that assumption, predict whether wild-caught results would differ from common garden results. Explain.

4. Did the invasion of Florida islands by *A. sagrei* lead to eco-evolutionary effects? Explain.

---

*Phenotypic plasticity refers to nonheritable phenotypic variation that occurs when individuals have different phenotypes depending on environmental conditions experienced during their lifetime. For example, plants may be taller or produce more seeds if they have ample nutrients than if they do not, while people may have larger muscles if they lift weights regularly than if they do not.

ADDITIONAL RESOURCES

This textbook provides resources such as **Web Extensions** and additional **Climate Change Connections** for instructors who can share them with students. Please contact your instructor, if you would like to access that content.

Life History

Nemo Grows Up: A Case Study

"Birds do it, bees do it, even educated fleas do it"[1]—they all produce offspring that perpetuate their species. But beyond that basic fact of life, the offspring produced by different organisms vary tremendously. A grass plant produces seeds a few millimeters long that can wait, buried in the soil, for years until conditions are right for germination. A sea star spews hundreds of thousands of microscopic eggs that develop adrift in the ocean. A rhinoceros produces one calf that develops in her womb for 16–18 months and can walk well several days after birth, but requires more than a year of care before it becomes fully independent (**FIGURE 7.1**).

Even this broad range of possibilities barely begins to describe the different ways in which organisms reproduce. In popular media, we humans often depict other animals as having family lives similar to ours. For example, in the animated film *Finding Nemo*, clownfish live in families with a mother, a father, and several young offspring. When Nemo the clownfish loses his mother to a predator, his father takes over the duties of raising him. But in a more realistic version of this story, after losing his mate, Nemo's father would have done something less predictable: he would have changed sex and become a female.

Actually, the correspondence between the movie and biology breaks down long before Nemo loses his mother. Clownfish spend their entire adult lives within a single sea anemone (**FIGURE 7.2**). Anemones are related to jellyfish, with a central mouth ringed by stinging tentacles. In what appears to be a mutually beneficial relationship, the anemone protects the clownfish by stinging their predators, but

---

[1] Lyrics from the song "Let's Do It" written by Cole Porter.

KEY CONCEPTS

CONCEPT 7.1 Life history patterns vary within and among species.

CONCEPT 7.2 There are trade-offs between life history traits.

CONCEPT 7.3 Organisms face different selection pressures at different life cycle stages.

CONCEPT 7.4 Life history patterns can be classified along several continua.

© Jiri Balek/Shutterstock.com

FIGURE 7.1 Offspring Vary Greatly in Size and Number Organisms produce a large range of offspring numbers and sizes. A rhinoceros produces a single calf that weighs 40–65 kg (90–140 pounds). On the other end of the spectrum, many plants produce hundreds to thousands of seeds that are less than a millimeter long and weigh as little as 0.8 µg (roughly one fifty-billionth the weight of a rhinoceros calf).

FIGURE 7.2 Life in a Sea Anemone Clownfish (*Amphiprion percula*) form hierarchical groups of unrelated individuals that live and reproduce among the tentacles of their anemone host (*Heteractis magnifica*).

? Predict the sex of each of these clownfish (assuming that they live together as a group of four fish in an anemone host). Explain your answer.

the clownfish themselves are not stung. The clownfish, in turn, may help the anemone by eating its parasites or driving away its predators.

Two to six clownfish typically inhabit a single anemone, but they are far from a traditional human family—in fact, they are usually not related to one another. The clownfish that live in an anemone interact according to a strict pecking order that is based on size. The largest fish in the anemone is a female. The next fish in the hierarchy, the second largest, is the breeding male. The remaining fish are sexually immature *nonbreeders*. If the female dies, as in Nemo's story, the breeding male undergoes a growth spurt and changes sex to become a female, and the largest nonbreeder increases in size and becomes the new breeding male.

The breeding male clownfish mates with the female and cares for the fertilized eggs until they hatch. The hatchling fish leave the anemone to live in the open ocean, away from the predator-infested reef. The young fish eventually return to the reef and develop into juveniles. Then they must find an anemone to inhabit. When a juvenile fish enters an anemone, the resident fish allow it to stay there only if there is room. If there is no room, the young fish is expelled and returns to the dangers of an exposed existence on the reef.

This life cycle, with its expulsions, hierarchies, and sex changes, is certainly as colorful as the fish that live it. But why do clownfish engage in these complicated machinations just to produce more clownfish? Organisms have arrived at a vast array of solutions to the basic problem of reproduction. As we will see, these solutions are often well suited for meeting the challenges and constraints of the environment where a species lives.

Introduction

Human history is a record of past events. Your personal history might consist of a series of details about the course of your life: your birth weight, when you started walking and talking, your adult height, and other relevant information about your development. Similarly, an individual organism's **life history** consists of major events related to its growth, development, reproduction, and survival.

In this chapter, we'll discuss traits that characterize the life history of an organism, including age and size at sexual maturity, amount and timing of reproduction, and survival and mortality rates. As we'll see, the timing and nature of life history traits, and therefore the life history itself, are products of adaptation to the environment in which the organism lives. Life history traits, such as reproductive capacity, are an important influence on the growth rate of populations (see Unit 3) and consequently the composition and diversity of communities (Unit 5). We'll also consider how biologists analyze life history patterns in order to understand the trade-offs, constraints, and selection pressures imposed on different stages of an organism's life cycle.

CONCEPT 7.1 Life history patterns vary within and among species.

LEARNING OBJECTIVES

7.1.1 Summarize the key stages that make up the life history of an organism.

7.1.2 Explain how genetics and the environment act as controls on life history traits.

7.1.3 Compare the benefits and costs associated with sexual versus asexual reproduction.

7.1.4 Describe how additional complexity in a life cycle, such as larval and adult forms, may benefit a species.

Life History Diversity

Studying variation in life history traits and analyzing the causes of that variation helps us to understand how life history traits interact with the environment and influence the potential growth rate of populations. In order to understand such analyses, it is helpful first to examine some of the broad life history patterns found within and among species.

Individuals within species differ in their life histories

Individual differences in life history traits are ubiquitous. Think about your own life experiences and those of your family and friends. Some members of your social group reached developmental milestones such as puberty earlier or later than others. Different women may have different numbers of children with different age gaps between them. Despite this variation, it is possible to make some generalizations about life histories in *Homo sapiens*: for example, women typically have one baby at a time and reproduction usually occurs between the ages of 15 and 45. Similar generalizations can be made for other species. The **life history strategy** of a species is the overall pattern in the timing and nature of life history events averaged across all the individuals in the species (**FIGURE 7.3**).

The life history strategy is shaped by the way the organism divides its energy and resources between growth, reproduction, and survival, and the associated trade-offs with the allocation patterns. Within a species, individuals often differ in how they divide their energy and resources among these activities. Such differences affect the ecological success of a species and may result from genetic variation, from differences in environmental conditions, or from a combination of both.

GENETIC DIFFERENCES   Some life history variation within species is determined genetically. Genetically influenced traits can often be recognized as those that are more similar within families than between them. Again, these kinds of traits are familiar in humans: for example, siblings are often similar in appearance and reach similar adult heights and weights. The same is true in other organisms. For example, in annual bluegrass (*Poa annua*), life history traits such as age at first reproduction, growth rate, and number of flowers produced are similar among sibling plants (Law et al. 1977). As with any other trait, heritable variation in life history traits is the raw material

on which natural selection acts. Selection favors individuals whose life history traits result in their having a better chance of surviving and reproducing than do individuals with other life history traits.

Much of life history analysis is concerned with explaining how and why life history patterns have evolved to their present states. Life histories are believed to be adapted to maximize **fitness** (the genetic contribution of an organism's descendants to future generations, determined both by the reproductive rate of the parent and the survival rates of both parent and offspring). However, no organism has a perfect life history—that is, one that results in the unlimited production of descendants. Instead, all organisms face *constraints* that prevent the evolution of a perfect life history. As we'll see in Concept 7.3, these constraints often involve ecological trade-offs in which an increase in the performance of one function (such as reproduction) can reduce the performance of another (such as growth or survival). Thus, although life histories often serve organisms well in the environments in which they have evolved, they are optimal only in the sense of maximizing fitness subject to constraints.

ENVIRONMENTAL DIFFERENCES   A single genotype may produce different phenotypes under different environmental conditions, a phenomenon known as **phenotypic plasticity** (see the Climate Change Connection in Chapter 6). Almost every trait shows some degree of plasticity, and life history traits are no exception. For example, most plants and animals grow at different rates depending on temperature. They do so because development

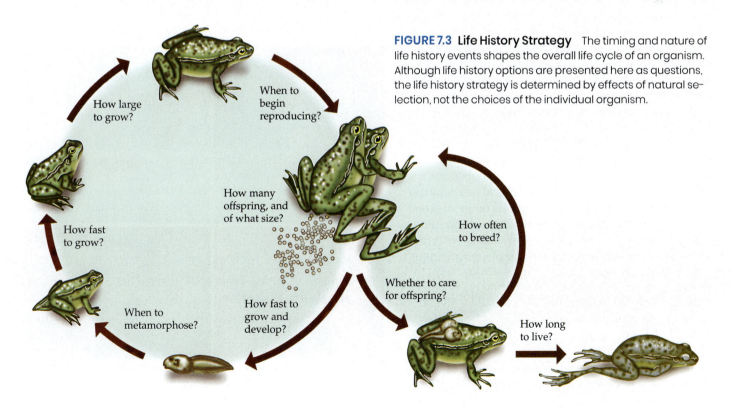

FIGURE 7.3 Life History Strategy   The timing and nature of life history events shapes the overall life cycle of an organism. Although life history options are presented here as questions, the life history strategy is determined by effects of natural selection, not the choices of the individual organism.

How large to grow?

When to begin reproducing?

How many offspring, and of what size?

How fast to grow?

How often to breed?

Whether to care for offspring?

When to metamorphose?

How fast to grow and develop?

How long to live?

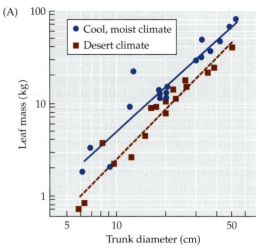

(A)

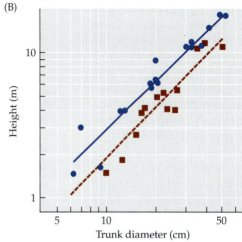

(B)

© Morales/AGE Fotostock

FIGURE 7.4 Plasticity of Growth Form in Ponderosa Pines
(A) Ponderosa pine trees (*Pinus ponderosa*) in cool, moist climates allocate more resources to leaf production than do trees in desert climates. (B) Desert trees are shorter than those grown in cooler climates, but for a given height, they have thicker trunks. (After R. M. Callaway et al. 1994. *Ecology* 75: 1474–1481.)

Use the solid (regression) line in (B) to estimate the trunk diameter of a tree that is 5 m tall and grows in a cool, moist climate versus the trunk diameter of a tree of the same height that grows in a desert climate.

typically speeds up as the temperature rises, then slows down again due to heat stress as the temperature approaches the organism's upper lethal temperature.

Changes in life history traits often translate into changes in adult morphology. Slower growth under cooler conditions, for example, may lead to a smaller adult size or to differences in adult shape. Callaway et al. (1994) showed that ponderosa pine (*Pinus ponderosa*) trees grown in cool, moist climates allocate more biomass to leaf growth relative to sapwood production than do those in warmer desert climates ("sapwood" refers to newly formed layers of wood that function in water transport). **Allocation** describes the relative amounts of energy or resources that an organism devotes to different functions. The result of allocation differences in ponderosa pines is that trees grown in different environments differ in adult shape and size. Desert trees are shorter and squatter, with fewer branches and leaves (**FIGURE 7.4**). As a result of having fewer leaves, they also lose less water and have lower photosynthetic rates per unit of ground area.

Phenotypic plasticity that responds to temperature variation often produces a continuous range of sizes. In other types of phenotypic plasticity, a single genotype produces discrete types, or **morphs**, with few or no intermediate forms. For example, populations

of spadefoot toad (*Spea multiplicata*) tadpoles in Arizona ponds contain two morphs: omnivore morphs, which feed on detritus and algae, and larger carnivore morphs, which feed on fairy shrimp and on other tadpoles (**FIGURE 7.5**). The differing body shapes of omnivores and carnivores result from differences in the relative growth rates of different body parts: carnivores have bigger mouths and stronger jaw muscles because of accelerated growth in those areas. Pfennig (1992) showed that omnivore

(A) Omnivore morph   (B) Carnivore morph

Omnivores feed on the pond bottom on detritus.

Carnivores feed in the water column on fairy shrimp.

All images courtesy of David Pfennig

FIGURE 7.5 Phenotypic Plasticity in Spadefoot Toad Tadpoles Spadefoot toad (*Spea multiplicata*) tadpoles can develop into small-headed omnivores (A) or large-headed carnivores (B), depending on the food they consume early in development. Later in development, omnivores and carnivores feed on different food sources that are located in different portions of their habitat.

tadpoles can turn into carnivores when fed on shrimp and tadpoles, and field studies show that the proportion of omnivore and carnivore morphs is affected by food supply. Carnivore tadpoles grow faster and are more likely to metamorphose before the ponds where they live dry up. As a result the rapidly growing carnivores are favored in ephemeral ponds. The more slowly growing omnivores are favored in ponds that persist longer, because they metamorphose in better condition and thus have better chances of survival as juvenile toads.

Examples of phenotypic plasticity such as the omnivore and carnivore morphs of the spadefoot toad may suggest situations where phenotypic plasticity is adaptive—that the ability to produce different phenotypes in response to changing environmental conditions increases the fitness of individuals. While that is often the case, adaptation must be demonstrated rather than assumed. For example, it may be adaptive for ponderosa pines to be stockier and have fewer leaves in hot, dry climates because these features can help reduce water loss. However, adaptation would have to be documented by measuring and comparing the survival and reproductive rates of stockier and taller trees in the desert environment. In some instances, phenotypic plasticity may be a simple physiological response, not an adaptive response shaped by natural selection. For example, as mentioned above, growth rate typically increases with temperature up to a point. This may occur because chemical reactions are slower at lower temperatures, and thus metabolism and growth are necessarily slower.

CLIMATE CHANGE CONNECTION

CLIMATE CHANGE AND THE TIMING OF SEASONAL ACTIVITIES The timing of seasonal life history activities can be of critical importance. For example, a bird that migrates north too early in the spring may starve if no food is available, while a plant that flowers when its pollinators are not present may fail to reproduce. As described in Concept 4.2, the timing of such seasonal events is affected by changing day length (photoperiod) and sometimes by other environmental cues such as temperature that also vary over the course of a year. As the climate has changed in recent decades, have species adjusted the times when they perform key seasonal activities?

Long-term data sets show that many species are initiating spring activities earlier than they once did in response to climate change. For example, as the climate warms, leaf production in plants, egg laying in birds, emergence from dormancy in insects, and arrival of migratory animals often occur earlier today than they did in the 1960s and 1970s.

In some cases, however, shifts in the timing of seasonal activities have not kept pace with climate change. One example involves seasonal change in fur coloration of the snowshoe hare (*Lepus americanus*). As winter approaches, the coat color of snowshoe hares changes from brown to white, providing camouflage against snow; the reverse coat-color change occurs in spring. As the climate has warmed, the length of time that the ground is covered by snow has decreased because snowfall now begins later in autumn and snowmelt occurs earlier in spring. If the timing of the fall coat-color change in snowshoe hares had kept pace with the delay in when snowfall begins, we would expect that snowshoe hares would molt to white later in the fall than they once did. Instead, however, the date and rate of the fall molt has not changed (Mills et al. 2013). As a result, the number of days in which a "camouflage mismatch" occurs has increased, making the hares easier for visually hunting predators to spot (**FIGURE 7.6**) and leading to increased mortality rates (Zimova et al. 2016). Mismatches in the timing of seasonal activities have also been found in caribou (*Rangifer tarandus*) and snow geese (*Chen caerulescens*): although the plants their young require for food are producing leaves earlier in the spring, neither species has adjusted the timing of reproduction. This has caused a decline in the reproductive success of both species because their young are not getting enough to eat. ☀

(A)

L. S. Mills et al. 2013. *Proc Natl Acad Sci USA* 110: 7360–7365

(B)

L. Scott Mills Research Photo

FIGURE 7.6 Camouflage Mismatch in Snowshoe Hares (A) Historically, snowshoe hares changed their color from brown to white at a time of year that matched the onset of snowfall, causing them to be well-camouflaged all winter. (B) With climate warming, snowfall now begins later in the year. However, the date of the fall coat-color change has remained the same, causing an increase in the number of days that snowshoe hares experience a camouflage mismatch.

FIGURE 7.7 Life Cycle of a Coral Reef-forming coral colonies grow by asexual reproduction before producing eggs and sperm. The sexually produced offspring establish new colonies.

? Would the larva shown in the diagram be genetically identical to the polyp to its left? Would two different larvae be genetically identical to each other? Explain.

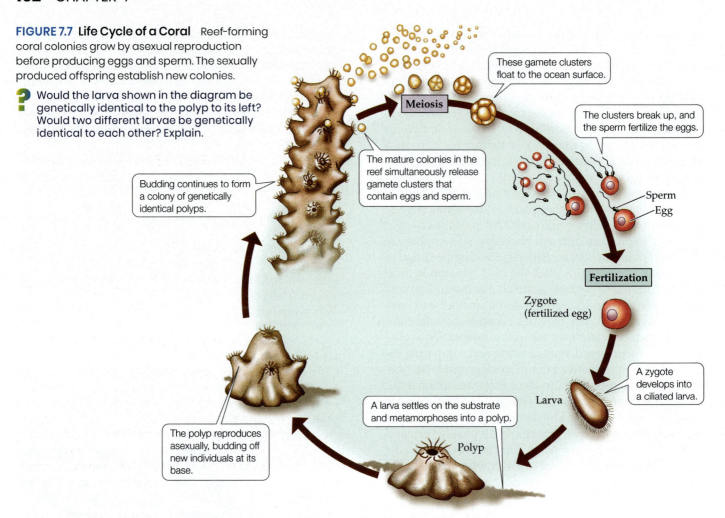

These gamete clusters float to the ocean surface.

Meiosis

The clusters break up, and the sperm fertilize the eggs.

The mature colonies in the reef simultaneously release gamete clusters that contain eggs and sperm.

Sperm
Egg

Budding continues to form a colony of genetically identical polyps.

Fertilization

Zygote (fertilized egg)

A zygote develops into a ciliated larva.

The polyp reproduces asexually, budding off new individuals at its base.

A larva settles on the substrate and metamorphoses into a polyp.

Larva

Polyp

Mode of reproduction is a basic life history trait

At the most basic level, evolutionary success is determined by successful reproduction. Despite this universal reality, organisms have evolved vastly different mechanisms for reproducing—from simple asexual splitting to complex mating rituals and intricate pollination systems.

ASEXUAL REPRODUCTION The first organisms to evolve on Earth reproduced asexually by binary fission (parental cell divides to produce two cells). The sexual reproductive processes of meiosis, recombination, and fertilization arose later. Today, all prokaryotes and many protists reproduce asexually. While sexual reproduction is the norm in multicellular organisms, many can also reproduce asexually. For example, after they are initiated by a (sexually produced) founding polyp, coral colonies grow by asexual reproduction (**FIGURE 7.7**). Each individual polyp in a colony is produced when a multicellular bud splits off from a parent polyp to form a new polyp; as a result, each polyp is a genetically identical copy, or *clone*, of the founding polyp. Once the colony has grown to a certain size and conditions are right, the polyps reproduce

sexually, producing offspring that develop into polyps that start their own new colonies of clones.

SEXUAL REPRODUCTION AND ANISOGAMY Most plants and animals reproduce sexually, as do many fungi and protists. Some protists, such as the green alga *Chlamydomonas reinhardtii* (**FIGURE 7.8A**), have two different *mating types*, analogous to males and females except that their gametes are the same size. The production of equal-sized gametes is called **isogamy**. In most multicellular organisms, however, the two types of gametes are different sizes, a condition called **anisogamy**. Typically, the eggs are much larger than the sperm and contain more cellular and nutritional provisions for the developing embryo. The sperm are small and may be motile (**FIGURE 7.8B**). As we'll see in Concept 8.3, differences between the sexes in gamete size can influence other reproductive characteristics, such as differences between the sexes in their mating behavior.

Although sexual reproduction is widespread, it has some disadvantages. Because meiosis produces haploid gametes that contain half the genetic content of the parent, a sexually reproducing organism can transmit only half of its genetic material to each offspring, whereas asexual

(A)

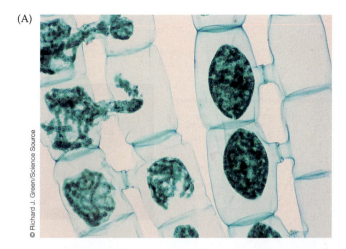

© Richard J. Green/Science Source

(B)

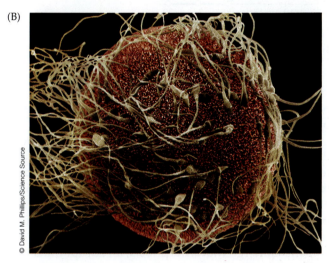

© David M. Phillips/Science Source

FIGURE 7.8 **Isogamy and Anisogamy** (A) An isogamous species: two gametes of the single-celled alga *Spirogyra* fusing. (B) An anisogamous species: fertilization of a human egg, showing the difference in size between egg and sperm.

reproduction allows transmission of the entire genome. Another disadvantage of sex is that recombination and the independent distribution of chromosomes into gametes (during meiosis) can disrupt favorable gene combinations, potentially reducing offspring fitness. Finally, the growth rate of sexually reproducing populations is only half that of asexually reproducing ones, all else being equal (**FIGURE 7.9**).

Given such disadvantages, why is sex so common? Sex has some clear benefits, including recombination, which promotes genetic variation and hence may increase the capacity of populations to evolve in response to environmental challenges such as drought or disease. In a test of this idea, Morran et al. (2011) examined the benefits of sex in the nematode worm *Caenorhabditis elegans*. Populations of *C. elegans* consist of males and hermaphrodites. The hermaphrodites can reproduce by self-fertilization (selfing) or by mating with males (outcrossing). In wild-type populations, outcrossing rates typically range from 1% to 30%. However, *C. elegans* can be manipulated genetically to form strains that always self-fertilize ("obligate selfers") or never self-fertilize ("obligate outcrossers"). The offspring of obligate selfers are very similar genetically to their parents, whereas the offspring of obligate outcrossers are more variable genetically; thus, these strains are well suited for testing the idea that sex is beneficial because it promotes increased levels of genetic variation.

Morran et al. exposed some *C. elegans* populations to a lethal bacterial pathogen, *Serratia marcescens*. In wild-type populations exposed to this pathogen, the rate of outcrossing increased, rising from an initial 20% to more than 80% over the course of 30 generations (**FIGURE 7.10A**). Moreover, *C. elegans* populations containing only obligate selfers were always driven to extinction by the pathogen, whereas wild-type and obligate-outcrossing populations

(A)

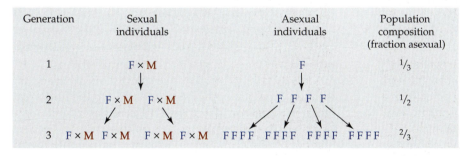

(B)

FIGURE 7.9 **The Cost of Sex** One cost of sex is referred to as the "cost of males." Imagine a population in which there are both sexual and asexual individuals. Assume that each sexual or asexual female can produce four offspring per generation, but half of the offspring produced by the sexual females are male and must pair with females to produce offspring. Under these conditions, the asexual individuals (A) will increase in number more rapidly

and (B) in less than 10 generations will constitute nearly 100% of the population.

? In generation 2 there are four sexual and four asexual individuals. How many sexual and asexual individuals are there in generation 3? How many of each will there be in generation 4? Explain your results in terms of the cost of males.

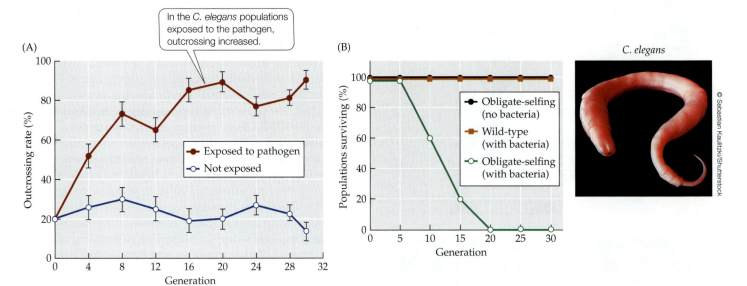

(A)

In the *C. elegans* populations exposed to the pathogen, outcrossing increased.

- Exposed to pathogen
- Not exposed

(B)

- Obligate-selfing (no bacteria)
- Wild-type (with bacteria)
- Obligate-selfing (with bacteria)

C. elegans

© Sebastian Kaulitzki/Shutterstock

FIGURE 7.10 Benefits of Sex in a Challenging Environment (A) Outcrossing rates were measured over time in wild-type populations of the nematode worm *Caenorhabditis elegans*. Some *C. elegans* populations were exposed to the bacterial pathogen *Serratia marcescens*, while others were not. Error bars show ± one standard error of the mean. (B) Percentage of replicate wild-type and obligate-selfing *C. elegans* populations surviving under different treatments. (After L. T. Morran et al. 2011. *Science* 333: 216–218.)

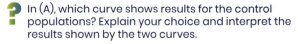 In (A), which curve shows results for the control populations? Explain your choice and interpret the results shown by the two curves.

always persisted (**FIGURE 7.10B**). Overall, these results support the hypothesis that the genetic variation generated by sexual reproduction is beneficial in a challenging environment. McDonald et al. (2016) obtained similar results in yeast, and showed that sex provided benefits by increasing the fixation of advantageous mutations while decreasing the fixation of deleterious mutations.

Life cycles are often complex

The small, early stages of many animal life cycles look and behave completely differently from adult stages. They frequently eat different foods and prefer different habitats. For example, coral reef fishes such as the damselfish *Chromis atripectoralis* start life as hatchlings only a few millimeters long. The hatchlings live and grow in the open ocean, feeding on planktonic algae. When they have grown to about a centimeter in length, they return to the reef and begin to eat larger food items. This life cycle may have evolved in response to high levels of predation on young fish that stay on the reef; young fish that spend more time growing in the open ocean may have better chances of survival. Complex life cycles can also lower competition between individuals

of the same species, since species at different ages use of different resources.

As corals (see Figure 7.7) and coral reef fishes both demonstrate, life cycles can involve stages that have different body forms or live in different habitats. A **complex life cycle** is one in which there are at least two distinct stages that differ in their habitat, physiology, or morphology. In many cases, the transitions between stages in complex life cycles are abrupt. For example, many organisms undergo **metamorphosis**, an abrupt transition in form from the larval to the juvenile stage that is sometimes accompanied by a change in habitat (exemplified by the frogs in Figure 7.3). As we will see in Concept 7.4, complex life cycles and metamorphosis often result when offspring and parents are subjected to very different selection pressures.

Because most vertebrates have simple life cycles that lack an abrupt transition between habitats or forms, we humans tend to think of metamorphosis as an exotic and strange process. However, complex life cycles and metamorphosis can be found even among vertebrates, including some fishes and most amphibians. Most marine invertebrates produce microscopic larvae that swim in the open ocean before settling to the bottom at metamorphosis. Many insects also undergo metamorphosis—from caterpillars to moths, grubs to beetles, maggots to flies, and aquatic larvae to dragonflies and mayflies. In fact, Werner (1988) calculated that of the 33 phyla of animals recognized at that time, 25 contained at least some subgroups that have complex life cycles. He also noted that about 80% of all animal species undergo metamorphosis at some time in their life cycle (**FIGURE 7.11**).

Complex life cycles occur in many types of algae and plants, reaching some of their most elaborate forms in

(A)   Larva   Adult

Develops into

Euroleon nostras

(B)

Develops into

Paracentrotus lividus

© H. Bellmann/F. Hecker/AGE Fotostock

Richard Garvey-Williams/Alamy Stock Photo

© Wim van Egmond/Science Source

© Damsea/Shutterstock.com

FIGURE 7.11 **The Pervasiveness of Complex Life Cycles** Most groups of animals include members that undergo metamorphosis. (A) Familiar examples are insects such as the antlion, which develops from a larva that lives in soil. (B) Most marine invertebrates have free-swimming larval stages, including echinoderms such as sea urchins.

these groups. Some algae and all plants have complex life cycles in which a multicellular diploid *sporophyte* alternates with a multicellular haploid *gametophyte*. The sporophyte produces haploid spores that disperse and grow into gametophytes, and the gametophyte produces haploid gametes that combine in fertilization to form zygotes that grow into sporophytes. This type of life cycle, called **alternation of generations**, has been elaborated on in different plant and algal groups. In mosses and a few other plant groups, the gametophyte is larger, but in most plants and some algae, the sporophyte is the dominant stage of the life cycle.

Over the course of evolution, complex life cycles have been lost in some species that are members of groups in which such cycles are considered the ancestral condition. The resulting simple life cycles are sometimes referred to as **direct development** because development from fertilized egg to juvenile occurs within the egg prior to hatching and no free-living larval stage occurs. For example, most species in one group of salamanders, the plethodontids, lack the gilled aquatic larval stage that is typical of salamanders. Instead, they lay their eggs on land, where they hatch directly into small terrestrial juveniles.

As we've seen, organisms vary greatly in key aspects of their life history strategies, such as when they reproduce, how many offspring they produce, and how much care is allotted to each offspring. How can we organize these diverse patterns into a coherent scheme?

CONCEPT 7.2   There are trade-offs between life history traits.

LEARNING OBJECTIVES

7.2.1   Illustrate how the number of offspring may affect the size of those offspring.

7.2.2   Explain how providing care to offspring may compromise other functions in adults.

7.2.3   Understand that allocating energy and resources to reproduction may affect parental growth, survival, and future reproduction.

Trade-Offs

Organisms have limited amounts of energy and resources that can be invested in growth, reproduction, and defense. As discussed in Concept 6.3, trade-offs occur when organisms allocate their limited energy or other resources to one structure or function at the expense of others. As we'll see, trade-offs among life history traits are common.

There is a trade-off between number and size of offspring

Many organisms show a trade-off between their investment in each individual offspring and the number of offspring they produce. Investment in offspring includes energy, resources, and time. In many cases, organisms that make a large investment in each offspring produce small numbers of large offspring, while organisms that make a small investment in each offspring produce large numbers of small offspring. As we'll see, parental investment can also affect offspring "quality," as when reduced investment per offspring increases the risk of offspring mortality.

LACK CLUTCH SIZE A classic example demonstrating the trade-off between how much investment goes into each offspring versus the number of offspring was first described by David Lack in 1947. Lack asserted that the number of eggs per reproductive bout (known as the clutch size) is limited by the maximum number of young that the parents can raise at one time, which in turn is related to the resource availability (prey and other factors needed to raise the young). If the parents rear fewer than this maximum number, they will reduce their genetic representation in future generations (fitness). If they attempt to rear more than this maximum number, their offspring may be more likely to die from starvation, predation, or other factors, again reducing the parents' fitness.

Lack made careful observations of the breeding biology of bird species, from the poles to the tropics. What struck him was that clutch size varied with latitude, with a greater number of offspring at higher latitudes. He hypothesized that the reason for larger clutches at higher latitudes was the longer periods of daylight during the breeding season. These longer days allowed parents more time for foraging, and they could therefore feed greater numbers of offspring.

The term "Lack clutch size" refers to the maximum number of offspring that a parent can successfully raise to maturity. Lack hypothesized that as a result of natural selection from the trade-off between numbers versus resource provisioning of young birds, the most productive clutch size would be found in natural populations. This hypothesis has generally been supported using experiments that manipulate the number of eggs in nests, and examining whether there are costs to unusually large clutch sizes. For example, Nager et al. (2000) artificially increased the number of eggs in clutches laid by the lesser black-backed gull (*Larus fuscus*). They did this by removing eggs from nests, which stimulated the females to lay more eggs. Nager et al. found that the increased clutch size resulted in a drop in the nutritional quality of later-produced eggs (specifically, these eggs had a lower lipid content). They also found that eggs from larger clutches had reduced survivorship to fledging (the point at which wing feathers are developed enough for flight) (**FIGURE 7.12**). Thus, in lesser black-backed gulls, production of larger clutches reduced both egg quality and survivorship to fledging.

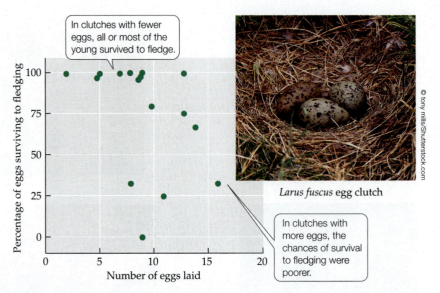

In clutches with fewer eggs, all or most of the young survived to fledge.

In clutches with more eggs, the chances of survival to fledging were poorer.

Larus fuscus egg clutch

© tony mills/Shutterstock.com

FIGURE 7.12 Clutch Size and Survival Lesser black-backed gulls typically lay three eggs in a clutch. However, when they are manipulated experimentally to produce larger clutches of eggs, their offspring have reduced chances of survival to fledging. (After R. G. Nager et al. 2000. *Ecology* 81: 1339–1350.)

TRADE-OFFS IN ORGANISMS WITHOUT PARENTAL CARE
Parental care like that provided by birds and some other vertebrates is relatively rare. In organisms that do not provide parental care, resources invested in *propagules* (such as eggs, spores, or seeds) are the main measure of reproductive investment. In this case, the size of the propagule is the primary measure of parental investment, and propagule size is traded off against the number of propagules produced in a reproductive bout. In plants, for example, the size of the seeds that a species produces is negatively correlated with the number of seeds it produces (**FIGURE 7.13**).

In some cases, the size–number trade-off also applies to variation within species. The western fence lizard (*Sceloporus occidentalis*), which is common throughout the coastal mountains of the western United States, does not provide parental care. Barry Sinervo (1990) found that lizard populations farther to the north laid more eggs per clutch (Washington: 12 eggs/clutch vs. California: 7 eggs/clutch) but laid smaller eggs (Washington: 0.40 g vs. California: 0.65 g) (**FIGURE 7.14**).

In order to determine the consequences of egg size for offspring performance, Sinervo raised fence lizard eggs in the laboratory. He artificially reduced the size of some of the eggs by using a syringe to remove some yolk from them. To control for any possible effects of this method on egg development, he inserted a syringe into some other eggs but did not remove any yolk. These eggs that had been poked, but not reduced, developed at the same rate as unmanipulated eggs, indicating that insertion of the syringe was not the cause of differences between unmanipulated and reduced eggs.

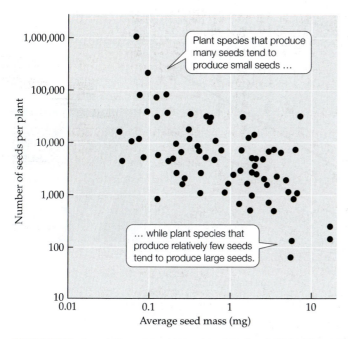

FIGURE 7.13 Seed Size–Seed Number Trade-Offs in Plants (After O. A. Stevens. 1932. *Am J Bot* 19: 784–794.)

Sinervo found that the reduced eggs developed faster than the unmanipulated eggs but produced smaller hatchlings. These small hatchlings grew faster than their larger siblings, but they were not able to sprint as fast to escape from predators. Many of the differences between the lizards hatched from the reduced eggs and

from the unmanipulated eggs echoed observed differences between populations with naturally differing egg sizes. Sinervo hypothesized that the differences between populations in egg and hatchling size may be the result of selection favoring faster sprint speeds in the south, where there may be more predators, or of selection favoring earlier hatching and faster growth in the north, where the growing season is shorter.

There are trade-offs between current reproduction and other life history traits

As we've seen, when parents produce more offspring, their investment per offspring may decline. Such a decline can have various effects on the offspring, including reduced survival (as in lesser black-backed gulls) and reduced size (as in western fence lizards). The allocation of resources to reproduction can also affect the parent by decreasing an individual's growth rate, its survival rate, and its potential for future reproduction.

For example, a trade-off between current reproduction and survival has been documented in studies that examine how life history traits differ among species. In one such study, Ricklefs (1977) observed a trade-off between annual fecundity (as measured by the number of offspring raised to maturity) and annual survivorship in birds (**FIGURE 7.15A**). Trade-offs between reproduction and survival have also been observed within a species. For example, in the fruit fly *Drosophila melanogaster*, males

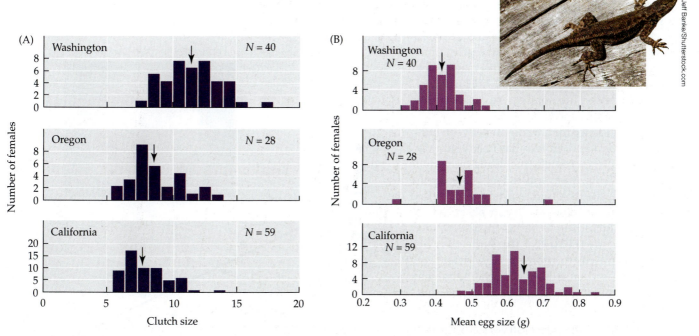

FIGURE 7.14 Egg Size–Egg Number Trade-Off in Fence Lizards Western fence lizards in northern populations produced (A) larger clutches and (B) smaller eggs than those in southern populations. The arrow points to the average for each population. (After B. Sinervo. 1990. *Evolution* 44: 279–294.)

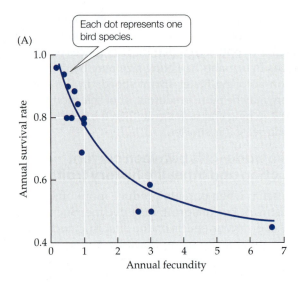

(A)

Each dot represents one bird species.

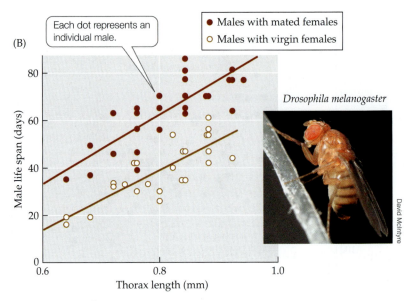

(B)

Each dot represents an individual male.

- ● Males with mated females
- ○ Males with virgin females

Drosophila melanogaster

David McIntyre

FIGURE 7.15 **Trade-Offs between Reproduction and Survival** (A) In a comparison of 14 different bird species, the annual survival rate declines as annual fecundity increases. (B) Life span versus size (thorax length in millimeters) for male *Drosophila* kept with eight virgin females or eight previously mated females. Regression lines represent average male life spans. (A after R. E. Ricklefs. 1977. *Am Nat* 111: 453–478; B after L. Partridge and M. Farquhar. 1981. *Nature* 294: 580–582.)

? In (B), what is the average life span of male flies with a 0.8-mm thorax kept with virgin females? How does this compare with that of males of the same size kept with previously mated females?

spend more time and energy courting unmated females than they spend courting recently mated females. Partridge and Farquhar (1981) tested whether such differences in courtship activity affected the longevity of male fruit flies. Males were kept with eight virgin females per day or with eight previously inseminated females per day. In the absence of sexual activity, a male's life span is correlated positively with his size, so Partridge and Farquhar also recorded the size of each male. Among males of any particular size, males kept with virgin females had a shorter life span than did males kept with inseminated females (**FIGURE 7.15B**), showing a cost (reduced life span) of sexual activity among males of this species.

Similarly, evidence for a trade-off between current reproduction and growth has been found in mollusks, insects, mammals (including humans), fishes, amphibians, and reptiles (see citations in Barringer et al. 2013). A trade-off between reproduction and growth has also been observed in many plants, including Douglas fir trees (*Pseudotsuga menziesii*) (**FIGURE 7.16**). Note that by allocating resources to reproduction instead of growth, an individual will reproduce at a smaller size than it would if it had continued to grow and reproduced at a later time (when it was larger). Small individuals often produce fewer offspring than do large individuals, so this observation suggests that allocating resources to current reproduction might decrease an individual's potential for future reproduction. This trade-off has also received empirical support, as you can explore in **ANALYZING DATA 7.1**.

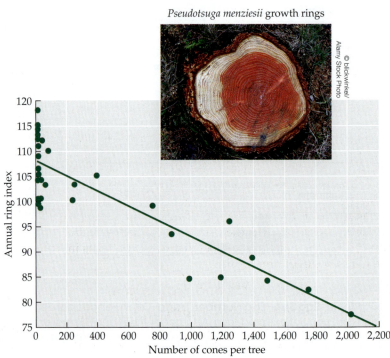

Pseudotsuga menziesii growth rings

© blickwinkel/ Alamy Stock Photo

FIGURE 7.16 **A Reproduction versus Growth Trade-Off** The thickness of annual growth rings (a measure of growth rate) declines in Douglas fir trees that produce many cones. (After S. Eis et al. 1965. *Can J Bot* 43: 1553–1559.)

ANALYZING DATA 7.1

Is There a Trade-Off between Current and Delayed Reproduction in the Collared Flycatcher?

Lars Gustafsson and Tomas Pärt (1990)* studied a population of collared flycatchers (*Ficedula albicollis*) on the Swedish island of Gotland. Gustafsson and Pärt monitored the survival and reproduction of each bird throughout its entire life. They noted that some females reproduced for the first time when they were 1 year old ("early breeders"), while others reproduced for the first time when they were 2 years old ("late breeders"). The average number of eggs laid by early breeders and late breeders are reported in the table.

	Average number of eggs	
Age (years)	Early breeders	Late breeders
1	5.8	–
2	6.0	6.3
3	6.1	7.0
4	5.7	6.6

1. Graph the average number of eggs (on the *y*-axis) versus age (on the *x*-axis) for both early breeders and late breeders.

2. Do the results suggest that it would be advantageous for birds to delay reproduction until they were 2 years old? Explain.

3. Do the results indicate that allocating resources to current reproduction can reduce an individual's potential for future reproduction? Explain.

4. These results were based on field observations. What are the limitations of such data? Propose an experiment to test whether there is a cost of reproduction in females that reduces their potential for future reproduction.

* Gustafsson, L., and T. Pärt. 1990. Acceleration of senescence in the collared flycatcher *Ficedula albicollis* by reproductive costs. *Nature* 347: 279–281.

CONCEPT 7.3 Organisms face different selection pressures at different life cycle stages.

LEARNING OBJECTIVES

7.3.1 Contrast the benefits and costs associated with small size in early life cycle stages.

7.3.2 Explain how adaptations at specific stages in a complex life cycle may benefit the species.

7.3.3 Compare the benefits of semelparity and iteroparity in the context of total lifetime reproduction of an organism.

Life Cycle Evolution

In Concept 7.1, we saw that an organism's size and form may vary greatly over the course of its life cycle. Each life cycle stage may have different habitat preferences, food preferences, and vulnerability to predation. These differences suggest that different morphologies and behaviors are adaptive at different life cycle stages. Differences in selection pressures over the course of the life cycle are responsible for some of the most distinctive patterns in the life histories of organisms.

Small size has benefits and drawbacks

Small, early life cycle stages can be particularly vulnerable to predation because there are many predators that are big enough to consume them (although for some predators, small prey may be more difficult to detect). These small stages may also be poor competitors for food and thus more susceptible to environmental perturbations that diminish food supply, because they have little storage capacity for energy and nutrients to help them withstand starvation. These vulnerabilities are typically counterbalanced by behavioral, morphological, and physiological adaptations. Furthermore, in some organisms, small, mobile early stages can perform essential functions that are not possible for large adult stages. Here, we examine how organisms protect small-sized life history stages and the important functions those stages can provide.

PARENTAL INVESTMENT In many organisms, the parents' main investment in their offspring is the provisioning of the eggs or embryos. Animals add yolk to their eggs, which helps their offspring survive and grow through the small, vulnerable stages of life. Female kiwis, for example, produce one yolk-rich egg at a time. The kiwi egg is so large that it makes up 15%–20% of the bird's body size (**FIGURE 7.17A**). During the month that it takes her to make the egg, the female kiwi eats about three times as much as when she is not producing an egg. Offspring of many invertebrate species with greater yolk volume develop more rapidly, and require less food during development, than those with less yolky eggs. Another pattern common among invertebrates is investment in energetically expensive egg coverings that protect the

(A) Egg   Head

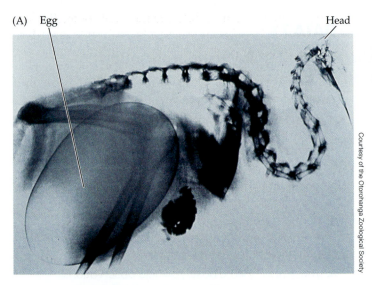

(B)

Courtesy of the Otorohanga Zoological Society

© Chien Lee/Minden Pictures

FIGURE 7.17 Parental Investment (A) This X-ray photograph shows the size of a kiwi egg in proportion to the female's body size. (B) A male horned land frog (*Sphenophryne cornuta*) carries its young on its back, from tadpole stage to small offspring, as shown here.

offspring during development. Plants provision the fertilized embryos in their seeds with *endosperm*, nutrient-rich material that sustains the developing embryo and often the young seedling. The starchy white part of corn kernels and the milk and meat of coconuts are examples of endosperm.

Another mechanism for protecting small, vulnerable offspring is parental care. Birds and mammals are the most familiar examples of parental care because they invest large amounts of time and energy in protecting and feeding their relatively helpless offspring. Some fishes, reptiles, amphibians (**FIGURE 7.17B**), and invertebrates also guard or brood their embryos and hatchlings, protecting them until they are big enough to be less vulnerable to starvation and predation.

DISPERSAL AND DORMANCY Although small offspring are vulnerable to many hazards, they are also well suited for several important functions, including dispersal and dormancy. **Dispersal**—the movement of organisms or propagules from their birthplace—is a key feature in the life history of all organisms. Even in organisms such as plants, fungi, and many marine invertebrates that are sessile or move very little as adults, the life cycle typically includes a stage in which dispersal occurs. The small pollen, seeds, spores, or larvae of these organisms can be carried long distances by water or wind or, in the case of pollen and seeds, by animals. In general, smaller propagules disperse more readily and can travel farther in a given amount of time.

Dispersal provides a number of potential advantages. For example, dispersal reduces competition among close relatives, and allows organisms to reach new areas where they can grow and reproduce. In some circumstances, dispersal can increase the chance of escaping regions of high mortality, as when pathogens and other natural enemies are abundant at the location from which organisms disperse.

The ability of an organism to disperse can also have important evolutionary consequences. For example, Hansen (1978) compared the fossil records of extinct marine snails with typical swimming larvae with those of species that had lost their swimming larval stages and developed directly into crawling juveniles. He found that the species without swimming larvae tended to have smaller geographic distributions and were more prone to extinction (**FIGURE 7.18**). Hansen attributed these differences to differences in dispersal ability. Species with swimming larvae would have been able to move greater distances and hence would have had more broadly distributed populations that were less vulnerable to random events that could lead to extinction.

Small size also makes eggs and embryos well suited to *dormancy*, a state of suspended growth and development in which an organism can survive unfavorable conditions. Many seeds are capable of long periods of dormancy before germination, which in extreme cases can last up to thousands of years. Most bacteria and some protists and animals can also undergo various forms of dormancy. The brine shrimp eggs that children purchase as "sea monkeys," for example, are in a dormant state that allows them to survive out of water, often for years. In general, small seeds, eggs, and embryos are better suited to dormancy than large multicellular organisms because they do not have to expend as much metabolic energy to stay alive. However, some animals do enter dormancy in mature stages in response to stressful environmental conditions (as described in Concept 4.2).

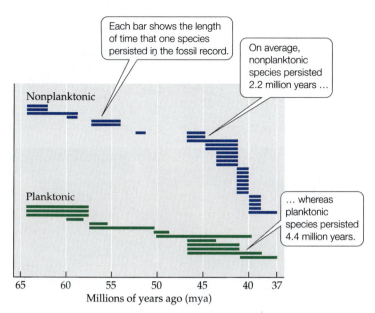

FIGURE 7.18 Developmental Mode and Species Longevity
Species of marine snails that undergo direct development without a swimming larval stage (nonplanktonic) have become extinct more rapidly than those with swimming larvae (planktonic). (After T. A. Hansen. 1978. *Science* 199: 885–887.)

Complex cycles may result from stage-specific selection pressures

Organisms with complex life cycles have multiple life stages, each adapted to its habitat and habits. This flexibility may be one of the reasons that complex life cycles are so common in so many groups of organisms. Because separate life history stages can evolve independently in response to size- and habitat-specific selection pressures, complex life cycles can minimize the drawbacks of small, vulnerable early stages.

LARVAL FUNCTION AND ADAPTATION Functional specialization of particular life stages is a defining feature of complex life cycles. Having multiple stages with largely independent morphological features can result in a pairing of particular functions with particular stages. Such a pairing can reduce some of the trade-offs that result from simultaneously optimizing multiple functions.

An example of this type of specialization occurs in many insects with complex life cycles. Such insects spend their entire larval stage in a very small area—sometimes on a single plant. Insect larvae such as caterpillars and grubs are specialized eating and growing machines. They spend almost all of their time taking in food and turning it into body mass, without forming many complex morphological structures other than mandibles (chewing mouthparts or "jaws"). Once they have accumulated sufficient mass, these larvae metamorphose into adult butterflies, moths, and beetles, whose main function is often to disperse, find a mate, and reproduce. In extreme cases,

such as mayflies, the adults are incapable of feeding and live only the few hours or days it takes them to reproduce.

Marine invertebrate larvae are also specialized for feeding, although they perform this function while dispersing on ocean currents. For example, the larvae of many mollusks (such as snails and clams) and echinoderms (such as sea urchins and sea stars) have intricate feeding structures that cover most of the larval body. These structures, called ciliated bands, are ridges covered in cilia that beat in coordinated patterns to catch tiny food particles and move them, like a conveyer belt, toward the mouth. The ciliated bands wind and fold their way around the larval bodies, many of which have extra lobes or arms that support and elongate the ciliated bands. In sea urchins, the longer the larval arms, and the longer the ciliated band, the more efficiently the larvae are able to feed (Hart and Strathmann 1994).

Other specialized larval structures can help to protect small life cycle stages from being eaten by other organisms. Examples include the toxin-bearing spines of some caterpillars, the head spines of crab larvae (**FIGURE 7.19**), and the setae or bristles of polychaete worm larvae, which deter some predators by making the larva a large and uncomfortable mouthful.

TIMING OF LIFE CYCLE SHIFTS Most organisms with complex life cycles use different habitats and food resources at different life stages. Such shifts can occur abruptly, as in organisms that undergo metamorphosis, but they can also occur more gradually. Regardless of the speed with which changes in habitat and food preferences occur, different-sized and different-aged individuals of the same species may have very different ecological roles. We'll use the term *niche shift* to refer to such size- or age-specific changes in an organism's ecological function or

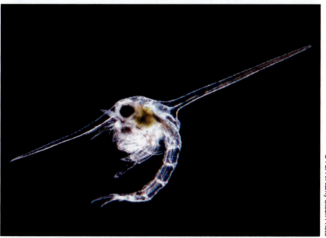

FIGURE 7.19 Specialized Defensive Structures in Marine Invertebrate Larvae The planktonic (floating) larvae of the sand crab *Corystes cassivelaunus* have defensive head spines that can make them difficult for fish to eat.

habitat. (As we'll see in Concept 9.5, an organism's *ecological niche* consists of the habitat and resource requirements that the organism needs to grow, survive, and reproduce.)

In species in which an abrupt metamorphosis occurs at the transition between life cycle stages, the organism spends relatively little time in vulnerable stages that are intermediate between larva and adult. In theory, there should be an optimal time to undergo metamorphosis, or any niche shift, that maximizes survival over the course of the life cycle. Thus, we might expect a niche shift to occur when the organism reaches a size at which conditions are more favorable for its survival or growth in the adult habitat than in the larval habitat.

Dahlgren and Eggleston (2000) tested this idea for the Nassau grouper (*Epinephelus striatus*), an endangered coral reef fish that spends its juvenile stages in and around large clumps of algae. Smaller juveniles spend their time hiding within the algae, whereas larger ones spend their time in rocky habitats near algal clumps. By tethering and enclosing juvenile fish of different sizes in the two habitats, Dahlgren and Eggleston were able to measure mortality and growth rates in each habitat. They found that the smaller juveniles were very vulnerable to predation in the rocky habitats, while the larger juveniles were less vulnerable and were able to grow faster in the rocky habitats. Thus, the niche shift in this species appears to be timed to maximize growth and survival, as predicted.

In some cases, the larval habitat may be so favorable for growth and survival that metamorphosis is delayed—or even eliminated altogether. For example, most salamanders have aquatic larvae that metamorphose into terrestrial adults, but some salamanders, such as the mole salamander *Ambystoma talpoideum*, can become sexually mature while retaining gills and remaining in the aquatic habitat (**FIGURE 7.20**). These aquatic, gilled adults are referred to as **paedomorphic**, which means that they result from a delay of some developmental events (loss of gills, development of lungs) relative to sexual maturation. Both aquatic paedomorphic adults and terrestrial metamorphic adults can exist in the same population of mole salamanders. The frequency of paedomorphosis in these mixed populations seems to depend on factors such as predation, food availability, and competition—all of which influence survival and growth in the aquatic habitat.

Some species reproduce only once, while others reproduce multiple times

An important life history trait influencing reproductive success is the number of reproductive events they have during their lifetime. **Semelparous** species (also known as monocarpic in plants) reproduce only once in a lifetime, whereas **iteroparous** species (also known as polycarpic in plants) have the capacity for multiple bouts of reproduction.

Many plant species complete their life cycle in a single year or less. Known as *annual plants*, such species are semelparous: after one season of growth, they reproduce

FIGURE 7.20 Paedomorphosis in Salamanders The mole salamander *Ambystoma talpoideum* can produce both (A) paedomorphic aquatic adults and (B) terrestrial metamorphic adults.

once and die. A more complex example of a semelparous plant is the century plant (a common name applied to several species in the genus *Agave*) of North American deserts. These plants grow vegetatively for up to 30 years before undergoing a single intensive bout of sexual reproduction. When it is ready to reproduce, a century plant produces a single stalk of flowers that is up to 6 m (20 feet) tall and towers over the rest of the plant. The plant produces a large quantity of seeds from this single reproductive event. The portion of the plant that produces the tall stalk of flowers dies after reproduction; hence, it is semelparous. At the genetic level, however, a century plant individual may not die when it flowers if it also reproduces asexually, producing genetically identical clones that surround the original plant (**FIGURE 7.21**). In this sense, some century plants are not semelparous after all—the clones survive after the flowering event and will eventually flower themselves.

A striking example of a semelparous animal is the giant Pacific octopus (*Enteroctopus dofleini*), which in its 3- to 5-year life span (relatively short for an octopus species) can reach about 8 m (25 feet) in length and weigh nearly 180 kg (400 pounds). The female of this marine invertebrate species lays a single clutch containing tens of thousands of fertilized eggs. She then broods the eggs for up to 6 months. During this time, the female does not feed at all; she is a constant presence over her eggs, cleaning and ventilating them. The female dies shortly after the eggs hatch, having exhausted herself in this intense period of parental investment. Other animals that exhibit semelparity include salmon, many spiders, and some insects such as butterflies.

This *Agave* produced the tall flowering stalk that extends behind it …

… but it also produced these genetically identical clonal offspring.

FIGURE 7.21 *Agave*: A Semelparous Plant? The *Agave* individual that produced the tall flowering stalk will die shortly after it flowers and so can be viewed as semelparous. But the individual that flowered also produced genetically identical clonal offspring. Thus, the genetic individual will live on after flowering, and in that sense it is not semelparous after all.

Why would an organism that lives multiple years only reproduce once in its lifetime? Theoretically, semelparous organisms gain an advantage in total lifetime reproductive output due to the conditions affecting the trade-off between reproduction and survival. Larger organisms have higher reproductive output, and reserving reproductive maturity to the end of the life cycle results in the highest reproductive output under certain conditions. If the probability of adult mortality is above a certain threshold, and the costs of reproduction are high even at low levels of reproductive output, then semelparity will result in higher lifetime reproductive output than iteroparity. Another reason that has been proposed for the evolution of semelparity is the benefit of producing massive amounts of offspring under high predation rates. Producing more offspring than can be consumed by predators allows some to escape and maintain the population.

Most organisms do not invest so heavily in single reproductive events. Iteroparous organisms engage in multiple bouts of reproduction over the course of a lifetime. Examples of iteroparous plants are long-lived trees such as

pines and spruces. Among animals, most large mammals are iteroparous. Of course, iteroparity can take a variety of forms, from plants that flower twice in a season and then die to trees that reproduce every year for centuries.

There are a multitude of different combinations of life history traits associated with reproduction and allocation of effort and resources. In the next section we consider some overarching conceptual models to match these life history patterns up with ecological conditions.

> **CONCEPT 7.4** Life history patterns can be classified along several continua.

LEARNING OBJECTIVES

7.4.1 Evaluate the environmental conditions that would favor the persistence of *r*-selected and *K*-selected species.

7.4.2 Describe the trade-offs in plant allocation described in Grime's competitive/stress/ruderal model.

7.4.3 Show how differences in species size or age can be accounted for in describing the allocation of energy and resources to reproduction and other life history stages.

Life History Continua

Ecologists have proposed several classification schemes for organizing patterns of life history traits in relation to the environment. Most of these schemes make broad generalizations about associated life history traits and attempt to place these associations along continua between two extremes that are shaped by the ecological conditions that influence mortality rates and resource availabilities. In this section, we examine the most prominent of these schemes and discuss how they relate to one another.

Live fast and die young, or slow and steady wins the race?

One of the best-known schemes for classifying life history diversity was also one of the first proposed. In 1967, Robert MacArthur and Edward O. Wilson coined the terms *r*-selection and *K*-selection to describe two ends of a continuum of life history patterns. The "*r*" in *r*-selection refers to the intrinsic rate of increase of a population, a measure of how rapidly a population can grow. The term **r-selection** refers to selection for high population growth rates. This type of selection can occur in environments where population density is low—for example, in recently disturbed habitats that are being recolonized. In this type of habitat, genotypes that can grow and reproduce rapidly will be favored over those that cannot. In contrast, **K-selection** refers to selection for slower rates of increase, which occurs in populations that are at or approaching *K*, the carrying capacity or stable population size for the environment in which they live (see Concepts 10.3 and

10.5 for in-depth discussions of *r* and *K*). *K*-selection occurs under crowded conditions, in which genotypes that can efficiently convert food into offspring are favored. By definition, *K*-selected populations do not have high population growth rates, because they are already near the carrying capacity for their environment and competition for resources can be intense.

One way to think of the *r*–*K* continuum is as a spectrum of population growth rates, from fast to slow. Organisms at the *r*-selected end of the continuum are often small and have short life spans, rapid development, early maturation, low parental investment, and high rates of reproduction. Examples of this "live fast, die young" end of the continuum include most insects, small short-lived vertebrates such as mice, and weedy plant species. In contrast, *K*-selected species tend to be long-lived, develop slowly, delay maturation, invest heavily in each offspring, and have low rates of reproduction. Examples of this "slow and steady" end of the continuum include large mammals such as elephants and whales, reptiles such as tortoises and crocodiles, and long-lived plant species such as oak and maple trees.

Like most classification schemes, the *r*–*K* continuum tends to emphasize the extremes. Most life histories are intermediate between these extremes, however, and hence the *r*–*K* approach is not informative in some situations. The distinction between *r*-selection and *K*-selection is perhaps most useful in comparing life histories in closely related species or species living in similar environments. For example, Braby (2002) compared three species of Australian butterflies in the genus *Mycalesis*. The species that occurs in the driest, least predictable habitats shows the most *r*-selected characteristics, including rapid development, early reproduction, production of many small eggs, and rapid population growth. In contrast, the two species found in more predictable, wet forest habitats have more *K*-selected characteristics.

Plant life histories can be classified based on habitat characteristics

In the late 1970s, Philip Grime (1977) developed a classification system specifically for plant life histories. The success of a plant species in a given habitat, he argued, is limited by two factors: stress and disturbance. Grime defined stress broadly as any external abiotic factor that limits vegetative growth (see Concept 4.1). Under this definition, examples of stress include extreme temperatures, shading, low nutrient levels, and water shortages. He defined disturbance broadly as any process that destroys plant biomass (see Concept 17.1). Under Grime's definition, disturbance can result from biotic sources such as outbreaks of herbivorous insects or abiotic sources such as fire.

If we consider that in a given habitat, stress and disturbance may each be either high or low, then there are four possible habitat types: high stress–high disturbance, low stress–high disturbance, low stress–low disturbance, and high stress–low disturbance. If we further consider that most habitats with high stress *and* high disturbance will not be suitable for plants, then there are three main habitat types to which plants may adapt. Grime developed a model for understanding the three plant life history patterns that correspond to these three habitat types: competitive (low stress–low disturbance), ruderal (low stress–high disturbance), and stress-tolerant (high stress–low disturbance) (**FIGURE 7.22**).

Grime defined competition between plants in a very specific manner as "the tendency of neighboring plants to utilize the same quantum of light, ion of a mineral nutrient, molecule of water, or volume of space." Under conditions of low stress and low disturbance, **competitive plants** that are superior in their ability to acquire light, minerals, water, and space should have a selective advantage.

FIGURE 7.22 Grime's CSR Model Grime categorized plant life histories within a triangle whose axes indicate the degree of competition, disturbance, and stress in the habitat type to which plants are adapted. Intermediate life history strategies are shown in the center of the triangle. (After J. P. Grime. 1977. *Am Nat* 111: 1169–1194.)

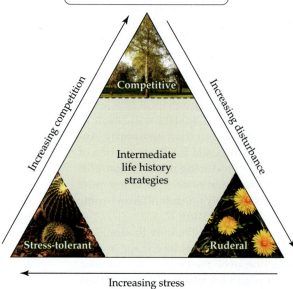

Competitive plants, such as birch trees, are adapted to low-stress, low-disturbance habitats.

Increasing competition

Increasing disturbance

Competitive

Intermediate life history strategies

Stress-tolerant plants, such as cacti, are adapted to high-stress, low-disturbance habitats.

Stress-tolerant

Ruderal

Ruderal plants, such as dandelions, are adapted to high-disturbance, low-stress habitats.

Increasing stress

Grime classified plants that are adapted to habitats with high levels of disturbance and low levels of stress as **ruderals**. The ruderal strategy generally includes short life spans, rapid growth rates, heavy investment in seed production, and seeds that can survive in the ground for long periods until conditions are right for rapid germination and growth. Ruderal species are often called "weedy" species and are adapted for brief periods of intense exploitation of favorable habitats after disturbance has removed competitors.

Finally, under conditions in which stress is high and disturbance is low, stress-tolerant plants are favored. Although stressful conditions may vary widely across habitats, Grime identified several features of **stress-tolerant plants**, including but not limited to slow growth rates, evergreen foliage, slow rates of water and nutrient use, high investment into defense from herbivores, and an ability to respond effectively to temporarily favorable conditions. Habitats favoring stress-tolerant plants might include places where water or nutrients are scarce or temperature conditions are extreme.

Grime's conceptual model posits that natural selection has resulted in three distinct yet very broad categories of life history strategies in plants. Although Grime focused on describing these three extreme strategies, he also recognized that intermediate strategies are commonly found. Indeed, various combinations of the three extreme strategies yield many possible intermediate strategies, such as competitive ruderals and stress-tolerant competitors, among others. However, the model also explicitly recognizes that there are trade-offs to the life history traits, and thus individual species can't be well adapted to all three of the evolutionary forces in the model.

Life histories can be classified independent of size and time

Unlike the classification schemes discussed above, an approach described by Charnov (1993) organizes life histories in a manner that removes the influence of size and time. As we saw in our discussion of the *r–K* continuum, size and time play a critical role in traditional classifications of life histories. For example, *r*-selected species are characterized as smaller and more short-lived than *K*-selected species. But if we could control for the effects of body size and life span, then we could ask whether closely related organisms experience similar selection pressures independent of those factors.

To illustrate this approach, we'll begin with the observation that the age of sexual maturity is positively correlated with life span in many species (Charnov and Berrigan 1990). Such a correlation is not surprising. Species with short life spans must mature in short periods, but the same is not true of species with long life spans, and as a result a positive correlation can arise automatically. One way to remove this effect of life span is to divide the average age of maturity of a species by its average

life span. This division yields a *dimensionless ratio*—that is, a ratio in which the units in the numerator (e.g., age of maturity in *years*) are identical to and hence cancel the units in the denominator (e.g., life span, also in *years*).

By removing the effects of variables such as size or (in our case) time, a dimensionless ratio allows ecologists to compare the life histories of very different organisms. Charnov and Berrigan compiled data for a wide range of bird, mammal, lizard, and fish species. To remove the effects of life span, they focused their analyses on the age of maturity: life span dimensionless ratio, which they denoted *c* (**FIGURE 7.23**). Their analysis revealed that *c* differed between ectothermic (fishes, lizards, and snakes) and endothermic (mammals and birds) organisms. For example, if we compare organisms with a given life span, the values of *c* indicate that it takes fishes three to six times longer to mature than mammals and birds, while it takes lizards and snakes two to four times longer. Such results can highlight major differences in the life histories of different groups of organisms, thus helping to make sense of life history variation.

While this dimensionless approach has some advantages over classification schemes that incorporate time and size, it also has potential disadvantages. Indeed, an emphasis on constant or "invariant" dimensionless life history parameters has been questioned by Nee et al. (2005), who argue that life history parameters can appear to be invariant simply as an artifact of the mathematical

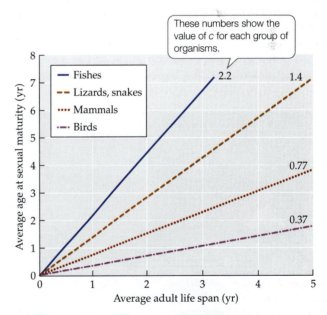

FIGURE 7.23 A Dimensionless Life History Analysis The average age at which females reach sexual maturity is plotted against the average female life span for different groups of organisms. The slope of each line yields the dimensionless ratio *c*: the average age of maturity divided by the average life span. (After E. L. Charnov and D. Berrigan. 1990. *Evol Ecol* 4: 273–275.)

? In groups of organisms for which *c* > 1, do most individuals live long enough to reproduce? Explain.

methods used to estimate them. Overall, there are many ways to organize the vast diversity of life history strategies. The classification scheme that is most useful in any given case will depend on the organisms and questions of interest. For example, the *r–K* continuum has a long history of use in relating life history characteristics to population growth characteristics, whereas Grime's scheme may be most appropriate for life history comparisons between groups of plants. Alternatively, dimensionless analyses may be most helpful when comparing life histories across broad ranges of taxonomy or size.

A CASE STUDY REVISITED

Nemo Grows Up

Why does a male clownfish that has lost his mate become a female rather than simply finding a new partner? As we have seen, large individuals often produce more offspring than do smaller individuals. In clownfish, the number of eggs an individual can produce is proportional to its body size. Thus, larger individuals can produce more eggs and presumably have a better chance of having some of their offspring survive. Smaller individuals are more easily able to make sperm cells, which are smaller and take fewer resources to produce. For these reasons, in clownfish and in many other animals, females are larger than males.

Changes in sex during the course of the life cycle, called **sequential hermaphroditism**, are found in 18 fish families and in many invertebrate groups (**FIGURE 7.24**). Researchers have hypothesized that these sex changes should be timed to take advantage of the maximum reproductive potentials of the different sexes at different sizes, and in some cases they appear to do so. This hypothesis helps to explain sex

changes in clownfish and the timing of those changes relative to size, but it leaves unanswered the question of how a hierarchy of clownfish is maintained within each anemone.

As a graduate student at Cornell University, Peter Buston set out to answer this question. He conducted experiments on a clownfish species, *Amphiprion percula*, that lives on reefs in Papua New Guinea. He found that each clownfish maintains the strict size hierarchy by remaining smaller than the fish ahead of it in line and bigger than the one behind it (**FIGURE 7.25**). If a fish grows to be too close in size to one of its anemone-mates, a fight results, which usually ends in the smaller fish being killed or expelled from the anemone. Buston suggested that the clownfish regulate their own growth to prevent such conflicts.

Buston also manipulated clownfish groups by removing the breeding males from anemones and measuring the growth of the remaining individuals. He found that the largest nonbreeder grew only enough to take the place of the breeding male; it avoided growing too big and threatening the female's dominance. Similarly, the next largest nonbreeder grew only enough to take the place of the fish that had become the breeding male, and so on. Thus, clownfish avoid conflict within their social groups by exerting remarkable control over their growth rates and reproductive status.

FIGURE 7.24 **Sequential Hermaphroditism** The moon wrasse (*Thalassoma lunare*) exhibits sequential hermaphroditism. Wrasses live among coral reefs in tropical and temperate seas. In some species, a change in sex, from female to male, may be accompanied by a change in color.

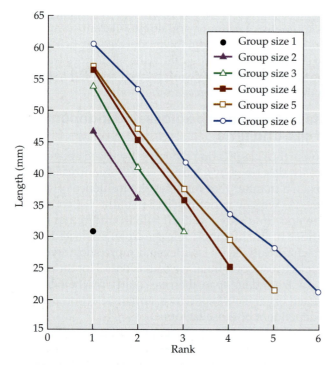

FIGURE 7.25 **Clownfish Size Hierarchies** Clownfish within an anemone regulate their growth to maintain a hierarchy in which each fish belongs to a distinct size class. Anemones may be home to between one and six fish, and the size of each fish is determined by that fish's rank and the size of the group in which it lives. (After P. M. Buston. 2003a. *Nature* 424: 145–146.)

🌱 CONNECTIONS *in* NATURE

TERRITORIALITY, COMPETITION, AND LIFE HISTORY
The physiology of clownfish growth regulation is not understood, but a more pressing ecological and evolutionary question is why the size hierarchy is maintained. What makes small clownfish bide their time as nonbreeders under the dominance of a single breeding female and male? The answer may lie in the clownfish's dependence on the protection of anemones for survival.

Clownfish are brightly colored, and they are poor swimmers. Outside the anemone's stinging tentacles, they are easy prey for larger fishes on the reef. Thus, expulsion from the anemone is often a death sentence. So the stakes are very high in conflicts between fish within an anemone: the loser will probably die without reproducing. This situation exerts strong selection pressure on the fish to avoid conflicts by regulating their growth. In evolutionary terms, growth regulation mechanisms have evolved because individuals that avoid growing to a size that leads to conflict with other fish have higher survival and reproductive rates (we described this process of adaptive evolution in Concept 6.2). Buston (2003b, 2004) demonstrated that remaining in an anemone as a nonbreeder is more advantageous than trying to leave the anemone and find a new one. Anemones are a limited resource for the clownfish, and those that bide their time once they find an anemone experience the highest lifetime fitness.

The scarcity of anemones also results in competition among clownfish at a key stage in their life history. As we have seen, hatchling clownfish disperse from their anemone and spend their early life stages in the open ocean. When they return to the reef, their survival depends on their choice of an anemone. The number of fish in an anemone is generally correlated with the anemone's size. However, Buston found that at any given time some anemones are undersaturated, meaning that they have room for more fish. If a juvenile fish is lucky enough to enter such an anemone, it is allowed to stay, and it enters the line of succession toward becoming a breeder. If the juvenile enters a saturated anemone, however, it is expelled, and it often dies before it can find another anemone. Similar settlement lotteries play out in many organisms that live in crowded habitats and compete for space. For example, in environments such as tropical rainforests, where many long-lived tree species compete for limited space and sunlight, the success of any one seed or seedling can depend on chance events, such as the death of a nearby large tree that creates a gap in the canopy (Denslow 1987). As we'll see in Concept 19.3, such settlement lotteries can play an important role in maintaining the diversity of species found in highly competitive environments. 🐟

SUMMARY

CONCEPT 7.1 Life history patterns vary within and among species.

7.1.1 Summarize the key stages that make up the life history of an organism.

- Life histories are diverse, varying between individuals of the same species as well as between species. The source of this variation may be genetic or environmental.

7.1.2 Explain how genetics and the environment act as controls on life history traits.

- Life histories are under genetic control and subject to natural selection. Environmental variation can also impact life histories through phenotypic plasticity.

7.1.3 Compare the benefits and costs associated with sexual versus asexual reproduction.

- Organisms may reproduce sexually or asexually. In many cases, the same organism can do both.

- There are advantages and disadvantages to sexual reproduction. The high levels of genetic variation resulting from sexual reproduction may be beneficial in challenging environments.

7.1.4 Describe how additional complexity in a life cycle, such as larval and adult forms, may benefit a species.

- Most organisms have complex life cycles with multiple stages that differ in size, morphology, or habitat.

CONCEPT 7.2 There are trade-offs between life history traits.

7.2.1 Illustrate how the number of offspring may affect the size of those offspring.

- There is a trade-off between offspring size and number, such that organisms tend to produce large numbers of relatively small offspring or small numbers of relatively large offspring.

7.2.2 Explain how providing care to offspring may compromise other functions in adults.

- Allocation of time and energy into caring for offspring lowers the ability of the adult to forage and may increase the risk of disease and predation on the adult.

7.2.3 Understand that allocating energy and resources to reproduction may affect parental growth, survival, and future reproduction.

- An individual's investment in current reproduction can result in a trade-off between reproduction and other life history traits, including survival, growth, and potential for future reproduction.

(Continued)

SUMMARY *(continued)*

CONCEPT 7.3 Organisms face different selection pressures at different life cycle stages.

7.3.1 Contrast the benefits and costs associated with small size in early life cycle stages.

- The small sizes of early life cycle stages make them vulnerable to predation and food shortages.

- Small life cycle stages are well suited to some important functions, such as dispersal and dormancy.

7.3.2 Explain how adaptations at specific stages in a complex life cycle may benefit the species.

- Complex life cycles allow life histories the flexibility to respond to differences in selection pressures on different life cycle stages.

7.3.3 Compare the benefits of semelparity and iteroparity in the context of total lifetime reproduction of an organism.

- Semelparous species reproduce only once in a lifetime, while iteroparous species reproduce multiple times.

CONCEPT 7.4 Life history patterns can be classified along several continua.

7.4.1 Evaluate the environmental conditions that would favor the persistence of *r*-selected and *K*-selected species.

- Environments subject to frequent disturbances and low population size favor species exhibiting *r*-selected traits. Where the environment is stable with population sizes near the carrying capacity, *K*-selected traits are favored.

7.4.2 Describe the trade-offs in plant allocation described in Grime's competitive/stress/ruderal model.

- Grime's triangular model categorizes plant life histories by the degree of competition, stress, and disturbance in the habitat type to which they are adapted.

7.4.3 Show how differences in species size or age can be accounted for in describing the allocation of energy and resources to reproduction and other life history stages.

- Charnov's dimensionless life history analysis attempts to remove the effects of size and time in order to compare life histories across a broad taxonomic range.

Review Questions

1. Many closely related animal species produce eggs of vastly different sizes. As we saw in Concept 7.3, one trade-off of producing larger eggs is that fewer eggs can be produced. Despite the apparent simplicity of this trade-off, it is still unclear why both strategies (many small eggs and few large eggs) are maintained in groups of closely related species. What are some other life history traits besides offspring number that might be correlated with egg size, and under what environmental conditions might those traits be advantageous? Can you think of any reasons why species that live in the same habitats continue to exhibit reproductive patterns that vary so widely?

2. Some animals exhibit both sexual and asexual reproduction, depending on the environmental conditions they experience. Rotifers are a classic example of this phenomenon. Females can produce diploid eggs by mitosis that hatch soon after release. In this manner, rotifer populations can double within hours. These same females, under other conditions, can produce haploid eggs that form males if unfertilized and form females if fertilized. What might be the reason for the maintenance in rotifers of both sexual and asexual reproduction?

3. The Nassau grouper is popular in Asia, where restaurant-goers can pick the grouper they want steamed for dinner from a selection of live fish swimming in tanks. Adult groupers can grow to large sizes (up to 3 feet long and 55 pounds), but those favored by restaurants are plate-sized juvenile and young adult fish. How would you expect the removal of these younger, smaller fish to affect the life history evolution of the remaining population? How might life history parameters such as age and size at reproduction and investment in growth versus reproduction evolve in response to fishing pressures?

Hone Your Problem-Solving Skills

California sheephead (*Semicossyphus pulcher*) are predatory fish that are born as females and change sex to males when they become large enough to defend territories from other males. Researchers measured the size at which sheephead became sexually mature ("size at maturation") and the size at which they changed sex from female to male ("size at sex change") for fish collected at Santa Catalina and San Nicolas Islands (Hamilton and Caselle 2015). The results are shown in the table. Little fishing occurred at these islands from 1940 to 1980. At Catalina, sheephead have been subjected to intensive fishing pressure from the early 1980s to the present. At San Nicolas, sheephead were subjected to intensive fishing from the early 1980s to 1998 but protected from fishing from 1999 on.

1. Do the data indicate that fishing by humans affects the sizes at which these fish reach sexual maturity and change sex? Explain.

2. Do size at maturation and size at sex change recover once fishing pressure is reduced? Explain which data you used to answer this question, and why.

| | Catalina | | San Nicholas | |
Year	Size at maturation (mm)	Six at sex change (mm)	Size at maturation (mm)	Size at sex change (mm)
1980	213	350	298	493
1998	198	256	211	294
2007	178	225	300	460

3. Fishing not only reduces the abundance of the targeted species, but also tends to remove larger individuals at higher rates than smaller individuals, leading to changes such as those shown in the table. Suppose larger sheephead produce more offspring than do smaller individuals (as is true for many species). Predict how protection from fishing will affect the average number of offspring produced per female and the abundance of the population over time.

ADDITIONAL RESOURCES

This textbook provides resources such as **Web Extensions** and additional **Climate Change Connections** for instructors who can share them with students. Please contact your instructor, if you would like to access that content.

Behavioral Ecology

Infanticide in Lion Packs: A Case Study

Lions are unique among cats in that they live in social groups called *prides*. A typical lion pride contains anywhere from 2 to 18 adult females and their cubs, along with a few adult males. The adult females form the core of the pride, and they are closely related: they are mothers, daughters, aunts, and cousins. The adult males in a pride may be closely related as well (e.g., brothers or cousins), or they may be a coalition of unrelated individuals that help one another.

The lions in a pride hunt cooperatively, and the females often feed, care for, and protect one another's cubs. But life in a pride has a dark side as well. Some males in the pride will attack and kill cubs in the pride (**FIGURE 8.1**), a behavior that seems both horrific and puzzling. Why do adult male lions do this? To shed light on this murderous behavior, let's consider some aspects of the life history of lions in more detail.

As young adults, male lions are driven from the pride into which they were born. A group of young males expelled from a pride may stay together to form a "bachelor pride." Bachelor prides may also consist of males from different prides that meet and begin to hunt together. By the time they are 4 or 5 years old, the young males in a bachelor pride are large and strong enough to challenge the adult males of an established pride. If their challenge is successful, the new males drive off the "dethroned" males, and they typically try to kill any young cubs that were recently fathered by those males. Although the females fight back, the new males often succeed in killing cubs.

If a female's cubs are killed, she becomes sexually receptive soon thereafter. In contrast, it can take up to 2 years for a female with cubs to resume sexual cycling. This delay in reproductive activity helps clarify the behavior of the incoming males. On average, incoming males remain with a pride for just 2 years before they

FIGURE 8.1 Killing the Cub The male African lion shown here is attempting to kill the juvenile offspring of another male; such attempts often succeed. Why might this behavior be evolutionarily adaptive for the murdering male?

© Laura Romin & Larry Dalton/ Alamy Stock Photo

© Roy Mangersnes/Minden Pictures

FIGURE 8.2 Females That Fight to Mate with Choosy Males Red phalarope (*Phalaropus fulicarius*) females (the two birds on the left) are larger and more colorful than the males of their species (on the right). In this species, the females fight over the right to mate with the males—and the males choose which females they will mate with.

are defeated and displaced by a new group of younger males. By killing cubs when he enters a pride, a new male increases the chance that he will reproduce before he is displaced by a younger male. As a result, incoming males that commit infanticide should leave more offspring than do males that do not commit infanticide. This logic suggests that infanticidal behavior by males is favored by natural selection, leading us to expect that it would be common in lion populations (which it is).

Infanticide is just one of the seemingly odd behaviors we see in animals. Fruit flies, for example, sometimes lay their eggs in food sources that contain high concentrations of ethyl alcohol, a toxic substance. Why do they do this? And why is it that the females of many species are more "choosy" than the males in selecting a mate—and yet in some species, such as the birds in **FIGURE 8.2**, the males are choosy and the females try to mate with as many males as possible? For answers, we turn to the strange and wonderful world of animal behavior.

Introduction

In nature, many of an animal's activities are centered on obtaining food, finding mates, or avoiding predators, all critical to the ecological success of a species. The behavioral decisions an animal makes often play key roles in its ability to meet these three critical needs. Consider the dilemma facing a young male lion deciding whether to challenge the adult males of a lion pride. An incorrect decision by the young male could lead to serious injury or death (if he is defeated in combat), or it could lead to a missed opportunity to join a pride and reproduce (if he delays combat unnecessarily). Likewise, a young trout that remains close to a hiding place while feeding may

increase its chance of escaping predators, but in so doing, it may forgo the opportunity to forage in areas that are rich with food but lacking in protective cover.

As these examples suggest, the behavioral decisions made by individuals have very real costs and benefits that affect their ability to survive and reproduce. These examples also highlight the fact that animal behaviors take place in an ecological setting: the behavioral decisions of the lion and the trout are made in the presence of competitors and predators. These behaviors affect survival and reproduction, and are central themes in the field of **behavioral ecology**, the study of the ecological and evolutionary basis of animal behavior.

Behavioral ecology is a dynamic field, broad in scope. In this chapter we emphasize three aspects of behavior particularly important in ecology: foraging behavior, mating behavior, and living in groups. Let's begin by taking a closer look at the types of questions that behavioral ecologists address in their research.

CONCEPT 8.1 Evolution is the basis for adaptive behavior.

LEARNING OBJECTIVES

8.1.1 Explain how natural selection can lead to the evolution of adaptive behaviors.

8.1.2 Illustrate how the environment can interact with genetics to influence behavior.

An Evolutionary Approach to Behavior

Researchers studying animal behavior can seek to answer questions using a variety of approaches focusing on different aspects of the behavior under study. You might ask, for example, why a robin hopping around your yard periodically tilts its head to the side. It turns out that robins do this because their sensory and nervous systems can detect the faint sounds of worms moving through the soil. Thus, one explanation for the robin's behavior might focus on how the required sensory equipment works. Furthermore, hunting by listening might enable a robin to detect otherwise hard-to-find prey. Hence, a second explanation of the robin's head-tilting behavior might focus on whether listening for worms increases the efficiency of foraging, thus enhancing the bird's survival and reproductive success. If so, then this behavior may have become common over time because it was favored by natural selection.

Notice that the first explanation we mentioned addresses a "how" question about behavior: it looks within an individual bird to explain how the head-tilting behavior functions. By focusing on events that take place during an animal's lifetime, this approach seeks to explain behaviors in terms of their immediate or **proximate causes**. In contrast, the second explanation addresses a "why" question about behavior: it examines the evolutionary and historical reasons for a particular behavior. By addressing previous events that influenced the features of an animal as we know it today, this approach seeks to explain behaviors in terms of their evolutionary or **ultimate causes**.

Although behavioral ecologists examine both proximate and ultimate causes in their research, they are primarily concerned with ultimate explanations of animal behaviors. We will follow their lead in this chapter, focusing on selected ultimate explanations for why animals behave as they do. We'll begin by examining how natural selection affects behavior.

Natural selection shapes animal behaviors over time

As we've seen in earlier chapters of this book, an individual's fitness—that is, its ability to survive and reproduce—depends in part on its behavior. Therefore, natural selection should favor individuals whose behaviors make them efficient at activities such as foraging, obtaining mates, and avoiding predators.

To explore this idea further, recall from Concept 6.1 that natural selection is not a random process. Instead, when natural selection operates, individuals with particular traits consistently leave more offspring than do other individuals *because of* those traits. If the traits that confer advantage are determined in part by genes, then individuals that have those traits will pass them to their offspring. In such cases, natural selection can cause **adaptive evolution**, a process in which traits that confer survival or reproductive advantages tend to increase in frequency over time.

Applying these ideas to heritable behaviors, we would predict that as an outcome of natural selection, individuals should exhibit behaviors that improve their chances of surviving and reproducing. As illustrated by the practice of infanticide by male lions—a behavior that increases a male's chance of reproducing before he is displaced by a younger male—animal behaviors are often consistent with this prediction. Further support comes from studies that have documented adaptive behavioral change as it took place.

For example, Silverman and Bieman (1993) reported an adaptive behavioral change in populations of the German cockroach (*Blattella germanica*) (**FIGURE 8.3**). In the 1980s, efforts to control this cockroach often used baits that combined an insecticide with a feeding stimulant, such as glucose. Initially, these baits were highly effective, killing the vast majority of the cockroaches that encountered them. Over time, however, a novel behavioral adaptation, glucose aversion, emerged in some cockroach populations. Cockroaches from these populations avoided feeding on glucose, causing the baits to become ineffective. This change in the feeding behavior of German cockroaches is heritable and is controlled by a single gene (Silverman

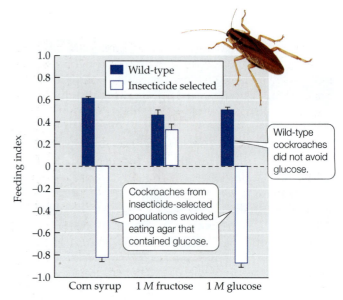

FIGURE 8.3 An Adaptive Behavioral Response Feeding behavior in two populations of the German cockroach (*Blattella germanica*), one of which (wild-type) had no prior exposure to insecticides, while the other had been exposed to insecticides. Cockroaches could choose to eat plain (unsweetened) agar, agar that contained one of three sources of sugar—fructose, glucose, or corn syrup (which contains both fructose and glucose)—or both. The diets the cockroaches selected were characterized by a feeding index ranging from 1.0 (indicating that 100% of their diet consisted of agar containing glucose) to −1.0 (indicating that 100% of their diet consisted of plain agar). Error bars show one standard error (SE) of the mean. (After J. Silverman and D. N. Bieman. 1993. *J Insect Physiol* 39: 925–933.)

? Give both a proximate and an ultimate explanation for glucose aversion in *B. germanica*.

and Bieman 1993). In particular, glucose aversion appears to result from mutations that affect taste receptor neurons. In individuals that exhibit glucose aversion, the presence of glucose activates taste receptor neurons that in other individuals are activated only by bitter substances (Wada-Katsumata et al. 2013).

The increase in the frequency of glucose aversion in populations of cockroaches exposed to baits containing glucose shows how natural selection under different environmental conditions can shape behaviors over time. But for selection to have this effect—and for ultimate explanations of behavior to be convincing—a behavior must be determined at least in part by genes. Because later sections of this chapter emphasize ultimate explanations of behavior, we turn now to a closer examination of this key underlying assumption: that animal behaviors are determined by genes.

Behaviors are determined by genes and by environmental conditions

Many characteristics of an animal, including aspects of its behavior, are influenced both by genes and by environmental conditions (see Concepts 6.2 and 7.1). Later in this chapter, we'll discuss how certain features of the environment, such as the presence of predators, can alter an animal's behavior. Here we'll focus primarily on genes, but it is essential to bear in mind that environmental conditions also affect most behaviors, even those that are strongly influenced by genes.

The glucose-aversion behavior of cockroaches that we have just discussed is heritable and appears to be controlled by a single gene. However, this behavior is a relatively simple one—a cockroach either avoids glucose or it does not. But is there evidence for a genetic basis to more complex behaviors?

Weber et al. (2013) examined the genetics of one such complex behavior, burrow construction in mice. They studied two closely related species, oldfield mice (*Peromyscus polionotus*) and deer mice (*P. maniculatus*). In the wild, oldfield mice build complex burrows with a long entrance tunnel and an escape tunnel, while deer mice build much simpler burrows (**FIGURE 8.4**). Most other *Peromyscus* species construct simple burrows, or no burrows at all. The complex burrows built by oldfield mice are unique, and they may be an adaptation to avoiding predators in open habitats that provide little protective cover. Although snakes and other predators might spot oldfield mice easily in such habitats, the length of the burrow entrance tunnel and the presence of an escape tunnel might help a mouse evade a pursuing predator.

Weber et al. wanted to evaluate the contribution of genes to the unique burrowing behavior of oldfield mice. To do this, they took advantage of the fact that oldfield mice and deer mice can interbreed to form viable and fertile hybrid offspring and that both species exhibit their usual field burrowing behaviors in a laboratory enclosure. They examined the burrowing behaviors of oldfield mice, deer mice, and two different types of hybrid offspring: first-generation (F_1) hybrids (offspring of matings between oldfield mice and deer mice) and later-generation backcross hybrids (offspring of matings between F_1 individuals and deer mice).

The results indicated that the complex burrowing behavior of oldfield mice is affected by several different DNA regions. As expected, all of the oldfield mice and none of the deer mice built escape tunnels. In addition, 100% of the F_1 hybrid mice built escape tunnels, and roughly 50% of the backcross mice built escape tunnels (**FIGURE 8.5**). These results and additional genetic mapping by Weber et al. indicate that a single chromosomal location, or *genetic locus*, controls whether the mice build escape tunnels, and that the genes for tunnel-building behavior are dominant. The complex burrow-building behavior of oldfield mice appears to have evolved as a combination of two simpler behaviors (construction of long entrance tunnels and construction of escape tunnels).

The study by Weber et al. is unusual in its use of both behavioral observations and genetic mapping to examine how genes affect a complex behavior of ecological

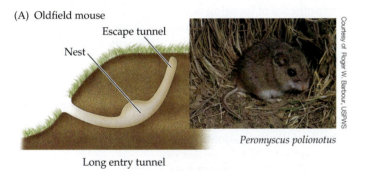

(A) Oldfield mouse

Escape tunnel
Nest
Long entry tunnel

Peromyscus polionotus

(B) Deer mouse

Nest
Short entry tunnel

Peromyscus maniculatus

FIGURE 8.4 Distinctive Mouse Burrows (A) The oldfield mouse (*Peromyscus polionotus*) constructs a complex burrow with a long entrance tunnel and an escape tunnel. (B) The deer mouse (*P. maniculatus*) constructs a simpler burrow, with a short entrance tunnel and no escape tunnel. (After E. Callaway. 2013. *Nature* 493: 284.)

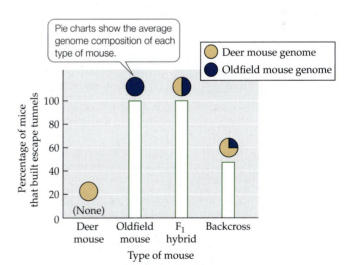

FIGURE 8.5 The Genetics of Escape Tunnel Construction The graph shows the proportions of deer mice, oldfield mice, F_1 hybrids, and backcross mice (i.e., offspring of a hybrid mouse and a deer mouse) that constructed burrows with escape tunnels. (After J. N. Weber et al. 2013. *Nature* 493: 402–405.)

? Do the colors shown in the pie charts match what you would expect based on the types of mice used in this study? Explain.

importance. Although relatively few studies have identified genes that affect other such behaviors, a wide range of behaviors are known to be heritable, and typically those behaviors are influenced by multiple genes (van Oers and Sinn 2013).

Overall, it is clear that genes affect many behaviors, but it is important to keep a few caveats in mind. In particular, it is usually a mistake to assume that behaviors are under the control of one or a few genes. It is also wrong to assume that an individual that has an allele associated with a certain behavior will always perform that behavior—like an inflexible robot under the strict control of its genes. Instead, two individuals with identical alleles may behave differently. Moreover, individuals often change their behavior when in different environments, and these changes can occur quickly. For example, when the COVID-19 pandemic hit in 2020, noise created by motorized vehicles in cities impacted by business and school shutdowns decreased substantially. In response to the quieter conditions, white-crowned sparrows (*Zonotrichia leucophrys*) in the San Francisco Bay area altered their song performance and decreased the volume of their songs by 30%. Birds were able to hear songs proclaiming territorial occupation from twice as far away compared to conditions under the previous noise levels prior to the pandemic (Dewberry et al. 2020). By assuming that genes affect behaviors and that natural selection has molded behaviors over time, we can make specific predictions about how animals will behave in particular situations. Even when these predictions turn out to be wrong, an evolutionary view of behavior provides a productive approach to the study of animal behavior that can help us understand how animals interact in nature.

> **CONCEPT 8.2**   Animals make behavioral choices that enhance their energy gain and reduce their risk of becoming prey.

LEARNING OBJECTIVES

8.2.1   Explain the theory of optimal foraging by outlining the factors that influence the net benefit of foraging.

8.2.2   Summarize what determines optimal foraging in an area with different food densities with reference to the marginal value theorem.

8.2.3   Describe how the presence of predators can impact foraging behavior.

Foraging Behavior

As we've seen, there are costs and benefits to the behavioral choices that animals make, which can influence the overall benefit of a behavior to animals. In this section, we'll consider the trade-offs faced by animals in their search for food and the potential for behavior to maximize the benefit they receive while foraging.

Optimal foraging theory addresses behavioral choices that enhance the rate of energy gain

Animals face several challenges when searching for food (Concept 5.4). The abundance of food and its nutritional content vary in space and time. Additionally, how easy the food items are to detect, capture, or subdue can impact an animal's success. These factors will influence the nutritional benefit and the amount of energy that must be expended to acquire food, and therefore its net benefit to the animal. For example, a lizard foraging for ants may find higher abundances of ants near the colony's tunnels and an abrupt decrease when moving away from the ant domicile. Ant species also vary in their concentrations of nitrogen and carbon and in their abilities to defend against predators. Feeding in areas with the most dense assembly of ant species with the highest nutrient concentrations and the greatest ease of capture would provide the greatest benefit to the foraging lizard.

Ecologists have theorized that foraging efficiency can be viewed as the energy benefit per amount of energy expended to find food. An optimal solution to the challenges posed above would result in the maximum energy gained per unit of energy expended. If energy is in short supply and impacts an animal's fitness, then selection should act on foraging behavior to increase efficiency. **Optimal foraging theory** was developed by several ecologists in the 1960s and 1970s as a framework to develop testable hypotheses of foraging behavior (MacArthur and Pianka 1966; Pyke 1984). This approach has been used for several aspects of foraging, focusing primarily on food patchiness, quality, and the ease of acquisition.

Optimal foraging theory provides an opportunity to use straightforward mathematical models to evaluate hypotheses associated with adaptive behavior. In simple mathematical terms, optimal foraging can be represented by the following equation:

$$E_{net} = E_{gross} / (h+s)$$

where E_{net} is the net energy obtained from a food source, E_{gross} is the total energy from a food source, h is the time spent handling the food source, and s is the time spent searching for the food. The net energy value is determined by the energy value of the food item as well as the energy expended in handling and searching for it. Food that requires substantial handling (e.g., a tough nut to crack or a fighting prey) will yield lower net energy than food that requires less handling. Likewise, food that is sparsely distributed will require greater searching time than food that is more abundant.

Simple conceptual models that describe the net amount of energy that an animal gets from its food can also be used to formulate predictions of optimal foraging (**FIGURE 8.6**). As an animal increases its effort to forage (the time and energy it spends searching for, capturing, subduing, and consuming food), the total amount of

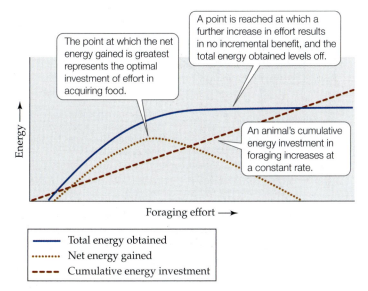

The point at which the net energy gained is greatest represents the optimal investment of effort in acquiring food.

A point is reached at which a further increase in effort results in no incremental benefit, and the total energy obtained levels off.

An animal's cumulative energy investment in foraging increases at a constant rate.

Energy →

Foraging effort →

— Total energy obtained
···· Net energy gained
--- Cumulative energy investment

FIGURE 8.6 Conceptual Model of Optimal Foraging
The net energy gained from foraging (dotted curve) equals the total energy obtained from the food acquired (solid curve) minus the cumulative energy invested in acquiring that food (dashed curve). This simple model can be used to test whether animals forage in a manner that results in the maximum benefit, based on estimates for the total energy obtained and the cumulative energy invested. (After G. Parker and J. Smith. 1990. *Nature* 348: 27–33.)

? Suppose you could estimate the net energy gained at different levels of foraging effort expended by lizards eating ants in the desert. How could you use that information to test whether the lizards foraged optimally?

energy obtained (blue curve) initially increases rapidly. At some point, however, a further increase in foraging effort provides relatively little additional energy, and the net energy gain begins to decrease. Several factors may cause this decrease, including a limitation on how much food the animal can carry or ingest.

While the models discussed here are simple, they provide a basis for making quantitative predictions about animal foraging behavior. More sophisticated models have been used to derive hypotheses that can be tested under field or laboratory conditions. An important component of these models is the currency (such as net energy gain) that is used to determine the benefit. Such models might incorporate, for example, net energy gained, time spent feeding, and risk of predation (Schoener 1971). If foraging behavior is an adaptation to limited food supplies, then we must be able to relate the benefit of that behavior to the survival and reproduction of the animal.

TESTS OF OPTIMAL FORAGING THEORY Research addressing optimal foraging has focused on diet selection, selection of patches to feed in, time spent in food patches, and prey movements (Pyke 2019). In one of the key early tests, John Krebs et al. (1977) devised a unique way to evaluate whether great tits (*Parus major*), a common bird found throughout much of Eurasia and northern Africa, selected prey types of greatest profitability (energy gained

per unit of handling time). They placed captive birds next to a moving conveyer belt carrying prey that differed in size (large and small mealworms) and in the time required to obtain them (each of the small mealworms was taped to the surface of the conveyer belt, thus requiring greater handling time, which decreased their profitability). By changing the proportions of the prey types and the distances between adjacent prey on the conveyer belt (variation in effort associated with *search time*), the researchers varied the profitability of the large and small mealworms. Using a model of optimal foraging and measurements of the times it took individual birds to subdue and consume the prey (*handling time*), they predicted how frequently the birds should select the large mealworms as encounter rates with the two prey types were varied. The birds consumed an increasing percentage of large mealworms as the relative profitability of those larger prey increased (**FIGURE 8.7**), just as the model predicted.

A field study by Meire and Ervynck (1986) focused on the diet selection of the Eurasian oystercatcher (*Haematopus ostralegus*), a shorebird that eats bivalves (e.g., clams and mussels). Oystercatchers must find a bivalve buried in the sand, lift it out, and open it before they can eat it. For bivalves below a certain size, the net energy

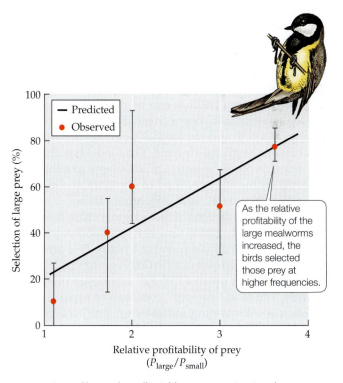

As the relative profitability of the large mealworms increased, the birds selected those prey at higher frequencies.

FIGURE 8.7 Effect of Profitability on Food Selection
Krebs et al. used an optimal diet selection model, along with measurements of prey handling time for individual birds, to predict the rate at which great tits (*Parus major*) would select large over small mealworms as their encounter rates with the two prey types were varied (expressed as the calculated ratio of profitabilities of the prey types), where profitability refers to the energy per unit of handling time. Error bars show ± one SE of the mean. (After J. B. Krebs et al. 1977. *Anim Behav* 25: 30–38.)

gain from this effort is small, setting a lower limit on the bivalve size selected by the oystercatchers. Bivalves above a certain size have thicker shells and require more effort to open, setting an upper limit on the bivalve size selected by the birds. Meire and Ervynck demonstrated that oystercatchers select prey of sizes that fall between these limits, which provide the most energy gain for the effort, despite the relatively low abundance of prey of these sizes.

THE MARGINAL VALUE THEOREM Another aspect of optimal foraging theory considers the habitat in which an animal forages as a heterogeneous landscape made up of patches containing different amounts of food. To optimize its energy gain, an animal should forage in the most profitable patches—those in which it can achieve the highest energy gain per unit of time. We can also consider the benefit obtained by a foraging animal from the perspective of time spent in a patch. Once the forager finds a profitable patch, its rate of energy gain is initially high, but that rate decreases and eventually becomes marginal as the forager depletes the food supply. Simple conceptual models predict that a foraging animal should stay in a patch until the time when the rate of energy gain in that patch has declined to the average rate for the habitat (known as the *giving-up time*), then depart for another patch. The giving-up time should also be influenced by the distance to other patches. Effort must be invested in traveling to another patch, so the animal may accept a lower rate of energy gain if the distance between patches is greater. This conceptual model, called the **marginal value theorem**, was initially developed by Eric Charnov (1976). It can be used to evaluate the influences of distance between patches, the quality of the food in a patch, and the animal's energy-extraction efficiency on the giving-up time. The model has also been extended to other "giving-up" problems in behavioral ecology, including how long to copulate and when to cease guarding a nest and seek other mates.

One of the predictions of the marginal value theorem is that the longer the travel time between food patches, the longer an animal should spend in a patch. This prediction was tested by Richard Cowie (1977) using a laboratory setup with great tits in a "forest" composed of wooden dowels. The food "patches" consisted of sawdust-filled plastic cups containing mealworms. The "travel time" among patches was manipulated by placing cardboard covers on top of the food cups and adjusting the ease with which they could be removed by the birds. Cowie used the marginal value theorem to predict the amount of time the birds should spend in the patches, based on the travel time between them. His results matched the model predictions fairly well (Cowie 1977) (**FIGURE 8.8**). Similar results have been obtained from other laboratory experiments as well as from studies using observational studies of foragers, such as Munger's 1984 study on the behavior of horned lizards (*Phrynosoma* spp.) foraging for ants in the Chihuahuan Desert.

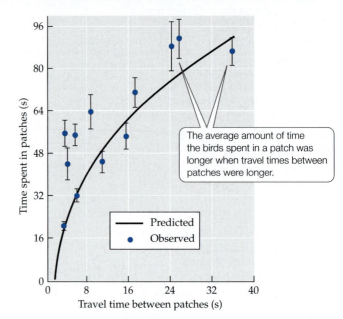

FIGURE 8.8 Effect of Travel Time between Patches In a laboratory experiment, Cowie used the marginal value theorem to predict how the travel time between patches would affect the average amount of time great tits (*Parus major*) spent in a patch. Error bars show ± one SE of the mean. (After R. J. Cowie. 1977. *Nature* 268: 137–139.)

The marginal value theorem also predicts that the time spent in food patches is positively related to their quality. Watanabe et al. (2014) studied foraging rates of Adélie penguins (*Pygoscelis adeliae*) on krill (*Euphausia crystallorophias*) in Antarctic waters. Using animal-borne video cameras and sensors to detect rapid head movement associated with capturing krill, they tested whether foraging penguins spent more time in denser patches of krill. Their results supported the marginal value theorem's prediction that in the short term (individual dives), the penguins spent more time in patches with higher krill abundance (**FIGURE 8.9**). This and other studies using methods that remotely track foraging animals and detect prey capture have led to model refinements that incorporate energetic costs of capturing prey in the consideration of time spent in patches (Foo et al. 2016).

Research testing optimal foraging theory has increasingly incorporated additional complexities into its models and predictions, including risk of predation and environmental stress, foraging experience of the animal, and interactions with other species (competition, predation, facilitation). Generally, optimal foraging theory has been successful at supporting qualitative predictions (descriptions of behavior construed as consistent with the theory, such as eating a more diverse diet under certain conditions) but less so with more quantitative ones (predictions based on measurements of foraging behavior, such as the average time spent feeding in a patch) (Pyke 2019). Optimal foraging theory best describes the foraging behavior of animals that feed on immobile prey and applies less well to animals feeding on mobile prey (Sih and Christensen 2001). In addition, the assumptions that energy is always in short supply

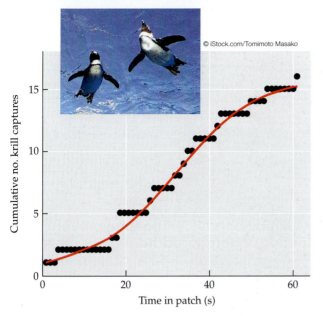

Cumulative no. krill captures vs Time in patch (s)

FIGURE 8.9 Foraging in Adélie Penguins The relationship between the amount of time spent in a patch of krill (*Euphausia crystallorophias*) by foraging Adélie penguins (*Pygoscelis adeliae*) versus the richness of the patch (as estimated by the rate of krill captures). The data support the prediction of the marginal value theorem that foragers will spend greater time in richer patches than in those with fewer prey. (After Y. Y. Watanabe et al. 2014. *Proc R Soc B* 281: 20132376.)

and that a shortage of energy dictates foraging behavior may not always be correct. Carnivores, in particular, may not lack for food resources to the degree assumed in optimal foraging models (Jeschke 2007). Furthermore, resources other than energy may be involved in the selection of food items, particularly nutrients such as nitrogen and sodium. And as we'll see next, additional considerations for foragers include the risk of predation and the defenses of prey.

Individuals often alter their foraging decisions when predators are present

As noted above, optimal foraging behavior may be compromised by the risk of exposure to predators of the foraging animal. An individual that is well fed but does not survive long enough to reproduce will not pass its genes to future generations. Foragers face trade-offs in which achieving one objective (such as eating) comes at the expense of another (survival). Recall from Concept 4.1 that trade-offs are traits or behaviors that are beneficial for one function but negatively influence traits or behaviors for another function.

FIGURE 8.10 Movement Responses of Male and Female Elk Results from a statistical analysis of daily movement patterns of male (A) and female (B) elk show that the probability of finding elk in grasslands drops when wolves arrive, then rises when wolves depart. (After S. Creel et al. 2005. *Ecology* 86: 3387–3397.)

? Compare and contrast how male and female elk respond to the presence of wolves.

Trade-offs that affect foraging decisions may be related to predators (an herbivore may avoid an area with ample food if predators are present), environmental conditions (in the desert, a foraging animal may retreat to a burrow or shade when temperatures become too hot), or physiological conditions (a hungry animal may tolerate greater risks when foraging than will a well-fed animal). Our focus here will be on how predators affect foraging decisions.

Creel et al. (2005) studied how the presence of wolves affected the foraging behavior of elk (*Cervus elaphus*) in the Greater Yellowstone Ecosystem. The researchers used GPS radio collars to track the daily movements of elk. On days when wolves were known to be present in the area, elk moved into wooded regions rather than the nearby grasslands where they usually preferred to forage. The wooded regions offered less food than the grasslands but more protective cover: in the grasslands they were more vulnerable to wolf predation. Results from a statistical analysis of elk movements provided additional evidence that elk moved into forests when wolves arrived and returned to grasslands when wolves departed (**FIGURE 8.10**).

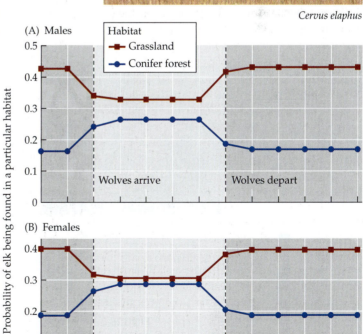

Cervus elaphus

Similar results have been found in aquatic environments. For example, Werner et al. (1983) examined how the presence of predators affected the foraging decisions of the bluegill sunfish (*Lepomis macrochirus*). Identical sets of sunfish in three size classes (small, medium, and large) were introduced on both sides of a divided pond; a predatory fish, the largemouth bass (*Micropterus salmoides*), was also introduced on one side of the pond. The sizes of sunfish and bass were selected such that smallest sunfish would be very vulnerable to the bass predators, while the largest ones would be too large for the bass to eat. Medium and large sunfish foraged in similar ways on both sides of the pond, with their habitat choice and diet matching predictions based on optimal foraging theory. The same was true for the small sunfish in the half of the pond lacking predators. In the presence of predators, however, they increased the time they spent foraging in vegetation, a habitat that provided greater cover from the bass but only one-third the rate of food intake available in more open habitats.

Researchers have also tested whether a perceived risk of predation can alter foraging patterns, even in the absence of actual predation. In one such study, Zanette et al. (2011) exposed the nests of some song sparrows (*Melospiza melodia*) to recordings of calls and sounds from their predators (such as raccoons, ravens, or hawks), while other nests were exposed to recordings of calls and sounds from nonpredators (such as seals or geese). (The researchers had protected all the nests from actual predators with electric fencing and netting.) Song sparrows exposed to recordings of predators fed their young fewer times per hour than did sparrows that heard recordings of nonpredators (**FIGURE 8.11**). Song sparrows that heard predators also built their nests in denser, thornier vegetation

and spent less time incubating their eggs than did sparrows exposed to recordings of nonpredators. We'll explore the consequences of such behavioral changes in Connections in Nature at the end of this chapter.

The song sparrow, elk, and sunfish examples are representative of hundreds of other studies showing that prey alter their foraging behavior in the presence of predators (Lima 1998; Verdolin 2006). As we'll see next, when predators are present, prey make other behavioral changes that can reduce their chance of being eaten.

Prey exhibit behaviors that can prevent detection or deter predators

Predators can exert strong selection on prey populations. As a result of such selection pressures, prey species have evolved a broad range of defenses against their predators. We'll focus here on antipredator behaviors; in Concept 13.2, we'll discuss other forms of defense, including physical defenses, toxins, and morphological forms of camouflage.

Antipredator behaviors include those that can help prey avoid being seen, detect predators, prevent attack, or escape once attacked (**FIGURE 8.12**). Behaviors that can help prey avoid being seen include hiding, remaining still when predators are nearby, and performing risky activities (such as foraging) during times of day when predators are not active. Other animals make themselves difficult to see by resembling objects that are less preferred by predators, such as portions of flower petals (in some caterpillars) or feces. With respect to detecting predators, prey often remain highly vigilant for predators, and some birds, lizards, and mammals can remain alert even while sleeping (see Figure 8.12B). There is also a wide variety of ways that prey seek to prevent attack once they are seen. For example, juvenile decorator crabs (*Libinia dubia*) attach to their bodies bits of an alga that local fishes find unpalatable, an action that was found to increase their rate of survival (Stachowicz and Hay 1999). Older decorator crabs that are too large for the fish to eat do not engage in this behavior. When threatened, some prey make sudden movements or display markings that confuse the predator, as illustrated by the display of eyespots shown in Figure 8.12C. Some prey send predators a signal, in effect conveying, "I see you, I'm faster than you, so don't bother to attack me." The stotting behavior of antelopes (see Figure 8.12D) is thought to be one such signal. Other examples of prey signaling to prevent attack include lizards that perform "push-ups" (indicating their overall physical condition) and ground squirrels

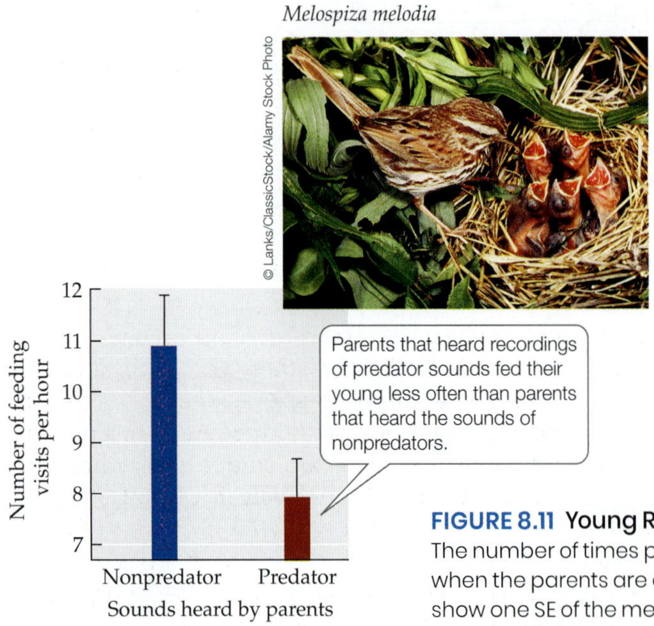

Melospiza melodia

Parents that heard recordings of predator sounds fed their young less often than parents that heard the sounds of nonpredators.

FIGURE 8.11 Young Receive Less Food When Parents Fear Predators The number of times per hour that song sparrow parents feed their offspring drops when the parents are exposed to recordings of sounds made by predators. Error bars show one SE of the mean. (After L. Y. Zanette et al. 2011. *Science* 334: 1398–1401.)

FIGURE 8.12 Examples of Antipredator Behaviors (A) Slug caterpillars (Family *Limacodidae*) are covered in protective, stinging hairs, making them unpalatable to predators. (B) Australian sea lions (*Neophoca cinerea*) can literally sleep with one eye open, with half their brain in a state of sleep while the other half remains alert for danger. (C) When threatened, the peacock butterfly (*Aglais io*) displays eyespots, a transformation that can startle predators. (D) A springbok (*Antidorcas marsupialis*) displays a stiff-legged jumping behavior known as stotting or pronking, which is thought to discourage predators from pursuing the small gazelle. (E) When captured, hognose snakes such as this eastern hognose snake (*Heterodon platirhinos*) play dead and emit an odor that smells like decaying meat; this behavior can deter predators that will not eat carrion.

that deliberately approach rattlesnakes, often within striking distance, while waving their tails from side to side (tail-flagging). Tail-flagging was found to be effective in deterring rattlesnakes from striking, and it increased the chance that a snake would abandon its ambush site (Barbour and Clark 2012).

If a predator attacks and captures (or is about to capture) its prey, the potential victim may resort to extreme behaviors. A hognose snake, for example, may play dead when captured, extruding its tongue and emitting a foul odor that resembles the smell of decaying meat, all the while keeping a close eye on its attacker (see Figure 8.12E). This behavior may work because many predators will not eat carrion. As a last resort, many prey defecate, urinate, or extrude other unpleasant substances, such as the large amounts of mucus secreted by a hagfish under attack (this mucus sometimes suffocates the predator). Other species detach parts of the body when threatened or grabbed. A gecko, for example, can drop its tail, which wriggles on the ground, distracting the predator. Some sea cucumbers take such evasive maneuvers to a unique level: when captured, they turn themselves partially inside out, startling the attacker and covering it with a tangled mass of internal organs. The sea cucumber then detaches those organs and swims away; later, it regrows the missing organs in a remarkable example of self-regeneration.

Having examined the foraging and antipredator behaviors of animals from an evolutionary perspective, we turn now to another key animal activity: sexual reproduction.

> **CONCEPT 8.3** Mating behaviors reflect the costs and benefits of parental investment and mate defense.

LEARNING OBJECTIVES

- **8.3.1** Describe examples of the behaviors utilized by animals to increase their access to mates.
- **8.3.2** Evaluate the benefits of being choosy with mate selection and the conditions that favor selectivity by females versus males.
- **8.3.3** Describe conditions under which different mating systems would be favored.

Mating Behavior

Males and females differ in their sexual organs and in other ways that are directly related to reproduction. But there are other, more puzzling differences between them. Males are often larger or more brightly colored than females, they may possess unusual weapons (such as large horns), or they may have gaudy ornaments, such as the extravagant plumage of a male Argus pheasant (*Argusianus argus*; **FIGURE 8.13**) or a male peacock. In addition, males and females often differ in their mating behavior. In many species, the males may fight, sing loudly, or perform strange antics to gain access to females (**FIGURE 8.14**). Furthermore, males may be willing to mate with any female who will have them. Females, on the other hand, rarely attempt to court males and typically are more choosy about who they will mate with. What causes such differences between the sexes?

FIGURE 8.13 A Male Shows Off The Argus pheasant (*Argusianus argus*) is native to the understory of the dense tropical forests of Southeast Asia. The males display their remarkable tail feathers as they attempt to attract and mate.

Differences between males and females can result from sexual selection

Charles Darwin (1859, 1871) concluded that the often-extravagant features of males did not provide a general advantage to members of a species, reasoning that if they did, both sexes would have them. He proposed instead that such features resulted from **sexual selection**, a process in which individuals with certain characteristics gain an advantage over others of the same sex *solely with respect to mating success*. We'll focus initially on sexual selection among males.

EVIDENCE FOR SEXUAL SELECTION Darwin pointed out that when individuals compete against others of their gender for mates, they typically use either force or displays. A male lion, for example, tries to repel his rivals by force, while a male pheasant or peacock tries to attract females to him (and away from other males) by displaying his beautiful tail feathers.

In species in which males fight over the right to mate with females, Darwin (1871) argued, the large size,

FIGURE 8.14 A Male Courtship Dance The male Victoria's riflebird (*Ptiloris victoriae*) of Australia accentuates the bright colors of his plumage in this courtship display. Part of this courtship display includes the "sky-pointing" behavior seen here.

strength, or special weapons of such males could have evolved by sexual selection. To make his case, Darwin began by pointing out that males often fought ferociously over females. He then described how males with the largest size, strength, and weaponry typically won such battles and therefore sired more offspring than other males. The large size, strength, or weaponry of the victors would then be passed on to their male descendants—causing these traits to become increasingly common over time. Multiple studies corroborate Darwin's argument. For example, in bighorn sheep, large rams with a long horn length typically defeat other males in battles over the right to mate with females and therefore sire more offspring than other rams (see the Case Study in Chapter 6). Since body size and horn length are heritable traits (Coltman et al. 2003; body size and horn length are also related to the age of the animal), the male offspring of the victors also tend to be large and strong. Over time, this process causes the body and horn growth rates of males to increase.

Darwin also thought that extravagant traits used by males to lure females could have arisen by sexual selection. For example, he wrote of his "conviction that the male Argus pheasant acquired his beauty gradually, through the preference of the females during many generations for the more highly ornamented males." But Darwin's hypothesis that female mating preferences could lead to the evolution of more highly ornamented or brightly colored males was experimentally tested by few researchers prior to Malte Andersson's classic 1982 study on the long-tailed widowbird (*Euplectes progne*).

Male long-tailed widowbirds are mostly black and have extremely long tail feathers, the longest of which reach 50 cm. In contrast, females are mottled brown and have short tails (approximately 7 cm). Like many other animals, male widowbirds establish **territories**, areas that they defend against other males. In the grasslands of Kenya, where Andersson studied these birds, male widowbirds establish and defend territories in which females can feed and build their nests.

To test whether female mating preferences could have driven the evolution of the long tails found in males, Andersson captured birds and subjected them to four treatments: (1) a control treatment in which the tails of the birds were not altered; (2) a second control treatment, in which the birds' tails were first cut at the midpoint and then glued back on; (3) a treatment in which the birds' tail lengths were shortened (cut to approximately 14 cm); and (4) a treatment in which the birds' tail lengths were increased (feathers cut from birds in treatment 3 were glued to the tails of these birds).

Andersson found that males with lengthened tails had higher mating success than control males or males with shortened tails (**FIGURE 8.15**). There were no differences among treatments in the courtship behavior of the males or the vigor with which they defended their territories. Overall, Andersson's results support the hypothesis that female mating preferences affect male mating success and hence may have selected for the extremely long tails of male widowbirds. Many other studies since have found similar results.

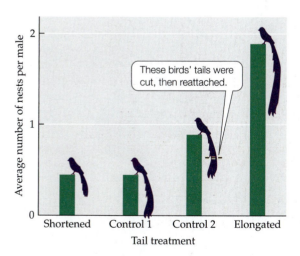

FIGURE 8.15 Males with Long Tails Get the Most Mates The mating success of male long-tailed widowbirds (*Euplectes progne*) depends on the length of their tails, as Malte Andersson discovered by experimentally altering the tails of wild birds. (After M. Andersson. 1982. *Nature* 299: 818–820.)

? Explain why Malte Andersson used the two types of controls described in the text.

BENEFITS TO CHOOSY FEMALES In some species, a male that attempts to convince a choosy female into mating with him may provide direct benefits to the female, such as gifts of food, help in rearing the young, or access to a territory that has good nesting sites, abundant food, or few predators. *Nuptial gifts*, usually offerings of food, are common in nature, and serve to enhance the chances that a female will select a male for mating. A bizarre form of nuptial gifts, which occurs in some spiders and mantid insects, involves *sexual cannibalism* or *sexual suicide* in which the male is consumed by the female after transferring his sperm to her. In redback spiders (*Latrodectus hasselti*) the male actually flips himself into the female's mandibles after successfully transferring sperm to her sperm receptacles (**FIGURE 8.16**). Why would the male do this, ending all future potential reproductive opportunities? Andrade (1996) discovered that a female redback spider was less likely to mate again if she had consumed her first mate, increasing the chances that the suicidal males would pass on their genes. Furthermore, the male's chances of survival and locating a female are very low (<20% for a single mate, <4% for subsequent matings), so the evolutionary cost of suicide for successful mating is low. The occurrence of sexual suicide is relatively

high in related spiders, indicating such a strategy is adaptive (Miller 2007).

In some cases there are few or no apparent direct benefits provided to choosy females in mate selection. Why do females prefer to mate with males that have certain features, such as the long tail of the widowbird in the example described above, variation in color, or a loud mating call? Several hypotheses suggest that the female receives indirect genetic benefits when she chooses such a male. This may occur when her choosiness enhances the probability that her sons will mate due to some associated trait that confers higher fitness, or simply that the male's greater attractiveness to females will be inherited by her sons, enhancing their reproductive success. Evidence in support of the link between sexual signals in males and associated fitness benefits comes from studies of the associations of male coloration and parasitic infections. For example, Molnár et al. (2013) studied how variation in throat and belly color of the European green lizard (*Lacerta viridis*) was associated with the degree of infection by blood parasites. Previous studies had shown that experimental alteration of the belly and throat coloration of European green lizards influences both female choice for mating

(A)

© Peter Yeeles/Alamy Stock Photo

(B) Male spider

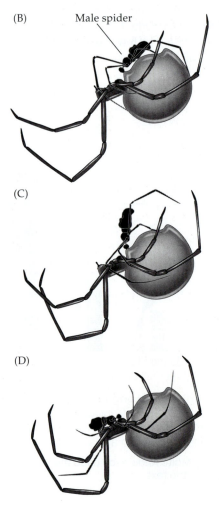

(C)

(D)

FIGURE 8.16 The ultimate gift for copulation (A) A female redback spider (*Latrodectus hasselti*). (B) A much smaller male spider aligns himself on the female's abdomen and inserts sperm into her reproductive tract. (C) The male spider then raises up and flips his body over into the jaws of the female (D) who consumes him, lowering the potential she will mate with another male. (B–D after L. M. Forster. 1992. *Australian J Zool* 40: 1–11.)

and the outcome of fights between males, with brighter colors favored. The researchers selected male European green lizards from populations in the wild and collected blood for assessment of protist parasitic infections. They also measured the coloration and brightness of the throat and belly using a spectrophotometer. The researchers found that nearly all (96%) of the male European green lizards they sampled were infected with blood parasites. However, the degree of infection varied among the lizards and was negatively correlated with the brightness of both the throat and belly (**FIGURE 8.17**). Since the immunological defense response to parasites has a genetic basis, female selection of mates with brighter throats and bellies would benefit her offspring, supporting the hypothesis of an indirect benefit to the females' choosiness.

Numerous studies indicate that choosy females receive a variety of direct and indirect benefits when they select their mates. Next, we'll examine a question raised in the opening pages of this chapter: Why are females usually more choosy than males about who they will mate with?

Gamete size, parental care, and ecological factors affect mating behavior

In addition to the differences we have discussed, females and males often differ in how much energy and resources they invest in their offspring. Such investments begin with the production of gametes and may continue in species in which the parents care for their offspring as they develop

into young adults. As we'll see, parental investments in offspring, along with ecological factors, can help us to understand the wide range of mating behaviors found in animal populations.

WHY ARE FEMALES USUALLY CHOOSIER THAN MALES?

One potential explanation for female choosiness comes from *anisogamy*: when there are large differences in size between the egg cells of a female and the sperm cells of a male (see Figure 7.8B). Because female gametes are so much larger than male gametes, a female typically invests more resources in producing a single gamete than does a male, including the nutritive material to support the developing embryo, and hence she has more at stake in each one. However, it should be noted that producing sperm also requires energy expenditure and can compromise fitness in males, and a comparison of single gametes may not provide an accurate view of the differential investment of males and females (Wedell et al. 2002).

In many species, females invest large amounts of resources as their offspring develop. For example, under natural conditions, a chicken hen incubates her eggs to keep them warm, and then cares for her chicks for several weeks after they hatch. The rooster does not provide any care for the young chicks. What is true for chickens is true for many other species as well: females spend more of their time and energy caring for their offspring than males do.

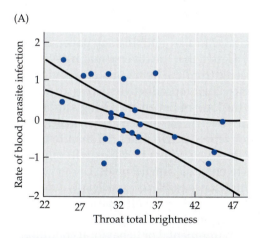

(A)

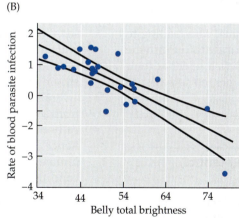

(B)

Filos.m/CC BY-SA 4.0

FIGURE 8.17 Skin Color Brightness Indicates Blood Parasites Male European green lizards (*Lacerta viridis*) collected from natural populations exhibit significant negative correlations between throat (A) and belly (B) brightness (reflection of light in visible wavelengths from 320 to 700 nm) and the amount of blood parasites (expressed as the residuals in the statistical model after accounting for other factors influencing infection rate). Center line is the correlation average, lines on either side are 95% confidence intervals. (After Molnár et al. 2013. *Naturwissenschaften* 100: 551–558.)

How do differences in gamete size and parental care relate to mating behavior? Since reproduction is costly, we would expect that in species in which females invest more time and energy in their offspring than males do, females will be choosy and males will compete for the right to mate with females. Moreover, where males invest relatively little per offspring produced, we would expect that males could produce more offspring during their lifetime than females could. This expectation often holds (**TABLE 8.1**). When the reproductive potential of males is higher than that of females, selection should favor different mating behaviors in males and females: it should be advantageous for a male to mate with as many females as possible, whereas a female should "protect" her investment by choosing to mate only with males that provide ample resources or that appear to be of high genetic quality.

As we've seen, observations in nature are often consistent with these predictions. But what about the exceptions, species in which females compete with one another to mate with males? Assuming that the mating behavior of such species has been shaped by natural selection, we would expect that where males provide more parental care than females, the males would be choosy in mate selection, leading to competition among females for the right to mate. Field observations generally support this. For example, in the red phalarope (*Phalaropus fulicarius*, a shorebird; see Figure 8.2), females are larger and more brightly colored than males and compete with other females to mate, aggressively pursuing males with circular flight patterns or swimming after them. After successful mating the female lays her eggs, then abandons the nest in search of another mate, leaving the male to incubate the eggs. Another example of male parental care leading to greater competition among females to mate is found in the pipefish *Syngnathus typhle*. Male pipefish select as their mates the largest, most highly ornamented females, as

they generally produce more eggs than other females do. The male pipefish carries the developing fertilized eggs in a special pouch that protects, aerates, and nourishes the fertilized eggs (Berglund and Rosenqvist 1993). A male does not mate while he is carrying the fertilized eggs, but during that time a female can produce additional eggs and mate with several other males. Thus, females have higher reproductive potential than males do, and (as predicted) they compete for the right to mate with males.

ECOLOGICAL FACTORS AND MATING BEHAVIOR As we saw in Concept 8.2, the foraging decisions of individuals are affected by ecological factors, such as a heterogeneous environment or the presence of predators. Not surprisingly, ecological factors can also affect decisions about mating. Female guppies, for example, mate less often and become less particular in their choice of mates (settling for less brightly colored males) when predators are present (Godin and Briggs 1996). Similar results have been found for many other species. Overall, the evidence shows that in fishes, birds, mammals, and other animals, an individual's decision to mate and its "choosiness" can be altered by such ecological factors as the number and spatial locations of potential mates, the quality of those mates, the availability of food, and the presence of predators or competitors.

Ecological factors can also influence the **mating system**, a term that refers to the number of mating partners that males or females have and the pattern of parental care. A rich variety of mating systems occur in nature (**TABLE 8.2**), and mating systems can vary not only among closely related species, but even among individuals within a population of a single species. How can we make sense of this variation? In a groundbreaking paper, Emlen and Oring (1977) argued that the diverse mating systems seen in nature result from the behaviors of individuals striving to maximize their reproductive success, or *fitness*.

Let's consider the logic of Emlen and Oring's approach from a male perspective. As mentioned earlier, males typically have greater reproductive potential than females; hence, the reproductive success of males will often be limited by access to potential female mates. Under certain conditions, this imbalance can lead to *polygyny*, a mating system in which one male mates with multiple females in a breeding season. As Emlen and Oring (1977) wrote, "Polygyny occurs if environmental or behavioral conditions bring about the clumping of females, and males have the capacity to monopolize them." For example, the availability of food or nest sites may affect where females are found. Whether females settle close to or far away from one another may determine whether a male can acquire and defend more than one mate (**FIGURE 8.18**).

Experimental studies in birds, fishes, and mammals have illustrated particular cases in which females clump together in high-resource areas—and the males then follow the females to those same areas. Moreover, in some

TABLE 8.1

Examples of the Reproductive Potential of Males and Females

Species	Maximum number of offspring produced during lifetime	
	Male	Female
Elephant seal	100	8
Red deer	24	14
Human	888	69

Source: N. B. Davies et al. 2012. *An Introduction to Behavioral Ecology*, 4th ed. Wiley-Blackwell: Oxford. Data from B. J. Le Boeuf and J. Reiter. 1988. In *Reproductive Success*, T. H. Clutton-Brock (Ed.), pp. 344–362. Chicago University Press: Chicago, IL; T. H. Clutton-Brock et al. 1982. *Red Deer: The Behavior and Ecology of Two Sexes*. Chicago University Press: Chicago, IL.

TABLE 8.2

Mating Systems

Mating system	Description
Monogamy	A male mates with only one female, and she with him. This pairing may last for one or more breeding seasons. In many cases, both parents care for the young.
Polygyny	One male mates with multiple females in a breeding season. The male may control access to these females directly (by fighting with other males) or indirectly (by controlling access to resources that females seek, such as food or good nesting sites). The female usually provides most or all of the parental care.
Polyandry	One female mates with multiple males in a breeding season. The female may defend these males directly (by fighting with other females) or indirectly (by controlling access to food or other resources). The male usually provides most or all of the parental care.
Promiscuity	Both males and females mate with multiple partners in a breeding season.

cases, field observations indicate that the availability of resources is correlated with both the locations of females and the mating system. For example, Martin and Martin (2007) found that the brushtail possum (*Trichosurus cunninghami*) was monogamous in a habitat where food and nest sites (and hence females) were widely separated,

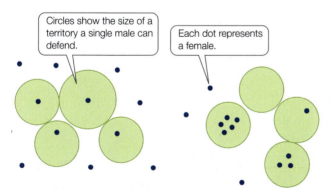

Low potential for polygyny: Females are spread evenly throughout the environment

High potential for polygyny: Females clump together in some regions of the environment

FIGURE 8.18 Ecological Factors Can Affect the Potential for Polygyny In this diagram, dots represent the locations of females and circles show the size of a territory that a male can defend.

but polygynous in a habitat where food and nest sites (and hence females) were located closer to one another. Similarly, Lukas and Clutton-Brock (2013) found that monogamy usually occurs in mammalian species where it is difficult for males to defend access to more than one breeding female—as hypothesized by Emlen and Oring's original paper.

CONCEPT 8.4 There are advantages and disadvantages to living in groups.

LEARNING OBJECTIVES

8.4.1 Describe the potential benefits and costs of species living in groups.

8.4.2 Explain how the balance between benefits and costs of living in groups determines the net impact to individual organisms and populations.

Living in Groups

Individuals of the same species often cluster together, forming groups. Familiar examples of such groups include herds of horses, prides of lions, schools of fish, and flocks of birds. How might the individuals in a group benefit from belonging to the group? And are there disadvantages to communal life that might limit the size of a group or prevent its formation altogether?

Benefits of group living include access to mates, protection from predators, and improved foraging success

Members of a group can enjoy higher reproductive success than solitary individuals. This is clear for males that hold high-quality territories, and it may also be true for females in such territories because they may gain access to good breeding sites or abundant supplies of food. Like the females of a lion pride, group members may also share the responsibilities of feeding and protecting the young, which can benefit the parents (who may have more time to obtain food for themselves) as well as the offspring (who may be both better fed and better protected).

Living in a group can provide other advantages as well, such as a reduced risk of predation. In some cases, the individuals in the group can band together to discourage attack (**FIGURE 8.19**). Moreover, predators are often detected sooner when they approach a group than when they approach a single individual. As a result, they are less likely to surprise their prey, which causes the predators' attack success rate to drop. For example, goshawks were successful in killing wood pigeons about 80% of the time when they attacked a single pigeon, but when they attacked pigeons in a large flock, they were detected sooner, and their success rate plunged (**FIGURE 8.20**).

FIGURE 8.19 A Formidable Defense A group of musk oxen that circles is a very difficult target for predators.

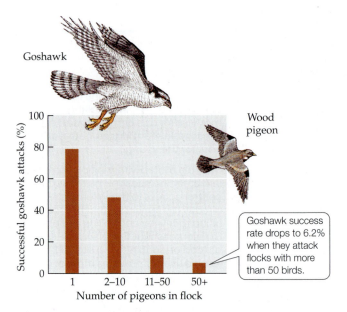

Goshawk

Wood pigeon

Goshawk success rate drops to 6.2% when they attack flocks with more than 50 birds.

FIGURE 8.20 Safety in Numbers The success rate of a goshawk attacking wood pigeon prey plummets when it attacks a large flock. (After R. E. Kenward. 1978. *J Anim Ecol* 47: 449–460.)

In other cases, group members do not cooperate against predators, yet individuals in a group still have a lower risk of predation than they would on their own. One reason for this is that as the number of individuals in a group increases, the chance of being the one attacked decreases, a phenomenon known as the **dilution effect**. In **ANALYZING DATA 8.1**, you can see whether the dilution effect applies to a marine insect attacked by fish predators. Furthermore, if group members respond to a predator by scattering in different directions, they may make it difficult for the predator to select a target, thus causing the predator's attack success rate to drop.

Group members may also experience improved foraging success. Two or more lions, for example, can bring down much larger prey than a single lion could tackle on its own. Furthermore, lions, killer whales, wolves, and many other predators may coordinate their attacks, such that the actions of one predator drive prey into the

---

ANALYZING DATA 8.1

Does the Dilution Effect Protect Individual Ocean Skaters from Fish Predators?

Individuals in a group may gain protection from predators because of the dilution effect: when a predator attacks, the larger the number of prey individuals in the group, the smaller the chance that any particular member of the group will be the victim.

Foster and Treherne* tested whether the dilution effect occurred when a predatory fish (*Sardinops sagax*) attacked groups of a marine insect, the ocean skater (*Halobates robustus*). A subset of their data is presented in the table, which shows the number of predator attacks (per 5 minutes) on ocean skater groups differing in size.

No. insects in group	No. groups observed	No. attacks (per 5 minutes per group)
1	3	15; 6; 10
4	2	16; 8
6	3	9; 12; 7
15	2	7; 10
50	2	15; 11
70	2	14; 7

1. Calculate the average number of attacks (per 5 minutes) for each group size. Do the fish predators show a clear preference for attacking small groups over large groups (or vice versa)? Explain.

2. For each group size, convert the average that you calculated for question 1 into the average number of attacks *per individual* (per 5 minutes). Is there a consistent relationship between the average number of attacks per individual (per 5 minutes) and group size? Explain.

3. Are these results consistent with the dilution effect?

*Foster, W. A., and J. E. Treherne. 1981. Evidence for the dilution effect in the selfish herd from fish predation on a marine insect. *Nature* 293: 466–467.

waiting jaws of another. Herbivores may also forage more effectively when in groups than when on their own by increasing the probability of finding high-quality patches of food resources.

Costs of group living include greater energy expenditures, more competition for food, and higher risks of disease

In one study of group living, a European goldfinch (*Carduelis carduelis*) in a flock of six birds consumed (on average) 20% more seeds per unit of time than did a bird feeding on its own, because goldfinches in a flock spent more time eating and less time scanning for predators than did European goldfinches feeding on their own (Glück 1987). But the increase in the number of seeds eaten per unit of time by a goldfinch in a flock has a downside: as the group size increases, group members deplete the available food more rapidly, which means the birds must spend more of their time flying between feeding sites (**FIGURE 8.21**). Traveling in search of food takes time and energy, and it can increase the risk of being spotted by predators.

Competition for food can also become more intense as the size of a group increases. As a result, a member of a large group may spend more time and energy fighting for food than would a member of a smaller group (or a solitary individual). In particular, in groups with a dominance hierarchy, subordinate group members can spend much of their time and energy on interacting with group members. For example, in a study on the cichlid fish (*Neolamprologus pulcher*), subordinates spent more of their energy on submissive behaviors (appeasing dominant group members) than they did on any other activity.

Finally, members of a large group may live closer together or come into contact with one another more often than do members of a small group. As a result, parasites and diseases often spread more easily in large groups than in small groups; we'll return to this topic in Concept 13.5.

Group size may reflect a balance between the costs and benefits of group living

If we apply the cost/benefit principles discussed in this chapter to group size, we might predict that groups should be of a size at which the benefits of belonging to a group exceed the costs. For example, using an approach similar to that introduced in Concept 8.2 for optimal foraging, we could predict that groups will have an "optimal" size—the size at which the net benefits received by its members are maximized. However, as shown in **FIGURE 8.22**, unless group members can prevent other individuals from joining the group once an optimal size is reached, the observed group size may be larger than the optimal size. In addition, it can be very difficult to measure all the benefits and costs of group living; it is particularly challenging to quantify both costs

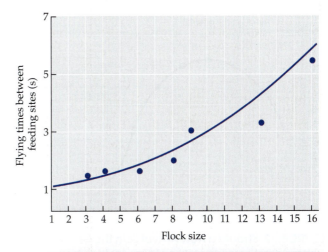

FIGURE 8.21 Traveling in a Group A study of European goldfinches (*Carduelis carduelis*) in groups of seven different sizes showed that the amount of time the birds spent flying between feeding sites increased with the size of the flock. (After E. Glück. 1987. *Ethology* 74: 65–79.)

? A goldfinch feeding in a flock eats more seeds per hour than does a goldfinch feeding alone. Can that benefit be compared directly with the cost shown in this figure? If not, what other information would you need to make this comparison?

and benefits with a single "currency," such as energy use or offspring production.

In general, an argument like that in Figure 8.22 suggests that it may be advantageous for individuals to belong to groups that are larger than the optimal size, but not so large that a new arrival would do better on its own. Such an intermediate-sized group might be large enough to reduce the risk of predation, but small enough to avoid running out of food. Using an overall measure of individual condition (level of stress as measured by fecal concentrations of the hormone cortisol), Pride (2005) found that ring-tailed lemurs in groups of intermediate sizes were less stressed than lemurs that belonged to smaller or larger groups. Similarly, Creel and Creel (1995) found that the per capita intake of food for Tanzanian wild dogs chasing prey was greatest for packs of intermediate sizes.

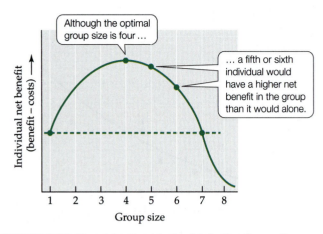

FIGURE 8.22 Should a New Arrival Join the Group? In this hypothetical example, the net benefit to an individual group member is maximized at the optimal group size of four. However, a new arrival would have a greater net benefit if it joined a group of size four than if it remained alone. Unless existing group members can prevent new arrivals from joining the group, new arrivals should continue to join until the group has reached a size of seven—at that point, the individual would do better on its own than in the group.

A CASE STUDY REVISITED

Infanticide in Lion Packs

Can an evolutionary perspective on behavior help us to understand infanticide? It turns out that the males of many species kill the young of their potential mates. For example, male langur monkeys (*Semnopithecus entellus*) kill the infants of females in their social group. This behavior appears to increase the reproductive success of the murderous males: DNA paternity analyses showed that infanticidal male langurs were not related to the infants they killed but were related to the females' subsequent offspring (Borries et al. 1999). Infanticide by males has been documented in dozens of other species, including horses, chimpanzees, bears, and marmots. Infanticide by males appears to be adaptive in many cases: it reduces the time that females spend between estrous cycles, thus enabling the males to sire more offspring than they otherwise could.

But in some species, females commit infanticide. For example, female giant water bugs (*Lethocerus deyrollei*) and female wattled jacanas (a type of shorebird, *Jacana jacana*) destroy the eggs or young of their own species. While gruesome, this behavior makes evolutionary sense: in these species, the males provide most or all of the parental care, and the females have higher reproductive potential than the males. Thus, as is true for male lions and langurs, the infanticidal behavior of female water bugs and jacanas appears to be adaptive: by killing the young,

the time before the male is willing to mate is shortened, thus potentially increasing her own reproductive success.

What about other puzzling behaviors mentioned in this chapter's Case Study? Recall that female fruit flies (*Drosophila melanogaster*) sometimes lay their eggs in foods that are high in ethyl alcohol. This behavior is not as strange as it first appears: evidence suggests that it provides a behavioral defense against the wasp *Leptopilina heterotoma*, which preys on the fruit flies. Females of this wasp lay their eggs on fruit fly larvae; when an egg hatches, the young wasp burrows through the body of the fly larva, consuming and killing it. A fruit fly larva infected by this wasp will preferentially choose to eat foods that are high in alcohol content, such as rotting fruit, as it helps defend it from the wasp young. Even though consuming foods containing high concentrations of alcohol harms the fruit fly larvae, the benefits of this action outweigh its costs: wasps are more susceptible to the harmful effects of the alcohol than the fruit flies, increasing the overall chances that the fruit fly larvae will survive. In addition, Kacsoh et al. (2013) showed that adult female fruit flies altered their egg-laying behavior in response to the presence of wasps. In the absence of wasps, the fruit flies laid about 40% of their eggs in high-alcohol foods, but when female wasps were present, the fruit flies laid over 90% of their eggs in high-alcohol foods. This behavior increased the survival of fruit fly larvae exposed to wasps (**FIGURE 8.23**), suggesting that the behavior can be viewed as a type of preventative medicine.

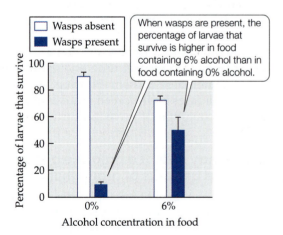

FIGURE 8.23 Fruit Flies Medicate Their Offspring Female fruit flies (*Drosophila melanogaster*) laid most of their eggs in foods containing alcohol when females of the wasp *Leptopilina heterotoma* were present. This behavior increased the percentage of fruit fly larvae that survived to adulthood. Error bars show one SE of the mean. (After B. Z. Kacsoh et al. 2013. *Science* 339: 947–950.)

? In the absence of wasps, what is the cost (in terms of reduced larval survival) of laying eggs on food containing 6% alcohol? When wasps are present, what is the benefit?

🌱 CONNECTIONS *in* NATURE

BEHAVIORAL RESPONSES TO PREDATORS HAVE BROAD ECOLOGICAL EFFECTS As you've seen, individuals often change their behavior in response to predators. For example, in Concept 8.2, we saw that when exposed to recordings of sounds made by predators, song sparrows fed their young less often, built their nests in less desirable areas, and spent less time incubating their eggs (Zanette et al. 2011). What were the consequences of these behavioral changes?

Zanette et al. found that when song sparrow parents altered their behavior in response to a perceived high risk of predation, their offspring lost body heat more rapidly (**FIGURE 8.24A**) and weighed less than did the offspring of sparrows exposed to recordings of nonpredators. These effects on individual offspring appear to have caused the number of offspring produced per year to decline (**FIGURE 8.24B**). Overall, the results of this study suggest that fear of predation alters the behavior of song sparrows in ways that decrease their reproductive success and may cause their population sizes to drop.

Behavioral responses to predators can also affect ecosystem processes, such as the decomposition of leaves and other plant litter in soil and subsequently the availability of nutrients (Concept 22.2). The presence of spider predators can lead to a series of events in their grasshopper prey that ultimately slowed the decomposition rates of plant litter. How did this happen? Hawlena et al. (2012) found this effect occurred indirectly via changes in foraging behavior. When the researchers raised grasshoppers in the presence of predators, the grasshoppers became physiologically stressed, one consequence of which was that they required more energy to maintain their basic body functions. This demand for additional energy appears to have altered their foraging behavior, leading the grasshoppers to increase their consumption of plants that are high in carbohydrates (and thus in energy) but low in nitrogen. Thus, grasshoppers stressed by predators had a higher carbon:nitrogen ratio in their bodies than did grasshoppers raised in the absence of predators. Although this change in nutrient content did not affect the decomposition of the grasshoppers' own bodies, it did decrease the

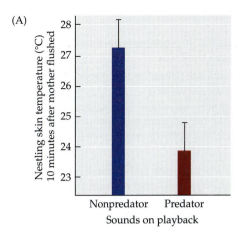

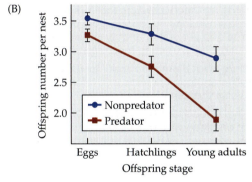

FIGURE 8.24 Costs of Fear (A) The rate at which young song sparrows lost body heat was higher for offspring of birds exposed to playbacks of predators than it was for offspring of parents exposed to playbacks of nonpredators. Error bars show one SE of the mean. (B) Fewer offspring were produced in nests exposed to playbacks of predators than in nests exposed to playbacks of nonpredators. Error bars show ± one SE of the mean. (After L. Y. Zanette et al. 2011. *Science* 334: 1398–1401.)

❓ Estimate the cost of fear on the number of offspring that survived to young adulthood.

decay of plant materials in the soil. This probably happened because the altered carbon:nitrogen ratio in the decomposing bodies of grasshoppers affected the carbon:nitrogen ratio in the soil, which in turn affected the community of soil microorganisms that decompose leaves and other plant matter. 🌿

SUMMARY

CONCEPT 8.1 Evolution is the basis for adaptive behavior.

8.1.1 Explain how natural selection can lead to the evolution of adaptive behaviors.

- Animal behaviors can be explained in terms of their immediate, or proximate, causes or in terms of their evolutionary, or ultimate, causes. Most research in behavioral ecology addresses ultimate causes.

- An individual's ability to survive and reproduce depends in part on its behavior. This observation suggests that natural selection will favor individuals whose behaviors make them efficient at activities such as foraging, obtaining mates, and avoiding predators.

(Continued)

8.1.2 Illustrate how the environment can interact with genetics to influence behavior.

- Animal behaviors are influenced by genes as well as by environmental conditions.

- By assuming that genes affect behaviors and that natural selection has shaped behaviors over time, we can predict how animals will behave in particular situations.

CONCEPT 8.2 Animals make behavioral choices that enhance their energy gain and reduce their risk of becoming prey.

8.2.1 Explain the theory of optimal foraging by outlining the factors that influence the profitability of foraging.

- Optimal foraging theory predicts that foraging animals will maximize their net energy gain per unit of feeding time and per unit of energy invested in seeking, capturing, and extracting food resources.

8.2.2 Summarize what determines optimal foraging in an area with different food densities with reference to the marginal value theorem.

- The marginal value theorem predicts that animals foraging in an area with patches of food resources of different densities will deplete rich patches first and leave them once the density of food is the same as the average of the entire area.

8.2.3 Describe how the presence of predators can impact foraging behavior.

- Individuals often alter their foraging decisions when predators are present. A perceived risk of predation can also alter foraging patterns, even in the absence of actual predation.

- Prey exhibit a wide range of behaviors that can help them avoid being seen by predators, detect predators, prevent attack, or escape once attacked.

CONCEPT 8.3 Mating behaviors reflect the costs and benefits of parental investment and mate defense.

8.3.1 Describe examples of the behaviors utilized by animals to increase their access to mates.

- Within a species, males are often larger or more brightly colored than females, or they may possess unusual weapons or have gaudy ornaments. Such differences between males and females of the same species can result from sexual selection.

8.3.2 Evaluate the benefits of being choosy with mate selection and the conditions that favor selectivity by females versus males.

- A female may receive indirect genetic benefits when she chooses to mate with a male that has certain features, such as a costly and unwieldy ornament. Females benefit when the males' features signal good genes that are passed on to both her sons and her daughters. Females may also receive indirect genetic benefits through her sons, who will themselves be attractive and produce many grandchildren.

- In most species, females invest more in their gametes and provide more parental care than males do. In this situation, males and females have different interests: it is to a male's advantage to mate with as many females as possible, whereas a female should "protect" her investment by mating with those males that provide the most resources or that appear to be of high genetic quality.

8.3.3 Describe conditions under which different mating systems would be favored.

- In the rare cases in which males typically provide more parental care than females do, it is the male who is the choosy partner.

- The rich variety of mating systems seen in nature result from the behaviors of individuals striving to maximize their reproductive success.

CONCEPT 8.4 There are advantages and disadvantages to living in groups.

8.4.1 Describe the potential benefits and costs of species living in groups.

- Benefits of group living include access to mates, protection from predators, and improved foraging success.

- Costs of group living include greater expenditures of energy, increased competition for food, and higher risks of disease.

8.4.2 Explain how the balance between benefits and costs of living in groups determines the net impact to individual organisms and populations.

- Group size may reflect a balance between the costs and benefits of group living; in some cases, this balance appears to have caused groups to be larger than the optimal size.

Review Questions

1. Distinguish between proximate and ultimate explanations of animal behavior.

2. Explain the links between the following: natural selection, heritable behaviors, adaptive evolution, and ultimate explanations of animal behaviors.

3. Describe how the presence of a predator may alter an individual's foraging decisions. Can fear of predators have similar effects, even in the absence of actual predators? Explain.

4. What is sexual selection? Summarize the evidence supporting the claim that differences between males and females can result from sexual selection.

5. Describe an example in which group living leads to both benefits and costs.

6. Consider two bird species that forage for insects that live in shrubs. The shrubs have a clumped, patchy distribution throughout their habitat. The two bird species have the same ability to locate, capture, and consume the insects. However, one species (species A) uses less energy to fly from patch to patch than the other species (species B). According to the marginal value theorem, which bird species should spend more time in each patch, and why?

Hone Your Problem-Solving Skills

Animals exhibit diverse mating systems, including monogamy, polygyny, polyandry, and promiscuity (see Table 8.2). Mating systems vary both among and within species. Gray wolves (*Canis lupus*) typically exhibit monogamy and live in cooperative breeding groups in which nonbreeding members help care for offspring. Reproductive success of the breeding pair also is tied to acquiring a high-quality territory. In monogamous species, such as gray wolves, how does pair bond duration affect reproductive success? Also, how might pair bond duration affect the likelihood of other mating strategies arising within groups? To address these questions, a researcher monitored 15 groups of gray wolves over 9 years in Idaho. He collected feces for DNA and pedigree analyses. Pedigree analyses allowed the determination of group size and composition; paternity; evidence of polygyny and polyandry; and duration of pair bonds. Pup survival was assessed between 3 and 15 months of age; pups rarely disperse before age 15 months, so apparent pup survival assumes that a pup's absence means it has died. The figure shows the relationship between pair bond duration and apparent pup survival. Error bars are standard errors.

1. Summarize the results shown in the figure. What demographic characteristic of breeding pairs is conflated with pair bond duration? Suggest two possible explanations for the relationship between pair bond duration and apparent pup survival.

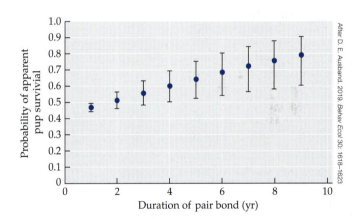

After D. E. Ausband, 2019, *Behav Ecol* 30: 1618–1623

2. Results shown are based on an analysis in which group size was held constant at six adults. Why would it be important to control for group size? Based on the results, can the researcher definitively conclude that increasing pair bond duration resulted in increasing pup survival in his study population?

3. Research with other species links social conflict with the emergence of different mating strategies within populations. Predict how pair bond duration affects the existence of polygyny and polyandry (considered together by the researcher as polygamy) within groups of gray wolves.

Population Distribution and Abundance

From Kelp Forest to Urchin Barren: A Case Study

Stretching over 1,600 km of the Pacific Ocean to the west of Alaska, the mountainous Aleutian Islands are often shrouded in fog and battered by violent storms. The islands have few large trees, and except for the eastern islands that once were connected to the mainland, they lack mainland terrestrial mammals such as brown bears, caribou, and lemmings. There is abundant marine wildlife in the surrounding waters, however, including seabirds, sea otters, whales, and a variety of fishes and invertebrates.

Although there are few trees on land, the nearshore waters of some Aleutian islands harbor fascinating marine communities known as kelp forests, made up of brown algae such as *Laminaria* and *Nereocystis*. Dense clusters of kelp rise from their holdfasts on the sea bottom toward the surface, producing what feels like an underwater forest (**FIGURE 9.1**). Other nearby islands do not have kelp forests. Instead, the bottoms of their nearshore waters are carpeted with sea urchins and support few kelp or other large algae. Areas with large numbers of urchins are called "urchin barrens" because they lack kelp forests. Why are some islands surrounded by kelp forests and others by urchin barrens?

One possibility is that islands with kelp forests differ from islands without kelp forests in terms of climate, ocean currents, tidal patterns, or physical features such as underwater rock surfaces. But no such differences have been found, leaving us to look for other reasons why some islands have kelp forests while others do not. Because urchins feed on algae and can eat vast quantities of it, investigators suspected that grazing by urchins might prevent the formation of kelp forests.

This hypothesis was tested in two ways. First, studies in the Aleutian Islands and elsewhere along the Alaskan coast consistently showed that kelp forests were not

FIGURE 9.1 Kelp Forests Depend on Sea Urchin Population Control The bull kelp *Nereocystis luetkeana* is one of several species that make up the kelp forests found off the coasts of some Aleutian islands. Research shows that the presence or absence of kelp forests near these islands is influenced by the population control of sea urchins by sea otters.

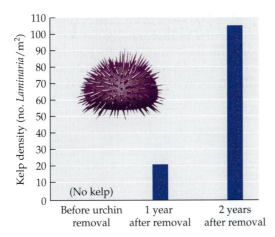

FIGURE 9.2 **Do Sea Urchins Limit the Distribution of Kelp Forests?** Mean densities of the kelp *Laminaria* in 50-m² plots increased dramatically after urchins were removed. (After D. O. Duggins. 1980. *Ecology* 61: 447–453.)

found in regions where there were many large urchins. Although such correlations did not prove that urchins suppress kelp forests, the fact that a number of studies found the same result suggested that urchins might determine where kelp forests are located. Second, the effect of urchins was tested in an experiment that measured change in kelp densities in several 50-m² plots containing urchins and in similar, nearby 50-m² plots from which urchins were removed (Duggins 1980). There were no kelp in any of the plots at the start of the experiment, and kelp densities remained at zero in the plots where urchins remained. In the plots from which urchins had been removed, however, the density of *Laminaria* rose to 21 individuals per square meter in the first year and reached 105 individuals in the second year (**FIGURE 9.2**). *Laminaria* is a dominant member of kelp forest communities, so these results suggested that kelp forests would grow in the absence of urchins.

These and other results indicated that the presence or absence of urchins is an answer to the question of why some Aleutian islands have kelp forests and others do not. But this answer just shifts the question from what determines the locations of kelp forests to what determines the locations of urchins. As we'll see, a more complete answer to our question about why kelp forests are found in some areas but not others turns out to depend on the voracious feeding habits of sea otters, which themselves may have become a meal of last resort for killer whales.

Introduction

In this chapter's Case Study, we focused on a fundamental ecological question: What determines the distribution and abundance of a species, in this case kelp? The **distribution** of a species is simply the geographic area where individuals of the species are present, while its **abundance** refers to the number of individuals of a species or population.

These two measures are highly related because the distribution of a species can be viewed as a map of all areas where the abundance of the species is greater than zero.

Determining the distributions and abundances of species, and the factors important to these patterns, can be challenging given that groups of individuals (or populations) often vary dramatically over space and time. Our ability to document this variability and predict these changes can serve as a "measuring stick" for how well we understand events in nature. In this chapter, we will focus on how and why the distribution and abundance of individuals within species and populations vary over the landscape and ways in which to measure that variation. In Chapters 10 and 11, we'll expand our view of populations by considering how they vary over time using both examples and population growth models. But first, we'll begin by describing aspects of populations and individuals, including estimations of abundance, in more detail.

CONCEPT 9.1 Populations are groups of individuals of the same species that vary in size over space and time.

LEARNING OBJECTIVES

9.1.1 Define the terms population, population size, and population density.

9.1.2 Compare the different ways in which individuals are defined, including the terms clones, genets, and ramets.

9.1.3 Compare the different methods used to measure the abundance of individuals within populations or species.

Populations and Individuals

A **population** is a group of individuals of the same species that live in the same area at the same time and interact with one another. To explore this definition further, what exactly do we mean by "interact"? In species that reproduce sexually, a population might be defined as the group of individuals that interact by interbreeding. In species that reproduce asexually, however, such as dandelions or the fish *Poecilia formosa*, a population must be defined by other kinds of interactions, such as competition for common sources of food. Our definition of a population also incorporates the area over which members of a species interact. If that area is known, as in a population of lizards that live on and move throughout a small island, we can report population abundance either as **population size** (the number of individuals in the population) or as **population density** (the number of individuals per unit of area). For example, if there were 2,500 lizards on an island of 20 hectares (ha), or roughly 50 acres, the population size would be 2,500 lizards, and the population density would be 125 lizards per hectare.

In some cases, the total area occupied by a population is not known. For example, when little is known about how far a sexually reproducing species or its gametes (e.g., plant pollen) can travel, it is difficult to estimate the area over which individuals interbreed frequently and hence represent a single population. For asexual species, similar problems are encountered when we try to estimate the area over which interactions other than interbreeding occur. When the area occupied by a population is not fully known, ecologists use the best available information about the biology of the species to delimit an area within which the size and density of the population can be estimated.

What are individuals?

As we've seen, it can be a challenge to determine the size or density of a population, because it is necessary to know how many individuals are present within the population. For some species, there is an added challenge—determining what constitutes an individual.

How can there be confusion over what an individual is? Consider the quaking aspen trees (*Populus tremuloides*) in **FIGURE 9.3**. Like many plant species, an individual aspen can produce genetically identical copies of itself, or **clones**. Aspens produce clones by forming new plants from root buds, while species such as clover and strawberries do so by forming new plants from buds located on horizontal stems, or "runners" (**FIGURE 9.4**). Among animals, many corals, sea anemones, and hydroids can form clones of genetically identical individuals, as can some frogs, fishes, lizards, and many insects. Some plant clones can grow to enormous sizes (e.g., covering 81 ha, or 200 acres, in aspen clones) or live for extremely long

periods (e.g., 43,000 years in the case of *Lomatia tasmanica*, a rare shrub found in Tasmania, Australia).

To cope with the complications that result from the formation of clones, biologists who study such organisms define individuals in several different ways. For example,

FIGURE 9.4 Plants and Animals That Form Clones
Many plants and animals reproduce asexually, thereby forming clones of genetically identical individuals. Examples of asexual reproduction include *budding* (in which clonal offspring detach from the parent), *apomixis* (in which clonal offspring are produced from unfertilized eggs; also known as *parthenogenesis*), and *horizontal spread* (in which clonal offspring are produced as the organism grows).

? How might groups of genetically identical individuals be identified in clones that form by budding? By apomixis? By horizontal spread?

FIGURE 9.3 Aspen Groves: One Tree or Many?
These quaking aspen (*Populus tremuloides*) growing in western Colorado could represent multiple different genetic individuals, each established from a seed. However, it is also possible that each of these aspens is part of one "tree," having been produced asexually from the root buds of a single genetic individual.

an individual can be defined as the product of a single fertilization event. Under this definition, a grove of genetically identical aspen trees is a single genetic individual, or **genet**. However, members of a genet are often physiologically independent of one another, and they may in fact compete for resources. Such actually or potentially independent members of a genet are called **ramets**. In strawberries, for example, a rooted plant is considered a ramet because it can persist even if it is not connected to the rest of its genet (see Figure 9.4). Whether we view a patch of strawberries or a grove of aspen trees as one individual or many depends on what we are interested in. If we are interested in evolutionary change over time, the genet level may be more appropriate. In contrast, if we are interested in how independent physiological units compete, the ramet level may be more appropriate.

The most direct way to determine how many individuals live in a population is to count all of them. This sounds simple enough, and it is possible in some cases, as for the lizards on one island, and other organisms that are confined to small areas, are easy to see, or do not move. But complete counts of organisms are often difficult or impossible. Consider the chinch bug (*Blissus leucopterus*), an insect that attacks crops such as corn and wheat. This insect can cover large areas and reach densities that exceed 5,000 individuals per square meter, making it impractical to count all the individuals in a population. In such cases, a variety of methods can be used to estimate abundance. Let's discuss some of those methods next.

Ecologists estimate abundance using a variety of methods

As just mentioned, many ecological studies require an estimate of a population's actual abundance, or **absolute population size**. For example, with species that are threatened or endangered, it is ideal to estimate their absolute

population sizes in order to keep track of possible further declines. In other cases, it may be sufficient to estimate the **relative population size**, the number of individuals in one time interval or place *relative to* the number in another. Estimates of relative population size are based on data that are presumed to be correlated with absolute population size but do not assess the actual number of individuals in the population. Examples of such data include the number of cougar tracks found in an area, the number of fish caught per unit of effort (e.g., per number of hooks trolled each day), or the number of birds observed while the observer walks a standard distance (or remains in one place for a standard time interval).

Relative population size estimates are usually easier and less expensive to obtain than are absolute estimates. While useful, estimates of relative population size must be interpreted carefully. The number of cougar tracks observed, for example, depends not only on cougar population density, but also on animal activity. Thus, if twice as many tracks were found in area A as in area B, we could not be confident that area A had twice as many cougars—there could be more or fewer than that, depending on whether cougars moved more frequently in one area than in another.

Methods for estimating abundance fall into three general categories: (1) area-based counts in which the number of individuals in a given area or volume are counted, (2) distance methods in which the distances of individuals from a line or a point are measured and then converted into estimates of the number of individuals per unit of area, and (3) mark–recapture studies, which involve releasing marked individuals and then recapturing them at a later time to estimate the total population size. You can learn more about these methods in **ECOLOGICAL TOOLKIT 9.1**, where we describe these approaches in more detail and provide examples.

ECOLOGICAL TOOLKIT 9.1

Estimating Abundance

Methods for estimating abundance fall into three general categories: area-based counts, distance methods, and mark–recapture studies. Many variations on these approaches have been developed, and a wide range of statistical techniques are available for analyzing abundance estimates obtained using each of them (Krebs 1999; Williams et al. 2002).

A. AREA-BASED COUNTS In an area-based count, as its name suggests, the number of individuals in a given area or volume are counted. Area-based counts are often used to estimate absolute population sizes of organisms that are sessile (e.g., plants) or can move

only short distances during the time it takes to count the individuals in a quadrat (e.g., sea urchins). Area-based counts can also be used to estimate the abundances of more mobile organisms, as when large mammals are observed in aerial surveys. Area-based counts of highly mobile organisms can provide estimates of relative population sizes; further information (such as the probability that an organism will be present but not seen when surveyed by air) may be required before such counts can be used to estimate absolute population sizes.

(Continued)

ECOLOGICAL TOOLKIT 9.1 (continued)

FIGURE A An Underwater Quadrat A marine biologist uses a square quadrat to count the numbers of individuals of different coral species found on a reef off the Caroline Islands, Micronesia.

Often area-based counts make use of a *quadrat* (**FIGURE A**), which is a sampling area (or volume) of any size or shape, such as a 0.25 × 0.25-m² square plot used to count small plants, a 0.1-ha plot used to count trees, or a soil core of a certain diameter and depth used to count soil organisms. The counts from multiple quadrats are then summed and averaged to estimate the number of individuals per unit of area (or volume), which can then be used to estimate the total population size. Suppose, for example, that an entomologist wants to estimate the population of chinch bugs in a 400-ha (ca. 1,000-acre) field of corn. If they counted chinch bugs in five 10 × 10-cm quadrats (i.e., five 0.01-m² quadrats), and their counts were 40, 10, 70, 80, and 50 chinch bugs, they would estimate that there were, on average,

$$\frac{(40+10+70+80+50)/5}{0.01} = 5{,}000 \quad \text{(9.1)}$$

chinch bugs per square meter. Thus, there would be an estimated 20 billion chinch bugs in the population (5,000 bugs/m² × 10,000 m²/ha × 400 ha = 20,000,000,000).

Area-based methods work well if individuals can be counted accurately within the quadrats, and if the quadrats provide a good representation of the entire area covered by the population. To help ensure that the latter condition is met, ecologists use as many quadrats as is feasible, and they often place those quadrats at locations selected at random from the entire area covered by the population. Quadrats can also be placed in a variety of other ways, such as at evenly spaced locations along a transect line or rectangular grid.

B. DISTANCE METHODS In distance methods, an observer measures the distances of individuals seen from a line or a point; these distances are then converted into estimates of the number of individuals per unit of area. In the **line transect** approach, an observer travels along a transect line. Each individual that the observer can see from the line is counted, and its perpendicular distance from the line is recorded (d_1, d_2, and d_3 in **FIGURE B**). The distances can then be used to estimate the absolute abundance; this conversion requires a *detection function*, which accounts for the probability of an individual at varying distances away from the transect. Other distance methods include *point sampling* techniques, in which the distance to the nearest (visible) individual is measured from a series of locations or "points"; as with line transect data, a detection function is used to convert these distances into estimates of the absolute population size (see Krebs 1999; Schwarz and Seber 1999).

C. MARK–RECAPTURE STUDIES Mark–recapture studies are used to estimate the absolute population size of mobile organisms; they are also used to obtain data on the survival or movement of individuals. With this method, a subset of the individuals in a population is captured, marked (as with a tag or dot of paint) so that they can be recognized at a later time, and released. After the marked individuals have been given enough time to recover and move throughout the population, individuals are captured a second time, and the proportion of marked individuals found in the second capture is used to estimate the total population size.

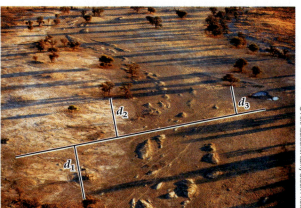

FIGURE B Counting Trees from a Line Transect The density of these camelthorn trees (*Acacia erioloba*) in Kgalagadi Transfrontier Park, South Africa, could be estimated using a line transect, as shown here.

ECOLOGICAL TOOLKIT 9.1 (continued)

Imagine, for example, that we capture 23 butterflies from a meadow, which we then mark and release (**FIGURE C**). A day later, we sample the meadow again, this time catching 15 butterflies, of which 4 are marked. In our first sample, we caught and marked $M_1 = 23$ butterflies from a total population of unknown size (N); thus, we initially caught a proportion M_1/N of the butterflies in the field. The second time butterflies were sampled, we caught $M_2 = 15$ butterflies, of which 4 were marked and hence were recaptured ($R = 4$).

Assuming that no butterfly births, deaths, or movements into or out of the meadow have occurred since our first sample, the proportion of marked individuals captured in our second sample (R/M_2) should equal the original proportion we caught, M_1/N. Thus, we have the equation

$$M_1/N = R/M_2 \qquad (9.2)$$

We can rearrange Equation 9.2 to estimate the total number of butterflies in the meadow as

$$N = (M_1 \times M_2)/R \qquad (9.3)$$

which in this case would equal $(23 \times 15)/4 = 86$.

Equation 9.3 assumes that (1) the population size does not change during the sampling period (no births, deaths, immigration, or emigration), (2) each individual has an equal chance of being caught, (3) marking does not harm individuals or alter their behavior (such as by making them harder to recapture), and (4) marks are not lost over time. A wide range of other mark–recapture methods have been developed to address cases in which one or more of these assumptions are violated (Krebs 1999; Schwarz and Seber 1999; Williams et al. 2002).

FIGURE C **Release of Marked Butterfly** To obtain mark-recapture estimates of butterfly abundance, ecologists tag and then release them back into their habitat (note tag on the left wing).

LEARNING OBJECTIVES

9.2.1 Describe the relationship between populations, metapopulations, and geographic ranges for species.

9.2.2 Compare the different dispersion patterns of populations.

9.2.3 Describe how the size and patchiness of geographic ranges vary among species and can be predicted using models.

Distribution and Abundance Patterns

The distribution and abundance patterns of species and populations vary in their spatial extent across the land-scape. An example of this variability is the distribution map for the herbaceous perennial *Clematis fremontii* (**FIGURE 9.5A**). *Clematis* has a patchy distribution across Missouri, Kansas, and Nebraska, where its populations are restricted to dry, rocky, treeless meadows or glades formed on particular limestone outcrops within the region. Populations, such as those of *Clematis*, rarely occur in isolation from one another and are usually connected through dispersal. **Dispersal** is simply the movement of individuals into (**immigration**) or out of (**emigration**) an existing population. A group of geographically isolated populations linked together by dispersal is known as a **metapopulation**. For example, a cluster of meadows might be considered a metapopulation if *Clematis* seeds from one meadow had the potential to disperse to another meadow. In Concept 9.4, we will discuss metapopulations in more detail. At larger spatial scales, the entire **geographic range**, or distribution, of a species might consist of one or multiple metapopulations, depending on the extent of the area occupied by a species.

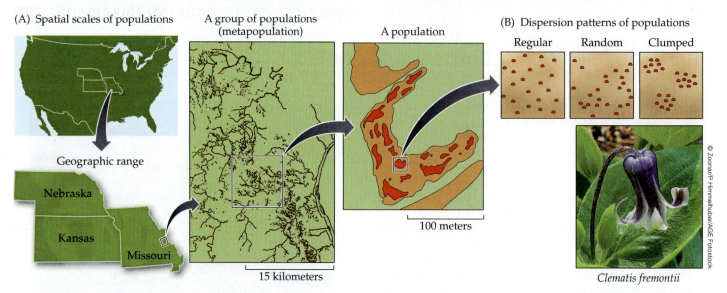

(A) Spatial scales of populations

A group of populations (metapopulation)

A population

(B) Dispersion patterns of populations

Geographic range

Nebraska

Kansas

Missouri

15 kilometers

100 meters

Regular Random Clumped

© Zoonar/P. Himmelhuber/AGE Fotostock

Clematis fremontii

FIGURE 9.5 Many Populations Have a Patchy Distribution
The distribution and abundance of the herbaceous perennial *Clematis fremontii* are patchy over different spatial scales. (A) Populations occur within limestone meadows. A group of populations makes up a metapopulation, and multiple metapopulations make up the geographic range (in this case, Missouri, Kansas, and Nebraska). (B) Individuals within a population show one of three different dispersion patterns. (A after R. O. Erickson. 1945. *Ann Mo Bot Gard* 32: 413–460.)

The aggregates of *C. fremontii* individuals found in the meadow populations provide an example of the **dispersion**, or spatial arrangement, of individuals within a population (**FIGURE 9.5B**). We can recognize three basic patterns in how the individuals of a population are positioned with respect to one another (see Figure 9.5A). In some cases, the members of a population have a **regular dispersion**, in which individuals are relatively evenly spaced throughout their habitat. In other cases, individuals show a **random dispersion**, similar to what occurs if individuals are positioned at locations selected at random. Finally, as in *C. fremontii*, individuals may be grouped together to form a **clumped dispersion**. In natural populations, clumped dispersions are more common than either regular or random dispersions. It is important to note that identifying dispersion patterns can depend on the spatial scale of the measured area. For example, clumped dispersion patterns might not be revealed except at large spatial scales.

The geographic ranges of species vary in size

As we discussed, the geographic range of a species is the entire region over which that species is found. Although there are no species that are found everywhere, there is considerable variation in the sizes of their geographic ranges. Examples of species with small geographic ranges include the Devil's Hole pupfish (*Cyprinodon diabolis*), which lives in a single desert pool (7 × 3 m across and 15 m deep). Many tropical plants also have small geographic ranges. This latter point was illustrated dramatically in 1978, when 90 new plant species were discovered on a single mountain ridge in Ecuador, each with a geographic range that was restricted to that ridge. We call such species **endemic** because they occur in one particular location and nowhere else on Earth.

Other species, such as coyotes, live over most of one continent (North America), while still others, such as gray wolves, live on small portions of several continents (North America and Eurasia). Relatively few terrestrial species are found on all or most of the world's continents. Notable exceptions include humans, Norway rats, and the bacterium *Escherichia coli*, which lives in the intestinal tracts of reptiles, birds, and mammals (including humans) and thus is found wherever its host organisms are found. Some marine species, including invertebrates with planktonic larvae (see Figure 7.11) and whales (see Figure 9.13), have large geographic ranges. But while range sizes vary greatly, the pattern in the oceans is similar to that on land, and for most marine species the geographic range is relatively small (Gaston 2003).

The geographic range of a species includes the areas it occupies during all of its life stages. It is particularly important to keep this fact in mind for species that migrate and for species whose biology is poorly understood. For example, if we wish to protect monarch butterfly populations, we must ensure that conditions are favorable for them in both their summer breeding grounds and their overwintering sites. In some cases, we understand an organism's range poorly because it has life stages that are hard to find or study; this is true of many fungi, plants, and insects. We may know under what conditions the adult organism lives, yet have no idea where or how other life stages live. In fact, that was long the case for the monarch butterfly. Biologists knew that each spring these butterflies arrived in eastern North America from the south, but it took almost 120 years (from 1857 to 1975) before their overwintering sites were discovered in mountains west of Mexico City.

The geographic ranges of species vary in patchiness

Even within the geographic range of a species, much of the habitat is not suitable for the species. As a result,

FIGURE 9.6 Abundance Varies throughout the Geographic Range of a Species The map shows abundances of the red kangaroo (*Macropus rufus*) throughout its range in Australia. These data were based on aerial surveys conducted from 1980 to 1982. (After G. Caughley et al. 1987. *Kangaroos: Their Ecology and Management in the Sheep Rangelands of Australia.* Cambridge University Press: Cambridge.)

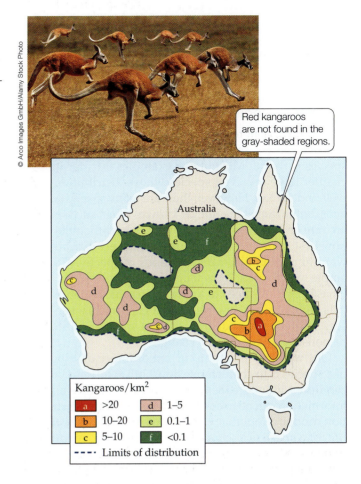

populations tend to have a patchy distribution. This observation holds at both large and small spatial scales. On land, for example, at the largest spatial scales, climate constrains where populations of a species are located (see Concept 3.1). At smaller spatial scales, factors such as topography, soil type, and the presence or absence of other species prevent populations from being spread evenly across the landscape.

As we saw with the *C. fremontii* example (see Figure 9.5), some species require a very particular habitat that is found only in portions of its geographic range; hence its populations have a highly patchy distribution. Other species tolerate a broader range of habitats, but their abundances still vary throughout their geographic range. The distribution of red kangaroos (*Macropus rufus*) in arid regions of Australia illustrates this point. The abundance of red kangaroos varies throughout their geographic range, which includes several regions of high density and several areas where red kangaroos are not found (**FIGURE 9.6**).

Finally, it is important to recognize that a population may exist in a series of habitat patches or fragments that are spatially isolated from one another but are linked by dispersal. Such a "patchy" population structure can result from features of the abiotic environment, as we saw with *Clematis*, but can also result from human actions. For example, the low-lying shrub-covered heathlands of England once covered large, continuous areas, but over the past 200 years the development of farms and urban areas has greatly reduced the extent of heath plant species (**FIGURE 9.7**). In some cases, this fragmentation results in patches that are so isolated that little dispersal can occur among them, thus breaking a single

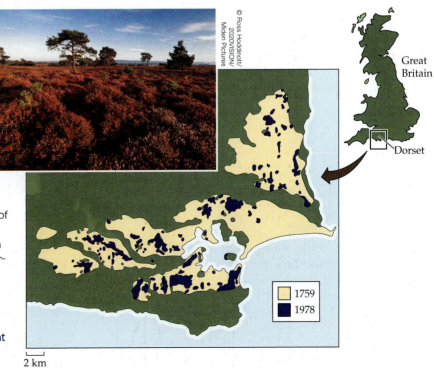

FIGURE 9.7 Fragmentation of Dorset Heathlands The low-lying shrub-covered heathlands of Dorset, England, reached their maximum extent in Roman times, 2,000 years before the present. From 1759 to 1978, the decline of this habitat type accelerated: the total area of heathlands shrank from 400 km² to less than 60 km², and the number of patches increased greatly. (After N. R. Webb and L. E. Haskins. 1980. *Biol Conserv* 17: 281–296.)

? How many patches of heathland were present in 1759? In 1978? Use your answers to estimate the average patch size in 1759 and 1978.

2 km

large population into a series of much smaller populations. In Concept 9.4, we will explore the occurrence and consequences of patchy population structures (*metapopulations*) in more detail.

Species distribution models can be used to predict a species' geographic range

As we have seen, to determine the geographic distribution of a species, scientists record all locations where the species is found. Most of our examples thus far in this chapter have involved species whose distributions are well understood. However, there are many species whose geographic ranges are not yet known. When such species are rare or in need of protection, it can be difficult to plan how best to protect them. Furthermore, ecologists often want to predict *future* distributions of species—for example, whether and how a pest species will spread after it has been introduced to a new geographic region. Scientists and policymakers face similar challenges when they seek to predict how distributions of species will shift in response to global climate change.

One way to predict the current or future distribution of a species is to characterize how both abiotic and biotic conditions influence their occurrence or abundance. Such information can be used to construct a **species distribution model**, a tool that predicts the geographic distribution of a species based on the environmental conditions at locations the species is known to occupy.

Investigators from the United States and Mexico used such an approach to predict the distributions of chameleons in Madagascar (Raxworthy et al. 2003). The researchers obtained information about vegetation cover (from satellite images), temperature, precipitation, topography (elevation, slope, aspect), and hydrology (water flow, tendency

to pool) from government and commercial sources. Values for these environmental variables were recorded for each of a series of 1×1-km^2 areas (referred to as grid cells) that covered all of Madagascar. Next, for 11 chameleon species, rules were developed that described the environmental conditions in which each of the species was most likely to be found; we'll refer to these rules as *habitat rules*.

There are many different ways to develop such habitat rules. The chameleon study used a computer program that compared the environmental conditions of grid cells selected at random from a map of Madagascar with the environmental conditions of grid cells where a chameleon species was known to occur. For example, initially a habitat rule might state that a species should be found in locations where the temperature ranges from 15°C to 25°C and the elevation ranges from 300 to 550 m. This rule might change at random to a temperature range of 15°C to 30°C and an elevation range of 300 to 500 m. If the new rule improves the ability of the program to predict where the species is actually found, it is retained, and other, less successful rules are discarded.

For the Madagascar chameleons, the accuracy of the distribution model developed was tested with chameleon location data that had not been entered into the program. The model performed well, correctly predicting where these chameleons lived 75% to 85% of the time. Next, the model was used to predict the geographic ranges of each of the 11 chameleon species—information that will be useful in efforts to protect chameleon habitat. Finally, the researchers investigated an interesting "error" in the model: there were several overlapping areas in which the model predicted that 2 or more of the 11 species would be found but in which no chameleons were known to occur (**FIGURE 9.8**). When

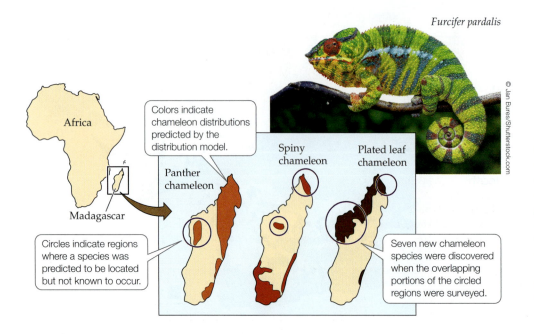

Furcifer pardalis

© Jan Bures/Shutterstock.com

FIGURE 9.8 Predicted Distributions of Madagascar Chameleons The predicted distributions of 3 of 11 species of chameleons are shown for the panther chameleon (*Furcifer pardalis*), the spiny chameleon (*F. verrucosus*), and the plated leaf chameleon (*Brookesia stumpffi*). All 11 of the predicted distributions proved accurate. (After C. J. Raxworthy et al. 2003. *Nature* 426: 837–841.)

two of these overlapping areas were surveyed, 7 previously unknown chameleon species were discovered. More intensive surveys conducted at the same time, but at sites outside these overlapping areas, found only 2 new species. Thus, the scientists were able to predict both the distributions of known chameleon species and the locations of habitats suitable for other chameleons, and the latter prediction led to the discovery of 7 new chameleon species.

CLIMATE CHANGE CONNECTION

EFFECTS OF CLIMATE CHANGE ON THE GEOGRAPHIC DISTRIBUTIONS OF SPECIES The waters along the east coast of Tasmania have warmed considerably since 1950 (**FIGURE 9.9A**). As this warming has occurred, the long-spined sea urchin (*Centrostephanus rodgersii*) has extended its range to the south (**FIGURE 9.9B**). The changes in the distribution of this urchin are consistent with the idea that climate change is the underlying cause: the larvae of *C. rodgersii* fail to develop properly in waters colder than 12°C, and the urchin has moved into new regions as waters in those locations have warmed to the point that they remain above that temperature. As *C. rodgersii* has expanded its range, it has established extensive urchin barrens in which all kelp have been removed by grazing (Ling 2008). Thus, through its effects on the geographic distribution of the long-spined sea urchin, ongoing climate change appears to be having

a profound effect on kelp ecosystems along the Tasmanian coast. As observed for the long-spined sea urchin, shifts in the geographic distributions of hundreds of other species have been linked to climate change (Parmesan and Yohe 2003). In some marine communities, range shifts driven by climate change have contributed to the rapid replacement of temperate species with species from subtropical or tropical regions, leading to the formation of entirely new communities (Wernberg et al. 2016). On land, many species in the Northern Hemisphere have expanded the northern edges of their ranges toward the pole, while the southern edges of their ranges have maintained relatively stable positions (Sunday et al. 2012). But range shifts do not always occur in this way, nor do they necessarily keep pace with ongoing climate change.

For example, Kerr et al. (2015) found that the geographic ranges of 67 species of bumblebees have shown rapid losses in the south and only a slow expansion in the north—as a result, their ranges are shrinking and the populations of some bumblebee species have declined as the climate has warmed. Moreover, even when the range expansion of one or more species keeps pace with climate change, such range shifts can have cascading and wide-ranging effects on other species (as illustrated by the decimation of kelp forests as the long-spined sea urchin expanded its range to the south). The exact nature of such cascading effects can be hard to predict, but it is clear that ongoing climate change will have major effects on ecosystems throughout the globe. ☀

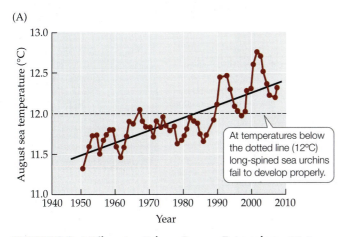

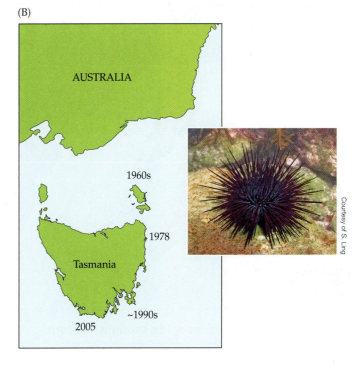

FIGURE 9.9 A Climate-Driven Range Extension Winter water temperatures along the east coast of Tasmania in August, the most important month for offspring production in long-spined sea urchins (A). The map in (B) shows the years in which long-spined sea urchins were first observed at points along the Tasmanian coast. (After S. Ling et al. 2009. *Proc Natl Acad Sci USA* 106: 22341–22345. © 2009 National Academy of Sciences, U.S.A.)

LEARNING OBJECTIVES

9.3.1 Describe the factors important to the suitability of habitat for populations and species.

9.3.2 Explain how the distribution and abundance of species can reflect their evolutionary and geologic history.

9.3.3 Describe the role of dispersal and migration in distributing organisms across the landscape.

Processes Important to Distribution and Abundance

It is clear from the chameleon example discussed earlier in this chapter that identifying the factors important to species presence is key to understanding their distribution and abundance patterns. Many factors can influence the distributions and abundances of organisms. We'll survey these factors by grouping them into three categories: habitat suitability, historical factors (such as evolutionary history and continental drift), and dispersal.

Habitat suitability determines distribution and abundance

Good and poor places to live exist for all species. A desert species is not likely to perform well in the Arctic, or vice versa. Even small differences among environments in survival and/or reproduction of individuals can cause differences in abundance. Thus, the distribution and abundance of a species are influenced strongly by the presence of appropriate habitat. But what factors make habitat suitable?

THE ABIOTIC ENVIRONMENT As we discussed in Unit 1, the climate and other aspects of the abiotic (non-living) environment, such as soil pH, salt concentration, and available nutrients, set limits on whether a habitat will be suitable for a particular species. Some species can tolerate a broad range of physical conditions, while others have more narrow requirements.

Creosote bush (*Larrea tridentata*), for example, has a broad distribution in North American deserts, ranging across much of the southwestern United States and northwestern and central Mexico (**FIGURE 9.10**). Creosote bush is very tolerant of arid conditions: it uses water rapidly when it is available, then shuts down its metabolic

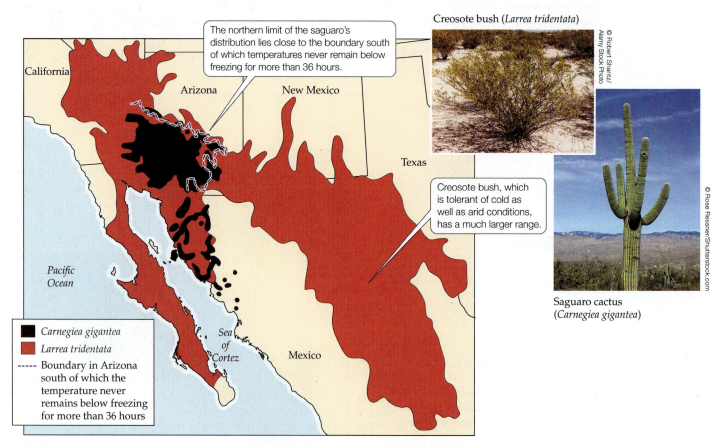

FIGURE 9.10 The Distributions of Two Drought-Tolerant Plants The geographic distribution of creosote bush (*Larrea tridentata*) is much larger than that of saguaro cactus (*Carnegiea gigantea*). (After T. W. Yang. 1970. *J Arizona Acad Sci* 6: 41–45; F. Shreve and I. L. Wiggins. 1951. *Vegetation and Flora of the Sonoran Desert*. Carnegie Institute of Washington: Washington, DC.)

processes during periods of extended drought. Creosote bush also tolerates cold, so its populations thrive in high-elevation deserts where winter temperatures can remain below freezing for several days.

The saguaro cactus (*Carnegiea gigantea*), on the other hand, has a more limited distribution. Like creosote bush, saguaro flourishes under arid conditions, but it achieves its drought tolerance in different ways. The saguaro cactus uses a unique photosynthetic pathway called crassulacean acid metabolism (see Concept 5.3), which allows it to reduce water loss. Furthermore, during wet periods, the saguaro stores water in its massive trunk and arms, saving it for use during times of drought. Saguaro cannot tolerate cold, however; it is killed when temperatures remain below freezing for 36 hours or more. The importance of saguaro's sensitivity to cold is revealed by its distribution: the northern limit of its distribution corresponds closely to a boundary north of which temperatures occasionally remain below freezing for at least 36 hours (see Figure 9.10).

THE BIOTIC ENVIRONMENT The biotic environment also has important effects on distributions and abundances of species. Obviously, species that depend on one another for their growth, reproduction, and/or survival have to live together at the same location. For example, all species require food resources, and thus habitat suitability will depend on the distribution and abundance of their food. An example of this is the Seychelles warbler (*Acrocephalus sechellensis*), an endangered songbird. In the 1950s, this bird nearly went extinct: its total world population was reduced to just 26 individuals located on Cousin Island in the Seychelles, a group of islands off the east coast of Africa. After the Seychelles warbler was legally protected in 1968, the Cousin Island population increased to about 300 birds, and the species was introduced successfully to two other islands.

Seychelles warblers are territorial: a breeding pair defends its territory against other birds of the species. But not all territories are equal: some are of higher quality than others because they provide more food (e.g., insects) (**FIGURE 9.11**). Birds that live in a high-quality territory live longer and produce more young. In addition, a breeding pair that lives in a high-quality territory often receives help rearing its young from offspring born in previous years. Because the high-quality sites attract offspring from previous years and are aggregated toward one end of the island, differences in territory quality make the dispersion of individuals in the population more clumped than it otherwise would be.

Organisms can also be excluded from an area by herbivores, predators, competitors, parasites, or pathogens, any of which can greatly reduce the survival or reproduction of members of a population. For example, the case study describes how the distribution of kelp forests is dependent on the presence of sea otters, which prey on sea

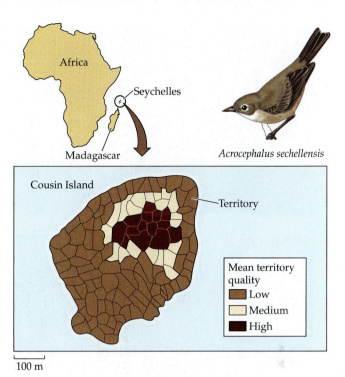

FIGURE 9.11 **Food Resources Affect Habitat Suitability** The mean quality of Seychelles warbler territories on Cousin Island across the years 1986 to 1990. Territories were grouped into three quality categories: high, medium, and low. High-quality territories were clustered inland; these territories had high vegetation cover, little wind, and abundant insects. Coastal areas had lower-quality territories because salt spray led to defoliation, which lowered insect abundance. (After J. Komdeur. 1992. *Nature* 358: 493–495.)

urchins, which are voracious herbivores on kelp. Another dramatic example of biotic control of species distribution and abundance is provided by the successful biological control of *Opuntia stricta*, an introduced cactus that spread rapidly to cover large areas in Queensland and New South Wales, Australia. The cactus was imported from the southern United States in 1839 and planted as hedge. Within 40 years, *O. stricta* had become a pest species, and by 1925 it covered 243,000 km^2. The cactus can grow up to 2 m high, and in many areas it covered the ground with dense, spiny thickets, making the rangelands it occupied useless (**FIGURE 9.12A**). In the hope of controlling the cactus, an Argentinean moth, *Cactoblastis cactorum*, known to feed on *Opuntia* was released in 1926 (**FIGURE 9.12B**). By 1931, the moths had spread widely and destroyed billions of cacti. Since 1940, the cactus has persisted in small numbers, but its distribution and abundance have been greatly reduced. Although the introduction of *C. cactorum* as a means of biological pest control appears to have been a great success, such introductions must be undertaken cautiously because they can lead to unintended consequences, such as damage to native species (Louda et al. 1997).

(A)

(B)

Reproduced with permission of the Department of Natural Resources, Queensland, Australia

FIGURE 9.12 Herbivores Can Limit Plant Distributions
In Australia, the moth *Cactoblastis cactorum* was used to control populations of an introduced cactus, *Opuntia stricta*. (A) A dense thicket of *O. stricta* 2 months before the release of the moth. (B) The same stand 3 years later, after the moth had killed the cacti by feeding on their growing tips.

INTERACTIONS BETWEEN ABIOTIC AND BIOTIC ENVIRONMENT   In reality, the distribution and abundance of species are a product of both the abiotic and biotic environment. For example, the quality of the territories of the Seychelles warbler depends not only on insect food resources but also the exposure to salt spray and wind (see Figure 9.11). In another example, the rocky intertidal barnacle *Semibalanus balanoides* cannot survive when exposed at low tide to summer air temperatures above 25°C, and it cannot reproduce if winter air temperatures during low tide do not remain below 10°C for 20 days or more. On the Pacific coast of North America, temperatures are such that *S. balanoides* could be found 1,600 km farther south than it currently is (**FIGURE 9.13**). But this barnacle is absent from the region shown in purple in Figure 9.13, presumably because competition from other species of barnacles prevents it from living in what would otherwise be suitable habitat. To the north, as temperatures become increasingly colder, a point is reached where *S. balanoides* outcompetes the other barnacles and maintains healthy populations. Thus, the

abiotic and biotic environments interact to determine where populations of this barnacle are found.

DISTURBANCE   The distributions of some organisms depend on regular forms of disturbance. A **disturbance** is an abiotic event that kills or damages some individuals and thereby creates opportunities for other individuals to grow and reproduce. Many plant species, for example, persist in an area only if there are periodic fires. If humans prevent fires, such species are replaced by other species that are not as tolerant of fires but are superior competitors in the absence of fires. Thus, a change in the frequency of fires can change the composition of ecological communities, as you can explore in **ANALYZING DATA 9.1**. Floods, windstorms, and droughts are other forms of disturbance that can harm some species but give others an advantage. We'll discuss the role of disturbance in more detail in Chapter 17.

Distribution and abundance reflect evolutionary and geologic history

Events in the evolutionary and geologic history of Earth have had a profound effect on where organisms live today. Why, for example, are polar bears (*Ursus maritimus*) found in the Arctic but not in Antarctica? Polar bears hunt on ice

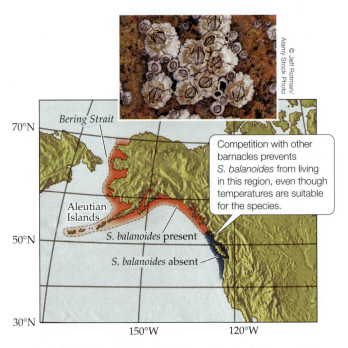

© Jeff Rotman/ Alamy Stock Photo

FIGURE 9.13 Joint Effects of Temperature and Competition on Barnacle Distribution   Although temperatures are suitable for the barnacle *Semibalanus balanoides* throughout the red- and blue-shaded regions, it is excluded from the southern region potentially by its competitors. In the red-shaded regions, temperatures are colder and *S. balanoides* is the superior competitor.

? Is global warming likely to increase or decrease the geographic range of *S. balanoides*? Explain.

ANALYZING DATA 9.1

Have Introduced Grasses Altered the Occurrence of Fires in Hawaiian Dry Forests?

Bush beardgrass (*Schizachyrium condensatum*), molasses grass (*Melinis minutiflora*), and several other non-native grasses were introduced by humans to Hawaii as forage for livestock. By 1969, introduced grasses had invaded the dry forests of Volcanoes National Park, Hawaii. These dry forests are open woodlands with an understory of shrubs; they contain few or no native grasses. Hughes et al. (1991)* provide data on fire occurrence (**TABLE A**) and on vegetation abundance in unburned and burned regions of dry forests in the park (**TABLE B**).

1. Using the data in Table A, calculate the frequency of fires and the average area burned before and after introduced grasses invaded Volcanoes National Park. What do your results suggest about how introduced grasses have affected the occurrence of fires in Volcanoes National Park?

2. Based on the data in Table B, does fire promote or limit the abundance of native trees and shrubs? How does fire affect introduced grasses?

*Hughes, F., P. M. Vitousek, and T. Tunison. 1991. Alien grass invasion and fire in the seasonal submontane zone of Hawai'i. *Ecology* 72: 743–746.

3. Introduced grasses recover quickly from fires, and they provide more fuel for future fires than do native trees and shrubs. Use this information to predict what may happen if a fire occurs in a Hawaiian dry forest after introduced grasses have invaded that forest. Do the events you've described help to explain the data in Tables A and B? Explain your reasoning.

Table A

Time frame	Number of fires	Total area burned
1928–1968	9	2.3 ha
1969–1988	32	7,800 ha

Table B

Vegetation type	Vegetation abundance index		
	Unburned	Burned once	Burned twice
Native trees and shrubs	112.3	5.2	0.7
Introduced grass	80.0	92.1	100.9

---

packs and eat seals, both of which abound in Antarctica. Part of the answer to our question can be found in the evolutionary history of these bears. Fossils and genetic evidence indicate that polar bears evolved from brown bears (*Ursus arctos*) in the Arctic (Lindqvist et al. 2010); hence *U. maritimus* is found in the Arctic because the species originated there. As for their absence from Antarctica, although polar bears can travel over 1,000 km in a year, it appears they cannot or will not cross the tropical regions that separate the Arctic from Antarctica. Thus, the distribution of polar bear populations is influenced by evolutionary history and dispersal as well as by the presence of suitable habitat.

Geologic history plays a key role in some curious distribution patterns that puzzled biologists for nearly 100 years. Consider Alfred Russel Wallace's observation that the animals of a region can differ considerably across relatively short geographic distances (Wallace 1860). The mammal communities of the Philippines, for example, are more similar to those in Africa (88% overlap at the family level) than they are to those in New Guinea (64% overlap), despite the fact that Africa is 5,500 km away and New Guinea is only 750 km away. No explanation for this and other similar observations could be found until the discovery of *continental drift*, the gradual movement of continents over time. This discovery led to the realization that the Philippines and New Guinea are on different tectonic plates and have been in close geographic contact for a relatively short period of time.

Dispersal is a process that distributes organisms across the landscape

Organisms differ greatly in their capacity for movement. In plants, for example, dispersal occurs when seeds move away from the parent plant. Although events such as storms can transport seeds long distances (hundreds of meters to many kilometers; see Cain et al. 2000), dispersal distances in plants are usually small (one to a few tens of meters). In some cases, typical seed dispersal distances are so small that they hardly count as movement. For example, seeds of the forest plant *Viola odorata* have been seen to disperse only 0.002 to 0.02 m (0.008–0.8 inches) when ants are not present; when ants are present, they may carry these seeds for a few meters. At the other end of the spectrum, some whale species travel tens of thousands of kilometers in a single year. Overall, the spatial extent of populations varies tremendously—from very small, in organisms that disperse little, to very large, in species that travel great distances.

FIGURE 9.14 Migration of North Pacific Humpback Whales Five separate populations (represented by different-colored arrows) of the North Pacific humpback whale (*Megaptera novaeangliae*) migrate between their winter breeding grounds off Mexico, Hawaii, and Japan and their summer feeding grounds in the Gulf of Alaska and the Northeast Pacific coast. (Map after https://hawaiihumpbackwhale.noaa.gov/explore/humpback_whale.html. Migration data from SPLASH Research.)

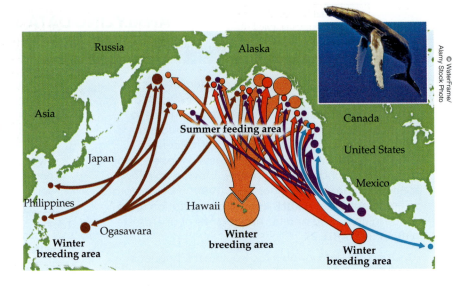

Whales also migrate, which is a specific type of dispersal in response to seasonal variation in resources. **Migration** involves round-trip movement and usually includes the entire population. For example, five separate North Pacific populations of the humpback whale migrate more than 4,800 km (~3,000 miles) between their winter breeding grounds in the south (Mexico, Hawaii, and Japan) and their summer feeding grounds in the north (Northeast Pacific coast and Gulf of Alaska) (**FIGURE 9.14**). A 2006 survey of North Pacific humpback whales determined that these populations have rebounded since 1966, when commercial whaling of these populations was banned.

As demonstrated by the polar bear's absence from Antarctica, a species' limited capacity for dispersal can prevent it from reaching areas of suitable habitat—a phenomenon known as **dispersal limitation**. In another example, the Hawaiian Islands have only one native terrestrial mammal, the Hawaiian hoary bat (*Lasiurus cinereus*), which was able to fly to the islands. No other land mammals have been able to disperse to Hawaii on their own, although cats, pigs, wild dogs, rats, goats, mongooses, and other mammals now thrive in Hawaii following their introduction to the islands by people.

Dispersal limitation can also occur on smaller spatial scales, preventing populations from expanding to nearby areas of apparently suitable habitat. An example of dispersal limitation was documented in a long-term study of the English bluebell (*Hyacinthoides non-scripta*). In 1960, 27 populations of 7 to 10 individuals each were established in apparently suitable forest habitat located near source populations (Van der Veken et al. 2007). Forty-five years later, only 11 (41%) of these experimental populations persisted, and most contained hundreds or thousands of individuals. These results suggested that dispersal limitation had prevented the bluebells from maintaining the majority of their original populations or creating new populations in areas nearby.

In the following section, let's consider in more detail how dispersal can create and maintain multiple populations and the role of these metapopulations in the conservation of endangered species.

> **CONCEPT 9.4** In metapopulations, sets of spatially isolated populations are linked by dispersal.

LEARNING OBJECTIVES

9.4.1 Describe how the rates of colonization and extinction of populations affect metapopulations.

9.4.2 Describe how the amount of suitable habitat and population isolation can affect metapopulation persistence or extinction.

Metapopulations

The checkered landscapes of the Dorset Heathlands demonstrate that the world is a patchy place (see Figure 9.7). The patchy nature of the landscape ensures that for many species, areas of suitable habitat do not cover large, continuous regions, but rather exist as a series of favorable sites that are spatially isolated from one another. As a result, the populations of a species are often scattered across the landscape, each in an area of favorable habitat but separated from one another by hundreds of meters or more. These seemingly isolated populations can be classified as a metapopulation when individuals (or gametes) occasionally disperse from one population to another. Literally, the term "metapopulation" refers to a population of populations, but it is usually defined in a more particular sense as a set of spatially isolated populations linked to one another by dispersal (**FIGURE 9.15**). In some metapopulations, certain populations are *sources* (from which the number of individuals that disperse to other populations is greater than

(A)

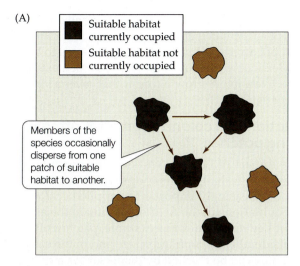

Suitable habitat currently occupied

Suitable habitat not currently occupied

Members of the species occasionally disperse from one patch of suitable habitat to another.

(B)

Courtesy of DigitalGlobe

FIGURE 9.15 The Metapopulation Concept A metapopulation is a set of spatially isolated populations linked by dispersal. (A) Seven patches of suitable habitat for a species are diagrammed, four of which are currently occupied and three of which are not. The area outside of these seven patches represents unsuitable habitat. (B) Satellite image of a group of lakes in northern Alaska that are sometimes connected to one another by temporary streams that form after the snow melts or after periods of heavy rainfall.

the number of migrants they receive) while other populations are *sinks* (which receive more immigrants than the number of emigrants they produce).

Metapopulations are characterized by repeated extinctions and colonizations

As ecologists have long recognized, populations of some species are prone to extinction for two reasons: (1) the patchiness of their habitat makes dispersal between populations difficult, and (2) environmental conditions can change in a rapid and unpredictable manner. Metaphorically, we can think of such populations as a set of "blinking lights" that wink on and off, seemingly at random, as patches of suitable habitat are colonized and the populations in those patches

then go extinct. Although the individual populations may be prone to extinction, the collection of populations—the metapopulation—persists because it includes populations that are going extinct and new populations established by colonization.

Building on this idea of random extinctions and colonizations, Richard Levins (1969, 1970) represented metapopulation dynamics in terms of the extinction and colonization of habitat patches:

$$\frac{dp}{dt} = cp(1-p) - ep \qquad (9.4)$$

where p represents the proportion of habitat patches that are occupied at time t, while c and e are the patch colonization and patch extinction rates, respectively.

In deriving Equation 9.4, Levins made a few assumptions, including the following: (1) there is a very large (infinite) number of identical habitat patches, (2) all patches have an equal chance of receiving colonists (hence the spatial arrangement of the patches does not matter), and (3) all patches have an equal chance of extinction.

As we'll discuss below, some of the assumptions of Levins's model are not realistic. Still, Equation 9.4 leads to a simple but fundamental insight: for a metapopulation to persist for a long time, the ratio e/c must be less than 1. This means that some patches will be occupied if the colonization rate is greater than the extinction rate. On the other hand, if the extinction rate is greater than the colonization rate (and hence, $e/c > 1$), the metapopulation will collapse and all populations in it will become extinct. Levins's groundbreaking approach focused attention on a number of key issues, such as how to estimate factors that influence patch colonization and extinction, the importance of the spatial arrangement of suitable patches, the extent to which the landscape between habitat patches affects dispersal, and the vexing problem of how to determine whether empty patches are suitable habitat or not. Levins's rule for persistence also has applied importance, as we will see shortly.

A metapopulation can go extinct even when suitable habitat remains

Human actions (such as land development) often convert large tracts of habitat into sets of spatially isolated habitat fragments (see Figure 9.7). Such **habitat fragmentation** can cause a species to have a metapopulation structure where it did not have one before. If land development continues and the habitat becomes still more fragmented, the metapopulation's colonization rate (c) may decrease because patches become more isolated and hence harder to reach by dispersal. Further habitat fragmentation also causes the patches that remain to become smaller; as a result, the extinction rate (e) may increase because smaller patches have smaller populations, which, as we have just seen, have a higher risk of extinction. Both of these trends (an increase in e and a decrease in c) cause the ratio e/c to increase. Thus, if too much habitat is removed, the ratio e/c may shift suddenly from less than 1 to greater than 1, thereby dooming all populations—and the metapopulation—to eventual extinction, even though some habitat remains.

The idea that all populations in a metapopulation might go extinct while suitable habitat remains was developed further in studies on the northern spotted owl (**FIGURE 9.16**). The northern spotted owl (*Strix occidentalis caurina*) is found in the Pacific Northwest region of North America. It lives in old-growth forest, where nesting pairs establish large territories that range in size from 12 to 30 km² (territories are larger in poor-quality habitat). Lande (1988) modified Levins's model to include a description of how owls might search for vacant "patches," which were interpreted as sites suitable for individual territories. Lande estimated that the entire metapopulation would collapse if the area covered by old-growth forest were reduced by logging to less than 20% of the total area of a large region. This result had a powerful impact: it illustrated how a species might go extinct if its habitat dropped below a critical threshold (in this case, 20% suitable habitat), and it contributed to the 1990 listing of the northern spotted owl as a threatened species in the United States. The importance of

FIGURE 9.16 The Northern Spotted Owl The northern spotted owl (*Strix occidentalis caurina*) thrives in old-growth forests of the Pacific Northwest; such forests include those that have never been cut, or have not been cut for 200 years or more.

conserving old-growth forest has been highlighted by the effects of a recent invader, the barred owl (*Strix varia*): the arrival of barred owls can cause spotted owls to become locally extinct within forest patches, but such extinctions are less likely in old-growth forests that cover a large area (Dugger et al. 2011).

Extinction and colonization rates often vary among patches

As the impact of Lande's work on the northern spotted owl suggests, the metapopulation approach has become increasingly important in applied ecology. But metapopulations in the field often violate the assumptions of Levins's model. For example, patches often differ considerably in population size and in the ease with which they can be reached by dispersal. As a result, extinction and colonization rates may vary greatly among patches. Therefore, most ecologists use more complex models (see Hanski 1999) when addressing practical questions in the field.

Consider the skipper butterfly *Hesperia comma*. In the early 1900s, this butterfly was found on grazed calcareous grasslands (i.e., grasslands growing in alkaline soils found on limestone or chalk outcrops) throughout a broad range of the United Kingdom. Starting in the 1950s, however, calcareous grasslands became overgrown because the numbers of cattle and other important grazers were reduced. As a result, *H. comma* populations began to decline. By the mid-1970s, the butterfly was found in only 10 restricted regions, a very small fraction of its original range.

Things began to pick up for the butterfly in the early 1980s. By this time, habitat conditions had improved because livestock had been reintroduced. Surveying these grasslands in 1982, Chris Thomas and Terésa Jones documented the locations of all patches containing *H. comma* populations and of all patches that appeared suitable for, but were not occupied by, *H. comma*. To determine the fate of each occupied and unoccupied patch over time, they surveyed the patches again in 1991 and noted which ones were occupied at that time. Their results highlight two important features of many metapopulations: isolation by distance and the effect of patch area.

Isolation by distance occurs when patches located far from occupied patches are less likely to be colonized than are nearby patches. In *H. comma*, distance from occupied patches had a strong effect on whether patches vacant in 1982 were colonized by 1991: few patches separated by more than 2 km from an occupied patch were colonized during that period (**FIGURE 9.17**). Patch area also affected the chance of colonization: the majority of colonized patches were at least 0.1 ha in size. Patch area may have affected colonization rates directly because small patches may be harder for the butterflies to find than large patches. Alternatively, *H. comma* might have colonized small patches, but then suffered extinction in those patches by 1991 due to problems associated with small population size; such

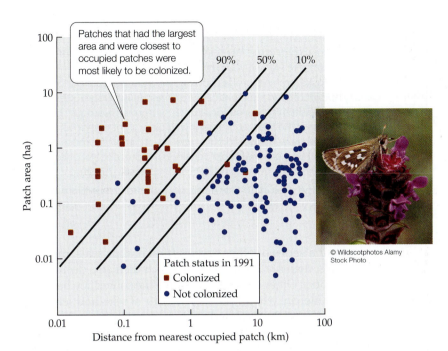

FIGURE 9.17 Colonization in a Butterfly Metapopulation Colonization of suitable habitat from 1982 to 1991 by the skipper butterfly *Hesperia comma* was influenced by patch area and patch isolation (distance to the nearest occupied patch). Each square or circle represents a patch of suitable habitat that was not occupied by *H. comma* in 1982. The lines show the combinations of patch area and patch isolation for which there was a 90%, 50%, and 10% chance of colonization (as calculated from a statistical analysis of the data). (After C. D. Thomas et al. 1992. *Oecologia* 92: 563–567; C. D. Thomas and T. M. Jones. 1993. *J Anim Ecol* 62: 472–481.)

? Based on these results, estimate the chance of colonization for a 1-ha patch located 1 km away from the nearest occupied patch.

patches would appear never to have been colonized because the sites were not sampled between 1982 and 1991.

Among patches occupied in 1982, Thomas and Jones found that the chance of extinction was highest in small patches (most likely because small patches tend to have small population sizes) and in patches that were far from another occupied patch. Isolation by distance can affect the chance of extinction because a patch that is near an occupied patch may receive immigrants repeatedly, which may increase the patch population size and make extinction less likely. This tendency for high rates of immigration to protect a population from extinction (by reducing the problems associated with small population size) is known as the **rescue effect** (Brown and Kodric-Brown 1977). In Chapter 24, we discuss in more detail the role of metapopulation dynamics in the conservation of species experiencing habitat fragmentation.

A CASE STUDY REVISITED

From Kelp Forest to Urchin Barren

When sea urchins graze kelp so heavily that kelp forests are replaced by urchin barrens, what happens next? We might expect that the urchins would starve because they have destroyed their food source. Field studies show that urchin barrens can persist for years on end, however, because urchins can use food sources other than kelp, including benthic diatoms, less preferred algae (including hard, encrusting forms that cover rock surfaces), and detritus. When food is extremely scarce, urchins can reduce their metabolic rate, reabsorb their sex organs (forgoing reproduction but increasing their chances of survival), and absorb dissolved nutrients directly from seawater.

As tough and resilient as urchins are, they are vulnerable to predation by sea otters (*Enhydra lutris*), which function as impressive urchin-eating machines. Otters need to eat large quantities of food each day because they have a high metabolic rate and they store little energy as fat. Urchins are a favorite food of otters, and since there are 20 to 30 otters per square kilometer around some Aleutian islands, the potential exists for otters to consume enormous quantities of urchins. These facts, coupled with the observation that urchins usually are common only where otters are absent, led investigators to suspect that otters might control the locations of urchins, and hence the locations of kelp forests.

To test this hypothesis, Estes and Duggins (1995) compared sites with and without otters, both in the Aleutian Islands and along the coast of southern Alaska. Confirming the results of previous studies, they found that sites where otters had been present for a long time usually had many kelp and few urchins, whereas sites without otters usually had many urchins and few kelp. Estes and Duggins also collected data from sites colonized by otters during the course of their study. At sites in southern Alaska, the arrival of otters had a rapid and dramatic effect: within 2 years, urchins virtually disappeared, and kelp densities increased dramatically (**FIGURE 9.18A**). At Aleutian Islands sites, however, kelp recovered more slowly after the arrival of otters (**FIGURE 9.18B**). At these sites, otters ate most of the large urchins, reducing urchin biomass by an average of 50%. However, in a twist that did not occur at the southern Alaska sites, the arrival of new urchin larvae (most likely from other sites via ocean currents) provided a steady supply of small urchins. These small urchins slowed the rate at which kelp forests replaced urchin barrens.

(A) Southern Alaska   (B) Aleutian Islands

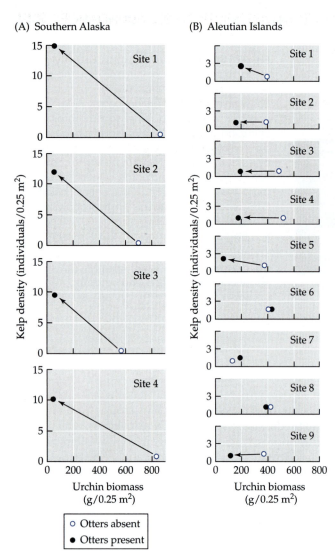

○ Otters absent
● Otters present

FIGURE 9.18 The Effect of Otters on Urchins and Kelp
Plots of kelp density versus sea urchin biomass measured at sites in southern Alaska and in the Aleutian Islands before and 2 years after the return of otters. (A) Two years after otters colonized four sites in southern Alaska, urchin biomass had declined considerably, and kelp density had increased substantially at all sites. (B) Two years after otters colonized nine sites in the Aleutian Islands, sea urchin biomass had declined at six of the sites, but kelp showed clear signs of recovery at only two of the sites. Arrows indicate a decline in urchin biomass and (at some sites) an increase in kelp density in the presence of otters. (After J. A. Estes and D. O. Duggins. 1995. *Ecol Mongr* 65: 75–100.)

 For the nine sites in (B), list the six sites where urchin biomass declined; also list the two sites where kelp density increased.

hunt a series of other species (first harbor seals, then fur seals, then sea lions), each of which then also declined in number. Other researchers dispute the connection between commercial whaling and the decline of seals and sea lions, suggesting that seal and sea lion populations declined for other reasons, such as a lack of food due to reduced fish populations in the open ocean (DeMaster et al. 2006). Whatever the cause, it was in the 1990s, when populations of harbor seals, fur seals, and sea lions had all declined to low levels, that killer whales were first seen attacking otters. Otters and killer whales had been observed in close proximity for decades, but within 10 years of the first known attack, otter populations crashed.

CONNECTIONS *in* NATURE

FROM URCHINS TO ECOSYSTEMS   Urchins, otters, and perhaps killer whales and people play important roles in determining the distribution of kelp. But does the presence or absence of kelp matter? Do kelp have strong effects on nearshore ecosystems?

Indeed they do. Kelp forests serve as nurseries for the young of many marine species and as havens from predators for the adults of still more species. Kelp forests are also among the most productive ecosystems in the world, rivaling tropical forests in the amount of new biomass they produce each year (up to 2,000 g of carbon/m²/year). Kelps grow from their base, and their tips are constantly "eroded" by wave action and other physical forces. Thus, much of their biomass ends up as floating bits of detritus, which provides food for suspension feeders such as barnacles and mussels that filter food from the water. As a result, barnacles and mussels grow more rapidly and are more abundant in kelp forests than in urchin barrens. Carbon-13 labeling studies (see Ecological Toolkit 5.1) have shown that the tissues kelp produce by photosynthesis provide a food source for a wide range of species (Duggins et al. 1989).

Overall, we can see that the effects of urchins on kelp, and of otters on urchins, do indeed matter: urchins and otters (and killer whales) all interact in ways that alter fundamental aspects of kelp forest ecosystems.

Historically, sea otters were abundant throughout the North Pacific, but by 1900 they had been hunted to near extinction for their fur. By 1911, when international treaties protected the sea otter, only about 1,000 otters remained—less than 1% of their early numbers. Scattered colonies of otters survived and gradually increased in size around some Aleutian islands, causing the observed pattern of kelp forests around some islands, and urchin barrens around others. In the 1990s, however, there was a sudden and unexpected decline in otter populations. Urchins made a comeback, and kelp densities were reduced (**FIGURE 9.19A–D**). The question then became, What caused the decline of sea otter populations in the 1990s?

James Estes and his colleagues have suggested that otters declined because of increased predation by the killer whale, *Orcinus orca* (**FIGURE 9.19E**). It is not known why killer whales began to eat more otters. Some researchers have argued that this change may have been part of a chain of events that began when commercial whaling drove populations of large whales to low numbers (Springer et al. 2003). According to this hypothesis, once their preferred prey (large whales) became rare, killer whales began to

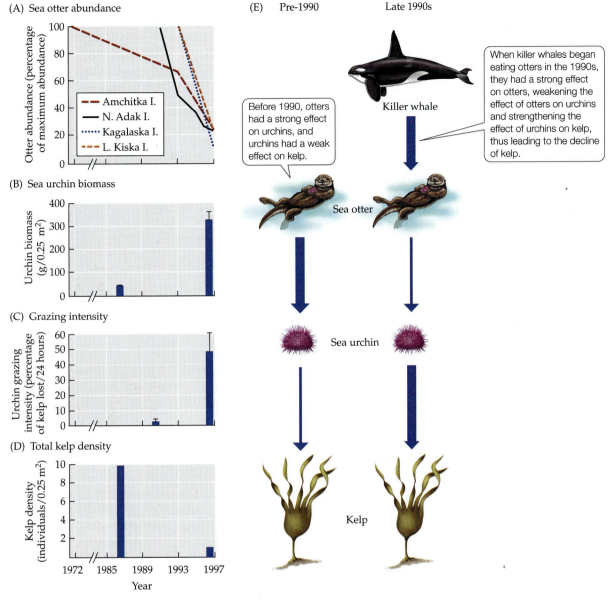

FIGURE 9.19 Killer Whale Predation on Otters May Have Led to Kelp Declines Declines in otter abundance over time (A) are associated with (B) a rise in urchin biomass, (C) an increase in the intensity of urchin grazing on kelp, and (D) a decrease in kelp density. (E) The proposed mechanisms for these changes. Strengths of the effects are indicated by the thicknesses of the arrows. Error bars in (B) and (C) show one standard error of the mean. (After J. A. Estes et al. 1998. *Science* 282: 5388.)

SUMMARY

CONCEPT 9.1 Populations are groups of individuals of the same species that vary in size over space and time.

9.1.1 Define the terms population, population size, and population density.

- A population is a group of individuals of the same species that live in the same area and at the same time and interact with one another.

- Population abundance can be defined as population size (the number of individuals in the population) or as population density (the number of individuals per unit of area).

9.1.2 Compare the different ways in which individuals are defined, including the terms clones, genets, and ramets.

(Continued)

SUMMARY (continued)

- In sexually reproducing species, the offspring is a single genetic individual.

- In asexually reproducing species, multiple genetically identical individuals, or clones, can be produced. The single genetic unit is known as the genet and the genetically identical individuals are known as ramets.

9.1.3 Compare the different methods used to measure the abundance of individuals within populations or species.

- The most direct way to determine the number of individuals in a population is to count all of them, obtaining their absolute population size. When this is not possible, the relative population size, or the number of individuals standardized for area or time, can be measured.

- Relative population size measures can include area-based counts, distance methods, or mark–recapture studies.

CONCEPT 9.2 Species vary in their distribution and abundance across their geographic range.

9.2.1 Describe the relationship between populations, metapopulations, and geographic ranges for species.

- Groups of individuals of the same species can be divided into populations, and groups of populations (metapopulations) are linked together by dispersal.

- At larger spatial scales, the entire geographic range, or distribution, of a species might consist of one or multiple metapopulations.

9.2.2 Compare the different dispersion patterns of populations.

- The dispersion of individuals within a population may be regular, random, or clumped. In the field, clumped dispersions are more common.

9.2.3 Describe how the size and patchiness of geographic ranges vary among species and can be predicted using models.

- Geographic ranges can be as small as a few meters to as large as multiple continents.

- Many species have a patchy distribution within their geographic range depending on the amount and variability of suitable habitat for that species.

- Species distribution models can be used to estimate the geographic range of a species when there are insufficient data on its distribution or when we want to predict the future locations of its populations.

CONCEPT 9.3 Species are limited in their distribution and abundance by habitat suitability, historical factors, and dispersal.

9.3.1 Describe the factors important to the suitability of habitat for populations and species.

- The suitability of habitat depends on both abiotic and biotic features of the environment, including factors that affect physiological tolerances, resources, and species interactions.

- Some species are thwarted or dependent on regular forms of disturbance, or abiotic events that kill or damage some individuals.

9.3.2 Explain how the distribution and abundance of species can reflect their evolutionary and geologic history.

- The distribution of species may be the result of their past evolutionary history, which has shaped their adaptations to the abiotic and biotic environment.

- The distribution of species may be the result of past geologic changes on Earth such as continental drift.

9.3.3 Describe the role of dispersal and migration in distributing organisms across the landscape.

- Dispersal, or the movement of individuals into (immigration) or out of (emigration) an existing

- population, can link or limit individuals among populations over small or large spatial scales.

- Migration involves round-trip movement, includes the entire population, and is often in response to seasonal variation in resources.

CONCEPT 9.4 In metapopulations, sets of spatially isolated populations are linked by dispersal.

9.4.1 Describe how the rates of colonization and extinction of populations affect metapopulations.

- The populations within metapopulations experience repeated colonization and extinction events.

- If the rate of population extinction is greater than the rate of population colonization, the metapopulation will eventually collapse.

9.4.2 Describe how the amount of suitable habitat and population isolation can affect metapopulation persistence or extinction.

- A metapopulation can be doomed to extinction, even when some suitable habitat remains, if the remaining habitat is not large enough to sustain individual populations or if the habitat is isolated by distance and is thus unable to receive immigrants from other populations (rescue effect).

Review Questions

1. Describe some of the complicating factors that can be encountered in studying a population.

2. No species is found everywhere on Earth. Why? Your answer should include an explanation of why organisms are not found in all places where you might expect them to thrive.

3. What is a species distribution model? Describe how such a model could be used to predict the future distribution of an organism that is spreading into a new geographic region.

4. Sea otters can eat 20% to 23% of their body weight in food each day. An average sea otter weighs 23 kg (roughly 50 pounds), and there are 20 to 30 otters per square kilometer where they are present. An average sea urchin weighs 0.55 kg. Assuming that the otters eat only sea urchins, use these data to calculate a conservative estimate of the number of sea urchins per square kilometer that an otter population would be expected to eat each year.

Hone Your Problem-Solving Skills

Climate change has enabled the long-spined sea urchin (*Centrostephanus rodgersii*) to expand its range poleward along the east coast of Tasmania, Australia (see Figure 9.9). To assess the impact of this range shift, Scott Ling (2008), then a graduate student at the University of Tasmania, estimated the number of taxa and total number of individuals in areas where (1) urchins had yet to colonize ("intact kelp beds"), (2) urchins were experimentally removed ("recovered kelp beds"), and (3) urchins had colonized and were not removed ("urchins present"). For each treatment, he placed four 0.25-m^2 quadrats on each of three patches. Data are shown in the table.

1. What is the total number of quadrats from which data were collected in each patch? In each treatment? Use this information to calculate mean values for the total number of taxa and the total number of individuals in each of the three treatments.

2. Suppose that a patch from which urchins had been removed (a recovered kelp bed) and a patch in which urchins were present each had an area of 40 m^2. Use your results from Question 1 to estimate the total number of individuals present in these two patches.

3. Interpret the results shown in the table.

Treatment	Patch	Total no. of taxa (in each of 4 quadrats)	Total no. of individuals (in each of 4 quadrats)
Intact kelp beds	1	61; 66; 44; 50	459; 402; 96; 179
	2	71; 87; 77; 66	497; 759; 560; 392
	3	69; 57; 90; 79	458; 188; 690; 533
Recovered kelp beds	1	84; 57; 79; 69	781; 341; 515; 771
	2	80; 86; 74; 61	401; 730; 429; 312
	3	72; 91; 69; 64	650; 1,132; 488; 419
Urchins present	1	11; 6; 9; 14	68; 36; 71; 47
	2	11; 22; 10; 15	38; 97; 33; 169
	3	12; 17; 13; 5	31; 134; 24; 14

ADDITIONAL RESOURCES

This textbook provides resources such as **Web Extensions** and additional **Climate Change Connections** for instructors who can share them with students. Please contact your instructor, if you would like to access that content.

10

Population Dynamics

KEY CONCEPTS

CONCEPT 10.1 Populations are dynamic entities that vary in size over time.

CONCEPT 10.2 The risk of extinction increases in populations that fluctuate in size and/or are small.

A Sea in Trouble: A Case Study

In the 1980s, the comb jelly *Mnemiopsis leidyi* (**FIGURE 10.1**) was introduced into the Black Sea, most likely by the discharge of ballast water from cargo ships. The timing of this invasion could hardly have been worse. At that time, the Black Sea ecosystem was already in decline due to increased inputs of nutrients such as nitrogen from sewage, fertilizers, and industrial wastes (and, as we'll see in this chapter's Connections in Nature, overfishing may also have contributed to the ecosystem's decline). The increased supply of nutrients had devastating effects across the northern Black Sea, where the waters are shallow (less than 200 m deep) and prone to problems that stem from **eutrophication** (an increase in the nutrient content of an ecosystem). As nutrient concentrations increased in these shallow waters, phytoplankton abundance increased, water clarity decreased, oxygen concentrations dropped, and fish populations experienced massive die-offs.

Such was the situation when *Mnemiopsis* arrived. This marine invertebrate species is a voracious predator of zooplankton, fish eggs, and young fish. Furthermore, *Mnemiopsis* continues to feed even when it is completely full, which causes it to regurgitate large quantities of prey stuck in balls of mucus. Small prey encased in mucus survive poorly. As a result, the negative effect of *Mnemiopsis* on its prey outstrips even its considerable ability to digest food.

Following its arrival in the Black Sea in the early 1980s, *Mnemiopsis* gradually increased in numbers. Then, in 1989, *Mnemiopsis* populations exploded (**FIGURE 10.2A**), reaching astonishing biomass levels (1,500 kg/m²) throughout the sea. The total biomass of *Mnemiopsis* in the Black Sea was estimated at 800 million tons (live weight) in 1989—far greater than the world's entire annual commercial fish catch, which has never exceeded 95 million tons.

The enormous numbers of *Mnemiopsis* present in 1989, and again in 1990, compounded the effects of the Black Sea's ongoing problems. *Mnemiopsis* ate huge quantities of zooplankton, causing their populations to crash (**FIGURE 10.2B**). Zooplankton eat phytoplankton, so *Mnemiopsis* indirectly caused phytoplankton populations to increase even more than they already had because of nutrient enrichment (**FIGURE 10.2C**). Upon their deaths, the phytoplankton and *Mnemiopsis* provided food for bacterial decomposers. Bacteria use oxygen as they decompose dead organisms, so as bacterial activity increased, oxygen concentrations in the water decreased, harming some fish populations. In addition, by devouring the food supplies

FIGURE 10.1 A Potent Invader The comb jelly *Mnemiopsis leidyi* was introduced from the east coast of North America to the Black Sea, wreaking havoc in its new ecosystem upon its arrival.

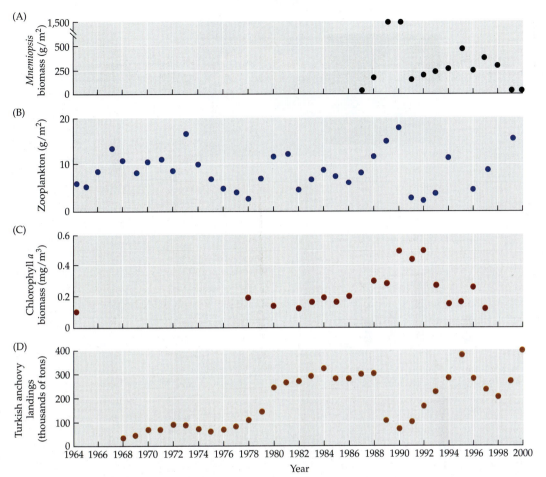

FIGURE 10.2 **Changes in the Black Sea Ecosystem** The graphs track long-term changes in four components of the Black Sea ecosystem: (A) mean biomass of the invasive species *Mnemiopsis leidyi* (first measured in 1987), (B) mean biomass of zooplankton, (C) mean biomass of chlorophyll *a* (an indicator of phytoplankton abundance), and (D) Turkish anchovy landings. (After A. E. Kideys. 2002. *Science* 297: 1482–1484.)

(zooplankton), eggs, and young of important commercial fishes such as anchovies, *Mnemiopsis* led to a rapid decline in fish catches (**FIGURE 10.2D**), causing extensive losses in the Turkish fishing industry.

The combined negative effects of nutrient enrichment and invasion by *Mnemiopsis* posed a serious threat to the Black Sea ecosystem. Although it covers a large surface area (over 423,000 km²), the Black Sea is nearly land-locked and exchanges little of its water each year with other ocean waters. In addition, the Black Sea is unusual in that only the top 150 to 200 m of its waters (~10% of its average depth) contain oxygen, which effectively makes the entire sea "shallow" for species that require oxygen. Its limited water exchange and anoxic deep waters make the Black Sea particularly vulnerable to the negative effects of nutrient enrichment.

Native Black Sea predators and parasites had failed to regulate *Mnemiopsis* populations. Thus, in the early 1990s, the future of the Black Sea ecosystem looked bleak. Fortunately, by the late 1990s, there were signs of improvement: *Mnemiopsis* and phytoplankton populations had fallen, paving the way for the recovery of the Black Sea. How did this happen?

Introduction

In the last chapter, we focused on how and why the distribution and abundance of populations and species vary across landscapes. But population abundance can also change over time, displaying different patterns of **population growth**. This is true whether abundance is measured on a small spatial scale, such as the number of plants found in a restricted area along a riverbank, or on a much larger spatial scale, such as the number of cod found in the North Atlantic Ocean. Some populations differ little in abundance over time and space; others differ considerably.

For example, Richard Root and Naomi Cappuccino (1992) studied abundances of 23 species of herbivorous insects that fed on tall goldenrod (*Solidago altissima*). They studied these insects for at least 12 consecutive years at each of 22 sites in the Finger Lakes region of New York (**FIGURE 10.3**). These sites were no more than 75 km (47 miles) apart; hence, in any given year, all the sites experienced roughly the same climate conditions. Nevertheless, insect abundances varied from one site to another and from one year to the next. For example, maximum abundances

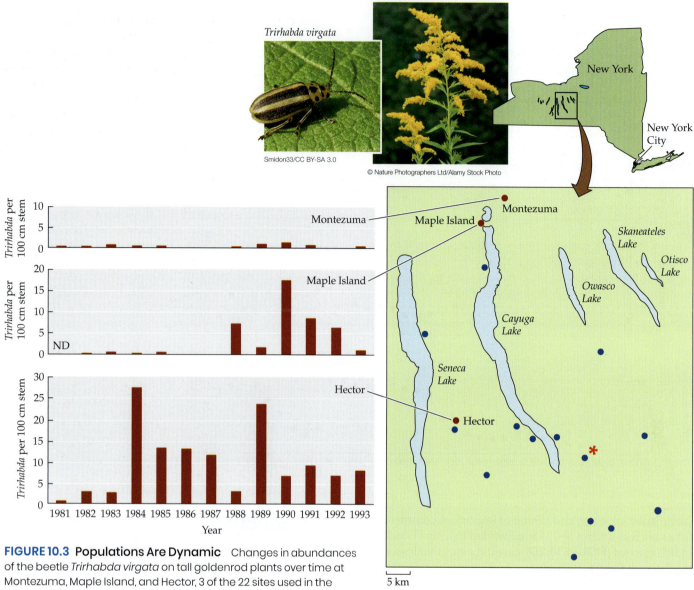

FIGURE 10.3 Populations Are Dynamic Changes in abundances of the beetle *Trirhabda virgata* on tall goldenrod plants over time at Montezuma, Maple Island, and Hector, 3 of the 22 sites used in the study. Five of these sites were located close to one another and are indicated on the map by an asterisk; all other sites are indicated by dots. (After R. B. Root and N. Cappuccino. 1992. *Ecol Monogr* 62: 393–420; additional data from R. B. Root, personal communication.)

❓ In what year or years did Trirhabda abundance vary greatly over space? Explain.

of the beetle *Trirhabda virgata* varied considerably in abundance (ranging from 0.03 to 10.1 insects per stem), both from one site to another and over time (see Figure 10.3).

Ecologists typically use the term *population dynamics* to refer to the ways in which population sizes change over time. In this chapter, we'll consider the dynamics of populations in more detail, placing special emphasis on the patterns of population growth and the risk of extinction for small populations. In Chapter 11, we'll narrow our discussion of population dynamics by using quantitative models to understand and measure population growth patterns. But first, we'll begin our discussion of population dynamics by surveying patterns of population growth in nature.

CONCEPT 10.1 Populations are dynamic entities that vary in size over time.

LEARNING OBJECTIVES

10.1.1 List the different patterns of population growth observed in nature.

10.1.2 Compare the patterns of exponential growth with that of logistic growth.

10.1.3 Describe population size fluctuations and the special case of population cycling.

Patterns of Population Growth

Most observed patterns of population growth can be grouped into four major types: exponential growth, logistic growth, population fluctuations, and regular population cycles (a special type of fluctuation). Bear in mind, however, that a single population could (and probably will) experience each of these four types of growth at different times. For example, as we will see shortly, a population may grow logistically yet fluctuate around the values expected in logistic growth.

Exponential growth can occur when conditions are favorable

Many organisms, such as giant puffball fungi and the desert shrub *Cleome droserifolia*, produce large numbers of offspring. In such cases, if even a fraction of those offspring survive to reproduce, the population can increase in size very quickly, showing a pattern of **exponential growth**, or a J-shaped pattern (see Figure 11.4). Exponential growth occurs when the rate of growth increases (or decreases) in proportion to the current number of individuals. Exponential growth cannot continue indefinitely, but when conditions are favorable, a population can increase exponentially for a limited time. Such periods of exponential growth can occur within the established range of a species, as when good weather occurs for several years running. They can also occur when a species reaches a new geographic area, either by dispersing on its own or with human assistance.

An example of how dispersal can lead to exponential growth is provided by the cattle egret subspecies *Bubulcus ibis ibis* (**FIGURE 10.4A**). These birds originally lived in the Mediterranean region and in parts of central and southern Africa. Since the late 1800s and early 1900s, however, they have colonized new regions on their own, including South America and North America. Typically, after the subspecies reached a new area, its population in that area increased exponentially as it became established in its new habitat (**FIGURE 10.4B**). For example, after the cattle

(A)

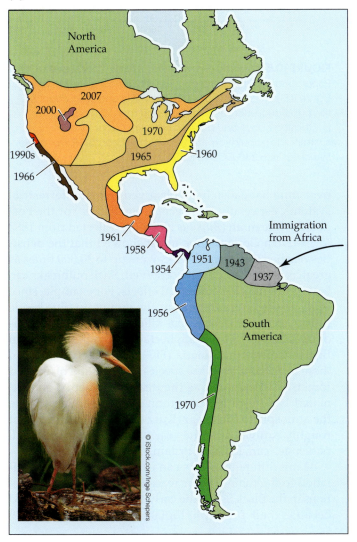

(B)

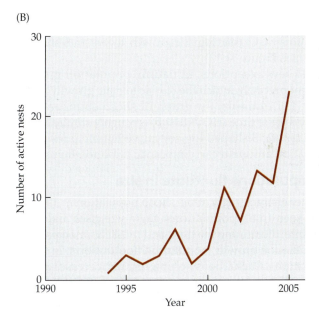

FIGURE 10.4 Colonizing the New World (A) The cattle egret subspecies *Bubulcus ibis ibis* dispersed from Africa to South America in the late 1800s. Once it established colonies in the northeastern region of South America, it then spread rapidly to other parts of South and North America. The contour lines and dates show the edges of the cattle egret's range at different times. (B) The number of active cattle egret nests observed annually within wetlands of the San Francisco Bay area after it first colonized this region in the early 1990s. (A after R. L. Smith. 1974. *Ecology and Field Biology*, 2nd ed. Harper & Row: New York; S. Osborn. 2007. In *The Birds of North America Online*, A. Poole [Ed.]. Cornell Lab of Ornithology: Ithaca, NY; B. after J. P. Kelly et al. 2007. *Waterbirds* 30: 455–478.)

egret colonized the San Francisco Bay area in the 1990s, its populations there grew exponentially for over a decade (Kelly et al. 2007). As in cattle egrets, species that successfully colonize new geographic regions on their own typically do so by long-distance events. Local populations in the new region may then increase in size—often growing exponentially—while also expanding (by relatively short-distance dispersal events) to occupy nearby areas of suitable habitat.

In logistic growth, the population approaches an equilibrium

Some populations appear to reach a relatively stable population size, or *equilibrium*, that changes little over time. When this occurs, the number of individuals first increases in size, then fluctuates by a relatively small amount around what appears to be the maximum sustainable population size. Such populations exhibit the second pattern of population growth, **logistic growth**. Logistic growth is a pattern in which numbers of individuals increase rapidly at first and then stabilize as the population reaches **carrying capacity**, or the maximum population size that can be supported indefinitely by the environment. Classic logistic growth shows an S-shaped curve (see Figure 11.13).

With few exceptions, in nature, population growth does not precisely match the classic logistic growth curve shape. For example, the graph of sheep abundance in Tasmania over time (**FIGURE 10.5**) is roughly similar to the characteristic S-shaped logistic curve but includes more fluctuation around a carrying capacity population size.

All populations fluctuate in size

Another characteristic of the sheep population in Tasmania is seen in all populations: their size rises and falls over time, illustrating the third and most common pattern of population growth, **population fluctuations**. In some populations, fluctuations occur as erratic increases or decreases in abundance from an overall mean (**FIGURE 10.6**). In other populations, fluctuations occur as deviations from a population growth pattern, such as exponential or logistic growth. If, for example, the growth of a population exactly matched a logistic curve, the population would not be said to fluctuate. But if population abundances rose above and fell below those expected in exponential growth (as in the cattle egret) and logistic growth (as in the Tasmanian sheep), the population would be said to fluctuate.

In some cases, population fluctuations are relatively small (as seen in Figure 10.5). In other cases, the number of individuals in a population can explode at certain times, causing a population **outbreak** (**FIGURE 10.7**). As we saw in Figure 10.2A, the biomass of the comb jelly *Mnemiopsis* increased 1,000-fold during a 2-year outbreak in the Black Sea. Rapid variations in population sizes

When sheep were first introduced to Tasmania, the population increased rapidly.

Later, population numbers fluctuated above and below a maximum sustainable population size.

FIGURE 10.5 Logistic Growth Rises First, Then Levels Off Population growth rarely matches a classic logistic curve (see Figure 11.13). For sheep introduced to the island of Tasmania, the population increased rapidly and then leveled off with fluctuation above and below a carrying capacity or maximum sustainable population size. (After J. Davidson. 1938. *Trans R Soc S Aust* 62: 342–346. CC BY-NC-SA 3.0.)

over time have also been observed in many terrestrial systems, especially in insects. Census data for the bordered white moth (*Bupalus piniarius*) collected from 1882 to 1940 in a German pine forest showed that the densities reached during outbreaks were up to 30,000 times as great as the lowest density observed. Such outbreaks can have wide-ranging ecological effects. For example, since 2000, an ongoing outbreak of the mountain pine beetle (*Dendroctonus ponderosae*) has killed hundreds of millions of trees across 18.1 million hectares (45 million acres) in British Columbia, Canada (**FIGURE 10.8**). The death of these trees has altered the species composition of affected forests. Furthermore, as the dead trees decay, an estimated 17.6 megatons of carbon dioxide is released into the atmosphere each year (Kurz et al. 2008)—an amount roughly equivalent to the yearly carbon emissions of all passenger cars in Great Britain.

Many different factors can cause the size of a population to fluctuate. The increase in zooplankton populations in the Black Sea in the early 1980s probably occurred because their prey (phytoplankton) had increased in

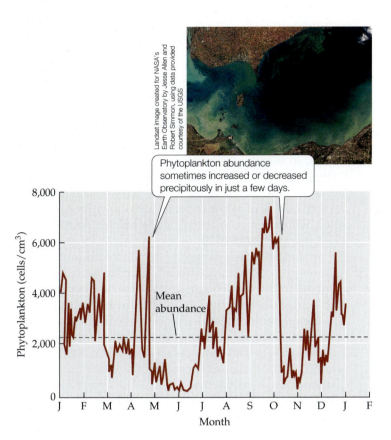

Landsat image created for NASA's Earth Observatory by Jesse Allen and Robert Simmon, using data provided courtesy of the USGS

Phytoplankton abundance sometimes increased or decreased precipitously in just a few days.

Mean abundance

FIGURE 10.6 Population Fluctuations Variation in phytoplankton abundance in water samples taken from Lake Erie during 1962, showing fluctuations above and below the overall mean abundance of 2,250 cells per cubic centimeter. The inset shows an October 2011 phytoplankton "bloom" (a rapid increase in phytoplankton numbers) in the lake. (After C. C. Davis. 1964. *Limnol Oceanogr* 9: 275–283.)

abundance (see Figure 10.2). Then, in 1991, zooplankton numbers plummeted, probably because of the spectacular increase in the abundance of their predator (*Mnemiopsis*) during the previous 2 years. The rapid changes in phytoplankton abundance in Lake Erie shown in Figure 10.6

could reflect changes in a wide range of environmental factors, including nutrient supplies, temperature, and predator abundance.

Analyzing the factors important to population fluctuations can also help identify the factors important to disease outbreaks. In 1993, dozens of people in the Four Corners region of the southwestern United States (where New Mexico, Arizona, Colorado, and Utah intersect) became sick with flu-like symptoms and shortness of breath, and 60% of them died within a few days of becoming ill. No one had seen this combination of symptoms before. An outbreak of a lethal, previously unknown disease appeared to be in progress, and there was no cure or successful treatment. The U.S. Centers for Disease Control (CDC) quickly identified the disease agent as a new strain of hantavirus carried by the deer mouse (*Peromyscus maniculatus*). Seeking more information about the new disease, now known as hantavirus pulmonary syndrome, or HPS, the CDC contacted ecologists who had been studying mouse populations in the Southwest. Examination of deer mouse specimens collected between 1979 and 1992 revealed that the virus had been present in the area for more than 10 years prior to the outbreak. Why, then, did the outbreak of HPS occur in 1993 and not before?

Photo by Chip Clark, Smithsonian

FIGURE 10.7 Populations Can Explode in Numbers When conditions are favorable, a population outbreak can occur in which the numbers of individuals increase very rapidly. The cockroaches covering the kitchen in this exhibit from the National Museum of Natural History represent the number that could have been produced by a single pregnant female in a few generations.

© Gunter Marx/Alamy Stock Photo

FIGURE 10.8 Consequences of an Insect Outbreak This aerial view shows the red foliage of lodgepole pine (*Pinus contorta*) trees killed by an outbreak of mountain pine beetles in British Columbia, Canada.

To address this question, ecologists used data on the abundances of *Peromyscus* species collected since 1989 at the nearby Sevilleta National Wildlife Refuge. These data showed that the densities of several *Peromyscus* species had increased 3- to 20-fold between 1992 and 1993. Next, a series of satellite images was used to develop an index of how much plant matter was available as food for *Peromyscus* at different times. When that index was compared with precipitation data, the results suggested that unusually high rainfall from September 1991 through May 1992 had led to enhanced plant growth in spring 1992 (**FIGURE 10.9**). In turn, the enhanced plant growth produced abundant food for rodents (seeds, berries, green plant matter, arthropods), which allowed mouse populations to increase in size by 1993—the year of the HPS outbreak.

Rodents shed hantavirus in their urine, feces, and saliva; hence, high mouse numbers, which led to increased mouse–human contact, were thought to be the cause of the 1993 outbreak. The actual risk to people varies greatly with location and depends on such factors as habitat type (which can influence mouse movements), microclimate (e.g., in arid regions, nearby areas often experience very different amounts of rainfall), and local food abundance. Overall, we now know enough about HPS to predict periods of

heightened risk to human populations, but more remains to be learned about whether these factors create predictable population cycles in mice and the hantavirus disease. Let's now turn to what ecologists know about the factors important in producing population cycles.

Some species exhibit population cycles

The fourth pattern of population growth is **population cycles**, in which alternating periods of high and low abundance occur after constant (or nearly constant) intervals of time. Such regular cycles have been observed in populations of small rodents such as lemmings and voles, whose abundances typically reach a peak every 3 to 5 years (**FIGURE 10.10**).

Population cycles are among the most intriguing patterns observed in nature. After all, what factors can cause numbers to fluctuate greatly over time, yet maintain a high degree of regularity? Possible answers to this question include both internal factors, such as hormonal or behavioral changes in response to crowding, and external factors, such as weather, food supplies, or predators. Gilg et al. (2003) used a combination of field observations and mathematical models to argue that the 4-year cycle of collared lemming (*Dicrostonyx groenlandicus*) abundance in Greenland is driven by predators, one of which, the stoat, specializes on eating lemmings (see Figure 10.10). Other investigators have suggested that cycles of the Norwegian lemming (*Lemmus lemmus*) are caused by interactions between lemmings and their food plants. Similarly, a number of studies (e.g., Korpimäki and Norrdahl 1998) have implicated predators as the driving force behind cycles of field voles in Scandinavia, but Graham and Lambin (2002), in a large-scale field experiment, showed that predator removal had no effect on field vole cycles in England.

One mechanism that may be important to population cycles is **delayed density dependence**. Delayed density dependence occurs when the number of individuals born in a given time period is influenced by the population densities or other conditions that were present several time periods ago. Consider a population of predators that reproduce more slowly than their prey. If there are few predators initially, the prey population may increase rapidly in size. As a result, the predator population may also increase, reaching a point at which there are many adult predators that survive well and produce a large number of offspring. However, if the resulting large population of predators eats so many prey that the prey population decreases sharply in size, there may be few prey available for the next generation of predators. In such a case, a mismatch in predator and prey numbers (high predator numbers, low prey numbers) occurs because

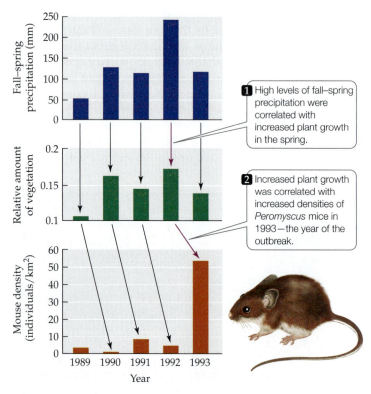

FIGURE 10.9 From Rain to Plants to Mice The outbreak of hantavirus pulmonary syndrome in the southwestern United States in 1993 may have been caused by a series of interconnected events. (After T. L. Yates et al. 2002. *BioScience* 52: 989–998.)

Figure annotations:

1 High levels of fall–spring precipitation were correlated with increased plant growth in the spring.

2 Increased plant growth was correlated with increased densities of *Peromyscus* mice in 1993—the year of the outbreak.

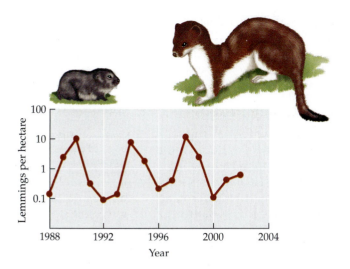

FIGURE 10.10 A Population Cycle In northern Greenland, collared lemming (*Dicrostonyx groenlandicus*, left) abundance tends to rise and fall every 4 years. In this location, the population cycle appears to be driven by predators, the most important of which is the stoat (*Mustela erminea*, right). In other regions, lemming population cycles may be driven by food supply. (After O. Gilg et al. 2003. *Science* 302: 5646.)

? Based on data from 1988 through 2000, how many lemmings per hectare would you have expected there to be in 2002? Explain your reasoning.

there is a time lag in the response of predator numbers to prey numbers. When such a mismatch takes place, the predators may survive or reproduce poorly and their numbers may drop. If prey numbers then increase (because there are now fewer predators), predator numbers may first rebound, then fall again because of the built-in time lag. Thus, in principle at least, it seems reasonable that a delay in the response of predators to prey density could cause predator numbers to fluctuate over time. You can explore how delayed density dependence can affect the population cycling in sheep blowfly (*Lucilia cuprina*) in **ANALYZING DATA 10.1**.

ANALYZING DATA 10.1

Does delayed density dependence produce cycles in blowfly populations?

In the 1950s, A. J. Nicholson (1957)* performed a series of pioneering laboratory experiments on delayed density dependence in the sheep blowfly *Lucilia cuprina*, an important agricultural pest of sheep. Before they can lay eggs, the females of this species need a protein meal. Once they have fed, the females attack living sheep by laying their eggs near the tail or near open wounds. Small white maggots hatch from those eggs and feed on dung attached to the skin or on exposed flesh. Eventually the maggots metamorphosize into adult flies. The sheep blowfly's full life cycle can be completed in as little as 7 days.

In the first of the two experiments, Nicholson tested for delayed density dependence by providing adult blowflies with unlimited food (ground liver) but restricting maggots to 50 g of food per day. Because adults had abundant food, each female was able to lay many eggs but when those eggs hatched, the lack of food caused many of the maggots to die before they reached adulthood (**FIGURE A**). In a second experiment, Nicholson recreated the same experimental conditions as the first experiment until roughly halfway through the

experiment (indicated by the dotted vertical line), when the food supply for adults was also limited (**FIGURE B**).

1. Based on the data, which of the four population growth patterns discussed under Concept 10.1 best characterizes the results shown in Figure A?

2. How does the pattern of population size change when food is limited for blowfly adults halfway through the second experiment (Figure B)?

3. Using data from the experiment, explain how delayed density dependence can cause populations to cycle.

4. Describe what you would expect to happen to the adult population of blowflies over time if both the adults and maggots were allowed to have unlimited food.

---

* Nicholson, A. J. 1957. The self-adjustment of populations to change. *Cold Spring Harbor Symposia on Quantitative Biology* 22: 153–173.

(Continued)

ANALYZING DATA 10.1 (continued)

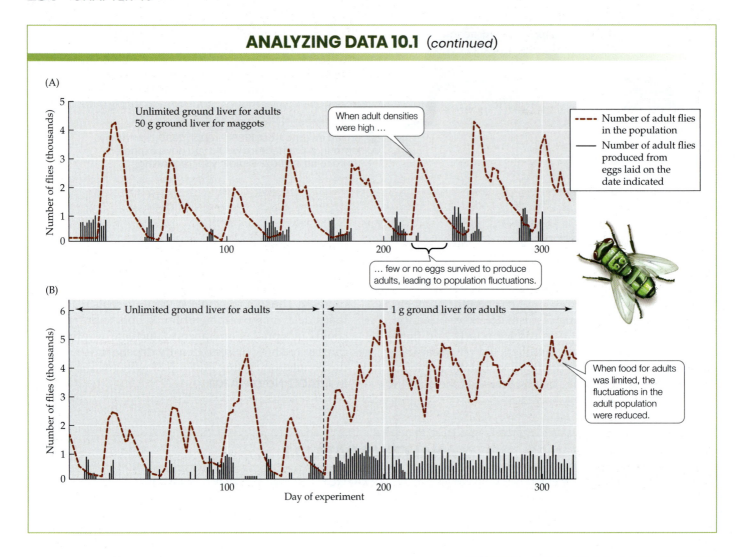

(A)

Unlimited ground liver for adults
50 g ground liver for maggots

When adult densities were high …

… few or no eggs survived to produce adults, leading to population fluctuations.

- - - Number of adult flies in the population
— Number of adult flies produced from eggs laid on the date indicated

(B)

Unlimited ground liver for adults ← | → 1 g ground liver for adults

When food for adults was limited, the fluctuations in the adult population were reduced.

Day of experiment

CLIMATE CHANGE CONNECTION

COLLAPSING POPULATION CYCLES AND CLIMATE CHANGE   Recent evidence suggests that population cycles may stop entirely if key environmental conditions change. For example, population cycles of lemmings (including the cycle shown in Figure 10.10), voles, and several insect herbivores have decreased in amplitude or ceased entirely in some high-latitude and high-elevation locations (Gilg et al. 2009; Johnson et al. 2010; Cornulier et al. 2013).

What factors can cause the collapse of a population cycle? Some evidence points to climate change as a possible cause. Lemmings, for example, thrive when warmth from the ground melts a thin layer of the snow cover, leaving a small gap between the ground and the snow. In some regions, warmer winter temperatures have caused the snow to melt and refreeze, preventing the formation of these gaps. As discussed in Gilg et al. (2009), a shortage of gaps has made it more difficult for lemmings to feed and has made lemmings easier for

their predators to catch. By holding lemming abundance in check (due to increased predation), these changes may have prevented lemming populations from increasing greatly in abundance every 3 to 4 years, thus halting the population cycles previously observed for this species (see Figure 10.10).

Climate warming also may have contributed to the collapse of vole population cycles throughout Europe and across different species (Cornulier et al. 2013). This hypothesis is reasonable since temperatures have increased and climate warming could affect populations of different species across Europe. However, vole cycles in some areas of Finland have continued despite regional warming, indicating that the effect of climate change may depend on the species or on the particular mechanisms that drive the cycles (Brommer et al. 2010). Moreover, the collapse of a population cycle can be caused by factors other than climate change. For example, Allstadt et al. (2013) concluded that the recent collapse of cycles in Canadian populations of the gypsy moth (*Lymantria dispar*) resulted from attack by a specialist pathogen rather than climate change. ☀

LEARNING OBJECTIVES

10.2.1 Justify why fluctuations in population growth rate can increase a population's risk of extinction.

10.2.2 List and describe the ways that chance events can drive small populations to extinction.

Population Extinction

Populations can be driven to extinction by many different factors, including changes in the environment, biological interactions, and human-caused events. In this section, we'll look at how the risk of extinction is affected by the fluctuations and sizes of populations.

Fluctuations in population size can increase the risk of extinction

Imagine a population that is increasing over time. If the population size fluctuates very little over time, then in most years the population will continue to increase in size. Under these circumstances, the population will face little or no risk of extinction. However, random variation in environmental conditions could cause the population size to change considerably from year to year. What are the implications of such fluctuations?

To show what happens when population size fluctuates, computer simulations were performed for three populations that were allowed to fluctuate at random. Examining the results in **FIGURE 10.11**, we see that two of the populations recovered from low numbers, but one went extinct. These results support what common sense tells us: fluctuations increase the risk of extinction. In part, this occurs because, for a given average population size, a population that fluctuates in size grows slower than one that does not vary and thus either takes longer to recover or may not recover at all. The take-home message is that when variable environmental conditions increase the extent to which a population's growth rate fluctuates over time, the risk of extinction also increases. This effect, however, is dependent on the size of the population: small populations are at particular risk.

Small populations are at much greater risk of extinction than large populations

The size of a population has a strong effect on its risk of extinction. For example, Jones and Diamond (1976) studied extinction in bird populations on the Channel Islands, located off the coast of California. By combining data from published articles (from 1868 on), museum records, unpublished field observations, and their own fieldwork, they showed that population size had a strong effect on the chance of extinction (**FIGURE 10.12**). They found that 39% of populations with fewer than 10 breeding pairs went

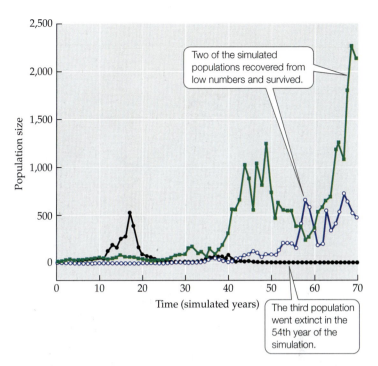

Two of the simulated populations recovered from low numbers and survived.

The third population went extinct in the 54th year of the simulation.

FIGURE 10.11 Fluctuations Can Drive Small Populations Extinct Simulated growth of three populations in which the population growth rate varied at random from year to year. This variation over time was intended to simulate random variation in environmental conditions. Each simulated population began with 10 individuals but ended with variable population sizes, including one population that went extinct.

extinct, whereas they observed no extinctions in populations with over 1,000 breeding pairs. Similar work by Pimm et al. (1988) showed that small populations can go extinct very rapidly: on islands off the coast of Britain, bird populations with 2 or fewer nesting pairs had a mean time to extinction of 1.6 years, while populations with 5 to 12 nesting pairs had a mean time to extinction of 7.5 years.

These findings for birds have been confirmed in many other groups of organisms. Overall, field data indicate that the risk of extinction increases greatly when population size is small. But what are the factors that place small populations at risk?

When populations are small, they reduce what is called **effective population size**, or the number of individuals that can contribute offspring to the next generation. A reduction in the effective population size can result in an *extinction vortex* in which smaller population sizes lead to further declines in population size, eventually resulting in extinction (**FIGURE 10.13**). How does the effective population size decline over time? There are three main categories of factors that place small populations at risk of extinction: genetic factors, demographic factors, and environmental factors. Let's consider each of them separately.

RISK FROM GENETIC FACTORS Small populations can encounter problems associated with genetic drift and

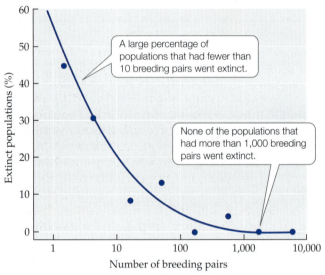

FIGURE 10.12 **Extinction in Small Populations** Among bird populations on the Channel Islands, the percentage of populations that went extinct declined rapidly as the number of breeding pairs in the population increased. (After H. L. Jones and J. M. Diamond. 1976. *Condor* 78: 526–549.)

? Assume that a population is at high risk (>30%) of extinction. Use the graph to estimate the total number of breeding pairs the population should have to reduce its risk of extinction to 5%.

inbreeding depression, processes that reduce genetic diversity and the fitness of individuals and the adaptability of populations (see Figure 10.13). Recall from Concept 6.2 that **genetic drift** is the process by which chance events influence which alleles are passed on to the next generation. Genetic drift can occur in many ways, including chance events that determine whether individuals reproduce or die. Imagine, for example, that an elephant walks through a population of ten small plants, 50% of which have white flowers (genotype *aa*) and 50% of which have red flowers (*AA*). If the elephant happens to crush more red-flowered than white-flowered plants, then by chance alone, there will be more copies of the *a* allele than of the *A* allele in the next generation. This scenario is just one

of many possible examples of how genetic drift can cause allele frequencies to change at random from one generation to the next.

Genetic drift has little effect on large populations, but in small populations it can cause losses of genetic variation over time. For example, if genetic drift causes the frequency of two alleles (e.g., *A* and *a*) to change at random in each generation, one allele may eventually increase to a frequency of 100% (reach *fixation*), while the other is lost (see Figure 6.7). Drift can reduce the genetic variation of small populations rapidly: for example, after ten generations, roughly 40% of the original genetic variation is lost in a population of ten individuals, while 95% is lost in a population of two individuals.

In addition, small populations can show a high frequency of **inbreeding** (mating between related individuals). Inbreeding is common in small populations because after several generations at a small population size, most of the individuals in the population will be closely related to one another (to see why, answer Review Question 3). Inbreeding tends to increase the frequency of homozygotes, including those that have two copies of a harmful allele. Thus, like genetic drift, inbreeding depression can lead to the loss of genetic diversity, reducing individual fitness and hence population growth rates.

The combined negative effects of genetic drift and inbreeding depression appear to have reduced the fertility of male lions that live on the floor of the Ngorongoro Crater, Tanzania (**FIGURE 10.14**). From 1957 to 1961, there were 60 to 75 lions living in the crater, but in 1962 an extraordinary outbreak of biting flies caused all but 9 females and 1 male to die. Seven males immigrated into the crater in 1964–1965, but no further immigration has occurred since that time. The population has increased in size since the 1962 crash. From 1975 to 1990, for example, the population fluctuated between 75 and 125 individuals. However, genetic analyses indicated that all these individuals were descendants of just 15 lions (Packer et al. 1991). In a population of 15 individuals, genetic drift and inbreeding depression have powerful effects. Those effects appear to be the reason why the crater population has less genetic variation and more frequent sperm abnormalities than the large population of lions found nearby on the Serengeti Plain. In such a situation, all is not necessarily lost: in some cases, populations in decline because of drift and inbreeding have been "rescued" by introducing a small number of individuals from other, more genetically diverse populations (see Figure 23.16).

RISK FROM DEMOGRAPHIC FACTORS A second factor important to the loss of genetic diversity, and ultimately extinction, in small populations is that of **demographic stochasticity**, or the fluctuations in population size as the result of chance differences among individuals in reproduction and survival (see Figure 10.13). For example, for an individual, survival and reproduction are

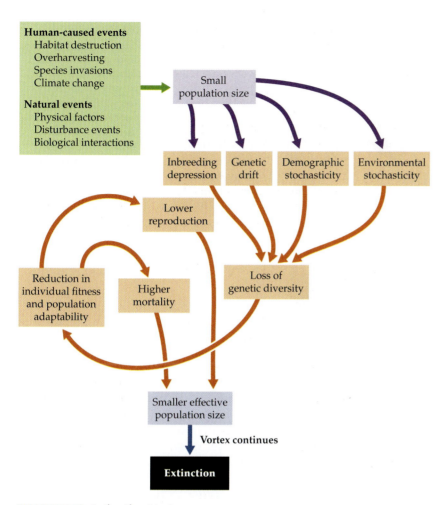

Human-caused events
Habitat destruction
Overharvesting
Species invasions
Climate change

Natural events
Physical factors
Disturbance events
Biological interactions

Small population size

Inbreeding depression

Genetic drift

Demographic stochasticity

Environmental stochasticity

Lower reproduction

Reduction in individual fitness and population adaptability

Higher mortality

Loss of genetic diversity

Smaller effective population size

Vortex continues

Extinction

FIGURE 10.13 Extinction Vortex Human-caused and natural events can reduce the effective population size of species, resulting in the loss of genetic diversity, and eventually leading to population- and species-level extinctions.

all-or-nothing events: an individual either survives or it does not, and it either reproduces or it does not. At the population level, we can transform such all-or-nothing events into a probability that survival or reproduction will occur. For example, if 70 out of 100 individuals in a population survive from one year to the next, then (on average) each individual in the population has a 70% chance of survival.

In a small population, however, demographic stochasticity can result in outcomes that differ from what such averages would lead us to expect. Consider a population of ten individuals for which previous data indicate that, on average, each individual has a 70% probability of surviving from one year to the next. However, many chance events—such as whether an individual is struck by a falling tree—can cause the percentage of individuals that actually do survive to be higher or lower than 70%. For example, if six of the ten individuals experienced (the ultimate) "bad luck" and died in chance mishaps, the observed survival rate (40%) would be much lower than the expected 70%. By affecting the survival and reproduction

of individuals in this way, demographic stochasticity can cause the size of a small population to fluctuate over time. In one year the population may grow, while the next it may decrease in size, perhaps so drastically that extinction results.

In contrast, when the population size is large, there is little risk of extinction from demographic stochasticity. The fundamental reason for this has to do with laws of probability. You are, for example, much more likely to receive zero heads if you toss a fair coin 3 times than if you toss the same coin 300 times. Similarly, when we consider the demographic fates of individuals, we can see that chance events are much more likely to cause reproductive failure or poor survival in small populations than in large populations. If each individual in a population has a 33% chance of producing zero offspring, then if there are two individuals in the population, there is an 11% chance $(0.33 \times 0.33 = 0.33^2 = 0.11)$ that no offspring will be produced—driving the population to extinction in one generation. Although demographic stochasticity could cause a population of 30 individuals to fluctuate in size (perhaps leading to eventual extinction), there is essentially no chance (0.33^{30}) that it could cause the population to go extinct in a single generation.

Demographic stochasticity is also one of several factors that can cause small populations to experience Allee effects. **Allee effects** occur when the population growth rate *decreases* as the population density decreases, perhaps because individuals have difficulty finding mates at low population densities (**FIGURE 10.15**). This phenomenon reverses the usual assumption that population growth rates tend to *increase* as population density decreases (see Figure 11.12). Allee effects can be disastrous for small populations. If demographic stochasticity or any other factor decreases the population size, Allee effects can cause the population growth rate to drop, which causes the population size to decrease even further in a downward spiral toward extinction.

RISK FROM ENVIRONMENTAL VARIATION   Finally, environmental stochasticity can cause declines in genetic diversity and ultimately extinction of small populations (see Figure 10.13). **Environmental stochasticity** refers to erratic or unpredictable changes in the environment. In the simulations described above (see Figure 10.11), we've already seen that (1) variation in environmental conditions that causes fluctuations in population growth rates can lead to population size fluctuations and thus an increased

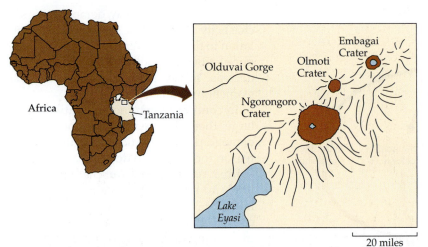

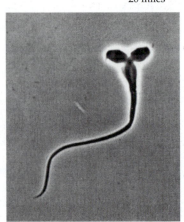

FIGURE 10.14 A Plague of Flies In 1962, the population of lions in the 260-km² (100-square-mile) Ngorongoro Crater of Tanzania was nearly driven to extinction by a catastrophic outbreak of biting flies similar to those on the face of this male. Lions became covered with infected sores and eventually could not hunt, resulting in many deaths. In the population that descended from the few survivors, genetic drift and inbreeding depression have led to frequent sperm abnormalities, such as this "two-headed" sperm.

risk of extinction, and (2) such environmental variation is more likely to cause extinction when the population size is small. Many species face such risks from environmental stochasticity. For example, census data on female grizzly bears (*Ursus arctos horribilis*) in Yellowstone National Park showed that the average population growth rate varied from year to year. Despite the fact that the population tends to grow in size, researchers using a mathematical model found that random variation in environmental conditions could place the Yellowstone grizzly population at high risk of extinction, especially if the population size were to drop to 40 females or less from its 1997 level of 99 females (**FIGURE 10.16**).

Environmental stochasticity differs from demographic stochasticity in a fundamental way. Environmental stochasticity refers to changes in the average birth or death rate of a population that occur from one year to the next. These year-to-year changes reflect the fact that environmental conditions vary over time, affecting all the individuals in a population: sometimes there are good years and sometimes there are bad years. In demographic stochasticity, the average (population-level) birth and death rates may be constant across years, but the actual fates of individuals differ because of the random nature of whether each individual reproduces or not, and survives or not.

Populations also face risks from extreme environmental events such as floods, fires, severe windstorms, or outbreaks of disease or natural enemies. Even though they occur rarely, such **natural catastrophes** can eliminate or drastically reduce the size of populations that otherwise would seem large enough to be at little risk of extinction. For example, disease outbreaks have resulted in mass mortality in populations of sea urchins (up to 98% of the individuals in some populations) and Baikal seals (killing about 2,500 of a population of 3,000 seals).

Environmental stochasticity also played a key role in the extinction of the heath hen (*Tympanuchus cupido cupido*). This bird was once abundant from Virginia to New England. By 1908, hunting and habitat destruction had reduced its population to 50 birds, all on the island of Martha's Vineyard, where a 1,600-acre reserve was established for its protection. Initially, the population thrived, increasing in size to 2,000 birds by 1915. A population of 2,000 may seem large enough to be nearly "bulletproof" against the problems that threaten small populations,

(A)

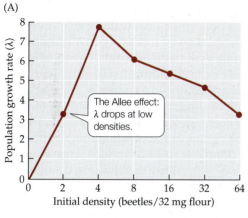

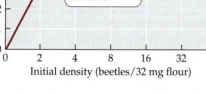

The Allee effect: λ drops at low densities.

(B)

Thunnus thynnus

(C)

Strigops habroptilus

(D)

Aconitum napellus

FIGURE 10.15 Allee Effects Can Threaten Small Populations Allee effects occur when the growth rate of a population decreases as population density decreases. (A) In laboratory experiments with the flour beetle *Tribolium*, population growth rates reached their lowest point at the lowest initial density. Allee effects can be important in animals such as (B) bluefin tuna (*Thunnus thynnus*), which form schools whose protective or early warning systems function poorly at small population sizes. Allee effects are also important in species in which individuals have difficulty finding mates at low population densities; there are many such species, including (C) kakapos (*Strigops habroptilus*) and (D) monkshood (*Aconitum napellus*). (A after F. Courchamp et al. 1999. *Trends Ecol Evol* 14: 405–410.)

including genetic drift and inbreeding, demographic stochasticity, and environmental stochasticity. However, a series of disasters struck between 1916 and 1920, including a fire that destroyed many nests, unusually cold weather, a disease outbreak, and a boom in the number of goshawks (a predator of heath hens). Because of the combined effects of these events, the heath hen population dropped to 50 birds by 1920 and never recovered. The last heath hen died in 1932.

With the benefit of hindsight, we can see that heath hens were vulnerable in 1915 because they all lived in a single population. More typically, members of a species are found in *metapopulations*, which are often isolated from one another by regions of unsuitable habitat. You can read more about metapopulation dynamics in Concept 9.4.

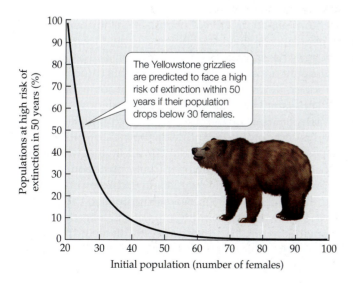

The Yellowstone grizzlies are predicted to face a high risk of extinction within 50 years if their population drops below 30 females.

FIGURE 10.16 Environmental Stochasticity and Population Size This graph plots the risk that the Yellowstone grizzly bear population will be close to extinction in 50 years against the population size (number of females). By studying 39 consecutive years of census data, researchers found that the average population growth rate of Yellowstone grizzlies could lead to explosive growth if it remained constant from year to year. The risk of extinction increased dramatically when random variation in environmental conditions dropped the population size to 40 females or less. (After W. F. Morris and D. F. Doak. 2002. *Quantitative Conservation Biology: Theory and Practice of Population Viability.* Oxford University Press/Sinauer: Sunderland, MA.)

A Sea in Trouble

In the late 1980s and early 1990s, the Black Sea ecosystem was under severe duress from the combined effects of eutrophication and invasion by the comb jelly *Mnemiopsis leidyi*, as described in the Case Study. Although *Mnemiopsis* numbers declined sharply in 1991, they rose steadily again from 1992 to 1995, and then remained high for several years—at about 250 g per square meter, which translates to over 115 million tons of *Mnemiopsis* throughout the Black Sea. The situation did not look promising. But by 1999, matters were different: the Black Sea was showing signs of recovery.

The events that set the stage for the recovery of the Black Sea actually began prior to the first onslaught of *Mnemiopsis*. In the mid- to late 1980s, the amounts of nutrients added to the Black Sea began to level off. From 1991 to 1997, nutrient inputs declined, probably because of hard economic times in former Soviet Union countries coupled with national and international efforts to reduce nutrient inputs. The reduction had rapid effects: after 1992, phosphate concentrations in the Black Sea declined, phytoplankton biomass began to fall, water clarity increased, and zooplankton abundance increased. *Mnemiopsis* still posed a threat, however, as evidenced by its high biomass and by falling anchovy catches from 1995 to 1998 (see Figure 10.2). Scientists and government officials were gearing up to combat the threat from *Mnemiopsis* when the problem was inadvertently solved by the arrival of another comb jelly, the predator *Beroe* (**FIGURE 10.17**).

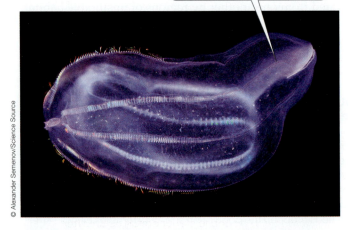

This *Beroe* (blue outline) has ingested an individual *Mnemiopsis*, which is seen glowing in the gut.

FIGURE 10.17 Invader versus Invader Another invasive comb jelly species, the predator *Beroe*, brought *Mnemiopsis* under control, thus contributing to the recovery of the Black Sea ecosystem.

Beroe arrived in 1997. Like *Mnemiopsis*, *Beroe* probably reached the Black Sea in the ballast water of ships from the Atlantic. *Beroe* feeds almost exclusively on *Mnemiopsis*. It is such an effective predator that within 2 years of its arrival, *Mnemiopsis* numbers plummeted (see Figure 10.2A). Following the sharp decline in *Mnemiopsis*, the Black Sea population of *Beroe* also crashed, presumably because it depended on *Mnemiopsis* for food. The fall of *Mnemiopsis* led to a rebound in zooplankton abundance (which had dropped again from 1994 to 1996) and to increases in the population sizes of several native jellyfish species. In addition, after the *Mnemiopsis* population crashed, there was an increase in the anchovy catch and in field counts of anchovy egg densities. Overall, the decline of *Mnemiopsis* helped to improve the condition of the Black Sea ecosystem, including the fisheries on which people depend for food and income.

 CONNECTIONS *in* NATURE

FROM BOTTOM TO TOP, AND BACK AGAIN The decrease in nutrient inputs by human activities and the control of *Mnemiopsis* by *Beroe* had rapid beneficial effects on the entire Black Sea ecosystem. The speed and magnitude of the ecosystem's recovery provide a source of hope, suggesting that it may be possible to solve large problems in other aquatic communities. Note, however, that ecologists rarely attempt to solve such problems by deliberately introducing new predators, such as *Beroe*, because such introductions can also have unanticipated negative effects.

The details of the fall and rise of the Black Sea ecosystem also illustrate two important types of causation in ecological communities: bottom-up and top-down controls. The fall of the Black Sea ecosystem began when increased nutrient inputs led to problems associated with eutrophication: increased phytoplankton abundance, increased bacterial abundance, decreased oxygen concentrations, and fish die-offs. The effect of adding nutrients to the Black Sea illustrates **bottom-up control**, which occurs when the abundance of a population is limited by nutrient supply or food availability. In this case, prior to nutrient enrichment, phytoplankton abundance—and thus the abundance of food for other organisms—was limited by the supply of nutrients.

Ecosystems are also affected by **top-down control**, which occurs when the abundance of a population is limited by predators. Recent evidence indicates that early steps in the decline of the Black Sea ecosystem were driven not only from the bottom up (by eutrophication), but also from the top down, by overfishing (Daskalov et al. 2007). Starting in the late 1950s, overfishing caused sharp drops in the abundances of predatory fishes. As predatory fish populations declined, their prey, planktivorous (plankton-eating) fishes, increased in number (**FIGURE 10.18A**). In turn, the increase in planktivorous fishes was associated with declining numbers of zooplankton and increasing

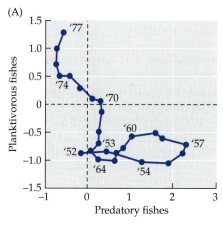 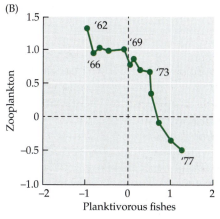 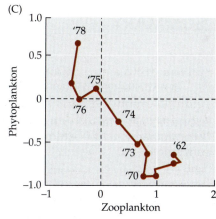

FIGURE 10.18 Ecosystem Changes in the Black Sea
Abundance indices of (A) planktivorous and predatory fishes, (B) zooplankton and planktivorous fishes, and (C) phytoplankton and zooplankton. In each graph, the organisms whose abundance is plotted on the *y*-axis are eaten by the organisms whose abundance is plotted on the *x*-axis. (Planktivorous fishes eat both zooplankton and phytoplankton, but they have a greater effect on zooplankton abundance than on phytoplankton abundance.) Numbers on the plots indicate years, beginning in 1952. In the abundance indices, data are standardized to have a mean of 0 and a variance of 1. (After G. M. Daskalov et al. 2007. *Proc Natl Acad Sci USA* 104: 10518–10523. © National Academy of Sciences, U.S.A.)

Referring to (A), describe predatory and planktivorous fish abundance from 1952 to 1957. Next, summarize how abundances of phytoplankton, zooplankton, planktivorous fishes, and predatory fishes changed in the 1970s. Finally, convert your summary of abundance changes in the 1970s into a chain of feeding relationships, where arrow thickness indicates the strength of each relationship (see Figure 9.19, in which similar chains are shown for Alaska). Is the chain you drew more similar to that in Alaska pre-1990 or that in Alaska in the late 1990s? Explain.

numbers of phytoplankton (**FIGURE 10.18B,C**), suggesting possible top-down control. Later, the arrival of the voracious predator *Mnemiopsis* also had a top-down effect, altering many key features of the ecosystem (e.g., zooplankton abundance, phytoplankton abundance, fish abundance). Top-down control also seems to have

influenced ecosystem recovery: it took another predator, *Beroe*, to rein in *Mnemiopsis*. In reality, as in the Black Sea, bottom-up and top-down controls interact to shape how ecosystems work. We'll return to bottom-up and top-down controls in Units 5 and 6, where we consider these important topics in more detail.

SUMMARY

CONCEPT 10.1 Populations are dynamic entities that vary in size over time.

10.1.1 List the different patterns of population growth observed in nature.

- Most observed patterns of population growth can be grouped into four major types: exponential growth, logistic growth, fluctuations, and regular cycles. These four patterns are not mutually exclusive, and a single population can (and probably will) experience each of them at different times.

10.1.2 Compare the patterns of exponential growth with that of logistic growth.

- Exponential growth occurs when the rate of growth increases (or decreases) in proportion to the current number of individuals and lasts for a limited time when conditions are favorable (or unfavorable).

- Logistic growth occurs when the rate of growth increases rapidly at first and then stabilizes as the population reaches carrying capacity (the maximum population size that can be supported indefinitely by the environment).

10.1.3 Describe population size fluctuations and the special case of population cycling.

- All populations show fluctuations in size, some of which are erratic and others of which are deviations from exponential or logistic growth. Some populations fluctuate greatly over time; others fluctuate relatively little.

- Some populations show regular population cycles, a special type of fluctuation in which alternating periods of high and low abundance occur after nearly constant intervals of time.

(Continued)

SUMMARY *(continued)*

- A time lag in population growth often occurs when a change in population size is delayed by density-dependent factors, such as growth, reproduction, and/or mortality.

CONCEPT 10.2 The risk of extinction increases in populations that fluctuate in size and/or are small.

10.2.1 Justify why fluctuations in population growth rate can increase a population's risk of extinction.

- For a given average population size, a population that fluctuates in size shows a slower growth rate than one that does not vary.

- A slower growth rate results in a smaller population size, which has a greater risk of extinction.

10.2.2 List and describe the ways that chance events can drive small populations to extinction.

- Small populations can be driven to extinction by chance events associated with genetic drift and inbreeding, demographic stochasticity, and environmental stochasticity.

- Genetic drift and inbreeding reduce the genetic variation of small populations, limiting the ability of populations to respond to environmental change and harmful alleles.

- Demographic stochasticity and environmental stochasticity produce unpredictable changes in population sizes, thus increasing the risk of extinction, especially for small populations.

Review Questions

1. Describe a factor that can cause a time lag or delay in the response of a natural population to a change in population density. How do such time lags affect changes in abundance over time?

2. Summarize how chance events can threaten small populations.

3. A population consists of four unrelated individuals, two females (F1 and F2) and two males (M1 and M2). Individuals live only 1 year, and they mate only once, producing two offspring (one female, one male) from each mating. Individuals avoid mating with relatives if possible.

   a. Starting with individuals F1, F2, M1, and M2 as the parent generation, can the first two generations of offspring be born to parents that are not related to each other? You may find it helpful to construct a diagram to illustrate the two generations of parents and their offspring.

   b. If the second generation of offspring become parents, how many of the matings in this third generation of parents can occur between unrelated individuals? Generalizing from your results, is inbreeding likely to be common or uncommon in small populations?

Hone Your Problem-Solving Skills

Svane (1984) studied population dynamics in the tunicate (or "sea squirt") *Ascidia mentula*, a filter-feeding marine invertebrate. The figure shows population densities over time for one of six study populations, and the table provides data (averaged across the six study populations) on how per capita birth and death rates are related to population density.

Population density (no. indiv/m²)	Birth rate (offspring/indiv/yr)	Death rate (deaths/indiv/yr)
38	0.40	0.38
60	0.39	0.32
75	0.56	0.51
105	0.47	0.36
180	0.44	0.48
230	0.30	0.32

1. Estimate the minimum and maximum density observed for the population whose densities are shown in the figure. Which of the four population growth patterns described in Concept 10.1 best represents the results shown in the figure? Explain.

2. Use the data in the table to graph the birth rate versus population density. To do this, first plot each data point and then draw an approximate best-fit curve that goes through those points. Are birth rates density dependent? Explain.

3. Use the data in the table to graph the death rate versus population density; to do this, first plot each data point and then draw an approximate best-fit curve that goes through those points. Are death rates density dependent? Explain.

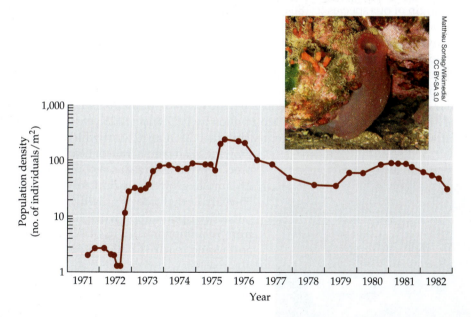

Matthieu Sontag/Wikimedia/ CC BY-SA 3.0

ADDITIONAL RESOURCES

This textbook provides resources such as **Web Extensions** and additional **Climate Change Connections** for instructors who can share them with students. Please contact your instructor, if you would like to access that content.

11

Population Growth and Regulation

KEY CONCEPTS

CONCEPT 11.1 Populations can grow exponentially when conditions are favorable, but exponential growth cannot continue indefinitely.

CONCEPT 11.2 Population size is determined by a combination of density-dependent and density-independent factors.

CONCEPT 11.3 The logistic equation incorporates limits to growth and shows how a population may stabilize at a maximum size, the carrying capacity.

CONCEPT 11.4 Life tables show how survival and reproduction vary with age or size structure, influencing population growth and size.

Human Population Growth: A Case Study

Viewed from space, Earth appears as a beautiful ball of blue and white in a vast sea of black. If we use satellite images to explore the surface of this beautiful ball in more detail, we find clear signs of human impacts across the globe. These signs range from the clear-cutting of forests, to the quilt-like patterns of agricultural fields, to the eerie red glow of fires burning out of control across the Amazon and other regions of the world (**FIGURE 11.1**).

People have a large effect on the global environment for two underlying reasons: our population has grown explosively, and so has our use of energy and resources. The human population crossed the 7.9 billion mark in 2022, more than double the 3 billion people alive in 1960 (**FIGURE 11.2**). Our use of energy and resources has grown even more rapidly. From 1860 to 1991, for example, the human population quadrupled in size, but our energy consumption increased 93-fold.

The addition of nearly 5 billion people since 1960 is remarkable. For thousands of years, the size of our population increased relatively slowly, reaching 1 billion for the first time in 1804. The time we took to reach the 1 billion mark puts the current growth of our population in perspective: it took roughly 200,000 years (from the origin of our species to 1804) for the human population to reach its first billion, but now we are adding 1 billion people every 13 years. When did we switch from relatively slow to explosive increases in the size of our population?

No one knows for sure, given that it is difficult to estimate population sizes from long ago. According to the best information we have, by 1550 there were roughly 500 million people alive, and the population was doubling every 275 years. By the time we reached our first billion in 1804, the human population was growing at a very rapid rate: it doubled from 1 to 2 billion by 1927, in just 123 years. Forty-seven years later, it had doubled again, reaching 4 billion in 1974, at which time it was growing at an annual rate of nearly 2%. To appreciate what that means, a population with a 2% annual growth rate doubles in size every 35 years. If that rate of growth could be sustained, our population would almost double from 7.9 billion in 2022 to 14.9 billion in 2054 and would reach 31 billion by 2090.

FIGURE 11.1 Amazon on Fire This late-night NASA satellite image of South America shows the vast area over which large, intense, and persistent fires are burning in this region of the world (red areas). Fire activity in the Amazon varies considerably from year to year, driven by changes in human activity and climate. The timing and location of fires in 2019 (when this photo was taken) suggest that they were associated with extensive land clearing that year rather than regional drought conditions.

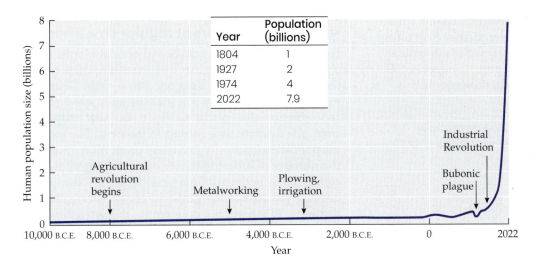

FIGURE 11.2 Explosive Growth of the Human Population The size of the human population increased relatively slowly until 1804, when the effects of the Industrial Revolution took hold. Since that time our population has increased in size to 7.9 billion people in 2022. (Based on estimates by the History Database of the Global Environment [HYDE] and the United Nations. Visualization from OurWorldinData.org. CC BY-SA 4.0/Max Roser. Retrieved from https://ourworldindata.org/world-population-growth.)

Year	Population (billions)
1804	1
1927	2
1974	4
2022	7.9

What do you think the world would be like with 31 billion people? Already, with 7.9 billion people, we have transformed the planet. However, it is unlikely that there will be 31 billion people on Earth in 2090. Over the last 50 years, the rate of human population growth has slowed considerably, from a high of 2.2% per year in the early 1960s to the present rate of 1.0% annually. Even so, the current rate translates into a human population that is increasing by about 80 million people per year, more than 215,000 people each day. Five countries—India, China, Pakistan, Nigeria, and the United States—account for almost half of this annual increase.

If the current annual growth rate of 1.0% were maintained, there would be more than 14 billion people on Earth in 2080. Can Earth support 14 billion people? Will there be that many people in 2080? Or will annual growth rates continue to fall? We'll return to these questions in the Case Study Revisited.

Introduction

Ecologists use population growth models to understand the ways in which populations change in abundance over time and what factors promote or limit population growth. What we learn from these models can surprise us. We may find, for example, that current methods used to protect an endangered species are inadequate. Such was the case for loggerhead sea turtles, a rare species whose young often die as they crawl to the sea after hatching from nests dug in the sand (**FIGURE 11.3**). Efforts to protect loggerhead turtles initially focused on protecting newborns. However, researchers found, using population growth models, that even if newborn survival could be increased to 100%, loggerhead turtle populations would still continue to decline. Fortunately, the researchers were able to use their population growth model to inform management techniques that resulted in more effective ways of protecting loggerhead turtles (see Ecological Toolkit 11.1).

As we have seen in Chapters 9 and 10, populations can change in size as a result of four processes: birth, death, immigration, and emigration. We can summarize the effects of these four processes on population size with the following equation:

$$N_{t+1} = N_t + B - D + I - E$$

where N_t is the population size at time t, B is the number of births, D is the number of deaths, I is the number of immigrants, and E is the number of emigrants between time t and time $t + 1$. As implied by this equation, populations are open and dynamic entities that can change from one time period to the next due to births and deaths. For simplification purposes, the population growth models we consider

FIGURE 11.3 Dash to the Sea These loggerhead sea turtle hatchlings have emerged from nests in the sand and must reach the sea to survive. On land, eggs and hatchlings face threats from predators, beach development, and artificial lighting (which can disrupt the hatchlings' sense of direction, preventing them from reaching the sea). Loggerhead turtles also face threats in the marine environment from predators, commercial fisheries (turtles can be caught accidentally in nets and traps), collisions with boats, and pollution.

© iStock.com/LagunaticPhoto

here do not include immigration or emigration. Let's use this basic information about population change over time to consider two common observed patterns of population growth that were described in Chapter 10: exponential population growth and logistic population growth.

CONCEPT 11.1   Populations can grow exponentially when conditions are favorable, but exponential growth cannot continue indefinitely.

LEARNING OBJECTIVES

11.1.1   Define geometric population growth.

11.1.2   Define exponential population growth.

11.1.3   Describe the characteristics of geometric and exponential growth.

Geometric and Exponential Growth

In general, populations can grow rapidly whenever individuals leave an average of more than one offspring over substantial periods of time. In this section, we describe **exponential growth** and the special case known as **geometric growth**, two related patterns of population growth that can lead to rapid increases in population size. For exponential growth, the organisms are assumed to reproduce continuously over time compared to geometric growth, in which the organisms reproduce in synchrony at discrete periods of time. Let's consider geometric growth first.

Populations grow geometrically when reproduction occurs at regular time intervals

Some species, such as cicadas and annual plants, reproduce in synchrony at regular time intervals. These regular time intervals are called *discrete time periods*. Geometric growth occurs when a population with synchronous reproduction changes in size by a constant proportion from one discrete time period to the next. The fact that the population grows by a constant proportion means that the number of individuals added to the population becomes larger with each time period (births > deaths). As a result, the population grows larger by ever-increasing amounts. When plotted on a graph, this growth pattern forms a J-shaped set of points, with each point representing the resulting population size after each time period (**FIGURE 11.4A**).

Mathematically, we can describe geometric growth as

$$N_{t+1} = \lambda N_t \tag{11.1}$$

where N_t is the population size after t generations or, equivalently, after t discrete time periods (e.g., t years if there is one generation per year), and λ is a constant whose value is determined by the per capita (or per individual) birth rate ($b = B/N$) minus the per capita death rate ($d = D/N$) over discrete time periods. In Equation 11.1, λ serves as a multiplier that allows us to predict the size of the population in the next time period. We'll refer to λ as the **geometric population growth rate**; λ is also known as the (per capita) **finite rate of increase**. We use

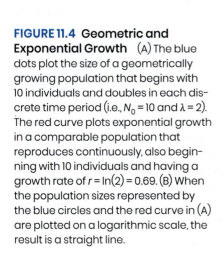

FIGURE 11.4   Geometric and Exponential Growth   (A) The blue dots plot the size of a geometrically growing population that begins with 10 individuals and doubles in each discrete time period (i.e., $N_0 = 10$ and $\lambda = 2$). The red curve plots exponential growth in a comparable population that reproduces continuously, also beginning with 10 individuals and having a growth rate of $r = \ln(2) = 0.69$. (B) When the population sizes represented by the blue circles and the red curve in (A) are plotted on a logarithmic scale, the result is a straight line.

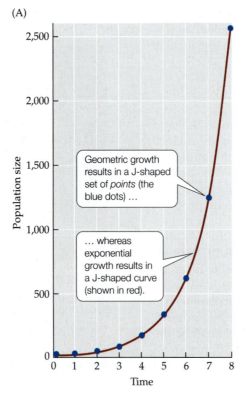

Geometric growth results in a J-shaped set of *points* (the blue dots) …

… whereas exponential growth results in a J-shaped curve (shown in red).

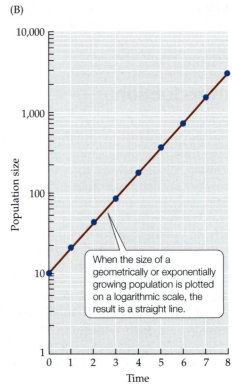

When the size of a geometrically or exponentially growing population is plotted on a logarithmic scale, the result is a straight line.

this terminology by convention, but it can be confusing: we can see from Equation 11.1 that when the population "growth" rate λ is between 0 and 1, the population does not grow, but rather decreases in size over time.

Geometric growth can also be represented by a second equation,

$$N_t = \lambda^t N_0 \qquad (11.2)$$

where N_0 is the initial population size (i.e., the population size at time = 0).

The two equations for geometric growth (Equations 11.1 and 11.2) are equivalent in that each can be derived from the other. Which one we use depends on what we are interested in. If we want to predict the population size in the next time period and we know λ and the current population size, either equation can be used. If we know the population size in both the current and previous time periods, we can rearrange Equation 11.1 by dividing N_{t+1} by N_t to get an estimate of λ. Finally, we can use Equation 11.2 to predict the size of the population after any number of discrete time periods. If $\lambda = 2$, for example, then after 12 time periods, a population that begins with $N_0 = 10$ individuals will have $N_{12} = 2^{12}N_0$ individuals, which (as we can determine by using a calculator with a y^x function) equals $4{,}096 \times 10$, or $40{,}960$.

Populations grow exponentially when reproduction occurs continuously

In contrast to the pattern described in the previous section, individuals in many species (including humans) do not reproduce in synchrony at discrete time periods; instead, they reproduce continuously over time. Exponential growth occurs when population size changes by a constant proportion at each instant in time (see the red curve in Figure 11.4A, representing continuous growth). Mathematically, exponential growth can be described by the following two equations:

$$\frac{dN}{dt} = rN \qquad (11.3)$$

and

$$N_t = N_0 e^{rt} \qquad (11.4)$$

where N_t is the population size at each instant in time, t.

In Equation 11.3, dN/dt represents the rate of change in population size at each instant in time; we see from the equation that dN/dt equals a constant rate (r; instantaneous birth rate (b) − instantaneous death rate (d)) multiplied by the current population size, N. Thus, the multiplier r provides a measure of how rapidly a population can grow; r is called the **exponential growth rate** or the (per capita) **intrinsic rate of increase**.

As we did for Equation 11.2, we can use Equation 11.4 to predict the size of an exponentially growing population at any time t, provided we have an estimate for r and know N_0, the initial population size. The "e" in Equation 11.4 is a constant, approximately equal to 2.718 ["e" is the base of the natural logarithm, $\ln(x)$]. We can calculate e^{rt} using the function e^x, which can be found on many calculators.

When plotted on a graph, the exponential growth pattern forms a curve that, like the geometric growth pattern, is J-shaped. Exponential growth and geometric growth are similar in that we can draw an exponential growth curve through the discrete points of a population that grows geometrically (see Figure 11.4A). Because exponential and geometric growth curves overlap, both types of growth are sometimes lumped together for simplicity and referred to as *exponential growth*.

Geometric and exponential growth curves overlap because Equations 11.2 and 11.4 are similar in form, except that λ in Equation 11.2 is replaced by e^r in Equation 11.4. Thus, if we want to compare the results of discrete time and continuous time growth models, we can calculate λ from r, or vice versa:

$$\lambda = e^r$$

$$r = \ln(\lambda)$$

where $\ln(\lambda)$ is the natural logarithm of λ, or $\log_e(\lambda)$. For example, if $\lambda = 2$ (as in Figure 11.4A), an equivalent value for r would be $r = \ln(2)$, which is approximately 0.69. **FIGURE 11.4B** illustrates a simple way to determine whether a population really is growing exponentially: plot the natural logarithm of population size versus time, and if the result is a straight line, the population is increasing exponentially.

Finally, look again at Equations 11.1 and 11.3. In Equation 11.1, which value of λ will ensure that the population does not change in size from one time period to the next? Similarly, in Equation 11.3, which value of r causes the population to remain fixed in size? The answers are $\lambda = 1$ (because then $N_{t+1} = N_t$) and $r = 0$ (because then the rate at which the population size changes is 0). When $\lambda < 1$ (or $r < 0$), the population will decline to extinction, whereas when $\lambda > 1$ (or $r > 0$), the population will increase geometrically (or exponentially) to form a J-shaped curve (**FIGURE 11.5**).

How can we estimate a population's growth rate (r or λ)? In one approach, Equation 11.4 is used to estimate the growth rate r at different points in time, as you can explore for the human population in **ANALYZING DATA 11.1**. There are a variety of other methods as well (see Caswell 2001), including estimating r (or λ) from life table data, as we will see in Concept 11.4.

FIGURE 11.5 How Population Growth Rates Affect Population Size Depending on the value of λ or r, a population with an exponential growth pattern will decrease in size, remain the same size, or increase in size.

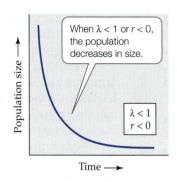

When $\lambda < 1$ or $r < 0$, the population decreases in size.

$\lambda < 1$
$r < 0$

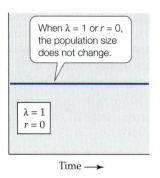

When $\lambda = 1$ or $r = 0$, the population size does not change.

$\lambda = 1$
$r = 0$

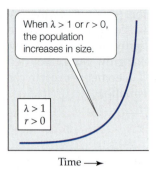

When $\lambda > 1$ or $r > 0$, the population increases in size.

$\lambda > 1$
$r > 0$

We can also use r to determine the **doubling time (t_d)** of a population, which is the number of years it will take the population to double in size. As interested readers can confirm (by solving Equation 11.4 for the time it takes a population to increase from its initial size, N_0, to twice that size, $2N_0$), doubling times can be estimated as

$$t_d = \frac{\ln(2)}{r} \qquad (11.5)$$

where r is the exponential growth rate.

Populations can grow rapidly because they increase by multiplication

Equations 11.1 and 11.3 show that populations increase by multiplication, not addition: at each point in time, the population changes in size according to the multiplier λ or r. As a result, populations have the potential to add large numbers of individuals rapidly whenever $\lambda > 1$ or $r > 0$. The principle at work here is the same one that applies to interest on a savings account. Even when the interest rate is low, you can earn a lot of money each year if you have

---

ANALYZING DATA 11.1

How Has the Growth of the Human Population Changed over Time?

Ecologists often use estimates of λ or r to determine how rapidly a population is growing (or declining) at various points in time. For a population that is growing exponentially, we can calculate such estimates by re-arranging Equation 11.4 to read

$$e^{rt} = \frac{N_t}{N_0}$$

where N_0 is the population size at the beginning of a time period, t is the length of the time period, and N_t is the population size at the end of the period. If we know t, N_0, and N_t, we can then estimate r:

$$r = \frac{\ln\left(\dfrac{N_t}{N_0}\right)}{t}$$

In this exercise, we'll use this technique and the data in the table to examine the growth rate of the world's human population at different points in time.

1. Calculate the exponential growth rates for the years shown in the table and graph your results. For example, from year 1 to year 400, the length of the time period, t, is $t = 400 - 1 = 399$, and we find that $r = [\ln(190 \text{ million}/170 \text{ million})]/399 = 0.1112/399 = 0.00028$.

Year (C.E.)	Population size	Exponential growth rate (r)
1	170 million	0.00028
400	190 million	?
800	220 million	?
1200	360 million	?
1550	500 million	?
1804	1 billion	?
1927	2 billion	?
1960	3 billion	?
1999	6 billion	?
2010	6.87 billion	?
2016	7.35 billion	?
2019	7.7 billion	(N/A)

2. If the human population continued to grow at the rate you calculated for 2016, how large would the population be in 2066?

3. What assumptions did you make in answering Question 2? Based on results for Question 1, is it likely that the human population will reach the size that you calculated for 2066? Explain.

a large amount deposited in the bank, because savings, like populations, grow by multiplication. Similarly, the fact that populations grow by multiplication means that even a low growth rate can cause the size of a population to increase rapidly.

Consider our own population. In this chapter's Case Study, we stated that the current annual growth rate of the human population was 1.0%, which is an $r = 0.01$. If we set the year 2022 as time $t = 0$, we have $N_0 = 7.9$ billion, the size of the human population in 2022. Plugging these values of r and N_0 into Equation 11.4, we calculate that the population size 1 year later should be $N_1 = 7.9 \times e^{0.01}$, which equals 7.98 billion people. Thus, in 2022, the human population was increasing by 80 million people per year (7.98 billion – 7.90 billion = 0.08 billion = 80 million). Since populations grow by multiplication, if r remained constant at 0.01 for an extended period of time, the yearly increments to the human population would become astronomical. For example, after 225 years, there would be 75 billion people, and our population would be increasing in size by almost a billion people *each year*.

Turning from humans to other species, what do field studies reveal about the growth rates of their populations? Some species, such as the woodland herb *Asarum canadense* (wild ginger), have maximum observed values of λ that are close to 1 ($\lambda = 1.01$ in young forests, $\lambda = 1.1$ in mature forests) (**FIGURE 11.6**). Similar values were observed for a population of 25 reindeer introduced to Saint Paul Island off the coast of Alaska in 1911. After 27 years, the population had increased from 25 to 2,046 individuals, which (when we solve for λ in Equation 11.2) yields $\lambda = 1.18$.

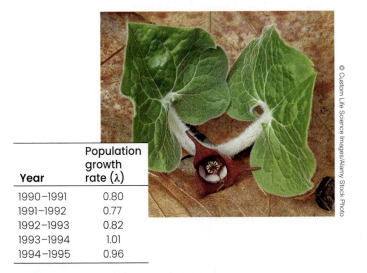

© Custom Life Science Images/Alamy Stock Photo

Year	Population growth rate (λ)
1990–1991	0.80
1991–1992	0.77
1992–1993	0.82
1993–1994	1.01
1994–1995	0.96

FIGURE 11.6 Some Populations Have Low Growth Rates   The growth rates of a population of wild ginger (*Asarum canadense*) in a young forest vary from year to year. The maximum growth rate in this forest is 1.01. However, growth rates are often less than 1.0, suggesting that the population will decline in size unless conditions improve. (Data from H. Damman and M. L. Cain. 1998. *J Ecol* 86: 13–26.)

Considerably higher annual growth rates have been observed for populations of many species, including western gray kangaroos ($\lambda = 1.9$), field voles ($\lambda = 24$), and rice weevils ($\lambda = 10^{17}$), which are insect pests of rice and other grains. Some bacteria, such as the mammalian gut inhabitant *Escherichia coli*, can double in number every 30 minutes, resulting in the unimaginably high annual growth rate of $\lambda = 10^{5,274}$.

Recall that when $\lambda > 1$ (or $r > 0$) for an extended period, populations increase exponentially in size, forming a J-shaped curve like that in Figure 11.4A. In natural populations, $\lambda > 1$ (or $r > 0$) when key factors in the environment are favorable for growth, survival, and reproduction. But can such favorable conditions last for long?

There are limits to the growth of populations

An argument from basic principles suggests that the answer to the question we just posed is no. Physicists estimate that the known universe contains a total of 10^{80} atoms. Yet if favorable conditions persisted for long enough, allowing λ to remain greater than 1, even populations of relatively slowly growing species would eventually increase to more than 10^{80} individuals. For example, based on *Asarum*'s growth rate of $\lambda = 1.01$ in young forests, a population that began with 2 plants would have more than 10^{82} plants after 19,000 years. For an extremely rapidly growing species such as *E. coli*, the numbers are even more absurd: it would take only 6 days for a population that began with a single bacterium to exceed 10^{80} individuals.

No population could ever come close to having 10^{80} individuals, because there would be no atoms with which to construct their bodies. Thus, exponential growth cannot continue indefinitely. While this is an extreme example (because other difficulties would be encountered long before there was a shortage of atoms), it illustrates a fundamental point: there are limits to population growth, which cause it to slow and eventually stop. We'll look at some of those limits in the following section.

> **CONCEPT 11.2**   Population size is determined by a combination of density-dependent and density-independent factors.

LEARNING OBJECTIVES

11.2.1 Define density-independent factors and describe how they affect population size and growth rate.

11.2.2 Define density-dependent factors and describe how they affect population size and growth rate.

Effects of Density

Although populations can show exponential growth under favorable conditions, conditions in nature are

(A)

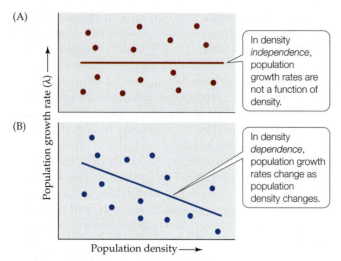

In density *independence*, population growth rates are not a function of density.

(B)

In density *dependence*, population growth rates change as population density changes.

FIGURE 11.7 Comparing Density Independence and Density Dependence Each point represents one population. (A) Density independence. (B) Density dependence. In this example, population growth rates decrease as population density increases.

rarely favorable for long. For example, Damman and Cain (1998) calculated the geometric growth rate (λ) in each of 5 years for a population of *A. canadense* located in a young forest. As mentioned above, the maximum growth rate was $\lambda = 1.01$. During the other 4 years, however, values for λ ranged from 0.77 to 0.96 (see Figure 11.6). Thus, far from threatening to overrun the planet with its offspring, we would expect this population to decline in the long run, unless conditions changed for the better λ.

What factors change population sizes and growth rates over time? One set of factors are referred to as **density-independent** factors, meaning that their effects on population size (N) or population growth rate (λ or r) are independent of the number of individuals in the population (**FIGURE 11.7A**). The other set of factors are known as **density-dependent** factors because their effects on population size or population growth rate are dependent on the number of individuals in the population (**FIGURE 11.7B**). Let's discuss density-independent factors first.

Density-independent factors can determine population size

In many species, year-to-year variation in weather leads to dramatic changes in abundance and hence in population growth rates. For example, Davidson and Andrewartha (1948) studied how weather in Adelaide, Australia, affected populations of the insect *Thrips imaginis*, a pest of roses. By correlating weather conditions with thrips population sizes over a 14-year period, they showed that yearly fluctuations in population size could be predicted accurately by an equation that used temperature and rainfall data (**FIGURE 11.8**).

The effects of climate can also change the birth or death rates of species more gradually over time, as is the case for forests across broad regions of the western United States.

CLIMATE CHANGE CONNECTION

EFFECTS OF CLIMATE CHANGE ON TREE MORTALITY RATES Over the course of several decades, mortality rates increased gradually in populations of coniferous forest trees across the western United States (**FIGURE 11.9**). These increases occurred in stands of seemingly healthy forest that had not been cut for more than 200 years, leading researchers to ask, "What is killing the trees?"

In seeking an answer to this question, Van Mantgem et al. (2009) ruled out several possible causes, including air pollution, forest fragmentation, changes in fire frequency, and within-stand increases in the intensity of competition. The researchers went on to note that during the time period covered by their study, regional temperatures in the western United States had increased at rates of 0.3°C to 0.5°C per decade. These rapid temperature increases were associated with declines in the snowpack, earlier spring snowmelt, and a lengthening of the summer dry period. These changes caused an increase in the trees' *climatic water deficit* (the amount by which a plant's annual evaporative demand for water exceeds available water). Previous studies had

FIGURE 11.8 Weather Can Influence Population Size Davidson and Andrewartha accurately predicted the mean number of thrips per rose observed in Adelaide, Australia, using an equation based on four weather-related variables. (After J. Davidson and H. Andrewartha. 1948. *J Anim Ecol* 17: 200–222.)

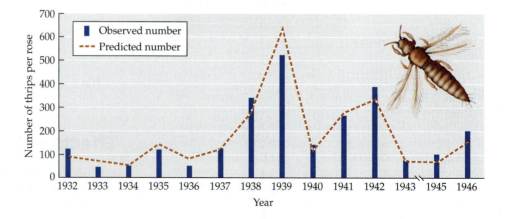

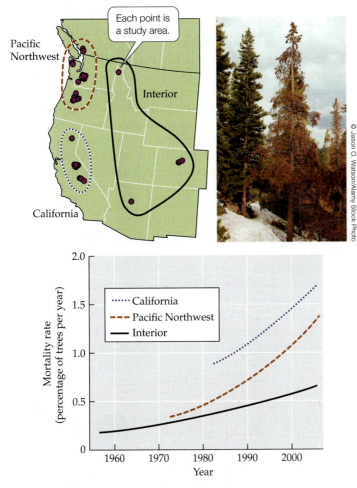

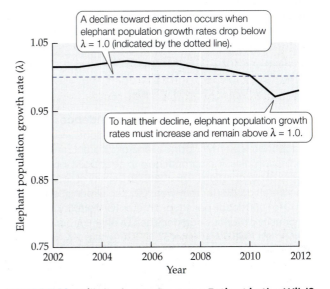

Each point is a study area.

Pacific Northwest

Interior

California

© Jason O. Watson/Alamy Stock Photo

FIGURE 11.9 Rising Tree Mortality Rates Trends in coniferous tree mortality rates for 76 study plots located in the three regions of the western United States shown on the map. (After P. J. van Mantgem et al. 2009. *Science* 323: 521–524.)

shown that tree mortality rates tend to increase when climatic water deficit increases (Bigler et al. 2007). Overall, van Mantgem et al.'s study suggests that the rise in tree mortality rates was driven by regional warming and the ensuing drought stress. Similarly, in the southwestern United States, warmer temperatures in the summer and reduced snowfall in the winter have produced "hotter droughts" that are associated with increases in the area burned by wildfires and the area affected by insect outbreaks—again causing tree mortality rates to rise (Williams et al. 2013). ☀️

Population size can also fluctuate as a consequence of biotic factors, such as hunting. For example, in 2002, a system for recording the cause of death of elephants was established in 45 sites across Africa. Wittemyer et al. (2014) combined that information with other demographic data to estimate how elephant population growth rates have changed over time (**FIGURE 11.10**). Their analyses indicated that across the African continent, elephant population growth rates have dropped below λ = 1.0,

primarily because of a rapid increase in illegal poaching (for tusk ivory) after 2009. For example, 100,000 elephants were killed for ivory over a 3-year period (2010–2012)—a level of illegal killing that cannot be sustained. To prevent elephants from becoming extinct in the wild, elephant population growth rates must increase and remain above λ = 1.0. For this to occur, new efforts must be taken to curb the rate of illegal killing and reduce the global demand for illegal ivory.

As these examples suggest, density-independent factors can have major effects on population size from one year to the next. In principle, such factors could account entirely for year-to-year fluctuations in the size of a population. But density-independent factors do not tend to increase the size of populations when they are small and decrease the size of populations when they are large. A factor that did consistently lead to such changes would cause the population growth rate to change as a function of density—that is, to be density *dependent*, not density independent.

Density-dependent factors regulate population size

Limited amounts of factors such as food or habitat can influence population size in a density-dependent manner, which means that they cause birth rates, death rates, or dispersal rates to change as the density of the population changes (see Figure 11.7B). As densities increase, it is common for birth rates to decrease, death rates to increase, and dispersal from the population (emigration) to increase—all of which tend to decrease population size. When densities decrease, the opposite occurs: birth rates tend to increase, and death and emigration rates decrease.

FIGURE 11.10 Will Elephants Become Extinct in the Wild? Population growth rates (λ) for 306 elephant populations show that elephants have been in decline across the African continent since 2010. (After G. Wittemyer et al. 2014. *Proc Natl Acad Sci USA* 111: 13117–13121.)

When one or more density-dependent factors cause population size to increase when numbers are low and decrease when numbers are high, **population regulation** is said to occur. Ultimately, when the density of any species becomes high enough, density-dependent factors decrease population size because food, space, or other essential resources are in short supply. Note that "regulation" has a particular meaning here, referring to the effects of factors that tend to increase λ or r when the population size is small and decrease λ or r when the population size is large. Density-independent factors can have large effects on population size, but they do not *regulate* population size because they do not consistently increase population size when it is small and decrease population size when it is large. Thus, by definition, only density-dependent factors can regulate population size.

Density dependence has been observed in many populations

Density dependence can often be detected in natural populations (**FIGURE 11.11**). For example, in a study that combined field observations with controlled experiments, Arcese and Smith (1988) examined the effect of population density on reproduction in the song sparrow (*Melospiza melodia*) on Mandarte Island, British Columbia. They found that the number of eggs laid per female decreased with density, as did the number of young that survived long enough to become independent of their parents (see Figure 11.11A). Because Mandarte Island is small and the birds were likely to suffer food shortages at high densities, Arcese and Smith predicted that if they provided food to a subset of nesting pairs when densities were high, the birds that were fed should be able to rear more young to independence. That is exactly what happened: nesting pairs that were fed reared nearly four times as many young to independence as did control birds that were not fed (see Figure 11.11A).

In addition to density-dependent reproduction, density-dependent mortality has been observed in many populations. For example, when Yoda et al. (1963) planted soybeans (*Glycine soja*) at various densities, they found that at the highest initial planting densities, many of the seedlings had died by 93 days of age (see Figure 11.11B). Similarly, in an experiment in which eggs of the flour beetle *Tribolium confusum* were placed in glass tubes (each with 0.5 g

(A)

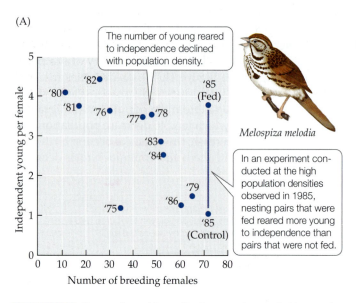

The number of young reared to independence declined with population density.

In an experiment conducted at the high population densities observed in 1985, nesting pairs that were fed reared more young to independence than pairs that were not fed.

Melospiza melodia

(B)

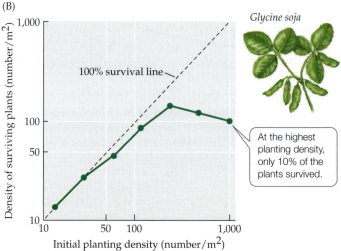

100% survival line

Glycine soja

At the highest planting density, only 10% of the plants survived.

(C)

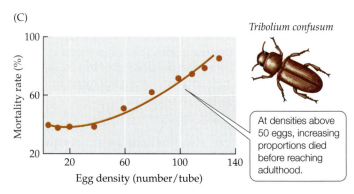

Tribolium confusum

At densities above 50 eggs, increasing proportions died before reaching adulthood.

FIGURE 11.11 Examples of Density Dependence in Natural Populations (A) Numbers of young song sparrows reared to independence on Mandarte Island at different densities of breeding females. The number next to each point indicates the year of observation (1975–1986). (B) Density of surviving soybeans 93 days after they were planted at densities ranging from 10 to 1,000 seeds per square meter. (C) Mortality rates in flour beetles at various egg densities. (A after P. Arcese and J. N. M. Smith. 1988. *J Anim Ecol* 57: 119–136; B after J. A. Yoda et al. 1963. *J Biol* 14: 107–120; C after T. S. Bellows, Jr. 1981. *J Anim Ecol* 50: 139–156.)

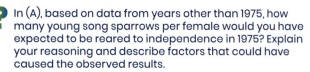 In (A), based on data from years other than 1975, how many young song sparrows per female would you have expected to be reared to independence in 1975? Explain your reasoning and describe factors that could have caused the observed results.

(A) *Poa annua*

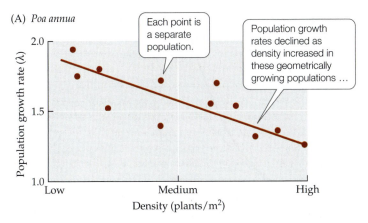

(B) *Daphnia pulex*

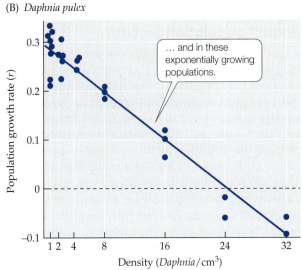

FIGURE 11.12 **Population Growth Rates May Decline at High Densities** Each point represents one population. (A) The geometric population growth rate (λ) of the grass *Poa annua* is density dependent, as is (B) the exponential growth rate (*r*) of the water flea *Daphnia pulex*. (A after R. Law. 1975. Unpublished PhD thesis. University of Liverpool; B after P. W. Frank et al. 1957. *Physiol Zool* 30: 287–305.)

? Are high-density populations increasing in size in (A)? In (B)? Explain.

of food), death rates increased as the density of eggs per tube increased—again revealing density dependence (see Figure 11.11C). Density dependence has also been detected in populations whose abundance is strongly influenced by factors usually considered to act in a density-independent manner, such as temperature or precipitation.

When birth, death, or dispersal rates show strong density dependence, population growth rates (λ or *r*) may decline as densities increase (**FIGURE 11.12**). Eventually, if densities become high enough to cause λ to equal 1 (or *r* to equal 0), the population stops growing entirely; if λ becomes less than 1 (or *r* < 0), the population declines. As we'll see in the next section, such density-dependent changes in the population growth rate can cause a population to reach a stable, maximum population size.

CONCEPT 11.3 The logistic equation incorporates limits to growth and shows how a population may stabilize at a maximum size, the carrying capacity.

LEARNING OBJECTIVES

11.3.1 Define logistic population growth and compare to exponential population growth.

11.3.2 Describe the growth patterns of the U.S. population.

Logistic Growth

Some populations exhibit **logistic growth**, a pattern in which abundance increases rapidly at first and then stabilizes at a population size known as the **carrying capacity**, the maximum population size that can be

supported indefinitely by the environment. The growth of such a population can be represented by an S-shaped curve (**FIGURE 11.13**). The growth rate of the population decreases as the population size nears the carrying capacity because resources such as food, water, or space begin to be in short supply. At the carrying capacity, the growth rate is zero, and hence the population size does not change.

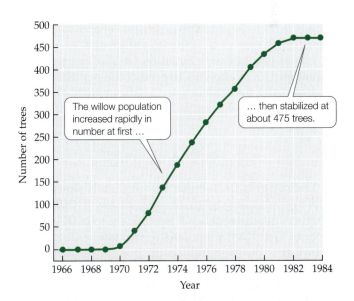

FIGURE 11.13 **An S-Shaped Growth Curve in a Natural Population** At a site in Australia, heavy grazing by rabbits had prevented willows from colonizing the area. The rabbits were removed in 1954, opening up new habitat for willows. When willows colonized the area in 1966, ecologists tracked the growth of their population. (After M. C. Alliende and J. L. Harper. 1989. *J Ecol* 77: 1029–1047.)

The logistic equation models density-dependent population growth

To see how the idea of a carrying capacity can be represented in a mathematical model of population growth, let's reconsider Figure 11.12. The data in both graphs show that population growth rates (r or λ) decreased approximately as a straight line as population densities increased. But r is assumed to be constant in the exponential growth equation, $dN/dt = rN$. As we've seen, a constant value of $r > 0$ allows for unlimited growth in population size. Thus, to modify the exponential growth equation to make it more realistic, we replace the assumption that r is constant with the assumption that r declines in a straight line as density (N) increases. When we do this, we obtain the *logistic equation*:

$$\frac{dN}{dt} = rN\left(1 - \frac{N}{K}\right) \tag{11.6}$$

where dN/dt is the rate of change in population size at time t, N is population density (also at time t), r is the (per capita) intrinsic rate of increase under ideal conditions, and K is the density at which the population stops increasing in size. K can be interpreted as the carrying capacity of the environment, and the term $(1 - N/K)$ can be viewed as the fraction of the carrying capacity that is available for population growth. As long as the population size is less than the carrying capacity (i.e., $N < K$), only a fraction of the available resources are being used and the population will continue to grow. As the population size approaches the carrying capacity, however, the fraction of resources available for individuals becomes smaller and the population growth slows and ultimately stops at K.

Just as you saw with Equation 11.4, we can rearrange Equation 11.6 to allow us to predict the population size at some later time, assuming logistic growth. When we do this, we get

$$N_t = \frac{K}{1 + \left[\left((K - N_0)/N_0\right)\right]e^{-rt}} \tag{11.7}$$

Logistic growth is similar to, but slightly slower than, exponential growth when densities are low (**FIGURE 11.14**). This occurs because when N is small, the term $(1 - N/K)$ is close to 1, and hence a population that grows logistically grows at a rate close to r. As the population density increases, however, logistic growth and exponential growth differ greatly. In logistic growth, the rate at which the population changes in size (dN/dt) approaches zero as the population size nears the carrying capacity, K. As a result, over time, the population size approaches K gradually, eventually remaining constant with K individuals in the population.

In Concept 10.1, we discussed the extent to which the growth of natural populations can be described by the

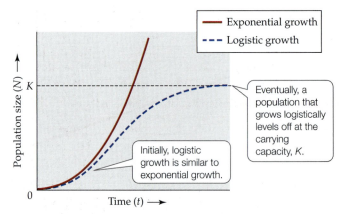

FIGURE 11.14 Comparison of Logistic and Exponential Growth Over time, logistic growth differs greatly from the unlimited growth of a population that increases exponentially.

? In the logistic equation, as the population size (N) becomes increasingly close to the carrying capacity, K, how does that affect the term $(1 - N/k)$? Why does this cause N to stop increasing in size?

S-shaped curve that results from the logistic equation; here, we examine efforts to fit the logistic equation to U.S. census data.

Can logistic growth predict the carrying capacity of the U.S. population?

In a groundbreaking paper published in 1920, Pearl and Reed examined the fit of several different mathematical models to U.S. census data for the period 1790–1910. Several of the approaches they tested did a good job of matching the historical data, but none included limits to the eventual size of the U.S. population. To address this shortcoming, they derived the logistic equation, which, unknown to them, had been first described in 1838 by the Belgian mathematician P. F. Verhulst. Pearl and Reed argued that the logistic equation provided a sensible way to represent population growth because it included limits to growth. When they fit the census data to the logistic curve, they obtained an excellent match, from which they estimated that the U.S. population had a carrying capacity of $K = 197,274,000$ people.

The logistic curve estimated by Pearl and Reed provides a good fit to U.S. population data through 1950. After that time, however, the actual population size differed considerably from Pearl and Reed's projections (**FIGURE 11.15**). By 1967, the carrying capacity (197 million) they had predicted had been surpassed. Pearl and Reed intended their estimate of the carrying capacity to represent the number of people that could be supported in the United States in a self-sufficient manner. They recognized that if conditions changed—for example, if agricultural productivity increased or if more resources were imported from other countries—the population could increase beyond 197 million. These and other changes have occurred, leading some ecologists and demographers to shift their focus from

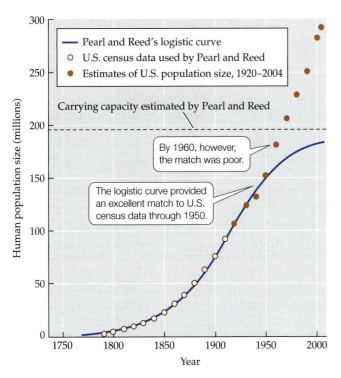

FIGURE 11.15 Fitting a Logistic Curve to the U.S. Population Size In 1920, Pearl and Reed fitted a logistic curve to U.S. census data for 1790–1910. They estimated the nation's carrying capacity (K) as 197 million people. (Data through 1910 from R. Pearl and L. J. Reed. 1920. *Proc Natl Sci Acad USA* 6: 275–288; other data from Statistical Abstracts, U.S. Census Bureau.)

Labels in figure:
- Pearl and Reed's logistic curve
- U.S. census data used by Pearl and Reed
- Estimates of U.S. population size, 1920–2004
- Carrying capacity estimated by Pearl and Reed
- By 1960, however, the match was poor.
- The logistic curve provided an excellent match to U.S. census data through 1950.

the number of humans that make up the carrying capacity to the total area of land required to support humans (the "ecological footprint," discussed in Connections in Nature).

> **CONCEPT 11.4** Life tables show how survival and reproduction vary with age or size structure, influencing population growth and size.

LEARNING OBJECTIVES

11.4.1 Justify the use of life tables to determine population growth and size.

11.4.2 Describe how age or size structure influences population growth and population size.

11.4.3 Compare the three types of survivorship curves.

11.4.4 Analyze life table data and calculate a net reproductive rate (R_0) and exponential growth rate (r).

Life Tables

Up to this point, we have assumed that individuals within a population do not vary in their birth (b) and death (d) rates. This is a big assumption given that we know real populations are made up of individuals of different ages,

sizes, and sexes, which vary in their capacity to reproduce and survive. Information about the varying patterns of reproduction and survival in a population is essential if we want to understand current population growth or predict future population sizes. A **life table** provides a summary of how survival and reproductive rates vary with the age, size, or life stage of the individuals within a population. These summaries can then be used to predict future population trends and develop strategies for managing populations of commercial or ecological value. Before we explore life tables in more detail, let's first consider how populations can differ in their age and survivorship structure.

Age or size structure influences how rapidly populations grow

Members of a population whose ages fall within a specified range are said to be part of the same *age class*. Age class 1, for instance, might include all individuals who are at least 1 year old but who are not yet 2 years old. Once individuals have been categorized in this way, a population can be described by its **age structure**: the proportions of the population in each age class. Imagine a population of a hypothetical organism in which all members die before they reach 3 years of age. In this population, every individual will be 0 ("newborns," which includes all individuals less than 1 year old), 1, or 2 years old. If there are 100 individuals in the population, and if 20 are newborns, 30 are 1-year-olds, and 50 are 2-year-olds, then the age structure will be 0.2 in age class 0, 0.3 in age class 1, and 0.5 in age class 2.

Age structure is a key feature of populations, in part because it influences whether a population increases or decreases in size. Consider two human populations of the same size and with the same survival and reproduction rates, but with different age structures. If one of the populations had many people older than 55, while the other had many people between ages 15 and 30, we would expect the second population to grow more rapidly than the first because it contained more individuals of reproductive age. Indeed, human populations that are growing rapidly typically have a greater percentage of people in younger age classes than do populations that are growing slowly or are in decline (**FIGURE 11.16**).

We have emphasized the importance of age because in many species, birth and death rates differ greatly among individuals of different ages. For other kinds of organisms, age is less important. In many plant species, for example, if conditions are favorable, a seedling may grow to full size relatively rapidly and reproduce at a young age. If conditions are not favorable, however, the plant may remain small for years and reproduce little or not at all; if conditions become favorable at a later time, the plant may then grow to full size and reproduce. For such species, whether an individual reproduces or not is more closely related to size than to age. When birth and death

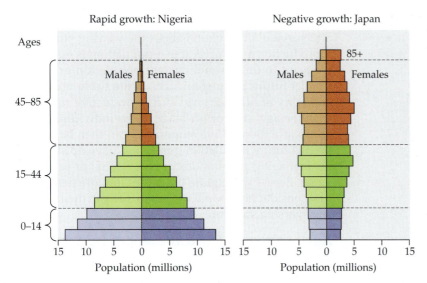

Nigeria		Japan
201	Population 2019 (millions)	126.2
5.7	Lifetime births per female	1.4
44	Percentage of population below age 15	12
3	Percentage of population over age 65	28
53(m) 55(f)*	Life expectancy at birth	81(m) 87(f)
67	Infant mortality (deaths per 1,000 births)	2
401.3	Projected population 2050 (millions)	109.9

*m = males; f = females
Source: Population Research Bureau.
https://www.prb.org/international/geography/nigeria and
https://www.prb.org/international/geography/japan.

FIGURE 11.16 Age Structure Influences Growth Rate in Human Populations Population pyramids for Nigeria and Japan show age structures that are typical of human populations with rapidly growing populations (Nigeria) and with growth rates that are negative or close to zero (Japan). The main reproductive ages (15–44) are shown in green. (After L. Roberts. 2011. *Science* 333: 540–543. Pyramids from United Nations, DESA, Population Division. World Population Prospects 2019/CC BY 3.0 IGO. https://population.un.org/wpp/Graphs/DemographicProfiles/Pyramid/392 and https://population.un.org/wpp/Graphs/DemographicProfiles/Pyramid/566.)

rates correlate poorly with age, or when age is difficult to measure, life tables based on the sizes or the life cycle stages (e.g., newborn, juvenile, adult) of individuals in the population can be constructed.

There are three types of survivorship curves

As discussed previously, different age classes of populations have different rates of survivorship. Survivorship data can be graphed as a **survivorship curve**. In such a curve, survivorship data are used to plot the numbers of individuals from a hypothetical cohort (typically, of 1,000 individuals) that will survive to reach different ages. Results from studies on a variety of species suggest that survivorship curves can be classified into three general types, which indicate the life stages at which high rates of mortality are most likely to occur (**FIGURE 11.17**). In populations with a **type I survivorship curve**, newborns, juveniles, and young adults all have high **survival rates**; death rates do not begin to increase until old age. Examples of populations with type I survivorship curves include U.S. females (**FIGURE 11.18**) and Dall mountain sheep (Figure 11.17A). In populations with a **type II survivorship curve**, individuals have an approximately constant chance of surviving from one age to the next throughout their lives. Some bird species have a type II survivorship curve (Figure 11.17B), as do mud turtles (after their second year), some fishes, and some plant species. Finally, in populations with a **type III survivorship curve**, individuals die at very high rates when they are young, but those that reach adulthood survive well later in life. Type III survivorship curves—the most common

type observed in nature—are typical of species that produce large numbers of young. Examples include giant puffballs, some plants, most insects, and many marine invertebrates, including the acorn barnacle *Balanus glandula* (Figure 11.17C). In this species, a population size of a million juveniles declines precipitously to 62 individuals after 1 year and to 2 individuals after 8 years.

We have discussed type I to III survivorship curves as if they were constant for each species, but that is not necessarily the case. Survivorship curves can vary among populations of a species, between males and females in a population, and among cohorts of a population that experience different environmental conditions (see Figure 11.18). In fact, by comparing birth and death rates in groups of individuals that experience different conditions, we can assess the effects of those conditions on populations. As we'll see in the next section, we can also use birth and death rates in a life table analysis to predict how the size and composition of a population will change over time.

Life tables can be based on age, size, or life cycle stage

The data that ecologists collect on the patterns of births and deaths for populations can be used to construct life tables. **TABLE 11.1** shows a life table using data from the acorn barnacle *B. glandula* on the shorelines of Scotland. This life table is known as a **cohort life table**, in which the fate of a group of individuals born during the same time period (a cohort) is followed from birth to death. The two columns on the left show the number of individuals surviving and the number of offspring produced at different

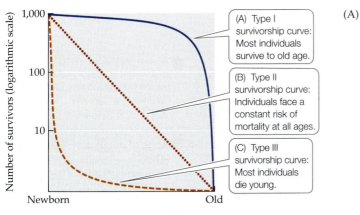

FIGURE 11.17 Three Types of Survivorship Curves
Ecologists recognize three general types of survivorship
curves. Survivorship curves are given for (A) the Dall mountain
sheep, (B) the song thrush, and (C) the acorn barnacle *Balanus
glandula*. Notice that the number of survivors has been plotted
on a logarithmic scale. (A,B after E. S. Deevey. 1947. *Q Rev Biol* 22:
283–314; C after J. H. Connell. 1970. *Ecol Monogr* 40: 49–78.)

? **What percentage of Dall mountain sheep survive
to age 11?**

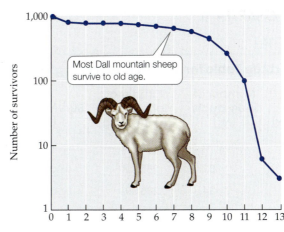

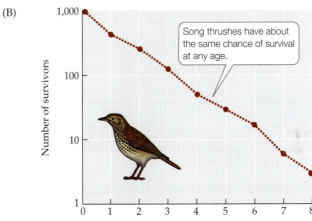

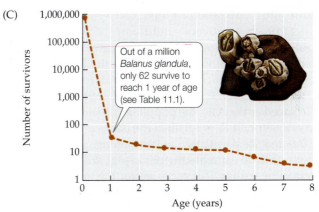

ages (*x*) through time. As the individuals within the cohort
die, N_x decreases from 1 million barnacles to only 2 barnacles after 8 years (see Figure 11.17C). The proportion of
individuals that survive, known as **survivorship (I_x)**, can
be calculated by simply dividing N_x by N_0, the number
of barnacles originally born into the cohort (represented
by age 0). In addition, we can calculate the **fecundity (F_x)**,
or the mean number of barnacle offspring produced per
surviving adult barnacle per age class, by dividing the
total number of offspring ($N_{x\ offspring}$) by the number of
individuals (N_x) that produce those offspring. Multiplying
survivorship (I_x) by fecundity (F_x) gives us the number of
offspring produced for individuals within a particular age
class within the population. The sum of these values for
all the age classes gives us the **net reproductive rate (R_0)**,

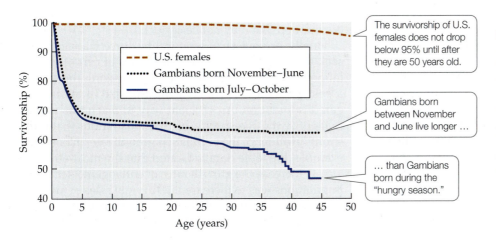

**FIGURE 11.18 Survivorship Varies
among Human Populations** In the
United States, survivorship (I_x) does not
drop greatly until old age. In Gambia, many
people die at much younger ages. (U.S.
data from E. Arias. 2015. National Vital Statistic Reports 64. National Center for Health
Statistics: Hyattsville, MD; Gambia data
from S. E. Moore et al. 1997. *Nature* 388: 434.)

? **The proportion of Gambians born in
the hungry season who live to age 45
is roughly the same as the
proportion of U.S. females who live
to what age (see Table 11.2)?**

TABLE 11.1

Cohort Life Table for Barnacle *Balanus glandula*[a]

Age (x)	Number of individuals (N_x)	Number of offspring (N_x offspring)	Survivorship (l_x)	Fecundity (F_x)	$l_x F_x$	$x\, l_x F_x$
0	1,000,000	0	1	0	0	0
1	62	285,200	0.000062	4,600	0.285	0.285
2	34	295,800	0.000034	8,700	0.296	0.592
3	20	232,000	0.000020	11,600	0.232	0.696
4	15	190,500	0.000015	12,700	0.191	0.764
5	11	139,700	0.000011	12,700	0.140	0.700
6	6	76,200	0,000006	12,700	0.076	0.456
7	2	25,400	0.000002	12,700	0.025	0.175
8	2	25,400	0.000002	12,700	0.025	0.200

1 To estimate the per capita growth rate, r, sum all l_x by F_x values to get the net reproductive rate, R_0.
$R_0 = \text{sum } (l_x F_x) = 1.27$

2 Then, sum all l_x by F_x by x values and divide by R_0 to get the generation time, G.
$G = \text{sum } (x\, l_x F_x)/R_0 = 3.05$

3 Finally, divide the natural log of R_0 by G to get the intrinsic rate of increase, r.
$r = (\ln R_0)/G = 0.08$

Source: Data from J.H. Connell. 1970. Ecol Monogr 40: 49–78.

[a]The life history of barnacles involves releasing larvae into the water column, where they feed and undergo a series of larval stages. Eventually they settle back on the rock and metamorphose into a juvenile barnacle, which grows into a reproductive adult.

which is simply the mean number of offspring produced per individual in the cohort, adjusted for survival:

$$R_0 = \text{sum } (l_x F_x) \qquad (11.8)$$

If R_0 is greater than 1.0, there is a net increase in offspring produced each generation, and assuming the birth and death rates do not change over time, the population should increase exponentially. If R_0 is less than 1.0, and individuals are not replaced as they die, the population declines eventually to extinction. If R_0 is 1.0, then the births and deaths balance out and the population will not change in size.

We can use R_0 to estimate the per capita growth rate, r, of a cohort by scaling R_0 to account for the generation time of the cohort. The **generation time (G)** is the average age of the parents of all the offspring produced within the cohort (see Table 11.1 for equation). To estimate r, we simply divide the natural log (ln) of R_0 by G and get

$$r = (\ln R_0) / G \qquad (11.9)$$

Note that R_0 equals λ when the generation time of the population is equal to 1.

Once we have calculated r from the life table, we can use it (or λ) in our population growth models (Equation 11.2 for geometric growth, Equation 11.4 for exponential growth, or Equation 11.7 for logistic growth) to predict population sizes in the future. In addition, other methods exist to calculate future population growth and size using life table data.

Cohort life tables follow individuals from birth to death as a function of calendar year or life stage (e.g., eggs, larvae, pupae, and adults in insects). This is relatively easy to do if the organisms are easily followed—for example, if they are sessile and short-lived, as we saw in the barnacle example. However, for organisms that are highly mobile or have long life spans (e.g., trees that live much longer than people), it is hard to observe the fate of individuals from birth to death. In some of these cases, a **static life table** can be used, in which the survival and reproduction of individuals of different ages during a single time period are recorded. To construct a static life table, one must be able to estimate the ages of the organisms under observation. Estimating ages is difficult in some species, but for others, reliable indicators of age are known, including annual growth rings in fish scales and tree wood and tooth wear in deer. Once ages have been estimated, age-specific birth rates can be determined by counting how many offspring the individuals of different ages produce. Age-specific survival rates can also be determined from a static life table (see Review Question 3), but only if we assume that survival rates have remained constant during the entire time that the individuals in the population have been alive—an assumption that may not be correct.

Finally, ecologists and natural resource managers can seek to change the birth or death rates of certain populations, with the ultimate goal of decreasing the size of a pest population or increasing the size of an endangered population. An efficient way to reach this goal is to identify the age-specific birth or death rates that most strongly influence the population growth rate. In one such example, life table data indicated that the most effective way to increase the growth rates of endangered sea turtle populations was to increase the survival rates of juvenile and mature turtles—a change from the common practice of protecting newborns (**ECOLOGICAL TOOLKIT 11.1**).

Extensive life table data exist for people

Many economic, sociological, and medical applications rely on human life table data. Life insurance companies, for example, use census data to construct static life tables that provide a snapshot of current survival rates; they use these data to determine the premiums they charge customers of different ages. Let's consider two examples of human life tables, one from the United States, the other from Gambia.

The U.S. Centers for Disease Control and Prevention periodically release reports that provide life table data

---

ECOLOGICAL TOOLKIT 11.1

Estimating Population Growth Rates in a Threatened Species

Loggerhead sea turtles (*Caretta caretta*) are large marine turtles that lay eggs in nests that adult females dig into sandy beaches. Newly hatched baby turtles weigh just 20 g (0.04 pounds) and have a shell length of 4.5 cm (1.8 inches). They reach adulthood after 20–30 years, at which point they can weigh up to 227 kg (500 pounds) and have a shell length of 122 cm (4 feet).

Loggerhead sea turtles have been listed as a threatened species under the U.S. Endangered Species Act since 1978. Many species eat loggerhead eggs or hatchlings, and the juveniles and adults are eaten by large marine predators such as tiger sharks and killer whales. Loggerheads also face threats from people, including the destruction of nesting sites by development, as well as commercial fisheries (in whose nets sea turtles can become trapped and drown).

Early efforts to protect loggerhead sea turtles focused on the egg and hatchling stages, which suffer extensive mortality and are relatively easy to protect. To evaluate this approach, Crouse et al. (1987) and Crowder et al. (1994) used life table data to determine how the existing exponential growth rate of $r = -0.05$ would change if new management practices improved the survival rates of turtles of various ages (**FIGURE A**). Their findings suggested that even if hatchling survival rates were increased by 90%, loggerhead populations would continue to decline. Instead, they found that the population growth rate was most responsive to increasing the survival rates of older juveniles and adults.

The results obtained by Crouse, Crowder, and their colleagues prompted the enactment of laws requiring turtle excluder devices (TEDs) to be installed in shrimp nets (**FIGURE B**). A TED functions as a hatch through which juvenile and adult sea turtles can escape when caught in a net. Shrimp nets were singled out because the data suggested that shrimping accounted for more loggerhead deaths (5,000–50,000 deaths per year) than all other human activities combined.

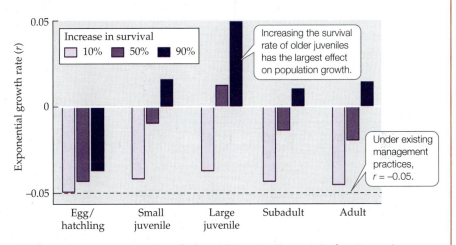

FIGURE A **Management Practices and Sea Turtle Population Growth Rates** Researchers used life table data to identify the age-specific death rates that most strongly influenced the population growth rate of loggerhead sea turtles. (After L. B. Crowder et al. 1994. *Ecol Appl* 4: 437–445.)

FIGURE B Turtle Excluder Device (TED)

Loggerheads are most easily counted when they nest, yet it takes 20–30 years for turtles to become sexually mature. As a result, it will be decades before we know whether TED regulations help turtle populations to increase in size. But early results are encouraging: the number of turtles killed in nets dropped substantially (up to 94%) after the TED regulations were implemented (Finkbeiner et al. 2011).

for people in the United States. Reports released in 2009 and 2015 provide information on the survivorship (l_x), fecundity (F_x), and life expectancy (expected number of years of life remaining) of U.S. females of different ages (**TABLE 11.2**). To make their interpretation easier, such data can be graphed, as in Figure 11.18, which plots l_x data for U.S. females. This curve shows that survival probabilities for U.S. females remain high for many years; in fact, as Table 11.2 reveals, these survival probabilities do not begin to drop sharply until around age 70.

The data from the United States are in stark contrast to data from Gambia, a country located on the west coast of Africa. Moore et al. (1997) analyzed birth and death records for 3,102 people born in three Gambian villages between 1949 and 1994. They found that the season of birth had long-term effects: individuals born during the "hungry

season" (July–October, when food stored from the previous year is in low supply) had lower survivorship as adults than did individuals born at other times of the year (see Figure 11.18). Their data also reveal large differences between the survivorship of people in Gambia and in the United States. For example, only 47% to 62% of Gambians (depending on their season of birth) survived to reach age 45, whereas 97% of U.S. females survived to that age.

A CASE STUDY REVISITED

Human Population Growth

Media reports often state that the human population is growing exponentially. As we saw in Figure 11.4, a simple way to determine whether a population is growing exponentially is to plot the natural logarithm of population size versus time. If a straight line results, the population is growing exponentially. When we plot the natural logarithm of human population size versus time for the last 2,000 years, however, we see that our population sizes deviate considerably from the straight line expected in exponential growth (**FIGURE 11.19**). In fact, as fast as exponential growth is, historically the human population has increased even more rapidly than that.

TABLE 11.2

Survivorship, Fecundity, and Years of Life Remaining by Age for U.S. Females

Age (yr), x	Survivorship, l_x	Fecundity, F_x	Expected no. of years of life remaining (at age x)
0	1.0	0.0	81.8
1	0.994	0.0	80.5
5	0.994	0.0	76.6
10	0.993	0.0	71.6
15	0.992	0.004	66.7
20	0.991	0.203	61.7
25	0.989	0.511	56.9
30	0.986	0.578	52.0
35	0.982	0.479	47.2
40	0.977	0.232	42.4
45	0.970	0.046	37.8
50	0.958	0.003	33.2
55	0.940	0.0	28.8
60	0.915	0.0	24.5
65	0.880	0.0	20.3
70	0.828	0.0	16.5
75	0.752	0.0	12.9
80	0.640	0.0	9.6
85	0.485	0.0	6.9
90	0.292	0.0	4.8
95	0.119	0.0	3.3
100	0.027	0.0	2.3

Sources: J. A. Martin et al. 2009. National Vital Statistics Reports 57. National Center for Health Statistics: Hyattsville, MD; E. Arias. 2015. National Vital Statistics Reports 64. National Center for Health Statistics: Hyattsville, MD.

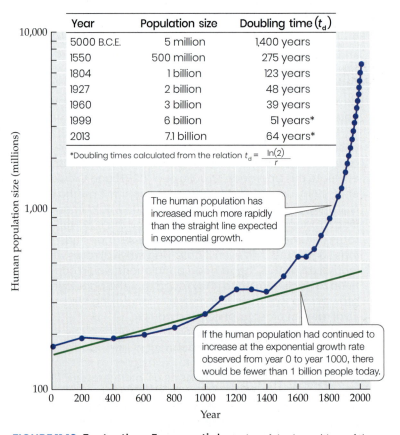

Year	Population size	Doubling time (t_d)
5000 B.C.E.	5 million	1,400 years
1550	500 million	275 years
1804	1 billion	123 years
1927	2 billion	48 years
1960	3 billion	39 years
1999	6 billion	51 years*
2013	7.1 billion	64 years*

*Doubling times calculated from the relation $t_d = \dfrac{\ln(2)}{r}$

The human population has increased much more rapidly than the straight line expected in exponential growth.

If the human population had continued to increase at the exponential growth rate observed from year 0 to year 1000, there would be fewer than 1 billion people today.

FIGURE 11.19 Faster than Exponential A plot of the logarithm of the human population size over the last 2,000 years differs dramatically from the straight line expected if it were growing exponentially.

The faster-than-exponential nature of human population growth is also evident from historical doubling times for the human population. Recall that in a population that grows exponentially, the doubling time remains constant. However, as shown in the inset of Figure 11.19, the doubling times observed for the human population dropped from roughly 1,400 years in 5000 B.C.E. to a mere 39 years in 1960—again indicating that historically, our population has increased more rapidly than expected of exponential growth.

Projecting into the future, we can predict how long it will take our population to double in size at current rates of growth. To do this, the doubling time is estimated from the relation $t_d = \ln(2)/r$ (see Equation 11.5), where r is the current growth rate of the human population. Such estimates have shown that the human population was growing most rapidly in the early 1960s, with a doubling time of 32 years. Since then, the doubling time has increased (because r has decreased), reaching 69 years in 2022.

The increase in the doubling time (and the decrease in r) over the past 5 decades indicates that the human population is now growing more slowly than expected in exponential growth. So, returning to the question we asked in the Case Study (whether there would be 14 billion people in 2080), the answer is probably not. U.S. Census Bureau projections indicate that population growth rates are likely to continue to fall over the next 40 years (**FIGURE 11.20**), leading to a predicted population size of 9.6 billion in 2050 (**FIGURE 11.21**). Extending that curve out to 2080 suggests that there will be roughly 10 billion people in that year. If these projections turn out to be correct, or nearly so, what will the future hold with that many people? Is 10 billion above the carrying capacity of the human population?

To answer these questions, we must determine the carrying capacity of the human population, but that is trickier than it may at first appear. Many researchers have estimated the human carrying capacity, obtaining values that range from fewer than 1 billion to more than 1,000 billion (see Cohen 1995). This large variation is due in part to the fact that many different methods—from logistic models to calculations based on crop production and human energy requirements—have been used. In addition, different researchers have made different assumptions about how people would live and how technology would influence our future, assumptions that have a large effect on the estimated carrying capacity.

For example, by using the ecological footprint approach described in this chapter's Connections in Nature, it has been estimated that Earth could support 1.5 billion people indefinitely if everyone used the amount of resources used by people in the United States in 2007 (Ewing et al. 2010). On the other hand, if everyone used the amount of resources used by people in India in 2007, the world could support over 13 billion people. Thus, as we suggested in this chapter's Case Study, issues of

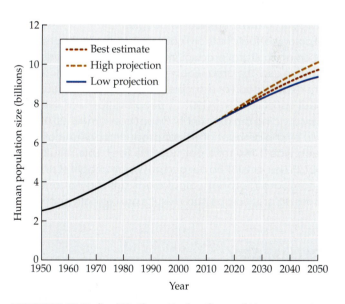

FIGURE 11.21 United Nations Projections of Human Population Size The human population is expected to increase to 9.7 billion by 2050; low and high projections range from 8.3 billion to 10.9 billion. (From United Nations, Department of Economic and Social Affairs, Population Division. 2019. *World Population Prospects 2019: Data Booklet* [ST/ESA/SER.A /424].)

❓ Using the best-estimate curve shown here and the annual growth rate estimated for the human population in 2050 (see Figure 11.20), approximately how large will our population be in 2051?

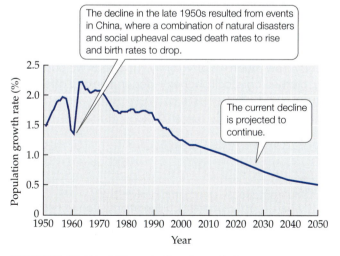

FIGURE 11.20 World Population Growth Rates Are Dropping Annual world population growth rates have declined since the early 1960s. (Data from U.S. Census Bureau, International Data Base, June 2011 update.)

❓ In 2050, will the human population still be increasing in size? Explain.

human population size and resource use are linked inextricably: more people means that more resources will be used, but the degree to which our growing population affects the environment depends on the amount of resources used by each person.

 CONNECTIONS *in* NATURE

YOUR ECOLOGICAL FOOTPRINT   When you turn on a light, purchase an appliance, drive a car, or eat fruit imported from another country, you may not think about the effects your actions have on the natural world. How, for example, does driving to the store to get groceries affect forests or coral reefs?

To answer this question, we must account for the resources required to support our actions. The grains we eat require farmland; the wood products we use require natural forests or plantations; the fish we eat require productive fishing grounds; the machines and appliances we purchase require raw materials and energy to build, as well as energy for their operation. Ultimately, every aspect of our economy depends on the land and waters of Earth. Recognizing this, William Rees proposed that we measure the environmental impact of a population as its **ecological footprint**, which is the total area of productive ecosystems required to support that population (Rees 1992). The ecological footprint approach turns the carrying capacity concept on its head: instead of asking how many people a given area can support, it asks how much area is required to support a given number of people.

Ecological footprints are calculated from national statistics on agricultural productivity, production of goods, and resource use. The area required to support these activities is then estimated. For example, the land required to support wheat consumption in 1993 by people in Italy was estimated by dividing the amount of wheat consumed (26,087,912 tons) by the amount of wheat produced per unit of land, which was 2.744 tons per hectare (ha). This resulted in 9,507,257 ha, or 0.167 ha per person (Wackernagel et al. 1999). To compare footprint calculations among nations and across different crops, such results are typically converted to *global hectares*, where a global hectare is defined as a hectare of world-average biological productivity (Kitzes and Wackernagel 2009).

Methods of calculating ecological footprints are still being refined, but results to date are sobering. In 2007, there were 11.9 billion global hectares of productive land available, and the ecological footprint of an average person was 2.7 global hectares (Ewing et al. 2010). These results suggest that Earth could have supported 4.4 billion people (11.9 billion ha/2.7 ha per person) for a long time. In fact, the human population in 2007 was 6.7 billion, more than a 50% overshoot of its carrying capacity. An overshoot of this magnitude indicates that in 2007, environmental resources were being used more rapidly than they could be regenerated, a pattern of use that cannot be sustained.

Similar calculations can be made for nations, cities, and even companies, schools, or individuals (see Review Question 4). In the United States, for example, the average ecological footprint was 8.0 global hectares per person in 2007. Since there were 1,330 million global hectares of productive land available in the nation, this calculation suggests that the carrying capacity of the United States in 2007 was 166 million people (1,330 million ha/8.0 ha per person); the actual population was 309 million, nearly double the carrying capacity.

Human use of resources changes from year to year, depending on population size, per capita rates of resource use, and technology (i.e., the efficiency of production). In addition, the total area of productive ecosystems available to support our activities changes over time because of factors such as gain or loss of farmland, destruction of natural habitat, pollution, and extinctions of species. As a result, our ecological footprint changes over time. People have now begun to use our changing footprint as a way to assess whether our current population size and resource use can be sustained. This approach highlights the fact that all of our actions—what we eat, how big a house we buy, how much we drive or fly, the goods we purchase (e.g., clothes, cars, cell phones)—depend on and affect the natural world.

SUMMARY

CONCEPT 11.1 Populations can grow exponentially when conditions are favorable, but exponential growth cannot continue indefinitely.

11.1.1 Define geometric population growth.

- Geometric growth occurs when a population with synchronous reproduction changes in size by a constant proportion from one discrete time period to the next.

- Mathematically, geometric growth is $N_{t+1} = \lambda N_t$, where N_t is the population size at time t and λ is the geometric population growth rate, a constant whose value is determined by per capita birth and death rates.

11.1.2 Define exponential population growth.

- Exponential growth occurs when a population with continuous reproduction changes in size by a constant proportion at each instant in time.

- Mathematically, population growth is $dN/dt = rN$, where N is the population size and r is the exponential population growth rate, a constant whose value is determined by instantaneous per capita birth and death rates.

11.1.3 Describe the characteristics of geometric and exponential growth.

- Populations undergoing geometric and exponential growth have the potential to increase rapidly in size because they grow by multiplication, not by addition.

- Similar to the interest on a savings account, populations undergoing geometric and exponential growth can grow rapidly even when the growth rate is low.

- All populations experience limits to growth, which ensure that geometric and exponential growth cannot continue indefinitely.

CONCEPT 11.2 Population size is determined by a combination of density-dependent and density-independent factors.

11.2.1 Define density-independent factors and describe how they affect population size and growth rate.

- Density-independent factors affect population size and growth rate independent of population density and include abiotic factors such as weather and climate and biotic factors such as hunting.

- In many species, density-independent factors play a major role in determining changes in population size.

11.2.2 Define density-dependent factors and describe how they affect population size and growth rate.

- Density-dependent factors affect population size and growth rate as a consequence of population density and include factors such as food or habitat.

- When population size becomes large enough, a lack of food, habitat, or other resources causes birth rates to decrease, death rates to increase, or dispersal to increase.

CONCEPT 11.3 The logistic equation incorporates limits to growth and shows how a population may stabilize at a maximum size, the carrying capacity.

11.3.1 Define logistic population growth and compare to exponential population growth.

- Logistic growth occurs when population density slows exponential population growth, stabilizing that growth at a maximum level, known as the carrying capacity (K) or the density at which the population stops increasing in size.

- Mathematically, the logistic growth equation ($dN/dt = rN (1 - N/K)$) is a modification of the exponential growth equation ($dN/dt = rN$), where the term ($1 - N/K$) represents the fraction of the carrying capacity that is available for population growth.

- Logistic growth is similar to, but slightly slower than, exponential growth when densities are low, but as densities increase, population growth slows, and eventually becomes zero, as it reaches K.

11.3.2 Describe the growth patterns of the U.S. population.

- Logistic population growth provides a close fit to the size of the U.S. population up to 1950; since that time, the growth rate of the U.S. population has been greater than expected in logistic growth and has not reached a carrying capacity.

CONCEPT 11.4 Life tables show how survival and reproduction vary with age or size structure, influencing population growth and size.

11.4.1 Justify the use of life tables to determine population growth and size.

- Birth and death rates vary with the age, size, or life stage of individuals within a population.

- Life tables provide a summary of how survival and reproductive rates vary with the age, size, or life stage of the individuals, allowing more accurate determinations of population growth rates and sizes.

11.4.2 Describe how age or size structure influences population growth and population size.

- Age or size structure influences the survival and reproductive rates of populations, thus affecting population growth and size.

- If one population has many older individuals, while the other has many younger individuals, the second population will likely grow faster because it contains more individuals of reproductive age and fewer individuals at the end of their life span.

11.4.3 Compare the three types of survivorship curves.

- In a population with a type I survivorship curve, most individuals survive to old age; death rates do not begin to increase until old age.

- In a population with a type II survivorship curve, individuals experience a constant chance of surviving from one age to the next throughout their lives.

- In a population with a type III survivorship curve—the most common type in nature—death rates are very high for young individuals, but adults survive well later in life.

11.4.4 Analyze life table data and calculate a net reproductive rate (R_0) and exponential growth rate (r).

- Cohort life tables can be constructed from data on the fates of individuals born during the same time period (age class) and used to calculate age-specific survivorship (l_x) and fecundity (F_x).

- Multiplying survivorship (l_x) by fecundity (F_x) and summing these values for all the age classes gives the net reproductive rate, R_0, or the mean number of offspring produced per individual, adjusted for survival.

- The per capita growth rate, r, can be estimated by dividing $\ln R_0$ by the generation time ($G = $ (sum $(x l_x F_x)/R_0$)).

- In highly mobile or long-lived organisms, a static life table may be constructed from data on the survival and fecundity of individuals of different ages during a single time period.

Review Questions

1. A population of insects triples every year. Initially, there were 40 insects.

   a. How many insects will there be after 4 years?

   b. How many insects will there be after 27 years? (Write your answer to this question as an equation.)

   c. The habitat of the insect is degraded such that the population growth rate (λ) changes from 3.0 to 0.75. If there were 100 insects in the population when its habitat became degraded, how many insects will there be after 3 years?

2. What is the distinction between factors that regulate population size and factors that determine population size?

3. For a field ecology project, you count the number of individuals of different ages found in a population during a single time period. There are 100 newborns, 40 1-year-olds, 15 2-year-olds, 5 3-year-olds, and 0 4-year-olds. These individuals produced the following number of offspring: newborns = 0 offspring, 1-year-olds = 100 offspring, 2-year-olds = 30 offspring, 3-year-olds = 25 offspring, 4-year-olds = 0 offspring.

   a. Use these data to create a static life table and calculate R_0 and r.

   b. Explain the difference between a static life table and a cohort life table.

4. Calculate your ecological footprint at https://www.footprintcalculator.org.

Hone Your Problem-Solving Skills

As discussed in this chapter, life table data can be used to estimate a population's growth rate (r). Here we'll calculate r and future population size using life table data collected for the grass *Poa annua*. These data were collected by marking 843 naturally germinating seedlings and then following their fates over time.

Age (x)	Number of individuals (N_x)	Number of offspring $(N_{x\ offspring})$	Survivorship (l_x)	Fecundity (F_x)	$l_x F_x$	$x l_x F_x$
0	843	0				
1	722	216,600				
2	527	326,740				
3	316	135,880				
4	144	30,240				
5	54	3,240				
6	15	450				
7	3	30				
8	0	0				

Source: Data in M. Begon et al. 1996. *Population Ecology: A Unified Study of Animals and Plants*, 3rd ed. Blackwell Science: Oxford. Adapted from R. Law. 1975. Unpublished PhD thesis. University of Liverpool.

1. Use the life table above and the method in Table 11.1 to fill in the columns and calculate R_0 and r.

2. Using the exponential growth model, calculate the population size of the grass species 10 years from now assuming $N_0 = 100$ individuals.

3. Now, using a logistic growth model, calculate the population size of the grass species 10 years from now assuming $N_0 = 100$ individuals and $K = 1,000,000$.

4. Comparing the population sizes after 10 years, does the grass population reach its carrying capacity?

ADDITIONAL RESOURCES

This textbook provides resources such as **Web Extensions** and additional **Climate Change Connections** for instructors who can share them with students. Please contact your instructor, if you would like to access that content.

Species Interactions

Predation

Snowshoe Hare Cycles: A Case Study

In 1899, a fur trader in northern Ontario reported to the Hudson's Bay Company that "Indians are bringing poor hunts. They have been starving all spring. Rabbits being scarce" (Winterhalder 1980). The "hunts" referred to were pelts of beavers and other fur-bearing animals trapped by members of the Ojibwa tribe, and the "rabbits" were actually snowshoe hares (*Lepus americanus*; **FIGURE 12.1**). Collectively, 200 years of such reports show that hare populations increased and decreased regularly. When hares were abundant, the Ojibwa had enough food to spend time trapping for pelts, which they then traded to the Hudson's Bay Company. But when hares were scarce, tribal members concentrated on gathering food, rather than trapping animals that provided pelts, but little meat.

Beginning in the early 1900s, wildlife biologists used the careful records of the Hudson's Bay Company to estimate abundances of snowshoe hares and their Canada lynx predators (*Lynx canadensis*). Both species exhibited regular population cycles, with abundances peaking about every 10 years and then falling to low levels (**FIGURE 12.2A**). Snowshoe hares constitute a major portion of the lynx diet, so it was not surprising that numbers of lynx should rise and fall with numbers of hares. But what drove the cyclic fluctuations in the hare population? Adding to the mystery, hare population sizes rose and fell in synchrony across broad regions of the Canadian forest, so explanations of hare cycles had to account for the large-scale synchrony as well.

One approach to finding the factors important to hare population cycles is to document the changes in birth, death, and dispersal rates that are associated with increasing or declining numbers of hares. Dispersal plays a relatively small role: it may alter local population sizes, but hares do not move far enough to account for the simultaneous changes in their abundance seen across broad geographic regions. In contrast, consistent patterns of birth and death rates have been found across different regions of Canada. Snowshoe hares can raise up to three or four

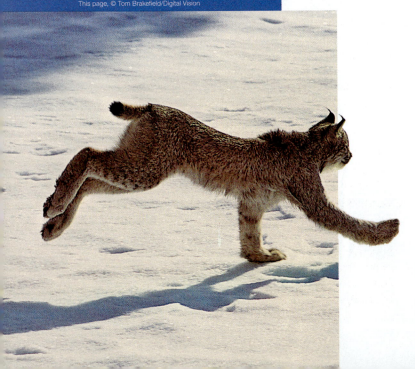

FIGURE 12.1 Predator and Prey
A snowshoe hare (*Lepus americanus*) flees from its specialist predator, the Canada lynx (*Lynx canadensis*).

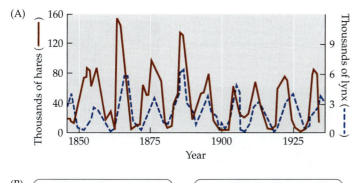

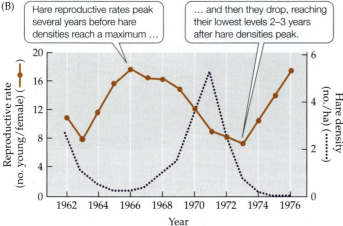

FIGURE 12.2 Hare Population Cycles and Reproductive Rates (A) Historical trapping data from the Hudson's Bay Company indicate that numbers of both hares and lynx fluctuated in a 10-year cycle. (B) The highest hare reproductive rates do not coincide with the highest hare densities. (A after D. A. MacLulich. 1936. *J R Astron Soc Can* 30: 233–246; B after J. R. Cary and L. B. Keith. 1979. *Can J Zool* 57: 375–390.)

❓ In (A), does the peak abundance of one species typically occur after the peak abundance of the other species? Describe the observed pattern and hypothesize why it might occur.

litters over the summer, with an average of five young per litter. Hare reproductive rates reach their highest levels (ca. 18 young per female) several years before hare density reaches a maximum. Reproductive rates then begin to fall, reaching their lowest levels 2–3 years after hare density peaks (**FIGURE 12.2B**). Hare survival rates show a similar pattern: they are highest several years before hare density peaks; then they fall and do not rise again until several years after hare density peaks.

Together, the changes over time in hare birth and survival rates drive the hare population cycle. But what causes these rates to change? Several hypotheses have been proposed, one of which focuses on food supplies. Large numbers of hares consume prodigious amounts of vegetation, and studies have shown that food can be limiting at peak hare densities (up to 2,300 hares/km^2). Two observations, however, indicate that food alone does not drive the hare cycle: first, some declining hare populations do not lack food, and second, the experimental addition of high-quality food does not prevent hare populations from cycling.

A second hypothesis focuses on predation. Many hares (up to 95% of those that die) are killed by predators such as lynx, coyotes, and birds of prey. In addition, lynx and coyotes kill more hares per day during the peak and decline phases of the hare cycle than during the increase phase. But questions remain. The killing of hares by predators explains the drop in survival rates as hare numbers decline, but by itself it does not explain (1) why hare birth rates drop during the decline phase of the cycle or (2) why hare numbers sometimes rebound slowly after predator numbers plummet. Nor does it explain other observations, such as why the physical condition of hares worsens as hares decrease in numbers. What other factors are at work?

Introduction

One certainty for all species on Earth is that they interact with other species. These interactions not only affect individuals, populations, and communities but often lead to evolutionary change. In this unit, we will focus on two species interactions, such as one species eating another or two species competing for the same limiting resources. In the next unit, we will consider the more complicated but real-life situation in which species interact with multiple species as part of an integrated community.

Ecologists characterize species interactions based on the effects species have on each other—positive (+), negative (–), or neutral (0)—and whether that effect is **trophic** (feeding) or **non-trophic** (competitive and facilitative) (**FIGURE 12.3**). In addition, species interactions may include a **symbiosis**, a positive or negative interaction in which one species lives in close physical and/or physiological contact. Finally, species interactions vary in their strength, a characteristic that we will consider in more detail in Concepts 14.1 and 16.3. Four broad categories into which species interactions fall are predation, competition, positive interactions, and amensalism.

1. **Predation** is a trophic interaction in which an individual of one species (a **predator**) kills and/or consumes individuals (or parts of individuals) of another species (its **prey**). Predation includes **carnivory**, in which the predator and prey are both animals; **herbivory**, in which the predator is an animal and the prey is a plant or algae; and **parasitism**, in which the predator (a **parasite**) lives symbiotically on or in the prey (its **host**) and consumes only certain tissues without necessarily killing the host. Some parasites are pathogens that cause disease in their hosts.

2. **Competition** is a non-trophic interaction in which two or more species overlap in the use of at least some of the same required limiting resources, negatively affecting their growth, reproduction, and/or survival. The limiting resource need not be food; species may compete for water, space, nesting sites, or sunlight.

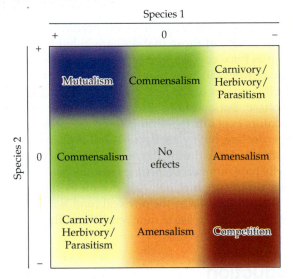

FIGURE 12.3 Types of Two Species Interactions
Interactions between two species can be grouped based on whether the effect of the interacting species (here denoted as species 1 and species 2) is negative (−), positive (+), or neutral (0). The strength of the two species' interaction increases from the center (no effect) to the corners. (After E. G. Pringle. 2016. *PLOS Biol* 14: e2000891.)

3. **Positive interactions** (also known as **facilitations**) are trophic or non-trophic interactions in which at least one species benefits from the interaction and none are harmed. Positive interactions include **mutualisms**, in which both species benefit from the interaction, sometimes in a highly dependent and symbiotic manner. **Commensalism** is a type of positive interaction in which one participant benefits but the other is unaffected; it includes a wide range of interactions that typically encompass the provisioning of resources that improve conditions for the commensal organism.

4. **Amensalisms** occur when one participant is harmed but the other is not affected. An example is a herd of animals trampling vegetation; the animals are not affected by this event, but the plants clearly are. Another example includes small understory plants that grow under trees. The understory plants are deprived of light but likely have no effect on the trees.

Over the course of this and the next chapter, we will consider the three broad categories of predation given above: carnivory, herbivory, and parasitism. These broad categories seem simple, and it is easy to think of examples: a lion that kills and eats a zebra, an insect that eats a plant leaf, a tapeworm that robs a dog of nutrients in its digestive tract. But the natural world defies such simple categorization. Consider those prototypical herbivores, sheep: they get most of their food from plants, but they have also been known to eat the helpless young of ground-nesting birds. Conversely, carnivores can act like herbivores: wolves, for example, will eat berries, nuts, and leaves. And some

FIGURE 12.4 Are Parasitoids Carnivores or Parasites?
Parasitoids such as the wasp *Aphidius colemani*, shown here depositing an egg into an aphid, can be considered unusual carnivores because during their lifetime they eat and slowly kill only one prey individual. Parasitoids can also be viewed as unusual parasites that eat all or most of their host, thereby killing it.

organisms do not fit neatly into any category. **Parasitoids** are insects that typically lay one or a few eggs on or in another insect (the host) (**FIGURE 12.4**). After they hatch from their eggs, the parasitoid larvae remain with the host, which they eat and usually kill. Parasitoids can be considered unusual parasites (because they consume most or all of their host, almost always killing it) or unusual carnivores (because over the course of their lives they eat only one individual, killing it slowly).

Despite these and other complications, we will approach the rich variety of trophic interactions in two chapters: this chapter will cover carnivory and herbivory, and Chapter 13 will focus on parasitism. We will begin by exploring some aspects of carnivores and herbivores that define and characterize their dietary preferences.

CONCEPT 12.1 Most carnivores have broad diets, whereas a majority of herbivores have relatively narrow diets.

LEARNING OBJECTIVES

12.1.1 Understand why carnivores tend to be generalists and have broad animal prey diets.

12.1.2 Know why herbivores tend to be specialists and have narrow plant prey diets.

Carnivore and Herbivore Dietary Preferences

Although they share some similarities, carnivores and herbivores differ from each other in many aspects. The most obvious difference is that carnivores invariably kill their prey (it is hard to eat only part of an animal without killing it), while herbivores usually do not kill the plants they eat, at least not immediately. Another difference is that animal prey can usually move away or hide from their predators but most plant prey cannot. Finally, even though plant prey are often more abundant, their body tissues have much lower nitrogen content, and thus are less nutritious, than animal prey (**FIGURE 12.5**). These three factors have important consequences for the dietary preferences of carnivores versus herbivores.

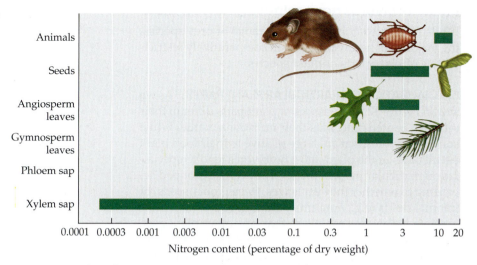

FIGURE 12.5 The Nitrogen Contents of Plants and Animals Differ Nitrogen is an essential component of any animal's diet. Body tissues of animals have much higher nitrogen content than those of plants. Of plant tissues, leaves tend to have the highest nitrogen content of any plant parts other than seeds. (After W. J. Mattson, Jr. 1980. *Ann Rev Ecol Syst* 11: 119–161.)

As we learned in Concept 8.2, optimal foraging and dietary preferences are thought to be dependent on two factors: (1) encounter rate, a function of search time, or the time it takes to search for and find prey, and (2) handling time, or the time it takes to subdue and consume the prey. If the encounter rate for prey is low, as would be the case for predators searching for mobile animal prey, then the prediction is that the predators should not be too narrow in their prey choices. As a consequence, these predators (carnivores) should be *generalists* and have fairly broad diets. On the other hand, if prey are relatively easy to search for but their handling times are longer, as is the case for immobile but less nutritious plants, then the predators (herbivores) should be *specialists* and have narrow diets. Let's consider these predictions in more detail below.

Many carnivores have broad diets

Most carnivores eat prey in relation to their availability without showing a preference for any particular prey species. This lack of preference is likely a result of a generalist strategy. A predator can be said to show a preference for a particular prey species if it eats that species more often than would be expected based on that prey's availability.

Some carnivores do show a strong preference for certain prey species. Lynx and coyotes, for example, eat more hares than would be expected based on their availability; even when hares constitute only 20% of the available food, they constitute 60%–80% of the diet of lynx and coyotes.

Some carnivores concentrate their foraging on whatever prey is most plentiful. When researchers provided guppies with two kinds of prey, fruit flies (floating on the water surface) and tubificids (aquatic worms found on the bottom), the guppies ate disproportionate amounts of whichever prey was most abundant (**FIGURE 12.6**). Predators like these guppies that focus on abundant prey tend to "switch" from one prey species to another. Such switching may occur because the predator forms a search image of the most common prey type and hence tends to orient toward that prey, or because learning enables it to become increasingly efficient at capturing the most common prey type. As we saw in Concept 8.2, in some cases predators switch from one type of prey to another in a manner consistent with the predictions of optimal foraging theory.

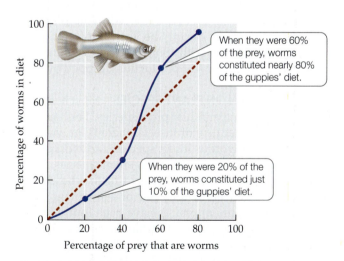

When they were 60% of the prey, worms constituted nearly 80% of the guppies' diet.

When they were 20% of the prey, worms constituted just 10% of the guppies' diet.

FIGURE 12.6 A Predator That Switches to the Most Abundant Prey Guppies focused their foraging efforts on whichever prey species was most common in their habitat: aquatic worms (tubificids) or fruit flies. The dashed red line indicates the results that would have been expected if the guppies had captured worms according to their availability instead of switching to whichever prey species was more abundant (solid blue line). (After W. W. Murdoch et al. 1975. *Ecology* 56: 1094–1105.)

Most herbivores have relatively narrow diets

While most predators eat a broad range of prey species, the majority of herbivores feed on comparatively restricted sets of plant parts or plant species.

SPECIALIZATION ON PARTICULAR PLANT PARTS As we saw in Figure 12.5, the tissues of plant parts differ in their nitrogen content and thus their nutritional value. While some herbivores that are large relative to their food plant eat all parts of the plant, most herbivores tend to specialize on particular parts of plants. They can be grouped according to whether they eat leaves, stems, roots, seeds, or internal fluids (e.g., nutrient-containing sap).

More herbivores eat leaves than any other plant part. Leaves are abundant, and they are available year-round in many places; leaves are also more nutritious than other plant parts (except for seeds) (see Figure 12.5). Herbivores that eat leaves range from large browsers, such as deer or giraffes, to grasshoppers and herbivorous fishes, to tiny "leaf miners" such as fly larvae that enter a leaf and eat it from the inside. By removing photosynthetic tissues, leaf-eating herbivores can reduce the growth, survival, or reproduction of their food plants.

Belowground herbivory can also have major effects on plants, as illustrated by the 40% reduction in growth observed in bush lupine plants after 3 months of herbivory by caterpillars of the root-killing ghost moth (*Hepialus californicus*). Similarly, herbivores that eat seeds can have large effects on plant reproductive success, sometimes reducing it to zero. The effects of herbivores that feed on internal fluids are not always obvious (because visible plant parts are not removed), but they too can be considerable. For example, Dixon (1971) showed that although the lime aphid (*Eucallipterus tiliae*) did not reduce aboveground growth in lime trees during the year of infestation, the roots of trees infested with aphids did not grow that year, and a year later, their leaf production dropped by 40%.

SPECIALIZATION ON PLANT SPECIES Most herbivores also specialize on particular plant species. This statement is true largely because of insects: there is an enormous number of herbivorous insect species, and most of them live on and eat only one (or a few) plant species. For example, most species of agromyzid flies, whose larvae are leaf miners, feed on only one or a few plant species (**FIGURE 12.7**). Similar results have been found for leaf-feeding beetles in the genus Blepharida: among 37 species of these beetles, 25 feed on a single plant species, 10 feed on 2–4 plant species, and only 2 feed on a relatively broad suite of plants (12–14 species) (Becerra 2007).

There are numerous examples of herbivores that eat many plant species, however. Grasshoppers, for example, graze on a broad range of plant species, and even among the leaf miners in Figure 12.7, several species eat more than ten different plants. Large browsers, such as deer, often switch from one tree or shrub species to another; in

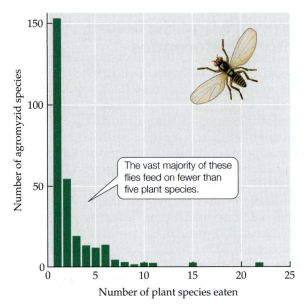

FIGURE 12.7 Most Agromyzid Flies Have Narrow Diets
The larvae of agromyzid flies are leaf miners that live inside leaves and feed on leaf tissue. (After K. A. Spencer. 1972. *Handbooks for the Identification of British Insects, Diptera, Agromyzidae*, Vol. X, Part 5G. Royal Entomological Society: London.)

❓ Using the data in the graph, make a rough estimate of the percentage of agromyzid fly species that feed on fewer than five host plant species.

addition, they eat all or most of the aboveground parts of many herbaceous plant species. The golden apple snail (*Pomacea canaliculata*) is a voracious generalist herbivore, capable of removing all the large plants from wetlands; the snail then survives by eating algae and detritus.

Now that we have considered diet preferences, we'll next focus on adaptations of predators and prey to either obtain food or avoid being eaten.

> **CONCEPT 12.2** Predation results in a wide range of capture and avoidance mechanisms.

LEARNING OBJECTIVES

12.2.1 Know the search and capture mechanisms of carnivores and the ways animal prey avoid being eaten.

12.2.2 Describe how plants avoid being eaten and how herbivores can overcome these defenses.

Mechanisms Important to Predation

Among the other challenges they face, all animals must obtain food while striving to avoid being eaten themselves. As we have seen in Concepts 5.4 and 8.2, this ongoing drama has resulted in the evolution of a dazzling array of morphological and behavioral mechanisms in both predator and prey. Let's consider some of them here.

Some carnivores move to search for and capture prey, while others sit and wait

Many carnivores forage throughout their habitat, moving about in search of prey. Examples of species that hunt in this way include wolves, sharks, and hawks. Other carnivores remain in one place and attack prey that move within striking distance (as do moray eels and some snakes, such as mambas and vipers) or enter a trap (such as a spider's web or the modified leaf of a carnivorous plant).

Many carnivores have unusual physical features that help them capture prey. The body form of the cheetah, for example, enables great bursts of speed that allow it to catch gazelles and other rapidly fleeing prey. In another example, most snakes can swallow prey that are considerably larger than their heads (**FIGURE 12.8**). Unlike those of other terrestrial vertebrates, the bones of a snake's skull are not rigidly attached to one another. This unique feature allows the snake to open its jaws to a seemingly impossible extent. Curved teeth mounted on bones that can move inward then help to pull prey items down the throat. A person with similar adaptations would be able to swallow a watermelon whole.

While some carnivores depend primarily on their physical structure, others subdue prey with poison (e.g., venomous spiders). Still others use mimicry: ambush bugs, scorpionfishes, and many other predators blend into their environment so well that prey may be unaware of their presence until it is too late. Some predators have inducible traits that improve their ability to feed on specific prey species. The predatory ciliate protist *Lembadion bullinum* has such an inducible offense: individuals gradually adjust their size to match the size of the available prey. Thus, if a ciliate is small but the available prey in its environment are large, the ciliate increases in size. Similarly, if a ciliate is large but available prey are small,

the ciliate decreases in size. Finally, some predators can detoxify or tolerate prey chemical defenses, as the following example shows.

The garter snake (*Thamnophis sirtalis*) is the only predator known to eat the toxic rough-skinned newt (*Taricha granulosa*). In some of its populations, the skin of this newt contains large amounts of tetrodotoxin (TTX), an extremely potent neurotoxin. TTX binds to sodium channels in nerve and muscle tissue, thus preventing nerve signal transmission and causing paralysis and death. A single newt can contain enough TTX to kill 25,000 mice—far more than enough to kill a person, as was tragically demonstrated in 1979 when a 29-year-old man died after eating a rough-skinned newt on a dare.

The garter snakes in some populations, however, can eat rough-skinned newts because they can tolerate TTX. These snakes have TTX-resistant sodium channels (Geffeney et al. 2005). Although these garter snakes are protected from the lethal effects of TTX, those individuals that can tolerate the highest concentrations of TTX move more slowly than less resistant individuals—a trade-off between tolerance for the poison and speed of locomotion. In addition, once they swallow a poisonous newt, the snakes are immobilized for up to 7 hours. During that time, the snakes are vulnerable to predation themselves and may also suffer from heat stress.

Escaping carnivores: Physical defenses, toxins, mimicry, and behavior

Many prey species have physical features that reduce their chances of being killed by predators. Such physical defenses include large size (e.g., elephants), a body plan designed for rapid or agile movement (e.g., gazelles), and body armor (e.g., snails and armored mammals such as the pangolin in **FIGURE 12.9A**).

Other species use poisons to defend themselves against predators. Species that contain powerful toxins are often brightly colored (**FIGURE 12.9B**). Such **warning** (aposematic) **coloration** can itself provide protection from predators, which may instinctually avoid prey that are brightly colored or may learn from experience not to eat them.

Other prey species use **mimicry** as a defense: by resembling less palatable organisms or physical features of their environment, they cause potential predators to mistake them for something less desirable to eat. There are many forms of mimicry. Some species have a shape or coloration that provides camouflage, allowing them to avoid detection by predators (**FIGURE 12.9C**); this form of mimicry is called **crypsis** (from *cryptic*, "hidden"). Other prey species use mimicry as a form

(A)

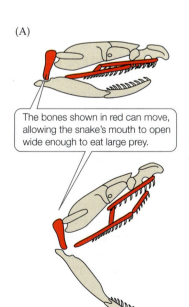

The bones shown in red can move, allowing the snake's mouth to open wide enough to eat large prey.

(B)

© Butterfly Hunter/Shutterstock.com

FIGURE 12.8 How Snakes Swallow Prey Larger Than Their Heads (A) Snakes have movable skull bones that allow them to swallow surprisingly large prey. (B) This golden tree snake (*Chrysopelea ornate*) eating a butterfly lizard larger than its head.

(A)

(B)

(C)

(D)

FIGURE 12.9 Adaptations to Escape Being Eaten Prey have evolved a wide range of mechanisms to escape from predators, including (A) physical features, such as the armor of the ground pangolin of South Africa (*Manis temmenickii*); (B) toxins, advertised by bright warning colors such as those of the nudibranch *Hypselodoris bullockii*; (C) crypsis, or camouflage, as in this female Saturniid moth (*Rhodinia fugax*), which blends in with the leaf litter on the forest floor; and (D) mimicry, as in this terrestrial flatworm (*Bipalium everetti*) that resembles a snake.

of "false advertising": their shape and coloration mimic those of a species that is fierce or that contains a potent toxin (**FIGURE 12.9D**). Finally, many prey species change their behavior when predators are present. When predators are abundant, snowshoe hares, like the elk described in Concept 8.2, forage less in open areas (where they are most vulnerable to attack). When threatened, musk oxen form a defensive circle, which makes them a difficult target (see Figure 8.19).

In some cases, there may be a trade-off between different types of defenses. For example, among four species of marine snails eaten by the green crab (*Carcinus maenas*), the species whose shells could be crushed most rapidly by crabs were the quickest to take refuge when crabs were detected, and vice versa (**FIGURE 12.10**). The exact negative correlation between resistance to crushing and predator avoidance behavior suggests that there may be a trade-off between a snail's physical and behavioral defenses.

Reciprocal plant–herbivore interactions

As we have learned, herbivores generally consume only parts of their food plant and usually do not kill them. Moreover, because most plants are not mobile and thus unable to escape herbivory in space, they employ defenses to reduce being eaten. Let's first consider plant responses to reduce herbivory and then ways that herbivores can respond.

REDUCING HERBIVORY: AVOIDANCE, TOLERANCE, AND DEFENSES Some plants avoid herbivory by producing great numbers of seeds in some years and few or no seeds in other years. For example, up to 100 years may pass between bouts of seed production, as in the mass flowering of bamboos in China. This phenomenon, known as masting, allows plants to hide (in time) from seed-eating herbivores, then overwhelm them by sheer numbers. Plants can also avoid herbivores in other ways, such as by producing leaves at times of the year when herbivores are scarce.

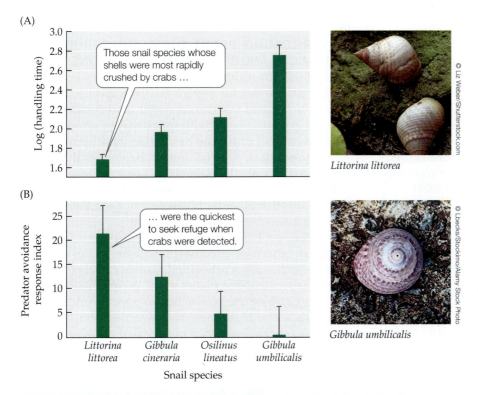

FIGURE 12.10 A Trade-Off in Snail Defenses against Crab Predation?
(A) Handling time taken by green crabs (*Carcinus maenas*) to manipulate and crush the shells of each of four snail species. (B) Index of the strength of the predator avoidance response of each of four snail species; larger values indicate a more rapid behavioral response to crabs. Error bars show one standard error of the mean. (After P. A. Cotton et al. 2004. *Ecology* 85: 1581–1584.)

Other plants have adaptive growth responses that allow them to compensate for, and hence tolerate, the effects of herbivory—at least up to a point. **Compensation** occurs when removal of plant tissues stimulates a plant to produce new tissues, allowing for relatively rapid replacement of the material eaten by herbivores. When *full compensation* occurs, herbivory causes no net loss of plant tissue. Compensation may occur when, for example, removal of leaf tissue decreases self-shading, resulting in increased plant growth, or when removal of apical buds (those at the end of a branch or shoot) allows lower buds to open and grow. Beech trees respond to simulated herbivory (clipping) by increasing both their leaf production and their photosynthetic rate. Similarly, moderate to high levels of herbivory may benefit field gentians (*Gentianella campestris*) under some circumstances (**FIGURE 12.11**). In this case, the timing of herbivory is critical: early in the growing season (up to July 20), the plant more than fully compensates for the lost tissue, but later in the season (July 28), it does not. If the amount of material removed from a field gentian—or any other plant—is large enough, however, or if insufficient resources are available for growth, the plant cannot fully compensate for the damage.

Finally, plants use an enormous array of structural and chemical defenses to ward off herbivores (Pellmyr et al. 2002; Agrawal and Fishbein 2006). A stroll through many plant communities makes this readily apparent: the leaves of many plants are tough, and many plant bodies are covered with spines, thorns, sawlike edges, or pernicious (nearly invisible) hairs that can pierce the skin like miniature porcupine quills. In some cases, such structures are an **induced defense** (stimulated by herbivore attack), as illustrated by individual cacti that increase their production of spines only after they have been grazed (Myers and Bazely 1991).

Plants also produce a wide variety of chemicals, called **secondary compounds**, that function to reduce herbivory. Some secondary compounds are toxic, protecting the plant from all but the relatively small number of herbivore species that can tolerate them. Others serve as chemical cues that attract predators or parasitoids to the plant, where they attack herbivores (Schnee et al. 2006).

Some plant species, such as oak trees, produce secondary compounds constantly, regardless of whether herbivores have attacked the plant. In other species, the production of secondary compounds is an induced defense. For example, when attacked by herbivores, a North

American tobacco species, *Nicotiana attenuata*, produces two induced defenses: toxic secondary compounds that deter herbivores directly, and volatile compounds that deter herbivores indirectly by attracting predators and parasitoids. Acting together, these defenses are very effective in reducing losses of tissue to herbivores. In one experiment, the application of compounds that are normally induced by herbivory to the stems of *N. attenuata* caused the numbers of a leaf-feeding herbivore on the plants to drop by more than 90% (Kessler and Baldwin 2001).

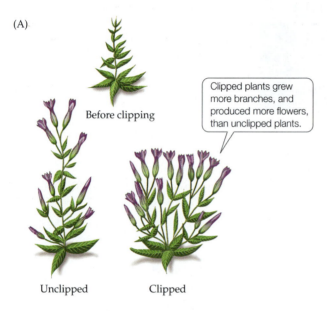

(A)

Before clipping

Clipped plants grew more branches, and produced more flowers, than unclipped plants.

Unclipped   Clipped

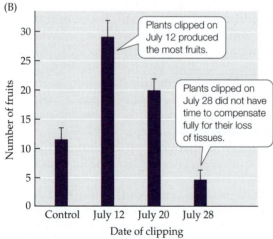

(B)

Plants clipped on July 12 produced the most fruits.

Plants clipped on July 28 did not have time to compensate fully for their loss of tissues.

FIGURE 12.11 Compensating for Herbivory Field gentians (*Gentianella campestris*) were clipped at different times during the growing season to simulate herbivory. (A) The shape and production of flowers in unclipped (control) and clipped plants. (B) Numbers of fruits produced by control plants and plants clipped on different dates. Error bars show one standard error of the mean. (After T. Lennartsson et al. 1998. *Ecology* 79: 1061–1072.)

? How many fruits would you expect to be produced by a field gentian that compensates fully for clipping? Explain your reasoning.

OVERCOMING PLANT DEFENSES: STRUCTURAL, CHEMICAL, AND BEHAVIORAL ADAPTATIONS The defenses used by plants prevent most herbivores from eating most plants. But for any given plant species, there are some herbivores that can cope with its defensive mechanisms. A plant covered with spines may be attacked by an herbivore that can avoid or tolerate those spines. Many herbivores have evolved digestive enzymes that enable them to disarm or tolerate plant chemical defenses. Such herbivores may gain a considerable advantage: they can eat plants that other herbivores cannot and thereby have access to an abundant food resource.

Some herbivores use behavioral responses to circumvent an otherwise effective plant defense. For example, some beetles use a behavioral response to cope with the defenses of tropical plants in the genus *Bursera*. These plants combine the production of toxic secondary compounds with a high-pressure delivery system: they store a toxic, sticky resin in a network of canals that runs through their leaves and stems (**FIGURE 12.12**). If an insect

(A)

(B)

Courtesy of Judith Becerra and D. Lawrence Venable

FIGURE 12.12 Plant Defense and Herbivore Counter-defense Some plants in the genus *Bursera* store toxic resin under high pressure in leaf canals. (A) When herbivores eat the leaves, they chew through these canals, causing the resin to be squirted up to 2 m from the leaf. (B) The larvae of some beetles in the genus *Blepharida* can disable this defense by chewing slowly through the canals, releasing the pressure in a gradual and harmless way.

herbivore chews through one of these canals, the resin squirts from the plant under high pressure and may repel or even kill the insect (the resin hardens after it is exposed to air, so if an insect is drenched in resin, it can be entombed). Yet some tropical beetles in the genus *Blepharida* have evolved an effective counterdefense (Becerra 2003). Their larvae chew slowly through the leaf veins where the resin canals are located, releasing the pressure so gradually that the resin does not squirt from the plant. It often takes a beetle larva more than an hour to "disarm" a leaf in this manner; once that job is done, the larva eats the leaf in 10–20 minutes.

Evolution can influence plant–herbivore interactions

The variety of antiherbivore defenses seen in plants suggests that herbivores represent a strong source of selection on plant populations. Several recent studies have tested this claim. For example, in an experiment lasting five plant generations, Züst et al. (2012) tested the hypothesis that aphid herbivores cause evolution by natural selection in populations of the annual plant *Arabidopsis thaliana*, a small plant in the mustard family that is often used in laboratory experiments and genetic studies. They began their experiment with equal mixtures of 27 different

Arabidopsis genotypes obtained from natural populations (**FIGURE 12.13A**). Typically, any one plant genotype expresses a subset of the full chemical arsenal of a species; collectively, however, the 27 genotypes used in this study were chosen to represent the full diversity of *Arabidopsis* chemical defenses.

Züst and colleagues found that feeding by aphids (two species were used, *Brevicoryne brassicae* and *Lipaphis erysimi*) reduced average plant size by up to 82% compared with a no-aphid (control) treatment, indicating that herbivory has a cost. However, they also found that the average sizes of plants exposed to the two species of aphids rose steadily over the course of the experiment (**FIGURE 12.13B**), suggesting that rapid evolution may have occurred in these populations. These increases in average plant size were associated with considerable changes in the genotypic composition of the plant populations. For example, ten plant genotypes were lost completely, and different aphid species selected for different plant genotypes. You can explore the extent to which different aphid species caused different plant genotypes to be favored by natural selection in **ANALYZING DATA 12.1**. Overall, Züst et al.'s results provide clear experimental evidence that herbivores can cause evolution by natural selection in plant populations.

(A)

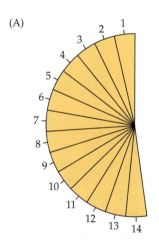 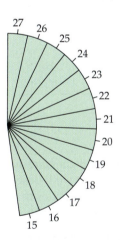

Plant genotypes that encode defensive compounds that have three-carbon side chains (3C defensive compounds)

Plant genotypes that encode defensive compounds that have four-carbon side chains (4C defensive compounds)

(B)

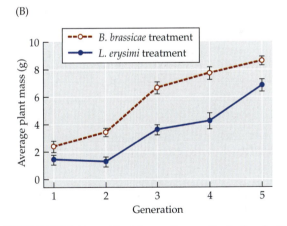

FIGURE 12.13 Does Herbivory Cause Evolution in Plant Populations? (A) This pie chart shows the equal proportions of 27 *Arabidopsis thaliana* genotypes used at the start of an experiment testing the hypothesis that herbivory by aphids caused evolution in experimental plant populations. Yellow indicates plant genotypes that encode defensive compounds that have three-carbon side chains (3C defensive compounds), while blue indicates plant genotypes that encode defensive compounds that have four-carbon side chains (4C defensive compounds). (B) The herbivory by aphids (two species were used, *Brevicoryne brassicae* and *Lipaphis erysimi*) caused the average mass of *A. thaliana* plants to increase from generation to generation, indicating an evolutionary response by plant populations. Error bars show ± one standard error of the mean. (After T. Züst et al. 2012. *Science* 338: 116–119.)

ANALYZING DATA 12.1

Do Different Herbivore Species Select for Different Plant Genotypes?

As a graduate student, Tobias Züst (Züst et al. 2012)* examined how herbivory by aphids affected evolution by natural selection in plant populations. Six replicate populations of the annual plant *Arabidopsis thaliana* were established for each of three experimental treatments: a control (no aphids), herbivory by the aphid *Brevicoryne brassicae*, and herbivory by the aphid *Lipaphis erysimi*. Each replicate population was initiated from 27 natural genotypes, and plants were grown at high densities (over 8,000 plants/m^2) in each of the three treatments.

At the start of the experiment, each replicate population contained equal proportions of the 27 plant genotypes (see Figure 12.13A). The experiment was conducted for five generations. At the end of the experiment, the frequencies of all surviving genotypes were determined.

The table shows the average plant genotype frequencies at the end of the selection experiment; in addition to the genotypes shown here, genotypes 12, 14, and 21 occurred at low frequencies (less than 1.5%) in one or two treatments. Other genotypes not shown in the table did not survive.

1. In total, how many plant populations were established in this experiment? In each of these populations, what was the initial frequency of each plant genotype?

2. Did evolution occur in the control populations? If so, what factor or factors may have caused evolution by natural selection in these populations? Explain your answers.

3. Did evolution occur in the populations exposed to aphid herbivores? If so, what factor or factors may have caused evolution by natural selection in these populations? Explain your answers.

4. Compare results for the *B. brassicae* treatment with those for the *L. erysimi* treatment, focusing on whether selection favored genotypes that code for 3C or 4C defensive compounds (see Figure 12.13A). To what extent do the plant genotypes favored by selection differ between these two treatments?

*Züst, T., C. Heichinger, U. Grossniklaus, R. Harrington, D. J. Kliebenstein, and L. A. Turnbull. 2012. Natural enemies drive geographical variation in plant defenses. *Science* 338: 116–119.

	Frequency (%) of surviving plant genotypes														
Treatment	1	2	3	4	5	6	8	9	10	15	16	22	25	27
Control	0.7	4.9	0	2.8	0	42.3	2.8	8.5	6.3	0	1.4	1.4	26.1	0.7
B. brassicae	2.8	3.5	0	0.7	0.7	0	0	9.9	3.5	1.4	2.1	1.4	67.4	2.8
L. erysimi	0.7	0	5.6	0	9.7	0	0	63.2	4.2	9.7	0	0.7	6.3	0

CONCEPT 12.3 Predator populations can cycle with their prey populations.

LEARNING OBJECTIVES

12.3.1 Know how to model predator and prey population cycling.

12.3.2 Describe the factors that affect whether predator–prey populations cycle.

Predator–Prey Population Cycles

We introduced population cycles in Chapter 10 (see Figure 10.10), and in the Case Study at the opening of this chapter, we described the most famous one of all: the hare–lynx cycle. Cyclic fluctuations in abundance are one of the most intriguing patterns in nature. After all, what could cause populations to change so considerably in size over time, yet in such a regular manner? We will return to the mechanisms that underlie the hare–lynx cycle in the Case Study Revisited, but first we'll describe some insights into the causes of population cycles that have come from models, experiments, and field observations of predator–prey interactions.

Predator–prey cycles can be modeled mathematically

One way to evaluate possible causes of population cycles is to investigate the issue mathematically. In the 1920s, Alfred Lotka and Vito Volterra independently represented the dynamics of predator–prey interactions with what is now called the **Lotka–Volterra predator–prey model**:

$$\frac{dN}{dt} = rN - aNP$$

$$\frac{dP}{dt} = baNP - mP$$

(12.1)

In these equations, N represents the number of prey individuals and P represents the number of predator individuals. The equation for change in the prey population over time (dN/dt) assumes that when predators are absent ($P = 0$), the prey population grows exponentially (i.e., $dN/dt = rN$, where r is the exponential growth rate; see Concept 11.1). When predators are present ($P \neq 0$), the rate at which they kill prey depends in part on how frequently predators and prey encounter one another. This frequency is expected to increase with the number of prey (N) and with the number of predators (P), so a multiplicative term (NP) is used in the equation for dN/dt. The rate at which predators kill prey also depends on the efficiency with which predators capture prey; this capture efficiency is represented by the constant a, so the overall rate at which predators remove individuals from the prey population is aNP.

Predators starve when there are no prey. Thus, the equation for change in the predator population over time (dP/dt) assumes that in the absence of prey ($N = 0$), the number of predators decreases exponentially with a mortality rate of m (i.e., $dP/dt = -mP$). When prey are present ($N \neq 0$), individuals are added to the predator population according to the number of prey that are killed (aNP) and the efficiency with which prey are converted into predator offspring (represented by the constant b). Thus, the rate at which individuals are added to the predator population is $baNP$.

We can determine the relationship between prey and predator populations by solving for the population growth equation of each species (Equation 12.1) when they stop changing in size (or reach an equilibrium). This approach involves determining the zero population growth isocline for both prey and predator. The zero population growth isocline (or simply **isocline**) is the condition in which the population size of the prey (or the predator) does not change in size for a given number of predators (or prey). For prey, their abundance does not change when $dN/dt = 0$, which occurs when $P = r/a$. Similarly, the abundance of predators does not change when $dP/dt = 0$, which occurs when $N = m/ba$.

Once we determine r/a and m/ba, we can then plot the isocline for both the prey (x-axis) and predators (y-axis) in graphical form. For the prey, the isocline will be a horizontal line originating at the value $P = r/a$ (**FIGURE 12.14A**). This line represents the number of predators needed to keep the prey population from changing (or at equilibrium). If the predator abundance is below the line, the prey population will increase in size. If the predator abundance is above this line, then the prey population will decrease in size. Similarly, for the predator, the isocline will be a vertical line originating at the value $N = m/ba$ (**FIGURE 12.14B**). This line represents the number of prey needed to maintain the predator population at zero growth. If the prey abundance is to the left of the line, the predator population will decrease in size. If the prey abundance is to the right of the line, then the predator population will increase in size.

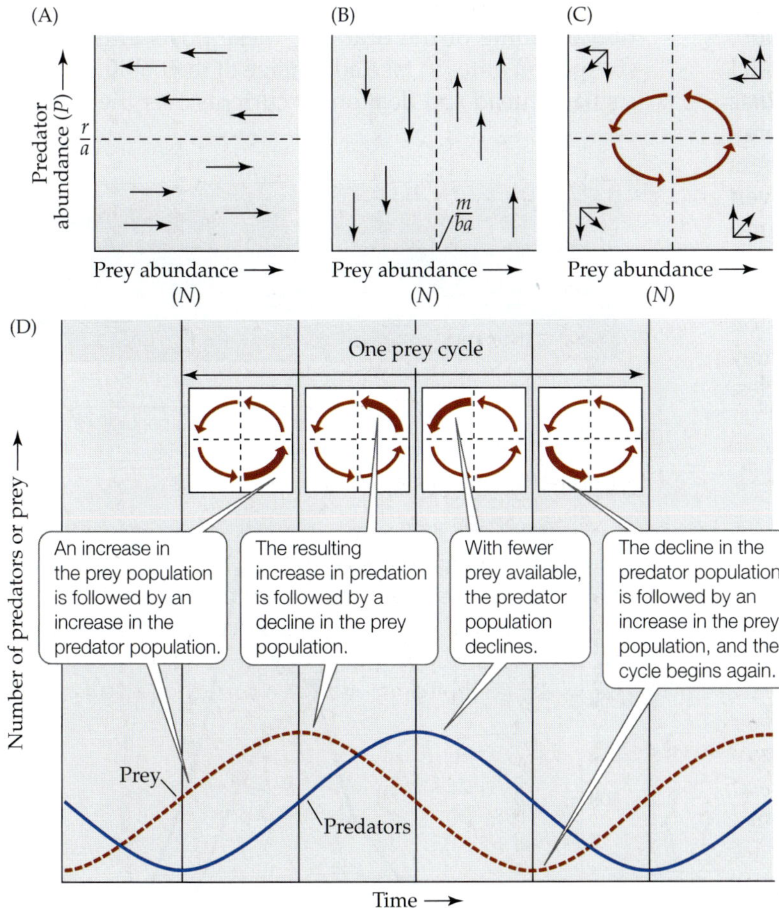

FIGURE 12.14 The Lotka–Volterra Predator–Prey Model Produces Population Cycles (A) Considering the prey population first, the abundance of prey does not change when $dN/dt = 0$, which occurs when $P = r/a$ (see Equation 12.1). (B) Similarly, considering the predator population, the abundance of predators does not change when $dP/dt = 0$, which occurs when $N = m/ba$. Combining the results in parts (A) and (B) shows that the combined abundances of predator and prey populations (represented by the red vectors) have an inherent tendency to cycle (C). These cycles are shown here in two ways: (C) by plotting the abundances of predators and prey populations together, and (D) by plotting the abundance of both predators and prey versus time; the four inset diagrams in (D) show the combined effect of prey and predator abundance. In (D), note that the predator abundance curve is shifted one-fourth of a cycle behind the prey abundance curve.

Combining the isoclines in Figure 12.14A,B shows that the isoclines cross at 90° angles and divide the graph into four regions (**FIGURE 12.14C**). We can then follow the population growth of both predator and prey in each of these regions and find that both cycle over time, with the predators lagging behind the prey by one-fourth of a cycle (**FIGURE 12.14D**). Starting in the lower right corner, both prey and predator populations are growing but the increasing numbers of predators cause the prey abundance to level off and eventually reach zero population growth at its isocline. As the populations move into the upper right corner, predator abundance is still increasing but prey abundance is in decline. This causes the predator population to slow its growth and eventually reach its isocline. Now the prey population has declined to the point that the predator population cannot sustain itself and it declines as well (upper left corner). Finally, in the lower left corner, the prey population rebounds because of low predator numbers and begins to increase. This increase eventually leads to an increase in predators when the cycle starts all over again.

The Lotka–Volterra predator–prey model thus yields an important result: it suggests that predator and prey populations have an inherent tendency to cycle because the abundance of one population is dependent on the abundance of the other population. The only condition in which the two populations do not cycle is when the predator and prey isoclines intersect. Here, by definition, both populations do not change in size. But the model also has a curious and unrealistic property: the *amplitude* of the cycle (the magnitude by which predator and prey numbers rise and fall) depends on the initial numbers of predators and prey. If the initial numbers shift even slightly, the amplitude of the cycle will change. More complex predator–prey models (e.g., Harrison 1995) still produce cycles but do not show this unrealistic dependence on initial population sizes. The same general conclusion emerges from all of these models, however: predator–prey interactions have the potential to cause population cycles.

Predator–prey cycles can be reproduced under laboratory conditions

Can the cycling behavior of predator–prey models be reproduced in the laboratory? Experiments show that such cycles can be difficult to achieve. When prey are easy for predators to find, predators typically drive prey to extinction, then go extinct themselves. Such was the case in C. B. Huffaker's experiments with the herbivorous six-spotted mite (*Eotetranychus sexmaculatus*) and the predatory mite, *Typhlodromus occidentalis*, that eats it (Huffaker 1958). In an initial set of experiments, Huffaker released 20 six-spotted mites on a tray with 40 positions, a few of which contained oranges, which these herbivorous mites could eat (**FIGURE 12.15A**). At first, the six-spotted mite population increased, in some cases reaching densities of 500 mites per orange. Eleven days after the start of the experiment, Huffaker released two predatory mites on the tray. Both prey and predator populations increased for a time, then declined to extinction (**FIGURE 12.15B**).

Huffaker observed that the prey persisted longer if the oranges were widely spaced—presumably because it took the predators more time to find their prey. He tested this idea in a follow-up experiment in which he increased the complexity of the habitat in the following way. First, he added strips of Vaseline that partially blocked the predatory mites as they crawled from one orange to another. Then he placed small wooden posts in an upright position on some of the oranges; these posts allowed the six-spotted mites to take advantage of their ability to spin a silken thread and float on air currents over the Vaseline

(A)

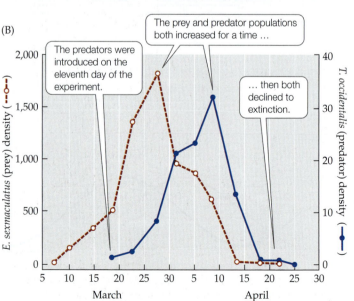

Courtesy of F. E. Skinner

(B)

FIGURE 12.15 In a Simple Environment, Predators Drive Prey to Extinction (A) C. B. Huffaker constructed a simple laboratory environment to test for conditions under which predators and prey would coexist and produce population cycles. He placed oranges in a few positions in an experimental tray to provide food for the herbivorous six-spotted mite (*Eotetranychus sexmaculatus*); the remainder of the positions contained inedible rubber balls. (B) When a predatory mite (*Typhlodromus occidentalis*) was introduced into this simple environment, it drove the prey to extinction, causing its own population to go extinct as well. (B after C. B. Huffaker. 1958. *Hilgardia* 27: 343–383.)

barriers. Thus, he altered the experimental environment to favor dispersal of the six-spotted mite and impede dispersal of the predatory mite. Under these conditions, the prey and the predators both persisted, illustrating a form of "hide-and-seek" dynamics that produced population cycles (**FIGURE 12.16**). The six-spotted mites dispersed to unoccupied oranges, where their numbers increased. Once the predators found an orange with six-spotted mites, they ate them all, causing both prey and predator numbers on that orange to plummet. In the meantime, however, some six-spotted mites dispersed to other portions of the experimental environment, where they increased in number until they too were discovered by the predators.

Predator–prey cycles can persist in the field

Natural populations of predators and prey can coexist and show dynamics similar to those of Huffaker's mites. Clumps of mussels off the coast of California, for example, can be driven to local extinction by predatory sea stars. However, mussel larvae float in ocean currents and hence disperse more rapidly than the sea stars. As a result, the mussels continually establish new clumps that flourish until they are discovered by sea stars. Thus, like the six-spotted mites in Huffaker's experiments, the mussels persist because portions of their population escape detection by predators for a time.

Field studies have also shown that predators influence population cycles in species such as southern pine beetles, voles, collared lemmings, snowshoe hares, and moose (Gilg et al. 2003; Turchin 2003). But predation is not the only factor that causes population cycles in these species. The supply of food plants for the herbivorous prey can also play an important role, and in some cases, social interactions are important as well. Thus, reality is not as simple as implied by the results of predator–prey models (in which cycles are maintained purely by predator–prey interactions). In the field, some population cycles may be caused by three-way feeding relationships—by the effects of predators and prey on each other, coupled with the effects of prey and their food plants on each other.

Whether their populations cycle or not, a variety of factors can prevent predators from driving prey to extinction. Such factors include habitat complexity and limited predator dispersal (as in Huffaker's mites), prey switching in predators (see Figure 12.6), spatial refuges (i.e., areas in which predators cannot hunt effectively), and evolutionary changes in the prey population.

In this section, we have seen how predation can alter the population size of predator and prey, resulting in population cycles. We turn next to how predators can have major effects on ecological communities.

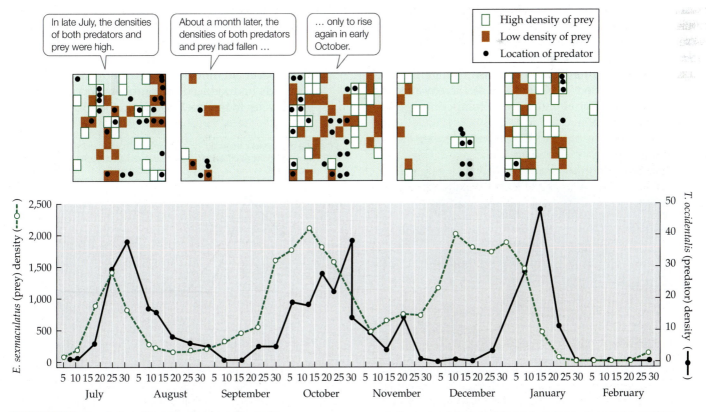

FIGURE 12.16 Predator–Prey Cycles in a Complex Environment Huffaker modified the simple laboratory environment shown in Figure 12.15 to create a more complex environment that aided the dispersal of the prey species but hindered the dispersal of the predator. Under these conditions, predator and prey populations coexisted, and their abundances cycled over time. The top panels show the locations within the environment of prey (shaded regions) and predators (circles) at five different points in time. (After C. B. Huffaker. 1958. *Hilgardia* 27: 343–383.)

CONCEPT 12.4 Predation can affect prey distribution and abundance, in some cases causing a shift from one community type to another.

LEARNING OBJECTIVES

12.4.1 Understand how predators can affect prey distribution and abundance.

12.4.2 Describe how carnivores and herbivores can alter communities in dramatic ways.

Effects of Predation on Communities

A general theme that runs through this book is that ecological interactions can affect the distributions and abundances of species, affecting communities and ecosystems. The community-level consequences of predation can be profound, in some cases causing major shifts in the types of organisms found at a given location.

All trophic interactions have the potential to reduce the growth, survival, or reproduction of the organisms that are eaten. These effects can be dramatic, as demonstrated in the case of a leaf-feeding beetle, *Chrysolina quadrigemina*, that rapidly reduced the density of Klamath weed, an invasive plant that is poisonous to livestock (**FIGURE 12.17**). Predators and parasitoids can also have dramatic effects when they are introduced as biological pest controls. In six cases, introductions of wasps that preyed on crop-eating insects decreased the herbivores' densities by more than 95%, thus greatly reducing the economic damage caused by those pests.

Predators and herbivores can also change the outcome of competition (see Concept 14.4), thereby affecting the distributions or abundances of competitor species.

In particular, inferior competitors may increase in abundance when they are in the presence of a predator that decreases the abundance or performance of a dominant competitor. Paine (1974) found such a result: he showed that the removal of a predatory sea star (*Pisaster*) led to the crowding out of large invertebrate species but one, the California mussel (*Mytilus californianus*). The mussel was a dominant competitor that, in the absence of the sea star, drove all the other large invertebrates to local extinction.

We turn now to examples of how trophic interactions can affect communities, first for carnivores and second for herbivores.

Carnivores can alter communities in dramatic ways

Anolis lizards are predators that eat a broad range of prey species, including spiders. Thomas Schoener and David Spiller studied the effects of lizard predators on their spider prey in the Bahamas. They selected 12 small islands and divided them into four groups of 3 islands each that were similar in size and vegetation. Initially, each group of 3 islands contained an island with lizards and 2 without. One of the latter 2 islands was then chosen at random to have two male and three female adult *Anolis sagrei* lizards introduced to it; the other island was left as a control where lizards were absent naturally.

The introduced lizards greatly reduced the distributions and abundances of their spider prey (Schoener and Spiller 1996). Before the experiment began, the numbers of spider species and the overall densities of spiders were similar among the 8 islands that lacked lizards. By the end of the experiment, however, the introduction of lizards to 4 islands had reduced the numbers and densities of spider species to the levels found on the 4 islands where

(A)

Courtesy of USDA ARS, European Biological Control Laboratory/Bugwood.org

(B)

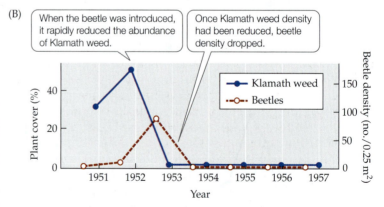

When the beetle was introduced, it rapidly reduced the abundance of Klamath weed.

Once Klamath weed density had been reduced, beetle density dropped.

FIGURE 12.17 A Beetle Controls a Noxious Rangeland Weed Klamath weed (*Hypericum perforatum*), which poisons cattle, once covered about 4 million acres of rangeland in the western United States. (A) This photograph, taken in 1949, shows a field completely covered with flowering Klamath weed. (B) The leaf-feeding beetle *Chrysolina quadrigemina* was introduced in 1951 in the hope of controlling Klamath weed. This graph tracks densities of beetles and of Klamath weed (as a percentage of plant cover)

in plots after the beetle's introduction. (B after C. Huffaker and C. Kennett. 1959. A ten-year study of vegetational changes associated with biological control of Klamath weed. *J Range Manage* 12: 69–82. doi:10.2307/3894934. Material used with permission of the publisher.)

? Explain how a plant community might change after *C. quadrigemina* reduced the density of Klamath weed.

The proportion of spider species that went extinct was nearly 13 times higher on islands where lizards were introduced than on islands without lizards.

FIGURE 12.18 Lizard Predators Can Drive Their Spider Prey to Extinction
The experimental introduction of lizards to small islands in the Bahamas greatly increased the rate at which their spider prey became extinct. Error bars show one standard error of the mean. The photograph shows Thomas Schoener on one of the study islands. (After T. W. Schoener and D. A. Spiller. 1996. *Nature* 381: 691–694.)

lizards were present naturally. The proportion of spider species that went extinct was nearly 13 times higher on islands where lizards were introduced than on islands without lizards (**FIGURE 12.18**). Similarly, the density of spiders was about 6 times higher on islands without lizards than on islands that had lizards (either naturally or experimentally). The introduction of lizards reduced the densities of both common and rare spider species, and most of the rare species went extinct. Similar experimental results have been obtained for beetles eaten by rodents and grasshoppers eaten by birds.

Schoener and Spiller's work on the effects of predatory lizard on spiders shows that the direct effects of a predator can greatly reduce the number of prey species in a community. In other cases, a predator that suppresses a dominant competitor can (indirectly) cause the number of species in a community to increase (as in the sea star and mussel example). Indirect effects of predators can also alter ecological communities by affecting the transfer of nutrients from one ecosystem to another, as the following study on arctic foxes illustrates.

In the late nineteenth and early twentieth centuries, humans introduced arctic foxes (*Vulpes lagopus*) to some of the Aleutian islands off the coast of Alaska. Other islands remained fox-free, either because foxes were never introduced there or because the introductions failed. Taking advantage of this inadvertent large-scale experiment, Croll et al. (2005) determined that, on average, the introduction of foxes to an island reduced the density of breeding seabird populations nearly 100-fold. The decrease in seabird numbers, in turn, reduced the input of guano (bird feces) to an island from roughly 362 to 6 g per square meter. Seabird guano, which is rich in phosphorus and nitrogen, transfers nutrients from the ocean (where seabirds feed) to the land. By reducing the amount of guano that fertilized the (nutrient-limited) plant communities on the islands, the introduction of foxes caused dwarf shrubs and forbs to increase in abundance at the expense of grasses. As a result, the introduction of foxes had the unexpected effect of transforming the island communities from grasslands to communities characterized by small shrubs and forbs.

Herbivores can alter communities in dramatic ways

Herbivores can have equally large effects. Lesser snow geese (*Chen caerulescens caerulescens*) migrate from their overwintering grounds in the United States to breed in salt marshes that border Canada's Hudson Bay. During the summer, the geese graze on marsh grasses and sedges. Historically, although the geese removed considerable plant matter, their presence benefited the marshes by adding nitrogen, which is a limiting resource for plant growth. As they eat, the geese defecate every few minutes, thereby adding nitrogen to the soil (nitrogen moves into the soil from goose feces more rapidly than it does from the decomposing leaves of marsh plants). The plants absorb the added nitrogen, which allows them to grow rapidly after being grazed. Overall, low to intermediate levels of grazing by geese lead to increased plant growth (Jefferies et al. 2003). For example, net primary production (NPP, measured as the amount of new aboveground plant growth) was higher in lightly grazed plots than in ungrazed plots (**FIGURE 12.19A**).

About 40 years ago, however, the situation described in the previous paragraph started to change. Beginning around 1970, lesser snow goose densities increased exponentially. This increase probably occurred because increased crop production near their overwintering sites provided the geese with a superabundant supply of food. The ensuing high densities of geese no longer benefited marsh plants. The geese completely removed the vegetation, drastically changing the community of marsh plant species (**FIGURE 12.19B**). Of an original 54,800 hectares (135,400 acres) of intertidal marsh in the Hudson Bay region, geese are estimated to have destroyed 35% (19,200 ha or 47,400 acres). An additional 30% (16,400 ha or 40,500 acres) of the original marsh has been badly damaged by the geese. Controlled hunts (from 1999 on) have slowed goose population growth; this strategy may eventually lead to marsh recovery.

In *The Origin of Species*, Darwin (1859) noted the speed with which Scotch fir trees replaced heaths after regions of heathland were enclosed to prevent grazing by cattle. When he observed heathlands grazed by cattle, "on looking closely between the stems of the heath, I found a multitude of seedlings and little trees, which had been perpetually browsed down by the cattle. In one square yard … I counted thirty-two little trees; and one of them, judging from the rings of growth, had during 26 years tried to raise its head above the stems of the heath, and had failed." Darwin concluded that seeds dispersed from trees located at the edge of the heath would germinate and overgrow the heath if not for grazing by cattle. Thus, the very existence of the heath community in that area depended on grazing.

Herbivores can also have pronounced effects in aquatic environments. The golden apple snail was introduced into Taiwan from South America in 1980 for local consumption and export. The snail escaped from cultivation and spread rapidly through Southeast Asia (**FIGURE 12.20**). Its spread caught the attention of researchers and government officials because it proved to be a serious pest of rice. The snail has also been found in Hawaii, the southern United States, and Australia and is expected to reach Bangladesh and India (Carlsson et al. 2004).

Most freshwater snails eat algae, but the golden apple snail prefers to eat aquatic plants, including those that float on the water surface and those that attach themselves to the bottom. However, as mentioned in Concept 12.1, golden apple snails are generalists, and if plants are not available, they can survive on algae and detritus. As a result, these snails are resilient and hard to get rid of.

As a first step toward assessing how the snail had affected natural communities, Nils Carlsson and colleagues

(A)

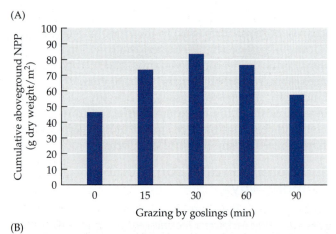

(B)

Courtesy of R. L. Jefferies (University of Toronto), a member of the Hudson Bay Project partly funded by NSERC, Canada

FIGURE 12.19 Snow Geese Can Benefit or Decimate Marshes (A) When lightly grazed (for a single 15- to 90-minute episode) by snow goose goslings, salt marsh plants increased their subsequent cumulative production of new biomass compared with no grazing, because of the nitrogen added by the defecating geese. (B) Heavy grazing by high densities of snow geese can convert salt marshes to mudflats, as seen by comparing this small remnant of marsh (protected from geese) with the surrounding mudflat (a former marsh that was grazed heavily by geese). (A after D. S. Hik and R. L. Jefferies. 1990. *J Ecol* 78: 180–195.)

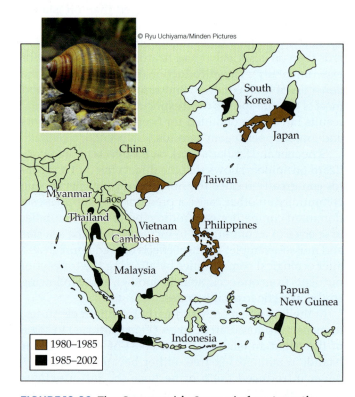

© Ryu Uchiyama/Minden Pictures

FIGURE 12.20 The Geographic Spread of an Aquatic Herbivore Since its introduction to Taiwan in 1980, the golden apple snail (*Pomacea canaliculata*) has spread rapidly across parts of Southeast Asia, threatening rice crops and native plant species. The map shows the regions the snail had occupied by 1985 and by 2002. (After J. O. L. Carlsson et al. 2004. *Ecology* 85: 1575–1580.)

surveyed 14 wetlands in Thailand with varying densities of snails. They found that wetland communities with high densities of snails were characterized by few plants, high nutrient concentrations in the water, and a high biomass of algae and other phytoplankton (**FIGURE 12.21**).

To test whether the trends observed in their survey could have been caused by the snail, Carlsson et al. (2004) placed 24 1 × 1 × 1-m enclosures in a wetland in which snail densities were low. To each enclosure, they added about 420 g of water hyacinth (*Eichhornia crassipes*), one of the most abundant plant species in many Southeast Asian wetlands. Next, they added 0, 2, 4, or 6 snails to the enclosures; there were six replicates of each of the four snail density treatments. Carlsson and colleagues then measured

the effects of the snails on plant biomass and phytoplankton biomass. Water hyacinth biomass increased in the enclosures where no snails were present but decreased in all the other enclosures. At the highest snail density tested (6 snails/m²), phytoplankton biomass increased.

The results of the survey and the experiment concur in suggesting that the golden apple snail can have an enormous effect on wetland communities, causing a shift from a wetland with clear water and many plants to a wetland with turbid water, few plants, high nutrient concentrations, and high phytoplankton biomass. It is likely that this shift occurs because the snails suppress plants directly (by eating them) and because they release the nutrients they obtain from the plants into the water, thus providing improved growth conditions for algae and other phytoplankton.

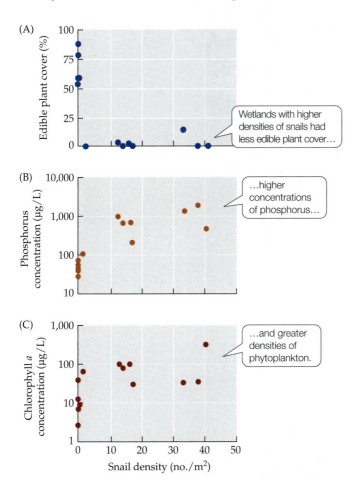

FIGURE 12.21 A Snail Herbivore Alters Aquatic Communities Nils Carlsson and colleagues measured characteristics of 14 natural wetlands in Thailand that differed in their densities of golden apple snails (*Pomacea canaliculata*). (A) Percentage of the wetlands covered by edible plant species. (B) Concentrations of phosphorus in the water. (C) Chlorophyll *a* concentrations (an indicator of phytoplankton biomass). Note the log scale in (B) and (C). Experiments conducted separately indicated that all the trends shown here could have been caused by the snail. (After J. O. L. Carlsson et al. 2004. *Ecology* 85: 1571–1580.)

In (B), compare the average total phosphorus concentration in wetlands without snails with that in wetlands with snails.

CLIMATE CHANGE CONNECTION

CLIMATE CHANGE AND SPECIES INTERACTIONS

Climate affects the physiology of organisms, the distribution and abundance of populations, and the outcome of interactions between species (see Chapter 2). As a result, changes in climate are expected to have wide-ranging effects on species interactions and thus ecological communities (Gilman et al. 2010). For example, in a review of over 600 articles, Tylianakis et al. (2008) found that climate change affected the strength and frequency of a wide range of ecological interactions (**FIGURE 12.22**).

Collectively, these results suggest that species interactions are likely to complicate efforts to predict how climate change will affect predator–prey relationships. For example, although the direct effects of climate on predators or herbivores might suggest that these species would extend their ranges in response to climate change, competition with other species could prevent this from occurring. If competition had this effect, then interactions with other organisms would have caused the actual distribution of a species under climate change to be smaller than its potential distribution (see Figure 4.3). In other circumstances, however, a very different result might be observed. For example, as discussed in Concept 25.2, changes in climate can lead to the formation of new types of communities that contain collections of species that differ from those found in current communities. In such a new community type, a predator might interact with new prey or hosts and hence might expand its geographic range farther than would otherwise be expected. In this case, over time, the actual distribution of the predator would become larger than its (predicted) potential distribution because of changes in community structure that resulted from climate change. Overall, the results of Tylianakis et al. (2008) and Gilman et al. (2010) indicate that ecological interactions will influence how future climate change will affect predator–prey interactions and many other species associated with these interactions.

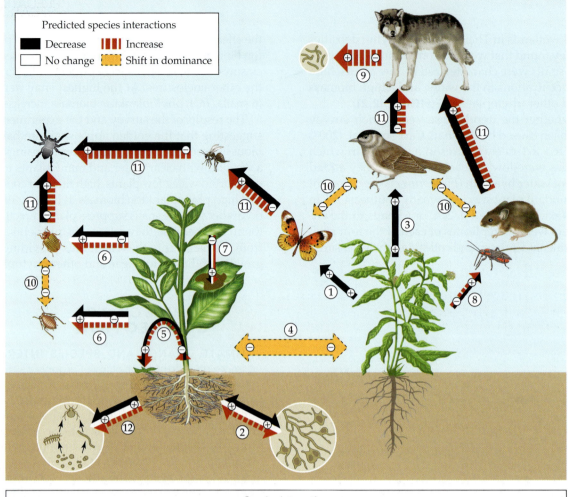

Predicted species interactions

- ■ Decrease
- ▥ Increase (red/striped)
- □ No change
- ▤ Shift in dominance (yellow/dotted)

Species interactions

1. Plant–pollinator
2. Plant–fungal mutualism
3. Plant–seed disperser
4. Plant–plant competition
5. Plant–hemiparasite
6. Plant–herbivore
7. Plant–pathogen
8. Plant–seed predator
9. Host–pathogen
10. Animal–animal competition
11. Predator–prey or parasitoid–host
12. Soil food web

FIGURE 12.22 Climate Change Alters Species Interactions
This diagram provides an overview of a literature review of how climate change is predicted to alter species interactions in terrestrial systems, including some with parasites, in studies that tested for the effects of increased temperature, changing rainfall patterns, or increased frequency of extreme weather events. Arrows with solid outlines indicate nutrient and energy flow; double-headed arrows with dotted outlines indicate competition. A + or – symbol within an arrow indicates benefit or cost to each participant. (After J. M. Tylianakis et al. 2008. *Ecol Lett* 11: 1351–1363.)

A CASE STUDY REVISITED

Snowshoe Hare Cycles

What is the cause of snowshoe hare population cycles? As we saw in the Case Study, neither the food supply hypothesis nor the predation hypothesis alone can explain these cycles. However, much of the variation in hare densities can be explained when we combine these two hypotheses—and add even more realism with a few new twists.

Charles Krebs and colleagues (1995) performed an experiment designed to determine whether food, predation, or their interaction caused population cycles in hares. The sheer scope of the experiment was impressive: the experimental treatments were performed in seven 1 × 1-km blocks of forest located in an isolated region of Canadian wilderness. Three blocks were not manipulated and were used as controls. Food for hares was added to two blocks (the "+Food" treatment). In 1987, an electric fence 4 km in length was constructed to exclude predators from one block of forest (the "–Predators" treatment). In the following year, a second 4-km fence was built; in the block of forest enclosed by this fence, food was added and predators were excluded (the "+Food/–Predators" treatment). The two fences (with a total length of 8 km) had to be monitored daily during the winter, when temperatures could plummet to –45°C (–49°F); this monitoring required so much time that the researchers could not replicate either fenced treatment. The survival rates and densities of hares in each block of forest were observed for 8 years.

(A)

Two fences

Photo by Tim Karels courtesy of Charles Krebs

(B)

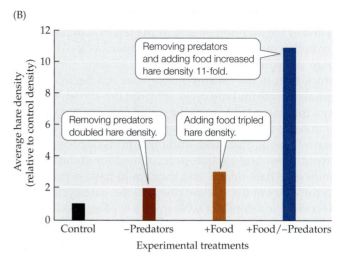

Removing predators and adding food increased hare density 11-fold.

Removing predators doubled hare density.

Adding food tripled hare density.

FIGURE 12.23 **Both Predators and Food Influence Hare Density** (A) This aerial photograph shows one of the 1-km² snowshoe hare study sites described in the text. (B) Average hare densities relative to their densities in control blocks of forest. (B after C. J. Krebs et al. 1995. *Science* 269: 1112–1115.)

Compared with the control blocks, hare densities were considerably higher in the +Food, the –Predators, and the +Food/–Predators blocks (**FIGURE 12.23**). The most pronounced effects were seen in the +Food/–Predators block, where, on average, hare densities were 11 times those in the control blocks. The strong effect of jointly adding food and removing predators suggests that hare population cycles are influenced by both food supply and predation.

This conclusion was supported by results from a mathematical model that examined feeding relationships across three levels: vegetation (the hares' food), hares, and predators (King and Schaffer 2001). Field data were used to estimate the model parameters, and the model's predictions were compared with the actual results for Krebs et al.'s four treatments. Although the match was not exact,

there was reasonably good agreement between the model and the results, again suggesting that both food and predators influence hare population cycles (**FIGURE 12.24**).

While much progress has been made in the study of snowshoe hare population cycles, some questions remain. We do not yet have a complete understanding of the factors that cause hare populations across broad regions of Canada to cycle in synchrony. Lynx can move from 500 to 1,100 km. If lynx move from areas with scarce prey to areas with abundant prey on a scale of hundreds of kilometers, their movements might be enough to cause geographic synchrony in hare cycles. In addition, large geographic regions in Canada experience a similar climate, and that may also affect the synchrony of hare population cycles.

(A)

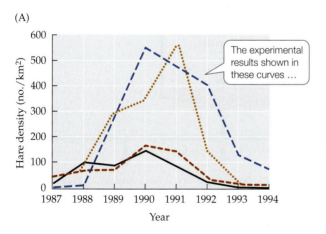

The experimental results shown in these curves …

(B)

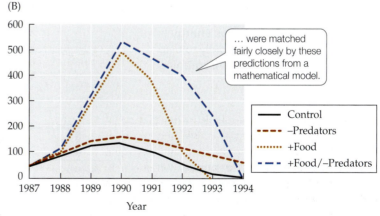

… were matched fairly closely by these predictions from a mathematical model.

FIGURE 12.24 **A Vegetation–Hare–Predator Model Predicts Hare Densities Accurately** The model assumes that hare population densities are influenced by feeding relationships across three levels: vegetation (the hares' food), hares, and predators. Parameters for the model were estimated from field data.

When the investigators compared the predictions of their model with the experimental results of Krebs et al. (1995), they found a reasonably good match between (A) the experimental results and (B) the model's predictions. (After A. A. King and W. M. Schaffer. 2001. *Ecology* 82: 814–830.)

Finally, the Krebs et al. experiment provided a test of whether the addition of food or the removal of predators (or both) could stop the hare population cycle. Although hare densities declined less in the +Food/−Predators block than in the control blocks, they did decline at the usual point in the hare cycle. Why did the +Food/−Predators treatment fail to stop the cycle? One possible reason is that the fences excluded lynx and coyotes but did not exclude owls, goshawks, and other birds of prey. Collectively, these bird predators accounted for about 40% of snowshoe hare deaths and thus could have contributed to the onset of the decline phase of the hare cycle in the +Food/−Predators block. Next we'll explore another possible explanation: stress caused by fear of predator attack.

🌱 CONNECTIONS *in* NATURE

FROM FEAR TO HORMONES TO POPULATION DYNAMICS
Predators not only affect their prey directly (by killing them) but also influence them indirectly (e.g., by altering their foraging behavior, as described for elk in Concept 8.2). Boonstra et al. (1998) tested snowshoe hares for another possible indirect effect of predators: fear. Their results hint at a fascinating way in which predation might influence the decline phase of the hare cycle.

When humans are in a dangerous situation, we often engage a set of fight-or-flight responses that can produce rapid and sometimes astonishing results (such as the ability to move unusually heavy weights). Snowshoe hares have a similar stress response. A hormone called cortisol stimulates the release of stored glucose into the blood, where it becomes available to the muscles; cortisol also suppresses body functions that are not essential for immediate survival, including growth, reproduction, and immune system function (**FIGURE 12.25**).

The stress response works well for immediate, or *acute*, forms of stress, such as an attack by a predator. Energy is provided to the muscles rapidly to help the animal deal with the threatening

situation. Shortly thereafter, the response is shut down by a negative feedback process. The stress response works less well for long-term, or *chronic*, stress, however. In such cases, the negative feedback signals are weak, and the stress response is maintained for a long time. A failure to "turn off" the stress response can have harmful effects, including decreased growth and reproduction and increased susceptibility to disease. Collectively, such effects can reduce a population's survival and reproductive rates.

When predators are abundant, as we have seen, they can cause up to 95% of snowshoe hare deaths. At such times, hares are at increased risk of encountering predators; hares would also be likely to see or hear predators killing other hares and to find the remains of hares that had been killed by predators. Reasoning that the fear provoked by such events could trigger chronic stress, Boonstra and colleagues measured the hormonal and immune responses of hares exposed in the field to high versus low numbers of predators. During the decline phase of the hare cycle (when hares are exposed to many predators), cortisol levels increased, blood glucose levels increased, reproductive hormone levels decreased, and overall body condition worsened—as expected for hares experiencing chronic stress (see Figure 12.25). Further experiments showed that a predator-induced increase in cortisol levels

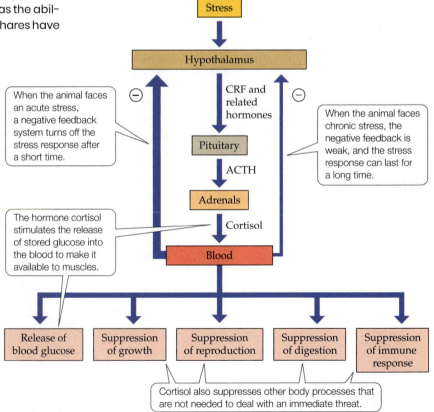

FIGURE 12.25 The Stress Response When an animal is stressed, the hypothalamus releases a hormone called CRF, which stimulates a cascade of reactions that affect a number of body processes. (After R. Boonstra et al. 1998. *Ecol Monogr* 79: 371–394.)

Stress → Hypothalamus → CRF and related hormones → Pituitary → ACTH → Adrenals → Cortisol → Blood

When the animal faces an acute stress, a negative feedback system turns off the stress response after a short time.

When the animal faces chronic stress, the negative feedback is weak, and the stress response can last for a long time.

The hormone cortisol stimulates the release of stored glucose into the blood to make it available to muscles.

Release of blood glucose | Suppression of growth | Suppression of reproduction | Suppression of digestion | Suppression of immune response

Cortisol also suppresses other body processes that are not needed to deal with an immediate threat.

led to a drop in the number and size of offspring produced by female snowshoe hares (Sheriff et al. 2009). In addition, female hares with high cortisol levels (caused by exposure to predators) transmitted high cortisol levels to their offspring, who also had reduced reproductive rates (Sheriff et al. 2010).

Overall, chronic stress induced by predation may help to explain some of the puzzling observations mentioned in the Case Study, including the drop in birth rates during the decline phase of the hare cycle and the slow rebound of hare numbers after predator numbers plummet. If future studies confirm the results of Boonstra et al. (1998) and Sheriff et al. (2009, 2010), their work will provide a clear example of how predation risk can alter the physiology of individual prey, thereby changing prey population dynamics and influencing predator–prey cycles.

SUMMARY

CONCEPT 12.1 Most carnivores have broad diets, whereas a majority of herbivores have relatively narrow diets.

12.1.1 Understand why carnivores tend to be generalists and have broad animal prey diets.

- Most carnivores eat prey in relation to their availability across a broad range of prey species.

- When the encounter rate and handling time for prey is low, as is the case for carnivores, they tend to be broad in their prey choices.

12.1.2 Know why herbivores tend to be specialists and have narrow plant prey diets.

- Many herbivores specialize on particular plant parts, such as leaves, roots, stems, seeds, or internal fluids, and specialize on one or a few plant species.

- Herbivores tend to be specialists because plants are relatively easy to search for but their handling times are longer, thus favoring specialization.

CONCEPT 12.2 Predation results in a wide range of capture and avoidance mechanisms.

12.2.1 Know the search and capture mechanisms of carnivores and the ways animal prey avoid being eaten.

- Many predators move about their habitat in search of prey to capture; others remain in one place and attack or trap prey that are within striking distance.

- In response to strong selection pressure exerted by carnivores, prey have evolved a rich variety of defensive mechanisms, including physical defenses, toxins, mimicry, or behavioral responses to escape predation.

12.2.2 Describe how plants avoid being eaten and how herbivores can overcome these defenses.

- Plants cope with herbivory via masting and other forms of avoidance, compensation (a form of tolerance), and secondary chemicals that deter herbivores.

- Although a plant's defensive mechanisms prevent most herbivores from eating it, some herbivores can overcome the plant's defenses by structural, chemical, or behavioral means.

CONCEPT 12.3 Predator populations can cycle with their prey populations.

12.3.1 Know how to model predator and prey population cycling.

- Results from mathematical predator–prey models, laboratory experiments, and field observations show that population cycles can be caused by trophic interactions.

- The Lotka–Volterra predator–prey model suggests that predator and prey populations have an inherent tendency to cycle because the abundance of one population is dependent on the abundance of the other population.

- The relationship between prey and predator populations can be determined by solving for the population growth equation of each species when they reach the zero growth isocline.

12.3.2 Describe the factors that affect whether predator–prey populations cycle.

- Whether predators and prey cycle may depend on several factors, including habitat complexity and dispersal rates, spatial refuges, and evolutionary changes.

CONCEPT 12.4 Predation can affect prey distribution and abundance, in some cases causing a shift from one community type to another.

12.4.1 Understand how predators can affect prey distribution and abundance.

- Predators can affect the distribution and abundance of their prey by decreasing their growth, reproduction, and/or survival.

12.4.2 Describe how carnivores and herbivores can alter communities in dramatic ways.

- Predation can dramatically alter the composition of ecological communities by directly affecting the abundance of dominant or multiple species of prey.

- Predators can also indirectly affect communities by controlling the abundance of dominant prey, which interact strongly with other species.

Review Questions

1. Compare and contrast the diet breadth of carnivores and herbivores, and hypothesize why these differences exist.

2. Summarize the effects that carnivores and herbivores have on their prey, as well as the effects that prey have on the carnivores and herbivores that eat them. Explain why these effects are pervasive and pronounced.

3. In this chapter, we claim that predation can have strong effects on ecological communities.
   a. Provide a logical argument to support this claim.
   b. Does the scientific evidence support or contradict this claim? Explain.

Hone Your Problem-Solving Skills

In the Case Study of this chapter, we considered a nearly 100-year record of the Hudson's Bay Company to estimate abundances of snowshoe hares and their Canada lynx predators (*Lynx canadensis*). Both species exhibited regular population cycles, with abundances peaking about every 10 years and then falling to low levels (see Figure 12.2A). Suppose for the purposes of this problem that you know the following components of the relationship between snowshoe hare and lynx:

$$r = 0.4, m = 0.9$$
$$a = 0.01, ba = 0.01$$

1. Using the values above, calculate the isoclines for prey and predator. Then, graph the isoclines for each species. Label all your axes and isoclines.

2. You are given the following abundance data for prey and predator for two different time periods where the prey or predators cross their respective isoclines:

   Time 1: 90 prey, 30 predators

   Time 2: 100 prey, 40 predators

   Plot the population abundances of both species for Time 1 and Time 2 on your graph. Now, given Lotka–Volterra model assumptions, determine the abundances for Time 3 (predator crosses its isocline) and Time 4 (prey crosses its isocline). Describe, and draw on the graph, what will happen to both populations over time.

3. Now, plot the abundance of both prey and predators over time, starting with Time 1 and ending at Time 5.

4. Suppose the population sizes of prey and predators at Time 1 were those values where the isoclines cross. Plot the abundances of both prey and predators over time starting with Time 1 and ending at Time 5. Describe what will happen to both populations over time.

ADDITIONAL RESOURCES

This textbook provides resources such as **Web Extensions** and additional **Climate Change Connections** for instructors who can share them with students. Please contact your instructor, if you would like to access that content.

Parasitism

Enslaver Parasites: A Case Study

In science fiction books and movies, villains sometimes use mind control or physical devices to break the will and control the actions of their victims. In these stories, people may be forced to perform strange or grotesque actions, or to harm themselves or others—all against their will.

Real life can be just as strange. Consider the hapless cricket shown in **FIGURE 13.1**. This cricket does something that a cricket ordinarily would never do: it walks to the edge of a body of water, jumps in, and drowns. Shortly afterward, a hairworm begins to emerge from the body of the cricket (see Figure 13.1). For the worm, this is the final step in a journey that begins when a terrestrial arthropod—such as a cricket—drinks water in which a hairworm larva swims. The larva enters the cricket's body and feeds on its tissues, growing from microscopic size into an adult that fills all of the cricket's body cavity except its head and legs. When fully grown, adult hairworms must return to the water to mate. After the adults mate, the next generation of hairworm larvae are released to the water, where they will die unless they are ingested by a terrestrial arthropod host.

Has the hairworm "enslaved" its cricket host, forcing it to jump into the water—an act that kills the cricket but is essential for the hairworm to complete its life cycle? The answer appears to be yes. Observations have shown that when crickets infected with hairworms are near water, they are much more likely to enter the water than are uninfected crickets (Thomas et al. 2002). Furthermore, in ten out of ten trials, when infected crickets were rescued from the water, they immediately jumped back in. Uninfected crickets do not do this.

Hairworms are not the only parasites that enslave their hosts. Maitland (1994) coined the term "enslaver parasites" for several fungal species that alter the perching behavior of their fly hosts in such a way that fungal spores can be dispersed more easily after the flies die (**FIGURE 13.2**). The fungus *Ophiocordyceps unilateralis* also manipulates the final actions of its host, the ant *Camponotus leonardi*. First, an infected ant climbs down from its home in the upper branches of trees and selects a leaf in a protected environment about 25 cm above the soil (Andersen et al. 2009). Then, just before the fungus kills it, the ant bites into the selected leaf with a "death grip" that will hold its body in place after it is dead. The fungus grows well in such protected environments, but it cannot survive where the ant usually lives—at the tops of trees, where the temperature and humidity are more variable. Thus, while the ant's final actions do not benefit the ant, they do allow the fungus to complete its life cycle in a favorable environment.

KEY CONCEPTS

CONCEPT 13.1 Parasites typically feed on only one or a few host species, but host species have multiple parasite species.

CONCEPT 13.2 Hosts have mechanisms for defending themselves against parasites, and parasites have mechanisms for overcoming host defenses.

CONCEPT 13.3 Host and parasite populations can evolve together, each in response to selection pressure imposed by the other.

CONCEPT 13.4 Hosts and parasites can have important effects on each other's population dynamics.

CONCEPT 13.5 Parasites can alter the outcomes of species interactions, thereby causing communities to change.

© Pascal Goetgheluck/Science Source

FIGURE 13.1 Driven to Suicide The behavior of this wood cricket (*Nemobius sylvestris*) was manipulated by the hairworm (*Paragordius tricuspidatus*) emerging from its body. By causing the cricket to jump into water (where it drowns), the parasite is able to continue its life cycle.

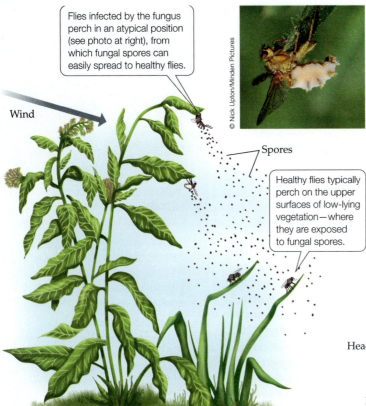

Flies infected by the fungus perch in an atypical position (see photo at right), from which fungal spores can easily spread to healthy flies.

Wind

Spores

Healthy flies typically perch on the upper surfaces of low-lying vegetation—where they are exposed to fungal spores.

FIGURE 13.2 Enslaved by a Fungus Shortly before they die from the infection, yellow dung flies infected by the fungus *Entomophthora muscae* move to the downwind side of a relatively tall plant and perch on the underside of one of its leaves. This position increases the chance that fungal spores released by *Entomophthora* will land on healthy yellow dung flies. (After D. P. Maitland. 1994. *Proc R Soc London* 258B: 187–193.)

Even vertebrates can be enslaved by parasites. Rats typically engage in predator avoidance behaviors in areas that show signs of cats. However, rats infected with the protist parasite *Toxoplasma gondii* behave abnormally: they do not avoid cats, and in some cases they are actually attracted to cats. While such a behavioral change can be a fatal attraction for the rat, it benefits the parasite because it increases the chance that the parasite will be transmitted to the next host in its complex life cycle—a cat.

How do some parasites enslave their hosts? Can the hosts fight back? More generally, what can these remarkable interactions tell us about host–parasite relationships?

Introduction

More than half of the millions of species that live on Earth are **parasites**, meaning that they live in or on other organisms and have a negative effect on those species (see Figure 12.3). To begin to understand how many parasites

there are, we need look no further than our own bodies (**FIGURE 13.3**). Our faces are home to mites that feed on exudates from the pores of our skin and on secretions at the base of our eyelashes. There are bacteria and fungi that grow on our skin and under our toenails. Arthropods such as lice may live on our heads, pubic regions, and other parts of our bodies. Moving inward, our tissues, organs, and body cavities can be infested with a rich variety of organisms, from viruses to bacteria to worms to fungi to protists.

Parasites consume the tissues or body fluids of the organism on or within which it lives, called its **host**; some parasites, called **pathogens**, cause diseases. Unlike carnivores but similar to herbivores, parasites typically harm, but do not immediately kill, the organisms they eat. The negative effects of parasites on their hosts vary widely, from mild to lethal. We see this variation in our own species, for which some parasites, such as the fungus that causes athlete's foot, are little more

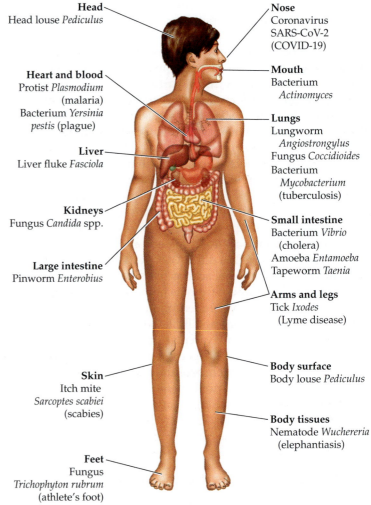

Head
Head louse *Pediculus*

Nose
Coronavirus SARS-CoV-2 (COVID-19)

Heart and blood
Protist *Plasmodium* (malaria)
Bacterium *Yersinia pestis* (plague)

Mouth
Bacterium *Actinomyces*

Lungs
Lungworm *Angiostrongylus*
Fungus *Coccidioides*
Bacterium *Mycobacterium* (tuberculosis)

Liver
Liver fluke *Fasciola*

Kidneys
Fungus *Candida* spp.

Small intestine
Bacterium *Vibrio* (cholera)
Amoeba *Entamoeba*
Tapeworm *Taenia*

Large intestine
Pinworm *Enterobius*

Arms and legs
Tick *Ixodes* (Lyme disease)

Skin
Itch mite *Sarcoptes scabiei* (scabies)

Body surface
Body louse *Pediculus*

Body tissues
Nematode *Wuchereria* (elephantiasis)

Feet
Fungus *Trichophyton rubrum* (athlete's foot)

FIGURE 13.3 The Human Body as Habitat for Symbionts Different parts of our bodies provide suitable habitat for a wide range of symbionts, many of which are parasites; only a few examples are shown here. Some of these organisms are pathogens that cause disease.

than a nuisance. Others, such as the protist *Leishmania tropica*, can cause disfigurement, and still others, such as the malaria bacterium *Plasmodium falciparum* or the coronavirus SARS-CoV2 (COVID-19), can kill. There is similar variation in the degree of harm caused by parasites that infect other species. Parasites vary in many other ways, as we'll see next as we examine their basic biology.

> **CONCEPT 13.1** Parasites typically feed on only one or a few host species, but host species have multiple parasite species.

LEARNING OBJECTIVES

13.1.1 Know why parasites are abundant and typically specialists.

13.1.2 Compare and contrast ectoparasites and endoparasites.

Parasite Natural History

Parasites vary in size from relatively large species (**macroparasites**), such as arthropods and worms, to species too small to be seen with the naked eye (**microparasites**), such as viruses, bacteria, protists, and unicellular fungi. But whether they are large or small, parasites typically feed on only one or a few host individuals over the course of their lives. Thus, defined broadly, parasites include herbivores, such as aphids or nematodes that feed on only one or a few host plants, as well as **parasitoids**, insects whose larvae feed on a single host, almost always killing it.

Most species are attacked by more than one parasite (**FIGURE 13.4**), and even parasites have parasites. Because parasites spend their lives feeding on one or a few host individuals, they tend to have a close relationship to the organisms they eat. For example, many parasites are closely adapted to particular host species, and many attack only one or a few host species. This specialization at the species level helps to explain why there are so many species of parasites—many host species have at least one parasite that eats *only* them. Overall, although the total number of parasite species is not known, a rough estimate is that 50% of the species on Earth are parasites (Windsor 1998).

Parasites are also specialized for living on or eating certain parts of the host's body. We'll focus next on this aspect of parasite specialization by describing both ectoparasites and endoparasites.

Ectoparasites live on the surface of their host

An **ectoparasite** lives on the outer body surface of its host (**FIGURE 13.5**). Ectoparasites include plants such as dodder and mistletoe that grow on, and obtain water and food from, another plant (see Figure 5.3). As described in Concept 5.1, such parasitic plants use modified roots

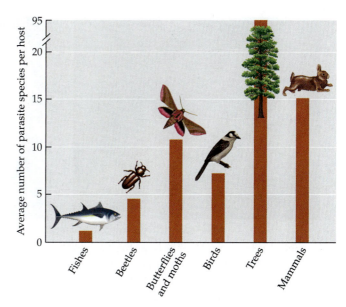

FIGURE 13.4 Many Species Are Host to More Than One Parasite Species In a study conducted in Britain, most host species were found to harbor more than one parasite species. The number of parasite species shown here for fishes, birds, and mammals includes only helminth worm parasites and hence is an underestimate of the actual number of parasite species found in these vertebrates. (After P. D. Stiling. 2002. *Ecology: Theories and Applications*, 4th ed. Prentice-Hall: Upper Saddle River, NJ.)

? Averaging across the six groups of organisms other than vertebrates (which we exclude because the data underestimate the true number of parasites), what is the average number of parasite species per host? Suppose the number of parasite species was determined for a previously unstudied host from one of the six groups. Is it likely that the number of parasites in that host would be close to the average you calculated? Explain.

called *haustoria* to penetrate the tissues of their host. Dodder cannot photosynthesize and hence depends on its host for both mineral nutrients and carbohydrates. In contrast, mistletoes are *hemiparasitic*: they extract water and mineral nutrients from their hosts, but since they have green leaves and can photosynthesize, they do not rely exclusively on their hosts for carbohydrates.

There are also many fungal and animal parasites that live on the surfaces of plants, feeding on their hosts' tissues or body fluids. More than 5,000 species of fungi attack important crop and horticultural plants, causing billions of dollars of damage each year. Some fungi that attack plants, including mildews, rusts, and smuts, grow on the surface of the host plant and extend their hyphae (fungal filaments) within the plant to extract nutrients from its tissues (see Figure 13.5A). Plants are also attacked by numerous animal ectoparasites, including aphids, whiteflies, and scale insects, which are found on stems and leaves, and nematode worms, beetles, and (juvenile) cicadas, which are found on roots. Animal ectoparasites that eat plants and live on their outer surfaces

(A)

(B)

FIGURE 13.5 Ectoparasites A wide range of parasites live on the outer surfaces of their hosts, feeding on host tissues. Examples include (A) the corn smut fungus (*Ustilago maydis*), seen here growing on an ear of corn, and (B) the velvet mite (*Trombidium* spp.), which in its larval form feeds parasitically on the blood of insects, such as this sawfly larva.

can sometimes be thought of both as herbivores (because they eat plant tissues) and as parasites (especially if they remain on a single host plant for much of their lives).

A similar array of fungal and animal ectoparasites can be found on the surfaces of animals. Familiar examples include *Trichophyton rubrum*, the fungus that causes athlete's foot, and fleas, mites, lice, and ticks, which feed on the tissues or blood of their hosts (see Figure 13.5B). Some of these parasites also transmit diseases to their hosts, including fleas that spread the plague and ticks that spread Lyme disease.

Endoparasites live inside their host

If we ignore the details of their shape, we can think of people and most other animals as being constructed in a similar way: their bodies consist of tissues that surround an open tube called the *alimentary canal*. The alimentary canal runs through the middle of the body, from the mouth to the anus. Parasites that live inside their hosts, called **endoparasites**, include species that inhabit the alimentary canal as well as species that live within host cells or tissues.

The alimentary canal provides an excellent habitat for parasites. The host brings in food at one end (the mouth) and excretes what it cannot digest at the other (the anus). Parasites that live within the alimentary canal often do not eat host tissues at all; instead, they rob the host of nutrients. A tapeworm, for example, has a *scolex*, a structure with suckers (and sometimes hooks) that it uses to attach itself to the inside of the host's intestine (**FIGURE 13.6A**). Once it is attached, the tapeworm simply absorbs food that the host has already digested. Tapeworms that infect humans can grow up to 5–10 m (16–33 feet) long; large tapeworms such as these can block the intestines and cause nutritional deficiencies.

Many other endoparasites live within the cells or tissues of animal hosts, causing a wide range of symptoms as they

(A)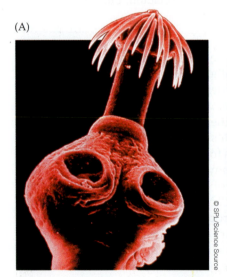

(B)

(C)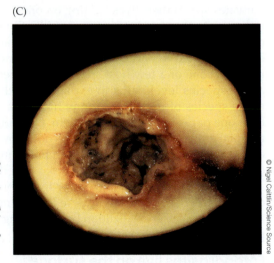

FIGURE 13.6 Endoparasites Many parasites live within the body of their host, feeding on the host's tissues or robbing it of nutrients. (A) The tapeworm *Taenia taeniaeformis* uses the suckers and hooks shown here to attach to the intestinal wall of its mammalian host, often a rodent, rabbit, or cat. Once attached, an adult can grow to over 5 m (16 feet) in length. (B) The bacterium *Mycobacterium tuberculosis* causes the lung disease tuberculosis, which kills 1–2 million people each year. (C) This section of a potato tuber shows the destruction wrought by *Erwinia carotovora*, a bacterium that causes soft rot. Affected areas become soft with decay and develop a distinctive foul odor.

reproduce or consume host tissues. Examples in humans include *Yersinia pestis*, the bacterium that causes the plague, and *Mycobacterium tuberculosis*, the bacterium that causes tuberculosis (TB) (**FIGURE 13.6B**). TB is a potentially fatal lung disease, aptly referred to as the "Captain of Death"; with the possible exception of malaria, it has killed more people than any other disease in human history. TB continues to kill 1–2 million adults each year (a number comparable to the roughly 1 million that currently die each year from AIDS). Another example is the COVID-19 pandemic caused by the coronavirus SARS-CoV2, which has caused nearly 7 million deaths since January 2020.

Plants too are attacked by a wide variety of endoparasites, including bacterial pathogens that cause soft rot in various plant parts, such as fruits (e.g., tomatoes) or storage tissues (e.g., potatoes; **FIGURE 13.6C**). Other plant pathogens include fungi that cause plant parts to rot from the inside out. Some bacteria invade plant vascular tissues, where they disrupt the flow of water and nutrients, causing wilting and often death. Plant pathogens can have large effects on natural communities, as illustrated by the protist *Phytophthora ramorum*, which causes sudden oak death, a disease that has recently killed more than a million oaks and other trees in California and Oregon (see also the chestnut blight in Figure 13.14).

Endoparasitism and ectoparasitism have advantages and disadvantages

There are advantages and disadvantages to living in or on a host (**TABLE 13.1**). Because ectoparasites live on the surface of their host, it is relatively easy for them or their offspring to disperse from one host individual to another. It is much more difficult for endoparasites to disperse to new hosts. Endoparasites solve this problem in a variety of ways. Some, like the enslaver parasites discussed in the Case Study at the opening of this chapter, alter the physiology or behavior of their host in ways that facilitate their

TABLE 13.1

Advantages and Disadvantages of Living in or on a Host

	Ectoparasitism	Endoparasitism
Advantages	Dispersal easier	Feeding easier
	Safer from host's immune system	More protection from external environment
Disadvantages	Feeding more difficult	Dispersal more difficult
	Greater exposure to external environment	Greater vulnerability to host's immune system
	Greater vulnerability to natural enemies	

dispersal. Other examples include the bacterium *Vibrio cholerae*, which causes cholera, and the amoeba *Entamoeba histolytica*, which causes amoebic dysentery. People with cholera and dysentery have diarrhea, a condition that increases the chance that the parasite will contaminate drinking water and thereby spread to new hosts. Other endoparasites have complex life cycles that include stages that are specialized for dispersing from one host species to another (see Figure 13.9).

Although dispersal is relatively easy for ectoparasites, there are costs to life on the surface of a host. Compared with endoparasites, ectoparasites are more exposed to natural enemies such as predators, parasitoids, and parasites. Aphids, for example, are attacked by ladybugs, birds, and many other predators, as well as by lethal parasitoids and by parasites such as mites that suck fluids from their bodies. Endoparasites, in contrast, are safe from all but the most specialized predators and parasites. Endoparasites are also relatively well protected from the external environment, and they have relatively easy access to food—unlike an ectoparasite, an endoparasite does not have to pierce the host's protective outer surfaces to feed. But living within the host does expose endoparasites to a different sort of danger: more exposure to the host's immune system. Some parasites have evolved ways to tolerate or overcome immune system defenses, as we will see in the following section.

> **CONCEPT 13.2** Hosts have mechanisms for defending themselves against parasites, and parasites have mechanisms for overcoming host defenses.

LEARNING OBJECTIVES

13.2.1 Describe the mechanisms organisms use for defending themselves against parasites.

13.2.2 List the mechanisms that parasites use to circumvent host defenses.

Defense and Counterdefenses

As we saw in Chapter 12, carnivores and herbivores exert strong pressure on their food organisms, and vice versa. The same is true of parasites and their hosts: hosts have ways to protect themselves against parasites, and parasites have countermeasures to circumvent host defenses.

Immune systems, biochemical defenses, and symbionts can protect hosts against parasites

Host organisms have a wide range of defensive mechanisms that can prevent or limit the severity of parasite attacks. For example, a host may have a protective outer covering, such as the skin of a mammal or the hard exoskeleton of an insect, that can keep ectoparasites from piercing its body or make it difficult for endoparasites to enter. Endoparasites that do manage to enter the host's body are often killed or rendered less effective by the host's immune system, biochemical defenses, or defensive symbionts.

IMMUNE SYSTEMS The vertebrate immune system includes specialized cells that allow the host to recognize microparasites to which it has been previously exposed; in many instances, the "memory cells" of the immune system are so effective that the host has lifelong immunity against future attack by the same microparasite species. Other immune system cells engulf and destroy parasites or mark them with chemicals that target them for later destruction.

Plants can also mount highly effective responses to invasion by parasites. Some plants have resistance genes, the different alleles of which provide protection against microparasites with particular genotypes; we will describe this defense system in more detail in Concept 13.3. Plants are not helpless, however, even when they lack alleles that provide resistance to a specific attacker. In such a case, the plant relies on a nonspecific immune system that produces antimicrobial compounds, including some that attack the cell walls of bacteria and others that are toxic to fungal parasites (**FIGURE 13.7**). The plant may also produce chemical signals that "warn" nearby cells of imminent attack, and still other chemicals that stimulate the deposition of lignin, a hard substance that provides a barricade against the invader's spread.

BIOCHEMICAL DEFENSES Hosts have ways of regulating their biochemistry to limit parasite growth. Bacterial and fungal endoparasites, for example, require iron to grow. Vertebrate hosts—including mammals, birds, amphibians, and fishes—have a protein called transferrin that removes iron from their blood serum (where parasites could use it) and stores it in intracellular compartments (where parasites cannot get to it). Transferrins are so efficient that the concentration of free iron in mammalian blood serum is only 10^{-26} M—so low that parasites cannot grow in vertebrate blood unless they can somehow outmaneuver the host. To do this, some parasites steal iron from the transferrin itself and use it to support their own growth.

Similar biochemical battles occur between plants and their parasites. As we saw in Concept 12.2, plants use a rich variety of chemical weapons to kill or deter the organisms that eat them. Plant defensive secondary compounds are so effective that some animals eat specific plants in order to treat or prevent parasite infections. For example, when parasitic flies lay eggs on the bodies of woolly bear caterpillars, the caterpillars switch from their usual food plant (lupines) to a diet of poisonous hemlock (Karban and English-Loeb 1997). The new diet does not kill the parasites, but it does increase the chance that the caterpillar will survive the attack and metamorphose into an adult tiger moth (*Platyprepia virginalis*). Chimpanzees infected with the nematode *Oesophagostomum stephanostomum* specifically seek out and eat a bitter plant that scientists have learned contains compounds that kill or paralyze the nematodes and can also deter many other parasites (Huffman 1997). Humans do essentially the same thing: we spend billions of dollars each year on pharmaceuticals that are based on compounds originally obtained from plants.

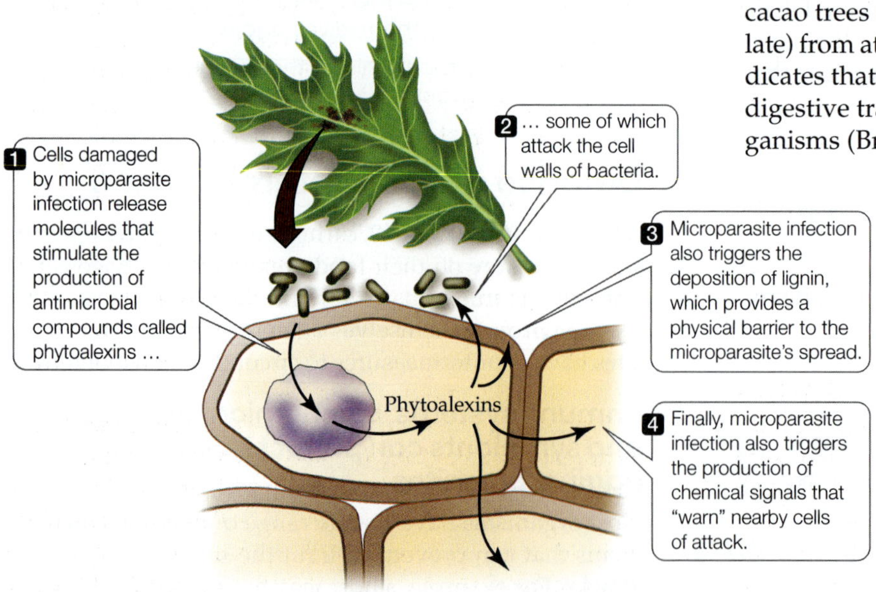

DEFENSIVE SYMBIONTS Some organisms are aided in their defense against parasites by mutualistic symbionts such as bacteria and fungi. For example, fungal symbionts living within leaves protect grasses and plants such as cacao trees (the source of the beans used to make chocolate) from attack by pathogens. Growing evidence also indicates that bacterial symbionts living within the human digestive tract can protect us against disease-causing organisms (Britton and Young 2012).

Many such "defensive symbionts" are heritable, meaning the symbiont is reliably transmitted from a host to its offspring. We might expect that hosts harboring heritable defensive symbionts should increase in frequency in a population when parasites are common—and indeed, that frequently happens. For example, in a laboratory experiment, the frequency of pea aphids (*Acyrthosiphon pisum*) harboring the bacterial symbiont *Hamiltonella defensa* increased rapidly in the presence of a lethal wasp parasite (Oliver et al. 2008). This was expected, because the symbiont is heritable and because pea aphids harboring the symbiont survived at higher rates than did

1. Cells damaged by microparasite infection release molecules that stimulate the production of antimicrobial compounds called phytoalexins …

2. … some of which attack the cell walls of bacteria.

3. Microparasite infection also triggers the deposition of lignin, which provides a physical barrier to the microparasite's spread.

4. Finally, microparasite infection also triggers the production of chemical signals that "warn" nearby cells of attack.

Phytoalexins

FIGURE 13.7 Nonspecific Plant Defenses Plants can mount a nonspecific defensive response that is effective against a broad range of fungal and bacterial microparasites.

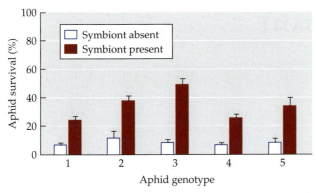

FIGURE 13.8 Protected by a Symbiont Pea aphids (*Acyrthosiphon pisum*) of five different genotypes were exposed to the pathogenic fungus *Pandora neoaphidis*. For each of these genotypes, some aphids were inoculated with the bacterial symbiont *Regiella insecticola*, while other aphids lacked the symbiont. Aphids harboring the symbiont survived at higher rates than did aphids lacking the symbiont. Error bars show one standard error of the mean. (After C. L. Scarborough et al. 2005. *Science* 310: 1781.)

pea aphids lacking the symbiont. In another study on pea aphids, the bacterial symbiont *Regiella insecticola* was found to protect against attack by a deadly fungal parasite (**FIGURE 13.8**). Defensive symbionts have also been shown to protect against attack by nematode parasites, as you can explore in **ANALYZING DATA 13.1**.

Parasites have mechanisms that circumvent host defenses

To survive and reproduce, a parasite must be able to tolerate or evade its host's defensive mechanisms. Aphids and other ectoparasites, for example, must be able to pierce the protective outer covering of the host, and they must be able to tolerate whatever chemical compounds are present in the host tissues or body fluids that they eat. Viewed broadly, the challenges faced by ectoparasites are similar to those faced by herbivores and carnivores as they attempt to cope with the toxins and physical structures that their prey use to defend themselves. We discussed such challenges in Concept 12.2, so here we focus on how endoparasites cope with defenses found inside the host.

COUNTERDEFENSES AGAINST ENCAPSULATION
Endoparasites face formidable challenges from host immune systems and related aspects of host biochemistry. Host species typically have a number of ways to destroy parasite invaders. In addition to the strategies we have already described, some hosts can cover parasites or parasite eggs with capsules that kill them or render them harmless, a process called *encapsulation*.

Some insects defend themselves against macroparasites using encapsulation. Insect blood cells can engulf small invaders, such as bacteria, but they cannot engulf large objects, such as nematodes or parasitoid eggs. However, some insects have *lamellocytes*, which are blood cells that can form multicellular sheaths (capsules) around large objects. When an insect mounts such an encapsulation defense, most or all of the attacking parasites may be destroyed. As a result, the parasites are under strong selection to develop a counterdefense.

For example, *Drosophila* fruit flies have an effective defense against wasp parasitoids: they encapsulate (and hence kill) their eggs. Parasitoid wasps that attack fruit flies avoid encapsulation in several different ways. When wasps in the genus *Leptopilina* lay their eggs inside a fruit fly host, they also inject virus-like particles into the host. These particles infect the host's lamellocytes and cause them to self-destruct, thus weakening the host's resistance and increasing the percentage of wasp eggs that survive (Rizki and Rizki 1990). Other parasitoid wasps, such as *Asobara tabida*, lay eggs covered with filaments. These filaments cause the eggs to stick to and become embedded in fat cells and other host cells, where they are not detected by circulating lamellocytes.

COUNTERDEFENSES INVOLVING HUNDREDS OF GENES
Some endoparasites have a complex set of adaptations that allows them to thrive inside their host. One such endoparasite is *Plasmodium falciparum*, a protist that causes malaria, a disease that kills half a million people each year (**FIGURE 13.9**). *Plasmodium*, like many endoparasites, has a complex life cycle with specialized stages that allow it to alternate between a mosquito and a human host. Infected mosquitoes contain one specialized *Plasmodium* stage, called a *sporozoite*, in their saliva. When an infected mosquito bites a human, sporozoites enter the victim's bloodstream and travel to the liver, where they divide to form another stage, called a *merozoite*. The merozoites penetrate red blood cells, where they multiply rapidly. After 48–72 hours, large numbers of merozoites break out of the red blood cells, causing the periodic chills and fever that are associated with malaria. Some of the offspring merozoites attack more red blood cells, while others transform into gamete-producing cells. If another mosquito bites the victim, it picks up some of the gamete-producing cells, which enter its digestive tract and form gametes. After fertilization occurs, the resulting zygotes produce thousands of sporozoites, which then migrate to the mosquito's salivary glands, where they await their transfer to another human host.

Plasmodium faces two potentially lethal challenges from its human host. First, red blood cells do not divide or grow, and hence they lack the cellular machinery needed to import nutrients necessary for growth. A *Plasmodium* merozoite inside a red blood cell would starve if it did not have a way to obtain essential nutrients. Second, after 24–48 hours, a *Plasmodium* infection causes red blood cells to have an abnormal shape. The human spleen recognizes and destroys such deformed cells, along with the parasites inside.

ANALYZING DATA 13.1

Will a Defensive Symbiont Increase in Frequency in a Host Population Subjected to Parasitism?

Although we would expect heritable defensive symbionts to increase in frequency in host populations subjected to parasitism, few studies have tested this hypothesis. Jaenike and Brekke (2011)* performed such a test, using laboratory populations of the fruit fly *Drosophila neotestacea*. These flies harbor a bacterial symbiont of the genus *Spiroplasma*, which protects flies from the nematode parasite *Howardula aoronymphium*. *Howardula* can sterilize female flies and reduce the mating success of male flies.

Jaenike and Brekke established five replicate populations in which flies were exposed every generation to the nematode parasite and five replicate populations in which the parasite was absent. Initially, each population had a 50:50 mixture of *Spiroplasma*-infected and uninfected adult flies. In a second experiment, the researchers established five replicate populations in which all flies were infected with *Spiroplasma* and five replicate populations in which all flies were uninfected. All populations in this second experiment were exposed to *Howardula* parasites (but not necessarily infected by *Howardula*) in the first generation only. Both experiments were run for seven fly generations. The results for each experiment are shown in the tables.

Experiment 1 Percentage of Fruit Fly Individuals Harboring *Spiroplasma* Symbionts

Treatment	Generation						
	1	2	3	4	5	6	7
Howardula absent	54	65	52	65	59	65	39
Howardula present	49	52	86	92	97	99	96

Experiment 2 Percentage of Fruit Fly Individuals Infected by the Nematode Parasite *Howardula*

Treatment	Generation						
	1	2	3	4	5	6	7
Spiroplasma absent	30	59	95	92	87	—	—
Spiroplasma present	25	15	7	2	1	0	0

By generation 6, all fly populations were extinct because *Howardula* had sterilized the flies it parasitized.

1. Plot the percentage of flies harboring *Spiroplasma* (*y*-axis) versus generation (*x*-axis) for both treatments in Experiment 1. Describe the hypothesis tested by this experiment. Which treatment represents the control? Do the results support the hypothesis?

2. Plot the percentage of flies infected by *Howardula* (*y*-axis) versus generation (*x*-axis) for both treatments in Experiment 2. Describe the hypothesis tested by this experiment. Which treatment represents the control? Do the results support the hypothesis?

3. Examine the graphs you made for Questions 1 and 2. Do the results indicate that there is a cost to flies for harboring *Spiroplasma*? Explain.

*Jaenike, J., and T. D. Brekke. 2011. Defensive endosymbionts: A cryptic trophic level in community ecology. *Ecology Letters* 14: 150–155.

Plasmodium addresses these challenges by having hundreds of genes whose function is to modify the host red blood cell in ways that allow the parasites to obtain food and escape destruction by the spleen (Hiller et al. 2004; Marti et al. 2004). Some of these genes cause transport proteins to be placed on the surface of the red blood cell, thereby enabling the parasite to import essential nutrients into the host cell. Other genes guide the production of unique knobs that are added to the surface of the red blood cell. These knobs cause the infected red blood cell to stick to other human cells, thereby preventing it from traveling in the bloodstream to the spleen, where it would be recognized as infected and then destroyed. The proteins on these knobs vary greatly from one parasite individual to another, making it difficult for the human immune system to recognize and destroy the infected cells.

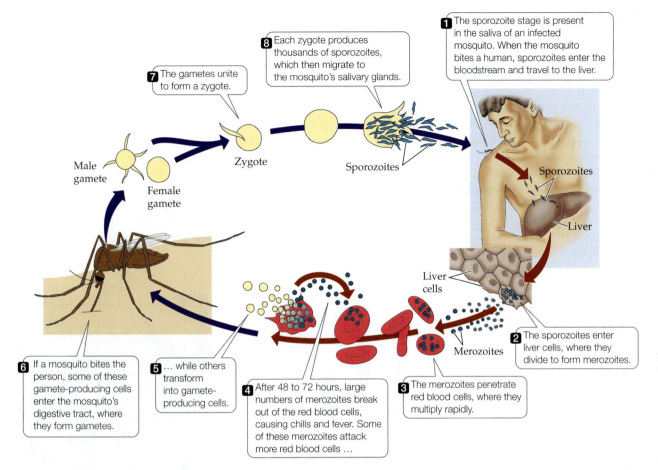

7 The gametes unite to form a zygote.

8 Each zygote produces thousands of sporozoites, which then migrate to the mosquito's salivary glands.

1 The sporozoite stage is present in the saliva of an infected mosquito. When the mosquito bites a human, sporozoites enter the bloodstream and travel to the liver.

Male gamete

Female gamete

Zygote

Sporozoites

Sporozoites

Liver

Liver cells

Merozoites

2 The sporozoites enter liver cells, where they divide to form merozoites.

6 If a mosquito bites the person, some of these gamete-producing cells enter the mosquito's digestive tract, where they form gametes.

5 … while others transform into gamete-producing cells.

4 After 48 to 72 hours, large numbers of merozoites break out of the red blood cells, causing chills and fever. Some of these merozoites attack more red blood cells …

3 The merozoites penetrate red blood cells, where they multiply rapidly.

FIGURE 13.9 Life Cycle of the Malaria Parasite The life cycle of the protist *Plasmodium falciparum* includes specialized stages that facilitate the dispersal of this endoparasite from one host to another. The sporozoite stage, for example, enables the parasite to disperse from an infected mosquito to a human host.

? Which stage in the life cycle enables the parasite to disperse from a human host to a mosquito?

CONCEPT 13.3 Host and parasite populations can evolve together, each in response to selection pressure imposed by the other.

LEARNING OBJECTIVES

13.3.1 Understand how host–parasite interactions can result in coevolution.

13.3.2 Know why host–parasite interactions can result in life history trade-offs.

Parasite–Host Coevolution

As we have just seen, *Plasmodium* has specific mechanisms that enable it to live inside a red blood cell. When both a parasite and its host possess such specific mechanisms, that observation suggests that the strong selection pressure that hosts and parasites impose on each other has caused their populations to evolve. Such changes have been directly observed in Australia, where the myxoma virus was introduced to control populations of the European rabbit (*Oryctolagus cuniculus*).

European rabbits were introduced to Australia in 1859, when 24 wild rabbits were released at a ranch in Victoria. Within a decade, rabbit populations had grown so large, and were consuming so much plant material, that they posed a threat to cattle and sheep pasturelands and wool production. Several control measures were enacted, including introductions of predators, shooting and poisoning of rabbits, and the building of fences to limit the spread of rabbits from one region to another (Fenner and Ratcliffe 1965). None of these methods worked: by the 1900s, hundreds of millions of rabbits had spread throughout much of the continent.

After years of investigation, Australian government officials settled on a new control measure: introduction of the myxoma virus. A rabbit infected with this virus may suffer from skin lesions and severe swellings, which can lead to blindness, difficulty with feeding and drinking, and death (usually within 2 weeks of infection). The virus is transmitted from rabbit to rabbit by mosquitoes. In 1950, when the virus was first used to control rabbit populations, 99.8% of infected rabbits died. In the ensuing decades, millions of rabbits were killed by the virus and the sizes of rabbit populations dropped dramatically throughout the Australian continent. Over time, however, rabbit populations evolved resistance to the virus, and the virus evolved to become less lethal (**FIGURE 13.10**). The myxoma virus is still used to control rabbit populations, but doing so requires a constant search for new, lethal virus strains to which the rabbit has not evolved resistance.

The increased resistance of the rabbit and the reduced lethality of the virus illustrate **coevolution**, which occurs when populations of two interacting species evolve together, each in response to selection pressure imposed by the other. The outcome of coevolution can vary greatly depending on the biology of the interacting species. In the European rabbit, selection favored the evolution of increased resistance to viral attack, as you might expect. In addition, viral strains of intermediate lethality predominated, perhaps because such strains allowed rabbits to live long enough for one or more mosquitoes to bite them and transmit the virus to another host (mosquitoes do not bite dead rabbits). In other cases of host–parasite coevolution, the parasite evolves counterdefenses to overcome host resistance mechanisms, as the following examples illustrate.

Selection can favor a diversity of host and parasite genotypes

As mentioned earlier, plant defense systems include a specific response that makes particular plant genotypes resistant to particular parasite genotypes. Such *gene-for-gene interactions* are well documented in a number of plant species, including wheat, flax, and *Arabidopsis thaliana*. Wheat has dozens of different genes for resistance to fungi such as wheat rusts (*Puccinia*). Different wheat rust genotypes can overcome different wheat resistance genes, however, and periodically, mutations occur in wheat rusts that produce new genotypes to which wheat is not resistant. Studies have shown that the frequencies of wheat

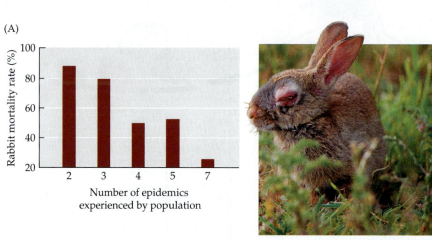

(A)

A rabbit infected with the myxoma virus

© Clifford Rhodes/Alamy Stock Photo

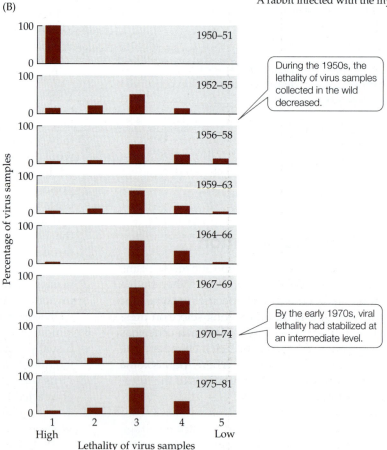

(B)

During the 1950s, the lethality of virus samples collected in the wild decreased.

By the early 1970s, viral lethality had stabilized at an intermediate level.

FIGURE 13.10 Coevolution of the European Rabbit and the Myxoma Virus (A) After the introduction of the myxoma virus to Australia, researchers periodically tested its lethality by collecting rabbits from a wild population and exposing them to a standard strain of the virus that killed 90% of naive (unselected) laboratory rabbits. Over time, mortality in those wild rabbits declined as the number of epidemics increased and the population evolved resistance to the virus. (B) The lethality of virus samples collected in rabbits in the wild also declined, as was determined when they were tested against a standard (unselected) line of rabbits. (A, extracted from P. J. Kerr and S. M. Best. 1998. Myxoma virus in rabbits. In *Genetic resistance to animal diseases*. M. Müller and G. Brem (Eds.). *Rev Sci Tech Off Int Epiz* 17(1): 256–268. Available at: http://dx.doi.org/10.20506/rst.17.1.1081. World Organisation for Animal Health at www.oie.int; B after R. M. May and R. M. Anderson. 1983. *Proc R Soc London* 219B: 281–313.)

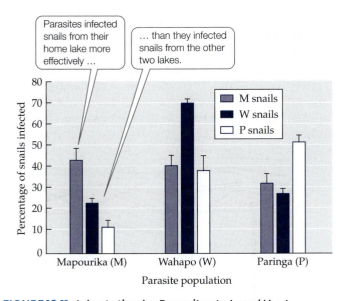

FIGURE 13.11 plot labels:
- Parasites infected snails from their home lake more effectively …
- … than they infected snails from the other two lakes.
- M snails
- W snails
- P snails
- Percentage of snails infected
- Mapourika (M) Wahapo (W) Paringa (P)
- Parasite population

FIGURE 13.11 **Adaptation by Parasites to Local Host Populations** The graph shows the frequencies with which *Microphallus* parasites from three lakes in New Zealand (Lake Mapourika, Lake Wahapo, and Lake Paringa) were able to infect snails (*Potamopyrgus antipodarum*) from the same three lakes.

Lake Mapourika

Error bars show one standard error of the mean. (After C. M. Lively. 1989. *Evolution* 43: 1663–1671.)

? Do snails with poor defenses against parasites from their own lake also have poor defenses against parasites from other lakes? Explain.

rust genotypes vary considerably over time as farmers use different resistant varieties of wheat. For example, a rust variety may be abundant in one year because it can overcome the resistance genes of the wheat varieties planted that year, yet less abundant the following year because it cannot overcome the resistance genes of the different wheat varieties planted that year.

Changes in the frequencies of host and parasite genotypes also occur in natural systems. In the lakes of New Zealand, a trematode worm (*Microphallus* sp.) parasitizes the snail *Potamopyrgus antipodarum*. The worm has serious negative effects on its snail hosts: it castrates the males and sterilizes the females. The parasite has a much shorter generation time than its host, and hence we might expect that it would rapidly evolve the ability to cope with the snail's defensive mechanisms. Lively (1989) tested this idea in an experiment that pitted parasites from each of three lakes against snails from the same three lakes. He found that parasites infected snails from their home lake more effectively than they infected snails from the other two lakes (**FIGURE 13.11**). This observation suggests that the parasite genotypes in each lake had evolved rapidly enough to overcome the defenses of the snail genotypes found in that lake.

The snails also evolved in response to the parasites, albeit more slowly. Dybdahl and Lively (1998) documented the abundances of different snail genotypes over a 5-year period in another New Zealand lake. The snail genotype that was most abundant changed from one year to the next. Moreover, roughly a year after a snail genotype was the most abundant one in the population, snails of that genotype had a higher-than-typical number of parasites. Together with Lively's earlier study (1989), these results suggest that parasite populations evolve to exploit the snail genotypes found in their local environment. Refining this idea further, Dybdahl and Lively hypothesized that as a result of evolution by natural selection, parasites would be able to infect snails with a common genotype at a higher rate than they could infect snails with a rare genotype. That is exactly what they found in a laboratory experiment (**FIGURE 13.12**). Hence, snail genotype frequencies may change from year to year because common genotypes are attacked by many parasites, placing them at a disadvantage and driving down their numbers in future years.

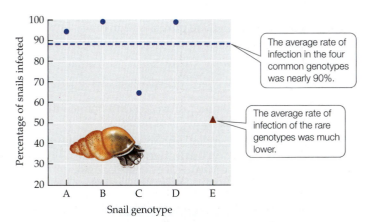

FIGURE 13.12 plot labels:
- Percentage of snails infected
- The average rate of infection in the four common genotypes was nearly 90%.
- The average rate of infection of the rare genotypes was much lower.
- A B C D E
- Snail genotype

FIGURE 13.12 **Parasites Infect Common Host Genotypes More Easily Than Rare Genotypes** In a laboratory experiment, Dybdahl and Lively compared rates of *Microphallus* infection in four common snail genotypes (A–D, represented by blue dots) and in a group of 40 rare snail genotypes (E, represented by a red triangle). The parasites and snails in this experiment were all taken from the same lake. (After M. F. Dybdahl and C. M. Lively. 1998. *Evolution* 52: 1057–1066.)

Host defenses and parasite counterdefenses both have costs

Parasites and hosts have such a powerful effect on each other that we might expect an ever-escalating "arms race" in which host resistance and parasite counterdefenses both get stronger and stronger over time. But such an outcome rarely occurs. In some cases—as in Dybdahl and Lively's snails and trematodes—host genotypes that are common decrease in frequency because they are attacked by many parasites, leading to an increase in the frequency of a previously rare genotype, and the arms race continually begins anew. An arms race may also stop because of trade-offs: a trait that improves a host's defenses or a parasite's counterdefenses may have costs that reduce other aspects of the organism's growth, survival, or reproduction.

Such trade-offs have been documented in a number of host–parasite systems, including *Drosophila* fruit flies and the parasitoid wasps that attack them (described in Concept 13.2). Alex Kraaijeveld and colleagues (2001) have shown that selection can increase both the frequency with which fruit fly hosts encapsulate wasp eggs (from 5% to 60% in five generations) and the ability of wasp eggs to avoid encapsulation (from 8% to 37% in ten generations). But they have also shown that there are costs to these defenses and counterdefenses. For example, fruit flies from lineages that can mount an encapsulation defense have lower larval survival rates when they compete for food with flies of the same species that cannot. Similarly, wasp eggs that avoid encapsulation by becoming embedded in host tissues take longer to hatch than do other eggs.

The evolutionary changes in host and parasite populations that we've discussed in this section reflect the profound effects these organisms have on each other. Next, we'll focus on some of the population consequences of host–parasite interactions.

> **CONCEPT 13.4** Hosts and parasites can have important effects on each other's population dynamics.

LEARNING OBJECTIVES

13.4.1 Understand how parasites can influence the population dynamics of hosts.

13.4.2 Explain how simple models of host–pathogen dynamics can be used to control the establishment and spread of diseases.

Host–Parasite Population Dynamics

As we've seen, parasites can reduce the survival, growth, or reproduction of their hosts—an observation that is illustrated clearly by the large drop in reproductive success that a sexually transmitted mite can inflict on its beetle host (**FIGURE 13.13**). At the population level, the harm that parasites cause host individuals translates into a reduction of the host population growth rate, λ (see Concept 11.1). As we will see in this section, the reduction in λ can be drastic: parasites may drive local host populations extinct or even reduce the geographic range of the host species.

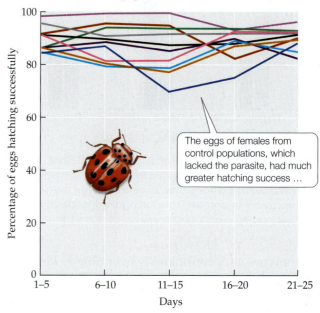

(A) Females from control populations

> The eggs of females from control populations, which lacked the parasite, had much greater hatching success …

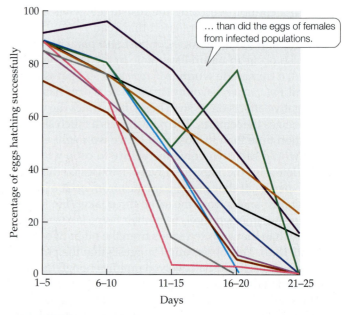

(B) Females from infected populations

> … than did the eggs of females from infected populations.

FIGURE 13.13 Parasites Can Reduce Host Reproduction Researchers infected experimental populations of the beetle *Adalia decempunctata* with a sexually transmitted mite parasite (*Coccipolipus hippodamiae*). Over the next 25 days, they monitored the proportions of the eggs laid by female beetles from (A) control and (B) infected populations that hatched. Each curve represents the eggs laid by a single female. (After K. M. Webberley et al. 2004. *J Anim Ecol* 73: 1–10.)

In other, less extreme cases, parasites may reduce host abundances or otherwise alter host population dynamics without causing the extinction of host populations.

Parasites can drive host populations to extinction

The amphipod *Corophium volutator* lives in North Atlantic tidal mudflats. *Corophium* is small (1 cm long) and often very abundant, reaching densities of up to 100,000 individuals per square meter. *Corophium* builds tubular burrows in the mud, from which it feeds on plankton suspended in the water and on microorganisms found in sediments near the burrow opening. It is eaten by a wide range of organisms, including migratory birds and trematode parasites. The parasites can reduce the size of *Corophium* populations greatly, even to the point of local extinction. For example, in a 4-month period, attack by trematodes caused the extinction of a *Corophium* population that initially had 18,000 individuals per square meter (Mouritsen et al. 1998).

Parasites can also drive host populations to extinction over a large geographic region. The American chestnut (*Castanea dentata*) once was a dominant member of deciduous forest communities in eastern North America (**FIGURE 13.14**), but the parasitic fungus *Cryphonectria parasitica* changed that completely. This fungal pathogen causes chestnut blight, a disease that kills chestnut trees. The fungus was introduced to New York City from Asia in 1904 (Keever 1953). By midcentury, the fungus had wiped out most chestnut populations, greatly reducing the geographic range of this once-dominant species.

Isolated chestnut trees still can be found in North American forests, and some of these trees show signs of resistance to the fungus. But it is likely that many of the standing trees simply have not yet been found by the fungus. Once the fungus reaches a tree, it enters the tree through a hole or wound in the bark, killing the aboveground portion of the tree in 2–10 years. Before they die, infected trees may produce seeds, which may germinate and give rise to offspring that live for 10–15 years before they are killed by the fungus in turn. Some infected trees also produce sprouts from their roots, but these are usually killed a few years after they appear aboveground. Efforts are under way to breed resistant chestnut varieties, but at present it is not known whether chestnut populations will ever recover from the onslaught of the chestnut blight fungus.

Parasites can influence host population cycles

Ecologists have long sought to determine the causes of population cycles. As we saw in Concept 12.3, such cycles may be caused by three-way feeding relationships—by the effects that predators and herbivorous prey have on each other, coupled with the effects that those prey and their food plants have on each other.

Population cycles can also be influenced by parasites. Consider the work of Peter Hudson and colleagues, who manipulated the abundances of parasites in red grouse (*Lagopus lagopus*) populations on moors in northern England. In this region, red grouse populations tend to crash every 4 years. Previous studies had shown that a parasitic nematode, *Trichostrongylus tenuis*, decreased the survival and reproductive success of individual red grouse. Hudson et al. (1998) investigated whether this parasite might also cause grouse populations to cycle.

(A)

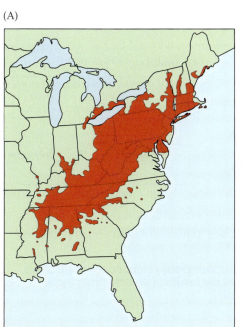

(B)

Some chestnut trees had a diameter up to twice the size of the one shown here.

FIGURE 13.14 Parasites Can Reduce Their Host's Geographic Range (A) The original distribution of the American chestnut (*Castanea dentata*) is shown in darker, shaded region. Although a few chestnut trees remain standing, a fungal parasite drove this once-dominant species virtually extinct throughout its entire former range. (B) Chestnuts were once important timber trees (note the two loggers shown in the photograph). (Range detail courtesy of Elbert L. Little, Jr. 1970. *Atlas of United States Trees.* U.S. Department of Agriculture, Forest Service, and other publications.)

The researchers studied changes in red grouse numbers in six replicate populations over the course of two population cycles. Long-term data on grouse population cycles indicated that these populations were likely to crash in 1989 and again in 1993. In two of the six study populations, the researchers treated as many grouse as they could catch in 1989 and 1993 with a drug that killed the parasitic nematodes. In two of the other study populations, grouse were caught and treated for parasites in 1989 only. The remaining two populations served as unmanipulated controls. Because each replicate population covered a very large area (17–20 km²), it was not possible to count red grouse directly. Instead, Hudson and colleagues used the number of red grouse shot by hunters as an index of the actual population size.

In the control populations, red grouse numbers crashed as predicted in 1989 and 1993 (**FIGURE 13.15**). Although parasite removal did not completely stop the red grouse population cycle, it did reduce the fluctuation in grouse numbers considerably; this was particularly true for the populations that were treated for parasites in both 1989 and 1993. Thus, the experiment provided strong evidence that parasites influence—and may be the primary cause of—red grouse population cycles.

As we've seen, parasites that cause diseases (pathogens) can greatly affect the population dynamics of both wild and domesticated plant and animal species. Pathogens also have large effects on human populations—so much so that they are thought to have played a major role in the rise and fall of civilizations throughout the course of human history (McNeill 1976; Diamond 1997). One example is the European conquest of North America, where up to 95% of the native population (19 million of the original 20 million) were killed by new diseases brought to the continent by European trappers, missionaries, settlers,

and soldiers. Even with such massive mortality, the conquest took roughly 400 years; without it, the conquest would certainly have taken longer, and might have failed. Pathogens continue to be a major source of human mortality today. Despite medical advances, millions of people die each year from diseases such as COVID-19, AIDS, tuberculosis, and malaria.

Simple models of host–pathogen dynamics suggest ways to control the establishment and spread of diseases

Considerable effort has been devoted to the development of mathematical models of host–pathogen population dynamics. These models often differ in three ways from those we have seen in earlier chapters. First, the host population is subdivided into categories, such as susceptible individuals, infected individuals, and recovered and immune individuals. Second, it is often necessary to keep track of both host and pathogen genotypes because, as we have seen, host genotypes may differ greatly in their resistance to the pathogen, and pathogen genotypes may differ greatly in their ability to cause disease. Third, depending on the pathogen, it may be necessary to account for other factors that influence its spread, such as (1) differences in the likelihood that hosts of different ages will become infected; (2) a latent period, in which a host individual is infected but cannot spread the disease; and (3) vertical transmission, the spread of the disease from mother to newborn, as can occur in AIDS.

Models that include all of these factors can be very complicated. Here we'll consider a simple model that does not incorporate most of these complicating factors, yet still yields a key insight: a disease will spread only if the density of susceptible hosts exceeds a critical **threshold density**.

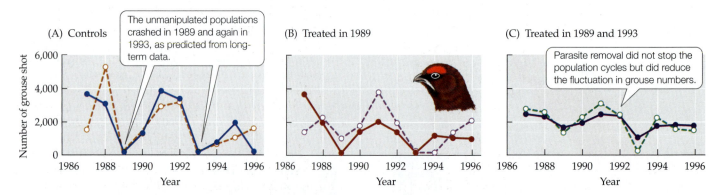

FIGURE 13.15 Parasite Removal Reduces Host Population Fluctuations Hudson et al. studied the effects of parasites on the cycling of six red grouse populations subjected to three treatments: (A) two control populations, (B) two populations treated for nematode parasites in 1989, and (C) two populations treated for parasites in 1989 and 1993. (After P. J. Hudson et al. 1998. *Science* 282: 2256–2258.)

? If parasite removal completely stopped the population cycles, how might the results in (C) differ from those actually obtained?

To develop a model that can be used to estimate the threshold density, we must determine how to represent the transmission of the disease from one host individual to the next. We'll denote the density of susceptible individuals by S and the density of infected individuals by I. For a disease to spread, infected individuals must encounter susceptible individuals. Such encounters are assumed to occur at a rate that is proportional to the densities of susceptible and infected individuals; here, we'll assume that this rate is proportional to the product of their densities, SI. Diseases do not spread with every such encounter, however, so we multiply the encounter rate (SI) by a transmission coefficient (β) that indicates how effectively the disease spreads from infected to susceptible individuals. Thus, an essential feature of the model—disease transmission—is represented by the term βSI.

The density of infected individuals increases when the disease is transmitted successfully (at the rate βSI) and decreases when infected individuals die or recover from the disease. If we set the combined death and recovery rate equal to m, these assumptions yield the equation

$$\frac{dI}{dt} = \beta SI - mI \qquad (13.1)$$

where dI/dt represents the change in the density of infected individuals at each instant in time.

A disease will become established and spread when the density of infected individuals in a population increases over time. This occurs when dI/dt is greater than zero, which, according to Equation 13.1, occurs when

$$\beta SI - mI > 0$$

We can rearrange this equation to get

$$S > \frac{m}{\beta}$$

Thus, a disease will become established and spread when the number of susceptible individuals exceeds m/β; this number of susceptible individuals is the threshold density, denoted by S_T. In other words,

$$S_T = \frac{m}{\beta}$$

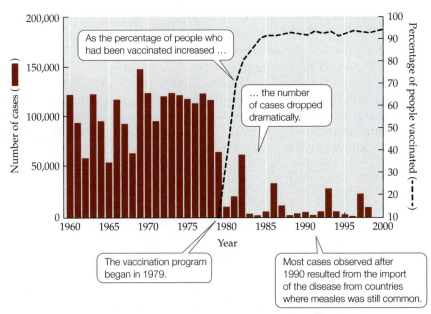

FIGURE 13.16 Vaccination Reduces the Incidence of Measles in Humans The results of a measles vaccination program in Romania show that lowering the density of susceptible individuals can control the spread of a disease. Measles often kills (especially in populations that are poorly nourished or that lack a history of exposure to the disease) and can cause severe complications in survivors, including blindness and pneumonia. (After P. M. Strebel and S. L. Cochi. 2001. *Nature* 414: 695–696.)

For some diseases that affect people or animals, the transmission rate β and the death and recovery rate m are known, permitting estimation of the threshold density.

CONTROLLING THE SPREAD OF DISEASES As Equation 13.1 suggests, to prevent the spread of a disease, the density of susceptible individuals must be kept below the threshold density (S_T). There are several ways of achieving this goal. People sometimes slaughter large numbers of susceptible domesticated animals to reduce their density below S_T and hence prevent disease spread. This is typically done when the disease in question can spread to humans, as in highly virulent forms of bird flu. In human populations, if an effective and safe vaccine is available, the density of susceptible individuals can be reduced below S_T by a mass vaccination program. Such programs work, as illustrated by the dramatic results of a measles vaccination program in Romania (**FIGURE 13.16**).

Other public health measures can also be taken to raise the threshold density, thereby making it more difficult for the disease to become established and spread. For example, the threshold density can be raised by taking actions that increase the rate at which infected individuals recover and become immune (thereby increasing m and hence increasing $S_T = m/\beta$). One way to increase the recovery rate is to improve the early detection and clinical

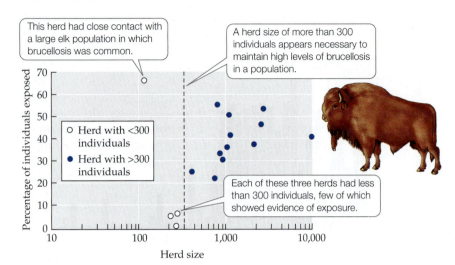

This herd had close contact with a large elk population in which brucellosis was common.

A herd size of more than 300 individuals appears necessary to maintain high levels of brucellosis in a population.

○ Herd with <300 individuals

● Herd with >300 individuals

Each of these three herds had less than 300 individuals, few of which showed evidence of exposure.

Percentage of individuals exposed

Herd size

FIGURE 13.17 Determining Threshold Population Densities for Disease Control The percentage of bison that showed evidence of previous exposure to brucellosis was monitored in six national parks in the United States and Canada. By plotting this percentage versus the size of each of 16 bison herds, researchers obtained a rough estimate of the threshold density for establishment of the disease (200–300 individuals, the upper bound of which is shown by the dashed line). (After A. Dobson and M. Meagher. 1996. *Ecology* 77: 1026–1036.)

treatment of the disease. The threshold density can also be raised if β, the disease transmission rate, is decreased. This can be achieved by quarantining infected individuals or by convincing people to engage in behaviors (such as hand washing or masks) that make it more difficult for the disease to be transmitted from one person to the next.

The same principles can be applied to wild populations. Dobson and Meagher (1996) studied bison populations to determine how best to prevent the spread of the bacterial disease brucellosis. Using data from previous studies in which 16 bison herds in six national parks in Canada and the United States had been tested for exposure to the disease, they found that the threshold density (S_T) for disease establishment appeared to be a herd size of 200–300 bison (**FIGURE 13.17**). This field-based estimate of S_T was very similar to the estimated threshold density of 240 individuals calculated from a model similar to Equation 13.1. Many of the herds in the six national parks had 1,000–3,000 individuals, so reducing herd sizes below a threshold value of 200–300 individuals would require implementing a vaccination program or killing large numbers of bison. An effective vaccine was not available, and killing many bison was not acceptable, either politically or ecologically (since herds as small as 200 individuals would face an increased risk of extinction). Thus, Dobson and Meagher concluded that it would be difficult to prevent the establishment of brucellosis in wild bison populations.

CONCEPT 13.5 Parasites can alter the outcomes of species interactions, thereby causing communities to change.

LEARNING OBJECTIVES

13.5.1 Know how parasites can affect the outcome of species interactions and community structure.

13.5.2 Describe how climate change might influence host–pathogen relationships.

Parasites Can Change Ecological Communities

The effects of parasites on their hosts can have ripple effects on communities: by reducing the performance of host individuals and the growth rates of host populations, parasites can change the outcome of species interactions, the composition of ecological communities, and even the physical environment in which a community is found.

Changes in species interactions

When two individual organisms interact with each other, the outcome of that interaction depends on many features of their biology. An individual predator that is young and healthy may be able to catch its prey—even though the prey organism literally "runs for its life"—whereas a predator that is old or sick may go hungry. Similarly, an individual that is in good condition may be able to compete effectively with others for resources, while an individual in poor condition may not.

Because they can affect host performance, parasites can affect the outcome of interactions between their hosts and other species. Thomas Park conducted a series of experiments on factors that influenced the outcome of competition between flour beetle species. In one of those experiments, Park (1948) examined how the protist parasite *Adelina tribolii* affected the outcome of competition experiments using two species of flour beetles, *Tribolium castaneum* and *T. confusum*. In the absence of the parasite, *T. castaneum* usually outcompeted *T. confusum*, driving it to extinction in 12 of 18 cases (**FIGURE 13.18**). The reverse was true when the parasite was present: *T. confusum* outcompeted *T. castaneum* in 11 of 15 cases. The outcome of competition was reversed because the parasite had a large negative effect on *T. castaneum* individuals, but virtually no effect on *T. confusum*. Parasites can also affect the outcome of competition in the field, as when the malaria

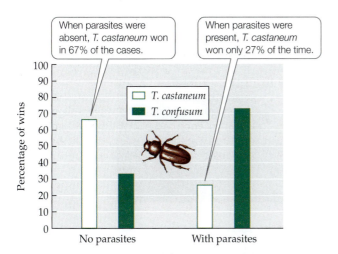

When parasites were absent, *T. castaneum* won in 67% of the cases.

When parasites were present, *T. castaneum* won only 27% of the time.

FIGURE 13.18 Parasites Can Alter the Outcome of Competition   Thomas Park performed competition experiments using populations of the flour beetles *Tribolium castaneum* and *T. confusum* that were or were not infected with a protist parasite. (After T. Park. 1948. *Ecol Monogr* 18: 267–307.)

parasite *Plasmodium azurophilum* reduced the competitive superiority of the lizard *Anolis gingivinus* over its smaller counterpart, *A. wattsi* (Schall 1992). Finally, parasites can alter the outcome of predator–prey interactions: by decreasing the physical condition of infected individuals, parasites may make predators less able to catch their prey, or prey less able to escape predation.

In the examples described in the previous paragraph, parasites affected the outcome of species interactions by altering the physical condition of their host. Parasites can also alter the outcome of species interactions by changing host behavior. For example, when infected by a parasite, the host may behave in an unusual manner that makes it more vulnerable to predation. There are numerous examples of this phenomenon, including protist parasites making rats less wary of cats, as described in the Case Study. Some worm parasites cause amphipods to move from sheltered areas to areas of relatively bright light, where the amphipods are more likely to be seen and eaten by fish or bird predators. In both of these cases, the parasite induces a change in host behavior that makes the host more likely to be eaten by a species that the parasite requires to complete its life cycle.

Changes in community structure

As we'll discuss in Chapter 16, ecological communities can be characterized by the number and relative abundances of the species they contain as well as by physical features of the environment. Parasites can alter and be altered by each of these aspects of communities.

In this chapter, we have seen several cases in which a parasite reduced the abundance, or even the geographic range, of its host, and we have also seen that parasites can change the outcome of species interactions. Such changes can have profound effects on the composition of communities. For example, a parasite that attacks a dominant competitor can suppress that species, causing the abundances of inferior competitors to increase. Such an effect was observed in six stream communities studied by Kohler and Wiley (1997). Prior to recurrent outbreaks of a fungal pathogen, the caddisfly *Glossosoma nigrior* was the dominant herbivore in each of the six communities. The fungus devastated *Glossosoma* populations, reducing their densities nearly 25-fold, from an average of 4,600 individuals per square meter to an average of 190 individuals per square meter. This drastic reduction in *Glossosoma* density allowed increases in the abundances of dozens of other species, including algae, grazing insects that ate algae, and filter feeders such as blackfly larvae. In addition, several species that previously were extremely rare or absent from the communities were able to establish thriving populations, thus increasing the diversity of the communities.

Parasites can also cause changes in the physical environment. This can happen when a parasite attacks an organism that is an *ecosystem engineer*, a species whose actions change the physical character of its environment, as when a beaver builds a dam (see Concept 16.3). As we learned earlier in Concept 13.4, the amphipod *Corophium volutator* can function as an ecosystem engineer in its tidal mudflat environment: in some circumstances, the burrows it builds hold the mud together, preventing the erosion of silt and causing the formation of "mud islands" that rise above the surface of the water at low tide. As described earlier, trematode parasites can drive local *Corophium* populations to extinction (**FIGURE 13.19A**). When this happens, erosion rates increase, the silt content of the mudflats decreases, and the mud islands disappear (**FIGURE 13.19B–D**). Along with these physical changes, in one instance, the abundances of ten large species in the mudflat community changed considerably in the presence of the parasite, including one species (a ribbon worm) that was driven to local extinction (K. N. Mouritsen, personal communication).

Finally, certain aspects of a community can be important in pathogen success and disease transmission. As we will learn in the Case Study in Chapter 19, the species diversity within a community can reduce the emergence and transmission of infectious diseases in wildlife and humans.

CLIMATE CHANGE CONNECTION

CLIMATE CHANGE AND DISEASE SPREAD   As we saw in Chapter 12, changes in climate are expected to have wide-ranging effects on species interactions and ultimately ecological communities (see Figure 12.22).

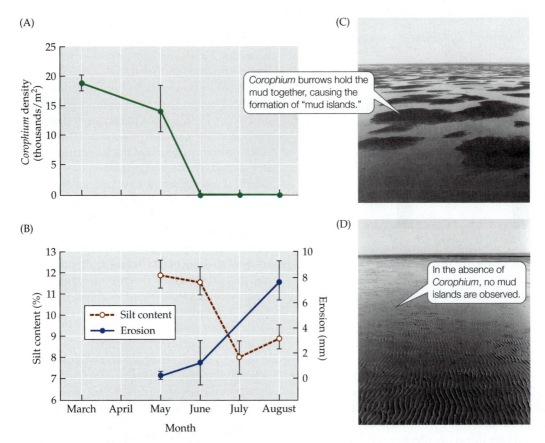

FIGURE 13.19 Parasites Can Alter the Physical Environment Infection of the amphipod *Corophium volutator* by a trematode parasite affects not only the host, but its entire tidal mudflat community. (A) The trematode can drive amphipod populations to local extinction. (B) In the absence of *Corophium*, the erosion rate increases and the silt content of the mudflats decreases. (C,D) The overall physical structure of the mudflats also changes [compare (C) with (D)]. Error bars show ± one standard error of the mean. (After K. N. Mouritsen et al. 1998. *J Mar Biol Assoc U.K.* 78: 1167–1180; K. N. Mouritsen and R. Poulin. 2002. *Parasitology* 124: S101–S117.)

For example, because mosquitoes and other *vectors* (organisms that transmit pathogens from one host to another) are often more active or produce more offspring under warm conditions, scientists have predicted that ongoing climate change may cause the incidence of some diseases to rise in human and wildlife populations (Epstein 2000; Harvell et al. 2002).

A growing body of evidence supports this prediction. In one such study, increases in ocean temperatures were strongly correlated with increases in coral diseases along Australia's Great Barrier Reef (Bruno et al. 2007). Similar results have been found in corals at other locations, as well as in a variety of amphibian and shellfish populations (Harvell et al. 2009).

Climate change is also expected to change the distributions of some pathogens and their vectors by changing the locations where conditions are suitable for those organisms. For example, González et al. (2010) found that climate change is likely to increase the risk of leishmaniasis in North America by increasing the geographic ranges of its reservoir species (rodents in the genus *Neotoma* that can harbor the pathogen) and its sand fly vectors (**FIGURE 13.20**). Similarly, the number of people at risk from malaria, cholera, and the plague may increase as global temperatures continue to warm (see citations in Ostfeld 2009).

Overall, these studies and others (see Tylianakis et al. 2008; Gilman et al. 2010) indicate that ecological interactions will influence how future climate change will affect the incidence of disease in humans and many other species. Likewise, Costello et al. (2009) outlined major threats to human health from direct and indirect effects of climate change on disease incidence, food and water insecurity, and extreme climate events (such as hurricanes and floods that create conditions that favor the spread of diseases). It is highly likely that climate change will also have direct and indirect effects on the incidence of disease in the populations of many species other than humans, contributing to the ongoing biodiversity crisis. ☀

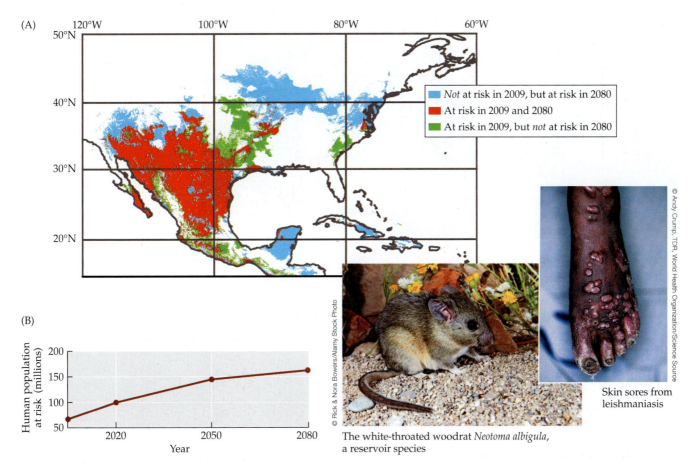

(A)

Legend:
- Not at risk in 2009, but at risk in 2080
- At risk in 2009 and 2080
- At risk in 2009, but not at risk in 2080

(B)

Skin sores from leishmaniasis

The white-throated woodrat *Neotoma albigula*, a reservoir species

FIGURE 13.20 Climate Change May Increase the Risk of Leishmaniasis in North America Leishmaniasis can cause severe skin sores, difficulty breathing, immune system impairment, and other complications that can lead to death. There are currently 1 million new cases each year. Leishmaniasis is caused by protists in the genus *Leishmania* and spread by sand flies (bloodsucking insects in the genera *Lutzomyia* and *Phlebotomus*). In addition to infecting humans, the pathogen can persist in several reservoir species (rodents in the genus *Neotoma*). (A) Change in the geographic regions in which people are predicted to be at risk from leishmaniasis due to the presence of at least one vector and reservoir species. (B) Change in numbers of people predicted to be at risk due to the presence of at least one vector and reservoir species. (After C. González et al. 2010. *PLOS Neglected Trop Dis* 4: 1–16.)

A CASE STUDY REVISITED

Enslaver Parasites

Returning to a question that we posed in the Case Study, how do enslaver parasites manipulate the behavior of their hosts? In some cases, we have hints of how they do this. Consider the tropical parasitoid wasp *Hymenoepimecis argyraphaga* and its host, the orb-weaving spider *Plesiometa argyra*. The larval stage of this wasp attaches to the exterior of a spider's abdomen and sucks the spider's body fluids. When fully grown, the wasp larva induces the spider to make a special "cocoon web" (**FIGURE 13.21**). Once the spider has built the cocoon web, the larva kills and eats the spider. The larva then spins a cocoon and attaches it to the cocoon web. As the larva completes its development within the cocoon, the cocoon web serves as a strong support that protects the larva from being swept away by torrential rains.

A parasitized spider builds normal webs right up to the night when the wasp induces it to make a cocoon web. This sudden change in the spider's web-building behavior suggested that the wasp might inject the spider with a chemical that alters its behavior. To test this idea, William Eberhard (2001) removed wasp larvae from spider hosts several hours before the time when a cocoon web would usually be made. Wasp removal sometimes resulted in the construction of a web that was very similar to a cocoon web, but more often resulted in the construction of a web that was intermediate in form but differed substantially from both normal and cocoon webs. In the days that followed the removal of the parasite, some spiders partially recovered the ability to make normal webs. These results are consistent with the idea that the parasite induces construction of a cocoon web by injecting a fast-acting chemical into the spider. The chemical appears to act in a dose-dependent manner; otherwise,

(A)

Unparasitized spiders build webs like this one.

A,B courtesy of William Eberhard

(B)

A wasp larva induces its spider host to build a "cocoon web" like this one …

… from which the larva hangs its cocoon.

Inset, © Anand Varma

FIGURE 13.21 Parasites Can Alter Host Behavior The parasitoid wasp *Hymenoepimecis argyraphaga* dramatically alters the web-building behavior of the orb-weaving spider *Plesiometa argyra*. (A) The web of an uninfected spider. (B) The "cocoon web" of a parasitized spider. (After W. G. Eberhard. 2001. *J Arachnol* 29: 354.)

we would expect spiders exposed to the chemical to build only cocoon webs, not webs that are intermediate in form. Spiders build cocoon webs by repeating the early steps of their normal web-building sequence a large number of times; thus, the chemical appears to act by interrupting the spiders' usual sequence of web-building behaviors.

Other enslaver parasites also appear to manipulate host body chemistry. In the Case Study, we described hairworm parasites that cause crickets to commit suicide by jumping into water. Thomas and colleagues (2003) have shown that the hairworm causes biochemical and structural changes in the brain of its cricket host. The concentrations of three amino acids (taurine, valine, and tyrosine) in the brains of parasitized crickets differ from those in crickets that have not been parasitized. Taurine, in particular, is an important neurotransmitter in insects, and it also regulates the brain's ability to sense a lack of water. Hence, it is possible that the parasite induces its host to commit suicide by causing biochemical changes in its brain that alter the host's perception of thirst.

The papers by Eberhard (2001) and Thomas et al. (2003) suggest that some parasites enslave their hosts by manipulating them chemically. But even in Eberhard's work,

which indicates that the wasp injects a chemical into its spider host, the chemical in question has not been found. If this chemical were known, it could be injected into unparasitized spiders; if those spiders constructed cocoon webs, we would have a clear understanding of how the parasite manipulates the spider.

Although a definitive chemical experiment such as this has yet to be performed, a similar genetic experiment was performed for spongy moths (*Lymantria dispar*) enslaved by a virus (Hoover et al. 2011). Spongy moths infected with this virus move to the tops of trees shortly before they die; after death, the bodies of the moths liquefy and release millions of infective viral particles. Uninfected spongy moths do not exhibit this climbing behavior before death. Based on previous work, Hoover and colleagues hypothesized that the expression of a particular viral gene (the *egt* gene) caused infected moths to move to treetops shortly before death. In a laboratory test of this hypothesis, they found that moths infected with the typical, or wild-type, virus strain died at higher positions than did moths infected by viruses from which the suspect gene had been removed (**FIGURE 13.22**)—strong evidence that they had succeeded in identifying the first known "enslaver gene."

FIGURE 13.22 A Parasite Gene That Enslaves Its Host Spongy moths infected by a virus (*Lymantria dispar* nucleopolyhedrovirus, or LdMNPV) climb to high locations before they die—a behavior that benefits the virus but not the moth. To test the hypothesis that a particular viral gene (the *egt* gene) affects this behavior, researchers reported the height at death of spongy moth caterpillars reared in cages and subjected to the following treatments: WT viruses (two different natural, or wild-type, viruses); EGT− viruses (two different experimental viruses from which the *egt* gene had been removed); and EGT+ viruses (two different experimental viruses from which the *egt* gene was first removed, then replaced). Error bars show one standard error of the mean. (After K. Hoover et al. 2011. *Science* 333: 1401.)

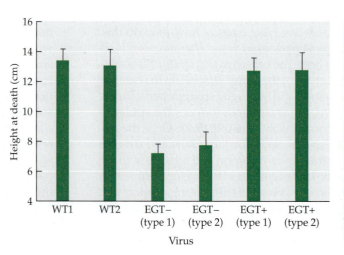

© Krishna.k/Shutterstock.com

❓ Explain why the researchers included the WT and EGT+ treatments.

CONNECTIONS *in* NATURE

FROM CHEMICALS TO EVOLUTION AND ECOSYSTEMS

Enslaver parasites that manipulate their hosts exert strong selection pressure on host populations, so resistance to the manipulations of enslaver parasites might be expected to evolve in host populations. For example, selection would favor host individuals with the ability to recognize and destroy chemicals that a parasite uses to alter host behavior. Likewise, enslaver parasites might be expected to evolve the ability to overcome host resistance mechanisms.

To date, we know of no such evidence of ongoing host–enslaver parasite coevolution. However, interactions between enslaver parasites and their hosts do provide evidence of previous evolutionary change. Like any parasite, an enslaver parasite has adaptations that allow it to cope with host defenses (otherwise it would not survive). More specifically, an enslaver parasite that uses a chemical to manipulate a specific host behavior is beautifully adapted to take advantage of the body chemistry of its host. Such evolutionary links between enslaver parasites and their hosts illustrate a central feature of both ecology and evolution: ecological interactions affect evolution, and vice versa, at times making it difficult to distinguish one from the other (see Concept 6.5). As we've seen in this chapter, the outcome of such ecological and evolutionary interactions can have profound effects on individuals, populations, communities, and ecosystems. As evolutionary change tips the balance back and forth, first in favor of the host, then in favor of the parasite, we can expect concomitant changes in the population dynamics of other species, such as those that compete with or eat the host or the parasite. Viewed in this way, communities and ecosystems are highly dynamic, always shifting in response to the ongoing ecological and evolutionary changes that are occurring within them.

SUMMARY

CONCEPT 13.1 Parasites typically feed on only one or a few host species, but host species have multiple parasite species.

13.1.1 Know why parasites are abundant and typically specialists.

- Parasites usually feed on only one or a few host individuals during the course of their lives.

- Parasites typically live on or in their hosts and feed on certain parts of the host's body.

- Most host species are attacked by more than one parasite and their parasites have parasites.

13.1.2 Compare and contrast ectoparasites and endoparasites.

- Some parasites are ectoparasites that live on the surface of their host; others are endoparasites that live within the body of their host.

- Advantages of ectoparasitism include ease of dispersal from one host individual to another and avoidance of the host's immune system.

- Advantages of endoparasitism include avoidance of natural enemies, ease of feeding, and protection from the external environment.

CONCEPT 13.2 Hosts have mechanisms for defending themselves against parasites, and parasites have mechanisms for overcoming host defenses.

13.2.1 Describe the mechanisms organisms use for defending themselves against parasites.

- Many host organisms have immune systems that allow them to recognize and defend against endoparasites.

- Biochemical conditions inside the host's body can also provide protection against parasites, as can defensive symbionts such as fungi and bacteria.

13.2.2 List the mechanisms that parasites use to circumvent host defenses.

- Parasites have a broad suite of mechanisms that allow them to circumvent host defenses, from relatively simple counterdefenses against encapsulation to more complex counterdefenses that involve hundreds of genes.

CONCEPT 13.3 Host and parasite populations can evolve together, each in response to selection pressure imposed by the other.

13.3.1 Understand how host–parasite interactions can result in coevolution.

- Host–parasite interactions show coevolution, in which populations of the host and parasite evolve together, each in response to selection pressure imposed by the other.

- Selection can favor a diversity of host and parasite genotypes. A rare host genotype may increase in frequency because few parasites can overcome its defenses; as a result, parasite genotypes that can respond to the new host genotype's defenses may also increase in frequency.

13.3.2 Know why host–parasite interactions can result in life history trade-offs.

- Host–parasite interactions can exhibit trade-offs in which a trait that improves host defenses or parasite counterdefenses has costs that reduce other aspects of the organism's growth, survival, or reproduction.

CONCEPT 13.4 Hosts and parasites can have important effects on each other's population dynamics.

13.4.1 Understand how parasites can influence the population dynamics of hosts.

(Continued)

SUMMARY (continued)

- Parasites can reduce the abundances of host populations, in some cases driving local host populations to extinction or changing the geographic distributions of host species.

- Evidence suggests that parasites can also cause cycling of hosts' populations.

13.4.2 Explain how simple models of host–pathogen dynamics can be used to control the establishment and spread of diseases.

- Some models of host–pathogen population dynamics subdivide the host population into susceptible individuals, infected individuals, and recovered and immune individuals; track different host and pathogen genotypes; and take into account factors such as host age, latent periods, and vertical transmission.

- A simple mathematical model of host–pathogen dynamics yields an important insight: for a disease to become established and spread, the density of susceptible hosts must exceed a critical threshold density.

- To control the spread of a disease, efforts may be made to lower the density of susceptible hosts (by slaughtering domesticated animals or undertaking vaccination programs) or to raise the threshold density (by increasing the recovery rate or decreasing the transmission rate).

CONCEPT 13.5 Parasites can alter the outcomes of species interactions, thereby causing communities to change.

13.5.1 Know how parasites can affect the outcome of species interactions and community structure.

- Parasites can affect the outcomes of interactions between their hosts and other species; for example, a host that is a dominant competitor may become an inferior competitor when infected by a parasite or a host may become more vulnerable to predation under these conditions.

- The effects of parasites can alter the composition of ecological communities and change features of the physical environment.

13.5.2 Describe how climate change might influence host–pathogen relationships.

- Ongoing climate change will likely cause the incidence of diseases to rise through both direct and indirect changes in the abundances and distributions of pathogens.

Review Questions

1. Define endoparasites and ectoparasites, giving an example of each. Describe some advantages and disadvantages associated with each of these two types of parasitism.

2. Given the effects that parasites can have on host individuals and host populations, would you expect that parasites could also alter the outcomes of species interactions and the composition of ecological communities? Explain.

3. **a.** Summarize the mechanisms that host organisms use to kill parasites or reduce the severity of their attack.

   b. With your answer to (a) as background material, do you think the following statement from a news report could be true?

   The parasite has a mild effect on a plant species in Australia, but after it was introduced for the first time to Europe, it had devastating effects on European populations of the same plant species.

   Explain your reasoning, and illustrate your argument with an example of how a plant defensive mechanism might work—or fail to work—in a situation such as this.

Hone Your Problem-Solving Skills

As we learned in Concept 13.5, leishmaniasis is a serious disease caused by protists that reside in and are spread by sand flies. The disease is found both in humans and in rodents in the genus *Neotoma* (see Figure 13.20). Suppose a control program is being designed to reduce new cases of leishmaniasis in humans that involves reducing the infection by leishmaniasis in *Neotoma* rodents. You are asked to consult on the program and answer the following questions.

1. Why would you expect that reducing the infection by leishmaniasis of *Neotoma* rodents would have an effect on similar infection in humans?

2. To reduce the spread of leishmaniasis in *Neotoma* rodents, the control program recommends reducing the threshold density, S_T, in rodent populations. What is meant by the concept of a threshold density for the establishment and spread of a disease? Why is this concept important?

3. Suppose the control program determines that the threshold density (S_T) of leishmaniasis in *Neotoma* populations is 5,000 individuals. Using the data in the table, determine which *Neotoma* rodent populations will need to be reduced, and by how much, to decrease leishmaniasis infection.

Population	1	2	3	4	5	6	7	8
Number of rodent individuals	9,000	1,200	5,500	4,000	8,000	1,000	4,000	10,000

4. What other measures besides reducing the host population sizes of both rodents and sand flies can be proposed to decrease the infection prevalence in humans?

ADDITIONAL RESOURCES

This textbook provides resources such as **Web Extensions** and additional **Climate Change Connections** for instructors who can share them with students. Please contact your instructor, if you would like to access that content.

Competition

© Chris Mattison/Alamy Stock Photo

Competition in Plants That Eat Animals: A Case Study

Despite repeated reports that plants could eat animals, early scientists were skeptical of those claims. Charles Darwin (1875) laid their doubts to rest by providing clear experimental evidence of carnivory by plants. Today, more than 600 species of plants that eat animals have been identified, including bladderworts, sundews, pitcher plants, and the well-known Venus flytrap.

Plants use a variety of mechanisms to eat animals. The Venus flytrap has modified leaves that look like fanged jaws yet attract insects with a sweet-smelling nectar (**FIGURE 14.1**). The inner surface of the leaf has touch-sensitive hairs; if an insect trips those hairs, the leaf snaps shut in less than half a second. Once the insect is captured, the trap tightens further, forming an airtight seal around its victim, which is then digested over the course of 5–12 days. Some plants can capture animals at truly blinding speeds. For example, aquatic bladderworts (*Utricularia* spp.) have a trapdoor that springs inward when touched, creating a suction that pulls in prey in less than half a millisecond.

Other plants lack moving parts yet still can eat animals. Consider the pitcher plants, which can use nectar or visual cues to lure insects into a pitcher-shaped trap. The inside of the pitcher often has downward-facing hairs, which make it easy for the insect to crawl in, but hard to crawl out. What's more, in many pitcher plants, once it is about halfway down, the insect encounters a layer of flaky wax. An insect that steps onto this wax is doomed: the wax sticks to its feet, causing it to lose its grip and tumble into a vat that contains either water (in which it drowns) or deadly digestive juices.

Why do some plants eat animals? The answer may relate to the subject of this chapter: competition. Because plants are immobile, competition for limiting resources such as nutrients or water can be intense. Many carnivorous plants are found in environments with nutrient-poor soils. Furthermore, evolutionary relationships among plants reveal that in nutrient-poor environments, carnivory has evolved multiple times, in a variety of independent plant lineages. Overall, these observations suggest that carnivory in plants is an adaptation for life in nutrient-poor environments—perhaps providing a way to avoid competing with other plants for soil nutrients. Hence, carnivorous plants could use animal prey as an alternative source of nutrients when competition is intense.

To test this idea, Stephen Brewer (2003) measured how the growth of the pitcher plant *Sarracenia alata* was affected when he cut off its access to prey (by covering the pitchers) and when he reduced noncarnivorous competitor plant species ("neighbors") by weeding and clipping. His results show that biomass in *Sarracenia* increased when neighbors were reduced (**FIGURE 14.2**), suggesting that competition had an important effect on growth. But further examination of Figure 14.2 reveals that

FIGURE 14.1 A Plant That Eats Animals Attracted to the plant's sweet-smelling nectar, this wasp is about to become a meal. Although the Venus flytrap typically captures insects, it can also feed on other animals, such as slugs and small frogs.

FIGURE 14.2 Competition Decreases Growth in a Carnivorous Plant To test the effects of competition on the carnivorous pitcher plant *Sarracenia alata*, the growth of control plants ("neighbors present") was compared with the growth of plants whose noncarnivorous competitors were weeded and clipped ("neighbors removed"). Neighbor removal increased plant growth, especially when animal prey were available. Error bars show one standard error of the mean. (After J. S. Brewer. 2003. *Ecology* 84: 451–462.)

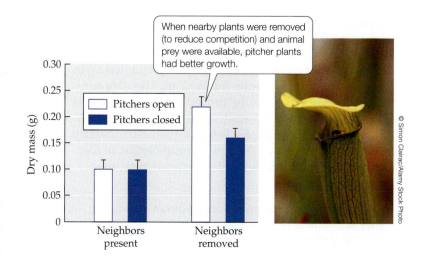

matters are not as simple as they may at first appear. Although the growth of pitcher plants with competition was expected to decline when they were deprived of prey, that did not happen. Instead, it seems that pitcher plants were only able to benefit from animal prey when neighbors were removed. Why is this so?

Introduction

In 1917, A. G. Tansley reported results from a series of experiments designed to explain the distribution in Britain of two species of bedstraw plants, *Galium hercynicum* and *G. pumilum* (then known as *G. sylvestre*). *Galium hercynicum* was restricted to acidic soils, *G. pumilum* to calcareous soils. Even in places where the two species grew within inches of each other, each remained confined to its characteristic soil type. In his experiments, Tansley found that when grown alone, each species could survive on both acidic and calcareous soils. However, when he grew the species together on acidic soils, only *G. hercynicum* survived, and if he grew them together on calcareous soils, only *G. pumilum* survived. Tansley concluded that the two species competed with each other and that when grown on its native soil type, each species drove the other to extinction.

Tansley's work on bedstraws is one of the first experiments ever performed on **competition**, a non-trophic interaction between individuals of two or more species in which all species are negatively affected by their shared use of a resource that limits their ability to grow, reproduce, or survive (see Figure 12.3). In this chapter, we specifically focus on **interspecific competition** (between individuals of different species) as opposed to **intraspecific competition** (between individuals of a single species),

as in *density-dependent growth*, a topic we addressed in Chapters 10 and 11.

What are some of the resources that species compete for? **Resources** are simply the components of the environment, such as food, water, light, and space, that are required by species. Food is an obvious example—when food is scarce, population growth rates fall unless individuals can successfully compete. In terrestrial ecosystems—especially arid ones—water is also a resource. But an organism does not need to consume a substance for it to be a resource. Plants "consume" light in the sense that they use it to fix carbon, and they can deplete the supply available to other plants by shading them. Space is also an important resource. Plants, algae, and sessile animals require space to attach and grow, and competition for space can be intense. Mobile animals also compete for space as they seek access to foraging grounds, territories to attract mates, or refuges from heat, cold, or predators.

Finally, the full set of resources, along with other biotic and abiotic requirements, is what is known as the ecological or **fundamental niche** of a species (**FIGURE 14.3A**).

FIGURE 14.3 The Concept of the Fundamental and the Realized Niche In this conceptual representation of species 1's use of two resources, (A) its fundamental niche is contained within the entire blue area, but (B) the use of resources in that area is limited by interactions with other species, which set the limits of its realized niche.

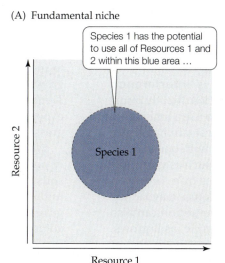

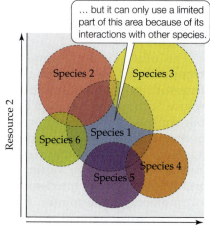

However, within the context of species interactions, no one species has exclusive access to all the resources within its fundamental niche. Thus, ecologists recognize a more restricted set of conditions that a species is limited to, in large part because of species interactions. These more restricted conditions form the **realized niche** of a species (**FIGURE 14.3B**). We will consider the niche concept later in Concept 14.2 and in Chapter 19 when we discuss resource partitioning. For now, let's begin our exploration of competition by considering some of its general characteristics.

> **CONCEPT 14.1** Competition can be direct or indirect, vary in its intensity, and occur between similar or dissimilar species.

LEARNING OBJECTIVES

14.1.1 Define the difference between exploitation competition and interference competition.

14.1.2 Analyze how and why competition can vary in its intensity.

14.1.3 Describe the importance of competition within communities.

General Features of Competition

The simple definition of competition as two or more species negatively affecting one another because of shared resources belies the complicated ways in which species actually compete with one another. Because resources are the mitigating factor in the interaction, and each species requires and obtains resources in different ways, the mechanisms used to compete and the intensity and ultimate outcome of competition can vary widely among species. Let's next consider some of the ways in which species compete.

Species may compete directly or indirectly

Species often compete indirectly through their mutual effects on the availability of a shared resource. Known as **exploitation competition**, this type of competition occurs simply because individuals reduce the supply of a shared resource as they use it. We have already considered an example of exploitation competition in the Case Study on pitcher plants.

The other kind of competition is **interference competition**, a case in which one species directly interferes with the ability of its competitors to use a limiting resource. Such interactions are perhaps most familiar in mobile animals, as when carnivores fight with one another over animal prey. Similarly, herbivores such as voles may aggressively exclude other vole species from preferred habitat, and members of warring ant colonies may kidnap and kill one another. Interference competition can also occur among sessile animals. For example, as it grows, the acorn barnacle *Semibalanus balanoides* often crushes or smothers nearby individuals of another barnacle species,

FIGURE 14.4 Interference Competition in Plants A formidable competitor, the kudzu vine (*Pueraria montana*) has grown over and completely covered these Georgia trees and shrubs, competing with them for light.

Chthamalus stellatus. As a result, *Semibalanus* directly prevents *Chthamalus* from living in most portions of the rocky intertidal zone (we'll describe competition between these barnacles in more detail in Concept 14.4).

Interference competition also occurs in plants, as when one plant species grows over another, reducing its access to light (**FIGURE 14.4**). There is also circumstantial evidence that interference competition can take the form of **allelopathy**, in which individuals of one species release toxins that harm individuals of another species. Although allelopathy appears to be important in some crop systems (Minorsky 2002; Belz 2007), there are few rigorous tests of the effects of allelopathy in competitive environments.

Competition can vary in intensity depending on resource availability and type

Plants can compete for aboveground resources, such as light, as well as for belowground resources, such as soil nutrients. Researchers have suggested that the relative importance of aboveground and belowground competition in plants might change depending on whether aboveground or belowground resources are more scarce: belowground competition, for example, might be expected to increase in intensity when the competing plants are growing in nutrient-poor soils. Scott Wilson and David Tilman (1993) tested this idea by performing transplant experiments with *Schizachyrium scoparium*, a perennial grass species native to their study site in Minnesota.

Wilson and Tilman selected a series of 5 × 5 m plots of natural vegetation growing in sandy, nitrogen-poor soils. For 3 years, they treated half of the plots with high-nitrogen fertilizer each year. This 3-year period gave the plant communities in the fertilized plots time to adjust to the experimentally imposed change in soil nitrogen levels. At the end of the 3-year period, they planted *Schizachyrium* individuals in all the plots.

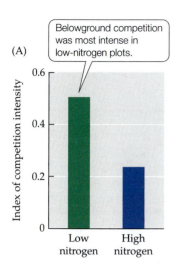

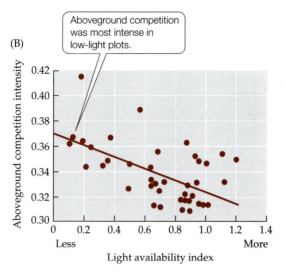

FIGURE 14.5 Resource Availability Affects the Intensity of Competition (A) In transplant experiments with the grass *Schizachyrium scoparium*, belowground competition between plant species for nutrients increased in intensity when soil nutrients were scarce. (B) Similarly, aboveground competition for light increased as light levels decreased. (After S. D. Wilson and D. Tilman. 1993. *Ecology* 74: 599–611.)

Once they were added to the high-nitrogen (fertilized) and low-nitrogen (unfertilized) plots, *Schizachyrium* individuals were grown under three treatments: (1) with neighbors present (competition), (2) with neighbor roots present but neighbor stems tied back (which prevented light competition with *Schizachyrium*), or (3) with neighbor roots and stems both removed (no competition). Wilson and Tilman found that while total competition (the sum of belowground and aboveground competition) did not differ between the low- and high-nitrogen plots, belowground competition was most intense in the low-nitrogen plots (**FIGURE 14.5A**). They also found that aboveground competition increased when light levels were low (**FIGURE 14.5B**). Thus, their work demonstrates that the intensity of competition can increase when the particular resource being competed for is scarce.

Competition is often asymmetrical

When two species compete for a resource that is in short supply, each obtains less of the resource than it could if the competitor were not present. Because competition reduces the resources available for the growth, reproduction, and survival of both species, the abundance of each species is reduced to some extent. In many cases, however, the effects of competition are unequal, or asymmetrical: one species is harmed more than the other. This asymmetry is especially clear in situations in which one competitor drives another to extinction.

For example, in a laboratory experiment, Tilman et al. (1981) examined competition for silica (SiO$_2$) between species of freshwater diatoms, which use silica to construct their cell walls. Tilman and colleagues grew two diatom species, *Synedra ulna* and *Asterionella formosa*, alone and in competition with each other. They measured how the population densities of the diatoms and silica concentrations in the water changed over time. When grown alone, each species reduced silica (the resource) to a low and approximately constant concentration; each species also

reached a stable population size (**FIGURE 14.6**). *Synedra* had a lower stable population size than *Asterionella*, and it reduced silica to lower levels than did *Asterionella*. When the two species competed with each other, *Synedra* drove *Asterionella* to extinction, apparently because it reduced silica to such low levels that *Asterionella* could not survive.

As this example suggests, before the inferior competitor goes extinct, the superior competitor typically loses potential resources to its competitor or invests energy in the competitive interaction. Hence, even when one species drives the other to extinction, both the superior and the inferior competitor are harmed to some extent. However, the effect of the superior competitor is still greater than the effect of the inferior competitor. Indeed, in general, there is a continuum in how strongly each competitor affects the other (**FIGURE 14.7**). Note that the two ends of this continuum do not represent competitive (–/–) interactions. Instead, such interactions are referred to as **amensalism**, –/0 interactions in which individuals of one species are harmed while individuals of the other species are not affected at all. Possible examples of amensal interactions include small woody plants that grow beneath towering trees, or corals in which the individuals of one species can grow over those of another, depriving them of light.

Competition can occur between closely or distantly related species

We've seen that competition can occur between pairs of closely related species, such as the diatom species studied by Tilman. Brown and Davidson (1977) examined whether competition also occurs between groups of more distantly related species. In particular, they suspected that rodents and ants might compete because both eat the seeds of desert plants, and the sizes of the seeds they eat overlap considerably (**FIGURE 14.8**).

Brown and Davidson established experimental plots (each about 1,000 m^2) in a desert region near Tucson, Arizona. Their experiment lasted 3 years and used

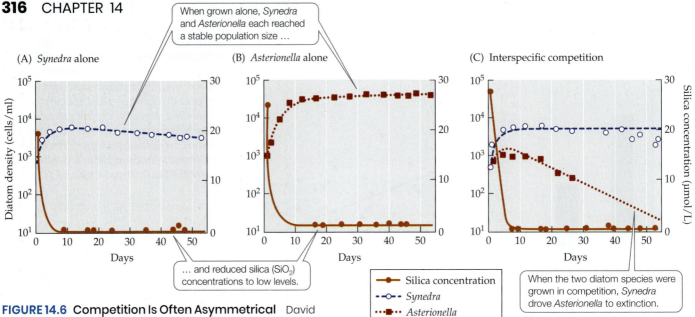

(A) *Synedra* alone

When grown alone, *Synedra* and *Asterionella* each reached a stable population size …

(B) *Asterionella* alone

(C) Interspecific competition

… and reduced silica (SiO₂) concentrations to low levels.

Silica concentration
--o-- *Synedra*
...■... *Asterionella*

When the two diatom species were grown in competition, *Synedra* drove *Asterionella* to extinction.

FIGURE 14.6 Competition Is Often Asymmetrical David Tilman and his colleagues demonstrated competition between two diatom species for silica by growing them alone and in competition with each other. *Synedra* (A) reduced silica concentrations to lower levels than did *Asterionella* (B) This result may explain why *Synedra* outcompeted *Asterionella* when the two species were grown together (C). (After D. Tilman et al. 1981. *Limnol Oceanogr* 26: 1020–1033.)

❓ Suppose a third diatom species reduced the concentration of silica to 5 µmol/L when grown alone. Predict what would happen if this species were grown in competition with *Asterionella*.

four treatments: (1) plots in which a ¼-inch wire mesh fence excluded seed-eating rodents and from which rodents within the fence were removed by trapping; (2) plots in which seed-eating ants were excluded by applying insecticides; (3) plots in which both rodents and ants were excluded by fencing, trapping, and insecticides; and (4) plots in which both rodents and ants were left undisturbed (control plots).

The results indicated that rodents and ants do compete for food. Relative to the control plots, the number of ant colonies increased by 71% in the plots from which rodents were excluded, and rodents increased by 18% in number and 24% in biomass in the plots from which ants were excluded. In the plots from which both rodents and ants were excluded (treatment 3), the density of seeds increased by 450% compared with all other plots. Treatments 1 (no rodents), 2 (no ants), and 4 (the control plots, with both rodents and ants present) all resulted in similar

densities of seeds. These results suggest that when either rodents or ants were removed, the group that remained ate roughly as many seeds as rodents and ants combined ate in the control plots. Thus, under natural conditions, each group would be expected to eat fewer seeds in the presence of the other group than it could eat when alone.

It is not surprising that species as different as ants and rodents compete. After all, people differ greatly from bacteria, fungi, and insects, yet we compete with these organisms for food in farm fields, in grain storage bins—even in our refrigerators. Overall, whether they are closely or distantly related, organisms can compete if they share the use of a limiting resource.

Competition for resources is common in natural communities

How important is competition in natural communities? To answer this question, results from many field studies must be compiled and analyzed. The findings of three such analyses indicate that competition has important effects on many species. For example, Schoener (1983) examined the results of 164 published studies on competition and found that of 390 species studied, 76% showed effects of competition under some circumstances, and 57% showed effects of competition under all circumstances tested. Connell (1983) examined the results of 72 studies and found that competition was important for 50%

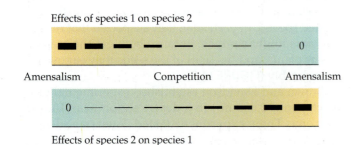

Effects of species 1 on species 2

Amensalism   Competition   Amensalism

0

Effects of species 2 on species 1

FIGURE 14.7 A Continuum of Competitive Effects Competition may affect members of both species equally, or the members of one species may be harmed more than are members of the other species. The thickness of the bars represents the strength of the competitive effects and the −/0 represents amensalism.

❓ Indicate the interactions that represent asymmetrical competition.

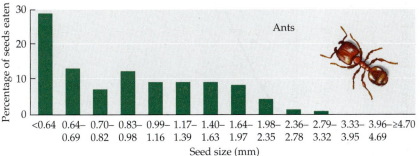

FIGURE 14.8 Ants and Rodents Compete for Seeds There is extensive overlap in the sizes of seeds eaten by ants and by rodents. Removal experiments showed that these two distantly related groups compete for this food source. (After J. H. Brown and D. W. Davidson. 1977. *Science* 196: 4292.)

of the 215 species studied. Gurevitch et al. (1992) took a different approach: they did not report the percentage of species for which competition was important, but rather analyzed the magnitude of competitive effects found for 93 species in 46 studies published between 1980 and 1989. They showed that competition had significant (though variable) effects on a wide range of organisms.

Surveys such as those by Schoener, Connell, and Gurevitch et al. face potential sources of bias, including investigators' failure to publish studies that show no significant effects and the tendency for investigators to study "interesting" species (i.e., those they suspect will show competition). Despite such potential sources of bias, the fact that hundreds of studies have documented effects of competition makes it clear that competition is common—though not ubiquitous—in nature. We explore the relative importance of competition to community structure in Chapter 19.

> **CONCEPT 14.2** Competing species are more likely to coexist when they use resources in different ways.

LEARNING OBJECTIVES

14.2.1 Define the competitive exclusion principle and explain how it differs from competitive coexistence.

14.2.2 Define and give examples of resource partitioning (or niche partitioning).

14.2.3 Describe how competition can lead to character displacement and resource partitioning.

Competitive Coexistence

As indicated above, ecologists have long thought that competition between species was important in communities. For example, although he often focused on competition within species, Darwin (1859) also argued that competition between species could influence both ecological and evolutionary processes. Darwin recognized that interspecific competition could lead to two possible outcomes. At one extreme, if a dominant species prevents another species from using essential resources, the inferior species may become locally extinct, a process known as **competitive exclusion**. We saw this result in the diatom example tested in a laboratory setting (see Figure 14.6). However, in reality, most species show some sort of **competitive coexistence**, or the ability to coexist with one another despite sharing limiting resources. Let's consider some general features of competition that lead to either competitive exclusion or competitive coexistence.

Competitors that use limiting resources in the same way cannot coexist

In the 1930s, the Russian ecologist G. F. Gause performed laboratory experiments on competition using three species of the single-celled protist group *Paramecium*. He constructed miniature aquatic ecosystems by growing paramecia in tubes filled with a liquid medium that contained bacteria and yeast cells as a food supply. He found that grown alone, *P. caudatum* and *P. bursaria* populations showed logistic growth and reached a stable carrying capacity (**FIGURE 14.9A**). When grown together, each species showed slower population growth, but they were able to coexist. However, when *P. aurelia* was grown in competition with *P. caudatum*, *P. aurelia* drove *P. caudatum* to extinction (**FIGURE 14.9B**). The difference in outcome, Gause suggested, was a consequence of *P. caudatum* and *P. aurelia* competing for bacteria as a food source, while *P. bursaria* avoided competition by eating yeast cells that settled to the bottom of the tubes. Thus, *P. caudatum* and *P. bursaria* partitioned their food resource in the presence of one another and were able to coexist as a result.

Experiments with a wide range of other species (e.g., algae, flour beetles, plants, and flies) have yielded similar results: one species drives the other to extinction unless the two species use the available resources in different ways. Such results led to the formulation of the **competitive exclusion principle**, which states that two species that use a limiting resource in the same way cannot coexist indefinitely. As we'll see next, field observations are consistent with this explanation of why competitive exclusion occurs in some situations but not others.

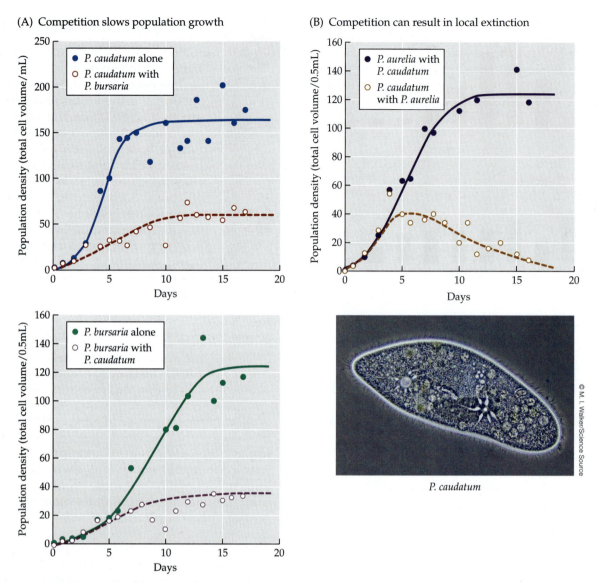

(A) Competition slows population growth

(B) Competition can result in local extinction

P. caudatum

FIGURE 14.9 Competition in *Paramecium* G. F. Gause grew three species of *Paramecium* in tubes filled with a liquid medium containing bacteria and yeast cells. (A) When grown alone, *Paramecium caudatum* and *P. bursaria* each showed logistic population growth and reached carrying capacity. When grown together, each species showed slower population growth, but they were able to coexist by feeding on different food items.

(B) When *P. caudatum* was grown with *P. aurelia*, it experienced local extinction. (A after G. F. Gause. 1935. *Vérifications Expérimentales de la Théorie Mathématique de la Lutte pour la Vie.* Hermann et Cie: Paris; B after G. F. Gause. 1934. *The Struggle for Existence.* Williams & Wilkins: Baltimore, MD.)

❓ Predict what would happen if *P. aurelia* and *P. bursaria* were grown together. Explain.

Competitors may coexist if they use resources differently

In natural communities, many species use the same limiting resources yet manage to coexist with one another. This observation does not violate the competitive exclusion principle, because a key point of that principle is that species must use limiting resources in the same way. Field studies often reveal differences in how species use limiting resources. Such differences are referred to as **resource partitioning** (or, sometimes, **niche partitioning**).

Thomas Schoener studied resource partitioning in four lowland *Anolis* lizard species that live on the West Indian island of Jamaica. Although these species all live together in trees and shrubs and eat similar foods, Schoener (1974) found differences among them in the height and thickness of their perches and in the time they spent in sun or shade. As a result of these differences, members of the different *Anolis* species competed less intensely than they otherwise would.

In a marine example, Maayke Stomp and her colleagues (2004) studied resource partitioning in two types of cyanobacteria collected from the Baltic Sea. The species identities of these cyanobacteria are unknown, so we will refer to them as BS1 and BS2 (standing for Baltic Sea 1 and Baltic

Sea 2). BS1 absorbs green wavelengths of light efficiently, which it uses in photosynthesis. However, BS1 reflects most of the red light that strikes its surface; hence, it uses red wavelengths inefficiently (and is red in color). In contrast, BS2 absorbs red light and reflects green light; hence, BS2 uses green wavelengths inefficiently (and is green in color).

Stomp and colleagues explored the consequences of these differences in a series of competition experiments. They found that each species could survive when grown alone under green or red light. However, when they were grown together under green light, the red cyanobacterium BS1 drove the green cyanobacterium BS2 to extinction (**FIGURE 14.10A**)—as might be expected, since BS1 uses green light more efficiently than does BS2. Conversely, under red light, BS2 drove BS1 to extinction (**FIGURE 14.10B**), as also might be expected. Finally, when grown together under "white light" (the full spectrum of light, including both green and red light), both BS1 and

BS2 persisted (**FIGURE 14.10C**). Taken together, these results suggest that BS1 and BS2 coexist under white light because they differ in which wavelengths of light they use most efficiently in photosynthesis.

Following up on their laboratory experiments, Stomp et al. (2007) analyzed the cyanobacteria present in 70 aquatic environments that ranged from clear ocean waters (where green light predominates) to highly turbid lakes (where red light predominates). As could be predicted from Figure 14.10, only red cyanobacteria were found in the clearest waters and only green cyanobacteria were found in highly turbid waters—but both types were found in waters of intermediate turbidity, where both green and red light were available. Thus, the laboratory experiments and field surveys conducted by the researchers suggest that red and green cyanobacteria coexist because they partition the use of a key limiting resource: the underwater light spectrum.

Competition can lead to character displacement and resource partitioning

When two species compete for resources, natural selection may favor individuals whose phenotype either (1) allows them to outcompete their competitors, resulting in competitive exclusion, or (2) allows them to partition their limiting resources, thus decreasing the intensity of competition. For example, when two fish species live apart from each other (each in its own lake), the two species may catch prey of similar size. If some factor (such as dispersal) were to cause members of these two species to live in the same lake, their use of resources would overlap considerably (**FIGURE 14.11A**). In such a situation, natural selection might favor individuals of species 1 whose morphology was such that they ate smaller prey, hence reducing competition with species 2; similarly, selection might favor individuals of species 2 that ate larger prey, hence reducing competition with species 1. Over time, such selection pressures could cause species 1 and species 2 to evolve to become different when they live together than when they live apart (**FIGURE 14.11B**). Such a process illustrates **character displacement**, which occurs when competition causes the phenotypes of competing species to evolve to become different over time.

Character displacement appears to have occurred in two species of finches on the Galápagos archipelago. Specifically, the beak sizes of the two species, and hence the

(A) Competition in green light

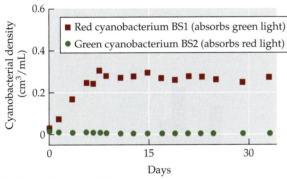

(B) Competition in red light

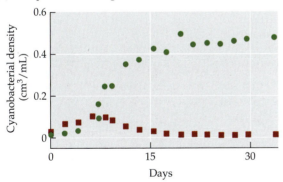

(C) Competition in white light

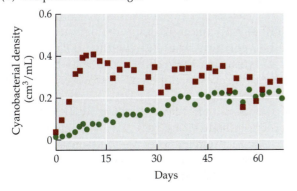

FIGURE 14.10 Do Cyanobacteria Partition Their Use of Light? Two types of cyanobacteria, BS1 and BS2, were grown together under (A) green light (550 nm), (B) red light (635 nm), and (C) "white" light (the full spectrum, which includes both green and red light). BS1 absorbs green light more efficiently than it absorbs red light; the reverse is true for BS2. Only BS1 persists when the two types are grown together under green light, and only BS2 persists when they are grown under red light. However, both types persist under white light, suggesting that BS1 and BS2 coexist by partitioning their use of light. (After M. Stomp et al. 2004. *Nature* 432: 104–107.)

(A)

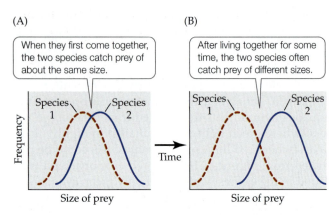

(B)

FIGURE 14.11 Character Displacement Competition for resources can cause competing species to change over time. Imagine that two fish species that once lived apart and tended to catch prey of about the same size are brought together in a single lake. (A) When the two species first come together, there is considerable overlap in the resources they use. (B) As the two species interact over time, the characteristics they use to obtain prey may evolve such that they tend to catch prey of different sizes.

sizes of the seeds the birds eat, are different on islands where both species live than on islands that have only one of the two species (**FIGURE 14.12**). Field observations suggest that these two finch species probably differ when they live together because of competition, not because of other factors, such as differences in food supplies (Schluter et al. 1985; Grant and Grant 2006).

Data suggestive of character displacement have also been observed in plants, frogs, fishes, lizards, birds, and crabs: in each of these groups, there are pairs of species that consistently differ more where they live together than where they live apart. Additional evidence is needed, however, if we are to make a strong argument that such differences result from competition (as opposed to other factors). Strong support for the role of character displacement can come from experiments designed to test whether competition occurs and has a selective effect on morphology. Such experiments were conducted on sticklebacks of the genus *Gasterosteus*, a group of fish species whose morphology varies most when different species live in the same lake (Schluter 1994). The results indicated that individuals whose morphology differed the most from that of their competitors had a selective advantage: they grew more rapidly than did individuals whose morphology was more similar to that of their competitors. Support for character displacement has also been found in field experiments with spadefoot toad tadpoles (Pfennig et al. 2007) and in laboratory experiments with the bacterium *Escherichia coli* (Tyerman et al. 2008). In each of these studies, experimental results suggest that competition caused the observed morphological differences—that is, that character displacement occurred—and the species were better able to partition their resources as a result.

Evidence for resource partitioning has been used as an explanation for the patterns of species diversity found

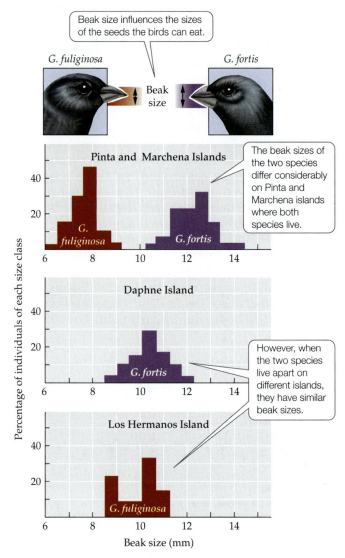

FIGURE 14.12 Competition Shapes Beak Size On islands harboring both *Geospiza fuliginosa* and *G. fortis*, competition between these two species of Galápagos finches may have had a selective effect on the sizes of their beaks. (After D. Lack. 1947. *Darwin's Finches*. Cambridge University Press: Cambridge.)

in communities, as we will see in Chapter 19. For now, let's next turn to mathematical models designed to predict whether the outcome of competition results in competitive exclusion or competitive coexistence.

CONCEPT 14.3 Competitive interactions can be modeled using the logistic equation.

LEARNING OBJECTIVES

14.3.1 Formulate the components of the Lotka–Volterra competition model, including the competition coefficient.

14.3.2 Diagram and interpret the four competitive outcomes of the Lotka–Volterra competition model.

The Lotka–Volterra Competition Model

Working independently of each other, A. J. Lotka (1932) and Vito Volterra (1926) both modeled competition by modifying the logistic equation. Recall from the discussion under Concept 11.3 that in the logistic equation, the rate at which a population changes in size (dN/dt) is

$$\frac{dN}{dt} = rN\left(1 - \frac{N}{K}\right)$$

or, alternatively,

$$\frac{dN}{dt} = rN\left(\frac{K - N}{K}\right)$$

where N is the population size, r is the intrinsic rate of increase (the maximum possible growth rate for the species, achieved only under ideal conditions), and K is the number at which the population stops increasing in size (which can be interpreted as the carrying capacity of the population).

As we have seen in Concept 14.2, competition deprives species of resources and hence reduces population growth rates. Thus, the presence of a competitor should reduce the growth rate of the original population. To incorporate the effects of the competitor species on one another, we can modify the logistic equation of each species by subtracting a **competition coefficient**, which is a constant used to indicate how strong the competitive effect of one species is on another. The new equations, known as the **Lotka–Volterra competition model**, can be written as

$$\frac{dN_1}{dt} = r_1 N_1 \left(\frac{K_1 - N_1 - \alpha N_2}{K_1}\right)$$

$$\frac{dN_2}{dt} = r_2 N_2 \left(\frac{K_2 - N_2 - \beta N_1}{K_2}\right)$$

(14.1)

In these equations, N_1 is the population density of species 1, r_1 is the intrinsic rate of increase of species 1, and K_1 is the carrying capacity of species 1; N_2, r_2, and K_2 are similarly defined for species 2. The competition coefficients (α and β) are constants that describe the effect of one species on the other: α is the effect of species 2 on species 1, and β is the effect of species 1 on species 2. For example, if $\alpha = 1$, then individuals of the two species have the same effect in depressing the growth of species 1. If $\alpha = 5$, each individual of species 2 decreases the growth of species 1 by the same amount as five additional individuals of species 1. Thus, the competition coefficient α is a measure of the effect, on a per-individual basis, of species 2 on the population growth of species 1, measured relative to the effect of species 1. Similar reasoning applies to β, which is the effect, on a per-individual basis, of species 1 on the population growth of species 2.

We can also think of α and β as "translation terms," each of which converts the number of individuals of one species into the number of individuals of the other species that has an equivalent effect on population growth rates. For example, if $\alpha = 3$, one individual of species 2 decreases the growth of species 1 by the same amount as would three individuals of species 1. Thus, if there are 100 individuals of species 2, it would take 300 individuals of species 1 to decrease the growth rate of species 1 by the same amount as do the 100 individuals of species 2 (i.e., $\alpha = 3$ and $N_2 = 100$, so it takes $\alpha N_2 = 3 \times 100 = 300$ individuals of species 1 to have an equivalent effect).

In the remainder of this section, we'll see how Equation 14.1 can be used to predict the outcome of competition; then we'll explore how competitive coexistence is affected by species interaction strength.

Predicting the outcome of competition

The outcome of competition can be predicted if we know how the population sizes of species 1 and species 2 are likely to change over time. For example, if the population size of species 2 is likely to increase while that of species 1 is likely to decrease to zero, then species 2 should drive its competitor to extinction, thus "winning" the competitive interaction. A computer can be programmed to solve Equation 14.1, thereby predicting the population sizes of species 1 and 2 at different times. Here, however, we'll use a graphical approach to examine the conditions under which each species would be expected to increase or decrease in population size.

We begin by determining when the population size of each competing species would stop changing in size. This approach, which we also used for the Lotka–Volterra predator–prey model (see Concept 12.3), is based on the idea that the population size (N) does not change when the population growth rate (dN/dt) equals zero (or reaches an equilibrium). For example, based on the Lotka–Volterra competition model (Equation 14.1), the population size of species 1 does not change when $dN_1/dt = 0$. When we set dN_1/dt equal to zero, we find that the population size of species 1 (N_1) does not change when

$$N_1 = K_1 - \alpha N_2 \qquad \textbf{(14.2)}$$

Likewise, the population size of species 2 (N_2) does not change when

$$N_2 = K_2 - \beta N_1 \qquad \textbf{(14.3)}$$

Notice that Equations 14.2 and 14.3 are straight lines, written with N_1 as a function of N_2 and N_2 as a function of N_1, respectively. Each of these lines is called the zero population growth isocline (or simply **isocline**), so named because a population does not increase or decrease in size for any combination of N_1 and N_2 that lies on these lines. For species 1, the abundance does not change when $dN_1/dt = 0$, which occurs when $N_2 = K_1/\alpha$ and $N_1 = K_1$. Similarly, for species 2, the abundance does not change when $dN_2/dt = 0$, which occurs when $N_1 = K_2/\beta$ and $N_2 = K_2$.

Once we determine K_1/α and K_2/β, we can then plot the isoclines for both species 1 (x-axis) and species 2 (y-axis) in graphical form. For species 1, the isocline will be a diagonal line originating at the value $N_2 = K_1/\alpha$ and ending at the value $N_1 = K_1$ (**FIGURE 14.13A**). This isocline represents the number of individuals of species 2 that would keep species 1's population from changing (or at equilibrium). For example, in Figure 14.13A, because a point to the right of the N_1 isocline represents more individuals than zero population growth will allow, the population size of species 1 will decrease until it reaches the isocline. This is true for the entire region shaded in blue: the population size of species 1 decreases for all points to the right of the N_1 isocline. In contrast, when the population size of species 1 is to the left of the N_1 isocline, the population size of species 1 increases. Similar reasoning applies to species 2's isocline, which can be plotted as the diagonal line originating at the value $N_1 = K_2/\beta$ and ending at the value $N_2 = K_2$ (**FIGURE 14.13B**). This isocline represents the number of individuals of species 1 that would keep species 2's population from changing (or at equilibrium). Here the population size of species 2 decreases in regions above the N_2 isocline and increases in regions below the N_2 isocline.

The graphical approach we have just described can be used to predict the end result of competition between species. To do this, we plot the N_1 and N_2 isoclines together. Because there are four possible ways that the N_1 and N_2 isoclines can be arranged relative to each other, we must make four different graphs. In two of these graphs, the isoclines do not cross, and competitive exclusion results: depending on which isocline is above the other, either species 1 (**FIGURE 14.14A**) or species 2 (**FIGURE 14.14B**) always drives the other to extinction. Note that in the regions shaded in blue, the population sizes of both species are greater than the population sizes on their isoclines, and hence both

species *decrease* in number (as indicated by the thick black arrows). Similarly, in the regions shaded in yellow, the population sizes of both species are less than those on their isoclines, and hence both species *increase* in number. In the regions shaded in light or dark gray, one species increases in number (because its population sizes are less than those on its isocline) while the other decreases until the species that increases reaches its carrying capacity (K) and the species that decreases reaches zero and becomes extinct.

Competitive exclusion also occurs in the third graph (**FIGURE 14.14C**), but which species "wins" depends on whether the changing population sizes of the two species first enter the region shown in dark gray (in which case, species 2 drives species 1 to extinction) or the region shown in light gray (in which case, species 1 drives species 2 to extinction). Finally, **FIGURE 14.14D** shows the only case in which the two species coexist, and hence competitive exclusion does not occur. Although in this case neither species drives the other to extinction, competition still has an effect: the final or equilibrium population size of each species (indicated by the box in the figure) is lower than its carrying capacity, as in Gause's experiments with *Paramecium* (see Figure 14.9A).

Researchers have used the graphical approach described in Figure 14.14 to predict the outcome of competition under different ecological conditions. For example, Livdahl and Willey (1991) used this approach to predict whether competition with a native species of mosquito could prevent the invasion of an introduced mosquito species. You can explore their results in **ANALYZING DATA 14.1**.

The strength of competitive interactions affects coexistence

Now that we've seen the four possible outcomes predicted by the Lotka–Volterra competition model, let's focus on the single case in which competitive coexistence occurs. We can use Figure 14.14D to show that coexistence occurs when the values of α, β, K_1, and K_2 are such that the following inequality holds:

FIGURE 14.13 Graphical Analyses of Competition The zero population growth isoclines from the Lotka–Volterra competition model can be used to predict changes in the population sizes of competing species. (A) The N_1 isocline. The change in population size of species 1 (indicated by black solid arrows) increases in the yellow region and decreases in the blue region. (B) The N_2 isocline. The change in population size of species 2 (indicated by red dashed arrows) increases in the yellow region and decreases in the blue region.

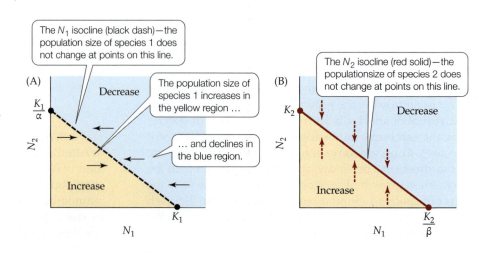

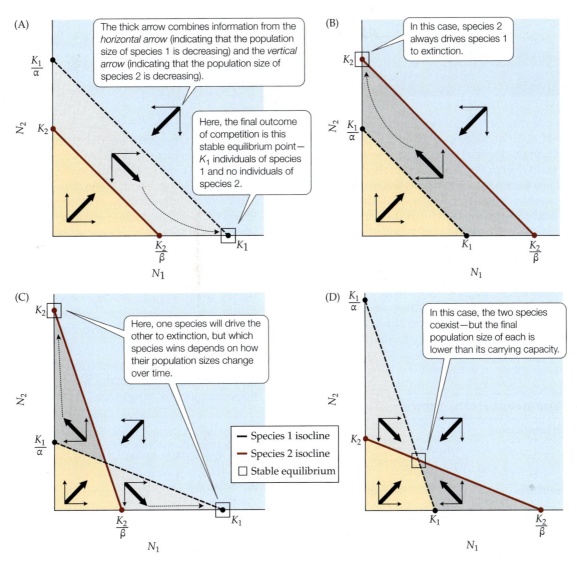

FIGURE 14.14 Outcome of Competition in the Lotka–Volterra Competition Model The outcome of competition depends on how the N_1 and N_2 isoclines are positioned relative to one another. (A) Competitive exclusion of species 2 by species 1; species 1 always wins. (B) Competitive exclusion of species 1 by species 2; species 2 always wins. (C) The two species cannot coexist; either species 1 or species 2 wins depending on population sizes of both species. (D) Species 1 and species 2 coexist. The box in each graph indicates a stable equilibrium point—a combination of population sizes of the two species that, once reached, does not change over time.

❓ In (B), if $K_2 = 1,000$ and if species 1 went extinct when $N_2 = 1,200$, how would the population size of species 2 change after the extinction of species 1?

$$\alpha < \frac{K_1}{K_2} < \frac{1}{\beta} \qquad (14.4)$$

To see what we can learn from this inequality, consider a situation in which the competing species are equally strong competitors, indicating that $\alpha = \beta$. If the two species are also very similar in how they use resources, an individual of species 1 will have nearly the same effect on the growth rate of species 2 as would an individual of species 2 (and vice versa). Thus, when the two species use resources in very similar ways and thus strongly compete, α and β should both be close to 1.

Suppose, for example, that $\alpha = \beta = 0.95$. If we substitute these values for α and β into Equation 14.4, we obtain

$$0.95 < \frac{K_1}{K_2} < 1.05$$

This result suggests that when species strongly compete, coexistence is predicted only when the two species also have similar carrying capacities.

In contrast, if the competing species do not compete strongly but rather differ greatly in how they use resources, α and β will be much lower than 1. To illustrate this case, suppose that $\alpha = \beta = 0.1$. In this situation, coexistence

ANALYZING DATA 14.1

Will Competition with a Native Mosquito Species Prevent the Spread of an Introduced Mosquito?

The mosquito *Aedes albopictus* breeds in small volumes of water, such as those in tree holes (cavities in trees that can hold water) and in abandoned tires. Introduced from Asia to North America in the 1980s, this species is a public health concern because it can transmit diseases such as dengue fever. Once in North America, *A. albopictus* colonized tree holes and tires, where it encountered thriving populations of several different native species of mosquitoes.

Livdahl and Willey (1991)* sought to predict the outcome of competition between *A. albopictus* and the native mosquito *A. triseriatus*, a predominant member of tree hole communities. To do this, they estimated competition coefficients and carrying capacities for *A. albopictus* and *A. triseriatus* mosquito larvae developing in water obtained from tree holes and from tires. Their results are shown in the table.

Water obtained from tree holes	Water obtained from tires
Competition coefficients	
$\alpha = 0.43$	$\alpha = 0.84$
$\beta = 0.72$	$\beta = 0.25$
Carrying capacities (no. individuals/100 ml water)	
$K_1 = 42.5$	$K_1 = 33.4$
$K_2 = 53.2$	$K_2 = 44.7$

1. Using Equation 14.1, designate *A. triseriatus* as species 1 and *A. albopictus* as species 2. Use the data in the table to plot the N_1 and N_2 isoclines (see Equations 14.2 and 14.3) for these two species competing in tree hole communities. Predict the equilibrium population density (no. individuals per 100 ml of water) for each species. Describe the likely outcome of competition between these two species in tree hole communities.

2. On a separate graph, plot the N_1 and N_2 isoclines for these two species competing in tires. Predict the equilibrium population density (no. individuals per 100 ml of water) for each species. Describe the likely outcome of competition between these two species in tires.

3. Is it likely that competition with the native species (*A. triseriatus*) will prevent the spread of the introduced species (*A. albopictus*)? Explain.

* Livdahl, T. P., and M. S. Willey. 1991. Prospects for an invasion: Competition between *Aedes albopictus* and native *Aedes triseriatus*. *Science* 253: 189–191.

---

is predicted even if the carrying capacity of one species is nearly 10 times that of the other species, namely

$$0.1 < \frac{K_1}{K_2} < 10$$

As you can demonstrate on your own, other values for the competition coefficients α and β yield similar results. Taken together, such analyses of the Lotka–Volterra competition model suggest the following refinement of the competitive exclusion principle: competing species are more likely to coexist (and hence competitive exclusion is *less* likely) when they do not compete strongly but rather use resources in different ways.

A variety of factors can influence how species divide their use of resources, thereby preventing one competitor from driving the other to extinction. As we'll see in the next section, some of these factors can alter the outcome of competition entirely, turning the inferior competitor into the superior one.

CONCEPT 14.4 The outcome of competition can be altered by predation, the physical environment, and disturbance.

LEARNING OBJECTIVES

14.4.1 Describe how herbivores or predators can change or reverse the outcome of competition.

14.4.2 Explain how the physical environment can affect the outcome of competition and distribution of species.

14.4.3 Explain how disturbances can allow coexistence in highly asymmetrical competitive interactions.

Altering the Outcome of Competition

The outcome of competition between species can be changed by a broad suite of factors, including features of the physical environment, disturbance, and interactions with other species. For example, a difference in abiotic conditions—as might occur from one place to another—can cause a competitive reversal, in which the species that was the inferior competitor in one habitat becomes the superior competitor in another. Cases in which the outcome of competition may be modified under different abiotic conditions include Tansley's bedstraws, described in the Introduction, and the acorn barnacle in North America (see Figure 9.13).

Interactions with other species can have similar effects on the outcome of competition between species. The presence of herbivores has been shown to reverse the outcome of competition between species of encrusting marine algae (Steneck et al. 1991) and between ragwort (*Senecio jacobaea*) and other plant species (**FIGURE 14.15**). Herbivores can have this effect if they prefer to feed on the superior competitor, thereby reducing the growth, reproduction, or survival of that species. What is true of herbivores is also true of predators, pathogens, and mutualists: an increase or decrease in the abundance of such

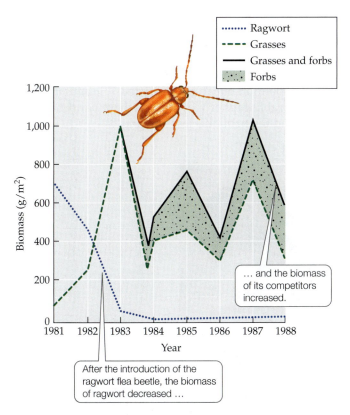

FIGURE 14.15 Herbivores Can Alter the Outcome of Competition Ragwort flea beetles are herbivores that feed on ragwort (*Senecio jacobaea*), an invasive plant species. The graph tracks the biomasses of ragwort, grasses, and forbs (broad-leaved herbaceous plants) at a site in western Oregon after the flea beetle was introduced there in 1980. The results show that in the absence of the flea beetle, ragwort was a superior competitor, but it declined precipitously when the beetle was introduced. (After P. McEvoy et al. 1991. *Ecol Appl* 1: 430–432.)

species can change the outcome of competition among the species with which they interact.

In later chapters, we'll explore many examples in which species interactions alter competitive outcomes—sometimes preventing a superior competitor from driving other species to extinction. Here, we'll focus on the effects of the physical environment and disturbance.

The physical environment can affect competition and ultimately the distribution of species

In a series of classic experiments, Joseph Connell (1961a,b) examined factors that influenced the local distribution, survival, and reproduction of two barnacle species, *Chthamalus* and *Semibalanus*. The larvae of barnacles drift through ocean waters, then settle on rocks or other surfaces (such as boat hulls), where they metamorphose into adults, forming a hard outer shell.

At Connell's study site along the coast of Scotland, the distributions of *Chthamalus* and *Semibalanus* larvae overlapped considerably: the larvae of both species were found throughout the upper and middle intertidal zones. However, adult *Chthamalus* were usually found only near the top of the intertidal zone, whereas adult *Semibalanus* were not found there but were found throughout the rest of the intertidal zone (**FIGURE 14.16**). What accounted for these differences in distribution?

To answer this question, Connell examined the effects of competition and of abiotic features of the environment, such as the risk of desiccation (drying out because of exposure to air, which is greatest in the upper intertidal zone). To test the importance of competition under different abiotic conditions, he chose some individual young barnacles of each species that had settled in each zone and removed all nearby members of the other species. For other focal individuals, he left nearby members of the other species in place. He found that competition with *Semibalanus* excluded *Chthamalus* from all but the top of the intertidal zone, where *Chthamalus* was able to thrive under reduced competition. As they grew, *Semibalanus* smothered (by growing on top of) or crushed the *Chthamalus* in the middle intertidal but not in the upper intertidal zone. Averaging across all regions of the intertidal zone, only 14% of *Chthamalus* survived their first year when faced with competition from *Semibalanus*, whereas 72% survived where Connell had removed *Semibalanus*. *Chthamalus* individuals that survived a year of competition with *Semibalanus* were small and reproduced poorly.

Semibalanus, in contrast, was not affected strongly by competition with *Chthamalus*. However, whether *Chthamalus* was removed or not, *Semibalanus* dried out and survived poorly near the top of the intertidal zone. Thus, *Semibalanus* appears to have been excluded from that zone by its sensitivity to desiccation rather than its interactions with *Chthamalus*.

FIGURE 14.16 Squeezed Out by Competition
Removal experiments at a field site in Scotland showed that competition mediated by the physical environment determines the local distribution of two species of barnacles, *Chthamalus stellatus* and *Semibalanus balanoides*. (After J. H. Connell. 1961b. *Ecology* 42: 710.)

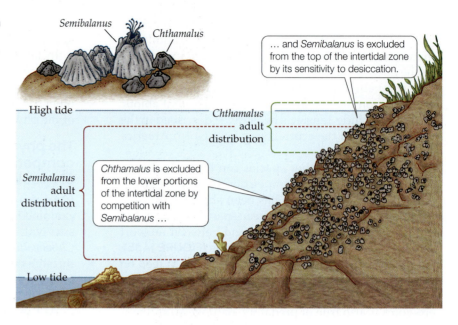

... and *Semibalanus* is excluded from the top of the intertidal zone by its sensitivity to desiccation.

High tide

Chthamalus adult distribution

Chthamalus is excluded from the lower portions of the intertidal zone by competition with *Semibalanus* ...

Semibalanus adult distribution

Low tide

Semibalanus   *Chthamalus*

As observed for Tansley's bedstraw plants and Connell's barnacles, competition can restrict the local distribution of a species to a particular set of environmental conditions—the bedstraws, for example, could be growing inches away from each other, but each species was restricted to a particular soil type. Competition has also been shown to prevent a wide range of species, including mammals, marine invertebrates, birds, and plants, from occupying geographic regions in which they would otherwise thrive.

In some cases, a "natural experiment"—a situation in nature that is similar in effect to a controlled removal experiment—provides evidence that competition can vary depending on environmental conditions and ultimately affect geographic distributions. Such a situation was found for chipmunks in the genus *Tamias* (previously known as *Neotamias* or *Eutamias*). These chipmunks live in forests on mountains in the southwestern United States, where mountain ranges are separated from one another by desert flatlands. Patterson (1980, 1981) studied the distributions of *Tamias* chipmunks and found that when

a species lived alone on a mountain range because it preferred those environmental conditions, it consistently occupied a broader range of habitats and elevations than when it lived with a competitor species (**FIGURE 14.17**). As in Connell's removal experiments, this result suggests that competition may have restricted *T. quadrivittatus* to less desirable higher elevations where conditions are colder and the habitat is less suitable.

T. quadrivittatus

FIGURE 14.17 A Natural Experiment on Competition between Chipmunk Species Observations of the distributions of *Tamias* chipmunks on mountain ranges in New Mexico suggest that competition may restrict the preferred habitats in which they live. In competition, *T. quadrivittatus* was restricted to less desirable colder elevations compared to *T. dorsalis*. Similar results were obtained for *Tamias* species living in Nevada. (After M. V. Lomolino et al. 2006. *Biogeography*, 3rd ed. Oxford University Press/Sinauer: Sunderland, MA.)

In mountain ranges where only one chipmunk species was present, that species occupied a broader range of elevations than it did where a competitor was present.

In mountain ranges where two chipmunk species were present, their distributions did not overlap greatly.

Tamias quadrivittatus

T. dorsalis

Organ Mountains   Mt. Taylor   Magdalena Mountains

CLIMATE WARMING, NON-NATIVE SPECIES, AND COMPETITION We have learned that competition often varies in its intensity, can be asymmetrical in its effects, can occur between closely or distantly related species, and can be modified by the physical and biotic environment. Given an understanding that species interactions can vary under a variety of contexts, ecologists are interested in exploring how climate change might modify competitive interactions and ultimately influence abundance and distribution of species. Nowhere is this more important than how climate change might influence the competitive interactions of non-native, invasive species. To explore the effects of climate change on the competitive interactions of invasive species, and how that will ultimately influence their distribution and abundance, experiments can be conducted to compare those interactions with and without simulated climate conditions. One such experiment examined how warming temperatures could alter the competitive interactions of the two intentionally introduced non-native beachgrass species along the U.S. Pacific Northwest coast (Biel and Hacker 2021). The European beachgrass *Ammophila arenaria* (**FIGURE 14.18A**) was introduced in the early 1900s and spread widely along the Pacific coast (Mexico to Canada), and the second species, the American beachgrass *A. breviligulata* (**FIGURE 14.18B**), was introduced in the mid-1930s and occurs only in the north

(Canada to northern Oregon). Over the last century, these two invasive species have dramatically converted coastal landscapes from sparsely vegetated and shifting sandy environments to highly vegetated hills of sand (**FIGURE 14.18C, D**). *Ammophila breviligulata* is a better competitor than *A. arenaria*, but given morphological and growth-form differences between the two species, *A. arenaria* captures more sand and builds taller and more protective dunes. The competition experiment was conducted by planting both species together and alone in large bags filled with sand that were artificially warmed with heating tape (**FIGURE 14.18E**). The researchers found that *A. breviligulata* had lower biomass when exposed to warming conditions, but *A. arenaria* showed neutral or positive responses to warming. Nevertheless, under either experimental treatment, *A. breviligulata* had strong negative effects on *A. arenaria*, while *A. arenaria* had weaker effects on *A. breviligulata*. Using the Lotka–Volterra competition model (see Concept 14.3), the researchers predicted that although *A. breviligulata* mostly excludes *A. arenaria*, elevated temperatures can increase the likelihood of species coexistence. Thus, under climate warming, the differences in physiological tolerance and the mediation of species interactions might result in an expansion of the northern distributional limit of *A. arenaria* but restriction in the southern limit of *A. breviligulata*. Given that beachgrass abundance can have direct effects on dune ecosystems, reductions in vigor and competitive ability from warming could ultimately alter coastal protection and biodiversity.

(A) (B) (C) (D) (E)

All photos courtesy of Sally Hacker

FIGURE 14.18 Implications of Climate Warming to Competition of Invasive Species Warming from climate change affects the competitive interactions of two non-native beachgrass species with consequences for dune ecosystems. (A) European beachgrass *Ammophila arenaria*. (B) American beachgrass *A. breviligulata*. (C) Dune covered in *A. breviligulata* in Oregon. (D) Dune covered in *A. arenaria*. (E) Beachgrass competition experiment that manipulated warming. (After R. Biel and S. D. Hacker. 2021. *Oecologia* 197: 757–770.)

FIGURE 14.19 Population Decline in an Inferior Competitor Lacking Disturbance In this graph, each point represents an observed change in density (*N*, the number of individuals per square meter) from one year (year *x*) to the next (year *x* + 1) at sites where sea palms are growing in competition with mussels and lack disturbance. These points can be used to estimate a *replacement curve* (solid blue line), which shows the extent to which sea palm individuals replace themselves over time without disturbance. The exact replacement curve (dashed red line) shows the densities at which the population size would not change from one year to the next. (After R. T. Paine. 1979. *Science* 205: 685–687.)

❓ Based on the observed replacement curve (the solid blue line), how many years would it take for a sea palm population to decline from 100 individuals to fewer than 20 individuals?

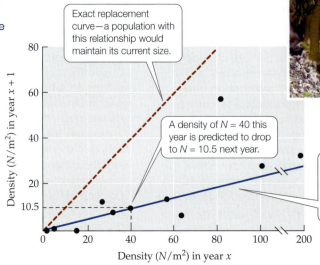

Exact replacement curve—a population with this relationship would maintain its current size.

A density of *N* = 40 this year is predicted to drop to *N* = 10.5 next year.

The observed replacement curve is well below the exact replacement curve. This indicates that the sea palm population is not replacing itself— instead it is declining toward extinction.

Postelsia palmaeformis

Peter Pearsall/U.S. Fish and Wildlife Service/CC BY 2.0

Disturbance can prevent competition from running its course

As we saw in Concept 9.3, a disturbance such as a fire or major storm may kill or damage some individuals while creating opportunities for others. Some species can persist in an area only if such disturbances occur regularly. Forests, for example, contain some herbaceous plant species that require abundant sunlight and are therefore found only in areas where wind or fire has created an opening in the tree canopy. Over time, a population of such plants is doomed: as trees recolonize the area, shade increases to the point at which the species are competitively excluded. Such species are called **fugitive species** because they must disperse from one place to another to take advantage of disturbances that open up resources and allow them to avoid competitive exclusion.

Robert Paine, a marine ecologist from the University of Washington, has described how periodic disturbance allows a fugitive algal species, the sea palm (*Postelsia palmaeformis*), to coexist with a competitively dominant species, the mussel *Mytilus californianus* (Paine 1979). The sea palm is a brown alga that lives in the intertidal zone and must attach itself to rocks to grow. It competes for attachment space with mussels. Although a sea palm can outcompete an individual mussel (by growing on top of

it), the sea palm is eventually displaced by other mussels that grow in from the side. Competition with mussels causes sea palm populations to decline over time (**FIGURE 14.19**). Hence, if competition ran its course, mussels would drive sea palm populations to extinction. That is exactly what happens on low-disturbance shorelines (with a mean rate of 1.7 disturbances per year), where waves only occasionally tear patches of mussels from the rocks. However, sea palms can persist in shoreline areas where high-energy waves remove mussels more frequently (with a mean rate of 7.7 disturbances per year), thereby creating temporary openings for sea palm individuals.

A CASE STUDY REVISITED

Competition in Plants That Eat Animals

In plants, competition for nutrients can be especially important (see Figure 14.5), but other resources, such as light and water, can also be in short supply. Carnivorous plants live in nutrient-poor soils, and their root systems are usually less well developed compared to their non-carnivorous neighbors. These observations suggest that carnivorous plants might be especially hard-hit by below-ground competition if they were denied access to their unique, alternative source of nutrients (animal prey).

Instead, contrary to what would be expected if competition for nutrients were important, *Sarracenia* was not especially hard-hit when neighbors were present and pitchers were deprived of prey (see Figure 14.2). In fact, when neighbors were present, pitcher plants had the same biomass regardless of whether they had access to prey. These results suggest that there was relatively little competition between *Sarracenia* and noncarnivorous plants for soil nutrients and that some other factor was driving the positive response these plants had to neighbor removal.

Further investigation revealed that competition for light appeared to be more important to pitcher plants. Brewer (2003) found that neighbors reduced the availability of light to *Sarracenia* by a factor of 10. When neighbors were removed, *Sarracenia* responded by greatly increasing its growth, especially if pitchers were open and the plants could capture prey (see Figure 14.2). Hence, *Sarracenia* responded to higher light levels when neighbors were removed by growing more rapidly—but only if prey were available to supply the extra nutrients they needed for such growth. Overall, it appears that pitcher plants compete with their neighbors for light but avoid competition for soil nutrients by eating animals and by using changes in light levels as a cue for growth.

 ## CONNECTIONS *in* NATURE

THE PARADOX OF DIVERSITY As we've seen, some field data show that superior competitors can drive inferior competitors extinct—which is exactly what the competitive exclusion principle states should happen whenever two or more species use the same set of limiting resources. Natural communities, however, contain many species that share the use of scarce resources without driving one another to extinction. Pitcher plants, for example, coexist with a diverse group of other species (**FIGURE 14.20**), even though they were predicted to be inferior competitors for soil nutrients. In the context of Brewer's experiments on pitcher plants, let's reconsider why superior competitors do not always drive inferior competitors to extinction.

The concept of resource partitioning suggests that a number of species could coexist in nutrient-poor environments if they avoided competition for scarce nutrients by acquiring them in different ways. This idea helped to motivate Brewer's study: he wanted to know whether differences in their means of nutrient acquisition could explain the coexistence of carnivorous and noncarnivorous plants. To find out, Brewer (2003) deprived carnivorous plants of their unique source of nutrients (animal prey), thus increasing the overlap between the ways in which carnivorous

FIGURE 14.20 Coexistence in a Nutrient-Poor Environment The pitcher plant *Sarracenia alata*, seen in the close-up at the left, co-exists with noncarnivorous plants that can compete with it for both nutrients and light.

and noncarnivorous plants acquired nutrients. If competition for nutrients was important, pitcher plants that were deprived of prey should have experienced more severe competitive effects, or they should have compensated for reduced nutrient intake by increasing their production of roots or pitchers. Neither of these outcomes occurred, so Brewer sought other explanations of species coexistence.

As we'll see in Concept 19.3, environmental variation provides a second mechanism for the coexistence of species in communities: if environmental conditions fluctuate over space or time (or both), species may coexist if different species are superior competitors under different environmental conditions. Tansley's bedstraw example (given in the Introduction) illustrates how differences in soils can alter the outcome of competition, thus promoting coexistence in environments that vary over space. With respect to variation over time, an inferior competitor may persist whenever competition fails to run its course. Consider a species such as the sea palm (see Figure 14.19), which competes poorly but tolerates disturbance well. Such a species may persist if a disturbance periodically "resets the clock" by decreasing the abundance of a superior competitor before that species drives the inferior competitor to extinction. Such a scenario may also apply to the pitcher plant *Sarracenia* The habitat in which it lives is prone to fire; pitcher plants tolerate fire well, and they use changes in light levels as a cue for growth. As a result, *Sarracenia* grows primarily when its competitors are reduced by fire. This growth strategy may allow it to escape competition for nutrients by reducing its demand for scarce nutrients when competition is potentially most intense (i.e., in years without fire) and increasing its demand for nutrients when competitors have been reduced (years with fire).

SUMMARY

CONCEPT 14.1 Competition can be direct or indirect, vary in its intensity, and occur between similar or dissimilar species.

14.1.1 Define the difference between exploitation competition and interference competition.

- The most common form of competition is exploitation competition, which occurs when species compete indirectly as they share the use of a limiting resource.

- Another form of competition, called interference competition, occurs when species compete directly for access to resources.

14.1.2 Analyze how and why competition can vary in its intensity.

- If resource levels become sufficiently low, the intensity of competition can increase.

- Competition is often asymmetrical, affecting one competitor more strongly than the other.

- Competition can occur between closely or distantly related species.

14.1.3 Describe the importance of competition within communities.

- Competition for resources is common—though not ubiquitous—in natural communities.

CONCEPT 14.2 Competing species are more likely to coexist when they use resources in different ways.

14.2.1 Define the competitive exclusion principle and explain how it differs from competitive coexistence.

- The competitive exclusion principle states that if competing species use the same limiting resource in the same way, they cannot coexist.

- In reality, most species show competitive coexistence, or the ability to coexist with one another despite sharing limiting resources.

14.2.2 Define and give examples of resource partitioning (or niche partitioning).

- Field studies have revealed many examples of resource partitioning, in which competing species use one or more shared resources in different ways.

14.2.3 Describe how competition can lead to character displacement and resource partitioning.

- In character displacement, competition causes the phenotypes of competing species to evolve to become different from each other over time, thereby reducing the intensity of competition and allowing resource partitioning.

CONCEPT 14.3 Competitive interactions can be modeled using the logistic equation.

14.3.1 Formulate the components of the Lotka–Volterra competition model, including the competition coefficient.

- Lotka and Volterra modeled the effects of interspecific competition by modifying the logistic equation.

- The Lotka–Volterra competition model includes a competition coefficient, which is a constant used to indicate how strong the competitive effect of one species is on another.

14.3.2 Diagram and interpret the four competitive outcomes of the Lotka–Volterra competition model.

- Graphical analysis of the Lotka–Volterra competition model allows four predictions for the outcome of competition—three lead to competitive exclusion, and one leads to competitive coexistence.

CONCEPT 14.4 The outcome of competition can be altered by predation, the physical environment, and disturbance.

14.4.1 Describe how herbivores or predators can change or reverse the outcome of competition.

- Herbivores or predators can change or reverse the outcome of a competitive interaction if they prefer to feed on the superior competitor.

14.4.2 Explain how the physical environment can affect the outcome of competition and distribution of species.

- The physical environment can affect the outcome of competition and ultimately the distribution of species.

14.4.3 Explain how disturbances can allow coexistence in highly asymmetrical competitive interactions.

- Periodic disturbances that remove a superior competitor can allow an inferior competitor to persist.

Review Questions

1. Plant species require nitrogen to grow and reproduce. Assume that a single application of high-nitrogen fertilizer is added to a low-nitrogen, sandy soil in which two plant species are found. Predict how the intensity of competition for soil nitrogen will change over time, and explain your prediction.

2. List four general features of competition described in Concept 14.1, and provide an example of each.

3. Suppose that each of 20 meadows contains a population of plant species 1, a population of plant species 2, or populations of both plant species. Species 1 and 2 are known to compete with each other. Each meadow is separated from the others by areas in which neither species 1 nor species 2 can grow or survive.

    a. List three possible reasons why the meadows contain different combinations of the two plant species.

    b. Describe an experiment that would help to evaluate one or more of the reasons you have listed.

Hone Your Problem-Solving Skills

As described in Concepts 14.2 and 14.3, laboratory experiments, field observations, and mathematical models have been used to explain why competing species coexist in some situations but not others.

1. From *each* of these three approaches, describe a result that helps to explain when competing species are likely to coexist. Taken together, do these three approaches to studying competition yield similar or different explanations for why coexistence occurs in some situations but not others?

2. If $\alpha = 0.8$, $\beta = 1.6$, $N_1 = 140$, and $N_2 = 230$, are individuals of species 1 or species 2 having a greater effect on the growth rate of species 2?

3. Based on graphical analyses of the Lotka–Volterra competition model, evaluate the following statement: if $\alpha < \beta$, species 1 will always drive species 2 to extinction. Explain your answer.

ADDITIONAL RESOURCES

This textbook provides resources such as **Web Extensions** and additional **Climate Change Connections** for instructors who can share them with students. Please contact your instructor, if you would like to access that content.

Mutualism and Commensalism

The First Farmers: A Case Study

Humans first began to farm about 10,000 years ago. Agriculture was a revolutionary development that led to great increases in the size of our population as well as to innovations in government, science, the arts, and many other aspects of human societies. But people were far from the first species to farm. That distinction goes to ants in the tribe Attini, a group of 210 species, most of which live in tropical forests of South America. These ants, known informally as the attines or fungus-growing ants, started cultivating fungi for food at least 50 million years before the first human farmers (**FIGURE 15.1**).

Like human farmers, the ant farmers nourish, protect, and feed on the species they grow, forming a relationship that benefits both the farmer and the crop. The attines cannot survive without the fungi they cultivate; many of the fungi depend on the ants as well. When a virgin queen ant leaves her mother's nest to mate and begin a new colony, she carries in her mouth some of the fungi from her birth colony. The fungi are cultivated in subterranean gardens (**FIGURE 15.2**). An ant colony may contain hundreds of gardens, each roughly the size of a football; these gardens can provide enough food to support 2–8 million ants. Some attines occasionally replace the fungi in their gardens with new, free-living fungi that they gather from surrounding soils. Other species, such as leaf-cutter ants in the genera *Atta* and *Acromyrmex*, do not cultivate fungi found in the environment. Instead, the fungi in their gardens come only from propagules passed from a parent ant colony to each of its descendant colonies.

As their name suggests, leaf-cutter ants cut portions of leaves from plants and feed them to the fungi in their gardens. Back at the nest, the ants chew the leaves to a pulp, fertilize them with their own droppings, and "weed" the fungal gardens to help control bacterial and fungal invaders. In turn, the cultivated fungi produce specialized structures, called *gongylidia*, on which the ants feed. The partnership between leaf-cutter ants and fungi has been called an "unholy alliance" because each partner helps the other to overcome the formidable defenses that protect plants from being eaten. The ants, for example, scrape a waxy covering from the leaves that the fungi have difficulty penetrating, while the fungi digest and render harmless the chemicals that plants use to kill or deter insect herbivores.

© Martin Dohrn/Minden Pictures

FIGURE 15.1 Collecting Food for Their Fungi Fungus-growing ants (*Atta cephalotes*) in Costa Rica carry leaf segments to their colony, where the leaves will be fed to the fungus (the gray material) the ants cultivate for food.

(A)

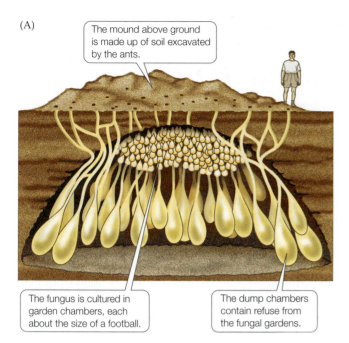

The mound above ground is made up of soil excavated by the ants.

The fungus is cultured in garden chambers, each about the size of a football.

The dump chambers contain refuse from the fungal gardens.

(B)

© Mark Moffett/Minden Pictures

FIGURE 15.2 The Fungal Garden of a Leaf-Cutter Ant (A) A diagrammatic representation of a large *Atta* leaf-cutter ant colony. (B) This photo shows a cutaway view of a garden chamber in a central Paraguay colony of the leaf-cutter ant *Atta laevigata*. Inside the chamber is a specialized structure called a gongylidia, which is produced by the cultivated fungus and eaten by the ants. (A after B. Hölldobler and E. O. Wilson. 1990. *The Ants*. Belknap Press of Harvard University Press: Cambridge, MA; modified from J. C. M. Jonkman, in Weber 1979.)

But all is not perfect in the gardens. Nonresident fungi, which themselves would benefit from ant cultivation, periodically invade leaf-cutter ant colonies. Furthermore, pathogens and parasites that attack the cultivated fungi occasionally outstrip the ants' ability to weed them out. What prevents such unwanted guests from destroying the gardens?

Introduction

Chapters 12, 13, and 14 emphasized interactions between species in which at least one member is harmed (predation, herbivory, parasitism, and competition). But life on Earth is also shaped by **positive interactions**, those in which one or both species benefit and neither is harmed. Most vascular plants, for example, form beneficial associations with fungi that improve the growth and survival of both species. In fact, fossil evidence indicates that the earliest vascular plants formed similar associations with fungi more than 400 million years ago (Selosse and Le Tacon 1998). These early vascular plants lacked true roots, so their interactions with fungi may have increased their access to soil resources and aided their colonization of land.

As this example suggests, positive interactions have influenced key events in the history of life as well as the growth and survival of organisms living today. As we'll see in this chapter, positive interactions can have wide-ranging effects on many species, thus shaping communities and influencing ecosystems. We will begin our study of positive interactions with definitions of some key terms and an overview of the scope of these interactions in ecological communities.

CONCEPT 15.1 In positive interactions, no species is harmed, and the benefits are greater than the costs for at least one species.

LEARNING OBJECTIVES

15.1.1 Compare mutualism and commensalism and give examples of their importance in communities.

15.1.2 Describe how positive interactions form and evolve over space and time.

15.1.3 Explain how positive interactions can vary in their strength under different physical environments.

Positive Interactions

There are two fundamental types of positive interactions: mutualism and commensalism. **Mutualism** is a mutually beneficial interaction between individuals of two or more species (+/+ relationship) (see Figure 12.3). **Commensalism** is an interaction in which individuals of one species benefit, while those of the other species do not benefit and are not harmed (+/0 relationship) (see Figure 12.3). Many ecologists refer to mutualism and commensalism collectively as **facilitation**.

In some cases, the species involved in a positive interaction form a **symbiosis**, a relationship in which individuals of the two species live in close physical and/or physiological contact with each other. Examples include the relationships between pea aphids and their bacterial symbionts (see Concept 13.2) and between humans and bacteria (we have a diverse set of bacteria living in our guts, many of which are beneficial). However, parasites also form symbiotic associations with their hosts (see Figure 13.3). Thus, symbiotic relationships can range from parasitism (+/−) to commensalism (+/0) to mutualism (+/+).

In mutualism and commensalism, the growth, reproduction, or survival of individuals of one or multiple species is increased by their interaction with other species (and no species is harmed). The benefits can take a variety of forms. A species may provide its partner with food, shelter, or a substrate to grow on; it may transport its partner's pollen or seeds; it may reduce heat or water stress; or it may decrease the negative effects of competitors, herbivores, predators, or parasites. In a mutualism, there can be costs to an organism that provide a benefit to its partner, as when supplying food to its partner reduces its own opportunity for growth. Nevertheless, the net effect of the interaction is positive because the benefits are greater than the costs for each of the partners.

In the remainder of this section, we will discuss some general observations that apply to both mutualism and commensalism; in Concept 15.2, we'll examine some characteristics that are specific to mutualism.

Mutualism and commensalism are ubiquitous

Mutualistic associations literally cover the land surface of Earth. For example, most vascular plant species, including those that dominate terrestrial ecosystems, form **mycorrhizae**, symbiotic associations between plant roots and various types of fungi that are usually mutualistic (**FIGURE 15.3**). About 80% of angiosperms (flowering plants) and all gymnosperms (e.g., conifers, cycads, and the ginkgo) form mycorrhizal associations. Mycorrhizae provide clear benefits to the plants, improving their growth and survival in a wide range of habitats (Smith and Read 2008; Booth and Hoeksema 2010). One way in which mycorrhizal fungi benefit plants is by increasing the surface area over which the plants can extract water and nutrients from the soil; in some cases, over 3 m of fungal filaments, known as *hyphae*, may extend from 1 cm of plant root. The fungi may also protect the plants from pathogens, while the plants typically benefit the fungi by supplying them with carbohydrates.

There are two major types of mycorrhizae (**FIGURE 15.4**). In **ectomycorrhizae**, the fungal partner typically grows between root cells and forms a mantle around the exterior of the root; hyphae in the mantle often extend varying distances into the soil. In **arbuscular mycorrhizae**, the

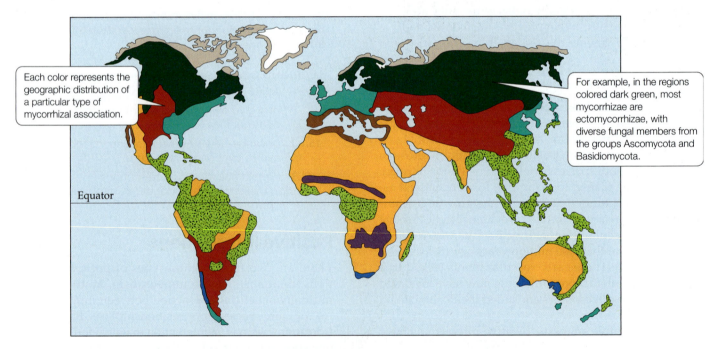

Each color represents the geographic distribution of a particular type of mycorrhizal association.

For example, in the regions colored dark green, most mycorrhizae are ectomycorrhizae, with diverse fungal members from the groups Ascomycota and Basidiomycota.

Equator

FIGURE 15.3 Mycorrhizal Associations Cover Earth's Land Surface Each color on the map shows the region in which one of eight major types of mycorrhizal associations is found (see Fitter 2005 to learn which fungi are involved in each of these eight mycorrhizal associations). Notice that the locations of the different types of mycorrhizal associations correspond fairly closely to the locations of major terrestrial biomes (see Figure 3.5A). (After A. H. Fitter. 2005. *J Ecol* 93: 231–243; based on D. J. Read. 1991. *Experientia* 47: 376; D. J. Read et al. 2004. *Can J Bot* 82: 1243–1263.)

? What types of plants are likely to be involved in the mycorrhizal association shown in the light green stippled areas? (Hint: Refer to Figure 3.5A.)

fungal partner also grows into the soil, and it grows between some root cells while penetrating the cell walls of others. Hyphae of arbuscular mycorrhizae that penetrate a root cell form a branched network, called an *arbuscule*. Because their hyphae can penetrate root cells, arbuscular mycorrhizae once were called "endomycorrhizae" (from the Greek *entos*, "in"). However, most researchers no longer use the term "endomycorrhizae," because the hyphae of some ectomycorrhizae can also penetrate root cells.

Mutualistic associations can be found in many other organisms and habitats. In the oceans, corals form mutualisms with symbiotic algae, as mentioned in Concept 3.3. The corals provide the algae with habitats, nutrients (nitrogen and phosphorus), and access to sunlight; the algae provide the corals with carbohydrates produced by photosynthesis. All the numerous invertebrate and vertebrate species that live in and on coral reefs depend directly or indirectly on the coral–alga mutualism. On land, mammalian herbivores such as cattle and sheep depend on bacteria and protists that live in their guts and help them metabolize otherwise indigestible plant material, such as cellulose. Similarly, insects rely on mutualisms with a number of other species, including plants (e.g., pollination mutualisms, fungi, protists; see **FIGURE 15.5**), and bacteria.

Commensalism, like mutualism, is everywhere—the ecological world is built on it. As we'll see in Concept 16.3, millions of species form +/0 relationships with so-called foundation species, which provide the habitat in which they live. In these relationships, a species that depends on the habitat provided by another species often has little or no effect on the species that provides that habitat. Examples include species that live on other species, such as lichens found on the bark of a tree or the harmless bacteria that grow on the surface of your skin. Many algae, invertebrates, and fishes found in marine kelp forests go locally extinct if the kelp are removed (see Case Study in Chapter 9); such species depend on the kelp for a home, but most of them do not harm or benefit the kelp. Likewise, although the numbers are quite uncertain, there may be more than a million insect species and thousands of understory plant species that live in tropical forests and nowhere else. These insects and small plants depend on the forest for their habitat, yet many have little or no effect on the trees that tower above them.

Positive interactions can be obligate or facultative and loosely structured

Mutualism and commensalism include a broad set of interactions, ranging from those that are *obligate* (that is, required for species) to those that are *facultative* (not required). The leaf-cutter ant–fungus mutualism discussed in the Case Study illustrates one end of this spectrum: the ants and the fungi they cultivate have a highly specific, obligate relationship in which neither partner can survive without the other,

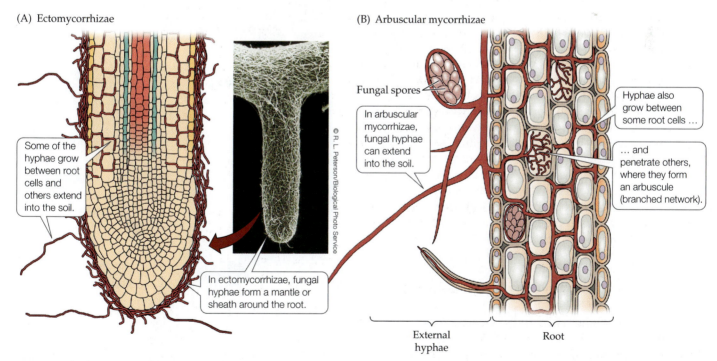

(A) Ectomycorrhizae

Some of the hyphae grow between root cells and others extend into the soil.

In ectomycorrhizae, fungal hyphae form a mantle or sheath around the root.

© R. L. Peterson/Biological Photo Service

(B) Arbuscular mycorrhizae

Fungal spores

In arbuscular mycorrhizae, fungal hyphae can extend into the soil.

Hyphae also grow between some root cells …

… and penetrate others, where they form an arbuscule (branched network).

External hyphae

Root

FIGURE 15.4 Two Major Types of Mycorrhizae Mycorrhizae can be classified as (A) ectomycorrhizae or (B) arbuscular mycorrhizae. In arbuscular mycorrhizae, hyphae can live between the root cells and penetrate the cell wall. (After A. D. Rovira et al. 1983. In *Inorganic Plant Nutrition* [*Encyclopedia of Plant Physiology*, new series, Vol. 15B], A. Läuchli and R. L. Bieleski [Eds.], pp. 61–93. Springer: New York.)

? Describe morphological features that distinguish ectomycorrhizae from arbuscular mycorrhizae.

Hypermastigote (*Barbulanympha* sp.)

Wood-eating cockroach (*Cryptocercus punctulatus*)

10 μm

Courtesy of Patrick Keeling

Courtesy of Kevin Carpenter and Patrick Keeling

FIGURE 15.5 A Protist Gut Mutualist This wood-eating cockroach (like other wood-eating insects, such as termites) would starve if gut mutualists such as the protist shown here (a hypermastigote) did not help it to digest wood. The hypermastigote can break down cellulose, a major structural component of wood that the cockroach cannot digest on its own.

and their interaction has led each partner to evolve unique features that benefit the other species.

Similarly, many tropical fig trees are pollinated by one or a few species of fig wasps. These relationships are mutually beneficial and obligate for both species in that neither species can reproduce without the other. Fig–fig wasp interactions also show clear signs of coevolution (Bronstein 1992). Fig flowers are contained within structures of fleshy stem tissue known as *receptacles* (**FIGURE 15.6**). In monoecious figs (those in which each tree has separate male and female flowers), the male and female flowers are located in different parts of the receptacle, and the male flowers mature after the female flowers. The forms of female flowers range from those with short styles to those with long styles.

A female fig wasp enters the receptacle, carrying pollen she collected from male flowers in another receptacle. Once inside, the wasp inserts her ovipositor through the styles of the female flowers to lay eggs in the ovaries (see Figure 15.6). She then deposits pollen on the stigmas of those flowers. The wasp pollinates both long-styled and short-styled flowers, and hence both flower types develop seeds. Perhaps because wasp ovipositors are not long enough to reach the ovaries of long-styled flowers, wasp larvae typically develop within short-styled flowers and feed on some of their seeds.

When the young wasps complete their development, they mate, the males burrow through the receptacle, and the females exit through the passageway the males have made. Before the females leave the receptacle, however, they visit male flowers (which are now mature), collect pollen from them, and store it in a specialized sac for use when they lay their eggs in another receptacle. The wasp's reproductive behavior is a remarkable example of a specialization that provides a benefit to another species.

Unlike the ant–fungus and fig–fig wasp mutualisms, many mutualisms and commensalisms are facultative. In desert environments, for example, the soil beneath an adult plant is often cooler and moister than the soil of an adjacent open area. These differences in soil conditions may be so pronounced that the seeds of many plant species can germinate and survive only in the shade provided by an adult plant; such adults are called *nurse plants* because they "nurse" or protect the seedlings. A single species of nurse plant may protect the seedlings of many different species. Desert ironwood (*Olneya tesota*), for example, serves as a nurse plant for 165 different species, most of which can also germinate and grow under other plant species. This situation is typical of facultative interactions: a species that requires "nursing" may be found under a variety of nurse plant species (and hence has a facultative relationship with each of them), and the nurse plant and the beneficiary species may evolve little in response to one another.

Facultative positive interactions also occur in forest communities. For example, large herbivores such as deer or moose may inadvertently eat the seeds of small herbaceous plants whose leaves they feed on. The seeds may

Receptacle

Female flowers

In this fig species, the receptacle contains both male and female flowers, but the female flowers mature 3–4 weeks earlier than the male flowers.

Male flowers

Wasp ovipositor

Stigma

Long-styled female flower

Short-styled female flower

Ovary

Some of the female flowers have short styles, while others have long styles.

FIGURE 15.6 Fig Flowers and the Wasp That Pollinates Them The receptacle and flowers of a typical monoecious fig tree, *Ficus sycomorus*. (After J. L. Bronstein. 1992. In *Insect-Plant Interactions*, Vol. 4, E. A. Bernays [Ed.], pp. 1–44. CRC Press: Boca Raton, FL.)

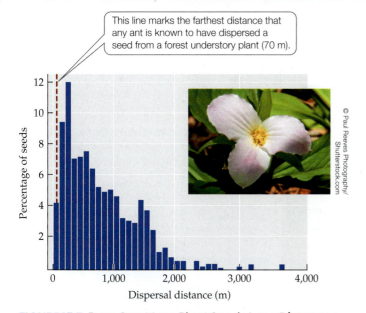

This line marks the farthest distance that any ant is known to have dispersed a seed from a forest understory plant (70 m).

FIGURE 15.7 Deer Can Move Plant Seeds Long Distances These estimates of the distances that white-tailed deer disperse the seeds of the forest understory plant *Trillium grandiflorum* are based on observations of deer movements and of the length of time that deer retain plant seeds in their digestive tracts (from the time they eat the seeds until they defecate them). Although *T. grandiflorum* seeds are also dispersed by ants, deer move the seeds much farther. (After M. Vellend et al. 2003. *Ecology* 84: 1067–1072.)

pass unharmed through the herbivore's digestive tract and be deposited with its feces, often far from the parent plant (**FIGURE 15.7**). As we saw in Concept 9.3, dispersal of offspring away from parents may be advantageous, so benefits may accrue to both the plant (whose seeds are dispersed) and the herbivore (which feeds primarily on leaves).

Positive interactions can cease to be beneficial under some circumstances

Interactions between two species can be categorized by determining for each species whether the outcome of the interaction is positive (benefits > costs), negative (costs > benefits), or neutral (benefits = costs). However, the costs and benefits experienced by interacting species can vary from one place and time to another (Bronstein 1994). Thus, depending on the circumstances, an interaction between two species may have either positive or negative outcomes.

Soil temperature, for example, influences whether a pair of wetland plant species interact as commensals or competitors (Callaway and King 1996). Some wetland plants aerate

FIGURE 15.8 From Benefactor to Competitor The growth of the small-flower forget-me-not *Myosotis laxa* under (A) low soil temperatures (11°C–12°C) and (B) high soil temperatures (18°C–20°C) in the presence and absence of the cattail *Typha latifolia* was measured by changes in three parameters: root length (left *y*-axis), root mass (right *y*-axis), and shoot mass (right *y*-axis). Error bars show one standard error of the mean. (After R. M. Callaway and L. King. 1996. *Ecology* 77: 1189–1195.)

? Under what conditions does *Myosotis laxa* best grow? Explain.

hypoxic soils by passively transporting oxygen through air channels in their leaves, stems, and roots. Oxygen leaked into the soil from the roots of such plants can become available to other plant species, thereby reducing the negative effects of the hypoxic soil conditions. In a greenhouse experiment, Ragan Callaway and Leah King grew the cattail *Typha latifolia*, a species that has extensive air channels, together with the small-flowered forget-me-not *Myosotis laxa*, a species that lacks air channels. They grew these plants under two different temperature regimes (11°C–12°C and 18°C–20°C) in pots filled with a mix of natural pond soil and peat, with the soil in the pots submerged under 1–2 cm of water to make it hypoxic. They also grew some pots of *Myosotis* without *Typha* under the same conditions.

At the low soil temperatures, the dissolved-oxygen content of the soil increased when *Typha* was present, but that did not happen at the high soil temperatures. How did these different oxygen levels affect the outcome of the *Myosotis–Typha* interaction? At the low soil temperatures, the growth of *Myosotis* roots and shoots increased when *Typha* was present (**FIGURE 15.8A**). At the high soil temperatures, however, *Myosotis* growth decreased when *Typha* was present (**FIGURE 15.8B**). Overall, these results suggest that at the low soil temperatures, *Typha* provided benefits to *Myosotis* (perhaps by aerating the soil), while at the high temperatures, *Typha* had a negative effect on *Myosotis*—just one example of how a change in environmental conditions can alter the outcome of an ecological interaction (other examples are discussed in Concepts 16.3 and 17.3 and in Bronstein 1994).

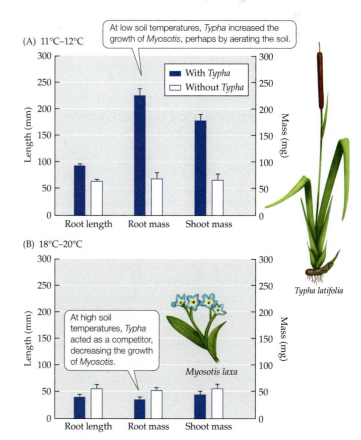

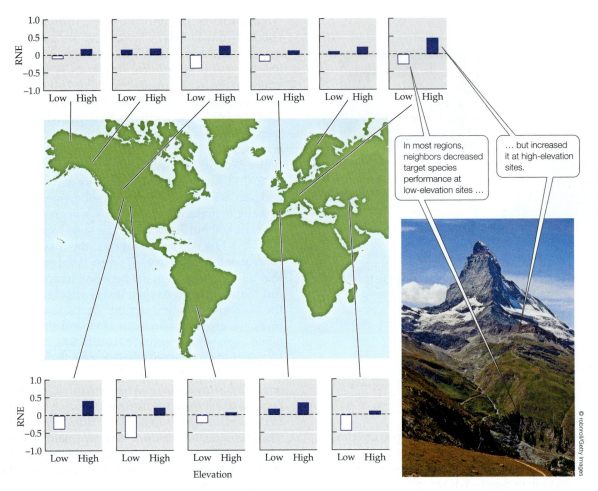

FIGURE 15.9 Neighbors Increase Plant Growth at High-Elevation Sites The relative neighbor effect (RNE, defined as the growth of the target plant species when neighboring plants are present minus its growth when neighbors are removed) of alpine plants was measured in plots at high and low elevations in 11 regions. Plant growth was measured as change in biomass (for most sites) or in leaf number. RNE values greater than zero (in blue) indicate that neighbors increased the growth of target species; RNE values less than zero (in white) indicate that neighbors decreased the growth of target species. (After R. M. Callaway et al. 2002. *Nature* 417: 844–848.)

Positive interactions may be more common in stressful environments

In recent decades, studies have shown that positive interactions are important in a number of ecological communities, such as oak woodlands, coastal salt marshes, and marine intertidal communities. Many of these studies have focused on how individuals of a target species are affected by nearby individuals of one or more other species. These effects can be assessed by comparing the performance of the target species when neighbors are present with its performance when neighbors are removed. Although results from such studies cannot be used to determine whether mutualism, commensalism, or competition is occurring (because two-way interactions are not examined), they do provide a rough assessment of whether positive interactions are common in ecological communities.

In one of the most comprehensive studies of this type, an international group of ecologists tested the effects that neighboring plants had on a total of 115 target plant species in 11 regions worldwide (Callaway et al. 2002). In 8–12 replicate plots for each treatment of each target species, neighbors were either left in place or removed from the vicinity of the target species. The "relative neighbor effect" (RNE, defined as the growth of the target species with neighbors present minus its growth when neighbors were removed) was then measured. The researchers found that RNE was generally positive at high-elevation sites, indicating that neighbors had a positive effect on the target species, but negative at low-elevation sites (**FIGURE 15.9**). In addition, neighbors tended to reduce the survival and reproduction of target species individuals at low-elevation sites, but to increase their survival and reproduction at high-elevation sites. Callaway et al. determined that the RNE was negatively related to the maximum temperature in the summer, suggesting that positive interactions were more common in colder, more stressful environments and

Woudloper/CC BY-SA 3.0

Arenaria bryophylla

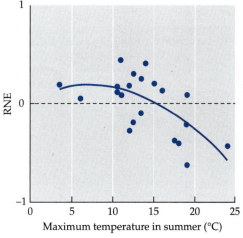

FIGURE 15.10 Neighbors Ameliorate Cold Temperatures in Alpine Plants

© imageBROKER/Alamy Stock Photo

Rhododendron ferrugineum

FIGURE 15.10 Neighbors Ameliorate Cold Temperatures in Alpine Plants
The relative neighbor effect (RNE, defined in Figure 15.9) of alpine plants changes from positive (above zero) to competitive (below zero) as temperature increases at lower elevations. (After R. M. Callaway et al. 2002. *Nature* 417: 844–848.)

competition was more common in warmer, less stressful environments (**FIGURE 15.10**). Similar results have been found in salt marsh communities (see Figure 16.13 and Figure 17.14) and intertidal communities (Bertness 1989; Bertness and Leonard 1997).

With this discussion of positive interactions as background, let's examine some of the characteristics that are unique to mutualism. Our discussion will place special emphasis on what can be learned from studies that document the costs and benefits of mutualistic interactions.

> **CONCEPT 15.2** Each partner in a mutualistic interaction acts in ways that serve its own ecological and evolutionary interests.

LEARNING OBJECTIVES

15.2.1 Categorize different types of mutualisms.

15.2.2 Justify why mutualisms are not altruistic.

Characteristics of Mutualism

In the previous section, we discussed some features that apply to both mutualism and commensalism: these two types of positive interactions are ubiquitous, they can evolve in many ways, and they can cease to be beneficial under some conditions. However, because mutualism is a reciprocal relationship in which both parties benefit, some of its characteristics differ from those of commensalism. A mutualism has costs as well as benefits, and if the costs exceed the benefits for one or both partners, their interaction will change. Before we describe the special characteristics of mutualism, however, we'll begin with a discussion of how mutualisms are classified.

Mutualisms can be categorized according to the benefits they provide

Mutualisms are often categorized by the types of benefits that the interacting species provide to each other, such as food or a place to live. As we'll see, one partner in a mutualism may receive one type of benefit (such as food) while the other receives a different benefit (such as a place to live). In such cases, the mutualism could be classified in two different ways.

There are many **trophic mutualisms**, in which a mutualist receives energy or nutrients from its partner. In the leaf-cutter ant–fungus mutualism described in this chapter's Case Study, each partner feeds the other. (Recall that the ant and the fungus also help each other to overcome plant defenses, so each also provides the other with an ecological service.) In other trophic mutualisms, one organism may receive an energy source while the other receives limiting nutrients. In mycorrhizae, for example, the fungus receives energy in the form of carbohydrates and the plant may get help in taking up water or a limiting nutrient such as phosphorus. An exchange of energy for limiting nutrients also occurs in the coral–alga symbiosis, in which the coral receives carbohydrates and the alga receives nitrogen.

In **habitat mutualisms**, one partner provides the other with shelter, a place to live, or favorable habitat. Alpheid (pistol) shrimps form a habitat mutualism with some gobies (fishes of genera *Cryptocentrus* and *Vanderhorstia*) in environments with abundant food but little protective cover. The shrimp digs a burrow in the sediments, which it shares with a goby, thus providing the fish with a safe haven from danger. For its part, the goby serves as a "seeing-eye fish" for the shrimp, which is nearly blind. Outside the burrow, the shrimp keeps an antenna on the

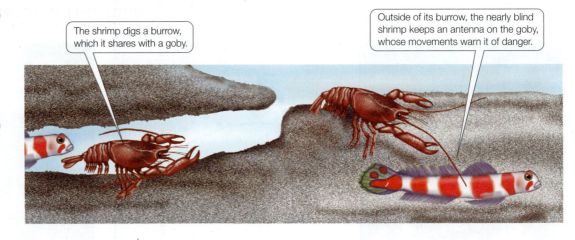

FIGURE 15.11

A Seeing-Eye Fish
In environments with little protective cover, a habitat mutualism between an alpheid (pistol) shrimp and a goby benefits both partners.

The shrimp digs a burrow, which it shares with a goby.

Outside of its burrow, the nearly blind shrimp keeps an antenna on the goby, whose movements warn it of danger.

fish (**FIGURE 15.11**); if a predator or some other form of disturbance causes the fish to move suddenly, the shrimp darts back into the burrow.

In other habitat mutualisms, a species may provide its partner with favorable habitat by altering local environmental conditions or by improving its partner's tolerance of existing conditions. The grass *Dichanthelium lanuginosum* grows next to hot springs in soils whose temperatures can be as high as 60°C (140°F). Regina Redman, Russell Rodriguez, and colleagues performed laboratory and field experiments in which this grass was grown with and without *Curvularia protuberata*, a symbiotic fungus that grows throughout the plant body (such fungi are called *endophytes*). In the laboratory, 100% of the grass plants that had the *Curvularia* endophyte survived intermittent soil temperatures of 60°C, while none of the plants without the endophyte survived (Redman et al. 2002). In field experiments in which soil temperatures reached up to 40°C (104°F), plants with endophytes had greater root and leaf mass than plants without endophytes. In soils above 40°C, the grass plants with endophytes continued to grow well, but all the plants without endophytes died. Thus, *Curvularia* increased the ability of its grass host to tolerate high soil temperatures. *Curvularia* is not alone: many other fungal endophytes can increase the tolerance of their host plants for soils that are high in temperature or salinity (Rodriguez et al. 2009), as can some mycorrhizal fungi (Bunn et al. 2009).

Mutualists are in it for themselves

Although both partners in a mutualism benefit, that does not mean that a mutualism has no costs for the partners. In the coral–alga mutualism, for example, the coral receives benefits in the form of energy, but it incurs the costs of supplying the alga with nutrients and space. Likewise, the alga gains limiting nutrients, but it provides the coral with energy that it could have used to support its own growth and metabolism. The costs of mutualism may be especially clear when one species provides the other with a "reward" such as food for a service such as pollination. For example, during flowering, milkweeds use up to 37% of the energy they gain from photosynthesis to produce the nectar that attracts insect pollinators such as honeybees.

For an ecological interaction to be a mutualism, the net benefits must exceed the net costs for both partners. However, neither partner in a mutualism is in it for altruistic reasons. Should environmental conditions change so as to reduce the benefits or increase the costs for one of the partners, the outcome of the interaction may change. This is especially true if the interaction is not obligate. Ants, for example, often form facultative relationships in which they protect other insects from competitors, predators, and parasites. In one such case, ants protect treehoppers from predators, and the treehoppers secrete honeydew (a sugar syrup substance), which the ants feed on (**FIGURE 15.12**). Treehoppers always secrete honeydew, so the ants always have access to this food source. However, in years when predator abundances are low, the treehoppers may receive

© Morley Read/Alamy Stock Photo

FIGURE 15.12 **A Facultative Mutualism** Ants often form facultative mutualisms with insects that secrete honeydew, a sugar syrup substance on which the ants feed. The ants shown here will protect these Ecuadorian treehoppers from predators and parasites in exchange for honeydew.

(A)

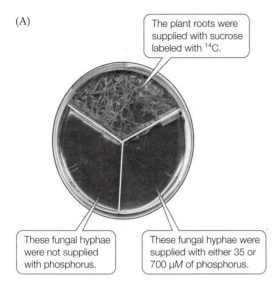

The plant roots were supplied with sucrose labeled with ^{14}C.

These fungal hyphae were not supplied with phosphorus.

These fungal hyphae were supplied with either 35 or 700 μM of phosphorus.

(B)

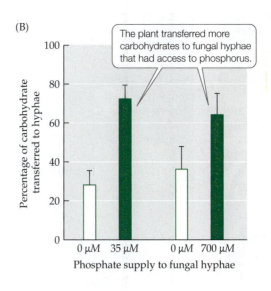

The plant transferred more carbohydrates to fungal hyphae that had access to phosphorus.

FIGURE 15.13 Rewarding Those Who Reward You Researchers tested the hypothesis that *Medicago truncatula* plants allocate more carbohydrates to those mycorrhizal fungi that provide them with higher concentrations of phosphorus, a key plant nutrient. (A) They used a split-plate design to separate the fungal hyphae into two groups. Some fungal hyphae lacked access to phosphorus, while other fungal hyphae were supplied with either 35 or 700 μM of phosphorus. (B) They then tracked the proportion of sucrose (labeled with ^{14}C) that the plant provided to each group of hyphae. Error bars show one standard error of the mean. (After E. T. Kiers et al. 2011. *Science* 333: 880–882.)

no benefit from the ants. In such years, the outcome of the interaction may shift from +/+ (a mutualism) to either +/0 (a commensalism) or +/− (parasitism), depending on whether the consumption of honeydew by ants reduces treehopper growth or reproduction.

Finally, under certain conditions, a mutualist may withdraw or modify the reward that it provides to its partner. In high-nutrient environments, for example, some plants reduce the carbohydrate rewards that they usually provide to mycorrhizal fungi. In such environments, the plant can obtain ample nutrients on its own, and hence the fungus is of little benefit. Thus, when nutrients are plentiful, the plant may cease to reward the fungus because the costs of supporting fungal hyphae are greater than the benefits the fungus can provide. Moreover, a recent study found that the plant *Medicago truncatula* can discriminate among mycorrhizal fungi, allocating more carbohydrate rewards to those fungal hyphae that are supplying the most nutrients (**FIGURE 15.13**). You can explore this relationship further in **ANALYZING DATA 15.1**, where you will examine whether the fungus also modifies its provision of nutrients to the plant depending on the rewards it receives from the plant.

Some mutualists have mechanisms to prevent overexploitation

As we've seen, there is an inherent conflict of interest between the partners in a mutualism: the benefit to each species comes at a cost to the other. In such a situation, natural selection may favor **cheaters**, individuals that increase their production of offspring by overexploiting their mutualistic partner. When one of the partners in a mutualism overexploits the other, it becomes less likely that the mutualism will persist. But mutualisms do persist, as the 50-million-year association between fungus-growing ants and the fungi they cultivate readily attests. What factors allow a mutualism to persist in spite of the conflict of interest between the partners?

One answer is provided by "penalties" imposed on individuals that overexploit a partner. If those penalties are high enough, they can reduce or remove any advantage gained by cheating. Olle Pellmyr and Chad Huth documented such a situation in an obligate, coevolved mutualism between the yucca plant *Yucca filamentosa* and its exclusive pollinator, the yucca moth *Tegeticula yuccasella* (Pellmyr and Huth 1994). Female yucca moths collect pollen from yucca plants with their unique mouthparts (**FIGURE 15.14A**). After collecting pollen, a female moth typically moves to another plant, lays eggs in the ovary of a flower, and then walks up to the top of the style. There, the moth deliberately places some of the pollen she carries on the stigma, thus pollinating the plant (**FIGURE 15.14B**). The larvae that hatch from the moth's eggs complete their development by eating some of the seeds, which then develop in the ovary of the flower.

The moth and the plant depend absolutely on each other for reproduction. However, the mutualism is vulnerable to overexploitation by moths that lay too many

Pollen grains

From O. Pellmyr. 2003. Ann Mo Bot Gard 90: 35

Courtesy of Olle Pellmyr

FIGURE 15.14 Yuccas and Yucca Moths *Yucca filamentosa* has an obligate relationship with its exclusive pollinator, the yucca moth *Tegeticula yuccasella*. (A) The female moth collects pollen from a yucca flower using specialized mouthparts. She may carry a load of up to 10,000 pollen grains, nearly 10% of her own weight. (B) The moth at the lower right of this photo is laying eggs in the ovary of a yucca flower; the moth at the top is placing pollen on the stigma.

eggs and hence consume too many seeds. Yuccas have a mechanism to prevent such overexploitation: they selectively abort flowers in which female moths have laid too many eggs (**FIGURE 15.15**). On average, yuccas retain 62% of the flowers that contain up to six moth eggs, but only 11% of the flowers that contain nine or more eggs. When the yucca aborts a flower, it does so before the moth larvae hatch from their eggs. Although the cue that determines flower abortion is not known, it is clear that it is a powerful mechanism for reducing overexploitation: all the moth larvae in an aborted flower die.

Few other clear cases of penalties for cheating have been documented, so we do not yet know whether such penalties are common in nature. Be that as it may, the yucca–yucca moth interaction illustrates the theme that runs throughout this section: the partners in a mutualism are not altruistic. Instead, the yucca takes actions that promote its own interests, and

---

ANALYZING DATA 15.1

Does a Mycorrhizal Fungus Transfer More Phosphorus to Plant Roots That Provide More Carbohydrates?

As seen in Figure 15.13, Kiers et al. (2011)* found that the plant *Medicago truncatula* transfers more carbohydrates to those fungal hyphae that have greater access to phosphorus. The researchers also tested the reciprocal interactions: whether the plant's mycorrhizal partner, the fungus *Rhizophagus irregularis* (previously known as *Glomus intraradices*), behaves in a similar manner, transporting more phosphorus to roots that have greater access to carbohydrates.

To do this, Kiers et al. used a split-plate experimental design similar to that in Figure 15.13. They provided fungal hyphae with radioactively labeled phosphorus (^{33}P) and monitored the transfer of phosphorus to plant roots differing in access to carbohydrates (sucrose). Some plant roots had no access to sucrose, while other plant roots were supplied with either 5 or 25 mM of sucrose. In the results shown in the figure, "dpm" refers to disintegrations per minute, a measure of radiation intensity; error bars show one standard error of the mean.

1. Draw and label a sketch of the split-plate experimental design, modeling your diagram on the photograph in Figure 15.13A.

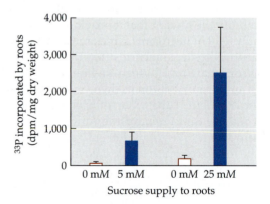

2. Interpret the results shown in the figure.

3. Compare the results in the figure here with those in Figure 15.13B. Does the plant or the fungus control the exchange of materials, or do both partners play a role? Explain.

---

*Kiers, E. T., and 14 others. 2011. Reciprocal rewards stabilize cooperation in the mycorrhizal symbiosis. *Science* 333: 880–882.

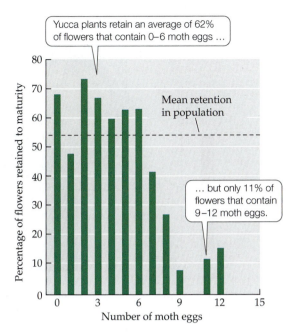

Yucca plants retain an average of 62% of flowers that contain 0–6 moth eggs …

Mean retention in population

… but only 11% of flowers that contain 9–12 moth eggs.

FIGURE 15.15 A Penalty for Cheating Yucca plants selectively abort flowers in which yucca moths have laid too many eggs. (After O. Pellmyr and C. J. Huth. 1994. *Nature* 372: 257–260.)

the yucca moth does the same. In general, a mutualism evolves and is maintained because its net effect is advantageous to both parties. If the net effect of a mutualism were to impair the growth, survival, or reproduction of one of the interacting species, the ecological interests of that species would not be served, and the mutualism might break down, at least temporarily. Should such a situation continue, the longer-term or evolutionary interests of that species might also fail to be served, and the mutualism might break down on a more permanent basis.

Although it is possible for a mutualism to break down, we've also seen that mutualism and commensalism are very common and that some of these interactions have been maintained for millions of years. Let's turn now to the ecological effects of these pervasive interactions.

CONCEPT 15.3 Positive interactions affect the abundances and distributions of populations as well as the structure of ecological communities.

LEARNING OBJECTIVES

15.3.1 Explain the consequences of positive interactions on the distribution and abundance of species.

15.3.2 Describe how positive interactions can increase species diversity in communities.

15.3.3 Illustrate ways in which positive interactions can affect ecosystem processes.

Ecological Consequences of Positive Interactions

So far in this chapter, we've considered features that are common to commensalism and mutualism as well as characteristics that are unique to mutualism. At various points, we've mentioned some ecological consequences of positive interactions, including increased survival rates and the provision of food and habitat. In this section, we'll take a closer look at how positive interactions affect populations of organisms and the communities in which they are found.

Positive interactions influence the abundances and distributions of populations

As examples earlier in this chapter suggest, mutualism and commensalism can provide benefits that increase the growth, reproduction, or survival of individuals in one or both of the interacting species—a point that was demonstrated recently for a defensive bacterial symbiont that increased the reproductive success of its fruit fly host (**FIGURE 15.16**). As a result, mutualism and commensalism can affect the abundances and distributions of the interacting species. To explore these issues further, we will first examine how an ant–plant mutualism affects the abundances of its members. We will then consider how mutualism and commensalism influence the distributions of organisms.

EFFECTS ON ABUNDANCE The effects of mutualism on abundance can be seen in the mutualistic relationship between ants in the genus *Pseudomyrmex* spp. and the bullhorn acacia (*Acacia* spp.). This plant has unusually large thorns, which provide a home for the ants (**FIGURE 15.17A**). The thorns have a tough, woody covering but a soft, pithy interior that is easy for the ants to excavate. A queen ant establishes a new colony by burrowing into a green thorn, removing some of its pithy interior, and laying eggs inside the thorn. As the colony grows, it eventually occupies all of the acacia's thorns.

The ants feed on nectar, which the plant secretes from specialized nectaries, and on modified leaflet tips called Beltian bodies, which are high in protein and fat (**FIGURE 15.17B**). The ants aggressively attack insect and even mammalian herbivores (such as deer) that attempt to eat the plant. The ants also use their mandibles to maul other plants that venture within 10–150 cm of their home acacia, thus providing the acacia with a competitor-free zone in which to grow (**FIGURE 15.17C**).

Do the services provided by the ants benefit the acacias? To find out, Dan Janzen removed ants from some acacia plants and compared the growth and survival of those plants with those of plants that had ant colonies.

FIGURE 15.16 A Symbiont Increases the Fertility of Its Host Bacteria in the genus *Spiroplasma* are obligate symbionts that live within the cells of their host, the fruit fly *Drosophila neotestacea*. The graph shows the number of eggs produced by laboratory-reared female flies that either had *Spiroplasma* symbionts (red bars) or did not have *Spiroplasma* symbionts (white bars), and that either were infected by the nematode parasite *Howardula* (parasitized) or were not infected by it (unparasitized). *Howardula* can sterilize female flies and reduce the mating success of male flies. Error bars show one standard error of the mean. (After J. Jaenike et al. 2010. *Science* 329: 212–215.)

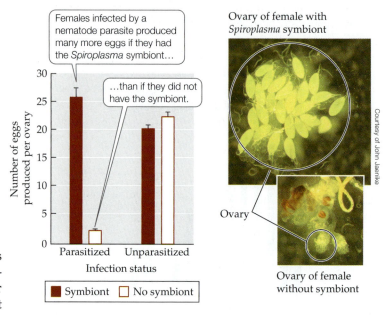

Females infected by a nematode parasite produced many more eggs if they had the *Spiroplasma* symbiont…

…than if they did not have the symbiont.

Ovary of female with *Spiroplasma* symbiont

Ovary

Ovary of female without symbiont

Courtesy of John Jaenike

The results were striking. On average, bullhorn acacias with ant colonies weighed over 14 times as much as acacias that lacked colonies; acacias with ants also had higher survival rates (72% vs. 43%) and were attacked by insect herbivores much less frequently (Janzen 1966).

If a bullhorn acacia lacks an ant colony, the repeated loss of its leaves and growing tips to herbivores often kills the plant in 6–12 months. The ants, in turn, depend on the acacias for food and a home, and they cannot survive without these plants. Thus, the ant–acacia mutualism has considerable effects on the abundance of each partner. Furthermore, both the ant and the plant have evolved unusual characteristics that benefit their partners. For example, *Pseudomyrmex* ants that depend on acacias are highly aggressive, remain active for 24 hours a day

(A)

© Mark Moffett/Minden Pictures

(B)

© Mark Moffett/Minden Pictures

Nectary

Beltian body

(C)

© Nicholas Smythe/Science Source

FIGURE 15.17 An Ant–Plant Mutualism (A) Acacia ants (*Pseudomyrmex spinicola*) tending to larvae and pupae inside an acacia thorn. (B) A nectary at the base of a leaf and Beltian bodies at the leaflet tips. (C) Ants have removed the plants that grew near this acacia, creating a competitor-free zone for the plant.

(patrolling the plant surface), and attack vegetation that grows near their home plants; *Pseudomyrmex* species that do not form mutualisms with acacias show none of these traits. Similarly, acacias that form mutualisms with ants have enlarged thorns, specialized nectaries, and Beltian bodies on their leaves; few nonmutualistic acacia species show these traits. Overall, both the ants and the acacias appear to have evolved in response to their partners, making the ant–acacia partnership an example of an obligate and coevolved mutualism.

EFFECTS ON DISTRIBUTION   There are literally millions of positive interactions in which one species provides another with favorable habitat and thus influences its distribution. Specific examples include corals that provide their algal symbionts with a home and fungal symbionts that enable plants to live in environments they otherwise could not tolerate (such as the example of the *Curvularia* fungi that enable the grass *Dichanthelium* to live in high-temperature soils). Of course, obligate mutualisms, such as the fig–fig wasp mutualism discussed earlier, have a profound influence on the geographic distribution of the interacting species because neither can live where its partner is absent.

It is very common for a group of dominant species, such as the trees in a forest, to determine the distributions of other species by physically providing the habitat on which they depend. Many plant and animal species are found only in forests. Such "forest specialists" either cannot tolerate the physical conditions of more open areas (such as a nearby meadow) or are prevented from living in those open areas by competition with other species. Similarly, at low tide in marine intertidal communities, many species (e.g., crabs, snails, sea stars, sea urchins, barnacles) can be found under the strands of seaweeds that are attached to the rocks. The seaweeds provide a moist and relatively cool environment that enables some species to live in higher regions of the intertidal zone than they otherwise could. Finally, many sandy and cobblestone beaches are stabilized by grasses such as *Ammophila breviligulata* and *Spartina alterniflora*. By holding the substrate together, these species enable the formation of entire communities of plants and animals.

Many forest specialists have little direct effect on the trees under which they live; hence, they have a commensalism with the trees of the forest. The same is true of many marine species that seek shelter under seaweeds and of many of the organisms that depend on substrate stabilization by grasses. In each of these cases, a positive interaction (often a commensalism) allows one species to have a larger distribution than it otherwise would.

Positive interactions can alter communities and ecosystems

The effects that commensalism and mutualism have on the abundances and distributions of species can affect other species interactions, which in turn can have consequences for community. For example, if a dominant competitor depends on a facilitator, loss of the facilitator may reduce the performance of that dominant species and increase the performance of other species—thus changing the mix of species in the community or their relative abundances. As we'll see, when the structure of a community changes, properties of the ecosystem may also change.

COMMUNITY DIVERSITY   Coral reefs are known for their astonishing beauty, and they are exceptional ecologically in that their fish communities are the most diverse vertebrate communities in the world. One of the most common interactions among these diverse coral reef fish is a service mutualism in which a small species (the "cleaner") removes parasites from a larger fish (the "client"). The cleaner often ventures into the mouth of the client (**FIGURE 15.18A**). What prevents the client from simply eating the cleaner?

The answer appears to be that the benefit a client receives from cleaning (parasite removal) is greater than

(A)

Cleaner fish

© WaterFrame/Alamy Stock Photo

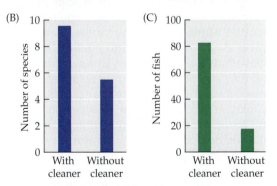

(B)

Number of species

With cleaner   Without cleaner

(C)

Number of fish

With cleaner   Without cleaner

FIGURE 15.18 Ecological Effects of the Cleaner Fish *Labroides dimidiatus*   (A) Looking for parasites, a cleaner fish places its head within the mouth of a much larger client, this sweetlips fish. The experimental removal of *L. dimidiatus* from small reefs within the Great Barrier Reef of Australia led to (B) a drop in the number of fish species found on the reefs and (C) a decrease in the total abundance of fish on the reefs. (B,C after A. S. Grutter et al. 2003. *Curr Biol* 13: 64–67.)

the energy benefit it could gain by eating the cleaner. In the Great Barrier Reef of Australia, individuals of the cleaner species *Labroides dimidiatus* were visited by an average of 2,297 clients each day, from which the cleaner removed (and ate) an average of 1,218 parasites per day (0.53 parasites per client). To determine whether the activities of cleaners were translated into a reduction in the number of parasites found on clients, Alexandra Grutter experimentally removed *L. dimidiatus* from three of five small reefs. After 12 days, on the reefs from which the cleaners had been removed, there were 3.8 times more parasites on *Hemigymnus melapterus* fish than on the control reefs. In follow-up studies, Grutter and colleagues (2003) examined the effect of *L. dimidiatus* on the number of species and the total abundance of fish found on coral reefs. The results were dramatic: removal and exclusion of *L. dimidiatus* for a period of 18 months caused large drops in both the number of fish species and the total abundance of fish found on the reefs (**FIGURE 15.18B,C**).

Grutter's work shows that a mutualism can have a major effect on the diversity of species found in a community. Most of the species lost from the reefs without cleaners were species that typically move among reefs, including some large predators. Large predators can themselves affect the diversity and abundance of species, so the removal of cleaner fish could also result in further, but difficult-to-predict, long-term changes to the community.

CLIMATE CHANGE CONNECTION

CLIMATE WARMING AND EXTINCTION OF PLANT–POLLINATOR NETWORKS As we have learned, plant–pollinator mutualisms are often highly specialized and have important benefits for both species involved in the interaction. These codependent species are also embedded within whole communities of plants and pollinators, connected through multiple mutualistic interactions. But what happens when climate change causes the extinction of one or more plant species within these networks, and what effects could that have on the extinction of pollinator species dependent on those plants? To explore this question, Jordi Bascompte, María García, and colleagues (2019) used species distribution models (see Concept 9.2 for use of species distribution models to predict the locations of chameleon species in Madagascar) to predict the extinctions and subsequent coextinctions of species due to warming temperatures within seven plant–pollinator networks across Europe, due to warming temperatures at two time horizons of 2050 and 2080 (**FIGURE 15.19A**). The researchers were interested in measuring the response of plant–pollinator

networks under two time horizons (2050 and 2080) and for two sources of extinction: (1) the direct effects of climate warming on plant species extinction rates and (2) the subsequent coextinction of pollinator species that depend on a mutualistic interaction with those plants. As **FIGURE 15.19B** reveals, there was a large latitudinal range in the rate of species extinctions, with the two Mediterranean networks showing higher numbers of species lost compared to the other northern networks for both 2050 and 2080. The differences were even greater when considering the additional coextinctions that were predicted to occur from the direct loss of plants to their pollinators (see Figure 15.19B). The results suggest that the Mediterranean plant–pollinator networks may be more vulnerable to climate change than the Eurosiberian networks. If so, the greater species loss in the Mediterranean networks could be the consequence of greater warming temperatures in southern regions but also a narrower distributional range of plants and pollinators in these networks, making it harder to "escape" to higher latitudes. This study suggests that it is important to consider the interdependencies among species when predicting extinction risk in the face of climate warming. ☀

SPECIES INTERACTIONS AND ECOSYSTEM PROPERTIES

Barbara Hetrick and colleagues (1989) performed greenhouse experiments in which the presence of mycorrhizal fungi altered the outcome of competition between two prairie grasses, big bluestem (*Andropogon gerardii*) and junegrass (*Koeleria macrantha*). They found that big bluestem dominated when mycorrhizal fungi were present and that junegrass dominated when they were not. In a natural prairie community of which big bluestem was a dominant member, when David Hartnett and Gail Wilson (1999) suppressed mycorrhizal fungi with a fungicide, the performance of big bluestem decreased. At the same time, the performance of a variety of other plant species, including both grasses and wildflowers, increased. Hartnett and Wilson suggested that big bluestem's dominance may have come from a competitive advantage conferred by its association with mycorrhizal fungi and that removal of those fungi removed that advantage and released the inferior competitors from the negative effects of competition.

Mycorrhizal associations can affect other features of ecosystems in addition to diversity, as shown in a 1998 study by Marcel van der Heijden, John Klironomos, and colleagues. In a large-scale field experiment, these scientists manipulated the number of species of mycorrhizal fungi (from 0 to 14 species) found in soils in which identical mixtures of the seeds of 15 plant species had been sown. After one growing season, plant dry weights

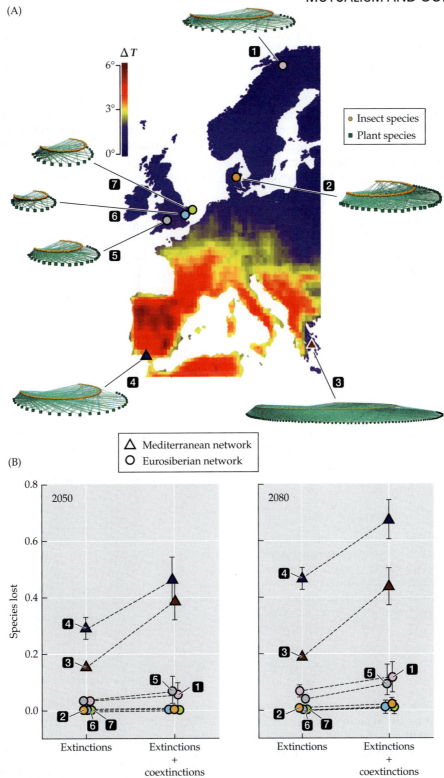

FIGURE 15.19 Plant–Pollinator Extinctions Predicted Under Climate Warming (A) Researchers measured the effects of the predicted increase in temperature (ΔT is the change in temperature in the warmest month from 2020 to 2080) on seven plant–pollinator networks (the green squares represent plants, the orange circles represent pollinator insects, and the solid lines represent interactions) across Europe. (B) Plant species extinctions and subsequent coextinctions of pollinators within each pollination network at two times in the future, 2050 and 2080. Error bars show one standard error of the mean. The different numbers identify different networks, with the triangles representing the Mediterranean networks and the circles representing the Eurosiberian networks. (After J. Bascompte et al. 2019. *Sci Adv* 5: eaav2539.)

(A)

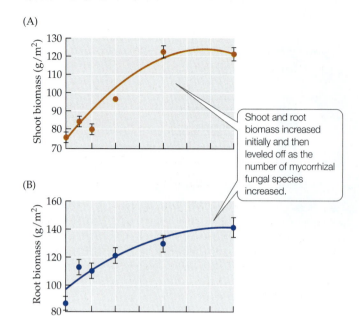

Shoot and root biomass increased initially and then leveled off as the number of mycorrhizal fungal species increased.

(B)

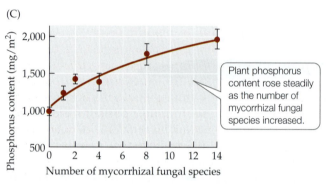

(C)

Plant phosphorus content rose steadily as the number of mycorrhizal fungal species increased.

FIGURE 15.20 **Mycorrhizal Fungal Species Richness Affects Ecosystem Properties** Researchers measured the effects of the number of mycorrhizal fungal species in the soil on (A) mean shoot biomass, (B) mean root biomass, and (C) mean phosphorus content in mixtures of 15 species of plants grown from seed in a field experiment. Error bars show ± one standard error of the mean. (After M. van der Heijden et al. 1998. *Nature* 396: 69–72.)

and phosphorus content were measured. Plant root and shoot biomass increased as the number of species of fungi increased (**FIGURE 15.20A,B**), as did the efficiency of phosphorus uptake by plants (**FIGURE 15.20C**). These results show that mycorrhizal fungal species richness can influence key features of ecosystems such as net primary production (measured as the amount of new plant growth over one growing season) and the supply and cycling of nutrients such as phosphorus.

A CASE STUDY REVISITED

The First Farmers

As Figure 15.2A illustrates, the fungal gardens of leaf-cutter ants represent a potentially enormous food resource for any species able to overcome the defenses of the resident ants. In particular, are there any parasites that specialize in attacking fungal gardens? As we saw in Chapter 13, roughly half the world's species are parasites, and many of them have remarkable adaptations for evading host defenses.

Although you might expect that the answer would be yes, for more than 100 years after the fungus-growing role of leaf-cutter ants was discovered (Belt 1874), no such parasites were known. That changed in the early 1990s, when Ignacio Chapela (Seifert et al. 1995) observed that leaf-cutter ant gardens were plagued by a virulent parasitic fungus of the genus *Escovopsis* (see also Currie et al. 1999a). This parasite can spread from one garden to the next, and it can rapidly destroy the gardens it invades, leading to the death of ant colonies. Leaf-cutter ants respond to *Escovopsis* by increasing the rate at which they weed their gardens (**FIGURE 15.21**) and, in some cases, by

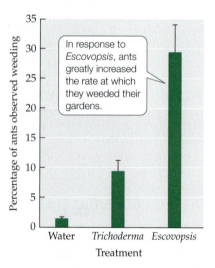

In response to *Escovopsis*, ants greatly increased the rate at which they weeded their gardens.

FIGURE 15.21 **A Specialized Parasite Stimulates Weeding by Ants** Currie and Stuart measured the frequency with which the leaf-cutter ant *Atta colombica* weeded its fungal gardens after colonies were exposed to water, *Trichoderma viride* (a generalist fungal parasite), and the specialized fungal parasite *Escovopsis*. Error bars show one standard error of the mean. (After C. R. Currie and A. E. Stuart. 2001. *R Soc* 268: 1471.)

 Suppose 2% of ants were observed weeding in colonies exposed to water, 20% in colonies exposed to *Trichoderma*, and 20% in colonies exposed to *Escovopsis*. Propose a hypothesis that might explain these results.

increasing how often they dose the garden with antimicrobial toxins, which they produce in specialized glands (Fernández-Marín et al. 2009).

The ants also enlist the help of other species in combatting *Escovopsis* (Currie et al. 1999b). On the underside of the ant's body lives a bacterium that produces chemicals that inhibit *Escovopsis*. The queen carries this bacterium on her body when she begins a new colony. While the ants clearly benefit from the use of these fungicides, what of the bacterium? Recent work (Currie et al. 2006) indicates that the bacterium also benefits: the ant provides it with both a place to live (it is housed in specialized structures called *crypts* that are located on the ant's exoskeleton) and a source of food (glandular secretions). Thus, the bacterium appears to be a third mutualist that benefits from, and contributes to, these unique fungal gardens.

 ## CONNECTIONS *in* NATURE

FROM MANDIBLES TO NUTRIENT CYCLING   While you have been reading this chapter, billions of pairs of leaf-cutter ant mandibles have been removing leaves from the forests of the Americas. The workers of a single colony can harvest as much plant matter each day as it would take to feed a cow. People have long known that leaf-cutter ants are potent herbivores. Weber (1966) describes reports—the earliest from 1559—of leaf-cutter ants destroying the crops of Spanish colonists, and they still plague farmers today. In tropical regions, these ants tend to increase in abundance after a forest is cut down. Anecdotal evidence suggests that the thriving ant colonies found in deforested areas are one of the reasons why farms in some tropical regions are often abandoned just a few years after trees have been removed to make room for them (other reasons relate to a point made in Chapters 3 and 22: some tropical soils are nutrient-poor).

In addition to their effects on human farmers, leaf-cutter ants introduce large amounts of organic matter into tropical forest soils. As a consequence, they affect the supply and cycling of nutrients in the forest ecosystem (a topic we will discuss in more detail in Concepts 22.2 and 22.3). Normally, nutrients in the leaf litter that falls to the forest floor enter the soil when the leaves decompose. Bruce Haines (1978) compared the amounts (g/m²) of 13 mineral nutrients contained in leaf litter with the amounts of the same nutrients found in aboveground areas where colonies of the leaf-cutter ant *Atta colombica* deposit their refuse (other *Atta* species deposit refuse belowground, as shown in Figure 15.2A). Averaged across the 13

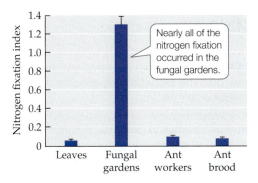

FIGURE 15.22 Nitrogen Fixation in Fungal Gardens When researchers measured nitrogen fixation activity in different parts of the colonies of leaf-cutter ants, they found that most of it was taking place in the fungal gardens. In addition, bacteria from genus *Klebsiella* were isolated from the fungal gardens and shown to fix nitrogen. Error bars show one standard error of the mean. (After A. A. Pinto-Tomás et al. 2009. *Science* 326: 1120–1123.)

nutrients, the ants' refuse areas contained about 48 times the nutrients found in the leaf litter. Plants respond to this concentration of nutrients by increasing their production of fine roots in the *Atta* refuse areas. Furthermore, the activities of leaf-cutter ants have the effect of tilling the soil near their nests, making it easier for plant roots to penetrate the soil (Moutinho et al. 2003). Moutinho and colleagues also found that the leaf material ants bring into their colonies fertilizes the soil, causing soils beneath ant colonies to be 3–4 times richer in calcium and 7–14 times richer in potassium than are soils 15 m away from the nest. Finally, recent evidence suggests that the fungal gardens tended by ants may also house nitrogen-fixing bacteria (**FIGURE 15.22**). These bacteria may be part of yet another mutualism found in the gardens—a mutualism that may prove to be an important source of nitrogen in tropical ecosystems.

The overall effects of leaf-cutter ants on the ecosystems in which they live are complex. In forest ecosystems, net primary production (NPP) is usually measured as new aboveground plant growth (see Concept 20.1); root growth is often ignored, since it is difficult to measure in trees. Although leaf-cutter ants reduce NPP by harvesting leaves, some of the other activities of ants (e.g., tillage, fertilization) may increase NPP. As a result, the net effect of the ants on the NPP of their ecosystem is difficult to estimate. While it may prove possible to disentangle such effects in future studies, there is no doubt that the ants and their partners have considerable effects on the ecosystems in which they are found.

SUMMARY

CONCEPT 15.1 In positive interactions, no species is harmed, and the benefits are greater than the costs for at least one species.

15.1.1 Compare mutualism and commensalism and give examples of their importance in communities.

- Mutualism is a mutually beneficial interaction between species, and commensalism occurs when one species benefits but the other is not affected.

- Mutualism and commensalism are ubiquitous interactions that are important in both terrestrial and aquatic communities.

15.1.2 Describe how positive interactions form and evolve over space and time.

- Mutualism and commensalism can evolve from other kinds of ecological interactions; for example, over time, a host–parasite interaction may evolve to become a mutualistic interaction.

- The costs and benefits of a positive interaction can vary from one place and time to another; as a result, a positive interaction may cease to be beneficial under some circumstances.

15.1.3 Explain how positive interactions can vary in their strength under different physical environments.

- Positive interactions may be more common in stressful environments.

CONCEPT 15.2 Each partner in a mutualistic interaction acts in ways that serve its own ecological and evolutionary interests.

15.2.1 Categorize different types of mutualisms.

- Mutualisms can be categorized by whether one partner provides the other with food (a trophic mutualism) or a place to live (a habitat mutualism).

15.2.2 Justify why mutualisms are not altruistic.

- The partners in a mutualism are in it for themselves—it is not an altruistic interaction.

- Some mutualists have mechanisms to prevent over-exploitation by cheaters.

CONCEPT 15.3 Positive interactions affect the abundances and distributions of populations as well as the structure of ecological communities.

15.3.1 Explain the consequences of positive interactions on the distribution and abundance of species.

- Positive interactions can provide benefits that increase the growth, reproduction, or survival of one or both of the interacting species.

- As a result of such demographic effects, positive interactions can determine the abundances and distributions of populations of the interacting species.

15.3.2 Describe how positive interactions can increase species diversity in communities.

- Species that are dependent on positive interactions can be lost from the community, causing species diversity to decline.

- If a dominant competitor depends on a facilitator, loss of the facilitator may reduce the performance of that dominant species and increase the performance of other species, thereby increasing species diversity.

15.3.3 Illustrate ways in which positive interactions can affect ecosystem processes.

- Positive interactions can provide resources that contribute to the overall production of an ecosystem.

Review Questions

1. Summarize the key features of positive interactions described in Concept 15.1.

2. Researchers who study mutualism do not think of it as an altruistic interaction. Explain why.

3. High water temperature is one of several stressors that can cause coral bleaching, in which a coral expels its algal mutualists and hence loses its color. If bleaching occurs repeatedly, coral death may result. Some corals are more sensitive than others to high water temperatures. If a reef containing a mixture of coral species were exposed to increasingly high water temperatures over a series of years, how might the community change over time?

Hone Your Problem-Solving Skills

The survival rates of small seedlings of the Scotch pine tree (*Pinus sylvestris*) growing in open areas and shaded beneath *Salvia* shrubs are given in the table below, as are mean values for certain abiotic conditions.

	Under *Salvia*	Open areas
Seedling survival		
Pinus sylvestris	55%	22%
Abiotic conditions (means)		
Light level (W/m^2)	473	917
Soil moisture (%)	12	9
Soil temperature (°C)	15	19

Source: J. Castro et al. 2002. *Restor Ecol* 10: 297–305.

1. How would you characterize the relationship between *Pinus sylvestris* seedlings and *Salvia* shrubs?

2. What factors may lead to the increased survival of *Pinus sylvestris* seedlings that live under *Salvia* shrubs?

3. How might the relationship between these species change over time?

ADDITIONAL RESOURCES

This textbook provides resources such as **Web Extensions** and additional **Climate Change Connections** for instructors who can share them with students. Please contact your instructor, if you would like to access that content.

Communities

16

The Nature of Communities

"Killer Algae!": A Case Study

In 1988, a French marine biology student dove into the crystal-clear water of the Mediterranean Sea and made an unusual discovery. On the seafloor, just below the cliffs on which stood the palatial Oceanographic Museum of Monaco, grew an unusual seaweed, *Caulerpa taxifolia* (**FIGURE 16.1**), a native of the warm tropical waters of the Caribbean. The student told Alexandre Meinesz, a leading expert on tropical algae and a professor at the University of Nice, about the unusual species. Over the following year, Meinesz confirmed its presence and determined that its feathery fluorescent green fronds, interconnected by creeping underground stems called rhizomes, carpeted an underwater area in front of the museum.

Meinesz was astonished because this species had never been seen in such cold waters, and it had certainly never reached the high densities he recorded. As it later turned out, earlier sightings from 1984 allowed Meinesz to calculate a spread of more than 1 hectare (ha) in 5 years. Over the next few months, he asked himself and his colleagues some important questions. First, how did the seaweed get to the Mediterranean in the first place, and how could it survive in temperatures as cold as 12°C–13°C (54°F–55°F), given that its normal temperature range is 18°C–20°C (64°F–68°F)? Second, did this species occur anywhere else in the Mediterranean, and was it spreading beyond the soft-sediment habitats found in front of the museum? Most importantly, at such high population densities, how was it interacting with native algae and seagrasses, both of which are critical habitats and sources of food for fish and invertebrate species?

A definitive answer to the second question came in July 1990, when the alga was found 5 km east of the museum, at a popular fishing location. Evidently, fragments had been caught on the gear and anchors of fishing vessels and transported to new sites of colonization. The find generated media coverage that included information on the toxicity of the seaweed, which produces a peppery secondary compound to deter the fish and invertebrate herbivores that abound in the tropics. The press sensationalized *Caulerpa*'s natural toxicity with headlines such as "Killer algae!"— a misleading title that suggested that the seaweed was toxic to humans (it is not). As the news spread, so did the sightings of *Caulerpa*. By 1991, 50 sightings had been reported in France alone. The fluorescent green alga indiscriminately colonized muddy, sandy, and rocky bottoms at a depth of 3–30 m. By 2000, the alga had moved from France to Italy, then to Croatia to the east and Spain to the west, eventually spreading as far as Tunisia (**FIGURE 16.2**). It had invaded thousands of hectares, despite frantic but futile efforts to remove it.

FIGURE 16.1 Invading Seaweed *Caulerpa taxifolia* rapidly invaded and dominated marine communities in the Mediterranean Sea.

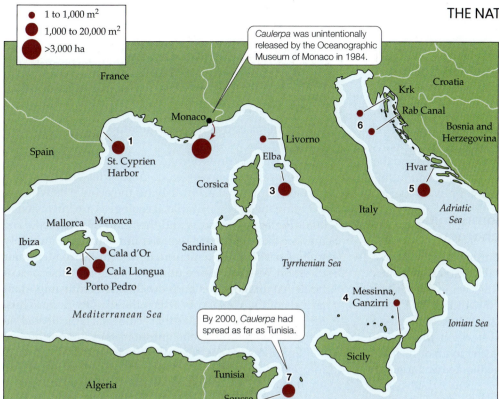

FIGURE 16.2 Spread of *Caulerpa* in the Mediterranean Sea *Caulerpa taxifolia* first invaded the waters off Monaco and France. By 2000, this algal species had reached Croatia and Tunisia. (After A. Meinesz et al. 2001. *Biol Inv* 3: 201. Based on A. Meinesz. 1997. *Le roman noir de l'algue "tueuse"*: Caulerpa taxifolia *contre la Méditerranée*. Belin Éditeur: Paris.)

Using the order of appearance on the map, describe the possible invasion pathways of *Caulerpa* within this region.

From the very beginning, Meinesz suspected that the answer to his first question lay with the museum. In 1980, a cold-resistant strain of *Caulerpa taxifolia* had been discovered and propagated in the tropical aquariums at the Wilhelma Zoo in Stuttgart, Germany. Cuttings were sent to other aquariums, including the one in Monaco, to be grown as aesthetically pleasing backdrops to tropical fish displays. The museum admitted to unintentionally releasing *Caulerpa* in the process of cleaning tanks but believed the alga would die in the cold waters of the Mediterranean.

Given that *Caulerpa* did not die, but instead quickly invaded and overtook shallow areas of the Mediterranean, scientists and fishermen alike wanted to understand how this abundant and fast-spreading seaweed would affect marine habitats and the fisheries dependent on them. How do interactions with one very abundant species influence the hundreds of other species with which it shares a community?

Introduction

We have emphasized throughout this book that species are connected with one another and with their environment. Ecology is, at its very essence, the study of these interconnections. In Unit 4, we looked at interactions between species as two-way relationships, with one species

eating, competing with, or facilitating another species. For ease of mathematical modeling, we considered these pairwise interactions in isolation, even though we have emphasized that, in reality, species experience multiple interactions. In this chapter, we will explore multiple-species interactions and how they shape the nature of communities. We will consider the various ways in which ecologists have defined communities, the metrics used to measure community structure, and the types of species interactions that characterize communities.

CONCEPT 16.1 Communities are groups of interacting species that occur together at the same place and time.

LEARNING OBJECTIVES

16.1.1 Describe the ways ecologists delineate communities.

16.1.2 Explain why ecologists use subsets of species to define communities and list some of the subsets used.

What Are Communities?

Ecologists define **communities** as groups of interacting species that occur together at the same place and time.

Interactions among multiple species and their physical environment give communities their character and function. Whether we are dealing with a desert, a kelp forest, or the gut of an ungulate, the existence of the community is dependent on the individual species that are present and on how they interact with one another and their physical surroundings. As we will see in this chapter and others in this unit, the relative importance of species interactions and the physical environment, which can vary among communities, is a major focus of research for community ecologists.

Ecologists often delineate communities by their physical or biological characteristics

The technical definition of a community given above is more theoretical than operational. In practical terms, ecologists often delineate communities using physical or biological characteristics as a guide (**FIGURE 16.3**). A community may be defined by the physical characteristics of its environment; for example, a physically defined community might encompass all the species in a hot springs, a mountain stream, or a desert. The biomes and aquatic biological zones described in Chapter 3 are based largely on the physical characteristics thought to be important in defining communities. Similarly, a biologically defined community might include all the species associated with a kelp forest, a freshwater bog, or a coral reef. This way of thinking uses the presence and implied importance of abundant species, such as kelp, wetland plants, or corals, as the basis for community delineation.

In most cases, however, communities end up being defined somewhat arbitrarily by the ecologists who are studying them. For example, if ecologists are interested in studying aquatic insects and their amphibian predators, they are likely to restrict their definition of the community to that particular interaction. Unless they broaden their question, researchers are unlikely to consider the roles of birds that forage in wetlands or other inherently important aspects of the wetland in which they are working. Thus, is it important to recognize that ecologists typically define communities based on the questions they are posing.

Regardless of how a community is defined, ecologists interested in knowing which species are present in a

(A) Desert

(B) Hot springs

(C) Kelp forest

(D) Coral reef

FIGURE 16.3 Defining Communities Ecologists often delineate communities based on their physical attributes or their biological attributes.

? Of the four communities shown in this figure, which are mostly defined by physical attributes and which are mostly defined by biological attributes?

FIGURE 16.4 Subsets of Species in Communities Ecologists may use subsets of species to define communities. These examples show three ways in which such subsets could be designated. (A) All the bird species in a community could be grouped together by taxonomic affinity. (B) All the species that use pollen as a resource could be grouped together as a guild. (C) All the plant species in a community that have nitrogen-fixing bacteria (e.g., legumes) could be placed in the same functional group.

(A) Taxonomic affinity

(B) Guild

(C) Functional group

community must contend with the difficult issue of accounting for them. Merely creating a species list for a community is a huge undertaking, and one that is essentially impossible to complete, especially if small or relatively unknown species are considered. Taxonomists have officially described about 1.9 million species, but we know from sampling studies of tropical insects and microorganisms that this number greatly underestimates the actual number of species on Earth, which could be closer to 9 million or even more. For this reason, and because of the difficulty of studying many species at one time, ecologists usually consider a subset of species when they define and study communities.

Ecologists may use subsets of species to define communities

One common way of subdividing a community is based on taxonomic affinity—that is, by groups of species classified together because of evolutionary lineage (**FIGURE 16.4A**). For example, a study of a forest community might be limited to all the bird species within that community (in which case an ecologist might speak of "the forest bird community"). Another useful subset of a community is a **guild**, a group of species that use the same resources, even though they might be taxonomically distant (**FIGURE 16.4B**). For example, some birds, bees, and bats feed on flower pollen, thus forming a guild of pollen-eating animals. Finally, a **functional group** is a subset of a community that includes species that function in similar ways but may or may not use similar resources (**FIGURE 16.4C**). For example, nitrogen-fixing plants (legumes) can be placed in the same functional group.

There are other subsets of communities that allow ecologists to organize species based on their *trophic*, or food-based, interactions (**FIGURE 16.5A**). Species can be organized in a **food web**, a representation of the trophic or energetic connections among species within a community. Food webs can be further organized into **trophic levels**, or groups of species that

have similar ways of interacting and obtaining energy. The lowest trophic level contains *primary producers*, which are autotrophs such as plants. The primary producers are fed on by organisms at the second level, the *primary consumers*, which are herbivores. The third level contains *secondary consumers*, which are carnivores, or animals that

FIGURE 16.5 Food Webs and Interaction Webs (A) Food webs describe trophic or energetic connections among species within a community. (B) Interaction webs include both trophic interactions (vertical arrows) and non-trophic (horizontal arrows) competitive and positive interactions.

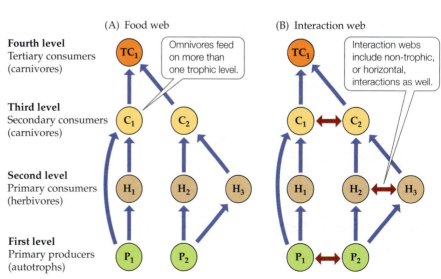

(A) Food web

Fourth level
Tertiary consumers (carnivores)

Omnivores feed on more than one trophic level.

Third level
Secondary consumers (carnivores)

Second level
Primary consumers (herbivores)

First level
Primary producers (autotrophs)

(B) Interaction web

Interaction webs include non-trophic, or horizontal, interactions as well.

eat animals. Secondary consumers are fed on in turn by *tertiary consumers*, also carnivores.

Traditionally, food webs have been used as a descriptive or idealized method of understanding the trophic relationships among the species in a community. Food webs tell us little, however, about the strength of those interactions or their importance in the community. In addition, the use of trophic levels can create confusion for a number of reasons: for example, some species span two trophic levels (e.g., corals can be classified as both carnivores and herbivores because they eat zooplankton and they have symbiotic algae), some species change their feeding status as they mature (e.g., amphibians can be herbivores as tadpoles and carnivores as adults), and some species are *omnivores*, feeding on more than one trophic level (e.g., some fish feed on both algae and invertebrates). Moreover, idealized food webs often do not include certain important resources and consumers that are common within communities. For example, all organisms that die without being consumed become organic matter known as *detritus* and can be consumed by *detritivores* (mostly fungi and bacteria) through a process known as *decomposition*; see Concept 22.2. Another example is symbionts, including parasites and mutualists, which are present at almost all trophic levels (see Chapters 13 and 15).

Another characteristic of food webs is that they do not include non-trophic interactions (so-called **horizontal interactions**, such as competition and some positive interactions), which, as we have seen in Unit 4, can also influence community character. The concept of an **interaction web** has been introduced to more accurately describe both the trophic (vertical) and non-trophic (horizontal) interactions among the species in a traditional food web (**FIGURE 16.5B**). Despite these drawbacks, the food web concept remains a strong one, if only for its visual representation of important consumer relationships within a community.

We will learn much more about food webs in Chapter 21. Next let's consider the important properties of communities that allow us to characterize them and to distinguish one from another.

> **CONCEPT 16.2** Species diversity and species composition are important descriptors of community structure.

LEARNING OBJECTIVES

16.2.1 Define and quantify species diversity and compare to biodiversity.

16.2.2 Define and graph rank abundance and species accumulation relationships.

16.2.3 Define species composition and explain why it is an important characteristic of communities.

Community Structure

We have seen that communities vary greatly in the number of species they contain. A tropical rainforest, for example, has many more tree species than a temperate rainforest, and a midwestern prairie has many more insect species than a New England salt marsh. Ecologists have devoted substantial effort to measuring this variation at a number of spatial scales. Species diversity and species composition are important descriptors of **community structure**, the set of characteristics that shape a community. Community structure is descriptive in nature but provides the necessary quantitative basis for generating hypotheses and experiments directed at understanding how communities work.

Species diversity is an important measure of community structure

Species diversity is the most commonly used measure of community structure. Even though the term is often used generally to describe the number of species within a community, it has a more precise definition. **Species diversity** is a measure that combines the number of species (species richness) and their abundances compared with those of the other species (species evenness) within the community. **Species richness** is the easiest metric to determine: one simply counts all the species in the community. **Species evenness**, which tells us about the commonness or rarity of species, requires knowing the abundance of each species relative to those of the other species within the community, a harder value to obtain. (See Ecological Toolkit 9.1 for methods of estimating abundances in terms of number, biomass, or percentage of cover.)

The contributions of species richness and species evenness to species diversity can be illustrated using a hypothetical example (**FIGURE 16.6**). Let's imagine two meadow communities, each containing four species of butterflies. Both communities have the same butterfly species richness, but their species evenness differs. In community A, one species constitutes 85% of the individuals in the community, while the other three species constitute only 5% of the individuals in the community; thus, species evenness is low. In community B, the number of individuals are evenly divided among the four species (25% each), so species evenness is high. In this case, even though each community has the same species richness (four species), community B has the higher species diversity because it has higher species evenness.

A number of species diversity indices can be used to describe species diversity quantitatively. By far the most commonly used is the **Shannon index**,

$$H = -\sum_{i=1}^{s} p_i \ln(p_i)$$

where

H = the Shannon index value

p_i = the proportion of individuals found in the *i*th species

ln = the natural logarithm

s = the number of species in the community

FIGURE 16.6 Species Richness and Species Evenness
These two hypothetical butterfly communities have the same number of species (species richness) but different relative abundances (species evenness). Species diversity, as measured using the Shannon index, is lower in community A (see Table 16.1).

> In community A, the abundance of one species (the blue butterfly) is high relative to the other species, so this community has low species evenness.

Community A

> In community B, each species has the same abundance, so this community has high species evenness.

Community B

The lowest possible value of H is zero. The higher a community's H value, the greater its species diversity. **TABLE 16.1** calculates the Shannon index for the two butterfly communities in Figure 16.6. These calculations show that community A has the lower Shannon index value (H), confirming mathematically that this community has lower species diversity than community B. Given that both communities have the same species richness, the difference in species diversity is driven by the lower species *evenness* in community A. You can practice calculating the Shannon index in **ANALYZING DATA 16.1**, which explores how an invasive plant affects community structure in central European grasslands.

As we mentioned earlier, the term "species diversity" is often used imprecisely to describe the number of species in a community without regard to the relative abundances of species or species diversity indices. For example, one commonly hears the assertion that "species diversity" is higher in tropical communities than in temperate communities, without any accompanying information about the actual relative abundances of species in the two community types. Another term that is

TABLE 16.1

Calculation of Species Diversity Using the Shannon Index for Communities A and B

> To calculate the Shannon Index (H), the natural logarithm (ln) is applied to p_i for each species (i)…

> …and then this value is multiplied by p_i once again.

Community A

Species	Abundance	Proportion (p_i)	ln (p_i)	p_i ln (p_i)
Blue	17	0.85	−0.163	−0.139
Yellow	1	0.05	−2.996	−0.150
Pink	1	0.05	−2.996	−0.150
Orange	1	0.05	−2.996	−0.150
Total	20	1.00		−0.589

> All the values are summed for all the species in the community and multiplied by −1 to get H.

$$H = -\sum_{i=1}^{s} p_i \ln(p_i) = 0.589$$

Community B

Species	Abundance	Proportion (p_i)	ln (p_i)	p_i ln (p_i)
Blue	5	0.25	−1.386	−0.347
Yellow	5	0.25	−1.386	−0.347
Pink	5	0.25	−1.386	−0.347
Orange	5	0.25	−1.386	−0.347
Total	20	1.00		−1.388

> Community B has higher species diversity than community A.

$$H = -\sum_{i=1}^{s} p_i \ln(p_i) = 1.388$$

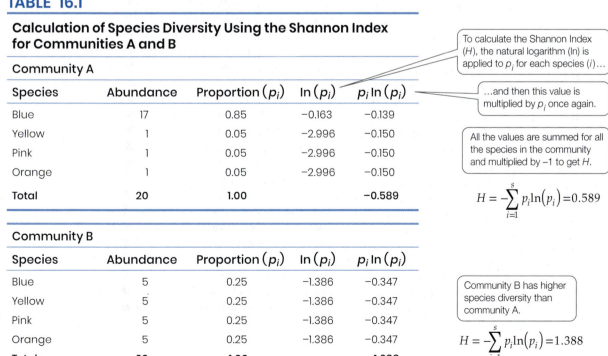

ANALYZING DATA 16.1

What Are the Effects of Invasive Species on Species Diversity?

Invasions of non-native species have been implicated in both increases and decreases of species diversity within communities. One study looked at the effects of 13 "neophyte" plant species (i.e., those introduced since 1500) on the species diversity of a variety of plant communities in the Czech Republic in Central Europe (Hejda et al. 2009).* To understand the importance of species invasions to species diversity, the researchers measured species richness and abundance (percent plant cover) in plots with similar site conditions that differed in whether they had been invaded or not (i.e., native) by particular invasive species. They then subtracted the species richness of the invaded plots from that of the native plots, averaged the resulting values, and obtained an average change in species richness (*y*-axis) for each species invasion (*x*-axis). The results are shown in **FIGURE A**. The researchers also calculated the Shannon index (*H*) for each of the plots and conducted the same analysis: they calculated an average change in species diversity (*y*-axis) for each invasive species (*x*-axis). These values are given in **FIGURE B**. Error bars show one standard error of the mean.

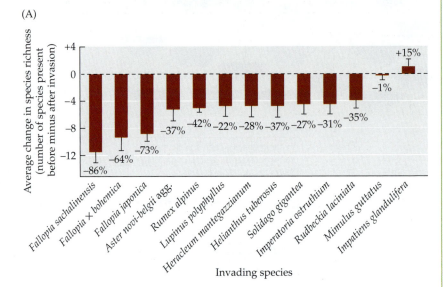

(A)

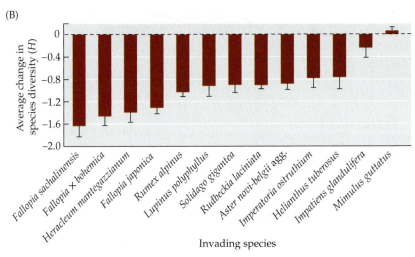

(B)

1. Based on the average changes in species richness in Figure A, how many invasive species probably had negative effects on species richness, how many probably had positive effects on species richness, and how many probably had no effect on species richness?

2. Above (or below) each bar in Figure A is the percentage change in species richness for that invader. What do these percentages tell you about the likely direction and strength of the effect of invasive species on native community richness? Compare the rank order of the magnitude of the average change in species richness from Figure A with that of the change in species diversity (*H*) in Figure B. Does the order differ between the two measures and, if so, why?

*Hejda, M., P. Pysek, and V. Jorosik. 2009. Impact of invasive plants on the species richness, diversity, and composition of invaded communities. *Journal of Ecology* 97: 393–403.

often used interchangeably with "species diversity" is "biodiversity." Technically, **biodiversity** is a term used to describe the diversity of important ecological entities that span multiple spatial scales, from genes to species to communities (**FIGURE 16.7**). Implicit in the term is the interconnectedness of genes, individuals, populations, species, and even community-level components of diversity. As we saw in Chapter 10, the genetic variation among

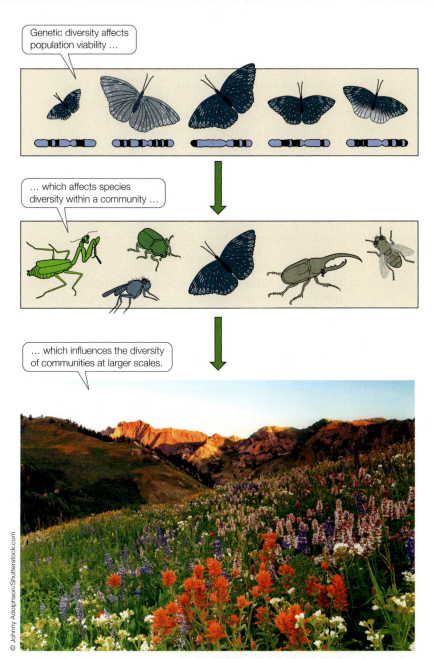

Genetic diversity affects population viability ...

... which affects species diversity within a community ...

... which influences the diversity of communities at larger scales.

FIGURE 16.7 Biodiversity Considers Multiple Spatial Scales Diversity can be measured at spatial scales that range from genes to species to communities. The term "biodiversity" encompasses diversity at all of these scales.

individuals within a population influences that population's *viability* (its chance of persistence). Population viability, in turn, has important consequences for species persistence, and ultimately for species diversity within communities. Moreover, the number of different kinds of communities in a region is critical to diversity at larger regional and latitudinal scales (see Figure 18.5). We will discuss the importance of spatial scale and biodiversity in chapters to come, but it is worth understanding some of the ways in which the term "diversity" is used, as a starting point for those later discussions.

Species within communities differ in their commonness or rarity

Although species diversity indices allow ecologists to compare different communities, graphical representations of species diversity can give us a more explicit view of the commonness or rarity of the species in communities. Such graphs, called **rank abundance curves**, plot the proportional abundance of each species (p_i) relative to the others in rank order, from most abundant to least abundant (**FIGURE 16.8**). If we use rank abundance curves to compare our two butterfly communities from Figure 16.6, we can see that community A has one abundant species (i.e., the blue butterfly) and three rare species (i.e., the yellow, pink, and orange butterfly species), whereas in community B, all the species have the same abundance.

These two patterns could suggest the types of species interactions that might occur in these two communities. For example, the dominance of the blue butterfly in community A might indicate that it has a strong effect on one or more of the other species in the community. In community B, where all the species have the same abundance, their interactions might be fairly equivalent, with no one species dramatically affecting the others. To test these hypotheses, we can design manipulative experiments to explore relationships between species abundances and the types of interactions that occur among the species in a community. As we will see in the next section, experiments of this kind typically involve adding or removing individuals of one species and measuring the responses of individuals of another species in the community to the manipulation.

Species diversity estimates vary with sampling effort and scale

Let's imagine that you are sampling your backyard for insect species. It makes sense that the more samples you collect, the more species you are likely to find. However, eventually you reach a point in your sampling effort at which any additional sampling will reveal so few new species that you could stop sampling and still have a good idea of the species richness of your backyard. That point of "no significant return" for your effort can be determined using a **species accumulation curve** (**FIGURE 16.9**). These curves are calculated by plotting species richness as a function of sampling effort. In other words, each data point on a species accumulation curve represents the total number of species and the sampling

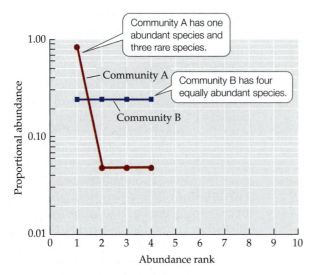

FIGURE 16.8 **Are Species Common or Rare?** Using rank abundance curves, we can see that the two hypothetical butterfly communities in Figure 16.6 differ in the commonness and rarity of the same four species.

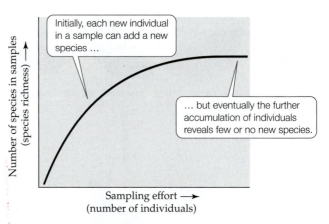

FIGURE 16.9 **When Are All the Species Sampled?** Species accumulation curves can help us determine when most or all of the species in a community have been observed. In this hypothetical example, the number of new species observed in each sample decreases after about half the individuals in the samples have accumulated.

effort up to that point. The more samples taken, the more individuals will be added, and the more species will be found. In theory, one could imagine, a threshold could be reached at which no new species would be added by additional sampling. In reality, this never occurs in natural systems, because new species are constantly being found.

Jennifer Hughes and colleagues (2001) used species accumulation curves to ask how communities differ in the relationship between species richness and sampling effort. Are there some very diverse communities in which we are unable to estimate species richness accurately despite intensive sampling? Hughes and colleagues calculated species accumulation curves for five different communities: a temperate forest plant community in Michigan, a tropical bird community in Costa Rica, a tropical moth community in Costa Rica, a bacterial community from the human mouth, and a bacterial community from tropical soils in the eastern Amazon (**FIGURE 16.10**). To compare the curves properly, given that the communities differed substantially in organismal abundance and species richness, the data sets were standardized by calculating for each data point the proportions of the total number of individuals and species that had been sampled up to that point. The results showed that the species richnesses of the Michigan forest plant and Costa Rican bird communities were adequately represented well before half the individuals were sampled. Human oral bacteria and Costa Rican moth communities had species accumulation curves that never completely leveled off, suggesting that their species richness was high and that additional sampling would be required to achieve an approximation of that richness. Finally, the eastern Amazon soil bacterial community had a linear species accumulation curve, demonstrating that each

new sample resulted in the observation of many new bacterial species. Based on this analysis, it is clear that the sampling effort for tropical bacteria was well below that needed to adequately estimate species richness in these hyperdiverse communities.

A comparison of species accumulation curves not only provides valuable insight into the differences in species richness among communities, but also demonstrates the influence of the spatial scale at which sampling is carried out. For example, if we were to sample the richness of bacteria in tropical soils at the same scale at which we sampled Costa Rican moths, the bacterial richness would be immense in comparison. But such comparisons do suggest that our ability to sample all the bacteria in the human mouth is roughly equivalent to our ability to sample all the moth species in a few hundred square kilometers of tropical forest. The work of Hughes et al. also reminds us how little we know about the community structural characteristics of rarely sampled assemblages, such as microbial communities.

Species composition tells us who is in the community

A final element of a community's structure is its **species composition**: the identity of the species present in the community. Species composition is an obvious but important characteristic that is not revealed in species diversity indices. For example, two communities might have the same species diversity value but have completely different members. For example, a study in Scottish pastures showed that two bacterial communities had diversity indices that were nearly identical but their taxonomic composition differed (McCaig et al. 1999). Five taxonomic groups of bacteria out of the 20

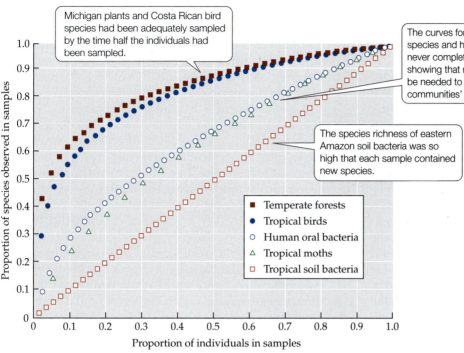

Michigan plants and Costa Rican bird species had been adequately sampled by the time half the individuals had been sampled.

The curves for Costa Rican moth species and human oral bacteria never completely leveled off, showing that more sampling would be needed to estimate those communities' species richness.

The species richness of eastern Amazon soil bacteria was so high that each sample contained new species.

- ■ Temperate forests
- ● Tropical birds
- ○ Human oral bacteria
- △ Tropical moths
- □ Tropical soil bacteria

FIGURE 16.10 **Communities Differ in Their Species Accumulation Curves** Hughes and colleagues found that communities of five different types varied greatly in the sampling effort that would be needed to estimate their species richness. The data sets were standardized by calculating for each data point the proportions of the total number of individuals and species that had been sampled up to that point. (After J. B. Hughes et al. 2001. *Appl Environ Microbiol* 67: 4399–4406.)

? Based on the graph, which of these communities would require more sampling to adequately estimate their species richness? Which would require very little additional sampling?

the researchers found were present in one or the other pasture, but not in both.

In many ways, community structure is the starting point for more interesting questions: How do species in the community interact with one another? Do some species play greater roles in the community than others? How is species diversity maintained? How does this information shape our view of communities in terms of conservation and the services they provide to humans? Let's move from the rather static view of communities as groups of species occurring together in space and time to a more active view of them as complex networks of species with connections and interactions that vary in strength, direction, and significance.

> **CONCEPT 16.3** Communities can be characterized by complex networks of direct and indirect interactions that vary in strength and direction.

LEARNING OBJECTIVES

16.3.1 Compare direct versus indirect species interactions and describe the effects they have on communities.

16.3.2 Explain how species interactions can vary in strength and direction, and how these attributes can be measured experimentally.

16.3.3 Compare foundation species, keystone species, and ecosystem engineers and the effects they have on communities under different environmental contexts.

Interactions of Multiple Species

The way we think about species interactions changes dramatically when we consider that they are embedded in a community of multiple interactors. Instead of a particular species experiencing a single, direct interaction with another species, we are now dealing with multiple species interactions that generate a multitude of connections—some direct, but many indirect (**FIGURE 16.11**). **Direct interactions** occur between two species and include trophic and non-trophic interactions—the interactions we explored in Unit 4. **Indirect interactions** occur when the relationship between two species is mediated by a third

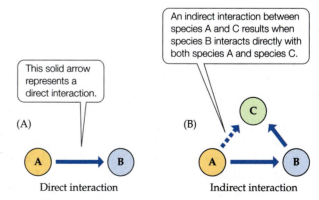

This solid arrow represents a direct interaction.

An indirect interaction between species A and C results when species B interacts directly with both species A and species C.

(A) Direct interaction

(B) Indirect interaction

FIGURE 16.11 **Direct and Indirect Species Interactions** (A) A direct interaction occurs between two species. (B) An indirect interaction (dashed arrow) occurs when the direct interaction between two species is mediated by a third species.

(or more) species. The simple addition of a third species to a two-species interaction creates many more effects, both direct and indirect, which have the potential to change the outcome of the original interaction dramatically.

A social interaction analogy fits well here. Consider Figure 16.11B. Let's say you are person A and you have a good friend (person B) with whom you interact well. Now, suppose this friend meets another person (person C) who dominates your friend's time. They go to dinner, the movies, and bowling—all things you and your friend enjoyed together—without you. At some point, this new friend might begin to interfere with your friendship, possibly compromising it to the point at which it becomes antagonistic. Sadly, the indirect effect of person C changes your friendship irreparably. You might say that "the friend of my friend is my enemy." Likewise, one could imagine gaining a friend indirectly, if your foe had a foe (in this case, "the enemy of my enemy is my friend"). The point is that simply adding another person to the social circle can change the outcome of your relationship completely. The same is true of species interactions when we view them in the community context, rather than as isolated entities.

Indirect species interactions can have large effects

Charles Darwin was one of the first to convey the importance of indirect interactions. In *The Origin of Species* (1859), Darwin set the scene by describing the role of bees in flower pollination, and hence in seed production, among native plants in the region of England where he lived. In the book, he established the hypothesis that the number of bees is dependent on the number of field mice, which prey on the combs and nests of bees. Mice, in turn, are eaten by cats, leading Darwin to muse, "Hence it is quite credible that the presence of a feline animal in large numbers in a district might determine, through the intervention first of mice and then of bees, the frequency of certain flowers in that district!" (Darwin 1859, p. 59).

It is only recently that the sheer number and variety of effects of indirect interactions have been documented (Menge 1995). In many cases, indirect effects are discovered almost by accident when species are experimentally removed to study the strength of a direct negative interaction such as predation or competition. A good example of this type of indirect effect comes in the form of an interaction web called a trophic cascade (**FIGURE 16.12A**). A **trophic cascade** occurs when the rate of consumption at one trophic level results in a change in species abundance or composition at lower trophic levels. For example, when a carnivore eats an herbivore (having a direct negative effect on the herbivore) and decreases its abundance, there may be an indirect positive effect on a primary producer that was eaten by that herbivore. One of the best-known examples is the indirect regulation of kelp forests by the sea otter (*Enhydra lutris*) through its direct interaction with sea urchins (*Strongylocentrotus* spp.) along the west coast of North America (see the Case Study Revisited in Chapter 9) (Simenstad et al. 1978). Two direct trophic interactions, those of sea otters feeding on sea urchins and sea urchins feeding on kelp, generate indirect positive effects, including that of sea otters on kelp (via their reduction of urchin abundance) and that of kelp on sea otters (via the food they provide for the urchins). Furthermore, the kelp can positively affect the abundances of other seaweeds, which serve as habitat

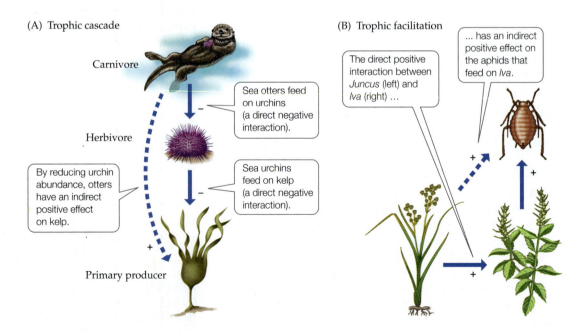

FIGURE 16.12 Indirect Effects in Interaction Webs (A) A trophic cascade occurs when a carnivore feeds on an herbivore and thus has an indirect positive effect on a primary producer that is eaten by that herbivore. (B) Trophic facilitation occurs when a consumer is indirectly helped by a positive interaction between its prey and another species.

(A) Trophic cascade

Carnivore

Sea otters feed on urchins (a direct negative interaction).

Herbivore

By reducing urchin abundance, otters have an indirect positive effect on kelp.

Sea urchins feed on kelp (a direct negative interaction).

Primary producer

(B) Trophic facilitation

The direct positive interaction between *Juncus* (left) and *Iva* (right) ...

... has an indirect positive effect on the aphids that feed on *Iva*.

and food for many marine invertebrates and fishes. The indirect effects generated in this simple food web are just as important as the direct effects in determining whether the ecosystem will be a kelp forest or an urchin barren (see Figure 9.18). We will explore the effects of indirect interactions on species diversity (Chapter 19), food webs (Chapter 21), and ecosystem management (Chapter 24) in more detail later in the book.

Indirect effects can also emerge from direct positive interactions called trophic facilitations. A **trophic facilitation** occurs when a consumer is indirectly helped by a positive interaction between its prey and another species (**FIGURE 16.12B**). An example of this type of indirect effect was demonstrated by Sally Hacker (Oregon State University) and Mark Bertness (Brown University), who studied salt marsh plant and insect interaction webs in New England. Their research showed that a commensal interaction between two salt marsh plants—a rush, *Juncus gerardii*, and a shrub, *Iva frutescens*—has important indirect effects on aphids feeding on *Iva* (Hacker and Bertness 1996).

To explore these findings in greater detail, let's first consider the commensal interaction between the two plant species. When *Juncus* was experimentally removed, the photosynthetic rate of *Iva* was lower (**FIGURE 16.13A**). In contrast, removing *Iva* had no effect on *Juncus*. In the absence of *Juncus*, soil salinity increased and oxygen content decreased considerably around *Iva*, suggesting that the presence of *Juncus* ameliorated harsh physical conditions for *Iva*. *Juncus*, by shading the soil surface and thus decreasing water evaporation from the surface of the marsh, decreases salt buildup. *Juncus* also has specialized tissue called aerenchyma, through which oxygen can move into the belowground parts of the plant, thus keeping it from "drowning" during daily high tides. Some of the oxygen "leaks" out of the plant and can be used by other neighboring plants, such as *Iva*.

To understand the importance of this direct positive interaction, Hacker and Bertness (1996) measured the population growth rate of aphids on *Iva* growing with and without *Juncus*. They found that aphids had a much harder time finding shrubs in the presence of the rush but that once they did, their population growth rates were significantly higher (**FIGURE 16.13B**). Using the exponential growth equation, they predicted that aphids would become locally extinct in the salt marsh without the indirect positive effects of *Juncus* (**FIGURE 16.13C**). It is clear from this example that interactions in trophic facilitation webs can have both positive effects (as when *Juncus* improves soil conditions for *Iva*) and negative effects (as when *Juncus* facilitates aphids that feed on *Iva*), but it is the sum total of these effects that determines whether the interaction is beneficial or not. Given that the ultimate fate of *Iva* without *Juncus* is death, the positive effects greatly outweigh the negative.

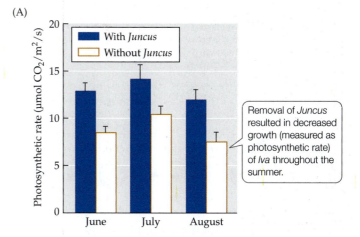

(A)

Removal of *Juncus* resulted in decreased growth (measured as photosynthetic rate) of *Iva* throughout the summer.

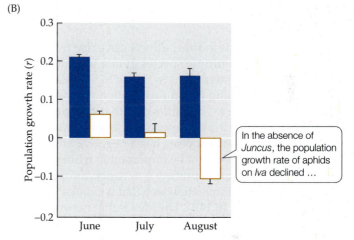

(B)

In the absence of *Juncus*, the population growth rate of aphids on *Iva* declined …

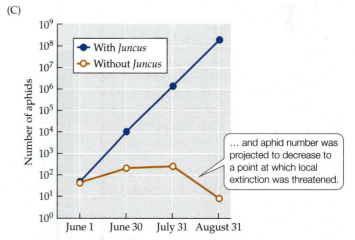

(C)

… and aphid number was projected to decrease to a point at which local extinction was threatened.

FIGURE 16.13 Results of Trophic Facilitation in a New England Salt Marsh Removal experiments demonstrated that aphids are indirectly facilitated by the rush *Juncus gerardii*, which has a direct positive effect on the shrub *Iva frutescens*, on which the aphids feed. (A) Photosynthetic rate of *Iva* with and without *Juncus*. (B) Growth rate of aphid populations with and without *Juncus*. (C) Projected numbers of aphids with and without *Juncus*. Error bars show one standard error of the mean. (After S. D. Hacker and M. D. Bertness. 1996. *Am Nat* 148: 559–575.)

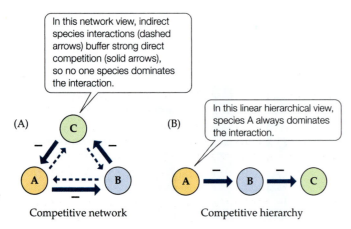

In this network view, indirect species interactions (dashed arrows) buffer strong direct competition (solid arrows), so no one species dominates the interaction.

In this linear hierarchical view, species A always dominates the interaction.

(A) Competitive network

(B) Competitive hierarchy

FIGURE 16.14 Competitive Networks versus Competitive Hierarchies

Finally, important indirect effects can arise from interactions among multiple species at one trophic level (i.e., the horizontal interactions in Figure 16.5B). Buss and Jackson (1979), looking for an explanation for the coexistence of competitors, hypothesized that **competitive networks**—competitive interactions among multiple species in which every species has a negative effect on every other species—might be important in maintaining species richness in communities. A network, as opposed to a hierarchy, is an interaction web that is circular rather than linear (**FIGURE 16.14A**). The idea is that networks of interacting species indirectly buffer strong direct competition, thus making competitive interactions weaker and more diffuse. So, for example, species A may have the potential to outcompete species B, and species B may have the potential to outcompete species C, but because species C also has the potential to outcompete species A, no one species dominates the interaction. This is clearly an example of the "enemy of my enemy is my friend" effect described earlier. All else being equal, a hierarchical view of competition, with species A outcompeting B and B outcompeting C (**FIGURE 16.14B**), always results in species A dominating the interaction.

Buss and Jackson tested this hypothesis using encrusting invertebrates and algae that live on the undersides of coral reefs in Jamaica (**FIGURE 16.15**). These species compete for space by growing over one another. The researchers collected samples at the margins between species, where one species overgrows another, for as many pairs of individuals of different species as possible to determine the proportion of wins (species on top) to losses (species on the bottom) for each interaction. Their results showed that every species both overgrew and was overgrown by at least one other species and that no one species consistently won competition for space at their borders. The species interacted in a circular network rather than a linear hierarchy. These observations demonstrate how competitive networks, by fostering diffuse and indirect interactions, can promote diversity in communities.

Species interactions vary greatly in strength and direction

It should be clear by now that species interactions in a community vary greatly in their strength and direction. Some species have a strong negative or positive effect on the community, while others probably have little or no effect. **Interaction strength**, the effect of one species on the abundance of another species, can be measured experimentally by removing individuals of one species (referred to as the *interactor species*) from the community and looking at the effect on individuals of the other species (the *target species*, as described in **ECOLOGICAL TOOLKIT 16.1**). If the removal of individuals of the interactor species results in a large decrease in individuals of the target species, we know that the interaction is strong and positive. However, if the abundance of the target species increases significantly after removal, we know that the interactor species has a strong negative effect on the target species.

The interaction strength "dynamic" (i.e., the relative proportion of strong to weak interactions or positive to negative interactions) is not well understood for any community, because of the numbers of species involved and the many indirect interactions that emerge. As you will see throughout Unit 5, however, we can get

Courtesy of Robert S. Steneck

FIGURE 16.15 Competitive Networks in Coral Reef Communities Encrusting invertebrates and algae compete for space on coral reefs by overgrowing one another, but no one species consistently "wins" this competition.

ECOLOGICAL TOOLKIT 16.1

Measurements of Interaction Strength

We can measure interaction strength by experimentally manipulating species interactions. The usual procedure involves the removal (or sometimes the addition) of individuals of one species involved in the interaction (the interactor species) and measurement of the response of the individuals of the other species (the target species). There are multiple ways to calculate interaction strength, but one simple way is to calculate the relative interaction intensity (RII) (Armas et al. 2004) using the following equation:

$$\text{Relative interaction intensity} = (C - E)/(C + E)$$

where

C = the number or biomass of target individuals in the presence of the interactor

E = the number or biomass of target individuals in the absence of the interactor

Interaction strength can vary depending on the environmental context in which the interaction is measured.

For example, Menge et al. (1996) measured the interaction strength of sea star (*Pisaster ochraceus*) predation on mussels (*Mytilus trossulus*) in wave-exposed versus wave-protected areas of the coastline at Strawberry Hill on the coast of Oregon (see **FIGURE**). Sea stars were excluded from some mussel beds by cages in both exposed and protected areas. At the end of the experiment, the numbers of mussels in the cages (E) were compared with the numbers of mussels in control plots (C) that had been exposed to sea star predation (see Figure). The results (see Figure) showed that interaction strength was greater in wave-protected than in wave-exposed areas. Sea stars probably cannot feed as efficiently when subjected to the crashing waves characteristic of wave-exposed areas. Thus, this study demonstrates the importance of environmental context (in this case, wave exposure) to the strength of species interactions. It also shows how those interactions can change over relatively small spatial scales (e.g., between wave-exposed and wave-protected areas of the Strawberry Hill rocky shore).

How Much Does Predation by Sea Stars Matter? It Depends
(Top left) The coastline at Strawberry Hill, Oregon. Plots with (bottom left) and without (middle) cages that excluded sea stars were set up in both wave-exposed and wave-protected areas along the rocky shore. (Bottom right) When mussels were counted and interaction strengths calculated, the results showed that interaction strength was greater in protected than in exposed areas. Error bars show one standard error of the mean. (Graph after B. A. Menge et al. 1996. *Food Webs: Integration of Patterns and Dynamics*, G. A. Polis and K. O. Winemiller [Eds.], pp. 258–274. Chapman & Hall: New York.)

All photos courtesy of Sally Hacker

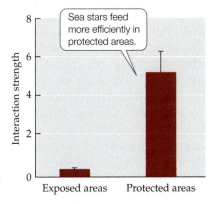

Sea stars feed more efficiently in protected areas.

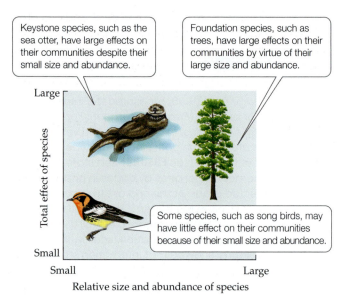

Keystone species, such as the sea otter, have large effects on their communities despite their small size and abundance.

Foundation species, such as trees, have large effects on their communities by virtue of their large size and abundance.

Some species, such as song birds, may have little effect on their communities because of their small size and abundance.

FIGURE 16.16 Foundation versus Keystone Species
Species that have large effects on their communities may or may not do so by virtue of their large size and abundance. Some species (lower left-hand corner) have little overall effect relative to their size and abundance, especially if they are redundant in the community. (After M. E. Power et al. 1996b. *BioScience* 46: 609–620.)

influencing temperature and moisture levels in the forest. The roots of the tree can increase weathering and aeration of the soil, and they can stabilize surrounding substrates. The tree's leaves fall to the forest floor, where they add moisture and nutrients to the soil and provide habitat for soil-dwelling invertebrates, seeds, and microorganisms. If the tree falls, it can become a "nurse log," providing space, nutrients, and moisture for tree seedlings. Thus, trees can have a large physical influence on the structure of a forest community, which obviously changes over time as trees grow, mature, and die.

Other strong interactors, so-called **keystone species**, have large effects not because of their abundance, but because of the vital roles they play in their communities. They differ from foundation species in that their effect is large in proportion to their size and abundance (see Figure 16.16). Keystone species usually influence community structure indirectly, via trophic means, as we saw in the case of sea otters (see Figure 16.12A and Chapter 9). Sea otters are considered keystone species because, by preying on sea urchins, they indirectly enhance the presence of kelp, which provides important habitat for many other species. We will consider the role of keystone species in

an idea of the importance of species to communities through both observations and experiments.

There are some large or abundant species, such as trees, that are likely to have large community-wide effects by virtue of providing habitat or food for other species. They may also be good competitors for space, nutrients, or light. These species, known as **foundation species** (Dayton 1971), have large effects on other species, and thus on the species diversity of communities, by virtue of their considerable size and abundance (**FIGURE 16.16**).

Some foundation species act by "bioengineering" their environment. These species, known as **ecosystem engineers** (Jones et al. 1994), are able to create, modify, or maintain physical habitat for themselves and other species. Consider the simple example of the trees mentioned above. Just like any other species, trees provide food for other organisms and compete for resources. However, trees also engineer their environment in subtle but important ways (**FIGURE 16.17**). The trunk, branches, and leaves of a tree provide habitat for a multitude of species, from birds to insects to lichens. The physical structure of the tree reduces sunlight, wind, and rainfall,

Tree leaves, branches, and trunk provide habitat for other species.

Trees affect temperature and moisture by reducing the effects of sun, wind, and rain.

Tree leaves fall to the forest floor, providing habitat for invertebrates, seeds, and microbes.

Fallen trees can serve as nurse logs, providing space, nutrients, and moisture for seedlings.

Tree roots aerate soil and anchor and bind rocks and soils, thus stabilizing the forest floor.

FIGURE 16.17 Trees Are Foundation and Ecosystem Engineering Species
Trees not only provide food for and compete with other species, but also act as ecosystem engineers by creating, modifying, or maintaining physical habitat for themselves and other species. (After C. G. Jones et al. 1997. *Ecology* 78: 1946–1957.)

more detail in Chapter 21 and the Case Study Revisited in Chapter 24.

There are also keystone species that act as ecosystem engineers. A great example is the beaver, a species in which just a few individuals can have dramatic effects on the landscape. Beavers dam streams with cut trees and woody debris. Very quickly, flooding ensues and sediment accumulates as the increasing number of woody obstacles slows the water flow. Eventually, the once swiftly flowing stream is replaced by a wetland, containing plants that can deal with flood conditions; plants that cannot do so, such as trees, are lost from the community. At the landscape level, by creating a mosaic of wetlands within a larger forested community, beavers can increase regional species diversity significantly (**FIGURE 16.18**). Naiman et al. (1988) showed that there was a 13-fold increase in wetland area in one region of Minnesota (from roughly 200 to 2,600 ha) when beavers were allowed to recolonize areas where they had been hunted nearly to extinction some 60 years previously.

Finally, it is worth mentioning that there are species that play only a small role in a community's structure and function. Rather than being keystone species or ecosystem engineers, these species are more like bit players: they contribute to the overall diversity of the community, but their presence or absence has little significance for the ultimate regulation of the community (see Figure 16.16). Some of these species may be **redundant**—that is, they may have the same function in the community as other species within a larger functional group (see Figure 16.4C). Their loss from a community might have little effect as long as other species within the same functional group remain present. We will discuss the role of species in community regulation in more detail in Chapter 19.

Environmental context can change the outcome of species interactions

As we have seen in this section, interactions among species can vary in strength and direction, and their outcomes are highly dependent on the influence of each of the species in the community. As we saw in Unit 4 and Ecological Toolkit 16.1, another important factor in the outcome of species interactions is the environmental context in which they occur. For example, under benign environmental conditions that are favorable for population growth, it makes sense that species will thrive and be limited by resources and will therefore engage in negative interactions such as competition or predation. Under harsh environmental conditions, species will naturally be more strongly limited by physical factors and will therefore interact either weakly or positively with other species.

This view of species interactions as *context dependent*, or changeable under different environmental conditions, is relatively new to ecology, but a number of important examples of context dependence exist. Most of these examples involve keystone or foundation species that play

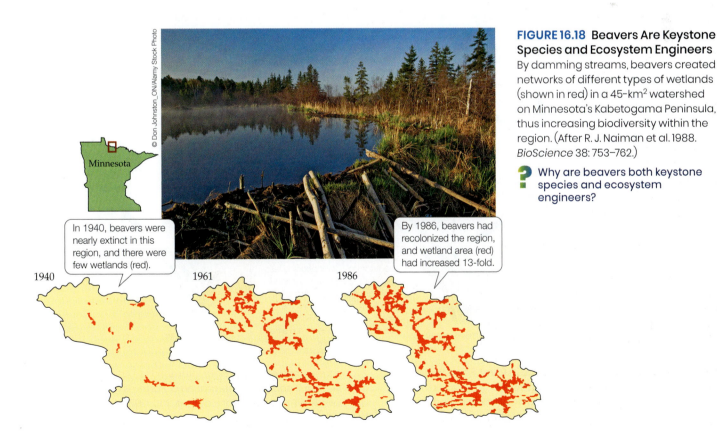

FIGURE 16.18 Beavers Are Keystone Species and Ecosystem Engineers By damming streams, beavers created networks of different types of wetlands (shown in red) in a 45-km² watershed on Minnesota's Kabetogama Peninsula, thus increasing biodiversity within the region. (After R. J. Naiman et al. 1988. *BioScience* 38: 753–762.)

? Why are beavers both keystone species and ecosystem engineers?

© Don Johnston_ON/Alamy Stock Photo

Minnesota

In 1940, beavers were nearly extinct in this region, and there were few wetlands (red).

By 1986, beavers had recolonized the region, and wetland area (red) had increased 13-fold.

1940

1961

1986

important roles in their communities in one context, but not in another. Mary Power, a professor at the University of California, Berkeley, who works on stream communities in Northern California, has shown that the role of fish predators (roach [*Hesperoleucas symmetricus*] and steelhead [*Oncorhynchus mykiss*]) changes from year to year. The role of these predators shifts from that of keystone species following winters of scouring floods to that of weak interactors during years with winter droughts and in places where flood control is operating (Power et al. 2008).

In the Northern California rivers where Power works, there is a natural winter flood regime that produces dramatic population cycles of the green filamentous alga *Cladophora glomerata*. In most years, scouring winter floods remove most of the inhabitants—particularly armored herbivorous insects such as large caddisfly larvae—from the river bottom. In the following spring, there are large blooms of *Cladophora*. Increased light and nutrients and the lack of invertebrate herbivores allow *Cladophora* mats to grow profusely over the rocks, producing filaments up to 8 m (26 feet) long. By midsummer, these mats detach from the rocks and cover large portions of the river, at which time midge larvae, which feed on the floating alga and use it to weave small homes,

increase in number. The midges are fed on by small fish and damselfly larvae, which in turn are eaten by steelhead and roach (a trophic cascade with four levels; **FIGURE 16.19A**). The steelhead and roach are able to decrease the size of the algal mats by eating small fish and damselfly larvae, which feed on midge larvae, which feed on the mats. The roach also feed on the algae directly, but only to a small degree.

During drought years, however, and in rivers where flood control is operating, flooding and scouring of the river bottom do not occur. In those years, *Cladophora* persists but does not form large, lush mats. Power and colleagues showed that this change was due to the presence of more armored herbivorous insects, which were not removed by floods and which ate the *Cladophora* while it was still attached to the rocks. This interaction led to declines in *Cladophora* and the loss of the detachment phase of the alga. The armored insects are much less susceptible to predation than the midges and thus are not controlled by higher trophic levels. In essence, the typical river food web with four trophic levels is converted into a two-level food web during drought years, and the steelhead and roach, which in flood years are keystone predators, become minor players in the food web (**FIGURE 16.19B**).

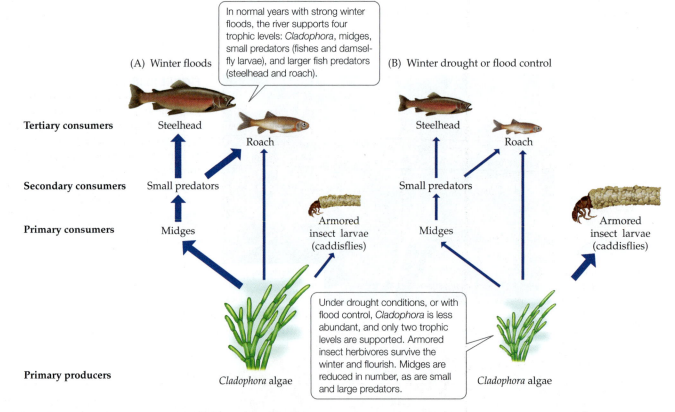

FIGURE 16.19 Context Dependence in River Food Webs Environmental changes alter the relative importance of different trophic levels in the Eel River of Northern California during winter floods (A) and winter droughts or flood control conditions (B). Wider arrows represent stronger interactions. (After M. E. Power et al. 1996a. In *Food Webs: Integration of Patterns and Dynamics*, G. A. Polis and K. O. Winemiller [Eds.], pp. 286–297. Chapman & Hall: New York.)

CLIMATE CHANGE CONNECTION

CONTEXT DEPENDENCE OF OCEAN ACIDIFICATION

One emerging environmental context that will affect communities is ocean acidification. Estimates show that oceans are absorbing about 48% of all atmospheric CO_2. Marine primary producers use some of the human-created CO_2 in photosynthesis, but the remainder reacts chemically with seawater, lowering its pH and causing oceans to become more acidic (see Concept 25.1). Ocean acidification can have negative effects on calcifying organisms such as corals, mollusks, and crustaceans, which rely on calcium carbonate for the accretion and maintenance of their external shells. But the negative effects of increasing CO_2 and acidification are not universal. For example, primary producers such as seagrasses, algae, and phytoplankton are known to increase their productivity under elevated CO_2. Although the effects of ocean acidification on the physiology of single species is a growing field of study, much less is known about how a world with a lower pH will affect the structure and function of communities. Moreover, along with the acidification of oceans, temperatures are increasing, creating multiple stressors. As a result, marine ecologists are asking, How will species interactions, both direct and indirect, be influenced by ocean acidification?

Christian Alsterberg and his colleagues (2013) considered how a food web in an estuarine community on the western coast of Sweden was influenced by ocean acidification and warming. The researchers chose to focus on single-celled microalgae living in the sediments and their interactions with macroalgae and consumers. Under normal conditions in this estuarine system, omnivores (i.e., a guild of medium-sized crustaceans and snails that feed on two trophic levels) increase the productivity of benthic microalgae in two ways: (1) they increase the light available to microalgae by feeding on macroalgae, and (2) they prey on herbivores of microalgae (i.e., a guild of small crustaceans, snails, and worms) (**FIGURE 16.20A**). To estimate the importance of ocean acidification and increased warming in a multispecies food web, the researchers conducted an experiment in which CO_2 and temperature, as well as omnivores, were manipulated in mesocosms (in this case, large aquariums). After a 5-week period, the experiment showed that with elevated CO_2 and temperature, macroalgae and herbivores increased in biomass. When omnivores

(A) Food web under ambient conditions

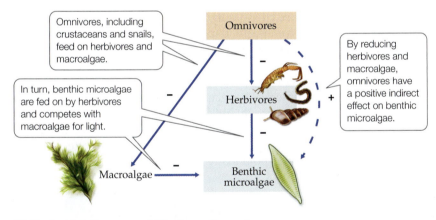

Omnivores, including crustaceans and snails, feed on herbivores and macroalgae.

In turn, benthic microalgae are fed on by herbivores and competes with macroalgae for light.

By reducing herbivores and macroalgae, omnivores have a positive indirect effect on benthic microalgae.

FIGURE 16.20 Food Webs in an Acidic and Warming World (A) The interaction web of species in an estuarine community off the western coast of Sweden. (B) The effects of ocean acidification and warming on the interaction web with (left) and without (right) omnivores. The biomass of benthic microalgae did not change with omnivores (left) but declined without them (right). Thicker arrows represent stronger interactions. (After C. Alsterberg et al. 2013. *Proc Natl Acad Sci USA* 110: 8603–8608.)

(B) Food web under ocean acidification and warming

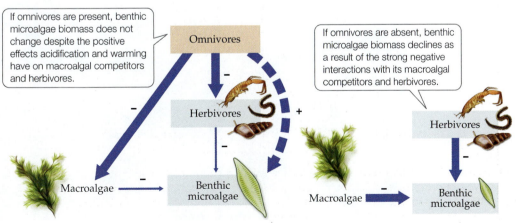

If omnivores are present, benthic microalgae biomass does not change despite the positive effects acidification and warming have on macroalgal competitors and herbivores.

If omnivores are absent, benthic microalgae biomass declines as a result of the strong negative interactions with its macroalgal competitors and herbivores.

were present, this greater biomass was consumed by omnivores, resulting in a stronger negative interaction between them and their food sources (**FIGURE 16.20B**). The stronger negative interaction lead to a stronger indirect positive effect for benthic microalgae. As a result, there was no change in benthic microalgae abundance despite the elevated CO_2 and temperature conditions. However, when omnivores were absent, macroalgae and herbivores were not kept in check, and their negative interactions with benthic microalgae grew even stronger, resulting in the decline of benthic microalgae. Thus, this experiment showed that consumers were able to modulate the negative effects of elevated CO_2 and temperature on primary producers in this estuarine system. This study highlights the importance of understanding how organisms, placed within the context of their food web communities, gain some resilience against the effects of climate change. ☀

In upcoming chapters, we will consider in much more detail the effects of physical and biological factors on the outcome of species interactions and ultimately on the diversity of communities.

A CASE STUDY REVISITED

"Killer Algae!"

The introduction of *Caulerpa taxifolia* into the Mediterranean Sea in the early 1980s set in motion a series of unfortunate events that resulted in large carpets of fluorescent green algae dominating formerly species-rich nearshore marine communities. *Caulerpa* thrived because humans facilitated its dispersal and its physiological tolerance. Even in the early stages of the invasion, Meinesz documented the seaweed in at least three types of communities, with different species compositions, on rocky, sandy, and muddy substrates. Together, these communities are home to several hundred species of algae and three marine flowering plants, as well as a number of animal species. Once *Caulerpa* arrived, native competitor and herbivore species were unable to keep it from spreading.

The invasion of *Caulerpa* has changed the ways in which native species interact with one another, and thus the structure and function of the native communities. One obvious consequence of the presence of *Caulerpa* is the decline of seagrass meadows dominated by *Posidonia oceanica* (**FIGURE 16.21**). This seagrass has been likened to an "underwater tree" because of its long life span and slow growth (patches grow to 3 m in diameter in 100 years). Just like forests, seagrass meadows support a multitude of species that use the vegetation as habitat. Research showed that *Posidonia* and *Caulerpa* have different growth cycles: *Posidonia* loses blades in summer, when *Caulerpa* is most productive. Over time, these asynchronous growth patterns result in *Caulerpa* overgrowing the existing seagrasses and establishing itself as the foundation species. Additional research has shown that *Caulerpa* acts as an ecosystem engineer, accumulating sediments around its roots more readily than *Posidonia*, which can change the species composition of the small invertebrates that live on the seafloor. Some surveys have revealed a significant drop in the numbers and sizes of fish using the communities invaded by *Caulerpa*, suggesting that these habitats may be less suitable for some commercially important species.

Future changes in Mediterranean seagrass meadows, and in the species dependent on them, will be difficult to predict, given the sheer number of species that are potentially affected by *Caulerpa*, the indirect effects that will be generated by changing interactions, and the relatively short time that has elapsed since the invasion began. A scientific approach, guided by a combination of theory and real-world observations, will be necessary if future predictions are to be made about the ultimate effect of *Caulerpa* on this potentially vanishing underwater community.

FIGURE 16.21 A Mediterranean Seagrass Meadow Native communities like this one, dominated by the seagrass *Posidonia oceanica*, can be replaced by invasive *Caulerpa taxifolia*. Compare this photograph with Figure 16.1.

CONNECTIONS *in* NATURE

STOPPING INVASIONS REQUIRES COMMITMENT Even though it may be too late to stop the invasion of *Caulerpa taxifolia* in the Mediterranean, the lessons learned there have been important in other regions of the world. In 2000, just as Meinesz was making progress in banning international trade of the alga, he received an e-mail from an environmental consultant in San Diego, California. While surveying eelgrass in a lagoon, she had noticed a large patch of what was later identified as *Caulerpa taxifolia*. Acting on Meinesz's recommendation, a team of scientists and managers from county, state, and federal agencies immediately assembled to design an eradication plan. This plan involved treating the alga with chlorine gas injected under tarps placed on top of algal patches. More than $1 million was initially budgeted for the project in 2000, but it eventually took 6 years and $7 million to eradicate the alga. The invasion was widely publicized, resulting in the discovery of another patch of *Caulerpa* in another lagoon near Los Angeles, which was also eradicated. The California experience is a rare success story only because immediate action was taken by scientists, managers, and policymakers to deal with the invasion before eradication became an ecological and fiscal impossibility.

To determine the origin of the *Caulerpa* that invaded California, molecular evidence was needed. This shift in the team's focus from communities to genes illustrates a point made in Chapter 1: ecologists must study interactions in nature across many levels of biological organization. The team sent specimens of the alga to geneticists at two universities, who analyzed the sequences of its ribosomal DNA and quickly determined that they were identical to those of *Caulerpa* from the Mediterranean, the Wilhelma Zoo (where the strain was first cultivated), and many other public aquariums around the world (Jousson et al. 2000). Unfortunately, it is still unknown how the species was introduced into the two California lagoons, but hypotheses range from amateur aquarists cleaning their tanks in the lagoons to an accidental release from aquariums on board a Saudi Arabian prince's yacht, which was being repainted in San Diego at about the time the alga probably arrived. Through the use of DNA analysis, it has been determined that the *Caulerpa* algae involved in subsequent invasions in Australia and Japan are genetically identical to the original German *Caulerpa taxifolia* strain. The molecular evidence makes it clear that the trade of this alga in aquarium circles poses a global threat to nearshore temperate marine environments. Legislation is now in place to ban the "killer alga" from a number of other countries where it has a good chance of invading successfully.

SUMMARY

CONCEPT 16.1 Communities are groups of interacting species that occur together at the same place and time.

16.1.1 Describe the ways ecologists delineate communities.

- Communities can be delineated by the characteristics of their physical environment or by biological characteristics, such as the presence of abundant species.

16.1.2 Explain why ecologists use subsets of species to define communities and list some of the subsets used.

- Ecologists often use subsets of species to define and study communities because it is impractical to count or study all the species within a community, especially if they are small or undescribed.

- Subsets of species used to study communities include taxonomic groups, guilds, functional groups, and food and interaction webs.

CONCEPT 16.2 Species diversity and species composition are important descriptors of community structure.

16.2.1 Define and quantify species diversity and compare to biodiversity.

- Species diversity, the most commonly used measure of community structure, is a combination of the number of species (species richness) and the abundances of those species relative to one another (species evenness).

- *Biodiversity* is a term used to describe the diversity of important ecological entities that span multiple spatial scales, from genes to species to communities.

16.2.2 Define and graph rank abundance and species accumulation relationships.

- Communities can differ in the commonness or rarity of their species. Rank abundance curves allow one to plot the proportional abundance of each species relative to the others, from most abundant to least abundant.

- Species accumulation curves can be used to determine when most or all of the species in a community have been observed. Species richness increases with increased sampling effort up to a certain point, at which additional samples reveal few or no new species.

16.2.3 Define species composition and explain why it is an important characteristic of communities.

- Species composition—the identity of the species present in a community—is an obvious but important characteristic of community structure that is not revealed in measures of species diversity.

(Continued)

SUMMARY (continued)

- Knowing species composition allows an understanding of the roles of those species within the community.

CONCEPT 16.3 Communities can be characterized by complex networks of direct and indirect interactions that vary in strength and direction.

16.3.1 Compare direct versus indirect species interactions and describe the effects they have on communities.

- Indirect species interactions, in which the relationship between two species is mediated by a third (or more) species, can have large effects on the outcomes of direct species interactions.

16.3.2 Explain how species interactions can vary in strength and direction, and how these attributes can be measured experimentally.

- Some species have a strong negative or positive effect on their communities, but others probably have little or no effect.

- Interaction strength, the effect of one species on the abundance of another species, can be measured experimentally by removing individuals of the *interactor species* and measuring the effect on the number of individuals or biomass of the *target species*.

16.3.3 Compare foundation species, keystone species, and ecosystem engineers and the effects they have on communities under different environmental contexts.

- Species that have large effects on their communities by virtue of their size and abundance are known as *foundation species*. Those that have large effects due to the roles they play in their communities are known as *keystone species*.

- Ecosystem engineers create, modify, or maintain physical habitat for themselves and other species.

- The environmental context, including climate change, can modify species interactions enough to change their outcome.

Review Questions

1. What is the formal definition of a community? Why is incorporating species interactions into that definition important?

2. Species diversity measurements take into account both species richness and species evenness. Why would these measurements be preferred to species richness alone? What do rank abundance curves add to one's knowledge about community structure?

3. Species vary in the strength of their interactions with other species. Species that interact strongly with other species include foundation species, keystone species, and ecosystem engineers. Describe the differences among these three types of species and give some examples. Can foundation and keystone species also be ecosystem engineers?

Hone Your Problem-Solving Skills

The interaction web shown in the figure is representative of one found in the rocky intertidal zone of Washington and Oregon. The arrows and their signs (+ or −) represent interactions that occur between species in the food web.

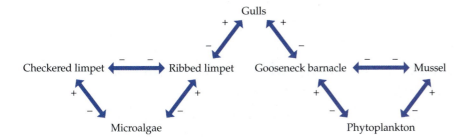

Suppose an ecologist wants to know the influence of gulls on other members of the food web. She conducts a gull removal experiment along a stretch of rocky shoreline to determine the interaction strength between gulls and other members of the food web (with the exception of phytoplankton). The results are shown in the table. Determine the interaction strength values between gulls and each of the five species given in the table, using the relative interaction intensity (RII) equation in Ecological Toolkit 16.1.

Target species	Control (C) number of individuals with gulls	Experimental (E) number of individuals without gulls
Ribbed limpet	10	100
Gooseneck barnacle	500	3,000
Checkered limpet	100	50
Mussel	3,000	2,500
Microalgae	500	100

1. Based on your relative interaction intensity (RII) calculations, which species of prey is most negatively affected by gulls? Which nonprey species is most affected by gulls, and what is that effect?

2. Determine the four indirect effects that occur when gulls are removed from the community. How does each species respond (i.e., does it increase, decrease, or not change in abundance) and through what type of interaction (i.e., herbivory or competition)?

3. Would the effect of gulls be stronger or weaker on microalgae and phytoplankton if the mussel and checkered limpet were excluded from the food web? Explain.

17

Change in Communities

© Steve Terrill/Getty Images

A Natural Experiment of Mountainous Proportions: A Case Study

The eruption of Mount St. Helens was a defining moment for ecologists interested in natural catastrophes. Mount St. Helens, located in Washington State, is part of the geologically active Cascade Range, located in the Pacific Northwest region of North America (**FIGURE 17.1**). The once frosty-topped mountain had a rich diversity of ecological communities. If you had visited Mount St. Helens in the summer, you could have seen alpine meadows filled with colorful wildflowers and grazing elk. At lower elevations, you could have hiked across the cool fern- and moss-covered forest floor under massive old-growth trees. You could have swum in the blue, clear water of Spirit Lake, or fished along its shores. But a few minutes after 8:30 a.m. on May 18, 1980, all that was living on Mount St. Helens would be gone. On the north side of the mountain, a huge magma-filled bulge had been forming for months. The bulge gave way that morning in an explosive eruption and the largest avalanche in recorded history.

Photos of the eruption show that mud and rock flowed down the face of Mount St. Helens and were deposited tens of meters deep in some areas (**FIGURE 17.2**). The wave of debris that passed over Spirit Lake was 260 m (858 feet) deep and decreased the lake's water depth by 60 m (200 feet). The bulk of the avalanche traveled 23 km (14 miles) in about 10 minutes to the North Fork Toutle River, where it scoured the entire valley, from floor to rim, with material from the volcano and left a truly massive pile of tangled vegetation at its tail end. In addition to the avalanche, the blast produced a cloud of hot air that burned forests to ash near the mountain, blew down trees over a large area, and left dead but standing trees stretching for miles away from the mountain. Ash from the explosion blanketed forests, grasslands, and deserts located hundreds of kilometers away.

The destruction that ensued on that day created whole new habitats on Mount St. Helens, some of which were completely devoid of any living organisms. At one extreme, there was the Pumice Plain, a large, gently sloping moonscape of a place below the volcano, that had been pelted with hot, sterilizing pumice (see Figure 17.2). This harsh and geologically monotonic environment lacked life, or even organic matter, of any form. All life in Spirit Lake was extinguished, and huge amounts of woody debris were deposited there, some of which still floats on top of the lake today. But, not surprisingly, given the large forests that had surrounded the mountain, the majority of the landscape consisted of downed or denuded trees covered with rock, gravel, and mud tens of meters deep in some places (see Figure 17.2).

FIGURE 17.1 Once a Peaceful Mountain Before the eruption on May 18, 1980, Mount St. Helens, in southwestern Washington State, had a diversity of communities, including alpine meadows, old-growth forests, and lakes and streams.

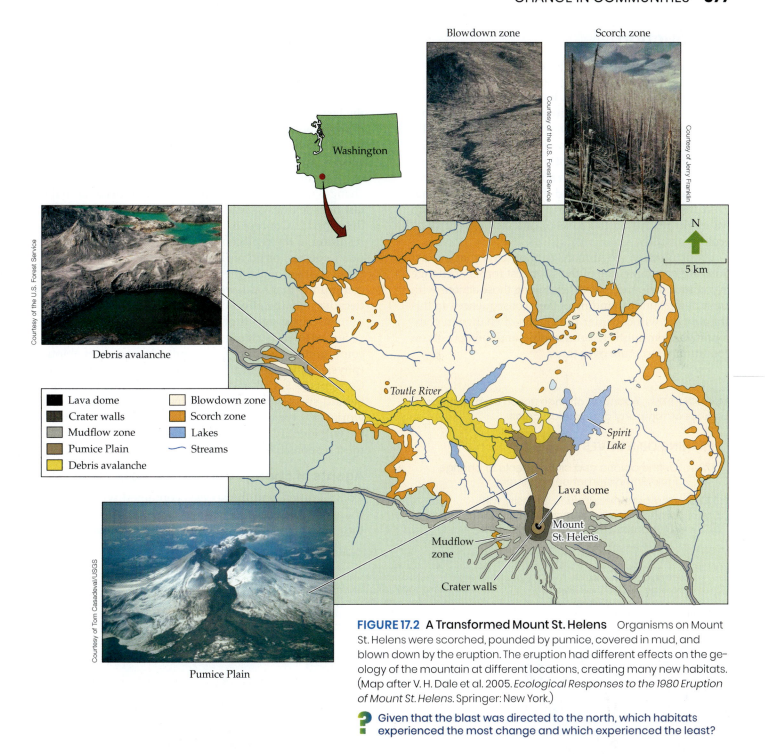

Blowdown zone

Scorch zone

Washington

N

5 km

Courtesy of the U.S. Forest Service

Courtesy of Jerry Franklin

Debris avalanche

Courtesy of the U.S. Forest Service

⬛ Lava dome	⬜ Blowdown zone
Crater walls	🟧 Scorch zone
Mudflow zone	🟦 Lakes
Pumice Plain	〜 Streams
🟨 Debris avalanche	

Toutle River

Spirit Lake

Lava dome

Mount St. Helens

Mudflow zone

Crater walls

Pumice Plain

Courtesy of Tom Casadevall/USGS

FIGURE 17.2 **A Transformed Mount St. Helens** *Organisms on Mount St. Helens were scorched, pounded by pumice, covered in mud, and blown down by the eruption. The eruption had different effects on the geology of the mountain at different locations, creating many new habitats.* (Map after V. H. Dale et al. 2005. *Ecological Responses to the 1980 Eruption of Mount St. Helens.* Springer: New York.)

❓ Given that the blast was directed to the north, which habitats experienced the most change and which experienced the least?

Compared with the Pumice Plain, this blowdown zone had some hope of a biological legacy buried under the piles of trees and ash.

Shortly after the eruption, helicopters delivered the first scientists to the mountain to begin studying what was essentially a natural experiment of epic proportions. A few lucky ecologists recorded the first observations of the sequence of biological changes that began soon after the eruption. Field excursions in the summers of 1980 and 1981 were organized, and valuable baseline data were collected. Now, more than 40 years later, hundreds of ecologists have studied the reemergence of life on Mount St. Helens. For many, the experience has been life-changing, and their careers have been consumed by research on this fascinating study system. Much of what has been learned has been unexpected and has changed the way we view the recovery of communities and the persistence of life on Earth.

Introduction

One constant that all ecologists can agree on is that communities are always changing. Some communities show more dynamism than others. For example, it is hard to imagine that desert communities, with their large, stoic cacti, have changed much over time. This is especially true if you compare deserts with, for example, high mountain streams or rocky intertidal zones, where species are coming and going on a regular basis. But community change is relative, and there is no question that even deserts change, though at a much slower pace than we might realize on the basis of one visit, or even one ecological study.

Unfortunately, we humans cannot deny that our actions are becoming one of the strongest forces of change in communities, and that we are taking those actions with an imperfect understanding of their consequences. In this chapter, we will consider the agents of change in communities, from subtle to catastrophic, and their effects on community structure over time.

CONCEPT 17.1 Agents of change act on communities across all temporal and spatial scales.

LEARNING OBJECTIVES

17.1.1 Define abiotic and biotic agents of change including disturbance and stress.

17.1.2 Compare disturbance intensity with disturbance frequency and describe their differential effects on communities.

Agents of Change

Let's imagine for a moment that you have the ability to look back in time and follow the change in a typical coral reef community in the Indian Ocean (**FIGURE 17.3**). Over the last few decades, you might have seen considerable change, both subtle and catastrophic. Subtle changes might include the slow rise to dominance of certain coral species, and the slow decline of others, due to the effects of competition, predation, and disease. More catastrophic changes might include the massive deaths of corals in the last decade due to bleaching (loss of symbiotic algae, as described in Concept 3.3) and the great tsunami of 2004, resulting in the replacement of some coral species with other species, or no replacement at all. Taken together, these changes make the community what it is today: a community that has fewer coral species than it did a few decades ago, the effect of a combination of natural and human-caused agents of change.

Succession is change in the species composition of communities over time as a result of a variety of abiotic (physical and chemical) and biotic agents of change. In Concepts 17.2, 17.3, and 17.4, we will consider the theory behind succession and examples that illustrate how it works in a variety of systems. But first, in this section, we will identify and define the agents of change that are most responsible for driving succession.

Agents of change can be abiotic or biotic

Communities, and the species contained within them, change in response to a number of abiotic and biotic factors (**TABLE 17.1**). We have considered many of these factors

Species interactions, such as competition, predation, and disease, can cause the gradual replacement of species over time.

Changes in abiotic conditions, such as sea level rise and warmer water temperatures, can cause physiological stress, coral bleaching, and eventually mortality.

Catastrophic disturbances, such as tsunamis and blast fishing, can cause massive injury and death in coral reefs.

FIGURE 17.3 Change Happens Coral reef communities in the Indian Ocean have experienced large changes over the last few decades. The agents of change have been both subtle and catastrophic, natural and human-caused.

TABLE 17.1

Examples of Abiotic and Biotic Agents of Stress, Disturbance, and Change in Communities

Agent of change	Examples
Abiotic factors	
Waves, currents	Storms, hurricanes, floods, tsunamis, ocean upwelling
Wind	Storms, hurricanes and tornados, wind-driven sediment scouring
Water supply	Droughts, floods, mudslides
Chemical composition	Pollution, acid rain, high or low salinity, high or low nutrient supply
Temperature	Freezing, snow and ice, avalanches, excessive heat, fire, sea level rise or fall
Volcanic activity	Lava, hot gases, mudslides, flying rocks and debris, floods, changes in weather or climate
Biotic factors	
Negative interactions	Competition, predation, herbivory, disease, parasitism, trampling, digging, boring

Source: Adapted, with additions, from W. P. Sousa. 2001. In *Marine Community Ecology*, M. D. Bertness et al. (Eds.), pp. 85–130. Oxford University Press/Sinauer: Sunderland, MA.

in previous chapters. In Unit 1, we learned that abiotic factors, in the form of climate, soils, nutrients, and water, vary over daily, seasonal, decadal, and even 100,000-year time scales. This variation has important implications for community change. For example, in Indian Ocean coral reef communities (see Figure 17.3), unusually high water temperatures driven by large-scale climate change have been implicated in recent losses of symbiotic algae from corals, resulting in coral bleaching. If the symbiotic algae do not return, the corals will eventually die, thus creating the conditions for species replacement. Likewise, increases in sea level can decrease the amount of light that reaches the corals. If light availability falls below the physiological limits of some coral species, they could slowly be replaced by more tolerant species, or even by macroalgae (seaweeds). Finally, increasing ocean acidification can dissolve the skeletons of corals, hindering their growth (see Chapter 25 and Chapter 16's Climate Change Connection for more information on climate change and ocean acidification). Because these abiotic conditions are constantly changing, communities are doing the same, at a pace consistent with their environment.

Abiotic agents of change can be placed into two categories, both of which can have either natural or human origins, but which differ in the effects they have on species: disturbances and stresses. A **disturbance** is an abiotic

event that physically injures or kills some individuals and creates opportunities for other individuals to grow or reproduce. Some ecologists also consider biotic events such as digging by animals to be disturbances. In our coral reef example, the 2004 tsunami can be viewed as a disturbance because the force of water passing over the reef injured and killed many coral individuals. Likewise, the outlawed practice of blast fishing, which involves using dynamite to stun or kill fish for easy collection, can cause massive injury and death in coral reefs. Even biotic events such as coral boring by snails or predation by parrot fishes can be considered disturbances because they remove coral tissue and weaken coral skeletons. *Stress*, on the other hand, occurs when some abiotic factor reduces the growth, reproduction, or survival of individuals and creates opportunities for other individuals. A stress in our coral reef might be the effect of warmer water temperatures or sea level rise on the growth, reproduction, or survival of corals. Examples of other stresses and disturbances are included in Table 17.1. Both disturbance and stress are believed to play critical roles in driving succession.

How do biotic factors influence community change? In Unit 4, we saw that species interactions, both negative and positive, can result in the replacement of one species with another through stress and disturbance. In our coral reef (see Figure 17.3), change might be driven by competition between, for example, platelike corals and branched corals, with the platelike forms eventually dominating over time. Coral diseases are another example of a species interaction that can initiate change in communities by causing particular coral species to grow more slowly or eventually die. Equally common agents of change are the actions of ecosystem engineers and keystone species (see Figures 16.16 and 16.17). Both types of species have large effects on other species that result in community change.

Finally, it is important to realize that abiotic and biotic factors often interact to produce change in communities. We can see this interaction in the case of ecosystem engineers such as beavers, which cause changes in abiotic conditions that in turn cause species replacement (see Figure 16.18). Similarly, abiotic factors such as wind, waves, or temperature can act by modifying species interactions, either positively or negatively, thus creating opportunities for other species. We have seen examples of this kind of effect on sea palms in the rocky intertidal zone (see Figure 14.18), plants in alpine regions (see Figure 15.9), and stream insects in Northern California (see Figure 16.19).

Agents of change vary in their intensity, frequency, and extent

As you might guess, the tempo of succession is largely determined by how often, at what magnitude, and to what areal extent agents of change act. For example, when the avalanche produced by Mount St. Helens ripped through the alpine community back in 1980, it produced a disturbance that was larger and more severe than any others that had occurred that year, that decade, or that century. The *intensity*, or severity, of that disturbance—the amount of damage and death it caused—was huge, both because of the massive physical force involved and because of the area covered. In contrast, the *frequency* of that kind of disturbance is low because such eruptive episodes are so rare (occurring once every few centuries). Extremely intense and infrequent events, such as the eruption of Mount St. Helens, are at the far end of the spectrum of disturbances organisms experience in communities (**FIGURE 17.4**). In this case, the entire community is affected, and recovery involves the complete reassembly of the community over time.

At the other end of the spectrum are weak and frequent disturbances that may have more subtle effects or affect a smaller area (see Figure 17.4). Prior to the eruption of Mount St. Helens, such disturbances might have included wind blowing down old trees living in the Douglas fir forests surrounding the mountain. These more frequent disturbances open up patches of resources that can be used by individuals of the same or different species. A mosaic of disturbed patches can promote species diversity in communities over time but may not lead to much successional change.

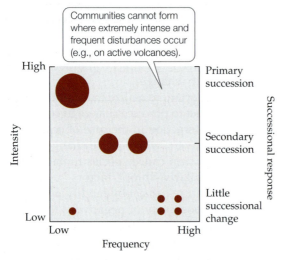

FIGURE 17.4 The Spectrum of Disturbance How much biomass is removed (the intensity, or severity, of disturbance) and how often it is removed (the frequency of disturbance) can influence the amount of change (represented by the size of the red circles) that occurs and the type of succession that is possible afterward (right side of the graph).

❓ Describe how the type of organism being studied might influence whether we classify a disturbance as being intense or frequent.

CLIMATE CHANGE CONNECTION

VOLCANOES AND CLIMATE AS AN AGENT OF CHANGE The eruption of Mount St. Helens not only created a massive disturbance to land, it also affected the atmosphere. An ash cloud roiled as high as 18 km (11 miles) and began to drift eastward in the atmospheric winds (see Figure 17.1). After the eruption, climate scientists Clifford Mass and Alan Robock (1982) were interested in whether the ash and gases propelled into the atmosphere affected local climate and, if so, for how long. Volcanic eruptions are known to have local to global effects on climate. For example, the 1883 eruption on the island of Krakatoa in Indonesia (see Figure 18.22), one of the most deadly and destructive volcanic events, injected a huge amount of sulfur dioxide (SO_2) gas into the stratosphere (80 km [50 miles] high), which was then carried by winds around the globe. The sulfur dioxide gas interacted with water vapor, producing sulfuric acid, which can partially reflect solar radiation. In the year following the eruption, the average temperature in North America dropped by 0.4°C (0.72°F). A similar decrease in global temperatures (0.5°C [1°F]) from the eruption of Mount Pinatubo occurred in the Philippines in 1992.

What did Mass and Robock (1982) find out about the eruption of Mount St. Helens on climate? The blanket of ash cooled Eastern Washington and Idaho by 8°C (15°F) during the day and warmed it by the same amount at night. But the local temperature effects lasted only one day as the volcanic dust cloud dissipated and moved east. In addition, because the volcano did not produce a high-elevation plume of sulfur dioxide gas, there was no long-term or global influence on climate. Thus, the consequences of volcanic explosions on the atmosphere, and the potential ensuing changes to local or global climate, vary considerably depending on the intensity, frequency, and extent of the eruption. ☀

Let's turn our attention from the intensity, frequency, and extent of agents of change to their consequences for community succession.

CONCEPT 17.2 Succession is the process of change in species composition over time as a result of abiotic and biotic agents of change.

LEARNING OBJECTIVES

17.2.1 Compare primary succession with secondary succession.

17.2.2 Discuss the early research and differential views of early ecologists with regard to succession.

17.2.3 Outline the multiple models of succession and compare their differences.

The Basics of Succession

At the most basic level, the term "succession" refers to the process by which the species composition of a community changes over time. Mechanistically, succession involves colonization and extinction due to abiotic and biotic agents of change. Even though studies of succession often focus on changes in vegetation, the roles of animals, fungi, bacteria, and other microbes are equally important.

Theoretically, succession progresses through various stages that include a **climax stage** (**FIGURE 17.5**). The climax is thought to be a stable end point that experiences little change until a particularly intense disturbance sends the community back to an earlier stage. As we will see in Concepts 17.3 and 17.4, there is some argument about whether succession can ever lead to a stable end point.

Primary succession and secondary succession differ in their initial stages

Ecologists recognize two types of succession that differ in their initial stages. The first type, **primary succession**, involves the colonization of habitats that are devoid of life (see Figure 17.5), either as a result of catastrophic disturbance, as we see on the Pumice Plain at Mount St. Helens, or because they are newly created habitats, such as volcanic rock. As you can imagine, primary succession can be very slow because the first arrivals (known as *pioneer* or *early successional* species) typically face extremely inhospitable conditions. Even the most basic resources needed to fuel life, such as soil, nutrients, and water, may be lacking. The first colonizers, then, tend to be species that are capable of withstanding great physiological stress and transforming the habitat in ways that benefit their further growth and expansion (and that of other species, as we will see).

The other type of succession, known as **secondary succession**, involves the reestablishment of a community in which most, but not all, of the organisms or organic constituents have been destroyed (see Figure 17.5). Agents of change that can create such conditions include fire, hurricanes, logging, and herbivory. Despite the catastrophic effect of the eruption on Mount St. Helens, there were many areas, such as the blowdown zone, where some organisms survived and secondary succession took place. As you might expect, the legacy of the preexisting species and their interactions with colonizing species can play a large role in the trajectory of secondary succession.

The early history of ecology is a study of succession

The modern study of ecology had its beginnings at the turn of the twentieth century. At that time, it was dominated by scientists who were fascinated with plant communities and the changes they undergo over time. One of these pioneers was Henry Chandler Cowles, who studied the successional sequence of vegetation in sand dunes on the shore of Lake Michigan (**FIGURE 17.6**). In this ecosystem, the dunes are continually growing as new sand is deposited at the shoreline. This new sand is blown onshore when shorelines are exposed during droughts. Cowles was able to infer the successional pattern along a dune by assuming that the plant assemblages farthest from the lake's edge were the oldest and that the ones nearest the lake, where new sand was being deposited, were the youngest. As you walked from the lake to the back of the dune, he believed, you were traveling forward in time and able to imagine what the areas you had just passed through would look like in centuries to come. The first stages were dominated by a hardy ecosystem engineer, American beach grass (*Ammophila breviligulata*). *Ammophila* (whose genus name literally means "sand lover") is excellent at trapping sand and creating hills, which provide refuge on their leeward side for plants less tolerant of the constant burial and sand scouring experienced on the beachfront.

Cowles (1899) made the assumption that the different plant assemblages—or "societies," as he called them—that he saw in different positions on a dune

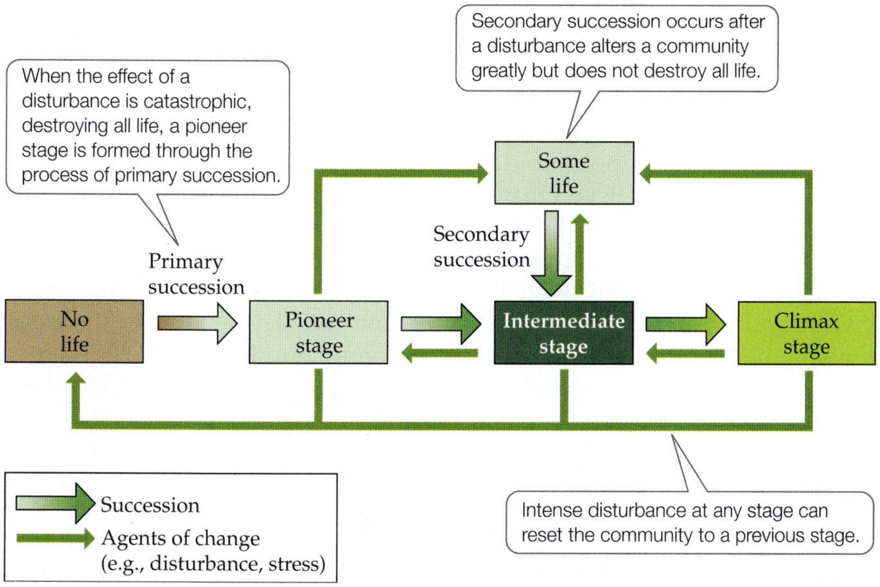

FIGURE 17.5 The Trajectory of Succession A simple model of succession involves transitions between stages driven by species replacements over time. Theoretically, these changes ultimately result in a climax stage that experiences little change. There is some argument, however, about whether succession can ever lead to a stable end point.

FIGURE 17.6 Space for Time Substitution (A) The portion of a dune nearest the shoreline on Lake Michigan is covered with *Ammophila*. (B) When Henry Chandler Cowles studied succession on these dunes, he assumed that the earliest successional stages occurred on the newly deposited sand at the front of the dune, and that later successional stages occurred at the back of the dune.

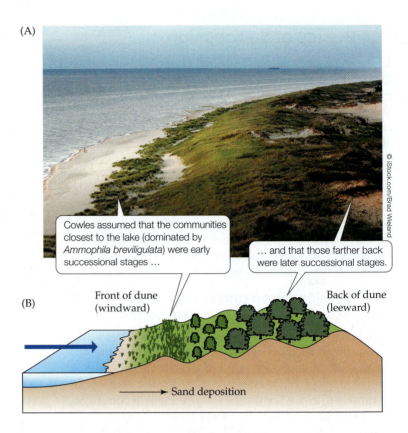

(A)

(B)

Cowles assumed that the communities closest to the lake (dominated by *Ammophila breviligulata*) were early successional stages …

… and that those farther back were later successional stages.

Front of dune (windward)

Back of dune (leeward)

Sand deposition

represented different successional stages. That assumption allowed him to predict how a community would change over time without actually waiting for the pattern to unfold, which would have taken decades to centuries. This idea, known as the "space for time substitution" (Pickett 1989), is used frequently as a practical way to study communities over time scales that exceed the life span of an ecologist. It assumes that time is the main factor causing communities to change and that unique conditions in particular locations are inconsequential. These are big assumptions, and they have fueled a debate about the predictability of community dynamics over time. We will discuss this debate in more detail in Concept 17.4, when we deal with alternative stable state theory.

Henry Cowles was not alone in his interest in plant succession. His peers included Frederick Clements and Henry Gleason, two men who had completely different and contentious views on the mechanisms driving succession (Kingsland 1991). Clements, who in 1907 wrote one of the first formal books on the new science of ecology, believed that plant communities were like "superorganisms," groups of species that worked together in a mutual effort toward some deterministic end. Succession was similar to the development of an organism, complete with a beginning (embryonic stage), middle (adult stage), and end (death). Clements (1916) thought that each community had its own predictable life history and, if left undisturbed, ultimately reached a stable end point. This "climax community" was composed of species that dominated and persisted over many years and provided the type of stability that could potentially be maintained indefinitely.

Gleason (1917) thought that viewing a community as an organism, with various interacting parts, ignored the responses of individual species to prevailing conditions. In his view, communities were not the predictable and repeatable result of coordinated interactions among species, but rather the random product of fluctuating environmental conditions acting on individual species. Each community was the product of a particular place and time and was thus unique in its own right.

Looking back, it is clear that Gleason and Clements had extreme views of succession. As we will see in the next section, we can find elements of both theories in the results of studies that have accumulated over the last century. First, however, it is important to mention one last ecologist, Charles Elton (**FIGURE 17.7A**), whose perspective on succession was shaped not only by those of the botanists who came before him, but also by his interest in animals. He wrote his first book, *Animal Ecology* (1927), in 3 months' time at the age of 26. The book addresses many important ideas in ecology, including succession. Elton believed that organisms and the environment interact to shape the direction succession will take. He presented an example from pine forests in England that were being subjected to deforestation. After the felling of the pines, the trajectory of succession varied depending on the moisture content of the environment (**FIGURE 17.7B**). Wetter areas developed into sphagnum bogs, while slightly drier areas developed into wetlands containing rushes and grasses. Eventually, these communities all became birch scrub, but then ultimately diverged into two types of forest. Through these observations, Elton demonstrated that the only way to predict the trajectory of succession was to understand the biological and environmental context in which it occurred.

Elton's greatest contribution to the understanding of succession was his acknowledgment of the role of animals.

(A)

Courtesy of Charles Elton

(B)

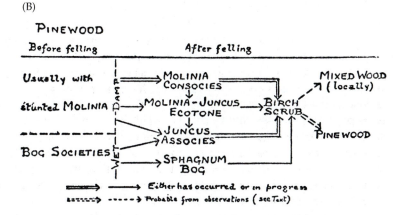

FIGURE 17.7 Elton's Context-Dependent View of Succession
(A) Charles Elton at the age of 25, a year before the publication of his first book, *Animal Ecology* (1927). (B) Elton's book contained this diagram of succession in pine forests after logging. The successional trajectory differed depending on the moisture content of a particular area: wetter areas became sphagnum bogs, while slightly drier areas became wetlands containing rushes (*Juncus*) and grasses (*Molinia*). Eventually, these communities all became birch scrub but then ultimately diverged into pine woods or mixed woods, again depending on moisture. (B from V. S. Summerhayes and P. H. Williams. 1926. *J Ecol* 14: 203–243.)

Up to that point, most ecologists believed that plants drove succession, while animals were passive followers. Elton provided many examples showing how animals could create successional patterns by eating, dispersing, trampling, and destroying vegetation in ways that greatly affected the sequence and timing of succession. We will review some examples of animal-driven succession in the next section, but it is clear that the observations and conclusions Elton made 90 years ago still hold today.

Multiple models of succession were stimulated by lack of scientific consensus

Fascination with the mechanisms responsible for succession, and attempts to integrate the controversial theories of Clements, Cowles, and Elton, led ecologists to use more scientifically rigorous methods to explore succession, including comprehensive reviews of the literature and manipulative experiments. Joseph Connell and his collaborator Ralph Slatyer (1977) surveyed the literature and proposed three models of succession that they believed to be important (**FIGURE 17.8**).

• The *facilitation model*, inspired by Clements, describes situations in which the earliest colonizers modify the environment in ways that ultimately benefit later-arriving species but hinder their own continued dominance. These early successional species have characteristics that make them good at colonizing open habitats, dealing with physical stress, growing quickly to maturity, and ameliorating the harsh physical conditions often characteristic of early successional stages. Eventually, however, a sequence of species facilitations leads to a climax community composed of species that no longer facilitate other species and are displaced only by disturbances.

• The *tolerance model* also assumes that the earliest colonizers modify the environment, but in neutral ways that neither benefit nor inhibit later species. These early successional species have life history strategies that allow them to grow and reproduce quickly. Later species persist merely because they have life history strategies such as slow growth, few offspring, and long life that allow them to tolerate increasing environmental or biological stresses that would hinder early successional species.

• The *inhibition model* assumes that early successional species modify the environment in ways that hinder later successional species. For example, these early colonizers may monopolize resources needed by subsequent species. This suppression of the next stage of succession is broken only when stress or disturbance decreases the abundance of the inhibitory species. As in the tolerance model, later species persist merely because they have life history strategies that allow them to tolerate environmental or biological stresses that would otherwise hinder early successional species.

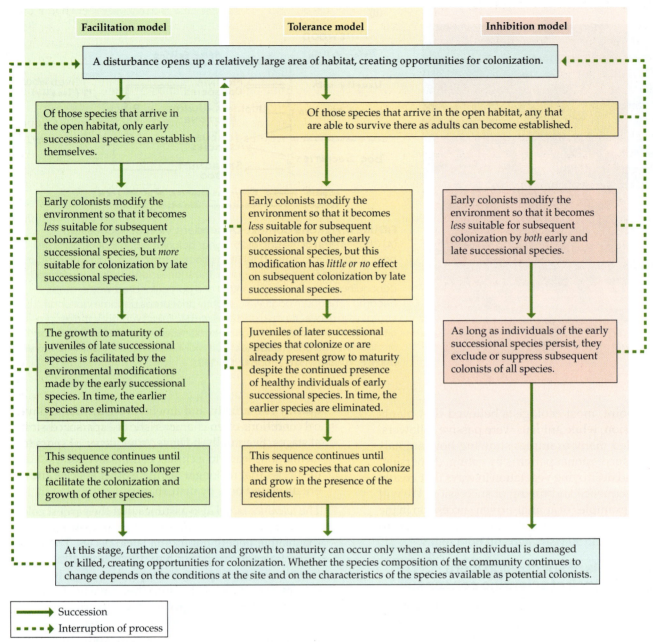

Facilitation model	Tolerance model	Inhibition model

A disturbance opens up a relatively large area of habitat, creating opportunities for colonization.

Facilitation model

Of those species that arrive in the open habitat, only early successional species can establish themselves.

Early colonists modify the environment so that it becomes *less* suitable for subsequent colonization by other early successional species, but *more* suitable for colonization by late successional species.

The growth to maturity of juveniles of late successional species is facilitated by the environmental modifications made by the early successional species. In time, the earlier species are eliminated.

This sequence continues until the resident species no longer facilitate the colonization and growth of other species.

Tolerance model

Of those species that arrive in the open habitat, any that are able to survive there as adults can become established.

Early colonists modify the environment so that it becomes *less* suitable for subsequent colonization by other early successional species, but this modification has *little or no* effect on subsequent colonization by late successional species.

Juveniles of later successional species that colonize or are already present grow to maturity despite the continued presence of healthy individuals of early successional species. In time, the earlier species are eliminated.

This sequence continues until there is no species that can colonize and grow in the presence of the residents.

Inhibition model

Early colonists modify the environment so that it becomes *less* suitable for subsequent colonization by *both* early and late successional species.

As long as individuals of the early successional species persist, they exclude or suppress subsequent colonists of all species.

At this stage, further colonization and growth to maturity can occur only when a resident individual is damaged or killed, creating opportunities for colonization. Whether the species composition of the community continues to change depends on the conditions at the site and on the characteristics of the species available as potential colonists.

→ Succession
┄┄▸ Interruption of process

FIGURE 17.8 Three Models of Succession Connell and Slatyer proposed three conceptual models—the facilitation, tolerance, and inhibition models—to describe succession. (After J. H. Connell and R. O. Slatyer. 1977. *Am Nat* 111: 982.)

CONCEPT 17.3 Experimental work on succession shows its mechanisms to be diverse and context dependent.

LEARNING OBJECTIVES

17.3.1 Analyze multiple studies to understand the diverse mechanisms involved in primary and secondary succession.

17.3.2 Summarize the results of experiments designed to determine the mechanisms of succession.

Mechanisms of Succession

More than 40 years have gone by since Connell and Slatyer wrote their influential theoretical paper on succession. Since that time, there have been a number of experimental tests of their three models. Those studies show that the mechanisms driving succession rarely conform to any one model, but instead are dependent on the community and the context in which experiments are conducted.

No one model fits any one community

To illustrate the types of successional mechanisms that have been revealed by experiments, we will focus on three studies: communities that form (1) after glacial retreat in Alaska, (2) after vegetation disturbance in salt marshes in New England, and (3) after wave disturbance in the rocky intertidal zone of the U.S. Pacific coast.

PRIMARY SUCCESSION IN GLACIER BAY, ALASKA One of the best-studied examples of primary succession occurs in Glacier Bay, Alaska, where the melting of glaciers has led to a sequence of community change that reflects succession over many centuries (**FIGURE 17.9**). Captain George Vancouver first recorded the location of glacial ice there in 1794, while exploring the west coast of North America. Over the last 200 years, the glaciers have retreated up the bay, leaving behind bare, broken rock (known as *glacial till*). John Muir, in his book *Travels in*

Alaska (1915), first noted how much the glaciers had melted since Vancouver's time. When he visited Glacier Bay in 1879, he camped among ancient tree stumps that had once been covered by ice and saw forests that had grown up in previously glaciated areas. He was impressed with the dynamic nature of the landscape and how the plant community responded to the changes.

Muir's book sparked the interest of William S. Cooper (1923a), who began his studies of Glacier Bay in 1915. A former student of Henry Chandler Cowles, Cooper saw Glacier Bay as an example of the "space for time substitution" so well documented by his advisor in the Lake Michigan dunes. He established permanent plots (Cooper 1923b) that have allowed researchers to observe the pattern of community change along the bay from Vancouver's time to today. This pattern is generally characterized by an increase in plant species richness and a change in plant species composition with time and distance from the melting ice front (**FIGURE 17.10**). In the first years after new habitat is exposed, a primary or **pioneer stage** develops, dominated by a few species that include lichens, mosses, horsetails, willows, and cottonwoods. Roughly 30 years after exposure, a second community develops, named the *Dryas* stage after the small shrub (*Dryas drummondii*) that dominates this community. In this stage, species richness increases, with willows, cottonwoods, alders (*Alnus sinuata*), and Sitka spruce (*Picea sitchensis*) sparsely distributed among the carpet of *Dryas*. After about 50 years (or

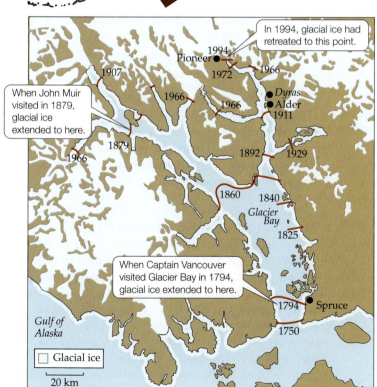

When John Muir visited in 1879, glacial ice extended to here.

In 1994, glacial ice had retreated to this point.

When Captain Vancouver visited Glacier Bay in 1794, glacial ice extended to here.

☐ Glacial ice

20 km

FIGURE 17.9 Glacial Retreat in Glacier Bay, Alaska
Over more than 200 years, the melting of glaciers has exposed bare rock to colonization and succession. (After F. S. Chapin et al. 1994. *Ecol Monogr* 64: 149–175.)

❓ Based on the locations of the glaciers over time, describe where the oldest and youngest communities are located.

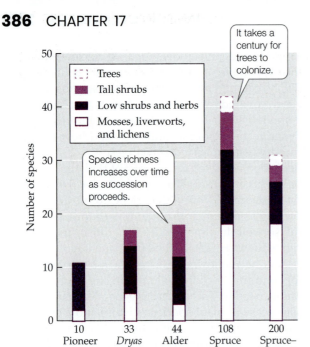

FIGURE 17.10 **Successional Communities at Glacier Bay, Alaska** Plant species richness has generally increased over the 200 years following glacial retreat. (After W. A. Reiners et al. 1971. *Ecology* 52: 55–69.)

some 20 km from the ice front), alders dominate, forming the third community, referred to as the alder stage. Finally, a century after glacial retreat, a mature Sitka spruce forest (the spruce stage) is in place, which fosters a diverse array of lichens, low shrubs, and herbs. Reiners et al. (1971) documented that 200 years after exposure, species richness decreases somewhat as Sitka spruce forests are transformed into forests of longer-living western hemlocks.

The mechanisms underlying succession in this system have been studied extensively by F. Stuart Chapin and colleagues (1994). They wondered, given the harsh physical conditions experienced by most species in the pioneer stage, whether the facilitation model could explain the pattern of succession observed by Cooper and Reiners et al. First, they analyzed the soils of the different successional stages. They found significant changes in soil properties that were coincident with the increases in plant species richness (**FIGURE 17.11**). Not only were there increases in soil organic matter and soil moisture in later stages of succession, but nitrogen increased more than fivefold from the alder stage to the spruce stage. (This increase resulted from the action of nitrogen-fixing bacteria associated with plant roots, which we'll describe in more detail in this chapter's Connections in Nature.) Chapin hypothesized that the assemblage of species at each stage of succession was having effects on the physical environment that largely shaped the pattern of community formation. The question remained, however, whether those effects were facilitative or inhibitory, and how they varied across the different successional stages.

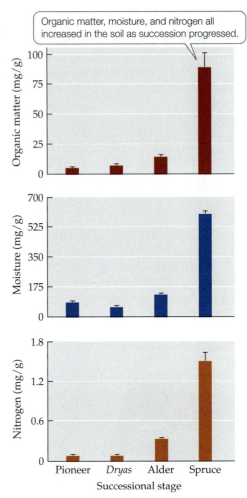

FIGURE 17.11 **Soil Properties Change with Succession** Chapin and colleagues studied the properties of the soils in each of four successional stages at Glacier Bay. Error bars show one standard error of the mean. (After F. S. Chapin et al. 1994. *Ecol Monogr* 64: 149–175.)

To test their facilitation hypothesis, Chapin et al. (1994) conducted manipulative experiments. They added spruce seeds to each of the successional stages and observed their germination, growth, and survival over time. These experiments, along with observations of unmanipulated plots, showed that neighboring plants had both facilitative and inhibitory effects on the spruce seedlings but that the directions and strengths of those effects varied with the stage of succession (**FIGURE 17.12**). For example, in the pioneer stage, spruce seedlings had a low germination rate, but a higher survival rate, than in later successional stages. In the *Dryas* stage, spruce seedlings had low germination and survival rates due to increases in seed predators, but those individuals that did survive grew better because of the presence of nitrogen fixed by symbiotic bacteria associated with *Dryas*. In the alder stage, a further increase in nitrogen (alders also host nitrogen-fixing bacteria) and an increase in soil organic matter had positive effects on spruce seedlings, but shading and seed predators led to overall low germination and

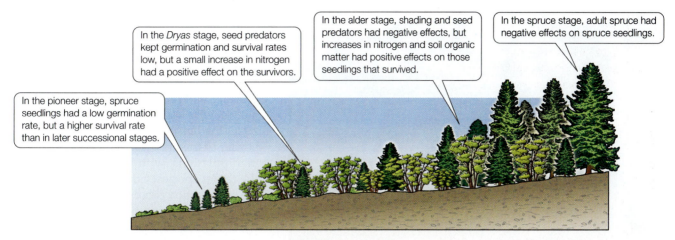

In the pioneer stage, spruce seedlings had a low germination rate, but a higher survival rate than in later successional stages.

In the *Dryas* stage, seed predators kept germination and survival rates low, but a small increase in nitrogen had a positive effect on the survivors.

In the alder stage, shading and seed predators had negative effects, but increases in nitrogen and soil organic matter had positive effects on those seedlings that survived.

In the spruce stage, adult spruce had negative effects on spruce seedlings.

Effects on spruce seedlings	Successional stage			
	Pioneer	*Dryas*	Alder	Spruce
Positive	Higher survival	Higher nitrogen level Higher growth	Higher soil organic matter Higher nitrogen level More mycorrhizae Higher growth	Higher germination

Negative	Lower germination	Lower germination
		Lower survival
		Higher seed predation
		  and seed mortality
			Lower germination
			Lower survival
			Higher seed predation
			  and seed mortality
			Root competition
			Competition for light
				Lower growth
				Lower survival
				Higher seed predation
				  and seed mortality
				Root competition
				Competition for light
				Lower nitrogen level

FIGURE 17.12 Both Positive and Negative Effects Influence Succession The relative contributions of positive and negative effects of other species on spruce seedling establishment changed across successional stages in Glacier Bay, Alaska. Positive effects equaled or outweighed negative effects in the first three stages, but the opposite was seen in the last spruce stage. (After F. S. Chapin et al. 1994. *Ecol Monogr* 64: 149–175.)

survival rates. In this stage, alders had a net positive effect on spruce seedlings that germinated before alders were able to dominate. Finally, in the spruce stage, the effects of large spruce on spruce seedlings were mostly negative and long-lasting. Growth and survival rates were low because of competition with adult spruce for light, space, and nitrogen. Interestingly, seed production by adults was enhanced, which led to relatively high seedling numbers merely as a consequence of the many more seeds available for germination.

Thus, in Glacier Bay, the mechanisms outlined in Connell and Slatyer's models were operating in at least some stages of succession. Early on, aspects of the facilitation model were seen as plants modified the habitat in positive ways for other plants and animals. Species such as alders had negative effects on later successional species unless they were able to colonize early, supporting the inhibition model. Finally, some stages—such as the spruce stage, in which dominance was a result of slow growth and long life—were driven by life history characteristics, a signature of the tolerance model.

SECONDARY SUCCESSION IN A NEW ENGLAND SALT MARSH What do other studies show with regard to Connell and Slatyer's three models? Mark Bertness and Scott Shumway studied the relative importance of facilitative versus inhibitory interactions in controlling secondary succession in a New England salt marsh. Salt marshes are characterized by different species compositions and physical conditions at different tidal elevations. The shoreline border of the marsh is dominated by the cordgrass *Spartina patens*, whereas dense stands of the black rush *Juncus gerardii* are found between the shoreline and the terrestrial border. A common natural disturbance in salt marsh habitats is the deposition of tidally transported dead plant material known as *wrack* (**FIGURE 17.13**). The wrack smothers and kills plants, creating bare patches where secondary succession takes place. Soil salinity is high in these patches because, without shading by plants, water evaporation increases, leaving behind salt deposits. The patches are initially colonized by the spike grass *Distichlis spicata*, an early successional species that is eventually outcompeted by *Spartina* and *Juncus* in their respective zones.

FIGURE 17.13 Wrack Creates Bare Patches in Salt Marshes A tidal deposit of wrack at Rumstick Cove, Rhode Island, where Bertness and Shumway conducted their research on secondary succession. This dead plant material smothers living plants, creating bare patches with high soil salinity.

Bertness and Shumway (1993) hypothesized that *Distichlis* could either facilitate or inhibit later colonization by *Spartina* or *Juncus* depending on the salt stress experienced by the interacting plants. To test this idea, they created bare patches in two zones of a marsh and

manipulated plant interactions shortly after the patches had been colonized (**FIGURE 17.14**). In the low intertidal zone (the *Spartina* zone, close to the shoreline), they removed *Distichlis* from half the newly colonized patches, leaving *Spartina*, and removed *Spartina* from the other half, leaving *Distichlis*. In the middle intertidal zone (the *Juncus* zone, closer to the terrestrial border of the marsh), they performed similar manipulations, with *Juncus* and *Distichlis* as the target species. Control patches, in which the colonization process was not manipulated, were maintained in both zones. In addition, they watered half the patches in each treatment group with fresh water to alleviate salt stress, and left half as controls.

After observing the patches for 2 years, Bertness and Shumway found that the mechanisms of succession differed depending on the level of salt stress experienced by the plants and the species interactions involved. In the low intertidal zone, *Spartina* always colonized and dominated the plots, whether or not *Distichlis* was present or watering occurred (Figure 17.14A). *Distichlis* was able to dominate only if *Spartina* was removed from the plots, so it was clearly inhibited by *Spartina*, the dominant competitor. In the middle intertidal zone, *Juncus* was able to colonize only if *Distichlis* was present or watering

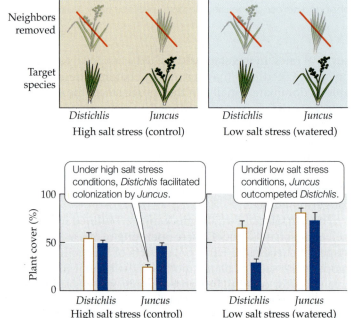

FIGURE 17.14 New England Salt Marsh Succession Is Context Dependent The trajectory of succession in salt marshes depends on soil salinity and the physiological tolerances of plant species. The kinds of interactions observed differed between the low intertidal zone (A) and the middle intertidal zone (B). Error bars show one standard error of the mean. (After M. D. Bertness and S. W. Shumway. 1993. *Am Nat* 142: 718–724.)

occurred (Figure 17.14B). Measurements of soil salinity confirmed that the presence of *Distichlis* helped to shade the soil surface, thus decreasing salt accumulation and reducing stress for *Juncus*. *Distichlis*, however, was able to colonize plots with *Juncus* only when salt stress was high—that is, under the control conditions. If plots were watered, *Distichlis* was easily outcompeted by *Juncus*.

These experimental manipulations confirmed that the mechanisms important to succession are context dependent. No single model is sufficient to explain the underlying causes of succession. In the middle intertidal zone, *Distichlis* was a strong facilitator of colonization by *Juncus*. Once this facilitation occurred, the balance was tipped in favor of *Juncus*, which outcompeted *Distichlis* (see Figure 17.14B). In the low intertidal zone, *Distichlis* and *Spartina* were equally able to colonize and grow in salty patches. If *Spartina* arrived first, it inhibited *Distichlis* colonization. If *Distichlis* arrived first, it persisted only if *Spartina* did not arrive and displace it (see Figure 17.14A).

PRIMARY SUCCESSION IN ROCKY INTERTIDAL COMMUNITIES

Our final examples come from an environment where succession has been studied extensively: the rocky intertidal zone. Here, disturbances are created mainly by waves, which can tear organisms from the rocks during storms or propel objects such as logs or boulders into them. In addition, stresses caused by low tides that expose organisms to high or low air temperatures can easily kill them or cause them to lose their attachment to the rocks. The resulting bare rock patches become active areas of colonization and succession.

Some of the first experimental work on succession in the rocky intertidal zone was done on boulder fields in Southern California by Wayne Sousa, a graduate student at the time. Sousa (1979) noticed that the algae-dominated communities on these boulders experienced disturbance every time the boulders were overturned by waves. When he cleared some patches on the boulders and observed succession in those patches over time, he found that the first species to colonize and dominate a patch was always the bright green alga *Ulva lactuca* (**FIGURE 17.15A**). It was followed by the red alga *Gigartina canaliculata*. To understand the mechanisms controlling this successional sequence, Sousa performed removal experiments on concrete blocks that he had allowed *Ulva* to colonize. He found that colonization by *Gigartina* was accelerated if *Ulva* was removed (**FIGURE 17.15B**). This result suggested inhibition was the main mechanism controlling succession, but a question remained: If *Ulva* is able to inhibit other seaweed species, why doesn't it always dominate? Through a series of further experiments, Sousa found that grazing crabs preferentially fed on *Ulva*, thus initiating a transition from the

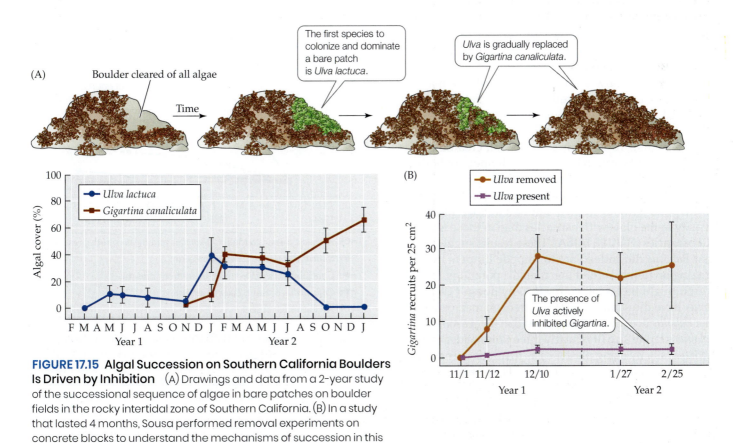

FIGURE 17.15 Algal Succession on Southern California Boulders Is Driven by Inhibition (A) Drawings and data from a 2-year study of the successional sequence of algae in bare patches on boulder fields in the rocky intertidal zone of Southern California. (B) In a study that lasted 4 months, Sousa performed removal experiments on concrete blocks to understand the mechanisms of succession in this ecosystem. Error bars show ± one standard error of the mean. (After W. P. Sousa. 1979. *Ecol Monogr* 49: 227–254.)

early *Ulva* stage to other mid-successional algal species. In turn, the mid-successional species were more susceptible to the effects of stress and parasitic algae than the late successional *Gigartina*. *Gigartina* dominated because it was the least susceptible to stress and consumer pressures.

This view of rocky intertidal succession as driven by inhibition was the accepted paradigm for many years. Facilitation and tolerance were thought to be much less important in a system where competition for space was strong. More recent work by Terence Farrell and others (e.g., Berlow 1997) demonstrated that the relative importance of inhibition is probably much more context dependent than previously thought. In the more productive rocky intertidal zone of the Oregon coast, the communities include many more sessile invertebrates, such as barnacles and mussels, than Sousa's communities of the Southern California coast, where seaweeds dominate. In the high intertidal zone of Oregon, Farrell (1991) found that the first colonizer of bare patches was a barnacle, *Chthamalus dalli*. It was replaced by another, larger barnacle species, *Balanus glandula*, which was then replaced by three species of macroalgae, *Pelvetiopsis limitata*, *Fucus gardneri*, and *Endocladia muricata*. A series of removal experiments showed that *Chthamalus* did not inhibit colonization by *Balanus*, but that *Balanus* was able to outcompete *Chthamalus* over time, thus supporting the tolerance model. Likewise, *Balanus* did not hinder macroalgal colonization, but in fact facilitated it, lending credibility to the facilitation model.

But why and how would *Balanus* facilitate macroalgal colonization? Farrell suspected that *Balanus* protected the algae in some way, possibly from desiccation stress or grazing by limpets (herbivorous marine snails). To test this idea, Farrell created experimental plots from which *Balanus*, limpets, or both were removed, then observed macroalgal colonization in those plots. He found that macroalgae colonized all of the plots without limpets but had a much higher density in the plots with barnacles than in those without barnacles (**FIGURE 17.16A**). These results suggested that *Balanus* did indeed act to impede limpets from grazing on newly settled macroalgal sporelings.

You might be asking yourself, Why doesn't *Chthamalus* have the same facilitative effect on macroalgae that *Balanus* does? Farrell suspected that the reason was *Balanus*'s larger size (it is nearly three times wider than *Chthamalus*). By using plaster casts to mimic barnacles that were slightly larger than *Balanus*, Farrell found that these barnacle mimics had an even more positive effect on macroalgal colonization than did smaller-sized live barnacles of either species (**FIGURE 17.16B**). It seems likely that the smaller and smoother *Chthamalus* does not retain as much moisture, or block as many limpets, as the larger and more sculpted *Balanus*—or the mimics, for that matter.

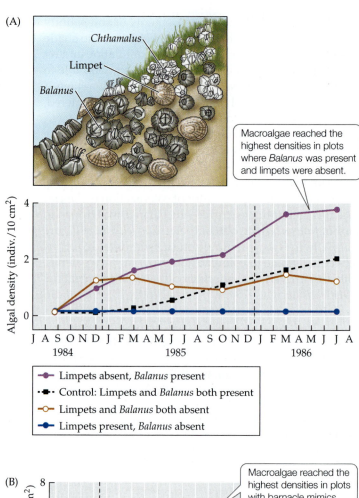

(A)

Macroalgae reached the highest densities in plots where *Balanus* was present and limpets were absent.

- — Limpets absent, *Balanus* present
- ▪▪ Control: Limpets and *Balanus* both present
- ○ Limpets and *Balanus* both absent
- ● Limpets present, *Balanus* absent

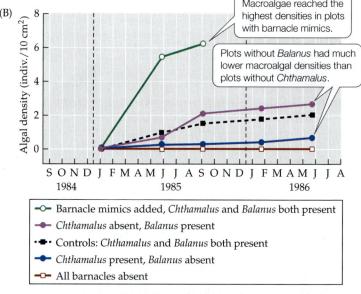

(B)

Macroalgae reached the highest densities in plots with barnacle mimics.

Plots without *Balanus* had much lower macroalgal densities than plots without *Chthamalus*.

- ○ Barnacle mimics added, *Chthamalus* and *Balanus* both present
- ● *Chthamalus* absent, *Balanus* present
- ▪ Controls: *Chthamalus* and *Balanus* both present
- ● *Chthamalus* present, *Balanus* absent
- □ All barnacles absent

FIGURE 17.16 Algal Succession on the Oregon Coast Is Driven by Facilitation (A) Changes in macroalgal densities over time were measured in plots from which *Balanus* barnacles, limpets, or both had been removed. The results suggested that *Balanus* facilitates macroalgae by reducing limpet grazing. (B) To understand the mechanisms of the facilitation, large barnacle mimics were added to some plots and compared with plots from which the real barnacle species—*Balanus*, *Chthamalus*, or both—had been removed. The results suggested that the larger the barnacle species, the better it protects macroalgae against limpet grazing and desiccation. (After T. M. Farrell. 1991. *Ecol Monogr* 61: 95–113.)

Experiments show facilitation to be important in early stages

A number of experimental studies like the ones we have just described, initially stimulated by Connell and Slatyer's 1977 paper, suggest that succession in any community is driven by a complex array of mechanisms (see **ANALYZING DATA 17.1**). No one model fits any one community; instead, each community is characterized by elements of all three of Connell and Slatyer's models. In most successional sequences, especially those in

---

ANALYZING DATA 17.1

What Kinds of Species Interactions Drive Succession in Mountain Forests?

We learned in Concept 17.3 that successional patterns are often the result of complex species interactions. Such interactions are exemplified in a study investigating the patterns of succession in mountain forests in Utah dominated by quaking aspen (*Populus tremuloides*) and subalpine fir (*Abies lasiocarpa*) (Calder and St. Clair 2012).* In some cases, aspen can form stable and self-sustaining populations, but more commonly these trees occur in mixed stands with firs. Observations show that aspen initiate the earliest stage of secondary succession in open meadows created by fire or deforestation, using root suckers (underground shoots that produce clonal plants; see Figure 9.4) to colonize open meadows. Over time, mixed aspen–fir stands are formed as the shade-tolerant firs become established and increase in abundance while aspen decline. The stands are eventually dominated by firs, which are more susceptible to fire than pure stands of aspen, thus increasing the chance of starting the successional cycle anew.

To understand the transition from one successional stage to another, Calder and St. Clair counted the aspen suckers and fir seedlings in four successional stages—meadow, aspen, mixed aspen–fir, and fir—with the results shown in **FIGURE A**. To test for the type of interaction important in the transition from one stage to another, the researchers then measured the mortality of canopy aspen and subalpine fir trees as a function of their distance from the nearest neighboring tree of the other species. These results are given in **FIGURE B**. Error bars show one standard error of the mean.

1. Based on the data in Figure A, what is the pattern of aspen abundance over the four successional stages? How does the abundance pattern of subalpine fir differ? Do these patterns of abundance of aspen and fir support the successional sequence described in the first paragraph above?

2. What type of interspecific interaction would you hypothesize could account for the difference between the number of fir seedlings in the aspen stage and in the meadow stage in Figure A? What type of interspecific interaction might explain the difference between the number of aspen suckers in the mixed and the fir stages?

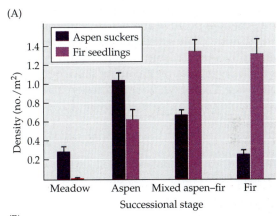

(A)

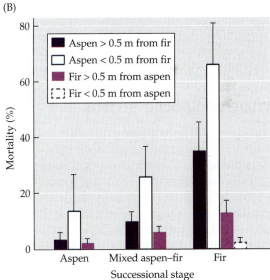

(B)

3. Now consider Figure B. What happens to fir trees when they live close (<0.5 m) to aspen trees? What happens to aspen when they live close (<0.5 m) to fir trees? Do these data support your previous hypotheses (from question 2) about the types of species interactions driving succession?

4. Which of Connell and Slatyer's three models (see Figure 17.8) best fits the results of this study? Why?

---

*Calder, W. J., and S. B. St. Clair. 2012. Facilitation drives mortality patterns along successional gradients of aspen–conifer forest. *Ecosphere* 3: 1–11. CC BY 3.0.

which a pioneer stage is exposed to physically stressful conditions, facilitative interactions are important drivers of early succession. Organisms that can tolerate and modify these physically challenging environments will thrive and facilitate other organisms that lack those capabilities. As succession progresses, slow-growing and long-lived species begin to dominate. Those species tend to be larger and more competitively dominant than early successional species. For this reason, one might expect competition to play a more dominant role than facilitation later in succession.

As succession proceeds, species richness typically increases (see Figure 17.10); thus, we must recognize that vast arrays of both positive and negative interactions are operating in mid- to late successional stages. We will learn more about the mechanisms responsible for controlling species diversity in Chapter 19, but let's turn our attention next to the question of whether succession always takes one predictable path, as Clements believed, or whether other paths are possible.

> **CONCEPT 17.4** Communities can follow different successional paths and display alternative states.

LEARNING OBJECTIVES

17.4.1 Define alternative stable states and stability using empirical data and models.

17.4.2 Describe how human activities have caused regime shifts.

Alternative Stable States

Up to this point, we have assumed that the trajectory of succession is repeatable and predictable. But what if, for example, a boulder in the rocky intertidal zone of Southern California turns over and, instead of a seaweed community forming, as Sousa (1979) observed, a sessile invertebrate community forms instead? Or what if *Dryas* never colonizes the till left behind by a glacier at Glacier Bay, but is replaced by a grass that competes with, rather than facilitates, later successional species such as Sitka spruce? Might spruce forests never develop? Possibly. There are cases in which different communities develop in the same area under similar environmental conditions. Ecologists refer to such alternative scenarios as **alternative stable states**. Richard Lewontin (1969) was one of the first to formally define and model alternative stable states in natural communities.

A community is said to have **stability**, or to be stable, when it remains in or returns to the original structure and function after some perturbation. How stable are natural communities? This question has perplexed ecologists for some time, partly because the notion of stability depends on spatial and temporal scale. At a small spatial scale, such as a 1-m² plot in a midwestern prairie, there might

be considerable change or instability over time. If all the plants were removed from the plot, it is unlikely that all the same species would recolonize that particular plot, and certainly not in the exact same locations. However, if a larger area is considered (e.g., a 100-m² plot), the chance of finding the same species increases. Similarly, if one followed the plot for a short time, the chance that its species composition would change would be low. But the longer you observed it, the more likely it would be that the community would change and thus appear unstable. With these caveats in mind, let's take a closer look at examples of communities that, once disturbed, do not revert to previous states, but instead show alternative stable states.

Alternative states are controlled by strong interactors

John Sutherland (1974) studied alternative states in marine fouling communities: the sponges, hydroids, tunicates, and other invertebrates that encrust ships, docks, and other hard surfaces in bays and estuaries. He suspended ceramic tiles from the dock at Duke University Marine Lab in Beaufort, North Carolina, in early spring and allowed them to be colonized by planktonic invertebrate larvae (**FIGURE 17.17A**). At the end of 2 years, even though a handful of species had colonized the tiles, most of them were dominated by a solitary tunicate species, *Styela*. Its dominance was not universal, however: *Styela* actually declined on Sutherland's tiles during the first winter and was replaced by the hydroid species *Tubularia*. This effect was due to the annual nature of *Styela*, which dies off in winter, and the tunicate quickly regained dominance the following spring when larvae started to settle.

By placing new tiles out periodically, Sutherland also showed that *Styela* was able to persist despite the existence of other potential colonizers. These colonizers fouled the new tiles but were unable to colonize those dominated by *Styela*. For this reason, Sutherland viewed this fouling community as stable. Within a few months, he also identified what he believed to be another stable fouling community, this one dominated by *Schizoporella*, an encrusting bryozoan (see Figure 17.17A). This community developed on new tiles suspended from the dock in late summer and was also impervious to colonization by other species, including *Styela*.

To understand what might be controlling these two alternative outcomes of succession, Sutherland submerged new tiles at the same spot on the dock but excluded fish predators from half of the tiles by surrounding them with cages (**FIGURE 17.17B**). After a year, Sutherland found that the tiles protected from fish predation had formed communities dominated by *Styela*, while those exposed to fish predation had formed communities dominated by *Schizoporella*. He also noticed that the abundances of both species on the tiles protected from predators were reversed when *Styela* began to die off in the winter. These results suggested that *Styela* is competitively dominant if left

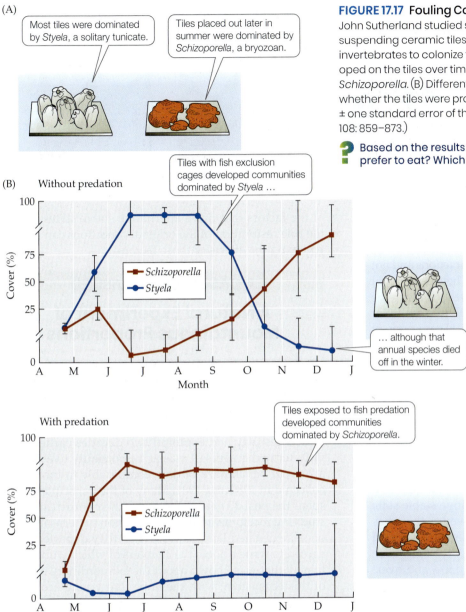

FIGURE 17.17 Fouling Communities Show Alternative States
John Sutherland studied succession in fouling communities by suspending ceramic tiles from a dock in North Carolina and allowing invertebrates to colonize them. (A) Two types of communities developed on the tiles over time, one dominated by *Styela* and another by *Schizoporella*. (B) Different communities developed depending on whether the tiles were protected from fish predation. Error bars show ± one standard error of the mean. (After J. P. Sutherland. 1974. *Am Nat* 108: 859–873.)

❓ Based on the results shown in (B), which fouling species did fish prefer to eat? Which species was the competitive dominant?

undisturbed but is outcompeted by *Schizoporella* when disturbed. Sutherland explained his original observations of *Styela* dominance by suggesting that fish predation was spotty and that the tunicates themselves, once they reached a certain large size, might have acted as a natural "cage" or predator exclusion mechanism.

Lewontin (1969) and Sutherland (1974) both believed that multiple stable states existed in communities and could be driven by the addition or exclusion of particularly strongly interacting species. If those species were missing or ineffective, communities could follow alternative successional trajectories that might never lead back to the original community type (state) but might instead form a new community type. We can visualize the theory behind alternative stable states by imagining a landscape in which different states are represented by valleys and in which a community is represented by a ball (**FIGURE 17.18A**). Just as the ball can move from one valley to another, the community can move from one state to another, depending on the presence or absence of strongly interacting species and how they affect the community (**FIGURE 17.18B**). For example, it may take only a slight change in the abundance of one or more dominant species to force the community (ball) into an alternative state (valley), or it may require complete removal of a species to cause this change. If we use Sutherland's work as an example, we can think of the *Styela* and *Schizoporella* community types as two different valleys. Whether the ball resides in the *Schizoporella* valley or the *Styela* valley depends on the presence of fish predators. Interestingly, in this system, the ball may not simply move back to the *Schizoporella* valley if access is restored to fish predators (**FIGURE 17.18C**). As Sutherland noted, *Styela* is able to escape predation once it reaches a certain size. Thus, this system might show **hysteresis**, an inability to shift back to the original community type even when the original conditions are restored.

Connell and Sousa (1983) were skeptical that Sutherland had demonstrated the existence of alternative stable states, for several reasons. First, they thought that his tile communities did not persist long enough, or have a spatial scale large enough, to be considered stable. If the tiles could be followed over multiple years, they asked, would they not all end up being dominated by one or the other species? In addition, they wondered whether the fouling communities could have been sustained outside of an

FIGURE 17.18 A Model of Alternative Stable States (A) A community is represented by a ball that moves within a landscape of community states (valleys). (B) Note that some valleys can be deeper than others, suggesting the magnitude of change (ΔX) needed to shift the community from one state to another. (C) Hysteresis occurs when reversal of the change (–ΔX) does not return the community to its original state. (After B. E. Beisner et al. 2003. *Front Ecol Environ* 1: 376–382.)

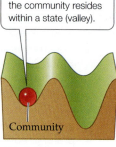

Stability occurs when the community resides within a state (valley).

Community

(A) Stability

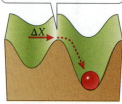

A change in some factor (ΔX) may cause the community to move to another state (valley).

ΔX

(B) Change

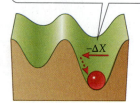

Reversal of the change (–ΔX) may not result in a return to the original conditions if the initial shift was sufficiently large.

–ΔX

(C) Hysteresis

experimental setting in which predators were removed. Their final argument, although it was not a criticism of Sutherland's study per se, was that alternative stable states could be driven only by species interactions and not by physical changes in the community. They argued that many of the examples Sutherland had used to bolster the importance of alternative stable states fell into the latter category. Their requirement that the physical environment not change is problematic because it excludes as drivers of succession all species that interact with other species by modifying their physical environment—that is, all ecosystem engineers. We know that ecosystem engineers can have strong effects on communities, so excluding them is unrealistic to most ecologists.

Human actions have caused communities to shift to alternative states

The stringent requirements suggested by Connell and Sousa had the effect of delaying alternative stable state research for two decades. Recently, however, there has been renewed interest in alternative stable states, spurred by the increasing evidence that human activities, such as habitat destruction, species introductions, and over-harvesting of wild species, are shifting communities to alternative states. We have already seen examples of such changes in several of the Case Studies in this book, including the change from kelp forests to urchin barrens due to the decline of sea otters (see Chapter 9), the crash of the anchovy fishery in the Black Sea due to the introduction of *Mnemiopsis* (see Chapter 10), and the invasion of the aquarium strain of *Caulerpa taxifolia* in the Mediterranean, Australia, Japan, and North America (see Chapter 16). These so-called *regime shifts* are caused by the removal or addition of strongly interacting species that maintain one community type over others. Ecologists are uncertain whether the results can be reversed or whether hysteresis will occur once communities have been "manipulated" by human activities and new regimes are in place. Will recolonization by sea otters rejuvenate kelp forests? Will the cessation of nutrient enrichment in the Black Sea revitalize the anchovy fishery? And will the removal of *Caulerpa* restore seagrass communities? These are all questions whose answers may be found in a better understanding of the factors that drive alternative stable states and of the role restoration of the original conditions can play in reversing the effects of those factors.

A Natural Experiment of Mountainous Proportions

On the twentieth anniversary of the eruption of Mount St. Helens, in 2000, a group of ecologists gathered on the once smoking and ash-covered volcano to participate in a week-long field camp. They gathered their gear, including tape measures, quadrat frames, and maps, and visited the same sites they had explored two decades earlier. This visit, termed a "pulse," was an opportunity to establish a 20-year benchmark of data comparable to those first collected in 1980 and 1981. Many of the participants had spent the past 20 years—for some, their entire careers—studying recolonization and succession patterns in those once-devastated landscapes. When they departed, they agreed to write a book, the chapters of which would contain all that was known about the extraordinary ecology of this ecosystem, with the hope that young ecologists would be motivated to continue the research and carry on their legacy. The book, *Ecological Responses to the 1980 Eruption of Mount St. Helens* (Dale et al. 2005), was published 5 years later.

What does the book tell us about succession on Mount St. Helens? First, the eruption created disturbances that varied in their effects depending on distance from the volcano and habitat type (e.g., aquatic vs. terrestrial). Although areas close to the summit, such as the Pumice Plain, were literally sterilized by the heat of the eruption, ecologists were surprised to discover how many species actually survived on the mountain (**TABLE 17.2**). Because the eruption occurred in spring, many species had been still dormant under the winter snows. Survivors included plants with underground buds or rhizomes, animals such as rodents and insects with burrows, and fish and other aquatic species in ice-covered lakes. In the blowdown zone, large trees and animals perished while smaller organisms survived in the protection of

TABLE 17.2

Surviving Organisms Found on Mount St. Helens within a Few Years after the Eruption

Disturbance zone	Mean vegetation cover (%)	Average number of plant species/m²	Animals					
			Small mammals	Large mammals	Birds	Lake fish	Amphibians	Reptiles
Pumice Plain	0.0	0.0	0	0	0	0	0	0
Mudflow zone	0.0	0.0	0	0	0	N/A	0	0
Blowdown zone			8	0	0	4	11	1
Pre-eruption clear-cut	3.8	0.0050						
Forest without snow	0.06	0.0021						
Forest with snow	3.3	0.0064						
Scorch zone	0.4	0.0039	0	0	0	2	12	1

Source: C. M. Crisafulli et al. 2005. In *Ecological Responses to the 1980 Eruption of Mount St. Helens*, V. H. Dale et al. (Eds.), pp. 287–299. Springer: New York, based on references cited there.

their larger neighbors. The opposite was true in areas outside the blowdown zone, where falling rocks and ash smothered smaller plants and animals, but not larger organisms.

A second important research discovery from Mount St. Helens is the role survivors have played in controlling the pace and pattern of succession. In many cases, these species were thrust into novel physical environments and species assemblages without time to adapt over evolutionary time scales. Some species thrived, while others fared poorly, but their adaptability and unpredictability were surprising. Unlikely alliances were formed that hastened succession in particular habitats. For example, newly formed and isolated ponds and lakes were colonized by amphibians much faster than had been thought possible (**FIGURE 17.19**). Scientists discovered that frogs and salamanders were using tunnels created by northern pocket gophers (*Thomomys talpoides*) to make their way from one pond to another across the arid landscape (Crisafulli et al. 2005). The gophers were particularly successful on Mount St. Helens, both because they survived the eruption in their tunnels and because grassy meadows—their preferred habitat—expanded greatly after the eruption. Interestingly, the gophers were also responsible for facilitating plant succession: their burrowing activity brought to the soil surface organic matter, seeds, and fungal spores buried deep under the volcanic rock and ash (Crisafulli et al. 2005) (**FIGURE 17.20**).

A third important discovery was the realization that multiple mechanisms were responsible for primary succession on Mount St. Helens. Facilitation on the Pumice Plain was exemplified by the dwarf lupine (*Lupinus lepidus*), the first plant to arrive there. Dwarf lupines trapped seeds and detritus and increased the nitrogen content of the soil though their symbiotic association with nitrogen-fixing bacteria (del Moral et al. 2005). The lupines, in turn, were inhibited by multiple insect herbivores, which essentially controlled the pace of primary succession (Bishop et al. 2005). Tolerance was evident in some primary successional habitats, where Douglas fir lived in concert with annual herbs. The diversity of strategies species used, and the resulting community

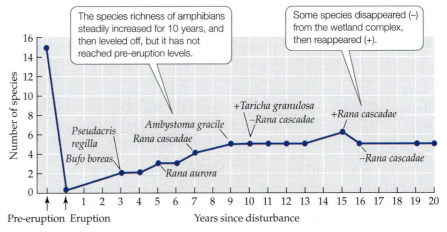

FIGURE 17.19 Rapid Amphibian Colonization Frog and salamander species rapidly colonized a wetland complex in the Pumice Plain on Mount St. Helens. (After C. M. Crisafulli et al. 2005. In *Ecological Responses to the 1980 Eruption of Mount St. Helens*, V. H. Dale et al. [Eds.], pp. 183–197. Springer: New York.)

FIGURE 17.20 **Pocket Gophers to the Rescue** The burrowing activity of northern pocket gophers, some of which survived the eruption underground, brought organic matter, seeds, and fungal spores to the soil surface, creating microhabitats, like this one in the Pumice Plain, where plants could grow.

compositions, never ceased to amaze ecologists, who up to that point had been guided mostly by the models of Connell and Slatyer (1977).

Despite decades of data and a treasure trove of novel discoveries, research on Mount St. Helens has only just begun. Will communities there follow paths of succession that lead to predictable and repeatable outcomes? Or will they form alternative states that are highly dependent on their historical legacies? Geologic studies suggest that Mount St. Helens erupts roughly every 300 years. The life span of its community succession thus greatly exceeds our own life span by hundreds of years, so we must be content with the limited knowledge we have gained from studying what is arguably the most interesting phase of succession on Mount St. Helens and with the hope that ecologists will continue their research there for years to come.

 ## CONNECTIONS *in* NATURE

PRIMARY SUCCESSION AND MUTUALISM We saw in Chapter 15 that positive relationships can alter communities, and that they may be particularly important in stressful environments. Primary succession in terrestrial environments illustrates both of these effects: some of the examples presented in this chapter involve plants that interact in a mutualistic way with symbiotic nitrogen-fixing bacteria. These bacteria form nodules in the roots of their plant hosts, where they convert nitrogen gas from the atmosphere (N_2) into a form that is usable by plants (ammonia [NH_4^+]). The plants provide the bacteria with sugars produced by photosynthesis. This interaction appears to be extremely important for plants and animals colonizing completely sterile environments. We have seen that *Dryas* and alders, both species that form tight mutualisms with nitrogen-fixing bacteria, were some of the first species to colonize the till left behind by glaciers at Glacier Bay, Alaska. Similarly, *Lupinus lepidus* was able to use the nitrogen produced by its bacterial symbionts to colonize the sterile Pumice Plain of Mount St. Helens after the eruption. Lupines were the major source of nitrogen for subsequent plants and herbivorous insects for many years. Thus, lupines and their symbiotic bacteria play a large role in controlling the rate of primary succession on Mount St. Helens.

The nitrogen-fixing bacteria involved in symbioses are extremely diverse. Only a few groups of bacteria live in root nodules; all the rest are associated with either the surfaces of roots or the guts of ruminants. The nodule-forming bacteria include the rhizobia, a taxonomic group associated with legumes (such as lupines), and actinomycetes of the genus *Frankia*, which are associated with woody plants such as alders and *Dryas*. Nodule formation involves a complex series of chemical and cellular interactions between the root and the bacteria (**FIGURE 17.21**). Free-living bacteria are attracted to root exudates that cause the microbes to attach to the roots and multiply. Sets of genes are activated in both bacterial and root cells that allow the bacteria to enter the root, the root cells to divide, and the nodule to be formed.

The enzyme involved in nitrogen fixing (nitrogenase) is highly sensitive to oxygen and requires anaerobic conditions. Thus, wherever nitrogen-fixing symbioses occur, there are structural components to the interaction (such as membranes within the root nodules) that produce anaerobic conditions. The bacteria, however, need oxygen to metabolize, so a hemoglobin protein known as leghemoglobin, which has a high affinity for oxygen, is produced in the nodules to deliver oxygen to the bacteria in an essentially anaerobic environment. The nodules often have an eerie pink color that is associated with the leghemoglobin. In addition, the nodule develops a specialized vascular system that supplies sugars to the bacteria and carries fixed nitrogen to the plant.

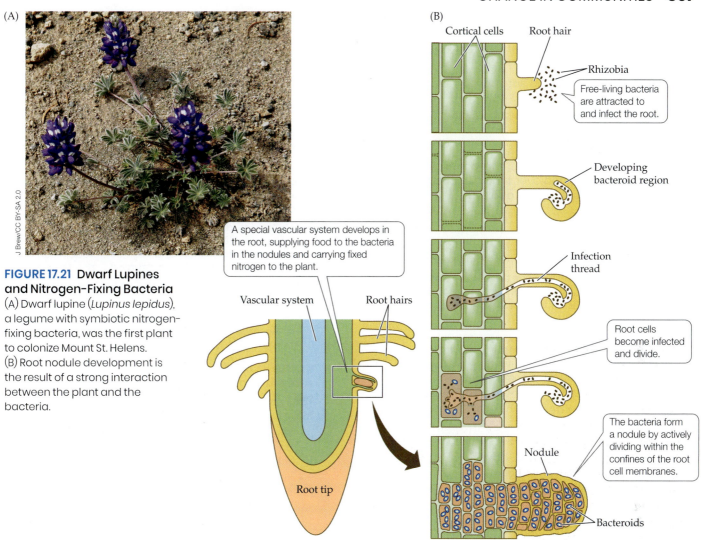

FIGURE 17.21 Dwarf Lupines and Nitrogen-Fixing Bacteria (A) Dwarf lupine (*Lupinus lepidus*), a legume with symbiotic nitrogen-fixing bacteria, was the first plant to colonize Mount St. Helens. (B) Root nodule development is the result of a strong interaction between the plant and the bacteria.

Maintaining a symbiosis with nitrogen-fixing bacteria is costly to plants. Estimates suggest that creating and maintaining the nodules alone costs a plant 12%–25% of its total photosynthetic output. Plants may be able to shoulder this cost, especially if it allows them to live in environments free of competitors and herbivores. But as they increase the nitrogen content of the soils in which they live, plants with symbionts make conditions better for other plant species as well—some of which are likely to be competitors. Thus, these plants face a trade-off between improving the environment for themselves and competing with other species, which makes their role in early successional environments important, if somewhat ironic.

SUMMARY

CONCEPT 17.1 Agents of change act on communities across all temporal and spatial scales.

17.1.1 Define abiotic and biotic agents of change including disturbance and stress.

- Agents of change include both abiotic and biotic factors.
- Abiotic agents of change can act as disturbances (injuring or killing organisms) or as stresses (reducing the growth, reproduction, or survival of organisms).

- Biotic agents of change include negative species interactions such as competition, predation, and trampling. Ecosystem engineers and keystone species are common agents of change.

17.1.2 Compare disturbance intensity with disturbance frequency and describe their differential effects on communities.

- Agents of change vary in their intensity, frequency, and areal extent. Disturbance intensity is how much

(Continued)

SUMMARY (*continued*)

biomass is removed and disturbance frequency is how often it is removed.

- Intense but infrequent disturbances produce entire community reassembly, whereas weak but frequent disturbances produce gradual community change.

CONCEPT 17.2 Succession is the process of change in species composition over time as a result of abiotic and biotic agents of change.

17.2.1 Compare primary succession with secondary succession.

- Succession involves a series of stages that theoretically include a stable end point, or climax stage.
- Primary succession involves the colonization of habitats that are devoid of life and can be slow because the early successional species typically face inhospitable conditions.
- Secondary succession involves the reestablishment of a community in which most, but not all, of the organisms or organic constituents have been destroyed. More hospitable conditions and preexisting species can lead to faster successional change.

17.2.2 Discuss the early research and differential views of early ecologists with regard to succession.

- Henry Chandler Cowles used a space for time substitution method to study succession in dunes.
- Early ecologists were fascinated with succession but disagreed about whether it proceeded in deterministic or random ways and whether plants were the main drivers.

17.2.3 Outline the multiple models of succession and compare their differences.

- In 1977 Connell and Slatyer proposed three models of succession, known as the facilitation model, tolerance model, and inhibition model.
- The facilitation model proposes that early colonizers modify the environment in positive ways for later species. The tolerance model proposes that early colonizers modify the environment but in ways that

neither benefit nor inhibit later species. The inhibition model proposes that early colonizers modify the environment in negative ways for later species.

CONCEPT 17.3 Experimental work on succession shows its mechanisms to be diverse and context dependent.

17.3.1 Analyze multiple studies to understand the diverse mechanisms involved in primary and secondary succession.

- Multiple studies of succession have shown that no one model fits any one community. Aspects of the facilitation, tolerance, and inhibition models can be seen in almost all systems studied.

17.3.2 Summarize the results of experiments designed to determine the mechanisms of succession.

- Generally, experiments show that facilitation tends to be important in early stages of succession, and competition in later stages of succession.

CONCEPT 17.4 Communities can follow different successional paths and display alternative states.

17.4.1 Define alternative stable states and stability using empirical data and models.

- Alternative stable states occur when different communities develop in the same area under similar environmental conditions.
- In communities that experience alternative states, succession is typically controlled by strongly interacting species and may show an inability to shift back to the original community even when the original conditions are restored (known as hysteresis).

17.4.2 Describe how human activities have caused regime shifts.

- Human activities have caused regime shifts in communities by changing the underlying biotic and/or abiotic conditions that determine community structure. The communities may or may not show hysteresis.

Review Questions

1. List some abiotic and biotic agents of change in communities. Describe the intensities and frequencies with which they are likely to act.

2. Describe the differences between primary and secondary succession and what those differences mean for colonizing species.

3. Connell and Slatyer proposed three separate models of succession: the facilitation model, the tolerance model, and the inhibition model. Choose a hypothetical community and describe the different circumstances that would be required to support each of the models.

4. Why is it hard to determine whether a community is stable? Do you think John Sutherland was able to demonstrate alternative stable states on his ceramic tiles? Why or why not?

Hone Your Problem-Solving Skills

The eruption of Mount St. Helens created a number of successional communities including primary, secondary, and climax stages. One study by Charles Crisafulli and colleagues (2005) considered the pace of recovery of small mammals in four of these successional communities: the Pumice Plain (primary succession), the blowdown zone (secondary succession), the scorch zone (secondary succession), and an undisturbed reference area 21 km away from the mountain (climax stage). To compare the recovery of small mammals in these successional communities, the researchers conducted extensive trapping between 1982 and 2000. Below is a graph of small mammal species richness over time in different successional communities, organized according to the degree of disturbance the communities experienced from the eruption:

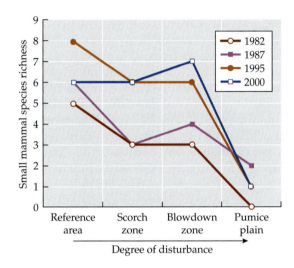

1. What happened to small mammal species richness over time and within the different successional habitats?

2. Did small mammals recover in all or some of the successional habitats, if you use the reference area as a standard for recovery? Describe what you think may have controlled recovery (or lack thereof).

3. The researchers recorded species composition of the small mammals they caught. They found that the deer mouse (*Peromyscus maniculatus*) was the only species present in all four communities. What does this suggest about the deer mouse and its role in succession?

4. The reference area, which served as a control for the small mammal surveys, increased in species richness over time. Give some plausible reasons for this increase.

18

Biogeography

The Largest Ecological Experiment on Earth: A Case Study

There is probably only one place on Earth where a person can hear the calls of 100 species of birds or smell the fragrances of 1,000 species of flowering plants or see the leaf patterns of 300 species of trees, all in 1 hectare (2.5 acres) of land. That place is the Amazon, where half the world's remaining tropical rainforests and species reside. Just 1 hectare (ha) of rainforest in the Amazon contains more plant species than all of Europe! Of course, not all of the species diversity of the Amazon is confined to the rainforest itself. The Amazon Basin contains the largest watershed in the world; one-fifth of all the fresh water on Earth falls on its slopes, collects in over 1,000 forested tributaries, and eventually flows into the Amazon River and out to sea. A trip to a fish market in Manaus, Brazil, would reveal the amazing diversity of aquatic life in these rivers (**FIGURE 18.1**). The number of fish species in the Amazon Basin exceeds that of the entire Atlantic Ocean.

Ironically, with this incredible species diversity can come devastating species losses when these ecosystems are disturbed. The main destructive force in the Amazon Basin has been deforestation, which began in earnest with the building of roads in the 1960s (Bierregaard et al. 2001). Before then, most of the region had no roads and was relatively isolated from the rest of society. Within 50 years, however, 20% of the rainforest has been converted to pastureland, towns, roads, and mines. Although this percentage might seem modest, it is deceiving, both because of the sheer number of species involved and because of the pattern of deforestation. Logging practices have caused extreme *habitat fragmentation*, sometimes resulting in a "fishbone" pattern in which thin linear fragments of rainforest are surrounded by strips of nonforested land. These fragments of forest can be thought of as isolated "islands" of forest within a "sea" of deforested habitat. As we will see, habitat fragmentation, by isolating species, can have serious consequences for species diversity.

The fragmentation of the Amazon rainforest motivated Thomas Lovejoy and his colleagues to initiate one of the largest and longest-running ecological experiments ever conducted. The Biological Dynamics of Forest Fragments Project (BDFFP) began in 1979, and Lovejoy seized a unique opportunity to find out what

© Alexandre Rotenberg/Alamy Stock Photo

FIGURE 18.1 Diversity Abounds in the Amazon
Freshwater fish caught in the Amazon River on display in a market in Manaus, Brazil.

was happening to the species diversity of the Amazon as logging eliminated more and more of the forest. He was guided by an elegant model in Robert MacArthur and Edward O. Wilson's 1967 book *The Theory of Island Biogeography*, which presents an explanation for the observation that more species are found on large islands than on small islands. By taking advantage of a Brazilian law requiring landowners to leave half of their land as forest, Lovejoy arranged to designate different-sized forest plots ("islands") that would be surrounded by either forested land (controls) or deforested land ("sea") (**FIGURE 18.2**). The control plots and fragments were designated before logging took place and were either 1, 10, 100, or 1,000 ha in size. Baseline data collected immediately after logging showed little difference in species diversity between control plots and fragments.

By the mid-1980s, the ecologists had a fully replicated experiment at a scale unimaginable in the past. For more than 40 years, the BDFFP has evolved from a study that asks the simple question, What is the minimum area of rainforest needed to maintain species diversity? to one that asks, What roles do the shape, configuration, and connectivity of forest fragments play in maintaining species diversity? How does the surrounding habitat influence that diversity? And what is the prognosis for the Amazon rainforest, one of the most deforested but species-rich terrestrial biomes on Earth?

Introduction

Looking out over a community such as a rocky intertidal zone on the Northern California coast, it is obvious that the locations of species on the shoreline are influenced not only by physical factors, such as tide height and wave action, but also by a variety of biological interactions. Sea stars eat sessile mussels in the low intertidal zone, thus limiting them to the higher intertidal zones. In those zones, the crevices between mussels provide habitat for many species that otherwise would be absent. Local conditions such as these are important regulators of species distributions. However, as important as these conditions appear to us, we must always be cognizant of the influence of processes operating at larger geographic scales. Oceanographic processes, such as currents and ocean upwelling, regulate the delivery of invertebrate larvae to rocky shorelines. At a global scale, oceanic circulation patterns control current direction. By limiting dispersal, those patterns can isolate species over ecological and evolutionary time. As a result, the local assemblage of species on the Northern California coast is ultimately based on a foundation of global and regional processes. In this chapter, we will consider the effects of these large-scale geographic processes on one of the most recognizable ecological patterns known: the distribution and diversity of species on Earth.

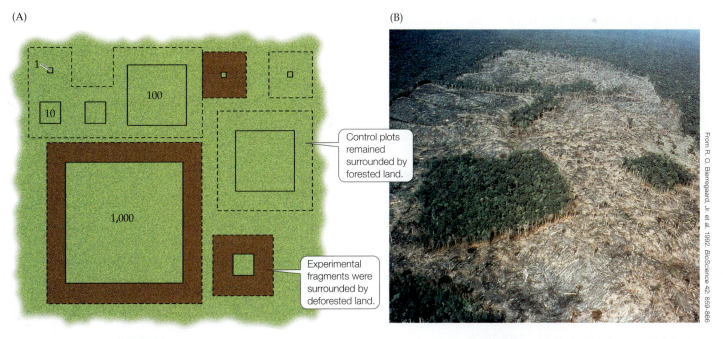

(A) (B)

Control plots remained surrounded by forested land.

Experimental fragments were surrounded by deforested land.

From R. O. Bierregaard, Jr. et al. 1992. *BioScience* 42: 859-866

FIGURE 18.2 Studying Habitat Fragmentation in Tropical Rainforests The Biological Dynamics of Forest Fragments Project (BDFFP) near Manaus, Brazil, was designed to study the effects of habitat fragment size on species diversity. (A) Plots of four sizes (1, 10, 100, and 1,000 ha) were designated before logging took place, then either isolated by logging or left surrounded by forest as controls. (B) Aerial photo of a 10-ha and a 1-ha fragment isolated in 1983. (A after R. O. Bierregaard, Jr. et al. 2001. *Lessons from Amazonia: The Ecology and Conservation of a Fragmented Forest*. Yale University Press: New Haven, CT.)

❓ Why didn't the experimental manipulation involve removing forest from the fragments?

LEARNING OBJECTIVES

18.1.1 Define biogeography and explain how patterns of species diversity and composition are connected across different spatial scales.

18.1.2 Outline the processes important to global-, regional-, and local-scale biogeography.

18.1.3 Analyze the relative importance of species pools versus local-scale processes in determining local community species diversity.

Biogeography and Spatial Scale

One of the most obvious ecological patterns on Earth is the variation in species composition and diversity among geographic locations. The study of this variation is known as **biogeography**. Pretend for a moment that you have a lifelong desire to see all the forest biomes on Earth. In this imaginary scenario, you have the ability to move from one geographic region on Earth to another. Think Google Earth, but with the ability to fly down into a community and see species up close. You start in the tropics at 4°S latitude and 60°W longitude and fly into the Amazon rainforest, the most species-rich forest on Earth (**TABLE 18.1**). At 20 m altitude, you fly through the middle of the humid forest, and as you travel over each hectare, you see new tree species (**FIGURE 18.3A**). You may have encountered half of them in the previous hectare, but at least half are completely new. The more area you cover, the more tree species you see. The richness is almost overwhelming, and the heat and humidity are stifling, so you decide to head north to drier climes.

You arrive at 35°N, 125°W. This is the southern coast of California, where the forests are oak woodland—a dry biome, as we learned in Chapter 3. Most of the trees and shrubs are evergreen, but they are not conifers. Instead, they are flowering plants with small, tough (*sclerophyllous*) leaves. The woodlands are interspersed with grasslands (**FIGURE 18.3B**). Flying down through the vegetation, you notice the many kinds of trees and shrubs, all with small leaves and thick bark. The woodland is aromatic because of the volatile oils contained in the shrubs and herbaceous plants. Plant species richness is high, but just a fraction of that in the Amazon (see Table 18.1).

It's still warm, so you decide to head north to 45°N, 123°W, where the forest is cool and very wet. You are in the Pacific Northwest region of North America, where the forests are dominated by large conifers. As you fly through, you notice the lushness of the forest, with its lichen-filled canopy and fern-covered floor (**FIGURE 18.3C**). Tree species richness in these lowland temperate evergreen forests is a fraction of that in the two previous forests you've visited (see Table 18.1). There are only a handful of tree species: Douglas fir, western hemlock, western red cedar, red alder, and big leaf maple. What these forests lack in species richness, however, is made up by their huge biomass.

You want to see the extremes in species richness, so your next stop is 64°N, 125°W, in the boreal forests of Canada. Flying over the cold landscape, you notice rows and rows of identical spruce trees, broken once in a while by large wetlands (**FIGURE 18.3D**). Dipping down into the canopy, you are struck by the dense and monotonous nature of the forest. It's dark down under those spruce boughs, but low-lying berry bushes are a reminder that light does penetrate the canopy, especially in the summer months. You continue to fly north, and the forests thin until the landscape is one long expanse of treeless tundra.

TABLE 18.1

Tree Species Richness in Different Forests around the World

Forest location/type	Latitude, longitude	Approximate tree species richness	Source
Amazon, Brazil	4°S, 60°W	1,300	Laurance 2001
Southern California, USA	35°N, 125°W	57	Allen et al. 2007
Pacific Northwest, USA			Franklin and Dyrness 1988
Douglas fir forest	45°N, 123°W	7	
Garry oak forest	45°N, 123°W	4	
Boreal forest, Canada	64°N, 125°W	2	Kricher 1998
New Zealand			Dawson and Lucas 2000
Beech forest	45°S, 170°E	20	
Flowering tree forest	35°S, 170°E	100	

Sources: Laurance 2001, Allen et al 2007, Franklin and Dyrness 1988, Kricher 1998, Dawson and Lucas 2000.

FIGURE 18.3 **Forests around the World** Forest biomes vary greatly in their species composition and species richness. (A) A tropical rainforest in Brazil. (B) Oak woodland in Southern California. (C) Lowland temperate evergreen forest in the Pacific Northwest. (D) Boreal spruce forest in Denali National Park, Alaska.

Your trip could end here, but you have always wanted to visit New Zealand, so you take the time to fly back to the Southern Hemisphere. New Zealand was separated from the ancient continent of Gondwana roughly 80 million years ago, and since that time, evolution has produced unique forests there (**FIGURE 18.4**). Roughly 80% of the species in New Zealand are **endemic**, meaning that they occur nowhere else on Earth. Dialed into 45°S, 170°E, on the South Island of New Zealand, you fly through the Southern Hemisphere equivalent of the Pacific Northwest. Instead of conifers, the forests are dominated by four species of southern beech trees with billowy layers of twisted branches (see Figure 18.4A). Below the canopy are "divaricating shrubs," whose multiple-angled branches give them a zigzag appearance. Plants with this growth form are found in highest abundance in New Zealand. Although temperate evergreen forests in the Northern and Southern Hemispheres are similar in some ways (e.g., each has low tree species richness compared with forests in the tropics),

they are made up of completely different species assemblages with very different evolutionary histories.

Even within New Zealand, over a distance that extends from 35°S to 47°S (a latitudinal distance identical to that from Southern California to British Columbia in the Northern Hemisphere), there are big differences in tree species richness and composition. The North Island is warmer (closer to the equator) than the South Island and has more diverse forests, consisting of many flowering tree species with a few tall emergent conifers (see Figure 18.4B). These forests have a tropical feel to them because of all the flowering trees and the multitude of vines and epiphytes (plants and lichens that live on larger plants). The tree ferns growing here are similar to those that were dominant 100 million years ago, during the age of the dinosaurs. One of the most extraordinary trees is the kauri (*Agathis australis*), which is among the largest tree species on Earth (interestingly, the largest is the giant sequoia, *Sequoiadendron giganteum*, which occurs at roughly the same latitude in the Northern

(A) South Island

(B) North Island

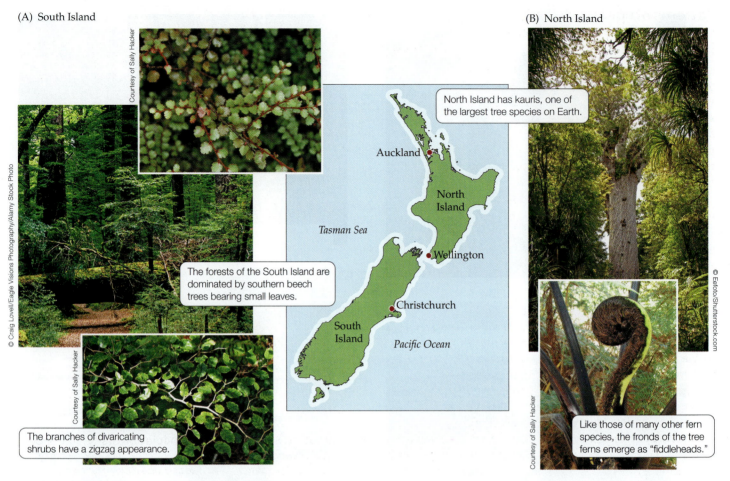

North Island has kauris, one of the largest tree species on Earth.

The forests of the South Island are dominated by southern beech trees bearing small leaves.

The branches of divaricating shrubs have a zigzag appearance.

Like those of many other fern species, the fronds of the tree ferns emerge as "fiddleheads."

FIGURE 18.4 Forests of the North and South Islands, New Zealand The two islands of New Zealand span a large latitudinal gradient (35°S–47°S) and thus have different forest types. (A) The forests of the South Island are dominated by beeches.

(B) The forests of the warmer North Island have greater tree species diversity and a different species composition than those on the South Island (see Table 18.1).

Hemisphere). Some kauri trees are 60 m (200 feet) high and 7 m (23 feet) in diameter. Unfortunately, like redwoods, kauris have been extensively logged, and they exist in a forest community in only two small reserves, 100 km² in total size. Given that old-growth stands of kauris take 1,000–2,000 years to generate, these forests are virtually irreplaceable. If we contrast the tree species richness of the forests characteristic of the North Island with those on the South Island, we find more than 100 tree species in the warmer northern forests, compared with the 10–20 species in the less diverse beech forests characteristic of the temperate south (see Table 18.1).

With our world forest tour at its end, what can we conclude about biogeographic patterns on Earth, assuming that forest communities are good global representatives?

- First, species richness and composition vary with latitude: the lower tropical latitudes have many more, and different, species than the higher temperate and polar latitudes.

- Second, species richness and composition vary from continent to continent, even where longitudes or latitudes are roughly similar.

- Third, the same community type or biome can vary in species richness and composition depending on its location on Earth.

As we will see in the rest of this chapter, these are reliable patterns that have been demonstrated over and over again for many regions of the world and many community types. What has puzzled naturalists for centuries is just what processes control these biogeographic patterns. Why are more species found in some areas than in other areas? Why do some regions harbor species assemblages that are not found anywhere else on Earth?

A number of hypotheses have been proposed to explain biogeographic variation in species diversity and composition. As we'll see, these hypotheses are highly dependent on the spatial scale at which they are applied.

Patterns of species diversity at different spatial scales are interconnected

On our world forest tour, we saw that patterns of species diversity and composition varied at global, regional, and local spatial scales. We can think of these spatial scales as interconnected in a hierarchical way, with patterns of species diversity and composition at one spatial scale setting the conditions for patterns at smaller spatial scales (Whittaker et al. 2001). Let's start with the largest spatial scale and work downward.

The *global scale*, as the term suggests, includes the entire world, a huge geographic area over which there are major variations with changes in latitude and longitude (**FIGURE 18.5A**). Species have been isolated from one another, often on different continents or in different oceans, by long distances and over long time periods. Differences in the rates of three processes—speciation, extinction, and dispersal—help determine differences in species diversity and composition at the global scale. We will consider these processes in more detail in the following section.

The **regional scale** encompasses smaller geographic areas in which the climate is roughly uniform and to which species are restricted by dispersal limitation (see Concept 9.3). The **regional species pool**, sometimes called the **gamma diversity** of the region, encompasses all the species contained within a region (**FIGURE 18.5B**). Earth's regions differ in species diversity and composition because of differences in the rates of speciation, extinction, and dispersal at the global scale, as mentioned above. The Amazon, for example, has many more species, and thus a larger species pool, than the Canadian boreal forest.

The physical geography of a region, such as the number, area, and distance from one another of mountains, valleys, deserts, islands, and lakes—referred to collectively as the *landscape*—is critical to within-region biogeography. Species diversity and composition vary within a region depending on how the landscape shapes the rate of extinction in, and the rates of immigration to and emigration from, local habitats (**FIGURE 18.5C**). Ecologists consider within-region biogeography in two related ways:

- The **local scale**, which is essentially equivalent to a community, reflects the suitability of the abiotic and biotic characteristics of habitats for species from the regional species pool once they reach those habitats through dispersal (**FIGURE 18.5D**). Species physiology and interactions with other species both influence

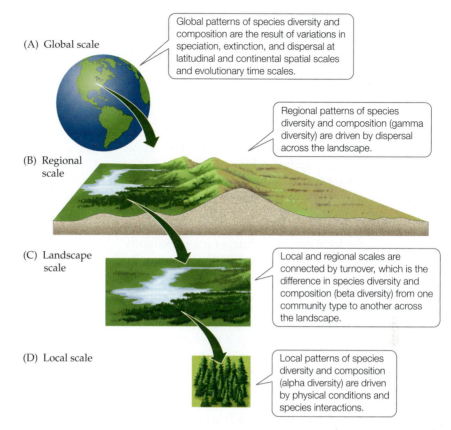

(A) Global scale

Global patterns of species diversity and composition are the result of variations in speciation, extinction, and dispersal at latitudinal and continental spatial scales and evolutionary time scales.

(B) Regional scale

Regional patterns of species diversity and composition (gamma diversity) are driven by dispersal across the landscape.

(C) Landscape scale

Local and regional scales are connected by turnover, which is the difference in species diversity and composition (beta diversity) from one community type to another across the landscape.

(D) Local scale

Local patterns of species diversity and composition (alpha diversity) are driven by physical conditions and species interactions.

FIGURE 18.5 Interconnected Spatial Scales of Species Diversity The arrows represent the relationships between, and processes important to, species diversity and composition at (A) global, (B) regional, (C) landscape, and (D) local scales.

species diversity at the local scale (sometimes called **alpha diversity**).

- The connection between local and regional scales of species diversity is expressed by a measurement known as **beta diversity**. Beta diversity tells us the change in species diversity and composition, or **turnover** of species, as one moves from one community type to another across the landscape (see Figure 18.5C).

Knowing how spatial scales are related to one another in a hierarchical way is important, but are there actual area values one could apply to local and regional spatial scales? For example, how much area does a region or locality encompass? The answer is highly dependent on the species and communities of interest. For example, Shmida and Wilson (1985) suggest that terrestrial plants might have a local scale of 10^2–10^4 m^2 and a regional scale of 10^6–10^8 m^2. But for bacteria, the local scale might be something more like 10^2 cm^2. As we will see, the actual area we use to define species diversity measurements can be critical to our interpretation of the processes controlling biogeographic patterns.

Local and regional processes interact to determine local species diversity

Figure 18.5 shows that patterns of species diversity, and the processes that control them, are interconnected across spatial scales. Given these interconnections, ecologists are interested in knowing just how much variation in species diversity at the local scale is dependent on larger spatial scales. The regional species pool provides the raw material for local species assemblages and sets the theoretical upper limit on species richness for communities in the region. But is local species richness also determined by local conditions, including species interactions and the physical environment?

One way we can consider this question quantitatively is by plotting the local species richness for a community against the regional species richness for that community (**FIGURE 18.6**). Three basic types of relationships can be seen in such plots. First, if local species richness and regional species richness are equal (slope = 1), then all the species within a region will be found in the communities of that region. Although this pattern is theoretically possible, we would not expect to find it in the real world, for the simple reason that all regions have varying landscape and habitat features that exclude some species from some communities (e.g., lowland tree species will not be found in alpine forests). Second, if local species richness is simply proportional to regional species richness (i.e., local species richness increases with increasing regional species richness, but the relationship is not 1:1), then we can assume that local species richness is largely determined by the regional species pool, with local processes such

as species interactions and physical conditions playing a more minor role. Finally, if local species richness levels off despite an increasing regional species pool, then local processes can be assumed to limit local species richness. The degree to which local richness levels off can tell us something about how important species interactions and physical conditions are in setting a *saturation point*—a limit on species richness—for communities.

Let's move away from these theoretical constructs and look at what real data show us about the relationship between local and regional species richness. Witman and colleagues (2004) considered this relationship for marine invertebrate communities living on subtidal rock walls at a variety of locations throughout the world (**FIGURE 18.7A**). At 49 local sites in 12 regions, they surveyed species richness in 0.25-m² plots on rock walls at a 10–15 m (33–50 feet) water depth. They then compared the local species richness values they found at the sites with regional species richness values from published lists of invertebrate species capable of living on hard substrates at similar depths. A plot of local versus regional species richness at all the sites (**FIGURE 18.7B**) showed that local species richness was always proportionally lower than regional species richness. Furthermore, local species richness never leveled off—that is, the communities never became saturated—at high regional richness values. Instead, regional species richness explained approximately 75% of the variation in local species richness. The results of this study suggest that regional species pools largely determine the number of species present in these marine invertebrate communities.

Does the lack of saturation detected in this study and others indicate that local processes are unimportant in determining local species richness? The answer is no, for at least two reasons. First, there was still considerable unexplained variation among local communities within regions, which could be attributable to the effects of local processes such as species interactions, abiotic conditions, or dispersal limitation (see Figure 19.4). Second, the effects of species interactions, in particular, are likely to be highly sensitive to the local spatial scale chosen. Although the small spatial scale of Witman and colleagues' study is probably appropriate for species interacting on subtidal rock walls, other studies have used inappropriate (usually too large) spatial scales that were unlikely to detect local effects. Nevertheless, the strong influence of regional-scale processes on local species richness suggests that both marine and terrestrial communities are likely to be much more susceptible to changes such as species invasions from outside their regions than previously thought.

In the remainder of this chapter, we will explore the factors controlling variation in species diversity at global and regional biogeographic scales. Chapter 19 will delve in more detail into the causes and consequences of species diversity differences at the local scale.

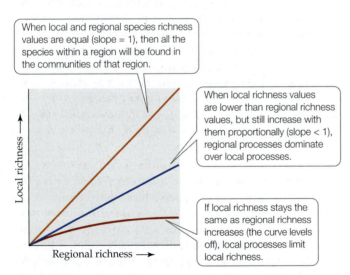

When local and regional species richness values are equal (slope = 1), then all the species within a region will be found in the communities of that region.

When local richness values are lower than regional richness values, but still increase with them proportionally (slope < 1), regional processes dominate over local processes.

If local richness stays the same as regional richness increases (the curve levels off), local processes limit local richness.

FIGURE 18.6 What Determines Local Species Richness? The relative influences of local and regional processes in a community can be determined by plotting local species richness against regional species richness. (After H. V. Cornell and J. H. Lawton. 1992. *J Anim Ecol* 61: 1–12.)

? Would you ever have a local to regional species richness relationship that had a slope of more than 1? Why or why not?

(A) Study sites

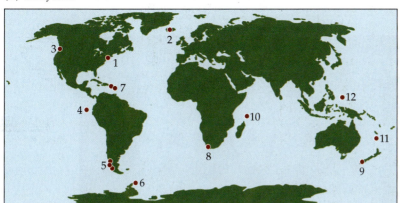

1	Gulf of Maine
2	Iceland
3	Northeastern Pacific
4	Galápagos Islands
5	Chilean Patagonia
6	Antarctic Peninsula
7	Eastern Caribbean
8	Southwestern Africa
9	Southwestern New Zealand
10	Seychelles
11	Norfolk Island
12	Palau

(B) Local versus regional species richness

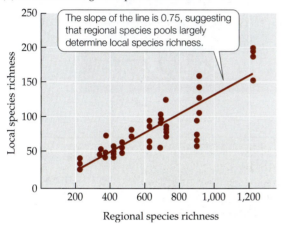

The slope of the line is 0.75, suggesting that regional species pools largely determine local species richness.

FIGURE 18.7 **Marine Invertebrate Communities May Be Limited by Regional Processes**   Among shallow subtidal marine invertebrate communities, regional species richness explains approximately 75% of the local species richness. (A) The 12 regions of the world where the 49 sampling sites were located. (B) A plot of local species richness against regional species richness. Each dot represents one of the 49 sampling sites. (After J. D. Witman et al. 2004. *Proc Natl Acad Sci USA* 101: 15664–15669. © National Academy of Sciences, U.S.A.)

CONCEPT 18.2   **Global patterns of species diversity and composition are influenced by geographic area and isolation, evolutionary history, and global climate.**

LEARNING OBJECTIVES

18.2.1 Describe the two major biogeographic patterns—biogeographic regions and latitudinal gradients in species diversity—at the global scale.

18.2.2 Explain the underlying forces thought to be important in creating biogeographic regions.

18.2.3 Outline the hypotheses proposed to explain the latitudinal gradient in species diversity pattern.

Global Biogeography

It must have been incredible to be a European scientific explorer 200 years ago. You would have left the safety of your home to travel by ship to a destination largely unknown. You would have had to endure seasickness, disease, accidents of all kinds, and years away from your family, friends, and colleagues. You might have had many years of financial debt to repay unless you were independently wealthy or could sell your collections. But you would have been the first scientist to document and collect animal and plant species of beauty, novelty, and rarity. It was under these circumstances that the science of biogeography was born and many important discoveries were made. Up to that point, European scientists had very little information about the natural history and ecology of other parts of the world; most was secondhand or anecdotal. What these early naturalists were able to bring back were specimens and, most of all, theories to help make sense of their observations.

Although not the first in his field, Alfred Russel Wallace (1823–1913) rightly earned his place as the father of biogeography (**FIGURE 18.8**). Inspired by naturalists such as Alexander von Humboldt, Charles Darwin, and Joseph Hooker, Wallace came on the scene with considerably less wealth or education, but his intellect and motivation more than made up for what he lacked in financial resources and training. Wallace is best known, along with Charles Darwin, as the co-discoverer of the principles of natural selection, although he has always stood in the shadow of Darwin in that regard. But his main contribution was the study of species distributions across large spatial scales.

Wallace left England for Brazil in 1848 and explored the Amazon rainforest for 4 years. On his way back, the ship he was traveling on burned in the middle of the Sargasso Sea, destroying all his specimens and most of his notes and illustrations. After 10 days in a lifeboat, he was rescued and made his way back to England, where he published an impressive six papers on his observations.

Even though he had vowed never to travel again, in 1852 Wallace left England for the Malay Archipelago (present-day Indonesia, the Philippines, Singapore, Brunei,

(A)

(B)

FIGURE 18.8 Alfred Russel Wallace and His Collections (A) A photograph of Wallace taken in Singapore in 1862, during his expedition to the Malay Archipelago. (B) Part of Wallace's rare beetle collection from the Malay Archipelago, found in an attic by his grandson in 2005. (C) A map of the Malay Archipelago illustrating Wallace's travels.

(C)

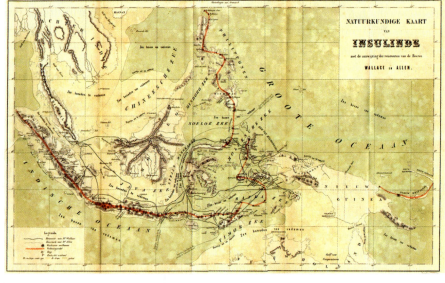

East Malaysia, and East Timor). It was here that he made the puzzling observation described in his 1869 book *The Malay Archipelago*: that the mammals of the Philippines were more similar to those in Africa (5,500 km away) than they were to those in New Guinea (750 km away). Wallace was the first to notice the clear demarcation between these two faunas, which came to be known as *Wallace's line*. It turns out, as we'll see shortly, that these separate groups of mammals evolved on two different continents that have come into close proximity only within the last 15 million years.

Wallace's biogeographic research culminated in the publication of a two-volume work called *The Geographical Distribution of Animals*, published in 1876. In this book, Wallace overlays species distributions on top of geographic regions and reveals two important global patterns:

- Earth's land masses can be divided into six recognizable **biogeographic regions** containing distinct biotas that differ markedly in species diversity and composition.

- There is a gradient of species diversity with latitude: species diversity is greatest in the tropics and decreases toward the poles.

These two patterns are necessarily interrelated; the latitudinal gradient is superimposed over the biogeographic regions. For ease of explanation, we'll begin by exploring the biogeographic regions described by Wallace and the underlying forces that created them. We will then consider some of the processes likely to be responsible for the latitudinal gradient in species diversity.

The biotas of biogeographic regions reflect evolutionary isolation

The six biogeographic regions described by Wallace are the Nearctic (North America), Neotropical (Central and South America), Palearctic (Europe and parts of Asia and Africa), Ethiopian (most of Africa), Oriental (India, China, and Southeast Asia), and Australasian (Australia, the Indo-Pacific, and New Zealand) (**FIGURE 18.9**). It is no coincidence that these regions correspond roughly to Earth's six major tectonic plates. These plates are sections of Earth's crust that move across Earth's surface through the action of currents generated deep within its molten rock mantle (**FIGURE 18.10**). Before scientists understood the processes driving the movement of these plates, they hypothesized that the continents drifted over Earth's surface; thus, the name **continental drift** was given to the early theory describing these movements. There are three major types of boundaries between tectonic plates. In areas known as *mid-ocean ridges*, molten rock flows out

of the seams between plates and cools, creating new crust and forcing the plates apart in a process called *seafloor spreading*. In some areas where two plates meet, known as *subduction zones*, one plate is forced downward under another plate. These areas are associated with strong earthquakes, volcanic activity, and mountain range formation. In other areas where two plates meet, the plates slide sideways past each other, forming a *fault*.

As a result of processes such as seafloor spreading and subduction, the positions of the plates, and of the continents that sit on them, have changed dramatically over geologic time. For our purposes, let's consider the movements of the major tectonic plates since the early Triassic period (251 million years ago), when all of Earth's land masses, a single continent named Pangaea, began to break up (**FIGURE 18.11A**). At that time, there was a mass extinction (see Figure 6.19), which eventually led to the rise of the first archosaurs (precursors to dinosaurs) and the cynodonts (precursors to mammals). About 100 million years ago, during the mid-Cretaceous period, Pangaea had split into Laurasia to the north and Gondwana to the south. During that time, dinosaurs were in their heyday and mammals were small and a relatively minor component of the fauna. The end of the Cretaceous period was marked by another mass extinction, which resulted in the disappearance of dinosaurs. By the early Paleogene period (60 million years ago), Gondwana had separated into the

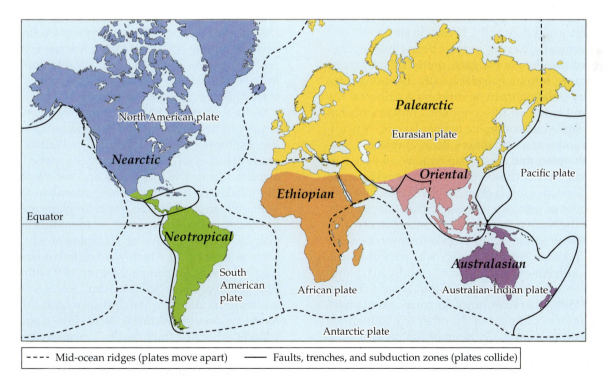

- - - - Mid-ocean ridges (plates move apart) ——— Faults, trenches, and subduction zones (plates collide)

FIGURE 18.9 Six Biogeographic Regions Wallace identified six biogeographic regions using the distributions of terrestrial animals. These six regions roughly correspond to Earth's major tectonic plates. (Based on A. R. Wallace. 1876. *The Geographical Distribution of Animals.* Harper and Brothers: New York.)

? Compare Wallace's 6 regions with the 11 biogeographic divisions shown in Figure 1.2. What types of data were used to expand the number of regions to 11?

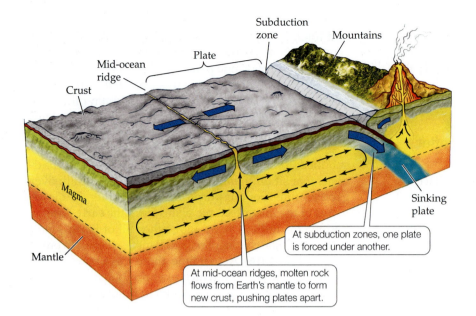

FIGURE 18.10 Mechanisms of Continental Drift Over geologic time, currents generated deep within Earth's molten rock mantle move sections of Earth's crust across its surface.

At subduction zones, one plate is forced under another.

At mid-ocean ridges, molten rock flows from Earth's mantle to form new crust, pushing plates apart.

present-day continents of South America, Africa, India, Antarctica, and Australia. Laurasia eventually split apart to form North America, Europe, and Asia. Most of these movements resulted in the separation of continents from one another, but some continents were brought together (**FIGURE 18.11B**). For example, North and South America joined at the Isthmus of Panama, India collided with Asia to create the Himalayas, Africa and Europe united at the Mediterranean Sea, and a land bridge formed between North America and Asia at the Bering Strait.

The movement of Earth's tectonic plates thus separated the terrestrial biota of Pangaea, united by geography and phylogeny, into biogeographically distinct groups of species by isolating them on different continents. The sequence and tempo of the continental movements has resulted in some biogeographic regions having very different flora and fauna than others. For example, the Neotropical, Ethiopian, and Australian regions, all once part of Gondwana, have been isolated for quite some time and have very distinctive forms of life. In other cases, however, distinct groups of species have been united. For example, the biota of the Nearctic region differs substantially from that of the Neotropical region despite their modern-day proximity. Because North America was part of Laurasia while South America was part of Gondwana, North and South America had no contact until about 6 million years ago. Within that time, however, many species have moved from one continent to the other (e.g., mountain lions, wolves, and the precursors of llamas spread to South America, while armadillos and opossums spread to North America), somewhat homogenizing the biotas of the two regions. Interestingly, there is also evidence that several families of terrestrial mammals went extinct once the two continents merged, suggesting that ecological coexistence was not possible for some species (Flessa 1975). Finally, the Nearctic and Palearctic, both part of ancient

Laurasia, have similarities in biota across what is now Greenland as well as across the Bering Strait, where a land bridge has intermittently allowed exchanges of species over the last 100 million years.

The legacy of continental movements can be found in a number of existing taxonomic groups as well as in the fossil record. The evolutionary separation of species due to barriers such as those formed by continental drift is known as **vicariance**. Tracing the threads of vicariance over large geographic areas and long periods provided important evidence for early theories of evolution. For example, as Wallace began to amass knowledge of the distributions of more and more species and make geographic connections between them, his ideas about the origin of species started to solidify. In an 1855 paper titled "On the law which has regulated the introduction of new species," he wrote, "Every species has come into existence coincident both in space and time with a pre-existing closely allied species." Despite the biogeographic evidence of evolutionary connections among species, it took a few more years for one mechanism of evolution (i.e., natural selection) and its role in the origin of new species to be formally proposed by both Wallace (1858) and Darwin (1859).

Before we move on, it is important to consider contemporary research that updates and expands on the biogeographic regions first identified by Wallace. One recent study (Holt et al. 2013) used phylogenetic information acquired from DNA analysis and recent observations of global species distribution patterns to test whether Wallace's original biogeographic regions are supported by modern data. The researchers identified more biogeographic regions (a total of 11), some of which were the same and others of which were different from Wallace's original 6 regions (compare Figure 18.9 with Figure 1.2). This new analysis suggests that additional isolation mechanisms beyond continental drift are responsible for the different regions. Interestingly, New Guinea and the Pacific Islands, separated from the Philippines by Wallace's line, emerge as a new biogeographic region, completely separate from the Australian or Oriental regions.

Another recent analysis of biogeographic regions involved mapping the distribution of species in the oceans. After all, the oceans make up 71% of Earth's surface area and, just as we have seen for continents, they are dynamic, in the sense that they are created, merged, and destroyed

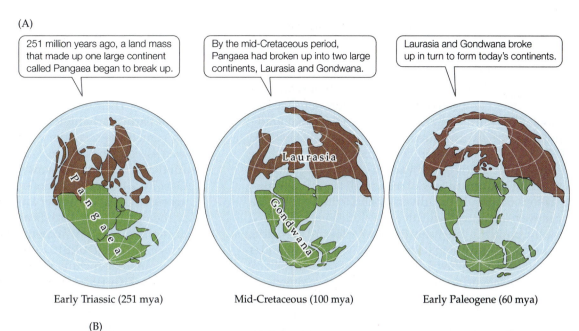

(A)

251 million years ago, a land mass that made up one large continent called Pangaea began to break up.

By the mid-Cretaceous period, Pangaea had broken up into two large continents, Laurasia and Gondwana.

Laurasia and Gondwana broke up in turn to form today's continents.

Early Triassic (251 mya)   Mid-Cretaceous (100 mya)   Early Paleogene (60 mya)

(B)

FIGURE 18.11 The Positions of Continents and Oceans Have Changed over Geologic Time The locations of continents and oceans have changed dramatically over the last 251 million years because of continental drift. (A) The breakup of Pangaea. (B) A summary of the movements that led to the configuration of the continents we know today. Larger, red arrows are labeled with the time (in millions of years) since land masses joined; Smaller, black arrows are labeled with the time since land masses separated. (After E. C. Pielou. 1979. *Biogeography.* Wiley: Hoboken, NJ.)

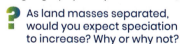 As land masses separated, would you expect speciation to increase? Why or why not?

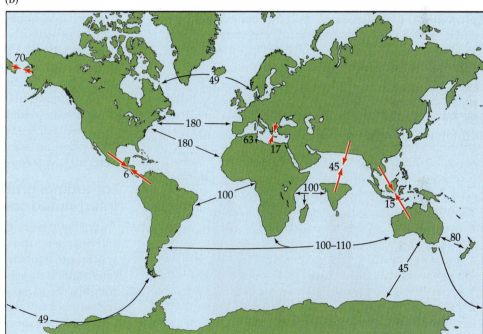

by the movements of Earth's tectonic plates (see Figure 18.11). The main question, then, is whether there are barriers to dispersal between oceans as there are between continents. Despite their appearance of connectivity, oceans do have significant impediments to the exchange of biotas: these impediments take the form of continents and currents; thermal, salinity, and oxygen gradients; and differences in water depth. Oceanographic discontinuities have isolated species from one another, allowed for evolutionary change, and created unique oceanic biogeographic regions (Briggs 2006). Unfortunately, delineation of marine biogeographic regions has been hindered by the extra complicating factor of water depth and by our basic lack of natural history and taxonomic knowledge of the deep oceans. One recent model by Adey and Steneck (2001) identifies 24 recognizable biogeographic regions for intertidal benthic marine macroalgae. Although it is hard to compare these macroalgal regions with terrestrial biogeographic regions, the analysis does suggest that the marine realm has much more biogeographic variation than previously realized.

Species diversity varies with latitude

If you recall our Google Earth–style tour of the globe in the previous section, it was clear that plant species diversity and community composition changed dramatically with latitude: species diversity was highest at tropical latitudes and decreased toward the poles. Wallace and other nineteenth-century European scientific explorers became keenly aware of this pattern as they collected thousands of species in the tropics and compared them with their more meager European collections. As more data have accumulated over the last 200 years, the latitudinal gradient in species diversity has been more firmly established (**FIGURE 18.12**). Willig and colleagues (2003) tallied the results of 162 studies on a variety of taxonomic groups extending over broad spatial scales (20° latitude or more) that considered whether diversity and latitude showed a negative relationship (with diversity decreasing toward the poles), a positive relationship (increasing toward the poles), a unimodal relationship (increasing toward mid-latitudes and then declining toward the poles), or no relationship. Negative relationships were by far the most common.

In addition to this undeniably strong latitudinal gradient, biogeographers have observed an important pattern of longitudinal variation. Gaston et al. (1995) measured the numbers of families along multiple transects running north to south and separated by 10° longitude. Families

of seed plants, amphibians, reptiles, and mammals all increased in number toward the equator and declined at higher latitudes in both the Northern and Southern Hemispheres. These researchers determined, however, that the number of families also depended on the longitude chosen. Their observations showed that there were areas of particularly high species richness at some locations, sometimes secondary to latitude. These areas are known as *biodiversity hot spots* in situations in which they are under threat from humans.

Of course, not all groups of organisms show decreases in species richness at higher latitudes; some groups display the opposite pattern. Seabirds, for example, have their highest diversity at temperate and polar latitudes (Harrison 1987) (**FIGURE 18.13A**). Seabirds of the Antarctic and subantarctic include penguins, albatrosses, petrels, and skuas (**FIGURE 18.13B**). In the Arctic and subarctic, auks replace penguins, and gulls, terns, and grebes are common. In the tropics and subtropics, seabird diversity declines: the seabird community there is composed mostly of pelicans, boobies, cormorants, and frigatebirds. This pattern of seabird diversity correlates well with marine productivity, which is substantially higher in temperate and polar oceans than in the tropics (see Figure 20.7). The same pattern of diversity has been observed in marine benthic communities, which also experience much higher productivity at higher latitudes.

As we will see, productivity differences are one possible cause of latitudinal gradients in species diversity. Let's turn now to some other possible explanations.

Latitudinal gradients have multiple, interrelated causes

As we have seen, global patterns of species richness are ultimately controlled by the rates of three processes: speciation, extinction, and dispersal. Let's assume here, for simplicity's sake, that the rate of species dispersal is roughly the same worldwide. We can then predict that the number of species at any particular location will reflect a balance, or equilibrium, between the rates of two fundamental processes: speciation and extinction. Subtracting the extinction rate from the speciation rate gives us the rate of species diversification: the net increase or decrease of species diversity over time. What ultimately controls this rate? Dozens of hypotheses have been proposed to explain species diversification with latitude, but there is very little agreement among biogeographers and ecologists. Part of the reason lies in the fact that there are multiple and confounding latitudinal gradients in area, evolutionary age, and climate that are correlated with species diversity gradients. In addition, because speciation and extinction occur at a global spatial scale and over evolutionary time scales, it is impossible to conduct manipulative experiments to isolate various factors and separate correlation from causation.

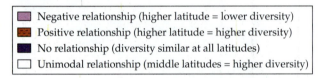

- ■ Negative relationship (higher latitude = lower diversity)
- ■ Positive relationship (higher latitude = higher diversity)
- ■ No relationship (diversity similar at all latitudes)
- □ Unimodal relationship (middle latitudes = higher diversity)

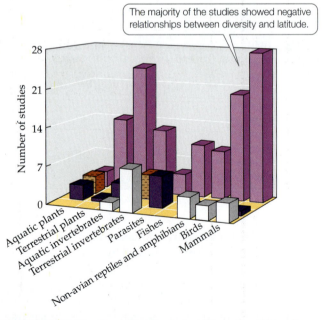

The majority of the studies showed negative relationships between diversity and latitude.

FIGURE 18.12 Studies of Latitude and Species Diversity Confirm Conventional Wisdom The relationship between species diversity and latitude (measured at 20° increments), tallied for a variety of taxonomic groups, shows that most are negative correlations (i.e., increasing species diversity with decreasing latitude). (After M. R. Willig et al. 2003. *Annu Rev Ecol Syst* 34: 273–309.)

(A) Global seabird diversity

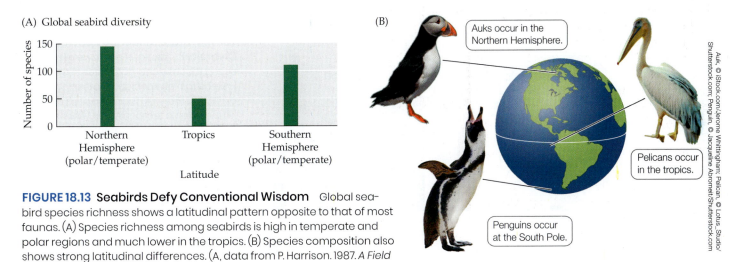

FIGURE 18.13 **Seabirds Defy Conventional Wisdom** Global sea-bird species richness shows a latitudinal pattern opposite to that of most faunas. (A) Species richness among seabirds is high in temperate and polar regions and much lower in the tropics. (B) Species composition also shows strong latitudinal differences. (A, data from P. Harrison. 1987. *A Field Guide to Seabirds of the World.* Penguin Random House: London.)

In an effort to summarize the most convincing ideas, Gary Mittelbach and colleagues (2007) suggested that hypotheses proposed to explain latitudinal gradients in species richness fall into three broad categories. The first category of hypotheses is based on the assumption that the rate of species diversification in the tropics is greater than that in temperate regions (**FIGURE 18.14A**). The second category of hypotheses suggests that the rates of diversification in the tropics and at higher latitudes are similar, but that the evolutionary time available for diversification has been much greater in the tropics (**FIGURE 18.14B**). The third category of hypotheses suggests that resources are more plentiful in the tropics because of higher productivity, and thus that species there have higher carrying capacities and a greater ability to

coexist (**FIGURE 18.14C**). Let's take a look at each category of hypotheses in more detail.

SPECIES DIVERSIFICATION RATE There are a number of hypotheses that seek to explain why species diversification might be higher in the tropics. One hypothesis relates diversification to geographic area and temperature. John Terborgh (1973) and Michael Rosenzweig (1992) proposed that terrestrial species diversity is highest in the tropics because the tropics have the largest land area (**FIGURE 18.15A**). Rosenzweig calculated that the region between 26°N and S has 2.5 times more land area than any other latitude range on Earth. This makes intuitive sense, given that this latitude range is at the middle, and thus at the widest part, of the planet. Equally interesting

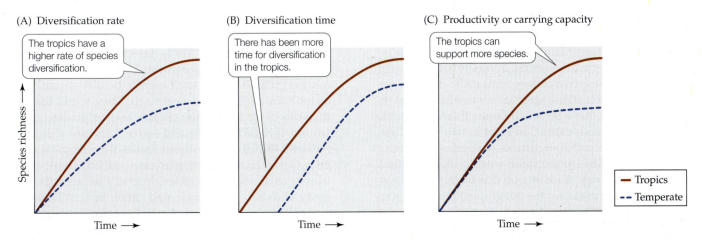

FIGURE 18.14 **Hypotheses Proposed to Explain the Latitudinal Gradient in Species Richness** (A) The tropics have a higher diversification rate (speciation rate – extinction rate) than temperate areas do, so they have accumulated species faster. (B) The tropics have had more time for diversification than temperate areas have, so they have accumulated more species. (C) Because their productivity is higher, the tropics have a higher carrying capacity than temperate areas, so more species can coexist there. (After G. G. Mittelbach et al. 2007. *Ecol Lett* 10: 315–331.)

(A) Land area

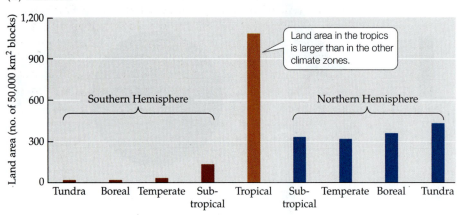

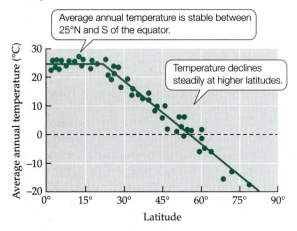

(B) Temperature

FIGURE 18.15 **Do Land Area and Temperature Influence Species Diversity?** Michael Rosenzweig hypothesized that two characteristics of the tropics lead to high speciation rates and low extinction rates: (A) their land area and (B) their stable temperatures. (After M. L. Rosenzweig. 1992. *J Mammol* 73: 715–730.)

SPECIES DIVERSIFICATION TIME   The second category of hypotheses, which proposes that latitudinal gradients in species diversity are influenced by evolutionary history, was first championed by Wallace (1878). He suggested that tropical regions, because they are thought to have been more climatically stable over time (see Figure 18.15B), could have considerably longer evolutionary histories than temperate or polar regions, where severe climate conditions (such as ice ages) might have disrupted species diversification. Thus, even if rates of speciation and extinction were the same worldwide, the tropics should have accumulated more species over time merely because species should have had more uninterrupted time to evolve there.

With these ideas in mind, we can consider another possibility: that most species actually originate in the tropics and then disperse to temperate regions during warmer periods of greater climate homogeneity. The idea that the tropics serve as a "cradle" for diversity was originally proposed by Stebbins (1974). Jablonski et al. (2006) recently examined this hypothesis by comparing modern marine bivalve faunas with marine bivalve fossils from as far back as 11 million years ago. They found that the majority of extant marine bivalve taxa originated in the tropics (**FIGURE 18.16A**) and spread toward the poles (**FIGURE 18.16B**), but without losing their tropical presence. Thus, in this particular case, we can think of the tropics as a cradle of species diversity because the majority of extant taxa originated there. But, as Jablonski and colleagues also pointed out, the tropics can serve as a "museum" as well as a cradle. If extinction rates in the tropics are low, then species that diversify there will tend to stay there "on display," if you will. Jablonski and colleagues suggested that the current loss of biodiversity in the tropics is likely to have profound effects because it not

are data showing that this very large area is also the most thermally homogeneous region on Earth (**FIGURE 18.15B**). A plot of average annual temperature against latitude by Terborgh showed that land temperatures are remarkably uniform over a wide area between 25°N and S, but then drop off rapidly at higher latitudes.

Why would a larger land area and more constant temperatures foster greater species diversity? Rosenzweig suggested that these two factors combine to decrease extinction rates and increase speciation rates in tropical regions. He argued that a larger and more thermally stable area should decrease extinction rates in two ways: first, by increasing the population sizes of species (assuming that their densities are the same worldwide), and thus decreasing their risk of extinction due to chance events, and second, by increasing the geographic ranges of species, and thus decreasing their chances of extinction by spreading the risk over a larger geographic area (see Concept 10.2). He further suggested that speciation should increase in larger areas because species should have larger geographic ranges, and thus should have a greater chance of reproductive isolation of populations (see Concept 6.4).

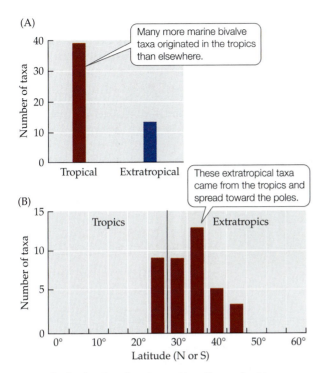

FIGURE 18.16 The Tropics Are a Cradle and a Museum for Speciation Extant and fossil marine bivalve taxa were examined to evaluate the hypothesis that longer evolutionary histories in the tropics contribute to the latitudinal gradient in species diversity. (A) Climate zones of first occurrence of marine bivalve taxa (based on families of fossils). (B) Range limits of modern marine bivalve taxa with tropical origins. (After D. Jablonski et al. 2006. *Science* 314: 102–106.)

❓ What is meant by the tropics being a cradle and a museum for diversity?

only compromises species richness today, but also could conceivably cut off the supply of new species to higher latitudes in the future.

PRODUCTIVITY The final category of hypotheses for the latitudinal gradient in species diversity that we will consider is based on resources—in particular, productivity. The productivity hypothesis, proposed as long ago as 1959 by G. E. Hutchinson, posits that species diversity is higher in the tropics because that is where productivity is highest, at least for terrestrial systems (see Figure 20.7). The thought is that higher productivity promotes larger population sizes because species will have higher carrying capacities. This higher productivity will lead to lower extinction rates, greater species coexistence, and overall higher species richness. The productivity hypothesis might explain why we see a reversal in the latitudinal gradient for some marine organisms, such as seabirds (see Figure 18.13), given that productivity is generally higher in temperate coastal marine habitats than

in tropical regions (see Figure 20.7). But we also know that some of the most productive habitats on Earth, such as estuaries, typically have very low species diversity. Suffice it to say, the productivity hypothesis is complex and unsatisfactory in many cases. In Chapter 19, we will consider how productivity influences diversity at local scales, where manipulative experiments can give us more insight into its effects.

CLIMATE CHANGE CONNECTION

LATITUDINAL GRADIENTS IN DIVERSITY UNDER CLIMATE CHANGE One way to explore the potential causes of latitudinal gradients in species diversity is to consider them over evolutionary time and with major changes in climate. We can ask, Did the fundamental pattern of increasing species diversity toward the tropics exist in the past, and if not, why? Philip Mannion and colleagues (2014) used the fossil record and fluctuations in past global temperatures as a window into past latitudinal and species diversity gradients and their potential causes. Their analysis showed that a tropical peak in species diversity has not been a universal pattern but has been restricted to particular intervals of time throughout the Phanerozoic when the Earth experienced colder, "icehouse" conditions (**FIGURE 18.17**). Likewise, they found that during warmer, "greenhouse" conditions, species diversity peaked in temperate latitudes, exhibiting a more unimodal relationship. These switches from temperate to tropical peaks in species diversity corresponded to transitions in greenhouse to icehouse climate conditions, lending support for the notion that polar to temperate glaciations could drive species to the tropics where extinctions would be lower. Conversely, during greenhouse conditions, the tropics might become too hot for many organisms, leading to increased extinction rates and dispersal out of the tropics. One might speculate, in fact, that with global warming, latitudinal gradients in species diversity could become shallower or more unimodal as warming causes species to disperse poleward or become increasingly extinct within tropical latitudes. ☀

As we have seen, biogeographic patterns have motivated and inspired some of the best and brightest scientists of modern times. Their fascination with the differences in the numbers and kinds of species at the global scale and their overwhelming drive to understand why these differences exist have resulted in some of the most influential scientific theories of all time, including that of the origin of species. In the next section, we will consider another important theory that strives to understand species diversity at smaller spatial scales.

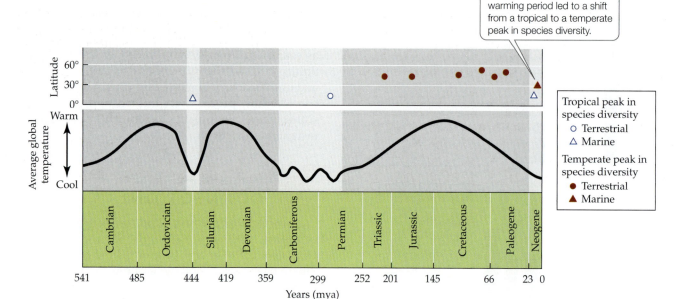

During the Neogene, a short warming period led to a shift from a tropical to a temperate peak in species diversity.

Tropical peak in species diversity
○ Terrestrial
△ Marine

Temperate peak in species diversity
● Terrestrial
▲ Marine

FIGURE 18.17 Latitudinal Species Diversity Gradients Vary with Climate The latitudinal species diversity gradients under fluctuating global temperature through the Phanerozoic. Tropical peaks in species diversity (open symbols) are restricted to cold, "icehouse" conditions, whereas temperate peaks in diversity (solid symbols) occur during warm, "greenhouse" intervals. Note that during the icehouse conditions of the Neogene, a short warming period during the Pleistocene interglacial period led to a peak in diversity at temperate latitudes. Circles are terrestrial examples, and triangles are marine examples. (After P. D. Mannion et al. 2014. *Trends Ecol Evol* 29: 42–50. CC BY 3.0)

CONCEPT 18.3 Regional differences in species diversity are influenced by area and distance, which determine the balance between immigration and extinction rates.

LEARNING OBJECTIVES

18.3.1 Graph and explain the species–area relationship and know why it differs between islands and mainland areas.

18.3.2 Explain regional species diversity for islands and island-like areas using the equilibrium theory of island biogeography.

Regional Biogeography

An important thread that runs through this chapter, and through biogeography generally, is the relationship between species richness and geographic area. We saw in the Case Study that large fragments of Amazon rainforest had greater species richness than smaller fragments. In our global tour of the world's forests, we saw that species diversity was greatest in the tropics (see Table 18.1), the climate zone whose geographic area is largest (see Figure 18.15A). This so-called **species–area relationship**, in which species richness increases with the area sampled, has been documented at a variety of spatial scales, from small ponds to whole continents. Most studies of species–area relationships have been targeted at regional spatial scales, where these relationships tend to be good predictors of differences in species richness.

Species richness increases with area and decreases with distance

In 1859, H. C. Watson plotted the first curve showing a quantitative species–area relationship—in this case, for plants within Great Britain (**FIGURE 18.18**) (Williams 1943). The curve starts with a small "bit" of the county of Surrey and expands to ever-increasing areas that eventually

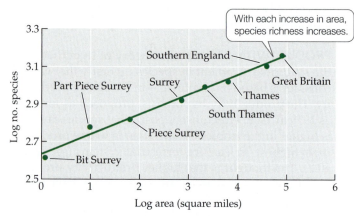

With each increase in area, species richness increases.

FIGURE 18.18 The Species–Area Relationship The first species–area curve, for British plants, was constructed by H. C. Watson in 1859. (After M. Rosenzweig. 1995. *Species Diversity in Space and Time.* Cambridge University Press: Cambridge; based on data in C. B. Williams. 1964. *Patterns in the Balance of Nature.* Academic Press: London; H. C. Watson. 1859. *Cybele Britannica: or British Plants and Their Geographical Relations* 4: 379. Longman and Company: United Kingdom.)

encompass all of Surrey, southern England, and finally Great Britain. With each increase in area, species richness increases until it reaches a maximum number bounded by the largest area considered. (**ECOLOGICAL TOOLKIT 18.1** and **ANALYZING DATA 18.1** provide further insight on how species–area curves are plotted and interpreted.)

Most species–area relationships have been documented for islands (**FIGURE 18.19**). Islands, in this case, include all kinds of isolated areas surrounded by a "sea" of dissimilar habitat (referred to as *matrix* habitat). So "islands" can include real islands surrounded by ocean, lake "islands" surrounded by land, or mountain "islands" surrounded by valleys. They can also include habitat fragments, like those produced by the deforestation of the Amazon (see Figure 18.2). Nonetheless, all of these islands and island-like habitats display the same basic pattern: large islands have more species than small islands.

In addition, because of the isolated nature of islands, species diversity on islands shows a strong negative relationship to distance from the main source of species. For example, Lomolino et al. (1989) found that mammal species richness on mountaintops in the American

---

ECOLOGICAL TOOLKIT 18.1

Species–Area Curves

Species–area curves are the result of plotting the species richness (S) of a particular sample against the area (A) of that sample. A linear regression equation estimates the relationship between S and A in the following manner:

$$S = zA + c$$

where z is the slope of the line and c is the y intercept of the line.

Because species–area data are typically nonlinear, ecologists transform S and A into logarithmic values so that the data fall along a straight line and conform to a linear regression model.

The figure shows species–area curves for plants on the Channel Islands (off the coast of France) and on the French mainland (Williams 1964). Log transformations were conducted on both the island and mainland data, the two data sets were plotted separately, and a linear model was used to estimate the best-fit curve for each of the data sets.

An important characteristic of species–area curves is evident in this figure: the steeper the slope of the line (i.e., the greater the z value), the greater the difference in species richness among the sampling areas. The Channel Islands have a much steeper slope than the French mainland areas, for the reasons outlined at the end of Concept 18.3.

Species–Area Relationships of Island versus Mainland Areas Species–area curves for plant species on the Channel Islands and in mainland France show that the slope of a linear regression equation (z) is greater for the islands than for the mainland areas. (After M. Rosenzweig. 1995. *Species Diversity in Space and Time*. Cambridge University Press: Cambridge; based on data in C. B. Williams. 1964. *Patterns in the Balance of Nature*. Academic Press: London.)

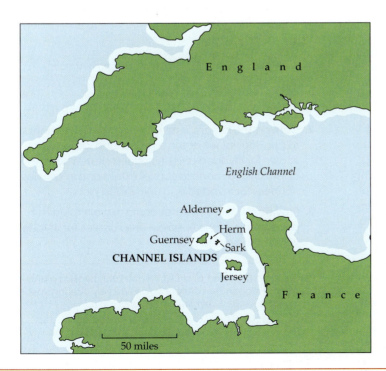

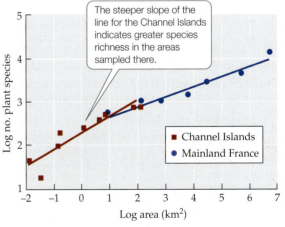

The steeper slope of the line for the Channel Islands indicates greater species richness in the areas sampled there.

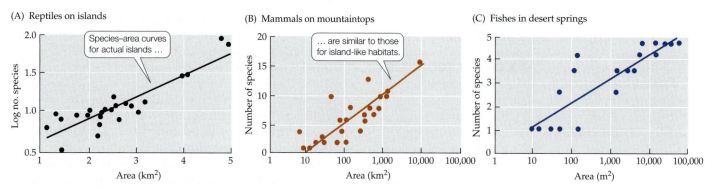

(A) Reptiles on islands

Species–area curves for actual islands …

(B) Mammals on mountaintops

… are similar to those for island-like habitats.

(C) Fishes in desert springs

FIGURE 18.19 **Species–Area Curves for Islands and Island-Like Habitats**
Species–area curves plotted for (A) reptiles on Caribbean islands, (B) mammals on mountaintops in the American Southwest, and (C) fishes living in desert springs in Australia all show a positive relationship between area and species richness. (A after S. J. Wright. 1981. *Am Nat* 118: 726–748; B after M. V. Lomolino et al. 1989. *Ecology* 70: 180–194; C after A. Kodric-Brown and J. H. Brown. 1993. *Ecology* 74: 1847–1855.)

ANALYZING DATA 18.1

Do Species Invasions Influence Species–Area Curves?

As we learned in Analyzing Data 16.1, the invasion of non-native species has been implicated in both increases and decreases of species diversity within communities. In the study we considered in that exercise, the majority of the non-native species had negative effects on species diversity at relatively small scales (16 m^2). Does this pattern hold as we increase the spatial scale over which we sample species diversity?

Kristin Powell and colleagues (2013)* considered this question by comparing the effect of native and non-native plants on forest communities at different spatial scales. They used species–area curves to plot the number of plant species versus the area sampled for three separate tree communities across the United States: tropical forests in Hawaii being invaded by the fire tree (*Morella faya*), oak–hickory forests in Missouri being invaded by Amur honeysuckle (*Lonicera maackii*), and hardwood hammock forests in Florida being invaded by the cerulean flax lily (*Dianella ensifolia*). In each of the forests, they identified multiple pairs of sites on opposite sides of an invasion front that had been ongoing for at least 30 years. At invaded sites, more than 90% of the plant cover was invaders, while the second site remained uninvaded. Powell et al.'s results for the Florida forest community are shown in the figure.

1. How do the slope (*z*) and y intercept (*c*) of the curve differ for invaded and uninvaded sites? What does this difference tell us about the effect of invaders on species richness at small versus large spatial scales?

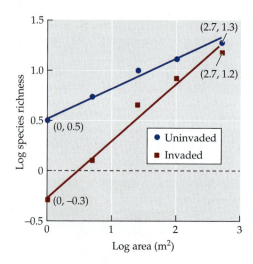

2. Convert the log area (m^2) and log species richness to non-log values at the smallest and largest spatial scales for invaded and uninvaded sites. What is the approximate range in spatial extent and in species richness for invaded and uninvaded plots?

3. Provide a hypothesis that could explain the difference between the species–area curves for invaded versus uninvaded areas.

* Powell, K. I., J. M. Chase, and T. M. Knight. 2013. Invasive plants have scale-dependent effects on diversity by altering species–area relationships. *Science* 339: 316–318.

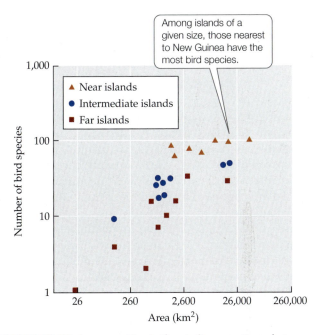

Among islands of a given size, those nearest to New Guinea have the most bird species.

▲ Near islands
● Intermediate islands
■ Far islands

FIGURE 18.20 Area and Isolation Influence Species Richness on Islands MacArthur and Wilson plotted species–area relationships for birds on islands of different sizes and at different distances from source populations (on New Guinea). (After R. H. MacArthur and E. O. Wilson. 1963. *Evolution* 17: 373–387.)

Southwest decreases as a function of the distance from the main source of species—in this case, two large mountain ranges in the region. This and other examples generally show that islands more distant from source populations, such as those in mainland areas or unfragmented habitats, have fewer species than islands of roughly the same size closer to source populations.

Almost always, however, island isolation and size are confounded. Robert MacArthur and Edward O. Wilson (1963) illustrated this problem by plotting the relationship between bird species richness and island area for a group of islands in the Pacific Ocean off New Guinea (**FIGURE 18.20**). Here, the islands varied in both size and degree of isolation from the mainland, but some patterns were evident. For example, if we compare islands of equivalent size, the island farthest from source populations (on New Guinea) has fewer bird species than the island closest to source populations.

Let's turn now to the question of how island area and isolation could together act to produce these commonly observed species diversity patterns.

Species richness is a balance between immigration and extinction

The Theory of Island Biogeography (1967) represented one of the most important breakthroughs in the science of biogeography since Wallace's time. The book was born out of the common interests of two scientists:

an ecologist, Robert MacArthur, and a taxonomist and biogeographer, Edward O. Wilson. Wilson, who had studied the biogeography of ants for his PhD thesis work, had made a few key observations about islands in the South Pacific, which he found himself discussing with MacArthur when they met at a scientific meeting (Wilson 1994). The first observation was that for every tenfold increase in island area, there was a rough doubling of ant species richness. The second was that as ant species spread from mainland areas to islands, the new species tended to replace the existing species, but there was no net gain in species richness. There appeared to be an equilibrium number of species on the islands, which was dependent on their size and distance from the mainland, but species composition on the islands could, and did, change over time.

MacArthur, a gifted mathematical ecologist, was just 31 years old when he and Wilson developed these observations into the beginnings of a simple but elegant theoretical regional biogeographic model. The model, published in their book 5 years later, became more commonly known as the **equilibrium theory of island biogeography**. The theory is based on the idea that the number of species on an island, or in an island-like habitat, depends on a balance between immigration or dispersal rates and extinction rates. The theory works something like this: Imagine an empty island open for colonization by species from mainland, or source, populations. As new species arrive on the island, by whatever means necessary, the island starts to fill up. The rate of immigration (the number of new species arriving) decreases over time as more and more species are added, eventually reaching zero when the entire pool of new species that could reach the island and be supported there is exhausted. But as the number of species on the island increases, there should also be an increase in the rate of extinction. This assumption makes sense according to the simple principle of balance mentioned above: with more species, there are more species extinctions. Additionally, as the number of species increases, the population size of each species should get smaller. Conceivably, this could occur for two reasons. First, competition may increase, thus decreasing the population sizes of species as they vie for the same space and resources. Second, predation may increase as more consumer species are added to the island. The result of either interaction is smaller population sizes and thus a greater risk of species extinction. If we plot the immigration rate against the extinction rate, the actual number of species on the island should fall where the two curves intersect, or where species immigration and extinction are in balance (**FIGURE 18.21**). This equilibrium number is the number of species that should theoretically "fit" on the island, irrespective of the turnover, or replacement of one species with another, that occurs on the island over time.

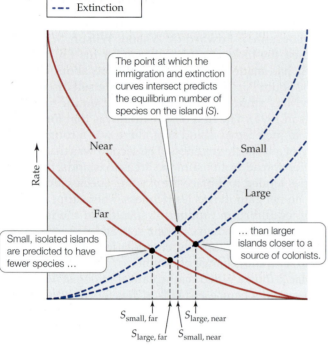

The point at which the immigration and extinction curves intersect predicts the equilibrium number of species on the island (S).

Small, isolated islands are predicted to have fewer species …

… than larger islands closer to a source of colonists.

$S_{small, far}$ $S_{large, near}$
$S_{large, far}$ $S_{small, near}$

Number of species present →

FIGURE 18.21 The Equilibrium Theory of Island Biogeography MacArthur and Wilson's theory emphasized the balance between species immigration rates and species extinction rates for islands of different sizes and at different distances from a source of colonizing species. (After R. H. MacArthur and E. O. Wilson. 1963. *Evolution* 17: 373–387.)

To understand the influence of island size and isolation on island species richness, MacArthur and Wilson simply adjusted their curves up or down to reflect their effects (see Figure 18.21). They assumed that island size mainly controls the extinction rate. They reasoned that small islands should have higher extinction rates than large islands, for the same two reasons described above, resulting in an extinction curve for small islands that is higher than that for large islands. Likewise, they reasoned that the distance of an island from the mainland mainly controls the immigration rate. Distant islands should have lower rates of immigration than islands near the mainland, resulting in an immigration curve for distant islands that is lower than that for islands near the mainland.

To test their theory, MacArthur and Wilson (1967) applied it to observations from the small volcanic island of Krakatau, between Sumatra and Java, which erupted violently in 1883, wiping out all life on the island (**FIGURE 18.22**). Surprisingly, animal and plant species

(A)

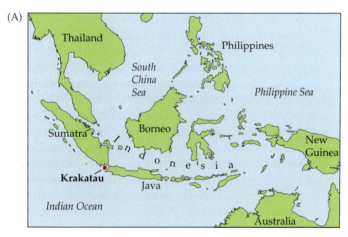

(B)

Parker & Coward, Britan. 1888

(C)

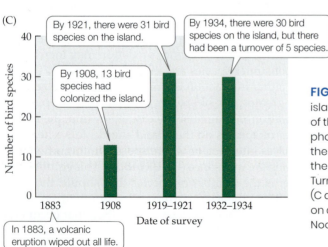

By 1921, there were 31 bird species on the island.

By 1934, there were 30 bird species on the island, but there had been a turnover of 5 species.

By 1908, 13 bird species had colonized the island.

In 1883, a volcanic eruption wiped out all life.

FIGURE 18.22 The Krakatau Test (A) The eruption of the small volcanic island of Krakatau, near Sumatra and Java, in 1883 provided a natural test of the equilibrium theory of island biogeography. (B) A drawing based on a photo taken from a ship a few hours before the major eruption. (C) By 1921, the number of bird species had reached 31, and in 1934, it was at 30— the equilibrium number predicted by MacArthur and Wilson's theory. Turnover, however, was five times higher than the theory had predicted. (C after R. H. MacArthur and E. O. Wilson. 1963. *Evolution* 17: 373–387; based on data from K. W. Dammerman. 1948. *The Fauna of Krakatau: 1883–1933.* Noord-Hollandsche Uitg.-Mig.)

began returning to what little remained of the island within a year of the explosion. MacArthur and Wilson used data from three surveys at various times since the eruption to calculate the immigration and extinction rates of birds on the island. Based on these rates, they predicted that the island should sustain roughly 30 bird species at equilibrium, with a turnover of 1 species. The data showed that bird species richness on the island had indeed reached 30 species within 40 years after the eruption and had remained close to that number thereafter. However, they also found that turnover was much higher, at 5 species. Whether this difference was due to a sampling error or a problem with the model is unknown, but this example motivated Wilson and others (e.g., the BDFFP researchers whose work is described in this chapter's Case Study) to start testing the model using manipulative experiments.

One of the best-known experiments to test the equilibrium theory of island biogeography was conducted by Daniel Simberloff and his advisor, Edward O. Wilson, on small mangrove islands and their arthropod inhabitants in the Florida Keys (Simberloff and Wilson 1969; Wilson and Simberloff 1969). These islands were scattered at various distances from large "mainland" mangrove stands (**FIGURE 18.23A**). After surveying species richness on the islands, Simberloff and Wilson manipulated a handful of them by fumigating them with an insecticide to remove all of their insects and spiders (**FIGURE 18.23B**). They then surveyed the defaunated islands over a year-long period (**FIGURE 18.23C**). By the end of the year, species numbers on the islands were similar to those before the defaunation; furthermore, the island closest to a source of colonists had the most species, and the farthest island had the least (**FIGURE 18.23D**). Interestingly, the farthest island had not quite regained its original species richness even after 2 years. All the islands showed considerable turnover of species, as might be expected for small islands where extinction rates are predicted to be high (see Figure 18.21).

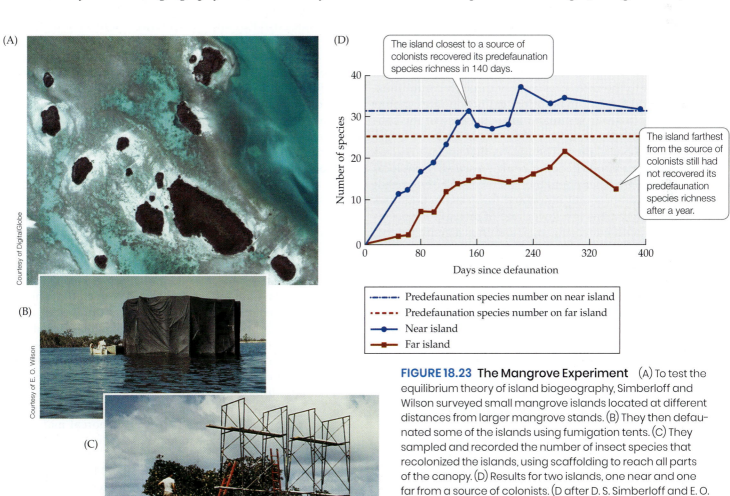

FIGURE 18.23 The Mangrove Experiment (A) To test the equilibrium theory of island biogeography, Simberloff and Wilson surveyed small mangrove islands located at different distances from larger mangrove stands. (B) They then defaunated some of the islands using fumigation tents. (C) They sampled and recorded the number of insect species that recolonized the islands, using scaffolding to reach all parts of the canopy. (D) Results for two islands, one near and one far from a source of colonists. (D after D. S. Simberloff and E. O. Wilson. 1969. *Ecology* 50: 278–296.)

The equilibrium theory of island biogeography holds true for mainland areas

Do the effects of area and isolation influence differences in species richness in mainland areas as well as on islands? As we saw in Watson's graph of plant species richness in Great Britain (see Figure 18.18), the species–area relationships observed on islands can also hold for mainland areas. How, then, does the biogeography of mainland areas differ from that of islands and island-like areas?

Let's consider a plot of plant species richness in mainland areas of France and on the Channel Islands in the English Channel (see Ecological Toolkit 18.1). Williams (1964) showed that plant species richness increases with area in both locations but that the slope of the line representing the increase is steeper for the Channel Islands than for the French mainland (i.e., the z value was greater on the islands). How can we interpret this difference? In mainland areas, just as on islands, species richness is theorized to be controlled by rates of immigration and extinction. In mainland areas, however, these rates are likely to be different from those on islands. Immigration rates should be greater in mainland areas because the barriers to dispersal are lower. Species can move from one area to the next, presumably through continuous, non-island habitat. In addition, extinction rates should be much lower in mainland areas because of the continual immigration of new individuals from the larger mainland population. The idea is that species will always have a good chance of being "rescued" from local extinction by other population members. The end result of these higher immigration and lower extinction rates in mainland areas is a lower rate of increase in species richness with increasing area, and thus a gentler slope, than in island areas.

We have seen over and over again in this chapter that geographic area has a large influence on species diversity at global and regional spatial scales. This effect takes on heightened significance as more habitats become "island-like" because of human influences. As we will see in the Case Study Revisited, the theory and practice of island biogeography is timely and relevant to the issues of conservation that we deal with today.

A CASE STUDY REVISITED

The Largest Ecological Experiment on Earth

One goal of ecologists is to understand the science behind the conservation of species threatened by habitat destruction and fragmentation. As we set aside more and more reserves to protect species diversity, the areas around those reserves continue to be changed by human activities, leaving many of them islands in a matrix of degraded habitat that is unsuitable for the species they contain. Thus, it is critical that we understand reserve design if we are to meet our conservation goals. When Lovejoy and his colleagues embarked, more than 40 years ago, on the Biological Dynamics of Forest Fragments Project (BDFFP) in the Amazon, one of their goals was to study the effects of reserve design on the maintenance of species diversity (Bierregaard et al. 2001). As it turned out, they learned that habitat fragmentation had even more negative and complicated effects than they had anticipated.

One of the first things they learned was that forest fragments needed to be large and close together to effectively maintain their original species diversity. For example, in a study of forest understory birds, Ferraz et al. (2003) found that even the largest fragments they surveyed (100 ha) lost 50% of their species within 12 years. Given that the regeneration time for these tropical rainforests ranges from several decades to a century, they projected that even fragments of 100 ha would be ineffective at maintaining bird species richness until forest regeneration could "rescue" species surviving within the fragments. The ecologists calculated that over 1,000 ha would be needed to maintain bird species richness until the forests could be regenerated, an area far greater than the average Amazon rainforest fragment in existence today (Gascon et al. 2000). If forest regeneration did not occur—as is likely when the land around a forest fragment is developed or used for agriculture—the fragment would have to contain 10,000 ha or more to maintain most of its bird species over more than 100 years of isolation (although even a fragment of that size could not sustain them all).

The researchers of the BDFFP were also surprised at how even minimal distances between fragments resulted in almost complete isolation of species. Clearings even 80 m (265 feet) wide hindered the recolonization of fragments by birds, insects, and arboreal (tree-dwelling) mammals (Laurance et al. 2002). It seemed that animals avoided entering the clearings for a number of interrelated reasons, the most obvious of which is that they have no innate reason to do so, having evolved within large, continuous, and climatically stable habitats that lacked the fragmentation imposed on them by deforestation. Moreover, even if some animals were inclined to venture into the clearings, specific requirements for their movement, such as trees for arboreal mammals, would not be present to facilitate their travel to other forest patches.

A second major finding of the BDFFP was that habitat fragmentation exposes the species within a fragment to a wide variety of potential hazards, including harsh environmental conditions, fires, hunting, predators, diseases, and invasive species. These *edge effects*, which occur at the transition between forest and nonforested

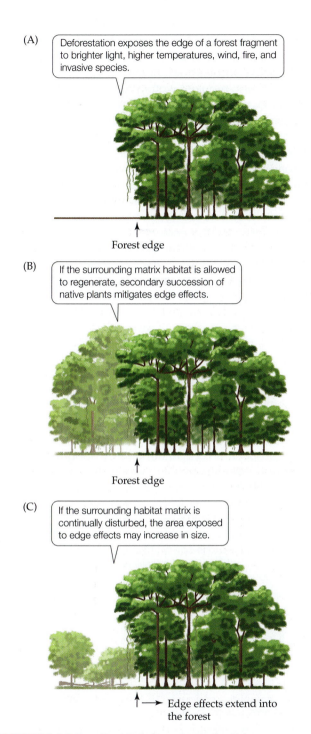

(A) Deforestation exposes the edge of a forest fragment to brighter light, higher temperatures, wind, fire, and invasive species.

Forest edge

(B) If the surrounding matrix habitat is allowed to regenerate, secondary succession of native plants mitigates edge effects.

Forest edge

(C) If the surrounding habitat matrix is continually disturbed, the area exposed to edge effects may increase in size.

Edge effects extend into the forest

FIGURE 18.24 Tropical Rainforests on the Edge
The BDFFP's research showed that deforestation subjects the forest fragments that remain to negative edge effects. (After C. Gascon et al. 2000. *Science* 288: 1356–1358.)

matrix habitat, can act together to increase local species extinctions. Trees, for example, can be killed or damaged by their sudden exposure to brighter light, higher temperatures, wind, fire, and diseases (**FIGURE 18.24**). Over time, depending on the surrounding matrix habitat, the

ultimate influences of edge effects are revealed. If the matrix habitat is left undisturbed, secondary succession occurs, as described in Chapter 17, reducing edge effects. If the matrix habitat continues to be disturbed, however, then the area subjected to edge effects may increase in size. For example, Gascon et al. (2000) describe forest fragments in the southern Amazon that are embedded in huge non-native sugarcane and *Eucalyptus* plantations where burning is used regularly for crop rotation. The burning keeps the forest edges in a constant state of disturbance. Fire-tolerant plant species, many of them non-native, become more common at the edges and act as conduits for more fires. This positive feedback loop ends up decreasing the effective size of the forest fragments and continually increasing the area subjected to edge effects. Some edge effects can extend a kilometer or more into a fragment, essentially influencing the entire area of a 1,000-ha fragment.

The results of the BDFFP have made an immense and sobering contribution to our understanding of forest fragmentation. As Laurance et al. (2002) point out, the BDFFP is a controlled experiment that probably provides a conservative estimate of species losses. The BDFFP has shown us that most of the forest fragments human activities are creating are too small to maintain all their original species; thus, habitat fragmentation is likely to result in the loss of many species. We'll see how the BDFFP's findings are being applied to reserve design and other conservation efforts when we discuss habitat fragmentation and edge effects in more detail in Concepts 24.2 and 24.3.

CONNECTIONS *in* NATURE

TROPICAL RAINFOREST DIVERSITY BENEFITS HUMANS Why do we care when species go extinct in a rainforest far away? As we will see in Concept 23.1, such extinctions raise ethical and aesthetic concerns similar to those that arise when great works of art or antiquities are lost to society. In addition, there are economic concerns about the loss of important *ecosystem services* produced by natural systems, which help sustain human health and well-being. For example, tropical deforestation raises concerns about losses of important foods and medicines that have their origins in rainforests. At least 80% of the developed world's diet originated in tropical rainforests, including corn, rice, squash, yams, oranges, coconuts, lemons, tomatoes, and nuts and spices of many kinds. Twenty-five percent of all commercial pharmaceuticals are derived from tropical rainforest plants, but less than 1% of tropical rainforest plants have been tested for their potential medical uses.

These statistics raise the question, How does the economic value of tropical rainforest plants used for nontimber purposes compare with the value of deforestation? It turns out that there have been very few economic analyses of this type. A few studies come from the Millennium Ecosystem Assessment (2005), a synthesis of studies on the use of the environment and its relationship to human needs, created by leading scientists from around the world. An example comes from Cambodia, where the total economic value of traditional forest products (e.g., fuelwood, rattan and bamboo, malva nuts, and medicines) was compared with that of unsustainable forest harvesting. The value of traditional forest products is four to five times greater ($700–$3,900 per hectare) than that of unsustainable forest harvesting ($150–$1,100 per hectare).

Recognition of the economic benefits of changing our resource management practices has only just begun. Why is this? Part of the answer lies in our not formally recognizing the economic value of the services provided to humans by species or whole communities. Tropical rainforests provide food, medicine, fuel, and a destination for tourists, all of which can be obtained without complete deforestation. Rainforests also regulate water flow, climate, and atmospheric CO_2 concentrations. Assigning a value to any of these important services is difficult compared with setting the market price of timber or agricultural products. For that reason, it is easier to justify the use of rainforest timber and land (and even some sustainable forest products) for private profit than to press for the conservation of rainforests based on the ecological services they provide to society in general. If private landowners are not given incentives to value the larger social benefits of ecological services, maximization of personal gain often drives their decisions. Given the importance of ecological services to our planet, we can no longer afford to ignore these economic trade-offs.

SUMMARY

CONCEPT 18.1 Patterns of species diversity and distribution vary at global, regional, and local spatial scales.

18.1.1 Define biogeography and explain how patterns of species diversity and composition are connected across different spatial scales.

- Biogeography is the study of variation in species diversity and composition among geographic locations.

- Patterns of species diversity and composition at the global, regional, and local spatial scales are connected to one another in a hierarchical way.

18.1.2 Outline the processes important to global-, regional-, and local-scale biogeography.

- Global-scale biogeography is the result of variations in speciation, extinction, and dispersal at latitudinal and continental spatial scales and evolutionary time scales.

- Regional-scale biogeography (gamma diversity) encompasses a smaller geographic area in which the climate is roughly uniform and the species contained therein are bound by dispersal limitation to that region.

- Local-scale biogeography (alpha diversity) is equivalent to a community and is determined by dispersal, physical conditions, and species interactions.

- Beta diversity is the change in species number and composition, or turnover of species, across the landscape from one local community to another.

18.1.3 Analyze the relative importance of species pools versus local-scale processes in determining local community species diversity.

- Studies show that regional species pools largely determine the numbers of species present in local communities but that local physical conditions and species interactions are also important.

CONCEPT 18.2 Global patterns of species diversity and composition are influenced by geographic area and isolation, evolutionary history, and global climate.

18.2.1 Describe the two major biogeographic patterns—biogeographic regions and latitudinal gradients in species diversity—at the global scale.

- Earth's land mass can be divided into biogeographic regions that vary markedly in species diversity and composition.

- For most taxonomic groups, species diversity is greatest in the tropics and declines at higher latitudes.

18.2.2 Explain the underlying forces thought to be important in creating biogeographic regions.

- The biotas of the biogeographic regions reflect an evolutionary history of isolation due to continental drift caused by the movements of Earth's tectonic plates.

- Tracing the threads of vicariance over large geographic areas and long time periods provided important evidence for early theories of evolution.

18.2.3 Outline the hypotheses proposed to explain the latitudinal gradient in species diversity pattern.

- A number of hypotheses, involving species diversification rate, species diversification time, and productivity, have been proposed to explain the latitudinal gradient in species diversity.

CONCEPT 18.3 Regional differences in species diversity are influenced by area and distance, which determine the balance between immigration and extinction rates.

18.3.1 Graph and explain the species–area relationship and know why it differs between islands and mainland areas.

- Species–area relationships show that species richness increases with the area sampled and decreases with distance from a source of species.

- Most species–area relationships have been documented for "islands," which include isolated areas surrounded by dissimilar habitat but also mainland areas.

- Species–area curves are steeper for islands compared to mainland areas because of the greater difference in species richness among isolated islands.

18.3.2 Explain regional species diversity for islands and island-like areas using the equilibrium theory of island biogeography.

- The equilibrium theory of island biogeography predicts that a balance between immigration and extinction rates controls species diversity on islands or in island-like areas.

- According to the theory, larger islands closer to a source of species have more species than smaller islands that are more distant from a source of species, because they have higher immigration rates and lower extinction rates.

Review Questions

1. Spatial scale is important to the biogeographic patterns of species diversity and composition that we see on Earth. Define the various spatial scales that are important to biogeography, and describe how they are related to or interconnected with one another.

2. Describe the factors that Alfred Russel Wallace believed created biogeographic regions on land and in the oceans.

3. Latitudinal gradients in species diversity and composition are strong global features of biogeography. Describe three hypotheses proposed to explain why species diversity is higher in the tropics and decreases toward the poles for the majority of taxonomic groups.

Hone Your Problem-Solving Skills

One study from the Biological Dynamics of Forest Fragments Project (BDFFP) considered the number of understory bird species living in different-sized forest fragments surrounded by deforested land (see the Ferraz et al. [2003] study in the Case Study Revisited). This study involved counting the number of bird species in the fragments at the start of the experiment and then over a handful of years afterward. A scaling factor was estimated to determine the time it takes to lose half of the bird species (t_{50}) in the different fragment sizes. Below is a table with the results of the study, organized by the fragment area:

Fragment area (ha)	Bird species richness initial count	t_{50} (yr)
1	83	5
10	92	8
100	113	12

Source: G. Ferraz et al. 2003. *Proc Natl Acad Sci USA* 100: 14069–14073.

1. Graph the initial numbers of bird species by fragment area. Do the fragments follow the species–area relationship?

(Continued)

2. Assuming that the species loss is linear over time, use the table to calculate the percentage loss of species per year in the 1-, 10-, and 100-ha fragments. Which fragment has the greatest species loss per year, and which has the least?

3. Now use the percentage loss per year to calculate the number of species in each of the fragment sizes 9 years after the start of the experiment. Graph the number of species by fragment area on the graph you developed for Question 1.

4. If you were to draw linear regressions for the species–area data points at the start of the experiment and 9 years after fragmentation, which species–area relationship would have the steepest slope (z)? Explain why.

ADDITIONAL RESOURCES

This textbook provides resources such as **Web Extensions** and additional **Climate Change Connections** for instructors who can share them with students. Please contact your instructor, if you would like to access that content.

Species Diversity in Communities

Can Species Diversity Suppress Human Diseases? A Case Study

On May 14, 1993, a 19-year-old cross-country track star, riding in the back seat of his family's car, began struggling to breathe. The family immediately stopped at a convenience store to call for help, and the young man was rushed to a hospital in Gallup, New Mexico. The ambulance crew tried to revive him, but he died soon after reaching the emergency room. A chest X-ray showed that his lungs were filled with fluid. The deputy medical investigator based in Gallup was called in, and over the course of 2 weeks, he determined that at least five other residents of the area, which included members of the Navajo Nation living in the Four Corners region (where New Mexico, Arizona, Colorado, and Utah intersect), had also mysteriously died in the same sudden manner. After interviewing families of the victims, the medical examiner determined that all had experienced flu-like symptoms and then acute respiratory distress as a result of their lungs being filled with fluid. The disease appeared to be infectious and viral.

By early June 1993, the Viral Special Pathogens Branch of the Centers for Disease Control and Prevention had determined that the culprit was a previously unknown species of hantavirus, a pathogen carried by rodents. It was given the name Sin Nombre virus (SNV) or "the nameless virus." Rodents shed the virus in their urine, feces, and saliva, which if aerosolized, can be inhaled by humans. It was subsequently determined that the new viral strain was carried by a species of deer mouse (*Peromyscus maniculatus*) whose populations had recently boomed in the Four Corners region (**FIGURE 19.1**). Research showed that deer mouse populations had increased 20-fold in some locations, triggering the transmission of SNV infections in humans (see Concept 10.1 and Figure 10.9).

Over the last 70 years, the number of emerging diseases affecting humans has substantially increased. Of these diseases, 62% are zoonotic—hosted by wildlife and infectious to humans. Diseases such as the Zika virus, Ebola virus, and avian influenza are all zoonotic diseases that have emerged over the last few decades. The most recent evidence for the coronavirus SARS-CoV-2 (COVID-19) is that it is a zoonotic disease, having been transmitted to humans from wild animals sold at the Huanan Seafood Wholesale Market in Wuhan, China. The factors that affect zoonotic disease emergence are complex and sometimes disease specific but often include human-caused events such as species invasions, climate change, pollution, and land use conversion. One seemingly unlikely factor, that of declining species diversity, is starting to be recognized as an important mechanism that may facilitate the emergence and transmission of zoonotic diseases.

© E.R. Degginger/Alamy Stock Photo

FIGURE 19.1 Deer Mice Trigger Hantavirus Infection in Humans Can the number of small-mammal species affect the transmission of hantavirus by the deer mouse?

It turns out that hantaviruses provide a nice model system for studying how the loss of species diversity within a community may affect disease emergence and transmission. A number of observational studies have linked hantavirus infection prevalence in deer mouse host populations with declining small-mammal species diversity. For example, in a field study in Oregon, the one variable that was significantly linked to SNV infection prevalence was small-mammal species diversity, with the prevalence of SNV rising from 2% to 14% as species diversity declined (Dizney and Ruedas 2009). A similar study in Utah came to the same conclusion. These researchers too found a negative correlation between small-mammal diversity and SNV infection prevalence in deer mice (Clay et al. 2009).

These observational studies are supported by an experimental study of hantaviruses in rodent communities of Panama. In their study, Gerardo Suzán and colleagues (2009) conducted a small-mammal removal experiment in replicate field plots, where zoonotic hantaviruses are native and common. Small-mammal diversity was reduced through trapping of species that were not host to the virus. They found that plots with reduced small-mammal diversity had an increase in rodent host individuals and also that more of those individuals were infected by hantavirus (**FIGURE 19.2**).

The observational and experimental evidence presented here point to the role of species diversity in buffering the transmission of zoonotic pathogens to wildlife and ultimately humans. But what explains the effect of species diversity in disease transmission? As we will see, the response of the host to changes in species diversity makes all the difference in the answer to this question.

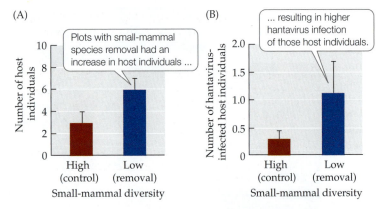

FIGURE 19.2 Disease Transmission Increased with Species Diversity Loss An experiment in Panama showed that plots with small-mammal diversity removal (low-diversity plots) increased in (A) the number of rodent host individuals and (B) the number of hosts infected with the hantavirus compared with the control (high-diversity plots). Error bars show one standard error (SE) of the mean. (After F. Keesing et al. 2010. *Nature* 468: 647–652.)

Introduction

Communities vary tremendously in the numbers and kinds of species they contain. In Chapter 18, our world-wide tour of forest communities demonstrated the wide variation in species diversity that occurs both globally and regionally. We saw that communities in the tropics (such as the Amazon rainforest) had many more tree species than those at higher latitudes (such as the forests of the Pacific Northwest or New Zealand). Moreover, we found that regional species pools had an important, but not an exclusive, influence on the number of species within a community.

In this chapter, we will focus on species diversity at the local scale. We will ask two important questions: First, what are the factors that control species diversity within communities? Second, what effects does species diversity have on the functioning of communities?

> **CONCEPT 19.1** Species diversity differs among communities as a consequence of regional species pools, abiotic conditions, and species interactions.

LEARNING OBJECTIVES

19.1.1 Describe how regional species' pools and dispersal abilities contribute to community membership.

19.1.2 Describe how local environmental conditions act as a "filter" for community membership.

19.1.3 Describe how species interactions may act to include species in, or exclude species from, communities.

Community Membership

If you looked across a landscape from the top of a mountain, you would see a patchwork of different communities—say, forests, meadows, lakes, streams, and marshes (**FIGURE 19.3**). You could be sure that each of those communities would have a different species richness and composition. The meadows would be dominated by a variety of grasses, herbs, and terrestrial insects. The lakes would be filled with various species of fish, plankton, and aquatic insects, and they might possibly harbor as many species as the meadows. Even though some species (such as amphibians) would be able to move from one community to another, the communities would still be highly distinct.

How do collections of species end up coming together to form different communities? One way to answer this question is to consider the factors that control species membership in communities. If you think about the sheer number of species that coexist within any community, it is clear that no one process is responsible for all the species we find there. As we saw in Concepts 9.3

FIGURE 19.3 A View from Above Looking at these mountains in Glacier National Park, Montana, it is easy to see that the landscape is made up of a patchwork of communities of different types.

and 18.1, the distributions and abundances of organisms are dependent on three interacting factors: (1) regional species pools and dispersal ability (species supply), (2) environmental conditions, and (3) species interactions. We can think of these three factors as "filters" that act to exclude species from (or include them in) particular communities (**FIGURE 19.4**). Let's briefly consider each of them in more detail.

Species supply is the "first cut" to community membership

In Concept 18.1, we saw that the regional species pool provides an absolute upper limit on the numbers and types of species that can be present within communities (see Figure 18.6). Not surprisingly, we saw that regions of high species richness tend to have communities of high species richness (see Figure 18.7). This relationship is due to the role of the regional species pool and, more specifically, the role of dispersal in "supplying" species to communities (see Figure 19.4A). Nowhere is the controlling effect of dispersal on community membership more evident than in the invasion of communities by non-native species.

As ecologists are beginning to learn, humans have greatly expanded the regional species pools of communities

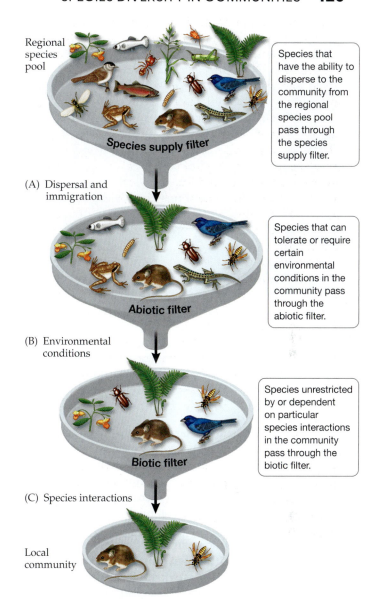

FIGURE 19.4 Community Membership: A Series of Filters Species end up in a local community by passing through a series of "filters" that determine community membership. Species are lost at each filter, so local communities contain a fraction of the species in the regional pool. In practice, all the filters work at the same time, rather than in series as the figure suggests. (After J. H. Lawton. 2000. In *Community Ecology in a Changing World*, O. Kinne [Ed.], Excellence in Ecology, Vol. 11. Ecology Institute: Luhe, Germany.)

❓ Would it make sense for the fish and frog species in the regional pool to be present in the local community shown in the figure? Explain.

by serving as vectors of dispersal. For example, we know that many aquatic species travel to distant parts of the world, which they could not otherwise reach, in the ballast water carried by ships (**FIGURE 19.5A**). Seawater is pumped into and out of ballast tanks, which serve to balance and stabilize cargo-carrying ships, all over the world. Most of the time, the water—along with the organisms it contains

(A)

© bjeayes/Getty Images

(B)

Courtesy of NOAA/GLERL

In the absence of natural enemies, invasive zebra mussels reach astounding—and destructive—population densities.

© aquafun/123RF

Dreissena polymorpha

FIGURE 19.5 Humans Are Vectors for Invasive Species (A) Large and fast oceangoing ships can carry marine species to all parts of the world in their ballast water. (B) The zebra mussel, a destructive invader of the inland waterways of the United States, was carried there from Russia in ballast water.

(from bacteria to planktonic larvae to fish)—is taken up and released close to ports, where some of the organisms have the opportunity to colonize nearshore communities. An estimated 10,000 marine species are transported in the ballast water of oceangoing vessels each day. Ballast water introductions have increased substantially over the past few decades because ships are larger and faster, so more species can be taken up and more survive the trip. In 1993, Carlton and Geller listed 46 known examples of ballast-water-mediated invasions in the previous 20 years. One species, the zebra mussel (*Dreissena polymorpha*), arrived in North America in the late 1980s in ballast water discharged into the Great Lakes (**FIGURE 19.5B**). As a non-native, invasive species, it has had community-changing effects on inland waterways and native species. Another example of a ballast water introduction with negative ecological consequences, which we learned about in the Case Study in Chapter 10, was the release of the comb jelly *Mnemiopsis leidyi* into the Black Sea. These introductions highlight the important role that dispersal plays in allowing a non-native species to gain a foothold in a community and potentially cause community-wide effects.

Next let's turn our attention to the role of local conditions, particularly the abiotic and biotic characteristics of communities that help determine their structure.

Environmental conditions play a strong role in limiting community membership

A species may be able to get to a community but fail to become a member of the community because it is physiologically unable to tolerate the environmental or abiotic conditions there (see Figure 19.4B). Such physiological constraints can be quite obvious. For example, if we return to our thought experiment of viewing a landscape from the top of a mountain, it is reasonable to assume that the abiotic attributes of the lakes we see make them good places for fishes, plankton, and aquatic insects, but not for terrestrial plants. Similarly, lakes might be good habitat for certain species of fish, plankton, and aquatic insect, but not for all of them. Some of these species depend on fast-flowing water and are thus restricted to streams. These differences among abiotic environments are obvious constraints (or requirements, depending on how you look at it) that largely determine where particular species can and cannot occur within a region. There are many examples throughout this book that demonstrate how physiological constraints can control the distributions and abundances of species—see, for example, the discussions of aspen (Concept 4.1), creosote bush and saguaro cactus (Concept 9.3), and the barnacle *Semibalanus balanoides* (Concept 9.3).

In our earlier discussion of species introductions by ballast water, it was clear that humans transport many more species than can actually survive in the new locations to which they are carried. For example, the majority of organisms released with ballast water find themselves in coastal waters that do not have the temperature, salinity, or light regimes they need to survive or grow. Luckily, many of these individuals die before they can become a threat to the native community. But ecologists know, based on examples such as the *Caulerpa taxifolia* invasion in the Mediterranean (see the Case Study in Chapter 16), that it is not wise to rely on physiological constraints to exclude potential invaders from a community. It may be that, with multiple introductions, particular individuals with slightly different physiological capabilities can survive and reproduce in an environment once thought uninhabitable by individuals of their species.

Who interacts with whom makes all the difference in community membership

Even if species can disperse to a community and cope with its potentially restrictive abiotic conditions, the final cut to community membership is coexistence with other species (see Figure 19.4C). Clearly, if a species depends on other species for its growth, reproduction, and survival, those other species must be present if it is to gain membership in a community. Equally important, some species may be excluded from a community by competition, predation, parasitism, or disease. For example, returning to our thought experiment, we might assume that lakes are suitable habitats for many fish species, but could those species all live together in one lake, given that resources are limiting? A simple view suggests that the best competitors or predators should dominate the lake, thus excluding weaker competitors and resulting in a low-diversity community. But we know that most communities are full of species that are actively interacting and coexisting. So what allows this coexistence? There are many important mechanisms that allow species to coexist, and we will spend the next two sections considering them. But first, let's ask how species might be excluded from communities by biological interactions—a question that is a bit different, but equally relevant.

The invasive species literature provides some of the best tests of whether species interactions can exclude species from communities. The failure of some non-native species to become incorporated into communities has been attributed to interactions with native species that exclude, or slow the population growth of, the non-native species—a phenomenon that ecologists call **biotic resistance**. Multiple studies in a variety of communities have shown that native herbivores have the ability to reduce the spread of non-native plants in substantial ways. Maron and Vila (2001) found that mortality of non-native plants due to native herbivores can be quite high (about 60%), especially at the seedling stage (up to 90% in some studies). But while native herbivores can kill individual non-native plants, it is still unknown how important native species are in completely excluding non-native species from a community. For example, Faithfull (1997) found that in Australia, adults and larvae of the native lucerne seed web moth (*Etiella behrii*) breed and feed on the seedpods of the invasive gorse shrub *Ulex europaeus*, but the plant still continues to spread (**FIGURE 19.6**). This lack of knowledge about biotic resistance may be an artifact of ecologists being more likely to study why a particular non-native species does or does not spread once it

Etiella behrii

© Arco Images GmbH/Alamy Stock Photo

FIGURE 19.6 Stopping Gorse Invasion? Herbivory by adults and larvae of the native lucerne seed web moth (*Etiella behrii*) has slowed, but not stopped, an invasion of the non-native gorse shrub *Ulex europaeus* (the plants with yellow flowers) in Australia.

becomes a provisional member of the community than to study all the cases in which it is unable to gain a foothold because of interactions with native species. It may also be true that most failed introductions of non-native species go completely undetected.

CLIMATE CHANGE CONNECTION

HOW ARE SPECIES INVASIONS ENHANCED BY CLIMATE CHANGE? There is growing evidence that climate change—and, in particular, rising temperatures—may facilitate the invasions of species that would be unable to survive under cooler conditions. As you might guess, climate change can play a role in mediating the ability of species to pass through the three filters described in Figure 19.4, thus potentially exacerbating the arrival, spread, impact, and management of invasive species.

Hellmann et al. (2008) outline five potential consequences of climate change for invasive species (**FIGURE 19.7A**). The first consequence arises when climate change alters the pathways (transport and introduction) of non-native species (see Figure 19.7A, consequence 1). Such alterations could occur if climate change better links areas that are geographically separate prior to climate change. For example, Sylvia Behrens Yamada and colleagues (2005) have shown that non-native European green crabs (*Carcinus maenas*)

(A)

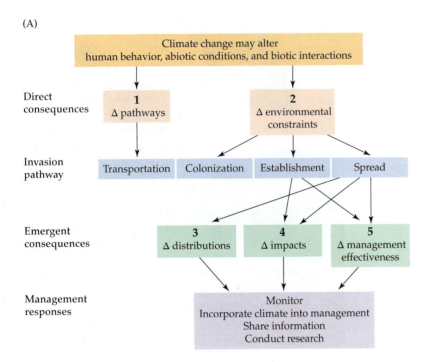

(B)

FIGURE 19.7 **The Five Consequences of Climate Change for Species Invasions** (A) Consequences 1 and 2 directly affect the invasion pathway for new non-native species. Consequences 3, 4, and 5 are emergent after an invader has become established and spread, and they have management implications. Delta (Δ) means "change in." (B) The European green crab (*Carcinus maenas*) has invaded estuaries along the U.S. Pacific coast. (A after J. J. Hellmann et al. 2008. *Conserv Biol* 22: 534–543.)

(**FIGURE 19.7B**), introduced to San Francisco Bay from the east coast of North America, were able to colonize northern Pacific estuaries during warm El Niño years. At these times, crab larvae were transported in stronger, warmer northward-flowing coastal currents to new estuarine locations in Oregon and Washington, where they were able to survive as adults. Thus, changes in coastal currents caused by global warming could create new pathways of dispersal for other non-native species.

A second consequence of climate change is the alteration of environmental constraints on non-native species that allows some species to overcome physiological or biotic constraints on their persistence outside of their native range (see Figure 19.7A, consequence 2). For example, in the green crab invasion mentioned above, it was predicted that green crabs would become locally extinct in northern estuaries once El Niño subsided, because of their intolerance of cold winter ocean temperatures (they are unable to molt and reproduce at temperatures below 10°C). In fact, the researchers found that green crabs persist as invaders where they experience occasional warm winters, during which they have much greater survival, growth, and reproduction.

A third consequence of climate change is the alteration of the distributions of existing invasive species (see Figure 19.7A, consequence 3). For non-native species that have gained a foothold outside of their biogeographic range, climate change could expand (or contract) their new range in dramatic ways. For example, one could imagine that with warming of estuarine water temperatures above 10°C, green crabs would not merely exist in

small populations but, instead, increase their numbers through enhanced survival and reproduction.

The fourth consequence of climate change occurs when the impacts of non-native species are altered (see Figure 19.7A, consequence 4). The impacts of the green crab under the climatic conditions of today are minimal. Hunt and Behrens Yamada (2003) observed very little overlap in the distribution and resource use of the green crab and the larger native red rock crab (*Cancer productus*). The red rock crab is dominant in the colder and more saline portions of estuaries, while the invasive green crab occurs in warmer and less saline areas. With climate change, increased temperatures or more rainfall could lead to warmer and less saline estuarine conditions, thus favoring green crabs over red rock crabs and having a greater impact on the estuarine community as a result.

The fifth and final consequence of climate change is its effects on the management of non-native species (see Figure 19.7A, consequence 5). Current management, whether it involves removal of invasive species or restoration of habitats impacted by these species, will need to adapt to changing climate in ways that maintain its control and efficacy. In the green crab example, management has been minimal beyond destroying individuals found in traps. If green crabs expand in population size and range in response to climate change, however, active management may be required to keep this invasive species from becoming a pest to shellfishery and aquaculture operations. As you can see, climate change can act on invasive species in a multitude of ways that may be hard to predict. ☀

Studying invasions gives us valuable insights into how species are included or excluded from communities, but how species coexist can be complicated. In the next two sections, we will consider theories of species coexistence and ultimately species diversity. We will start by revisiting the concept of **resource partitioning** (also known as **niche partitioning**), which relies on ecological and evolutionary "compromises" that result in divergence in resource use as a mechanism for coexistence (see Concept 14.2). We will then explore alternative theories and studies that consider the importance of disturbance, stress, predation, and even positive interactions to the coexistence of species and, ultimately, the species diversity of communities.

CONCEPT 19.2 Resource partitioning is theorized to reduce competition and increase species diversity.

LEARNING OBJECTIVES

19.2.1 Define resource partitioning.

19.2.2 Outline observations, experiments, and models that support resource partitioning as a mechanism of species coexistence.

Resource Partitioning

A simple model of resource partitioning envisions each type of resource available in a community as varying along a "resource spectrum." This spectrum could represent, for example, different nutrients, prey sizes, or habitat types; note that such a spectrum represents the *variability* of an available resource, not the amount. We can assume that the resource use of each species falls somewhere along this spectrum and overlaps with the resource uses of other species to varying degrees (**FIGURE 19.8A**).

The assumption is that the more overlap, the more competition between species, with the extreme being complete overlap and competitive exclusion. The less overlap, the more partitioning of resources has occurred, and the less strongly the species will compete with one another.

Using this guiding theory, we can consider some of the ways in which resource partitioning might result in higher species richness in some communities than in others. First, species richness could be high in some communities because species show a high degree of partitioning along the resource spectrum (**FIGURE 19.8B**). More species could be "packed" into a community if the overlap in resource use among the species is low, leading to less competition and ultimately higher species richness. This lower overlap could be due to the evolution of specialization or character displacement (see Figure 14.12), which may reduce competition over time. Second, species richness could be high in some communities because the resource spectrum is broad (**FIGURE 19.8C**). Presumably, a broader resource spectrum would make a greater diversity of resources available to be used by a wider variety of species, resulting in higher species richness.

At this point, let's turn our attention away from models and take a look at some real communities to see how resource partitioning might work in practice.

Early studies suggested that resource partitioning was the main mechanism of coexistence

As we learned earlier from the two-species studies of Gause (1934a) on *Paramecium* (see Concept 14.2) and Connell (1961a,b) on barnacles (see Concept 14.4), species that compete with each other may coexist by using slightly different resources. Robert MacArthur, whose work on the equilibrium theory of island biogeography we described in Concept 18.3, played a pioneering role in understanding how this principle might be applied to whole communities, where multiple species interactions are occurring all at once.

MacArthur studied warblers, small and brightly colored birds that co-occur in the forests of northern North America. The idyllic New England forests that MacArthur

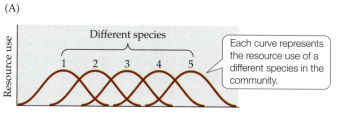

(A)

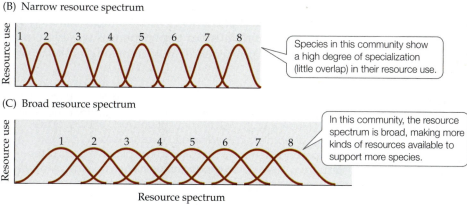

(B) Narrow resource spectrum

(C) Broad resource spectrum

Resource spectrum

FIGURE 19.8 Resource Partitioning Species coexistence within communities may depend on how the species divide resources. (A) The principle of resource partitioning along a resource spectrum. (B,C) Two characteristics of communities that can result in higher species richness. (After J. Hill and R. Hill. 2001. *Prog Phys Geogr* 25: 326–354.)

 Which panel shows the most resource partitioning? Which shows the least?

studied are home to an array of warbler species, *Setophaga* spp. (renamed from *Dendroica* spp.) that migrate from the tropics each spring to breed and feed on insects. Through a series of detailed natural history observations in the summers of 1956 and 1957 in Maine and Vermont, MacArthur (1958) recorded the feeding habits, nesting locations, and breeding territories of five species of warblers to find out how they might coexist in the face of very similar resource needs.

MacArthur began mapping the locations of warbler activity in tree canopies and found that the warblers were using different parts of the habitat in different ways (**FIGURE 19.9**). For example, yellow-rumped (*S. coronata*) warblers fed from the middle parts of trees to the forest floor, while bay-breasted (*S. castanea*) and black-throated green (*S. virens*) warblers fed more in the middle of a tree,

both inside and toward the outside of the tree canopy. Blackburnian (*S. fusca*) and Cape May (*S. tigrina*) warblers both fed on the outside tops of trees, often catching their prey in midflight. MacArthur found that the nesting heights of the five warbler species also varied, as did their use of breeding territories. Taken together, these observations supported his hypothesis that the warblers, although using the same habitat and food resources, were able to coexist by partitioning those resources in slightly different ways. MacArthur's work, which was part of his PhD thesis, earned him the prestigious Mercer Award, bestowed each year for the best paper in ecology.

MacArthur, along with his brother John MacArthur (MacArthur and MacArthur 1961), extended these ideas about resource partitioning in a study of the relationship between bird species diversity (calculated using the Shannon index; Concept 16.2) and foliage height diversity (a measure of the number of vegetation layers in a community that serves as an indication of habitat complexity, also calculated using the Shannon index). They found a positive relationship between the two in 13 tropical and temperate bird habitats from Panama to Maine (**FIGURE 19.10**). Interestingly, bird species diversity was not related to plant diversity per se, beyond the effects of foliage height diversity, suggesting that tree species identity was less important than the structural complexity of the habitat.

Another important resource partitioning study comes from phytoplankton communities. In Concept 14.1, we learned about David Tilman and colleagues' (1981) study of two species of diatoms that competed for silica (which diatoms use to build their cell walls). When the two species were grown together in a laboratory environment with limited supplies of silica, one outcompeted and excluded the other (see Figure 14.6). How, then, do diatom species coexist in nature? Tilman (1977) proposed what has become known as the **resource ratio hypothesis**, which posits that species coexist by using resources in different ratios or proportions. He predicted that diatoms would be able to coexist, despite using the same set of limiting nutrients, by acquiring those nutrients in different ratios. By growing two diatom species, *Cyclotella* and *Asterionella*, in laboratory environments that differed in their ratios of silica (SiO_2) to phosphorus (PO_4), Tilman found that *Cyclotella* was able to dominate only when the ratio of silica to phosphorus was low (approximately 1:1). When the ratio of silica to phosphorus was high (more like 1,000:1), *Asterionella* outcompeted *Cyclotella*. Only

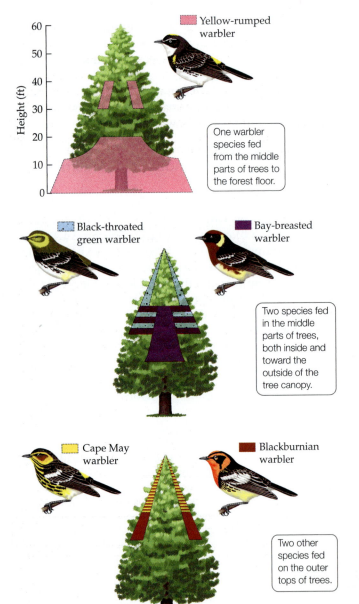

FIGURE 19.9 Resource Partitioning by Warblers
Robert MacArthur studied the habitat and food choices of five species of warblers in New England forests. He found that the warblers partition resources by feeding in different parts of the same trees. The colored shaded areas in each tree diagram represent the parts of trees where each warbler species fed most often. (After R. H. MacArthur. 1958. *Ecology* 39: 599–619.)

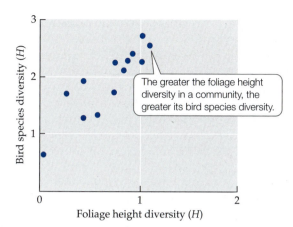

FIGURE 19.10 Bird Species Diversity Is Higher in More Complex Habitats MacArthur and MacArthur plotted bird species diversity against foliage height diversity (a measure of habitat complexity) for 13 different communities. Both kinds of diversity were calculated for each community, using the Shannon index (H). (After R. H. MacArthur and J. W. MacArthur. 1961. *Ecology* 42: 594–598.)

when the ratios of silica and phosphorus were limiting to both species (in the range of 100:1 to 10:1) could they coexist. Even though both species needed the same set of nutrients, it was the way in which they partitioned those resources that allowed them to coexist.

Outside of a laboratory setting, this type of partitioning would work best if resources naturally varied within the environment. What is the support for this possibility in the field? In a detailed survey, Robertson and colleagues (1988) mapped resource distribution in an abandoned agricultural field in Michigan that had been colonized by grassland plants. They found considerable variation in soil nitrogen and moisture at spatial scales of a meter or less (**FIGURE 19.11**). These patches of nitrogen and water resources did not necessarily correspond to topographic differences, and they were not correlated with each other. If we were to overlay the nitrogen map on the water map, we would find even smaller patches corresponding to different proportions of these two resources. Some of the best evidence of resource partitioning in plants comes from experiments that manipulate species richness and measure productivity, as we will explore in more detail in Concept 19.4.

The theory of resource partitioning relies on the assumption that species have evolved mechanisms for using resources in different, but complementary, ways, thus increasing their ability to coexist. As we learned in our discussion of species interactions in Unit 4, there are numerous other processes that can alter the outcome of species interactions and allow coexistence. In the next section, we will consider how those processes control species diversity at the local scale.

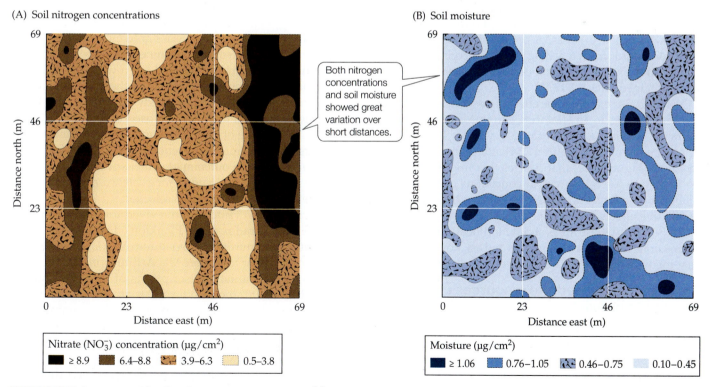

FIGURE 19.11 Resource Distribution Maps Mapping of (A) nitrogen concentrations and (B) soil moisture in an abandoned agricultural field revealed considerable small-scale variation ($\mu g/cm^2$ = micrograms per square centimeter). (From G. P. Robertson et al. 1988. *Ecology* 69: 1517–1524.)

CONCEPT 19.3 Processes such as disturbance, stress, predation, and positive interactions can mediate resource availability, thus promoting species diversity.

LEARNING OBJECTIVES

19.3.1 Describe the role of disturbance, stress, and predation in mediating coexistence and promoting species diversity.

19.3.2 Define and give examples of the intermediate disturbance hypothesis and its variations, including those that consider predation and positive interactions.

19.3.3 Define and give examples of lottery or neutral models.

Resource Mediation and Species Diversity

We have seen in previous chapters that disturbance, stress, and predation can modify species interactions and allow for species coexistence. We saw that when two species are competing with each other for the same resource, as in the case of the sea palms and mussels competing for space in the rocky intertidal zone (see Concept 14.4), coexistence can be achieved if the population growth of the dominant species is disrupted. In that example, mussels are the dominant competitors, and sea palms can coexist with them only where the mussels are disturbed frequently enough by wave action to allow the sea palms to acquire space. In this and many other examples in this book, as long as disturbance, stress, or predation keeps the dominant competitor from reaching its own carrying capacity, competitive exclusion cannot occur, and coexistence will be maintained (**FIGURE 19.12**).

We have also explored the effect of positive interactions between species in ameliorating extreme conditions and allowing coexistence. For example, we saw in the cases of salt marsh plants (Figure 17.14) and plants at high elevations (Figure 15.9) that species that might normally be unable to tolerate stressful conditions can maintain viable populations under those conditions because of the facilitative effects of other species.

Let's expand these ideas about modification of species interactions to whole communities and ask how processes that mediate resources influence species diversity.

Processes that mediate resources can allow species to coexist

There is an old adage among ecologists that goes something like this: "If you think it's a new idea, check Darwin. He probably proposed it first." In fact, when it comes to theories that explain coexistence, Darwin was the first to formally recognize disturbance as a mechanism for the maintenance of species diversity. In *The Origin of Species* (1859, p. 55), he noted the following results after an impromptu experiment in which he left a meadow on his property undisturbed by

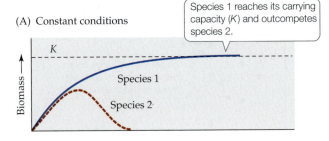

(A) Constant conditions

Species 1 reaches its carrying capacity (*K*) and outcompetes species 2.

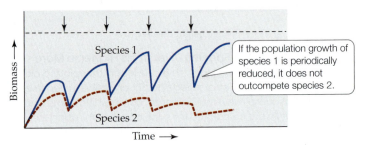

(B) Variable conditions

If the population growth of species 1 is periodically reduced, it does not outcompete species 2.

FIGURE 19.12 The Outcome of Competition under Constant and Variable Conditions (A) Under constant conditions, species 1 (the dominant competitor) outcompetes species 2 when it reaches its own carrying capacity (*K*). (B) If disruptive processes such as disturbance, stress, or predation (represented by the arrows) reduce the population growth of species 1, it will not reach its carrying capacity and will not outcompete species 2, thus allowing coexistence. (After M. Huston. 1979. *Am Nat* 113: 81–101.)

mowing: "Out of twenty species growing on a little plot of mown turf (three feet by four) nine species perished, from the other species being allowed to grow freely." Without mowing, the dominant competitors in the meadow community competitively excluded weedy plants and cut species richness nearly in half. Darwin used this example, along with a multitude of others, to support the argument that nature applies limits to the tendency of species to increase in abundance and outcompete other species. His hypothesis was that species struggle for existence, a necessary first piece to his theory of natural selection.

In 1961, G. E. Hutchinson revived this idea in a paper titled "The Paradox of the Plankton." Hutchinson, an influential community ecologist from Yale University (and major professor to Robert MacArthur), provided one of the first mechanistic descriptions of how coexistence could be maintained under fluctuating environmental conditions. He focused on phytoplankton communities in temperate freshwater lakes (**FIGURE 19.13**). The simple idea behind Hutchinson's model was the seeming paradox of the presence of 30–40 species of phytoplankton given the relatively limited resources at their disposal. He reasoned that all of the phytoplankton compete for the same array of resources, including carbon dioxide, nitrogen, phosphorus, sulfur, and trace elements, which are likely to be evenly distributed in lakes. How could so

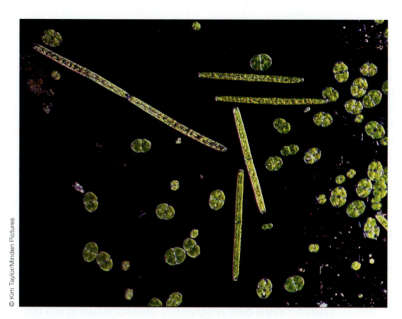

FIGURE 19.13 Paradox of the Plankton Phytoplankton from a freshwater lake. How can so many species coexist using the same set of basic resources? G. E. Hutchinson suggested that the answer is the influence of environmental variation over time.

many species manage to coexist with so few resources and in such a structurally simple environment as a lake? Hutchinson hypothesized that the conditions in the lake changed seasonally and over longer periods and that those changes kept any one species from outcompeting the others. As long as conditions in the lake changed before the competitively superior species eliminated others, coexistence would be possible.

Hutchinson's model has two components that interact to control coexistence among species. One is the time required for one species to competitively exclude another species (t_c), which depends on the population growth rates of the two competing species. The second is the time it takes for environmental variation to act on the population growth of the two competing species (t_e). Hutchinson predicted that when competitive exclusion occurs more rapidly than environmental conditions can change ($t_c \ll t_e$), coexistence cannot be achieved. One could imagine this occurring in communities where there is little environmental change or where the dominant competitor has very rapid growth rates. Conversely, in a fluctuating environment to which the competitors are adapted (where $t_c \gg t_e$), environmental variation does not affect the competitive interactions, and competitive exclusion occurs. One could imagine this pattern in environments with frequent, low-intensity environmental fluctuations and long-lived species. Hutchinson argued that it is only when the time it takes for competitive exclusion to occur is roughly equal to the time it takes for environmental variation to interrupt the competitive interaction (when $t_c = t_e$) that competitive exclusion is thwarted and coexistence occurs. Hutchinson argued that

this condition is likely to be met often in lake phytoplankton communities; otherwise, very few species, rather than tens of species, would coexist.

Hutchinson proposed the idea that competitive exclusion is rare in nature, but he did not test it. It was Robert Paine's work in the rocky intertidal zone of the west coast of North America in the late 1960s that provided some of the most rigorous and convincing evidence that coexistence could be maintained by disruptive processes such as predation or disturbance. Paine (1966) manipulated population densities of *Pisaster*, a predatory sea star that feeds preferentially on the mussel *Mytilus californianus*. In plots from which *Pisaster* was removed, species richness decreased as mussels outcompeted barnacles and other competitively inferior species. In plots where *Pisaster* was present, species richness was enhanced. There are several important aspects to Paine's work, including the keystone species concept and the effects of indirect interactions, but we will consider those aspects in more detail in Concept 21.4 when we discuss food webs. For now, let's concentrate on an idea that arose from the work of Darwin, Hutchinson, and Paine: the intermediate disturbance hypothesis.

The intermediate disturbance hypothesis considers species diversity under variable conditions

The **intermediate disturbance hypothesis** was proposed to explain how gradients in disturbance (although we can easily include stress and predation in this model) affect species diversity in communities (**FIGURE 19.14**). This hypothesis was first formally proposed by Joseph Connell, Paine's contemporary and an author of the classic work on barnacle competition (see Figure 14.16). Connell (1978) recognized that the level of disturbance (its frequency and intensity; see Figure 17.4) experienced by a particular community could have dramatic effects

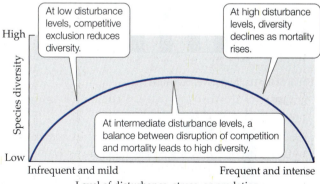

At low disturbance levels, competitive exclusion reduces diversity.

At high disturbance levels, diversity declines as mortality rises.

At intermediate disturbance levels, a balance between disruption of competition and mortality leads to high diversity.

High / Low — Species diversity

Infrequent and mild — Frequent and intense
Level of disturbance, stress, or predation

FIGURE 19.14 The Intermediate Disturbance Hypothesis Species diversity is expected to be greatest at intermediate levels of disturbance, stress, or predation. (After J. H. Connell. 1978. *Science* 199: 1302–1310.)

on its species diversity. He hypothesized that species diversity would be greatest at intermediate levels of disturbance and lowest at high and low levels of disturbance. Why would this be the case? At low levels of disturbance, competition would regulate species diversity because dominant species would be free to exclude competitively inferior species. At high levels of disturbance, on the other hand, species diversity would decline because many individuals would die and some species would become locally extinct as a result. At intermediate levels of disturbance, species diversity would be maximized simply by the balance between disruption of competition and mortality due to disturbance.

The intermediate disturbance hypothesis is highly amenable to testing. One such test was carried out by Wayne Sousa (1979a), who studied succession in intertidal boulder fields in Southern California (see Figure 17.15). In a different but related study, Sousa measured the rate of disturbance of communities living on the boulders and documented their species richness (**FIGURE 19.15**). Small

boulders were rolled over frequently by waves and thus constituted highly disturbed environments for the marine algae and invertebrate species that lived on them. The opposite was true for large boulders, which rarely experienced wave forces large enough to dislodge them. Intermediate-sized boulders, of course, were rolled over at intermediate frequencies. After 2 years, Sousa found that most of the small boulders had only one species (early successional species: the macroalga *Ulva* or the barnacle *Chthamalus*), while the greatest percentage of the large boulders had two species (late successional species: the macroalga *Gigartina canaliculata* and others). The greatest percentage of the intermediate-sized boulders had four species, but some had up to seven species (a mixture of early, mid, and late successional species). Sousa's study is just one of many that have demonstrated the highest diversity at intermediate disturbance levels.

There have been several elaborations on the intermediate disturbance hypothesis

The intermediate disturbance hypothesis is a simple model that relies on variation in disturbance levels to explain species diversity in communities. A handful of ecologists have used it as a foundation for adding more complexity and realism to their theories. One of the first to elaborate on the model was Michael Huston (1979), who acknowledged the effect of disturbance on competition but reasoned that a second process, competitive displacement, could be an important mediating factor. **Competitive displacement** occurs when the best competitor uses limiting resources that the weaker competitor requires, ultimately causing a decline in the weaker competitor's population growth to the point of extinction. Huston's **dynamic equilibrium model** considers how the frequency or intensity of disturbance and the rate of competitive displacement combine to determine species diversity (**FIGURE 19.16**). Like Hutchinson's model, Huston's model predicts maximum species diversity when the level of disturbance and the rate of competitive displacement are roughly equivalent (hence the term *equilibrium* in the model name). Species diversity will be highest when the frequency or intensity of disturbance and the rate of competitive displacement are both at low to intermediate levels (see Figure 19.16, point LL). Moreover, species diversity will be lowest either when disturbance is high and competitive displacement is low (point HL) or when competitive displacement is high and disturbance is low (point LH). When both processes are high and roughly similar (point HH), we expect

Large boulders/low level of disturbance

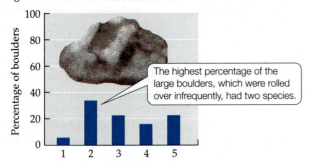

The highest percentage of the large boulders, which were rolled over infrequently, had two species.

Intermediate boulders/intermediate level of disturbance

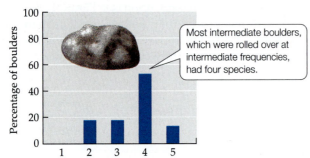

Most intermediate boulders, which were rolled over at intermediate frequencies, had four species.

Small boulders/high level of disturbance

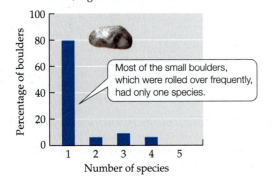

Most of the small boulders, which were rolled over frequently, had only one species.

Number of species

FIGURE 19.15 A Test of the Intermediate Disturbance Hypothesis Marine intertidal communities were surveyed on boulders that differed in the level of disturbance they experienced from being rolled over by wave action. (After W. P. Sousa. 1979a. *Ecology* 60: 1225–1239.)

? Which size boulder had the lowest species richness, and why?

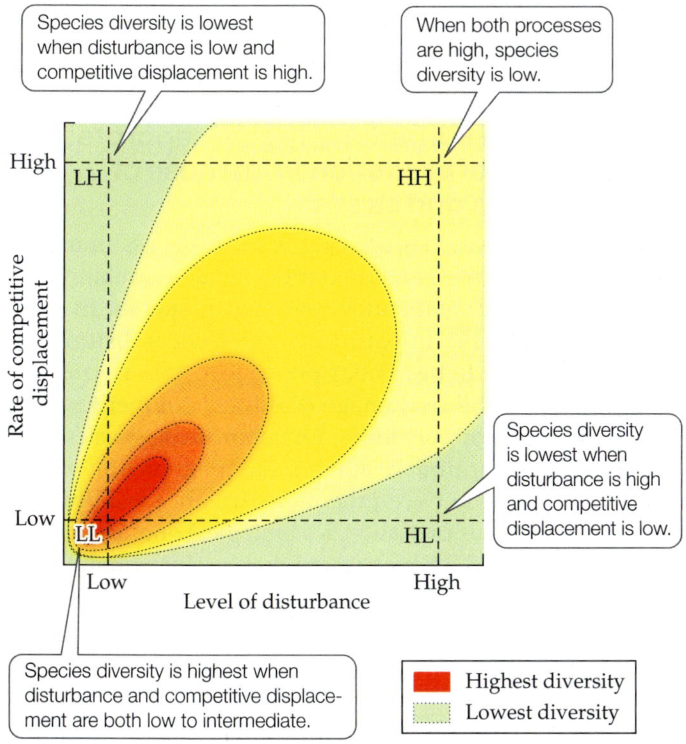
Species diversity is lowest when disturbance is low and competitive displacement is high.
When both processes are high, species diversity is low.
Species diversity is lowest when disturbance is high and competitive displacement is low.
Species diversity is highest when disturbance and competitive displacement are both low to intermediate.
High
Low
Rate of competitive displacement
LH
HH
LL
HL
Low
High
Level of disturbance
Highest diversity
Lowest diversity

FIGURE 19.16 The Dynamic Equilibrium Model The dynamic equilibrium model predicts that species diversity will be highest when the frequency and intensity of disturbance and the rate of competitive displacement are both low to intermediate. (After M. Huston. 1979. *Am Nat* 113: 81–101.)

species diversity to be relatively low because both high mortality and competitive displacement will be acting to reduce species diversity. Perhaps because of its added complexity, there have been few observational or experimental studies of the dynamic equilibrium model. One example comes from an observational study of riparian wetlands in Alaska by Pollock et al. (1998).

Another elaboration of the intermediate disturbance hypothesis comes from Hacker and Gaines (1997), who incorporated positive interactions into their model. If we think back to the previous chapters in this unit, we learned that species interactions are highly context dependent, varying in direction and strength depending on certain physical and biological factors. Theory and experiments both suggest that positive interactions should be more common under relatively high levels of disturbance, stress, or predation—all circumstances in which associations among species could increase their growth and survival. Hacker and Gaines reasoned that positive interactions might be particularly important in promoting species diversity at intermediate to high levels of disturbance (or stress or predation) for two reasons (**FIGURE 19.17**). First, at high levels of disturbance, positive interactions should increase the survival of individuals of the interacting species through both the amelioration of harsh conditions and associational defenses. Second, at intermediate levels of disturbance, species will be released from competition and thus are more likely to engage in positive interactions, an effect that should further increase species diversity.

Hacker and Gaines used studies of a New England salt marsh to support their theory. In this community, there is a strong gradient of physical stress due to saltwater inundation. The highest stress occurs closest to the shoreline, where the tides inundate the plants most frequently. A survey of plants, insects, and spiders across the marsh revealed three distinct intertidal zones, each with a different species composition, and showed that the middle intertidal zone had a higher species richness than either the high or low intertidal zone (**FIGURE 19.18A**). The researchers then conducted transplant experiments in which all the plant species were moved to all three zones, with or without the most abundant plant of their own zone: the tall shrub *Iva frutescens* in the high intertidal zone, and the rush *Juncus gerardii* in the middle and low intertidal zones (Bertness and Hacker 1994; Hacker and Bertness 1999). The results revealed that competition with *Iva* in the high intertidal zone led to the competitive exclusion of most plant species transplanted there, whether or not *Juncus* was also present. In the low intertidal zone, physiological stress was the main factor in controlling population numbers, as many individuals died whether *Juncus* was present or absent. In the middle intertidal zone, however, *Juncus* facilitated other plant species. Without *Juncus*, mortality was 100% for most species by the end of the summer. The mechanism of facilitation, described in Concept 16.3, was amelioration of both hypoxia and salt stress by *Juncus*. Additionally, as we saw

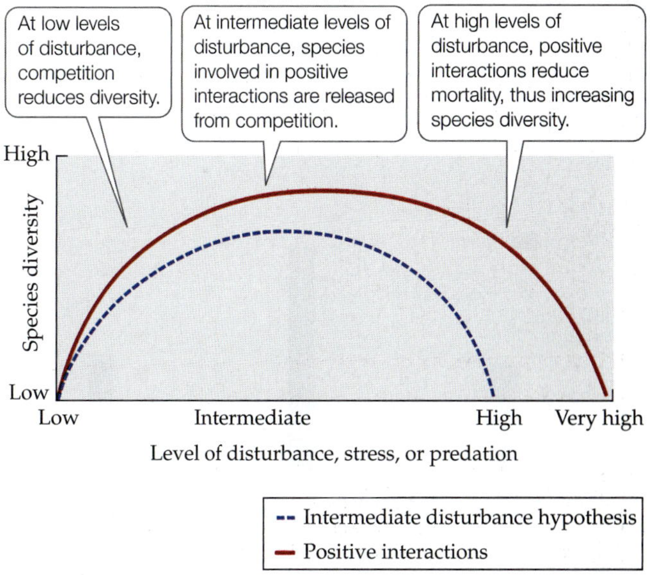
At low levels of disturbance, competition reduces diversity.
At intermediate levels of disturbance, species involved in positive interactions are released from competition.
At high levels of disturbance, positive interactions reduce mortality, thus increasing species diversity.
High
Low
Species diversity
Low
Intermediate
High
Very high
Level of disturbance, stress, or predation
Intermediate disturbance hypothesis
Positive interactions

FIGURE 19.17 Positive Interactions and Species Diversity The intermediate disturbance hypothesis has been elaborated to include positive interactions. (After S. D. Hacker and S. D. Gaines. 1997. *Ecology* 78: 1990–2003.)

in that discussion, *Juncus* indirectly facilitates an aphid herbivore that depends on *Iva* for survival (see Figures 16.12 and 16.13). It turns out that such indirect interactions affect a number of insect herbivores that feed on a variety of other plants facilitated by *Juncus* in the marsh. Hacker and Gaines (1997) concluded, based on these studies, that positive interactions are critically important in maintaining species diversity, especially at intermediate levels of physical stress (**FIGURE 19.18B**). They recognized that physical stress in the middle intertidal zone of the New England salt marsh both decreased the competitive effect of *Iva* and increased the facilitative effect of *Juncus* (and its indirect effects on insects), thus providing ideal conditions for enhanced species coexistence and diversity.

The Menge–Sutherland model separates the effects of predation from those of disturbance and stress

The intermediate disturbance hypothesis assumes that disturbance, stress, and predation all have similar effects on interspecific competition, and thus on species diversity (see Figure 19.14). In particular, it considers disturbance and predation to be similar processes—that is, processes that act to kill or damage dominant competitors and thereby create opportunities for subordinate species. This equating of disturbance and predation ignores an important difference between them: disturbance is a physical process, whereas predation is a biological one. Menge and Sutherland (1987) have argued that because predation is a biological interaction, it is independently affected by physical disturbance and stress and thus should be considered separately.

The Menge–Sutherland model predicts that predation should be relatively important in maintaining species richness at low levels of stress (or disturbance), at which predators can most easily feed on, and thus limit the abundance of, competitively dominant species (**FIGURE 19.19**). As stress increases, the effect of predation decreases as predators become less able to inflict damage on their prey at lower trophic levels. These prey, which are predicted by the model to be more tolerant of physical stress or disturbance, are more likely to compete for resources, especially at intermediate levels of stress or disturbance. But as environmental stress increases to high levels, both predation and competition become less important as more and more species are excluded from the community by their physiological limitations. As with the intermediate disturbance hypothesis, the influences of positive interactions, which are especially important at either extreme of predation or physical stress, have since been incorporated into the Menge–Sutherland model (Bruno et al. 2003), leading to conclusions similar to those of Hacker and Gaines (1997) (see Figure 19.17).

Another important factor that Menge and Sutherland considered in their model was the influence of a particular kind of dispersal known as *recruitment*, defined as the addition of young individuals to a population.

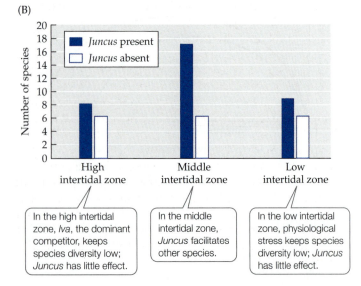

(A)

High intertidal zone	Middle intertidal zone	Low intertidal zone
Plants:	**Plants:**	**Plants:**
Atriplex patula	*Atriplex patula*	*Distichlis spicata*
Iva frutescens (tall)	*Distichlis spicata*	*Juncus gerardii*
Juncus gerardii	*Iva frutescens* (stunted)	*Limonium nashii*
Solidago sempervirens	*Juncus gerardii*	*Salicornia europaea*
Insects:	*Limonium nashii*	**Insects:**
Conocephalus spartinae	*Salicornia europaea*	*Coleophora caespititiella*
Hippodamia convergens	*Solidago sempervirens*	*Coleophora cratipennella*
Trirhabda bacharidis	**Insects:**	*Conocephalus spartinae*
Spiders:	*Coleophora caespititiella*	*Erynephala maritima*
Pardosa littoralis	*Coleophora cratipennella*	**Spiders:**
	Conocephalus spartinae	*Pardosa littoralis*
	Erynephala maritima	
	Hippodamia convergens	
	Microrhopala vittata	
	Trirhabda bacharidis	
	Uroleucon ambrosiae	
	Uroleucon pieloui	
	Spiders:	
	Pardosa littoralis	

(B)

In the high intertidal zone, *Iva*, the dominant competitor, keeps species diversity low; *Juncus* has little effect.

In the middle intertidal zone, *Juncus* facilitates other species.

In the low intertidal zone, physiological stress keeps species diversity low; *Juncus* has little effect.

FIGURE 19.18 Positive Interactions: Key to Diversity in Salt Marsh Communities? (A) Surveys of plant and arthropod species diversity in a New England salt marsh show diversity to be greatest in the middle intertidal zone. (B) Experiments suggest that the high diversity of plants and arthropods in this zone is controlled by the direct and indirect effects of the facilitating rush species *Juncus gerardii* as well as by a decrease in the effect of the dominant competitor, *Iva frutescens*, due to physical stress. (After S. D. Hacker and S. D. Gaines. 1997. *Ecology* 78: 1990–2003.)

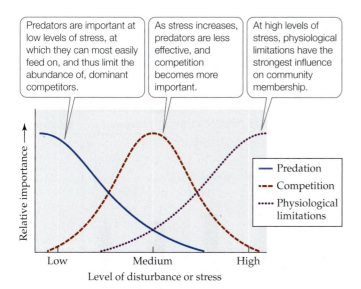

Predators are important at low levels of stress, at which they can most easily feed on, and thus limit the abundance of, dominant competitors.

As stress increases, predators are less effective, and competition becomes more important.

At high levels of stress, physiological limitations have the strongest influence on community membership.

— Predation
--- Competition
···· Physiological limitations

FIGURE 19.19 **The Menge–Sutherland Model** Menge and Sutherland's model of influences on community diversity is similar to the intermediate disturbance hypothesis (see Figure 19.14), but it accounts for the effect of predation separately from that of stress or disturbance. (After B. A. Menge and J. P. Sutherland. 1987. *Am Nat* 130: 730–757.)

They predicted that if recruitment was low, competition might not be particularly important in determining species diversity, because resources would be less likely to be limiting. Instead, the interplay between predation under benign environmental conditions and physical stress under extreme conditions would be the most influential factor regulating community membership. If recruitment increased, however, the role of competition would also increase, ultimately resulting in predictions similar to those in Figure 19.19. Thus, Menge and Sutherland suggest that dispersal (in the form of recruitment) can be another important influence on species diversity and species composition, as shown in Figure 19.4 and demonstrated in **ANALYZING DATA 19.1**.

The intermediate disturbance hypothesis and the Menge–Sutherland model assume that there is an underlying competitive hierarchy among species—that is, that some species are much stronger competitors than others and thus dominate communities if they are not kept in check by disruptive processes. What happens if we assume that there is no competitive hierarchy among species? If species have equivalent effects on one another, then the ability of any one species to live in a community will depend more on chance than on "conflict resolution." Let's spend a moment discussing this alternative theory of species diversity.

Lottery and neutral models rely on equality and chance

A final group of models proposed to explain species coexistence are so-called **lottery models** and **neutral models**

(Sale 1977; Chesson and Warner 1981; Hubbell 2001). As their names suggest, these models emphasize the role of chance in the maintenance of species diversity. Lottery and neutral models assume that resources in a community made available by the effects of disturbance, stress, or predation are captured at random by recruits from a larger pool of potential colonists. For this mechanism to work, species must have fairly similar interaction strengths and population growth rates, and they must have the ability to respond quickly, by dispersing, to disturbances that free up resources. If there is a large disparity in competitive abilities among species, the dominant competitor will have a greater chance of obtaining resources and eventually monopolizing them. In lottery and neutral models, the equal chance of all species to obtain resources is what allows species coexistence.

Lottery and neutral models have most often been applied to highly diverse communities. Peter Sale (1977, 1979) conducted one of the earliest and best-known tests of the lottery model on fishes of the Great Barrier Reef of Australia. Fish species diversity on this reef ranges from 1,500 species in the north to 900 species in the south. On any one small patch of reef (about 3 m, or 10 feet, in diameter), up to 75 species might be recorded. In the reef ecosystem, there is strong habitat fidelity and severe space limitation, and many individual fish spend their entire adult lives in roughly the same spot on the reef. Given these conditions, Sale asked the obvious question: How could so many species coexist in such a small space for so long?

Sale reasoned that only a portion of the coexistence among these fishes could be explained by resource partitioning, because the species tended to have very similar diets. He noted that vacant sites or territories were highly desirable and were made available rather unpredictably by the deaths of individual occupants (due, for example, to predation, disturbance, starvation, or disease). To look at this system in more detail, Sale observed losses of occupants and recruitment to newly vacated sites among three species of territorial pomacentrid fishes (*Eupomacentrus apicalis*, *Plectroglyphidodon lacrymatus*, and *Pomacentrus wardi*). He found the pattern of occupation to be random (**FIGURE 19.20**)—the identity of the species that had previously occupied a site had no bearing on which species was recruited to that site when it became vacant. One species, *P. wardi*, both lost and occupied sites at a greater rate than the other two species, but this had no effect on its overall ability to coexist with the other two species. Sale noted that one important component of this lottery system is that fishes produce many, highly mobile juveniles that can saturate a reef and take advantage of open space made available (as described for clownfish in Connections in Nature for Chapter 7). As Sale put it, "The species of a guild are competing in a lottery for living space in which larvae are tickets and the first arrival at a vacant site wins that site" (Sale 1977, p. 351).

ANALYZING DATA 19.1

How Do Predation and Dispersal Interact to Influence Species Richness?

A prominent theme in this chapter is that processes such as disturbance, stress, and predation can mediate resource availability, thus promoting species coexistence and species diversity. Another important theme in this and the previous chapter is that regional species pools and the dispersal abilities of species can play important roles in supplying new species to communities. What happens when we combine these concepts in an attempt to explain the factors important to species coexistence within local communities? That was the goal of research on zooplankton communities conducted by Jonathan Shurin (2001),* who explored the effects of predation and dispersal on the species diversity of local zooplankton communities. He used experimental ponds made from plastic cattle watering tanks, which he stocked with a diversity of local zooplankton to create individual zooplankton communities. Next, he imposed one of four predation treatments on each pond: (1) no predators, (2) fish predators only (juvenile bluegill sunfish, *Lepomis macrochirus*), (3) insect predators only (the backswimmer bug *Notonecta undulata*), and (4) both fish and insect predators. Finally, Shurin applied a second type of treatment: either the ponds received dispersers of a large number of zooplankton species from the regional pool (which Shurin repeatedly added to the ponds at low densities throughout the experiment), or they received no dispersers. The experiment ran over a summer, after which time Shurin counted the number of zooplankton species in each of the pond communities. His results are shown in the graph.

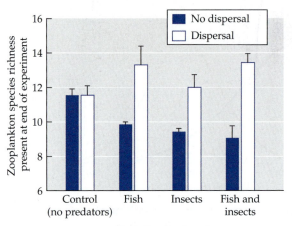

1. How did predation alone affect the species richness of zooplankton within the ponds? Give a plausible explanation for why this occurred. Did fish and insect predators have different effects on local species richness?

2. How does species richness change with the addition of zooplankton dispersal into the ponds? Without knowing anything about the species composition of the ponds, can you say what these results suggest about the dual effects of predation and dispersal on local species richness?

3. Suppose an additional treatment, that of doubling the number of predators, was added to this experiment. Suppose the results showed a decline in zooplankton richness (let's say six species without dispersal and ten species with dispersal). What would these results suggest about the role of dispersal in pond communities subjected to heavy predation? Considering the entire range of predation intensity, from none to intermediate to heavy, do the results fit the intermediate disturbance hypothesis? Why or why not?

---

*Shurin, J. B. 2001. Interactive effects of predation and dispersal of zooplankton communities. *Ecology* 82: 3404–3416.

The role of chance in maintaining species diversity, especially in unpredictable environments, has intuitive appeal. As long as species win the lottery every once in a while, they will continue to reproduce (i.e., buy more tickets) and be able to enter the lottery once again. It is easy to see how this mechanism might be particularly relevant in highly diverse communities such as tropical rainforests and grasslands, where so many species overlap in their resource requirements. Its relevance decreases, however, in communities where species have large disparities in interaction strength. In those communities, it appears that the "great equalizers" are processes that decrease competitive exclusion, such as disturbance,

stress, or predation, or increase inclusion, such as positive interactions.

Ecologists are a long way from agreeing on any one theory to explain why certain species end up coexisting in space and time. Instead, they continue to strive for generalities while recognizing that the relative importance of different mechanisms of species diversity may depend on the characteristics of the community in question.

Up to this point in the chapter, we have focused on the causes of species diversity at the community level. We have asked, Why and how does species diversity differ among communities? In the next section, we will shift gears and instead ask what might be considered the flip

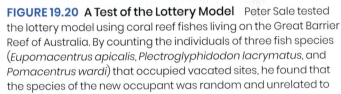

FIGURE 19.20 A Test of the Lottery Model Peter Sale tested the lottery model using coral reef fishes living on the Great Barrier Reef of Australia. By counting the individuals of three fish species (*Eupomacentrus apicalis*, *Plectroglyphidodon lacrymatus*, and *Pomacentrus wardi*) that occupied vacated sites, he found that the species of the new occupant was random and unrelated to the species that had previously occupied the site. The drawings represent the original occupants of vacated sites, and the colored arrows pointing to each drawing show the number of individuals of another species (straight arrows) or the same species (circular arrows) that took over those sites when they became vacant. (Data from P. F. Sale. 1979. *Oecologia* 42: 159–177.)

side of that question. We want to know, given the variation in species diversity among communities (and the current losses of species diversity due to human activities), whether species diversity matters. In other words, what do species do in communities? Does species diversity have functional significance?

> **CONCEPT 19.4** Many experiments show that species diversity affects community function.

LEARNING OBJECTIVES

19.4.1 Describe the relationships between species diversity and ecosystem functions from observations and experiments.

19.4.2 Compare the hypotheses given to explain species diversity and ecosystem function relationships.

The Consequences of Diversity

In the Case Study at the opening of this chapter, we saw lower hantavirus prevalence in small-mammal communities with higher species diversity than in those with lower species diversity (see Figure 19.2). These results support the notion that species diversity can control certain ecological functions of a community. These **community functions**, or processes that control community structure, are numerous and include not only disease suppression, but also plant productivity, water quality and availability, atmospheric gas exchange, and even resistance to disturbance (and recovery afterward). Many of these functions of communities provide valuable ecosystem services to humans, such as food and fuel production,

water purification, oxygen and carbon dioxide exchange, and protection from catastrophic events such as floods or tsunamis (see Concept 23.1). The Millennium Ecosystem Assessment (2005), a synthesis of studies produced under the auspices of the United Nations, details the importance of these ecosystem services to humans. The assessment predicts that if the current losses of species diversity continue, the world's human populations will be severely affected by the loss of the services those species, and the communities in which they live, provide.

What evidence underlies these dire predictions? Recent research has attempted to look at the connections between species diversity and community function, not only to seek basic insights into community ecology, but also because of concerns over species losses and the services that may be affected as a result.

Some relationships between species diversity and community function are positive

The consequences of species diversity to communities were first proposed by both Robert MacArthur (1955) and Charles Elton (1958), who theorized that species richness should be positively related to **community stability**. A community is thought to have stability when it remains, or returns, to its original structure and function after some perturbation (see Concept 17.4). The diversity-stability theory remained "conventional wisdom" until the mid-1970s, when it was tested mathematically using food web models that varied in species richness and complexity (considered in more detail in Concept 21.4). But it was not until 40 years later that the theory was first tested experimentally.

David Tilman and colleagues used a set of experimental plots on abandoned agricultural land at Cedar Creek, Minnesota, to explore the relationship between plant species richness and measures of community function (**FIGURE 19.21A**). In the first study, Tilman and Downing (1994) noticed that some of their experimental plots at Cedar Creek seemed to be responding to a drought differently than others. A survey of their plots showed that plots with higher species richness were better able to withstand the drought than plots with lower species richness (but the same density of plants) (**FIGURE 19.21B**). Drought-induced total plant biomass decrease was less in species-rich plots than in species-poor ones, resulting in a positive, curvilinear relationship between species richness and drought resistance (measured as the difference between biomass before and after the drought). Tilman and Downing reasoned that a curvilinear relationship would be expected if additional species beyond some threshold (the point at which the curve levels off; roughly 10–12 species in this study) had little additional effect on drought resistance. These species could be considered redundant in the sense that they had essentially the same effects on drought resistance as other species. Tilman and Downing suggested, however, that once the number of species in a plot declined below that threshold, each additional species lost from the plot would result in a progressively greater negative effect of drought on the community.

To test this idea more rigorously, Tilman et al. (1996) conducted a well-replicated experiment in which species diversity was directly manipulated. In the same prairie ecosystem, a series of plots that differed in plant species richness, but not in the number of individual plants, was created by randomly selecting sets of species from a pool of 24 species. Each plot was provided with the same amounts of water and nutrients. When biomass in the plots was measured after 2 years of growth, the results confirmed the curvilinear effect of species richness on biomass (**FIGURE 19.21C**) and additionally showed that nitrogen was more efficiently used as species richness increased.

There is debate over diversity-function relationships and their explanations

Although experiments documenting the relationships between species diversity and community function continue to increase in their sophistication, ecologists have debated over the generality of the relationships and their underlying mechanisms. Naeem and colleagues (1995) summarized at least three possible relationships between species diversity and community function and their corresponding hypotheses. Two variables distinguish these hypotheses: the degree of overlap in the ecological functions of species, and variation in the strength of the ecological functions of species.

The first hypothesis, known as the **complementarity hypothesis**, proposes that as species richness increases, there will be a linear increase in community function (**FIGURE 19.22A**). In this case, each species added to the

(A)

Courtesy of G. David Tilman

(B)
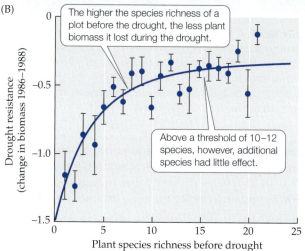

The higher the species richness of a plot before the drought, the less plant biomass it lost during the drought.

Above a threshold of 10–12 species, however, additional species had little effect.

(C)
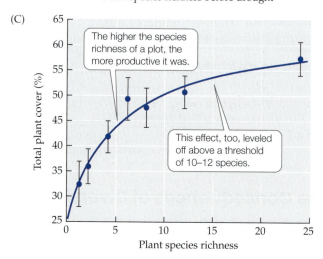

The higher the species richness of a plot, the more productive it was.

This effect, too, leveled off above a threshold of 10–12 species.

FIGURE 19.21 Species Diversity and Community Function (A) Tilman and colleagues used their prairie plots at the Cedar Creek site in Minnesota to test the effects of species richness on community function. (B) First, they measured the effects of a drought on plant biomass in plots that varied in species richness. (C) They then created plots that varied in species richness, though all had the same density of individual plants, and measured biomass in those plots after 2 years of growth. Error bars show ± one standard error (SE) of the mean. (B after D. Tilman and J. A. Downing. 1994. *Nature* 367: 363–365; C after D. Tilman et al. 1996. *Nature* 379: 718–720.)

(A) Complementarity hypothesis

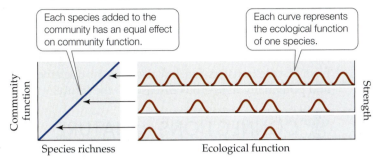

Each species added to the community has an equal effect on community function.

Each curve represents the ecological function of one species.

(B) Redundancy hypothesis

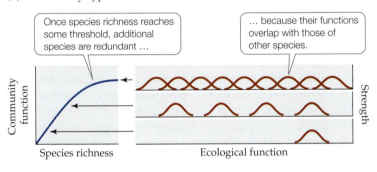

Once species richness reaches some threshold, additional species are redundant …

… because their functions overlap with those of other species.

(C) Idiosyncratic hypothesis

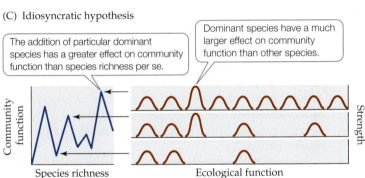

The addition of particular dominant species has a greater effect on community function than species richness per se.

Dominant species have a much larger effect on community function than other species.

FIGURE 19.22 Hypotheses on Species Richness and Community Function At least three possible relationships between species diversity and community function and their corresponding hypotheses have been proposed. Two variables distinguish these hypotheses: the degree of overlap in the ecological functions of species, and variation in the strength of the ecological functions of species. (After G. Peterson et al. 1998. *Ecosystems* 1: 6–8.)

community will have a unique and equally incremental effect on community function. We might expect this type of pattern if we assume that species are equally partitioning their functions within a community. For example, as more and more species are added to the community, each of their unique individual functions will accumulate and increase the overall community function.

The second hypothesis, known as the **redundancy hypothesis**, relies on assumptions similar to those of the complementarity hypothesis, but it places an upper limit on the effect of species richness on community function (**FIGURE 19.22B**). This model best fits the results of Tilman and colleagues described above (see Figure 19.21), in

which the functional contribution of additional species reaches a threshold. This threshold is reached because as more species are added to the community, there is overlap in their function—essentially, there is redundancy among species. In this model, species can be thought of as belonging to certain functional groups (see Figure 16.4C). As long as all the important functional groups are represented, the actual species composition of the community is of little importance to its overall function.

The third hypothesis, known as the **idiosyncratic hypothesis**, proposes that the ecological functions of some species have stronger effects than others do and that they vary dramatically (**FIGURE 19.22C**). Some species have a large effect on community function, while other species have a minimal effect. The addition of dominant species to a community will therefore have a large effect on community function, producing a curve with an idiosyncratic shape, as shown in Figure 19.22C. If communities are assembled in such a way that there are only a few dominant species (e.g., keystone or foundation species; see Figure 16.16), then one would expect community function values to vary dramatically with species richness—that is, there would be peaks and valleys in community function values, depending on whether the dominant species were present or not. As species richness increases, however, the chance that the dominant species will be present also increases. As a result, the variation in community function values should eventually stabilize.

Although these models provide a theoretical foundation for understanding how species contribute to community function, testing them is logically challenging because of the number of species involved and the variety of community functions that could be considered. In many ways, these models and tests are at the heart of modern community ecology, not only because they tell us something about how communities work, but also because they may be able to tell us what the future holds for communities that are both losing (by extinction) and gaining (by invasions) species through human influence.

A CASE STUDY REVISITED

Can Species Diversity Suppress Human Diseases?

The potential value of understanding how species diversity controls community function is limitless when we consider the services communities provide to humans. As we have seen, these services are numerous and diverse. One potential service that has been overlooked until recently is the role species diversity plays in infectious disease emergence and transmission. As we saw in the Case

Study at the opening of this chapter, Suzán and colleagues (2009) showed that plots with reduced small-mammal diversity increased in both host rodent species abundance and the number of SNV-infected rodent individuals (see Figure 19.2). How can species diversity have this effect on disease transmission? Several hypotheses have been proposed. First, if the species that are lost within the community compete with or prey on the host species, then their loss can lead to an increase in the population density of the host and the pathogen. Second, it might be that hosts in more species-diverse situations are simply more likely to come into contact with individuals of other species than their own species (conspecifics), reducing the probability of transmission. Finally, it may be that more diverse communities allow hosts to build up greater resistance to diseases because those hosts are exposed to similar pathogens in other species within the community.

The research to date on the effects of species diversity on hantavirus transmission best supports the first two hypotheses. In the case of the experimental plots in Panama, the data support the first hypothesis; there was an increase in the number of rodent individuals that led to an increase in the number of SNV-infected hosts (see Figure 19.2). Presumably, as the number of small-mammal competitors declined, the rodent host species were able to take advantage of greater resources and their numbers increased. More host individuals then led to greater hantavirus disease transmission. However, the results from the observational studies in Utah and Oregon showed a different pattern that more closely supports the second hypothesis. In those studies, the lower small-mammal diversity increased infection prevalence by simply increasing the encounter rate, rather than density, of individuals of the same host species.

Disentangling the effect of higher density from the effect of reduced species diversity can be difficult. One study, using the trematode parasite *Schistosoma mansoni* and its snail host, manipulated species richness while keeping density constant (Johnson et al. 2009). The researchers showed that the presence of other snail species reduced parasite transmission even when the density of the host remained constant. In this case, the multispecies treatments reduced the encounter rate of the snail host with its trematode parasite by providing alternative but suboptimal host species. Other studies have shown that which species are lost within a community can make a difference in disease transmission, supporting principles of the idiosyncratic hypothesis (see Figure 19.22C). It is clear that the number of examples of species diversity loss and disease transmission is increasing, but the generalities that can be drawn from these examples are still unfolding.

By applying basic principles of ecology to zoonotic disease transmission, we can see that we cannot underestimate the role of species diversity in regulating community integrity. We must consider what might seem like inconsequential and esoteric details, such as the number of species that coexist within communities. In this case, species richness makes all the difference, not only in protecting humans from disease transmission, but also in thwarting emerging and potentially dangerous diseases in the future.

CONNECTIONS *in* NATURE

MANAGING PATHOGENS BY MANAGING BIODIVERSITY

As more evidence accumulates that changes in biodiversity can trigger infectious diseases, there is interest in managing for these outbreaks. Management can come in many forms depending on the pathogen in question. Beyond the obvious recommendation that genetic and species diversity be maintained within ecosystems, there are other management suggestions that can help provide early warning signs or reduce the risk of emerging pathogens altogether.

First, it is critical to survey potential "emergence hot spots" where land use changes and agricultural intensification reduce diversity and have the potential to trigger endemic wildlife pathogens, potentially causing them to jump to new host species, including livestock and humans. In fact, research shows that almost half of the zoonotic diseases that have emerged since 1940 have occurred in regions where major changes in land use, agriculture, or wildlife hunting practices have occurred (Jones et al. 2008). In addition, taking care to reduce disease transmission of captive wildlife should be a special focus, given the recent evidence that coronavirus SARS-CoV-2 (COVID 19) was transmitted to humans from captive wild animals sold at a market in China.

Second, the research also suggests that another 20% of diseases emerging since the 1940s have arisen through the widespread use of antibiotics and the production of resistant strains of microbes. Antibiotics are thought to select for resistant microbes both by eliminating the diversity of nonresistant microbial strains and by eliminating species that suppress those strains. The observation that a more diverse microbiome can suppress strains that are resistant to antibiotics suggests that avoiding the overuse of these pharmaceuticals in medicine and agriculture is critical in preventing emerging diseases.

Finally, managing emerging diseases will involve considering the complex ways that factors such as climate change, invasive species, and pollution interact with biodiversity loss to increase the emergence and transmission of diseases. Despite the many questions that remain, it is clear that managing for biodiversity is a critical component in protecting human populations from potential disease epidemics.

SUMMARY

CONCEPT 19.1 Species diversity differs among communities as a consequence of regional species pools, abiotic conditions, and species interactions.

19.1.1 Describe how regional species' pools and dispersal abilities contribute to community membership.

- The regional species pool and the dispersal abilities of species play important roles in supplying species to communities.

- Humans have greatly expanded the regional species pools of communities by serving as vectors for the dispersal of non-native species.

19.1.2 Describe how local environmental conditions act as a "filter" for community membership.

- Local environmental or abiotic conditions act as a strong "filter" for community membership because species must be physiologically adapted to those conditions of the community.

19.1.3 Describe how species interactions may act to include species in, or exclude species from, communities.

- When a species depends on other species for its growth, reproduction, and survival, those other species must be present if it is to gain membership in a community.

- Species may be excluded from communities by competition, predation, parasitism, or disease. When non-native species are excluded, the community has biotic resistance to the potential invasion.

CONCEPT 19.2 Resource partitioning is theorized to reduce competition and increase species diversity.

19.2.1 Define resource partitioning.

- Resource partitioning theory predicts that species must use resources slightly differently if they are to avoid competitive exclusion.

19.2.2 Outline observations, experiments, and models that support resource partitioning as a mechanism of species coexistence.

- One model of resource partitioning states that the less overlap there is among species in their use of resources, along a resource spectrum, the more species can coexist in the community.

- The resource ratio hypothesis posits that species that use the same set of resources are able to partition them by using them in different proportions.

CONCEPT 19.3 Processes such as disturbance, stress, predation, and positive interactions can mediate resource availability, thus promoting species diversity.

19.3.1 Describe the role of disturbance, stress, and predation in mediating coexistence and promoting species diversity.

- If disturbance, stress, or predation keeps dominant competitors from reaching their carrying capacity, competitive exclusion will not occur and coexistence will be maintained.

19.3.2 Define and give examples of the intermediate disturbance hypothesis and its variations, including those that consider predation and positive interactions.

- The intermediate disturbance hypothesis states that intermediate levels of disturbance, stress, or predation promote species diversity by reducing competitive exclusion. At low levels of disturbance, competitive exclusion reduces species diversity, and at high levels of disturbance, high mortality reduces species diversity.

- The dynamic equilibrium model predicts that species diversity will be highest when the level of disturbance and the rate of competitive displacement are roughly equivalent.

- Positive interactions can promote species diversity, particularly at intermediate to high levels of disturbance, stress, or predation.

- The Menge-Sutherland model is similar to the intermediate disturbance hypothesis except that it separates the effect of predation from that of physical disturbance.

19.3.3 Define and give examples of lottery or neutral models.

- Lottery and neutral models assume that resources in a community made available by disturbance, stress, or predation are captured at random by recruits from a larger pool of potential colonists, whose chances for capturing resources are equal.

CONCEPT 19.4 Many experiments show that species diversity affects community function.

19.4.1 Describe the relationships between species diversity and ecosystem functions from observations and experiments.

- Evidence suggests that species diversity can control numerous functions of communities, including productivity, soil fertility, water quality and availability, atmospheric gas exchange, and responses to disturbance.

- Some manipulative experiments in different communities have shown that as species diversity increases, so does community function.

19.4.2 Compare the hypotheses given to explain species diversity and ecosystem function relationships.

- Hypotheses proposed to explain the positive relationship between species diversity and community function fall into three general categories, which include different assumptions about the degree to which individual species vary in their contribution to community function.

Review Questions

1. Suppose you are an ecologist studying prairie grassland communities in Minnesota. As you are doing your fieldwork, grass seeds with hooked spines attach themselves to your shoes. You then travel to New Zealand to study the grasslands on the South Island. When you enter the customs area in the Auckland airport, the officers in charge ask if you have visited a natural area or farm recently. You say yes, and they tell you to take off your shoes and wait while they disinfect them with bleach. Given what you know about the mechanisms important to community membership, is it worth the time and money required to clean all that footwear before allowing it into New Zealand?

2. We know that species diversity varies greatly among communities. Describe how some of the models proposed to explain this variation differ in their explanations of the mechanisms involved.

3. Suppose you are studying a tropical rainforest community in Panama. You obtain a 50-year data set for the forest that records both the mortality of adult trees and the emergence of new tree seedlings. As you analyze the data, you try to determine whether there is a pattern of species replacement, in which individuals of one species generally replace one another in the same sites, rather than individuals of other species establishing. After much work, you are convinced that no pattern of replacement exists in this forest—instead, sites are colonized in an entirely random fashion, with no one species having an advantage. What general set of models of species diversity best describes your observations, and why?

4. Recent experimental work in communities has shown positive relationships between species diversity and community function. We learned that there is considerable debate about the relationships and their controlling mechanisms and that at least three hypotheses have been developed to explain them. Below are three graphs (A, B, and C) of species richness–community function relationships that vary in the shapes of their curves. Describe which hypothesis best fits each curve, and why.

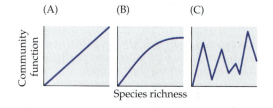

Hone Your Problem-Solving Skills

In the Case Study of this chapter, we explored the results of a study that considered the relationship between species diversity and disease transmission of the Sin Nombre virus in deer mouse populations in Oregon (Dizney and Ruedas 2009). Suppose a similar study is conducted in six parks in or around a city that vary in their degree of human disturbance. The researchers record the number of small-mammal species, the density of deer mice, and the number of deer mouse individuals infected with the Sin Nombre virus. Below is a table with the results of this hypothetical study, organized by the degree of human disturbance at the trapping locations:

Park	Degree of human disturbance (lowest to highest)	Small-mammal species richness	Deer mouse density	Sin Nombre virus infection prevalence
1	1	12	5	0.045
2	2	14	6	0.045
3	3	16	6.5	0.04
4	4	13	6.5	0.04
5	5	9	5.5	0.08
6	6	6	6	0.12

1. Graph the relationship between degree of human disturbance and small-mammal species richness. What species diversity model do your data best describe, and why?

2. Now graph the relationship between species richness and deer mouse density. What does the graph show? Do the data support the theory that species richness is a consequence of resource partitioning in this small-mammal community?

3. Graph both the relationship of species richness with Sin Nombre virus infection prevalence in the deer mouse and the relationship of deer mouse density with Sin Nombre virus infection prevalence in the deer mouse. Which factor seems to be more important in spreading infection in deer mice—species richness or deer mouse density? Explain why this might be so, based on theory.

Ecosystems

UNIT 6

20

Production

KEY CONCEPTS

CONCEPT 20.1 Energy in ecosystems originates with primary production by autotrophs.

CONCEPT 20.2 Net primary production is constrained by both physical and biotic environmental factors.

CONCEPT 20.3 Global patterns of net primary production reflect climate constraints and biome types.

CONCEPT 20.4 Secondary production is generated through the consumption of organic matter by heterotrophs.

Life in the Deep Blue Sea, How Can It Be? A Case Study

Ecologists once considered the deep sea to be the marine equivalent of a desert. The physical environment at depths between 1,500 and 4,000 m (5,000–13,000 feet) did not seem conducive to life as we knew it. It is completely dark, so photosynthesis is not possible. The water pressure reaches values 300 times greater than those at the surface of the ocean, similar to the pressure used to crush cars at a junkyard. Organisms living on the floor of the deep sea were thought to obtain energy exclusively from the sparse rain of dead material falling from the upper layers of the ocean where sunlight is sufficient for phytoplankton to carry out photosynthesis. Most of the known deep-sea organisms were detritus feeders such as echinoderms (e.g., sea stars), mollusks, crustaceans, and polychaete worms.

Our view of deep-sea life was changed dramatically in 1977, when an expedition led by Robert Ballard of the Woods Hole Oceanographic Institution used the submersible craft *Alvin* to dive to a mid-ocean ridge near the Galápagos archipelago (**FIGURE 20.1**). The team aboard the *Alvin* was in search of the deep-sea hot springs thought to occur along mid-ocean ridges. These ridges lie at the junctions of tectonic plates, where the seafloor spreads as the plates are pushed apart by molten rock rising from Earth's mantle (see Figure 18.10). Because mid-ocean ridges are volcanically active, geologists and oceanographers had hypothesized that seawater seeping into cracks in the ocean floor near the ridges would be superheated by pockets of magma, chemically transformed, and ejected as hot springs. These hot springs were considered potential sources of chemicals for the ocean system as well as sources of heat. Despite their hypothesized existence, no such hot springs had ever been located.

Ballard's group did indeed find hot springs, known as *hydrothermal vents*. However, this geochemical finding paled in comparison with their biological discovery: the areas around the hydrothermal vents were teeming with life. Dense assemblages of tube worms (e.g., *Riftia* spp.), giant clams (e.g., *Calyptogena* spp.), shrimps, crabs, and polychaete worms were found in the areas surrounding the vents (**FIGURE 20.2**). The density of organisms was unprecedented for the deep, dark seafloor.

The discovery of these diverse and productive hydrothermal vent communities posed an immediate question: How did the organisms obtain the energy needed to sustain themselves in such abundance?

FIGURE 20.1 Black Smoker Vent A hydrothermal vent emits superheated water as hot as 400°C, rich in iron sulfide, known as a "black smoker." Despite the high temperature and toxic nature of the water, abundant life surrounds these features.

FIGURE 20.2 Life around a Hydrothermal Vent Mussels in the genus *Bathymodiolus* are scattered near a hydrothermal vent, with several crabs lacking pigmentation in their carapaces.

The rate at which dead organisms from the upper zones of the ocean accumulate on the seafloor is very low (0.05–0.1 mm/year). The newly formed areas of seafloor where the vents are located are only decades old, and thus the amount of organic material that would have accumulated should not be enough to sustain these high densities of organisms. Photosynthesis in the surface waters therefore did not appear to be the energy source supporting these hydrothermal vent communities. Additionally the water emitted from the hydrothermal vents is rich in poisonous sulfides as well as heavy metals such as lead, cobalt, zinc, copper, and silver, constituting another serious challenge for life in the ocean vents.

Hydrothermal vent communities thus pose two mysteries: First, what is the source of energy that sustains them, and second, how do the organisms tolerate the high concentrations of potentially toxic sulfides in the water? As we shall see, the answers to these two questions are intimately related.

Introduction

In 1942, the journal *Ecology* published a controversial paper by Raymond Lindeman, describing the nature of energy flow in aquatic ecosystems. Lindeman had studied the energy relationships among the organisms and nonliving components in lakes in Minnesota. Rather than grouping its component plants, animals, and bacteria according to their taxonomic categories, Lindeman grouped them into categories based primarily on how they obtained their energy (**FIGURE 20.3**). His views on the importance of the energy base of the system—an "ooze" of particulate and dissolved dead organic matter—and on the efficiency of energy transfer among the system's

biological components were groundbreaking. Lindeman's treatment of energy flow in the lake was considered too theoretical at the time, and his paper was initially rejected. The editors later reconsidered their rejection of the paper after Lindeman's mentor, the prominent limnologist G. E. Hutchinson, advocated its acceptance. Lindeman's paper was among the first in the area of ecosystem science, and it is now considered a fundamental paper in the discipline.

The term **ecosystem** was coined by A. G. Tansley, a plant ecologist, to refer to all of the components of an ecological system, biotic and abiotic, that influence the flow of energy and elements (Tansley 1935). The "elements" considered in ecosystem studies are primarily nutrients, but they also include pollutants; the movements of elements through ecosystems are the topic of Chapter 22. The ecosystem concept is now well established and has become a powerful tool for integrating ecology with other disciplines such as geochemistry, hydrology, and atmospheric science.

In Chapter 5, we described the physiological basis for the capture of energy through photosynthesis and chemosynthesis by autotrophs, and we explained how heterotrophs obtain that energy by consuming autotrophs. In this chapter, we return to the topic of energy as we explore how energy enters ecosystems, how it is measured, and what controls rates of energy flow through ecosystems.

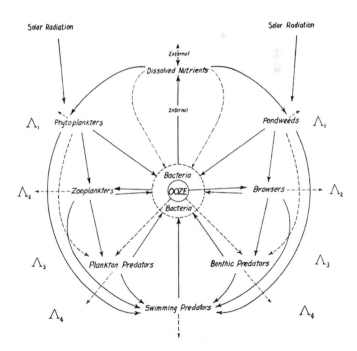

FIGURE 20.3 Energy Flow in a Lake Raymond Lindeman's diagram describes the movement of energy among groups of organisms at Cedar Bog Lake, Minnesota. Note the general functional categories of organisms Lindeman used, as well as the central position of "ooze" (organic matter) in the diagram. The subscripts next to the uppercase Greek lambdas represent trophic levels. (From R. L. Lindeman. 1942. *Ecology* 23: 399–418.)

LEARNING OBJECTIVES

20.1.1 Explain how variation in the leaf area index influences gross primary production through both carbon gains and losses.

20.1.2 Describe the relationship between gross primary production and net primary production and how this relationship can change during succession.

20.1.3 Summarize why it is important to measure net primary production and describe some of the ways ecologists measure net primary production at large spatial scales.

20.1.4 Illustrate how the relationship between net primary production and net ecosystem exchange is influenced by the mass of autotrophs versus the mass of heterotrophs.

Primary Production

The generation of chemical energy by autotrophs, known as **primary production**, is derived from the uptake of carbon during photosynthesis and chemosynthesis (see Chapter 5). Chemosynthesis can be the main source of energy in some rare circumstances, as we will see at the end of this chapter. However, the majority of energy on Earth comes from photosynthesis, and thus primary production derived from photosynthesis will be the focus of this section. Primary production represents an important energy transition: the conversion of light energy from the sun into chemical energy that can be used by autotrophs and consumed by heterotrophs. Primary production is the source of energy for all organisms, from bacteria to humans; even the fossil fuels we use today are derived from primary production. Primary production also accounts for the largest movement of carbon dioxide between Earth and the atmosphere, and it is therefore an important influence on global climate through its effect on greenhouse gas concentrations (see Chapters 2 and 25).

The energy assimilated by autotrophs is stored as carbon compounds in plant and phytoplankton tissues; therefore, carbon (C) is the currency used for the measurement of primary production. The *rate* of primary production is sometimes referred to as *primary productivity*.

Gross primary production is total ecosystem photosynthesis

The amount of carbon taken up by the autotrophs in an ecosystem is called **gross primary production (GPP)**. The GPP in most terrestrial ecosystems is equivalent to the total of all plant photosynthesis.

The GPP of an ecosystem is controlled by climate through its influence on rates of photosynthesis, as we saw in Concept 5.2, and by the leaf area of the plants, expressed as the **leaf area index**, the amount of leaf area over an area of ground. The leaf area index varies among biomes, from less than 0.1 in tundra (i.e., less than 10% of the ground surface has leaf cover) to 12 in boreal and tropical forests (i.e., on average, there are 12 layers of leaves between the canopy and the ground). Shading of the leaves below the uppermost layer increases with the addition of each new leaf layer, so the incremental gain in photosynthesis for each added leaf layer decreases (**FIGURE 20.4**). Eventually, the respiratory costs associated with building and maintaining additional leaf layers outweigh the photosynthetic benefits. Plants generally match their leaf area index to the climate conditions and the supplies of resources, particularly water and nutrients, in order to maximize carbon gain.

A plant uses approximately half of the carbon it fixes by photosynthesis in cellular respiration to support biosynthesis and cellular maintenance. All living plant tissues lose carbon via respiration, but not all of them acquire carbon via photosynthesis. Thus, plants that have a large proportion of nonphotosynthetic stem tissue, such as trees and shrubs, tend to have higher overall respiratory carbon losses than herbaceous plants such as grasses and forbs. Plant respiration rates increase with increasing temperatures, and as a result, respiratory carbon losses are higher in tropical forests than in temperate and boreal forests.

Net primary production is the energy remaining after respiratory losses

Not all of the carbon taken up in photosynthesis is available for growth and other functions in plants. As noted above, some carbon is lost in respiration. Carbon not used in respiration is available for growth and reproduction, storage, and defense against herbivory. The carbon available for these functions is determined by the balance between GPP and autotrophic respiration, and is called **net primary production (NPP)**:

$$NPP = GPP - respiration \qquad (20.1)$$

The NPP of a terrestrial ecosystem is the amount of energy captured by autotrophs that results in an increase in living plant matter, or **biomass**. NPP is the energy left over for plant growth, plant reproduction, defense, and consumption by herbivores and detritivores. It also represents the total net input of carbon into ecosystems.

Plants respond to varying environmental conditions by allocating carbon to the growth of different tissues. The allocation of carbon within a plant varies according to the species, the availability of resources, and the climate. Allocation of carbon to photosynthetic tissues is an investment in potential future NPP, but the demands of the plant for other resources, particularly water and nutrients, as well as biological interactions such as herbivory, influence whether that investment pays off.

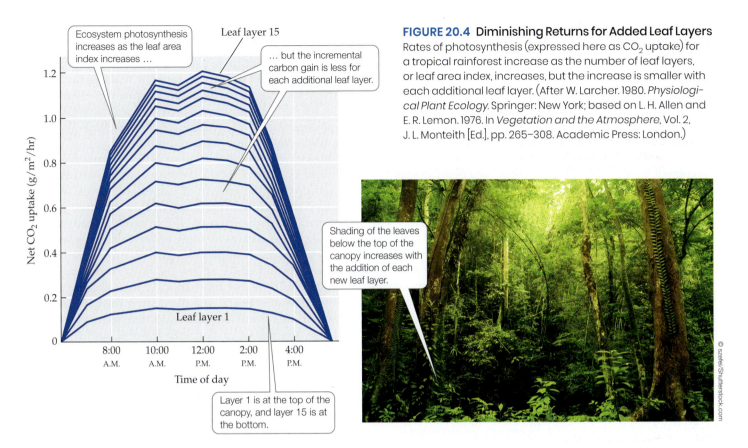

FIGURE 20.4 Diminishing Returns for Added Leaf Layers Rates of photosynthesis (expressed here as CO_2 uptake) for a tropical rainforest increase as the number of leaf layers, or leaf area index, increases, but the increase is smaller with each additional leaf layer. (After W. Larcher. 1980. *Physiological Plant Ecology*. Springer: New York; based on L. H. Allen and E. R. Lemon. 1976. In *Vegetation and the Atmosphere*, Vol. 2, J. L. Monteith [Ed.], pp. 265–308. Academic Press: London.)

A plant's allocation of NPP to the growth of leaves, stems, and roots is generally balanced so as to maintain supplies of water, nutrients, and carbon to match the plant's requirements. For example, plants growing in desert, grassland, and tundra ecosystems are regularly exposed to shortages of water or nutrients. Plants in these ecosystems may allocate a greater proportion of NPP to root growth, relative to the growth of shoots (leaves and stems), than plants growing in ecosystems with higher soil water and nutrient availability (**FIGURE 20.5**). This greater allocation to root growth facilitates their acquisition of the resources that are in short supply. In contrast,

plants growing in dense communities, with neighbors that may shade them, may allocate NPP preferentially to stems and leaves in order to capture more sunlight for photosynthesis. In other words, plants tend to allocate the most NPP to those tissues that acquire the resources that limit their growth.

Allocation of NPP to storage compounds such as starch and carbohydrates provides insurance to compensate for losses of tissues to herbivores, disturbances such as fire, and weather events such as frost. These compounds are usually stored in the stems of woody plants or in belowground stems and roots of herbaceous plants. Where levels

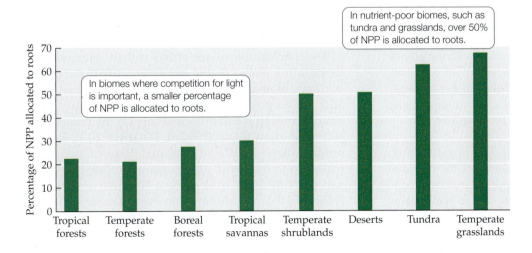

FIGURE 20.5 Allocation of NPP to Roots The proportion of NPP that plants allocate to roots varies with the resources available to them. (After B. Saugier et al. 2001. In *Terrestrial Global Productivity*, J. Roy, B. Saugier, and H. A. Mooney [Eds.], pp. 543–557. Academic Press: San Diego, CA.)

? In addition to low supplies of resources in the soil, what other factors might favor greater allocation of NPP to tissues below the soil surface?

of herbivory are high, plants may allocate a substantial amount of NPP (up to 20%) to defensive secondary compounds, such as tannins or terpenes, that inhibit grazing.

NPP changes during ecosystem development

As ecosystems develop during primary or secondary succession (see Concept 17.2), NPP changes as the abundance of plants and associated leaf area index, the ratio of photosynthetic to nonphotosynthetic tissue, and plant species composition all change. Disturbance and succession can therefore influence gains or losses of CO_2 from ecosystems, thereby affecting atmospheric CO_2 concentrations.

Most ecosystems have their highest NPP at mid-successional stages. Several factors contribute to this pattern, including the tendency for the proportion of photosynthetic tissues, plant diversity, and nutrient supply to be highest at mid-successional stages. In forest ecosystems, the leaf area index and the photosynthetic rates of leaves decrease in old-growth stands, lowering GPP and thus NPP. In some grasslands, such as the tallgrass prairies of the central United States, the accumulation of dead leaves near the ground surface and the development of a closed upper canopy of leaves decrease light availability to short plants, lowering the photosynthetic carbon gain of the ecosystem. However, the decrease in NPP over time is far less pronounced in grasslands than in forest ecosystems. Although NPP may decrease in late successional stages, lowering the uptake of CO_2 from the atmosphere, these old-growth ecosystems contain large pools of stored carbon and nutrients and provide habitat for late successional animal species.

NPP can be estimated by a number of methods

There are several reasons why it is important to be able to measure NPP in an ecosystem. As we have seen, NPP is the ultimate source of energy for all organisms in an ecosystem and thus determines the amount of energy available to support that ecosystem. It varies tremendously over space and time. Year-to-year variation in NPP provides a metric for examining ecosystem health, because changes in primary productivity can be symptomatic of stresses such as high temperature, drought, or acid rain. Finally, as noted earlier, NPP is intimately associated with the global carbon cycle, and it is therefore an important influence on global climate change (see Chapters 2 and 25). For all these reasons, scientists have put great effort into improving techniques for estimating NPP over the past 3 decades.

TERRESTRIAL ECOSYSTEMS Methods for estimating NPP in forest and grassland ecosystems are the best developed because of the economic importance of these ecosystems for wood and forage production as well as the high amount of carbon that they store. Traditional techniques include measuring the increase in plant biomass during the growing season by harvesting plant tissues in experimental plots. In forests, the radial growth of wood must be included in estimates of NPP. In the tropics, plants may continue to grow throughout the year, and tissues that die decompose rapidly, making the use of harvest techniques problematic. Despite these shortcomings, harvest techniques still provide reasonable estimates of aboveground NPP, particularly if corrections are made for tissue loss to herbivory and mortality.

Measuring the allocation of NPP to growth belowground is more difficult because root growth is more dynamic than the growth of leaves and stems, and the soil makes it difficult to observe this dynamic growth pattern. The proportion of NPP in roots exceeds that in aboveground tissues in some ecosystems: in grassland ecosystems, for example, root growth may be twice that of aboveground leaves, stems, and flowers combined. The finest roots *turn over* more quickly than shoots; that is, more roots are "born" and die during the growing season than stems and leaves. In addition, roots may exude a large amount of carbon into the soil, and they may transfer carbon to mycorrhizal or bacterial symbionts. Therefore, harvests for measuring root biomass must be more frequent, and additional correction factors must be used when estimating belowground NPP. Proportional relationships correlating aboveground to belowground NPP have been developed for some forest and grassland ecosystems so that measurements of aboveground NPP can be used to estimate whole-ecosystem NPP. The use of *minirhizotrons*, underground viewing tubes outfitted with video cameras, has led to advances in the understanding of belowground production processes (**FIGURE 20.6**).

The labor-intensive and destructive nature of harvest techniques makes them impractical for estimating NPP over large areas or in biologically diverse ecosystems. Several nondestructive techniques have been developed that allow more frequent estimation of NPP over much larger spatial scales, although with lower precision than harvest techniques. Some of these techniques, which include remote sensing and frequent atmospheric CO_2 measurements, provide a quantitative index rather than an absolute measure of NPP. Some techniques use a combination of data collection and modeling of plant physiological and climate processes to estimate the actual fluxes of carbon associated with NPP.

The concentration of the photosynthetic pigment chlorophyll in a plant canopy provides a proxy for photosynthetic biomass that can be used to estimate GPP and NPP. Chlorophyll concentrations can be estimated using remote sensing techniques that rely on the reflection of solar radiation (**ECOLOGICAL TOOLKIT 20.1**). Remote sensing allows NPP to be measured frequently, at spatial scales up to the entire globe, using satellite-based sensors (**FIGURE 20.7**). Indicators of NPP that are based on chlorophyll concentrations can overestimate NPP if the vegetation is not physiologically active, as in boreal forests in winter, but remote sensing generally provides the best estimate for NPP at regional to global scales.

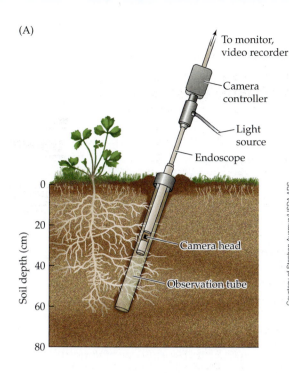

FIGURE 20.6 **A Tool for Viewing Belowground Dynamic** (A) Minirhizotrons allow researchers to observe the dynamics of root growth and death belowground. (B) A view of roots from a minirhizotron tube installed in a bog ecosystem in northern Minnesota. Small-diameter roots from ericaceous shrubs can be seen in the foreground against a background of decomposing *Sphagnum* mosses and peat.

NPP can also be estimated from direct measurements of its components: GPP and plant respiration. This approach may involve measuring the change in CO_2 concentration in a closed system, which can be created by placing a chamber around stems and leaves, whole plants, or whole stands of plants. For example, Howard Odum estimated NPP for a tropical forest in Puerto Rico by enclosing a stand of trees inside a 200 m^2 × 20 m tall clear plastic "tent" (Odum and Jordan 1970). The emissions of CO_2 to the atmosphere in such a closed system are from respiration by the plants and heterotrophs, including microorganisms in the soil and animals in the forest. Uptake of CO_2 from the atmosphere results from photosynthesis. Thus, the net change in CO_2 inside the system results from the balance between GPP and total respiratory release by the plants and the heterotrophs. This net exchange of CO_2

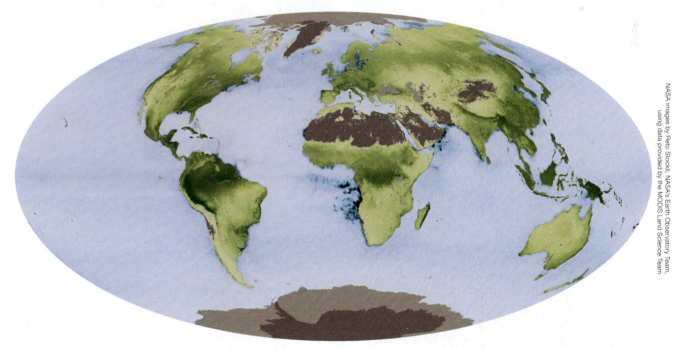

FIGURE 20.7 **Remote Sensing of NPP** Global NPP estimated using a satellite-based sensor [Moderate Resolution Imaging Spectroradiometer (MODIS)]. Note the latitudinal patterns in NPP corresponding to climate zones.

? In addition to zones of upwelling, what other coastal zones have high rates of NPP as indicated in this map?

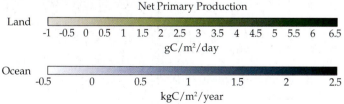

ECOLOGICAL TOOLKIT 20.1

Remote Sensing

When sunlight strikes an object, it is absorbed or scattered in such a way that the amount and quality of the light that reflects off of the object is changed. For example, when sunlight strikes a clear lake, about 5% of the visible light is reflected, while a light-colored sandy soil, such as might be found in a desert, reflects back as much as 40%. The amount of light reflected depends on the wavelengths of the light: different kinds of objects absorb or reflect some wavelengths more than others. The atmosphere scatters more blue wavelengths than red or green, and therefore the sky appears blue to our eyes. The lake, however, appears blue because most of the red and green light is absorbed by the water before it can be scattered back to our eyes. Lakes with high concentrations of phytoplankton appear green because much of the blue light is absorbed by the phytoplankton, leaving only the green light to be scattered back to our eyes.

Remote sensing is a technique that takes advantage of light reflection and absorption to estimate the density and composition of objects on Earth's surface, in its waters, and in its atmosphere. Ecologists use remote sensing to estimate NPP by taking advantage of the unique reflectance pattern of chlorophyll-containing plants, algae, and bacteria (**FIGURE A**). Because chlorophyll absorbs visible solar radiation in blue and red wavelengths, it has a characteristic *spectral signature* with greater reflection of green wavelengths. In addition, vegetation absorbs more light of red wavelengths than does bare soil or water.

Ecologists can measure the reflection of specific wavelengths from a land or water surface and estimate NPP using several indices that have been developed. One of the most commonly used indices is the *normalized difference vegetation index*, or NDVI, which uses differences between visible-light and near-infrared reflectance to estimate the density of chlorophyll:

$$NDVI = \frac{(NIR - red)}{(NIR + red)} \quad (20.4)$$

where NIR is the near-infrared wavelength band (700–1,000 nm) and red is the red wavelength band (600–700 nm). Note that the spectral signature of vegetation in Figure A shows a large difference between reflectance of red and near-infrared wavelengths relative to the spectral signatures of water and soil, which gives vegetation a high NDVI value and water and soil low NDVI values. The NDVI is coupled with estimates of the efficiency of light absorption to estimate photosynthetic CO_2 uptake.

Remote sensing of light reflectance from Earth's surface and atmosphere can be done at large spatial

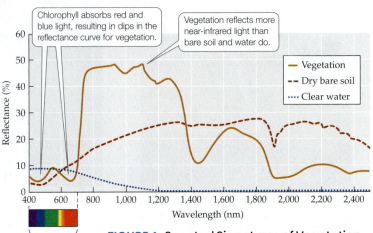

Chlorophyll absorbs red and blue light, resulting in dips in the reflectance curve for vegetation.

Vegetation reflects more near-infrared light than bare soil and water do.

Visible spectrum

FIGURE A Spectral Signatures of Vegetation, Clear Water, and Bare Soil   Note the low reflectances of blue and red wavelengths for vegetation. (After A. R. Huete. 2004. In *Environmental Monitoring and Characterization*, J. F. Artiola et al. [Eds.], pp. 183–206. Academic Press: Amsterdam.)

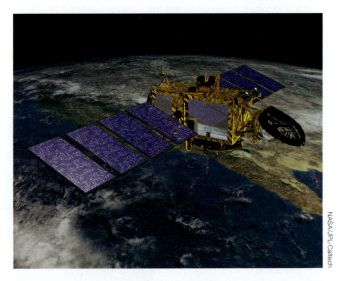

NASA/JPL-Caltech

FIGURE B Remote Sensing by Satellite   Remote sensing instruments mounted on satellites can measure the reflectance of solar radiation from Earth to provide ecologists with large-scale measurements of NPP and other phenomena.

scales using satellites (**FIGURE B**), which transmit their measurements to receiving stations. Depending on the spatial resolution of the surface measurement and the number of wavelengths measured, satellite remote sensing can generate massive amounts of data that need to be processed. Advances in computing power have enhanced the spatial and temporal capabilities of remote sensing, making it a powerful tool for measuring NPP as well as deforestation, desertification, atmospheric pollution, and many other phenomena of interest to ecologists.

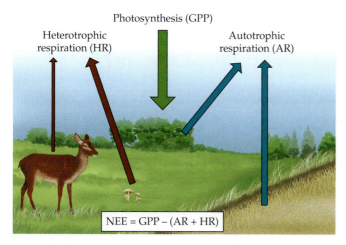

FIGURE 20.8 Components of Net Ecosystem Exchange (NEE) Net ecosystem exchange includes all of the components of an ecosystem that either take up CO_2 (autotrophs, through photosynthesis) or release CO_2 (both autotrophs and heterotrophs).

Photosynthesis (GPP)

Heterotrophic respiration (HR)

Autotrophic respiration (AR)

$$NEE = GPP - (AR + HR)$$

is called **net ecosystem exchange** (**NEE**) (**FIGURE 20.8**). Heterotrophic respiration must be subtracted from NEE to obtain NPP; as a result, NEE provides a more refined estimate of ecosystem carbon storage than NPP. Carbon movement into and out of ecosystems, such as carbon lost through leaching from the soil or through disturbances (e.g., fire or deforestation; **ANALYZING DATA 20.1**), can influence estimates of NEE and NPP.

Another noninvasive approach to estimating NEE uses frequent measurements of CO_2 and microclimate at various heights throughout a plant canopy and into the open air above the canopy. The movement of air in these zones is complex and can be modeled as rotating eddies of air, much like the eddies in flowing streams. These eddies can be modeled using high-frequency measurements at different heights. This technique, known as *eddy covariance* or *eddy correlation*, takes advantage of the gradient in CO_2 concentration between the plant canopy and the atmosphere that develops because of photosynthesis and respiration. During the day, when plants are photosynthetically active, the concentration of CO_2 is lower in the plant canopy than in the air above the plant canopy. At night, when photosynthesis shuts down but respiration continues, the CO_2 concentration in the canopy is higher than that in the atmosphere. Instrument-bearing towers established in forest, shrubland, and grassland canopies have been used to measure the NEE of CO_2 over long periods (**FIGURE 20.9**). Depending on the tower height, eddy covariance can provide an integrated NEE for up to several square kilometers of the surrounding area. A network of eddy covariance sites across the globe

(A)

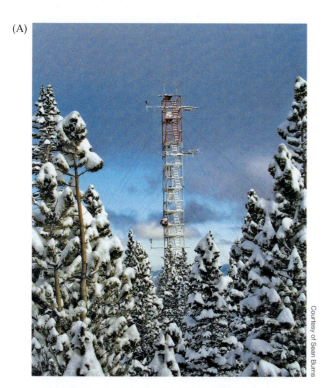

Courtesy of Sean Burns

(B)

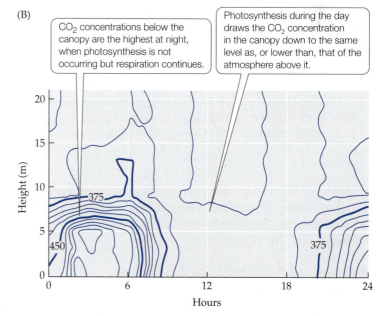

CO_2 concentrations below the canopy are the highest at night, when photosynthesis is not occurring but respiration continues.

Photosynthesis during the day draws the CO_2 concentration in the canopy down to the same level as, or lower than, that of the atmosphere above it.

FIGURE 20.9 Eddy Covariance Estimates of NEE (A) A tower projecting above a subalpine forest on Niwot Ridge, Colorado. Instruments attached to the tower measure the microclimate (temperature, wind speed, radiation) and atmospheric CO_2 concentrations at frequent intervals. These measurements are used to estimate net ecosystem exchange of CO_2. (B) Concentrations of CO_2 (in parts per million) from the ground surface to above the canopy in a boreal forest in Siberia, measured over the course of a 24-hour period in the summer. Average canopy height was 16 m. (B after D. Y. Hollinger et al. 1998. *Agr For Meteorol* 90: 291–306.)

❓ What would the daily pattern of CO_2 concentrations look like during the summer in a community made up primarily of cacti?

ANALYZING DATA 20.1

Does Deforestation Influence Atmospheric CO_2 Concentrations?

We know that on a yearly basis trees take up large amounts of CO_2 from the atmosphere, converting it through photosynthesis to fixed carbon. We also know that occasionally large numbers of trees succumb to fire, insect predation, diseases, and human activities. What effect, if any, might this deforestation have on atmospheric concentrations of CO_2? Two studies shed light on this question.

Over the past decade, mountain pine beetles (*Dendroctonus ponderosae*) killed millions of trees throughout western North America. Kurz et al. (2008)[*] studied the effects of a massive beetle infestation in British Columbia, Canada. The team measured and estimated NPP and heterotrophic respiration before and after the outbreak. Use their data (below) to answer Questions 1 and 2.

	NPP[a]	Heterotrophic respiration[a]
Before outbreak	440	408
After outbreak	400	424

[a] In g C/m²/yr

1. Prior to the mountain pine beetle outbreak, was the forest taking up more CO_2 than it was releasing? In other words, was the forest a sink or a source of CO_2 for the atmosphere?

2. Was the forest a sink or a source of atmospheric CO_2 following the outbreak? Would you expect this trend in net carbon exchange with the atmosphere to change over the next 100 years?

Trees are also being lost at a high rate from the tropical rainforest biome, in this case because of land use change (see Concept 3.1). The ongoing conversion of tropical rainforest to pasture by humans is altering the NEE of this biome. In a study that compared the NEE of a tropical pasture with that of second-growth tropical rainforest in Panama, Wolf et al. (2011)[†] obtained the following data, which you can use to answer Questions 3 and 4.

	GPP[a]	Total respiration[a] (autotrophic + heterotrophic)
Pasture	2,345	2,606
Second-growth forest	2,082	1,640

[a] In g C/m²/yr

3. What is the NEE for the tropical pasture? For the second-growth forest?

4. As noted in Table 20.1, today the tropical forest biome accounts for 35% of terrestrial NPP. The NEE of Earth's total land surface accounts for a net uptake of 3 petagrams ($3 \times 1,015$ grams) of carbon each year. Given these considerations, use the NEE figures you obtained for Question 3 to determine how much less annual global carbon uptake there would be if half of the existing tropical forest were converted to pasture. (Assume that the numbers from the Wolf et al. study represent the average conditions for undisturbed tropical forest and tropical pasture.)

[*] Kurz, W. A., and 7 others. 2008. Mountain pine beetle and forest carbon feedback to climate change. *Nature* 452: 987–990.

[†] Wolf, S., W. Eugster, C. Potvin, B. L. Turner, and N. Buchmann. 2011. Carbon sequestration potential of tropical pasture compared with afforestation in Panama. *Global Change Biology* 17: 2763–2780.

(FLUXNET: fluxnet.org) has been established to help researchers better understand the uptake and fate of carbon in terrestrial ecosystems and how carbon uptake is influenced by climate.

AQUATIC ECOSYSTEMS The dominant autotrophs in both freshwater and marine ecosystems are phytoplankton, including algae and cyanobacteria. These organisms have much shorter life spans than terrestrial plants, so the biomass present at any given time is very low compared with NPP; therefore, harvest techniques are not used to estimate NPP for phytoplankton, although they can be used for seagrasses and macroalgae. One approach to estimating NPP involves measuring the rates of photosynthesis and respiration in water samples collected in bottles and incubated at the collection site with light (for photosynthesis) and without light (for respiration). Although there are errors associated with the artificial environment of the bottles, as well as the inclusion of respiration by heterotrophic bacteria and zooplankton, this technique is used widely in freshwater and marine ecosystems.

Remote sensing of chlorophyll concentrations in the oceans using satellite-based instruments provides good estimates of marine NPP (see Figure 20.7). As described for terrestrial remote sensing, indices based on absorption and reflection of light of different wavelengths are used to indicate how much light is being absorbed by chlorophyll, which is then related to NPP by using a light utilization coefficient, a term that incorporates the efficiency of light absorption into photosynthetic CO_2 uptake.

As Figure 20.7 shows, there can be as much as a 50-fold difference in NPP between Arctic and tropical ecosystems. In the following section we will investigate the role of abiotic and biotic factors that influence differences in NPP among ecosystems.

> **CONCEPT 20.2** Net primary production is constrained by both physical and biotic environmental factors.

LEARNING OBJECTIVES

20.2.1 Contrast the direct and indirect influences of climate on the amount of terrestrial net primary production through its influence on photosynthesis and resource supply.

20.2.2 Describe how plant allocation and growth rates can influence the relationship between climate and net primary production.

20.2.3 List the nutrients that commonly limit the net primary production of freshwater and marine ecosystems.

Environmental Controls on NPP

As we have seen, NPP varies substantially over space and time. Much of this variation is associated with differences in climate, such as the latitudinal gradients in temperature and precipitation discussed in Concept 2.3. In this section, we explore the factors that constrain rates of NPP.

NPP in terrestrial ecosystems is controlled by climate

Variation in terrestrial NPP at the continental to global scales correlates with variation in temperature and precipitation. NPP increases as average annual precipitation increases up to a maximum of about 2,400 mm per year, after which it decreases in some ecosystems (e.g., highland tropical forests), but not in others (e.g., lowland tropical forests) (**FIGURE 20.10A**). NPP may decrease at very high precipitation levels for several reasons. Cloud cover over long periods lowers available sunlight. High amounts of precipitation leach nutrients from soils, and high soil water content results in hypoxic conditions that cause stress for both plants and decomposers.

NPP increases with average annual temperature (**FIGURE 20.10B**). This does not mean, however, that ecosystem carbon storage (NEE, discussed earlier) does the same. The loss of carbon from ecosystems due to respiration of heterotrophic organisms also increases at warmer temperatures, so NEE may potentially decrease. Several lines of evidence suggest that climate change over the past decades has changed NEE in some ecosystems. For example, tundra sites that were once carbon *sinks* (with GPP greater than carbon loss due to respiration) are

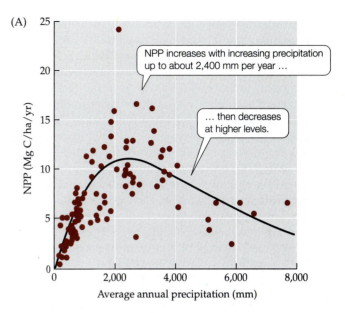

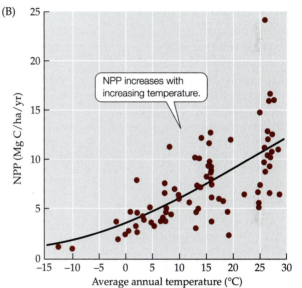

FIGURE 20.10 Global Patterns of Terrestrial NPP Are Correlated with Climate The graphs show the relationships between NPP and (A) precipitation and (B) temperature in terrestrial ecosystems worldwide. (Mg = 10^6 g.) (After E. A. G. Schuur. 2003. *Ecology* 84: 1165–1170.)

now carbon *sources* (with respiratory carbon loss greater than GPP). These changes are increasing CO_2 losses to the atmosphere.

These correlations of NPP with climate suggest that NPP is directly linked to water availability and temperature. Such links make sense when we consider the direct influence of water availability on photosynthesis via the opening and closing of stomates and the influence of temperature on the enzymes that facilitate photosynthesis (see Concept 5.2). In deserts and in some grassland ecosystems, water availability has a clear, direct influence on NPP. In other ecosystems where water limitation is not as severe, the causal connection between precipitation and NPP is less clear.

The links between climate and NPP may also be indirect, mediated by factors such as climate effects on nutrient availability or the particular plant species found within an ecosystem. How can we detect whether the influence of climate on NPP is direct or indirect? Several approaches, both observational and experimental, have been used. William Lauenroth and Osvaldo Sala examined how NPP in a short-grass steppe ecosystem responded to year-to-year variation in precipitation (Lauenroth and Sala 1992). They also examined the average annual NPP and precipitation across several grassland ecosystems at different locations in the central United States. When they compared the correlations between NPP and precipitation in their two analyses, they found that NPP increased more as precipitation increased for the site-to-site comparison than for the comparison among years in the short-grass steppe (**FIGURE 20.11**). They attributed the difference in the response of NPP to precipitation to variation in plant species composition among the grasslands. Some grass species have a greater inherent capacity to increase growth than others in response to enhanced water availability, associated with greater ability to produce new shoots and flowers. Lauenroth and Sala also suggested that there was a time lag in the response to increased precipitation in the short-grass steppe ecosystem; that is, the increase in NPP in response to an increase in precipitation did not occur in the same year, but was delayed one to several years. Within the grassland biome, differences in the abilities of species to respond to climate variation can contribute to site-to-site variation in NPP, influencing the correlation between climate and NPP among sites.

Experimental manipulations of water, nutrients, carbon dioxide, and plant species composition have been used to examine the direct influence of those factors on NPP. The results of numerous experiments indicate that nutrients, particularly nitrogen, control NPP in terrestrial ecosystems. For example, William Bowman, Terry Theodose, and their colleagues used a fertilization experiment in alpine communities of the southern Rocky Mountains to determine whether the supply of nutrients limits NPP (Bowman et al. 1993). They knew that spatial differences in NPP among alpine communities were correlated with differences in soil water availability, as in the grassland ecosystems described above. Bowman and colleagues' fertilization experiment was performed in two communities, a nutrient-poor dry meadow and a more nutrient-rich wet meadow. They sought to determine whether the supply of nutrients influenced NPP and, if so, whether the response differed between the two communities. They added nitrogen or phosphorus or both nitrogen and phosphorus to different plots in both communities, and they maintained plots with no nutrient additions as controls. Their results indicated that the supply of nitrogen limited NPP in the dry meadow, while nitrogen and phosphorus both limited NPP in the wet meadow (**FIGURE 20.12**). An additional experiment indicated that the addition of water to the dry meadow did not increase NPP, despite the positive relationship between NPP and soil moisture across the communities. These results suggest that the correlation between soil moisture and NPP in these alpine communities does not indicate a direct causal relationship, but rather is determined by the effect of soil moisture on nutrient supply through its effects on decomposition and movement of nutrients in the soil (described in Concept 22.2).

Closer examination of Figure 20.12B shows that the increase in NPP was not uniform across all plant species groups. The dominant plant type of the alpine dry meadow (*Kobresia myosuroides*) did not increase its biomass as much as the less common sedge and grass species. The change in NPP in the dry meadow occurred largely as a result of a change in plant species composition within the experimental plots. This was not the case in the wet meadows, where the dominant sedges increased their growth more than the subdominant forb species. These results are consistent with the general trend of results from many fertilization experiments, which indicate that plant species from resource-poor communities have lower growth responses to fertilization than species from resource-rich communities. This apparent contradiction is the result of differences in the capacity of plant species to respond to fertilization. Plants of resource-poor communities tend to have low intrinsic growth rates, a characteristic that lowers their resource requirements. Plants of resource-rich

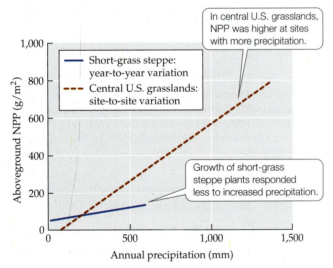

FIGURE 20.11 The Sensitivity of NPP to Changes in Precipitation Varies among Grassland Ecosystems The relationship between aboveground NPP and precipitation is shown for a short-grass steppe ecosystem and for several grassland ecosystems of different types at different sites in the central United States. (After W. K. Lauenroth and O. E. Sala. 1992. *Ecol Appl* 2: 397–403.)

(A)

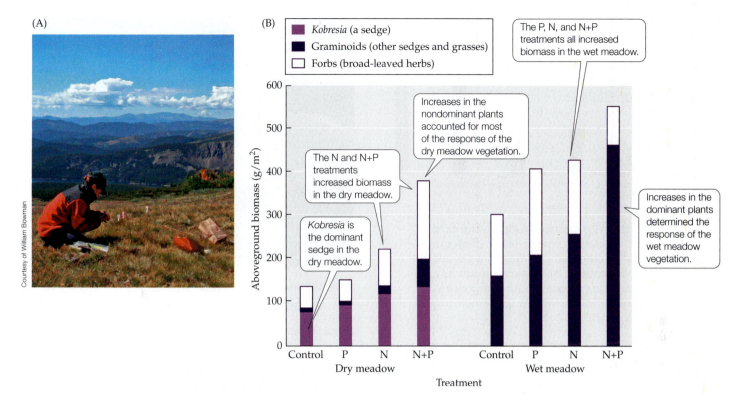

Courtesy of William Bowman

(B)

FIGURE 20.12 Nutrient Availability Influences NPP in Alpine Communities (A) Fertilized plots in an alpine dry meadow community in the Colorado Rocky Mountains, dominated by sedges, forbs, and grasses (see Figure 3.11). (B) Fertilization of plots in a resource-poor dry meadow and a resource-rich wet meadow with nitrogen (N), phosphorus (P), and both N and P showed that nutrient availability limits NPP. (B after W. D. Bowman et al. 1993. *Ecology* 74: 2085–2098.)

 In which community would you expect a higher proportion of belowground NPP? Would the allocation to belowground NPP change in response to fertilization?

communities tend to have higher growth rates, which make them better able to compete for resources, particularly light. Although NPP increases in nutrient-poor communities when they are fertilized, the change in plant species composition that occurs in many such experiments indicates that plant species composition can determine the intrinsic capacity of an ecosystem to increase its NPP when resources are increased (**FIGURE 20.13**). This study provides an example of the important roles that community dynamics can play in ecosystem function.

NPP is often limited by nutrients in non-desert terrestrial ecosystems. Some general differences among terrestrial ecosystem types have emerged from resource manipulation experiments and measurements of plant and soil chemistry. In lowland tropical rainforests, NPP is often limited by the supply of phosphorus, since the relatively old, leached tropical soils in which they grow are low in available phosphorus relative to other nutrients. Other nutrients, such as calcium and potassium, can also limit production in lowland tropical ecosystems. Montane tropical ecosystems, and most temperate and Arctic ecosystems, are limited by the supply of nitrogen, and occasionally by phosphorus. Even in some desert ecosystems, NPP is co-limited by water and nitrogen.

NPP in aquatic ecosystems is controlled by nutrient availability

The primary producers in lake ecosystems are phytoplankton and rooted macrophytes. NPP in lake ecosystems is often limited by the supply of both phosphorus and nitrogen, as we know not only from the results of experimental manipulations, but also from unintentional "experiments" set in motion by wastewater discharges into lakes (see Figure 22.18). A common approach to determining the response of NPP in lakes to changes in nutrient supply is to incubate translucent or open-top containers, sometimes referred to as "limnocorrals," of lake water, amended with one or more nutrients, in the lake (**FIGURE 20.14**). The NPP response is measured by changes in chlorophyll concentrations or numbers of phytoplankton cells.

One of the most convincing studies of the effect of nutrients on NPP in lakes was a series of whole-lake fertilization experiments by David Schindler (Schindler 1974). The experiments were initiated in 1969 in the Experimental Lakes Area in Ontario, a series of 58 small lakes set aside for experimental manipulations. Concern over declining water quality in the lakes of North America and Europe motivated Schindler and his colleagues to

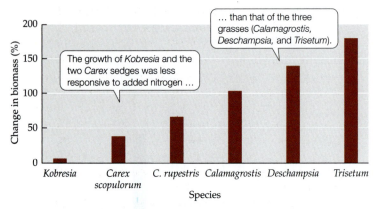

FIGURE 20.13 Growth Responses of Alpine Plants to Added Nitrogen The effect on plant growth of low to high nitrogen levels (with all other nutrients maintained at optimal concentrations) indicated that alpine plant species vary substantially in their ability to increase growth in response to an increase in nitrogen availability. (After W. D. Bowman and C. J. Bilbrough. 2001. *Plant Soil* 233: 283–290.)

FIGURE 20.14 Limnocorrals A researcher snorkels within a contained area (limnocorral) in McKinley Lake near Cordova, Alaska, subjected to experimental fertilization to examine the effects of nutrients on NPP.

establish several experiments to determine whether inputs of nutrients in wastewater were involved in the dramatic increases in the growth of phytoplankton that had been observed. They added nitrogen, carbon, and phosphorus to all or half of several individual lakes. The results of these experiments provided strong evidence for phosphorus limitation of NPP (**FIGURE 20.15**). Massive increases in the abundances of cyanobacteria were responsible for the increase in NPP in response to phosphorus addition. Evidence for nitrogen limitation of NPP in high-elevation lakes, based on small-scale fertilization experiments and measurements of the ratio of nitrogen to phosphorus in the water, also exists (Elser et al. 2007).

NPP in streams and rivers is often low, and the majority of the energy in those ecosystems is derived from terrestrial organic matter (see Concept 21.1). Water movement limits the abundance of phytoplankton, except where the water velocity is relatively low. In Concept 3.2, we introduced the *river continuum concept*, which describes the increasing importance of in-stream NPP as the river flows downstream. Most of the NPP in streams and rivers comes from photosynthesis by macrophytes and algae attached to the bottom in shallow areas where there is enough light for photosynthesis. Suspended sediment in rivers can limit light penetration; thus, turbidity often controls NPP. Nutrients, particularly nitrogen and phosphorus, can also limit NPP in streams and rivers.

Marine NPP is usually limited by nutrient supply, but the specific limiting nutrients vary among marine ecosystem types. Estuaries, the zones where rivers empty into the ocean (see Concept 3.3), are rich in nutrients relative to other marine ecosystems. Variation

FIGURE 20.15 Response of a Lake to Phosphorus Fertilization Experimental Lake 226 was divided into two sections as part of David Schindler's experiments on the effects of nutrient availability on NPP.

in NPP among estuaries is correlated with variation in nitrogen inputs from rivers. Agricultural and industrial activities have increased riverine inputs of nitrogen into estuaries, which have caused periodic "blooms" of algae. These blooms have been implicated in the development of "dead zones"—areas of high fish and zooplankton mortality—in over 400 nearshore ecosystems worldwide.

NPP in the open ocean is derived primarily from phytoplankton, including a group referred to as the *picoplankton*, consisting of cells smaller than 1 μm. Picoplankton contribute as much as 50% of the total marine NPP. Smaller contributions come from floating mats of seaweeds such as *Sargassum*. Near the coast, kelp forests may have leaf area indices and rates of NPP as high as those of tropical forests. "Meadows" of seagrasses such as eelgrass (*Zostera* spp.) are also important contributors to NPP in shallow nearshore zones.

In much of the open ocean, NPP is limited by nitrogen. In the equatorial Pacific Ocean, however, detectable concentrations of nitrogen can be found in the water even when peak NPP occurs, suggesting that some other factor limits NPP. John Martin and colleagues measured the concentrations of nutrients in the open waters of the Pacific and performed bottle incubation experiments with added nutrients. They found that adding iron to the bottles increased NPP (Martin et al. 1994). Based on this evidence that iron limits NPP in some ocean regions, Martin suggested that windblown dust from Asia, an important source of iron for the open ocean, could play an important role in the global climate system through its influence on marine NPP, and thus on atmospheric CO_2 concentrations. During glacial periods, large areas of the continents lacking vegetative cover could have

contributed aeolian dust that would have fertilized the ocean. As NPP in marine ecosystems increased, those ecosystems might have taken up more CO_2 from the atmosphere, reducing its atmospheric concentration and serving as a positive feedback to cool the climate further. Martin suggested that these findings might be applied to address global warming, saying at the time, "Give me half a tankerload of iron, and I'll give you an Ice Age." He recommended the use of large-scale experiments to investigate the influence of iron on ocean NPP. Unfortunately, Martin died in 1993, before his ambitious experiments could be carried out.

Martin's colleagues subsequently performed the first of several experiments in 1993, adding iron sulfate to surface waters of the equatorial Pacific west of the Galápagos archipelago. This experiment was alternatively referred to as IronEx I or the "Geritol solution"[1] to global climate change. During IronEx I, a 64-km^2 area was fertilized with 445 kg of iron, which resulted in a doubling of phytoplankton biomass and a fourfold increase in NPP (**FIGURE 20.16**). Three other iron fertilization experiments were subsequently performed, one in 1995 (IronEx II), which produced a tenfold increase in phytoplankton biomass; a second in 1999 in the Southern Ocean; and the last in 2002, also in the Southern Ocean. While the iron limitation hypothesis has been strongly supported by these and other experiments, fertilizing large areas of the ocean is unlikely to provide a solution to increasing atmospheric CO_2 concentrations and global climate change. Some of the CO_2 taken up by phytoplankton is

[1] Geritol is a dietary supplement once widely believed to help cure "iron-poor, tired blood."

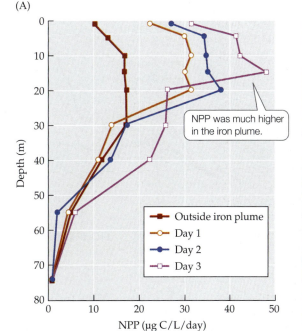

(A)

(B)

FIGURE 20.16 Effect of Iron Fertilization on Marine NPP
IronEx I released a plume of iron into the equatorial Pacific Ocean to study the effects of iron fertilization on NPP. (A) This vertical profile shows primary production at various depths outside and inside the iron plume on three specific days: 1, 2, and 3 days following the release of the iron. (B) Researchers deploy a pump to add iron to the ocean. (A after J. H. Martin et al. 1994. *Nature* 371: 123–129.)

NPP was much higher in the iron plume.

Outside iron plume
Day 1
Day 2
Day 3

eventually re-emitted to the atmosphere via respiration by zooplankton and bacteria that consume the phytoplankton. In addition, the iron is lost relatively quickly from the surface photic zone, sinking to deeper layers where it is unavailable to support phytoplankton photosynthesis and growth. Iron fertilization on a large scale could also have detrimental effects on ocean biodiversity and could create large dead zones similar to those generated by nitrogen inputs into estuaries.

The development of remote sensing and eddy covariance techniques has improved our ability to discern global patterns of NPP. We'll examine those patterns in the next section.

CONCEPT 20.3 Global patterns of net primary production reflect climate constraints and biome types.

LEARNING OBJECTIVES

20.3.1 Compare the net primary production of terrestrial and marine ecosystems at a global scale by considering their spatial coverage to evaluate their overall contribution to global net primary production.

20.3.2 Predict which biomes contribute the greatest amount of net primary production at a global scale based on the climatic factors that control net primary production.

Global Patterns of NPP

Which biomes and marine biological zones have the highest NPP and, as a consequence, the greatest effect on atmospheric CO_2 dynamics? Knowing how NPP varies at a global scale is key to understanding how biotic factors affect the global carbon cycle and how future changes in biomes could affect climate change (see Concept 25.1).

Initial estimates of global NPP were based on compilations of plot-level measurements from different biomes, scaled up using estimates of the spatial distributions of those biomes. These estimates were subject to error associated with the uncertainty of the actual area covered by each biome type, as well as with the potential for overestimating NPP if undisturbed, old-growth study plots were selected to represent a biome. Remote sensing data now give us rapid direct measurements of NPP, providing an estimate of Earth's capacity to take up CO_2 and its response to climate variation and climate change.

Terrestrial and oceanic NPP are nearly equal

Chris Field and colleagues estimated total planetary NPP to be 105 petagrams (1 Pg = 10^{15} g) of carbon per year, based on remote sensing data collected over multiple years (Field et al. 1998). They determined that 54% of this carbon is taken up by terrestrial ecosystems, while the remaining 46% is taken up by primary producers in the oceans. Their estimate of oceanic NPP (which comes to 48 Pg C/year) was considerably higher than previous estimates. Despite the similar contributions of land and oceans to total global NPP, the average rate of NPP on the land surface (426 g C/m²/year) is higher than that in the oceans (140 g C/m²/year). The lower rate in the oceans is compensated for by the greater percentage (70%) of Earth's surface they cover.

Most of the surface of both oceans and land is dominated by areas with relatively low NPP (see Figure 20.7). The highest rates of NPP on land are found in the tropics (**FIGURE 20.17**). This pattern results from latitudinal variation in climate and in the length of the growing season. Higher latitudes have shorter growing seasons and sparse short-statured plants, and low temperatures constrain nutrient supply by lowering decomposition rates, which in turn limits NPP. Tropical zones have long growing seasons and high rates of precipitation, promoting high rates of NPP. NPP declines to the north and south of the tropics at about 25°, reflecting the increasing aridity associated with the high-pressure zones generated by the descending air of the Hadley cells (see Concept 2.2). Another peak in terrestrial NPP occurs at the northern mid-latitudes, where the temperate forest biome is found. NPP in the mid- to high latitudes shows

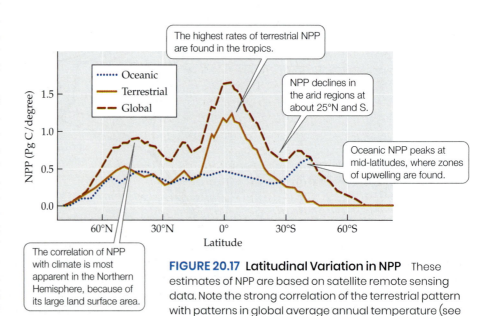

FIGURE 20.17 Latitudinal Variation in NPP These estimates of NPP are based on satellite remote sensing data. Note the strong correlation of the terrestrial pattern with patterns in global average annual temperature (see Figure 2.14) and precipitation (see Figure 2.16). (After C. B. Field et al. 1998. *Science* 281: 237–240.)

strong seasonal trends, with peaks in summer and declines in winter. In contrast, seasonal trends in the tropics are often slight and are associated with wet–dry cycles.

Oceanic NPP peaks at the mid-latitudes between 40° and 60° (see Figure 20.17). These peaks are associated with zones of upwelling, areas where ocean currents bring nutrient-rich deep water to the surface (see Concept 2.2). High NPP is also associated with estuaries at these latitudes. Seasonal trends in NPP occur in the oceans, but their magnitude is less than on the land surface.

Differences among biomes in NPP reflect climate and biotic variation

It is not surprising that NPP varies among biomes, since biomes are associated with latitudinal climate variation. For example, the high NPP in the tropics is associated with tropical forests, grasslands, and savannas. The low NPP at high latitudes is associated with boreal forests and tundra. Tropical forests and savannas contribute approximately 60% of terrestrial NPP and around 37% of global NPP (**TABLE 20.1**). In the oceans, zones of upwelling have high rates of NPP, but they cover less than 5% of the ocean surface. Although they cover less area than the open ocean, shallow oceans account for almost half of oceanic NPP. Despite its low rate of NPP, the vast area of the open ocean accounts for the majority of oceanic NPP and approximately 40% of total global NPP.

As noted in Concept 20.1, much of the variation in NPP among terrestrial biomes is associated with differences in leaf area index. Similarly, the complex structure of ocean macrophyte communities such as kelp beds accounts for their high rate of NPP (see Table 20.1) In addition, the length of the growing season varies markedly among terrestrial biomes, from year-round in some tropical ecosystems to 100 days or less in tundra. Variation associated with different plant growth forms (e.g., grasses vs. shrubs vs. trees) is also important but contributes less to variation among biomes than do growing season and leaf area index. Variation in NPP among aquatic ecosystems, as we saw in Concept 20.2, is primarily related to variation in inputs of nutrients.

What happens to all of this NPP? In the next section, we will introduce some of the concepts associated with secondary production. We will cover energy flow among organisms and its consequences for population growth, community dynamics, and ecosystem function in Chapter 21.

TABLE 20.1

Variation in NPP among Terrestrial Biomes and Oceanic Provinces

	NPP (g C/m²/yr)	Total NPP (Pg C/yr)	Percentage of global NPP
Biome			
Tropical forest	2,500	21.9	22.7
Tropical savanna	1,080	14.9	15.4
Temperate forest	1,550	8.1	8.4
Temperate grassland	750	5.6	5.8
Boreal forest	390	2.6	2.7
Temperate shrubland	500	1.4	1.4
Tundra	180	0.5	0.5
Desert	250	3.5	3.6
Crops	610	4.1	4.2
Total terrestrial		62.6	64.8
Oceanic Province			
Oligotrophic (e.g., open ocean)	91	14.5	15.0
Mesotrophic (e.g., shallow ocean)	132	15.7	16.3
Eutrophic (e.g., upwelling zone, coral reef)	422	2.8	2.9
Macrophytes (e.g., kelp beds, sea grass)	1,500	1.0	1.0
Total oceanic		34.0	35.2

Source: B. Saugier et al. 2001. In *Terrestrial Global Productivity*, J. Roy et al. (Eds.), pp. 543–557. Academic Press: San Diego, CA.

CONCEPT 20.4 Secondary production is generated through the consumption of organic matter by heterotrophs.

LEARNING OBJECTIVES

20.4.1 Describe how the diet of a heterotroph can be determined by observation and isotopic analysis.

20.4.2 Evaluate how ingestion, egestion, and respiration influence the amount of net secondary production.

Secondary Production

Energy that is derived from the consumption of organic compounds produced by other organisms is known as **secondary production**. Organisms that obtain their energy in this manner are known as *heterotrophs*, and they include archaea, bacteria, fungi, animals, and even a few plants (see the Case Study in Chapter 14).

Heterotrophs are classified according to the type of food they consume. The most general categories, introduced in Concept 5.4, are **herbivores**, which consume plants and algae; **carnivores**, which consume live animals; and **detritivores**, which consume dead organic matter (*detritus*). Organisms that consume live organic matter from both plants and animals are called **omnivores**. Further refinement of feeding preferences is sometimes incorporated into the terminology used to describe heterotrophs; insect eaters, for example, are referred to as *insectivores*. Here we briefly introduce some concepts of secondary production. In Chapter 21, we will discuss secondary production in more detail and in relation to the amount and efficiency of energy transfer between trophic levels, the controls on the magnitude of secondary production in terrestrial and aquatic ecosystems, and the concept of food webs.

Heterotroph diets can be determined from the isotopic composition of food sources

As we saw in Concept 5.4, chemistry is an important determinant of the benefit heterotrophs get from their food. In particular, the ratios of carbon to nutrients such as nitrogen and phosphorus influence the growth rates and reproductive output of heterotrophs and thus their secondary production. We will revisit this topic in Concept 21.2.

Determining what heterotrophs eat may be as simple as watching them feed. Such observations, however, may be a time-consuming and imprecise exercise. Another option is examining their fecal material, which can also be imprecise and not a pleasant task. An alternative method of determining a heterotroph's diet involves measuring stable isotopes (see Ecological Toolkit 5.1). The ratios of naturally occurring stable isotopes of carbon ($^{13}C/^{12}C$), nitrogen ($^{15}N/^{14}N$), and sulfur ($^{34}S/^{32}S$) differ among potential food items. Measurements of the isotopic composition of a heterotroph and its potential food sources can identify the food sources that make up its diet (Peterson and Fry 1987).

Isotopic measurements of preserved bone specimens have been used to study the diets of extinct animals as well as modern ones. One mystery in feeding ecology that was solved using isotopic measurements was the diet of European cave bears (*Ursus spelaeus*). Cave bears went extinct about 25,000 years ago, during the peak of the last Ice Age. Cave bears were much larger than the temperate-zone bears of today, as much as triple the size of the modern grizzly bear (*Ursus arctos horribilis*) of North America. Examination of the teeth and the jaw structure of cave bears led some mammalogists to hypothesize that they were primarily herbivores. However, the fact that plants are a poor-quality food, as noted in Concept 5.4, led to skepticism about an herbivorous diet adequately sustaining such a massive bear. G. V. Hilderbrand and colleagues measured the C and N isotope composition of bone samples provided by museums from across the world (Hilderbrand et al. 1996). The samples included cave bears and the herbivores that occurred alongside

them (woolly rhinoceros, woolly mammoth, horse, and aurochs, an ancestor of modern cattle). Hilderbrand and colleagues found that bones of cave bears had an isotopic composition different from that of Pleistocene herbivores (**FIGURE 20.18**). Using information about the isotopic composition of food sources, the researchers estimated that the average diet of cave bears consisted of 58% meat (range from 41% to 78%). This finding refuted the hypothesis that cave bears were primarily herbivores, indicating that the bulk of their diet was meat. In this and other studies, isotopic measurements have provided a useful tool for determining the diets of animals that is more accurate and integrative, and less time-consuming, than other techniques. See the Hone Your Problem-Solving Skills for this chapter to see how isotopic analysis can be used to compare grizzly bear diets in different environments.

Net secondary production is equal to heterotroph growth

Not all of the organic matter consumed by heterotrophs is incorporated into heterotroph biomass. Some is used in respiration, and some is egested (lost in urine and feces). **Net secondary production** is the balance among ingestion, respiratory loss, and egestion:

$$\frac{\text{net secondary}}{\text{production}} = \text{ingestion} - \text{respiration} - \text{egestion} \quad \textbf{(20.2)}$$

Net secondary production by a heterotroph depends on the quality of its food, related to its digestibility and nutrient content. In addition, the physiology of the

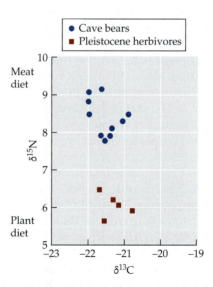

FIGURE 20.18 Isotopic Composition and Diet Carbon and nitrogen isotope composition of bones of museum specimens of cave bears and herbivores from about 20,000 years ago. The isotopic compositions are expressed as ratios of heavier to lighter isotopes compared with a standard. Higher numbers mean more of the heavier isotope. (After G. V. Hilderbrand et al. 1996. *Can J Zool* 74: 2080–2088.)

heterotroph influences how effectively its food intake is channeled into growth. Animals with high respiration rates (e.g., endotherms) have less energy left over to allocate to growth.

Net secondary production in most terrestrial ecosystems is a small fraction of NPP, because of predation on herbivores, plant defenses, and the low nutrient content of many plants, as we'll see in Chapter 21. Net secondary production represents a greater fraction of NPP in aquatic ecosystems than it does in terrestrial ecosystems. The majority of net secondary production in most ecosystems is associated with detritivores, primarily bacteria and fungi.

A CASE STUDY REVISITED

Life in the Deep Blue Sea, How Can It Be?

In this chapter, we have emphasized the importance of photosynthetic autotrophs as the source of energy for ecosystems, since the vast majority of the energy that enters ecosystems is derived from visible solar radiation. Here and in Chapter 5, however, we have alluded to another source of energy for ecosystems: chemosynthesis. Some bacteria can use chemicals such as hydrogen sulfide (H_2S and related chemical forms, HS^- and S^{2-}) as electron donors to take up carbon dioxide and convert it into carbohydrates:

$$S^{2-} + CO_2 + O_2 + H_2O \rightarrow SO_4^{2-} + (CH_2O)_n \quad \textbf{(20.3)}$$

Bacteria that provide energy for ecosystems via chemosynthesis are known as *chemoautotrophs*. The existence of chemoautotrophic bacteria was known for at least a century before the discovery of hydrothermal vents, but their role in providing energy for the vent communities was uncertain.

Initially, hypotheses suggested that the high velocity of water flow around the hydrothermal vents helped direct organic matter from the photic zone toward the filter-feeding invertebrates. However, several lines of evidence suggested that chemoautotrophs were the major source of energy for these ecosystems. First, the carbon isotopic ratios ($^{13}C/^{12}C$) in the bodies of the vent invertebrates were different from those of phytoplankton in the photic zone (see Ecological Toolkit 5.1). Second, the tube worms collected from the vents (*Riftia* spp.) lacked mouths and digestive systems. These gutless tube worms also had structures called trophosomes, made up of highly vascularized tissues with specialized cells containing large amounts of bacteria (**FIGURE 20.19**). Elemental sulfur was found in the trophosomes, suggesting that sulfides were being chemically transformed in the tube worms' bodies. Enzymes associated with the Calvin cycle, the biochemical pathway used by autotrophs to synthesize carbohydrates (see Concept 5.2), as well as enzymes involved in sulfur metabolism were found in

the trophosomes. Furthermore, the clams and other mollusks collected from the vent communities lacked some of the critical tissues for filter feeding, and they also had large amounts of bacteria in specialized tissues, as well as enzymes associated with the Calvin cycle.

All of this evidence pointed to the conclusion that deep-sea hydrothermal vent communities derive their energy from chemoautotrophic bacteria. These bacteria also aid in detoxifying the sulfides in the water, which would normally inhibit aerobic respiration. Many of the abundant ocean vent organisms have symbiotic relationships with the bacteria—that is, they house the chemoautotrophs in their bodies, often in specialized structures. Is this interaction a mutualistic symbiosis of the kind described in Chapter 15? The tube worms and clams housing the bacteria benefit by obtaining carbohydrates to fuel their metabolic processes, growth, and reproduction, as well as from detoxification of the sulfides. Do the bacteria derive any benefit from the invertebrates? The answer is yes: the invertebrates provide them with a chemical environment unlike that found in the surrounding water, supplying them with more

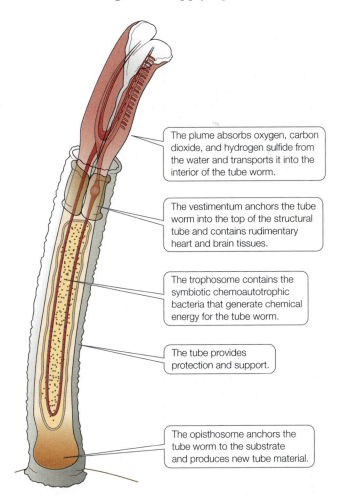

The plume absorbs oxygen, carbon dioxide, and hydrogen sulfide from the water and transports it into the interior of the tube worm.

The vestimentum anchors the tube worm into the top of the structural tube and contains rudimentary heart and brain tissues.

The trophosome contains the symbiotic chemoautotrophic bacteria that generate chemical energy for the tube worm.

The tube provides protection and support.

The opisthosome anchors the tube worm to the substrate and produces new tube material.

FIGURE 20.19 *Riftia* **Anatomy** *Riftia* tube worms have a number of specialized structures that make them well adapted to their hydrothermal vent environment.

carbon dioxide, oxygen, and sulfides than they could obtain if they were free-living in the water or the sediments surrounding the vent. The symbiosis between the bacteria and the invertebrates is therefore mutualistic, resulting in higher productivity than if the organisms lived separately.

 ## CONNECTIONS *in* NATURE

ENERGY-DRIVEN SUCCESSION AND EVOLUTION IN HYDROTHERMAL VENT COMMUNITIES

Hydrothermal vent environments are dynamic, born with the eruption of new hot springs, which eventually cease to emit sulfide-laden water as the subsurface water channels are altered and the underlying magma cools (Van Dover 2000). When the hot springs no longer emit water, and the sulfide in the seawater has been consumed, the communities surrounding the vents collapse as their energy source disappears and the physical substrate falls apart. The life span of vent communities varies from approximately 20 to 200 years. Studies of colonization and development in these communities over the past 3 decades have provided insights into succession in marine communities in general (see Chapter 17 for a general discussion of succession).

Succession in hydrothermal vent communities is relatively rapid and can be observed by periodically revisiting specific vents (**FIGURE 20.20**). Although the logistic difficulty and expense of such investigations has limited the number of observations, some general trends have emerged. The rates of colonization and development of hydrothermal vent communities are higher when they are closer to other existing vent communities, as we might predict based on the theory of island biogeography (see Concept 18.3). Because the community's energy is derived from chemosynthesis, colonization begins with chemoautotrophic bacteria, sometimes in numbers large enough to cloud the water. Tube worms are often the first invertebrates to arrive. Clams and other mollusks are thought to be stronger competitors for sites with optimal temperatures and water chemistry, and over time they increase in abundance at the expense of the tube worms. A few scavengers and carnivores, such as crabs and lobsters, are found in the developing community, although at low abundances. As the tube worm and bivalve populations decline with the drop in sulfide input when water flow from the vent decreases, the abundance of scavenger organisms increases until the energy available in the form of detritus is gone.

FIGURE 20.20 Succession in Hydrothermal Vent Communities Species composition and abundances in a hydrothermal vent community change over time following the eruption of a hot spring. (From T. M. Shank et al. 1998. *Deep-Sea Res II* 45: 465–515.)

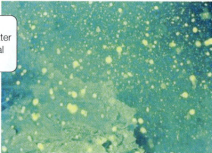

April 1991
Bacterially generated sediments cloud the water within weeks of the initial eruption of a vent.

March 1992
The site has been colonized by tube worms in the genus *Tevnia* (bottom right).

December 1993
Larger tubeworms in the genus *Riftia* dominate the site.

October 1994
Continued dominance by *Riftia*.

November 1995
A decrease in the temperature of the vent water has increased its iron concentration, resulting in iron oxide precipitation that gives *Riftia* individuals a rusty appearance.

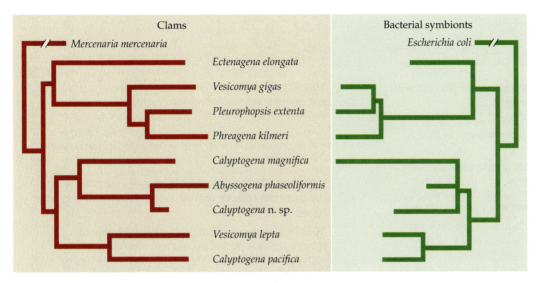

FIGURE 20.21 **Coevolution of Vent Clams and Their Symbiotic Bacteria** The phylogenetic trees of vesicomyid clams collected from hydrothermal vents and their accompanying chemo-autotrophic bacterial symbionts show remarkable parallels, suggesting that these species have coevolved. (After A. S. Peek et al. 1998. *Proc Natl Acad Sci USA* 95: 9962–9966. © 1998 National Academy of Sciences, U.S.A.)

The pattern of succession in hydrothermal vent communities is subject to the same random factors that influence succession in other habitats: the order of arrival of organisms at a site can influence the long-term dynamics of the community (see Concept 17.4). Neighboring vent communities found in the same area of a mid-ocean ridge may show different stages of succession, associated with the stages of hot spring development, as well as different trajectories of succession due to differences in the organisms present. Thus, collections of hydrothermal vents within the same general area are a mosaic of communities at different successional stages, similar to those in terrestrial forest patches, albeit separated by greater distances than the patches within a forest.

The unique nature of the energy supply in hydrothermal vent communities would suggest strong evolutionary divergence between the organisms that inhabit the vents and their nearest non-vent relatives (see Concept 6.4). Where phylogenetic relationships between the vent organisms and their non-vent relatives have been worked out, the divergence is indeed deep, usually at the level of genus, family, or order. Since the discovery of hydrothermal vents, approximately 500 new vent species have been described; of these species, about 90% are endemic to hydrothermal vents. However, large areas of mid-ocean ridges potentially containing hydrothermal vents have yet to be explored.

The close association between the chemoautotrophic bacteria and their invertebrate hosts suggests the potential for a coevolutionary relationship of the type described in Concept 15.1. Have the invertebrates and their chemosynthetic bacterial symbionts evolved in concert? To address this question, Andrew Peek and colleagues compared the evolutionary relationships (phylogenetic trees; see Figure 6.16) of vent-dwelling clams in the family Vesicomyidae with those of their symbiotic bacteria (Peek et al. 1998). Clams in this family transfer bacteria to their offspring in the cytoplasm of their eggs. Peek and colleagues collected eight species of clams in three genera from hydrothermal vent communities at latitudes ranging from 18°N to 47°N and at depths ranging from 500 to 6,370 m. Ribosomal DNA taken from the clams and the bacteria was used to construct the phylogenetic trees. The two trees showed remarkable congruence (**FIGURE 20.21**), providing strong evidence that speciation in the clams and in their bacterial symbionts has occurred synchronously. Other vent groups lack this apparent coevolutionary relationship, however. For example, three different species of tube worms found in different geographic locations have been found to contain the same species of sulfur-oxidizing bacteria.

Some researchers have suggested that hydrothermal vents are a potential site for the origin of life on Earth. The reducing (i.e., electron-donating) geochemical environment of hydrothermal vents is conducive to the abiotic synthesis of amino acids, which would have been required for the development of living systems. Although amino acids are not stable in ocean water under the high pressures and temperatures found at some deep-sea hydrothermal vents, there are vents with lower temperatures at shallower depths where amino acid genesis could (and does) occur. As Cyndy Lee Van Dover (2000) so eloquently stated, "Vent water may be the ultimate soup in the sorcerer's kettle."

SUMMARY

CONCEPT 20.1 Energy in ecosystems originates with primary production by autotrophs.

20.1.1 Explain how variation in the leaf area index influences gross primary production through both carbon gains and losses.

- Gross primary production (GPP) is the total amount of carbon fixed by the autotrophs in an ecosystem. The GPP of a terrestrial ecosystem is determined by the rate of photosynthesis and the leaf area index.

20.1.2 Describe the relationship between gross primary production and net primary production and how this relationship can change during succession.

- Net primary production (NPP) is equal to GPP minus autotrophic respiration.

- NPP changes during succession because of changes in leaf area index and in the proportions of photosynthetic and nonphotosynthetic plant tissues.

20.1.3 Summarize why it is important to measure net primary production and describe some of the ways ecologists measure net primary production at large spatial scales.

- Researchers have developed diverse approaches to measuring NPP at different spatial and temporal scales.

20.1.4 Illustrate how the relationship between net primary production and net ecosystem exchange is influenced by the mass of autotrophs versus the mass of heterotrophs.

- Net ecosystem exchange is the balance between carbon gained from photosynthesis and carbon lost from both autotrophs and heterotrophs.

CONCEPT 20.2 Net primary production is constrained by both physical and biotic environmental factors.

20.2.1 Contrast the direct and indirect influences of climate on the amount of terrestrial net primary production through its influence on photosynthesis and resource supply.

- Variation in terrestrial NPP is associated with variation in temperature and precipitation, both of which affect resource availability and the types and abundances of plants.

20.2.2 Describe how plant allocation and growth rates can influence the relationship between climate and net primary production.

- The intrinsic growth rates of different plant species influence spatial variation in NPP and its response to variation in resource availability.

20.2.3 List the nutrients that commonly limit the net primary production of freshwater and marine ecosystems.

- NPP in aquatic ecosystems is controlled by the supply of nutrients, particularly phosphorus and nitrogen.

CONCEPT 20.3 Global patterns of net primary production reflect climate constraints and biome types.

20.3.1 Compare the net primary production of terrestrial and marine ecosystems at a global scale by considering their spatial coverage to evaluate their overall contribution to global net primary production.

- Terrestrial and oceanic NPP contribute nearly equal proportions of global NPP.

20.3.2 Predict which biomes contribute the greatest amount of net primary production at a global scale based on the climatic factors that control net primary production.

- The majority of terrestrial NPP occurs in the tropics.

- Although zones of upwelling and coastal zones have the highest rates of NPP, the open ocean accounts for the majority of oceanic NPP because of its larger area.

- Differences in NPP among terrestrial biomes reflect differences in leaf area index and in the length of the growing season.

CONCEPT 20.4 Secondary production is generated through the consumption of organic matter by heterotrophs.

20.4.1 Describe how the diet of a heterotroph can be determined by observation and isotopic analysis.

- Heterotrophs derive energy from the consumption of live or dead organic matter.

- Heterotroph diets can be determined by measuring and comparing the ratios of stable isotopes in the tissues of the feeding organism and in those of its potential food sources.

20.4.2 Evaluate how ingestion, egestion, and respiration influence the amount of net secondary production.

- Net secondary production is the energy ingested by heterotrophs minus the energy used in respiration and egested in feces and urine.

Review Questions

1. Why is it important to know how much primary production occurs in ecosystems?

2. Plants allocate the energy they acquire through photosynthesis to different functions, including growth and metabolism. The allocation to growth can go preferentially to particular organs, such as leaves, stems, roots, or flowers. How would you expect the allocation of energy among plant organs to change as the amount of terrestrial ecosystem NPP increased? Explain why you would expect the allocation pattern you describe.

3. Some ecologists interested in the underlying factors that control variation in NPP in tundra measured the growth of plants at several locations over multiple years. They also measured air and soil temperatures, wind speed, solar radiation, and soil moisture. When they analyzed all of their data, they found that the best correlation between NPP and any of the physical environmental factors they had measured was with soil temperature. The researchers concluded that NPP in tundra is controlled by the effect of soil temperature on root growth. Is this conclusion correct?

4. What are some of the benefits and drawbacks associated with measuring NPP using (a) harvest techniques and (b) remote sensing?

Hone Your Problem-Solving Skills

In addition to examining the cave bear food preferences described in this chapter, Hilderbrand and colleagues wanted to see whether grizzly bears living near coastal areas relied more heavily on fish in their diet than bears from inland areas that did not have access to salmon (Hilderbrand et al. 1996). They measured the isotopic composition of bone and hair samples from grizzly bears from the western United States killed before 1931, prior to the construction of dams that restricted salmon runs up the Columbia River basin. Meat is more enriched in ^{15}N than plants, so bears with a carnivorous diet have higher ratios of $^{15}N/^{14}N$, expressed as a higher $\delta^{15}N$, than bears with a diet of plant tissues. Meat food sources have $\delta^{15}N$ values of 12 to 15, while $\delta^{15}N$ values of plant food sources are between 7 and 8. Marine sources of meat (e.g., fish) have a higher proportion of ^{13}C (higher $^{13}C/^{12}C$ ratio expressed as a higher $\delta^{13}C$) than terrestrial food sources (e.g., small mammals). Marine food sources have $\delta^{13}C$ values of −17.5 to −18.5, while terrestrial food sources have a $\delta^{13}C$ of between −20 and −22.

1. Using the data in the table, construct a graph that tests the hypothesis that grizzly bears of coastal areas consume more fish than bears from inland areas. Use $\delta^{13}C$ on the x-axis (going from more negative to less negative) and $\delta^{15}N$ on the y-axis.

2. The availability of salmon for consumption by bears has declined since 1931 because of lower numbers of fish spawning in rivers.

   a. What changes in isotopic composition would you expect if the bears compensated for the lower salmon availability by consuming a higher proportion of terrestrial plant material in their diet?

   b. What changes would you expect in the isotopic composition if they compensated by eating more terrestrial mammals?

Individual	$\delta^{13}C$	$\delta^{15}N$
Coastal population		
1	−17.9	12.6
2	−18.2	13.0
3	−18.6	12.2
4	−19.1	11.5
5	−19.0	11.5
River basin population		
1	−18.9	11.3
2	−19.1	10.8
3	−19.5	10.5
4	−19.6	10.0
5	−20.0	10.0
Inland population		
1	−20.0	8.9
2	−20.1	8.2
3	−20.6	8.9
4	−20.8	8.2
5	−21.2	7.8

21

Energy Flow and Food Webs

KEY CONCEPTS

CONCEPT 21.1 Trophic levels describe the feeding positions of groups of organisms in ecosystems.

CONCEPT 21.2 The amount of energy transferred from one trophic level to the next depends on food quality and on consumer abundance and physiology.

CONCEPT 21.3 Changes in the abundances of organisms at one trophic level can influence energy flow at multiple trophic levels.

CONCEPT 21.4 Food webs are conceptual models of the trophic interactions of organisms in an ecosystem.

Toxins in Remote Places: A Case Study

The Arctic is considered one of most pristine regions on Earth. Human effects on its environment are thought to be slight relative to those in the temperate and tropical zones, where the vast majority of humans live. Thus, the Arctic is one of the last places one would expect to find high levels of pollutants in living organisms.

In the mid-1980s, Eric Dewailly, a toxicologist, was studying concentrations of polychlorinated biphenyls (PCBs) in the breast milk of mothers in southern Quebec. PCBs belong to a group of chemical compounds called persistent organic pollutants (POPs) because they remain in the environment for a long time. POPs originate from industrial and agricultural activities and from the burning of industrial, medical, or municipal wastes. Exposure to PCBs has been linked to increased incidence of cancer, impaired ability to fight infections, decreased learning ability in children, and lower birth weights in newborns.

Dewailly was seeking a human population from a pristine area that could be used as a control in his study. He enlisted the help of some Inuit mothers from Arctic Canada. The Inuit are primarily subsistence hunters, and they have no developed industry or agriculture that would provide direct exposure to POPs (**FIGURE 21.1**). Dewailly therefore assumed that Inuit mothers would have few or no PCBs in their breast milk, providing a benchmark against which to compare populations in more industrialized areas.

What Dewailly found was startling: the Inuit women had concentrations of PCBs in their breast milk that were seven times higher than those in women of southern Quebec (**FIGURE 21.2**) (Dewailly et al. 1993). These alarming findings were reinforced by the work of Harriet Kuhnlein, who at the same time found that approximately two-thirds of the children from an Inuit community in northeastern Canada had PCB levels in their blood that exceeded Canadian health guidelines (Kuhnlein et al. 1995). More extensive surveys found that POPs were widespread in Inuit populations. As many as 95% of the people in Inuit communities of Greenland had blood levels of PCBs that exceeded health standards (Pearce 1997).

FIGURE 21.1 Subsistence Hunting Inuit hunters skin a seal they have successfully hunted in a remote, very sparsely populated Arctic region.

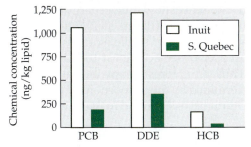

FIGURE 21.2 Persistent Organic Pollutants in Canadian Women The breast milk of Inuit mothers from Arctic Canada was found to contain substantially higher concentrations of polychlorinated biphenyls (PCBs) and two other POPs—dichlorodiphenyldichloroethylene (DDE, a pesticide similar to DDT) and hexachlorobenzene (HCB, an agricultural fungicide)—than that of mothers from southern Quebec. (After E. Dewailly et al. 1993. *Environ Health Perspect* 101: 618–620.)

How were these toxins finding their way into the Arctic environments where the Inuit live? The POPs that were found in the tissues of Inuit populations occur in gaseous form at most environmental temperatures. Produced in lower-latitude industrial areas, these compounds enter the atmosphere under warm temperatures, but when carried by winds into the colder regions of the Arctic, they condense into liquid forms and fall from the atmosphere, sometimes in snowflakes. The manufacture and use of most POPs has been banned in North America since the 1970s. Some developing countries continue to produce POPs, however, and they are important sources of the compounds found in Arctic regions. Although emissions of POPs have decreased, these compounds may remain in Arctic snow and ice for many decades, being released slowly during snowmelt every spring and summer.

While the source of the POPs was known, the high concentrations of these compounds in the Inuit were a mystery. The concentrations of POPs in their drinking water were not high enough to explain this phenomenon. One hint came from the correlation between the levels of the toxins in people and their preferred diets. Communities that had traditionally relied on marine mammals for their food tended to have the highest levels of POPs, while communities that consumed herbivorous caribou

had lower levels. We will discover the ecological basis for this difference as we trace the flow of energy and materials through ecosystems in this chapter.

Introduction

To begin our discussion of energy flow in ecosystems, let's move from the Arctic to a much warmer place: a North American desert. Deserts contain diverse assemblages of plants, animals, and microorganisms. This diversity is reflected in the variation in the sizes, shapes, and physiology of the animals making up the desert fauna, from termites in the soil to grasshoppers in the plant canopy to hawks in the sky. What links these animals together in the context of ecological functioning isn't necessarily their physical appearances or their evolutionary relationships. Rather, their ecological roles are determined by what they eat and by what eats them—that is, by their feeding, or *trophic*, interactions. In other words, the influence an organism has on the movement of energy and nutrients through an ecosystem is determined by the type of food it consumes as well as by what consumes it. For example, grasshoppers and scorpions are both arthropods, with similar morphology and physiology, yet their ecological effects on energy flow through the desert ecosystem are quite different. In the context of energy flow, grasshoppers are more similar to mule deer than to scorpions. Grasshoppers and mule deer are both generalist herbivores that consume a variety of desert plant species. The scorpion, by contrast, is a carnivorous arthropod feeding primarily on insects and thus has an ecological role more similar to that of a kestrel than to that of a grasshopper.

In this chapter, we continue the discussion of energy that we began in Chapter 20, describing its flow through ecosystems and the factors that control its movement through different trophic levels. We will also look at the feeding relationships in an ecosystem as an intricate web of interactions among species, a view that has important implications for energy flow and ecosystem function as well as for species interactions and community dynamics (topics that were covered in Units 4 and 5).

CONCEPT 21.1 Trophic levels describe the feeding positions of groups of organisms in ecosystems.

LEARNING OBJECTIVES

21.1.1 Describe how energy flow among trophic levels in an ecosystem is related to the food selection of consumers.

21.1.2 Explain how both primary production and detritus can be at the base of food chains.

21.1.3 Evaluate how terrestrial detrital energy inputs from outside an ecosystem (allochthonous) would change in a river from its source to where it reaches the ocean.

Feeding Relationships

In Chapter 20, we introduced Ray Lindeman's simplified approach to categorizing groups of organisms in an ecosystem according to how they obtain energy (see Figure 20.3). Rather than grouping them by their taxonomic identity, he grouped them into categories based on how they obtained energy in the ecosystem. In this section, we'll take a closer look at these feeding categories.

Organisms can be grouped into trophic levels

Each feeding category, or **trophic level**, is based on the number of feeding steps by which it is separated from autotrophs (**FIGURE 21.3**). The first trophic level consists of the autotrophs, the primary producers that generate chemical energy from sunlight or inorganic chemical compounds. The first trophic level also generates most of the dead organic matter in an ecosystem, which also provides energy for organisms in the ecosystem. In our desert ecosystem, the first trophic level includes all of the plants, which we lump together to form a single group, regardless of their taxonomic identity. In Lindeman's lake ecosystem (see Figure 20.3), the first trophic level was composed primarily of dead organic matter, which Lindeman poetically referred to as "ooze," as well as autotrophs such as phytoplankton and pondweeds. The second trophic level consisted of the herbivores that consume autotroph biomass—which in our desert ecosystem would include grasshoppers and mule deer—as well as organisms that consume dead organic matter, called *detritivores*. The remaining trophic levels (third and up) contain the carnivores that consume animals at the trophic level below them. The primary carnivores constituting the third trophic level in our desert ecosystem would include small birds and scorpions, while examples of the secondary carnivores making up the fourth trophic level would be foxes and birds of prey. Most ecosystems have four or fewer trophic levels.

Some organisms do not fit conveniently into the trophic levels defined here. Coyotes, for example, are opportunistic feeders, consuming vegetation, mice, other carnivores, and old leather boots. In trophic studies, heterotrophs that feed at multiple trophic levels are called **omnivores**.[1] Such heterotrophs defy our attempt to group organisms into simple feeding categories. However, their diets can be partitioned to reflect how much energy they consume within each trophic level (Pimm 2002). This partitioning is facilitated by the use of stable isotopes to trace food sources (Post 2002a) (see Ecological Toolkit 5.1 and Concept 20.4). Thus, omnivores occupy intermediate trophic levels as determined by the proportions of the foods they consume. Omnivory is common in many ecosystems.

All organisms are either consumed or end up as detritus

All organisms in an ecosystem are either consumed by other organisms at higher trophic levels or enter the pool of dead organic matter, or *detritus* (Lindeman's "ooze" in Figure 20.3 or, as songwriter Tom Waits put it, "We're all gonna be just dirt in the ground"). In most terrestrial ecosystems, a relatively small proportion of the biomass is consumed, and most of the energy flow passes through detritus (**FIGURE 21.4**). Because most of this energy flow occurs in the soil, we are not always aware of its magnitude and importance. Dead plant, microbial, and animal matter and feces are consumed by a multitude of detritivores (primarily bacteria, archaea, and fungi) in a process known as *decomposition*. We will describe decomposition in more detail in Chapter 22 in the context of nutrient cycling. Since detritus is part of the first trophic level, detritivores are placed with herbivores in the second trophic level. Although autotroph-based and detritus-based

---

[1] This definition contrasts with our earlier definition of omnivores in Chapter 20 as heterotrophs that consume both plants and animals

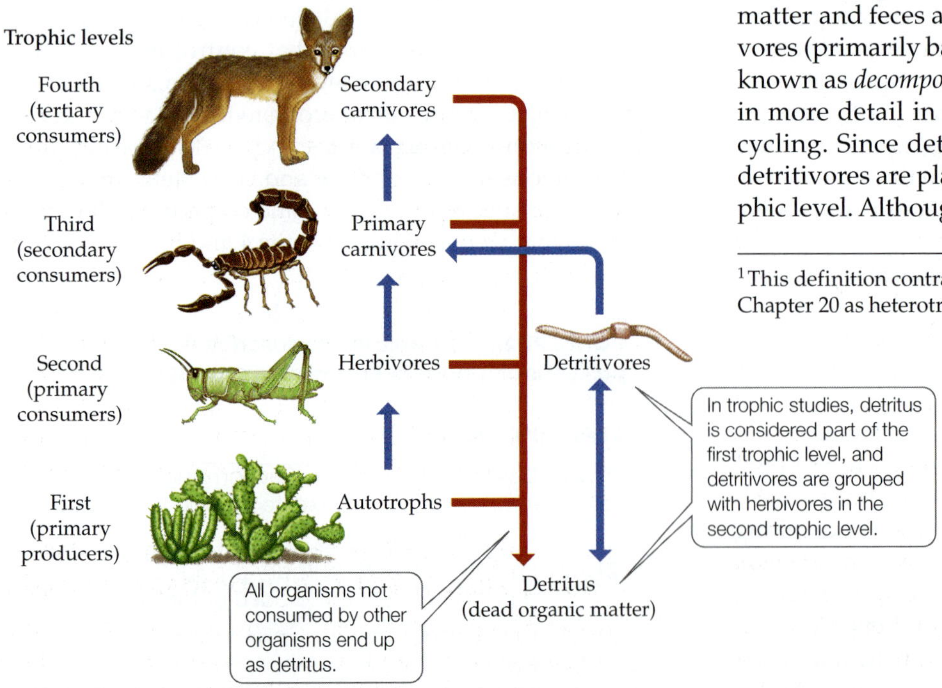

Trophic levels

Fourth (tertiary consumers) — Secondary carnivores

Third (secondary consumers) — Primary carnivores

Second (primary consumers) — Herbivores | Detritivores

First (primary producers) — Autotrophs

Detritus (dead organic matter)

All organisms not consumed by other organisms end up as detritus.

In trophic studies, detritus is considered part of the first trophic level, and detritivores are grouped with herbivores in the second trophic level.

FIGURE 21.3 Trophic Levels in a Desert Ecosystem Each trophic level is characterized by the number of feeding steps by which it is removed from autotrophs (primary producers).

Mycena interrupta

Bohadschia argus

FIGURE 21.4 Ecosystem Energy Flow through Detritus Detritus is consumed by a multitude of organisms, including fungi such as *Mycena interrupta* in Myrtle Forest and Leopard sea cucumber (*Bohadschia argus*) in the Great Barrier Reef.

trophic levels are sometimes considered separately, they are tightly linked through primary production, nutrient cycling, and the many organisms that acquire energy from both plants and detritus.

Energy flow through detritus is important in both terrestrial and aquatic ecosystems. Detritus in terrestrial ecosystems comes primarily from plants within the ecosystem. On the other hand, a large proportion of the input of detritus into stream, lake, and estuarine ecosystems is derived from terrestrial organic matter, which is considered external to the aquatic ecosystem. External energy inputs are referred to as **allochthonous inputs**, while energy produced by autotrophs within the system is known as **autochthonous energy**. Allochthonous inputs into aquatic ecosystems include plant leaves, stems, wood, and dissolved organic matter. These inputs fall into the water from adjacent terrestrial ecosystems or flow in via groundwater. Allochthonous inputs tend to be more important in stream and river ecosystems than in lake and marine ecosystems. For example, Bear Brook, a headwater stream in New Hampshire, receives 99.8% of its energy as allochthonous inputs; the rest is net primary production (NPP) derived from benthic algae and mosses in the stream (Fisher and Likens 1973). In contrast, autochthonous energy accounts for almost 80% of the energy in nearby Mirror Lake (Jordan and Likens 1975). Allochthonous energy is often of lower quality, however, because of the chemical composition of the carbon compounds that enter the system. As a result, the fraction of allochthonous energy that is actually used as an energy source is lower than the inputs would suggest (Pace et al. 2004). The importance of autochthonous energy inputs usually increases from the headwaters toward the middle reaches of a river, in concert with decreases in water velocity and increases in nutrient concentrations, as suggested by the river continuum concept (described in Concept 3.2).

As this aquatic example shows, grouping organisms into trophic levels makes it easier to trace the flow of energy through an ecosystem. That flow is the topic to which we'll turn next.

CONCEPT 21.2 The amount of energy transferred from one trophic level to the next depends on food quality and on consumer abundance and physiology.

LEARNING OBJECTIVES

21.2.1 Describe how the relationship between biomass and energy is influenced by the life span of primary producers.

21.2.2 Summarize the factors that may influence why a greater proportion of net primary production is consumed in aquatic ecosystems relative to terrestrial ecosystems.

21.2.3 Evaluate how consumer thermal physiology, body size, diet preference, and digestive specialization can determine trophic efficiency.

Energy Flow between Trophic Levels

The second law of thermodynamics states that during any transfer of energy, some energy is dispersed as unusable energy because of the tendency toward an increase in disorder (entropy). Thus, we can expect that the cumulative available energy will decrease with each trophic level as we move from the first trophic level upward. We know from our discussion of primary production in Chapters 5 and 20 that autotrophs and heterotrophs lose chemical energy through cellular respiration, lowering the amount of energy available to the organisms that consume them. In this section, we will examine more closely the factors influencing energy movement between trophic levels.

Energy flow between trophic levels can be depicted using energy or biomass pyramids

A common approach to conceptualizing trophic relationships in an ecosystem is to construct a stack of rectangles, each of which represents the amount of energy or biomass within one trophic level. When assembled from lower to higher trophic levels, these rectangles form a

trophic pyramid. By portraying the relative amounts of energy or biomass at each trophic level, these pyramids show us how energy flows through the ecosystem.

As we have noted, some of the biomass at each trophic level is not consumed, and a proportion of the energy at each trophic level is lost in the transfer to the next trophic level. Therefore, the rectangles in a trophic energy pyramid always decrease in size as we move from one trophic level to the one above it. In terrestrial ecosystems, energy and biomass pyramids are usually similar because biomass is typically a good proxy for energy (**FIGURE 21.5A**). In aquatic ecosystems, however, the high consumption rates and the relatively short life spans of the primary producers (mainly phytoplankton) can, in some cases, result in a biomass pyramid that is inverted relative to the energy pyramid

(**FIGURE 21.5B**). In other words, the biomass of heterotrophs may be greater at any given time than the biomass of autotrophs. However, the *energy* produced by the autotrophs is still greater than that produced by the heterotrophs.

This tendency toward inverted biomass pyramids is greatest where productivity is lowest, such as in nutrient-poor regions of the open ocean (**FIGURE 21.5C**). The higher proportion of primary consumer biomass relative to producer biomass in these nutrient-poor regions results from a more rapid turnover of phytoplankton, which have higher growth rates and shorter life spans than the phytoplankton of more nutrient-rich waters. Phytoplankton in nutrient-poor regions thus provide a greater energy supply per unit of time (Gasol et al. 1997). In addition, detritus makes a higher proportional contribution to energy flow in these nutrient-poor waters than in nutrient-rich waters.

Energy flow between trophic levels differs among ecosystem types

What factors determine the amount of energy that flows from one trophic level to the next? In Concept 20.2, we evaluated the factors that influence NPP in terrestrial and aquatic ecosystems, emphasizing abiotic factors such as climate and nutrient availability as well as differences in the inherent ability of autotroph species to produce biomass. It would be reasonable to assume that the flow of energy to higher trophic levels is associated with the amount of NPP at the base of the food web. As we will see, however, the situation is not quite so simple. The proportion of each trophic level consumed by the one above it; the nutritional content of autotrophs, detritus, and prey; and the efficiency of energy transfers also play roles in determining the flow of energy between trophic levels.

A comparison of the proportions of autotroph biomass consumed in terrestrial and in aquatic ecosystems provides some insight into the factors that influence energy flow between trophic levels. When viewed from space, some parts of Earth's terrestrial surface appear green, while the ocean appears blue. Why is the land surface green and the ocean blue? Furthermore, in Concept 20.2, we saw that very productive lakes (e.g., those that are experimentally fertilized; see Figure 20.15) can appear green. What these green areas have in common is primary productivity that far exceeds rates of herbivory. Herbivores on land consume a much lower proportion of autotroph biomass than do herbivores in most aquatic ecosystems. On average, about 13% of terrestrial NPP is consumed (range 0.1%–75%), while in aquatic ecosystems, an average of 35% of NPP is consumed (range 0.3%–100%) (Cebrian and Lartigue 2004).

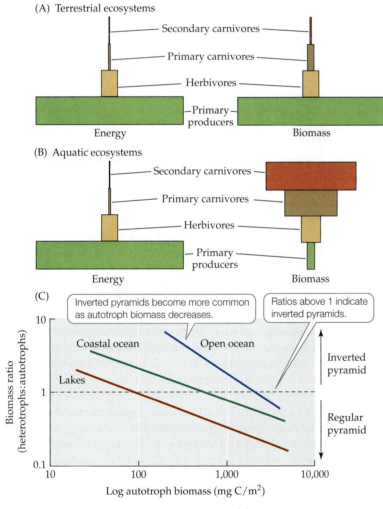

FIGURE 21.5 Trophic Pyramid Schemes (A) In terrestrial ecosystems, energy and biomass pyramids are usually similar. (B) In many aquatic ecosystems, the biomass pyramid is inverted relative to the energy pyramid. (C) Inverted biomass pyramids in aquatic ecosystems are most common in nutrient-poor waters with low autotroph biomass. (A,B after F. S. Chapin et al. 2002. *Principles of Terrestrial Ecosystem Ecology.* Springer-Verlag: New York. C after J. M. Gasol et al. 1997. *Limnol Oceanogr* 42: 1353–1363.)

There is a positive relationship between NPP and the *amount* of biomass consumed by herbivores (**FIGURE 21.6**). This relationship, which holds within most ecosystem types, would seem to suggest that herbivore production is limited by the amount of food available. Why, then, is the *proportion* of autotroph biomass consumed in terrestrial ecosystems relatively low? If herbivore production is limited by the supply of energy and nutrients from plants, why don't terrestrial herbivores consume a greater proportion of the biomass that is available?

Several hypotheses have been proposed to explain the lower proportion of autotroph biomass consumed in terrestrial ecosystems. First, Hairston and Hairston (1993) have argued that the population growth of herbivores is more constrained by predation in aquatic ecosystems than in terrestrial ecosystems because of the better-developed higher trophic levels in aquatic ecosystems. Predator removal experiments such as those described in Concept 12.4 and Concept 21.4 demonstrate that predators can effectively influence autotroph biomass through their influence on the abundance of herbivores.

Second, defenses against herbivory, such as the secondary compounds and structural defenses described in Concept 13.2, lower the amount of autotroph biomass that is consumed. Plants of resource-poor environments, such as desert and tundra, tend to be more strongly defended against herbivory than plants from resource-rich environments. This greater allocation to defense may explain why the proportion of plant biomass consumed

is lower in resource-poor terrestrial environments. Unicellular algae make up the bulk of autotroph biomass in aquatic ecosystems, and they generally lack the chemical and structural defenses of their multicellular terrestrial counterparts.

Third, the chemical composition of phytoplankton makes them more nutritious for herbivores than terrestrial plants are. Terrestrial plants contain nutrient-poor structural materials such as stems and wood, which are typically absent in aquatic autotrophs. Herbivores typically require large amounts of nutrients such as nitrogen and phosphorus to meet their demands for structural growth, metabolism, and protein synthesis. The ratio of nutrients to carbon (with carbon representing energy) is thus an important measure of food quality. Carbon:nutrient ratios differ markedly between autotrophs in terrestrial and in freshwater ecosystems. Freshwater phytoplankton have carbon:nutrient ratios closer to those of herbivores than terrestrial plants do (Elser et al. 2000) and thus better meet the nutritional needs of the herbivores that eat them. Each of these factors—predation, plant defenses, and food quality—contributes to differences in the proportion of NPP consumed among ecosystems and, in particular, the greater consumption of autotroph biomass in aquatic ecosystems (Shurin et al. 2006).

The efficiency of energy transfer varies among consumers

Not all of the food energy consumed by a heterotroph gets incorporated into heterotroph biomass. Some is lost to respiration and excretion. We can use the concept of *energy efficiency*, defined as the output of energy per unit of energy input, to characterize the transfer of energy between trophic levels. In studies of energy transfer in trophic systems, the concept of **trophic efficiency** is used, defined as the amount of energy at one trophic level divided by the amount of energy at the trophic level immediately below it. Trophic efficiency incorporates the proportion of available energy that is consumed (consumption efficiency), the proportion of ingested food that is assimilated by the consumer (assimilation efficiency), and the proportion of assimilated food that goes into producing new consumer biomass (production efficiency) (**FIGURE 21.7**).

As we saw above, not all of the biomass available at one trophic level is consumed by the next trophic level. The proportion of the available biomass that is ingested is the **consumption efficiency**. Consumption efficiency is typically higher in aquatic ecosystems than in terrestrial ecosystems (see Figure 21.6). Consumption efficiencies also tend to be higher for carnivores than for herbivores, although a systematic survey comparing the two groups has not been done.

Once biomass is ingested by the consumer, it must be assimilated by the digestive system before the energy it contains can be used to produce new biomass. The

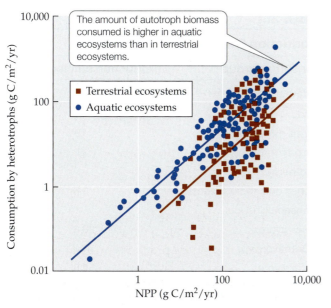

FIGURE 21.6 Consumption of Autotroph Biomass Is Correlated with NPP The amount of autotroph biomass consumed increases with increasing available NPP in both terrestrial and aquatic ecosystems. (After J. Cebrian and J. Lartigue. 2004. *Ecol Monogr* 74: 237–259.)

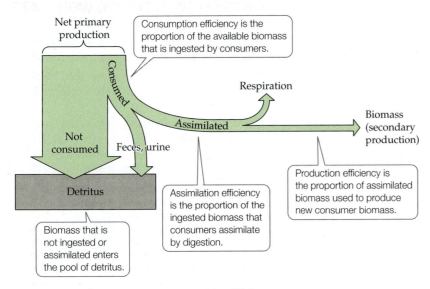

FIGURE 21.7 Energy Flow and Trophic Efficiency The proportion of energy transferred between trophic levels depends on efficiencies of consumption, assimilation, and production. (After F. S. Chapin et al. 2002. *Principles of Terrestrial Ecosystem Ecology.* Springer-Verlag: New York.)

? How do the trends in consumption efficiency vary in Figure 21.6? What does this variation suggest about differences in consumption efficiency in aquatic versus terrestrial ecosystems?

proportion of the ingested food that is assimilated is the **assimilation efficiency**. Food that is ingested but not assimilated is lost to the environment as feces, entering the pool of detritus, or as urine. Assimilation efficiency is determined by the quality of the food (its chemical composition) and the physiology of the consumer.

The quality of the food available to herbivores and detritivores is generally lower than that of the food available to carnivores. Plants and detritus are composed of relatively complex carbon compounds, such as cellulose, lignins, and humic acids, that are not easily digested. In addition, plants and detritus have low concentrations of nutrients. Animal bodies, on the other hand, have carbon:nutrient ratios that are usually very similar to those of the animals consuming them and so are assimilated more readily. Assimilation efficiencies of herbivores and detritivores vary between 20% and 50%, while those of carnivores are about 80%.

How thoroughly food is digested is influenced by the consumer's thermal physiology and the complexity of the consumer's digestive system. Endotherms tend to digest food more completely than ectotherms, due to higher thermal stability and a tendency to have a more developed digestive system, and therefore have higher assimilation efficiencies. Additionally, some herbivores have mutualistic symbionts that help them digest cellulose. For example, as described in Concept 5.4, ruminants have a modified stomach chamber that contains bacteria and protists that increase the breakdown of cellulose-rich foods. This mutualistic symbiosis, coupled with a longer period of digestion, gives ruminants higher assimilation efficiencies than nonruminant herbivores.

Assimilated food can be used to produce new biomass in the form of consumer growth and production of new consumer individuals (reproduction). However, a portion of the assimilated food must be used for respiration associated with maintenance of existing molecules and tissues as well as with construction of new biomass (see Concept 5.4). The proportion of the assimilated food that is used to produce new consumer biomass is **production efficiency**.

Production efficiency is strongly related to the thermal physiology and size of the consumer. Endotherms allocate much of their assimilated food to metabolic production of heat and therefore have less energy left over to allocate to growth and reproduction than ectotherms do (**TABLE 21.1**). Thus, ectotherms have considerably higher production efficiencies than endotherms. Body size in endotherms is an important determinant of heat loss and thus of production efficiency. If body shape and insulation (fat, feathers, and fur) are held constant, then, as animal body size increases, the surface area-to-volume ratio decreases. Thus, a small endotherm, such as a shrew, will lose a greater proportion of its internally generated heat across its body surface than a large endotherm, such as a grizzly bear, and thus a small endotherm will tend to have a lower production efficiency than a large endotherm.

TABLE 21.1

Production Efficiencies of Consumers

Consumer group	Production efficiency (%)
Endotherms	
Birds	1.3
Small mammals	1.5
Large mammals	3.1
Ectotherms	
Fishes and social insects	9.8
Nonsocial insects	
Herbivores	38.8
Detritivores	47.0
Carnivores	55.6
Non-insect invertebrates	
Herbivores	20.8
Detritivores	36.2
Carnivores	27.6

Sources: F. S. Chapin et al. 2002. *Principles of Terrestrial Ecosystem Ecology.* Springer-Verlag: New York; W. F. Humphreys. 1979. *J Anim Ecol* 48: 427–454.

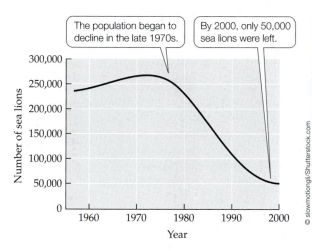

The population began to decline in the late 1970s.

By 2000, only 50,000 sea lions were left.

FIGURE 21.8 Steller Sea Lion Population Decline in Alaska The population of sea lions in the Gulf of Alaska and the Aleutian Islands decreased by about 80% over 25 years. (After A. W. Trites and C. P. Donnelly. 2003. *Mamm Rev* 33: 3–28; based on A. W. Trites and P. A. Larkin. 1996. *Aquat Mamm* 22: 153–166; A. W. Trites, unpublished data.)

Eumetopias jubatus

Trophic efficiencies can influence population dynamics

Changes in food quantity and quality, and the resulting changes in trophic efficiency, can determine the consumer population sizes that can be sustained, as well as the health of the individuals in consumer populations. Here we'll examine the potential contribution of changes in food quality to the decline in numbers of Steller sea lions (*Eumetopias jubatus*) in Alaska.

From the late 1970s into the 1990s, the total population of Steller sea lions in the Gulf of Alaska and the Aleutian Islands decreased by about 80%, from approximately 250,000 sea lions in 1975 to 50,000 in 2000 (**FIGURE 21.8**). The population in the eastern part of the range has recovered some since that time, and the species was taken off the endangered list in 2013. However, the population in the western part of its range, along the southern coast of Alaska, has continued to decline. Andrew Trites and C. P. Donnelly reviewed the available information to try to determine possible causes for this decline (Trites and Donnelly 2003). They found that individual sea lions collected during the period of decline were smaller than individuals within the same age classes collected before the start of the decline. There was also a reduction in the number of pups born per female during this period, which resulted in a shift in the age structure toward older individuals. No evidence was found for outbreaks of disease or parasites. The smaller body sizes and declining birth rates suggested that there were fewer prey available, or that the available prey were not providing sufficient nourishment to sustain the sea lions—in other words, that trophic efficiency had declined. Additional data indicated that the sea lions were obtaining prey—primarily fish—as regularly as they had before the decline. Nursing females in the declining population were actually spending less time hunting for the same amount of fish as nursing females in other populations that were not declining. Therefore, the availability of prey, or the sea lions' ability to capture it, did not appear to be limiting their growth and reproduction.

Trites and Donnelly considered the possibility that changes in the species of prey fish available had contributed to the decline of the Steller sea lions. They and others suggested that the decline might be related to declining prey quality, an idea they referred to as the "junk food hypothesis." Prior to the decline, the diet of the sea lions had been primarily herring, a fish that is relatively rich in fats, along with small amounts of pollock, cod, salmon, and squid. During the period of the population decline, the sea lions' diet shifted away from herring toward a greater proportion of pollock and cod (**TABLE 21.2**). This change in diet reflected a shift toward cod dominance of the fish community from the 1970s through the 1990s. The causes of the change in fish community composition

TABLE 21.2

Proportion of Steller Sea Lion Scats and Stomachs Containing Five Prey Categories

Years	Gadids (cod, pollock, hake)	Salmon	Small schooling fish (herring, capelin, eulachon, sand lance)	Cephalopods (squid)	Flatfish (flounder, sole)
1990–1993	85.2	18.5	18.5	11.1	13.0
1985–1986	60.0	20.0	20.0	20.0	5.0
1976–1978	32.1	17.9	60.0	0.0	0.0

Source: R. L. Merrick et al. 1997. *Can J Fish Aquat Sci* 54: 1342–1348.

are uncertain but may be associated with overfishing, oil spills, disease, and long-term climate change. The proportions of fat and energy per mass of pollock and cod are approximately half those of herring. Captive Steller sea lions raised on a diet of herring and then switched to a diet of pollock lose body mass and fat, even with an unlimited supply of pollock.

Based on their review of the available information, Trites and Donnelly concluded that nutritional stress was the most likely cause of the decline in the Steller sea lion population. The amount of prey available to the sea lions did not appear to have changed, but changes in the quality of that prey, and associated changes in trophic efficiency, contributed to the decline in the population through their effects on individual growth rates and birth rates. Others have suggested that the decline in Steller sea lion numbers may also be linked to changes in the trophic structure of the North Pacific (Springer et al. 2003). As described in the Case Study Revisited in Chapter 9, massive harvesting of great whales by humans in the mid-twentieth century may have forced their predators, killer whales, to hunt other prey, including Steller sea lions. As we describe in the next section, such "top-down" effects of predators on prey can have important consequences for energy flow in ecosystems.

> **CONCEPT 21.3** Changes in the abundances of organisms at one trophic level can influence energy flow at multiple trophic levels.

LEARNING OBJECTIVES

21.3.1 Compare the factors that influence energy flow in an ecosystem between bottom-up and top-down perspectives.

21.3.2 Describe how changes in the abundance of organisms at the fourth trophic level may impact rates of primary production through a trophic cascade.

21.3.3 Explain how the size of an ecosystem, the rates of disturbance, and the amount of primary production can influence the number of trophic levels.

Trophic Cascades

There are two possible ways to look at the control of energy flow through ecosystems. First, the amount of energy that flows through trophic levels may be determined by how much energy enters an ecosystem via NPP, which in turn is related to the supply of resources (as we saw in Concepts 20.2 and 21.2). The greater the NPP entering the ecosystem, the more energy can be passed on to higher trophic levels. This view, which is often referred to as "bottom-up" control of energy flow, holds that the resources that limit NPP determine energy flow through an ecosystem (**FIGURE 21.9A**). Alternatively, energy flow may be governed by rates of consumption (as well as

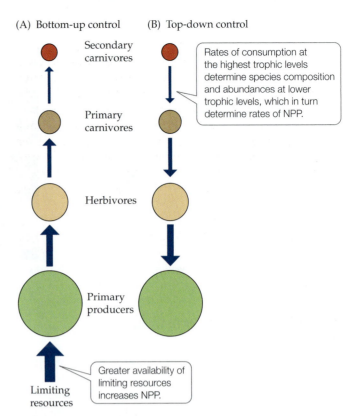

FIGURE 21.9 Bottom-Up and Top-Down Control of NPP Production in an ecosystem can be viewed as being controlled (A) by limiting resources or (B) by controls exerted on the species composition and abundances of autotrophs by consumption at higher trophic levels.

other, nonconsumptive interactions such as competition and facilitation, as discussed in Concept 16.3) at the highest trophic levels, which influence abundances and species composition at multiple trophic levels below them. This view is often referred to as "top-down" control of energy flow (**FIGURE 21.9B**). In reality, both bottom-up and top-down controls are operating simultaneously in ecosystems, but the top-down view has important implications for the effects of trophic interactions on energy flow in ecosystems.

Trophic interactions can trickle down through multiple trophic levels

Changes in abundance or species composition at one trophic level can lead to important changes in abundance and species composition at other trophic levels. For example, an increase in the rate of predation by a carnivore at the fourth trophic level on carnivores at the third trophic level would lead to a lower rate of consumption of herbivores at the second trophic level. More herbivory would decrease the abundance of autotrophs and would therefore lower rates of NPP. Nonconsumptive species interactions, such as competition, can have similar top-down effects on abundance and species composition at lower trophic levels, as we'll see shortly. Such a series of changes in abundance and species composition is referred to as a **trophic cascade**.

Our understanding of trophic cascades comes primarily from aquatic ecosystems, although there are examples from terrestrial ecosystems as well. Several generalizations have been drawn from studies of these interactions. First, trophic cascades are most often associated with changes in the abundance of top specialist predators. Second, omnivory may act to buffer the effects of trophic cascades through the consumption of prey at multiple trophic levels. Finally, trophic cascades have been hypothesized to be most important in relatively simple, species-poor ecosystems. However, several recent experiments have demonstrated trophic cascades in ecosystems with relatively high species diversity.

AN AQUATIC TROPHIC CASCADE There are several studies of trophic cascades associated with near extinctions of native species. A classic example is the interaction among sea otters (*Enhydra lutris*), sea urchins, and killer whales on the west coast of North America, which was discussed in the Case Study Revisited in Chapter 9. Additionally there are numerous examples of trophic cascades associated with the intentional or unintentional introduction of non-native species. One such example resulted from the release of brown trout (*Salmo trutta*), a popular sport fish, into streams and lakes of New Zealand. The stocking of Kiwi waters by European settlers began in the 1860s, and by 1920 an estimated 60 million fish had been released throughout New Zealand. Native fish populations have declined as a result, and some species have disappeared from streams now dominated by trout.

Alexander Flecker and Colin Townsend (1994) investigated the influence of the brown trout on the species composition of its prey (primarily stream insects) and associated effects on primary production in the Shag River. Brown trout were originally released into the Shag River in 1869 by the "Otago Acclimatisation Society" to make settlers feel more at home. The Shag River is one of a small number of streams in New Zealand that still holds both native fish and trout in the same sections. Native fish species include the common river galaxias (*Galaxias vulgaris*). The morphology and feeding behavior of galaxias are similar to those of trout, as indicated by the common name for the galaxias, Maori trout.

Flecker and Townsend compared the effects of brown trout and galaxias on stream invertebrate species composition and abundance as well as on primary production by algae. To manipulate fish presence and absence, they constructed artificial stream channels adjacent to the natural channel, made of 5-m lengths of half-cylinders of PVC pipe. The PVC channels had mesh on the ends that kept fish in or out but allowed free movement of stream invertebrates and algae. The researchers placed clean gravel and stone cobbles in the bottoms of the channels to provide a substrate for the invertebrates and algae. The channels were allowed to accumulate algae and invertebrates for 10 days before the fish were added. Three treatments were initiated: channels with introduced brown trout, channels with galaxias, and channels with no fish (controls). Eight fish of similar size and mass were used for each fish

species addition. The experiment was run for 10 days, after which samples were collected to determine invertebrate species composition and abundance and algal biomass.

Flecker and Townsend had expected brown trout to decrease invertebrate diversity more than the native galaxias, but the effect of fish on invertebrate diversity was relatively small and did not differ between the two fish species. The brown trout, however, reduced total invertebrate density by approximately 40% relative to the control channels, while galaxias resulted in a smaller reduction (**FIGURE 21.10A**). The abundance of algae increased with both fish, but the effect was greater in the channels with trout (**FIGURE 21.10B**). Flecker and Townsend suggested that the effect on algal

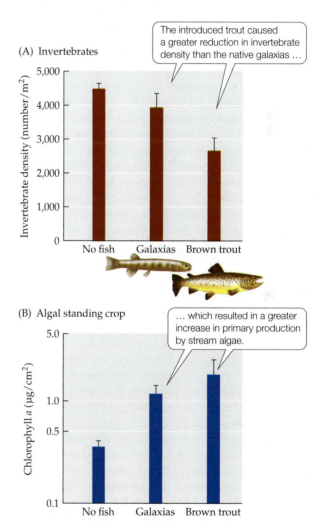

FIGURE 21.10 An Aquatic Trophic Cascade Flecker and Townsend used artificial stream channels to study the effects of non-native brown trout and a native fish (galaxias) on stream invertebrates and algae in the Shag River, New Zealand. (A) Effects on invertebrate density. (B) Effects on algal biomass, as estimated using chlorophyll concentrations in stream water. Error bars show one SE of the mean. (After A. S. Flecker and C. R. Townsend. 1994. *Ecol Appl* 4: 798–807.)

? What factor other than overall consumption rate might explain why the presence of brown trout results in a larger increase in primary production than the presence of native galaxias?

biomass was the result of a trophic cascade in which fish predation not only reduced the density of stream invertebrates, but also caused them to spend more time in refugia on the stream bottom rather than feeding on algae. The trout had a much greater effect on invertebrate density, and thus on primary production, than the native galaxias. These results suggested that trophic cascades associated with the stocking of non-native fish for sport may have consequences not just for native biodiversity, but for the functioning of stream ecosystems as well.

A TERRESTRIAL TROPHIC CASCADE   As mentioned earlier, trophic cascades have been most commonly observed in aquatic ecosystems, where they are more frequent and their effects are stronger than in terrestrial ecosystems (Shurin et al. 2002). Terrestrial ecosystems are generally thought to be more complex than aquatic ecosystems. In addition, it was believed that a decrease in the abundance of one species in a terrestrial ecosystem was more likely to be compensated for by an increase in the abundances of similar species that were not being consumed as heavily. Thus, trophic cascades were considered unlikely in diverse terrestrial ecosystems such as tropical forests.

Lee Dyer and Deborah Letourneau (1999a) tested the effects of a potential trophic cascade on the production of *Piper cenocladum* trees in the understory of a lowland tropical rainforest in Costa Rica. *Piper cenocladum* is a relatively common component of the understory in these forests and is eaten by dozens of different herbivore species. Ants of the genus *Pheidole* live in chambers in the petioles of the leaves of the *Piper* trees. The ants eat food bodies provided by the trees, and they also consume herbivores that attack the trees. These ants, in turn, are eaten by beetles of the genus *Tarsobaenus*. Thus, four distinct trophic levels exist in this system (**FIGURE 21.11**). Dyer and Letourneau had previously noted that plant biomass was lower, and rates of herbivory were higher, when densities of *Tarsobaenus* beetles were high. They performed experiments to test whether a trophic cascade involving the beetles, ants, and herbivores influenced the production of the *Piper* trees and how strong that influence was, compared with that of bottom-up factors such as light and soil fertility.

Dyer and Letourneau established experimental plots in the understory by planting uniform-sized cuttings of *Piper* trees. They treated two groups of plots with an insecticide to kill any ants present, then added *Tarsobaenus* beetle larvae to one of those groups of plots. This procedure established three groups of treatment plots: one group of insecticide-treated plots with beetles, one control group of insecticide-treated plots without beetles, and one control group of untreated plots. In the plots with beetles, the insecticide treatment facilitated the establishment of the beetles by preventing ant attacks on the beetle larvae. In addition, half of the plots were on a relatively fertile soil type, and the other half were on a relatively infertile soil

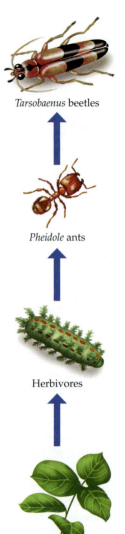

FIGURE 21.11 A Terrestrial Trophic Cascade   Trophic interactions in the understory ecosystem of a lowland tropical rainforest in Costa Rica. *Piper cenocladum* trees are consumed by herbivores but provide shelter for *Pheidole* ants, which consume herbivores attacking the trees. *Pheidole* ants are consumed by *Tarsobaenus* beetles. Both ants and beetles also consume food bodies produced by the trees. (After L. A. Dyer and D. K. Letourneau. 1999a. *Proc Natl Acad Sci USA* 96: 5072–5076. © 1999 National Academy of Sciences, U.S.A.)

Tarsobaenus beetles

Pheidole ants

Herbivores

Piper cenocladum

type. Natural light levels in the plots were also varied such that half of the plots were assigned to a high-light treatment and half to a low-light treatment. Dyer and Letourneau maintained these treatments for 18 months and measured herbivory and leaf production within each of the plots.

If the production of the *Piper* trees was limited primarily by resource supply (bottom-up control), then the addition of the *Tarsobaenus* beetles would be expected to have little effect on *Piper* leaf production. Soil fertility and light levels would be expected to have greater effects on leaf production if these effects were more important than the influence of the trophic cascade associated with beetles, ants, and herbivores (top-down control). Dyer and Letourneau found, however, that the trophic cascade was the only significant influence on leaf production. The addition of the predatory beetles decreased ant abundance fivefold, increased rates of herbivory threefold, and decreased leaf area per tree to half that in the control plots (**FIGURE 21.12**). This experiment provided convincing evidence of a trophic cascade affecting the production of the *Piper* trees. It should be noted, however, that the lack of an effect of soil fertility and light in the control treatments, which had low rates of herbivory, indicates that the resource(s) that actually limit production may not have been manipulated in this experiment. An additional experiment that used more controlled manipulation of light levels and soil nutrients, rather than relying on variation in natural levels, found significant effects of these resources on *Piper* production, but it also found a continued strong effect of herbivory

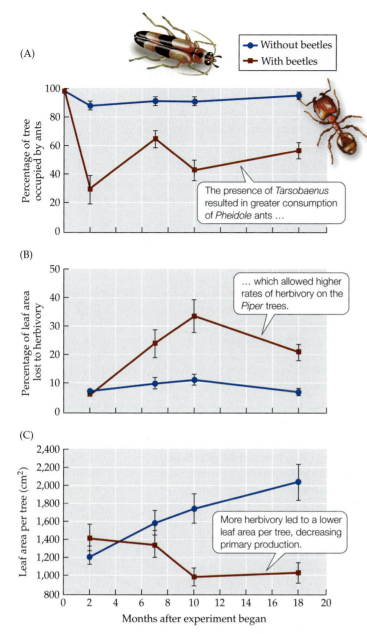

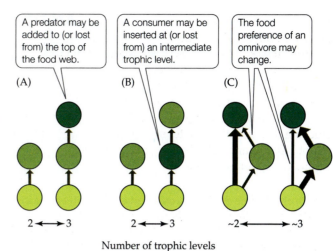

FIGURE 21.13 Changes in the Number of Trophic Levels Circles represent species at different trophic levels, and the thickness of the arrows represents the amount of energy flowing between species pairs. Differences among ecosystems in the number of trophic levels may occur because of (A) the addition or loss of a consumer at the top level, (B) the insertion or loss of a consumer at an intermediate level, or (C) a change in the preferred feeding level of an omnivore. (After D. M. Post and G. Takimoto. 2007. *Oikos* 116: 775–782.)

FIGURE 21.12 Effects of a Trophic Cascade on Production A trophic cascade in a tropical rainforest understory ecosystem (see Figure 21.11) was shown to have important effects on (A) predation, (B) herbivory, and (C) production. Error bars show ± one SE of the mean. (After L. A. Dyer and D. K. Letourneau. 1999a. *Proc Natl Acad Sci USA* 96: 5072–5076. © 1999 National Academy of Sciences, U.S.A.)

(Dyer and Letourneau 1999b). Thus, it is clear that trophic cascades do occur in diverse terrestrial ecosystems, although they may require strong interactions between specialist predators and their prey.

What determines the number of trophic levels?

What determines the variation in the number of trophic levels that occur among different ecosystems, and why do so few ecosystems have five or more trophic levels? These questions are not simply academic. Through trophic cascades, the number of trophic levels in an ecosystem can influence movements of energy and nutrients as well as the potential for toxins in the environment to become concentrated at higher trophic levels, as we will see in this chapter's Case Study Revisited. Change in the number of trophic levels may be due to the addition or loss of a predator at the top of the food web, the insertion or loss of a predator in the middle of the food web, or changes in omnivore feeding preference for foods at different trophic levels (**FIGURE 21.13**).

Several interacting ecological factors can control the number of trophic levels in ecosystems (Post 2002b). First, the amount of energy entering an ecosystem through primary production has been proposed as a determinant of the number of trophic levels. Because a relatively large amount of energy is lost in the transfer from one trophic level to the next, the more energy there is entering a system, the more is potentially available to support viable populations of higher-level predators (see **ANALYZING DATA 21.1**). However, this explanation appears to be important primarily in ecosystems with low resource availability. Second, the frequency of disturbances or other agents of change, such as disease outbreaks, can determine whether populations of higher-level predators can be sustained. Because lower trophic levels are required to sustain higher trophic levels, there is a longer time lag for the reestablishment of the higher trophic levels following a disturbance. If

ANALYZING DATA 21.1

Does the Identity of Organisms Influence Energy Flow between Trophic Levels?

Ecologists have noted that individuals and populations of some species (known as keystone species; see Concept 16.3) influence energy flow between trophic levels more than others. In particular, we've seen several examples in which invasive species have greatly altered energy transfers as well as diversity within communities. Attention has largely been focused on the behavioral characteristics of a species, such as how effective individuals of a species are at hunting or grazing, or its population dynamics (e.g., whether a population exhibits exponential growth; see Concept 10.3). Additionally, the thermal physiology and sizes of the species making up a trophic level can influence how much energy makes it from one trophic level to the next.

Using information from the text and Table 21.1, provide a rough estimate of how much energy would make it to the second, third, and fourth trophic levels in the following simplified food chains. Start with 100 units of energy in the autotrophic base of each of these food chains

(i.e., plants or algae). Assume the production efficiencies for endotherms do not vary according to diet.

1. Plants → non-insect invertebrate herbivores → small mammals → large mammals

2. Algae → aquatic non-insect invertebrate herbivores → insect predators →fish

3. Plants → large mammal herbivores → large mammal predators → large mammal predators

4. Plants → insect herbivores → insect predators → insect predators

5. Remembering that the transfer of energy between trophic levels can influence the number of trophic levels an ecosystem can sustain, and that greater energy transfer usually enhances the establishment of higher tropic levels, which of the hypothetical food chains in Questions 1–4 would be *most* likely, and which *least* likely, to sustain the highest trophic level?

---

disturbances occur frequently, then higher trophic levels may never become established, no matter how much energy is entering the system (Pimm and Lawton 1977). While some support for this hypothesis exists, the ability of some organisms to adapt to frequent disturbances and the potential for rapid colonization of disturbed sites (see Concept 17.1) result in a smaller effect of disturbance on trophic level number than expected. Finally, the area of an ecosystem can influence the number of trophic levels. Larger ecosystems support larger population sizes, which are less prone to local extinction (see Concept 11.3). Larger ecosystems also have more habitat heterogeneity and thus tend to have higher species diversity.

Support for the effect of ecosystem size on the number of trophic levels is derived primarily from studies of lakes and oceanic islands, ecosystems with discrete boundaries. For example, Gaku Takimoto and colleagues (2008) tested the relative effects of disturbance and island size on the number of trophic levels on 36 islands in the Bahamas. The effect of disturbance was tested by examining 33 of the smaller islands that were either exposed to (19 islands) or protected from (14 islands) storm surges. The number of trophic levels was estimated using isotopic ratios of carbon and nitrogen (as described in Concept 20.4) in tissues from the top predators, spiders and lizards. Takimoto and colleagues found that exposure to storm surges had no effect on the number of trophic levels. However, disturbance did influence the identity of the top predators: orb spiders were more frequently the top predators on exposed islands, and *Anolis* lizards were at the apex of the food web on protected islands. Island

size, however, was strongly correlated with the number of trophic levels (**FIGURE 21.14**), providing evidence that ecosystem size can influence the number of trophic levels in a terrestrial ecosystem.

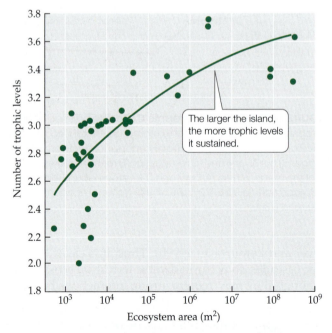

FIGURE 21.14 Ecosystem Size Is Correlated with the Number of Trophic Levels   On islands in the Bahamas, Takimoto and colleagues found that as island size increased, the number of trophic levels also increased. (After G. Takimoto et al. 2008. *Ecology* 89: 3001–3007.)

We turn our attention next to a more detailed investigation of trophic relationships in ecosystems as we cross the disciplinary boundaries of ecosystem ecology and community ecology (the topic of Unit 5) to examine how energy flow can influence the diversity and stability of communities and ecosystems.

LEARNING OBJECTIVES

21.4.1 Explain how food webs are helpful for envisioning ecosystem energy flow, and outline the factors that compromise their accuracy in portraying the full extent of interactions among organisms.

21.4.2 Describe how the use of interaction strengths can aid in the construction of more accurate food web models.

21.4.3 Summarize how ecologists have viewed the relationship between the complexity of food webs and the stability of associated communities and ecosystem processes.

Food Webs

Ever since Charles Darwin, in *The Origin of Species* (1859), described "a tangled bank, clothed with many plants of many kinds, with birds singing on the bushes, with various insects flitting about, … dependent upon each other in so complex a manner," the interdependence of species has been a central concept in ecology. When we examine these links among species with a focus on feeding relations, they can be described by a **food web**, a diagram showing the connections among organisms and the food they consume. For the desert ecosystem we considered at the start of this chapter, we can construct a simplified food web showing that plants are consumed by insects and ground squirrels and that these herbivores are food for scorpions, eagles, and foxes (**FIGURE 21.15A**). In this way, we can begin to understand qualitatively how energy flows from one component of this ecosystem to another, and how that energy flow may influence changes in population sizes and the species composition of communities.

Food webs are complex

The desert food web in Figure 21.15A is far from complete. Depending on our purposes, we may wish to add other organisms and links to the food web, providing additional complexity. For example, the scorpion consumes insects such as the grasshopper, but like the grasshopper, it may be food for birds such as shrikes and owls (**FIGURE 21.15B**). As we continue to add more and more organisms to the food web, we add complexity, such that the food web may take on the appearance of a "spaghetti

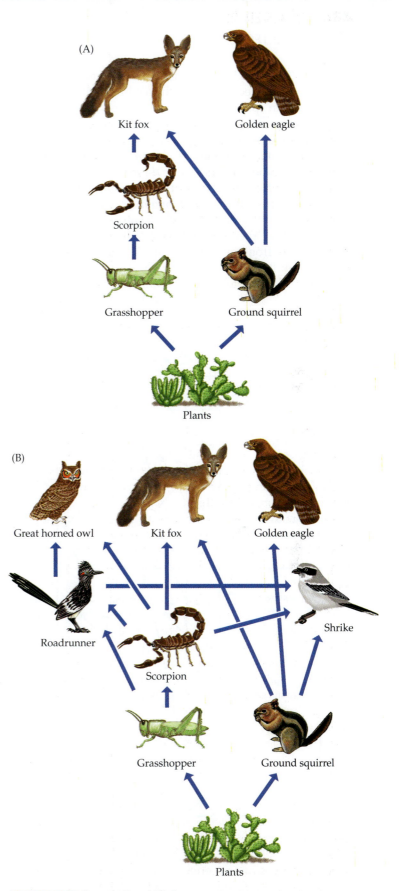

FIGURE 21.15 Desert Food Webs Food webs may be simple or complex depending on their purpose. (A) A simple six-member food web for a representative North American desert. (B) Addition of more participants to the food web adds realism, but the inclusion of additional species adds complexity.

diagram" (**FIGURE 21.16**). In order to add greater realism, it is important to recognize that the feeding relationships of animals can span multiple trophic levels (omnivory) and may even include cannibalism (half-circle arrows in Figure 21.16) (Polis 1991).

Although food webs are useful conceptual tools, even a simplified food web is a static description of energy flow and trophic interactions in a temporally dynamic ecosystem. Actual trophic interactions can change over time (Wilbur 1997). Some organisms alter their feeding patterns as they age. Maturing frogs, for example, make the transition from omnivorous aquatic tadpoles to carnivorous adults. Some animals, such as migratory birds, are relatively mobile and are thus components of multiple food webs. Furthermore, most food webs fail to account for additional biological interactions among organisms that influence population and community dynamics, such as pollination mutualisms. (In community studies, this problem may be addressed by the use of interaction webs, as described in Figure 16.5.) The critically important roles of microorganisms are often ignored as well, despite their processing of a substantial amount of the energy moving through ecosystems. What are food webs good for, then? Despite these apparent shortcomings, food webs are important conceptual tools for understanding the dynamics of species interactions and energy flow in ecosystems and hence the community and population dynamics of their component organisms.

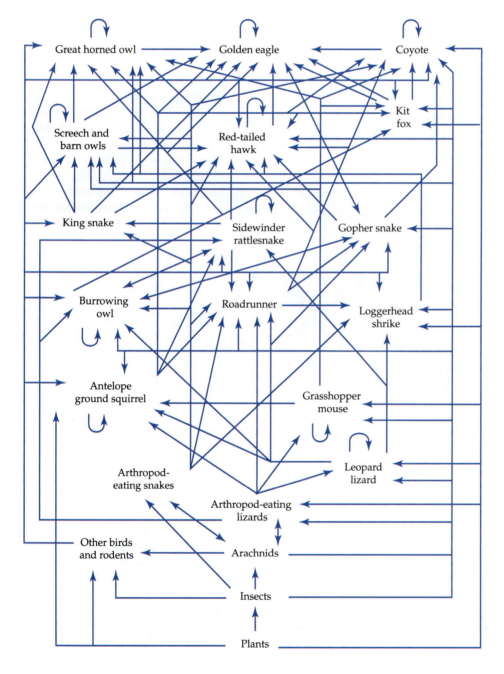

FIGURE 21.16 Food Webs Can Be Complex In this North American desert food web, complexity overwhelms any interpretation of interactions among the members. Even this food web, however, lacks the majority of the trophic interactions in the ecosystem. (From G. A. Polis. 1991. *Am Nat* 138: 123–155.)

? How many of the organisms or feeding groups depicted in this food web consume both plants and animals as food sources? What does this suggest about the frequency of omnivory in this food web?

The strengths of trophic interactions are variable

As indicated in the quote from Darwin above and in earlier chapters, a core concept of ecological thought is that "everything is connected to everything else." However, the links among the species in an ecosystem vary in their importance to energy flow and species population dynamics; in other words, not all connections are equally important. Some trophic relationships play larger roles than others in dictating how energy flows through the ecosystem. *Interaction strength* is a measure of the effect of the population of one species on the population size of another species (see Ecological Toolkit 16.1). Determining interaction strengths is an important goal of ecologists because it helps us simplify the "spaghetti" in a complex food web by focusing attention on those links that are most important for basic research and conservation (Berlow 1999).

How are interaction strengths determined? Several approaches have been used. Removal experiments, like those described in Concept 16.3 to determine competition or facilitation, can be employed, but performing such experimental removals to quantify every link in a food web would be logistically overwhelming. Therefore, much current ecological research is devoted to discovering simpler, less direct measures that can still give us a reliable estimate of the relative importance of different links. For example, simple food webs can be coupled with observations of the feeding preferences of predators and of changes in the population sizes of predators and prey over time to provide an estimate of which interactions are the strongest. Similarly, comparisons of two or more food webs in which a predator or prey species is present in some but absent in others may provide evidence for the relative importance of links. Predator and prey body sizes have been used to predict the strengths of predator–prey interactions because feeding rate is known to be related to metabolic rate, which in turn is governed by body size. The best estimates of interaction strengths in food webs often come from a combination of these approaches.

A series of classic studies examining interaction strengths in food webs was performed in rocky intertidal zones of the Pacific Northwest by Robert Paine. Paine (1966) had observed that the diversity of organisms in rocky intertidal zones declined as the density of predators decreased. He reasoned that some of those predators might be playing a greater role than others in controlling the diversity of these communities. One of Paine's critical observations was that one mussel species (*Mytilus californianus*) had the ability to overgrow and smother many of the other sessile invertebrate species that compete with it for space. Paine hypothesized that predators might play a key role in maintaining diversity in this community by consuming these mussels and preventing them from competitively excluding other species.

To test these hypotheses, Paine conducted an experiment in Washington State in which he removed the top predator in the system, the sea star *Pisaster ochraceus*, from experimental plots. *Pisaster* feeds primarily on bivalves and barnacles and to a lesser extent on other mollusks, including chitons, limpets, and a predatory whelk (*Nucella* sp.) (**FIGURE 21.17**). Following the continuous manual removal of *Pisaster* from 16-m² plots, acorn barnacles (*Balanus glandula*) became more abundant, but with time, they were crowded out by mussels (*Mytilus*) and gooseneck barnacles (*Pollicipes* spp.). After 2½ years, the number of species in the community had decreased from 15 to 8. Even 5 years after the experiment began, when sea stars were no longer being removed, dominance by the mussels continued, as individual mussels had grown to sizes that prevented predation by sea stars, and diversity remained lower in the experimental plots than in adjacent control plots (Paine et al. 1985). Experimental removals of higher-level predators in other intertidal zones, including one in New Zealand, which shares no species with the intertidal zone of the Pacific Northwest, resulted in similar reductions in diversity. Predators in these intertidal ecosystems are thus key to maintaining species diversity by preventing competitive exclusion. Such species are more important in food webs than their numbers would indicate.

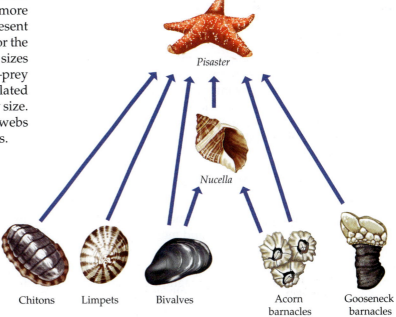

FIGURE 21.17 An Intertidal Food Web This food web from the rocky intertidal zone of Mukkaw Bay, Washington State, was used by Robert Paine to investigate the strength of the interaction between the sea star *Pisaster ochraceus* and its prey. (After R. T. Paine. 1966. *Am Nat* 100: 65–75.)

The experimental research of Paine and others was an encouraging advance in ecology because it demonstrated that, despite the potential complexity of trophic interactions among species, patterns of energy flow and community structure might be governed by a small subset of those species. Paine called animals like it *Pisaster keystone species*, defining them as species that have a greater influence on energy flow and community composition than their abundance or biomass would predict (see Figure 16.16). The keystone species concept has become an important focus in ecology and conservation biology because it implies that protecting such species may be critical for protection of the many other species that depend on them (as we'll see in Concept 23.5). Many keystone species are predators at higher trophic levels, which tend to have large effects on prey populations relative to their own abundance.

Some species act as keystone species in only part of their geographic range, suggesting that interaction strengths are dependent on the environmental context. Several studies, including those described in Figure 16.19 and Ecological Toolkit 16.1, have found context-dependent variation in the degree to which species behave as keystone species. Thus, while the keystone species concept is intuitively simple, predicting when and where a particular species will behave as a keystone species remains a challenge.

One reason it remains difficult to predict the strength of trophic interactions is that the ecological importance of a keystone predator such as *Pisaster* manifests itself not only through one strong link, such as that between *Pisaster* and mussels, but also through strong indirect effects (see Figure 16.11), such as the effects *Pisaster* has on other species by reducing the abundance of mussels. If *Pisaster* consumed only the species that are inferior competitors for space (such as barnacles), it would not play a keystone role in the rocky intertidal community. Thus, predicting the effects of species losses on the remaining community requires an understanding of not only the strengths of individual links, but also the strengths of chains of indirect effects.

Does complexity enhance stability in food webs?

Ecologists have pondered whether more complex food webs—those with more species and more links among them—are more stable than simpler food webs with lower diversity and fewer links. Stability, in this context, is usually evaluated by the magnitude of changes in the population sizes of the organisms in the food web over time. Stability may also be expressed through ecosystem processes such as primary production. As we saw in Chapter 11, large oscillations in population size over time increase the susceptibility of species to local extinction. Thus, a less stable food web means a greater

potential for extinction of its component species. The question of stability is taking on ever greater importance with increasing rates of biodiversity loss and non-native species invasions worldwide, which have significant implications for ecosystem function.

Early proponents of the idea that food web complexity increases stability based their arguments on observations of real trophic interactions as well as on intuition. Ecologists such as Charles Elton and Eugene Odum argued that simpler, less diverse food webs should be more easily perturbed, experience larger changes in species population densities, and experience greater species losses as a result. More rigorous mathematical analyses of food webs, however, provided a contrary view. Robert May (1973) used food webs made up of random assemblages of organisms to demonstrate that food webs with higher diversity are less stable than those with lower diversity. The instability in May's models resulted from accentuation of population fluctuations by strong trophic interactions: the more interacting species there were, the more likely that their population fluctuations would reinforce one another, leading to the extinction of one or more of the species.

May's work overturned the notion that more complex systems are inherently more stable than simpler ones. Yet anyone visiting a tropical rainforest or a coral reef can attest to the fact that highly diverse, productive, and complex communities do persist in nature. Therefore, much ecological research has been devoted to discovering the factors that allow naturally complex food webs to be stable. More recent models, for example, have incorporated distributions of interaction strengths more closely resembling those observed in nature. These models and experiments suggest that, while more complex systems are not necessarily more stable, some natural food webs may have a particular structure or organization that allows increased species diversity to have a stabilizing effect. Other studies suggest that the buffering influence of weak interactions (McCann et al. 1998; Neutel et al. 2002) and of behavioral or evolutionary changes in prey choice (Kondoh 2003) can help to reduce the population fluctuations associated with complex food webs. Additionally, the identity of the species in a food web is important to its behavior, with some species exerting a disproportionally greater influence on stability, and others being more likely to go extinct (Lawler 1993).

How diversity at one trophic level affects the stability of populations at other trophic levels has also been of interest to ecologists, particularly in the context of biodiversity loss (as we will see in Concept 23.3). Elton (1958) proposed that plant diversity influences diversity at higher trophic levels, with greater plant diversity stabilizing animal populations. We saw in Concept 19.4 that plant production is often higher in more diverse communities and that more diverse plant communities are better able

FIGURE 21.18 **Plant Diversity and Stability in Food Webs** Greater plant diversity enhanced the stability of arthropod communities in experimental plots. The potential mechanisms of this effect include greater and more stable plant biomass. Plant diversity, which is associated with greater habitat complexity, may be associated with greater abundance and diversity of predators, which may lead to greater top-down influences on herbivores and plants (trophic cascades). In addition, plant diversity enhances the diversity of the arthropod community as a whole, enhancing portfolio effects, which keep overall abundance stable. (After N. M. Haddad et al. 2011. *Ecol Lett* 14: 42–46.)

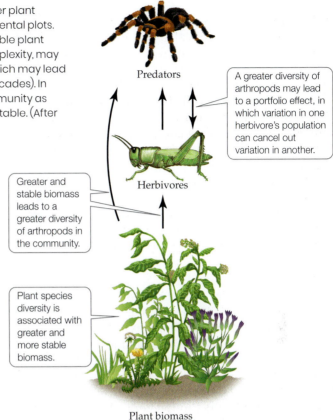

Predators

A greater diversity of arthropods may lead to a portfolio effect, in which variation in one herbivore's population can cancel out variation in another.

Herbivores

Greater and stable biomass leads to a greater diversity of arthropods in the community.

Plant species diversity is associated with greater and more stable biomass.

Plant biomass

to recover from disturbances. Do these properties convey greater stability to higher trophic levels? Nick Haddad and colleagues set out to test this hypothesis, using experimental prairie plots established by David Tilman at the Cedar Creek Ecosystem Science Reserve in Minnesota (Haddad et al. 2011). They studied the abundance and species composition of arthropod (primarily insect and spider) communities in plots with 1, 2, 4, 8, or 16 plant species over an 11-year period. A total of 733 different arthropod species were sampled during this period. These arthropods were divided into communities by their feeding preferences, which included detritivores, herbivores, predators, and parasitoids. Stability was evaluated by the amount of change in the abundances of individuals within populations and communities.

Haddad et al. found that, in general, the arthropod communities were more stable in the plots with higher plant diversity. However, not all arthropod communities exhibited the same relationship between plant diversity and stability. Populations of specialist herbivores (those that eat one or a few species of plants) had lower stability with increases in plant diversity. In contrast, the community of all herbivores showed greater stability with increasing plant diversity. The researchers suggested that the underlying mechanisms by which plant diversity influences arthropod community stability include greater and more stable plant biomass and increased diversity in the arthropod communities (**FIGURE 21.18**). Higher plant diversity was linked to greater predator abundance and diversity through its influence on habitat diversity. These predators may exert top-down effects on herbivore and plant abundances (trophic cascades). Haddad et al. also suggested that community stability is enhanced by a *portfolio effect*, in which variation in the population of one species cancels out variation in another such that overall abundance in the community remains the same. Greater diversity among the arthropods would lead to a greater probability of the portfolio effect. The researchers concluded that plant diversity in the prairie ecosystem provides services to humans not only in the form of potential biofuels, but also by keeping arthropod communities more stable and preventing outbreaks of insects that can be problematic for crops and forests.

A CASE STUDY REVISITED

Toxins in Remote Places

Knowledge of how energy flows through the trophic levels of ecosystems is key to understanding the environmental effects of persistent organic pollutants like those described in this chapter's Case Study. Some chemical compounds taken up by organisms, either directly from the environment or by consumption with their food, can become concentrated in their tissues. For a variety of reasons, these compounds are not metabolized or excreted, so they become progressively more concentrated in the body over the organism's lifetime, a process known as **bioaccumulation**. Bioaccumulation can lead to increasing tissue concentrations of these compounds in animals at successively higher trophic levels as animals at each trophic level consume prey with higher concentrations of the compounds. This process is known as **biomagnification** (**FIGURE 21.19**). The POPs we discussed at the beginning of this chapter are particularly susceptible to these processes.

The potential dangers associated with bioaccumulation and biomagnification of POPs were well publicized by Rachel Carson's book *Silent Spring*, published in 1962,

Cormorant, © Brian Lasenby/Shutterstock; Carp, © Vladimir Wrangel/Shutterstock

FIGURE 21.19

Bioaccumulation and Biomagnification Levels of mercury (a toxic heavy metal) show bioaccumulation and bio-magnification in a Czech pond ecosystem. (After P. Houserová et al. 2007. *Environ Pollut* 145: 185–194.)

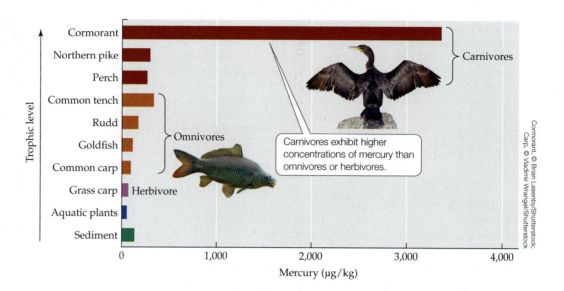

Carnivores exhibit higher concentrations of mercury than omnivores or herbivores.

in which she described the devastating effects that pesticides, particularly DDT, were having on nontarget bird and mammal populations. DDT was thought of as a "miracle" insecticide during the 1940s and 1950s, when it was widely used to control a variety of crop and garden pests and disease vectors. However, DDT was also accumulating in higher-level predators as a result of biomagnification, and it contributed to the near extinction of some birds of prey, including the peregrine falcon and the bald eagle. In *Silent Spring*, Carson described the persistence of DDT in the environment, its accumulation in the tissues of consumers, including humans, and its health hazards. Because of Carson's careful documentation and her ability to convey her message in a well-crafted manner that could be appreciated by the general public, *Silent Spring* led to increased scrutiny of the use of chemical pesticides, which eventually resulted in a ban on the manufacture and use of DDT in the United States.

The concept of biomagnification led researchers to suspect that the high concentrations of POPs found in the Inuit resulted from their position at the highest trophic levels of the Arctic ecosystem. This suspicion was reinforced by comparisons of the concentrations of toxins among different Inuit communities. The highest concentrations of toxins were found in communities that consumed marine mammals such as whales, seals, and walruses—animals that occupy the third, fourth, or fifth trophic levels. Inhabitants of communities where herbivorous caribou (at the second trophic level) were a more important part of the diet had lower concentrations of toxins. The Inuit preference for foods rich in fatty tissues, such as whale blubber (muktuk), poses a problem as well because many POPs are preferentially stored in the fatty tissues of animals.

Although emissions of some POPs and other pollutants are declining globally as awareness of their effects increases and regulations are put in place, the potential for long-term persistence of these compounds in the Arctic

environment means that their effects may not disappear any time soon (Pearce 1997). While the cold temperatures and relatively low light levels in the Arctic limit the chemical breakdown of POPs, their concentrations have gradually decreased in lake sediments. There has also been a gradual decline in the concentrations of some POPs and heavy metals in the blood of Inuit individuals, but new emerging POPs and mercury continue to be a concern for public health. While switching to alternative food sources might seem to be a potential solution to the problem, the cultural identity of the Inuit is strongly associated with their hunting traditions and their diet, and they would be unlikely to make such a switch easily.

CONNECTIONS *in* NATURE

BIOLOGICAL TRANSPORT OF POLLUTANTS Pollutants have been reported in almost all environments on Earth—even Antarctic ice holds trace amounts of DDT and lead emitted from the burning of leaded gasoline. Animals in many remote areas have high concentrations of industrial and agricultural toxins in their tissues. Fish in isolated alpine lakes of the Canadian Rockies, for example, contain high concentrations of POPs, which have been associated with condensation of these compounds in snowfields and glaciers above the lakes (Blais et al. 1998). As suggested in the Case Study Revisited, the concentrations of these pollutants are related to the trophic positions of the animals: consumers at the highest trophic levels, such as polar bears, seals, and birds of prey, contain the highest concentrations. The widespread nature of this problem underscores the notion that ecosystems are connected by the movements of energy and materials among them. Ecological processes in one ecosystem can have effects on other ecosystems through these movements (Polis et al. 2004).

The movement of POPs and other human-made toxins is usually associated with atmospheric transport from low

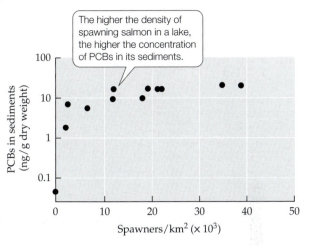

FIGURE 21.20 Biological Pumping of Pollutants   Spawning salmon act as biological pumps, concentrating toxins from the oceans in their bodies and transporting them en masse to freshwater ecosystems. (After E. M. Krümmel et al. 2003. *Nature* 425: 255–256.)

to high latitudes. However, the behaviors of animals can also influence the movement of POPs. Salmon, for example, have been shown to transport nutrients from marine to freshwater and terrestrial ecosystems during their spawning runs. At reproductive maturity, salmon leave the ocean and move up rivers in large numbers, as described in the Case Study in Chapter 2. From the rivers, they move into freshwater lakes and streams, where they spawn and then die. The potential exists for salmon to move toxins, as well as nutrients, from the oceans to freshwater ecosystems via this spawning behavior.

E. M. Krümmel and colleagues studied the potential for spawning sockeye salmon (*Oncorhynchus nerka*) to act as a "fish pump" for pollutants by moving PCBs from the ocean to remote lakes in Alaska (Krümmel et al. 2003). Salmon occupy the fourth trophic level, and thus, through bioaccumulation and biomagnification, they accumulate PCBs in their body fat at concentrations more than 2,500 times higher than those found in seawater. Krümmel and

colleagues collected sediment cores from eight lakes in southwest Alaska that had different densities of spawning salmon (ranging from 0 to 40,000 spawners/km²) and measured PCBs in the sediments. They found that the concentrations of PCBs were strongly correlated with the density of spawners (**FIGURE 21.20**). Lakes that did not have visits from spawning fish had concentrations of PCBs similar to expectations based on atmospheric transport alone. The lake with the highest density of spawning fish (40,000 per km²) had PCB concentrations that were six times higher than the levels associated with atmospheric transport. A similar study found that DDT, other POPs, and mercury are transported by northern fulmars (*Fulmarus glacialis*, pelagic fish-eating seabirds) from the ocean to small ponds near their nesting colonies (Blais et al. 2005). These examples demonstrate how the behaviors of some species (spawning in fish, colonial nesting in birds) can exacerbate problems of pollution associated with biomagnification in ecosystems.

SUMMARY

CONCEPT 21.1 Trophic levels describe the feeding positions of groups of organisms in ecosystems.

21.1.1 Describe how energy flow among trophic levels in an ecosystem is related to the food selection of consumers.

- An organism's trophic level is determined by the number of feeding steps by which it is removed from the first trophic level, which contains autotrophs and detritus.

- Omnivores feed at multiple trophic levels, although their diets can be partitioned to reflect their consumption at each level.

21.1.2 Explain how both primary production and detritus can be at the base of food chains.

- All organisms eventually end up as food for other organisms or as detritus.

- Energy from both primary production and detritus supports the second trophic level of herbivores and detritivores, respectively.

21.1.3 Evaluate how terrestrial detrital energy inputs from outside an ecosystem (allochthonous) would change in a river from its source to where it reaches the ocean.

(Continued)

SUMMARY (continued)

- Allochthonous sources of energy dominate in the higher portions of river systems, with autochthonous energy sources becoming increasingly important as the rivers become larger and flow velocity decreases.

CONCEPT 21.2 The amount of energy transferred from one trophic level to the next depends on food quality and on consumer abundance and physiology.

21.2.1 Describe how the relationship between biomass and energy is influenced by the life span of primary producers.

- Trophic energy and biomass pyramids portray the relative amounts of energy and biomass at different trophic levels.

21.2.2 Summarize the factors that may influence why a greater proportion of net primary production is consumed in aquatic ecosystems relative to terrestrial ecosystems.

- The high turnover of autotroph biomass in aquatic ecosystems can result in biomass pyramids that are inverted relative to energy pyramids.

- The proportion of autotroph biomass consumed in terrestrial ecosystems tends to be lower than that in aquatic ecosystems.

21.2.3 Evaluate how consumer thermal physiology, body size, diet preference, and digestive specialization can determine trophic efficiency.

- The efficiency of energy transfer from one trophic level to the next is determined by food quality and the physiology of consumers.

CONCEPT 21.3 Changes in the abundances of organisms at one trophic level can influence energy flow at multiple trophic levels.

21.3.1 Compare the factors that influence energy flow in an ecosystem between bottom-up and top-down perspectives.

- Changes in the numbers and types of consumers at higher trophic levels can influence primary production through influences on the consumption of herbivores.

21.3.2 Describe how changes in the abundance of organisms at the fourth trophic level may impact rates of primary production through a trophic cascade.

- Trophic cascades tend to be more apparent in aquatic ecosystems than in terrestrial ecosystems,

but they have been demonstrated in complex terrestrial ecosystems as well.

21.3.3 Explain how the size of an ecosystem, the rates of disturbance, and the amount of primary production can influence the number of trophic levels.

- The number of trophic levels that can be sustained in an ecosystem is determined by the size of the ecosystem, the amount of energy entering the ecosystem through primary production, and the frequency of disturbances.

CONCEPT 21.4 Food webs are conceptual models of the trophic interactions of organisms in an ecosystem.

21.4.1 Explain how food webs are helpful for envisioning ecosystem energy flow, and outline the factors that compromise their accuracy in portraying the full extent of interactions among organisms.

- Food webs are diagrams that portray the diverse trophic interactions among species in an ecosystem.

21.4.2 Describe how the use of interaction strengths can aid in the construction of more accurate food web models.

- Although trophic interactions are extremely complex, food webs can be simplified by focusing on the strongest interactions among the component organisms.

- Keystone species have greater effects on energy flow and community composition than their abundance or biomass would predict.

- Indirect effects of a predator on a target prey species, including its effects on other species that compete with, facilitate, or modify the environment of the target species, can offset or reinforce the direct effects of predation on the target species. These indirect effects may have stabilizing effects on inherently unstable food webs.

21.4.3 Summarize how ecologists have viewed the relationship between the complexity of food webs and the stability of associated communities and ecosystem processes.

- Although early considerations of food web complexity predicted that greater complexity confers more stability to food webs, models indicated the opposite. Recent modeling incorporating interactions' strengths more closely resembles patterns found in nature.

Review Questions

1. Suppose one population of coyotes (population A) demonstrates a greater degree of omnivory than another population (population B). Population A relies on a diet that includes road-killed animal carcasses, plants, and rotten food from dumpsters, while population B has a steady diet of small rodents. Which population should have a higher assimilation efficiency, and why?

2. Mammals in temperate terrestrial and temperate marine ecosystems occupying similar trophic levels may have different production efficiencies. Assuming similar food

quality, food abundance, and food capture rates, explain why the production efficiencies of these mammals would differ between a marine ecosystem and a terrestrial ecosystem. (Hint: Consider how the mammals maintain their body heat, as well as the temperature variation of their environments as described in Chapter 2.)

3. Which ecosystem would you expect to have a greater total amount of energy passing through its trophic levels: a lake or a forest adjacent to the lake? Which of these ecosystems would have a higher proportion of NPP moving through all of its trophic levels, the forest or the lake?

Hone Your Problem-Solving Skills

Generalist herbivores, which are often insects, consume a greater number of plant species than specialist herbivores do. Specialist herbivores are often protected against predation by acquiring protective chemicals from their plant food sources. Thus, dietary specialization may have consequences for the impact of herbivory on the flow of energy and nutrients in ecosystems.

1. Would you expect that a trophic cascade would have a greater or lesser effect on herbivory and NPP if only specialist herbivores were present? Assume a high diversity of plant species. Provide your answer in the form of a prediction, and describe an experiment in which you could test this hypothesis.

2. Michael Singer and colleagues investigated the influence of predatory birds on caterpillars and the subsequent effect on plant damage through herbivory in a deciduous forest ecosystem (Singer et al. 2014). They manipulated the presence and absence of birds (third trophic level) using exclosures, manipulated the proportion of specialist and generalist caterpillars (second trophic level), and measured abundances of the caterpillars and levels of damage to trees. **FIGURE A** shows the impact of bird predation on damage to the trees, expressed as relative to controls with no bird predation (zero point). Negative values indicate less herbivore damage to trees as a result of the bird predation; positive values indicate more herbivore damage to trees as a result of bird predation. **FIGURE B** shows the effect of predation on abundance of herbivores according to whether they are generalists (G) or specialists (S). Error bars in Figure B show ± one SE of the mean.

(A)

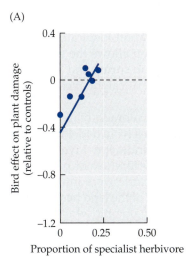

(B)

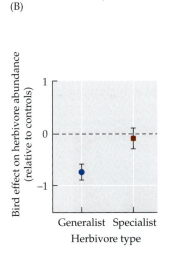

a. How do these results support or refute the hypothesis you derived in Question 1?

b. What factors would have contributed to the observed result?

22

Nutrient Supply and Cycling

A Fragile Crust: A Case Study

The Colorado Plateau in western North America includes vast expanses of isolated mountains, intricately folded sandstone formations, and deeply cut, multicolored canyons. One of the most unusual features found in this rugged and beautiful region, however, occurs at a very small scale: its patchy cover of dark, convoluted soil (**FIGURE 22.1**). On closer examination, the soil looks like a miniature landscape of hills and valleys, covered with black, dark green, and white splotches resembling lichens. The comparison is apt, because this crust on the soil surface, known simply as a **biological soil crust** (or **biocrust**), is composed of a mix of hundreds of species of cyanobacteria, lichens, and mosses (Belnap 2003). Approximately 70% of the soils on the Colorado Plateau, a geographic province that covers parts of Utah, Arizona, Colorado, and New Mexico, have some biocrust development. Similar crusts, containing a surprisingly similar suite of species, cover approximately 12% of terrestrial ecosystems globally (Rodriguez-Caballero et al. 2018). The crusty nature of the soil is largely the work of filamentous cyanobacteria, which create a sheath of mucilaginous material as they move through the soil after a rain. When the soil dries out, the cyanobacteria withdraw to deeper soil layers, leaving behind the sheathing material, which helps bind the coarse soil particles together (**FIGURE 22.2**).

The soils of the Colorado Plateau are exposed to tremendous climate variation and strong erosive forces (Belnap 2003). Surface temperatures can range from –20°C (–4°F) in winter to 70°C (158°F) in summer. High evapotranspiration rates often dry out the soils, and the sparseness of the vegetation allows the strong surface winds to carry away fine soil particles. Precipitation in spring and summer often occurs as brief, intense thunderstorms. Biocrusts are critical for anchoring the soil in place in the face of high winds and torrential rains.

Although the Colorado Plateau is sparsely populated, humans have had a large and lasting effect on its landscape. Livestock grazing has been an important use of public lands in the region since cattle were introduced there in the 1880s. Most of the land has been affected to some degree by grazing, which has resulted in the trampling of biocrusts and overgrazing of vegetation. Until recently, grazing was the most important human-associated disturbance

Courtesy of William Bowman

FIGURE 22.1 Biological Soil Crust on the Colorado Plateau Biological soil crusts are a common feature in the deserts of the Colorado Plateau. The surface topography and coloration of the crust are clearly visible in this photo.

(A)

(B)

Both images courtesy of Jayne Belnap/USGS

15 μm

150 μm

FIGURE 22.2 **Cyanobacterial Sheaths Bind Soil into Crusts** (A) Cyanobacterial strands surround themselves with a sheath of mucilaginous material as they move through the soil. (B) The sheaths left behind by the cyanobacteria help to bind soil particles together and protect soils from erosional loss.

in the region. However, an increasing number of off-road vehicles have been using the region. Each year around 3 million people visit Moab, Utah, a town with a year-round population of 5,400. Many of these visitors use all-terrain vehicles, motorcycles, mountain bikes, or hiking to visit the surrounding red rock country. The majority of these users of the desert backcountry obey federal and local laws, staying on designated trails and roads. Unfortunately, some users drive their vehicles off designated roads and across soils covered with biocrusts. The extraction of fossil fuels has also increased substantially over the past several decades with the advent of new technologies (e.g., hydraulic fracturing, or "fracking").

While the spatial extent of soil surface disturbance associated with off-road vehicle use, fossil fuel extraction, and livestock grazing has not been well quantified, it is clear that a large part of the landscape has been disturbed to some degree during the past 150 years, and that the rate of disturbance is increasing. The recovery of biocrusts following disturbance is extremely slow in arid environments: decades are required for the reestablishment of the cyanobacteria and up to centuries for recolonization by lichens and mosses (Belnap and Eldridge 2001).

What are the implications of the loss of biocrusts for the functioning of desert ecosystems? How important are they to the supply of nutrients in those ecosystems? Given the long-term nature of disturbances associated with livestock grazing across the Colorado Plateau, can we still find areas that can serve as controls for studies of the disturbance that has already occurred?

Introduction

In addition to energy, all organisms require specific chemical elements to function and grow. Organisms get these elements by absorbing them from the environment or by consuming other organisms, living or dead. Iron, for example, is needed by all organisms for several important metabolic functions, but how those organisms get their iron and where it comes from vary substantially. Phytoplankton in the Atlantic Ocean may take up iron that came from dust that blew in from the Sahara Desert. Lions in African savannas get their iron from the prey they kill and consume. Aphids get their iron in the sap they suck from a plant, whereas the plant takes up water containing dissolved iron from the soil. The ultimate source of all of this iron, however, is solid minerals in Earth's crust, which are subjected to chemical transformations as they move through the different physical and biological components of ecosystems.

The study of the physical, chemical, and biological factors that influence the movements and transformations of elements is known as **biogeochemistry**. An understanding of biogeochemistry is important for determining the availability of **nutrients**, which are defined as the chemical elements an organism requires for its metabolism and growth. Nutrients must be present in certain chemical forms to be available for uptake by organisms. The rate at which physical and chemical transformations occur determines the supply of nutrients, which in turn influences organismal function and population growth. Biogeochemistry also encompasses the study of non-nutrient elements that can serve as tracers in ecosystems and of pollutant compounds, such as persistent organic pollutants and heavy metals, that cause environmental damage. Biogeochemistry is a discipline that integrates contributions from soil science, hydrology, and atmospheric science as well as ecology.

In this chapter, we will consider the biological, chemical, and physical factors that control the supply and availability of nutrients in ecosystems. We will emphasize nutrient requirements and acquisition by autotrophs because they in turn are the principal source of nutrients for heterotrophs. We will describe what nutrients are most important, their sources, and how they enter ecosystems, and review some of the important chemical and biological transformations that constitute the cycling of nutrients in ecosystems. In Concept 25.1, we will consider the global-scale cycling of some of these elements.

> **CONCEPT 22.1** Nutrients enter ecosystems through the chemical breakdown of minerals in rocks or through fixation of atmospheric gases.

LEARNING OBJECTIVES

22.1.1 Describe the basic roles of nutrients in organisms, and differentiate between the ways in which microorganisms, plants, and animals obtain them.

22.1.2 Summarize the steps involved in the breakdown of minerals and the subsequent release of nutrients.

22.1.3 Describe the physical and chemical properties of soil that influence its fertility.

22.1.4 Explain the processes that fix carbon and nitrogen, converting them into usable forms for organismal function and growth.

Nutrient Requirements and Sources

All organisms, from bacteria to blue whales, share similar nutrient requirements. How those nutrients are obtained, the chemical forms of those nutrients that are taken up, and the relative amounts of those nutrients that are required vary greatly among organisms. All nutrients, however, come from common sources: inorganic mineral forms that are present in Earth's crust or as gases in the atmosphere.

Organisms have specific nutrient requirements

An organism's nutrient requirements are related to its physiology. The amounts and specific nutrients needed therefore vary according to the organism's mode of energy acquisition (autotrophs vs. heterotrophs), mobility, and thermal physiology (ectotherms vs. endotherms). Mobile animals, for example, generally have

higher rates of metabolic activity than plants or bacteria, and they therefore have higher requirements for nutrients such as nitrogen (N) and phosphorus (P) to support the biochemical reactions associated with movement. Differences in nutrient requirements are reflected in the chemical composition of organisms (**TABLE 22.1**). Carbon is often associated with structural compounds in plant cells and tissues, while nitrogen is largely found in enzymes. Accordingly, the ratios of carbon to nitrogen (C:N) in organisms can indicate the relative concentrations of biochemical machinery in cells. Animals and microorganisms typically have lower C:N ratios than plants: for example, humans and bacteria have C:N ratios of 6.0 and 3.0, respectively, whereas those of plants range from 10 to 40. This difference is one reason why herbivores must consume more food than carnivores to acquire enough nutrients to meet their nutritional demands.

The nutrients essential for all plants, and the functions associated with them, are presented in **TABLE 22.2**. Some plant species have specific requirements for other nutrients not found in Table 22.2. For example, many, but not all, C_4 and CAM plants (see Concept 5.3 for discussion of these photosynthetic pathways) require sodium, while

TABLE 22.1

Elemental Composition of Organisms (as Percentage of Dry Mass)

Element (symbol)	Bacteria (in general)	Plant (corn, *Zea mays*)	Animal (human, *Homo sapiens*)
Oxygen (O)	20	44.43	14.62
Carbon (C)	50	43.57	55.99
Hydrogen (H)	8	6.24	7.46
Nitrogen (N)	10	1.46	9.33
Silicon (Si)		1.17	0.005
Potassium (K)	1–4.5	0.92	1.09
Calcium (Ca)	0.01–1.1	0.23	4.67
Phosphorus (P)	2.0–3.0	0.20	3.11
Magnesium (Mg)	0.1–0.5	0.18	0.16
Sulfur (S)	0.2–1.0	0.17	0.78
Chlorine (Cl)		0.14	0.47
Iron (Fe)	0.02–0.2	0.08	0.012
Manganese (Mn)	0.001–0.01	0.04	—
Sodium (Na)	1.3	—	0.47
Zinc (Zn)		—	0.01
Rubidium (Rb)		—	0.005

Sources: (Bacteria) S. Aiba et al. 1973. *Biochemical Engineering*. 2nd ed. Academic Press: New York; (Plant and Animal) E. Epstein and A. J. Bloom. 2005. *Mineral Nutrition of Plants: Principles and Perspectives*. 2nd ed. Oxford University Press/Sinauer: Sunderland, MA; based on E. C. Miller. 1938; P. B. Hawk and B. L. Oser. 1965 (cited within).

Note: Dashes indicate a negligible amount of an element; blank spaces indicate that the element has not been measured.

TABLE 22.2

Plant Nutrients and Their Principal Functions

Nutrients	Principal functions
Carbon, hydrogen, oxygen	Components of organic molecules
Nitrogen	Component of amino acids, proteins, chlorophyll, nucleic acids
Phosphorus	Component of ATP, NADP, nucleic acids, phospholipids
Potassium	Ionic/osmotic balance, pH regulation, regulation of guard cell turgor
Calcium	Cell wall strengthening and functioning, ionic balance, membrane permeability
Magnesium	Component of chlorophyll, enzyme activation
Sulfur	Component of amino acids, proteins
Iron	Component of proteins (e.g., heme groups), oxidation–reduction reactions
Copper	Component of enzymes
Manganese	Component of enzymes, activation of enzymes
Zinc	Component of enzymes, activation of enzymes, component of ribosomes, maintenance of membrane integrity
Nickel	Component of enzymes
Molybdenum	Component of enzymes
Boron	Cell wall synthesis, membrane function
Chlorine	Photosynthesis (water splitting), ionic and electrochemical balance

Sources: F. B. Salisbury and C. Ross. 1992. *Plant Physiology*, 4th ed. Wadsworth: Belmont, CA; H. Marschner. 1995. *Mineral Nutrition of Higher Plants*. Academic Press: San Diego, CA.

most plants do not. In contrast, sodium is an essential nutrient for all animals, critical for maintaining pH and osmotic balances. Cobalt is required by some plants that host nitrogen-fixing symbionts (discussed later in this section). Selenium is toxic to most plants, but a small number of plants growing on soils rich in selenium may require it. In contrast, selenium is an essential nutrient for animals and bacteria.

Plants and microorganisms usually take up nutrients from their environment in relatively simple, soluble chemical forms, from which they synthesize the larger molecules needed for their metabolism and growth. Animals, on the other hand, typically take up their nutrients through the consumption of living organisms or detritus, obtaining their nutrients in larger, more complex chemical compounds. Animals break down some of these compounds and resynthesize new molecules; others are absorbed intact and used directly in biosynthesis. For example, 9 of the 20 amino acids that are essential for metabolism in humans and other mammals must be absorbed intact, since we cannot synthesize them ourselves.

Minerals and atmospheric gases are the ultimate sources of nutrients

All nutrients are ultimately derived from two abiotic sources: minerals in rocks and gases in the atmosphere. Over time, as nutrients are taken up and incorporated by organisms, they accumulate in ecosystems in organic forms (i.e., in association with carbon and hydrogen molecules). Nutrients may be cycled within an ecosystem, repeatedly passing through organisms and the soil or water in which the organisms live. They may even be cycled internally within an organism, stored or mobilized for use as its needs for specific nutrients change. Here we describe the inputs of nutrients into ecosystems from minerals and the atmosphere. In the following sections, we will complete the steps that constitute nutrient cycling within an ecosystem.

MINERAL SOURCES OF NUTRIENTS   The breakdown of minerals in rock supplies ecosystems with nutrients such as potassium, calcium, magnesium, and phosphorus. *Minerals* are solid substances with characteristic chemical properties, derived from a multitude of geologic processes. *Rocks* are collections of different minerals. Nutrients and other elements are released from minerals in a two-step process known as **weathering**. The first step, **mechanical weathering**, is the physical breakdown of rocks. Expansion and contraction processes, such as freeze–thaw and drying–rewetting cycles, act to break rocks into progressively smaller particles. Gravitational mechanisms (such as landslides) and the growth of plant roots also contribute to mechanical weathering. Mechanical weathering exposes greater amounts of surface area of mineral particles to **chemical weathering**, in which the minerals are subjected to chemical reactions that release soluble forms of nutrients.

Weathering is one of the processes involved in soil development. **Soil** is formally defined as a mix of mineral particles; solid organic matter (detritus, primarily decomposing plant matter); water containing dissolved organic matter, minerals, and gases (the *soil solution*); and organisms. Soils have several important properties that influence the delivery of nutrients to plants and microorganisms. One property is their texture, which is defined by the sizes of the particles that make up the soil. The coarsest soil particles (0.05–2 mm) are referred to as **sand**. Intermediate-sized particles (0.002–0.05 mm) are called **silt**. Fine soil particles (<0.002 mm), known as **clays**, have weak negative charges on their surfaces that can hold on to cations and exchange them with the soil solution. As a result, clay particles serve as a reservoir of nutrient cations such as Ca^{2+}, K^+, and Mg^{2+}. A soil's ability to hold these cations and exchange them with the soil solution, referred to as its **cation exchange capacity**, is determined by the amounts and types of clay the soil contains. Soil texture also influences the soil's water-holding capacity and thus the movement of nutrients in the soil solution. Soils with a high proportion of sand have a large volume of spaces between particles. These spaces (called *macropores*) allow water to drain through the soil and limit the amount of water it can hold.

Another important property of a soil influencing its texture and chemistry is its **parent material**, the rock or mineral material that was broken down by weathering to form that soil. Parent material for soil is usually the underlying bedrock but may also include thick layers of sediment deposited by glaciers (known as **till**), by wind (**loess**), or by water (**alluvium**). The chemistry and structure of the parent material, along with the physical environment, determine the rate of weathering and the amount and types of nutrients released, and they thus influence the fertility of the soil. Limestone, for example, is high in the nutrient cations Ca^{2+} and Mg^{2+}. Soils derived from more acidic parent material, such as granite, have lower concentrations of these elements. In addition, the higher acidity (lower pH) of soils derived from granite lowers the availability of nitrogen and phosphorus to plants.

The chemistry and pH of the parent material exert an important influence on the abundance, growth, and diversity of plants in ecosystems. For example, Laura Gough and colleagues (2000) demonstrated that variation in the acidity of the parent material is associated with differences in plant species richness among Arctic ecosystems in Alaska. They surveyed Arctic vegetation across natural gradients in soil acidity associated with the differential distribution of calcium-rich loess, which has lower acidity than other parent materials. They found that the

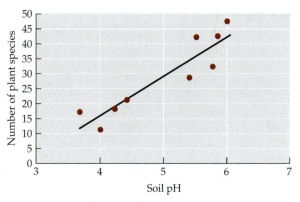

FIGURE 22.3 Species Richness Increases with Decreasing Soil Acidity Vascular plant species richness in the Alaskan Arctic tundra varies with soil acidity. The gradient in soil acidity is primarily due to differences in parent material: less acidic soils (with higher pH) are associated with greater loess deposits. (After L. Gough et al. 2000. *J Ecol* 88: 54–66.)

number of plant species increased as acidity decreased (**FIGURE 22.3**). This variation in diversity was attributed to the negative effects of soil acidity on nutrient availability as well as its inhibitory effects on plant establishment.

Over time, soils undergo changes associated with weathering, accumulation and chemical alteration of organic matter, and **leaching** (the movement of dissolved organic matter and fine mineral particles from upper to lower layers). These processes form **horizons**, layers of soil distinguished by their color, texture, and permeability (**FIGURE 22.4**). Variations in soil horizons are used by soil scientists to characterize different soil types.

Climate influences the rates of many of the processes associated with soil development, including weathering, biological activity (such as the input of organic matter from net primary production [NPP] and its decomposition in the soil), and leaching. In general, these processes occur most rapidly under warm, wet conditions. Thus, the

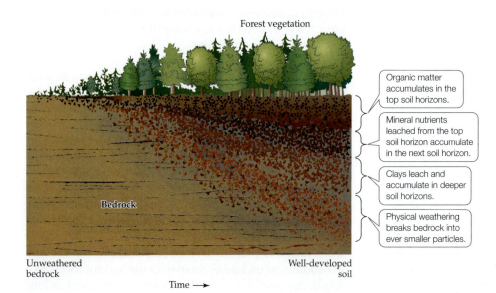

Forest vegetation

Organic matter accumulates in the top soil horizons.

Mineral nutrients leached from the top soil horizon accumulate in the next soil horizon.

Clays leach and accumulate in deeper soil horizons.

Physical weathering breaks bedrock into ever smaller particles.

Bedrock

Unweathered bedrock

Well-developed soil

Time →

FIGURE 22.4 Development of Soil Horizons Soils develop over time as parent material is weathered and broken up into ever finer soil particles, increasing amounts of organic matter accumulate in the soil, and materials are leached and deposited in deeper soil layers. The rate of soil development is dependent on the climate, the parent material, and the organisms associated with the soil. (After N. C. Brady and R. R. Weil. 2001. *The Nature and Property of Soils*. Prentice-Hall: Upper Saddle River, NJ.)

? Given what you've learned about primary production in Chapter 20 and about the climate factors that determine weathering and soil development in this chapter, what do you think the horizons of a desert soil would look like?

soils of lowland tropical forest ecosystems, which have experienced high rates of weathering and leaching for a long time, are poor in mineral-derived nutrients such as calcium and magnesium. A high proportion of the nutrients in lowland tropical forest ecosystems are found in the living biomass of trees, in contrast to most other terrestrial ecosystems, in which these nutrients are mostly found in the soil. When lowland tropical forests are cleared and burned to make way for pastures or cropland, most of the nutrients are lost in smoke and ash and through soil erosion following the fires. As a result, these ecosystems may become severely nutrient-impoverished, and it may take them centuries to return to their previous state. Soils in higher-latitude ecosystems have lower leaching rates and are usually richer in mineral-derived nutrients.

Organisms—primarily plants, bacteria, and fungi—influence soil development by contributing organic matter, which is an important reservoir of nutrients such as nitrogen and phosphorus. Organisms also increase rates of chemical weathering through the release of organic acids (from plants and detritus) and CO_2 (from metabolic respiration). Thus, rates of biological activity have a strong influence on the development of soils.

ATMOSPHERIC SOURCES OF NUTRIENTS The atmosphere is composed of 78% nitrogen (as dinitrogen gas, N_2), 21% oxygen, 0.9% argon, increasing amounts of carbon dioxide (0.041%, or 416 parts per million, in 2021), and other trace gases—some natural, others pollutants derived from human activities. The atmosphere is the ultimate source of carbon and nitrogen for ecosystems. These nutrients become biologically available when they are taken up from the atmosphere and chemically transformed, or *fixed*, by organisms. They may then be transferred from organism to organism before returning to the atmosphere.

Carbon is taken up by autotrophs as CO_2 through photosynthesis and chemosynthesis. (These processes were described in Concept 5.2, and the global cycling of carbon is discussed in Concept 25.1.) Carbon compounds store energy in their chemical bonds, and they are important structural components of autotrophs (e.g., cellulose) as well.

Although the atmosphere is a huge reservoir of nitrogen, it is in a chemically inert form (N_2) that cannot be used by most organisms because of the high energy required to break the triple bond between the two atoms. The process of taking up N_2 and converting it into chemically available forms is known as **nitrogen fixation** (see Connections in Nature in Chapter 17). Biological nitrogen fixation is accomplished with the aid of the enzyme *nitrogenase*, which is synthesized by around 150 species of bacteria. Some of these nitrogen-fixing bacteria are free-living, and others are partners in mutualistic symbiotic relationships (see Figure 15.21). Nitrogen-fixing symbioses include associations between plant roots and soil bacteria, most notably between legumes and bacteria in the family Rhizobiaceae. Legumes "host" rhizobia in special root structures called nodules and supply them with carbon compounds as an energy source to meet the high energy demands of nitrogen fixation (**FIGURE 22.5**; see also Figure 17.21). In return for supplying the rhizobia with room and board, the plant gets nitrogen fixed by the bacteria. Other examples of nitrogen-fixing symbioses include associations between woody plants such as alders and bacteria in the genus *Frankia* (called actinorhizal associations), associations between *Azolla* ferns and cyanobacteria, lichens that include fungal and nitrogen-fixing symbionts, and termites with nitrogen-fixing bacteria in their guts. Humans also fix atmospheric nitrogen when they manufacture synthetic fertilizers using the Haber–Bosch process, in which ammonia is produced from atmospheric nitrogen and hydrogen under high pressures and temperatures using an iron catalyst. The Haber–Bosch process requires substantial energy input in the form of fossil fuels.

(A)

(B)

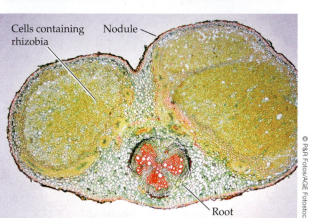

FIGURE 22.5 Legumes Form Nitrogen–Fixing Nodules
(A) These swollen nodules on the roots of a red clover (*Trifolium pratense*) plant contain nitrogen–fixing bacteria. (B) Cells inside this soybean root nodule (yellow in this micrograph) are filled with rhizobia.

Natural nitrogen fixation also requires a large amount of energy. It consumes as much as 25% of the photosynthetic energy obtained by plants with nitrogen-fixing symbiotic partners. Thus, nitrogen fixation provides these plants with a source of nitrogen, but it represents a trade-off with other energy-demanding processes such as growth, defense, and reproduction. Allocation of energy to nitrogen fixation rather than to growth lowers the ability of nitrogen-fixing plants to compete for resources other than nitrogen. Nitrogen fixation is particularly important during the early stages of primary succession, as we saw in Chapter 17.

In addition to carbon and nitrogen, the atmosphere contains fine soil particles (dust) and a collection of suspended solid, liquid, and gaseous particles known as **aerosols**. Some of this particulate matter enters ecosystems when it falls from the atmosphere because of gravity or in precipitation, a process known as **atmospheric deposition**. Atmospheric deposition represents an important natural source of nutrients for some ecosystems. Aerosols containing cations derived from sea spray, for example, may be an important source of nutrients in coastal areas. Atmospheric deposition of dust originating in the Sahara is an important input of iron into the Atlantic Ocean and of phosphorus into the Amazon Basin. On the other hand, some ecosystems have been negatively affected by atmospheric deposition associated with human industrial and agricultural activities. Acid rain, for example, is an atmospheric deposition process that has been associated with declines in forest ecosystems in the eastern United States and Europe (as we will see in Concept 25.3).

Now that we've seen how nutrients enter ecosystems, let's follow their movements within ecosystems as they are taken up and transformed. The next two sections will focus on terrestrial ecosystems; we will take a closer look at nutrient cycling in aquatic ecosystems in the final section.

CONCEPT 22.2 Chemical and biological transformations in ecosystems alter the chemical form and supply of nutrients.

LEARNING OBJECTIVES

22.2.1 Describe why decomposition is a critical process in the supply of nutrients in ecosystems.

22.2.2 Evaluate the biological and physical factors that influence rates of decomposition.

22.2.3 Explain how microbial processes may alter the chemical forms of nutrients and make them either more or less available to plants.

Nutrient Transformations

Once they have entered an ecosystem, nutrients are subjected to further modifications as a result of uptake by organisms and chemical reactions that alter their form and influence their movement and retention within the ecosystem. Foremost among these transformations is the decomposition of organic matter, which releases nutrients back into the ecosystem.

Decomposition is a key nutrient recycling process

As detritus (dead plants, animals, and microorganisms and egested waste products) builds up in an ecosystem, it becomes an increasingly important source of nutrients, particularly nitrogen and phosphorus, which often limit production of terrestrial and aquatic ecosystems. Nutrients in detritus are made available by **decomposition**, the process by which detritivores break down organic matter to obtain energy and nutrients (**FIGURE 22.6**). Decomposition releases nutrients as soluble organic and inorganic compounds that can be taken up by other organisms.

Organic matter in soil is derived primarily from plant matter, which comes from above and below the soil surface. Fresh, undecomposed organic matter on the soil surface is known as **litter** and is typically the most abundant substrate for decomposition. The litter is used by animals, protists, bacteria, and fungi as a source of energy and nutrients. As animals such as earthworms, termites, and nematodes consume the litter, they break it up into progressively finer particles. This physical fragmentation enhances the chemical breakdown of the litter by increasing its surface area.

An important final step in decomposition is the chemical conversion of organic matter into inorganic nutrients (i.e., nutrients that are not associated with carbon molecules), which is known as **mineralization**. It is the result of the breakdown of organic macromolecules in the soil by enzymes released by heterotrophic microorganisms. Because plants often rely on inorganic nutrients, ecologists use measurements of mineralization to estimate rates of nutrient supply. An understanding of the abiotic and biotic controls on decomposition and mineralization is key to understanding nutrient availability to autotrophs.

Rates of decomposition are greatly influenced by climate. Decomposition, like other biologically mediated processes, proceeds most rapidly at warm temperatures. Soil moisture also controls rates of decomposition by influencing the availability of water and oxygen to detritivores. Dry soils may not provide enough water for these organisms, and wet soils have low oxygen concentrations, which lower aerobic respiration and the rate of biological activity. Therefore, the activity of detritivores is highest at intermediate soil moistures and warm temperatures (**FIGURE 22.7**).

Some nutrients are consumed by detritivores during decomposition, so not all of the nutrients released during mineralization become available for uptake by autotrophs. The amounts of nutrients that are released from

FIGURE 22.6 Decomposition Decomposition of organic matter in the soil provides an important input of nutrients into terrestrial ecosystems. Similar steps occur in freshwater and marine ecosystems.

❓ How would the use of a nonselective pesticide (i.e., one that does not target any specific animals) to control insect herbivores affect the rate of decomposition in a lawn ecosystem?

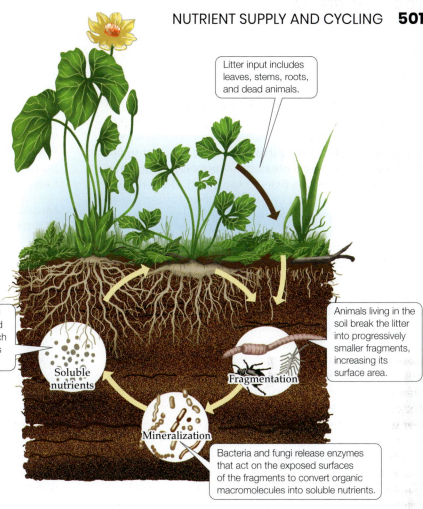

Litter input includes leaves, stems, roots, and dead animals.

Animals living in the soil break the litter into progressively smaller fragments, increasing its surface area.

Fragmentation

Small organic compounds and inorganic nutrients are released into the soil solution, from which they can be taken up by plants and microorganisms.

Soluble nutrients

Mineralization

Bacteria and fungi release enzymes that act on the exposed surfaces of the fragments to convert organic macromolecules into soluble nutrients.

organic matter during decomposition depend on the nutrient requirements of the decomposers and the amount of energy the organic matter contains. These factors can be approximated by the ratio of carbon (representing energy) to nitrogen (since nitrogen is the nutrient most often in short supply for detritivores) in the organic matter. A high C:N ratio in organic matter will result in a low net release of nutrients during decomposition, since heterotrophic microbial growth is more limited by nitrogen supply than by energy. For example, most heterotrophic microorganisms require approximately 10 molecules of carbon for every molecule of nitrogen they take up. About 60% of the carbon they take up is lost through respiration. Therefore, the optimal C:N ratio of organic matter for microbial growth is about 25:1. Organic matter with a C:N ratio greater than 25:1 would result in all of the nitrogen being taken up by the microbes during decomposition. Decomposition of organic matter with a C:N ratio of less than 25:1 would result in some nitrogen being released into the soil and made available for plants.

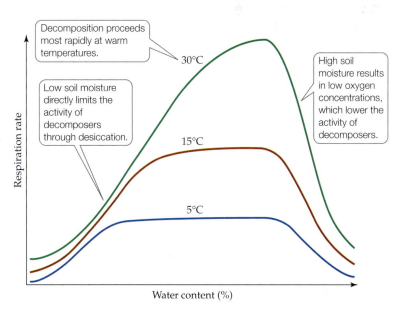

Decomposition proceeds most rapidly at warm temperatures.

Low soil moisture directly limits the activity of decomposers through desiccation.

High soil moisture results in low oxygen concentrations, which lower the activity of decomposers.

30°C

15°C

5°C

Respiration rate

Water content (%)

FIGURE 22.7 Climate Controls the Activity of Decomposers Changes in soil microbial respiration, used as an estimate of decomposition, are modeled as a function of soil moisture at different temperatures. (After E. A. Paul and F. E. Clark. 1996. *Soil Microbiology and Biochemistry.* Academic Press: San Diego, CA; F. L. Bunnell and D. E. N. Tait. 1974. In *Soil Organisms and Decomposition in Tundra.* Tundra Biome Steering Committee: Stockholm, Sweden.)

Not all of the carbon in litter is equally available as an energy source for decomposers: the chemistry of that carbon determines how rapidly the material can be decomposed. **Lignin**, a structural carbon compound that strengthens plant cell walls, is difficult for soil microorganisms to break down and thus decomposes very slowly (**FIGURE 22.8** and **ANALYZING DATA 22.1**). The rate of nutrient release from plant material containing high lignin concentrations, such as oak or pine leaves, is lower than that from material with low lignin concentrations, such as maple and aspen leaves. In addition, plant litter may contain secondary compounds, chemical compounds not used directly for growth (examples include those described in Concepts 5.4 and 12.2 associated with defense against herbivores and excess light), that can lower nutrient release during decomposition. Secondary compounds slow decomposition by inhibiting the activity of heterotrophic microorganisms and the enzymes they release into the soil or, in some cases, by stimulating their growth, leading to greater microbial uptake of nutrients.

By varying the chemistry of their litter, as well as the amount of litter they produce, plants can influence decomposition rates in the soil. Lowering decomposition rates lowers the fertility of the soil. What is the consequence for a plant of decreasing its own nutrient supply? For plants that have inherently slow growth rates, lowering soil fertility may protect them from competitive exclusion by neighbors with higher growth and resource uptake rates. Low soil nutrient concentrations can therefore be perpetuated through plant chemistry in a way that benefits the plants themselves (Van Breemen and Finzi 1998).

Microorganisms modify the chemical form of nutrients

Microorganisms in soil and freshwater and marine ecosystems transform some of the inorganic nutrients released during the process of mineralization. These transformations are particularly important in the case of nitrogen, since they can determine its availability to plants and the rate at which it is lost from the ecosystem (see Figure 22.11). Certain chemoautotrophic bacteria, known as *nitrifying bacteria*, convert ammonia (NH_3) and ammonium (NH_4^+) released by mineralization into nitrate (NO_3^-) by a process called **nitrification**. Nitrification occurs under aerobic conditions, so it is limited primarily to terrestrial environments. Under hypoxic conditions, some bacteria use nitrate as an electron acceptor, converting it into N_2 and nitrous oxide (N_2O, a potent greenhouse gas) by a process known as **denitrification**. These gaseous forms of nitrogen are lost to the atmosphere and thus represent a loss of nitrogen from ecosystems.

Plant ecologists and physiologists once believed that nitrogen availability to plants was dependent solely on the supply of inorganic nitrogen—nitrate and ammonium. Therefore, soil fertility has traditionally been estimated using measurements of these inorganic forms of nitrogen.

During the 1990s, much effort was invested in understanding what controls nitrogen mineralization rates, particularly in ecosystems where fertilization experiments had indicated that nitrogen availability limits primary production and influences community diversity. Measurements of inorganic nitrogen production in forest and grassland soils generally came close to estimates of the amount taken up by plants. However, rates of inorganic nitrogen supply in Arctic and alpine ecosystems were substantially lower than what plants were actually taking up. These apparent shortfalls in nitrogen supply led to the realization that some plants were using organic forms of nitrogen to meet their nutritional requirements. Earlier work in marine ecosystems had shown that phytoplankton could take up amino acids directly from water, and mycorrhizae had been shown to take up organic nitrogen from the soil and supply it to plants. However, Terry Chapin and colleagues (1993) and Ted Raab and colleagues (1996) demonstrated that some plant species, primarily sedges, take up organic forms of nitrogen

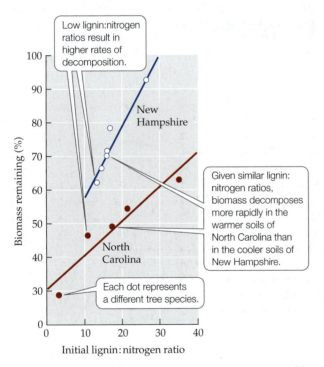

FIGURE 22.8 Lignin Decreases the Rate of Decomposition The rate of decomposition of leaf litter, expressed as the percent of biomass remaining, decreases as the ratio of lignin to nitrogen in the litter increases. This ratio varies among forest tree species. Note, however, that climate also has an important influence on decomposition rates. (After J. M. Melillo et al. 1982. *Ecology* 63: 621–626.)

ANALYZING DATA 22.1

Does Lignin Always Inhibit Decomposition?

We've learned that lignin, a structural compound found in leaves and stems, can lower rates of decomposition because it is a poorer carbon substrate for microorganisms. However, not all organic matter degradation is biotic. In arid ecosystems, for example, sunlight can break down organic matter on the surface of soils, and it can be more important than biological decomposition (Austin and Vivanco 2006). How does lignin influence the abiotic decomposition associated with photodegradation? Lignin absorbs more solar radiation than cellulose, and thus it might potentially *increase* abiotic decomposition. To test this hypothesis, Amy Austin and Carlos Ballaré (2010)* did a field experiment examining how the concentration of lignin influenced decomposition via both abiotic photodegradation and biotic activity. They used uniform cellulose sheets (filter paper) with a dilute solution of nutrients added to mimic leaf litter substrate. They added varying amounts of lignin to the sheets and then subjected them to conditions of mainly abiotic or biotic decomposition, by filtering light (biotic) or keeping the substrates isolated from the soil (abiotic). The mass loss from each sheet was measured to estimate the rate of decomposition. The results of Austin and Ballaré's experiment are presented in the table.

Biotic decomposition		Abiotic decomposition	
Lignin concentration (%)	Mass loss (%/day)	Lignin concentration (%)	Mass loss (%/day)
0	0.29	0	0.01
5	0.15	5	0.07
8	0.13	9	0.10
13	0.11	14	0.13
17	0.10		

1. Use the data in the table to plot the relationship between lignin concentration and mass loss for both biotic and abiotic decomposition.

2. What can you conclude about the influence of lignin on abiotic versus biotic decomposition? How does your conclusion support the general hypothesis that plant tissues high in lignin decompose more slowly than plant tissues low in lignin?

3. Under what kinds of environmental conditions and in what types of biomes would you expect the assumption that lignin will lower decomposition *not* to hold true?

* Austin, A. T., and C. L. Ballaré. 2010. Dual role of lignin in plant litter decomposition in terrestrial ecosystems. *Proceedings of the National Academy of Sciences USA* 107: 4618–4622.

without mycorrhizae. Arctic sedges may take up as much as 60% of their nitrogen in organic form. Organic nitrogen uptake has been observed in plants in other ecosystems as well, including boreal forests, salt marshes, savannas, grasslands, deserts, and rainforests. Thus, the mineralization step in decomposition may not be as important for plant nutrition as has been commonly thought (Schimel and Bennett 2004).

The use of soluble organic nitrogen by plants has important implications for competition among plants and between plants and soil microorganisms. There is evidence to support the hypothesis that plants in some Arctic and alpine communities avoid competition through the preferential uptake of specific forms of nitrogen—an example of resource partitioning (described in Concept 14.2).

Robert McKane and colleagues (2002) examined the forms of nitrogen taken up by several plant species growing together in the Arctic tundra of northern Alaska. For each species, they measured uptake of inorganic and organic forms of nitrogen, as well as the depth in the soil at which nitrogen was taken up and the time of year when it was taken up. They found that all three factors (form of nitrogen, depth of uptake, and timing of uptake) differed among species. Furthermore, the researchers found that the dominant plants in the community tended to be the species that used the form of nitrogen that was most abundant in the soil (**FIGURE 22.9**). Thus, the ability of a species to dominate a community where nitrogen limits growth may be determined in part by its ability to take up a specific form of nitrogen.

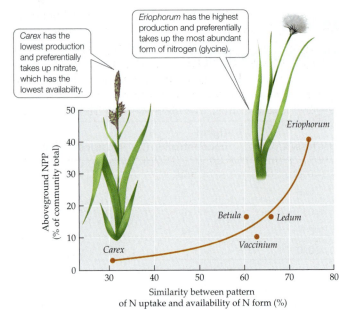

Carex has the lowest production and preferentially takes up nitrate, which has the lowest availability.

Eriophorum has the highest production and preferentially takes up the most abundant form of nitrogen (glycine).

FIGURE 22.9 Community Dominance and Nitrogen Uptake Dominance of a species in a plant community in the Alaskan Arctic tundra (measured by proportional contribution to the community's total NPP) is related to the similarity between the plant's preferred form of nitrogen (ammonium, nitrate, or glycine, a small amino acid) and the availability of that form in the soil. (After R. B. McKane et al. 2002. *Nature* 415: 68–71.)

Plants can recycle nutrients internally

Leaves, fine roots, and flowers contain the highest nutrient concentrations of any plant organ. During seasonal leaf senescence, nutrients and nonstructural carbon compounds (such as starch and carbohydrates) in perennial plants are broken down into simpler, more soluble chemical forms and moved into stems and roots, where they are stored. This phenomenon is most obvious in mid- to high-latitude ecosystems as chlorophyll molecules in the leaves of deciduous species are broken down to recover their nitrogen and other nutrients, while other pigments, such as carotenoids, xanthophylls, and anthocyanins, remain, providing the autumnal splendor we humans enjoy. Some of the fall coloration is due to an increase in pigment production, possibly to protect the leaves from high light levels or from herbivores. When growth resumes in spring, the nutrients are transported to growing tissues for use in biosynthesis. Plants may resorb as much as 60%–70% of the nitrogen and 40%–50% of the phosphorus in their leaves before they fall. This recycling reduces their need to take up "new" nutrients in the following growing season.

As we've traced the chemical transformations of nutrients in terrestrial ecosystems, we've seen that they move through various components of those ecosystems as they are transformed. In the next section, we'll look at those movements in more detail and trace the fates of the nutrients as they move through an ecosystem.

CONCEPT 22.3 Nutrients cycle repeatedly through the components of ecosystems.

LEARNING OBJECTIVES

22.3.1 Describe the key processes involved in nutrient cycling in ecosystems.

22.3.2 Summarize what determines the mean residence time of nutrients in ecosystems.

22.3.3 Evaluate how the processes that determine the loss of nutrients from an ecosystem would change during succession and why the loss rates of specific nutrients may vary through time.

Nutrient Cycles and Losses

In the previous section, we saw how nutrients undergo biological, chemical, and physical transformations as they are taken up by organisms and released through decomposition, ultimately returning to their original forms (or similar ones). This movement of nutrients within ecosystems is known as **nutrient cycling** (**FIGURE 22.10**).

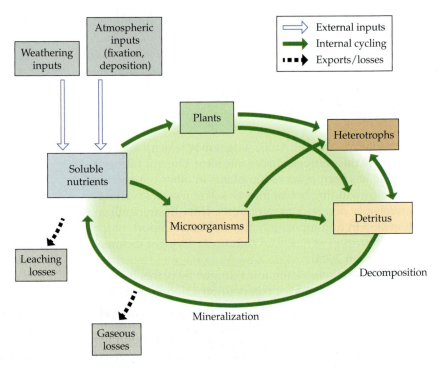

FIGURE 22.10 Nutrient Cycles A generalized nutrient cycle, showing the movements of a nutrient among the components of an ecosystem and the potential pathways for inputs and losses.

For example, we've traced the path of nitrogen into and through an ecosystem, starting with nitrogen-fixing microorganisms, as it is converted into chemical forms that can be used by plants. The plants incorporate the nitrogen into organic compounds (e.g., proteins and enzymes), which may end up being consumed by heterotrophs. Eventually plants, heterotrophs, and microorganisms all end up as detritus. Inorganic and organic nitrogen released from the detritus by decomposition is taken up again by plants and microorganisms, thereby completing the nitrogen cycle (**FIGURE 22.11**).

Nutrients cycle at different rates according to element identity and ecosystem type

The time it takes a nutrient molecule to cycle through an ecosystem, from uptake by organisms to release to subsequent uptake, can vary substantially depending on the element in question and the ecosystem where the cycle is occurring. In general, nutrients that limit primary production are cycled more rapidly than nonlimiting nutrients. For example, nitrogen and phosphorus may cycle through the photic zone of the open ocean over a period of hours or days, while zinc may cycle over geologic time scales associated with sedimentation, mountain building, and erosional processes. Nutrient cycling rates also vary with climate because of the effects of temperature and moisture on the metabolic rates of the organisms associated with production, decomposition, and chemical transformations of nutrients.

Biogeochemists measure rates of nutrient cycling by estimating the **mean residence times** of elements in some component of an ecosystem:

$$\text{mean residence time} = \frac{\text{total pool of element}}{\text{rate of input}}$$

The mean residence time is the amount of time an average molecule of an element spends in a pool before leaving it. The **pool** of an element is the total amount found within a physical or biological component of the ecosystem, such as soil or biomass. The inputs include all possible sources of the element for that ecosystem component. This approach to estimating mean residence time assumes that pools of nutrients do not change over time and that the mean residence time reflects the overall rate of nutrient cycling. It is most commonly used for estimating rates of nutrient turnover in soil organic matter, which reflect rates of nutrient input and subsequent decomposition. Decomposition rates, as we have seen, are related to climate and the chemistry of plant litter.

Given that both inputs of plant litter and decomposition rates control the mean residence times of nutrients in soil, and that both are subject to climatic control, what differences would we expect to see among ecosystems with similar plant growth forms (e.g., forests) in different climates? Relative to boreal and temperate forests, tropical forests have higher net primary productivity, and therefore higher litter input rates. Does this difference result in differences in the mean residence times of nutrients? A comparison of mean residence times for organic matter and for several nutrients indicates that nutrient pools in the soils of tropical forests are much smaller than those in boreal forests (**TABLE 22.3**). The turnover rates of nitrogen and phosphorus are more than 100 times faster in tropical forest soils than in boreal forest soils.

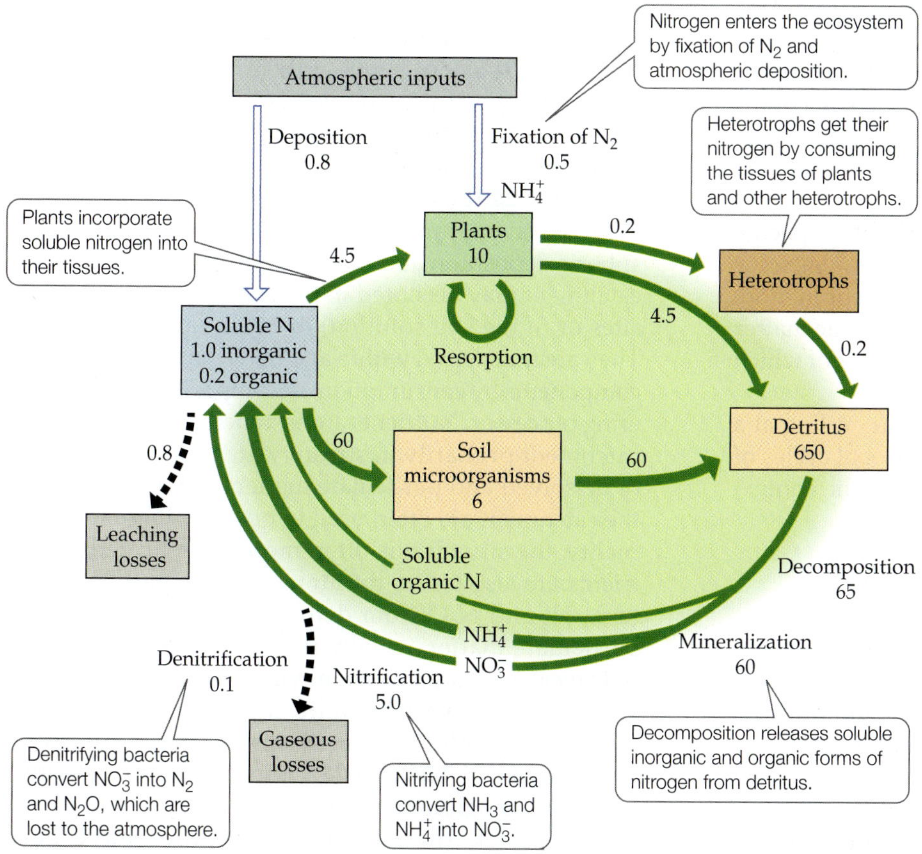

FIGURE 22.11 Nitrogen Cycle for an Alpine Ecosystem, Niwot Ridge, Colorado
Boxes represent pools of nitrogen, measured in grams per square meter; arrows represent flows of nitrogen, measured in grams per square meter per year. Note the large amount of nitrogen passing through soil microorganisms, which indicates a high turnover rate for nitrogen in this relatively small pool. (After W. D. Bowman and T. R. Seastedt. 2001. *Structure and Function of an Alpine Ecosystem, Niwot Ridge, Colorado.* Oxford University Press: New York.)

TABLE 22.3

Mean Residence Times of Soil Organic Matter and Nutrients in Forest and Shrubland Ecosystems

Ecosystem type	Mean resident time (yr)					
	Soil organic matter	N	P	K	Ca	Mg
Boreal forest	353	230	324	94	149	455
Temperate coniferous forest	17	18	15	2	6	13
Temperate deciduous forest	4	5	6	1	3	3
Chaparral	4	4	4	1	5	3
Tropical rainforest	0.4	2	2	1	1.5	1

Source: W. H. Schlesinger. 1997. *Biogeochemistry: An Analysis of Global Change*, 2nd ed. Academic Press: San Diego, CA, and references within.

Temperate forests and chaparral have turnover rates that fall in between but are closer to those in the tropics.

The main reason for this trend in mean residence times is that the influence of climate on rates of decomposition is greater than its influence on primary productivity. Boreal forest soils often have permafrost layers, which cool the soils and lower rates of biological activity. The permafrost also blocks the percolation of water through the soil, creating wet, anoxic soil conditions. Furthermore, the litter produced by boreal forest trees is rich in secondary compounds that slow rates of decomposition in the soil.

The variation in mean residence times among specific nutrients is related to their chemical properties (e.g., solubility). Some nutrients, such as potassium, occur in more soluble forms, and thus are lost from soil organic matter more quickly than others such as nitrogen, some of which is found as insoluble organic compounds.

In Chapter 25, we will return to nutrient cycling at a much larger spatial scale as we consider global cycles of carbon, nitrogen, phosphorus, and sulfur in the context of human alterations of these cycles.

Catchment studies measure losses of nutrients from ecosystems

What determines how long nutrients remain in an ecosystem? The retention of nutrients within an ecosystem is related to their uptake into its biological and physical pools and to the stability of their forms. Nitrogen, for example, is more stable as part of an insoluble organic molecule, such as a protein, than as nitrate, which is more easily leached from the soil. Nutrients are lost from an ecosystem when they move below the rooting zone by leaching, and from there into groundwater and streams. Nutrients are also lost to the atmosphere as gases or small particles and by conversion into chemical forms that cannot be used by organisms.

In our consideration of nutrient inputs into and losses from ecosystems, we have been referring to ecosystems as if they had definitive spatial units, but what defines the boundaries of an ecosystem? Ecologists studying terrestrial ecosystems commonly focus on a single drainage basin. This unit of study, which is called a **catchment**, includes the terrestrial area that is drained by a single stream and its tributaries (**FIGURE 22.12**). By measuring the inputs and outputs of elements in a catchment and calculating the balance between them, ecologists can make inferences about the use of nutrients in the ecosystem and their importance to ecosystem processes such as primary production.

FIGURE 22.13 presents a conceptual model of a catchment. Nutrient inputs into the catchment include atmospheric deposition and fixation. Nutrients that enter the catchment may be stored in the soil (on cation exchange sites or in the soil solution) or taken up by organisms. They are transferred within and between these ecosystem components by consumption, decomposition, and weathering processes. Nutrients are assumed to be lost from the catchment primarily in stream water, so measurements of dissolved and particulate matter in streams draining the catchment are often used to quantify these losses. In reality, the situation is often more complicated, as nutrients are also lost to the atmosphere in gaseous forms (e.g., N_2 and N_2O from denitrification) and as coarse particulate matter, usually fragmented litter (e.g., bits of leaves), and organisms moving out of the ecosystem. However, measurement of the input–output balance of different nutrients, using methods such as those described in **ECOLOGICAL TOOLKIT 22.1**, is instructive for determining their biological importance.

The best-known catchment studies have been performed at the Hubbard Brook Experimental Forest in New Hampshire (Likens and Bormann 1995), which is considered to be representative of the northern deciduous forests of the United States. Continuous monitoring of the Hubbard Brook catchment began in 1963 under the direction of Herb Bormann and Gene Likens, whose studies have served as models for a number of

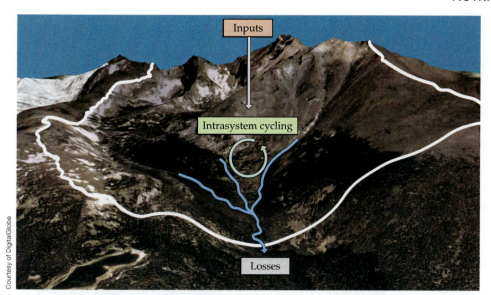

FIGURE 22.12 Catchments Are Common Units of Ecosystem Study A drainage basin (known as a catchment or watershed) associated with a single stream system (blue lines), with boundaries determined by topographic divides (outlined in white), is a unit commonly used in terrestrial ecosystem studies to measure inputs and outputs of nutrients. This catchment is the upper Hunters Creek basin, draining the south side of Longs Peak in Rocky Mountain National Park.

? What assumptions are made in this simple input–output model of a catchment that may not be realistic? (Hint: Compare this figure with Figure 22.13.)

other catchment-level studies. These studies are providing information about the roles of organisms and soils in nutrient retention, how ecosystems respond to disturbances such as logging and fire, and long-term trends in nutrient flows associated with acid rain and climate change.

Long-term ecosystem development affects nutrient cycling and constraints on primary production

As terrestrial ecosystems develop on new substrates (e.g., in primary succession on new volcanic flows), soil weathering, nitrogen fixation, and the buildup of organic matter in the soil determine the supply of nutrients available to plants. Early in ecosystem development, there is little organic matter in the soil, so supplies of nitrogen derived from decomposition are low. Supplies of mineral nutrients derived from weathering are also low, but higher relative to the supply of nitrogen. Accordingly, nitrogen availability should be an important constraint on primary production and plant community composition early in primary succession (see Chapter 17). As the pool of nitrogen in soil organic matter increases, its limitation of primary production should decrease.

Phosphorus enters ecosystems through the weathering of a single rock mineral (apatite), and its supply is high relative to that of nitrogen early in succession. As the supply of phosphorus from weathering is exhausted over time, however, decomposition becomes increasingly important as a source of

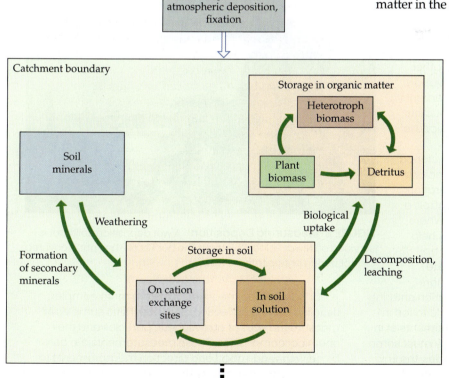

FIGURE 22.13 Biogeochemistry of a Catchment This conceptual model depicts the major pathways of nutrient movement into, through, and out of a catchment. (After G. E. Likens and F. H. Bormann. 1995. *Biogeochemistry of a Forested Ecosystem*. Springer: New York.)

ECOLOGICAL TOOLKIT 22.1

Instrumenting Catchments

Measuring the inputs of nutrients into catchments via atmospheric deposition, as well as their losses in stream water, requires knowing the concentrations of the elements in water as well as the volume of water entering and leaving the catchment (i.e., the amount of precipitation from precipitation gauges and stream flow from weirs, **FIGURE A**). The product of the two, concentration times volume, gives the total amount of the element entering or leaving the catchment. These values are usually averaged over periods ranging from a week to a year to provide input–output balances of specific elements.

Atmospheric deposition includes (1) elements captured in precipitation when it falls to the surface (wet deposition) and (2) particles, including aerosols and fine dust, that are transferred to the surface by gravitational fallout or air movement (dry deposition). Total atmospheric deposition can be sampled by placing buckets above the surrounding vegetation to collect the deposited material. However, buckets make good perches for birds, which may deposit their own contribution to ecosystem nutrient input inside the bucket, albeit at much higher concentrations than those found in most other parts of the catchment. This problem can be avoided by placing spiky projections around the edge of the bucket to prevent birds from landing on it. Another problem is that open buckets lose water to evaporation, increasing the concentration of the elements inside. Furthermore, in windy, cold climates, buckets are not good collectors or holders of snow because of their aerodynamics.

Wet deposition collectors have been developed that open to the atmosphere only during precipitation events and then close to prevent evaporation (**FIGURE B**). A moisture-sensitive surface controls a switch that opens and closes the collector. Where snow and wind occur together, windscreens help to prevent loss of snow from the bucket and enhance the capture of the deposition. Separate precipitation gauges may also be used to more accurately estimate the volume of precipitation entering the ecosystem. At regular intervals, the precipitation in the bucket or collector is analyzed for the elements of interest using chemical analyses that typically meet some government standard (e.g., in the United States, the Environmental Protection Agency provides these standards). In many developed nations, networks of wet deposition samplers have been established to provide spatial estimates of atmospheric deposition (e.g., the National Atmospheric Deposition Program in the United States: nadp.sws.uiuc.edu; see Figure 25.19).

Courtesy of William Bowman

FIGURE A Measuring Water Flow A weir on Fool Creek in the Fraser Experimental Forest, Colorado.

Courtesy of Jennifer Morse

FIGURE B Measuring Deposition A wet deposition collector is serviced on Niwot Ridge, Colorado. The bucket on the right is covered except during precipitation events.

Dry deposition measurements are more complex, usually involving collection of atmospheric samples to measure the sizes of atmospheric particles and their chemical composition. These measurements are combined with wind speed and direction measurements to estimate movements of elements to the surface. Because of the greater difficulty of the sampling and the larger uncertainties, dry deposition is measured less frequently than wet or bulk (total) deposition. In some areas, however, such as deserts or Mediterranean-type ecosystems, dry deposition is the largest component of total deposition.

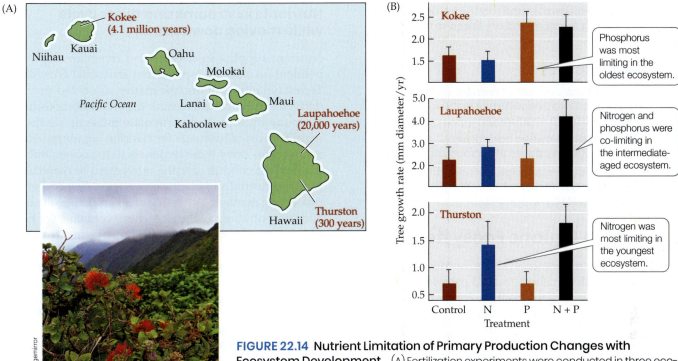

FIGURE 22.14 Nutrient Limitation of Primary Production Changes with Ecosystem Development (A) Fertilization experiments were conducted in three ecosystems of different ages in the Hawaiian Islands: Thurston (300 years old), Laupahoehoe (20,000 years old), and Kokee (4.1 million years old). Vegetation at all three sites is dominated by a single tree species, Ohi'a (*Metrosideros polymorpha*). (B) Ohi'a growth rates in response to fertilization treatments with nitrogen (N), phosphorus (P), and both (N + P) in the three ecosystems. The more an added nutrient increased tree growth, the more limiting it was assumed to be. Note the differences in the ranges of the *y* axes. Error bars show one SE of the mean. (A after T. E. Crews et al. 1995. *Ecology* 76: 1407–1424; B after P. M. Vitousek and H. Farrington. 1997. *Biogeochemistry* 37: 63–75; Thurston data from P. M. Vitousek et al. 1993. *Biogeochemistry* 23: 197–215; Kokee data from D. Herbert et al. 1999. *Ecology* 80: 908–920.)

phosphorus for plants. In addition, soluble phosphorus may combine with iron, calcium, or aluminum to form secondary minerals that are unavailable as nutrients, a process known as **occlusion**. The amount of phosphorus in occluded forms increases over time, further reducing its availability. As a result, phosphorus should become more limiting to primary production during later stages of succession (Walker and Syers 1976).

These observations of changes in nutrient cycling during ecosystem development provide a hypothetical framework for considering how those changes should influence the specific nutrients that limit primary production. Nitrogen should be most important in determining rates of primary production early in succession, nitrogen and phosphorus should both be important at intermediate stages of succession, and phosphorus should be most important late in succession. This hypothesis was tested in the Hawaiian Islands by Peter Vitousek and his colleagues. The movement of the Pacific tectonic plate over millions of years has given rise to the chain of volcanoes that form these islands. The oldest islands are in the northwestern part of the chain, the youngest in the southeast (**FIGURE 22.14A**). Vitousek's group studied Hawaiian ecosystems on soils

with ages ranging from 300 years to over 4 million years to determine which nutrients were most limiting to primary production. Their study was aided by the similarity of the vegetation and climate at each of the study sites. Vitousek and colleagues added nitrogen, phosphorus, or both nitrogen and phosphorus to plots in three ecosystems of different ages and measured the effects of these treatments on the growth of the dominant tree, Ohi'a (*Metrosideros polymorpha*). Consistent with their hypothesis, nitrogen was most limiting to tree growth in the youngest ecosystem, while phosphorus was most important in the oldest ecosystem (Vitousek and Farrington 1997) (**FIGURE 22.14B**). Nitrogen and phosphorus added in combination increased tree growth in the intermediate-aged ecosystem. In contrast to these tropical soils, the soils of ecosystems in temperate, high-latitude, and high-elevation zones are often subjected to major disturbances (e.g., large-scale glaciation, landslides) and are less likely to reach ages at which phosphorus becomes limiting.

Nutrients lost from terrestrial ecosystems often end up in streams, lakes, and oceans. They are a critical source of nutrients for those aquatic ecosystems, but they can have negative effects as well, as we'll see in the next section.

CONCEPT 22.4 Freshwater and marine nutrient cycles occur in a moving medium and are linked to terrestrial ecosystems.

LEARNING OBJECTIVES

22.4.1 Describe the concept of nutrient spiraling in moving waters and summarize the factors that control the spiral lengths.

22.4.2 Differentiate between natural and anthropogenic causes of eutrophication.

22.4.3 Explain why seasonal lake turnover and upwelling are important to enhancing nutrient supply in lakes and oceans.

Nutrients in Aquatic Ecosystems

In freshwater and marine ecosystems, nutrient transformations and transfers have the added complexity of occurring in a moving aqueous medium. Inputs of nutrients from outside the ecosystem are much more important than in terrestrial ecosystems. Furthermore, oxygen concentrations are often lower than in terrestrial ecosystems, constraining biological activity and the biogeochemical processes associated with it.

Nutrients in streams and rivers cycle while moving downstream

Nutrient supplies in streams and rivers are highly dependent on external inputs from terrestrial ecosystems. Terrestrial inputs of organic matter, dissolved nutrients derived from chemical weathering and decomposition in surrounding soils, and particulate minerals are the primary sources of nutrients for riverine organisms. Rivers and streams carry these materials to the ocean, but they are not just conduits for the movement of material between terrestrial and marine ecosystems. Biogeochemical processing in moving stream water can change the forms and concentrations of the elements it contains. For example, denitrification and biological uptake in streams and rivers may result in significant losses of nitrogen during transport in stream water. These processes may explain why rivers export less nitrate from regions receiving high amounts of nitrogen pollution than would be expected (**FIGURE 22.15A**). Both denitrification and biological uptake are enhanced when detritus is abundant on the stream bottom (**FIGURE 22.15B**).

Nutrients in rivers and streams are cycled repeatedly as the water flows downstream. Dissolved inorganic forms

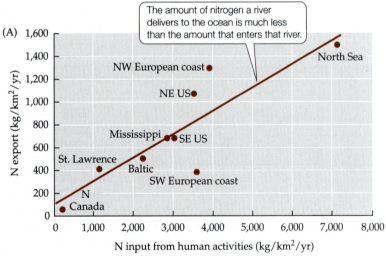

(A)

The amount of nitrogen a river delivers to the ocean is much less than the amount that enters that river.

x-axis: N input from human activities (kg/km^2/yr)
y-axis: N export (kg/km^2/yr)

Labeled points: North Sea, NW European coast, NE US, Mississippi, SE US, St. Lawrence, Baltic, SW European coast, N Canada

FIGURE 22.15 Rivers Are Important Modifiers of Nitrogen Exports Nitrogen that enters rivers from terrestrial ecosystems is not simply carried to the ocean. (A) The rates of nitrogen exports to the North Atlantic Ocean from major drainage basins are correlated with rates of nitrogen inputs into rivers by human activities. The export rates, however, are substantially lower than the input rates because of biogeochemical processing of the nitrogen in the rivers (notice the difference between the scales in the *x* and *y* axes). (B) Denitrification and biological uptake are two of the main processes that lower the export of nitrogen from drainage basins. Both processes are enhanced when benthic detritus is high. DON, dissolved organic nitrogen. (A after R. W. Howarth et al. 1996. *Biogeochemistry* 35: 75–139; B after E. S. Bernhardt et al. 2005. *BioScience* 55: P219–P230.)

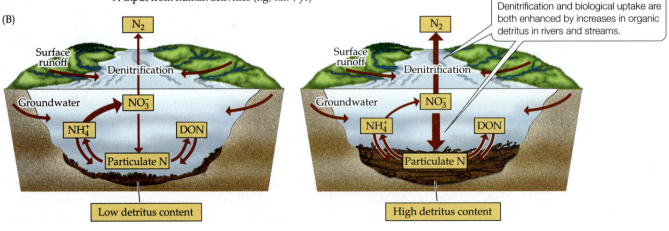

(B)

Denitrification and biological uptake are both enhanced by increases in organic detritus in rivers and streams.

Left diagram: Surface runoff, Denitrification, Groundwater, N$_2$, NO$_3^-$, NH$_4^+$, DON, Particulate N — **Low detritus content**

Right diagram: Surface runoff, Denitrification, Groundwater, N$_2$, NO$_3^-$, NH$_4^+$, DON, Particulate N — **High detritus content**

of nutrients are taken up by organisms, including fungi, bacteria, and phytoplankton, which incorporate them into organic molecules. These organisms may be consumed by others and pass through a food web, eventually entering the pool of stream detritus. Following decomposition of the detritus, the mineralized nutrients are released back into the water in dissolved inorganic forms. This repeated uptake and release in association with the movement of water can be thought of as nutrient "spiraling" (Newbold et al. 1983) (**FIGURE 22.16**). The time it takes for a full nutrient spiral to occur (i.e., from uptake and incorporation into organic forms to release in inorganic forms) is related to the amount of biological activity in the stream, the water velocity, and the chemical form of the nutrient. These variables have important impacts on the retention of nutrient pollutants (nitrate and phosphate) in rivers and can therefore impact the formation of dead zones in estuaries (see Concept 20.2). Greater biological retention (longer nutrient spirals) helps to buffer the impacts of nutrient pollution to downstream sources (lakes, estuaries). The turnover of nitrate in rivers tends to increase downstream, as indicated by increasing spiral lengths, while phosphate is retained equally well upstream and downstream (Ensign and Doyle 2006).

Nutrients in lakes cycle efficiently in the water column

Lake ecosystems receive inputs of nutrients from streams, by atmospheric deposition and nitrogen fixation, and as litter falling from adjacent terrestrial ecosystems. Biological demand for nutrients is highest in the photic zone, where phytoplankton are suspended in the water column, and in the shallow zones at the margins of the lake, where rooted aquatic plants are found. Phosphorus and sometimes nitrogen limit primary production in lakes. Nutrient transfers between trophic levels, like energy transfers (see Figure 21.5C), are very efficient in lakes. Some detritus is decomposed and mineralized in the water column and in sediments in the shallow zones, providing an internal input of nutrients. Nitrogen fixation by cyanobacteria

occurs in the photic zone, particularly when demand for nitrogen by organisms is greater than for phosphorus. Rates of nitrogen fixation in lake ecosystems are similar to those in terrestrial ecosystems.

Over time, nutrients are progressively lost from the photic zone of a lake. Dead organisms sink through the water column and are deposited in the sediments of the benthic zone. These sediments are characterized by hypoxic conditions that limit decomposition, and by a reducing chemical environment that may change the chemical form of some nutrients. Iron, for example, is often reduced from Fe^{3+} to Fe^{2+}, contributing to the dark color of lake sediments. Denitrification is also promoted by the low oxygen concentrations in the sediments, and bacteria may reduce sulfate (SO_4^{2-}) to hydrogen sulfide (H_2S).

Decomposition in the benthic sediments cannot provide nutrients to the photic zone unless there is mixing of the water column. In stratified temperate-zone lakes, as we saw in Concept 2.5, this mixing occurs in fall and spring, when the lake's water becomes isothermal throughout and wind facilitates its turnover. This seasonal turnover brings dissolved nutrients from the bottom water to the surface layers, along with detritus that may be subsequently decomposed by bacteria. Mixing of water layers is less common in tropical lakes, so external inputs of nutrients may be more important for maintaining production in those lakes.

Lake ecosystems are often classified according to their nutrient status. Nutrient-poor waters with low primary productivity are referred to as **oligotrophic**, while nutrient-rich waters with high primary productivity are referred to as **eutrophic**. **Mesotrophic** waters are intermediate in nutrient status between oligotrophic and eutrophic waters. The nutrient status of a lake is the result of natural processes associated with climate and with lake size and shape. For example, lakes in high mountain areas are typically oligotrophic because of their short growing season, low temperatures, and tendency to be deep with a low surface-area-to-volume ratio, which constrains the rate of nutrient input relative to the volume. In contrast, shallow lakes at lower elevations or in the tropics tend to be eutrophic because of their warmer temperatures and higher nutrient availability.

The nutrient status of a lake tends to shift naturally from oligotrophic to eutrophic over time. This process, known as **eutrophication**, occurs as sediments accumulate on the lake bottom (**FIGURE 22.17**). As the lake becomes shallower, its summer temperatures become warmer, more decomposition occurs, nutrient pools and the amount of mixing increase, and the lake becomes more productive. Human activities have accelerated the process

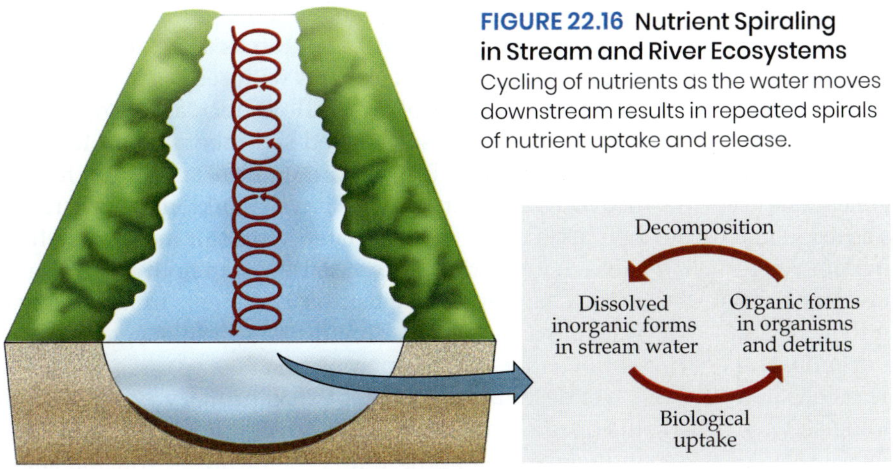

FIGURE 22.16 Nutrient Spiraling in Stream and River Ecosystems
Cycling of nutrients as the water moves downstream results in repeated spirals of nutrient uptake and release.

Decomposition

Dissolved inorganic forms in stream water

Organic forms in organisms and detritus

Biological uptake

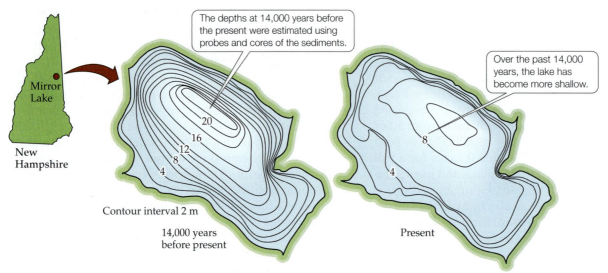

FIGURE 22.17 Lake Sediments and Depth Sediments accumulate at the bottom of a lake over time, making it progressively shallower and leading to eutrophication. Changes in the depth contours of Mirror Lake in New Hampshire show the accumulation of sediments there over the past 14,000 years. (After M. B. Davis et al. 1985. In *An Ecosystem Approach to Aquatic Ecology: Mirror Lake and Its Environment*, G. E. Likens [Ed.], pp. 345–366. Springer: New York.)

of eutrophication in many lakes through discharges of sewage, agricultural fertilizers, and industrial wastes containing high concentrations of nitrogen and phosphorus. For example, the water of Lake Tahoe, on the border between Nevada and California, has lost much of its clarity because of increased inputs of phosphorus and nitrogen from streams, groundwater, and surface runoff from neighboring communities. Water clarity, which is used as an indicator of a lake's nutrient status, is primarily determined by the density of plankton in the water column. It can be measured using a Secchi disk, a black-and-white circular plate that is lowered gradually into the water; the maximum depth at which the disk can be seen is referred to as the *depth of clarity*. Over the past 3 decades, the average depth of clarity in Lake Tahoe has risen by 10 m (Murphy and Knopp 2000). The rate at which water clarity has been declining has slowed since 2000, partly because of lower amounts of precipitation as well as lower concentrations of nutrient pollutants in streams draining into the lake.

Anthropogenic eutrophication can be reversed if the discharge of wastes into surface waters is decreased. A classic example of such a reversal occurred in the 1960s and 1970s in Lake Washington, near Seattle. Treated sewage, containing high concentrations of phosphorus, was released into Lake Washington beginning in the late 1940s as neighborhoods and accompanying sewage treatment plants were built near the shore of the lake. Decreases in water clarity were noted during the 1950s, corresponding to increases in phytoplankton densities and blooms of cyanobacteria. Public concern grew, and local governments debated what action to take. A prominent local limnologist, W. T. Edmondson, believed that the problem was associated with phosphorus inputs from the treated sewage, which included wastewater from washing machines containing phosphorus-laden detergents. Based on Edmondson's advice, Seattle stopped its sewage input into Lake Washington completely by 1968. Increases in water clarity were soon noted, and by 1975, the lake was considered recovered from eutrophication (**FIGURE 22.18**). Edmondson's recommendation was crucial to the lake's recovery, and the case contributed to the current U.S. and European Union restrictions on the use of phosphates in detergents.

Imports and upwelling are important sources of nutrients in marine ecosystems

Rivers join marine ecosystems in estuaries (described in Concept 3.3). In these zones where fresh water meets seawater, salinity—and thus water density—is variable. This variation influences the mixing of waters and the chemical forms of some nutrients. For example, phosphorus bound to soil particles may be released in a form more easily available to phytoplankton as a result of changes in pH when river water mixes with seawater.

As the velocity of water flow decreases toward the mouth of a river, suspended sediments begin to settle out of the water. These sediments are substrates for detritivores and nutrients for phytoplankton in the estuary. Estuaries are often associated with salt marshes, which are rich in nutrients because they trap both riverine and ocean sediments. Like benthic sediments in lakes, estuarine and salt marsh sediments have low oxygen concentrations that limit decomposition.

As described in Concept 20.2, primary production in the open ocean is limited by several nutrients, including nitrogen, phosphorus, and, in some areas, iron and silica.

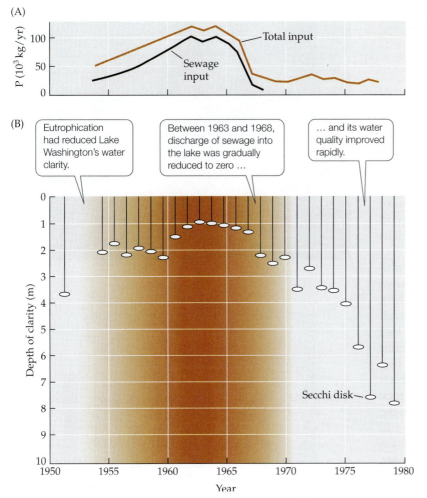

(A)

(B)

Eutrophication had reduced Lake Washington's water clarity.

Between 1963 and 1968, discharge of sewage into the lake was gradually reduced to zero …

… and its water quality improved rapidly.

Secchi disk

FIGURE 22.18 **Lake Washington: Reversal of Fortune** Inputs of treated sewage between the 1940s and the 1960s caused eutrophication in Lake Washington; cessation of sewage inputs between 1963 and 1968 increased lake clarity. (A) Phosphorus inputs. (B) Measurements of water clarity made with a Secchi disk. (After W. T. Edmondson and A. H. Litt. 1982. *Limnol Oceanogr* 27: 272–293.)

❓ While the story of Lake Washington seems to be a clear "experimental" demonstration of pollution influencing the nutrient status of a lake, what would make it an even more convincing example?

Seawater has relatively high concentrations of magnesium, calcium, potassium, chloride, and sulfur. Sources of nitrogen in marine ecosystems include inputs from rivers and atmospheric deposition as well as tight internal cycling through decomposition. Rates of nitrogen fixation by cyanobacteria in the oceans are lower than those in freshwater lakes, possibly because these organisms are limited by molybdenum, which is a component of the nitrogenase enzyme. Phosphorus, iron, and silica enter the marine ecosystem primarily in dissolved and particulate form in rivers; a smaller but important contribution comes from atmospheric deposition of dust. Inputs from both of these terrestrial sources are increasing as a result of human activities, including large-scale desertification and deforestation.

Deep deposits of sediments (up to 10 km, or 6 miles thick!) have accumulated in the benthic zones of the open ocean. These deposits, which consist of a mix of ocean-derived detritus and terrestrial erosional sediments, are important potential sources of nutrients. Sulfate reduction and denitrification occur in these anoxic sediments, and some decomposition and mineralization of organic matter also occur there. Bacteria have been found as deep as 500 m in these sediments. The deep ocean layers are dense due to cold temperatures and high salt concentrations (see Concept 2.2) and usually don't mix with the surface waters. Mixing of deep, nutrient-rich waters with nutrient-poor surface waters does occur in zones of upwelling, where ocean currents bring deep waters to the surface (**FIGURE 22.19**). These zones of upwelling are highly productive and thus are important areas for commercial fisheries.

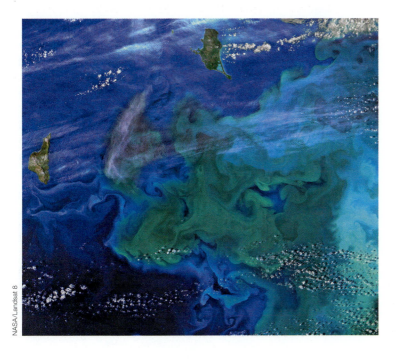

FIGURE 22.19 **Zones of Upwelling Enhance Nutrient Supply for Marine Ecosystems** Phytoplankton blooms (green areas), fed by upwelling of nutrient-rich deep ocean water, can be seen off the coast of the Pribilof Islands (Alaska) in this satellite image.

A CASE STUDY REVISITED

A Fragile Crust

We've seen that nutrient supplies for plants in terrestrial ecosystems are dependent on the weathering of rock minerals and the decomposition of detritus in the soil, as well as on the fixation of atmospheric nitrogen. How might the loss of biocrusts from desert soils influence these processes? As this chapter's Case Study explained, the crusts prevent erosional losses of soil by helping to bind soil particles together. The activity of the organisms that make up the crusts may also influence nutrient inputs and, in turn, the productivity of the desert ecosystem, as well as its capacity to withstand the desert climate.

Jason Neff and colleagues conducted a study to evaluate the effects of cattle grazing on soil erosion and nutrient availability on the Colorado Plateau (Neff et al. 2005). They selected three study sites in Canyonlands National Park: one that had never been grazed and two that had been grazed historically but were closed to grazing after 1974 (30 years of recovery). Cattle grazing in the park first occurred in the 1880s, and most of its soil surface has been affected. The ungrazed study site was surrounded by rock formations that prevented the movement of cattle into the area. The study sites all had the same parent material and similar plant communities and were located within 10 km of one another. Biocrusts were present at all three sites, although those at the historically grazed sites had clearly been damaged, as they appeared less well developed than those at the site that had never been grazed.

Samples of soil and bedrock were collected from each of the sites, and the textures and nutrient contents of the soils were compared. In addition, the retention of fine dust from the atmosphere was estimated by measuring the magnetic properties of the soil. Dust blown in from distant areas contains higher amounts of iron oxides than the native soil, so the more dust present, the stronger the magnetic signal. Retention of this dust is important because it is a source of mineral nutrients. In addition, loss of this dust indicates the potential for erosional loss of the native soil as well.

Neff and colleagues found that the historically grazed soils had less fine-textured soil, and substantially less magnesium and phosphorus, than the ungrazed soils (**FIGURE 22.20**). They attributed these differences to greater retention of dust and lower loss of native soils with better-developed biocrusts. The crusts may also enhance rates of weathering by altering pH, by increasing the rates of chemical reactions that release mineral nutrients, and by increasing water retention in the soil. Soils in the historically grazed sites also contained 60%–70% less carbon (from organic matter) and nitrogen than those in the ungrazed sites. These differences were also related to the development of biocrusts. Although a crust had begun to recover at the historically grazed sites, comparison with the

Courtesy of Nichole Barger

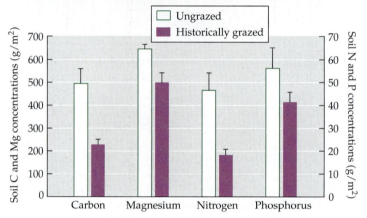

FIGURE 22.20 Loss of Biocrusts Results in Smaller Nutrient Supplies Historically grazed soils in Canyonlands National Park contained less carbon, magnesium, nitrogen, and phosphorus than soils that had never been grazed. Error bars show one SE of the mean. (Graph after J. C. Neff et al. 2005. *Ecol Appl* 15: 87–95.)

ungrazed site showed that the cumulative loss of carbon and nitrogen from the soils during the period of grazing was high. The cyanobacteria in biocrusts fix atmospheric N_2 (Belnap 2003), which represents an important input of a nutrient that may limit plant growth in the absence of water limitation during the spring growing season. In addition, crust-covered soils absorb more solar radiation and retain more water than soils without crusts, creating conditions more conducive to decomposition and mineralization.

 CONNECTIONS *in* NATURE

NUTRIENTS, DISTURBANCE, AND INVASIVE SPECIES By increasing nutrient supplies and stabilizing soils, biocrusts enhance primary production. Plants growing in association with the crusts have higher growth rates, and contain more nutrients, than plants growing in soils without crusts. Plant cover also increases in the presence of biocrusts. Furthermore, biocrusts have been shown to lower the germination

FIGURE 22.21 Scourge of the Intermountain West Large areas of the intermountain West of North America are now dominated by cheatgrass (*Bromus tectorum*), an invasive species that increases fire frequencies, outcompetes native plants for resources, and spreads rapidly across the landscape.

and survival rates of invasive plants (Havrilla et al. 2019) (see Chapter 23). Thus, the destruction of crusts by cattle grazing has had multiple ecological effects.

Are the negative effects of cattle grazing on soil stability and nutrient availability that Neff and colleagues observed in Canyonlands National Park common in other areas? The answer lies in part with the long-term history of grazing and climate in North America. Prior to Euro-American settlement, soils in much of the intermountain West did not experience the amount of grazing by native animals that occurred in other areas, such as the Great Plains, where large herds of bison roamed (see the Case Study in Chapter 3 and the discussion in Concept 3.2). A combination of aridity and long-term development of biocrusts may have given the soils of the Colorado Plateau an especially low tolerance for heavy grazing.

In the grasslands of the intermountain West, the combination of soil disturbance and loss of biocrust has created a situation conducive to the spread of non-native species—most notably cheatgrass (*Bromus tectorum*; **FIGURE 22.21**), a native of Eurasia. Cheatgrass has had profound effects on the ecology of much of western North America. Cheatgrass is a spring annual that sets seed, dies, and dries out by early summer. This life history increases the amount of dry, combustible vegetation that is present during the summer. As a result, cheatgrass has increased the frequency of fires, which now occur about every 3–5 years, compared with more natural fire frequencies of 60–100 years. Native grasses and shrubs cannot recover from such frequent fires, so cheatgrass increases its dominance under these conditions. Cheatgrass is an effective competitor for soil resources, and it also lowers rates of nitrogen cycling by producing litter with a C:N ratio higher than those of native species (Evans et al. 2001). This combination of increasing fire frequency, increasing competition, and changes in nutrient cycling has led to decreases in native species richness in many parts of the intermountain grasslands.

SUMMARY

CONCEPT 22.1 Nutrients enter ecosystems through the chemical breakdown of minerals in rocks or through fixation of atmospheric gases.

22.1.1 Describe the basic roles of nutrients in organisms, and differentiate between the ways in which microorganisms, plants, and animals obtain them.

- The nutrient requirements of organisms are specific to their physiology and thus differ between autotrophs and heterotrophs.

- Autotrophs absorb nutrients in simple, soluble forms from their environment, while heterotrophs obtain them in more complex forms by consuming prey or detritus.

22.1.2 Summarize the steps involved in the breakdown of minerals and the subsequent release of nutrients.

- The physical and chemical breakdown of minerals (weathering) releases soluble nutrients.

22.1.3 Describe the physical and chemical properties of soil that influence its fertility.

(Continued)

SUMMARY (continued)

- Soils are made up of mineral particles, detritus, dissolved organic matter, water containing dissolved minerals and gases, and organisms.

22.1.4 Explain the processes that fix carbon and nitrogen, converting them into usable forms for organismal function and growth.

- Carbon and nitrogen enter ecosystems through fixation of atmospheric gases by autotrophs and by bacteria, respectively.

CONCEPT 22.2 Chemical and biological transformations in ecosystems alter the chemical form and supply of nutrients.

22.2.1 Describe why decomposition is a critical process in the supply of nutrients in ecosystems.

- Decomposition of organic matter releases the nutrients it contains in soluble forms that can be reused by plants and microorganisms.

22.2.2 Evaluate the biological and physical factors that influence rates of decomposition.

- Moisture and temperature are key controls on decomposition in terrestrial ecosystems, with the highest rates occurring in warm temperatures and intermediate moisture conditions.

- Decomposition rates are influenced by the ratio of carbon to nutrient concentration, along with the chemistry of the carbon of the detritus, which determines how easily it can be degraded and used as an energy source.

22.2.3 Explain how microbial processes may alter the chemical forms of nutrients and make them either more or less available to plants.

- Modification of the chemical forms of nutrients, particularly nitrogen, by microorganisms influences the nutrients' availability to organisms or loss from the ecosystem.

- Plants recycle nutrients by reabsorbing them from senescing tissues and remobilizing them when growth commences again.

CONCEPT 22.3 Nutrients cycle repeatedly through the components of ecosystems.

22.3.1 Describe the key processes involved in nutrient cycling in ecosystems.

- Nutrient cycling rates are controlled primarily by the rate of decomposition, which in turn is controlled by climate and the chemistry of plant litter.

22.3.2 Summarize what determines the mean residence time of nutrients in ecosystems.

- The mean residence time of nutrients in ecosystems is related to nutrient uptake rates, associated with biological demand, and the rate of turnover of organic matter, associated with decomposition.

22.3.3 Evaluate how the processes that determine the loss of nutrients from an ecosystem would change during succession and why the loss rates of specific nutrients may vary through time.

- Changes in the relative amounts of nutrients supplied by weathering and decomposition determine the specific nutrients that limit primary production at different stages of ecosystem development.

- Losses of nutrients from terrestrial ecosystems can be estimated by measuring nutrient outputs in stream water.

CONCEPT 22.4 Freshwater and marine nutrient cycles occur in a moving medium and are linked to terrestrial ecosystems.

22.4.1 Describe the concept of nutrient spiraling in moving waters and summarize the factors that control the spiral lengths.

- The cycling of nutrients in streams and rivers can be thought of as a spiral of repeated biological uptake and incorporation into organic forms followed by release in inorganic forms.

22.4.2 Differentiate between natural and anthropogenic causes of eutrophication.

- Increases in the supply of nutrients limiting NPP, associated with both natural and anthropogenic causes, can lead to the enhancement of algal biomass in lakes and estuaries.

22.4.3 Explain why seasonal lake turnover and upwelling are important to enhancing nutrient supply in lakes and oceans.

- In lakes, nutrients are cycled between the water column and the benthic sediments.

- Imports of nutrients from rivers and terrestrial ecosystems support production in marine ecosystems.

Review Questions

1. Describe the processes involved in the transformation of solid minerals in rock into soluble nutrients in soil. What biological factors can influence the rate of this transformation?

2. Why is nitrogen often in short supply relative to other nutrients required by plants, despite being the most abundant element in the atmosphere? How does the supply of nitrogen change during terrestrial ecosystem development?

3. Which factor is more important in controlling the mean residence times and pools of nutrients in soil organic matter in terrestrial ecosystems: the rate of input (i.e., primary productivity) or the rate of decomposition? Would you expect to find larger nutrient pools in the soils of a tropical forest than in those of a boreal forest, given that primary productivity is higher in the tropics?

4. Why might you expect nutrient input from terrestrial and stream ecosystems to be more important in tropical lakes than in temperate-zone lakes?

Hone Your Problem-Solving Skills

How would disturbances and subsequent succession (see Chapter 17) influence nutrient cycling and nutrient losses in an ecosystem? As discussed above, nutrient losses are in part related to uptake by plants (see Figure 22.13). Uptake in turn is related to the rates of plant growth (NPP). As a result the losses of nutrients during succession are related to patterns of plant growth. The lowest nutrient losses should correspond to the highest growth rates.

1. Based on patterns of community replacement during succession discussed in Chapter 17, what pattern of nutrient loss from a catchment would you hypothesize following a disturbance in a forest? Consider how nutrient loss would change just after the disturbance, into the intermediate stages of succession, and finally into an old-growth community made up of long-lived mature trees.

2. Would the patterns of nutrient loss that you hypothesized above be the same for all nutrients? Would nutrients that limit NPP show the same patterns as nutrients that are not limiting to growth?

3. Peter Vitousek (1977) used a catchment approach to study nutrient retention by spruce–fir forests in the White Mountains of New Hampshire at different stages of secondary succession following logging. He measured nutrient loss in streams draining catchments of different successional stages. In order to evaluate the hypothesis that late successional communities would be more "leaky" than intermediate-stage communities, he examined the ratio of old-growth losses to losses during intermediate successional stages. He did this for several elements, some potentially more limiting to NPP than others. His results are shown in the table (μeq/L, microequivalents/liter). Do the data support Vitousek's hypothesis regarding changes in nutrient losses between intermediate and late successional forest communities? Do all of the elements show the same patterns of loss for intermediate and late successional communities? How do the differences in the patterns of element losses relate to their importance to plant growth?

	Mean growing-season stream water concentrations (μeq/L) ± SE		
	Late successional	Intermediate successional	Ratio late:intermediate
NO_3^-	53 ± 5	8 ± 1.3	6.62
K^+	13 ± 1	7 ± 0.5	1.81
Mg^{2+}	40 ± 4.9	24 ± 1.6	1.66
Ca^{2+}	56 ± 4.5	36 ± 2.5	1.56
Cl^-	15 ± 0.3	13 ± 0.3	1.16
Na^+	29 ± 2.6	28 ± 0.9	1.03

ADDITIONAL RESOURCES

This textbook provides resources such as **Web Extensions** and additional **Climate Change Connections** for instructors who can share them with students. Please contact your instructor, if you would like to access that content.

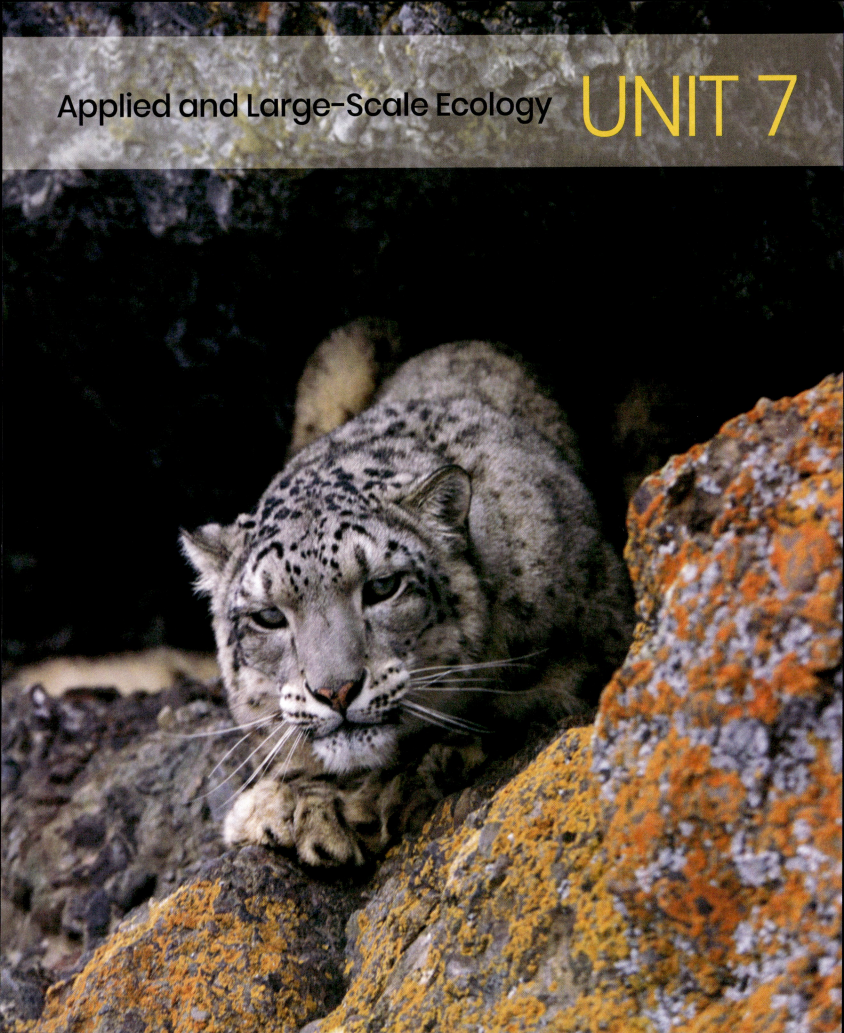

23

Conservation Biology

Can Birds and Bombs Coexist? A Case Study

How could the chaos and destruction of preparing for battle be beneficial for conservation efforts? Although it may seem strange, bombing for decades on the Fort Bragg military base in the North Carolina Sandhills has inadvertently protected thousands of acres of longleaf pine savanna, aiding efforts to save the endangered red-cockaded woodpecker (**FIGURE 23.1**).

For a century, the forests of Fort Bragg have been used for military training exercises, degraded by off-road vehicles and earth-moving equipment, and set on fire by explosives. These destructive activities take place in the midst of a once common but now rare ecosystem—one that, ironically, survives in large part as a result of the military presence. How can this be? First, pine savanna depends on fire for its persistence, so the fires that result from explosions benefit rather than harm the ecosystem. Second, the designation of large blocks of forest land for military use has kept them from being converted to farmland, forest plantations, and residential uses.

While some longleaf pine savanna has been preserved at Fort Bragg and other military bases, overall, this ecosystem has been reduced to only 3% of the more than 35 million hectares (>134,000 square miles) it once covered (**FIGURE 23.2**). Various factors have contributed to its decline, including rapid growth of the human population; the clearing of land for large plantations where other tree species, such as loblolly pine, are grown; and fire suppression. With the decline of the longleaf pine savanna ecosystem, several plant, insect, and vertebrate species that depend on it have also undergone substantial declines.

One of these species is the red-cockaded woodpecker (*Picoides borealis*), a small insectivorous bird that requires large tracts of open pine savanna. Once numbering around 1.3 million breeding pairs and associated helpers (known as groups living in clusters), the species currently stands at about 7,500 clusters. Whereas other woodpeckers nest in dead snags, red-cockaded woodpeckers require mature, living pine trees, especially the longleaf pine (*Pinus palustris*), for their nesting cavities.

Periodic fires historically helped to maintain longleaf pine savanna. Without those fires, the longleaf pine community soon undergoes succession. As an understory of young oaks and other hardwoods grows up, red-cockaded woodpeckers abandon their nesting cavities, apparently because of a decrease in food resources. In the past, the birds would move to parts of the forest that had been more recently burned, but as the area of suitably mature longleaf pines declines, there are fewer and fewer places for the birds to go. This loss of habitat has reduced the woodpecker's populations, making them vulnerable to the problems associated with small, isolated populations that we discussed in Concept 11.3. There

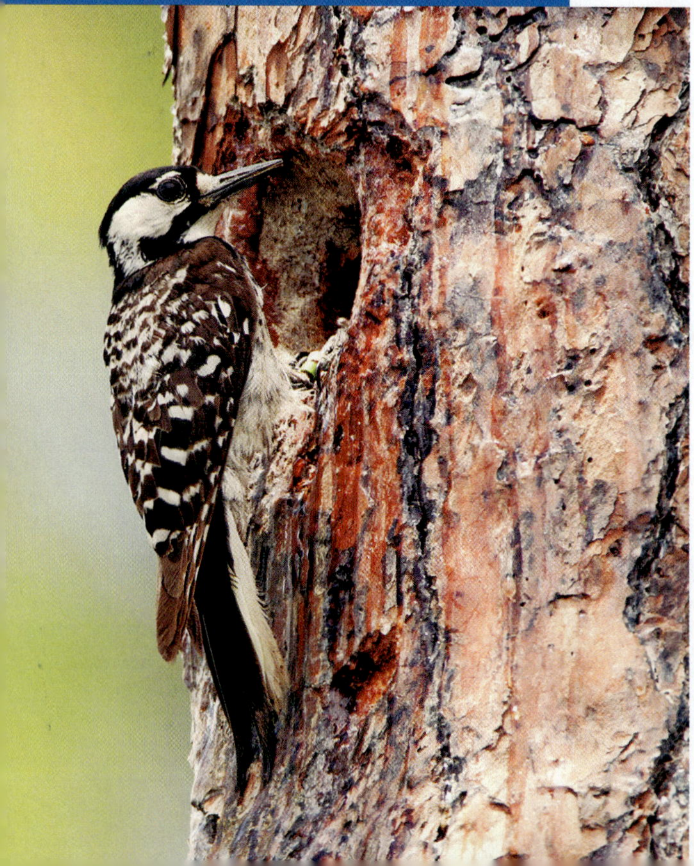

FIGURE 23.1 The Red-Cockaded Woodpecker: An Endangered Species
A female red-cockaded woodpecker (*Picoides borealis*) approaches her nest cavity. This species was once abundant throughout the pine savannas (communities dominated by grasses intermixed with pine trees) of the United States but has been severely reduced in numbers by the loss of its required habitat.

(A)

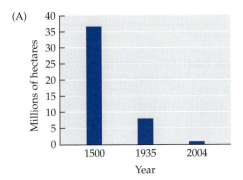

FIGURE 23.2 Decline of the Longleaf Pine Savanna Community (A) The estimated area covered by longleaf pine savanna at different times. The cover of this community has not changed from 2004 to the present. (B) As seen in this photograph from the southeastern United States, longleaf pine (*Pinus palustris*) savanna consists of open forest with a grass understory. (A after D. H. Van Lear et al. 2005. *For Ecol Manage* 211: 150–165; based on data from C. C. Frost. 1993. *Tall Timbers Fire Ecol Conf* 18: 17–44; W. G. Wahlenberg. 1946. Longleaf pine: Its use, ecology, regeneration, protection, growth, and management. C.L. Pack Forestry Foundation & USDA Forest Service; USDA Natural Resource Conservation Service, Longleaf Pine Initiative: Washington, DC. Accessed November 13, 2019. https://www.nrcs.usda.gov/wps/portal/nrcs/detailfull/national/home/?&cid=nrcsdev11_023913.)

? Estimate the hectares of longleaf pine savanna that existed in 1500, 1935, and 2004. Was the annual loss of longleaf pine savanna greater from 1500 to 1935, or from 1935 to 2004?

(B)

© iStock.com/cturtletrax

is evidence of genetic inbreeding among the birds, and in 1989, Hurricane Hugo killed 70% of the birds in one population.

The recent history of the red-cockaded woodpecker reflects that of thousands of other imperiled species around the world that have experienced gradual population declines with extensive loss of habitat, to critically low numbers. Species that require a specific habitat that is degraded by human activities will experience reductions in populations until, in some cases, they vanish. What can be done to protect species such as the red-cockaded woodpecker? Do we have a responsibility to protect existing biodiversity and to restore some of what has been lost? If so, how can we best allocate our limited resources to be most effective in our conservation efforts?

Introduction

Over the last few centuries, as the human population has grown and disproportionately increased its use of resources, many species have lost their habitats through direct destruction or through changes in their biological or physical properties. These changes have precipitated a major increase in the rate of species extinctions and loss of diversity. The 2021 analysis of the Red List of Threatened Species, compiled by the International Union for Conservation of Nature and Natural Resources (IUCN), estimates that 41,459 species are threatened with extinction—28% of the species that have been evaluated and almost 2% of all

described species worldwide (**TABLE 23.1**). This number is certainly an underestimate, as only 7% of the species that have been described have been evaluated, and many species have yet to be taxonomically described.

Ecologists play an important role in measuring the losses of species and their underlying causes. As we'll see in this chapter and the next, ecologists are also one part of a diverse team working to find ways to slow the decline of species and their habitats. We'll begin by introducing you to the field of biology dedicated to reversing those declines: conservation biology.

CONCEPT 23.1 Conservation biology is an integrative discipline that applies the principles of ecology to the protection of biodiversity.

LEARNING OBJECTIVES

23.1.1 Describe the different biological levels of diversity associated with conservation biology.

23.1.2 Evaluate the reasons why preserving biodiversity may be important.

23.1.3 Assess the factors that prompted the rapid development of the field of conservation biology.

Conservation Biology

The preservation of longleaf pine savanna at the Fort Bragg military base (described in the Case Study) and on other federal and state lands, coupled with legal protection and extraordinary human effort, has led to stabilization and slow recovery of the numbers of red-cockaded woodpeckers, leading to the U.S. Fish and Wildlife Service proposing a move from endangered to threatened status in 2020. As we'll see in the Case Study Revisited, this slow recovery has required expertise from biological disciplines such as population biology, genetics, and pathology as well as contributions from disciplines outside biology, including law, economics, political science, communications, and sociology. It has also required working

TABLE 23.1

Global Summary of the Number of Documented Imperiled Species

Group	Estimated number of described species	Number of species evaluated by 2022	Number of threatened species by 2022	Estimated percentage of species threatened in 2022
Vertebrates				
Mammals	6,577	5,969	1,337	26%
Birds	11,162	11,162	1,409	13%
Reptiles	11,690	10,150	1,845	21%
Amphibians	8,463	7,316	2,515	41%
Fishes	36,248	24,356	3,548	*
Subtotal	74,140	58,953	10,654	
Invertebrates				
Insects	1,053,578	12,161	2,291	*
Molluscs	84,528	9,017	2,384	*
Crustaceans	80,122	3,197	745	*
Corals	5,574	846	232	*
Arachnids	110,615	441	251	*
Velvet worms	210	11	9	*
Horseshoe crabs	4	4	2	100%
Others	157,543	904	152	*
Subtotal	1,492,174	26,581	6,066	
Plants				
Mosses	21,925	282	165	*
Ferns and allies	11,800	747	288	*
Gymnosperms	1,113	1,046	436	42%
Flowering plants	369,000	59,222	23,551	*
Green algae	12,382	16	0	*
Red algae	7,480	58	9	*
Subtotal	433,700	61,371	24,449	
Fungi & Protists				
Lichens	17,000	86	62	*
Mushrooms, etc.	120,000	511	222	*
Brown algae	4,485	15	6	*
Subtotal	141,485	612	290	*
Total	**2,131,499**	**147,517**	**41,459**	

Source: IUCN. 2022. Summary Statistics Table 1a. In: The IUCN Red List of Threatened Species. Version 2022-1. https://www.iucnredlist.org/resources/summary-statistics. Downloaded on September 5, 2022.

Note: "Imperiled" includes the IUCN Red List categories "critically endangered," "endangered," and "vulnerable." Some groups have been more thoroughly evaluated (mammals, birds) for conservation status than other groups, for which only a small percentage of described species have been evaluated. For those groups, there may be a bias toward completing assessments of imperiled species and making assessments of more common species a lower priority, and thus the estimate of the percent of species is not included. That only 1% of described species are shown as imperiled is an artifact of incomplete evaluation, as the percentage is believed to be much higher. An asterisk (*) indicates insufficient coverage for accurate estimate.

with farmers, landowners, the U.S. military, and the business community. Arriving at a successful management approach required not only data collection and analysis, but also creativity and the ability to work with a wide variety of people with interests and concerns (stakeholders) in the longleaf pine savanna ecosystem. Such an integrative approach is characteristic of conservation biology.

Conservation biology is the scientific study of the amount of biodiversity (including genetic diversity, species richness, and landscape diversity), how human activities are impacting it, and how best to maintain it and prevent its loss. **Biodiversity** includes genetic diversity within a species, the number of species, and the diversity of communities across landscapes (see Figure 16.7). Conservation biology applies many of the ecological principles and tools that you have studied in this book to the halting or reversal of biodiversity declines. Later in this chapter, we will look at the reasons why biodiversity is declining and at the strategies conservation biologists use to address conservation problems. But first let's consider why it is so important to prevent and reverse declines in biodiversity.

Protecting biodiversity is important for both practical and moral reasons

People rely on nature's diversity. In addition to the hundreds of domesticated species that sustain us, we make abundant use of wild species for food, fuel, and fiber. We harvest wild species for medicines, building materials, spices, and decorative items. Many people rely on these natural resources for their livelihoods. As discussed in Concept 19.4, the natural functioning of biological communities provides valuable services to humans. All of us are dependent on a wide range of these **ecosystem services**, such as water purification, generation and maintenance of soils, pollination of crops, climate regulation, and flood control (Costanza et al. 2014). These life-sustaining functions are themselves dependent on the integrity of natural communities and ecosystems. Furthermore, our emotional health is improved by time spent surrounded by nature's beauty and complexity. Spiritually, we go to natural ecosystems for solace, wonder, and insight.

But beyond our physical dependence on biodiversity, do we have some moral obligation to the other species that inhabit Earth? For many people, biodiversity has inherent value and warrants protection simply for that reason. For others, religious or spiritual beliefs lead to a sense of stewardship, or to the view that other species have a right to exist just as we do. Still others, however, do not share these views and see natural resources primarily as commodities that benefit human society.

The field of conservation biology arose in response to global biodiversity losses

Scientists have long been aware that human activity affects the abundances and distributions of organisms. In the nineteenth century, Alfred Russel Wallace, the "father of biogeography" whose work we described in Concept 18.2, foresaw the current biodiversity crisis, warning in 1869 that humanity was at risk of obscuring the record of past evolution by bringing about

extinctions. In the United States, there was a rising public outcry over the rapid decline of bison in the West, the stunning harvest to extinction of the passenger pigeon (**FIGURE 23.3**), the extensive use of bird feathers in ladies' hats, and other assaults on animal populations.

Ecologists in the United States in the first half of the twentieth century were divided over how strongly they could advocate for the preservation of nature while still maintaining scientific objectivity (Kinchy 2006). Before 1945, the Ecological Society of America frequently lobbied Congress for the establishment of national parks or for better management of existing parks. In 1948, however, the society decided to separate "pure" science from advocacy, and the Ecologists' Union branched off as an independent entity focused on the preservation of nature. In 1950, this offshoot organization changed its name to The Nature Conservancy, rising in prominence as a nonprofit organization that integrates science with advocacy and on-the-ground conservation work (Burgess 1977).

Conservation biology emerged as a scientific discipline in the early 1980s as ecologists and other scientists saw the need to apply their knowledge to the preservation of species and ecosystems. The Society for Conservation Biology, founded in 1985, arose in response to the biodiversity crisis. The emergence of professional journals dedicated to conservation biology during the 1980s and 1990s, and an ongoing increase in the number of academic programs for the training of graduate students and professionals,

FIGURE 23.3 The Passenger Pigeon: From Great Abundance to Extinction The passenger pigeon (*Ectopistes migratorius*), once one of the most abundant birds in North America, was hunted extensively in the nineteenth century. The last passenger pigeon died in the Cincinnati Zoo in 1914. The ecological effects of its extinction on the eastern deciduous forest, coincident with the loss of the American chestnut (see Concept 13.4), are difficult to estimate but are presumed to be considerable.

demonstrate the growing acceptance of and need for this specialized discipline.

Conservation biology is a value-based discipline

The methods of science call for objectivity—an assurance that the collection and interpretation of data are unbiased by preconceived ideas. Yet science is not free of human values, and it inevitably takes place within a larger social context. Conservation biologists have had to come to terms with the implicit and explicit values that are part of their work. From the founding of the Society for Conservation Biology, the designation of the discipline as "mission driven" (Soulé and Wilcox 1980; Meine et al. 2006) and "crisis-oriented" (Soulé 1985) explicitly revealed the values behind the science.

Many ecologists have chosen to speak up or refocus their research programs as they have come to understand the biological consequences of the changes taking place on the planet. For example, in 1986, Dan Janzen, a tropical biologist who had largely committed himself until then to studying tropical plant–insect interactions, wrote that "if biologists want a tropics in which to biologize, they are going to have to buy it with care, energy, effort, strategy, tactics, time, and cash." Such motivation does not necessarily detract from the objectivity of the scientific studies done by conservation biologists, as they understand that conserving biodiversity will require decisions based on sound and credible analyses, and weighing the trade-offs associated with conservation versus resource extraction. Furthermore, those analyses are subjected to rigorous scientific review by other scientists, who may challenge or even refute their conclusions.

In the next section, we'll meet one ecologist who put the values of conservation biology into practice. Then we'll examine the extent and causes of the current declines in biodiversity.

> **CONCEPT 23.2** Biodiversity is declining globally.

LEARNING OBJECTIVES

23.2.1 Differentiate between the current anthropogenically enhanced rate of extinction and the long-term background extinction rate.

23.2.2 Describe the pathway to species extinction from changes in population growth to the disappearance of the species.

Declining Biodiversity

The tropical botanist Alwyn Gentry devoted his life to identifying, classifying, and mapping the immense diversity of plants found in Central and South America. He also became an eyewitness to plant species extinctions as the region underwent rapid deforestation. It was not uncommon for him to identify a new *endemic* plant species (i.e., a species that occurs in a particular geographic region and nowhere else) during an expedition to Ecuador or Peru, only to return to the same spot a few years later to find the forest cleared and the species gone (Dodson and Gentry 1991) (**FIGURE 23.4**). Gentry worked with a

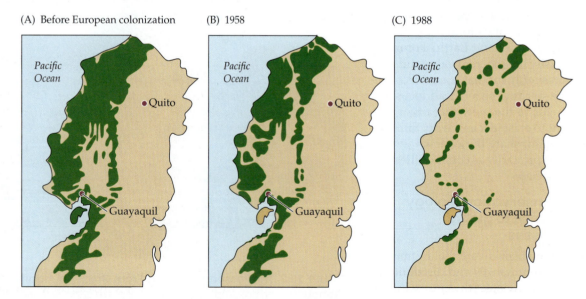

FIGURE 23.4 **Loss of Forest Cover in Western Ecuador**
Between 1958 and 1988, a growing human population and government policies intended to stimulate economic development led to rapid deforestation in western Ecuador. Green indicates forest cover. The extensive loss of forest habitat in this region is estimated to have resulted in the loss of more than 1,000 endemic species. (After C. H. Dodson and A. H. Gentry. 1991. *Ann Mo Bot Gard* 78: 273–295. Permission granted by Missouri Botanical Garden Press, St. Louis.)

growing sense of urgency to identify rare species in order to protect them from this fate. His death in a plane crash in the Ecuadorian forest in 1993, while doing an aerial survey of land proposed for conservation, cut this work short and was an enormous loss to conservation biology.

Gentry was just one of many taxonomists who have been finding and describing species while witnessing their rapid disappearance due to habitat destruction, disease, or climate change. Extinctions of barely known tropical species continue despite our decades-long recognition of the problem. Through greater efforts to explore Earth's ecosystems, ecologists are gaining knowledge of the world's biota and tabulating new species at a faster rate, but threats to those species are keeping pace with such gains in our knowledge about them.

The rate at which Earth is losing species is accelerating

How rapidly are species being lost? That is a difficult question to answer, in part because we do not know how many species exist that remain unknown to us. Most studies have estimated that there are about 5 million to 10 million eukaryotic species on Earth, but there may be as few as 3 million or as many as 50 million (Scheffers et al. 2012) or even more, particularly with greater consideration of microbial diversity (Locey and Lennon 2016).

Despite this uncertainty, extinction rates can be estimated using several indirect measures (May et al. 1995; May 2011). For example, estimates of extinction rates from the fossil record can be used to establish a "background" extinction rate with which current rates can be compared. For the best-known taxonomic groups, the mammals and birds, paleontologists have estimated that the background extinction rate is around one extinction every 200 years, which is equivalent to an average species life span of 1 million to 10 million years. By contrast, there was about one extinction per year among the mammals and birds over the twentieth century, which is equivalent to an average species life span of only 10,000 years. Thus, overall, the rate of extinction in the twentieth century was 100 to 1,000 times higher than the background rate estimated from the fossil record (Jablonski 2005).

A second method for estimating extinction rates uses the species–area relationship discussed in Concept 18.3. In particular, the relationship between number of endemic species and area is used to estimate the number of species that would be driven to extinction by a given amount of habitat loss (Kinzig and Harte 2000). In a third approach, biologists have used changes over time in the assessed conservation statuses of species (e.g., a shift from endangered to critically endangered) to forecast rates of extinction (Smith et al. 1993). Finally, a fourth approach is based on the rates of population decline or range contraction of common species (Balmford et al. 2003). All of these methods have uncertainties affecting their estimates of extinction rates, yet they are the best ways we have to document losses of biodiversity.

It can also be difficult to ascertain when a species is definitely extinct. Many species are known from a single specimen or location, and the logistics of relocating them can be daunting. Even an exhaustive hunt for a very rare species can fail to detect some remnant populations. Declaring a species extinct, however, has been known to stimulate biologists' search efforts, recently aided by the use of drones. Since the publication of a flora of Hawaiian plants in 1990, for example, 35 species listed there as extinct have been rediscovered, though only a few individuals have been found. The joy of their rediscovery is compromised by the realization that these extremely small populations cannot serve the same ecological functions as more substantial populations, and that 8% of Hawaii's native flora of 1,342 species is now considered extinct (Wagner et al. 1999).

Although humanity's growing ecological footprint (see Connections in Nature in Chapter 10) has accelerated the rates of biodiversity loss over the last century, people have had substantial effects on Earth's biota for millennia (see the Case Study in Chapter 3). David Steadman (1995) described how bones found on Pacific islands revealed the prehistoric extinction of up to 8,000 species of birds (of which perhaps 2,000 species were endemic flightless rails) after these islands were colonized by Polynesians. Most of these species were island endemics, and in some cases the extinctions encompassed entire ecological guilds (**FIGURE 23.5**). Ecologists can only speculate about the roles the lost frugivores and nectarivores played in maintaining endemic tree populations. Steadman's findings remind us that extinctions do not only eliminate individual species, but can also cause large changes in ecological communities.

Extinction is the end point of incremental biological decline

In 1954, Andrewartha and Birch wrote that "there is no fundamental distinction to be made between the extinction of a local population and the extinction of a species other than this: that the species becomes extinct with the extinction of the last local population." Sometimes the populations of a species gradually erode away, and sometimes they vanish in a spectacular collapse, as in the case of the passenger pigeon described earlier.

Conservation biologists have approached the process of biological decline and extinction in numerous ways. For example, as we saw in Concept 11.3, small populations are particularly vulnerable to genetic, demographic, and environmental stochasticity, each of which can reduce the population growth rate and increase the risk of local extinction as the population size declines further. Known as an **extinction vortex**, this pattern can doom a population to eventual extinction once its size drops below a certain point. With this in mind, Caughley (1994) argued

(A) All species

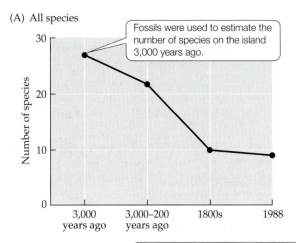

Fossils were used to estimate the number of species on the island 3,000 years ago.

(B) Guilds

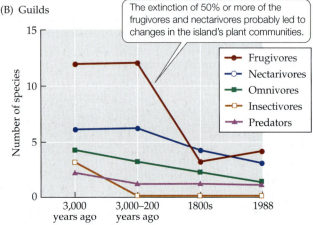

The extinction of 50% or more of the frugivores and nectarivores probably led to changes in the island's plant communities.

- Frugivores
- Nectarivores
- Omnivores
- Insectivores
- Predators

FIGURE 23.5 Humans Have Been Causing Extinctions for Millennia Trends over time in (A) the total number of bird species and (B) the number of species classified by feeding guild found in the Pacific island 'Eua in the nation of Tonga. Prehistoric extinctions (3,000–200 years ago) occurred on many Pacific islands as a result of hunting and the introduction of rats, dogs, and pigs. (After D. W. Steadman. 1995. *Science* 267: 1123–1131.)

? Speculate on reasons why losses of birds that feed on fruit (frugivores) or nectar (nectarivores) may have affected the island's plant communities. (Hint: See the discussion of mutualism in Concepts 15.1 and 15.2.)

that it is important to determine the causes of population declines in particular species, with the aim of identifying actions that could counteract these declines before the extinction vortex takes hold.

Ecologists may also study the declines of species using a spatial context by tracking changes in their geographic ranges. Ceballos and Ehrlich (2002) examined patterns of range contraction in 173 declining mammal species worldwide. They found that, collectively, these species had lost 68% of their range area over the past 100–200 years, with the greatest losses in Asia (83%). In a similar study, Channell and Lomolino (2000) examined patterns of range contraction in 309 declining species. They found that a decline often moves through the historical range

of a species like a wave, from one end to the other; this could occur, for example, if an invasive species entered the range at one edge and then spread through the range, eliminating the declining species population by population. Such a pattern contrasts with a retreat from all edges of the range into its center, which would probably occur if effects of small population size were prevailing.

When populations are lost from an ecological community, there are consequences not only for the declining species, but also for its predators, prey, competitors, and mutualistic partners. The loss of bird pollinators, for example, can reduce the reproductive success of plants that depend on those pollinators (**FIGURE 23.6**), causing plant densities to drop as well (Anderson et al. 2011; Galetti et al. 2013). The resulting changes at the community level may bring about secondary extinctions and ultimately affect ecosystem processes. Examples from earlier chapters include the local extinctions and other changes caused by the loss or removal of such species as the amphipod crustacean *Corophium volutator* (see Concept 13.4), the marsh plant *Juncus gerardii* (see Concept 16.3), and the sea star *Pisaster ochraceus* (see Concept 21.4). Modeling results also suggest that while food webs can be resilient to species removal, the loss of certain species can trigger a cascade of secondary extinctions. As might be expected, the stronger the interactions of a species in the food web, the greater the effect of its removal (Solé and Montoya 2001). Overall, both empirical and modeling results demonstrate that incremental species loss can have broad ecological consequences.

Earth's biota is becoming increasingly homogenized

Organisms are naturally mobile, which influences their dispersal across their geographic ranges, although they are still subject to dispersal barriers such as oceans and mountain ranges. Over the last century, however, people have moved over Earth's surface at an unprecedented rate, carrying organisms with them and greatly enhancing rates of introductions of new species to all parts of the globe (**FIGURE 23.7**).

While the introduction of non-native species can increase local diversity, they generally have negative effects on native species diversity. For example, introductions of non-natives can contribute to the range contractions of native species whose numbers may already be in decline because of habitat loss and other factors. Typically, the greatest "losers" among the native species tend to be specialists—those with morphological, physiological, or behavioral adaptations to a particular habitat—while the "winners" tend to be generalists with less stringent habitat requirements. The spread of non-native species and native generalists, coupled with declining abundances and distributions of native specialists, is part of a growing **taxonomic homogenization** of Earth's biota (Olden et al. 2004). In rare circumstances, non-native species can

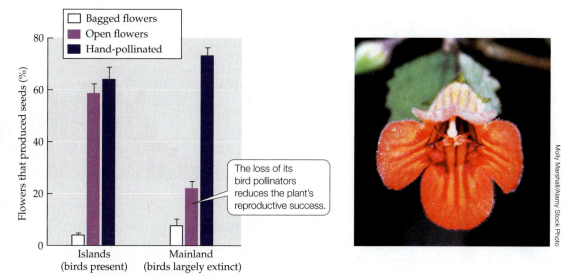

FIGURE 23.6 Loss of Bird Pollinators Reduces Reproductive Success in a New Zealand Shrub Birds that pollinate the shrub *Rhabdothamnus solandri* are nearly extinct on the New Zealand mainland, but densities of these birds remain high on nearby islands. Researchers recorded the percentage of *R. solandri* flowers that reproduced successfully (produced seeds) on island and mainland sites for each of three treatments: bagged flowers (which allowed only self-pollination), open flowers (which allowed bird pollination), and open flowers that were hand-pollinated. Error bars show one SE of the mean. (After S. H. Anderson et al. 2011. *Science* 331: 1068–1071.)

? Identify the control and experimental treatments in this study, and explain what can be learned from each of the three treatments.

provide conservation benefits, such as habitat or food for rare species (Schlaepfer et al. 2011). Non-native tamarisk shrubs (*Tamarix* spp.), for example, provide nesting habitat for the endangered southwestern willow flycatcher (*Empidonax traillii extimus*). Species introductions have also increased regional biodiversity in many parts of the world (**FIGURE 23.8**), although the value of increasing diversity through the increase in non-natives is questionable, as it typically comes at a cost to native diversity.

Island biotas are particularly vulnerable to both invasion and extinction. The decline of island endemics is often accelerated by the introduction of more cosmopolitan species. In a survey of American Samoa, Robert Cowie (2001) found just 19 of the 42 species of land snails historically known

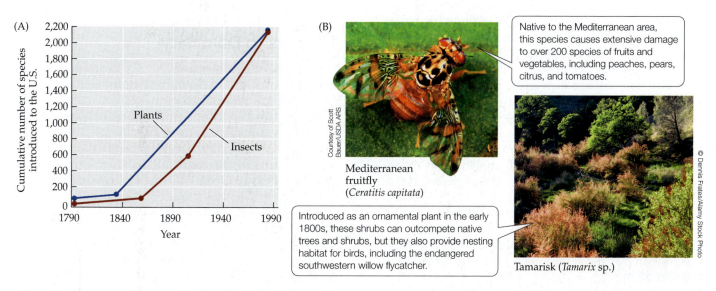

FIGURE 23.7 Species Introductions Are Increasing Globally The number of non-native species that have become established in the United States has increased about fivefold over the past century for various organisms, including molluscs, fishes, terrestrial vertebrates, and (A) plants and insects. Similar patterns are seen in many other countries. Photographs in (B) show two examples of introduced species. (After U. S. Congress, Office of Technology Assessment. 1993. *Harmful Non-Indigenous Species in the United States*. U.S. Government Printing Office: Washington, DC, based on contractor reports done for OTA.)

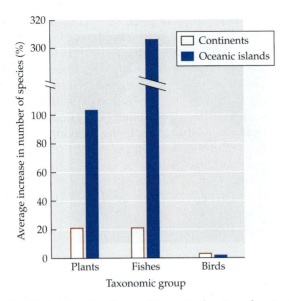

FIGURE 23.8 Introductions of Non-Native Species Can Increase Regional Biodiversity The introduction of non-native species to new regions has led to sizable increases in the numbers of species found on oceanic islands and within continental regions for plants and fishes, but not for birds. (After D. F. Sax and S. D. Gaines. 2003. *Trends Ecol Evol* 18: 561–566.)

❓ The introduction of non-native plants to new regions is associated with a decrease in the global diversity of plants. Explain how that can be true given the results shown in this figure.

from that island group, plus 5 species not previously found there but which he presumed were native. He also found that there were 12 non-native species present on the islands. These non-natives occurred in high abundances, representing about 40% of the individuals collected (there was also one abundant native species). Cowie concluded that most native species were declining in abundance, while many non-natives were increasing. Furthermore, the predators contributing to the declines of native land snail species were also non-natives, such as the predatory snail *Euglandina rosea* and the house mouse (*Mus musculus*). Cowie has found this trend toward homogenization of land snail faunas to be widespread among Pacific islands.

Homogenization has also been observed among the freshwater fishes of the United States, largely as the result of widespread introductions of game fishes. Rahel (2000) quantified the homogenization of U.S. fish faunas by examining the change in the number of species shared between all possible pairs of the 48 conterminous states. He found that, on average, pairs of states shared 15 more species than they did at the time of European colonization (**FIGURE 23.9**).

On a global scale, it is clear that biodiversity is being lost as a result of humanity's impact on the planet. Let's look in more detail at the reasons for these losses, and then consider what steps can be taken to counteract them.

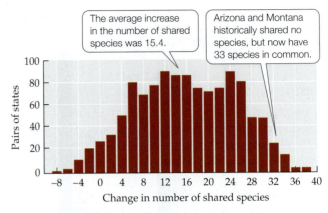

FIGURE 23.9 U.S. Fish Faunas Are Undergoing Taxonomic Homogenization The numbers of fish species shared by pairs of the 48 conterminous U.S. states have increased since European settlement. (After F. J. Rahel. 2000. *Science* 288: 854–856.)

CONCEPT 23.3 Primary threats to diversity include habitat loss, invasive species, overexploitation, pollution, disease, and climate change.

LEARNING OBJECTIVES

23.3.1 Compare the most important threats to diversity over the past century with the threats that are increasing the most in importance.

23.3.2 Describe the causes of diversity losses associated with habitat loss and degradation.

23.3.3 Explain the underlying mechanisms determining how invasive species, overexploitation, and water and air pollution lead to diversity loss.

Threats to Diversity

Understanding the causes of diversity losses is a first step toward reversing them. Multiple factors are likely to contribute to the decline and eventual extinction of any particular species. For example, while the last Pyrenean ibex (*Capra pyrenaica pyrenaica*) was killed in 2000 by a falling tree, declines in its populations following the fourteenth century resulted from overexploitation and competition with domesticated livestock, leading to its eventual extinction (Perez et al. 2002).

Multiple causes of diversity loss are also apparent in higher taxonomic groups. For example, over 1,223 mammal species (25% of those for which adequate data are available) are currently threatened with extinction (see Table 23.1). Globally, the primary threats facing mammals are loss of habitat, overexploitation, accidental mortality (e.g., road kills), and pollution—but the relative importance of these factors differs between terrestrial and marine mammals (**FIGURE 23.10**). Some mammals are threatened by additional factors, such as disease. As we'll see, this scenario, in which multiple types of threats contribute to the decline and extinction of a taxon, is common.

Habitat loss and degradation are the most important threats to diversity

The next time you fly in an airplane over Earth's surface, look down and ask yourself, "What species lived here before these farms and cities were here? Where do the species native to this place live now, and how do they move about?" From 30,000 feet above the landscape, you will find yourself face to face with the source of the diversity crisis: the scale of the human impact on the planet. Earth has been modified greater than 70% of its land surface (Winkler et al. 2021), and all marine ecosystems have been affected by humans (Halpern et al. 2008). One species, *Homo sapiens*, is now appropriating about 25% of Earth's primary production (Haberl et al. 2014).

The influence of human activities on natural habitat is the most important factor contributing to global declines in diversity (Sax and Gaines 2003). There are areas of extreme human influence, such as agricultural regions and certain coastal waters, and areas of little human influence, such as deserts and some polar seas. Overall, however, most of the lands and waters of Earth are at least moderately affected by humans (see Figure 3.5B). Addressing the loss, fragmentation, and degradation of habitat caused by human activities is central to conservation work. **Habitat loss** refers to the outright conversion of habitat to another use, such as urban development or agriculture, while **habitat fragmentation** refers to the breaking up of once continuous habitat into a series of habitat patches amid a human-dominated landscape. **Habitat degradation** refers to changes that reduce the quality of the habitat for many, but not all, species. Concepts 24.2 and 24.3 will address habitat fragmentation and its effects in detail; in this and the following sections, we'll cover habitat loss and habitat degradation.

On a continental scale, the extent of loss of some habitats is staggering (see Figure 24.12). Similar losses can be observed on more local scales, as in the forests of western Ecuador (see Figure 23.4). Another example is provided by the Atlantic forest of Brazil (Ranta et al. 1998). This moist tropical forest has many endemic species, perhaps because it has been isolated from the Amazon rainforest for millions of years. Of South America's 904 mammal

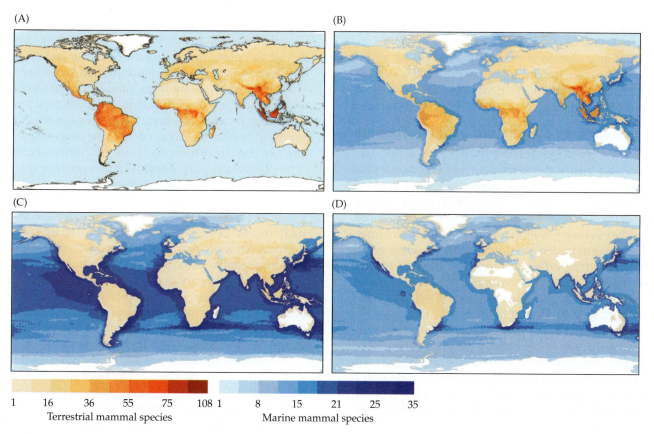

FIGURE 23.10 Threats to Mammal Species Globally, 22% of mammal species are threatened by extinction. These maps show the numbers of terrestrial and marine mammal species in various parts of the globe that are negatively affected by (A) habitat loss, (B) overexploitation, (C) accidental mortality, and (D) pollution. (From J. Schipper et al. 2008. *Science* 322: 225–230.)

? Contrast the threats to land mammals with those to marine mammals.

species, 73 are endemic to this forest, and 25 of those endemics are threatened with extinction. The forest's location also coincides with that of 70% of Brazil's human population. As a result, more than 92% of this habitat has been cleared to make room for agriculture and urban development, and what remains has been highly fragmented, pushing many species to endangerment.

How has the loss of Atlantic forest habitat affected diversity? Brooks and colleagues (1999) asked why there have been no reports of extinctions among birds of this region. They offered three possible explanations, which may apply to patterns of biological decline in other regions as well. First, the birds may be adjusting to living in forest fragments. Second, the most vulnerable species might have gone extinct before they were known to biologists. Their third explanation, which they see as the most plausible, is that the time lag between deforestation and extinction has not yet played out. By 2021 five to seven species of birds were reported to have gone extinct, with nine more listed as critically endangered (Develey and Phalan 2021). Unless drastic measures are taken, additional species are doomed to extinction. Moreover, the loss of bird species will have negative effects on other species. As bird populations in the Brazilian Atlantic forest have dwindled to low numbers, reductions in seed size and seedling survival have been observed in plant populations that depend on these birds for seed dispersal (Galetti et al. 2013).

Habitat degradation is extremely widespread, and it has diverse causes, including invasive species, overexploitation, and pollution. We'll turn now to one of those causes, invasive species.

Invasive species can displace native species and alter ecosystem properties

As discussed earlier, the introduction of non-native species generally has negative effects on diversity. Here, we'll consider how declines in diversity can be caused by the arrival of these **invasive species**: non-native, introduced species that sustain growing populations and have large effects on communities. Worldwide, 20% of endangered vertebrates are imperiled as a result of invasive species (MacDonald et al. 1989), with as many as 75% of vertebrate species on islands going extinct as a result of invasive mammals (McCreless et al. 2016).

Invasive species are of particular concern where they compete with, prey on, or change the physical environment of endangered native species. The effect of the Eurasian zebra mussel (*Dreissena polymorpha*) on the freshwater mussel species of North America is a prime example (see Figure 19.5). North America is the center of diversity for freshwater mussels (bivalves of the order Unionoida), with 297 species, a third of those in the world. Prior to the invasion of the zebra mussel in the late 1980s, North American freshwater mussels were already in trouble. Most of these species are globally imperiled, many are endemic and thus naturally rare,

and all are threatened by water pollution and river channelization. Competition with zebra mussels has brought about steep declines in populations of native freshwater mussels (60%–90%), including some regional extinctions (Strayer and Malcom 2007).

Invasive predators can also contribute to extinctions. In Lake Victoria, introduction of the Nile perch (*Lates niloticus*) has reduced the diversity and abundance of the native cichlid fishes, a group that shows *adaptive radiation* (a phenomenon discussed in Concept 6.4), with many species in specific habitats. Historically, about 600 species of cichlids had been recorded, most of which were endemic to Lake Victoria. The Nile perch is a large predator, and its introduction into the lake in the early 1960s as a food source for human populations has contributed to the extinction of roughly 200 cichlid species. Before the introduction, the cichlids made up 80% of the biomass of fish in the lake; the Nile perch now accounts for 80% of the biomass. As is often the case, more than one factor is driving the cichlids' decline: pollution and overfishing augment the negative effect of predation by the Nile perch (Seehausen et al. 1997).

In many ecosystems, habitat loss and degradation have increased vulnerability to invasion by non-native species, which in turn may lead to consequences that further degrade the ecosystem. The tropical dry forest of Hawaii, for example, harbors more than 25% of Hawaii's threatened plant species. The area of tropical dry forest has been reduced by 90% since human settlement. The arrival of invasive feral hogs, rats, and plants has made a bad situation worse. In addition to outcompeting and displacing local plants, invasive grasses are an excellent source of fuel for fires. As a result, the frequency of fires has increased (see Analyzing Data 9.1), furthering the decline of Hawaiian dry forests but favoring the spread of the fire-adapted, introduced grasses.

Ecosystem properties such as nitrogen cycling (see Figure 22.11) can be altered by some invasive species. One such species is kudzu (*Pueraria montana*), an invasive vine that covered more than 3 million ha (7.4 million acres) in the southeastern United States at its peak spread. This species disrupts communities by outcompeting other plants for light (see Figure 14.4). In addition, kudzu can fix up to 235 kg of nitrogen per hectare per year, an amount that far exceeds the atmospheric deposition of nitrogen in the eastern United States (7–13 kg N/ha/year).

To examine the extent to which nitrogen fixation by kudzu affects the nitrogen cycle, Hickman et al. (2010) measured the nitrogen mineralization rate in plots with and without kudzu (as discussed in Concept 22.2, the nitrogen mineralization rate provides an estimate of the rate at which nitrogen is supplied to plants). On average, nitrogen mineralization rates increased more than sevenfold in plots invaded by kudzu (**FIGURE 23.11**), indicating a large effect on soil nitrogen supply. In addition, more than twice as much of the gas nitric oxide (NO) was

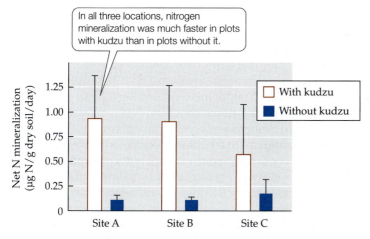

FIGURE 23.11 Invasive Species Can Alter the Nitrogen Cycle At three sites in Georgia, net nitrogen mineralization rates (an index of how rapidly nitrogen cycling occurs in an ecosystem) were much higher in soils supporting kudzu than in soils with native vegetation. Error bars show one SE of the mean. (After J. E. Hickman et al. 2010. *Proc Natl Acad Sci USA* 107: 10115–10119.)

released from the soil in plots invaded by kudzu as in plots lacking kudzu (see **ANALYZING DATA 23.1** to test whether NO emissions at one of the study sites differ statistically between plots with kudzu and plots lacking kudzu). In the atmosphere, NO participates in chemical reactions that produce ground-level ozone, a pollutant that affects human health and agricultural production (see Concept 25.4). Modeling results suggest that kudzu has the potential to increase the number of high-ozone event days by as many as 7 days per summer across broad regions of the southeastern United States (Hickman et al. 2010). Additionally, the greater supply of soil nitrogen fosters greater spread of invasive plants.

As we saw in the Case Study of the invasive alga *Caulerpa taxifolia* in Chapter 16, control or eradication of invasive species is difficult, labor-intensive, and expensive, but at times it may be warranted in the interest of protecting economically or culturally valuable native

---

ANALYZING DATA 23.1

Do Nitric Oxide Emissions Differ Statistically between Plots with and without Kudzu?

Hickman et al. (2010)* examined the impact of the invasive species kudzu (*Pueraria montana*) on nitric oxide (NO) emissions at three study sites in Georgia. NO is an important contributor to pollutant ozone formation. At each site, NO emissions were recorded from four plots with kudzu and four plots lacking kudzu.

Data from one study site are presented in the table. In this exercise, you will perform a statistical test (the *t*-test) to determine whether NO emissions in plots invaded by kudzu are significantly different from NO emissions in plots lacking kudzu.

Nitric Oxide Emissions (ng N/cm²/hr)	
Plots with kudzu	Plots lacking kudzu
4.1	2.0
1.7	0.9
6.1	1.1
2.8	0.9

1a. What is the sample size (*n*) for plots with kudzu and plots without kudzu?

1b. Using the definitions provided below, calculate the mean ($\bar{x}$) and standard deviation (*s*) of NO emissions for plots invaded by kudzu and for plots lacking kudzu. What do your results suggest?

2. The *t*-test provides a standardized way to determine whether the means of two treatments differ enough from one another to be considered

"significantly different." The *t*-test is based on calculation of the *T* statistic, defined in the Definitions list. Calculate the *T* statistic using the data provided above.

3. Determine the "degrees of freedom" and "*p* value" associated with the value you obtained for *T*. Interpret the results of your *t*-test.

DEFINITIONS

Mean: For *n* data points $x_1, x_2, x_3, \ldots, x_n$, the (arithmetic) mean ($\bar{x}$) equals

$$\bar{x} = \frac{(x_1 + x_2 + x_3 + \ldots + x_n)}{n} = \frac{1}{n}\sum_{i=1}^{n} x_i$$

Standard deviation: For *n* data points $x_1, x_2, x_3, \ldots, x_n$, the standard deviation (*s*) equals

$$s = \sqrt{s^2} = \sqrt{\frac{1}{n-1}\sum(x_i - \bar{x})^2}$$

T statistic: When comparing the means of two samples, each of size *n*, the *T* statistic equals

$$T \frac{\bar{x}_1 - \bar{x}_2}{\sqrt{\frac{1}{n}\left(s_1^2 + s_2^2\right)}}$$

---

* Hickman, J. E., et al. 2010. Kudzu (*Pueraria montana*) invasion doubles emissions of nitric oxide and increases ozone pollution. *Proceedings of the National Academy of Sciences U.S.A.* 107: 10115–10119.

species or natural resources. The best strategy for combating invasive species is to prevent their arrival through careful screening of biological materials at international borders. But once potentially invasive species are present, control measures are best implemented immediately; constant vigilance and quick action are key to minimizing their effects (Simberloff 2003).

Overexploitation of species has large effects on ecological communities

Overexploitation, the harvest of wild organisms at a rate that exceeds their replacement, can also lead to loss of diversity. For example, many of the world's people obtain their food, at least in part, directly from a natural ecosystem. The problem is that as the human population increases and natural habitats shrink, the harvesting of many species from the wild has become unsustainable. Globally, overexploitation is contributing to the imperilment of species of fishes, mammals, birds, reptiles, and plants. Overexploitation has been the cause of the probable extinction of at least one primate, Miss Waldron's red colobus monkey (*Piliocolobus waldroni*), a species endemic to Ghana and Côte d'Ivoire whose last confirmed sighting was in 1978 (Oates et al. 2000; McGraw 2005).

The effects of overexploitation on tropical forests have been substantial, resulting in what Kent Redford (1992) has called an "empty forest." This phrase refers to forests that look healthy in satellite images, but in which the abundances and diversity of large vertebrates have decreased. The increased accessibility of forests as roads are built through them facilitates this overharvesting of wildlife, as does the widespread availability of guns. The enormous quantity of "bushmeat" being taken from tropical forests is sobering. Redford has calculated that 13 million mammals are killed each year in the Amazon rainforests of Brazil by rural hunters, and it is estimated that in western and central Africa, 1 million tons of forest animals are taken annually for food (Wilkie and Carpenter 1999). Vast numbers of animals are also captured from tropical forests, coral reefs, and other ecosystems and then exported legally to other countries. For example, government records indicate that from 2000 to 2006, 1.5 billion animals, most of which were for the pet trade, were imported to the United States alone (Smith et al. 2009).

In the oceans, rapid and steep declines have taken place in both the abundances (**FIGURE 23.12**) and sizes (**FIGURE 23.13**) of top-level predators (Myers and Worm 2003). For every ton of fish caught by commercial

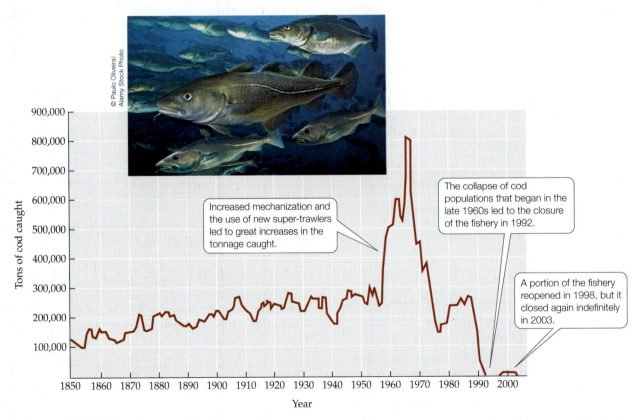

FIGURE 23.12 The Collapse of the Cod Fishery Changes over time in the amount of cod (*Gadus morhua*) caught off the coast of Newfoundland, Canada. Overharvesting led to the collapse of cod populations, which still have not recovered. (After Millennium Ecosystem Assessment. 2005. *Ecosystems and Human Well-Being: Biodiversity Synthesis.* World Resources Institute: Washington, DC.)

? Based on data prior to 1950, roughly how many tons of cod could have been harvested in a sustainable manner? Explain.

trawlers, 1–4 tons of other marine life may be brought aboard by the nets. Some organisms may survive the experience and be released back into the sea; the rest make up what is called *bycatch*. The bycatch of certain threatened species, such as marine mammals, seabirds, and marine turtles, has received attention from fisheries managers, and in some cases, losses have been reduced through changes in gear design (see Ecological Toolkit 10.1). But bycatch remains common, and concern has been raised about the ecological effects of this unnecessary mortality on marine food webs (Lewison et al. 2004). In addition, repeated trawling on the coastal sea bottom has affected benthic species such as corals and sponges and has thereby degraded benthic habitat for many other species. Studies indicate that habitat recovery following trawling is very slow (National Research Council 2002).

Whenever a species has market value, it is likely to be overharvested. This results in an unfortunate confluence between human behavior (i.e., greed) and declining animal and plant populations, when economically valuable threatened species are subjected to an "anthropogenic Allee effect" (see Figure 11.15) due to more aggressive search and harvesting strategies. Many scientists and policymakers argue that the best approach to protecting overexploited species is to determine the levels of harvest that will be sustainable and to establish regulatory mechanisms to permit only those levels to be taken. In one example of how this could be done, Bradshaw and Brook (2007) describe management options that provide revenue from meat and trophy hunting of the wild banteng (*Bos javanicus*), a member of the cattle genus, yet do not jeopardize the prospects for the recovery of this rare species.

Pollution, disease, and climate change erode the viability of populations

Human developments and activities result in air and water pollution and climate change, and are causing declines in populations of many species. Additionally, the emergence of new diseases and the transmission of diseases from domesticated animals into wildlife is having

(A)

(B)

FIGURE 23.13 Overharvesting Has Led to a Decline in the Sizes of Top Marine Predators Photographs of trophy fish caught on charter fishing boats based in Key West, Florida, in (A) 1957 and (B) 2007. In commercial and recreational fisheries, the largest fish are often the preferred prey. (C) The total length of trophy fish declined more than 50% between 1960 and 2007. Error bars show ± one SE of the mean. (C after L. McClenachan. 2009. *Conserv Biol* 23: 636–643.)

(C)

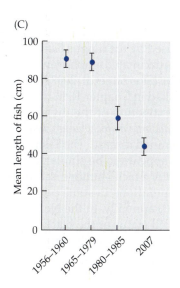

detrimental impacts on biodiversity. The effects of all these factors exacerbate declines in species already reduced by habitat loss, invasive species, or overexploitation.

Pollutants released by human activities are omnipresent in air and water. These pollutants become contributors to habitat degradation and diversity loss where they are present at levels that cause physiological stress. We will see in Concept 25.3 how some of these pollutants degrade habitats, reduce populations, and threaten the persistence of species.

One example of an emerging pollution threat is the growing concentration of persistent endocrine-disrupting contaminants (EDCs), particularly in the marine environment. As we saw in the Case Study Revisited in Chapter 21, persistent organic pollutants such as DDT, PCBs, flame retardants, and organophosphates from agricultural pesticides, some of which are EDCs, end up in marine food webs, where they are bioaccumulated and biomagnified, particularly in top predators. The number of chemicals found in marine mammals, the number of individuals affected, and the concentrations found have risen markedly in the last 40 years (Tanabe 2002). The orcas of British Columbia have been described as "fireproof killer whales" because of the extremely high levels of flame-retardant chemicals (polybrominated diphenyl ethers, or PBDEs) found in their bodies (**FIGURE 23.14**). These EDCs have been observed to interfere with reproduction, neurological development, and immune function in mammals (Ross 2006). Through their influence on hormones EDCs can turn males into females in some fish species, including the endangered pallid sturgeon (*Scaphirhynchus albus*) in the Mississippi River. Such problems for species already at low numbers do not improve the outlook for their future.

Disease has also contributed to the decline of many endangered species. In a striking example, an emerging disease caused by the fungus *Batrachochytrium dendrobatidis* has decimated amphibian populations around the globe (Skerratt et al. 2007) (see also the Case Study Revisited in Chapter 1). In the 1930s, the final decline to extinction of the thylacine, or Tasmanian wolf (*Thylacinus cynocephalus*), was hastened by an undetermined disease. The Tasmanian devil (*Sarcophilus harrisii*) appears to be similarly threatened because of the spread of a facial tumor disease (Hawkins et al. 2006), with populations in some parts of Tasmania decreasing by 50% annually. In the North American prairie, the endangered status of the black-footed ferret (*Mustela nigripes*) was exacerbated by canine distemper (Woodroffe 1999).

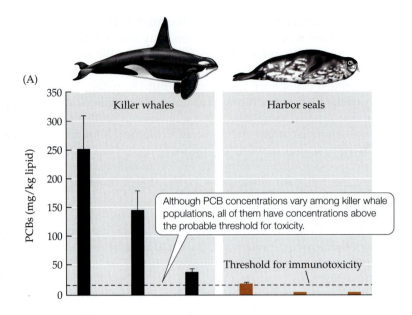

Although PCB concentrations vary among killer whale populations, all of them have concentrations above the probable threshold for toxicity.

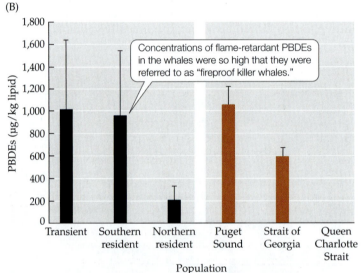

Concentrations of flame-retardant PBDEs in the whales were so high that they were referred to as "fireproof killer whales."

FIGURE 23.14 Persistent Organic Pollutants That Disrupt the Endocrine System Are a Growing Threat to Marine Mammals In British Columbia, the concentrations of PCBs (A) and PBDEs (B) found in killer whales (*Orcinus orca*) and harbor seals (*Phoca vitulina*) are very high. Error bars show one SE of the mean. (After P. S. Ross. 2006. *Can J Fish Aquat Sci* 63: 224–234, based on data from P. S. Ross et al. 2000. *Mar Pollut Bull* 40: 504–515; P. S. Ross et al. 2004. *Environ Toxicol Chem* 23: 157–165; S. Rayne et al. 2003. *Environ Sci Technol* 36: 2847–2854; P. S. Ross, unpublished data.)

CLIMATE CHANGE CONNECTION

IMPACTS ON DIVERSITY Although hundreds of species have shifted their distributions to higher latitudes or elevations in response to global warming (Parmesan 2006), only a few cases are known at this time in which species are imperiled directly by climate change (e.g., Bramble Cay melomys, a small rodent [Waller et al. 2017]).

However, the number of extinctions associated with climate change is expected to increase (Thomas et al. 2004; Wiens 2016). Throughout this book, we've emphasized that climate change has influenced and will continue to influence diversity in multiple ways. Warmer temperatures can directly impact physiological activity and behavior, influencing reproduction and mortality of individuals. We saw an example of this with changes in the length of time lizards can be active, alterations due to climate warming. As a consequence of climate change, the probability of population extinctions increases because of constraints on the amount of time the lizards can forage, which may explain local extinction of some lizard populations in Mexico (Sinervo et al. 2010). Climate change may affect how species interact and the intensity of those interactions, as exemplified by some aquatic ecosystems which have experienced increases in food web connections in a warmer world (Woodward et al. 2010). Changes in the type (antagonistic vs. facilitative) and intensity of biotic interactions make prediction of the fates of species in a warmer world challenging.

We saw in Concept 12.3 that the distribution of organisms and diversity in communities can be influenced by predation. If predators and prey respond differently to climate change, the influence of predation on diversity can be positive or negative, depending on which species is more sensitive. If prey are more sensitive to warmer temperatures than predators, then climate change will accentuate the negative impact of predation on diversity. This hypothesis was supported in the rocky intertidal zone by Christopher Harley (2011). Using a combination of experiments and observational studies employing variation in both space and time to examine variation in climate, Harley demonstrated a decrease in diversity of shellfish communities (barnacles and mussels) in the rocky intertidal zone consistent with greater predation and a restriction of habitat associated with climate change. The main predator, a sea star, was less sensitive to warming than barnacle and mussel species. The reduction in habitat increased the susceptibility of the prey species to predation, contributing to the local extinction of some species under warmer conditions. We will explore climate change in greater depth in Chapters 24 and 25. ☀

As the human population passed the 7 billion mark, our impact on the environment had already caused all of the world's biomes to be affected by the threats we have just described. However, the importance of these threats varies among biomes (**FIGURE 23.15**). Habitat loss is greater in the tropics than in the polar zones, for example, but climate change is having more of an effect in the polar zones than in the tropics. What can conservation biologists offer as solutions to these threats from so many fronts?

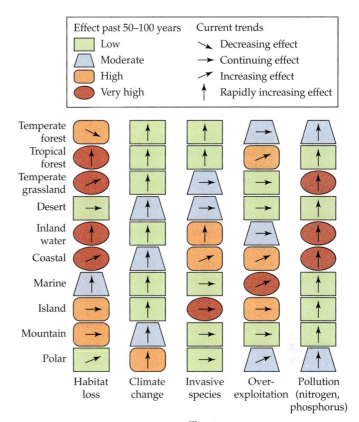

FIGURE 23.15 Different Biomes Face Different Principal Threats The effects of different types of threats on different biomes over the past 50–100 years were examined as part of the Millennium Ecosystem Assessment, an international collaboration among more than 1,000 ecologists commissioned by the United Nations. The color and shape of each box indicates the effect of the threat to date; the direction of the arrow indicates the trend in that threat. (After Millennium Ecosystem Assessment. 2005. *Ecosystems and Human Well-Being: Biodiversity Synthesis.* World Resources Institute: Washington, DC.)

❓ At a global scale, what factors have been the most important threats to diversity over the past decades, and what factors are projected to be the most important in the future? How do these current and future threats differ between terrestrial and marine biological zones?

CONCEPT 23.4 Conservation biologists use many tools and work at multiple scales to manage declining populations.

LEARNING OBJECTIVES

23.4.1 Describe how advances in molecular genetics have assisted with assessing genetic diversity in populations.

23.4.2 Evaluate the use of demographic models in projecting the fates of endangered species.

23.4.3 Explain why ex situ conservation may be the best approach to saving a species once it appears destined for extinction.

Approaches to Conservation

Where should we put our focus in preventing species loss—on the species or the habitat? Conservation biologists have debated this question and have generally concluded that protecting habitat is of primary importance but that understanding species is also important. There is no real dichotomy here, as we must understand the biology of a threatened species in order to identify and preserve its habitat. The U.S. Endangered Species Act functions through the listing of particular species threatened with extinction, but for each of those species, it mandates the identification and protection of critical habitat. Worldwide, many other laws protecting biodiversity take a similar approach.

Chapter 24 will describe how the principles of ecology are applied to protecting habitat and how conservation biologists work to manage ecosystems and landscapes. In this section, we will look at the variety of ways in which conservation biologists work to understand and protect biodiversity at the level of genes, populations, and species.

Genetic analyses are important conservation tools

As we saw in Concepts 6.2 and 11.3, small populations are more likely to go extinct than large populations due to genetic drift and inbreeding, lowering genetic variation and increasing the frequency of deleterious alleles. A decrease in genetic variation can limit the extent to which a population can evolve in response to environmental change. An increase in the frequency of deleterious alleles is also of concern because it can cause birth or survival rates to drop, thereby decreasing the population growth rate.

By increasing the risk of extinction in these ways, genetic problems resulting from small population sizes can hinder efforts to conserve a species. In some cases, conservation biologists have addressed this threat head-on by attempting the "genetic rescue" of populations that otherwise would appear doomed to extinction. Such an effort was used to help preserve the Florida panther (*Puma concolor coryi*), a subspecies of puma (pumas are also called panthers, cougars, and mountain lions). By the early 1990s, the number of panthers in Florida had decreased to fewer than 25 individuals. Compared with other puma populations, the Florida panther population had low genetic diversity and a high frequency of problems such as heart defects, kinked tails, poor sperm quality, and adult males in which one or both testes failed to descend properly. Models similar to those discussed in Concept 11.3 indicated a 95% chance that the population would become extinct within 20 years.

In 1995, to rescue the Florida panther from genetic decline and likely extinction, biologists captured eight female pumas from populations in Texas and released them in southern Florida. They selected females from Texas because historically gene flow occurred between the Florida and Texas puma populations. The results were striking (Johnson et al. 2010): panther numbers tripled by 2007, levels of genetic variation doubled, and the frequency of genetic abnormalities decreased substantially (**FIGURE 23.16**). Increases in panther numbers no doubt were aided by other conservation efforts, including habitat protection and the construction of highway underpasses to reduce mortality from collisions with vehicles, but it is clear that genetic restoration has contributed to the recovery of the Florida panther. The population size has continued to increase, reaching around 200 individuals by 2020. Another example of successful genetic

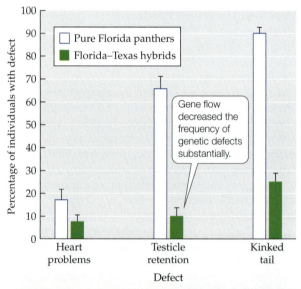

FIGURE 23.16 Genetic Rescue of the Florida Panther With depleted genetic diversity, frequent genetic defects, and a precariously small population size (fewer than 25 individuals), the Florida panther (*Puma concolor coryi*) seemed doomed to extinction in the early 1990s. The gene flow that resulted from the translocation of eight females from *P. concolor* populations in Texas helped to reverse these trends. Error bars show one SE of the mean. (After W. E. Johnson et al. 2010. *Science* 329: 1641–1645.)

rescue includes the case of the greater prairie chicken (see Concept 6.2).

Genetic rescue is not without risk, however. Introducing populations from other locations to help increase population sizes of endangered species can potentially introduce genes that are maladaptive to the new location and can have the opposite effect than what is desired. For example, ibex (*Capra ibex*) from the Middle East were introduced into the Tatra Mountains of Czechoslovakia in the 1950s to help rescue declining local populations of ibex (Templeton 1986). Unfortunately, the introduced ibex mated in the fall, rather than in the winter like the local populations. As a result, the young were born in the winter, when food was scarce, rather than in the spring, and the rescue effort failed when the population could no longer sustain itself due to low survival rates of young ibex.

As the Florida panther example suggests, genetic analyses can inform conservation decisions by revealing the genetic diversity present in a species and, in extreme cases, by guiding efforts to rescue a population or species from problems stemming from genetic decline. Genetic techniques can also be used in forensic applications related to conservation biology. For example, molecular genetic analyses permitted the identification of illegally harvested whale species in meat that was sold in Japan and labeled as either dolphin or (Southern Hemisphere) minke whale, both of which are legal to hunt (Baker et al. 2002). Cycads have also been genetically "fingerprinted," allowing tracking of these highly valuable and frequently poached plants (Little and Stevenson 2007). In **ECOLOGICAL TOOLKIT 23.1**, we explore how such "forensic conservation biology" is done and how it was

---

ECOLOGICAL TOOLKIT 23.1

Forensics in Conservation Biology

As we saw in Concept 23.3, overexploitation of wildlife can lead to population declines across entire continents and throughout the world's oceans. In some cases, conservation biologists or wildlife authorities may know that individuals from protected populations have been captured or killed, but without further information they cannot determine the extent or source of such illegal harvests. This lack of information can make laws that protect threatened species difficult to enforce. Fortunately, in some species, molecular genetic techniques can be used to monitor the extent of illegal harvesting or trace the source of illegally harvested wildlife products.

As an example, consider the trade in ivory. High demand for ivory led to the widespread slaughter of African elephants (*Loxodonta africana*), causing their numbers to drop from 1.3 million to 600,000 individuals between 1979 and 1987. As a response to this problem, an international ban on ivory trade was established in 1989. Initially the ban was successful, but soon an illegal ivory trade sprang up, leading to further declines in elephant populations.

The illegal trade in ivory proved hard to combat because even if a shipment was intercepted, it could be difficult to identify where the tusks had come from. In June 2002, more than 5,900 kg (>13,000 pounds) of ivory were confiscated in Singapore—the largest seizure of ivory since the 1989 ban (**FIGURE A**). Law enforcement officials suspected that these tusks came from elephants killed in multiple regions of Africa. Were they correct?

As in some human forensic cases, DNA evidence was used to answer this question. First, DNA was obtained from tusks seized in the June 2002 raid. As you may recall

FIGURE A Ivory from the 2002 Seizure in Singapore

Courtesy of the Center for Conservation Biology, University of Washington

from your introductory biology class, the polymerase chain reaction (PCR) can be used to amplify (i.e., produce many copies of) specific regions of DNA that often differ from one individual to another. Such highly variable DNA segments can then be visualized in a computer scan, as shown in **FIGURE B**. By amplifying several of these highly variable segments, researchers can create a "DNA profile" that characterizes an individual's genetic makeup.

To locate the source of the confiscated ivory, Samuel Wasser and colleagues amplified seven highly variable DNA segments and used them to produce a DNA profile for each of 37 of the confiscated tusks. The place of origin of each tusk was then estimated by comparing

(Continued)

ECOLOGICAL TOOLKIT 23.1 *(continued)*

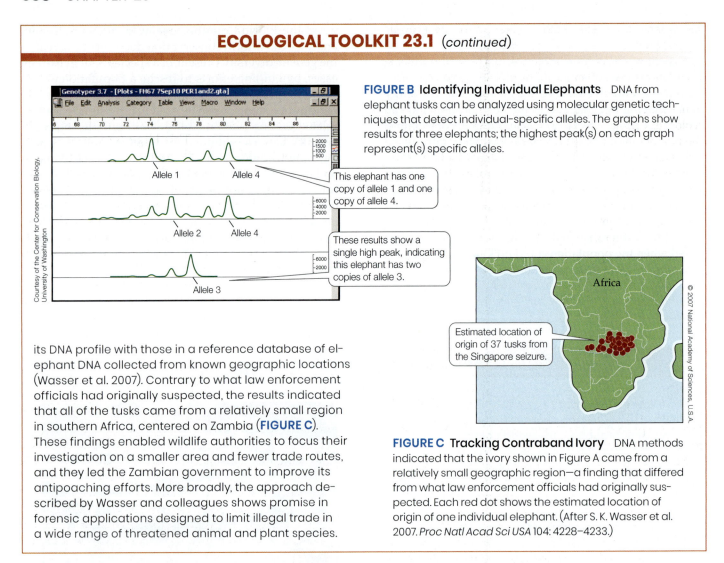

FIGURE B **Identifying Individual Elephants** DNA from elephant tusks can be analyzed using molecular genetic techniques that detect individual-specific alleles. The graphs show results for three elephants; the highest peak(s) on each graph represent(s) specific alleles.

This elephant has one copy of allele 1 and one copy of allele 4.

These results show a single high peak, indicating this elephant has two copies of allele 3.

Estimated location of origin of 37 tusks from the Singapore seizure.

Africa

Courtesy of the Center for Conservation Biology, University of Washington

© 2007 National Academy of Sciences, U.S.A.

its DNA profile with those in a reference database of elephant DNA collected from known geographic locations (Wasser et al. 2007). Contrary to what law enforcement officials had originally suspected, the results indicated that all of the tusks came from a relatively small region in southern Africa, centered on Zambia (**FIGURE C**). These findings enabled wildlife authorities to focus their investigation on a smaller area and fewer trade routes, and they led the Zambian government to improve its antipoaching efforts. More broadly, the approach described by Wasser and colleagues shows promise in forensic applications designed to limit illegal trade in a wide range of threatened animal and plant species.

FIGURE C **Tracking Contraband Ivory** DNA methods indicated that the ivory shown in Figure A came from a relatively small geographic region—a finding that differed from what law enforcement officials had originally suspected. Each red dot shows the estimated location of origin of one individual elephant. (After S. K. Wasser et al. 2007. *Proc Natl Acad Sci USA* 104: 4228–4233.)

---

used to track the source of a large shipment of contraband elephant ivory.

The availability of molecular genetic tools has enhanced our ability to understand the genetic problems faced by small populations and has helped us to address some of those problems. Let's turn next to some of the ways we can approach conservation at the population level.

Demographic models can guide management decisions

In Chapters 10 and 11, we introduced population demographic characteristics such as birth and death rates, and models that use them to project the growth of populations. Demographic models have been used to address the following questions, among others: Is the growth rate of the Yellowstone grizzly bear population high enough to allow it to persist? At what life stages are loggerhead sea turtles most vulnerable to predation, and what management decisions would be most expedient to ensure their continued viability? How much old-growth forest habitat must be preserved to ensure the persistence of the northern spotted owl?

There are hundreds of quantitative demographic models in use, tailored to the specific biological traits of particular species. The quantitative approach most widely used for projecting the potential future status of populations is referred to as **population viability analysis (PVA)**. This approach allows ecologists to assess extinction risks and evaluate management options for populations of rare or threatened species (Morris and Doak 2002). PVA is a process by which biologists can calculate the likelihood that a population will persist for a certain amount of time under various scenarios. A variety of PVA models have been developed, ranging from relatively simple stage- or age-based demographic models like those described in Concept 10.2 to more complex, spatially explicit models that can take actual landscape features and dispersal of individuals from multiple populations into account.

PVA provides conservation biologists with the probabilities that certain outcomes will occur, given assumptions about future conditions (e.g., changes in threats or in management efforts). Thus, PVA is a tool with which ecologists can synthesize data collected in the field, assess the risk of extinction for a population, identify particularly vulnerable age or stage classes, determine how many animals to release or how many plants to propagate to ensure the establishment of a new population, or determine what might be a safe number of animals to harvest (Beissinger and Westphal 1998).

PVA has been used to make a wide variety of decisions about how best to manage rare species. In Florida, the fire regime that would best serve population growth in the rare plant *Chamaecrista keyensis* was determined through PVA simulations of burns at different times of year and at different intervals (Liu et al. 2005). In Australia, the forest-cutting practices that would best serve the persistence of two endangered arboreal marsupial species, the greater glider and Leadbeater's possum, were determined through extensive PVA modeling coupled with long-term monitoring to verify the accuracy of the data going into the model (Lindenmayer and McCarthy 2006). Such analyses have played a critical role in management decisions for a number of species.

Some conservation biologists, however, caution against excessive reliance on conclusions based on the results of PVA. They point to the high level of uncertainty in the dynamics of small populations, the paucity of demographic and environmental data for many endangered species, and the high probability that a model will leave critical factors unaccounted for. To be used effectively, PVA models need to be constantly refined and revisited by different researchers to check their validity against field observations, just as management strategies must be checked and adjusted for effectiveness (Beissinger and Westphal 1998).

Ex situ conservation is a last-resort measure to rescue species on the brink of extinction

When remaining populations of a species fall below a certain size, direct, hands-on action may be called for. Such actions can include the introduction of individuals into threatened populations (as in the Florida panther) or extensive habitat manipulations intended to improve the chance that individuals will reproduce successfully (as in the red-cockaded woodpecker, as we will see in the Case Study Revisited). In some cases, however, the only hope for preserving a species may be to take some or all of the remaining individuals out of their habitat—ex situ—and allow them to multiply in sheltered conditions

under human care with the hope of later returning some individuals to the wild.

Ex situ conservation efforts have played a major role for around 25% of the endangered vertebrate species whose numbers have increased in recent years (Hoffmann et al. 2010). The rescue of the California condor (*Gymnogyps californianus*) is a leading example of this strategy (**FIGURE 23.17**). This large bird once ranged throughout much of North America, and by the nineteenth century, it was still distributed from British Columbia to Baja California. The condor population declined steeply between the 1960s and 1980s, however, reaching a low of 22 birds by 1982. The species became extinct in the wild in 1987, when the last birds were captured and brought to an ex situ facility in California for breeding (Ralls and Ballou 2004).

There are now over 450 California condors, some in the wild and some remaining in captivity. Increasing the population to this point has required careful genetic analysis, hand rearing of some chicks, and cooperation among zoos, managers of natural areas, hunters, and ranchers. Maintaining the current population will require continued input of individuals reared ex situ into wild populations, though the ultimate goal is to establish self-sustaining condor populations in the wild. One of the greatest remaining threats is lead poisoning from ammunition found in the carrion condors eat, which has prevented this goal from being met (Finkelstein et al. 2012). Other barriers to the condor's recovery include the negative health effects of ingesting plastic and other trash, West Nile virus, and genetic drift. Given all these risks and costs, is the recovery of the California condor worth all the effort that has gone into it? Without that effort, the species would now be extinct.

Ex situ conservation programs are taking place in zoos, special breeding facilities, botanical gardens, and aquaria all over the world. Such programs have allowed many species at risk of extinction to increase their numbers sufficiently to permit reintroduction into the wild. While ex situ programs play important roles in keeping our most threatened species from extinction, as well as in publicizing the plight of those species, they are expensive, and they can introduce a host of problems, such as exposure to disease, genetic adaptation to captivity, and behavioral changes (Snyder et al. 1996). Furthermore, as the case of the California condor shows, it can be difficult to restore self-sustaining populations in the wild. Could the funds dedicated to ex situ efforts be better spent on managing species in the wild or on securing land for the establishment of new protected areas—that is, for in situ conservation? Sometimes the answer is no, usually when populations have been reduced to critical levels or when not enough suitable habitat is available. But the question must always be asked.

(A)

(B)

(C)

(D)

A–C courtesy of the U.S. Fish and Wildlife Service

Courtesy of Gary Kramer/U.S. Fish and Wildlife Service

FIGURE 23.17 Ex Situ Conservation Efforts Can Rescue Species from the Brink of Extinction Ex situ efforts to save the California condor (*Gymnogyps californianus*) involve multiple steps. (A) To reduce inbreeding and increase the number of eggs that hatch successfully, a U.S. Fish and Wildlife Service biologist removes eggs from the wild (to be taken to an ex situ breeding facility) and replaces them with one egg from the San Diego Zoo. (B) At the San Diego Zoo, condor chick "Hoy" is being fed by a condor-feeding puppet to avoid its becoming acclimated to humans. (C) Two condors at the time of their release (spring 2000). The instrument in the right foreground is a scale from which condor weight can be read by telescope when a bird perches on it. (D) This adult, with a wingspan of 9 feet, was bred in captivity and later released.

CONCEPT 23.5 Prioritizing species helps maximize the biodiversity that can be protected with limited resources.

LEARNING OBJECTIVES

23.5.1 Summarize why the rarity of a species may be both helpful and misleading as an indicator of its endangerment.

23.5.2 Describe the potential benefits of using a surrogate species to help conserve habitat and other species found in that habitat.

Ranking Species for Protection

Conservation efforts can succeed. Indeed, an analysis concluded that conservation actions have reduced the rate of loss of threatened vertebrates by over 20% (Hoffmann et al. 2010). But such successes are outweighed by the severity of ongoing threats. How do we allocate the limited resources that are available for species conservation? Do we protect those species that are most threatened, or do we focus on those that play a substantial ecological role? And how should conservation biologists and policymakers decide which areas are critical to protect?

The rarest and the most rapidly declining species are priorities for protection

Many species have become rare as a result of anthropogenic threats we outlined earlier in this chapter. Other species may have always been rare. In either case, having a measure of how threatened a species is permits us to focus our efforts on those species that are most threatened: the rarest and the most rapidly declining. We may be able to postpone attending to species that are naturally low in abundance but not particularly threatened.

What do we mean by rarity, and how do we determine just how rare something is? To clarify the different concepts of rarity, we can use a matrix that sorts out

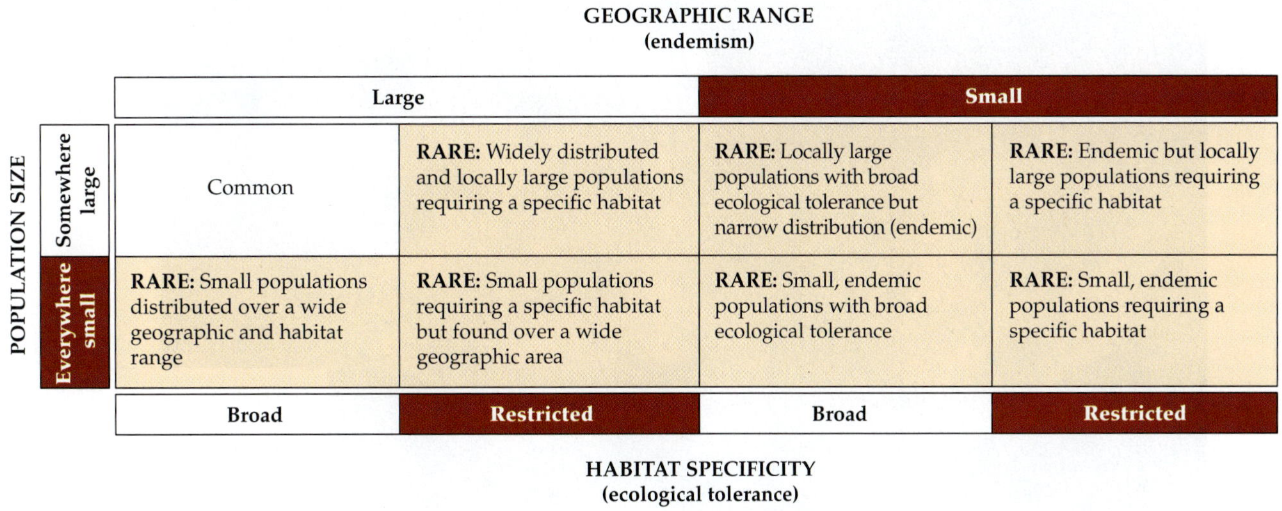

GEOGRAPHIC RANGE
(endemism)

		Large		Small	
POPULATION SIZE	**Somewhere large**	Common	**RARE:** Widely distributed and locally large populations requiring a specific habitat	**RARE:** Locally large populations with broad ecological tolerance but narrow distribution (endemic)	**RARE:** Endemic but locally large populations requiring a specific habitat
	Everywhere small	**RARE:** Small populations distributed over a wide geographic and habitat range	**RARE:** Small populations requiring a specific habitat but found over a wide geographic area	**RARE:** Small, endemic populations with broad ecological tolerance	**RARE:** Small, endemic populations requiring a specific habitat
		Broad	Restricted	Broad	Restricted

HABITAT SPECIFICITY
(ecological tolerance)

FIGURE 23.18 Seven Forms of Rarity Appropriate conservation measures for a rare species depend on the size of its geographic range, the sizes of its populations, and its habitat specificity. (After D. Rabinowitz. 1981. In *The Biological Aspects of Rare Plant Conservation*, H. Synge [Ed.], pp. 205–217. John Wiley & Sons Ltd: New York.)

whether a species has a wide or a narrow geographic range, whether it is broad or restricted in its habitat specificity, and whether its local populations tend to be small or large (**FIGURE 23.18**). There are some rare species, for example, that exist over a wide geographic area and are relatively broad in their habitat requirements, yet tend to occur in very small populations. Other rare species inhabit specific habitats within a narrow geographic range, but may have large populations in those specific locations (Rabinowitz et al. 1986). Conservation of these different types of rare species requires different approaches. Some species require small reserves to protect well-established populations; others require management practices that create habitat conditions suitable for a rare but geographically widespread species.

An important scientific assessment of the conservation status of species began in 1963 with the establishment of the IUCN Red List (see Table 23.1). A parallel effort was developed in the United States by The Nature Conservancy, which established the Natural Heritage Program (now NatureServe) in the early 1970s in order to assess the conservation status of U.S. species. Both organizations have developed a ranking structure that indicates how threatened a species is and an assessment protocol to determine its rank. The assessment protocol takes into account not only numbers of populations or individuals, but also the total geographic area the species occupies, the rate of its decline, and the threats it faces. Because of the challenge of creating a system that can be applied equally well to a skipper butterfly, a cycad, or a shark, and because the information available on rare species is often incomplete, both systems allow assessors to choose among different sets of criteria to decide whether a species is critically endangered, endangered, vulnerable, or under some lesser level of threat.

Such assessments of conservation status can be used to locate clusters of threatened species and thus identify areas that are critical to protect (**FIGURE 23.19**). They are frequently consulted when development projects are planned, and they are important for keeping the public aware of the degree of threat faced by Earth's biota. These databases are dynamic in that they can change as scientific information is updated. The conservation status assigned to a species can be downgraded if its numbers increase or upgraded if its numbers decline.

Protection of surrogate species can provide protection for other species with similar habitat requirements

If we protect the habitat that is necessary for the red-cockaded woodpecker, as described in the Case Study, will we simultaneously provide protection for the gopher tortoise, Bachman's sparrow, Michaux's sumac, and other rare species that are dependent on the longleaf pine savanna ecosystem? Species may become conservation priorities not only because of their own conservation status, but also because of their capacity to serve as **surrogate species** whose conservation will serve to protect many other species with overlapping habitat requirements. Some surrogate species can help us garner public support for a conservation project; examples of such **flagship species** include charismatic animals such as the mountain gorilla (**FIGURE 23.20**). Other surrogate species are referred to as **umbrella species**, which we select with the assumption that protection of their habitat will serve as an "umbrella" to protect many other species with similar habitat requirements. Umbrella species are typically species with large area requirements, such as grizzly bears, or habitat specialists, such as the red-cockaded woodpecker, but

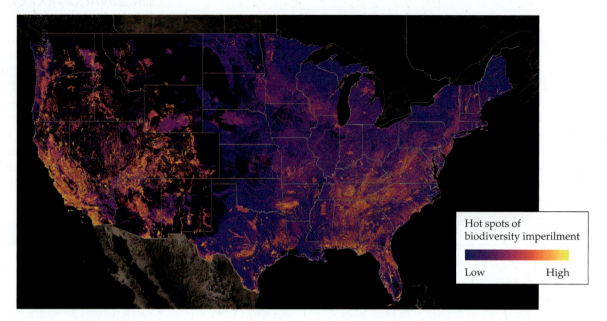

FIGURE 23.19 Hot Spots of Imperilment The compilation of NatureServe data on the location of imperiled species and their geographic ranges in the United States has permitted the identification of the critical areas to protect. California, Hawaii, the Florida Panhandle, and the southern Appalachian Mountains are "hot spots" of imperilment—they have high concentrations of imperiled species due to their high rates of endemism. (Copyright 2022, NatureServe and the NatureServe Network, 2550 South Clark Street, Suite 930, Arlington VA 22202, USA. All Rights Reserved)

they may also include animals that are relatively easy to count, such as butterflies (Fleishman et al. 2000). Some researchers prefer to choose not just one species, but several **focal species**, selected for their different ecological requirements or susceptibility to different threats, with the realization that by thus casting a broader net, we improve our chances of covering regional biodiversity with protection.

Methods have been devised and criteria established to allow for strategic selection of the one or several surrogate species that will best serve conservation aims (Favreau et al. 2006). Conservation biologists recognize, however, that surrogate species approaches are not without problems, and that the distribution or habitat requirements of any one species cannot capture all the conservation targets we may have.

FIGURE 23.20 A Flagship Species The mountain gorilla (*Gorilla beringei beringei*) is critically endangered in its highland forest habitat of central Africa. Only two populations remain in the wild, with a combined total of only 300 mature animals. Threats to their persistence include loss of habitat, hunting, and disease transmission from humans.

Can Birds and Bombs Coexist?

As the longleaf pine ecosystem lost 97% of its area over the last several hundred years, the biological traits of the red-cockaded woodpecker that had worked well in the extensive pine savannas of the past turned out to be detrimental in its changing environment. Prime woodpecker habitat became fragmented, consisting of islands of usable habitat in an unsuitable landscape. As a result, the woodpecker's unusual habit of excavating cavities in living trees—a process that usually takes a year or more to complete—made the availability of cavities a limiting factor for woodpecker populations.

Jeff Walters and his colleagues tested the hypothesis that a lack of high-quality habitat was limiting the woodpecker's population growth, by constructing artificial nest cavities, placing them in clusters, and observing woodpecker behavior. They tried this strategy for two reasons. First, they put cavities in clusters of trees because red-cockaded woodpeckers are *cooperative breeders* (males born in previous years help their parents raise young, forming the woodpecker groups described at the start of the chapter) and each bird in a group must have its own cavity. Second, the birds typically abandon cavity clusters after several years' use, primarily because of cavity entrance enlargement by other species, or mortality of cavity trees, so there is a continual demand for cavity clusters (Harding and Walters 2002). The artificial cavity clusters constructed by the researchers were rapidly colonized, mostly by helper birds from the vicinity and young dispersing birds (Copeyon et al. 1991; Walters et al. 1992).

These results suggested that people could increase numbers of red-cockaded woodpeckers by going out with drill, wood, wire, and glue and installing groups of cavities within living longleaf pines (**FIGURE 23.21**). Indeed, these activities have proved a boon to woodpecker recovery. Aided by the construction of artificial cavities, the population of red-cockaded woodpeckers at Fort Bragg increased from 238 breeding groups in 1992 to 368 breeding groups in 2006. Cavity construction has also contributed to increased abundances of red-cockaded woodpeckers at other military bases, including Eglin Air Force Base (Florida), Fort Benning (Georgia), Fort Polk (Louisiana), Fort Stewart (Georgia), and Marine Corps Base Camp Lejeune (North Carolina). Similar successes have occurred at sites other than military bases. For example, when Hurricane Hugo hit the South Carolina coast in 1989, the population of red-cockaded woodpeckers in Francis Marion National Forest, previously home to 344 breeding groups, was severely reduced. The hurricane killed 63% of the birds, and another 18% died the following winter (Hooper et al. 2004). Within 2 years of the storm, however, national forest workers had installed 443 artificial cavities. This strategy averted a severe population decline; by 1992, the population had recovered to 332 clusters. Across the southeastern United States the number of clusters of red-cockaded woodpeckers has increased from less than 1,500 in the early 1990s to over 7,800 in 2020 as a result of the recovery efforts. However, many of the clusters occur in areas susceptible to hurricanes, which appear to be increasing in intensity due to climate change.

Now that managers have identified cavity construction and maintenance as a critical factor for the recovery of red-cockaded woodpeckers, they are obliged by the Endangered Species Act to continue doing it. This strategy is labor-intensive and expensive, but for now it is necessary for the red-cockaded woodpecker's continued existence. How long can we sustain this effort? Will we reach a point at which there is enough longleaf pine savanna that the woodpeckers will be able to maintain their own numbers without human assistance? We do not know the answers to these questions.

In the decades during which Walters and others have been researching the red-cockaded woodpecker, they have used many of the tools described in this chapter. Models of population dynamics have facilitated the identification of vulnerable stages in the woodpecker's life cycle. Genetic studies and modeling have focused attention on the threat of inbreeding. Field studies have demonstrated the need for prescribed burning to maintain the community

(A)

(B)

Both courtesy of the U.S. Fish and Wildlife Service

FIGURE 23.21 Installation of Artificial Nest Cavities Has Allowed Populations of Red-Cockaded Woodpeckers to Increase (A) A researcher with the U.S. Fish and Wildlife Service cores a tree to help determine its suitability for an artificial nest cavity. (B) Installing the artificial nest cavity.

structure required by the woodpeckers. Economic and sociological analyses have led to the development of a "safe harbor" program that makes endangered species management more acceptable to private landowners.

 ## CONNECTIONS *in* NATURE

SOME BURNING QUESTIONS As we saw in Chapter 3, recurrent fires promote the establishment of savanna. Hence, to maintain red-cockaded woodpecker populations and the longleaf pine savannas on which they depend, fire is key—whether it is ignited naturally, accidentally by military training exercises, or intentionally under controlled conditions. As with other regular forms of disturbance (see Concept 9.2), differences in the frequency of fires can affect the distributions and abundances of species, and those changes, in turn, along with changes in the physical environment, can affect the cycling of nutrients and water. Because fire affects communities at so many levels, prescribed burning is used as a management tool for conserving species in numerous ecosystems where fire has been a regular natural disturbance (**FIGURE 23.22**).

But the use of fire as a management tool can have unintended and undesirable ecological outcomes where non-native invasive species are present. In some longleaf pine savannas in Florida, openings resulting from burning have provided favorable habitat for the establishment of cogongrass (*Imperata cylindrica*), an invasive plant from Asia. The presence of this grass, in turn, causes fires to burn hotter and more extensively. The consequences of these hotter fires are increased mortality of longleaf pine seedlings and native wiregrass, favorable conditions for further infiltration of cogongrass, and a resulting threat to the high levels of native plant diversity found in the understory of the longleaf pine savanna (Lippincott 2000). Land managers are faced with a dilemma: to burn or not to burn? The right question is more likely to be when to burn, and how often.

Adding people to the burning landscape further complicates matters. Throughout the southeastern United States, prescribed burns are taking place in a complex landscape where patches of forest are adjacent to people's homes and businesses. Convincing the public that these fires are necessary has required considerable outreach and public education. In the North Carolina Sandhills, the days for

FIGURE 23.22 Prescribed Burning Is a Vital Management Tool in Some Ecosystems Throughout the United States prescribed control burning, such as that shown here, helps to maintain a natural disturbance regime that maintains the high plant biodiversity characteristic of the understory in pine savanna ecosystems. Many threatened species, including the red-cockaded woodpecker, rely on regular burning for their persistence.

prescribed burns are chosen not only for safe conditions, but also with regard to wind direction so as to minimize the amount of smoke in population centers.

Here, as elsewhere, recognition of people as an integral component of the landscapes that must harbor all of nature's diversity has been a vital piece of the conservation picture. Establishing protected natural areas as sanctuaries for wildlife is an important part of the solution to the biodiversity crisis, but we must also do what we can to ensure that the vast majority of Earth's surface outside of protected areas is able to sustain both people's livelihoods and habitat for other species. This is a difficult challenge that will involve education, cooperation, legislation, and many creative approaches.

© Operation 2021/Alamy Stock Photo

SUMMARY

CONCEPT 23.1 Conservation biology is an integrative discipline that applies the principles of ecology to the protection of biodiversity.

23.1.1 Describe the different biological levels of diversity associated with conservation biology.

- Conservation biology is the scientific study of phenomena that affect the maintenance, loss, and restoration of biodiversity.
- Biodiversity includes genetic diversity, species diversity, and landscape diversity.

23.1.2 Evaluate the reasons why preserving biodiversity may be important.

- Biodiversity is important to human society because of our reliance on natural resources and ecosystem services that depend on the integrity of natural communities and ecosystems.

23.1.3 Assess the factors that prompted the rapid development of the field of conservation biology.

- With the growing awareness of accelerating losses of global biodiversity, ecologists saw a need for a separate discipline that would apply the principles of ecology to the preservation of species and ecosystems.
- Conservation biology is a scientific discipline instilled with the value of biodiversity.

CONCEPT 23.2 Diversity is declining globally.

23.2.1 Differentiate between the current anthropogenically enhanced rate of extinction and the long-term background extinction rate.

- Earth is losing species at an accelerating rate, 100 to 1,000 times higher than the background rate, largely because of humanity's growing footprint on the planet.

23.2.2 Describe the pathway to species extinction from changes in population growth to the disappearance of the species.

- Extinction is the end point of incremental biological decline as species lose individuals and populations and become increasingly vulnerable to the problems of small populations.
- Earth's biota is becoming increasingly homogenized because of a rise in generalist species and a decline in specialist species.

CONCEPT 23.3 Primary threats to diversity include habitat loss, invasive species, overexploitation, pollution, disease, and climate change.

23.3.1 Compare the most important threats to diversity over the past century with the threats that are increasing the most in importance.

- Habitat degradation, fragmentation, and loss are the most important threats to diversity.

23.3.2 Describe the causes of diversity losses associated with habitat loss and degradation.

- Habitat degradation and habitat loss lower the sizes of populations, increasing the probability that genetic and stochastic factors will result in local extinction of species.

23.3.3 Explain the underlying mechanisms determining how invasive species, overexploitation, and water and air pollution lead to diversity loss.

- Invasive species degrade local habitats by preying on or competing with native species and by altering ecosystem properties.
- Overexploitation of selected species has large effects on communities and ecosystems.
- Other factors that erode the viability of populations and contribute to losses of diversity include air and water pollution, diseases, and global climate change.

CONCEPT 23.4 Conservation biologists use many tools and work at multiple scales to manage declining populations.

23.4.1 Describe how advances in molecular genetics have assisted with assessing genetic diversity in populations.

- Genetic analyses have been used to understand and manage genetic diversity within rare species, as well as in forensic analyses of illegally harvested organisms.

23.4.2 Evaluate the use of demographic models in projecting the fates of endangered species.

- Population viability analysis (PVA) is an approach that uses demographic models to assess extinction risks and evaluate proposed management actions.

23.4.3 Explain why ex situ conservation may be the best approach to saving a species once it appears destined for extinction.

- Ex situ conservation, which involves taking organisms from the wild into human care, is a last-resort measure to rescue species on the brink of extinction.

CONCEPT 23.5 Prioritizing species helps maximize the biodiversity that can be protected with limited resources.

23.5.1 Summarize why the rarity of a species may be both helpful and misleading as an indicator of its endangerment.

- Conservation biologists identify those species of the highest priority for protection—the rarest and the most rapidly declining species—by assessing numbers of individuals and populations, total geographic area occupied, rates of decline, and the degree of threat faced.

23.5.2 Describe the potential benefits of using a surrogate species to help conserve habitat and other species found in that habitat.

- Identification of surrogate species can provide protection for other species with similar habitat requirements.

Review Questions

1. What are the principal threats to biodiversity? Describe some examples in which multiple threats have contributed to decline of a species.

2. Describe tools that conservation biologists use to protect biodiversity at the level of genes and populations.

3. What is the difference between a species determined to be endangered by the Natural Heritage/NatureServe program and one that is listed as endangered under the U.S. Endangered Species Act? What are the consequences for management of each?

4. Identify five imperiled species that live in your region: a plant, a mammal, a bird, a fish, and an invertebrate. Are any of these species endemic to your region? For each species you have identified, try to find out whether it was rare prior to human settlement of the region. What threats does this species face today? What is being done to protect this species? Based on the ecological knowledge you have gained, what questions do you think should be researched to aid in the recovery of the species? (Much of this information is available at www.natureserve.org.)

Hone Your Problem-Solving Skills

With the realization of declining abundances of fish, greater attention has been given to demographic approaches to sustainable harvests. Here we present a hypothetical example of such an approach. A species of commercial fish lives for 4 years, with harvestable fish occurring only in the 2- and 3-year-old age classes. The growth rate of the population can be modeled using a life table approach such as the one found in Tables 10.3 and 10.4. In the absence of harvest, the fish exhibits the following demographic properties:

Age (x)	Survival rate (S_x)	Survivorship (l_x)	Fecundity (F_x)
0	0.40	1.00	0
1	0.60	0.40	0
2	0.80	0.24	2
3	0.70	0.19	4
4	0	0.13	2

1. Starting with the following distribution of individuals in age cohorts, plot or tabulate the population growth for a population of this fish species for 10 years: $n_0 = 30$, $n_1 = 30$, $n_2 = 20$, $n_3 = 10$, $n_4 = 10$ (initial total of 100 individuals).

2. Project the same population growth if 60% of the 2- and 3-year-old cohorts are harvested for commercial use, by adjusting the survival rates. Is this level of harvest sustainable—that is, will the population be able to maintain a relatively stable size under these conditions indefinitely?

3. Now project the population growth if 40% of the 2- and 3-year-old cohorts are harvested for commercial use, and estimate whether this rate of harvest is sustainable. Try different combinations of fecundity and survival within the different age cohorts to consider how these factors influence the sustainability of populations under pressure from commercial harvest.

ADDITIONAL RESOURCES

This textbook provides resources such as **Web Extensions** and additional **Climate Change Connections** for instructors who can share them with students. Please contact your instructor, if you would like to access that content.

Landscape Ecology and Ecosystem Management

Wolves in the Yellowstone Landscape: A Case Study

The Greater Yellowstone Ecosystem (GYE) both symbolizes the soul of the American wilderness and encapsulates the challenges of managing public lands. The landscape is shaped by a unique complex of natural elements: volcanic eruptions, geothermal activity, glaciers, and repeated fires, sometimes on a massive scale. In addition, the interplay between large herbivores and their predators has the potential to impact the landscape, including the vegetation and landforms. All of these factors have contributed to the mosaic of forests, meadows, grasslands, lakes, and rivers that characterize the GYE.

Prior to the early twentieth century, wolves were important predators in the GYE, but extermination programs led to their local extinction. After 70 years of absence, wolves were reintroduced into the GYE between 1995 and 1997 from populations in Canada and northwestern Montana. Wolves hunt among a wide diversity of ungulates and other prey (**FIGURE 24.1**). The reintroduction of wolves was the culmination of years of research effort and hotly contested policy debate, with strong objection from some residents of the region. Twenty years later, its ecological consequences have proved to be multifaceted and profound, and public opinion has become generally more favorable. Wolf reintroduction is perceived as restoring an important natural element to the GYE.

But how "wild" and natural is the GYE? Larger in area than the state of West Virginia, the GYE includes two national parks and seven national forests as well as other public and private lands (**FIGURE 24.2**). The region is actively managed by more than 25 different state and federal agencies as well as private corporations, nongovernmental organizations, and private landowners. Decisions about the use of its land and natural resources are complex and often uncoordinated, yet when considered together, these decisions determine which species will or will not be sustained by the ecosystem (Parmenter et al. 2003).

Despite its fragmented management, the GYE is often perceived as one of the most biologically pristine regions in North America. It sustains seven species of native ungulates and five large carnivore species. Understanding how these predator and prey populations interact, and how their abundances affect the whole ecosystem, has been a persistent challenge to ecologists who study the GYE, particularly in light of a century of management of

KEY CONCEPTS

CONCEPT 24.1 Landscape ecology examines spatial patterns and their relationship to ecological processes.

CONCEPT 24.2 Habitat loss and fragmentation decrease habitat area, isolate populations, and alter conditions at habitat edges.

CONCEPT 24.3 Biodiversity can best be sustained by large reserves connected across the landscape and buffered from areas of intense human use.

CONCEPT 24.4 Ecosystem management is a collaborative process with the maintenance of long-term ecological integrity as its core value.

FIGURE 24.1 A Top Predator Returns A showdown between a pack of wolves (*Canis lupus*) and an American bison (*Bison bison*). After nearly 70 years of absence, wolves were reintroduced in 1995 to Yellowstone National Park, where they are now the main predators of ungulate herbivores including bison, moose, and elk.

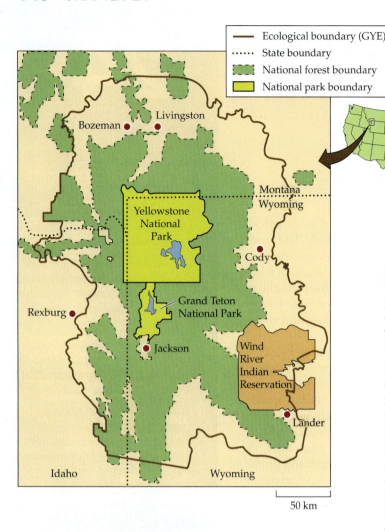

FIGURE 24.2 The Greater Yellowstone Ecosystem The Greater Yellowstone Ecosystem contains Yellowstone and Grand Teton National Parks, seven different national forests, and land managed by the Bureau of Land Management, as well as private lands. (After A. Parmenter et al. 2003. *Ecol Appl* 13: 687–703.)

Legend:
— Ecological boundary (GYE)
···· State boundary
▬ National forest boundary
▬ National park boundary

50 km

Introduction

In this chapter we will broaden the spatial scale of our view of ecology to take a landscape perspective. This broader view is facilitated by a powerful assemblage of tools that permit us to monitor the environment in multiple dimensions and at many scales. For example, the emergence of aerial photography gave ecologists a technique to examine "the big picture." Similarly, our ability to acquire images of Earth through remote sensing, from drones to satellites, has permitted the interpretation of many large-scale ecological patterns, including global patterns of net primary production (see Ecological Toolkit 20.1). The use of geographic information systems (GIS), methods used to visualize and analyze spatial data, has become standard in landscape planning efforts, whether for urban development or for conservation (**ECOLOGICAL TOOLKIT 24.1**). In the field, handheld global positioning systems (GPS) have permitted ecologists to document precise locations and integrate them with other landscape variables through GIS. Radiotelemetry and animal-borne GPS tracking devices have greatly enhanced our ability to follow animal movements and migration patterns, again with the help of GIS. Our ability to analyze all this information is constantly growing, thanks to better computers and new statistical methods of spatial analysis.

We saw in Concept 23.3 that habitat loss, fragmentation, and degradation are primary causes of the current declines in biodiversity. In this chapter, we'll see how the tools and methods of landscape ecology are used to assess the occurrence and possible causes of biodiversity declines at the landscape and ecosystem scales. Because protected natural areas are at the heart of conservation strategies, we will also consider how conservation biologists identify and design them to maximize their effectiveness. Finally, we'll examine how ecosystem management integrates ecological principles with social and economic information to help guide decisions about land and water use.

wildlife populations. After wolves were eradicated in the mid-1920s, there were concerns that elk were overgrazing meadows in the northern part of the park. The elk population was regulated from the 1920s to the late 1960s by exporting animals to elk farms and by culling. In 1968, a new policy of "natural regulation" without human intervention in population growth was implemented. The elk population nearly quadrupled over a 30-year period and suppressed the plants they fed on. The reintroduction of wolves has not only reduced the elk population but has also affected the populations of many other species. How?

To start to answer that question, let's go back to the 1950s, when ecologists noticed that beavers had become scarce in Yellowstone National Park. Gradually, it became clear that the cause was increased elk herbivory on the beavers' preferred food plants, willow and aspen. But a whole suite of other species depend on beaver ponds for their own persistence, and their abundances had declined along with the beavers'. The decision to eradicate wolves did not anticipate these ecological changes to the Yellowstone ecosystem. How can ecologists help managers of nature reserves make decisions that will take future consequences into account?

Geographic information systems (GIS) are computer-based systems that allow the storage, analysis, and display of data pertaining to specific geographic areas. The data used in GIS are derived from multiple sources, including aerial photographs, satellite imagery, and ground-based field studies (**FIGURE A**). Examples of such data include rainfall, elevation, and vegetation cover at specific locations. Each of these and many other variables may be used in a particular application of GIS—but whatever variables are used, the data are keyed to or referenced by spatial or geographic coordinates so that they can be assembled into a multilayered map.

Layers of mapped data can be put together in ways that help to address particular questions. We'll illustrate this process with an approach often used in conservation biology, called *gap analysis*. The acronym GAP refers to the Gap Analysis Program, a U.S. Geological Survey program whose mission is to help prevent biodiversity decline by identifying species and communities that are not adequately represented on existing conservation lands.

The lark bunting (*Calamospiza melanocorys*) is one such species. It depends on prairie habitat for its breeding grounds, but much of this habitat has been destroyed by conversion to agriculture. As a result, populations of the lark bunting have been declining by an average of 1.6% per year over the past 40 years, making it a species of conservation concern (U.S. Fish and Wildlife Service 2008).

For the lark bunting, or any other species, gap analysis is a two-step process. First, data on vegetation cover (see the top GIS layer in Figure A) and on other environmental conditions required or preferred by the lark bunting are used to predict its geographic distribution (the second GIS layer in Figure A). Next, that predicted distribution is compared with a third GIS layer showing the locations of conservation lands and protected areas. By combining these two layers, we can calculate that only a small percentage of the bird's distribution is protected (**FIGURE B**). Such information is critical to decisions about what lands should be protected to prevent future losses of biodiversity.

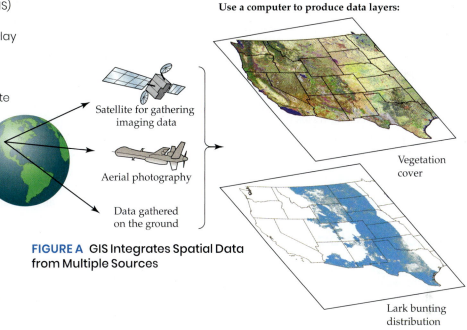

Use a computer to produce data layers:

Satellite for gathering imaging data

Aerial photography

Data gathered on the ground

Vegetation cover

Lark bunting distribution

FIGURE A GIS Integrates Spatial Data from Multiple Sources

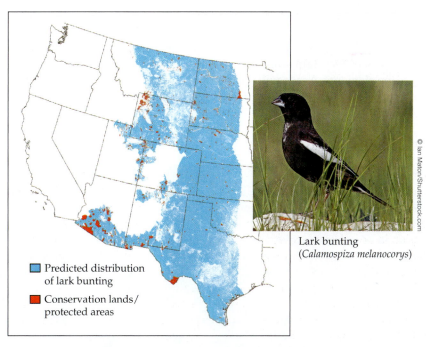

■ Predicted distribution of lark bunting

■ Conservation lands/ protected areas

Lark bunting (*Calamospiza melanocorys*)

© Ian Maton/Shutterstock.com

FIGURE B A Conservation Gap   Less than 3% of the lark bunting's predicted distribution is in protected areas. (Maps courtesy of U.S.G.S. Gap Analysis Project. https://doi.org/10.5066/F7V122T2 and https://doi.org/10.5066/F7ZS2TM0.)

CONCEPT 24.1   Landscape ecology examines spatial patterns and their relationship to ecological processes.

LEARNING OBJECTIVES

24.1.1   Describe the elements that make up a landscape and illustrate how they can influence ecological processes such as dispersal and ecosystem function.

24.1.2   Show how landscape structure can be evaluated using the number and areas of the elements that make up the landscape.

24.1.3   Compare the benefits and drawbacks associated with using coarse-scale versus fine-scale characterization of a landscape.

24.1.4   Describe how disturbances can affect and be affected by the landscape structure.

Landscape Ecology

Landscape ecology examines how landscape patterns are influenced by ecological processes as well as how these spatial patterns influence ecological processes. Landscape ecologists are interested in the spatial arrangement of different *landscape elements* across Earth's surface. Examples of landscape elements include surface features such as areas covered (patches) by forest, meadows, or lakes. At smaller spatial scales, individual creosote bushes in a desert, or areas of a certain soil type, could be considered landscape elements. As we will see, the spatial pattern of landscape elements can influence what species live in an

area, as well as the dynamics of ecological processes such as disturbance and dispersal of organisms.

A landscape is a heterogeneous area composed of a dynamic mosaic of interacting ecosystems

A **landscape** is an area in which at least one element is spatially heterogeneous (varies from one place to another) (**FIGURE 24.3**). Landscapes can be heterogeneous either in what they are composed of—for example, twelve different vegetative cover types versus only three—or in the way their elements are arranged—such as many small patches arranged regularly over the landscape versus a few large patches. Ecologists often refer to this composite (or pattern) of heterogeneous elements that make up a landscape as a **mosaic**.

Landscapes often include multiple ecosystems. The different ecosystems that make up a landscape are dynamic and continually interacting with one another. These interactions may occur through the flow of water, energy, nutrients, or pollutants between ecosystems.

There is also biotic exchange between habitat patches in the mosaic as individuals or their gametes (e.g., pollen) move between them (Forman 1995). For such movement to occur, patches of the same habitat type must be connected to one another, or the surrounding habitat (the *matrix*) must be of a type through which dispersal is possible (**FIGURE 24.4**). In Australia, for example, rats regularly leave patches of forest habitat to forage in adjacent macadamia nut plantations (a part of the surrounding

(A)

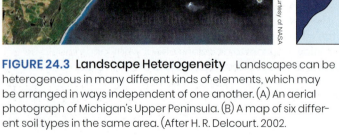

(B)

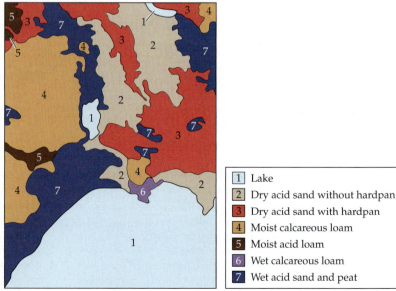

1	Lake
2	Dry acid sand without hardpan
3	Dry acid sand with hardpan
4	Moist calcareous loam
5	Moist acid loam
6	Wet calcareous loam
7	Wet acid sand and peat

FIGURE 24.3   Landscape Heterogeneity   Landscapes can be heterogeneous in many different kinds of elements, which may be arranged in ways independent of one another. (A) An aerial photograph of Michigan's Upper Peninsula. (B) A map of six different soil types in the same area. (After H. R. Delcourt. 2002.

In *Learning Landscape Ecology*, S. E. Gergel and M. G. Turner [Eds.], pp. 62–82. Springer: New York.)

?   In part (B), which landscape element covers the least area?

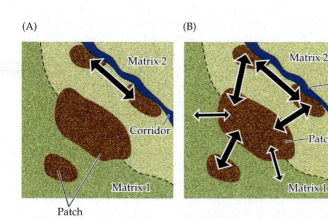

FIGURE 24.4 Movements across the Landscape
Movements between adjacent landscape elements may occur frequently (thicker arrows) or rarely (thinner arrows). (A) Exchange between patches of the same type occurs frequently if a corridor that allows movement connects the patches. (B) Exchange between patches of the same type occurs frequently, but exchange with the matrix occurs only rarely. (After A. M. Hersperger. 2006. *Landscape Urban Plann* 77: 227–239.)

? Do organisms move more freely across the matrix in (A) or in (B)? Explain.

matrix). As a result, nut losses to rats along plantation edges adjacent to forests are greater than along edges adjacent to grasslands or agricultural fields (White et al. 1997).

Next, let's focus in more detail on two aspects of landscape heterogeneity: how it is described, and the scale at which it is studied.

DESCRIBING LANDSCAPE HETEROGENEITY The heterogeneity that we see in landscapes can be described in terms of composition and structure. **Landscape composition** refers to the kinds of elements or patches in a landscape, as well as to how much of each kind is present. These elements are defined by the investigator and are influenced by the source of data used and goals of the analysis. In an example from Yellowstone National Park, researchers working on the history of landscape dynamics designated five different age classes of lodgepole pine forest using ground-based fieldwork, aerial photographs,

and GIS (Tinker et al. 2003). The composition of the landscape in **FIGURE 24.5** can thus be quantified by counting the kinds of elements in the mapped area (five in this case), by calculating the proportion of the mapped area covered by each kind of element, or by measuring the diversity and dominance of the different landscape elements much as one does for species, using a measure such as the Shannon index (described in Concept 16.2).

Landscape structure is the physical configuration of the different elements that compose the landscape. In Figure 24.5, we can see that some parts of the landscape contain large contiguous blocks of older forest, while other parts are more fragmented and contain smaller patches of forest with a variety of different ages. Landscape ecologists quantify landscape structure primarily by addressing whether the landscape is characterized by large or small patches, how aggregated or dispersed the patches are, whether the patches are simple or complicated in their shape, and how fragmented the landscape is (With 2019). Quantitative analyses of landscape structure allow us to compare one landscape with another and to relate landscape patterns to ecological processes and to the dynamics of landscape change. For example, Tinker and colleagues (2003) were able to use the measures of landscape structure that they derived for Yellowstone to compare the natural, fire-caused fragmentation within the park with fragmentation caused by clear-cutting in adjacent national forests. Logging created greater heterogeneity relative to the landscape primarily impacted by fire, with important implications for differences in population and community processes between the two landscape management types.

Only remnants of older forest remain following recent fires.

Stand age classes
1–25	248–323
26–158	324–560
159–247	

Yellowstone Lake

Large stands of older trees are found in this unburned region.

16 km

FIGURE 24.5 Landscape Composition and Structure This map of lodgepole pine (*Pinus contorta* var. *latifolia*) forest in Yellowstone National Park shows five different age classes of forest. Structural complexity varies across the landscape, as seen in the varying degree of natural fragmentation. (From D. B. Tinker et al. 2003. *Landscape Ecol* 18: 427–439.)

THE IMPORTANCE OF SCALE Consideration of scale is an important aspect of landscape ecology and can vary depending on the size of the area viewed. A landscape may be heterogeneous at a scale important to a tiger beetle, but homogeneous to a warbler or moose. The scale at which we choose to study a landscape determines the results we will obtain. Part of landscape ecology, therefore, is dedicated to understanding the implications of scale.

Scale, the spatial or temporal dimension of an object or process, is characterized by both grain and extent. **Grain**, which is the size of the smallest homogeneous unit of study (such as a pixel in a digital image), determines the resolution at which we view the landscape (**FIGURE 24.6A**). The selection of grain will affect the quantity of data that is used in analysis: using a large-grained approach may be appropriate when one is looking at patterns at a regional to continental scale. **Extent** refers to the area or time period encompassed by a study. Consider how differently we might describe the composition of a landscape depending on how we define its spatial extent. Panel 4 of **FIGURE 24.6B**, for example, shows little late successional whitebark pine, while panel 6 contains a considerable area of it (Turner et al. 2001). There may be natural or human-created boundaries that determine the extent of a study, or they may be defined by the researcher.

Ecosystem and landscape studies considering questions impacted by scale must also determine how processes scale up or down. For example, a researcher studying carbon exchange at the landscape level needs to know how leaf-based measurements of CO_2 exchange scale up to the whole plant, the ecosystem, and ultimately the mosaic of ecosystems that make up the landscape. This example shows the importance of connecting processes across different scales. Ecologists have developed methods to analyze how patterns and phenomena at one scale affect those occurring at either larger or smaller scales (see Levin 1992).

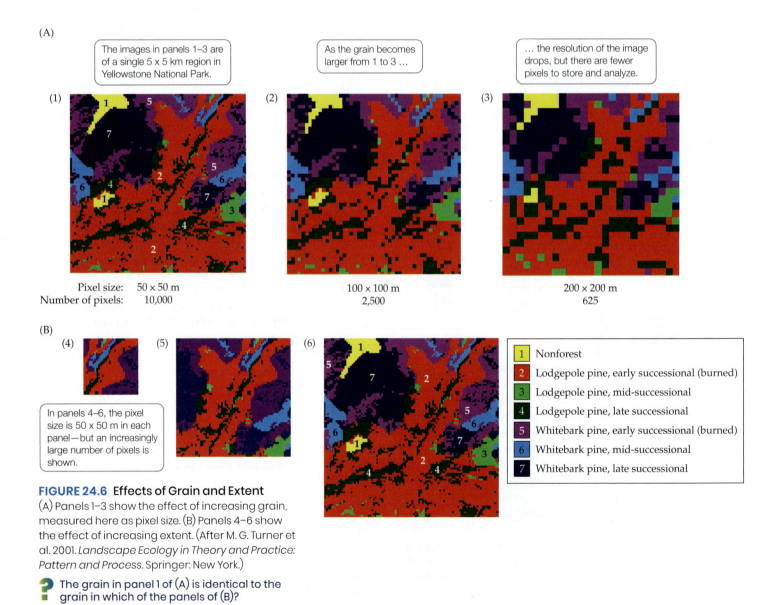

(A)

The images in panels 1–3 are of a single 5 x 5 km region in Yellowstone National Park.

As the grain becomes larger from 1 to 3 …

… the resolution of the image drops, but there are fewer pixels to store and analyze.

Pixel size: 50 x 50 m	100 x 100 m	200 x 200 m
Number of pixels: 10,000	2,500	625

(B)

In panels 4–6, the pixel size is 50 x 50 m in each panel—but an increasingly large number of pixels is shown.

1	Nonforest
2	Lodgepole pine, early successional (burned)
3	Lodgepole pine, mid-successional
4	Lodgepole pine, late successional
5	Whitebark pine, early successional (burned)
6	Whitebark pine, mid-successional
7	Whitebark pine, late successional

FIGURE 24.6 Effects of Grain and Extent
(A) Panels 1–3 show the effect of increasing grain, measured here as pixel size. (B) Panels 4–6 show the effect of increasing extent. (After M. G. Turner et al. 2001. *Landscape Ecology in Theory and Practice: Pattern and Process.* Springer: New York.)

? The grain in panel 1 of (A) is identical to the grain in which of the panels of (B)?

Landscape patterns affect ecological processes

Landscape structure plays an important role in ecological dynamics. For example, it can affect whether and how animals move and can therefore influence rates of pollination, dispersal, or consumption. Mickaël Henry and his associates studied the movements of the fruit-eating bat *Rhinophylla pumilio* in a tropical forest in French Guiana that had been fragmented by the construction of a reservoir. Using landscape metrics that quantified the degree of patch (the degree to which landscapes allow movement between patches) at their sampling sites, they found that more isolated forest fragments were less likely to be visited by bats, even if they contained abundant food resources (Henry et al. 2007). Thus, the landscape structure affected bat foraging behavior. Furthermore, because frugivorous bats disperse plant seeds, it is also likely that the landscape structure affected the dispersal of the plants that the bats fed on.

Landscape structure also influences biogeochemical cycling. Ecosystem ecologists have identified biogeochemical "hot spots" where chemical reaction rates are higher than in the surrounding landscape. Many such hot spots are found at the interfaces between terrestrial and aquatic ecosystems (McClain et al. 2003), but other factors may also play a part. For example, Kathleen Weathers and her colleagues found that inputs of sulfur, calcium, and nitrogen from atmospheric deposition were higher at forest edges than in forest interiors, primarily as a result of greater interception of airborne particles by the denser and more complex vegetation typically found at a forest edge. The fragmented forests that typically surround urban areas are therefore more likely to be influenced by inputs of atmospheric pollutants and nutrients than intact forests. This finding has implications for soil microbial dynamics, plant growth and diversity, and animal communities in the edges of these fragments (Weathers et al. 2001). We will discuss other such "edge effects" in Concept 24.2.

Habitat patches typically vary in both quality and resource availability. This variation can affect the diversity and population densities of the species inhabiting each patch, the time animals spend foraging in a patch (see Concept 8.2), and the movement of organisms between patches (see Concept 9.3). Patch boundaries, connections between patches, and the matrix between patches can also affect population dynamics, both within and among patches. For example, Schtickzelle and Baguette studied the movement patterns of the bog fritillary butterfly (*Proclossiana eunomia*) across fragmented landscapes in Belgium (**FIGURE 24.7**). Where patches of suitable butterfly habitat were aggregated, female butterflies crossed readily from patch to patch. However, where the habitat was more fragmented and there was a wider distance of matrix to cross, the butterflies were less likely to leave a patch (Schtickzelle and Baguette 2003).

FIGURE 24.7 **The Bog Fritillary Butterfly** The travel patterns of these butterflies (*Proclossiana eunomia*) are influenced by features of the surrounding landscape. Butterflies will hesitate to leave the patches they inhabit if there is not another suitable habitat patch nearby, but they will traverse a matrix of unsuitable habitat when the next patch is close.

While ecological processes are influenced by landscape patterns, landscape patterns are in turn influenced by ecological processes. Large grazing mammals, for example, often shape the landscapes they inhabit. The effects of moose (*Alces alces*) on Isle Royale in Lake Superior have been studied through the use of exclosures since the 1940s. These studies have shown that high rates of browsing by moose decrease plant growth, not just directly through the removal of biomass, but also indirectly by decreasing nitrogen supply via lower mineralization and litter decomposition rates (Concept 22.2). Moose browsing also shifts the tree species composition toward spruce, which in turn influences rates of biogeochemical processes (Pastor et al. 1988). The moose are thus both responding to and shaping the landscape. At a broader scale, landscape patterns interact with larger-scale disturbances, as we will see next.

Disturbance both creates and is influenced by landscape heterogeneity

Landscapes are dynamic. Changes can occur gradually as a result of shifts in climates and moving continents. Change can also be sudden, such as large disturbances, which is our focus in this section. For example, forests and prairies may burn over large areas, or floods bring sudden inputs of sediment into river ecosystems. We saw in Chapter 17 that disturbances can influence community composition. Landscape ecologists have asked, in turn, whether particular landscape patterns slow or accelerate the spread of disturbances or increase or decrease an ecosystem's vulnerability to disturbances.

FIGURE 24.8 Disturbances Can Shape Landscape Patterns The fires that burned through nearly one-third of Yellowstone National Park in the summer of 1988 resulted in a complex mosaic of burned and unburned patches. Areas that appear black in this aerial view of Madison Canyon were burned by intense crown fires, and brown patches were burned by severe ground fires, both of which killed most or all of the vegetation.

Courtesy of Jim Peaco/NPS Photo

An opportunity to examine the influence of landscape patterns on the spread of fire occurred after the 1988 forest fires that burned nearly one-third of the 898,000 hectares (ha), or 2.2 million acres, of Yellowstone National Park. Fires this extensive are thought to have occurred in the northern Rockies at 100- to 500-year intervals over the past 10,000 years. The 1988 fires burned through forest stands of different ages and species compositions, leaving a complex mosaic of patches that were burned at different intensities (**FIGURE 24.8**). The type and arrangement of these patches will influence the landscape composition for centuries (Turner et al. 2003). Here, a disturbance—fire—was a primary force shaping the landscape pattern of the future. At the same time, the fire was also responding to the existing landscape structure through its influence on burn probability. This reciprocal interaction between landscape pattern and disturbance is a common one.

Human actions have greatly altered the nature and extent of landscape-level disturbance. Some places have experienced greater human disturbance for a longer period of time than others. People first settled and cleared the areas with the most fertile soils, subjecting these ecosystems to the earliest human disturbance. Areas close to human settlements were converted to agriculture or had logging and hunting earlier than outlying areas. These disturbance patterns can be detected in ecological communities even centuries after people have left the land and it has reverted to forest (Butzer 1992).

Such *landscape legacies* can influence communities in subtle ways for centuries. In central France, Etienne Dambrine and his colleagues (2007) found that forest plant communities on the sites of recently uncovered Roman farming settlements still reflected the impacts of those disturbances 1,600 years later (**FIGURE 24.9**).

Roman ruins in France

Courtesy of Laure Laut

FIGURE 24.9 Landscape Legacies In central France, the legacy of Roman farming settlements, abandoned for nearly two millennia, is still reflected by higher plant species richness in the forest that replaced them. More plant species were found closer to the center of settlement sites, including more species that prefer a higher soil pH. The y axis represents departure from the mean calculated for plots 100–500 m from the settlement. (After E. Dambrine et al. 2007. *Ecology* 88: 1430–1439.)

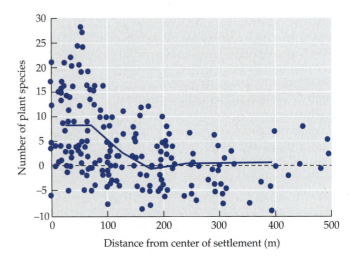

These researchers studied plant diversity in the forest at various distances from the Roman ruins. Dambrine and colleagues found that plant species richness increased in the vicinity of the ruins. An examination of soil properties revealed that this increase was primarily a consequence of higher soil pH and phosphorus, associated with the remnants of the lime mortar used in Roman buildings and from Roman agricultural practices. How many other ecosystems on Earth might display the signatures of human activities long since abandoned in their current community structure?

Disturbance, whether natural or human-caused, is an important factor shaping the landscape. Some current human activities are creating disturbances with far-reaching ecological effects, as we'll see in the next section.

FIGURE 24.10 The Islands of Lago Guri An aerial view of Lago Guri, Venezuela. This lake was formed when 4,300 km² (1.1 million acres) of forested land were inundated by a hydroelectric dam, leaving isolated islands of tropical forest.

> **CONCEPT 24.2** Habitat loss and fragmentation decrease habitat area, isolate populations, and alter conditions at habitat edges.

LEARNING OBJECTIVES

24.2.1 Describe the impacts of habitat fragmentation that lead to loss of diversity in landscapes.

24.2.2 Evaluate why fragmentation is more likely to impact higher trophic levels relative to plants and herbivores.

24.2.3 Explain how edges between habitat patches and the matrix in a fragmented landscape influence the physical environment and how this in turn impacts ecological processes such as dispersal and habitat use.

Habitat Loss and Fragmentation

In 1986, a massive hydroelectric project in the Caroni River valley of Venezuela inundated a large area of uneven terrain to create a reservoir known as Lago Guri (**FIGURE 24.10**). The result was the formation of scores of islands of tropical dry forest surrounded by water. This change in the landscape presented an opportunity for John Terborgh and his students and colleagues to study the effects of fragmentation in a tropical dry forest ecosystem. They found that small- and medium-sized islands were lacking the top predators found on the mainland, primarily wild cats (ocelots, jaguars, and pumas), raptors, and large snakes (Terborgh et al. 2006). As a result, generalist herbivores, seed predators, and predators of invertebrates were 10 to 100 times more abundant on the islands than in the remaining intact forest. Species that increased in abundance included leaf-cutter ants, birds, rodents, frogs, spiders, howler monkeys, porcupines, tortoises, and lizards. The increased abundances of these species had a dramatic effect on the vegetation of these islands: tree recruitment decreased and tree mortality increased because of high rates of herbivory, primarily by leaf-cutter ants (**FIGURE 24.11**). What lessons can we take from this "experiment" that apply to other fragmented ecosystems?

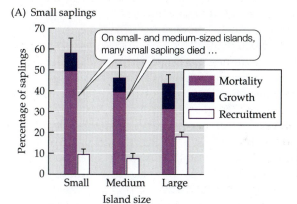

(A) Small saplings

On small- and medium-sized islands, many small saplings died …

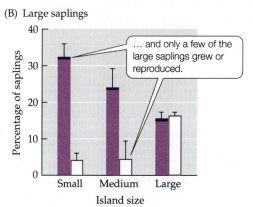

(B) Large saplings

… and only a few of the large saplings grew or reproduced.

FIGURE 24.11 Effects of Habitat Fragmentation by Lago Guri The high abundances of herbivores on small and medium-sized islands in Lago Guri caused a dramatic decline in sapling establishment and survival. The bars show the percentages of (A) small saplings and (B) large saplings in study plots that left their size class through either mortality or growth to a larger size, as well as the number of saplings recruited to each size class, over a 5-year period. Error bars show one SE of the mean. (After J. Terborgh et al. 2006. *J Ecol* 94: 253–263.)

Habitat loss and fragmentation are among the most widespread and important human-caused changes occurring in Earth's landscapes (**FIGURE 24.12**). When large blocks of habitat are cleared of forests, flooded by dam construction, divided by roads, or converted to human land uses, there are several consequences for the landscape and the species living there. The first is the direct loss of habitat area. Reductions in the amount of suitable habitat available have contributed to the declines of thousands of species, including the red-cockaded woodpecker (see the Case Study in Chapter 23). Second, as the remaining habitat becomes divided into smaller and smaller patches, it is increasingly degraded and influenced by edge effects, as the Biological Dynamics of Forest Fragments Project showed (see the Case Study in

Chapter 18). Third, fragmentation results in the spatial isolation of populations, making them vulnerable to the problems of small populations described in Concept 10.3.

The process of habitat loss and fragmentation may take place over many decades. A typical pattern begins with a clearing in a forest, which is then widened bit by bit until only isolated habitat fragments remain (**FIGURE 24.13**). Roads are often catalysts of habitat conversion, though human access along rivers can also serve to accelerate deforestation. The principal drivers of habitat fragmentation are conversion of land for agriculture and urban expansion.

Habitat fragmentation is a reversible process. The northeastern United States, for example, has more forest cover than it did a century ago as a result of agricultural

1620

Old-growth forest is shown in red.

1926

Each dot represents 25,000 acres of old-growth forest.

1990

© danilo donadoni/Marka/AGE Fotostock

Tulip tree in old-growth forest, Great Smoky Mountains National Park

FIGURE 24.12 Loss and Fragmentation of U.S. Old-Growth Forests Beginning in 1620, vast regions of old-growth forest (also known as ancient or virgin forest) in the United States were cut down to provide lumber and to make room for agriculture, housing, and other forms of development. (Adapted from A. Gould et al. 2001. In *Global Systems Science: A New World View*. The Lawrence Hall of Science. University of CA, Berkeley. © The Regents of the University of California. http://www.globalsystemsscience. org/studentbooks/anwv/ch3/. Based on C. O. Paullin. 1932. *Atlas of Historical Geography of the United States*. Carnegie Institution of Washington and the American Geographical Society of New York: Washington, DC, and New York; R. Findley and J. P. Blair. 1990. *Nat Geogr* 178: 106–136.)

FIGURE 24.13 **The Process of Habitat Loss and Fragmentation** Historically intact habitats are gradually reduced with increased human presence. These contemporaneous photographs (taken from different locations) illustrate a process that typically takes decades to complete. (A) An intact eucalyptus forest in Western Australia. (B) Areas within the forest have been cleared for grazing. (C) The forest has become further fragmented over time. (D) Only a few remnants of forest remain.

abandonment—but it will take centuries before these young forests contain as many species as were found in the old-growth forests that once covered the region. Furthermore, the global trend is toward net loss of forests (FAO and UNEP 2020) and toward increasingly fragmented forest, grassland, and riverine ecosystems.

Fragmented habitats are biologically impoverished relative to intact habitats

When habitat is fragmented, some species go locally extinct within many of the fragments. Reasons for the loss of species include loss of habitat, which can involve changes in climate, shelter, and nesting sites; lower resources; and the impact of genetic and stochastic factors on small populations. Mutualisms may be disrupted if pollinators are missing or as mycorrhizal fungi fail to persist in a particular fragment. Local extinction or decline is not inevitable; indeed, some species flourish under the changed conditions that follow fragmentation.

Fragmentation often leads to losses of top predators, giving rise to trophic cascades (see Concept 21.3), sometimes with large consequences for the remaining community as we saw with the Lago Guri example. Another example of such a cascade that has implications for human health is the growing risk of Lyme disease as a result of forest fragmentation in the northeastern United States. Brian Allan and colleagues found that forest fragments of less than 2 ha (5 acres) were very densely populated with white-footed mice (*Peromyscus leucopus*). Fragments of that size did not support substantial predator populations, and the mice had few competitors there. White-footed mice are the most important reservoir of *Borrelia burgdorferi*, the spirochete bacterium that causes Lyme disease. Ticks are the vector of this disease. Tick nymphs collected in these small fragments were significantly more likely to carry the disease, and occurred at higher densities, than nymphs in larger fragments (**FIGURE 24.14**). The outcome—an increased risk of human infection with Lyme disease—is ultimately a result of the biological impoverishment of habitat fragments (Allan et al. 2003).

Edge effects change abiotic conditions and species abundances in fragments

As intact habitat is fragmented, an abrupt boundary between two dissimilar patch types is created. The total length of habitat boundary, or *edge*, increases as fragmentation

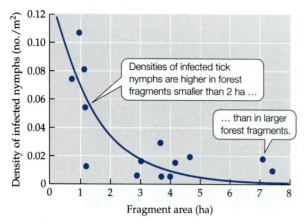

FIGURE 24.14 Habitat Fragmentation Can Have Consequences for Human Health The loss of predators from small forest fragments in New York State has led to elevated populations of white-footed mice in those fragments. As a result, densities of tick nymphs infected with the spirochete bacterium that causes Lyme disease are higher than in larger forest areas. (After B. F. Allan et al. 2003. *Conserv Biol* 17: 267–272.)

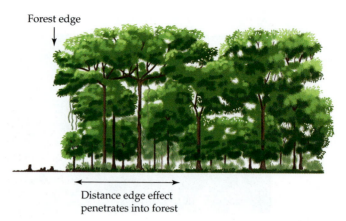

FIGURE 24.15 Edge Effects Deforestation creates new forest edges, exposing trees that once were surrounded by forest to edge effects such as increased light levels, higher temperatures, greater wind speeds, decreased soil moisture, and invasion of disturbance-adapted plants and animals. Some edge effects penetrate a few tens of meters into the forest fragment, while others penetrate hundreds of meters (see Analyzing Data 24.1).

increases. **Edge effects** are the diverse abiotic and biotic changes that are associated with habitat boundaries (**FIGURE 24.15**). The effect of edge formation is a change in the physical environment over a certain distance into the remaining fragment. As a result, biological interactions and ecological processes can change as well, as you can explore in **ANALYZING DATA 24.1**. The impact of the edge on the environment changes over time, so we can separate the immediate responses to fragmentation and edge formation from the responses that develop later (see Figure 18.24).

Analyzing Data 24.1 and the Case Study Revisited in Chapter 18 describe edge effects seen in large-scale experiments in Brazil. The effects of abiotic changes at a forest edge were also illustrated by a study of microclimates 10 to 15 years after the clear-cutting of an old-growth Douglas fir forest in the Pacific Northwest (Chen et al. 1995). Edges were generally characterized by higher temperatures, higher wind speeds, and more light penetration. Daily temperature extremes were also greater at the edges because more heat was lost from the forest edge at night than in the interior forest. The biotic consequences of these abiotic changes included higher rates of decomposition, more windthrown trees and thus more woody debris on the forest floor, and greater seedling survival of some tree species (Pacific fir) and the same or lower in others (Douglas fir and western hemlock).

Habitat edges provide habitat diversity that can facilitate or inhibit ecological functions. Edges can either enhance or inhibit dispersal of organisms. Dispersal of organisms can be greater for some species and lower for others at edges. Novel species interactions may take place at the junctions of two ecosystems. Some species may benefit from foraging in one habitat and reproducing in another. Invasive species are commonly more abundant in habitat edges,

influencing the population dynamics for native species (Fagan et al. 1999). For example, birds adapted to the forest interior often have lower breeding success when their nests are close to habitat edges. This lower breeding success can result from higher rates of egg predation by raccoons, crows, and other predators as well as higher rates of nest parasitism, especially by cowbirds. In the tallgrass prairie of Wisconsin, Johnson and Temple (1990) studied the reproductive success of five species of ground-nesting birds. They found that the closer nests were to a wooded edge of the prairie habitat, the greater the probability of nest predation by medium-sized predators and of nest parasitism by cowbirds, and the lower the rate of reproductive success. Similar patterns have been observed in other prairies, in Scandinavian forests, in eastern deciduous forests, and in the tropics (Paton 1994). Some biologists have characterized edges as "biological traps" as a result of the increased risks that some species face there (Battin 2004).

Fragmentation alters evolutionary processes

Throughout this book we've emphasized the interactions between ecological and evolutionary processes, particularly through adaptation and impacts on population dynamics. Human alteration of landscapes adds a twist to these interactions, which we emphasized in our consideration of conservation biology in the previous chapter. What are the evolutionary consequences when populations of species are restricted to smaller and more isolated patches?

Chapters 10 and 23 considered the genetic and demographic problems of small, isolated populations. The evolutionary consequences of fragmentation were studied by Marcel Goverde and his colleagues by watching bumblebee behavior in the Jura Mountains of Switzerland (Goverde et al. 2002). Their experimental plots included

ANALYZING DATA 24.1

How Far Do Edge Effects Penetrate into Forest Fragments?

When an intact forest is first fragmented, abiotic conditions change near the edge of the patch of forest that remains, giving rise to biotic changes (see Figure 24.15). In a landmark study on edge effects, William Laurance and his colleagues (2002)* synthesized 22 years of data from the Biological Dynamics of Forest Fragments Project, the world's largest ecological experiment (see the Case Study in Chapter 18). The graph shows some of the changes they measured in Amazon rainforest fragments.

1. According to the graph, how far from the edge must a tree be located if it is not to experience an increase in wind disturbance?

2. If the tree mortality effect penetrated 300 m on each side of an 800 × 800-m forest fragment, tree mortality would increase in what percentage of the fragment's area?

3. Are edge effects such as those shown here likely to cause other changes (not shown) in species interactions, community structure, or ecosystem processes? Explain.

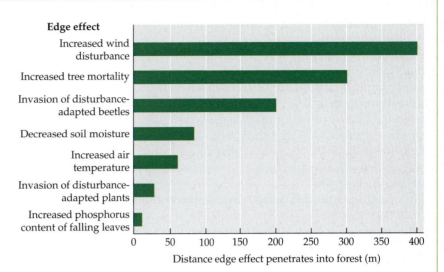

* Laurance, W. F., and 10 others. 2002. Ecosystem decay of Amazonian forest fragments: A 22-year investigation. *Conservation Biology* 16: 605–618.

---

meadow fragments of different sizes (created by mowing) and control plots in unfragmented meadow habitat. The researchers studied the foraging behavior of bumblebees as they visited the flowers of wood betony (*Stachys officinalis*), which were common in both experimental fragments and control plots. The bees visited fragments less frequently than they visited control plots, and once there, they tended to stay longer in the fragments. Ultimately, these two changes in bumblebee behavior resulted in a lower probability of pollination and an increased likelihood of inbreeding for the wood betony in the fragments, resulting in an altered evolutionary trajectory for those plants.

In many other cases, habitat fragmentation has been shown to increase rates of inbreeding and genetic drift for those species confined to fragments. For example, Keller and Largiadèr (2003) found significant genetic divergence between populations of the flightless ground beetle *Carabus violaceus* that had been isolated by roads. Habitat fragmentation can also alter selection pressures on organisms. Where plant populations become small and isolated, their chances of encountering their pollinators, their pathogens, their herbivores, their dispersers, and their competitors may all be reduced, with subsequent ecological and evolutionary consequences. Similar effects

have been observed in animals, whose breeding systems and survival patterns can be altered in small fragments (Barbour and Litvaitis 1993).

The evolutionary implications of habitat fragmentation continue to be important topics for study. As we'll see in the next section, however, the evolutionary dynamic associated with fragmentation is only one part of what must be considered in designing nature reserves that will work well to maintain biodiversity in landscapes increasingly modified by humans.

CONCEPT 24.3 Biodiversity can best be sustained by large reserves connected across the landscape and buffered from areas of intense human use.

LEARNING OBJECTIVES

24.3.1 List the factors that constitute a suitable core natural area that sustains diversity in a landscape.

24.3.2 Describe the importance of buffer zones around a core natural area in designing nature reserves.

24.3.3 Evaluate the importance of corridors in the context of environmental change.

Designing Nature Reserves

You may have a favorite national park, such as Everglades in Florida, Grand Canyon in Arizona, Bialowieski in Poland, or the Great Barrier Reef in Australia. How did these places get to be national parks? What were they before they were parks? Are they the best possible sites for maintaining biodiversity in their regions? Now consider how well the area around you is functioning to sustain native species. Your view is undoubtedly shaped by where you are right now, by what the human history of your area is, and by how effective past conservation work there has been. We turn now to an examination of the ways in which people can work to improve the likelihood of the persistence of species native to their region.

To counteract habitat loss, conservation planners worldwide are working to locate and design protected areas where species' populations can persist. The identification and preservation of core natural areas, buffer zones surrounding them, and habitat corridors connecting them is key to maintaining and allowing the growth of populations. In some cases, as we'll see, degraded ecosystems can be restored as viable habitat for wild species.

Core natural areas should promote species persistence

The principles of landscape ecology and conservation biology have come together to guide biologists in selecting the most vital lands for conservation. The design of new nature reserves focuses on **core natural areas**, where the conservation of biodiversity and ecological integrity take precedence over other values or uses (Noss et al. 1999). Populations that are able to maintain themselves in core areas may serve as sources of individuals for populations outside the protected areas. Ideally, core areas also provide enough land to meet the large habitat area requirements of top predators.

Madagascar is a large island that is a global priority for conservation. It has a rich biota and many endemic species, including more than 70 species of lemurs, a group of primates found only on Madagascar. The biota of Madagascar is seriously imperiled, as only 15% of the island's original forest remains. Efforts are under way to put more of its land into conservation. In designing a new national park in northeastern Madagascar, Claire Kremen and her colleagues examined both the biological and the socioeconomic circumstances of the region. Their design (**FIGURE 24.16**) was based on a core natural area that extended across

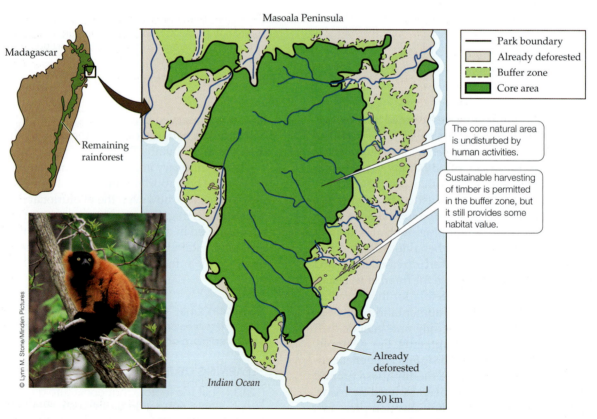

FIGURE 24.16 Designing Masoala National Park Masoala National Park, in northeastern Madagascar, was established after careful planning that took both ecological and socioeconomic concerns into account. It preserves habitat for many threatened species, including the red ruffed lemur (*Varecia variegata rubra*), which is endemic to this region of Madagascar. This map was simplified from more complex maps generated by using GIS techniques to analyze satellite imagery. (After C. Kremen et al. 1999. *Conserv Biol* 16: 605–618.)

several elevational and climate zones, encompassing a range of vegetation types. The proposed core area encompassed habitat for all of the region's rare species of butterflies, birds, and primates, and it had as yet been little affected by deforestation. The researchers excluded areas close to villages that had already been fragmented and where hunting had negatively affected animal populations (Kremen et al. 1999). The Masoala National Park, which opened in 1997, is now the largest national park in Madagascar at 211,230 ha (over 521,000 acres). With proper management, the park will give the unique biodiversity of this region an improved chance of being maintained in perpetuity.

Core natural areas are best when they are large and uncut by roads or even by trails. Large core areas facilitate maintaining large population sizes and help to minimize the problems of small populations and susceptibility to local extinction (see Concept 10.3). Thus, not all protected areas qualify as core natural areas. Many do not fully serve the purpose of protecting the whole biota from human interference. Most national parks in the United States were not established with the conservation of biodiversity as their primary mission, but rather to preserve scenery, often on land that was not considered useful for anything else. Conservation planners recognize that many countries do not have the short-term motivation of carving out large areas of land to be solely dedicated to biodiversity conservation given the requirements of supporting an economy based on resource extraction.

In the design of nature reserves, some spatial configurations are theoretically better than others for fostering the persistence of biodiversity (**FIGURE 24.17**). The principal concepts are that greater area, connectivity among reserves, and buffer zones promote greater biodiversity. While large areas with a greater proportion of core area should enhance diversity, multiple reviews indicate that several small reserves contain more species than a large reserve with the same total area (Fahrig 2019). The underlying reasons for this observation are unclear, but the realization that multiple small preserves can protect as much if not more biodiversity holds promise for the design of future reserves in the face of rapidly declining biodiversity and intact habitat. There are additional situations where small or disconnected reserves may be desirable. For example, diseases may spread less easily between isolated smaller reserves than within a large reserve.

In many settings where conservation is being accomplished, either the landscape or the social context

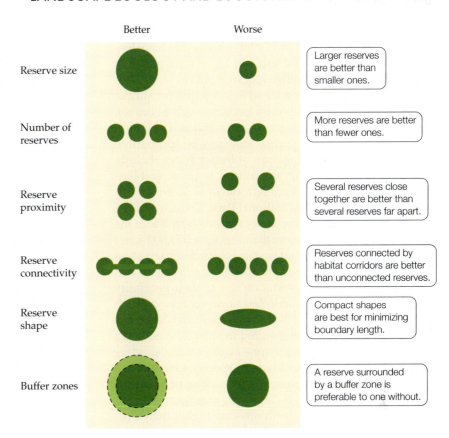

FIGURE 24.17 Guiding Principles for Designing Nature Reserves Some spatial configurations are preferable than others for fostering biodiversity. (After C. L. Shafer 1997. Terrestrial nature reserve design at the urban/rural interface. In M. W. Schwartz [ed.], *Conservation in Highly Fractured Landscapes*, pp. 345–347. Chapman and Hall: New York, and R. B. Primack and A. A. Sher. 2016. *An Introduction to Conservation Biology*. Sinauer and Associates: Sunderland MA.)

? Explain the underlying reasons why the design on the left is better than the one on the right for conservation as related to reserve size, proximity of multiple reserves, reserve connectivity, and reserve shape.

may not realistically permit adhering to the principles described above (Williams et al. 2005). There are many smaller reserves that have been established with the conservation of a single species or ecological community as their main objective. Such **biological reserves**, even if they are small, are nevertheless an important part of our conservation efforts. Critically situated smaller reserves may be the best available option, particularly where human population density is high and large reserves are unfeasible.

Core natural areas should be buffered by compatible land uses

Due to many constraints, relatively small areas of land are most commonly designated as core natural areas. If we are to conserve the majority of the world's species, however, areas outside of the core areas will have to be able to provide adequate habitat for biodiversity persistence (Soulé and Sanjayan 1998). We can augment the effectiveness of protected areas by surrounding them with **buffer zones** (see Figure 24.17), large areas that have less stringent controls on land use, yet are at least partially compatible with the resource requirements of many species. Such lands can be managed in ways that permit the extraction of resources used by humans, such as timber, fiber, wild fruits, nuts, and medicines, but still maintain some habitat value. Activities that may be compatible with the conservation function of buffer zones include selective logging, grazing, agriculture, tourism, and limited residential development (Groom et al. 1999).

In the plan for Masoala National Park, Kremen and her colleagues included a buffer zone on the eastern side of the park, which consisted of more than 71,000 ha (175,000 acres) of forest land designated for sustainable timber harvesting (see Figure 24.16). The researchers first identified areas that were at high risk of deforestation due to their proximity to villages. They then established how much wood each family, and thereby each village, consumed, and calculated how much area would be required to meet this need on a sustainable basis. The buffer zone augments the effective area of the park for many lowland species, even though they may be subjected to some level of hunting or collection.

On a cautionary note, buffer zones may serve as *population sinks* (areas where death rates are higher than birth rates) for some species, as animals that stray out of core areas and into buffer zones become vulnerable to hunting, vehicle collisions, or other sources of mortality. In Peru, slash-and-burn agriculture, in which plants are cut and then burned to enhance release of their nutrients, is commonly practiced just outside nature reserves. Wild animals such as agoutis, armadillos, and tapirs often damage farmers' crops. As a result, these animals are targeted by hunters, and such hunting has altered the relative abundances of mammals in the forest (Naughton-Treves et al. 2003). In other cases, however, buffer zones do not appear to act as population sinks and provide an effective transition zone between the core habitat and developed areas outside the reserve. An analysis of data from 785 animal species found that buffer zones can allow populations to persist in habitat fragments that might otherwise be too small or too isolated to support viable populations (Prugh et al. 2008). The key to success boils down to simple demography: if a buffer zone provides a threatened species with habitat in which birth rates are higher than death rates, it can aid conservation goals.

If we can succeed in establishing core areas for protection surrounded by sparsely inhabited buffer zones, have we done all that is necessary for conservation? Recall that landscape connectivity is another important consideration in reserve design.

Corridors can help maintain biodiversity in a fragmented landscape

Habitat corridors—narrow patches that connect blocks of habitat—have become a staple of urban, suburban, and rural planning (**FIGURE 24.18**). Connectivity among habitat patches might lessen the impact of fragmentation on small populations by enhancing dispersal through corridors of habitat that link them together.

When designing Masoala National Park discussed earlier, Kremen and her colleagues looked at the larger landscape and anticipated corridors that would be important in the future. Many of Masoala's target species are found in areas northwest of the park that lie between Masoala and two important protected areas to the north. The park plan included three corridors to those protected areas. The researchers developed this part of the plan by examining maps, but out of expediency, they did not actually do studies of animal movements (Kremen et al. 1999).

The intended function of habitat corridors is to prevent the isolation of populations in fragments. But do we know that corridors actually help to overcome this isolation? And is the effectiveness dependent on body size? For example, do corridors work for beetles as well as for wolves? Is a stream corridor in the suburbs providing necessary landscape connectivity for some species? At the continental scale, could we link the GYE to the Yukon through habitat corridors, as some have proposed? Experimental and observational studies of corridors' utility have shown mixed results.

Nick Haddad and his colleagues established a test of the utility of corridors at the Savannah River Ecology Laboratory in South Carolina. They set up patches of early successional habitat in a matrix of pine forest, some of

FIGURE 24.18 A Habitat Corridor Wildlife can cross this highway overpass over the A1 highway in the Netherlands.

them connected by corridors, and observed the movements of organisms between patches (**FIGURE 24.19**). Their results showed that the corridors did indeed serve to facilitate the movement of butterflies, pollen, and bird-dispersed fruits (Tewksbury et al. 2002).

Other studies, however, have found no benefits of corridors, and still others have found negative effects (reviewed in Haddad et al. 2014). For example, in the same experimental system at the Savannah River Ecology Laboratory, predation on indigo bunting (*Passerina cyanea*) bird nests was higher in patches connected by corridors (Weldon 2006). There are also concerns that corridors could facilitate the movement of pathogens (Hess 1994) or invasive species (Simberloff and Cox 1987). However, in general, corridors have been found to be effective for facilitating conservation of diversity (Resasco 2019).

Ecological restoration can increase biodiversity in degraded landscapes

What if habitat corridors are lacking and organisms' ability to move is impaired by an unsuitable matrix of degraded habitat? This was the case in Guanacaste Province on the Pacific coast of Costa Rica, where Santa Rosa National Park, in a lowland area of tropical dry forest, was largely separated by 35 km of cattle pasture and forest fragments from the upland forest habitat of the nearby mountains.

Tropical ecologist Dan Janzen knew that many insects, birds, and mammals needed to migrate between these lowland and upland forests. He also saw that the tropical dry forest that he had spent his career studying was fast disappearing. Janzen's effort to reverse this trend became one of the largest and most ambitious ecological restoration projects ever undertaken in the Neotropics. Now covering some 120,000 ha (300,000 acres) of land and 28,000 ha (70,000 acres) of marine reserve, the Area de Conservación Guanacaste (ACG) includes three national parks, a protected corridor linking them, and the surrounding agricultural areas. The ACG is home to some 230,000 species, or 65% of the species in Costa Rica (Daily and Ellison 2002).

Within the ACG, cattle ranches have occupied much of the land between the three parks for decades. Janzen has launched an effort to restore 75,000 ha (185,000 acres) of these pasturelands to the original forest types. His strategies include planting trees, suppressing fires, and limiting hunting to maintain mammalian and avian seed dispersers. Fire suppression is necessary to halt fires that burn readily in pastures covered in jaragua grass (*Hyparremia rufa*), an invasive plant introduced from Africa. Grazing will be maintained for some time in some areas to suppress the jaragua grass.

Ecological restoration is being applied in many other ecosystems, with varying degrees of success. To be successful, restoration ecologists must assess the current

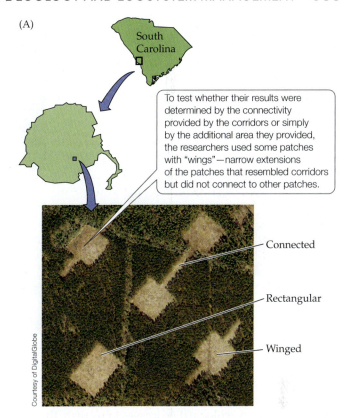

To test whether their results were determined by the connectivity provided by the corridors or simply by the additional area they provided, the researchers used some patches with "wings"—narrow extensions of the patches that resembled corridors but did not connect to other patches.

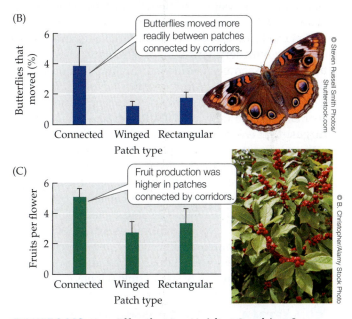

Butterflies moved more readily between patches connected by corridors.

Fruit production was higher in patches connected by corridors.

FIGURE 24.19 How Effective Are Habitat Corridors? (A) Nick Haddad and his colleagues tested the effectiveness of habitat corridors by creating experimental patches of early successional habitat within a pine forest and creating corridors between some of the patches. They then observed (B) movements of the common buckeye butterfly (*Junonia coenia*) between patches and (C) fruit production (which provides evidence of pollination) in winterberry (*Ilex verticillata*) in patches. Error bars in (B) and (C) show one SE of the mean. (After J. J. Tewksbury et al. 2002. *Proc Natl Acad Sci USA* 99: 12923–12926. © 2002 National Academy of Sciences U.S.A.)

(A)

Three years after restoration, densities of both adult and juvenile oysters were much higher in the restored habitat than in nearby unrestored habitat.

(B)

Before restoration

After restoration

Photos from D. M. Schulte et al. 2009. *Science* 325: 1124–1128

FIGURE 24.20 Dramatic Effects of an Ecological Restoration Project Native oyster populations have collapsed worldwide as a result of habitat loss and overharvesting. (A) In an ecological restoration experiment that began in 2004, oyster reefs were constructed in nine protected areas along the Great Wicomico River in Virginia. Three years later, native oyster populations had recovered dramatically across the 35-ha restoration project. Error bars show one SE of the mean. (B) Oyster habitat before and after restoration. The object on the right in each photograph is a robotic arm that can be used to pick up an individual oyster. (A after D. M. Schulte et al. 2009. *Science* 325: 1124–1128.)

ecosystem in the context of a desired ecosystem state (e.g., native biodiversity, function) and then apply their understanding of ecological processes to recreate the desired type of ecosystem. Anthony Bradshaw, a founder of restoration ecology, referred to this process as the "acid test" of ecology: "Each time we undertake restoration we are seeing whether, in the light of our knowledge, we can recreate ecosystems that function, and function properly" (Bradshaw 1987).

In some cases, such as the recovery of native oyster populations highlighted in **FIGURE 24.20**, results quickly suggest that we've passed this acid test. But in others, such as Janzen's efforts to restore tropical dry forests in Guanacaste, the process is likely to take decades if not centuries. In the next section, we will look more closely at how ecological principles are applied in making decisions about how to manage natural resources sustainably. Additionally, in the context of environmental change, particularly the spread of invasive species and climate change, it is unlikely that conserving and restoring communities to a perceived "original state" is possible (Hobbs et al. 2009).

CONCEPT 24.4 Ecosystem management is a collaborative process with the maintenance of long-term ecological integrity as its core value.

LEARNING OBJECTIVES

24.4.1 Evaluate how collaborative ecosystem management may lead to better solutions to preserving diversity than strictly science-based decisions.

24.4.2 Describe why iterative adjustments to land and marine reserve management policies are needed to help improve their effectiveness.

Ecosystem Management

In 1989, a convoy of logging trucks and about 300 loggers made the journey to a packed public hearing in Olympia, Washington, to defend their jobs. There was talk that the northern spotted owl (*Strix occidentalis caurina*; see Figure 9.15) could be listed as a threatened species under the U.S. Endangered Species Act, which would place its old-growth forest habitat off-limits to logging. Tempers were flaring among loggers and others supported by the timber industry. "When it comes to choosing between owls and the welfare of families, the hell with the owl as far as I'm concerned," said a state politician. At times, some of the testimony was drowned out by the honking of truck horns. The contentious debate about the logging of forests in the Pacific Northwest was reduced to "owls versus jobs" and resulted in bumper sticker and T-shirt slogans, vandalism by both sides, and the exchange of many angry words.

Some people recognized that there might be a better way to make decisions about the use of natural resources. The conflict in the Pacific Northwest was in part the outcome of a long history of top-down management of natural resources with a focus on resource production and extraction. In 1995, a federal Interagency Ecosystem Management Task Force was formed to develop alternatives to this approach (DellaSala and Williams 2006). Through selective withdrawal of old-growth forests from timber harvest, greater protection for the northern spotted owl has been accomplished. However, population recovery has been hampered by range expansion of larger barred owls into the Pacific Northwest, which compete with northern spotted owls for food. The recovery plan for northern spotted owls now includes both habitat protection and removal of barred owls.

Approaches to managing natural resources have become more collaborative over time

Through most of the twentieth century, management of natural resources on U.S. public lands was focused on maintaining individual resources of economic value, whether timber, deer or ducks for hunting, or scenery for visitors. This focus remained at the core of many land management policies until Congress passed the Multiple-Use Sustained-Yield Act of 1960. By the late 1980s, natural resource agencies had gradually expanded their missions to include "multiple use," in recognition that different people had different interests and that it was possible to manage public lands to meet diverse, and at times competing, demands. This was frequently done through spatial compartmentalization of uses, as when different blocks of land were designated as timber extraction zones, recreation zones, or wilderness areas.

Since the 1980s, with our greater awareness of the necessity of preserving biodiversity, our goals for land management have shifted again. The **ecosystem management** approach has emerged as a way to expand the scope of management to include protection of all native species and ecosystems while focusing on the sustainability of the whole system, not just the sustainability of resources of interest.

What is ecosystem management? Most simply stated, it is "managing ecosystems so as to assure their sustainability" (Franklin 1996). A committee of the Ecological Society of America arrived at a less simple but more comprehensive consensus definition in 1996: "Ecosystem management is management driven by explicit goals, executed by policies, protocols, and practices, and made adaptable by monitoring and research based on our best understanding of the ecological interactions and processes necessary to sustain ecosystem structure and function" (Christensen et al. 1996). This definition emphasizes sustainability but also recognizes the need for setting goals and using science to evaluate and adjust management practices over time.

The conflict in the late 1980s over old-growth forests in the Pacific Northwest was a stimulus to ecologists, government land managers, industry, and citizens to seek a less confrontational way to make decisions. Since that time, more collaborative decision making has been combined with better incorporation of science to arrive at management plans that not only attempt to sustain both biodiversity and people's livelihoods, but also are responsive to changing conditions. The focus of ecosystem management is on particular *ecoregions*, delineated by natural boundaries rather than political boundaries: a watershed, a mountain range, a stretch of coastline. The full range of *stakeholders*—people with some interest in the project—becomes involved in decision making for the ecoregion, joined together by the common goal of maintaining its ecological integrity and economic viability.

Ecosystem management sets sustainable goals, implements policies, monitors effectiveness, and adjusts as necessary

Ecosystem management is a process, one that may be implemented in different ways for different projects. Most ecosystem management projects begin with the gathering of scientific data to define the nature of the problems in the ecosystem. That information is then used to set sustainable goals. To meet those goals, a set of actions is needed, many of which may require adapting new policies. Once a new policy is implemented, the ecosystem is monitored to gauge whether that action brings about the desired result. Adjustments to the policies are then made as needed. In this iterative process, known as **adaptive management** (**FIGURE 24.21**), management actions are seen as experiments, and future management decisions are determined by the outcome of present decisions.

Monitoring is a vital component of adaptive management. For example, Mark Boyce developed a model predicting elk and wolf population dynamics in Yellowstone National Park following the reintroduction of wolves described in this chapter's Case Study that could be used to inform management of elk hunting by humans. Boyce and his colleague Nathan Varley took an adaptive management approach by updating this model based on demographic data from the first 10 years of wolf presence. Both the original and updated models provided good estimates of elk and wolf populations, and indicated stabilizing density-dependent control of elk populations due to predation and regulated elk hunting (Varley and Boyce 2006). This approach will be important for determining

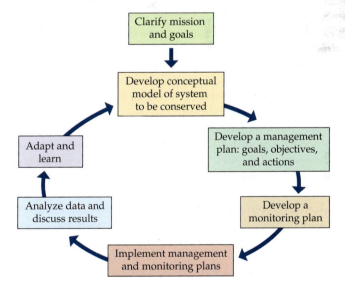

FIGURE 24.21 Adaptive Management Is a Vital Component of Ecosystem Management Adaptive management is a systematic way of learning from past management actions and adjusting future decisions accordingly. (After R. Margoluis and N. Salafsky. 1998. *Measures of Success: Designing, Managing, and Monitoring Conservation and Development Projects*. Island Press: Washington, DC.)

acceptable hunting levels for elk and for future adjustments in response to changing circumstances.

Although it is extremely useful, ecosystem management has limitations and drawbacks. One drawback is that it may take a long time to reach a consensus—yet averting an environmental crisis may require that preventive actions be taken quickly. There is also potential for continued conflict generated by those who simply want to disrupt the process, even when extensive efforts at stakeholder involvement have been made. In some instances, the intentional and effective spread of false information, a struggle for power among different government agencies, the presence of corruption, or the unmet needs of the people in local communities can produce situations that may not lend themselves to participatory governance.

Humans are an integral part of ecosystems

Human actions affect natural ecosystems, and human economies are affected by supplies of natural resources. Ecosystem managers must manage natural resources and biodiversity across large landscapes, as well as devise plans that protect both natural ecosystems and human economies. Ecosystem management incorporates human social and economic factors as fundamental parts of the decision-making process, along with legal requirements and, of course, ecological integrity (**FIGURE 24.22**). The integration of these different components is seen as necessary to achieve a successful management outcome.

As we have seen, people need natural ecosystems for many reasons, ranging from the economic to the spiritual. Ecosystem management incorporates education of the public about their reliance on ecosystem services as part of its mission. It also engages the public in helping to solve those problems that degrade the ecosystem services that they rely on.

Any conservation plan that excludes the human component will not be accepted, ultimately, by the stakeholders. The plan for Masoala National Park took the needs of the people living around the park into consideration. Conservation planners not only calculated their need for wood and provided for them in a buffer zone designated for managed forestry, but also surveyed the region for tree species that would have value in an export market and included them in an economic plan for future use. The idea was to remove economic pressure for park resources by identifying ways that people could support themselves and increase their incomes using resources outside the park. In addition, Kremen's team worked in conjunction with local people and with the Malagasy government to develop the plan, recognizing the importance of local acceptance of any proposal they made. In the end, the park plan provided for the economic needs of the people, by identifying forest resources that could be used to enrich the region, as well as for the habitat requirements of all the taxa included in the planners' analysis. While some problems have arisen with time, such as illegal hunting

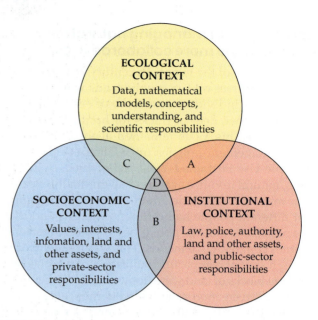

FIGURE 24.22 Humans Are an Integral Part of Ecosystem Management Ecosystem management integrates interests derived from ecological, institutional, and socioeconomic contexts. The letters represent the overlap of the three contexts: A, zone of regulatory or management authority; B, zone of social obligations; C, zone of informal decisions (as opposed to legal requirements); D, zone of win–win–win partnerships. (From Dennis A. Schenborn, personal communication.)

and logging within the park, the original conservation goals have generally been achieved (Kremen 2014).

A CASE STUDY REVISITED

Wolves in the Yellowstone Landscape

The reintroduction of wolves into the GYE in 1995 reflected a shift to using an ecosystem management approach in decision making. It was a bold step that followed years of study and preparation. That it happened at all reflects a quantum shift in human attitudes toward nature over the last century. In the late 1800s and early 1900s, wolves were feared and reviled. They were perceived primarily as a threat to people and livestock. Wolves were hunted to extinction in the area of Yellowstone National Park by the late 1930s and throughout virtually all of the conterminous United States not long thereafter.

The removal of a top predator can alter the landscape substantially, in part because herbivores whose populations were once controlled by the predator may increase in number and negatively affect vegetation dynamics, such as we've seen in several examples of trophic cascades. In Yellowstone, the growth and reproduction of riparian tree species, such as cottonwoods, aspens, and willows, declined after wolves were removed (Ripple and Beschta 2007). A possible reason was that the trees experienced heavy browsing by herbivores such as elk, which roamed freely along rivers and streams once the wolves were gone. How strong is the support for this explanation?

Many observations are consistent with this idea. The reintroduction of wolves began in the winter of 1995–1996, when 31 wolves captured in Canada were released into the park. Their numbers increased rapidly; by 2004, there were about 250 wolves in the park. Following the reintroduction, populations of elk, the wolves' principal prey, have declined by 50%. Elk were initially naive and very vulnerable to wolf predation, but they have since modified their behavior, showing a preference for foraging in places that provide high visibility (see Figure 8.10). Furthermore, cottonwoods, aspens, and willows have begun to recover in some areas. In some cases, the early signs of recovery appeared to be concentrated in areas where elk face a high risk of predation, such as locations where visibility is poor, escape routes are lacking, or ambush sites are common. Thus, elk may be avoiding areas where they are most vulnerable to attack by wolves, allowing trees in those areas to recover—and possibly leading to a series of other, cascading effects (**FIGURE 24.23**).

However, some studies have questioned whether a trophic cascade like that shown in Figure 24.23 is occurring. In an experimental test of the hypothesis that elk forage less in areas with wolves, leading to the recovery of woody species in those areas, Kauffman et al. (2010) found that aspen survival was not affected by the presence of wolves. Similarly, Creel and Christianson (2009) found that willow consumption by elk was more strongly affected by snow conditions than by the presence of wolves. Contrary to expectation, willow consumption actually increased when wolves were present. While the reintroduction of wolves may have affected willow and aspen abundance, it may be because predation by wolves has decreased the size of the elk population, not because fear of predation has led to changes in elk foraging behavior. Whatever the reason for the link between wolf introduction and increased willow and aspen cover, the reintroduction of wolves provides a wonderful opportunity to test hypotheses about how heterogeneity of a large landscape can be influenced by its component organisms.

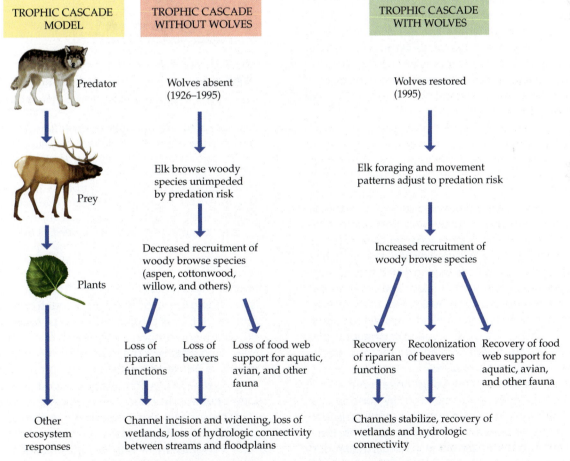

FIGURE 24.23 A Trophic Cascade Hypothesis Wolves are top predators, and their reintroduction to the Greater Yellowstone Ecosystem (GYE) has the potential to cause cascading trophic effects. According to the hypothesis shown here, elk now avoid those sites where they are most vulnerable to predation, and trees and shrubs are now returning to those sites after decades of suppression by elk. Researchers are actively testing this and other hypotheses about effects of wolves in the GYE. (After W. J. Ripple and R. L. Beschta. 2004. *BioScience* 54: 755–766.)

CONNECTIONS *in* NATURE

FUTURE CHANGES IN THE YELLOWSTONE LANDSCAPE If riparian trees continue to increase in abundance in the GYE, a series of linked effects (like those described in Concepts 16.3 and 21.3) may ensue. In some locations, increased numbers of willows have slowed stream flow and increased sedimentation rates (Beschta and Ripple 2006). The increased growth of riparian tree species is also expected to provide shade and habitat for migratory birds and for trout, which prefer shade-cooled waters. More riparian bird species have been observed under similar conditions in Alberta (Hebblewhite et al. 2005). As populations of willows, a preferred food for beavers, have increased, new beaver colonies have appeared. In turn, the dams built by the beavers have changed patterns of water flow, creating marshlands that may favor the return of otters, ducks, muskrats, and mink. The willow regrowth has also helped reverse the degradation of rivers and streams associated with the heavy grazing of streambank vegetation prior to wolf reintroduction (Beschta and Ripple 2019).

Other even more fundamental changes may be taking place in the Yellowstone ecosystem. Recall from Chapters 2, 3, and 4 that climate is the single most important determinant of where species live. With rising concentrations of greenhouse gases in the atmosphere, climate warming is occurring and will continue in the coming century (see Chapter 25). Will Yellowstone be able to maintain its current biological diversity in the face of global climate change?

A modeling study shows what the vegetation of the region surrounding Yellowstone National Park may look like under a doubling of current atmospheric CO_2 concentrations, which may happen within a century (**FIGURE 24.24**). Generally, the projections are for higher temperatures, no changes in precipitation, and more frequent fires. Based on these projected changes in the physical environment, the model predicts upslope and northward migrations of many species. These migrations will cause shifts in forest communities, with some species declining within the park and others increasing their range to include the park. Species currently rare in or absent from the GYE that may increase substantially there include gambel oak, western red cedar, and ponderosa pine. A near elimination of whitebark pine is predicted to occur as suitable habitat for that species shifts to the north (Bartlein et al. 1997).

The loss of whitebark pine would have a number of other ecological impacts. This tree is a keystone species that produces large, fatty, and nutritious nuts, an important food source for Clark's nutcracker, as well as for black and grizzly bears. Clark's nutcracker, in turn, is the primary disperser of the whitebark pine's seeds (Tomback 1982). One consequence of warmer winters during the past few decades has been an expansion of the range of the mountain pine beetle (*Dendroctonus ponderosae*) to high-elevation pine forests, including those where whitebark pine grows (Logan and Powell 2001). This beetle has devastating

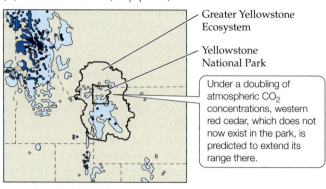

(A) Western red cedar (*Thuja plicata*)

Greater Yellowstone Ecosystem

Yellowstone National Park

Under a doubling of atmospheric CO_2 concentrations, western red cedar, which does not now exist in the park, is predicted to extend its range there.

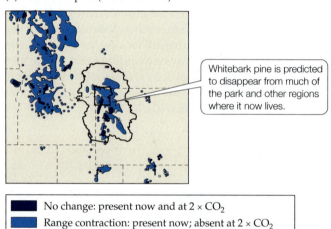

(B) Whitebark pine (*Pinus albicaulis*)

Whitebark pine is predicted to disappear from much of the park and other regions where it now lives.

■ (dark navy)	No change: present now and at $2 \times CO_2$
■ (blue)	Range contraction: present now; absent at $2 \times CO_2$
■ (light blue)	Range expansion: not present now; present at $2 \times CO_2$

FIGURE 24.24 Projected Effects of Climate Change in the Northern Rockies Shifts in the distributions of some principal tree species in the northern Rocky Mountains are projected by a model of a future climate driven by twice the current atmospheric CO_2 concentrations. These shifts include (A) the increased distribution of western red cedar, which is currently uncommon in the region, and (B) the near disappearance of whitebark pine. (After P. J. Bartlein et al. 1997. *Conserv Biol* 11: 782–792.)

effects on whitebark pine (**FIGURE 24.25**). Whitebark pine is also being attacked throughout much of its North American range by the fungus blister rust *Cronartium ribicola*, an introduced pathogen (Tomback and Achuff 2010). The combined effects of the mountain pine beetle and blister rust have caused an extensive die-off of whitebark pine, and this die-off has potentially reduced the occurrence of Clark's nutcracker in some areas (McKinney et al. 2009). Loss of whitebark pine also means loss of a food source for grizzly bears. Thus, it appears that climate change and introduced disease are having a major influence on whitebark pine populations, and that these effects have the potential to be transferred to wildlife, such as grizzly bears.

As we've seen in this chapter, landscape ecology and the use of tools such as remote sensing and GIS can elucidate current patterns of biodiversity and help us to predict future ones. Over the past century we have put much effort

FIGURE 24.25 Warm Winters Have Promoted a Devastating Insect Outbreak Once excluded from whitebark pine forests by cold winter temperatures, the mountain pine beetle has expanded its range as temperatures have warmed in recent decades. These beetles have contributed to the death of millions of whitebark pines, which turn red and subsequently gray when they die (as in this forest in the southern Rocky Mountains). In July 2011, the U.S. Fish and Wildlife Service announced that it will list whitebark pine as a candidate species under the Endangered Species Act.

into selecting, establishing, and undertaking management of new protected areas, but now we need to ask how well those areas will maintain their species in a warmer world. If biodiversity losses are projected under climate change, are there steps we can take now that can improve habitat connectivity, create or improve buffer zones around core natural areas, or restore degraded areas to greater ecological integrity? Or will we need to move species to new areas of suitable habitat, especially if they cannot migrate quickly enough to keep up with climate change?

With a growing human population and growing demands on ecosystems, these challenges will be considerable. Ecologists will have the critical role of providing the scientific information needed to make decisions about how we proceed as a society. The future of untold numbers of species relies on how effective we can be at this task.

SUMMARY

CONCEPT 24.1 Landscape ecology examines spatial patterns and their relationship to ecological processes.

24.1.1 Describe the elements that make up a landscape and illustrate how they can influence ecological processes such as dispersal and ecosystem function.

- A landscape is a heterogeneous area made up of a dynamic mosaic of different components that interact through the exchange of materials, energy, and organisms.

24.1.2 Show how landscape structure can be evaluated using the number and areas of the elements that make up the landscape.

- Landscapes are characterized by their composition— the elements that constitute them, including patches, mosaic, and corridors—as well as by their structure— how those elements are arranged on the landscape.

- Landscape patterns influence ecological processes by determining how easily organisms can move among elements as well as by influencing ecosystem properties such as rates of biogeochemical cycling.

24.1.3 Compare the benefits and drawbacks associated with using coarse-scale versus fine-scale characterization of a landscape.

- The scale that is used to characterize a landscape should match the needs of the research question or management issue.

24.1.4 Describe how disturbances can both affect and be affected by the landscape structure.

- Landscape patterns both shape and are shaped by disturbances.

CONCEPT 24.2 Habitat loss and fragmentation decrease habitat area, isolate populations, and alter conditions at habitat edges.

24.2.1 Describe the impacts of habitat fragmentation that lead to loss of diversity in landscapes.

- Habitat fragments are biologically impoverished compared with the intact habitat from which they were derived.

24.2.2 Evaluate why fragmentation is more likely to impact higher trophic levels relative to plants and herbivores.

- Once a patch of habitat is isolated by fragmentation, organisms reliant on that habitat may be isolated or may be able to cross the intervening matrix to some extent.

- The isolation of populations and the shifts in communities that result from habitat fragmentation alter ecological and evolutionary processes.

24.2.3 Explain how edges between habitat patches and the matrix in a fragmented landscape influence the physical environment and how this in turn impacts ecological processes such as dispersal and habitat use.

- The edges of habitat fragments have different abiotic conditions, and thus have different population dynamics, than interior habitats.

CONCEPT 24.3 Biodiversity can best be sustained by large reserves connected across the landscape and buffered from areas of intense human use.

(Continued)

SUMMARY (continued)

24.3.1 List the factors that constitute a suitable core natural area that sustains diversity in a landscape.

- The ideal spatial configuration for a core natural area is large, compact, and connected to or close to other protected natural areas.

24.3.2 Describe the importance of buffer zones around a core natural area in designing nature reserves.

- Core natural areas should be surrounded by buffer zones where human economic uses that are compatible with biodiversity conservation are allowed.

24.3.3 Evaluate the importance of corridors in the context of environmental change.

- Habitat corridors are instrumental in facilitating the movements of organisms between natural areas.

- Ecological restoration allows areas that have been degraded to support native species and ecosystem processes once again.

CONCEPT 24.4 Ecosystem management is a collaborative process with the maintenance of long-term ecological integrity as its core value.

24.4.1 Evaluate how collaborative ecosystem management may lead to better solutions to preserving diversity than strictly science-based decisions.

- Collaboration among all stakeholders is key to arriving at effective management plans.

24.4.2 Describe why iterative adjustments to land and marine reserve management policies are needed to help improve their effectiveness.

- Ecosystem management is a process of setting sustainable goals, developing and implementing land use management policies, monitoring the effectiveness of prior decisions, and adapting plans accordingly.

- Humans are an integral part of ecosystems. Conservation plans that include the economic and social well-being of local human populations are more likely to succeed over the long term.

Review Questions

1. How are islands of terrestrial habitat that result from habitat fragmentation like actual islands surrounded by water? How are they different? How do the principles of island biogeography (see Concept 18.3) apply to habitat islands in a fragmented landscape?

2. Are habitat corridors just long, skinny habitat patches? Describe how a corridor is like the habitat blocks it is meant to join and how it is different from them, and outline the implications for organisms using the corridor. Do you think corridors are beneficial? Do you think they are necessary? Why?

3. The western boundary of Yellowstone National Park, where it borders a national forest, is visible from space (check it out with Google Earth at 44°21'45"N, 111°05'50"W). Explain this observation in terms of the contrasting missions of a national park and a national forest. What are the ecological implications of these contrasting institutional functions for biodiversity? For ecosystem management?

Hone Your Problem-Solving Skills

One of the criteria used in determining the design of nature reserves is the shape of the habitat patch, which gives an indication of the amount of edge versus area of core habitat. Some species have greater needs for edge habitat, while others are negatively impacted by edges. Consider **FIGURE A**, indicating the sensitivity of three bird species to the shape of a habitat patch, estimated as the ratio of perimeter (length) to area.

1. Each of the three bird species is likely to do better in either design 1 or design 2 shown in the maps of habitat patches (**FIGURE B**). What is your prediction?

2. Describe some of the ecological considerations that might positively or negatively impact the response of a species to a habitat edge. Consider factors such as food preference, predation risk, physical environment, and reproduction.

(A)

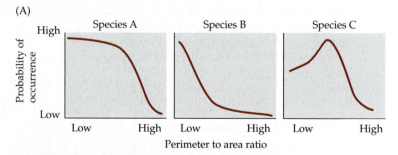

(B)

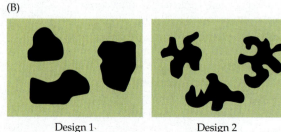

Design 1   Design 2

Global Ecology

Dust Storms of Epic Proportions: A Case Study

Dust is usually a subtle nuisance for most city dwellers, a reminder of neglect and lax housekeeping. Living in islands of asphalt and concrete, most urbanites see little bare soil, let alone clouds of blowing dust in the sky. Yet in late spring of 1934, a massive dust storm shrouded the U.S. cities of Chicago and New York in a dark haze never seen before by their residents. People choked on the dust, and it burned their eyes. Twelve million tons of dust fell on Chicago—4 pounds for each resident—and an estimated 350 million tons of dust were carried by the storm to the Atlantic Ocean. As frightening as this event was to city dwellers, farmers in the southern Great Plains had suffered through multiple years of frequent severe dust storms throughout the 1930s (**FIGURE 25.1**). During this period, many people in that region, known as the Dust Bowl, suffered from an often-fatal dust-induced pneumonia similar to the black lung disease that was killing coal miners.

Since the mid-1990s, widespread dust storms have impacted parts of Asia, including China, South Korea, and Japan. Residents of Beijing, China, have had experiences similar to the residents of Chicago and New York who faced massive, unexpected dust storms. An April 2006 storm dropped more than 300,000 tons of dust on Beijing. Residents were encouraged to stay indoors to avoid inhaling the dust and getting it in their eyes. Many of those brave enough to venture out wore surgical face masks to protect their lungs. Some residents lined their windows and doors with rags in an attempt to keep the dust out of their houses and apartments. More intense and frequent dust storms have occurred in the Middle East in the past decade. One storm in August 2015 was so bad that ports and airports throughout the region had to close. Several deaths and thousands of injuries were attributed to the dust. In addition to the direct health impact of dust on human respiratory systems, it can also spread disease such as meningitis.

Large dust storms in urban areas are perceived as rare events, potentially linked to unsustainable land use practices such as overgrazing or farming on marginal lands. In the examples mentioned above, farming and grazing in arid areas had increased prior to the dust storms. There is evidence, however, that massive dust storms occur at regular, but infrequent, intervals irrespective of human activities, moving large amounts of soil across whole continents. Over the past century, these events have been associated with prolonged droughts. The urban dust storms in the United States

KEY CONCEPTS

CONCEPT 25.1 Elements move among geologic, atmospheric, oceanic, and biological pools at a global scale.

CONCEPT 25.2 Earth is warming because of anthropogenic emissions of greenhouse gases.

CONCEPT 25.3 Anthropogenic emissions of sulfur and nitrogen cause acid deposition, alter soil chemistry, and affect the health of ecosystems.

CONCEPT 25.4 Losses of ozone in the stratosphere and increases in ozone in the troposphere both pose risks to organisms.

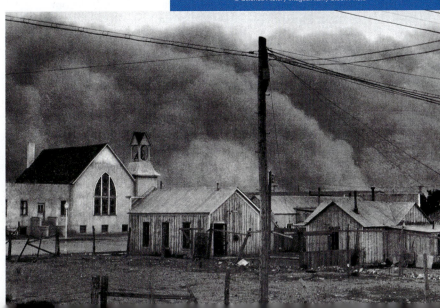

© Science History Images/Alamy Stock Photo

FIGURE 25.1 A Massive Dust Storm A wall of dust approaches the town of Clayton, New Mexico, on May 29, 1937. This storm was one of several "black dusters" that swept through the Dust Bowl during the 1930s.

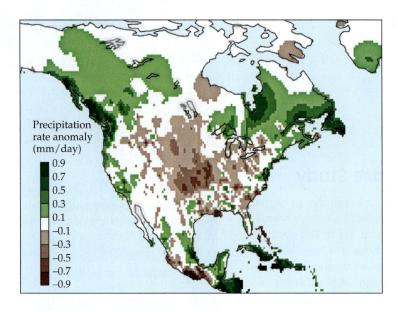

FIGURE 25.2 Drought in the Southern Plains During the 1930s, the southern Great Plains of the United States experienced the driest weather on record. The drought, in combination with loss of vegetation cover, created conditions conducive to dust input into the atmosphere. The values shown are anomalies (differences between averages for the period 1932–1939 and long-term averages). (After B. I. Cook et al. 2009. *Proc Natl Acad Sci USA* 106: 4997–5001. © 2009 National Academy of Sciences, U.S.A.)

during the 1930s were associated with a decade-long drought in the Dust Bowl (**FIGURE 25.2**). Similarly, the Beijing dust storms of the past two decades have been associated with drought in Mongolia. The increase in the Middle East dust storm frequency has been attributed to climate change and diversion of rivers for agriculture.

Dust in the atmosphere is made up of soil particles blown from regions that lack vegetative cover to protect their soils from the wind. As discussed in Chapters 4 and 22, soils are important as sources of nutrients, determinants of terrestrial moisture availability, and habitat for organisms. Therefore, the redistribution of soils from one area to another has the potential to cause ecological change. How widespread are these ecological effects? What role have humans played in the dust storms of the past century? As we will see in this chapter, the movement of dust is an important component in the movement of elements at the global scale.

Introduction

In Chapter 22, we reviewed the cycling of nutrients within ecosystems associated with biological uptake and decomposition. The movements of these biologically important elements are linked at a global scale that transcends ecological boundaries at the ecosystem and biome scales. Ecological processes at the ecosystem scale (e.g., net primary production, decomposition) influence global phenomena (e.g., climate via greenhouse gas emissions and uptake). In addition, the realization that humans are increasingly changing the physical and chemical environment at a global scale has fostered a greater awareness of ecology at these larger spatial scales. Emissions of pollutants, dust, and greenhouse gases into the atmosphere have caused widespread environmental problems, including climate change, acid precipitation, eutrophication, and loss of stratospheric ozone. A major focus of global ecology is therefore the study of the extensive environmental effects of human activities.

The first part of this chapter will cover the global-scale cycles of chemical elements, which are influenced by, but distinct from, the ecosystem-scale cycles covered in Chapter 22. Knowledge of these cycles is important for understanding global environmental change. Humans have had profound effects on these element cycles; for example, human activities now dominate the global nitrogen cycle (Fowler et al. 2015). We will review the environmental changes associated with anthropogenic effects in the remaining sections.

> **CONCEPT 25.1** Elements move among geologic, atmospheric, oceanic, and biological pools at a global scale.

LEARNING OBJECTIVES

25.1.1 Summarize the major pools and fluxes associated with global-scale cycles of carbon, nitrogen, phosphorus, and sulfur.

25.1.2 Describe why anthropogenic perturbations to the global carbon cycle are important mediators of environmental change, even though the fluxes are relatively small compared to net primary production and respiration.

25.1.3 Evaluate why changes to individual element cycles at the global scale have implications for the cycling of other elements.

Global Biogeochemical Cycles

In this section, we will follow the biogeochemical cycling of carbon, nitrogen, phosphorus, and sulfur at the global scale. These particular elements are emphasized both because of their importance to biological activity and because of their roles as pollutants. The cycles are discussed in terms of *pools*, or reservoirs—the amounts of elements within components of the biosphere—and *fluxes*, or rates of movement, between pools. For example, terrestrial plants constitute a pool of carbon, while photosynthesis represents a flux—in this case, the movement of carbon from the atmospheric pool to the terrestrial plant pool.

Carbon cycles dynamically at the global scale

Carbon (C) is critically important for life because of its role in energy transfer and the construction of biomass (see Concept 5.2 and Concept 20.1). At a global scale, C that is actively cycling is relatively dynamic, moving between atmospheric, terrestrial, and oceanic pools relatively quickly (over weeks to decades). It is important that we understand the global C cycle because changes in the fluxes of C among these pools are influencing Earth's climate system. Carbon in the atmosphere occurs primarily as carbon dioxide (CO_2) and methane (CH_4). As we saw in Chapter 2, both of these greenhouse gases influence atmospheric absorption of infrared radiation and its reradiation from Earth's surface. Thus, any changes in the atmospheric concentrations of these gases can have profound effects on the global climate, as we will see later in this chapter.

There are four major global pools of C: atmosphere, oceans, land surface (including soils and vegetation), and sediments and rock (**FIGURE 25.3**) (Schlesinger and Bernhardt 2020). The largest of these pools is the combination of sediments and rock, which contain 99% of global C. The C in this pool is found primarily in the form of carbonate minerals and organic compounds. It is the most stable of

the major pools, taking up and releasing C on geologic time scales.

The oceanic pool consists of two main components: surface waters (to depths of 75–200 m), where most marine biological activity occurs, and deeper, colder waters. Carbon dioxide dissolves in ocean water because of the concentration gradient between the atmosphere (which has a higher concentration) and the ocean (which has a lower concentration). There is relatively little mixing between ocean surface waters and deeper waters, although C is transferred between them by the sinking of detritus and carbonate shells of marine organisms and by the downwelling of polar ocean currents described in Concept 2.2. Most of the C in the oceans (>90%) is in the deeper waters. Some flux from this deep ocean pool occurs when upwelling brings carbon-enriched water to the surface, releasing CO_2 into the atmosphere.

The terrestrial pool, which includes vegetated and nonvegetated land surfaces and their associated soils, is the largest pool of biologically active C. The soil pool contains approximately twice as much C as the vegetation pool. The terrestrial pool exchanges C with the atmospheric pool primarily through photosynthetic uptake of CO_2 by plants and respiratory CO_2 release by plants and heterotrophs. Prior to the Industrial Revolution that began in the early nineteenth century, the exchanges between these two pools were roughly equal, with no net change in atmospheric CO_2.

As a result of the rapid growth of the human population over the past 200 years and associated industrial and

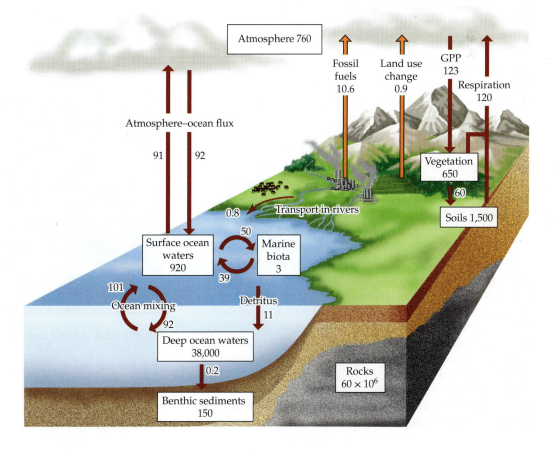

FIGURE 25.3 The Global Carbon Cycle Boxes represent major pools of C, measured in petagrams (1 Pg = 10^{15} g). Arrows represent major fluxes of C, measured in petagrams per year; anthropogenic fluxes are shown in orange. Note that the largest fluxes are terrestrial gross primary production (GPP) and respiration. (After W. H. Schlesinger and E.S. Bernhardt. 2013. *Biogeochemistry: An Analysis of Global Change*, 3rd eds. Academic Press: Cambridge, MA; F. S. Chapin et al. 2002. *Principles of Terrestrial Ecosystem Ecology*. Springer-Verlag: New York; P. Ciais et al. 2013. In *Climate Change 2013: The Physical Science Basis*, T. F. Stocker et al. [Eds.], pp. 466–570. Cambridge University Press: Cambridge.)

? How would deforestation influence the magnitude of carbon fluxes?

Atmosphere 760

Fossil fuels 10.6 · Land use change 0.9 · GPP 123 · Respiration 120

Atmosphere–ocean flux

91 92

Vegetation 650

60

Soils 1,500

Transport in rivers

0.8

50

Surface ocean waters 920 · Marine biota 3

39

Ocean mixing

101

Detritus 11

92

Deep ocean waters 38,000

0.2

Rocks 60 × 10^6

Benthic sediments 150

agricultural development, there has been an increase in the release of C to the atmosphere from the terrestrial pool. This **anthropogenic** (human-generated) release of C is the result of land use change—mainly forest clearing for agricultural development—and the burning of fossil fuels. Prior to the mid-nineteenth century, deforestation was the largest contributor to anthropogenic C release to the atmosphere. Removing the forest canopy warms the soil surface, increasing rates of decomposition and heterotrophic respiration. Burning of the trees also releases CO_2, as well as small amounts of carbon monoxide (CO) and CH_4, into the atmosphere. During the last half of the twentieth century, deforestation for agricultural development shifted from the mid-latitudes of the Northern Hemisphere to the tropics.

The rate of anthropogenic emission of C into the atmosphere has continued to increase in recent decades. In 1970, anthropogenic CO_2 emissions added C to the atmosphere at a rate of 4.1 petagrams (1 Pg = 10^{15} g) per year; by 2018, this rate had almost tripled to 11.5 Pg C per year. Today, burning of fossil fuels accounts for approximately 92% of the anthropogenic C flux to the atmosphere; the remaining 8% is associated with deforestation. Approximately half of these anthropogenic CO_2 emissions are taken up by the oceans and terrestrial biota. This proportion will decrease with time, however, as the uptake of CO_2 by terrestrial and marine ecosystems will not keep pace with the rate of emissions to the atmosphere (IPCC 2021).

Emissions of CH_4 to the atmosphere from the terrestrial pool have also increased in the past two centuries as a result of human activities. Although atmospheric concentrations of CH_4 are much lower than those of CO_2, even small increases in CH_4 could influence the global climate because it is 25 times more effective as a greenhouse gas per molecule than CO_2. Methane is emitted naturally by anaerobic methanogenic archaea that live in wetlands and shallow marine sediments. Methanogenic archaea in the rumens of ruminant animals are also a source of atmospheric CH_4. Anthropogenic emissions of CH_4 have doubled since the early nineteenth century as a result of the processing and burning of fossil fuels, agricultural development (primarily rice, which is grown in flooded fields), burning of forests and crops, and livestock production (IPCC 2021). As a result, atmospheric CH_4 concentrations have more than doubled over the past two centuries.

The process of photosynthesis is sensitive to the concentration of CO_2 in the atmosphere. As a result, photosynthesis has the potential to increase as anthropogenic CO_2 emissions increase, primarily in plants with the C_3 photosynthetic pathway (see Concept 5.3). Experiments have shown, however, that for some herbaceous plants, these increases may be short-term because the plants may acclimate to elevated CO_2 concentrations (Pastore et al. 2019). For other plants, such as forest trees, increases in photosynthetic rates may be more sustained.

It is extremely important that we understand the response of forest ecosystems to elevated CO_2 concentrations. Because much of terrestrial net primary production (NPP), and thus C uptake, occurs in these ecosystems, their response will have a profound effect on the fate of anthropogenic CO_2 emissions. However, it is difficult to manipulate atmospheric CO_2 concentrations experimentally in an intact forest. In one successful approach, called free-air CO_2 enrichment, or FACE, researchers inject CO_2 into the air through vertical pipes surrounding stands of trees while monitoring the atmospheric concentration of CO_2 within the experimental stands. The rate of CO_2 injection is controlled to maintain a relatively constant elevated level.

One of the first FACE experiments was initiated by Evan DeLucia and his colleagues to investigate the ecosystem effects of elevated CO_2 concentrations in a young loblolly pine (*Pinus taeda*) forest in North Carolina (DeLucia et al. 1999) (**FIGURE 25.4**). The experiment was initiated in 1997, when three plots exposed to elevated CO_2 concentrations and three control plots exposed to ambient CO_2 concentrations were established. The researchers used measurements of tree basal area to estimate aboveground NPP and repeated collections of soil cores to estimate fine root growth and belowground NPP. DeLucia and his colleagues found that the elevated CO_2 concentrations increased the overall NPP of the forest by 25%. Input of C into the soil, from both aboveground litter and belowground fine root turnover, also increased. The results of this experiment indicated that forests may be an important sink for anthropogenic CO_2. However,

FIGURE 25.4 A FACE Experiment The circles visible in this aerial photo are free-air CO_2 enrichment (FACE) treatment rings in a loblolly pine (*Pinus taeda*) forest in the Duke Forest in North Carolina. Carbon dioxide is released from plastic pipes surrounding treatment plots at a rate calculated to raise the CO_2 concentration to 200 ppm above ambient atmospheric CO_2 concentrations.

DeLucia et al. suggested that their young forest stand represents the upper limit of potential CO_2 uptake and that older forests, and forests with lower water and nutrient supplies, may not have as great a capacity to take up CO_2. Results from other FACE experiments in forest ecosystems have found similar responses to elevated CO_2 concentrations (an average increase in NPP of 23%; Norby et al. 2005). The greater productivity observed in forests in the Northern Hemisphere over the past five decades may be related in part to elevated atmospheric CO_2 concentrations (Graven et al. 2013), verifying the predictions of the FACE experiments.

Changing atmospheric CO_2 concentration directly alters the acidity (pH) of the oceans by affecting the rate at which CO_2 diffuses into seawater. Greater diffusion of CO_2 into seawater enhances the formation of carbonic acid, which lowers the pH of the seawater:

$$CO_2 + H_2O \rightleftharpoons H_2CO_3 \text{ (carbonic acid)} \rightleftharpoons H^+ + HCO_3^-$$
$$\text{(bicarbonate)} \rightleftharpoons 2 H^+ + CO_3^{2-} \text{ (carbonate)}$$

During the past century, ocean acidity has increased by about 30%. Further increases have been forecast using model simulations incorporating the expected increases in anthropogenic CO_2 emissions over the twenty-first century (Bopp et al. 2013). The predicted increases will have two negative effects on marine organisms that form protective external shells from calcium carbonate, including corals, mollusks, and many plankton. First, the increase in acidity will dissolve the existing shells of the organisms. Second, lower concentrations of carbonate in seawater will decrease the organisms' ability to synthesize shells (Feely et al. 2004; Orr et al. 2005). Between 1990 and 2009, the rate of formation of calcium carbonate by corals on Australia's Great Barrier Reef declined by 14%, an amount consistent with observed decreases in the pH of seawater (**FIGURE 25.5**) (De'ath et al. 2009). Both effects will increase mortality and lower the abundances of marine organisms that rely on calcium carbonate, altering the diversity and function of marine ecosystems. Under the current CO_2 emissions rate the majority of coral reefs will not be able sustain their carbonate structure by 2100 (Cornwall

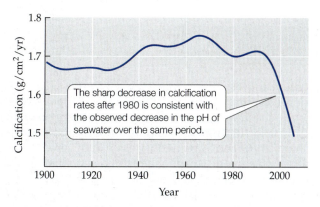

FIGURE 25.5 Rates of Calcification of Corals on Australia's Great Barrier Reef, 1900–2005 The sharp decline in calcification rates after 1980 is associated with the combined effects of decreasing pH and increasing ocean water temperature. (After G. De'ath et al. 2009. *Science* 323: 5910.)

et al. 2021). (Make your own prediction of the future of ocean pH and its effect in **ANALYZING DATA 25.1**.)

Atmospheric concentrations of C have changed dynamically throughout Earth's history in association with geologic and climate changes. Concentrations of CO_2 have ranged from greater than 3,000 parts per million (ppm) 60 million years ago to less than 200 ppm 140,000 years ago. Over the past 400,000 years, variations in the concentrations of CO_2 and CH_4, as measured in tiny bubbles preserved in polar ice, have followed glacial–interglacial cycles. The lowest CO_2 concentrations during this time were associated with glacial periods (**FIGURE 25.6**). Over most of the past 12,000 years, atmospheric CO_2 concentrations remained relatively stable, varying between 260 and 280 ppm. Since the mid-nineteenth century, however, CO_2 concentrations have increased at a rate faster than at any other time over the past 400,000 years (IPCC 2013), reaching values of 415 ppm in 2021. Even if we dramatically decreased our CO_2 emissions starting today, atmospheric CO_2 concentrations would remain elevated for a long time to come because of a time lag (decades to centuries) in oceanic uptake. The influence of CO_2 and CH_4 on climate change will be discussed later in this chapter.

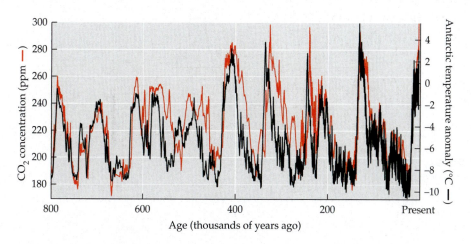

FIGURE 25.6 Changes in Atmospheric CO_2 Concentrations over Time Atmospheric CO_2 concentrations have varied with temperature over the past 800,000 years. These gas concentrations were measured in bubbles trapped in Antarctic ice; temperatures were estimated using oxygen isotopic analyses (see Ecological Toolkit 5.1). CO_2 concentrations in 2019 are 408 ppm. (After D. Lüthi et al. 2008. *Nature* 453: 379–382.)

How Much Will Ocean pH Drop in the Twenty-First Century?

Ocean acidification is one of the consequences of increased anthropogenic CO_2 emissions. There is already substantial evidence that the pH of ocean waters is declining (**FIGURE A**). Using information about the chemistry of seawater and diffusion of CO_2 from the atmosphere, marine geochemists have projected that the pH of the ocean will have decreased to 7.9 by the year 2050, and by 2100 it will be 7.75, assuming "business as usual" CO_2 emissions (a continued increase in the rate of emissions growth) during the twenty-first century (IPCC 2013).*

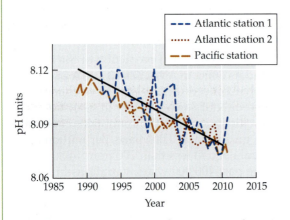

FIGURE A Measured Trend in Ocean pH for Two Stations in the Atlantic Ocean and One in the Pacific Ocean (After IPCC. 2013. *Climate Change 2013: The Physical Science Basis*. Cambridge University Press: Cambridge.)

1. Using the data in Figure A, derive a simple linear mathematical relationship or draw a graph to come up with your own prediction of the ocean pH in the years 2050 and 2100. How well do your predictions match up with the IPCC's predictions based on seawater chemistry and continued increases in atmospheric CO_2? Your answer should give you a higher estimated ocean pH than the one predicted in the IPCC report. What might account for this discrepancy?

2. The decrease in ocean pH is already affecting the calcification rates of marine organisms, as indicated for corals in Figure 25.5. To get a view of what may occur with an even more CO_2-rich future, Uthicke et al. (2013)[†] studied the abundance and diversity of foraminiferans (zooplankton that form carbonate shells) in sediments around natural CO_2 seeps in the ocean (**FIGURE B**). Using your own and the IPCC's prediction of change in ocean pH from Question 1 and the relationships shown in Figure B, estimate the percentage decrease in abundance (density) and species richness of foraminiferans from 2000 (pH = 8.10) to 2050 and from 2000 to 2100.

---

* IPCC. 2013. *Climate Change 2013: The Physical Science Basis*. Cambridge University Press: Cambridge. Available at the IPPC website: www.ipcc.ch/report/ar5/wg1.

[†] Uthicke, S., P. Momigliano, and K. E. Fabricius. 2013. High risk of extinction of benthic foraminifera in this century due to ocean acidification. *Scientific Reports* 3: 1–5.

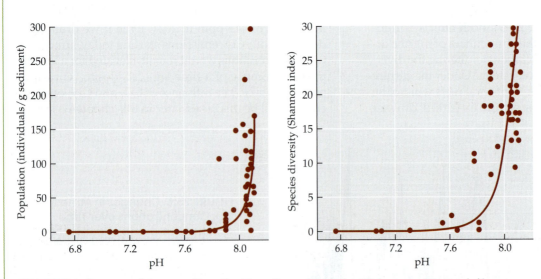

FIGURE B Influence of Ocean pH on the Density and Species Richness of Foraminiferans near Natural CO_2 Seeps (After S. Uthicke et al. 2013. *Sci Rep* 3:1–5.)

Biological fluxes dominate the global nitrogen cycle

Nitrogen (N) plays a key role in biological processes as a constituent of proteins and enzymes, and it is one of the resources that most commonly limits primary production, as we saw in Concept 20.2. Thus, cycles of N and C are tightly coupled through the processes of photosynthesis and decomposition.

The largest pool of N (>90%) is atmospheric dinitrogen gas (N_2) (**FIGURE 25.7**). This form of N is very stable chemically and cannot be used by most organisms, with the important exception of nitrogen-fixing bacteria, which are able to convert it to more chemically usable forms, as described in Concept 22.1. These fixed chemical compounds are referred to as *reactive* N because, unlike N_2, they can participate in chemical reactions in the atmosphere, soils, and water. Terrestrial N_2 fixation by bacteria provides approximately 128 teragrams (1 Tg = 10^{12} g) of reactive N per year (Cleveland et al. 1999; Galloway et al. 2004) and supplies 12% of the annual biological demand (Schlesinger and Bernhardt 2020). The remaining 88% is met by uptake of N from the soil in forms released by decomposition. Oceanic N_2 fixation contributes another 120 Tg to the biosphere annually. Geologic pools associated with sediments containing

organic matter represent a much smaller fraction of global N than of global C, but some N-rich sedimentary sources may be important in some sites (Morford et al. 2016).

Although the pools of N at land and ocean surfaces are relatively small, they are very active biologically, and they are held tightly by internal ecosystem cycling processes. Fluxes from these pools are small relative to the rates of internal cycling, usually less than 10% (Chapin et al. 2002). The natural flux of N between terrestrial and oceanic pools that occurs via rivers is tiny, but it plays an important biological role by enhancing primary production in estuaries and salt marshes. Denitrification, a microbial process that occurs in anoxic soils and in the ocean (described in Concept 22.2), results in movement of N (as N_2 and as N_2O, a greenhouse gas, also known as laughing gas) from terrestrial and marine ecosystems into the atmosphere. Oceanic and terrestrial ecosystems also lose N through burial of organic matter in sediments and through burning of biomass.

Human activities have altered the global N cycle tremendously—even more than they have altered the global C cycle. Anthropogenic fluxes are now the dominant components of the N cycle (Galloway et al. 2004; Canfield et al. 2010) (**FIGURE 25.8**). The rate of fixation of atmospheric N_2 by humans now exceeds the rate of

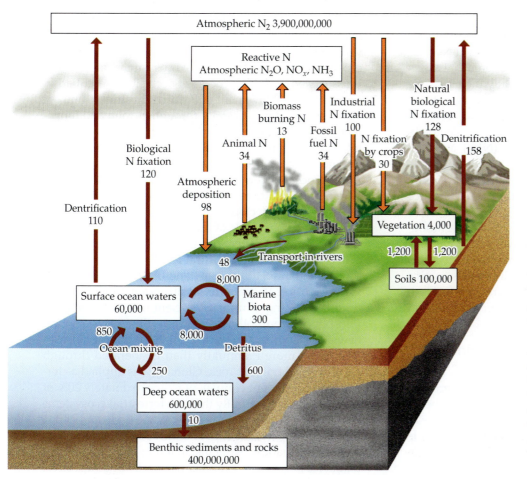

FIGURE 25.7 **The Global Nitrogen Cycle** Boxes represent major pools of N, measured in teragrams (1 Tg = 10^{12} g). Arrows represent major fluxes of N, measured in teragrams per year; anthropogenic fluxes are shown in orange. The percentage of the total atmospheric N pool made up of reactive N is minuscule (it is also difficult to quantify because it is very dynamic). (After W. H. Schlesinger and E.S. Bernhardt. 2013. *Biogeochemistry: An Analysis of Global Change*, 1st and 3rd eds. Academic Press: Cambridge, MA; F. S. Chapin et al. 2002. *Principles of Terrestrial Ecosystem Ecology*. Springer-Verlag: New York. Data from various sources including C. Cleveland et al. 1999. *Global Biogeochem Cycles* 13: 623–645; J. N. Galloway et al. 2004. *Biogeochemistry* 70: 153–226.)

Given its small size, why is the reactive pool of N of such great interest?

FIGURE 25.8 Changes in Anthropogenic Fluxes in the Global Nitrogen Cycle Increases in fertilizer production through the Haber–Bosch process, the growing of nitrogen-fixing crops, and combustion of fossil fuels have all contributed to the tremendous increase in biologically available (reactive) N. (After J. N. Galloway et al. 2004. *Biogeochemistry* 70: 153–226.)

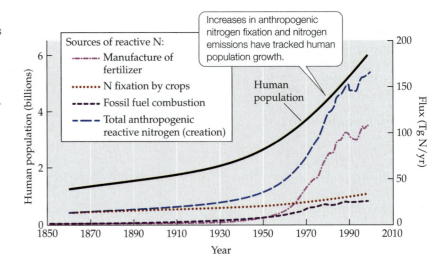

natural terrestrial biological fixation. Emissions of N associated with industrial and agricultural activities are causing widespread environmental changes, including acid precipitation, as we'll see in Concept 25.3. Three major processes account for these anthropogenic effects. The first is the manufacture of agricultural fertilizers by the Haber–Bosch process, described in Concept 22.1. Approximately 80% of the N in human tissues is derived from fixation of N_2 by this process. Second, growing N-fixing crops such as soybeans, alfalfa, and peas has increased biological N_2 fixation. Flooding of agricultural fields for other crops, such as rice, has increased N_2 fixation by cyanobacteria. Finally, anthropogenic emissions of certain gaseous forms of N have greatly increased the concentrations of these compounds in the atmosphere. Unlike N_2, these compounds, which include oxygenated N compounds (NO, NO_2, HNO_3, and NO_3^-, collectively referred to as NO_x, and N_2O), ammonia (NH_3), and peroxyacetyl nitrate (PAN), can undergo chemical reactions in the atmosphere and are potentially available for biological uptake. Fossil fuel combustion is the primary source of these nitrogenous gas emissions. Other contributors include biomass burning associated with deforestation, denitrification and volatilization (conversion to gaseous form) of fertilizers, and emissions from livestock feedlots and human sewage treatment plants. All of these reactive forms of N are returned to terrestrial and marine ecosystems through the process of atmospheric deposition (described in Concept 22.1).

The global phosphorus cycle is dominated by geochemical fluxes

Phosphorus (P) limits primary production in some terrestrial ecosystems—particularly those with old, well-weathered soils, such as tropical lowland forests—and in many freshwater and some marine ecosystems. Biologically available phosphorus is derived from the weathering of certain minerals, and decomposition. Phosphorus is added to crops as a fertilizer globally. Phosphorus availability can

also control the rate of biological N_2 fixation because that process has a high metabolic demand for P. Consequently, the C, N, and P cycles are linked to one another through photosynthesis and NPP, decomposition, and N_2 fixation.

Unlike C and N, P has essentially no atmospheric pool, with the exception of dust (**FIGURE 25.9**). Gaseous forms of P are extremely rare. The largest pools of P are in terrestrial soils and marine sediments. The largest fluxes of P occur in internal ecosystem cycles, which form a tight recycling loop between biological uptake by plants and microorganisms and release by decomposition. Typically, very little of the P cycling through terrestrial and aquatic ecosystems is lost. In terrestrial ecosystems, most P loss is associated with the process of occlusion (described in Concept 22.3). Movement of P from terrestrial to aquatic ecosystems occurs primarily through erosion and movement of particulate organic matter—mainly from plants—into streams. Much of the P transported from terrestrial to marine ecosystems (about 90%) is lost when it is deposited in deep ocean sediments. Ultimately, P in sediments in both marine and terrestrial ecosystems is cycled again in association with tectonic uplift and weathering of rocks, which occurs on a scale of hundreds of millions of years.

Anthropogenic effects on the global P cycle are associated with use of agricultural fertilizers, discharges of sewage and industrial wastes, and increases in terrestrial surface erosion. Phosphorus fertilizers are usually derived from the mining of uplifted ancient marine sediments. Phosphorus from soils and marine sediments is a nonrenewable resource, subject to depletion. Mining releases four times more P annually than is liberated through natural weathering of rock. Globally, P is applied as fertilizer in an amount equivalent to approximately 20% of the P that cycles naturally through terrestrial ecosystems (Schlesinger and Bernhardt 2020). While occlusion of P in the soil minimizes the flux of anthropogenic P from terrestrial to aquatic ecosystems, that flux still has great potential for negative environmental effects. One such effect is eutrophication in lakes, as described in Concept 22.4.

FIGURE 25.9 The Global **Phosphorus Cycle** Boxes represent major pools of P, measured in teragrams (Tg); arrows represent major fluxes of P, measured in teragrams per year. The major anthropogenic flux (P fertilization of crops) is shown in orange. (After W. H. Schlesinger and E.S. Bernhardt, 2013. *Biogeochemistry: An Analysis of Global Change*, 1st and 3rd eds. Academic Press: Cambridge, MA; F. S. Chapin et al. 2002. *Principles of Terrestrial Ecosystem Ecology*. Springer-Verlag: New York. Data from various sources cited within.)

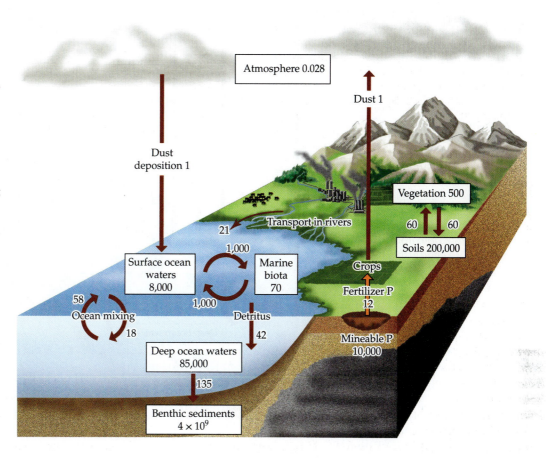

Biological and geochemical fluxes both determine the global sulfur cycle

Sulfur (S) is a constituent of some amino acids, but it is rarely, if ever, in short supply for organismal growth. Sulfur plays important roles in atmospheric chemistry. As with the C, N, and P cycles, anthropogenic changes to the global S cycle have important negative environmental consequences, primarily through the generation of acid precipitation.

The major global pools of S are in rocks, sediments, and the ocean, which contains a large pool of dissolved sulfate (SO_4^{2-}) (**FIGURE 25.10**). Fluxes of S among these global pools can occur in gaseous, dissolved, or solid forms. Weathering of S-containing minerals, mainly sedimentary pyrite, releases soluble forms of S that may enter the atmosphere or oceans. There is a net movement of S from the terrestrial pool to the oceanic pool, associated with transport in rivers and in atmospheric dust. Volcanic eruptions emit substantial amounts of sulfur dioxide (SO_2) into the atmosphere. Because they are episodic events, however, the amount of S emitted to the atmosphere by volcanic eruptions, on a time scale of centuries, is approximately the same as the amount blown into the atmosphere as dust from bare soils. Oceans release S to the atmosphere as small particles of windborne ocean spray and as gaseous emissions associated with microbial activity. Bacteria and archaea in anaerobic soils also emit S-containing gases such as hydrogen sulfide (H_2S). Most gaseous S compounds in the atmosphere undergo oxidation to SO_4^{2-} and H_2SO_4 (sulfuric acid), which are removed relatively quickly by precipitation.

Anthropogenic emissions of S to the atmosphere, which include gaseous and particulate forms (e.g., dust, aerosols), have quadrupled since the Industrial Revolution. Most of these emissions are associated with the burning of S-containing coal and oil and the smelting of metal-containing ores. What goes up must come down in the form of atmospheric deposition, usually within the same region from which it was emitted, but not always. Long-distance transport of fine dust occurs episodically in association with droughts and major wind events, as described in this chapter's Case Study. Increases in erosion associated with clearing of vegetation and overgrazing have contributed to anthropogenic input of S into the atmosphere as dust. Transport of S in rivers has doubled over the past 200 years (Schlesinger and Bernhardt 2020).

Human activities have resulted in changes in all four of the global biogeochemical cycles we have just described, and as we have noted, some of those changes have had important environmental effects. Let's turn our attention to those effects next.

FIGURE 25.10 **The Global Sulfur Cycle** Boxes represent major pools of S, measured in teragrams (Tg). Arrows represent major fluxes of S, measured in teragrams per year; anthropogenic fluxes are shown in orange. (After W. H. Schlesinger and E.S. Bernhardt, 2013. *Biogeochemistry: An Analysis of Global Change*, 1st and 3rd eds. Academic Press: Cambridge, MA; F. S. Chapin et al. 2002. *Principles of Terrestrial Ecosystem Ecology*. Springer-Verlag: New York. Data from various sources cited within.)

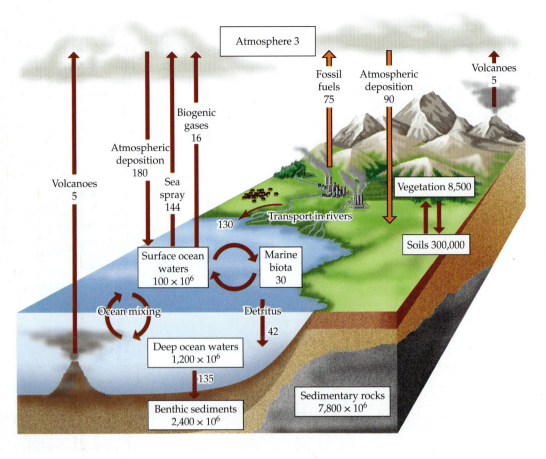

LEARNING OBJECTIVES

25.2.1 Relate the observed increase in global temperature over the past century to the potential causes for the change in climate.

25.2.2 Describe the expected and observed responses of organisms' geographic distributions resulting from climate change.

25.2.3 Evaluate why current climate–community distribution relationships may not adequately predict the future composition of communities under climate change.

25.2.4 Determine the factors that may increase the susceptibility of species to extinction due to climate change.

Global Climate Change

Throughout this book, we have emphasized the role that climate plays in ecological processes, including the distributions and physiological performance of organisms, the rates of resource supply, and the outcomes of biological interactions such as competition. Thus, changes in climate—particularly changes in the frequency of extreme events such as extensive droughts, violent storms, or extreme high and low temperatures—have profound effects on ecological patterns and processes. Because they are disturbances that result in significant mortality within populations, these extreme events are often critical in determining the geographic ranges of species.

As we learned in Concept 2.1, *weather* is the current state of the atmosphere around us at any given time. *Climate* is the long-term description of weather, including both average conditions and the full range of variation. Climate *variation* occurs at a multitude of time scales, from the daily changes associated with daytime solar heating and nighttime cooling, to seasonal changes associated with the tilt of Earth's axis, to decadal changes associated with interactions between ocean currents and the atmosphere (such as the Pacific Decadal Oscillation, described in the Case Study Revisited in Chapter 2). Climate change, on the other hand, refers to *directional* change in climate.

Evidence of climate change is substantial

Climate change is distinguished from climate variation by the presence of significant directional trends lasting at least three decades. Based on analyses of records from numerous climate-monitoring stations, atmospheric scientists have determined that Earth is currently experiencing significant climate change (IPCC 2021) (**FIGURE 25.11A**). Between 1880 and 2020, the average annual global surface temperature increased 0.97°C ± 0.2°C (1.8°F ± 0.4°F), with the greatest change occurring in the past 50 years. This

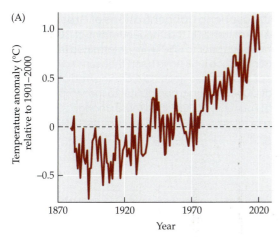

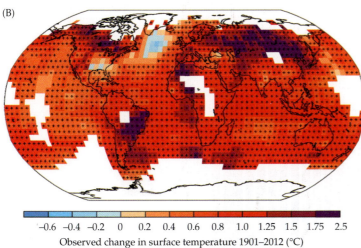

Observed change in surface temperature 1901–2012 (°C)

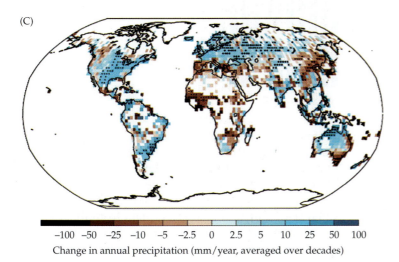

Change in annual precipitation (mm/year, averaged over decades)

FIGURE 25.11 Changes in Global Temperature and Precipitation
(A) Average annual global temperature anomalies (relative to the average global temperature for 1961–1990) between 1880 and 2021, averaged from numerous air and sea surface temperature records and normalized to sea level. (B) Regional trends in average annual temperatures for 1901–2012. (C) Trends in global precipitation from 1951 to 2010. (A, data from NOAA National Centers for Environmental Information, *Climate at a Glance: Global Time Series.* https://www.ncdc.noaa.gov/cag/; B,C from IPCC. 2013. *Climate Change 2013: The Physical Science Basis.* Cambridge University Press: Cambridge.)

rapid rise in global temperature is unprecedented in the past 10,000 years, although temperature changes at similar rates may have occurred at the onset and end of some glacial cycles (see Figure 25.6). The first two decades of the twenty-first century were the warmest decades of the previous 1,000 years, and 2016 was the warmest year since recordkeeping started. In association with this warming trend, there has been a widespread retreat of mountain glaciers, thinning of the polar ice caps, and thawing of permafrost. Sea level is rising at a rate greater than any estimated from the past 3,000 years (Kopp et al. 2016), posing a serious threat to coastal communities.

This warming trend has been heterogeneous across the globe, with most regions warming, others not changing significantly, and some even cooling (**FIGURE 25.11B**). The warming trend has been greatest in the middle to high latitudes of the Northern Hemisphere. Warming in the Arctic is four times greater than what has been recorded for Earth as a whole (Rantanen et al. 2022). Changes in terrestrial precipitation have also occurred, with more precipitation in portions of the high latitudes of the Northern Hemisphere and drier weather in the subtropics and tropics (**FIGURE 25.11C**). The frequencies of some extreme weather events, such as hurricanes (including massive storms such as Hurricane Katrina in 2005 and Hurricane Sandy in 2012) are increasing, fueled by warmer sea temperatures, droughts, and heat waves (IPCC 2013).

What are the causes of the observed climate change?

As we saw in Chapter 2, climate change may result from changes in the amount of solar radiation absorbed by Earth's surface or in the amount of absorption and reradiation of infrared radiation by gases in the atmosphere. Changes in absorption of solar radiation may be associated with variation in the amount of radiation emitted by the sun, in Earth's position relative to the sun, or in the reflection of solar radiation by clouds or surfaces with high reflectivity (albedo), such as snow and ice.

The warming of Earth by atmospheric absorption and reradiation of infrared radiation emitted by Earth's surface is known as the **greenhouse effect** (see Figure 2.4). This phenomenon is associated with radiatively active **greenhouse gases** in the atmosphere, primarily water vapor, CO_2, CH_4, and N_2O. The effectiveness of these gases in absorbing radiation depends on their concentrations in the atmosphere as well as their chemical properties. Water vapor contributes the most to the greenhouse effect, but its atmospheric concentration varies greatly from region to region, and changes in its average concentration have been small. Of the remaining greenhouse gases,

(A)

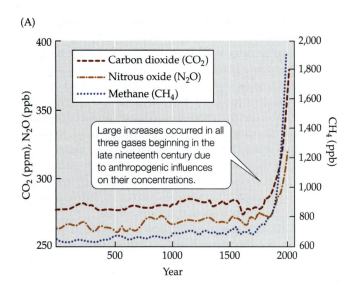

FIGURE 25.12 **Atmospheric Concentrations of Greenhouse Gases** (A) Long-term trends in the concentrations of CO_2, CH_4, and N_2O. Concentrations prior to 1958 were determined from ice cores; concentrations since 1958 have been measured directly. (B) Contributions of greenhouse gases to warming (radiative forcing), showing that CO_2 is the main contributor to the temporal change. (A after P. Forster et al. 2007. In *Climate Change 2007: The Physical Science Basis*, S. Solomon et al. [Eds.], pp. 129–234. Cambridge University Press: Cambridge; B from GlobalChange.gov.)

(B)

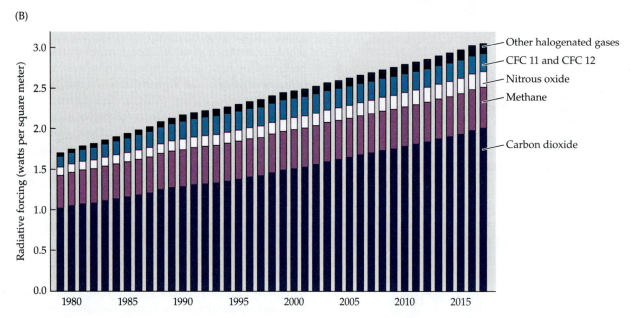

which tend to be more evenly distributed in the atmosphere, CO_2 contributes the most to greenhouse warming, followed by CH_4 (which has about 30% of the effect of atmospheric CO_2, although on a per molecule basis it is 30 times more effective than CO_2) and N_2O (which has about 10% of the effect of CO_2, but is around 280 times more effective than CO_2 on a per molecule basis).

As we saw in our discussion of the global biogeochemical cycles of C and N, atmospheric concentrations of CO_2, CH_4, and N_2O are increasing substantially, primarily as a result of fossil fuel combustion and land use change (**FIGURE 25.12**). Are increases in anthropogenic emissions of these greenhouse gases responsible for global climate change? To evaluate the underlying causes of climate change, its potential effect on ecological and socioeconomic systems, and our options for limiting climate change associated with human activities, the World Meteorological Organization and the United Nations Environment

Programme established the Intergovernmental Panel on Climate Change (IPCC) in 1988. The IPCC convenes panels of experts in atmospheric and climate science to evaluate trends in climate and the probable causes for any changes observed. These experts use a combination of sophisticated modeling and analysis of data from the scientific literature to evaluate potential underlying causes of observed climate change, as well as to predict future climate change scenarios. The IPCC releases assessment reports periodically to enhance the understanding of climate change among scientists, policymakers, and the general public. In recognition of their efforts to spread "knowledge about man-made climate change," the IPCC was awarded the Nobel Peace Prize in 2007.

In its third assessment report, released in 2001, the IPCC concluded that the majority of the observed global warming is attributable to human activities (**FIGURE 25.13**). While this conclusion is still occasionally debated in the

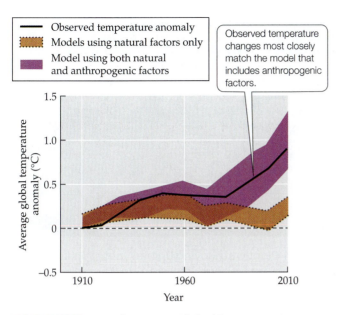

Legend:
— Observed temperature anomaly
Models using natural factors only
Model using both natural and anthropogenic factors

Observed temperature changes most closely match the model that includes anthropogenic factors.

FIGURE 25.13 Contributors to Global Temperature Change IPCC scientists compared observed global temperature changes between 1910 and 2010 with the results of computer models. The models predicted the temperature changes that would have been expected in that period due to natural climatological factors only, including variation in solar radiation and in atmospheric concentrations of aerosols from volcanic eruptions, and due to both natural and anthropogenic factors, including emissions of greenhouse gases and sulfate aerosols. These comparisons suggest that anthropogenic factors have played a large role in the observed warming. (After IPCC. 2013. *Climate Change 2013: The Physical Science Basis.* Cambridge University Press: Cambridge.)

political arena, it is backed by the majority of the world's leading atmospheric scientists. The certainty of an anthropogenic cause of climate change has increased with each new IPCC report, with the 2013 report stating, "It is extremely likely (95%–100% probability) that human influence has been the dominant cause of the observed warming since the mid-20th century." Paul Crutzen, who was awarded the Nobel Prize in Chemistry for his work on stratospheric ozone loss, suggested that we have entered a new geologic period, which he called the *Anthropocene epoch* (*anthropo*, "human"; *cene*, "recent"; *epoch*, "geologic age") to indicate the extensive impact of humans on our environment, particularly the climate system (Crutzen and Stoermer 2000).

Will the climate continue to grow warmer? The IPCC's models project an additional increase in average global temperature of 1.1°C–4.8°C over the twenty-first century (IPCC 2021). The range of variation in this estimate is associated with uncertainties about future rates of anthropogenic greenhouse gas emissions and about the behavior of the terrestrial–atmospheric–oceanic system. Model simulations incorporating different economic development scenarios have predicted vastly different

future rates of emissions. Aerosols in the atmosphere represent another source of uncertainty in the models' predictions. Aerosols, which reflect solar radiation, have a cooling effect on global temperatures. For example, emissions of large amounts of aerosols associated with major volcanic eruptions have had notable cooling effects at a global scale. While some aerosols have been increasing in the atmosphere (e.g., dust, in association with land use change and desertification), others have been decreasing (e.g., SO_4^{2-}, due to decreasing anthropogenic SO_2 emissions). Water in the atmosphere may play contradictory roles: clouds may have a cooling effect, while water vapor, which may increase because of greater evapotranspiration, may increase greenhouse warming. Despite these uncertainties in predicting the magnitude of future climate warming, the probability that global temperatures will continue to rise is extremely high. Even if anthropogenic CO_2 emissions were halted completely, global temperatures would continue to rise for decades or even centuries due to the reduced capacity of the ocean to absorb heat (Frölicher et al. 2014).

Ecological responses to climate change are occurring

As noted earlier, global warming of 0.97°C has occurred since 1880. Several physical environmental changes have occurred over the same period, including the retreat of glaciers, increased melting of sea ice, and a rise in sea level. Have ecological systems also responded to this warming? Numerous reports of biological changes are consistent with recent global warming (Parmesan 2006; Walther 2010). These changes include earlier migration of birds, local extinction of amphibian and reptile populations, bleaching of coral reefs, fish die-offs in lakes, and earlier spring greening of vegetation.

Although changes in species' range limits are more difficult to link directly to climate change, numerous changes in the geographic ranges of species have been noted that are consistent with warming. For example, Georg Grabherr and colleagues studied the vascular plant communities found on summits of mountains in the European Alps. They compared the current species richnesses of those communities with records dating back to the eighteenth and early nineteenth centuries (Grabherr et al. 1994). They found a consistent trend of increasing species richness resulting from the upward movement of species from lower elevations onto the summits (**FIGURE 25.14**). Bird species are shifting their ranges to higher elevations in the Andes, with lowland species expanding their ranges, but species at the highest elevations declining in abundance through time (Freeman et al. 2018). Similarly, Camille Parmesan and colleagues recorded a northward shift in the ranges of European nonmigratory butterfly species (Parmesan et al. 1999). Of the 35 species examined, 63% had shifted their ranges northward, while only 3% had shifted their ranges southward. More than half of the plant and animal species

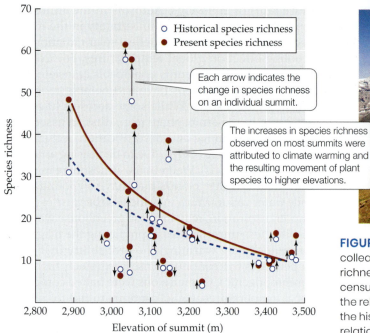

Each arrow indicates the change in species richness on an individual summit.

The increases in species richness observed on most summits were attributed to climate warming and the resulting movement of plant species to higher elevations.

FIGURE 25.14 Plants Are Moving Up the Alps Grabherr and colleagues compared historical records of vascular plant species richness on the summits of mountains in the European Alps with censuses taken in the early 1990s. The dashed blue curve indicates the relationship between species richness and summit elevation in the historical records, while the solid red curve indicates the present relationship. (After V. H. Grabherr et al. 1994. *BioScience* 53: 469–480.)

that have been investigated have shown geographic range shifts that are consistent with climate change (Parmesan and Yohe 2003).

Climate change may be causing populations of some species to go extinct. Barry Sinervo and colleagues (2010) found that 12% of Mexico's *Sceloporus* lizard populations had gone extinct between 1975 and 2009. Recall from Concept 23.2 that population extinctions are potentially the initial steps toward the extinction of a species. The extinctions of the lizard populations corresponded more closely to increases in temperature than to losses of habitat. Surprisingly, warming in the spring was better correlated with the extinctions than extreme temperatures during the summer. Sinervo and colleagues concluded that the warmer spring temperatures limited the lizards' foraging time during the breeding season. Ectothermic lizards must move into the shade and remain there to avoid overheating when temperatures become too warm (see Concept 4.2), and during that time they cannot seek out food. The observation that the probability of extinction was greatest at low-elevation and low-latitude sites, where the animals were most likely to be at the limits of their thermal tolerance, was consistent with this explanation.

Sinervo and colleagues also used a model of lizard thermal physiology to evaluate current and future worldwide effects of climate change on lizard populations. They estimated that climate change has already resulted in extinction of 4% of lizard populations worldwide. Using projections of future climate change, they suggested that 39% of the world's lizard populations, and 20% of its lizard species, may go extinct by 2050.

Migratory animals may also be adversely affected by climate change (Root et al. 2003). For example, migrating marine species, including whales and fish, may need to make longer journeys because of substantial changes in the distributions of their prey species as ocean temperatures warm. Some migratory bird species that breed in England and North America have been arriving at nest sites as much as 3 weeks earlier than they did 30 years ago because of warmer spring temperatures and faster snowmelt. However, plants and invertebrate prey species have responded faster to climate change than the migrating birds, resulting in a mismatch between bird arrival and prey availability. On the other hand, longer breeding seasons may increase the number of offspring produced by some bird species, particularly in high-latitude ecosystems.

Changes in community composition may also be indicators of climate change. These effects may be particularly apparent in some sessile marine species. Concepts 3.3 and 17.1 have described the effects of rising water temperatures on corals and the resulting changes in coral reef communities. Changes in the abundances of marine foraminiferans—a type of zooplankton—also reflect global climate trends during the past century (Field et al. 2006). Foraminiferan species have characteristic shells that allow them to be identified in marine sediments. Cores collected from benthic sediments can be examined to determine changes in the species composition of foraminiferans over time. Because the environmental tolerances of different species are known, these changes provide a means of reconstructing marine environments of the past. Following the mid-1970s, an increase in tropical and subtropical foraminiferan species, and a decrease in temperate and

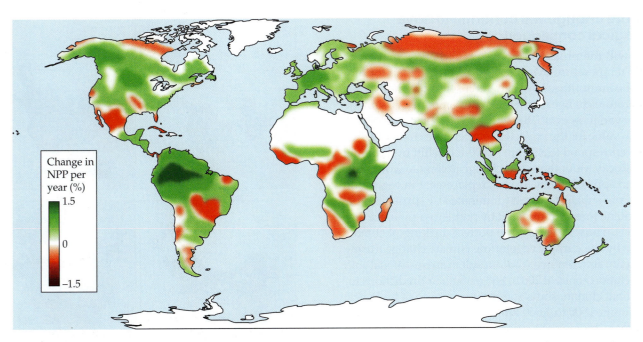

FIGURE 25.15 Changes in Terrestrial NPP Nemani and colleagues calculated changes in net primary production (NPP) between 1982 and 1999, expressed here as percentage change per year. The trend toward increased NPP in tropical regions of South America shown here was reversed in the first decade of the twenty-first century due to prolonged drought. (From R. R. Nemani et al. 2003. *Science* 300: 1560–1563.)

polar species, occurred in the eastern North Pacific Ocean, indicating a warming of ocean waters there.

Climate change is impacting forest composition in western North America through changes in the frequency and intensity of bark beetle attacks (Anderegg et al. 2015) and forest fires (Dennison et al. 2014; Higuera et al. 2021). Longer frost-free seasons are allowing mountain pine beetles (*Dendroctonus ponderosae*) in some regions to transition from completing one life cycle per year to two, greatly enhancing their population growth and potential outbreaks (Mitton and Ferrenberg 2012). In addition, the beetles are found at higher altitudes and latitudes than in the past, where they are attacking trees lacking defenses to the beetles. As a result of climate change effects on weather and fuel moisture content, forest fires have doubled since 1984 (Abatzoglou and Williams 2016). Climate change will continue to enhance forest fires until fuels become the limiting factor for their occurrence.

Changes in global NPP also indicate biological responses to climate change. Ramakrishna Nemani and colleagues used remote sensing data to examine global patterns of NPP over an 18-year period (1982–1999) (Nemani et al. 2003). They found that global NPP increased 6% during the study period, or 0.3% per year (**FIGURE 25.15**). Tropical ecosystems exhibited the largest increase in NPP, which was associated with increases in solar radiation due to less cloud cover in the tropics during the study period. During the first decade of the twenty-first century, however, the trend toward increasing NPP was reversed.

The decrease in global NPP during this decade was attributed to major droughts, particularly in the Southern Hemisphere (Zhao and Running 2010).

There has been a notable decrease in net ecosystem exchange (see Concept 20.1) in terrestrial high northern latitudes during the past 30 years, which has coincided with some areas of the Arctic switching from a net uptake of CO_2 from the atmosphere (acting as a *sink*) to a net export of CO_2 (acting as a *source*) (Oechel et al. 1993; Zhang et al. 2017). Large amounts of C are stored in the soils of boreal and tundra ecosystems as a result of low-temperature constraints on decomposition and the long-term buildup of carbon since the last glacial maximum. Warming during the twentieth century, however, increased the rate of CO_2 export from Arctic soils, such that losses now exceed gains from NPP. Warming of these high-latitude terrestrial ecosystems could provide a positive feedback to climate change by enhancing their emissions of CO_2 and CH_4. The rates of CO_2 loss from Arctic ecosystems have decreased since the early 1990s, possibly due to changes in rates of nutrient cycling and physiological and compositional changes in the plants (Oechel et al. 2000). However, greater loss of CO_2 due to heterotrophic consumption of soil organic matter during winter may offset the gains from greater summer productivity (Webb et al. 2016). During this same time period NPP of Arctic oceans has increased due to decreases in ice cover, warming, and greater reception of solar radiation and increased inputs of nutrients from rivers and coastal erosion (Terhaar et al. 2021).

Climate change will continue to have ecological consequences

What will the projected 1.1°C–4.8°C change in average global temperature over the next 80 years mean for biological communities? We can get a sense of what such a temperature change might mean by comparing it with the climate variation associated with elevation in mountains. A median value for the projected temperature change (2.9°C) would correspond to a 500-m (1,600-foot) shift in elevation. In the Rocky Mountains, this change in climate would correspond approximately to a full replacement in vegetation zone, from subalpine forest (dominated by spruce and fir) to montane forest (dominated by ponderosa pine; see Figure 3.11). Thus, if we assume perfect tracking of climate change by the current vegetation, climate change during the twenty-first century would result in an elevational shift in vegetation zones of 200 to 860 m. Similar predictions for latitudinal climate shifts suggest movement of biological communities 500 to 1,000 km toward the poles.

Climate–biome correlations, such as those described in Concepts 3.1 and 4.1, are useful as a demonstration of what could happen with climate change, but it would be naive to use them to predict what will actually happen to biological communities. We know that biological composition is influenced by a multitude of factors, including climate—particularly climate extremes—as well as species interactions, the dynamics of succession, dispersal ability, and barriers to dispersal (as described in Unit 5). Because the ongoing climate change will continue to be rapid relative to the climate changes that have shaped current biological communities, it is unlikely that the same assemblages of organisms will form the communities of the future.

Paleoecological records reinforce the suggestion that novel communities may emerge with climate change by showing that some plant communities of the past were quite different from modern plant communities. Jonathan Overpeck and colleagues used pollen records to reconstruct large-scale vegetation changes since the most recent glacial maximum in eastern North America (18,000 years ago) (Overpeck et al. 1992). They found not only that community types had made latitudinal shifts as the climate warmed, but also that community types without modern analogs existed under climate regimes that were unique and no longer present (**FIGURE 25.16**). Overpeck and his colleagues concluded that future vegetation assemblages would follow similar trends, given the predicted rapid rate of global warming and the potential for the emergence of unique climate patterns with no current analogs.

The rate of climate change will require rapid evolutionary change or the ability to disperse to new environments. Climate Change Connection 6.1 presents evidence that organisms with rapid life cycles have undergone evolutionary change in response to climate change. For more long-lived species, evolutionary responses are less likely, and thus for those species dispersal may be the only way to avoid extinction. Organisms' dispersal abilities, and

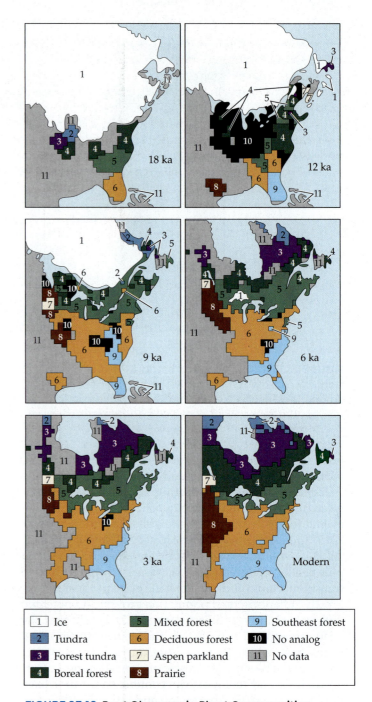

1	Ice	**5**	Mixed forest	**9**	Southeast forest
2	Tundra	**6**	Deciduous forest	**10**	No analog
3	Forest tundra	**7**	Aspen parkland	**11**	No data
4	Boreal forest	**8**	Prairie		

FIGURE 25.16 Past Changes in Plant Communities
Vegetation types in eastern North America have changed since the last glacial maximum, 18,000 years ago (ka = thousand years ago). Vegetation composition was determined from pollen preserved in sediments. (From J. T. Overpeck et al. 1992. *Geology* 20: 1071–1074.)

? What factors may have led to the development of vegetation types different from those found in North America today following retreat of the continental glacier?

barriers to their dispersal associated with anthropogenic habitat fragmentation, will be important constraints on their responses to climate change. Plant dispersal rates are, on average, much slower than the predicted rate of climate change. In order to track the projected change in climate over the twenty-first century, plant species populations will need to move 5 to 10 km per year. Plant species that have animal-dispersed seeds, and which can establish viable populations and grow to reproductive maturity in a relatively short time, may be able to disperse rapidly enough to keep pace with climate change. However, this kind of dispersal strategy is common mainly in ruderal (weedy) herbaceous plants. Shrubs and trees have much slower rates of dispersal; as a result, there may be significant time lags in their response to climate change.

For most animals, mobility is not a problem, but their habitat and food requirements are intimately associated with the presence of specific vegetation types. In addition, barriers to dispersal may prevent organisms of all kinds from migrating in response to climate change. Dams, for example, may prevent fish from moving to water with more suitable temperatures. Fragmentation of habitat by human development may pose significant barriers to dispersal for some species (see Concept 24.2). Without habitat corridors through which they can disperse, species face a greater probability of local extinction in the face of climate change. Projections of the risk of extinction due to climate change, based on multiple published studies, indicate as many as 17% of Earth's species could be lost (Urban 2015).

In addition to affecting the geographic ranges of species, climate change will affect ecosystem processes, such as NPP, decomposition, and nutrient cycling and retention. Both photosynthesis and respiration are sensitive to temperature, and because their balance determines NPP, the direct effects of climate warming on NPP may be relatively minor. As indicated in Concept 20.2, however, variation in NPP is related to water and nutrient availability and vegetation type, all of which may be affected by climate change. Changes in precipitation patterns and evapotranspiration rates resulting from climate change may strongly influence both water and nutrient availability. Because of the heterogeneity of climate change, and of the resulting changes in vegetation types, both increases and decreases in NPP may occur. Thus, the effect of climate change on NPP will probably not be uniform. The effect of warming on nutrient supplies will be most pronounced in mid- to high-latitude terrestrial ecosystems, where low temperatures constrain rates of nutrient cycling and soils have large pools of nutrients. As a result, climate change may lead to increases in NPP in some temperate forest ecosystems.

Biological indicators of global climate change are diverse, and they are increasing over time. Experiments, modeling, and comparisons with historical and paleoecological records provide clues to how Earth's biota will respond to climate change. Substantial uncertainties in predicting the effects of climate change still exist, however, many of which are associated with other environmental changes that are occurring at the same time. In the next section, we'll look at two such anthropogenic changes that are having profound effects on ecosystems: emissions of sulfur and nitrogen into the atmosphere.

CONCEPT 25.3 Anthropogenic emissions of sulfur and nitrogen cause acid deposition, alter soil chemistry, and affect the health of ecosystems.

LEARNING OBJECTIVES

25.3.1 Describe the causes of acid deposition and the mechanisms by which it affects ecosystems.

25.3.2 Assess how atmospheric nitrogen deposition can be both beneficial and detrimental to organismal and community function.

Acid and Nitrogen Deposition

The negative effects of air pollution have been known since at least the time of the ancient Greeks, when laws protected the quality of air, as indicated by its odor (Jacobson 2002). Since the Industrial Revolution, air pollution has mainly been associated with urban industrial centers, power plants, and oil and gas refineries. These stationary sources of atmospheric pollutants mainly affect the areas immediately adjacent to them and are usually considered regional rather than global problems. During the twentieth century, however, effective emissions dispersal strategies (e.g., tall smokestacks), widespread industrial development, and greater emissions of pollutants from mobile sources, such as automobiles, have increased the spatial extent of air pollution tremendously.

Fossil fuel combustion, agriculture, and urban and suburban development have influenced fluxes of N and S to an even greater degree than fluxes of C. Emissions of N and S into the atmosphere have resulted in two related environmental issues: acid precipitation and N deposition. Emissions of N and S are only a subset of the multiple types of air pollution, but they are among the most far-reaching. Sites affected by acid precipitation and N deposition now include national parks and wilderness areas (**FIGURE 25.17**).

Acid precipitation causes nutrient imbalances and aluminum toxicity

The detrimental effects of acidic air pollution on nearby vegetation, buildings, and human health have been known for several centuries, although their mechanisms were not well understood. In England during the mid-nineteenth century, industrial processes that released acidic compounds into the atmosphere, primarily hydrochloric acid, were implicated as a major source of harmful pollution (Jacobson 2002). Legislation was enacted in 1863 to reduce these acidic emissions. Despite such legislation, acid precipitation continued to be a problem throughout the nineteenth and twentieth centuries in large industrialized urban centers. During the 1960s, awareness of the widespread effects of acid precipitation, including its effects on nearby "pristine" ecosystems and agriculture, increased. In particular, damage to forests and mortality among aquatic organisms in northern Europe, parts of Asia, and northeastern North America prompted greater attention to acid precipitation.

USFS/Pacific Southwest Region 5

FIGURE 25.17 Air Quality Monitoring in the Sierra National Forest Air samples are collected regularly to monitor temporal changes in air chemistry and particulates. Air quality in national parks and wilderness areas, such as the Sierra National Forest, has been compromised by emissions of pollutants, including NO_x and sulfate aerosols. These pollutants not only lower visibility, but also pose a health hazard to the organisms that come into contact with them, including humans.

Sulfuric acid (H_2SO_4) and nitric acid (HNO_3) are the main acidic compounds found in the atmosphere. As we saw in Concept 25.1, sulfuric acid forms in the atmosphere from the oxidation of gaseous S compounds. Likewise, nitric acid originates from the oxidation of other NO_x compounds. Sulfuric and nitric acids can dissolve in water vapor and fall to the ground with precipitation (*wet deposition*). Naturally occurring precipitation has a slightly acidic pH of 5.0 to 5.6 due to the natural dissolution of CO_2 and formation of carbonic acid. Acid precipitation has a pH range from 5.0 to 2.0. Acidic compounds may also be deposited on Earth's surface when they form aerosols too large to be suspended or when they attach to the surfaces of dust particles (*dry deposition*).

Research has focused on determining the causes of the environmental degradation associated with acid precipitation, including increased mortality of plants and amphibians and decreased diversity. Initially, the acidity was considered the main culprit. In most cases, however, rainfall and surface waters did not have a low enough pH to cause the observed biological responses. An exception is found in regions at high latitudes or high elevations that develop a seasonal snowpack. During winter, acidic compounds accumulate in the snow. When temperatures increase in spring, water percolates through the snowpack, leaching out all the accumulated soluble compounds. The first meltwater of spring is therefore more acidic than the precipitation that fell during winter. This *acid pulse* has the potential to be toxic to sensitive organisms in soils and streams, including microorganisms, invertebrates, amphibians, and fish.

The vulnerability of organisms in soils, streams, and lakes to inputs of acid precipitation is determined by the ability of their chemical environment to counteract the acidity, known as its **acid neutralizing capacity**. The acid neutralizing capacity of soils and water is usually associated with their concentrations of base cations, including Ca^{2+}, Mg^{2+}, and K^+. Soils derived from parent material with high concentrations of these cations, such as limestone, are better able to neutralize acid precipitation than those derived from more acidic parent material, such as granite.

The detrimental effects of acid precipitation on plants and aquatic organisms are associated with biogeochemical reactions in the soil that decrease nutrient supplies and increase concentrations of toxic metals. As H^+ percolates through the soil, it replaces Ca^{2+}, Mg^{2+}, and K^+ at cation exchange sites on the surfaces of clay particles (see the description of cation exchange in Concept 22.1). These cations are released into the soil solution and can then leach out of the rooting zone of plants. The loss of these base cations leads to a decrease in soil pH, or *soil acidification*. Deficiencies in Ca and Mg, sometimes in combination with other stresses, were associated with large-scale mortality of trees in European forests during the 1970s and 1980s (**FIGURE 25.18**). In advanced stages of soil acidification, the metal cations aluminum (Al^{3+}) and manganese (Mn^{3+}) are released into the soil from cation exchange sites. Aluminum and manganese are toxic to plant roots,

Lovecz

FIGURE 25.18 Air Pollution Has Damaged European Forests The high tree mortality seen in this spruce forest in the Jizera Mountains, Czech Republic, is associated with acid precipitation and the resulting nutrient imbalance, particularly losses of base cations. Extensive forest decline occurred in Germany and northern Czechoslovakia (now part of the Czech Republic) in the 1970s and 1980s.

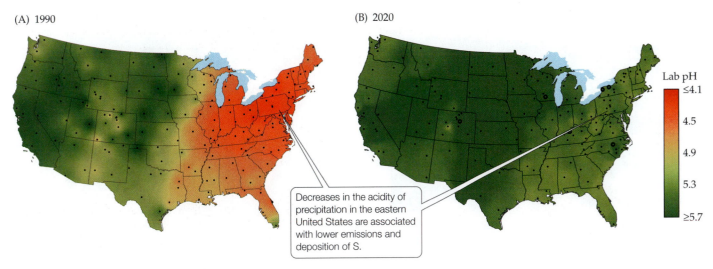

(A) 1990

(B) 2020

Lab pH
≤4.1
4.5
4.9
5.3
≥5.7

Decreases in the acidity of precipitation in the eastern United States are associated with lower emissions and deposition of S.

FIGURE 25.19 Decreases in Acid Precipitation The pH of precipitation in different parts of the United States as measured in (A) 1990 and (B) 2020, estimated based on measurements made at sampling points indicated by the dots. (From National Atmospheric Deposition Program/National Trends Network.)

soil invertebrates, and aquatic organisms, including fish. The combination of increasing acidity in precipitation and increasing aluminum concentrations in terrestrial runoff has been linked to fish die-offs in lakes and streams in northern Europe and eastern North America.

The realization that acid precipitation was negatively affecting the biota of forest and lake ecosystems prompted enhanced monitoring of atmospheric deposition and, eventually, laws to limit acidic emissions. Restrictions on emissions of S in North America and Europe have resulted in significant reductions in the acidity of precipitation (**FIGURE 25.19**). Forests are recovering from the effects of acid precipitation in central Europe, thanks to legislation limiting S emissions as well as decreased industrial activity in the former Soviet Union. Stream chemistry measurements also reflect the reduced acidity of precipitation and the recovery of aquatic ecosystems. Acid precipitation remains a problem, however, in some countries that have experienced rapid industrial development, such as China and India, though steps are being taken to reduce the emissions of acidic compounds.

Nitrogen deposition: Too much of a good thing can be bad

As we have seen, anthropogenic emissions of reactive N into the atmosphere from fossil fuel burning and agricultural activities have greatly altered global N cycles. Reactive N can fall back to Earth (via dry and wet deposition) after being transported away from the emission source in the atmosphere. Globally, anthropogenic emissions and deposition of reactive N compounds are more than three times greater now than they were in 1860 (Galloway et al. 2004, 2008) (**FIGURE 25.20**). Emissions and deposition of reactive N are expected to double between 2000 and 2050 as industrial development increases to keep pace with the

human population. Greater deposition of N will increase the supply of N for biological activity, but this abundance will come with an environmental cost.

The role of N as a determinant of rates of primary production was described in Concept 20.2. Nitrogen plays an important role in photosynthesis, which forms the base of the food webs that provide energy to all other organisms. Considerable benefit to humanity has accrued from the manufacture of N fertilizers and their widespread application to crops since the early twentieth century. We might expect, therefore, that an increased supply of N would facilitate plant growth and greater overall production in a N-limited ecosystem. Primary production has indeed increased in some ecosystems as a result of increased N deposition (e.g., forests in Scandinavia; Binkley and Högberg 1997). Nitrogen deposition may be partly responsible for the greater uptake of atmospheric CO_2 by terrestrial ecosystems observed in the Northern Hemisphere (Thomas et al. 2010).

Although primary production is increasing in some ecosystems because of N deposition, there is also strong evidence that N deposition is associated with environmental degradation, loss of biodiversity, decreases in primary production, and acidification of soils and surface waters. While N limits primary production in many terrestrial ecosystems, the capacity of vegetation, soils, and soil microbes to take up greater N inputs can be exceeded. This condition, known as *nitrogen saturation*, has a number of effects on ecosystems (Aber et al. 1998) (**FIGURE 25.21**). Greater concentrations of inorganic N compounds (NH_4^+ and NO_3^-) in the soil lead to enhanced rates of microbial processes (nitrification and denitrification) that release N_2O, a potent greenhouse gas. Nitrate (NO_3^-) is easily leached from soils and can move into groundwater, eventually entering aquatic ecosystems.

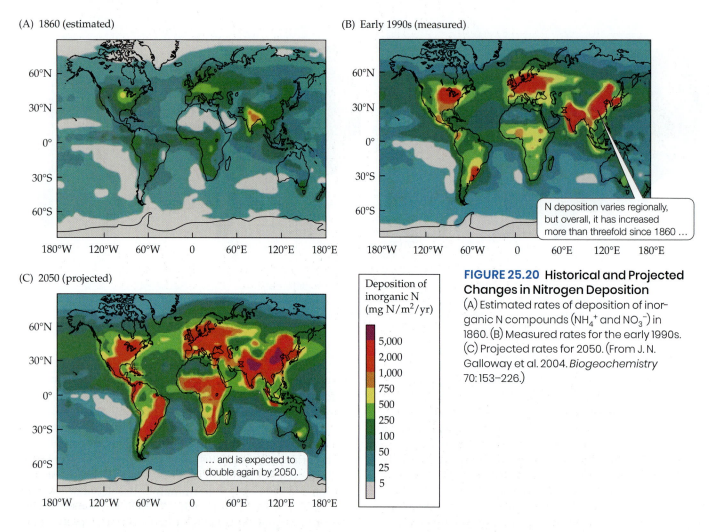

(A) 1860 (estimated)

(B) Early 1990s (measured)

N deposition varies regionally, but overall, it has increased more than threefold since 1860 …

(C) 2050 (projected)

… and is expected to double again by 2050.

Deposition of inorganic N
(mg N/m²/yr)

5,000
2,000
1,000
750
500
250
100
50
25
5

FIGURE 25.20 Historical and Projected Changes in Nitrogen Deposition (A) Estimated rates of deposition of inorganic N compounds (NH_4^+ and NO_3^-) in 1860. (B) Measured rates for the early 1990s. (C) Projected rates for 2050. (From J. N. Galloway et al. 2004. *Biogeochemistry* 70:153–226.)

When NO_3^- moves through the soil, it carries cations, including K^+, Ca^{2+}, and Mg^{2+}, in solution to maintain a charge balance. As in the case of acid precipitation, losses of these cations can lead to nutrient deficiencies and eventually to acidification of soils. Very high concentrations of NO_3^- in surface waters and ground water near agricultural areas has been linked to "blue baby"

syndrome, a dangerous condition in which an infant's ability to take up oxygen is compromised.

Most aquatic ecosystems are limited by P, so the biological uptake of anthropogenic NO_3^- that enters them from terrestrial ecosystems may be relatively small (although there is greater biological processing of N than expected; see Figure 22.15). Riverine transport of N to

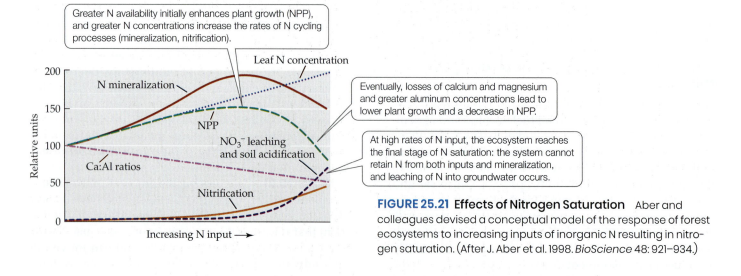

Greater N availability initially enhances plant growth (NPP), and greater N concentrations increase the rates of N cycling processes (mineralization, nitrification).

Leaf N concentration

N mineralization

NPP

NO_3^- leaching and soil acidification

Ca:Al ratios

Nitrification

Increasing N input ⟶

Eventually, losses of calcium and magnesium and greater aluminum concentrations lead to lower plant growth and a decrease in NPP.

At high rates of N input, the ecosystem reaches the final stage of N saturation: the system cannot retain N from both inputs and mineralization, and leaching of N into groundwater occurs.

FIGURE 25.21 Effects of Nitrogen Saturation Aber and colleagues devised a conceptual model of the response of forest ecosystems to increasing inputs of inorganic N resulting in nitrogen saturation. (After J. Aber et al. 1998. *BioScience* 48: 921–934.)

(A)

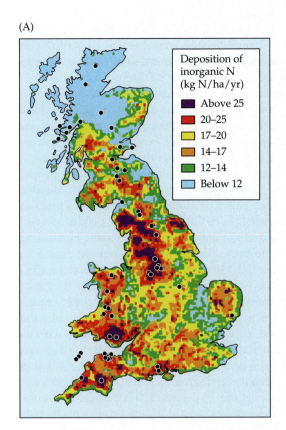

(B)

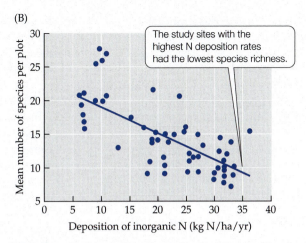

FIGURE 25.22 **Nitrogen Deposition Lowers Species Diversity**
(A) Inorganic N deposition in Great Britain. Dots on the map indicate the study sites where plant species richness in grassland ecosystems was measured. (B) Correlation between rates of inorganic N deposition and plant species richness. (After C. J. Stevens et al. 2004. *Science* 303: 1876–1879.)

nearshore marine ecosystems has increased as inputs of N fertilizer have increased (Howarth et al. 1996). Primary production in estuarine and marsh communities is often limited by N, and thus the influx of N from terrestrial sources into these ecosystems has resulted in eutrophication (described in Concept 22.4). Eutrophication results in heavy algal growth, which can create hypoxic conditions in the bottom waters of nearshore ecosystems. The resulting high inputs of organic matter lead to high rates of decomposition by microorganisms, which consume most of the available oxygen. The resulting hypoxic conditions are lethal for most marine life, including fish. Hypoxic conditions may occur over large areas, creating "dead zones." Dead zones of up to 18,000 km² form annually in the Gulf of Mexico, and over 400 dead zones form in locations around the world, including the Baltic Sea, the Black Sea, and the Chesapeake Bay.

In nutrient-poor ecosystems, many plants have adaptations that lower their nutrient requirements, which also lower their capacity to take up additional inputs of N. As a result, N inputs may cause faster-growing species to outcompete the species adapted to low-nutrient conditions. Acidification and increases in soluble aluminum may lead to declines in intolerant species. Eventually, this increased competition and toxicity can lead to lower diversity and alteration of community composition. In the Netherlands, species-rich heath communities adapted to low-nutrient conditions have been replaced by species-poor grassland communities as a result of very high rates of N deposition (Berendse et al. 1993). In Great Britain, Carly Stevens and colleagues surveyed grassland communities across the country with a range of N deposition rates (**FIGURE 25.22A**). At 68 sites, they measured the mean plant species richness in multiple study plots, along with several environmental variables, to try to explain the variation in plant diversity among the sites. The environmental variables included nine soil chemical factors, nine physical environmental variables, grazing intensity, and the presence or absence of grazing enclosures (Stevens et al. 2004). Of the 20 possible factors that may have influenced differences in species richness among the study sites, the amount of N deposition explained the greatest amount of variation (55%): higher inputs of N were associated with lower species richness (**FIGURE 25.22B**). The results of this study are supported by a similar large-scale study in the United States that found at least 25% of the sites surveyed had reduced species richness in association with greater N deposition (Simkin et al. 2016). In general, rare species appear to be most at risk for loss from plant communities (Suding et al. 2005). High rates of N deposition also facilitate the successful spread of some invasive plant species at the expense of native species (Dukes and Mooney 1999).

The ecological effects of S and N result when atmospheric deposition returns anthropogenic emissions to Earth's surface. In the next section, we'll describe some anthropogenic compounds that exert negative effects while remaining in the atmosphere.

LEARNING OBJECTIVES

25.4.1 Describe how the release of chlorofluoro-carbons poses a serious threat to organisms, including humans.

25.4.2 Explain how ozone in the stratosphere can benefit life on Earth but threaten diversity and ecosystem functioning when it occurs near the ground surface.

Atmospheric Ozone

Ozone is good for biological systems, but only when it is not in close contact with them. In the upper atmosphere (the *stratosphere*), ozone provides a shield that protects Earth from harmful ultraviolet radiation. When in contact with organisms in the lower atmosphere (the *troposphere*), however, ozone can harm them. Detrimental changes in ozone concentrations have occurred in both the stratosphere (losses) and the troposphere (increases) as a result of anthropogenic emissions of air pollutants.

Loss of stratospheric ozone increases transmission of harmful radiation

About 2.3 billion years ago, when prokaryotes first evolved the capacity to carry out photosynthesis, oxygen began to accumulate in Earth's atmosphere, leading to a series of changes that facilitated the evolution of greater physiological and biological diversity. The increase in atmospheric oxygen (in the form of O_2) also led to the formation of a layer of ozone (O_3) in the stratosphere (at an altitude 6–30 miles [10–50 km] above Earth's surface). This ozone layer acts as a shield protecting Earth's surface from high-energy ultraviolet-B (UVB) radiation (0.25–0.32 µm). UVB radiation is harmful to all organisms, causing damage to DNA and photosynthetic pigments in plants and bacteria, impairment of immune responses, and cancerous skin tumors in animals, including humans.

Stratospheric ozone concentrations change seasonally as a result of changes in atmospheric circulation patterns, particularly in the polar zones, where they decline in spring. British scientists measuring ozone concentrations in the Antarctic were the first to record an unusually large decrease in springtime stratospheric ozone concentrations starting in 1980. Springtime minimum ozone concentrations decreased by as much as 70% between 1980 and 1995 (**FIGURE 25.23**). There was also a concomitant increase in the area of the Antarctic region experiencing a decrease in ozone, called the ozone hole. An **ozone hole** is defined as an area with an ozone concentration of less than 220 Dobson units (= 2.7×10^{16} molecules of ozone) per square centimeter; prior to 1979, average annual ozone concentrations had never been recorded below this level. Ozone decreases have been recorded between 25°S and the South Pole. Similar reductions in ozone have been recorded in the Arctic (from 50°N to the North Pole), although the magnitude of the decrease has not been as great (thus conferring the name **Arctic ozone dent**, since ozone concentrations have not dropped below 220 Dobson units).

The decrease in stratospheric ozone was predicted in the mid-1970s by Mario Molina and Sherwood Rowland, who discovered that certain chlorinated compounds, particularly chlorofluorocarbons (CFCs), could destroy ozone molecules. CFCs were developed in the 1930s for use as refrigerants and were later found to be useful as propellants in spray cans dispensing hair spray, paint, deodorants, and many other products. By the 1970s, as much as a million metric tons of CFCs were being produced every year. Molina and Rowland (1974) found that CFCs did not degrade in the troposphere and could remain there for a very long time (50–140 years). From the troposphere, CFCs can move slowly into the stratosphere, where they react with other compounds, particularly in the polar regions during winter, to produce reactive chlorine molecules that destroy ozone. Other anthropogenic compounds with the same effect include carbon tetrachloride, used as a solvent and to fumigate grain, and methyl chloroform, used as an industrial solvent and degreaser. A single reactive chlorine atom has the potential to destroy 100,000 ozone molecules. Thus, the danger posed by chlorinated compounds to the stratospheric ozone layer was clear to Molina and Rowland.

The amount of UVB radiation at Earth's surface increases as concentrations of stratospheric ozone decrease (Madronich et al. 1998). These increases in UVB have been most notable in the Antarctic region, which has experienced an increase in UVB radiation of as much as 130% during spring. Increases have also been recorded in the Northern Hemisphere, including a 22% increase at mid-latitudes during spring.

These increases in UVB radiation at Earth's surface have coincided with an increasing incidence of skin cancer in humans, which is now approximately 10 times more common than it was in the 1950s. UVB radiation had an important role in the evolution of pigmentation in humans (Jablonski 2004). The production of melanin, a protective skin pigment, was evolutionarily selected for in humans living at low latitudes, where ozone levels are naturally lowest and the highest levels of UVB radiation reach Earth's surface. As humans migrated away from equatorial Africa into colder climates with less sunlight, however, high amounts of melanin in the skin limited production of vitamin D, resulting in selection for lower melanin production in peoples of higher latitudes. As these lighter-skinned humans have subsequently migrated into environments with higher UVB radiation, to which their complexions are not adapted, they have increased their risk of skin cancers. This is also true for populations at high latitudes in the Southern Hemisphere, including Australia, New Zealand, Chile, Argentina, and South Africa, where exposure to UVB is enhanced by stratospheric ozone loss. Concern is particularly great in Australia, where nearly 30% of the population has been diagnosed with some form of skin cancer.

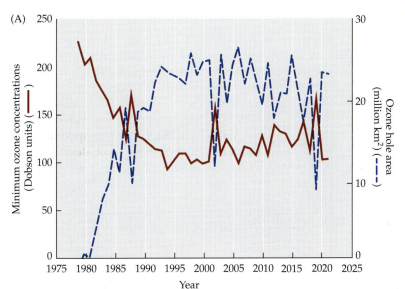

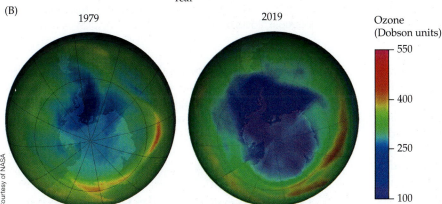

FIGURE 25.23 **The Antarctic Ozone Hole** (A) Since 1980, there has been a dramatic decrease in springtime ozone concentrations over the Antarctic region, with concentrations dropping below the threshold for ozone hole status (220 Dobson units) for a large proportion of the region after 1984. (B) Average ozone concentrations over Antarctica for the month of September in 1979 and 2019 demonstrate the dramatic decrease that occurred during this period. The lowest ozone concentrations are shown in dark blue. (A, data from ozonewatch.gsfc.nasa.gov.)

Substantial evidence exists to indicate that increasing UVB radiation has important ecological effects (Caldwell et al. 1998; Paul and Gwynn-Jones 2003). UVB radiation damages membranes and nucleic acids (DNA and RNA) of organisms, impacting metabolic function and increasing susceptibility to other stresses and disease. Sensitivity to UVB radiation varies among the species within a community, and as a result, changes in community composition are likely to result from increased UVB radiation. The potential for detrimental UVB effects due to stratospheric ozone loss is greatest at high latitudes and at high elevations (>3,000 m, or 9,800 feet) because of lower atmospheric filtering of UV radiation.

The realization of the rapid decreases in stratospheric ozone concentrations, and of their probable anthropogenic cause, resulted in several international conferences on ozone destruction in the 1980s. At these conferences, the Montreal Protocol, an international agreement calling for the reduction and eventual end of production and use of CFCs and other ozone-degrading chemicals, was developed. The Montreal Protocol has been signed by more than 150 countries. Atmospheric concentrations of CFCs have remained the same or, in most cases, declined since the Montreal Protocol went into effect in 1989 (**FIGURE 25.24**). A progressive recovery of the ozone layer is expected to occur over several decades, since the slow mixing of the troposphere, with the long-lived CFCs it still contains, and

the stratosphere will result in a time lag before stratospheric ozone concentrations rise. The trends in stratospheric ozone concentrations shown in Figure 25.23 indicate ozone destruction is slowly declining in response to lower emissions of CFCs, and a full recovery of the ozone layer is not expected until 2050. An estimated 280 million cases of skin cancer and 1.6 million skin cancer deaths have been avoided as a result of the Montreal Protocol.

Tropospheric ozone is harmful to organisms

Ninety percent of Earth's ozone is found in the stratosphere. The remaining 10% occurs in the troposphere. Tropospheric (including ground-level) ozone is generated by a series of reactions involving sunlight, NO_x, and volatile organic compounds such as hydrocarbons, carbon monoxide, and methane. In some regions, natural vegetation can be an important source of volatile organic compounds, which include terpenes (which give pines their characteristic odor) and isoprene. Under natural atmospheric conditions, the amount of ozone produced in the troposphere is very small, but anthropogenic emissions of ozone precursor molecules have greatly increased its production. Air pollutants that produce ozone can travel long distances, and thus tropospheric ozone production is a widespread concern.

Tropospheric ozone is environmentally damaging for two main reasons. First, ozone is a strong oxidant; that

(A)

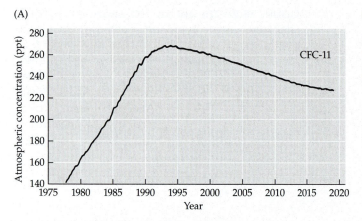

(B)

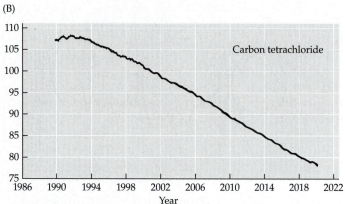

FIGURE 25.24 **Progress against the Ozone Killers** Measurements of atmospheric concentrations of ozone-destroying chlorinated compounds, in parts per trillion (ppt), at five monitoring locations across the globe show that several of them have declined since the signing of the Montreal Protocol in 1989. (Data from NOAA/Earth System Research Laboratory/Global Monitoring Division/HATS Flask Sampling Program. https://www.esrl.noaa.gov/gmd/hats/flask/flasks.html.)

is, the oxygen in it reacts easily with other compounds. Ozone causes respiratory damage and is an eye irritant in humans and other animals. An increase in the incidence of childhood asthma has been linked to exposure to ozone. Ozone damages the membranes of plants and can decrease their photosynthetic rates and growth. Ozone also increases the susceptibility of plants to other stresses, such as low water availability. Decreases in crop yields have been associated with exposure to ozone. Characteristic symptoms of ozone pollution have been found in plants near urban areas since the 1940s and 1950s (e.g., in the San Gabriel Mountains near Los Angeles and in the northern Alps in Italy), but more recently, symptoms have been noted in national parks and wilderness areas farther from sources of pollution. For example, plants in the Sierra Nevada of California are negatively affected by ozone generated in the Central Valley and the San Francisco and Los Angeles urban areas (Bytnerowicz et al. 2003). Growth rates of trees in forests of the eastern United States are as much as 10% lower than they would be in the absence of ozone (Chappelka and Samuelson 1998).

Second, ozone is a greenhouse gas that can contribute to global climate change. Ozone has a short life span in the atmosphere relative to other greenhouse gases, however, and its concentration can vary greatly from place to place. Thus, the effect of anthropogenic ozone on climate change is difficult to estimate.

Strategies to limit tropospheric ozone production have focused on lowering anthropogenic emissions of NO_x and volatile organic compounds. In most developed countries, efforts to lower emissions of ozone-producing compounds have met with success. In the United States, for example, emissions of volatile organic compounds dropped by 50% between 1970 and 2004, emissions of NO_x dropped by more than 30% (U.S. EPA 2005), and tropospheric ozone concentrations are decreasing near large urban areas (Cooper et al. 2014). Regulation of emissions of ozone-producing compounds has not been as strict in some developing countries, however. Ozone is a serious air pollutant in urban and agricultural regions of China and India.

A CASE STUDY REVISITED

Dust Storms of Epic Proportions

We've seen throughout this chapter that many aspects of global ecology—such as greenhouse gases and climate change, emissions and deposition of N and S, and stratospheric destruction and tropospheric production of ozone—involve transport and chemical processes in the atmosphere. The movements of dust described in this chapter's Case Study are also influenced by atmospheric processes, including rainfall patterns and wind. We've also seen that humans change the environment at a global scale through emissions of greenhouse gases and pollutants into the atmosphere. Land use change, which alters the amount and type of vegetation cover, generally influences the environment at a more local scale. However, land use change in arid zones that are subject to periodic severe droughts can have global-scale effects by enhancing the amount and spread of dust into the atmosphere.

During the early part of the twentieth century, the southwestern Great Plains of the U.S. was opened up for agricultural development. The natural vegetation of the region consisted of drought- and grazing-tolerant grasses. Bison, which had grazed the land for centuries, were replaced by cattle in the late nineteenth century. Economic demand for wheat, due to losses of agricultural lands in Europe during World War I, and the recent population expansion into the southern Great Plains encouraged the development of agriculture. Although this area was known to experience periodic droughts, farmers, encouraged by the notion that "rain follows the plow" and by recent technological developments in farming, cultivated large areas of land, plowing under the native prairie grasses and replacing them with wheat. For a while, the weather was conducive to agriculture, and the farmers prospered. However, the 1930s brought prolonged severe drought. Fields dried up, and with no protective network of roots to hold it together, the soil began to blow away. Major dust storms carried the soil across the North American continent and all the way to the Atlantic Ocean. The Dust Bowl event is still considered the worst environmental disaster the United States has ever experienced (Egan 2006).

Similar circumstances in Asia enhanced the severity of dust storms there. Deforestation, the development of agriculture in marginal zones, overgrazing, and the drainage of the Aral Sea for irrigation have all been implicated in the increased severity of dust storms following the mid-1990s (Wang et al. 2004).

While dust storms in urban areas are a rarity, large-scale dust storms regularly occur in desert regions (**FIGURE 25.25**). However, both the American Dust Bowl and Asian examples suggest that while dust storms are a natural phenomenon, a combination of agricultural development of marginal lands and severe drought exacerbates these events (Cook et al. 2009). At a global scale, extreme droughts and land use change contribute one-third to one-half of the inputs of dust into the atmosphere (Tegen and Fung 1995). Desert regions, such as the Gobi and Sahara–Sahel regions, have expanded at their margins because of land use change since the 1970s, increasing the global impact of dust storms. For example, Asian dust has been detected in the European Alps, traveling two-thirds of the way around the globe in approximately a week (Grousset et al. 2003). On a geologic time scale, major periods of dust redistribution occur in association with the recession of large ice sheets during interglacial periods, as evidenced by the distribution of loess soils, some hundreds of meters thick, across North America and Europe (**FIGURE 25.26**).

(A)

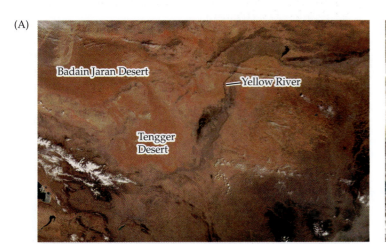

April 7, 2006

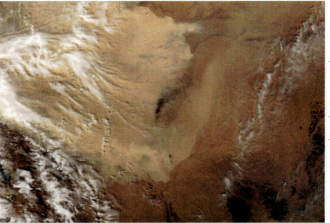

April 10, 2006

Both courtesy of the MODIS Rapid Response Team at NASA GSFC

(B)

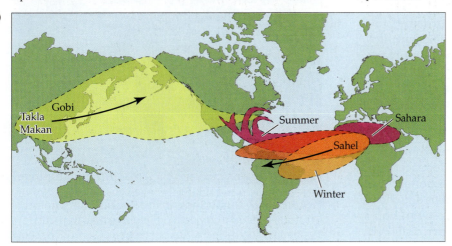

FIGURE 25.25 Desert Origins of Global Dust Storms Deserts are sources of dust that may travel large distances and have important ecological impacts in distant regions. (A) The photo on the left is a satellite image of the Gobi desert in early April 2006. The photo on the right shows the same region 3 days later, obscured by a massive dust storm. (B) Sources of the dust deposited in the Caribbean region include the deserts of North America and Asia. The main directions of dust flow are indicated by arrows. (B adapted from illustration by Betsy Boynton in V. H. Garrison et al. 2003. *BioScience* 53: 469–480.)

(A)

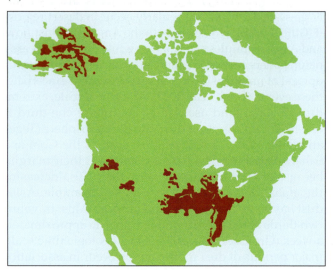

(B)

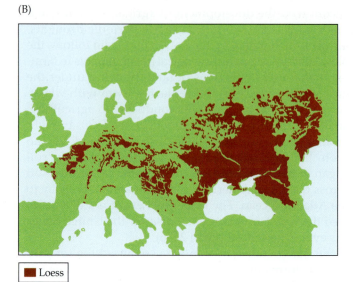

 Loess

FIGURE 25.26 **Distribution of Loess Soils** As continental glaciers receded following the most recent glacial maximum, wind carried substantial amounts of loose soil from the exposed areas. Large areas of (A) North America and (B) Europe were covered with deep layers of this material, which developed into loess soils. (A after D. R. Muhs. 2007. In *Encyclopedia of Quaternary Science*, pp. 2075–2086. Elsevier: New York; B after D. Haase et al. 2007. *Quat Sci Rev* 26: 1301–1312.)

🌱 CONNECTIONS *in* NATURE

DUST AS A VECTOR OF ECOLOGICAL IMPACTS The ecological effects of dust removal and deposition are not fully understood, but one of the best-studied effects is the movement of nutrients (as described in Chapter 22) at spatial scales ranging from a few meters to continents and oceans (Field et al. 2010). Dust deposition of nutrients can have important consequences for primary production and the global carbon cycle. The supply of iron (Fe) from dust deposition is important for oceanic primary production (Mahowald et al. 2005), as we saw in Concept 20.2. Dust from the Asian storms described earlier has been associated with algal blooms in the Pacific, and inputs of cations from African dust are important to primary production in tropical forests in the Amazon (Okin et al. 2004). In contrast, the removal of surface soils by wind can lead to lower production due to losses of organic matter and fine mineral particles, which are important for nutrient supply and retention. Dust may also be important in long-distance transport of pathogens (Garrison et al. 2003) and pollutants (Jaffe et al. 2003) and may influence disease dynamics (as described in Concept 13.5).

The ecological effects of dust movement can be both direct and indirect. Nutrient input and loss are examples of its direct effects. An example of an indirect effect occurs in the southwestern United States when dust transported from the Colorado Plateau falls in the Rocky Mountains and alters the timing of snowmelt. As noted in the Case Study in Chapter 22, grazing and recreational vehicle use have disturbed biological soil crusts in arid lands of the Colorado Plateau, increasing their erodibility and dust input into the atmosphere. Most of the dust is swept away in spring storms, and some ends up deposited in snow on the Rockies (**FIGURE 25.27**). The dust increases the amount of sunlight absorbed by the land surface, warming the snow and causing accelerated melting. Earlier snowmelt has the potential to increase the length of the growing season for plants growing in areas with deep snow cover. However, rather than stimulating earlier growth of plants in areas that melt sooner, accelerated snowmelt delays the initiation of growth and flowering of alpine plants, which wait to grow when air

FIGURE 25.27 **Dusty Snow in the Rockies** Dust from the Colorado Plateau is carried by spring storms to the Rocky Mountains, where it increases absorption of sunlight by snow and accelerates its melting. Earlier snowmelt has important implications for mountain ecosystems and regional hydrology.

Courtesy of the Center for Snow and Avalanche Studies, Silverton, CO temperatures are suitable. This delay results in greater synchrony of greening up of alpine plants, possibly leading to greater competition (Steltzer et al. 2009). In contrast, earlier snowmelt in lower-elevation subalpine meadows triggers some plants to initiate growth immediately, exposing them to potentially killing frosts (Inouye 2008). The surrounding

subalpine forests may experience water shortages when snowmelt occurs earlier, which may lower their NPP (Hu et al. 2010). The ecological impacts of dust, both direct and indirect, remind us that ecological phenomena occur at a global scale, have widespread importance, and testify to the role of humans in intensifying their effects.

SUMMARY

CONCEPT 25.1 Elements move among geologic, atmospheric, oceanic, and biological pools at a global scale.

25.1.1 Summarize the major pools and fluxes associated with global-scale cycles of carbon, nitrogen, phosphorus, and sulfur.

- The global carbon cycle includes large fluxes of CO_2 between the atmosphere and Earth's land surface associated with photosynthesis and respiration and, within the last 160 years, anthropogenic emissions of CO_2 and CH_4.

- Global fluxes of nitrogen are associated with biological uptake and chemical transformations. Anthropogenic nitrogen fixation and emissions now dominate the global nitrogen cycle.

- The global cycles of phosphorus and sulfur include both geochemical and biological fluxes.

- Anthropogenic fluxes of phosphorus associated with mining and industrial emissions of sulfur far exceed natural fluxes associated with weathering.

25.1.2 Describe why anthropogenic perturbations to the global carbon cycle are important mediators of environmental change, even though the fluxes are relatively small compared to net primary production and respiration.

- Atmospheric concentrations of CO_2 and CH_4 are increasing because of burning of fossil fuels, deforestation, and agricultural development.

- Elevated atmospheric CO_2 concentrations may increase terrestrial plant growth and the acidity of the oceans, causing ecological changes.

25.1.3 Evaluate why changes to individual element cycles at the global scale have implications for the cycling of other elements.

- Individual changes in global fluxes of C, N, P, or S can affect the fluxes of the other elements through impacts on primary production and environmental degradation.

CONCEPT 25.2 Earth is warming because of anthropogenic emissions of greenhouse gases.

25.2.1 Relate the observed increase in global temperature over the past century to the potential causes for the change in climate.

- Elevated levels of CO_2, CH_4, N_2O, and other greenhouse gases in the atmosphere have warmed Earth, particularly since the 1950s. This warming trend is expected to continue throughout the twenty-first century.

25.2.2 Describe the expected and observed responses of organisms' geographic distributions resulting from climate change.

- Large changes in species distributions, community composition, and ecosystem processes are expected as a result of global climate change.

- Recent changes in the geographic ranges of species and in carbon source–sink relationships have been attributed to climate change.

25.2.3 Evaluate why current climate–community distribution relationships may not adequately predict the future composition of communities under climate change.

- Individual species will vary in their response to climate change due to variation in dispersal rates, abilities to acclimatize and adapt, and changes in biotic interactions, resulting in a rearrangement of community composition.

25.2.4 Determine the factors that may increase the susceptibility of species to extinction due to climate change.

- Climate change will increase extinction rates of species with low dispersal rates or facing dispersal barriers due to habitat fragmentation, and species that cannot adapt or acclimatize to the new climate.

CONCEPT 25.3 Anthropogenic emissions of sulfur and nitrogen cause acid deposition, alter soil chemistry, and affect the health of ecosystems.

25.3.1 Describe the causes of acid deposition and the mechanisms by which it affects ecosystems.

- Sulfuric and nitric acids form in the atmosphere from compounds emitted by human activities. These compounds are subsequently deposited on Earth's surface as acid precipitation.

- Acid precipitation causes nutrient imbalances and aluminum toxicity in soils.

25.3.2 Assess how atmospheric nitrogen deposition can be both beneficial and detrimental to organismal and community function.

- Atmospheric deposition of reactive nitrogen compounds can increase productivity in some ecosystems, but it may also lead to soil acidification, eutrophication and dead zones in nearshore aquatic ecosystems, losses of species diversity, and increases in invasive species.

(Continued)

CONCEPT 25.4 Losses of ozone in the stratosphere and increases in ozone in the troposphere both pose risks to organisms.

25.4.1 Describe how the release of chlorofluorocarbons poses a serious threat to organisms, including humans.

- Anthropogenic emissions of chlorinated compounds have led to a loss of stratospheric ozone since the 1980s, particularly at high latitudes, and thus to an increase in the levels of harmful ultraviolet-B radiation reaching Earth's surface.

25.4.2 Explain how ozone in the stratosphere can benefit life on Earth but threaten diversity and ecosystem functioning when it occurs near the ground surface.

- Reactions involving volatile organic compounds, many of which are of anthropogenic origin, generate ozone in the troposphere, where it can harm organisms.

Review Questions

1. What are the major biological influences on the global carbon cycle? How have human influences during the past two centuries affected the fluxes of CO_2 associated with these biological influences (i.e., other than by fossil fuel burning) and, subsequently, atmospheric CO_2 concentrations?

2. Terrestrial animals are capable of migrating to regions where the climate is optimal for their function. Despite animals' mobility, ecologists are still predicting that as the climate changes, many animal species will experience local extinctions. Explain why animals' responses to climate change will depend on factors other than physiological tolerances and dispersal rates.

3. How can ozone in the atmosphere be both good and bad for organisms?

Hone Your Problem-Solving Skills

Forests are important to the global carbon cycle, taking up a substantial amount of CO_2 from the atmosphere via photosynthesis. As we discussed in Concept 20.2, the production of forests is often limited by the supply of N, and thus greater C uptake may occur with elevated N deposition.

1. How much additional C has been taken up and stored as wood if N deposition has added an average of 15 kg N per hectare per year to temperate deciduous forests for 20 years? Assume that 10% of the N deposition has been taken up and used for increased plant growth, that the C:N ratio in wood is 500:1, and that the temperate deciduous biome makes up 13 million km^2 (1 hectare [ha] = 0.01 km^2).

2. Make the same calculation for the boreal forest biome, with a N deposition rate of 5 kg N per hectare per year for 20 years, the same N uptake amount and C:N ratio of the wood, and an areal coverage of 19 million km^2 for this biome.

3. Calculate the annual sum of the C taken up and stored as wood from your answers to Questions 1 and 2, and compare that to the amount of anthropogenic C emitted in Figure 25.3.

Appendix
Some Metric Measurements Used in Ecology

Measures of	Unit	Equivalents	Metric → English conversion
Length	meter (m)	base unit	1 m = 39.37 inches = 3.28 feet
	kilometer (km)	1 km = 1000 (10^3) m	1 km = 0.62 miles
	centimeter (cm)	1 cm = 0.01 (10^{-2}) m	1 cm = 0.39 inches
	millimeter (mm)	1 mm = 0.1 cm = 10^{-3} m	1 mm = 0.039 inches
	micrometer (μm)	1 μm = 0.001 mm = 10^{-6} m	
	nanometer (nm)	1 nm = 0.001 μm = 10^{-9} m	
Area	square meter (m^2)	base unit	1 m^2 = 1.196 square yards
	hectare (ha)	1 ha = 10,000 m^2	1 ha = 2.47 acres
Volume	liter (L)	base unit	1 L = 1.06 quarts
	milliliter (ml)	1 ml = 0.001 L = 10^{-3} L	1 ml = 0.034 fluid ounces
	microliter (μl)	1 μl = 0.001 ml = 10^{-6} L	
Mass	gram (g)	base unit	1 g = 0.035 ounces
	kilogram (kg)	1 kg = 10^3 g	1 kg = 2.20 pounds
	teragram (Tg)	1 Tg = 10^{12} g	
	petagram (Pg)	1 Pg = 10^{15} g	
	milligram (mg)	1 mg = 10^{-3} g	
	microgram (μg)	1 μg = 10^{-6} g	
	picogram (pg)	1 pg = 10^{-12} g	
Temperature	degree Celsius (°C)	base unit	°C = $\frac{5}{9}$(°F − 32)
			0°C = 32°F (water freezes)
			100°C = 212°F (water boils)
			20°C = 68°F ("room temperature")
Pressure	Megapascal (MPa)		1 MPa = 145 psi (pounds per square inch)
Energy	joule (J)		1 J ≈ 0.24 calorie = 0.00024 kilocalorie*

*A calorie is the amount of heat necessary to raise the temperature of 1 gram of water 1°C.
The kilocalorie, or nutritionist's calorie, is what we commonly think of as a calorie in terms of food.

Answers

Chapter 1

Answers to Figure Legend Questions

FIGURE 1.4 Estimating from the graph, about 88% of tadpoles in the control group survived, and 0% of them had deformities. Since there were 35 tadpoles in the control group, this indicates that 31 (0.88 × 35) of the tadpoles in the control group survived, and none had deformities.

FIGURE 1.5 The results for cages from which *Ribeiroia* was excluded show that pesticides acting alone do not cause frog deformities. The results for cages exposed to *Ribeiroia* show that pesticides do affect frogs, since the percentage of frogs with deformities was higher in ponds where pesticides were present. However, the results do not indicate how pesticides caused that effect.

FIGURE 1.6 By comparing results from the controls with results from treatments in which pesticides were added, the investigator could test whether addition of a pesticide affected either the immune system response (number of eosinophils) of the tadpoles or the number of *Ribeiroia* cysts per tadpole. The intent of the "solvent control" was to check for possible effects of the solvent in which the pesticide was dissolved.

FIGURE 1.11 Producers take up nutrients such as nitrogen from the environment and use them for growth (step 1). The nitrogen in the producer's body may then be transferred to a series of consumers: to an herbivore that eats the plant, a carnivore that eats the herbivore, a second carnivore that eats the first, and so on (step 2). Eventually, however, the nitrogen is returned to the physical environment when the dead body of the organism containing it is broken down by decomposers (step 3).

Answers to Analyzing Data 1.1 Questions

1. Lakes with trout have lower densities of frogs than do lakes without trout, suggesting that the introduction of trout may have reduced frog density. However, while data from this study show that frog densities are correlated to the presence or absence of trout, they do not show that the trout *caused* frog densities to decline. To test whether the introduction of trout caused frog densities to decline, the researchers would need to perform a controlled experiment.

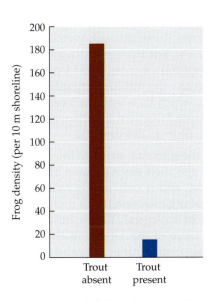

2. Results from control lakes that contain trout can be compared to results from trout-removal lakes: If the introduction of trout caused frog densities to decline, frog densities should increase in trout-removal lakes, whereas they should not change in the control lakes that still contain trout. Additionally, if the introduction of trout is the primary factor causing a decline in frog densities, frog densities also should not change very much in control lakes that never contained trout. Thus, if frog densities do not change very much in control lakes that never contained trout, such a result would strengthen the argument that changes in frog density observed in trout-removal lakes was due to the removal of trout, not to other unmeasured variables.

3. a. For the 1-year period that ends just prior to when trout began to be removed from Lakes 1, 2, and 3, frog densities were close to zero per 10 m of shoreline in each of these lakes.

   b. For the 1-year period that started one year after the removal of trout began, average frog densities were: 1.5 (Lake 1), 0.9 (Lake 2), and 0.9 (Lake 3) frogs per 10 m of shoreline. These data indicate that the removal of trout caused frog densities to increase within one year of beginning to remove trout from these lakes.

4. a. The experimental results suggest that introduced trout caused frog densities to decline.

   b. The experimental results suggest that frog populations can recover once trout are removed.

Answers to Review Questions

1. The phrase "connections in nature" is meant to evoke the fact that interactions among organisms and between organisms and their environment cause events in nature to be interconnected. As a result of such connections, an action that directly affects one part of an ecological community may cause unanticipated effects in another part of the community. Various examples related to amphibian deformities and population declines illustrate such connections and their indirect effects. For example, it appears that the addition of fertilizers to ponds has led to the following chain of events: the fertilizer stimulates increased algal growth, which then leads to increased snail abundance, increased *Ribeiroia* abundance, and hence more frequent amphibian deformities.

2. Ecology is the scientific study of interactions between organisms and their environment. The scope of ecology is broad, and it may address virtually any level of biological organization (from molecules to the biosphere). Most ecological studies, however, emphasize on one or more of the following levels: individuals, populations, communities, or ecosystems. Thus, if ecologists studied the effects of a particular gene, they probably would emphasize how the gene affected interactions in nature—they might, for example, study how a gene affected the ability of an organism to cope with its environment, or how a gene affected interactions among species. Compared with a geneticist or cell biologist, an ecologist would be less likely to emphasize either the gene itself or its effects on the workings of a cell, and more likely to study how the gene affected interactions in nature that occur at the individual, population, community, or ecosystem levels.

3. The scientific method summarizes the process of scientific inquiry. The four key steps in this inquiry process are: (1) observe nature and ask a question about those observations; (2) use previous knowledge or intuition to develop hypotheses (possible answers) to those questions; (3) evaluate different hypotheses by performing experiments, collecting new observations, or analyzing results from quantitative models; and (4) use the results from the approaches taken in (3) to modify the hypotheses, pose new questions, or draw conclusions about the natural world. An essential feature of many scientific investigations is a controlled experiment in which results from an experimental group (that has the factor being tested) are compared with results from a control group (that lacks the factor being tested).

Answers to Hone Your Problem-Solving Skills Questions

1. The five tanks with no atrazine serve as the control. By comparing results from control tanks to results from tanks with atrazine, an investigator could test whether the presence of atrazine affected one or more of the six variables measured in the experiment (phytoplankton abundance, attached algae abundance, water clarity, eosinophil number, tadpole survival, and number of *Ribeiroia* cysts).

2. Compared to the controls, when atrazine is added phytoplankton abundance decreases more than three-fold, the abundance of attached algae increases, and water clarity increases. To interpret these results, note that atrazine may have caused phytoplankton abundance to drop, which would cause water clarity to increase (because fewer phytoplankton were suspended in water), and that, in turn, would cause more sunlight to reach the algae attached to rocks, causing their abundance to increase.

3. Compared to the controls, when atrazine is added the number of eosinophils decreases more than two-fold, tadpole survival drops from 72% to 45%, and the number of *Ribeiroia* cysts increases more than four-fold. Atrazine may have impaired the tadpole's immune response, thereby causing the number of *Ribeiroia* cysts to increase, which would harm the tadpoles and cause their survival to drop.

4. The addition of atrazine to a pond could cause phytoplankton abundance to drop, thereby increasing the sunlight available to attached algae, hence increasing the growth of attached algae. Snails eat attached algae, so an increase in the abundance of those algae could cause snail abundance to increase, and that, in turn, could cause *Ribeiroia* abundance to increase (because *Ribeiroia* depends on snails to complete its life cycle). Atrazine also impairs the tadpole's immune response. Overall, since atrazine increases *Ribeiroia* abundance and impairs the tadpole's immune response, that could cause the number of *Ribeiroia* cysts to increase and tadpole survival to decrease.

Chapter 2

Answers to Figure Legend Questions

FIGURE 2.4 An increase in atmospheric greenhouse gases would increase the flux of infrared radiation back to Earth's surface and would have a warming effect on Earth's climate. Atmospheric aerosols reflect incoming solar radiation, so an increase in these particles would have a cooling effect on Earth's climate.

FIGURE 2.15 The larger a continent, the greater the seasonal temperature changes there. Because water has a higher heat capacity than land, seasonal temperature changes increase with distance from the ocean. Higher latitudes experience greater seasonal changes in radiation, for reasons we will explore in Concept 2.5.

FIGURE 2.18 Winds in the tropics blow from east to west, so the east-facing aspect would have the highest precipitation, and the west-facing slope would be in the rain shadow.

FIGURE 2.22 Seasonal changes in lake stratification would be unlikely in tropical lakes because seasonal changes in air temperature, and therefore water temperature, would be small.

FIGURE 2.26 In 11 out of 19 (58%) cases the cool phase of PDO corresponds with a higher-than-average catch. In 15 out of 22 cases (68%) the warm phase of PDO corresponds to a lower-than-average catch of salmon.

Answers to Analyzing Data 2.1 Questions

1. Greater solar radiation would be absorbed by the dark green crops. Given incoming radiation of 470 W/m², light-colored grasses reflect 122 W/m² (26% of 470) and absorb 345 W/m². With irrigated crops, 85 W/m² (18% of 470) is reflected and 385 W/m² is absorbed. Thus, with approximately 40 W/m² greater heat absorption, the change in albedo alone would result in warming.

2. The greater surface roughness of the crop plants would cause greater heat loss (approximately 40 W/m²) due to convective transport of pockets of warm air from the surface to the upper atmosphere.

3. Higher leaf area coupled with greater soil moisture in the irrigated crop system would result in higher evapotranspiration. As a result, more heat is lost from the surface to the atmosphere via latent heat flux by the irrigated cropland relative to the short-grass steppe.

4. The total difference in heat lost associated with the land use change from grassland to irrigated crop is 60 W/m² − 40 W/m² = 20 W/m². The greater total heat loss by the irrigated crop relative to the short-grass steppe would result in cooler temperatures.

Answers to Review Questions

1. Extreme environmental conditions, such as high and low temperatures or droughts, are important determinants of mortality in organisms. As a result, distributions of species often reflect extreme environmental conditions more than average conditions. The timing of changes in the physical environment is also important, as exemplified by the response of vegetation to the timing of precipitation, which is not reflected in average annual conditions.

2. Differences in the intensity of solar radiation across Earth's surface establish latitudinal gradients of surface heating. Greater heating in the tropics results in rising air currents, which establish large-scale atmospheric circulation cells, called Hadley cells. The warm rising air also promotes high amounts of precipitation on the tropics. Polar cells form where cold, dense air descends at the poles. Between the Hadley and polar

cells are the Ferrell cells, driven by the movement of the Hadley and polar cells and the exchange of energy between equatorial and polar air masses. The temperate zone is found at mid-latitudes in association with the Ferrell cells.

3. Salinization is a progressive increase in soil salinity due to surface evapotranspiration of water. Desert areas have high rates of evapotranspiration and little precipitation to leach salts to deeper soil layers. Some desert soils also have impervious soil layers underlying the surface layer that impede leaching, increasing the potential for salinization.

Answers to Hone Your Problem-Solving Skills Questions

1.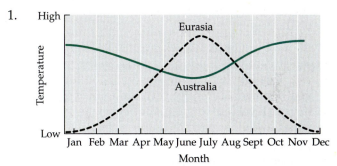

2.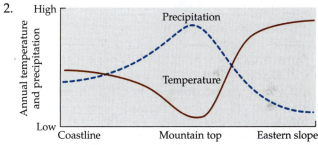

3.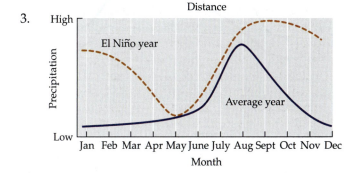

Chapter 3

Answers to Figure Legend Questions

FIGURE 3.4 Grasslands and shrublands might occur in areas with combinations of precipitation and temperature usually associated with forests or savannas due to disturbances such as fire or deforestation by humans or an outbreak of herbivory. These factors would limit successful establishment of trees, which would normally crowd out grasses and shrubs.

FIGURE 3.5 A comparison of Figures 3.5A and B shows that the greatest human impacts have occurred in grassland and deciduous forest biomes of North America and Eurasia (principally due to agricultural development). Note that in the Indian subcontinent and in South America, human impacts have occurred primarily in the tropical seasonal forest biome.

FIGURE 3.11 Both east- and west-facing slopes would have distinct biological zonation associated with gradients of temperature and precipitation, but precipitation would be lower on the east-facing slope due to the rain-shadow effect. As a result, a forest community on the west-facing slope might be replaced by a shrub or grassland community at the same elevation on the east-facing slope.

FIGURE 3.14 Oxygen levels would be highest where the stream velocity is the fastest, in the main channel. This is where organisms with the highest oxygen demands, typically fish, are found. The lowest oxygen levels are found in the benthic and hyporheic zones, where organisms must be able to tolerate hypoxic conditions.

Answers to Analyzing Data 3.1 Questions

1.

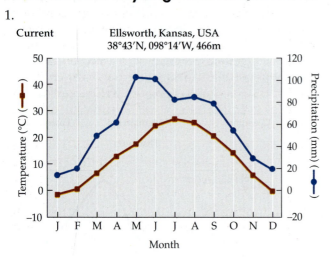

2.

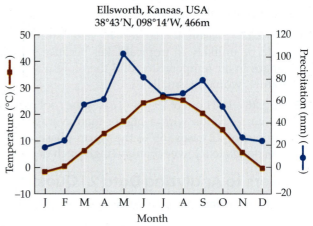

3. A decrease in precipitation during the summer growing season, coupled with warmer temperatures, results in a period of water stress in July, as indicated by the crossing of the temperature and precipitation lines. The occurrence of water stress in the summer and higher winter precipitation are more characteristic of the temperate shrubland biome, as shown in the exemplary climate diagram. With an increase in average annual temperature, the climate averages for Ellsworth cross the boundary between temperate grasslands and temperate shrublands.

4. Grazing and fire frequency also play roles in determining the occurrence of the grassland and shrubland biomes. If fires continue to be a part of the landscape, greater frequency due to warmer, drier conditions may allow grasslands to persist, as frequent fires promote grasses more than shrubs. Grazing may also help promote the persistence of grasslands rather than shrublands, as grasses are more tolerant of grazing.

Answers to Review Questions

1. Plant growth forms are good indicators of the physical environment, particularly climatic and soil conditions. Because plants are immobile as adults (seeds can move), they have evolved morphological features that allow them to cope with their physical environment, including its extremes. Leaf life span (evergreen vs. deciduous leaves), for example, reflects the fertility of the soil. Some biomes, such as grasslands, can also be indicators of disturbances such as grazing or fire. Animals can be important features of and controls on biome distribution, but their mobility renders them less useful as indicators of biomes.

2. Biomes are associated with the major climatic zones described in Chapter 2. Tropical rainforests are associated with a tropical climate characterized by high annual precipitation with only slight seasonal variations in the amount of precipitation. As the seasonality of rainfall becomes more pronounced further north and south from the tropics, regular dry periods occur, giving rise

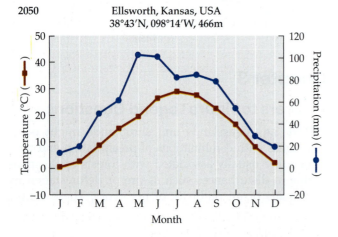

to the seasonal tropical forest biome. High-pressure zones associated with Hadley cells create extremely dry zones that promote the desert biome. Seasonality of both temperature (cool winters, warm summers) and precipitation in the temperate climatic zone gives rise to grassland (wet summers, dry winters) and shrubland (wet winters, dry summers) biomes. Temperate deciduous forests occur where seasonal temperature changes are moderate and both summer and winter are moist. Moving toward the polar climatic zone, winter temperatures and precipitation decrease, and the period of subfreezing winter temperature increases, marking the transition to the boreal and tundra biomes.

3. According to the river continuum concept, water velocity, stream bed particle size, and input of detritus from riparian vegetation all decrease as rivers move downstream. As a result, the importance of the surrounding terrestrial ecosystems as sources of energy for stream organisms tends to decrease downstream. Stream insects include more shredders near the source of a stream and more collectors in the lower portions. Attached plants and free-floating algae become more abundant downstream.

4. Light penetration varies according to the depth and clarity of the water. Where there is enough light for photosynthesis (the photic zone), photosynthetic organisms provide food for consumers, increasing the abundance of those organisms. The stability of the substrate determines whether organisms can anchor themselves or bury themselves in sand. Nearshore zones with rocky substrata tend to have the most abundant organisms and the most diverse communities. Photosynthetic organisms are more sparse in nearshore zones with sandy bottoms and below the photic zone in the open ocean.

Answers to Hone Your Problem-Solving Skills Questions

1. At the base of the mountains on the western slope, the biome type would be temperate evergreen forest (using the 12°C annual average temperature and 120 cm annual average precipitation). Using the environmental lapse rate of 4.5°C per 1,000 m, the annual average temperature will drop to −1.5°C by the summit ridges and peaks. With an annual average precipitation of 180 cm, this puts the trajectory of biome (vegetation) change through temperate deciduous forest, boreal forest, tundra, and finally into a gray area of ice and snow year-round. In fact, several coniferous forest bands are encountered, but the deciduous forest biome analog is generally missing.

Descending on the east slope, the temperature warms more quickly. Using an environmental lapse rate of 6.5°C per 1,000 m, the average annual temperature at the base of the mountains (2,700 m lower) would be 16°C. With an annual average precipitation of 50 cm,

shrubland vegetation would occur at the base of the mountains. Between the alpine ridges and the shrubland at the base, vegetation zones of tundra, boreal forest (subalpine forest), deciduous forest, grassland, and shrubland would be encountered.

2. The starting point on the western slope would have average annual temperature and precipitation of 16°C and 84 cm, respectively, with future climate change projections. The summit ridges would have annual average temperature and precipitation of 2.5°C and 126 cm with climate change projections. The vegetation transition from the base of the western slope to the summits would include grassland, deciduous forest, and boreal (subalpine) forest.

3. With projected climate change, annual average temperature and precipitation at the base of the western slopes of the Cascades would be 20°C and 35 cm. Descending the eastern slopes, the transect would encounter boreal (subalpine) forest, deciduous forest, grassland, shrubland, and finally desert.

Chapter 4

Answers to Figure Legend Questions

FIGURE 4.4   The southern limit of aspen's range tends to be associated with survival of drought conditions, which are becoming more frequent in the center of the continent. As a result, the southern range limit of aspen may move to the north. At the northern limit of aspen, the effects of low temperatures on its survival and reproduction tend to limit its distribution. Climate warming may offset this effect, and aspen may move northward in the future.

FIGURE 4.9   Cooling of leaves is important in any biome where leaf temperatures may rise to levels that are stressful, including many temperate and tropical biomes. However, a steady supply of water is needed to support transpirational cooling, which would be the case in tropical biomes and subtropical biomes during the rainy season.

FIGURE 4.10   Cooling mechanisms that do not use water, such as leaf pubescence or increasing convective heat loss, may be more important to cooling in deserts than in moister habitats such as the tropics, where the water supply is sufficient for transpirational cooling.

FIGURE 4.15   Moving between sun and shade influences the energy balance of the crocodile. The crocodile gains energy, particularly by solar radiation, when it moves to a sunny location. Moving into the shade results in net energy loss to the surrounding environment (losses > gains). If the riverbank on which the crocodile basks is warmer than its body, then it gains heat energy from the rock via conduction. A cooler rock in the shade will receive heat energy by conduction from the crocodile's body.

FIGURE 4.21 Closing stomates during midday lowers transpiration by increasing the resistance to water loss. Opening the stomates later in the afternoon when the air is cooler exposes the leaf to a concentration gradient of water from the plant to the air that is lower than at midday. As a result, transpirational water loss is less than it would be during the hotter part of the day.

FIGURE 4.25 The rate of water loss for each animal is given by slope of the line. If the external environment (light, temperature, humidity) is kept relatively constant, then the gradient of water potential from the animal to the air is the same, and the resistance modifies the actual water loss. Differences in the slopes therefore reflect differences in resistance to water loss.

Answers to Analyzing Data 4.1 Questions

1. Red represents the red squirrel, and blue represents the wolf. The larger animal (wolf) would have thicker fur with a greater insulative value than the red squirrel would. Longer fur in smaller mammals inhibits their mobility.

2. The circles represent the summer values for fur, and the triangles represent the winter values. Because the wolf is larger, its fur length varies more to adjust for seasonal changes in air temperature. The red squirrel may rely on torpor to survive the cold winter.

Answers to Review Questions

1. Plants as a group exhibit slightly greater tolerances of temperature extremes than ectotherms (see Figure 4.7), and both of these groups have tolerances much greater than those of endothermic animals. Plants and ectotherms, most of which do not generate heat internally, are more reliant on tolerance as a strategy for adapting to tissue temperature variation, while endotherms rely on avoidance of temperature extremes through internal heat generation and behavior, such as seasonal migration. Plants can exhibit avoidance of temperature extremes through leaf deciduousness.

2. a. Transpiration is an evaporative cooling mechanism that allows the plant to lower its leaf temperature below the air temperature. However, transpiration also results in water loss from the plant. If the water is not replaced, because the soil is too dry or the water loss is too rapid, the plant will experience water stress, and the rates of its physiological processes, such as photosynthesis, will decrease.

   b. Dark-colored animals may be able to warm themselves more effectively, but they may also be more visible to their predators or prey. In many cases, it appears that camouflage is more important than the ability to absorb sunlight effectively.

3. The principal ways in which plants determine their resistance to water loss are by adjusting the degree of opening of their stomates and by the thickness of the outer cuticle. Arthropods have cuticles that are extremely resistant to water loss. Similarly, skin thickness in amphibians, birds, and mammals affects their resistance to water loss. Reptiles have particularly thick skin, often overlain by scales, that provides a very effective barrier to water movement into the atmosphere. Note, however, that increasing the resistance of a barrier to water loss requires trade-offs with evaporative cooling as well as gas exchange.

Answers to Hone Your Problem-Solving Skills Questions

1. The most leaf pubescence would be expected for the population from the driest site (Death Valley), the least pubescence for the wettest site (Superior), with the amount for Oatman intermediate but probably closer to that for Death Valley, based on the magnitude of annual average rainfall. The same order would be expected for seasonal changes in pubescence (acclimatization): Death Valley > Oatman > Superior.

2. A quantitative expression of the answers from question 1 should show highest absorption in the plants from the Superior population, lowest absorption in the Death Valley population, and intermediate absorption in the Oatman population. If seasonal acclimatization is occurring, this will be indicated by lower absorption of radiation during the driest part of the year.

3. The results generally support the hypothesis that the Death Valley population has the most pubescence and lowest absorption of solar radiation, the Superior population has the least pubescence and highest absorption of solar radiation, and the Oatman population is intermediate for pubescence and absorption of solar radiation. While acclimatization occurs in all three populations during the drying cycle, the magnitude of the change in leaf absorption of solar radiation is roughly the same for each population.

Chapter 5

Answers to Figure Legend Questions

FIGURE 5.7 The light saturation level would be lower than the maximum light level the plant experiences because the energy invested in achieving a higher light saturation level might not pay off. The plant experiences the maximum light level for only short periods of time, and the increase in CO_2 taken up during those short periods might not pay for the additional machinery (e.g., chlorophyll, enzymes) needed to increase the light saturation level.

FIGURE 5.10 At low carbon dioxide and high oxygen concentrations, the photorespiratory carbon dioxide loss can exceed photosynthetic carbon dioxide gain. This is because oxygen is taken up to a greater extent than carbon dioxide by rubisco when the ratios of oxygen to carbon dioxide increase.

FIGURE 5.14   Extrapolation of the line used to fit the data to the x axis indicates that the proportion of the grass flora that is C_4 drops to zero when the growing-season minimum temperature is around 4°C–5°C. This would correspond to an average growing-season temperature of 9°C–10°C, which is at or above the growing-season temperatures for boreal forests and tundra shown in the climate diagrams. This result agrees well with the observed lack of C_4 plants in these biomes.

ECOLOGICAL TOOLKIT 5.1   CAM plants exhibit a wider range of $\delta\,^{13}C$ values because some are facultative CAM plants. At some times they use C_3 photosynthesis, but during drier periods they use CAM photosynthesis. The $\delta\,^{13}C$ of their tissues would reflect a mixing of C taken up using both of these photosynthetic pathways.

Answers to Analyzing Data 5.1 Questions

Note: Numerical answers may vary slightly due to differences in interpolation from the graph.

1. a. High-light grown plant

Gains: $(2.5\ \mu mol/m^2/s \times 7200\ s) + (32\ \mu mol/m^2/s \times 36{,}000\ s) + (2.5\ \mu mol/m^2/s \times 7200\ seconds) = 1{,}188{,}000\ \mu mols\ CO_2/m^2$ or $1.188\ mols\ CO_2/m^2$

Losses: $3\ \mu mol/m^2/s \times 36{,}000\ s = 108{,}000\ \mu mols\ CO_2/m^2$ or $0.108\ mols\ CO_2/m^2$

Total daily balance for the high-light grown plant: $+1.08\ mols\ CO_2/m^2$

Low-light grown plant

Gains: $(2.5\ \mu mol/m^2/s \times 7200\ s) + (5\ \mu mol/m^2/s \times 36{,}000\ s) + (2.5\ \mu mol/m^2/s \times 7200\ seconds) = 216{,}000\ \mu mols\ CO_2/m^2$ or $0.216\ mols\ CO_2/m^2$

Losses: $2\ \mu mol/m^2/s \times 36{,}000\ s = 72{,}000\ \mu mols\ CO_2/m^2$ or $0.072\ mols\ CO_2/m^2$

Total daily balance for the high-light grown plant: $+0.144\ mols\ CO_2/m^2$

b. High-light grown plant

Gains: $(-2\ \mu mol/m^2/s \times 7200\ s) + (2.5\ \mu mol/m^2/s \times 36{,}000\ s) + (-2\ \mu mol/m^2/s \times 7200\ seconds) = 61{,}200\ \mu mols\ CO_2/m^2$ or $0.0612\ mols\ CO_2/m^2$

Losses: $3\ \mu mol/m^2/s \times 36{,}000\ s = 108{,}000\ \mu mols\ CO_2/m^2$ or $0.108\ mols\ CO_2/m^2$

Total daily balance for the high-light grown plant: $-0.047\ mols\ CO_2/m^2$

Low-light grown plant

Gains: $(0\ \mu mol/m^2/s \times 7200\ s) + (2.5\ \mu mol/m^2/s \times 36{,}000\ s) + (0\ \mu mol/m^2/s \times 7200\ seconds) = 90{,}000\ \mu mols\ CO_2/m^2$ or $0.090\ mols\ CO_2/m^2$

Losses: $2\ \mu mol/m^2/s \times 36{,}000\ s = 72{,}000\ \mu mols\ CO_2/m^2$ or $0.072\ mols\ CO_2/m^2$

Total daily balance for the high-light grown plant: $+0.018\ mols\ CO_2/m^2$

2. The higher light saturation point in the high-light grown plants contributed significantly to the more positive carbon balance relative to the low-light grown plants when exposed to high-light conditions. Gains in carbon uptake were substantially higher in the high-light grown plants than in the low-light grown plants at high-light conditions. However, in low-light conditions, the lower light compensation point and nighttime respiration rates allowed the low-light grown plant to maintain a positive carbon balance, whereas the high-light grown plant had a negative carbon balance.

3. Low-light grown plants have lower concentrations of enzymes to support photosynthesis, and thus will have lower respiratory rates and lower carbon loss at night.

Answers to Review Questions

1. Autotrophy is the use of sunlight (photosynthesis) or inorganic chemicals (chemosynthesis) to fix CO_2 and synthesize energy storage compounds containing carbon–carbon bonds. Photosynthesis occurs in archaea, bacteria, protists, algae, and plants. Heterotrophy is the consumption of organic matter to obtain energy. The organic matter includes both living and dead organisms. Living organisms vary in their mobility, and their consumers (predators) have adapted ways to improve their efficiency in capturing their food (prey). Dead organic matter can be eaten and digested internally by multicellular heterotrophs or externally broken down by enzymes excreted into the environment and then absorbed by archaea, bacteria, and fungi.

2. CAM plants open their stomates to take up CO_2 at night, when the humidity of the air is higher than it is during the day. They store CO_2 in the form of a four-carbon organic acid, then release it to the Calvin cycle during the day. The storage of CO_2 allows the stomates to be closed during the day, when the potential for transpirational water loss is greater.

3. Live animals are a higher-quality food source, but they are rarer and thus harder to find, and they may have defense mechanisms that require expenditure of energy to overcome. Plant detritus is abundant in many ecosystems, so little energy needs to be expended in locating it, but its food quality is low.

Answers to Hone Your Problem-Solving Skills Questions

1. The photosynthesis rate for the C_3 plant increases from 32 to 39 $\mu mol/m^2/s$, or an increase of 22% with the increase in CO_2 concentrations. The photosynthesis rate for the C_4 plant does not increase at all—the photosynthesis rate is CO_2 saturated above an atmospheric concentration of about 200 ppm. If the increase in photosynthesis results in greater growth of the C_3 plants

but not the C_4 plants, the abundance of the C_3 plants may increase at the expense of the C_4 plants, which would decrease in abundance.

2. The observed increase in photosynthesis is greater than predicted for plants of both photosynthetic pathways but unexpectedly so for the C_4 plants, for which no increase was expected based on the modeled response. Reasons may be related to benefits to all plants in water savings, due to lower transpiration rates from stomatal closure, and thus less water stress lowering photosynthesis rates. Additionally, plants may be able to acclimatize to the elevated CO_2 and more effectively invest in enzymes to increase photosynthesis rates as CO_2 concentrations increase. Finally, the photosynthetic CO_2 response shown in the model may not be representative for all species. Some C_4 plants may have higher CO_2 saturation points than what is shown in the figure.

Chapter 6

Answers to Figure Legend Questions

FIGURE 6.6 The "Before selection" and "After selection" data show that nearly all fly larvae in galls less than 17 mm in diameter were killed by wasps. A much greater proportion of larvae in the largest galls survived, suggesting that wasps provide a stronger source of selection than do birds.

FIGURE 6.7 When the simulation began, each population had 9 *A* alleles and 9 *a* alleles. At generation 20, 8 populations still had both alleles. Eventually, it is likely that the *A* allele would either reach fixation (a frequency of 100%) or be lost from each of those 8 populations.

FIGURE 6.14 No. The added risk of mortality due to reproduction is represented by the difference between the solid blue curve (females that reproduced) and the red dashed curve (females that did not reproduce). That added risk decreases for females 3–7 years old, then rises for females 8–13 years old (and remains roughly constant thereafter).

FIGURE 6.23 If evolutionary changes in plant genotype did not affect moth abundance, we would expect that predicted and observed moth abundance would not be correlated to one another. If that were the case, the graph should look like this:

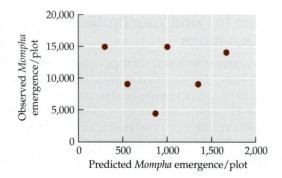

FIGURE 6.24 From the graph we can estimate that in 1832, the initial frequencies were 0.52 for genotype *AA*, 0.31 for genotype *Aa*, and 0.17 for genotype *aa*. Likewise, we can estimate that in 1923, the final frequencies were 0.73 for genotype *AA*, 0.22 for genotype *Aa*, and 0.05 for genotype *aa*. Using the approach for genotype frequencies described in the footnote in Concept 6.1, we can calculate that the frequency of the *a* allele was about 0.33 in 1832 and about 0.16 in 1923. Thus, the frequency of the *a* allele declined by more than 50% in about 100 years.

Answers to Analyzing Data 6.1 Questions

1. Releasing moths at densities and proportions similar to those observed in the field helped to remove potential complicating factors; this makes the experiment more realistic and the results easier to interpret. For example, some predators prefer to attack abundant prey, so if moths had been released at unusually high densities, predators might have devoted more effort to catching the moths than they typically do, thus making the results of the experiment more difficult to interpret.

2. We can see from the table from the Analyzing Data exercise for this chapter that in 2002 about 13% (101/807) of the moths that Majerus released were dark in color. Over time, that percentage dropped—from 13% in 2002 to 10% in 2003, 7% in 2004, 7% in 2005, 4% in 2006, and 2% in 2007. Because the proportions of dark moths that Majerus released were similar to those he observed in the field, this indicates that dark-colored moths were declining in frequency in the area where he conducted his experiment.

3. In every year but one (2006), the percentage of released dark-colored moths that were eaten was higher than the percentage of released light-colored moths that were eaten. Since the trees in the region in which the experiment was conducted were light in color, this result supports the hypothesis that natural selection caused the frequency of dark-colored moths to decline over time.

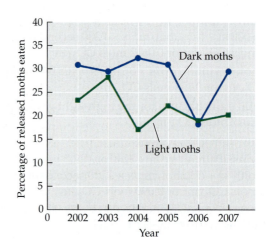

Answers to Review Questions

1. Natural selection acts as a sorting process, favoring individuals with some heritable traits over individuals with other heritable traits. As a result, the frequency of the favored traits in a population may increase over time. When this occurs, the frequencies of alleles that determine the favored traits also increase over time, and hence the population has evolved. But the individuals in the population do not evolve—each individual either has the trait favored by selection or it does not.

2. By consistently favoring individuals with one heritable trait over individuals with other heritable traits, natural selection can lead to a steady increase in the frequency of alleles that determine the favored trait. Although gene flow and genetic drift can also cause the frequency of alleles that determine an advantageous trait to increase over time, each of these processes can also do the reverse—that is, they can promote an increase in the frequency of disadvantageous alleles. Gene flow, for example, can transfer disadvantageous alleles to a population, thereby impeding adaptive evolution. Similarly, the random fluctuations in allele frequencies that result from genetic drift can promote an increase in the frequency of a disadvantageous allele. Hence, natural selection is the only evolutionary mechanism that consistently causes adaptive evolution.

3. Patterns of evolution over long time scales result from large-scale processes such as speciation, mass extinction, and adaptive radiation. The fossil record shows us that life on Earth has changed greatly over time, as seen in the rise and fall of different groups of organisms (for example, the rise of the amphibians and their later fall as reptiles became the dominant group of terrestrial vertebrates). Such changes in the diversity of life are due in part to speciation, the process by which one species splits to form two or more species. The rise and fall of different groups of organisms is also determined by mass extinctions and adaptive radiations. By removing large proportions of the species on Earth and hence altering the patterns of evolution observed after the extinction event, a mass extinction forever changes the evolutionary history of life. Similarly, by promoting an increase in the number of species in a group of organisms, an adaptive radiation shapes the patterns of evolution observed over long time scales.

4. Evolution occurs as organisms interact with one another and with their environment. Hence, evolution occurs partly in response to ecological interactions, and those interactions help to determine the course of evolution. The reverse is also true: as the species in a biological community evolve, the ecological interactions among those species change. Thus, ecology and evolution have joint effects because they both depend on how organisms interact with one another and their environments.

5. Rutter was concerned that by focusing harvesting efforts on the largest fish (because those fish are worth the most money), people would alter the fish population in ways that harm its future viability. In particular, by comparison to cattle, he is pointing out that it is a mistake to keep only the smallest individuals to breed. From an evolutionary perspective, Rutter was warning that fishing practices would cause the frequency of alleles favoring large size in fish to decrease over time, thus causing inadvertent and undesirable evolutionary change. Indeed, as we saw in the Case Study, harvesting-induced evolution is affecting fish populations today in ways that match his concerns.

Answers to Hone Your Problem-Solving Skills Questions

1. For *A. carolinensis* lizards that were either caught in the wild or reared in a common garden, the average toepad area of lizards from uninvaded islands was lower than the average toepad area of lizards from invaded islands.

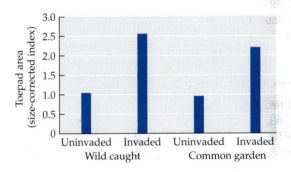

2. If toepad area differences resulted from evolution, individuals caught in the wild on uninvaded islands would differ genetically from individuals caught on invaded islands—and the same would be true for individuals reared from eggs collected on uninvaded versus invaded islands. Hence, if changes in *A. carolinensis* toepad area were caused by evolution, wild-caught results and common garden results should be similar.

3. If phenotypic plasticity was the primary cause of differences in toepad area, genes that affect toepad area would not differ between individuals living on uninvaded islands versus invaded islands. Thus, individuals reared from eggs collected on uninvaded versus invaded islands would also be similar genetically. In a common garden in which those (genetically similar) eggs were reared under identical conditions, toepad area should not change depending on whether the eggs were collected on uninvaded versus invaded islands. Hence, if changes in *A. carolinensis* toepad area were caused by phenotypic plasticity, wild-caught results and common garden results should differ from one another.

4. Since wild-caught results and common garden results were similar, this suggests that changes in toepad area resulted primarily from evolution, not phenotypic plasticity. Because an ecological event (invasion by a competitor species, *A. sagrei*) drove these evolutionary changes, this indicates that the invasion did lead to eco-evolutionary effects.

Chapter 7

Answers to Figure Legend Questions

FIGURE 7.2 Starting with the fish on the top left and proceeding clockwise, the genders are male, smallest nonbreeder, female, and largest nonbreeder. We can be confident of these predictions because the largest fish is female, the next largest a male, and the rest are sexually immature nonbreeders.

FIGURE 7.4 A 5-m-tall tree growing in a cool, moist climate is estimated to have a trunk diameter between 10 and 20 cm (the log scale makes it difficult to provide a precise estimate, but it is probably close to 15 cm), while a 5-m-tall tree growing in a desert climate is estimated to have a trunk diameter between 20 and 30 cm (probably close to 22 cm). To illustrate how these estimates are obtained: if you follow the line that moves horizontally to the right from the 5-m mark on the *y* axis, that line intersects the blue curve (the regression line for a cool, moist climate) at a point whose trunk diameter is about 15 cm.

FIGURE 7.7 The larva would be genetically identical to the polyp because both result from the same zygote (which in turn was produced when a sperm cell fertilized an egg cell). Two different larvae, however, would not be genetically identical because each resulted from a different fertilization event.

FIGURE 7.9 In generation 3 there are 8 sexual and 16 asexual individuals, while in generation 4 there will be 16 sexual and 64 asexual individuals. Note that the number of sexual individuals is increasing half as rapidly as the number of asexual individuals. This occurs because half of the offspring produced by sexual females are males (and males do not give birth to offspring). As a result, from one generation to the next, the number of sexual individuals doubles whereas the number of asexual individuals quadruples.

FIGURE 7.10 The blue line shows the results for the control populations. In this study, the experimental populations were exposed to a bacterial pathogen while the control populations were not. The results show that the outcrossing rate remained roughly constant in the control populations whereas it increased dramatically in the experimental populations, indicating that increased levels of outcrossing are favored by selection in populations exposed to pathogens.

FIGURE 7.15 For males with a thorax length of 0.8 mm, those kept with virgin females had an average life span of about 40 days while those kept with previously mated females had an average life span of about 63 days.

FIGURE 7.23 No. When $c > 1$, the average age of sexual maturity is greater than the average life span. For this to occur, the majority of individuals must die before they are old enough to reproduce.

Answers to Analyzing Data 7.1 Questions

1.

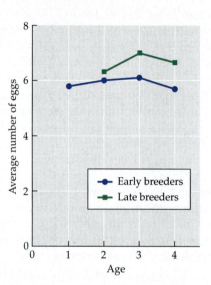

2. No. To see why, we can calculate how many eggs (on average) birds produced over the four years. Birds that reproduced in their first year had laid an average total of 23.6 eggs by the end of the fourth year, whereas birds that delayed reproduction until their second year had laid only 19.9 eggs in the same amount of time.

3. Yes. On a year-to-year basis, early breeders produced fewer eggs each year in years 2, 3, and 4 than did late breeders. This suggests that allocating resources to reproducing in their first year can reduce an individual's potential for reproducing in years 2–4.

4. Factors other than whether a bird reproduced in its first year may influence its reproductive success in years 2–4. An experimental approach to control for such factors would assign birds at random to the different treatments used in the experiment. There are several ways to test whether females experience a cost to reproduction that reduces their potential for future reproduction. For example, females could be assigned at random to one of the following three treatments: a control in which the number of eggs they laid was not altered; an experimental treatment in which extra eggs were added to their nest (increasing the female's costs of caring for eggs laid); and a second experimental treatment in which eggs were removed from the nest (reducing the female's costs).

Answers to Review Questions

1. In many plants and marine invertebrate animals, dispersal is negatively correlated with propagule size: smaller propagules can disperse farther than larger ones. In invertebrate animals, smaller egg size is also correlated with longer development times and increased reliance on food (rather than yolk provided in the egg) to complete development. However, in some vertebrates (for example, the fence lizards in Sinervo's study), smaller egg sizes actually lead to more rapid development to hatching. In both cases, the correlation between egg size and development time is striking, and the pattern that is favored varies with environmental conditions (e.g., temperature, rates of predation on larvae, etc.). An important reason why species that live in the same habitats may still exhibit different reproductive patterns is that different strategies may be favored in different years, depending on the particular environmental conditions. For example, in years with abundant food availability, a small-egg strategy may be favored, as offspring can acquire resources readily from the environment. However, in years when food is limited, a large-egg strategy may be advantageous due to its decreased reliance on external energy sources.

2. Asexual reproduction allows even a single individual to quickly increase the population size and allows a single highly successful genotype to dominate the population. The primary benefit of sex is the recombination of genetic material through the merging of unique genotypes, allowing potentially beneficial new combinations of genes to be introduced. The maintenance of both sexual and asexual reproduction allows rotifers to quickly increase the size of the reproductive population under beneficial environmental conditions while maintaining sufficient genetic variation to evolve in response to new environmental challenges.

3. Removal of small to medium-sized fish might produce selection for rapid growth through the size ranges that are favored by the fishery. This might lead to reproduction at older ages and larger sizes if there is a trade-off between growth and reproduction. Fish that are selected to grow quickly would allocate fewer resources to reproduction at smaller sizes so that they could allocate more resources to growth. Unfortunately, this is not the only effect of the Nassau grouper fishery. Because of heavy overfishing for both small and large fish and methods that target fish when they come together in large groups to spawn, Nassau grouper populations have declined precipitously.

Answers to Hone Your Problem-Solving Skills Questions

1. Intensive fishing began at Catalina and San Nicolas Islands in the early 1980s. At Catalina Island, where fishing pressure continued from the 1980s through 2007, the size at which sheephead became sexually mature decreased from 213 mm in 1980 to 178 mm in 2007; the size at which sheephead changed sex decreased from 350 mm to 225 mm during the same time period. At San Nicolas Island, fishing also appears to have affected size at maturation and size at sex change from 1980 to 1998, the time period during which intensive fishing occurred.

2. We can answer this question using data at San Nicolas Island because the sheephead population at that location was subjected to intensive fishing from the 1980s through 1998, but then (starting in 1999) the population was protected from fishing. At that island, size at maturation and size at sex change declined from 1980 to 1998. However, by 2007 (8 years after protection from fishing began), both the size at maturation and the size at sex change had increased substantially—indicating that size at maturation and size at sex change can recover once fishing pressure is reduced.

3. Protection from fishing should have an immediate effect of increasing population abundance (since fewer fish are being killed by humans). In addition, protection from fishing causes the size at maturation and the size at sex change to increase in size. As a result, the size of fertile females will increase over time, causing the number of offspring produced per female to increase over time (since larger individuals are assumed to produce more offspring). This change in the number of offspring produced per female should cause population abundance to increase more rapidly than it otherwise would.

Chapter 8

Answers to Figure Legend Questions

FIGURE 8.3 A proximate explanation for glucose aversion could describe how in cockroaches that exhibit this behavior, glucose activates taste neurons that in other individuals are activated only by bitter substances. An ultimate explanation for glucose aversion would be based on the fact that cockroaches exhibiting this behavior are more likely to survive than are other individuals (when exposed to baits containing glucose and insecticides).

FIGURE 8.5 Yes, the pie charts for the deer mouse and oldfield mouse each have a solid color (tan for the deer mouse, blue for the oldfield mouse), indicating that 100% of the genome of each mouse is composed of markers specific to its species. For the F_1 hybrids, 50% of the genome is from deer mice (as indicated by the tan half of the pie chart) and 50% of the genome is from oldfield mice (as indicated by the blue half of the pie chart). Backcross individuals represent offspring between F_1 hybrids (50% deer mouse genome and

50% oldfield mouse genome) and deer mice (100% deer mouse genome). Thus, we would expect that on average, 75% of their genome would be from deer mice and 25% of their genome would be from oldfield mice—and, as expected, the pie chart for backcross individuals is 75% tan in color and 25% blue in color.

FIGURE 8.6 Under conditions like those in which the relationship between net energy gained and foraging effort was estimated, you could test whether the effort lizards invested in acquiring food was similar to that which would maximize their net energy gained.

FIGURE 8.10 When wolves arrive, the probability that a female is found in grassland decreases whereas the probability a female is found in conifer forest increases; when wolves depart, the reverse is true. Similar patterns are observed for males, but males are less likely to change their behavior in response to the arrival of wolves than are females. For example, males are more likely to remain in grassland when wolves are present than are females.

FIGURE 8.15 In the first control, the tails of birds were not altered; results from this control can be compared to results from experimental treatments in which the tail lengths of birds were either shortened or lengthened. The second control (in which a portion of the tail was removed and then glued back on) was included so that Andersson could determine whether cutting a bird's tail had unintended effects.

FIGURE 8.21 This benefit cannot be compared directly to the cost shown in the figure because the benefit is in terms of food intake per hour, while the cost is in terms of increased flying times. To make this comparison you would need to use a common currency, such as the amount of energy gained from the increased food intake versus the amount of energy used during the increased flying times.

FIGURE 8.23 In the absence of wasps, laying eggs on food containing 6% alcohol causes larval survival to drop by about 18% (from 90% in food without alcohol to 72% in food with 6% alcohol). In the presence of wasps, larval survival increases by about 40% (from 10% in food without alcohol to 50% in food with 6% alcohol).

FIGURE 8.24 About 2.9 offspring per nest survived to young adulthood in nests that were not exposed to predator playbacks, whereas about 1.9 offspring per nest survived to young adulthood in nests exposed to predator playbacks. These results indicate that the "cost of fear" was a reduction of 1 offspring per nest.

Answers to Analyzing Data 8.1 Questions

1. The average number of attacks (per 5 minutes) is 10.3 for a single individual (a group of size 1); 12 for a group

of 4; 9.3 for a group of 6; 8.5 for a group of 15; 13 for a group of 50; and 10.5 for a group of 70. These results indicate that the predator does not have a strong preference for attacking either small or large groups—the risk of attack is similar for groups of all sizes.

Number of insects in group	Number of attacks (per 5 minutes)
1	10.3
4	12.0
6	9.3
15	8.5
50	13.0
70	10.5

2. To determine the average number of attacks per individual (per 5 minutes), we must divide the results we found in Question 1 by the number of individuals in the different groups. Thus we have:

Number of insects in group	Number of attacks per individual (per 5 minutes)
1	10.3
4	3.0
6	1.6
15	0.56
50	0.26
70	0.15

These results show that the average number of attacks per individual (per 5 minutes) declines dramatically with group size.

3. Yes, these results are consistent with the dilution effect: as the size of a group increases, an individual's chance of being eaten decreases.

Answers to Review Questions

1. A proximate explanation of a behavior would look within the organism to explain *how* the behavior occurs, focusing on events that serve as the immediate causes of the behavior. In contrast, an ultimate explanation of a behavior would seek to explain *why* the behavior occurs by examining the evolutionary reasons for the behavior.

2. Natural selection is a process in which individuals with certain traits consistently survive and reproduce at higher rates than do individuals with other traits. An animal's behaviors can affect its ability to survive and reproduce. Therefore, natural selection should favor individuals whose behaviors make them efficient at performing such activities as foraging,

obtaining mates, and avoiding predators. If the behaviors that confer advantage are heritable, then an animal will pass its advantageous behaviors to its offspring. When this is so, adaptive evolution can occur, in which the frequency of the advantageous behavior in a population increases over time. In cases where we demonstrate that natural selection has favored (or continues to favor) a particular heritable behavior, we can provide an ultimate explanation of the behavior by focusing on the evolutionary and historical reasons for why the behavior occurs.

3. A foraging animal often faces tradeoffs in which its ability to obtain food comes at the expense of other important activities, such as avoiding predators. When this occurs, individuals often alter their foraging decisions. Foragers may, for example, choose areas that provide less food but greater protective cover from predators. Fear of predators can have similar effects. For example, song sparrows exposed to playbacks of sounds made by predators (but no actual predators) fed their young less often, built their nests in denser, thornier vegetation, and spent less time incubating their eggs than did sparrows exposed to playbacks of nonpredators.

4. Sexual selection is a process in which individuals with certain characteristics have a consistent advantage over other members of their sex solely with respect to mating success. Charles Darwin pointed out that when sexual selection occurs, individuals typically use force or charm to gain access to mates. Often, the males compete with one another for the right to mate with females, while the females choose among the competing males; in some cases, the reverse occurs, and females compete for the right to mate with choosy males. Observational, genetic, and experimental evidence indicate that the large size, strength, or special weaponry of the males of many species result from sexual selection; such evidence also indicates that extravagant traits used to charm members of the opposite sex can result from sexual selection. Specific examples mentioned in the chapter include genetic evidence that the large body size and full curl of horns of male bighorn sheep result from sexual selection, along with Malte Andersson's classic experiments showing that sexual selection can explain the extremely long tails of male widowbirds.

5. In one example of how group living has both benefits and costs, goldfinches in a flock consumed more seeds per unit of time than did solitary birds. However, as the size of the flock increased, food supplies were depleted more rapidly, causing the birds to spend more time flying between feeding sites; traveling between feeding sites is energetically expensive and can lead to an increased risk of predation.

6. The greater expenditure of energy required by species B to fly between patches would dictate that it needs to spend longer in each patch in order to meet the assumptions of the marginal value theorem. Because its overall rate of energy gain in the habitat is lower, due to greater amount of energy it expends in traveling between patches, species B should deplete each patch to a greater degree before leaving it than species A.

Answers to Hone Your Problem-Solving Skills Questions

1. In the study population of gray wolves, pair bond duration ranged from 1 to 9 years and was positively associated with apparent pup survival (i.e., as the duration of pair bonds increased, apparent pup survival also increased). Age of breeders is conflated with the duration of the pair bond. One possible explanation for the pattern shown in the figure is that males and females with pair bonds of long durations are likely older and more experienced parents and are therefore better at coordinating parental care; this could result in increased pup survival. The pattern shown in the figure might also reflect longer territory occupancy by older pairs than younger pairs, which would presumably lead to more efficient use of resources and greater pup survival.

2. The presence and number of helpers will affect group size in gray wolves. Help provided by non-breeding group members could benefit the breeding pair by allowing them to spend more time foraging, which could influence their reproductive success. Similarly, help provided by non-breeding group members could benefit young of the breeding pair through the provision of extra food and protection, which also could influence the reproductive success of the breeding pair. Thus, because group size can influence reproductive success, it is important to control for group size in the analysis of the relationship between pair bond duration and apparent pup survival, which is one measure of reproductive success. The researcher has shown a correlation between pair bond duration and apparent pup survival. However, correlation does not imply causation. Other factors not measured by the researcher could have influenced both pair bond duration and apparent pup survival and produced the relationship shown in the figure.

3. Given that social conflict in other species is positively associated with the emergence of different mating strategies in populations, one would expect pair bond duration to be negatively associated with a prevalence of other mating strategies, such as polygyny and polyandry, in groups of gray wolves. In other words, as pair bond duration increased, the prevalence of other mating strategies within groups would be expected to decrease. This pattern is exactly what the researcher found for his study groups of gray wolves.

CHAPTER 9

Answers to Figure Legend Questions

FIGURE 9.4 In clones that form by budding or apomixis, identification of groups of genetically identical individuals may require the use of genetic analyses. In clones that form by horizontal spread, groups of individuals that are still connected to one another could be marked; however, to tell whether members of two such groups were in fact genetically identical would again require genetic analyses.

FIGURE 9.7 There were 7 habitat patches in 1759 and about 86 patches in 1978. Thus, in 1759, the average patch size was $400 \text{ km}^2/7 = 57.1 \text{ km}^2$. Patch sizes were much smaller in 1978: the average at that time was $60 \text{ km}^2/86 = 0.7 \text{ km}^2$.

FIGURE 9.12 Because it may compete poorly with other barnacle species in relatively warm waters, *S. balanoides* is currently excluded from the region shaded purple on the map. Thus, by warming northern waters, global warming will probably decrease the geographic range of *S. balanoides*.

FIGURE 9.16 The chance of colonization is between 50% and 90%.

FIGURE 9.17 Urchin biomass declined at Sites 1, 2, 3, 4, 5, and 9; kelp density increased at Sites 1 and 5.

Answers to Analyzing Data 9.1 Questions

1. During the 41-year period before introduced grasses had invaded the park, the fire frequency was 0.22 fires per year with an average burn size of 0.26 ha per fire. In the 20-year period after introduced grasses had invaded the park, the fire frequency was 1.6 fires per year with an average burn size of 243.8 ha per fire. These data suggest that the introduction of non-native grasses has resulted in a sevenfold increase in the frequency and a nearly 1000-fold increase in the scope of fires on Hawaii.

2. The data in Table B indicate that fire reduces the abundance of native trees and shrubs, while it increases the abundance of introduced grasses.

3. If a fire occurs in a Hawaiian dry forest after introduced grasses are present, the introduced grasses should recover quickly and provide fuel for later fires. We would predict that this fuel would make it more likely that a second fire would occur; in addition, should a second fire occur, the increased fuel levels would probably cause it to burn with greater intensity than the first fire. Thus there is the potential for a "fire cycle" in which a fire causes the abundance of introduced grasses to increase, and also makes future fires both more likely and more intense, leading to further increases in introduced grasses and further declines in native trees and shrubs. Such a fire cycle is consistent with data in Table A: after introduced grasses arrived, fires occurred more often and covered larger areas. A fire cycle is also consistent with data in Table B: introduced grasses were least abundant in unburned areas and most abundant in areas burned twice.

Answers to Review Questions

1. Complicating factors discussed in the text include (1) limited knowledge about the dispersal capabilities of the organism under study, (2) the fact that populations may have a patchy structure, and (3) the fact that individuals may be hard to define. The first two factors—limited information about dispersal and patchy populations—can make it difficult to determine the area within which individuals interact, and hence what constitutes a population. The third factor—difficulty in defining individuals—applies to the many organisms that reproduce asexually to form clones. In such organisms, it can be hard to determine what an individual is, thus making it difficult to estimate abundance.

2. The simplest reason that no species is found everywhere is that much of Earth does not provide suitable habitat. There can, in turn, be many reasons why portions of Earth are not suitable for a particular species. For example, the abiotic or biotic conditions of an environment may limit the growth, survival, or reproduction of the species, as may disturbance or the interaction between abiotic and biotic conditions. Furthermore, a species may be absent from environments where we would expect it to thrive because of dispersal limitation or historical factors (including evolutionary history and continental drift).

3. A species distribution model is a tool that predicts the environmental conditions occupied by a species based on the conditions present in its current distribution. Species distribution models can be used to predict the future distribution of an introduced species by collecting as much information as possible about environments where the species currently is found. Those data are then used to construct a species distribution model, which in turn is used to identify currently unoccupied locations that are likely to provide suitable habitat for the species. For such predictions to accurately reflect the future spread of the organism, information also must be gathered about its dispersal capabilities.

4. For a conservative estimate, assume there are 20 otters per square kilometer, each of which eats 20% of its body weight in food each day. Since urchins, on average, weigh 0.55 kg each, a kilogram of urchins consists of roughly $1/0.55 = 1.82$ urchins. Thus, the number of urchins per square kilometer that an otter population would be expected to eat each year is as follows: (20 otters/km^2) × (0.2 × 23 kg/otter/day) × (365 days/year) × (1.82 urchins/kg) = 61,116 urchins/km^2/year.

Answers to Hone Your Problem-Solving Skills Questions

1. Four quadrats were used in each patch. Each treatment had 3 patches, so there was a total of 12 quadrats used in each treatment. The mean values for each treatment are:

Treatment	Total no. taxa	Total no. individuals
Intact kelp beds	68.1	434.4
Recovered kelp beds	73.8	580.8
Urchins present	12.1	63.5

2. In the recovered kelp beds, there was an average of 580.8 individuals per quadrat or 580.8 individuals per 0.25 m². In the entire patch (which had an area of 40 m²), this suggests that we would have a total of

$$\frac{580.8 \text{ individuals}}{0.25 \text{ m}^2} \times 40 \text{ m}^2 = 92{,}928 \text{ individuals in the patch}$$

Likewise, in areas where urchins were present, there was an average of 63.5 individuals per quadrat or 63.5 individuals per 0.25 m². In the entire patch (which had an area of 40 m²), this suggests that we would have a total of

$$\frac{63.5 \text{ individuals}}{0.25 \text{ m}^2} \times 40 \text{ m}^2 = 10{,}160 \text{ individuals in the patch}$$

3. Urchins had large impacts on species diversity and overall abundance. Compared to intact or recovered kelp beds, the presence of urchins had very large effects, reducing the total number of taxa by 6–7-fold and the total number of individuals in the patch by about 8-fold. The total number of taxa and the total number of individuals in recovered kelp beds were similar to the total number of taxa and the total number of individuals in intact kelp beds; this indicates that when protected from urchins, previously degraded patches can recover.

CHAPTER 10

Answers to Figure Legend Questions

FIGURE 10.3 There was considerable variation in abundance from one field site to another in many of the years. In 1984 and 1989, for example, abundance was high at Hector but low at the other two locations.

FIGURE 10.10 From 1988 to 2000, the collared lemming population exhibited regular cycles, reaching peak abundance every 4 years. Because abundances peaked at about 10 lemmings per hectare in 1990, 1994, and 1998, we would have expected the next peak to occur in 2002, again at about 10 lemmings per hectare. However, the actual abundance in 2002 was less than 1 lemming per hectare.

FIGURE 10.11 In (A), abundance rises and falls in a regular manner, reaching a peak about every 40 days; thus, this curve shows regular population cycles. In (B), to the left of the dotted vertical line, the results are again consistent with a regular population cycle that reaches peak abundance every 40 days. After food for adults is limited, however, the regular population cycle no longer occurs. Instead, abundance rises and then fluctuates around a roughly stable population size. This pattern that can be viewed as illustrating either population fluctuations or logistic growth (with fluctuations).

FIGURE 10.13 About 100 breeding pairs would be needed for the risk of extinction to drop to 5%.

FIGURE 10.19 From 1952 to 1957, the abundance of predatory fish increased while the abundance of planktivorous fish showed little change. In the 1970s, predatory fish abundance dropped, planktivorous fish abundance increased, zooplankton abundance dropped, and phytoplankton abundance increased. Overall, the chain of feeding relationships for the Black Sea in the 1970s is more similar to that in Alaska pre-1990 than to that in Alaska in the late 1990s. In both cases, the organisms at the base of the food chain (phytoplankton in the Black Sea, kelp in Alaska) were only weakly controlled by their grazers (zooplankton in the Black Sea, urchins in Alaska), which in turn were strongly controlled by the organisms that ate them (planktivorous fish in the Black Sea, otters in Alaska).

Answers to Analyzing Data 10.1 Questions

1. For years 2–6 (respectively), the five missing values for the table are 1.22, 0.87, 1.17, 1.02, and 1.13.

2. If λ remained constant and equal to 1.02, when $N_0 = 1{,}000$ the population size at year 7 would be: $N_7 = N_0 (1.02)^7 = 1{,}149$. This predicted value is higher than the

observed value of 1,069, suggesting that the observed variation in λ decreased the growth of the population.

3. The geometric mean of the yearly population growth rates equals 1.00945.

4. Using the geometric mean calculated in Question 2 as our estimate of λ, we have: $N_7 = N_0 (1.000945)^7$ = 1,068. This value is lower than that calculated in Question 1 (1,149) and almost identical to the value in the table (1,069).

5. When environmental conditions vary, it is likely that the growth rate of a population will also vary over time. The results in Questions 1–3 suggest that using the arithmetic mean of such variable population growth rates will over-estimate the population size, whereas using the geometric mean would be more accurate. Because the arithmetic mean is known to over-estimate actual population sizes, in that sense it would be wrong to use the arithmetic mean to describe the growth of a population in a variable environment.

Answers to Review Questions

1. There are many built-in time lags in the responses of populations to changes in density. For example, the amount of available food may increase or decrease between the time the parent generation feeds and the time its offspring are born. In such a situation, the number of offspring produced may be more closely related to the previous conditions than to the conditions at the time of their birth. As a result of such time lags, the population may experience delayed density dependence, which may cause it to fluctuate in abundance over time.

2. Small populations can be threatened by chance events associated with genetic factors, demographic stochasticity, environmental stochasticity, and natural catastrophes. Genetic factors that increase the risk of extinction in small populations include genetic drift and inbreeding, both of which can increase the frequencies of harmful alleles. Demographic stochasticity results from chance events related to the reproduction and survival of individuals; such events can cause population growth rates to drop, as might occur if considerably more females than males happened to die in a small population, leaving few females to produce the next generation of offspring. Environmental stochasticity refers to unpredictable variation in environmental conditions; such variation can cause population growth rates to vary dramatically from year to year, increasing the chance of extinction in small populations. Finally, natural catastrophes can cause sudden reductions in population size, subjecting a population to increased risks from genetic factors, demographic stochasticity, and environmental stochasticity.

3. a. Yes, as illustrated by the two generations of parents and offspring in the diagram. KEY: FC = female child; MC = male child; FG = female grandchild, MG = male grandchild

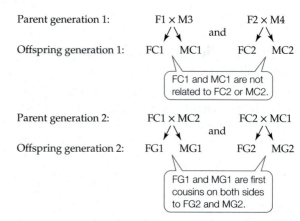

b. No, all of the individuals in the second generation of offspring are related to one another. As illustrated by this example, inbreeding is likely to be common in small populations.

Answers to Hone Your Problem-Solving Skills Questions

1. Densities in this population ranged from a minimum of fewer than 2 individual/m² to a maximum of nearly 300 individuals/m². Since the densities of this population show considerable variation over time (and do not cycle in a regular manner), the growth of this population is best described by the third pattern described in Concept 10.1, population fluctuations.

2. The graph shows that birth rates initially increase with density, indicating that Allee effects may occur in the study populations; for population densities greater than 100 individuals/m², birth rates decrease as density increases. Since birth rates change as a function of population density, birth rates are density dependent.

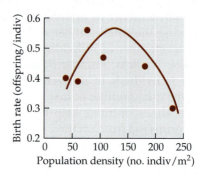

3. The graph shows that death rates do not change as a function of population density; thus, death rates are not density dependent.

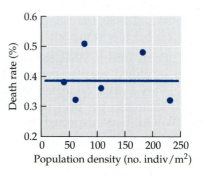

CHAPTER 11

Answers to Figure Legend Questions

FIGURE 11.11 Since there were about 35 breeding females in 1975, results from previous years suggest that roughly 4 young per female should have been reared to independence. In fact, fewer than 1.5 young per female were reared to independence, suggesting that conditions on the island were different in 1975 than in other years (there could have been a drought or a disease outbreak, among many other possibilities).

FIGURE 11.12 High-density populations are increasing in density in (A) because λ is greater than 1 in those populations. In contrast, in (B) the high-density populations are decreasing in size because r is less than zero in those populations.

FIGURE 11.14 As N becomes close to K, the term $(1 - N/K)$ becomes increasingly close to zero; this causes the population growth rate, dN/dt, to become close to zero. A population with a growth rate of zero does not increase in size; hence, as N approaches K, the population stops increasing in size.

FIGURE 11.17 100 sheep survive to age 11; thus, 10% $(100/1,000)$ of sheep survive from birth to age 11.

FIGURE 11.18 About 47% of Gambians born in the hungry season live to age 45; a similar percentage (48.5%) of U.S. females live to be 85 years old.

FIGURE 11.20 The graph shows that the human population is projected to have an annual growth rate of 0.5% in 2050. This rate is greater than zero, so the human population will still be increasing in size in 2050.

FIGURE 11.21 The best-estimate curve indicates there will be 9.6 billion people in 2050, and Figure 11.20 indicates that our annual growth rate will be 0.5% at that time. Hence, from 2050 to 2051, we would expect to add about 48 million (9,600,000,000 × 0.005) to our population. Thus, the human population size in 2051 would be about 9,048,000,000.

Answers to Analyzing Data 11.1 Questions

1.

Year	r
1	0.00028
400	0.00037
800	0.00123
1200	0.00094
1550	0.00252
1825	0.00660
1930	0.0135
1960	0.0178
1999	0.0123
2010	0.0113
2016	0.0155
2019	(N/A)

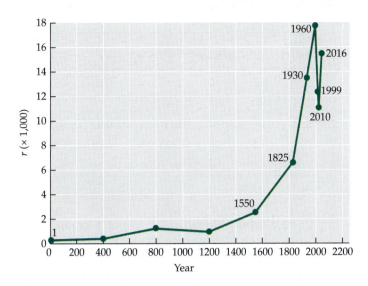

2. Based on a value of $r = 0.0155$ and a population size of 7.35 billion in 2016, we can use Equation 11.4 to estimate the population size in 2066:

$$N_{2066} = N_{2016} \times e^{rt} = 7.35 \times e^{(0.0155 \times 50)} = 15.9 \text{ billion people}$$

3. The calculations in Question 2 assume that the human population is growing exponentially and that the exponential growth rate, r, remains constant and equal to 0.0155 from 2016 to 2066. However, the answer to Question 1 indicates that r reached a maximum value (0.0178) in 1960 and has declined since that time, although it has risen from 2010 to 2016. If r continues to decline from the 1960 value, it is unlikely that the human population will reach 15.9 billion in 2066.

Answers to Review Questions

1. a. 3,240

   b. Substituting the values N0 = 40, $\lambda = 3$, and t = 27, we have
   $$N_t = N_0\lambda^t = 40 \times 3^{27}$$

   c. In this case, we have the values $N_0 = 100$, $\lambda = 0.75$, and $t = 3$, which we plug into the relation
   $$N_t = N_0\lambda^t = 100 \times (0.75)^3 = 42.19$$

2. Factors that regulate population size are density dependent: when N (the number of individuals in a population) is below some level, they cause the population size to increase, whereas when N is above some level, they cause the population size to decrease. Even if density-independent factors, such as year-to-year variations in temperature or rainfall, are the primary cause of year-to-year changes in abundance, those factors do not regulate population size.

3. a.

Age (x)	N_x	N_x offspring	l_x	F_x	$l_x F_x$	$x l_x F_x$
0	100	0	1	0	0	0
1	40	100	0.4	2.5	1	1
2	15	30	0.15	2	0.3	0.6
3	5	25	0.05	5	0.25	0.75
4	0	0	0	0	0	0

$R_0 = 0.25, r = 0.05$

b. In a cohort life table, the fate of a group of individuals born during the same time period (a cohort) is followed from birth to death. This type of life table is often used for sessile or relatively immobile organisms that do not have long life spans but is less useful for organisms that are highly mobile or long-lived. For those organisms, a static life table may be used, in which the survival and fecundity of individuals of different ages are observed during a single time period.

4. Each student will calculate their own answer.

Answers to Hone Your Problem-Solving Skills Questions

1.

Age (x)	Number of individuals (N_x)	Number of offspring ($N_{x \, offspring}$)	Survivorship (l_x)	Fecundity (F_x)	$l_x F_x$	$x l_x F_x$
0	843	0	1.000	0	0	0
1	722	216,600	0.856	300	256.94	256.94
2	527	326,740	0.625	620	387.59	775.18
3	316	135,880	0.375	430	161.19	483.56
4	144	30,240	0.171	210	35.87	143.49
5	54	3240	0.064	60	3.84	19.22
6	15	450	0.018	30	0.53	3.20
7	3	30	0.004	10	0.04	0.25
8	0	0	0.000	0	0.00	0.00

$R_0 = 846, r = 3.39$

2. $N_t = N_0 e^{rt} = 100 \times e^{3.39} \times {}^{10} = 5.27 \times 10^{16}$ individuals.

3. $N_t = 1,000,000/1 + [((1,000,000-100)/100)] \, e^{3.39} \times {}^{10} = 1,000,000/1 = 1,000,000$ individuals.

4. Yes, it easily reaches its carrying capacity of 1 million individuals after 10 years. Without density dependent growth, there would be 5.27×10^{16} individuals.

Chapter 12

Answers to Figure Legend Questions

FIGURE 12.2   The peak abundance of lynx usually occurs after the peak abundance of hares. One reason this might occur is that as hare abundance rises, the increased availability of food enables the lynx to produce more offspring; however, these offspring are not born immediately, so the rise in lynx abundance lags behind the rise in hare abundance.

FIGURE 12.7   To answer this question, we must use the data in the graph to determine the total number of agromyzid fly species and the number of agromyzid fly species that feed on fewer than five host plant species. We can do this using the scale on the y axis, which indicates that a bar that is 2.15 cm in height represents 50 fly species. Measuring all 13 bars on the graph, we find that their heights sum to 12.05 cm; this indicates that in total, there are about 280 fly species (280 = 12.05 cm/2.15 cm × 50). Similarly, the heights of the four bars representing fly species that feed on fewer than five host plant species sum to 10.4 cm, indicating that about 242 fly species feed on fewer than five host plant species. Thus, about 86% of agromyzid fly species feed on fewer than five host plant species.

FIGURE 12.11   On average (based on the height of the bar graph), the control plants produced about 11 or 12 fruits per plant. This indicates that a plant that compensated fully for clipping would also produce 11 or 12 fruits.

FIGURE 12.17   The density of other plants in the community would probably increase after herbivory by *C. quadrigemina* reduced the density of Klamath weed. Because Klamath weed was originally a dominant member of the community, it is likely that the community would change considerably after introduction of the beetle.

FIGURE 12.21   In the absence of snails, wetlands had phosphorus concentrations of less than 100 µg/L. When snails were present, phosphorus concentrations were usually much greater than 100 µg/L; for example, in the seven wetlands with snail densities greater than 10 snails per square meter, the average phosphorus concentration was close to 1,000 µg/L. Thus, the presence of snails is associated with an increase in the phosphorus concentration of these wetlands.

Answers to Analyzing Data 12.1 Questions

1. A total of 18 plant populations were established in this experiment. In each of these populations, the initial frequency of each plant genotype was $1/27 = 0.037$ or 3.7%.

2. If evolution had not occurred in the control populations, we would have expected all 27 plant genotypes to survive and their frequencies to change little from their initial values of 3.7% for each genotype. This was not the case: many genotypes did not survive (and hence had a final frequency of 0%), while others increased dramatically in frequency. Genotype 6, for example, reached a frequency of 42.3% when grown in the control environment. Genotype 6 may have been particularly well suited to the growing conditions experienced in the control populations, where plant genotypes were grown at high densities and in soil that may have differed from the soil of their home environments. Such changes in environmental conditions could have caused natural selection to occur in the control populations.

3. Plant genotype frequencies also changed in populations exposed to aphid herbivores, with many genotypes being driven to extinction while others increased dramatically in frequency. Hence, evolution occurred in these populations as well. Plant populations exposed to aphid herbivores could have experienced multiple sources of selection, such as novel environmental conditions (e.g., high plant densities and different soil from those found in their home environments) as well as the consequences of feeding by aphid herbivores.

4. In the *B. brassicae* treatment, 75% of the surviving plants encoded 4C defensive compounds; one of these, genotype 25, was the most common surviving genotype (67.4% of the surviving plants had this genotype). In contrast, in the *L. erysimi* treatment, 83% of the surviving plants had 3C genotypes, the most common of which was genotype 9, at 63.2%. Although a few genotypes performed reasonably well in both treatments (e.g., genotypes 9 and 25), overall, the outcome of selection differed considerably between treatments. These results suggest that natural selection by different herbivore species can favor different plant genotypes.

Answers to Review Questions

1. Most carnivores have a broad diet in that they eat a wide range of prey species. Although a substantial number of herbivores can eat many different plant species, the majority of herbivores are insects, most of which feed on just one or a few plant species. This difference is hypothesized to be due to the differences carnivores and herbivores experience related to encountering and handling their food. Carnivores are mostly generalists because their encounter rates are low for mobile prey, and thus they should not be too narrow in their prey choices. Herbivores are specialists because they have relatively high encounter rates with their immobile prey, but their handling times are longer because plants are less nutritious food.

2. A prey individual that cannot evade a carnivore is killed and eaten. While herbivores do not typically kill their food plants, they do have powerful negative effects on the plants on which they feed. As a result of this strong selection pressure that carnivores and herbivores exert on their food organisms, prey species have evolved a wide range of defensive mechanisms that increase the chance that they will not be eaten. Animals must eat if they are to survive, so there is also strong selection pressure on them to overcome the defenses of their prey. These effects are pervasive because all organisms must obtain food—setting in motion the conflicts just described. The effects are pronounced because there is such strong selection for both defensive and counterdefensive mechanisms.

3. a. Evidence described in this and preceding chapters indicates that predation can have a powerful effect on the abundance and distribution of prey species, and this can affect communities in dramatic ways.

   b. The scientific evidence strongly supports this claim. As described in this chapter, in many cases the effects of carnivory and herbivory have been so pronounced that they have altered ecological communities greatly, in some cases causing a shift from one community type to another. For example, arctic foxes feeding on seabirds, lesser snow geese feeding on marsh grasses, and aquatic snails feeding on large aquatic plants had such effects.

Answers to Hone Your Problem-Solving Skills Questions

1. $dP/dt = 0$, $N = m/ba$, which is $N = 0.90/0.01 = 90$ prey

   $dN/dt = 0$, $P = r/a$, which is $P = 0.40/0.01 = 40$ predators

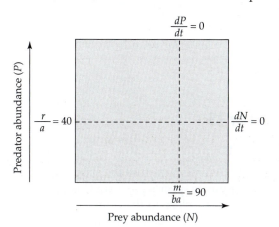

2. Given Lotka–Volterra model assumptions, the two populations will cycle with no change in the amplitude of the cycle. Thus, the values for Time 3 will be 90 prey and 50 predators and for Time 4 will be 80 prey and 40 predators.

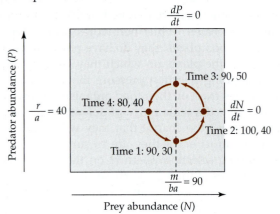

3.

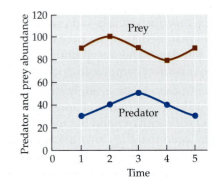

4. The populations will not cycle but will remain at equilibrium over time.

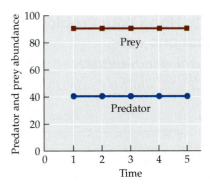

Chapter 13

Answers to Figure Legend Questions

FIGURE 13.4 Averaging across the six groups, there are about 21 parasite species per host. This average would probably not be close to the number of parasite species found in a previously unstudied host from one of the six groups of organisms. A reason for this is that in five of the groups (all but the trees, which had an average of 95), the average number of parasites per host

is fewer than 12. Thus, we might expect that 95 parasite species would be found in another tree, 7 parasite species would be found in another wasp, etc.—but we would not expect to find 21 parasite species in a host from any of the six groups.

FIGURE 13.9 The gamete-producing cells enable the parasite to disperse from a human host to a mosquito.

FIGURE 13.11 No. For example, with an infection rate of 70%, the Lake Wahapo snails are very poorly defended against parasites from their own lake, but they are reasonably well defended against parasites from both other lakes. Similarly, Lake Paringa snails are poorly defended against parasites from their own lake (infection rate = 51%), but they are well defended against parasites from Lake Mapourika (infection rate = 11%).

FIGURE 13.15 If the cycles stopped completely, we would not expect the numbers in both of the treated populations to drop in 1989 and again in 1993—the same years that the control populations were predicted to drop based on long-term data on population cycles in red grouse.

FIGURE 13.22 The WT and EGT+ treatments represent two types of controls. The WT treatments are unmanipulated controls; results from these controls can be compared with results from the EGT–experimental treatments. The EGT+ controls can be used to check whether the procedures used to remove (and insert) the *egt* gene have inadvertent effects. Hence, in the EGT+ controls, the gene is removed and then reinserted—if these experimental procedures do not have inadvertent effects, results from these controls should be similar to results from the WT controls. In fact, this is what was found.

Answers to Analyzing Data 13.1 Questions

1. This experiment tests the hypothesis that the symbiont *Spiroplasma* is more common in fruit flies harboring the nematode parasite *Howardula*. The "*Howardula* absent" treatment serves as the control. The frequency of *Spiroplasma* fluctuated in the control but did not rise

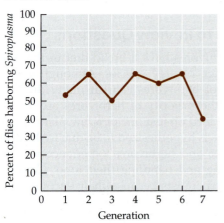

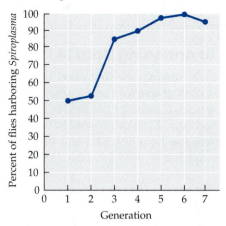

Howardula present

or fall consistently over time. In contrast, the frequency of *Spiroplasma* in the "*Howardula* present" treatment rose from its initial value of 50% to more than 95% by generation 5, supporting the hypothesis.

2. This experiment tests the hypothesis that the presence of the symbiont *Spiroplasma* protects fruit flies from the nematode parasite, *Howardula*. The "*Spiroplasma* absent" treatment serves as the control. In control populations, by generation 3, 95% of fruit flies were infected by the nematode parasite; all control

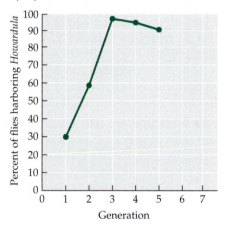

Spiroplasma absent

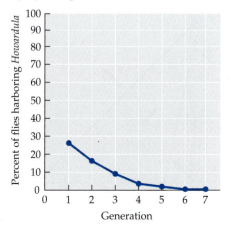

Spiroplasma present

populations declined to extinction by generation 6 (because, without the symbiont, the parasite sterilizes flies that it infects). In contrast, the frequency of *Howardula* dropped steadily in the "*Spiroplasma* present" treatment, reaching 0% by generation 6. These results support the hypothesis that the symbiont can protect fruit flies from attack by *Howardula*.

3. We would predict that if there was a large cost for harboring the symbiont, the frequency of flies harboring the symbiont would decline in the absence of the parasite. This did not occur (see the graph in Question 1 for the "*Howardula* absent" treatment), suggesting that the flies experience few costs for harboring the symbiont.

Answers to Review Questions

1. Ectoparasites live on the surface of their host, whereas endoparasites live inside the body of their host. Examples of ectoparasites include plants such as dodder and fungi such as rusts and smuts; examples of endoparasites include tapeworms and bacterial pathogens such as Mycobacterium tuberculosis. Ectoparasites can disperse more easily from one host individual to the next than can endoparasites; however, ectoparasites are at greater risk from natural enemies than are endoparasites.

2. Parasites can greatly reduce the growth, reproduction, or survival of host individuals, thereby reducing the growth rate of host populations. As a result, we would expect that parasites could also alter both the outcomes of species interactions and the composition of ecological communities. For example, if two plant species compete for resources and one typically outcompetes the other, a parasite that reduces the performance of the superior competitor may cause a competitive reversal in which the inferior competitor becomes the superior competitor. Such changes in the outcome of species interactions can cause changes in the relative abundances of the interacting species, thus altering the ecological community.

3. a. Host organisms have a wide range of defensive mechanisms that include a protective outer covering, an immune system that kills or limits the effectiveness of the parasite, and biochemical conditions inside the host's body that reduce the ability of the parasite to grow or reproduce.

   b. The statement could be true if the plant populations in Australia possessed specific defensive features that limited the ability of the parasite to grow or reproduce, yet the populations in Europe lacked such adaptations. Among many other possible examples, plants in the Australian populations might possess a specific allele that enabled them to kill or disable the parasite—hence causing the parasite to have mild effects there—whereas plants in the European populations might lack this allele, making them more vulnerable to parasite attack.

Answers to Hone Your Problem-Solving Skills Questions

1. The rodents serve as an alternative, or reservoir, host for the disease. Thus, if leishmaniasis infection can be reduced within rodents, fewer sand flies will carry the disease and fewer humans will be infected.

2. A given disease will become established and spread in a given host population only if the density of susceptible hosts exceeds a critical threshold density (S_T). The concept of a threshold density has considerable medical and ecological importance because it indicates that a disease will *not* spread if the density of susceptible hosts can be held below the threshold density.

3. The following populations will need to be reduced to S_T, which is 5,000 individuals: population 1 (decrease by 4,000 individuals), population 3 (decrease by 500 individuals), population 5 (decrease by 3,000 individuals), and population 8 (decrease by 5,000 individuals).

4. The threshold density can be raised by taking actions that increase the rate at which infected individuals recover and become immune (thereby increasing m and hence increasing $S_T = m/\beta$). This can be accomplished by early detection and clinical treatment of the disease. The threshold density can also be raised if β, the disease transmission rate, is decreased. This can be achieved by controlling the vector (sand flies) and/or changing the behaviors of the host (e.g., using bug spray).

Chapter 14

Answers to Figure Legend Questions

FIGURE 14.6 It is likely that *Asterionella* would drive the third diatom species to extinction. *Asterionella* reduces the concentration of silica to about 1 μmol/L when grown alone (see part B of this figure). This concentration is much lower than the concentration of silica (5 μmol/L) that results when the third diatom species is grown alone—suggesting that the third diatom species would not have enough silica and hence could not survive if it was grown in competition with *Asterionella*.

FIGURE 14.7 All of the interactions shown should be circled except the two on the ends (which represent amensalism, not competition) and the one in the middle (in which each competitor has an equal effect on the other).

FIGURE 14.9 *Paramecium aurelia* feeds mainly on floating bacteria, while *P. bursaria* feeds mainly on yeast cells. Because they rely on different food sources, it is likely that both species would persist if they were grown together.

FIGURE 14.14 The population size of species 2 would decrease to 1,000 because its abundance would be above its carrying capacity.

FIGURE 14.18 Two years. The observed replacement curve indicates that if a population begins with 100 individuals (in "year 0"), it will have about 22 individuals in the next year (year 1). A population that has 22 individuals in year 1 will have fewer than 10 individuals in year 2.

Answers to Analyzing Data 14.1 Questions

1. To draw each isocline, we first need to solve for the population sizes of species 1 and 2 when the other is at zero. For species 1, we use Equation 14.2 and find that $N_1 = 0$ when $N_2 = K_1/\alpha$ or $N_2 = 42.5/0.43 = 98.8$. For species 2, we use Equation 14.3 and find that $N_2 = 0$ when $N_1 = K_2/\beta$ or $N_1 = 53.2/0.72 = 73.9$.

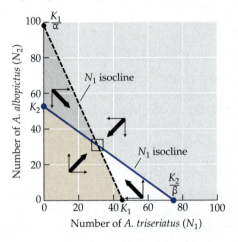

Graphing these two isoclines, we find that in tree holes, the likely outcome of competition is that the two species will coexist. The equilibrium population density for *A. triseriatus* (species 1) is 28.3, and the equilibrium population density for A. albopictus (species 2) is 32.8; these densities can be estimated from the graph or calculated algebraically by setting the N_1 and N_2 isoclines equal to one another.

2. To draw each isocline, we first need to solve for the population sizes of species 1 and 2 when the other is at zero. For species 1, we use Equation 14.2 and find that $N_1 = 0$ when $N_2 = K_1/\alpha$ or $N_2 = 33.4/0.84 = 39.8$. For species 2, we use Equation 14.3 and find that $N_2 = 0$ when $N_1 = K_2/\beta$ or $N_1 = 44.7/0.25 = 178.8$.

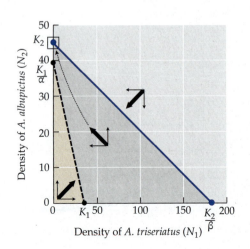

In tires, the likely outcome of competition is that A. albopictus (species 2) will drive the native species A. triseriatus (species 1) to extinction. The equilibrium population density for A. triseriatus (species 1) is zero, and the equilibrium population density for A. albopictus (species 2) equals $K_2 = 44.7$; these densities can be estimated from the graph.

3. In tree hole communities, the two species are predicted to coexist, whereas in tires, the introduced species is predicted to drive the native species to extinction. Hence, it is unlikely that competition with the native species, A. triseriatus, will prevent the spread of the introduced mosquito, A. albopictus.

Answers to Review Questions

1. Immediately after application of the fertilizer, it is likely that the intensity of competition for nitrogen will drop because soil nitrogen levels will not be as limited in supply. However, as the added nitrogen is used up by the plants (and leached from the soil by rainfall), soil nitrogen levels will decrease, and the intensity of competition for nitrogen will increase.

2. Four general features of competition, with an example of each, are (1) competition can be direct (as in allelopathy) or indirect (as in both pitcher plants and bedstraws), (2) competition is often asymmetrical (as in Tilman's diatoms), (3) competition can occur between closely (as in bedstraws) or distantly (as in ants and rodents) related species, and (4) competition is a common feature of natural communities.

3. a. Possible reasons why these meadows harbor one or the other (or both) of these two plant species include the following: (1) both species could persist at all locations, but one species (or the other) has yet to disperse to some meadows; (2) the physical conditions of the meadows differ such that in some meadows species 1 is favored, while in others species 2 is favored, and in still others, the species can partition resources such that both persist; (3) the abundances of herbivores or pathogens that feed on species 1 or 2 may vary between the meadows, causing the outcome of competition to differ from meadow to meadow; (4) the rates of a periodic disturbance such as fire may differ among meadows (if one of the species is an inferior competitor but is more tolerant of fire).

   b. Addition and removal experiments would help to evaluate these possible explanations for the observed distributions of the species. For example, in meadows where only species 1 is found, individuals of species 2 could be planted next to some individuals of species 1, but not others. Similarly, in meadows where both species are found, removal experiments could be performed in which individuals of species 1 would be removed from the vicinity of some species 2 individuals, but not others (and vice versa).

Answers to Hone Your Problem-Solving Skills Questions

1. Results from laboratory experiments, field observations, and mathematical models all suggest that competing species are more likely to coexist when they use resources in different ways. For example, in Gause's experiments with *Paramecium*, *P. caudatum* coexisted with *P. bursaria*, most likely because one species fed primarily on bacteria, the other on yeast. Likewise, in the case of four species of *Anolis* lizards that lived together on Jamaica and ate similar food, Schoener's field observations indicated that these species used space in different ways (an example of resource partitioning). Finally, graphical analysis of the Lotka–Volterra competition model indicates that competing species can coexist when the inequality shown in Equation 14.4 holds. That inequality is more likely to hold when competing species use resources in very different ways (e.g., when α and β are not close to 1).

2. Because $\beta = 1.6$ and there are 140 individuals of species 1, it would take $1.6 \times 140 = 224$ individuals of species 2 to reduce its own growth rate by the same amount that the 140 individuals of species 1 do. Therefore, because there are 230 individuals of species 2 present, species 2 is having a slightly greater effect on its own growth rate than is species 1.

3. The statement is not correct. For example, if $\alpha = 0.5$ and $\beta = 1$, Equation 14.4 predicts that both species will persist when $0.5 < K_1/K_2 < 1$. Thus, for example, if $K_1 = 100$ and $K_2 = 150$, both species should persist when $\alpha = 0.5$ and $\beta = 1$. (The statement can be shown to be false in many other ways; for example, in Figure 14.14B, values for α, β, K_1, and K_2 can be selected such that species 2 always drives species 1 to extinction, even though $\alpha < \beta$.)

Chapter 15

Answers to Figure Legend Questions

FIGURE 15.3 The stippled light green regions are similar to the regions in which tropical rainforests are found. Thus, the plants in this mycorrhizal association are likely to be tropical rainforest trees and other plants found in the rainforest biome.

FIGURE 15.4 Ectomycorrhizae form a mantle around the root, while arbuscular mycorrhizae can penetrate the cell wall of a root cell and form an arbuscule (a branching network of hyphae).

FIGURE 15.8 *Myosotis laxa* grows best under colder conditions of 11°C–12°C with cattail neighbors present.

FIGURE 15.20 These results would suggest that although ants increase their frequency of weeding when parasites are present, they do not discriminate among parasites.

Answers to Analyzing Data 15.1 Questions

1.

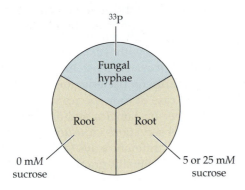

2. The results in the figure show that the fungus transferred more phosphorus to plant roots that had greater access to sucrose.

3. Both partners play a role. As shown in Figure 15.13, the plant transfers more carbohydrates to fungi that have access to phosphorus. Similarly, as shown in the figure here, the fungus transfers more phosphorus to plants that have greater access to carbohydrates.

Answers to Review Questions

1. Commensalism and mutualism share a number of characteristics: they are both very common, they can evolve in many ways, and they can cease to be beneficial if conditions change such that the costs of the interaction exceed its benefits. In addition, some evidence indicates that positive interactions may be particularly common in stressful environments. Positive interactions can also differ from one another in that they can range from obligate and coevolved to facultative and not coevolved relationships.

2. When a species in a mutualistic interaction provides its partner with a benefit, that action comes at a cost to the species providing the benefit. If circumstances change such that the costs of the interaction are greater than the benefits to one of the species, that species may cease to provide benefits to its partner, or it may penalize its partner. The fact that mutualists may stop providing benefits to their partners when it is not advantageous for them to do so has convinced researchers that mutualism is not an altruistic interaction.

3. Initially, we could expect a decrease in the growth or reproduction of the coral species that are most sensitive to high water temperatures. If high temperatures continued long enough to cause repeated bleaching, it is likely that these more sensitive species would begin to suffer heavy mortality. As a result of the decreased growth, reproduction, and survival of these sensitive species, the species composition of the reef would change: those coral species that were better able to tolerate high water temperatures would constitute an increasingly high percentage of the corals found in the reef. Such changes in the composition of the coral reef community might also affect other species; for example, a fish that depended on an increasingly rare coral for shelter or food might also decline in abundance. As water temperatures continued to rise, other, less sensitive corals might also experience negative effects. Eventually, if temperatures continued to rise, the abundance of all corals in the reef might decline, as would the abundance of the many species that depend on the reef.

Answers to Hone Your Problem-Solving Skills Questions

1. Scotch pine tree seedlings survive at much higher rates when growing under *Salvia* shrubs than when growing in open areas; thus, *Salvia* appear to serve as nurse plants for Scotch pine tree seedlings.

2. Scotch pine seedlings growing under Salvia shrubs experience lower light levels, lower soil temperatures, and higher soil moisture content than do Scotch pine seedlings growing in open areas. Any of these three (correlated) factors could contribute to the increased survival of pine seedlings that grow beneath Salvia; controlled experiments would be needed to separate their effects.

3. As a Scotch pine tree increased in size, it could begin to compete with (and hence harm) the *Salvia* shrub that once served as its nurse plant.

Chapter 16

Answers to Figure Legend Questions

FIGURE 16.2 Based on the dates of Caulerpa sightings, it is likely the seaweed spread from Monaco west to Spain and east to Sicily, Italy, at about the same time (1992, 1993). It was restricted to these locations for 2 years, but then traveled to the eastern coast of Italy, and from there to the island of Hvar (1995), from which it spread to the northern islands of Croatia (1996). Finally, it was sighted much later in Tunisia (2000), even though Tunisia is closer to Sicily than is Croatia. This may have been because there was less boat traffic to Tunisia than to Croatia, thus lowering the chance of invasion, or it may have been due to a lack of recognition until 2000 that the seaweed was present.

FIGURE 16.3 The desert and hot springs communities are defined by physical attributes of their environment, whereas the kelp forest and coral reef communities are defined by biological attributes of their environment, particularly by the presence and importance of abundant species (i.e., kelp and corals, respectively).

FIGURE 16.10 The tropical soil bacterial community requires much more sampling because each sample contains new species, thus producing a linear species accumulation curve. The sampling in the temperate forest plant and tropical bird communities was sufficient to identify a large majority of the species in these communities, and thus more sampling would not be needed. This is clear from the leveling off of the species accumulation curves once all the samples were analyzed. Finally, although the human oral bacterial and tropical moth communities showed some leveling off of their species accumulation curves, new species were being found even once all the samples were analyzed. Thus, they also need more sampling to adequately capture their species richness.

FIGURE 16.18 Beavers act as ecosystem engineers by damming streams with cut trees and woody debris. This behavior creates a flooded area, which accumulates sediment and eventually becomes dominated by marsh vegetation. At a landscape scale, by creating a mosaic of wetlands within a larger forest community, the beavers' actions enhance regional species diversity. Thus, beavers can also be classified as keystone species because they have such a large effect on diversity relative to their size and abundance.

Answers to Analyzing Data 16.1 Questions

1. The number of invasive species that likely caused negative effects on species richness is 11. The number of invasive species that likely caused positive effects on species richness is 1. The number of invasive species that likely had no effect is also 1. The percentage change in species richness suggests that most invasive species had strong to intermediate negative effects on species richness. Only two invasive plants had neutral or positive effects, and these effects were weak.

2. The order of the magnitude of the change in species diversity (H) did differ between the two measures, with some species having higher species richness but lower species diversity than other species. This suggests that the proportional abundance of the species within the plots (evenness), which is used along with species richness to calculate species diversity, also changed with the invasion. In some cases, evenness increased, and in others, it decreased.

Answers to Review Questions

1. A community is a group of interacting species that exist together at the same place and time. Interactions among multiple species and their physical environment give communities their character and function.

2. Species richness is the number of species in a community, but that measure tells us nothing about the relative abundances of those species. If two communities had a similar number of species, but great differences in species evenness (as in Figure 16.6), species richness would not reflect this difference, but species diversity indices would. Rank abundance curves (as in Figure 16.8) allow hypotheses to be generated about how those species may be interacting in the community based on their abundances.

3. Foundation species have a large effect on other species due to their large size and high abundance. For example, kelp and trees have a large influence on species diversity by virtue of providing their communities with habitat, food, and other services that are directly related to their size. Keystone species have a large effect despite their small size and low abundance, because of the important role they play in their communities. For example, sea otters have large effects on their communities by preying on herbivores (sea urchins), which, in turn, eat primary producers (kelp). This indirect interaction can allow primary producers to have higher abundances. Finally, ecosystem engineers are able to create, modify, or maintain physical habitat for themselves and other species. Trees and kelp are examples of ecosystem engineers that are foundation species, and beavers are an example of a keystone species that is also an ecosystem engineer.

Answers to Hone Your Problem-Solving Skills Questions

1. The per capita interaction strength (IS) values between gulls and the species listed in the table are the following: ribbed limpet IS = ln (10/100)/10 = –0.23, gooseneck barnacle IS = ln (500/3,000)/10 = –0.18, checkered limpet IS = ln (100/50)/10 = 0.07, mussel IS = ln (3,000/2,500)/10 = 0.02, microalgae IS = ln (500/100)/10 = 0.16. Of the prey species, ribbed limpet experience the greatest negative affect of gull predation. Of the nonprey species, microalgae have the greatest positive interaction with gulls because they indirectly benefit from the fact that their herbivore, the ribbed limpet, is eaten by the birds.

2. Indirect effect 1: Removing gulls decreases the abundance of mussels because of increased competition with the gooseneck barnacle. Indirect effect 2: Removing gulls decreases the abundance of the checkered limpet because of increased competition with the ribbed limpet. Indirect effect 3: Removing gulls decreases the abundance of microalgae because of increased herbivory by the ribbed limpet. Indirect effect 4: Although the experiment cannot test for the effect on phytoplankton, it is likely that removing gulls decreases the abundance of phytoplankton because of increased herbivory by the gooseneck barnacle.

3. The effect of gulls would be even more positive. That's because gulls, by eating the ribbed limpet and gooseneck barnacle, reduce the competition for the mussel and checkered limpet (the interaction strength measurements for those species indicated that they indirectly benefited from their interaction with gulls). By reducing competition, excluding the mussel and checkered limpet allows those nonprey species to increase in abundance and prey more heavily on microalgae and phytoplankton.

Chapter 17

Answers to Figure Legend Questions

FIGURE 17.2 The most destruction occurred immediately below the mountain, where a huge magma-filled bulge exploded and released rock and mud down the north side of the mountain. An area later known as the Pumice Plain, formed by the hot, pelting pumice rock, experienced the most destruction. The massive wave of debris from the explosion was funneled down the North Fork Toutle River, removing most life along the way. Spirit Lake was also completely destroyed because of its location within the path of the avalanche. Other areas, such as the south side of the mountain and the locations farther from the explosion (mudflow zone and blowdown zone), experienced blowdown of all trees, but some life remained, especially underground. Finally, the least destruction occurred in the scorch zone, where trees were denuded but remained standing.

FIGURE 17.4 Whether a disturbance is intense or frequent will depend on the susceptibility of the organisms involved and their ability to respond to the disturbance. The intensity and frequency of disturbance for an insect population will be quantitatively different from that for an elephant population. The same disturbance—let's say, a tree falling in a forest—could cause major destruction for the insect population living on that tree while having little effect on the elephant population, even if an elephant were struck by the tree. Of course, the insect population would recover much faster than an elephant population might.

FIGURE 17.9 The oldest communities are located in the areas that have been exposed the longest since glacial retreat, such as the mouth of the bay. Here, succession has been able to proceed for over 200 years and has allowed the formation of mature spruce forests. As the glacial retreat becomes more recent, the communities become younger, such that the youngest, pioneer community is located closest to the glacier.

FIGURE 17.17 The fish preferred to eat the tunicate *Styela*, because when the tiles were protected from fish predation, it was the species that dominated. When fish predation was allowed, the bryozoan *Schizoporella* dominated, suggesting that it was unpalatable to the fish. This

experiment suggests that *Styela* is the dominant competitor over *Schizoporella* in the absence of predation.

Answers to Analyzing Data 17.1 Questions

1. Aspen suckers colonize all the successional stages but are most abundant in aspen stands and least abundant in meadow and fir stands. In contrast, subalpine fir seedlings are most abundant in the mixed aspen–fir and fir stands and least abundant in the meadow and aspen stands. The data show that aspen are the first to colonize meadows, establishing aspen stands that are then colonized by firs. As firs increase and form mixed and fir-dominated stands, aspen decline and fir seedlings increase. This pattern supports the successional sequence described in the introductory paragraph of this Analyzing Data exercise.

2. The most consistent hypothesis is that fir seedlings are facilitated by aspen, because their densities are highest in aspen stands but lowest in meadows. However, competition seems to drive aspen out in later stages, because they decline in mixed aspen–fir and fir-dominated stages.

3. Fir trees have lower mortality when they live close to aspen than when they live farther away, suggesting that they are facilitated by aspen. However, aspen trees show greater mortality close to firs than farther away, suggesting that they compete with firs and are eventually excluded from the community. These results support the previous hypotheses from Question 2.

4. This study best fits the facilitation model inspired by Frederick Clements and later described by Connell and Slatyer. In this model, only certain species, such as aspen, can establish themselves in early successional habitats. This might be a consequence of clonal growth. In time, species such as aspen modify the habitat in such a way that they facilitate later successional species such as firs, which may be intolerant of full sun. As firs grow and mature, they continue to be facilitated by aspen but eventually displace them through competition for resources. In this system, firs are expected to dominate and exclude aspen until they are disturbed by fire or humans, resetting the system back to meadow.

Answers to Review Questions

1. Abiotic and biotic agents of change include those listed in Table 17.1. Intense disturbances such as hurricanes, tsunamis, fires, and volcanic eruptions can cause major damage but are relatively infrequent. Other agents of change, such as sea level rise, competition, or parasitism, may not cause major damage initially but may be frequent or constant and have dramatic effects over time. Still others, such as predation, may be relatively frequent but not very intensive, thus forming patches of available resources.

2. Primary succession involves the colonization of habitats devoid of life. Species colonizing these habitats must deal with stressful conditions and transform their habitats to create soils, nutrients, and food. Secondary succession involves the reestablishment of a community in which most, but not all, of the organisms have been destroyed. Under these conditions, colonizing species benefit from the biological legacy of the preexisting species, but they are likely to face more competition for resources than the species involved in primary succession.

3. A hypothetical community might be a newly cleared vacant lot in an unnamed city. The facilitation model would be supported if the first species to arrive were stress-tolerant and had the ability to modify their habitat in positive ways. In this case, those early species would facilitate the growth of later species, which would be better competitors but less stress tolerant. Over time, these later species would dominate as they outcompeted the facilitating species. The tolerance model assumes that the earliest species modify the environment, but in ways that neither help nor hinder later species. Later species are merely those that live longer and tolerate stressful conditions longer than early species. Finally, the inhibition model would be supported if the early species created conditions that benefited themselves but inhibited later species. Only through the removal of those inhibitory early species—for example, via disturbance or stress—would later species be able to displace them.

4. It is hard to know whether a community is stable because stability depends on the spatial and temporal scale at which the community is observed. All communities fluctuate and change over time, but how long must we wait for a community to return to some original state before we assume it is stable? There is no single answer. Although Sutherland did observe the formation of alternative communities on his tiles when predators were manipulated, did he follow the communities long enough, and at a large enough spatial scale, to show stability? Again, it depends on how you define "stability," leaving us with an unresolved question.

Answers to Hone Your Problem-Solving Skills Questions

1. In 1982, the Pumice Plain had no surviving species, while the reference area had the highest species richness (five species). The blowdown and scorch zones both had intermediate species richness (three species). Species richness generally increased over time, even in the reference area. By 2000, the Pumice Plain had one species, the blowdown zone had seven species, and the scorch zone and reference area had six species each.

2. Small mammals recovered in the two least disturbed successional habitats (blowdown and scorch zones) but not in the Pumice Plain. The pattern of species richness seen in the blowdown and scorch zones was likely due to their being secondary successional communities and thus having more resources available to them after the eruption. In the primary successional community of the Pumice Plain, the habitats and resources were nonexistent and had to be reestablished over time. Thus, it makes sense that this community could not support more than one small mammal species.

3. This suggests that the deer mouse has a life history that allows it to live in primary, secondary, and climax successional communities. It is likely able to disperse widely, grow quickly, and reproduce often—all characteristics of an early successional, pioneer species. The deer mouse is also likely to be an opportunistic and generalist species, living in a variety of habitats and feeding on a variety of food items.

4. It may be that some small mammal species were affected by the eruption, even though the trapping site was 21 km away from the mountain. Over time, these species recovered and were present in the reference community. In addition, it may be that the researchers did not trap individuals of a particular species some of the years, thus underestimating species richness. Alternatively, it may be that the animals became habituated to the traps because they contained bait. This could have resulted in individuals of rare species being caught more often over time.

Chapter 18

Answers to Figure Legend Questions

FIGURE 18.2 The goal of the study was to look at the effect of fragmentation on species diversity in the *remaining* forest fragments rather than considering the direct effects of deforestation itself.

FIGURE 18.6 No, there could never be more local species than would be contained within a region, because the spatial scale of the region is larger than that of the local community.

FIGURE 18.9 Holt et al. (2013) used phylogenetic information acquired from DNA analysis and more recent global species distribution patterns to test whether Wallace's original biogeographic regions were supported by modern data collection.

FIGURE 18.11 One would expect speciation to increase as land masses separate because species would become reproductively isolated from one another, thus increasing the chance that they would follow different evolutionary trajectories. The separation of species in this way is known as vicariance.

FIGURE 18.16 The idea that the tropics serve as a cradle is meant to suggest that it is a place in which species arise or "are born." The reference to the tropics as a museum is meant to suggest that it is a place in which species are protected from extinction and thus are "on display" for a long time.

Answers to Analyzing Data 18.1 Questions

1. There was a steeper slope (z) and lower y intercept (c) for the species–area relationship of invaded communities compared with uninvaded communities. These results suggest that invaders have strong negative effects on species richness at the smallest spatial scales and little or no effect at large spatial scales.

2. To convert log x values to x values, solve for 10^x. The approximate range of area values is from (at the smallest scale) 1 m² to (at the largest scale) 500 m². The approximate range in species richness for invaded plots is 0.6 (smallest scale) to 16 species (largest scale). For uninvaded plots, it is 3 (smallest scale) to 20 species (largest scale).

3. One hypothesis is that the invaded areas turn into island-like habitat where native species occur within a sea of invaders. As we saw in the example in Ecological Toolkit 18.1, island-like systems tend to have steeper slopes and lower y intercepts than mainland-like habitats. The equilibrium theory of island biogeography posits that smaller areas and those more distantly connected to the sources of species will have higher extinction rates and lower immigration rates and thus fewer species per given area. At the largest spatial scale, however, even though invasions may have negative effects on native species, the area may be large enough that immigration of species from uninvaded areas may rescue species from extinction. At some point, though, if large enough areas are invaded and/or immigration from uninvaded areas ceases, one could see a decline in species richness.

Answers to Review Questions

1. The largest spatial scale is the global scale, which covers the entire world, over which there are major differences in species diversity and composition with latitude and longitude. These patterns are controlled by speciation, extinction, and dispersal. The next scale down is the regional scale, defined by areas of uniform climate and by species that are bound by dispersal limitation to the region. Within a region, species diversity and composition depend on dispersal and extinction rates across the landscape. The regional species pool (also called gamma diversity) has an important influence on the species present at the next scale down, the local scale (also called alpha diversity). The relationship between regional and local species richness can help us to determine the extent to which the regional species pool or the local effects of species interactions and physical conditions determine local species richness.

2. Wallace identified six terrestrial biogeographic regions, which represent distinct biotas that vary in species diversity and composition. Wallace believed that these biogeographic regions reflect the evolutionary isolation of species due to the movements of the continents. Thus, the ancestors of many modern species may have occurred together in the evolutionary past, but since Pangaea began breaking up into the continents we know today, they have evolved separately. Recent research suggests that the biogeographic regions are more subdivided than previously thought, suggesting more isolation than simply the movements of the continents. There are also impediments to dispersal within oceans, such as currents, thermal gradients, differences in water depth, and the continents themselves, so it is assumed that the oceans could be divided into biogeographic regions, but that effort has received considerably less attention.

3. The three main hypotheses focus on (1) species diversification rate, (2) species diversification time, and (3) productivity. The first hypothesis proposes that both the large geographic land area and the thermal stability of the tropics might promote higher speciation rates and lower extinction rates, thereby increasing the population sizes and geographic ranges of species. Speciation rates should increase because larger geographic ranges should lead to greater reproductive isolation. Extinction rates should decrease because larger population sizes should lower the risk of extinction due to chance events while larger species ranges should spread extinction risk over a larger area.

The second hypothesis suggests that the tropics have had a longer evolutionary history than the temperate or polar zones because of their greater climatic stability. This stability may have allowed more species to evolve without the interruption of severe climatic conditions that would have hindered speciation and increased extinctions in the temperate and polar zones.

The third hypothesis suggests that the high productivity of the tropics increases species diversity by promoting larger population sizes, which should lead to lower extinction rates and overall higher species richness.

Answers to Hone Your Problem-Solving Skills Questions

1. Yes, the data follow the species–area relationship.

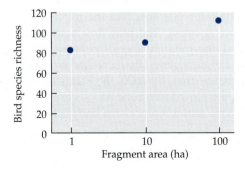

2. The percentage loss of species per year can be calculated using the t_{50} scaling factor. Divide the percentage loss (50%) by the number of years to reach that loss to get the percentage loss of species per year. Thus,

1 hectare: 50%/5 years = 10% species loss per year

10 ha: 50%/8 years = 6.25% species loss per year

100 ha: 50%/12 years = 4.17% species loss per year

The 1 ha fragments have the greatest species loss, and the 100 ha fragments have the least.

3. The number of species in the fragments 9 years after the start of the experiment can be calculated using the following equation: initial species number – (initial number of species × percentage species loss per year × 9 years). Thus,

1 ha: 83 species – (83 species × 10% loss per year × 9 years) = 8 species

10 ha: 92 species – (92 species × 6.25% loss per year × 9 years) = 40 species

100 ha: 113 species – (113 species × 4.17% loss per year × 9 years) = 71 species

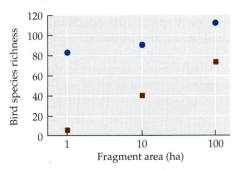

4. The fragments that had 9 years of isolation would have the steepest species–area slope. Once fragmentation occurs, the fragments act more like islands. Thus, the smallest fragments had greater species loss than the largest fragments because they had higher extinction rates and lower immigration rates. In addition, the smaller the fragment, the greater percentage of edge habitat, which is more hazardous for species, and thus extinction risk increases even more.

Chapter 19

Answers to Figure Legend Questions

FIGURE 19.4 No, it does not make sense that the fish and frog species should be present in the local community given in the figure, because that community contains terrestrial species. The abiotic filter should have excluded any aquatic species from this terrestrial community.

FIGURE 19.8 (B) shows the most resource partitioning (least overlap in resource use). (A) and (C) show the least resource partitioning (most overlap in resource use).

FIGURE 19.15 The lowest species richness occurred on the small boulders (their maximum richness was 4), which rolled over more frequently and thus experienced more disturbance compared with the other boulders.

Answers to Analyzing Data 19.1 Questions

1. The predation treatments all caused a decline in species richness compared with the control ponds without predation. Thus, it appears that predation caused the local extinction of zooplankton species. The two species of predators (fish and insect), either alone or together, did not differ in their effects on zooplankton richness.

2. Dispersal of zooplankton increased local species richness in the ponds but only if predation was present. If predators were not present, local species richness was similar with or without dispersal. The results suggest that dispersal can have a positive effect on local species richness but presumably only if resources are freed up by predation, thus increasing species coexistence.

3. The results suggest that the effect of predation on local species richness can be so intense that the process of dispersal is inadequate to "rescue" the community from species loss. Yes, the results fit the intermediate disturbance hypothesis but only if dispersal is incorporated into the model.

Answers to Review Questions

1. Yes. Community membership is dependent on dispersal, environmental factors, and biological interactions. Given all the introductions of non-native species that have occurred worldwide, it is clear that "getting there" has been an important constraint on the entrance of species into communities. In this particular case, the seeds on the ecologist's shoes are physically and biologically adapted to prairie grassland communities and are thus prime candidates for successful introduction into New Zealand grassland communities.

2. Resource partitioning is the idea that coexistence among species is possible if the species in a community use its resources in slightly different ways. Other models, such as the intermediate disturbance hypothesis, rely on population fluctuations due to disturbance, stress, or predation as the mechanism of coexistence. These models suggest that as long as populations of species never reach their carrying capacities, competitive exclusion will not occur, and coexistence will be possible. Lottery or neutral models assume that resources made available by disturbance, stress, or predation are captured at random by recruits from a larger pool of colonists, all of which have an equal chance of obtaining those resources.

3. Lottery and neutral models best support the tropical rainforest data set. These models assume that resources made available by the deaths of individuals are captured at random by recruits from a larger pool of colonists such that no one species has an advantage, and that species diversity is maintained as a result.

4. Species diversity–community function relationships can differ depending on two variables: the degree of overlap in the ecological functions of species, and variation in the strength of ecological functions of species. Graph A is best described by the complementarity hypothesis, which proposes that as species richness increases, there will be a linear increase in community function. This linear relationship occurs because each species added to the community has a unique and equally incremental effect on community function. Graph B is best described by the redundancy hypothesis, in which there is an upper limit on the effect of species richness on community function. This curvilinear relationship occurs because the unique functional contributions of species reach a threshold due to their overlap. Graph C best describes the idiosyncratic hypothesis, which suggests that the strengths of the effects of species' functions vary dramatically. Dominant species have a large effect on community function such that when they are present, they increase community function, but when they are absent, it declines. This produces a variable species richness and community function pattern.

Answers to Hone Your Problem-Solving Skills Questions

1. The model best describes the intermediate disturbance hypothesis, which shows a unimodal relationship between species richness and disturbance, stress, or predation. At low levels of disturbance, species diversity is low because dominant species are free to exclude competitively inferior species. At high levels of disturbance, species diversity declines because many species may become locally extinct as mortality increases. At intermediate levels of disturbance, species diversity is maximized simply by the balance between competition and mortality.

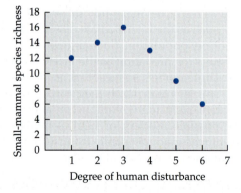

2. The graph shows that deer mouse density does not change with species richness. The data suggest that resource partitioning is not an important factor in the community. If the small-mammal species were partitioning resources, you would expect that where there is high species richness, there would also be lower densities of all species, including the deer mouse.

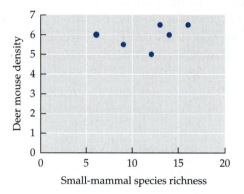

3. The graphs below show that Sin Nombre virus infection prevalence in the deer mouse is positively related to small-mammal species richness loss. There is no clear relationship with deer mouse density. The results suggest that when deer mouse hosts live in more species diverse communities, they are more likely to come into contact with individuals of other species than their own species (conspecifics), thus reducing the probability of transmission.

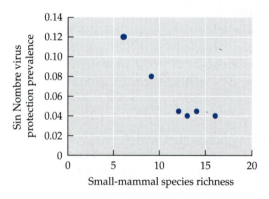

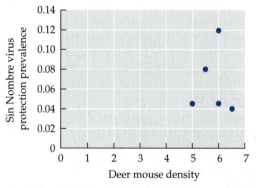

Chapter 20

Answers to Figure Legend Questions

FIGURE 20.5 Greater allocation of NPP to belowground tissues can be an adaptation to disturbances, such as fire, or to herbivory. Allocation of NPP to storage compounds allows more rapid recovery and higher survival rates following disturbance or loss of tissues to herbivory.

FIGURE 20.7 Estuaries also have high NPP due to the inputs of nutrients brought in by rivers. These nutrient subsidies include organic matter from both terrestrial and aquatic ecosystems as well as agricultural runoff.

FIGURE 20.9 Cacti are CAM plants (see Chapter 5), which open their stomates and take up CO_2 during the night when air temperatures are cooler and humidities are higher. The daily pattern of atmospheric CO_2 concentrations would be reversed from what is shown for the boreal forest, with lower concentrations at night and higher concentrations during the day.

FIGURE 20.12 The proportional allocation to belowground NPP would be greater in the more nutrient-poor community, the dry meadow. Greater allocation to roots enhances the uptake of the resources that most limit NPP, whereas light is more likely to be limiting in the more nutrient-rich wet meadow. Allocation to belowground NPP would decrease in response to fertilization.

Answers to Analyzing Data 20.1 Questions

1. Whether an ecosystem is a carbon sink (takes up more C than it releases) is determined by net ecosystem exchange (NEE). NEE is equal to NPP minus heterotrophic respiration. Prior to the beetle outbreak, NEE was equal to 440 g C/m²/year – 408 g C/m²/year = a net uptake (sink) of 32 g C/m²/year.

2. Following the beetle outbreak, NEE was 400 g C/m²/year – 424 g C/m²/year = –24 g C/m²/year, or a net source of 24 g C/m²/year. As tree regrowth occurs during secondary succession, the forest will again revert to a net sink of C, so the trend will reverse over the next 100 years.

3. NEE is equal to GPP minus the total (autotrophic and heterotrophic) respiration. For the pasture, NEE is equal to 2,345 g C/m²/year – 2,606 g C/m²/year = –262 g C/m²/year (net source), and for the second-growth forest, NEE is equal to 2,082 g C/m²/year – 1,640 C/m²/year = 442 C/m²/year (net sink). Thus, despite higher GPP in the pasture than in the second-growth forest, the higher respiratory losses in the pasture result in a net loss of C from the system.

4. Currently tropical rainforests account for around 3 Pg C/yr times 0.35 (35%) = 1.05 Pg C/yr. Converting half of the tropical rainforests to pasture would result in a decrease of NEE to 0.5 (–262 g C/m²/year) + 0.5 (442 g C/m²/year) = 90 g C/m²/year. This is an 80% reduction in NEE by tropical rainforests, or a 28% reduction in C uptake by the terrestrial land surface. Note that this scenario is a gross oversimplification of what would actually happen, and does not take into account biotic and functional variation among tropical rainforests and pastures.

Answers to Review Questions

1. Primary production is the source of the energy entering an ecosystem, and it therefore determines the amount of energy available to support that ecosystem. Primary production also results in the exchange of carbon between the atmosphere and the biosphere and thus is important in determining the atmospheric concentration of CO_2, an important greenhouse gas. Finally, primary production is a measure of the functioning of an ecosystem and provides a biological indicator of the ecosystem's response to stress.

2. As NPP increased in a terrestrial ecosystem, the leaf area index would increase along with overall plant biomass. The amount of shading would increase as the leaf area index increased, and light would become increasingly limiting to growth. To compensate, plants would allocate more energy to stems and less to roots so as to increase their height and overtop neighbors in order to acquire more light.

3. The researchers found a correlation between NPP and soil temperature, and they assumed that the causal link was through the effect of soil temperature on root growth. While this assumption may be correct, the researchers failed to show the causal link conclusively, which would require careful experimentation, or at least more thorough measurements of the effect of soil temperature on the factors that can influence plant growth. For example, soil temperature can affect the rate of decomposition of organic matter in the soil, and thus the availability of nutrients, which may influence growth rates.

4. a. Harvest techniques are simple and don't require high-tech equipment. However, harvesting can be labor-intensive, may fail to account for production that is lost to herbivores or decomposition, and is impractical at large scales.

   b. Remote sensing provides estimates of NPP at larger spatial scales and can be used at frequent intervals. However, remote sensing is expensive and requires handling of massive amounts of data. Because it is based on absorption of light by chlorophyll, remote sensing can potentially overestimate NPP if a plant canopy is physiologically inactive.

Answers to Hone Your Problem-Solving Skills Questions

1. Based on the isotopic composition of the bear tissues and of their food sources, grizzly bears living in inland areas consume less meat than coastal grizzlies, with a high proportion of their diet consisting of terrestrial plants. Grizzlies from along the coast of

southeast Alaska had the highest consumption of meat, derived primarily from marine sources, indicating fish makes up a large part of their diet. The population of grizzlies from the Columbia River drainage had an intermediate proportion of meat in their diet, with slightly less derived from marine sources.

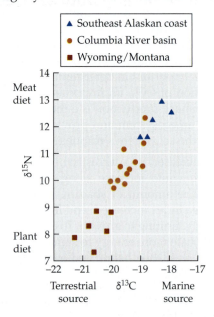

2. a. If bears switched from a diet of primarily fish to plants, the composition of N isotopes in bone and hair samples would shift to less enriched in ^{15}N and lower δ^{15}N values. The composition of C would be less enriched in ^{13}C and there would be a lower (more negative) $\delta\,^{13}$C.

   b. If bears switched from consuming mostly fish to mostly terrestrial mammals, the composition of C isotopes would be less enriched in ^{13}C and there would be a lower $\delta\,^{13}$C. N isotope composition would not change appreciably.

Chapter 21

Answers to Figure Legend Questions

FIGURE 21.7 Figure 21.6 shows that overall consumption efficiency in aquatic ecosystems is higher than in terrestrial ecosystems, as the line fitting the aquatic ecosystem data lies above the line fitting the terrestrial ecosystem data, indicating that a greater percentage of the NPP is being consumed.

FIGURE 21.10 Brown trout might preferentially feed on predators that are more effective in controlling insect herbivores than are the predators that galaxias feed on. As a result, the effect of the brown trout on algal abundance would be greater than the effect of the galaxias.

FIGURE 21.16 Eight of the 21 species or feeding groups (38%) eat both plants and animals, and most of the others

eat at more than one trophic level, indicating that omnivory is very common in this desert food web.

Answers to Analyzing Data 21.1 Questions

1. Plants (100); non-insect invertebrate herbivores (100 × 0.209 = 20.9); small mammals (20.9 × 0.015 = 0.31); large mammals (0.31 × 0.031 = 0.01)

2. Algae (100); aquatic insect herbivores (100 × 0.209 = 20.9); insect predators (20.9 × 0.556 = 11.62); fish (11.62 × 0.098 = 1.14)

3. Plants (100); large mammal herbivores (100 × 0.031 = 3.1); large mammal predators (3.1 × 0.031 = 0.10); large mammal predators (0.10 × 0.031 = 0.003)

4. Plants (100); insect herbivores (100 × 0.388 = 38.8); insect predators (38.8 × 0.556 = 21.57); insect predators (21.57 × 0.556 = 11.99)

5. The trophic chains in numbers 2 and 4 have substantially greater energy available to support a fifth trophic level than do the other trophic chains, due to the higher production efficiencies of their component ectothermic consumers. In contrast, the trophic chains in numbers 1 and 3 include larger endotherms, with much lower production efficiencies, and it is unlikely that they could sustain a fifth trophic level.

Answers to Review Questions

1. Population B should have a higher assimilation efficiency due to the higher food quality of its diet. The garbage and plant component of population A's diet is higher in materials that are difficult to digest, and its C:N ratio is also lower than that of population A's rodent diet. Thus, the amount of food assimilated would be greater in population B.

2. The seasonal and diurnal temperature variations in these animals' environments are different and should result in different production efficiencies. The marine environment is more thermally stable, and thus the marine mammals should need to invest less energy in coping with temperature changes than the mammals in the terrestrial ecosystem. As a result, the marine mammals should be able to invest more energy in growth and reproduction.

3. The forest would have a greater total amount of energy flowing through its trophic levels because a greater amount of energy would enter that ecosystem at the first trophic level. However, a larger proportion of the energy entering the lake ecosystem would pass through its higher trophic levels due to its higher consumption and production efficiencies.

Answers to Hone Your Problem-Solving Skills Questions

1. If only specialist herbivores were present, they would consume only a few or one species of plants.

Furthermore, if the herbivores were chemically defended, their consumption by predators would be limited. Given a diverse plant community, we would expect a lower impact on herbivory and NPP with a trophic cascade involving specialist herbivores than if there were generalist herbivores. This prediction could be tested by varying the presence/absence, or the proportions, of specialist and generalist herbivores and the abundances of predators at the third or fourth trophic level. The response variables would include abundance of the herbivores and the amount of plant consumption.

2. a. The results support the hypothesis that a trophic cascade would influence herbivory and NPP less with specialist herbivores than with generalist herbivores. At a mix of about 25% specialist herbivores, there is little influence of a trophic cascade on herbivory in the deciduous forest ecosystem under study.

   b. This appears to be primarily due to lower consumption of herbivores rather than to less consumption of plants due to specialization.

Chapter 22

Answers to Figure Legend Questions

FIGURE 22.4   Primary production is low and plants are sparse in desert ecosystems, so the amount of soil organic matter should also be low. Wetting–drying events should enhance mechanical weathering of soils, producing a range of soil particle sizes. However, without a protective covering, winds may remove some of the finest particles, as we describe in the Case Study Revisited in Chapter 22. The low amount of precipitation and plant growth should limit the development and depth of distinct soil horizons.

FIGURE 22.6   Pesticides applied to plants can wash into the organic surface layers of soils, where they can kill both herbivorous animals and soil detritivores. The loss of these animals would effectively lower the rate of decomposition and would thereby decrease soil fertility.

FIGURE 22.12   The simple input–output model depicted in the figure assumes that elements enter the ecosystem primarily through deposition and leave it in stream water. As noted in Figure 22.13, other modes of input and output occur, including inputs through N_2 fixation, outputs in groundwater, and gaseous losses (e.g., denitrification).

FIGURE 22.18   The study of eutrophication in Lake Washington is very convincing, but it lacks an appropriate control. Therefore, it is correlational; that is, it shows a quantitative link between depth of clarity and phosphorus inputs, but that link isn't necessarily causal. Appropriate controls might have included another lake that didn't have sewage inputs, or a lake that continued to have inputs of phosphorus-laden sewage

during the time sewage inputs to Lake Washington were halted. (Experiments with appropriate controls have demonstrated beyond a doubt that inputs of phosphorus in sewage entering lakes do cause eutrophication.)

Answers to Analyzing Data 22.1 Questions

1.

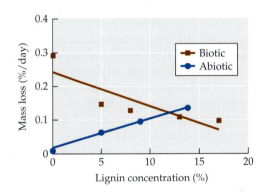

2. The results indicate that when plant litter is exposed to light, tissues higher in lignin degrade faster than those with lower lignin concentrations. Thus the inhibitory influence of lignin on biological degradation may be at least partially offset by the stimulatory effect of photodegradation.

3. The assumption that lignin will lower decomposition rate would not be expected to hold true in environments in which the influence of photodegradation is greatest: dry, high-light environments such as deserts, shrublands, grasslands, and some tundra ecosystems.

Answers to Review Questions

1. The transformation of minerals in rock involves both the physical breakdown (mechanical weathering) and chemical alteration (chemical weathering) of the minerals. Mechanical weathering occurs through expansion and contraction of solid materials due to freezing–thawing or drying–rewetting cycles, gravitational forces such as landslides, and pressure exerted by plant roots. Mechanical weathering exposes the surfaces of mineral particles to chemical weathering. Weathering is a soil-building process, leading to the development of ever finer mineral particles and greater release of the nutrients in the minerals. The release of CO_2 and organic acids into the soil from organisms and detritus enhances the rate of chemical weathering.

2. The original source of nitrogen for plants is dinitrogen gas (N_2) in the atmosphere, but they cannot use it unless it is converted to other forms by the process of nitrogen fixation. Only bacteria can carry out nitrogen fixation, which is an energetically expensive process. Some plants, such as legumes, have symbiotic relationships with nitrogen-fixing bacteria. As ecosystems develop, nitrogen builds up in the pool of detritus and

is converted into soluble organic and inorganic forms through decomposition. Some of the nitrogen released by decomposition is consumed by microorganisms, lowering the supply available to plants.

3. While both primary production and decomposition influence the buildup of organic matter and associated nutrients in the soil, decomposition is more sensitive to climatic controls than is primary production. The mean residence time of nutrients is therefore more strongly controlled by decomposition. Low soil temperatures in boreal forests result in very long mean residence times. High rates of decomposition limit the buildup of soil organic matter in tropical forests, and the mean residence times of nutrients such as nitrogen and phosphorus are two orders of magnitude lower than those in boreal forests.

4. Nutrient transfers between trophic levels are efficient in both tropical and temperate-zone lakes, but organic matter is progressively lost from the surface layers in both systems, falling into the sediments in the benthic zone, where oxygen concentrations, and thus decomposition rates, are low. In the temperate zone, some of these nutrient-rich sediments are brought back to the surface layers during seasonal turnover of water, where they decompose, providing nutrients to support production. Turnover is largely absent in tropical lakes, which are therefore more dependent on external inputs of nutrients from streams and terrestrial ecosystems.

Answers to Hone Your Problem-Solving Skills Questions

1. NPP should increase following the disturbance, reaching a maximum somewhere during the intermediate stages of succession, and then decrease at late stages as the forest matures and consists of old-growth stands of trees. As a result, nutrient losses should be lowest during the intermediate stages of succession, highest just following the disturbance, and intermediate late in succession.

2. Nutrient losses should vary according to their importance to plant growth. Limiting nutrients, such as N, will be retained more with lower losses than nutrients that are not limiting growth. Elements that are not taken up by plants should be lost at the same rate throughout succession.

3. The results support Vitousek's hypothesis regarding the patterns of nutrient loss between intermediate and late stages of succession. For nutrient elements, losses are generally higher in late successional communities than in intermediate-stage communities. In particular, N is retained much more than the other elements, suggesting it is probably the nutrient limiting growth of the plants. Elements such as Na and Cl, which have little or no importance to most plants, are lost at the same rates in intermediate and late stages of succession.

Chapter 23

Answers to Figure Legend Questions

FIGURE 23.2 The bar graphs indicate there were about 36 million ha in 1500, 8 million ha in 1935, and 1 million ha in 2004. The annual rate of loss appears to have been greater from 1935 to 2004 (7 million ha lost over 69 years, or approximately 100,000 ha lost per year) than from 1500 to 1935 (28 million ha lost over 435 years, or approximately 64,000 ha lost per year).

FIGURE 23.5 As discussed in Chapter 15, the seeds of many plant species are dispersed by animals that eat their fruit; hence the extinction of many frugivores may have reduced the ability of such plant species to disperse their seeds. Likewise, as also discussed in Chapter 15, many plants are pollinated by animals that visit flowers to collect nectar. Hence, the loss of nectarivores may have reduced the reproductive success of some plant species.

FIGURE 23.6 The "open flower" treatment is the control; results for this treatment indicate the percentage of flowers that currently can produce seeds on island and mainland sites. One experimental treatment was to bag flowers; results from this treatment show the percentage of flowers that produce seed in the absence of bird pollinators and all other means of pollination except self-pollination. A second experimental treatment was to hand-pollinate flowers; results from this treatment show the percentage of flowers that produce seeds when pollination is not limiting (as should be true when bird pollinators are abundant).

FIGURE 23.8 The difference between this statement and the results in the figure (which show that the introduction of non-native plant species can cause regional plant diversity to increase) is due to a difference in scale. When the introduction of non-native plant species causes the global extinction of one or more plant species, global plant diversity will decline even though regional plant diversity increases.

FIGURE 23.10 Habitat loss is the most important factor affecting terrestrial mammals; overharvesting is also an important threat. In contrast, accidental mortality and pollution are the most important threats affecting marine mammals.

FIGURE 23.12 Individual answers may vary but should include a line of reasoning similar to the following: Although there was year-to-year fluctuation in the cod harvest, overall the catch increased from roughly 100,000 tons caught in 1850 to roughly 300,000 tons caught in 1950. Because the harvest was maintained at these levels for 100 years, this suggests that at about 200,000 tons could have been caught in a sustainable manner.

FIGURE 23.15 Over the past decades habitat loss and pollution have been the primary causes of the loss of biodiversity from terrestrial, aquatic, and coastal habitats, while over-exploitation (hunting and harvesting) has been the largest factor influencing biodiversity in marine biological zones. Looking to the future (arrows) climate change and pollution are forecast to be the largest threats to all biological zones, with habitat loss continuing as a concern as well.

Answers to Analyzing Data 23.1 Questions

1. a. The sample size is $n = 4$ for plots with kudzu and for plots lacking kudzu.

    b. In plots with kudzu, $\bar{x}_1 = 3.68$ and $s_1 = 1.89$. In plots lacking kudzu, $\bar{x}_2 = 1.23$ and $s_2 = 0.53$.

    These results indicate that plots with kudzu have higher NO emissions than do plots lacking kudzu.

2.
$$T = \frac{3.68 - 1.23}{\sqrt{\frac{\left(1.89^2 + 0.53^2\right)}{4}}} = \frac{2.45}{\sqrt{\frac{3.853}{4}}} = 2.5$$

3. The degrees of freedom is df = 6, and the (two-tailed) p value of the test is $p = 0.047$. This result indicates that NO emissions in plots with kudzu differ significantly from NO emissions in plots lacking kudzu.

Answers to Review Questions

1. The principal threats to biodiversity are habitat loss, degradation, and fragmentation; the spread of invasive species; overharvesting; and climate change. For some species, disease poses a threat, and for others, particularly aquatic species, pollution is a particular threat. Many freshwater mussel species of North America are threatened both by pollution and by the invasion of the zebra mussel. The Pyrenean ibex was driven extinct by hunting, climate change, disease, and competition with domesticated species. Many other examples are possible.

2. DNA profiling (see Ecological Toolkit 23.1) and other genetic analyses are used to understand and manage genetic diversity in rare species; genetic approaches are also used in forensic studies of illegally harvested organisms. Conservation biologists use population viability analysis (PVA) models to assess extinction risk and evaluate options for managing rare species. Finally, ex situ conservation can be used to rescue species on the brink of extinction, as illustrated by ongoing efforts to protect the California condor.

3. The classification system set up by Natural Heritage/NatureServe documents each species' conservation status from a biological perspective, while a listing under the U.S. Endangered Species Act is a legal designation. While federally endangered species would generally also be considered globally rare by Natural Heritage/NatureServe, the reverse does not necessarily hold true: many extremely rare or threatened species are not on the federal endangered species list. The Endangered Species Act (ESA) provides legal protection for listed species, and it requires the designation of critical habitat and the development and implementation of a recovery plan for those species. In contrast, Natural Heritage/NatureServe can only recommend the protection of species.

4. Answers to this question will depend on where students are located and what species they identify. The objective of this question is to make students aware of species of conservation concern, threats to biodiversity, and efforts that are under way to protect species in their own region. It also invites them to identify research needs and to think about scientific approaches to conservation.

Answers to Hone Your Problem-Solving Skills Questions

1.

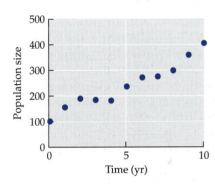

2. At this level of harvesting the population would decline through time and would not be sustainable.

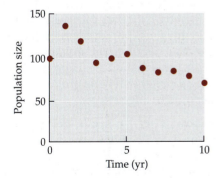

3. At this level of harvest the population size remains the same and is thus sustainable.

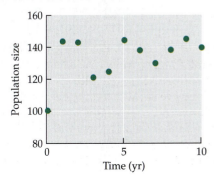

Chapter 24

Answers to Figure Legend Questions

FIGURE 24.3   Wet calcareous loam.

FIGURE 24.4   Organisms move more freely across the matrix in (B). We can infer this because exchange occurs between habitat patches separated by matrix in (B) whereas it does not occur in (A) (unless patches are connected to one another by a corridor).

FIGURE 24.6   It is identical to the grain in all three panels of part (B)—they each have a pixel size of 50×50 m.

FIGURE 24.17   *Reserve size*: A reserve that covers a small area typically harbors small populations—and small populations are at greater risk than larger ones from genetic factors (genetic drift and inbreeding), demographic stochasticity, environmental stochasticity, and natural catastrophes (see Chapter 10). In addition, a smaller proportion of the area is exposed to edge effects in a large reserve than in a small reserve; in a very small reserve, the entire area may be exposed to edge effects. *Number of reserves*: Although the total protected area is the same for both designs, in the design on the right each reserve is small in area and hence is likely to be at risk from problems associated with small populations. *Reserve proximity*: When several reserves are close to one another, individuals can move more freely between them. These movements help to prevent each reserve from experiencing problems associated with small population sizes. *Reserve connectivity*: Habitat corridors enable organisms to cross boundaries or landscape elements that otherwise might isolate each reserve from the other reserves (thereby exposing each reserve to problems associated with small population sizes). *Reserve shape*: When two reserves of equal area are compared, the reserve with a more compact shape (the best possible shape being a circle) will have proportionately less of its area exposed to edge effects.

Answers to Analyzing Data 24.1 Questions

1. The edge effect of increased wind disturbance penetrates 400 m into the forest; thus, for a tree not to experience increased wind disturbance it must be more than 400 m from the edge.

2. The total area of the forest is 800 m $\times$ 800 m = 640,000 m^2. Since we assume that the tree mortality effect penetrates 300 m on each side of the forest, the only region that does not experience a rise in tree mortality is a 200-$\times$ 200-m section in the center of the forest. This central section has an area of 40,000 m^2. Thus, the area experiencing a rise in tree mortality is 640,000 m^2 – 40,000 m^2 = 600,000 m^2, or 93.75% of the forest's total area.

3. The edge effects shown in the graph include changes to the abiotic conditions (such as increased wind disturbance and increased air temperature) and changes to aspects of the biotic environment (such as invasion of disturbance-adapted beetles and plants). By changing both abiotic and biotic components of the environment, other aspects of the environment not shown in the graph are likely to change as well. We would expect, for example, that the changing abiotic conditions could cause the abundance of some species originally present to decline, while others might increase. As we have seen throughout the textbook, such changes in abundance could lead to further changes in species interactions, community structure, and ecosystem processes (such as nutrient cycling).

Answers to Review Questions

1. Habitat islands resemble actual islands in the way that they spatially isolate populations of some species from one another, with potential demographic and genetic consequences. They differ from islands, however, in that the matrix between habitat fragments may be more or less permeable to some species, so that movement between habitat fragments may be constrained, but may still occur with some frequency. As we saw in Chapter 18, the principles of island biogeography apply to habitat islands in that there is immigration to fragments, extinction within fragments, and some equilibrium level of species diversity. Larger habitat islands can sustain greater species diversity than smaller fragments.

2. In a sense, corridors are long, skinny habitat patches. Animals may nest in them, plants will germinate in them if conditions are right, and predation and competition occur in them. But they are likely to be biologically impoverished relative to larger habitat blocks because of the effects of their narrow dimensions on their abiotic and biotic properties. They are likely to resemble edge habitat in experiencing more light, more rapid biogeochemical cycling, and more predation than larger habitat blocks. They may be more vulnerable to invasive species, and they may permit movement of diseases between habitat blocks. Nevertheless, they are generally beneficial, at least for some species, in allowing movement of organisms across a fragmented landscape.

3. National forests and national parks have different management objectives. The difference in the resulting land uses is visible from space, in the form of a clear line separating clear-cut patches of the Targhee National Forest from the uncut forests of Yellowstone National Park. National forests permit the harvesting of timber, which is generally not permitted in national parks. Timber harvesting makes for a patchy forest of different-aged stands, which may support a

different group of species than is found in a national park and may favor early successional species over old-growth–associated species. While both national parks and national forests have a mandate to protect biodiversity, national parks must balance these aims with recreation and visitor needs, while national forests must include timber production needs in their mission as well. Under an ecosystem management approach, the emphasis would be regional, and so the national forest and national park administrations would be working together to achieve conservation goals set by consensus.

Answers to Hone Your Problem-Solving Skills Questions

1. Species A is relatively insensitive to shape until the ratio of perimeter to area is very high, and therefore species A may be found in both reserve designs. In contrast, species B is more sensitive to a large amount of edge, so it has a higher probability of occurrence in patches with lower perimeter to area ratio, and thus design 1 would be best. Species C does best with intermediate perimeter to area ratio and thus would do better with design 2, which has more edge than design 1.

2. Greater food availability at the edges, such as the occurrence and abundance of food plants or prey that thrive in edge environments, would enhance the occurrence of a species there, while lower food availability would have the opposite effect. Lower diversity of species in patches with extensive edges may decrease food availability. Edges may enhance detection of predators or prey and also provide greater habitat for escape of predation. The physical environment at edges may be more extreme than at the core habitat in the patches, excluding some species. Some species may need extensive core habitat to protect and rear their young and so avoid patches with large amounts of edge.

Chapter 25

Answers to Figure Legend Questions

FIGURE 25.3 Deforestation would immediately lower the flux of carbon from the atmosphere to the land surface due to photosynthesis, but would increase the flux from the land surface to the atmosphere due to respiration. In other words, the deforested land would change from a sink to a source of atmospheric CO_2. Cutting the trees removes the most important autotrophs in the system. It also supplies carbon (from roots and woody debris) to soil heterotrophs and warms up the soil, both of which increase respiratory C emissions to the atmosphere.

FIGURE 25.7 Reactive N is chemically and biologically active, as the name implies. As a result, the pool of reactive N is a potential source of nutrients for organisms. In addition, it can influence soil chemistry and the health of organisms, as we will see later in the chapter. N_2, on the other hand, is chemically inert and must be converted to other chemical forms by nitrogen fixation to be used by organisms.

FIGURE 25.16 In Chapters 3, 16, and 17 we discussed several factors that determine the makeup of vegetation assemblages. These factors include physiological tolerances, biotic interactions such as competition and herbivory, and dispersal ability. Following deglaciation, combinations of temperature and precipitation different from any found today occurred in parts of North America, which resulted in unique combinations of plants relative to those that occur today. In addition, by differentially consuming specific plant species, particular species of herbivores can have an effect on vegetation types. As noted in the Case Study in Chapter 3, the animals that occurred at this time were quite different from those found today, including sloths, mastodons, and camels. Finally, the rates at which different species dispersed into the newly exposed substrate would have influenced the composition of the vegetation.

Answers to Analyzing Data 25.1 Questions

1. There is around a 0.05 drop in pH over the two-decade period of observation. Thus, between 2000 (pH = 8.10) and 2100, the pH should drop about 0.25 units (10 decades × 0.025 pH units/decade), for an estimated ocean pH of 7.85. The IPCC estimate is lower, due in part to the assumption of a continued increase in the rate of anthropogenic CO_2 emissions from fossil fuels.

2. Both the IPCC and empirically derived estimates for ocean pH in 2050 and 2100 are around 7.9 and 7.75, respectively. The results in Figure B indicate around a 90% decrease in abundance and a 75% decrease in species richness by 2050, and extinction of foraminiferans by 2100.

Answers to Review Questions

1. The two major biological influences on the global carbon cycle are photosynthesis, which takes up CO_2 from the atmosphere, and respiration, which releases CO_2 back to the atmosphere. Prior to the Industrial Revolution, uptake by photosynthesis and release by respiration were roughly equal at a global scale, and thus there was no net flux associated with Earth's biota. However, increasing human population growth rates resulted in increasing deforestation and agricultural development, which in turn resulted in greater decomposition and heterotrophic respiration due to warming of the soil surface. As a result, atmospheric CO_2 concentrations increased. Deforestation was the primary reason for increasing atmospheric CO_2 concentrations until the early part of the twentieth century.

2. While animals can respond to climate change by moving, their habitats cannot. Animals are dependent on plants to provide their food (or food for their prey). Climate change will be so rapid that evolutionary responses will not be possible for most species of plants, and the dispersal rates of most plant species are too slow to track the predicted climate changes. Dispersal may be inhibited by fragmentation of dispersal corridors due to land-use change. Loss of habitat will therefore result in decreased population growth for some animals. Additionally, migrating animals may respond to climate change more slowly than nonmigratory species. As a result, prey species may be less abundant or absent when these animals arrive at their destination.

3. The effect of atmospheric ozone on organisms depends on where in the atmosphere it is found. Ozone in the stratosphere acts as a shield against high-energy ultraviolet-B radiation, which is harmful to organisms. In contrast, ozone in the troposphere damages organisms that come in direct contact with it. Ozone in the troposphere also acts as a greenhouse gas, contributing to global climate change.

Answers to Hone Your Problem-Solving Skills Questions

1. 15 kg N/ha/yr for 20 years = 300 kg N/ha. Spread over 13,000,000 km^2 (1.3×10^9 ha), this is 3.9×10^{11} kg N. If 10% of this is taken up, then there is 3.9×10^{10} kg N incorporated into plant biomass. With a 500:1 ratio of C:N, that would be 7.8×10^7 kg C or 7.8×10^{10} g C.

2. 5 kg N/ha/yr for 20 years = 100 kg N/ha. Spread over 19,000,000 km^2 (1.9×10^9 ha), this is 1.9×10^{11} kg N. If 10% of this is taken up, then there is 1.9×10^{10} kg N incorporated into plant biomass. With a 500:1 ratio of C:N, that would be 3.8×10^7 kg C or 3.8×10^{10} g C.

3. On an annual basis, the greater C uptake would equal (7.8×10^{10} g C + 3.8×10^{10} g C)/ 20 years is equal to 5.8 $\times 10^9$ g C/yr. Annually, anthropogenic emissions are 10.4 Pg C, or 1.04×10^{16} g C, so the increased sequestration due to N deposition would be only a 0.00006% increase in C uptake.

Glossary

Numbers in brackets refer to the chapter(s) where the term is introduced.

A

abiotic   Of or referring to the physical or nonliving environment. *Compare* biotic. [1]

absolute population size   The actual number of individuals in a population. *Compare* relative population size. [9]

abundance   The number of individuals of a species that are found in a given area; abundance is often measured by population size or population density. [9]

acclimatization   An organism's adjustment of its physiology, morphology, or behavior to lessen the effect of an environmental change and minimize the associated stress. [4]

acid neutralizing capacity   The ability of the chemical environment to counteract acidity, usually associated with concentrations of base cations, including Ca^{2+}, Mg^{2+}, and K^+. [25]

acidity   A measure of the ability of a solution to behave as an acid, a compound that releases protons (H^+) to the water in which it is dissolved. *Compare* alkalinity. [2]

adaptation   (1) A physiological, morphological, or behavioral trait with an underlying genetic basis that enhances the survival and reproduction of its bearers in their environment. (2) *See* adaptive evolution. [1, 4]

adaptive evolution   A process of evolutionary change in which traits that confer survival or reproductive advantages tend to increase in frequency in a population over time. [6, 8]

adaptive management   A component of ecosystem management in which management actions are seen as experiments and future management decisions are determined by the outcome of present decisions. [24]

adaptive radiation   An event in which a group of organisms gives rise to many new species that expand into new habitats or new ecological roles in a relatively short time. [6]

aerosols   Solid or liquid particles suspended in the atmosphere. [22]

age structure   The proportion of individuals of a population in each age class. [11]

albedo   The amount of solar radiation reflected by a surface, usually expressed as a percentage of the incoming solar radiation. [2]

alkalinity   A measure of the ability of a solution to behave as a base, a compound that takes up protons (H^+) or releases hydroxide ions (OH^-). *Compare* acidity. [2]

Allee effect   A decrease in the population growth rate (r or λ) as the population density decreases. [10]

allele   One of two or more forms of a gene that result in the production of different versions of the protein that the gene encodes. [6]

allelopathy   A mechanism of competition in which individuals of one species release chemicals that harm individuals of other species. [14]

allocation   The relative amounts of energy or resources that an organism devotes to different functions. [7]

allochthonous inputs   Inputs produced outside the ecosystem. [21]

alluvium   Sediment deposited by flowing water. [22]

alpha diversity   Species diversity at the local or community scale. *Compare* beta diversity, gamma diversity. [18]

alternation of generations   A complex life cycle, found in many algae and all plants, in which there is both a multicellular diploid form, the sporophyte, and a multicellular haploid form, the gametophyte. [7]

alternative stable states   Different community development scenarios, or community states, that are possible at the same location under similar environmental conditions. [17]

amensalism   A species interaction in which individuals of one species are harmed while individuals of the other species do not benefit and are not harmed (–/0 relationship). [12, 14]

anisogamy   Production of two types of gametes of different sizes. *Compare* isogamy. [7]

anthropogenic   Of, relating to, or caused by humans or their activities. [25]

arbuscular mycorrhizae   Mycorrhizae in which the fungal partner grows into the soil, extending some distance away from the plant root, and also grows between some root cells while penetrating others. *Compare* ectomycorrhizae. [15]

Arctic ozone dent   An area of the stratosphere over the Arctic region where ozone concentrations are low but have not dropped below 220 Dobson units. [25]

assimilation efficiency   The proportion of ingested food that is assimilated by an organism. [21]

atmospheric deposition   The movement of particulate and dissolved matter from the atmosphere to Earth's surface by gravity or in precipitation. [22]

atmospheric pressure   The pressure exerted on a surface due to the mass of the atmosphere above it. [2]

autochthonous energy   Energy produced within the ecosystem. [21]

autotroph   An organism that converts energy from sunlight or from inorganic chemical compounds in the environment into chemical energy stored in the carbon–carbon bonds of organic compounds. *Compare* heterotroph. [5]

avoidance A response to stressful environmental conditions that lessens their effect through some behavior or physiological activity that minimizes an organism's exposure to the stress. *Compare* tolerance. [4]

B

behavioral ecology The study of the ecological and evolutionary basis of animal behavior. [8]

benthic zone The bottom of a body of water, including the surface and shallow subsurface layers of sediment. [3]

beta diversity The change in species diversity and composition, or turnover of species, from one community type to another across the landscape. *Compare* alpha diversity, gamma diversity. [18]

bioaccumulation A progressive increase in the concentration of a substance in an organism's body over its lifetime. [21]

biocrust *See* biological soil crust.

biodiversity The diversity of important ecological entities that span multiple spatial scales, from genes to species to communities. [16, 23]

biogeochemistry The study of the physical, chemical, and biological factors that influence the movements and transformations of chemical elements. [22]

biogeographic region A portion of Earth containing a distinct biota that differs markedly from the biotas of other biogeographic regions in its species composition and diversity. [18]

biogeography The study of variation in species composition and diversity among geographic locations. [18]

biological reserve An often small nature reserve established with the conservation of a single species or ecological community as the main conservation objective. [24]

biological soil crust A crust on the soil surface composed of a mix of species of cyanobacteria, lichens, and mosses; also called a biocrust. [22]

biomagnification A progressive increase in the tissue concentrations of a substance in animals at successively higher trophic levels that results as animals at each trophic level consume prey with higher concentrations of the substance due to bioaccumulation. [21]

biomass The mass of living organisms, usually expressed per unit of area. [20]

biome A large-scale terrestrial biological community shaped by the regional climate, soil, and disturbance patterns where it is found, usually classified by the growth form of the dominant plants. [3]

biosphere The highest level of biological organization, consisting of all living organisms on Earth plus the environments in which they live; located between the lithosphere and the troposphere. [1, 3]

biotic Of or referring to the living components of an environment. *Compare* abiotic. [1]

biotic resistance Interactions of the native species in a community with non-native species that exclude or slow the growth of those non-native species. [19]

bottom-up control Limitation of the abundance of a population by nutrient supply or by the availability of food. *Compare* top-down control. [10]

boundary layer A zone close to a surface where a flow of fluid, usually air, encounters resistance and becomes turbulent. [4]

buffer zone A portion of a nature reserve surrounding a core natural area where controls on land use are less stringent than in the core natural area, yet land uses are at least partially compatible with many species' resource requirements. *Compare* core natural area. [24]

C

C₃ photosynthetic pathway
A biochemical pathway involving the uptake of CO_2 by the enzyme ribulose 1,5 bisphosphate carboxylase/oxygenase (rubisco) and synthesis of sugars by the Calvin cycle. *Compare* C_4 photosynthetic pathway, crassulacean acid metabolism. [5]

C₄ photosynthetic pathway
A biochemical pathway involving the daytime uptake of CO_2 by the enzyme phosphoenol pyruvate carboxylase (PEPcase) in mesophyll cells; the carbon is then transferred as a four-carbon acid to the bundle sheath cells, where CO_2 is released to the Calvin cycle for sugar synthesis. *Compare* C_3 photosynthetic pathway, crassulacean acid metabolism. [5]

Calvin cycle The biochemical pathway used by photosynthetic and chemosynthetic organisms to fix carbon and synthesize sugars. [5]

carnivore An animal predator that kills and consumes tissues or fluids of live animals. [12, 20]

carnivory A trophic species interaction in which the predator is an animal (carnivore) and the prey is an animal. [12]

carrying capacity The maximum population size that can be supported indefinitely by the environment, represented by the term K in the logistic equation. [10, 11]

catchment The area in a terrestrial ecosystem that is drained by a single stream; a common unit of study in terrestrial ecosystem studies; also called a watershed. [22]

cation exchange capacity
A soil's ability to hold nutrient cations such as Ca^{2+}, K^+, and Mg^{2+} and exchange them with the soil solution, determined by the clay content of the soil. [22]

character displacement A process in which competition causes the phenotypes of competing species to evolve to become more different over time, thereby easing competition. [14]

cheater In a mutualism, an individual that increases its production of offspring by overexploiting its mutualistic partner. [15]

chemical weathering The chemical breakdown of soil minerals leading to the release of soluble forms of nutrients and other elements. *Compare* mechanical weathering. [22]

chemosynthesis The use of energy from inorganic chemical compounds to fix CO_2 and produce carbohydrates using the Calvin cycle; also called chemolithotrophy. [5]

clay Fine soil particles (<2 μm) that have a semicrystalline structure and weak negative charges on their surfaces that can hold onto cations and exchange them with the soil solution. [22]

climate The long-term description of weather, based on averages and variation measured over decades. *Compare* weather. [2]

climate change Directional change in climate over a period of three decades or longer. [1, 25]

climate envelope   The range of climate variables, including temperature, humidity, precipitation, and solar radiation, that are associated with a species geographic distribution. [4]

climax stage   The last stage of succession that is thought to be stable until disturbances or stresses shift the community back to earlier successional stages. [17]

cline   A pattern of gradual change in a characteristic of an organism over a geographic region. [6]

clone   A genetically identical copy of an individual. [9]

clumped dispersion   A dispersion pattern in which individuals are grouped together. *Compare* random dispersion, regular dispersion. [9]

coevolution   The evolution of two interacting species, each in response to selection pressure imposed by the other. [13]

cohort life table   A life table in which the fate of a group of individuals born during the same time period (a cohort) is followed from birth to death. [11]

commensalism   A species interaction in which individuals of one species benefit while individuals of the other species do not benefit and are not harmed (+/0 relationship). [12]

community   A group of interacting species that occur together at the same place and time. [1, 15, 16]

community function   The set of processes that shape community structure, including primary production, atmospheric gas exchange, or resistance to disturbance. [19]

community stability   *See* stability. [19]

community structure   The set of characteristics that shape a community, including the number, composition, and abundance of species. [16]

compensation   An adaptive growth response of plants to herbivory in which removal of plant tissues stimulates the plant to produce new tissues. [12]

competition   A non-trophic interaction in which individuals of the same species (intraspecific) or different species (interspecific) are harmed by their shared use of a resource that limits their ability to grow, reproduce, or survive (−/− relationship). [12, 14]

competition coefficient   A constant used in the Lotka–Volterra competition model to describe the extent to which an individual of one competing species decreases the per capita growth rate of the other species. [14]

competitive coexistence   The ability of two or more species to coexist with one another despite competing for the same limiting resources. [14]

competitive displacement   A process in which the best competitor uses limiting resources that the weaker competitor requires ultimately causing a decline in the weaker competitor's population growth to the point of extinction. [19]

competitive exclusion   *See* competitive displacement and competitive exclusion principle. [14]

competitive exclusion principle   The principle that two species that use a limiting resource in the same way cannot coexist indefinitely. [14]

competitive networks   Sets of competitive interactions involving multiple species in which every species negatively interacts with every other species, thus promoting species coexistence. [16]

competitive plants   In Grime's triangular model, plants that are superior competitors under conditions of low stress and low disturbance. *Compare* ruderals, stress-tolerant plants. [7]

complementarity hypothesis   A hypothesis proposing that as the species richness of a community increases, there is a linear increase in the positive effects of those species on community function. *Compare* idiosyncratic hypothesis and redundancy hypothesis. [19]

complex life cycle   A life cycle in which there are at least two distinct stages that differ in their habitat, physiology, or morphology. [7]

conduction   The transfer of sensible heat through the exchange of kinetic energy between molecules due to a temperature gradient. *Compare* convection. [2]

conservation biology   The scientific study of phenomena that affect the maintenance, loss, and restoration of biodiversity. [23]

consumer   An organism that obtains its energy by eating other organisms or their remains. *Compare* producer. [1]

consumption efficiency   The proportion of the biomass available at one trophic level that is ingested by consumers at the next trophic level. [21]

continental climate   The climate typical of terrestrial areas in the middle of large continental land masses at high latitudes, characterized by high variation in seasonal temperatures. *Compare* maritime climate. [2]

continental drift   The slow movement of tectonic plates (sections of Earth's crust) across Earth's surface. [18]

controlled experiment   A standard scientific approach in which an experimental group (that has the factor being tested) is compared with a control group (that lacks the factor being tested). [1]

convection   The transfer of sensible heat through the exchange of air and water molecules as they move from one area to another. *Compare* conduction. [2]

convergence   The evolution of similar growth forms among distantly related species in response to similar selection pressures. [3]

core natural area   A portion of a nature reserve where the conservation of biodiversity and ecological integrity takes precedence over other values or uses. *Compare* buffer zone. [24]

Coriolis effect   The apparent deflection of air or water currents when viewed from a rotating reference point such as Earth's surface. [2]

crassulacean acid metabolism (CAM)   A photosynthetic pathway in which CO_2 is fixed and stored as an organic acid at night, then released to the Calvin cycle during the day. *Compare* C_3 photosynthetic pathway, C_4 photosynthetic pathway. [5]

crypsis   A defense against predators in which prey species have a shape or coloration that provides camouflage and allows them to avoid detection. [12]

D

decomposition   The physical and chemical breakdown of detritus by detritivores, leading to the release of nutrients as simple, soluble organic and inorganic compounds that can be taken up by other organisms. [22]

delayed density dependence   A pattern of population growth in which delays in the effect of population density on population size can contribute to population fluctuations. [10]

demographic stochasticity Chance events associated with whether individuals survive or reproduce. [10]

denitrification A process by which certain bacteria convert nitrate (NO_3^-) into nitrogen gas (N_2) and nitrous oxide (N_2O) under hypoxic conditions. [22]

density-dependent A factor that causes birth rates, death rates, or dispersal rates to change as the density of a population changes. *Compare* density-independent. [11]

density-independent A factor whose effects on birth and death rates are independent of population density. *Compare* density-dependent. [11]

desertification Degradation of formerly productive land in arid regions resulting in loss of plant cover and acceleration of soil erosion. [3]

detritivore A heterotroph that consumes detritus. [20]

detritus Freshly dead or partially decomposed remains of organisms. [3, 5]

dilution effect A phenomenon in which the chance that any particular member of a group is the one attacked (as by a predator) decreases as the number of individuals in the group increases. [8]

direct development A simple life cycle that goes directly from fertilized egg to juvenile without passing through a free-living larval stage. [7]

direct interaction An interaction that occurs between two species, such as predation, competition, or a positive interaction. *Compare* indirect interaction. [16]

directional selection Selection that favors individuals with one extreme of a heritable phenotypic trait. *Compare* disruptive selection, stabilizing selection. [6]

dispersal The movement of organisms or propagules. [7, 9]

dispersal limitation A situation in which the limited capacity for dispersal of a species prevents it from reaching areas of suitable habitat. [9]

dispersion The spatial arrangement of individuals within a population. [9]

disruptive selection Selection that favors individuals with a phenotype at either extreme over those with an intermediate phenotype. *Compare* directional selection, stabilizing selection. [6]

distribution The geographic area where individuals of a species are present. [9]

disturbance An abiotic event that kills or damages some individuals and thereby creates opportunities for other individuals to grow and reproduce. [9, 17]

dormancy A state in which little or no metabolic activity occurs. [4]

doubling time (t_d) The number of years it takes a population to double in size. [11]

downwelling The sinking of deep ocean water from the surface. [2]

dynamic equilibrium model An elaboration of the intermediate disturbance hypothesis proposing that species diversity is maximized when the level of disturbance and the rate of competitive displacement are roughly equivalent. [19]

E

ecological footprint The total area of productive ecosystems required to support a population. [11]

ecology The scientific study of interactions between organisms and their environment. [1]

ecosystem All the organisms in a given area as well as the physical environment in which they live; an ecosystem can include one or more communities. [1, 20]

ecosystem management An approach to habitat management in which scientifically based policies and practices guide decisions on how best to meet an overarching goal of sustaining ecosystem structure and function for long periods. [24]

ecosystem services Natural processes that sustain human life and that depend on the functional integrity of natural communities and ecosystems. [23]

ecotype A population with adaptations to unique local environmental conditions. [4]

ectomycorrhizae Mycorrhizae in which the fungal partner typically grows between plant root cells and forms a mantle around the exterior of the root. *Compare* arbuscular mycorrhizae. [15]

ectoparasite A parasite that lives on the surface of another organism. *Compare* endoparasite. [13]

ectotherm An animal that regulates its body temperature primarily through energy exchange with its external environment. *Compare* endotherm. [4]

edge effects Abiotic and biotic changes that are associated with an abrupt habitat boundary such as that created by habitat fragmentation. [24]

effective population size The number of individuals in a population that can contribute offspring to the next generation. [10]

El Niño Southern Oscillation (ENSO) An oscillation of pressure cells and sea surface temperatures in the equatorial Pacific Ocean that causes widespread climatic variation and changes in upwelling currents. [2]

emigration The movement of individuals out of an existing population. [9]

endemic A species that occurs in a particular geographic location and nowhere else on Earth. [9, 18]

endoparasite A parasite that lives inside the body of its host organism. *Compare* ectoparasite. [13]

endotherm An animal that regulates its body temperature primarily through internal metabolic heat generation. *Compare* ectotherm. [4]

environmental science An interdisciplinary field of study that incorporates concepts from the natural sciences (including ecology) and the social sciences (e.g., politics, economics, ethics), focused on how people affect the environment and how we can address environmental problems. [1]

environmental stochasticity Erratic or unpredictable changes in the environment. [10]

epilimnion The warm surface layer of water in a lake, lying above the thermocline, that forms during the summer in some lakes of temperate and polar regions. *Compare* hypolimnion. [2]

equilibrium theory of island biogeography A theory proposing that the number of species on an island or in an island-like habitat results from a dynamic balance between immigration rates and extinction rates. [18]

eutrophic Nutrient-rich; characterized by high primary productivity. *Compare* oligotrophic, mesotrophic. [22]

eutrophication A change in the nutrient status of an ecosystem from nutrient-poor to nutrient-rich; such changes occur naturally, but they may also be caused by nutrient inputs that result from human activities. [10, 22]

evapotranspiration   The sum of water loss through evaporation and transpiration. [2]

evolution   (1) Change in allele frequencies in a population over time. (2) Descent with modification; the process by which organisms gradually accumulate differences from their ancestors. [6]

evolutionary tree   A branching diagram that represents the evolutionary history of a group of organisms. [6]

exploitation competition
An interaction in which species compete indirectly through their mutual effects on the availability of a shared resource. *Compare* interference competition. [14]

exponential growth   The change in the number of individuals within a population over time that is characterized by continuous birth and death at each instance in time. *Compare* geometric growth. [10, 11]

exponential growth rate (*r*)
A constant whose value is determined by the per capita (or per individual) birth rate minus the per capita death rate at each instance in time; also called the intrinsic rate of increase. *Compare* geometric population growth rate. [11]

extent   In landscape ecology, the area or time period encompassed by a study; together with grain, extent characterizes the scale at which a landscape is studied. *Compare* grain. [24]

extinction vortex   A pattern in which a small population that drops below a certain size becomes even more vulnerable to the problems that threaten small populations and hence may decrease even further in size, perhaps spiraling toward extinction. [23]

F

facilitation   *See* positive interaction. [12, 15]

fecundity   The average number of offspring produced by a female while she is of age x (denoted F_x in a life table). [11]

Ferrell cell   A large-scale, three-dimensional pattern of atmospheric circulation in each hemisphere, located at mid-latitudes between the Hadley and polar cells. [2]

finite rate of increase   *See* geometric population growth rate. [11]

fitness   The genetic contribution of an organism's descendants to future generations. [7]

fixation   (1) The uptake of the gaseous form of a compound, including CO_2 in photosynthesis and N_2 in nitrogen fixation, by organisms for use in metabolic functions. [5] (2) With respect to the genetic composition of a population, an allele frequency of 100%. [6]

flagship species   A charismatic species that may be emphasized in conservation efforts because it helps to garner public support for a conservation project. [23]

focal species   One of a group of species selected as a priority for conservation efforts, chosen because its ecological requirements differ from those of other species in the group, thereby helping to ensure that as many different species as possible receive protection. [23]

food web   A diagram showing the connections between organisms and the food they consume. [16, 21]

foundation species   A species that has large, community-wide effects on the habitat or food of other species by virtue of its size or abundance. [16]

fugitive species   A species that can persist in an area only if disturbances occur regularly and must therefore disperse from one place to another as environmental conditions change. [14]

functional group   A subset of a community that includes species that function in similar ways, but do not necessarily use the same resources. *Compare* guild. [16]

fundamental niche   The full set of resources, along with other biotic and abiotic requirements, that are suitable for a species excluding the negative interactions with other species. [14]

G

gamma diversity   Species diversity at the regional scale; the regional species pool. *Compare* alpha diversity, beta diversity. [18]

gene flow   The transfer of alleles from one population to another via the movement of individuals or gametes. [6]

generation time (*G*)   The average age of the parents of all the offspring produced within the cohort. [11]

genet   A genetic individual, resulting from a single fertilization event; in organisms that can reproduce asexually, a genet may consist of multiple, genetically identical parts, each of which has the potential to function as an independent physiological unit. *Compare* ramet. [9]

genetic drift   A process in which chance events determine which alleles are passed from one generation to the next, thereby causing allele frequencies to fluctuate randomly over time; the effects of genetic drift are most pronounced in small populations. [6, 10]

genotype   The genetic makeup of an individual. [6]

geographic range   The entire geographic region over which a species is found. [9]

geometric growth   The change in the number of individuals within a population over time that is characterized by birth and death from one discrete time period to the next. *Compare* exponential growth. [11]

geometric population growth rate (λ)
A constant whose value is determined by the per capita (or per individual) birth rate minus the per capita death rate at each discrete time period; also called the finite rate of increase. *Compare* exponential growth rate. [11]

grain   In landscape ecology, the size of the smallest homogeneous unit of study (such as a pixel in a digital image), which determines the resolution at which a landscape is observed; together with extent, grain characterizes the scale at which a landscape is studied. *Compare* extent. [24]

gravitational potential   The energy associated with gravity. [4]

greenhouse effect   The warming of Earth by gases in the atmosphere that absorb and reradiate infrared energy emitted by Earth's surface. [2, 25]

greenhouse gases   Atmospheric gases that absorb and reradiate the infrared radiation emitted by Earth's surface, including water vapor (H_2O), carbon dioxide (CO_2), methane (CH_4), and nitrous oxide (N_2O). [2, 25]

gross primary production (GPP)
The amount of energy that autotrophs capture by photosynthesis and chemosynthesis per unit of time. *Compare* net primary production. [20]

guild A subset of a community that includes species that use the same resources, whether or not they are taxonomically related. *Compare* functional group. [16]

H

habitat corridor A relatively narrow patch that connects blocks of habitat and often facilitates the movement of species between those blocks. [24]

habitat degradation Anthropogenic change that reduces the quality of habitat for many, but not all, species. [23]

habitat fragmentation The breaking up of once continuous habitat into a complex pattern of spatially isolated habitat patches. [9]

habitat loss The outright conversion of an ecosystem to another use by human activities. [23]

habitat mutualism A mutualism in which one partner provides the other with shelter, a place to live, or favorable habitat. [15]

Hadley cell A large-scale, three-dimensional pattern of atmospheric circulation in each hemisphere in which air is uplifted at the equator and subsides at about 30°N and S. [2]

heat capacity The amount of energy required to raise the temperature of a substance. [2]

herbivore An animal predator that consumes, or partially consumes, the tissues or internal fluids of living plants or algae. [12, 20]

herbivory A trophic species interaction in which the predator is an animal (herbivore) and the prey is a plant or alga. [12]

heterotroph An organism that obtains energy by consuming energy-rich organic compounds made by other organisms. *Compare* autotroph. [1, 5]

hibernation Torpor lasting several weeks during the winter; a strategy that is possible only for animals that have access to enough food and can store enough energy reserves. [4]

horizons Layers of soil distinguished by their color, texture, and permeability. [22]

horizontal interactions Non-trophic interactions, such as competition and some positive interactions, that occur within a trophic level. [16]

host An organism on or within which an herbivore, parasite, or mutualist lives and feeds. [12, 13]

hypolimnion The densest, coldest water layer in a lake, lying below the thermocline. *Compare* epilimnion. [2]

hyporheic zone The portion of the substrate below and adjacent to a stream bed where water movement still occurs, either from the stream or from groundwater moving into the stream. [3]

hypothesis A possible answer to a question developed using previous knowledge or intuition. *See also* scientific method. [1]

hypoxic Of or relating to a condition of oxygen depletion, usually below a level that can sustain most animals. [2]

hysteresis The inability of a community that has undergone change to shift back to the original community type, even when the original conditions are restored. [17]

I

idiosyncratic hypothesis A hypothesis proposing that as the species richness of a community increases, community function will vary idiosyncratically as the result of some species having stronger effects on the community than others. *Compare* complementarity hypothesis and redundancy hypothesis. [19]

immigration The movement of individuals into an existing population. [9]

inbreeding Mating between related individuals; inbreeding tends to increase the frequency of homozygotes, including those that have two copies of a harmful allele. [10]

indirect interaction An interaction in which the relationship between two species is mediated by a third (or more) species. *Compare* direct interaction. [16]

induced defense In plant–herbivore interaction, a defense against herbivory, such as production of a secondary compound, that is stimulated by herbivore attack. [12]

interaction strength A measure of the effect of one species (the interactor) on the abundance of another species (the target species). [16]

interaction web A concept that describes both the trophic (vertical) and non-trophic (horizontal) interactions among the species in a traditional food web. [16]

interference competition An interaction in which species compete directly by performing antagonistic actions that interfere with the ability of their competitors to use a resource that both require, such as food or space. *Compare* exploitation competition. [14]

intermediate disturbance hypothesis A hypothesis proposing that species diversity in communities should be greatest at intermediate levels of disturbance (or stress or predation) because competitive exclusion at low levels of disturbance and mortality at high levels of disturbance should reduce species diversity. [19]

interspecific competition An interaction in which individuals of different species are harmed by their shared use of a resource that limits their ability to grow, reproduce, or survive (–/– relationship). *Compare* intraspecific competition. [14]

intertidal Referring to the portion of the shoreline that is affected by the rise and fall of the tides. [3]

Intertropical Convergence Zone (ITCZ) The zone of maximum solar radiation, atmospheric uplift, and precipitation within the tropical zone. [2]

intraspecific competition An interaction in which individuals of the same species are harmed by their shared use of a resource that limits their ability to grow, reproduce, or survive (–/– relationship). *Compare* interspecific competition. [14]

intrinsic rate of increase *See* exponential growth rate. [11]

invasive species An introduced species that survives and reproduces in its new environment, sustains a growing population, and has large effects on the native community. [23]

isocline The set of abundances for which the population growth rate (dN/dt) of one of the species involved in a species interaction is zero. [12, 14]

isogamy The production of equal-sized gametes. *Compare* anisogamy. [7]

isolation by distance A metapopulation pattern in which habitat patches located far from occupied patches are less likely to be colonized than are nearby patches. [9]

iteroparous Having the capacity to reproduce multiple times in a lifetime. *Compare* semelparous. [7]

K

K-selection In the *r–K* continuum used for classifying life history strategies, the selection pressure for slower rates of increase faced by organisms that live in environments where population densities are high (at or near the carrying capacity, *K*). *Compare* *r*-selection. [7]

keystone species A strong interactor species that has an effect on energy flow and community structure that is disproportionate to its small size, abundance, or biomass. [16]

L

land use change The alteration of terrestrial surface, including vegetation and landforms, by human activities such as agriculture, forestry, or mining. [3]

landscape An area that is spatially heterogeneous in one or more features of the environment, such as the number or arrangement of different habitat types; a landscape typically includes multiple ecosystems. [1, 24]

landscape composition In landscape ecology, the kinds of elements or patches comprised by a landscape and how much of each kind is present. *Compare* landscape structure. [24]

landscape ecology The study of landscape patterns and the effects of those patterns on ecological processes. [24]

landscape structure In landscape ecology, the physical configuration of the different compositional elements of a landscape. *Compare* landscape composition. [24]

lapse rate The rate at which atmospheric temperature decreases with increasing distance from the ground. [2]

latent heat flux Heat transfer associated with the phase change of water, such as evaporation, sublimation, or condensation. [2]

leaching The vertical movement of dissolved matter and fine mineral particles from upper to lower layers of soil. [22]

leaf area index The area of leaves per unit of ground area (a dimensionless number, since it is an area divided by an area). [20]

lentic Of or referring to still water. *Compare* lotic. [3]

life history The major events relating to an organism's growth, development, reproduction, and survival; these events include the age and size of first reproduction, the amount and timing of reproduction, and longevity. [7]

life history strategy The overall pattern in the timing and nature of life history events, averaged across all the individuals of a species. [7]

life table A summary of how survival and reproductive rates in a population vary with the age of individuals; in species for which age is not informative or is difficult to measure, life tables may be based on the size or life cycle stage of individuals. [11]

lignin A structural compound that strengthens plant tissues. [22]

line transect A straight line from which an observer can count the number of individuals and estimate their abundance per unit of area. [9]

litter Fresh, undecomposed organic matter on the soil surface. [22]

littoral zone The nearshore zone of a lake where the photic zone reaches to the bottom. [3]

local scale A spatial scale that is essentially equivalent to a community. [18]

loess Sediment deposited by wind. [22]

logistic growth The change in the number of individuals within a population over time that is rapid at first, then decreases as the population approaches the carrying capacity of its environment. [10, 11]

lotic Of or relating to flowing water. *Compare* lentic. [3]

Lotka–Volterra competition model A modified form of the logistic equation used to model interspecific competition. [14]

Lotka–Volterra predator–prey model A modified form of the logistic equation used to model predator–prey interaction cycles. [12]

lottery model A hypothesis proposing that species diversity in communities is maintained by a "lottery" in which resources made available by the effects of disturbance, stress, or predation are captured at random by recruits from a larger pool of potential colonists. [19]

lower critical temperature The environmental temperature at which the heat loss of an endotherm triggers an increase in metabolic heat generation. [4]

M

macroparasites Relatively large parasite species, such as arthropods and worms. *Compare* microparasites. [13]

macrophyte A rooted or floating aquatic vascular plant. [3]

marginal value theorem A conceptual optimal foraging model proposing that an animal should stay in a food patch until the rate of energy gain in that patch has declined to the average rate for the habitat, then depart for another patch. [8]

maritime climate The climate typical of coastal terrestrial regions that are influenced by an adjacent ocean, characterized by low daily and seasonal variation in temperature. *Compare* continental climate. [2]

mass extinction An event in which a large proportion of Earth's species are driven to extinction worldwide in a relatively short time. [6]

mating system The number of mating partners that males or females have and the pattern of parental care in which they engage. [8]

matric potential The energy associated with attractive forces on the surfaces of large molecules inside cells or on the surfaces of soil particles. [4]

mean residence time The amount of time an average molecule of an element spends in a pool before leaving it. [22]

mechanical weathering The physical breakdown of rocks into progressively smaller particles without chemical change. *Compare* chemical weathering. [22]

mesotrophic Having a nutrient status that is intermediate between oligotrophic and eutrophic, usually used in reference to lakes. *Compare* eutrophic, oligotrophic. [22]

metamorphosis An abrupt transition from a larval to a juvenile life cycle stage that is sometimes accompanied by a change in habitat. [7]

metapopulation A set of spatially isolated populations linked to one another by dispersal. [9]

microparasites Parasite species too small to be seen with the naked eye, such as bacteria, protists, and fungi. *Compare* macroparasites. [13]

migration The round-trip movement of an entire population. [9]

mimicry A defense against predators in which prey species resemble less palatable organisms or physical features of their environment, causing potential predators to mistake them for something less desirable to eat. [12]

mineralization The chemical conversion of organic matter into inorganic compounds. [22]

morphs Discrete phenotypes with few or no intermediate forms. [7]

mosaic The composite or pattern of the heterogeneous features of the environment in a landscape. [24]

mutation Change in the DNA of a gene. [6]

mutualism A mutually beneficial interaction between individuals of two or more species (+/+ relationship). [12]

mycorrhizae Symbiotic associations between plant roots and various types of fungi that are usually mutualistic. [15]

N

natural catastrophe An extreme environmental event such as a flood, severe windstorm, or outbreak of disease that can eliminate or drastically reduce population size. [10]

natural selection The process by which individuals with certain heritable characteristics tend to survive and reproduce more successfully than other individuals because of those characteristics. [1, 6]

nekton Swimming organisms capable of overcoming water currents. *Compare* plankton. [3]

net ecosystem exchange (NEE) The combined fluxes of CO_2 into and out of an ecosystem principally by net primary production and autotrophic and heterotrophic respiration. [20]

net primary production (NPP) The amount of energy per unit of time that producers capture by photosynthesis and chemosynthesis, minus the amount they use in cellular respiration. *Compare* gross primary production. [1, 20]

net reproductive rate (R_0) The mean number of offspring produced by an individual in a population during its lifetime. [11]

net secondary production The balance between heterotroph energy gains through ingestion and heterotroph energy losses by cellular respiration and egestion. [20]

neutral model *See* lottery model. [19]

niche partitioning *See* resource partitioning. [14, 19]

nitrification A process by which certain chemoautotrophic bacteria, known as nitrifying bacteria, convert ammonia (NH_3) and ammonium (NH_4^+) into nitrate (NO_3^-) under aerobic conditions. [22]

nitrogen fixation The process of taking up nitrogen gas (N_2) and converting it into chemical forms that are more chemically available to organisms. [22]

North Atlantic Oscillation An oscillation in atmospheric pressures and ocean currents in the North Atlantic Ocean that affects climatic variation in Europe, in northern Asia, and on the east coast of North America. [2]

non-trophic A species interaction that does not involve feeding such as competition and facilitation. [12]

nutrient A chemical element required by an organism for its metabolism and growth. [22]

nutrient cycle The cyclic movement of nutrients between organisms and the physical environment. [1, 22]

O

occlusion A process by which soluble phosphorus combines with iron, calcium, and aluminum to form insoluble compounds (secondary minerals) that are unavailable to organisms as nutrients. [22]

oligotrophic Nutrient-poor, characterized by low primary productivity. *Compare* eutrophic, mesotrophic. [22]

omnivore (1) An organism that feeds on both plants and animals. [20] (2) In trophic studies, an organism that feeds on more than one trophic level. [21]

optimal foraging A theory proposing that animals will maximize the amount of energy acquired per unit of feeding time. [8]

osmotic adjustment An acclimatization response to changing water availability or salinity in terrestrial and aquatic environments that involves changing the solute concentration, and thus the osmotic potential, of the cell. [4]

osmotic potential The energy associated with dissolved solutes. [4]

outbreak An extremely rapid increase in the number of individuals in a population. [10]

ozone hole An area of the stratosphere with an ozone concentration of less than 220 Dobson units (= 2.7 × 10^{16} molecules of ozone) per square centimeter; found primarily over the Antarctic region. [25]

P

Pacific Decadal Oscillation (PDO) A long-term oscillation in sea surface temperatures and atmospheric pressures in the North Pacific Ocean that has widespread climatic effects. [2]

paedomorphic Resulting from a delay of a developmental event relative to sexual maturation. [7]

parasite An organism that lives in or on a host organism and feeds on its tissues or body fluids. [12, 13]

parasitism A trophic species interaction in which a predator (parasite) lives and feeds on or in its prey (host) without necessarily killing it. [12]

parasitoid An insect that lays one or a few eggs on or in a host organism (itself usually an insect), which the resulting larvae remain with, eat, and almost always kill. [12, 13]

parent material The rock or sediments that are broken down by weathering to form mineral particles in soil. [22]

pathogen A parasite that causes disease. [12, 13]

pelagic zone The open water column of a lake or ocean. [3]

permafrost A subsurface soil layer that remains frozen year-round for at least 3 years. [3]

phenotype The observable characteristics of an organism. [6]

phenotypic plasticity The ability of a single genotype to produce different phenotypes under different environmental conditions. [7]

photic zone The surface layer of a lake or ocean where enough light penetrates to allow photosynthesis. [3]

photorespiration A chemical reaction in photosynthetic organisms in which the enzyme rubisco takes up O_2, leading to the breakdown of sugars, the release of CO_2, and a net loss of energy. [5]

photosynthesis A process that uses sunlight to provide the energy needed to take up CO_2 and synthesize sugars. [5]

physiological ecology The study of the interactions between organisms and the physical environment that influence their survival and persistence. [4]

phytoplankton Photosynthetic plankton. *Compare* zooplankton. [3]

pioneer stage The first stage of primary succession. [17]

plankton Small, often microscopic organisms that live suspended in water; although many plankton are mobile, none can swim strongly enough to overcome water currents. *Compare* nekton. [3]

polar cell A large-scale, three-dimensional pattern of atmospheric circulation in which air subsides at the poles, moves toward the equator when it reaches Earth's surface, and is replaced by air moving through the upper atmosphere from lower latitudes. [2]

polar zone The major climatic zone above 60°N and S. [2]

pool The total amount of a nutrient or other element found within a component of an ecosystem. [22]

population A group of individuals of the same species that live within a particular area and interact with one another. [1, 9]

population cycles A pattern of population fluctuations in which alternating periods of high and low abundance occur after nearly constant intervals of time. [10]

population density The number of individuals per unit of area. [9]

population fluctuations The most common pattern of population growth, in which population size rises and falls over time. [10]

population growth The change in the number of individuals within a population over time. [10]

population regulation A pattern of population growth in which one or more density-dependent factors increase population size when numbers are low and decrease population size when numbers are high. [11]

population size The number of individuals in a population. [9]

population viability analysis (PVA) Projection of the potential future status of a population through use of demographic models; a PVA approach is often used to estimate the likelihood that a population will persist for a certain amount of time in different habitats or under different management scenarios. [23]

positive interaction A trophic or non-trophic species interaction in which one or both species benefit and neither is harmed. *See also* mutualism, commensalism. [12, 15]

predation A trophic interaction in which an individual of one species, a predator, consumes individuals (or parts of individuals) of another species, its prey. [12]

predator An organism that consumes other organisms (or parts of organisms), referred to as its prey. [12]

prey An organism eaten by a predator. [12]

pressure (or turgor) potential The energy associated with the exertion of pressure; has a positive value if pressure is exerted on the system and a negative value if the system is under tension. [4]

primary production The rate at which chemical energy in an ecosystem is generated by autotrophs, derived from the fixation of carbon during photosynthesis and chemosynthesis. *Compare* secondary production. *See also* gross primary production, net primary production. [20]

primary succession Succession that involves the colonization of habitats devoid of life. *Compare* secondary succession. [17]

producer An organism that can produce its own food by photosynthesis or chemosynthesis; also called a primary producer or autotroph. *Compare* consumer. [1]

production efficiency The proportion of assimilated food that is used to produce new consumer biomass. [21]

proximate cause An immediate, underlying cause that is based on internal features of an organism and can be used to explain how a behavior (or other characteristic of the organism) occurs. *Compare* ultimate cause. [8]

pubescence The presence of hairs on the surface of an organism. [4]

R

r-selection In the *r–K* continuum used for classifying life history strategies, the selection pressure for high population growth rates faced by organisms that live in environments where population densities are usually low. *Compare* K-selection. [7]

rain-shadow effect The effect a mountain range has on regional climate by forcing moving air upward, causing it to cool and release precipitation on the windward slopes, resulting in lower levels of precipitation and soil moisture on the leeward slope. [2]

ramet An actual or potential physiologically independent member of a genet that may compete with other members for resources. *Compare* genet. [9]

random dispersion A dispersion pattern in which individuals are positioned at random. Compare clumped dispersion, regular dispersion. [9]

rank abundance curve A graph that plots the proportional abundance of each species in a community relative to the others in rank order, from most abundant to least abundant. [16]

realized niche The part of a fundamental niche that a species occupies as a result of species interactions. *Compare* fundamental niche. [14]

recombination Rearrangements of genetic material during sexual reproduction that result in the production of offspring that have combinations of alleles that differ from those in either of their parents. [6]

redundant A species that may have the same function in the community as other species within a larger functional group. [16]

regional scale A spatial scale that encompasses a geographic area where the climate is roughly uniform, and the species contained therein are often restricted to that region by dispersal limitation. [18]

regional species pool All the species contained within a region; sometimes called gamma diversity. [18]

regular dispersion A dispersion pattern in which individuals are relatively evenly spaced throughout their habitat. *Compare* clumped dispersion, random dispersion. [9]

relative population size An estimate of population size based on data that are related to the absolute population size, but which can be compared from one time period or place to another. *Compare* absolute population size. [9]

replication The performance of each treatment of a controlled experiment, including the control, more than once. [1]

rescue effect A metapopulation pattern in which high rates of immigration protect a population from extinction. [9]

resistance Any force that impedes the movement of compounds such as water or gases such as carbon dioxide along an energy or concentration gradient; its inverse is conductance. [4]

resource A feature of the environment, such as food, water, light, and space, that is required for growth, reproduction, or survival. [14]

resource partitioning The use of limiting resources by different species in a community in different ways. [14, 19]

resource ratio hypothesis A hypothesis proposing that species can coexist in a community by using the same resources, but in differing proportions. [19]

ruderals In Grime's triangular model, plants that are adapted to environments with high levels of disturbance and low levels of stress. *Compare* competitive plants, stress-tolerant plants. [7]

S

salinity The concentration of dissolved salts in water. [2]

salinization A process by which high rates of evapotranspiration in arid regions result in a progressive buildup of salts at the soil surface. [2]

sand The coarsest soil particles (0.05–2 mm). [22]

savanna A vegetation type dominated by grasses with intermixed trees and shrubs. [3]

scale The spatial or temporal dimension at which ecological observations are collected. [1, 24]

scientific method An iterative and self-correcting process by which scientists learn about the natural world, consisting of four steps: (1) observe nature and ask a question about those observations; (2) develop possible answers to that question (hypotheses); (3) evaluate competing hypotheses with experiments, observations, or quantitative models; (4) use the results of those experiments, observations, or models to modify the hypotheses, pose new questions, or draw conclusions. [1]

secondary compound A chemical compound in plants not used directly in growth, and often used in such functions as defense against herbivores or protection from harmful radiation. [12]

secondary production Energy in an ecosystem that is derived from the consumption of organic compounds produced by other organisms. *Compare* primary production. [20]

secondary succession Succession that involves the reestablishment of a community in which some, but not all, of the organisms have been destroyed. *Compare* primary succession. [17]

semelparous Reproducing only once in a lifetime. *Compare* iteroparous. [7]

sensible heat flux The transfer of heat through the exchange of energy by conduction or convection. [2]

sequential hermaphroditism A change or changes in the sex of an organism during the course of its life cycle. [7]

sexual selection A process in which individuals with certain characteristics have an advantage over others of the same sex solely with respect to mating success. [8]

Shannon index The index most commonly used to describe species diversity quantitatively. [16]

silt Intermediate-sized soil particles, often ranging in size between 0.05 and 0.002 mm. [22]

soil A mix of mineral particles, detritus, dissolved organic matter, water containing dissolved minerals and gases (the soil solution), and organisms that develops in terrestrial ecosystems. [22]

speciation The process by which one species splits into two or more species. [6]

species accumulation curve A graph that plots species richness as a function of the total number of individuals that are present with each additional sample. [16]

species–area relationship The relationship between species richness and area sampled. [18]

species composition The identity of the species present in a community. [16]

species distribution model A tool that predicts the geographic distribution of a species based on the environmental conditions at locations where the species is known to occupy. [9]

species diversity A measure that combines the number of species (species richness) in a community and their relative abundances compared with one another (species evenness). [16]

species evenness The relative abundances of different species compared to one another in a community. [16]

species richness The number of species in a community. [16]

stability When a community retains, or returns to, its original structure and function after some perturbation. [17]

stabilizing selection Selection that favors individuals with an intermediate phenotype. *Compare* directional selection, disruptive selection. [6]

static life table A life table that records the survival and reproduction of individuals of different ages during a single time period. [11]

stomate A pore in plant tissues, usually leaves, surrounded by specialized guard cells that control its opening and closing. [4]

stratification The layering of water in oceans and lakes due to differences in water temperature and density with depth. [2]

stress An abiotic factor that results in a decrease in the rate of an important physiological process, thereby lowering the potential for an organism's growth, reproduction, or survival; the condition caused by such a factor. [4]

stress-tolerant plants In Grime's triangular model, plants that are adapted to conditions of high stress and low disturbance. *Compare* competitive plants, ruderals. [7]

subsidence A sinking (downward) movement of air in the atmosphere, usually over a broad area, leading to the development of a high-pressure cell. *Compare* uplift. [2]

succession The process of change in the species composition of a community over time as a result of abiotic and biotic agents of change. [17]

surrogate species A species selected as a priority for conservation with the assumption that its conservation will serve to protect many other species with overlapping habitat requirements. [23]

survival rate The proportion of individuals that survive over time. [11]

survivorship (l_x) The proportion of individuals that survive from birth (age 0) to age x (denoted l_x in a life table). [11]

survivorship curve A graph based on survivorship data (l_x) that plots the numbers of individuals from a hypothetical cohort (typically, of 1,000 individuals) that will survive to reach different ages. [11]

symbiosis A relationship in which two species live in close physical and/or physiological contact with each other. *See also* host, symbiont. [12, 15]

T

taxonomic homogenization The worldwide reduction of biodiversity resulting from the spread of non-native and native generalists coupled with declining abundances and distributions of native specialists and endemics. [23]

temperate zone The major climatic zone between 30° and 60°N and S. [2]

territory An area that an animal defends against intruders. [8]

thermocline The zone of rapid temperature change in a lake beneath the epilimnion and above the hypolimnion. [2]

thermoneutral zone The range of environmental temperatures over which endotherms maintain a constant basal metabolic rate. [4]

threshold density The minimum number of individuals susceptible to a disease that must be present in a population for the disease to become established and spread. [13]

tides Patterns of rising and falling of ocean water generated by the gravitational attraction between Earth and the moon and sun. [3]

till Layers of sediment deposited by glaciers. [22]

tolerance The ability to survive stressful environmental conditions. *Compare* avoidance. [4]

top-down control Limitation of the abundance of a population by consumers. *Compare* bottom-up control. [10]

torpor A state of dormancy in which endotherms drop their lower critical temperature and associated metabolic rate. [4]

trade-off An organism's allocation of its limited energy or other resources to one structure or function at the expense of another. [6]

trophic A species interaction that involves feeding such as carnivory, herbivory, parasitism, and facilitation. [12]

trophic cascade A change in the rate of consumption at one trophic level that results in a series of changes in species abundance or composition at lower trophic levels. [16, 21]

trophic efficiency A measure of the transfer of energy between trophic levels, consisting of the amount of energy at one trophic level divided by the amount of energy at the trophic level immediately below it. [21]

trophic facilitation An interaction in which a consumer is indirectly facilitated by a positive interaction between its prey or food plant and another species. [16]

trophic interaction An interaction in which a predator consumes a prey. [12]

trophic level A group of species that obtain energy in similar ways, classified by the number of feeding steps by which the group is removed from primary producers, which are the first trophic level. [16, 21]

trophic mutualism A mutualism in which one or both of the mutualists receives energy or nutrients from its partner. [15]

trophic pyramid A common approach to conceptualizing trophic relationships in an ecosystem in which a stack of rectangles is constructed, each of which represents the amount of energy or biomass within one trophic level. [21]

tropical zone The major climatic zone between 25°N and S, encompassing the equator; also called the tropics. [2]

tropics *See* tropical zone. [2]

turgor pressure Pressure that develops in a plant cell when water moves into it, following a gradient in water potential. [4]

turnover (1) The mixing of the entire water column in a stratified lake when all the layers of water reach the same temperature and density. (2) The change in species diversity and composition from one community type to another across the landscape; *See* beta diversity. [2, 18]

type I survivorship curve A survivorship curve in which newborns, juveniles, and young adults all have high survival rates and death rates do not begin to increase greatly until old age. [11]

type II survivorship curve A survivorship curve in which individuals experience a constant chance of surviving from one age to the next throughout their lives. [11]

type III survivorship curve A survivorship curve in which individuals die at very high rates when they are young, but those that reach adulthood survive well later in life. [11]

U

ultimate cause The underlying evolutionary or historical reason for a particular behavior (or other characteristic of an organism). *Compare* proximate cause. [8]

umbrella species A surrogate species selected with the assumption that protection of its habitat will serve as an "umbrella" to protect many other species; often a species with large or specialized habitat requirements or one that is easy to count. [23]

uplift The rising of warm, less dense air in the atmosphere due to heating of Earth's surface. *Compare* subsidence. [2]

upwelling The rising of deep ocean waters to the surface. [2]

V

vicariance The evolutionary separation of species due to a barrier that results in the geographic isolation of species that once were connected to one another. [18]

W

warning coloration A defense against predators in which prey species that contain powerful toxins advertise those toxins with bright coloration; also called aposematic coloration. [12]

water potential The overall energy status of water in a system; the sum of osmotic potential, gravitational potential, turgor pressure, and matric potential. [4]

weather The temperature, humidity, precipitation, wind, and cloud cover at a particular time and place. *Compare* climate. [2]

weathering The physical and chemical processes by which rock minerals are broken down, eventually releasing soluble nutrients and other elements. [22]

Z

zooplankton Nonphotosynthetic plankton. *Compare* phytoplankton. [3]

Literature Cited

Numbers in brackets indicate the chapter(s) where this reference is cited.

A

Abatzoglou, J. T., and P. Williams. 2016. Impact of anthropogenic climate change across western U.S. forests. *Proceedings of the National Academy of Sciences USA* 113: 11770–11775. [25]

Aber, J., and 9 others. 1998. Nitrogen saturation in temperate forest ecosystems: Hypotheses revisited. *BioScience* 48: 921–934. [25]

Adey, W. H., and R. S. Steneck. 2001. Thermogeography over time creates biogeographic regions: A temperature/space/time-integrated model and an abundance-weighted test for benthic marine algae. *Journal of Phycology* 37: 677–698. [18]

Agrawal, A. A., and M. Fishbein. 2006. Plant defense syndromes. *Ecology* 87: S132–S149. [12]

Agrawal, A. A., M. T. J. Johnson, A. P. Hastings, and J. L. Maron. 2013. A field experiment demonstrating plant life-history evolution and its eco-evolutionary feedback to seed predator populations. *American Naturalist* 181: S35–S45. [6]

Ahlenius, H. 2007. UNEP/GRID-Arendal. http://maps.grida.no/go/graphic/world-ocean-thermohaline-circulation1. [2]

Ahrens, C. D. 2005. *Essentials of Meteorology*. Thomson Brooks/Cole: Boston, MA. [2]

Aiba, S., A. E. Humphrey, and N. F. Mills. 1973. *Biochemical Engineering*. 2nd ed. Academic Press: New York. [22]

Allan, B. F. and 15 others. 2009. Ecological correlates of risk and incidence of West Nile virus in the United States. *Oecologia* 158: 699–708. [1]

Allan, B. F., F. Keesing, and R. S. Ostfeld. 2003. Effect of forest fragmentation on Lyme disease risk. *Conservation Biology* 17: 267–272. [24]

Allen, E. B., P. J. Temple, A. Bytnerowicz, M. J. Arbaugh, A. G. Sirulnik, and L. E. Rao. 2007. Patterns of understory diversity in mixed coniferous forests of Southern California impacted by air pollution. *Scientific World Journal* 7: 247–263. [18]

Alliende, M. C., and J. L. Harper. 1989. Demographic studies of a dioecious tree. I. Colonization, sex, and age structure of a population of *Salix cinerea*. *Journal of Ecology* 77: 1029–1047. [11]

Allin, E. F., and J. A. Hopson. 1992. Evolution of the auditory system in Synapsida ("mammal-like reptiles" and primitive mammals) as seen in the fossil record. In *The Evolutionary Biology of Hearing*, D. B. Webster, R. R. Fay, and A. N. Popper (Eds.), pp. 587–614. Springer: New York. [6]

Allstadt, A. J., K. J. Haynes, A. M. Liebhold, and D. M. Johnson. 2013. Long-term shifts in the cyclicity of outbreaks of a forest-defoliating insect. *Oecologia* 172: 141–151. [10]

Alpert, P. 2006. Constraints of tolerance: Why are desiccation-tolerant organisms so small or rare? *Journal of Experimental Biology* 209: 1575–1584. [4]

Alroy, J. 2015. Current extinction rates of reptiles and amphibians. *Proceedings of the National Academy of Sciences USA* 112: 13003–13008. [1]

Alsterberg, C., J. S. Eklöf, L. Gamfeldt, J. N. Havenhand, and K. Sundbäck. 2013. Consumers mediate the effects of experimental ocean acidification and warming on primary producers. *Proceedings of the National Academy of Sciences USA* 110: 8603–8608. [16]

AmphibiaWeb. 2019. University of California, Berkeley, CA, USA. Accessed 25 September 2019. https://amphibiaweb.org/declines/declines.html. [1]

Anderegg, W. R. L., and 16 others. 2015. Tree mortality from drought, insects, and their interactions in a changing climate. *New Phytologist* 208: 674–683. [25]

Andersen, S. B., and 7 others. 2009. The life of a dead ant: The expression of an adaptive extended phenotype. *American Naturalist* 174: 424–433. [13]

Anderson, J. T., D. W. Inouye, A. M. McKinney, R. I. Colautti, and T. Mitchell-Olds. 2012. Phenotypic plasticity and adaptive evolution contribute to advancing flowering phenology in response to climate change. *Proceedings of the Royal Society of London* 279B: 3843–3852. [6]

Anderson, R. C. 1970. Prairies in the Prairie State. *Transactions of the Illinois State Academy of Science* 63: 214. [6]

Anderson, S. H., D. Kelly, J. J. Ladley, S. Molloy, and J. Terry. 2011. Cascading effects of bird functional extinction reduce pollination and plant density. *Science* 331: 1068–1071. [23]

Andersson, M. 1982. Female choice selects for extreme tail length in a widowbird. *Nature* 299: 818–820. [8]

Andrade, M. C. B. 1996 Sexual selection for male sacrifice in the Australian redback spider. *Science* 271: 70–72. [8]

Andrewartha, H. G., and L. C. Birch. 1954. *The Distribution and Abundance of Animals*. University of Chicago Press: Chicago, IL. [23]

Arcese, P., and J. N. M. Smith. 1988. Effects of population density and supplemental food on reproduction in song sparrows. *Journal of Animal Ecology* 57: 119–136. [11]

Arias, E. 2015. *United States Life Tables, 2011*. National Vital Statistics Reports, Vol. 64, No. 11. National Center for Health Statistics: Hyattsville, MD. www.cdc.gov/nchs/data/nvsr/nvsr64/nvsr64_11.pdf. [11]

Armas, C., R. Ordiales, and F. I. Pugnaire. 2004. Measuring plant interactions: A new comparative index. *Ecology* 85(10): 2682–2686.

Armitage, K. B., D. T. Blumstein, and B. C. Woods. 2003. Energetics in hibernating yellow belly marmots (*Marmota flaviventris*). *Comparative Biochemistry and Physiology* 134A: 101–114. [4]

Asner, G. P., T. K. Rudel, T. M. Aide, R. Defries and R. Emerson. 2009. A contemporary assessment of change in humid tropical forests. *Conservation Biology* 6: 1386–1395. [3]

Ausband, D. E. 2019. Pair bonds, reproductive success, and rise of alternative mating strategies in a social carnivore. *Behavioral Ecology* 30: 1618–1623. [8]

Austin, A. T., and C. L. Ballaré. 2010. Dual role of lignin in plant litter decomposition in terrestrial ecosystems. *Proceedings of the National Academy of Sciences USA* 107: 4618–4622. [22]

Austin, A. T., and L. Vivanco. 2006. Plant litter decomposition in a semiarid ecosystem controlled by photodegradation. *Nature* 442: 555–558. [22]

Avers, C. J. 1985. *Molecular Cell Biology*. Addison-Wesley: Boston, MA. [5]

B

Baker, C. S., M. L. Dalebout, G. M. Lento, and N. Funahashi. 2002. Gray whale products sold in commercial markets along the Pacific coast of Japan. *Marine Mammal Science* 18: 295–300. [23]

Balanyá, J., J. M. Oller, R. B. Huey, G. W. Gilchrist, and L. Serra. 2006. Global genetic change tracks global climate warming in *Drosophila subobscura*. *Science* 313: 1773–1775. [6]

Balmford, A., R. E. Green, and M. Jenkins. 2003. Measuring the changing state of nature. *Trends in Ecology and Evolution* 18: 326–330. [23]

Barbier, E. B., S. D. Hacker, C. Kennedy, E. W. Koch, A. C. Stier and B. R. Silliman. 2011. The value of estuarine and coastal ecosystem services. *Ecological Monographs* 81: 169–193. [3]

Barbour, M. A., and R. W. Clark. 2012. Ground squirrel tail-flag displays alter both predatory strike and ambush site selection behaviours of rattlesnakes. *Proceedings of the Royal Society of London* 279B: 3827–3833. [8]

Barbour, M. S., and J. A. Litvaitis. 1993. Niche dimensions of New England cottontails in relation to habitat patch size. *Oecologia* 95: 321–327. [24]

Barnosky, A. D., P. L. Koch, R. S. Feranec, S. L. Wing and A. B. Shabel. 2004. Assessing the causes of Late Pleistocene extinctions on the continents. *Science* 306: 70–75. [3]

Barringer, B. C., W. D. Koenig, and J. M. H. Knops. 2013. Interrelationships among life-history traits in three California oaks. *Oecologia* 171: 129–139. [7]

Bartlein, P., J. C. Whitlock, and S. L. Shafer. 1997. Future climate in the Yellowstone National Park region and its potential impact on vegetation. *Conservation Biology* 11: 782–792. [24]

Bascompte, J., M. García, R. Ortega, E. L. Rezende, and R. Pironon. 2019. Mutualistic interactions reshuffle the effects of climate change on plants across the tree of life. *Science Advances* 5: eaav2539. [15]

Battin, J. 2004. When good animals love bad habitats: Ecological traps and the conservation of animal populations. *Conservation Biology* 18: 1482–1491. [24]

Beall, C. M. 2007. Two routes to functional adaptation: Tibetan and Andean high-altitude natives. *Proceedings of the National Academy of Sciences USA* 104: 8655–8660. [4]

Becerra, J. X. 2003. Synchronous coadaptation in an ancient case of herbivory. *Proceedings of the National Academy of Sciences USA* 100: 12804–12807. [12]

Becerra, J. X. 2007. The impact of herbivore–plant coevolution on plant community structure. *Proceedings of the National Academy of Sciences USA* 104: 7483–7488. [12]

Beck, B. B. 1980. Animal Tool Behavior: The Use and Manufacture of Tools by Animals. Garland STPM Press: New York. [5]

Begon, M., M. Mortimer, and D. J. Thompson. 1996. *Population Ecology: A Unified Study of Animals and Plants*. 3rd ed. Blackwell Science: Oxford. [11]

Behrens Yamada, S., B. R. Dumbauld, A. Kalin, C. E. Hunt, R. Figlar-Barnes, and A. Randall. 2005. Growth and persistence of a recent invader *Carcinus maenas* in estuaries of the northeastern Pacific. *Biological Invasions* 7: 309–321. [19]

Beisner, B. E., D. T. Haydon, and K. Cuddington. 2003. Alternative stable states in ecology. *Frontiers in Ecology and the Environment* 1: 376–382. [17]

Beissinger, S. R., and M. I. Westphal. 1998. On the use of demographic models of population viability in endangered species management. *Journal of Wildlife Management* 62: 821–841. [23]

Bell, M. A., M. P. Travis, and D. M. Blouw. 2006. Inferring natural selection in a fossil threespine stickleback. *Paleobiology* 32: 562–577. [6]

Bellows, T. S. Jr. 1981. The descriptive properties of some models for density dependence. *Journal of Animal Ecology* 50: 139–156. [11]

Belnap, J. 2003. The world at your feet: Desert biological crusts. *Frontiers in Ecology and Environment* 1: 181–189. [22]

Belnap, J., and D. Eldridge. 2001. Disturbance and recovery of biological soil crusts. In *Biological Soil Crusts: Structure, Function, and Management*, J. Belnap and O. L. Lange (Eds.), pp. 363–383. Springer: Berlin [22]

Belt, T. 1874. *The Naturalist in Nicaragua*. Murray: London. [15]

Belz, R. G. 2007. Allelopathy in crop/weed interactions: An update. *Pest Management Science* 63: 308–326. [14]

Benkman, C. W. 1993. Adaptation to single resources and the evolution of crossbill (*Loxia*) diversity. *Ecological Monographs* 63: 305–325. [5]

Benkman, C. W. 2003. Divergent selection drives the adaptive radiation of crossbills. *Evolution* 57: 1176–1181. [5]

Benton, M. J., and B. C. Emerson. 2007. How did life become so diverse? The dynamics of diversification according to the fossil record and molecular phylogenetics. *Palaeontology* 50: 23–40. [6]

Berendse, F., R. Aerts, and R. Bobbink. 1993. Atmospheric nitrogen deposition and its impact on terrestrial ecosystems. In *Landscape Ecology of a Stressed Environment*, C. C. Vos and P. Opdam (Eds.), pp. 104–121. Chapman & Hall: London. [25]

Berger, L., A. A. Roberts, J. Voyles, J. E. Loncore, K. A. Murray and L. F. Skerratt. 2016. History and recent progress on chytridioimycosis in amphibians. *Fungal Ecology* 19: 89–99. [1]

Berglund, A., and G. Rosenqvist. 1993. Selective males and ardent females in pipefishes. *Behavioral Ecology and Sociobiology* 32: 331–336. [8]

Berlow, E. L. 1997. From canalization to contingency: Historical effects in a successional rocky intertidal. *Ecological Monographs* 67: 435–460. [17]

Berlow, E. L. 1999. Strong effects of weak interactions in ecological communities. *Nature* 398: 330–334. [21]

Bernhardt, E. S., and 13 others. 2005. Can't see the forest for the stream? In-stream processing and terrestrial nitrogen exports. *BioScience* 55: 219–230. [22]

Bertness, M. D. 1989. Interspecific competition and facilitation in a northern acorn barnacle population. *Ecology* 70: 257–268. [15]

Bertness, M. D., and G. H. Leonard. 1997. The role of positive interactions in communities: Lessons from intertidal habitats. *Ecology* 78: 1976–1989. [15]

Bertness, M. D., and S. D. Hacker. 1994. Physical stress and positive associations among marsh plants. *American Naturalist* 144: 363–372. [19]

Bertness, M. D., and S. W. Shumway. 1993. Competition and facilitation in marsh plants. *American Naturalist* 142: 718–724. [17]

Beschta, R. L., and W. J. Ripple. 2006. River channel dynamics following extirpation of wolves in northwestern Yellowstone National Park, USA. *Earth Surface Processes and Landforms* 31: 1525–1539. [24]

Beschta, R. L., and W. J. Ripple. 2019. Can large carnivores change streams via a trophic cascade? *Ecohydrology* 12: e2048. [24]

Biel, R. G., and S. D. Hacker. 2021. Warming alters the interaction of two invasive beachgrasses with implications for range shifts and coastal dune functions. *Oecologia* 197: 757–770.

Bierregaard, R. O. Jr., C. Gascon, T. E. Lovejoy, and R. C. G. Mesquita. 2001. *Lessons from Amazonia: The Ecology and Conservation of a Fragmented Forest*. Yale University Press: New Haven, CT. [18]

Bigler, C., D. G. Gavin, C. Gunning, and T. T. Veblen. 2007. Drought induces lagged tree mortality in a subalpine forest in the Rocky Mountains. *Oikos* 116: 1983–1994. [11]

Binkley, D., and P. Högberg. 1997. Does atmospheric deposition of acidity and nitrogen threaten Swedish forests? *Forest Ecology and Management* 92: 119–152. [25]

Birkeland, C. 1997. Symbiosis, fisheries and economic development on coral reefs. *Trends in Ecology and Evolution* 12: 364–367. [3]

Bishop, J., W. F. Fagan, J. D. Schade, and C. M. Crisafulli. 2005. Causes and consequences of herbivory on prairie lupine (*Lupinus lepidus*) in early primary succession. In *Ecological Responses to the 1980 Eruption of Mount St. Helens*, V. H. Dale, F. J. Swanson, and C. M. Crisafulli (Eds.), pp. 151–161. Springer: New York. [17]

Björkman, O. 1981. Responses to different quantum flux densities. In *Physiological Plant Ecology I: Encyclopedia of Plant Physiology*, O. L. Lange et al. (Eds.), pp. 57–101. Springer: Berlin. [5]

Blais, J. M., D. W. Schindler, D. C. G. Muir, L. E. Kimpe, D. B. Donald, and B. Rosenberg. 1998. Accumulation of persistent organochlorine compounds in mountains of western Canada. *Nature* 395: 585–588. [21]

Blais, J. M., L. E. Kimpe, D. McMahon, B. E. Keatley, M. L. Mattory, M. S. V. Douglas, and J. P. Smol. 2005. Arctic seabirds transport marine-derived contaminants. *Science* 309: 445. [21]

Blaustein, A. R. and P. T. J. Johnson. 2003. Explaining frog deformities. *Scientific American* 288: 60–65. [1]

Boonstra, R., D. Hik, G. R. Singleton, and A. Tinnikov. 1998. The impact of predator-induced stress on the snowshoe hare cycle. *Ecological Monographs* 79: 371–394. [12]

Booth, M. G., and J. D. Hoeksema. 2010. Mycorrhizal networks counteract competitive effects of canopy trees on seedling survival. *Ecology* 91: 2294–2302. [15]

Bopp, L., and 11 others. 2013. Multiple stressors of ocean ecosystems in the 21st century: Projections with CMIP5 models. *Biogeosciences* 10: 6225–6245. [25]

Borland, A. M., H. Griffiths, C. Maxwell, M. S. J. Broadmeadow, N. M. Griffiths, and J. D. Barnes. 1992. On the ecophysiology of the Clusiaceae in Trinidad: Expression of CAM in *Clusia minor* L. during the transition from wet to dry season and characterization of three endemic species. *New Phytologist* 122: 349–357. [5]

Borries, C., K. Launhardt, C. Epplen, J. T. Epplen, and P. Winkler. 1999. DNA analyses support the hypothesis that infanticide is adaptive in langur monkeys. *Proceedings of the Royal Society of London* 266B: 901–904. [8]

Bouzat, J. L., H. A. Lewin, and K. N. Paige. 1998. The ghost of genetic diversity past: Historical DNA analysis of the greater prairie chicken. *American Naturalist* 152: 1–6. [6]

Bowman, W. D., T. A. Theodose, J. C. Schardt, and R. T. Conant. 1993. Constraints of nutrient availability on primary production in two alpine communities. *Ecology* 74: 2085–2098. [20]

Bowman, W. D., and C. J. Bilbrough. 2001. Influence of a pulsed nitrogen supply on growth and nitrogen uptake in alpine graminoids. *Plant and Soil* 233: 283–290. [20]

Bowman, W. D., and T. R. Seastedt. 2001. *Structure and Function of an Alpine Ecosystem, Niwot Ridge, Colorado*. Oxford University Press: New York. [22]

Braby, M. F. 2002. Life history strategies and habitat templets of tropical butterflies in north-eastern Australia. *Evolutionary Ecology* 16: 399–413. [7]

Bradshaw, A. D. 1987. Restoration: An acid test for ecology. In *Restoration Ecology: A Synthetic Approach to Ecological Restoration*, W. R. Jordan, M. E. Gilpin and J. D. Aber (Eds.), pp. 23–30. Cambridge University Press: Cambridge. [24]

Bradshaw, C. J. A., and B. W. Brook. 2007. Ecological–economic models of sustainable harvest for an endangered but exotic megaherbivore in northern Australia. *Natural Resource Modeling* 20: 129–156. [23]

Bradshaw, W. E., and C. M. Holzapfel. 2001. Genetic shift in photoperiodic response correlated with global warming. *Proceedings of the National Academy of Sciences USA* 98: 14509–14511. [6]

Breshears, D. D. and 12 others. 2005. Regional vegetation die-off in response to global-change-type drought. *Proceedings of the National Academy of Sciences USA* 10: 15144–15148. [2]

Brewer, J. S. 2003. Why don't carnivorous pitcher plants compete with non-carnivorous plants for nutrients? *Ecology* 84: 451–462. [14]

Briggs, J. C. 2006. Proximate sources of marine biodiversity. *Journal of Biogeography* 33: 1–10. [18]

Britton, R. A., and V. B. Young. 2012. Interaction between the intestinal microbiota and host in *Clostridium difficile* colonization resistance. *Trends in Microbiology* 20: 313–319. [13]

Brommer, J. E., H. Pietiainen, K. Ahola, P. Karell, T. Karstinen and H. Kolunen. 2010. The return of the vole cycle in southern Finland refutes the generality of the loss of cycles through "climatic forcing." *Global Change Biology* 16: 577–586. [10]

Bronstein, J. L. 1992. Seed predators as mutualists: Ecology and evolution of the fig/pollinator interaction. In *Insect-Plant Interactions* (Vol. 4), E. A. Bernays (Ed.), pp. 1–44. CRC Press: Boca Raton, FL. [15]

Bronstein, J. L. 1994. Conditional outcomes in mutualistic interactions. *Trends in Ecology and Evolution* 9: 214–217. [15]

Brooks, T., J. Tobias, and A. Balmford. 1999. Deforestation and bird extinctions in the Atlantic forest. *Animal Conservation* 2: 211–222. [23]

Brown, J. H. and A. Kodric-Brown. 1977. Turnover rates in insular biogeography: Effect of immigration on extinction. *Ecology* 58: 445–449. [9]

Brown, J. H., and D. W. Davidson. 1977. Competition between seed-eating rodents and ants in desert ecosystems. *Science* 196: 880–882. [14]

Bruno, J. F., and 7 others. 2007. Thermal stress and coral cover as drivers of coral disease outbreaks. *PLOS Biology* 5: 1220–1227. [13]

Bruno, J. F., J. J. Stachowicz, and M. D. Bertness. 2003. Inclusion of facilitation into ecological theory. *Trends in Ecology and Evolution* 18: 119–125. [19]

Bullock, S., H. Mooney and E. Medina, Eds. 1995. *Seasonally Dry Tropical Forests*. Cambridge University Press: Cambridge. [3]

Bunn, R., Y. Lekberg, and C. Zabinski. 2009. Arbuscular mycorrhizal fungi ameliorate temperature stress in thermophilic plants. *Ecology* 90: 1378–1388. [15]

Burgess, R. L. 1977. The Ecological Society of America: Historical data and preliminary analyses. In *History of American Ecology*, F. N. Egerton (Ed.), pp. 1–24. Arno Press: New York. [23]

Buss, L. W. ,and J. B. C. Jackson. 1979. Competitive networks: Nontransitive competitive relationships in cryptic coral reef environments. *American Naturalist* 113: 223–234. [16]

Buston, P. M. 2003a. Size and growth modification in clownfish. *Nature* 424: 145–146. [7]

Buston, P. M. 2003b. Forcible eviction and prevention of recruitment in the clown anemonefish. *Behavioral Ecology* 14: 576–582. [7]

Buston, P. M. 2004. Territory inheritance in clownfish. *Proceedings of the Royal Society of London* 271B: S252–S254. [7]

Butzer, K. W. 1992. The Americas before and after 1492: An introduction to current geographical research. *Annals of the Association of American Geographers* 82: 345–368. [24]

Bytnerowicz, A., M. J. Arbaugh, and R. Alonso. 2003. *Ozone Pollution in the Sierra Nevada: Distribution and Effects on Forests*. Elsevier: Amsterdam. [25]

C

Cain, M. L., B. G. Milligan, and A. E. Strand. 2000. Long-distance seed dispersal in plant populations. *American Journal of Botany* 87: 1217–1227. [9]

Calder, W. J., and S. B. St. Clair. 2012. Facilitation drives mortality patterns along successional gradients of aspen-conifer forest. *Ecosphere* 3: 1–11. [17]

Caldwell, M. M., L. O. Björn, J. F. Bornman, S. D. Flint, G. Kulandaivelu, A. H. Teramura, and M. Tevini. 1998. Effects of increased solar ultraviolet radiation on terrestrial ecosystems. *Journal of Photochemistry and Photobiology: Biology* 46B: 40–52. [25]

Callaway, R. M., and 12 others. 2002. Positive interactions among alpine plants increase with stress. *Nature* 417: 844–848. [15]

Callaway, R. M., and L. King. 1996. Temperature-driven variation in substrate oxygenation and the balance of competition and facilitation. *Ecology* 77: 1189–1195. [15]

Callaway, R. M., E. H. DeLucia, and W. H. Schlesinger. 1994. Biomass allocation of montane and desert ponderosa pine: An analog for response to climate change. *Ecology* 75: 1474–1481. [7]

Canfield, D. E., A. N. Glazer, and P. G. Falkowski. 2010. The evolution and future of Earth's nitrogen cycle. *Science* 330: 192–196. [25]

Carlsson, J. O. L., C. Brönmark, and L.-A. Hansson. 2004. Invading herbivory: The golden apple snail alters ecosystem functioning in Asian wetlands. *Ecology* 85: 1575–1580. [12]

Carlton, J. T., and J. B. Geller. 1993. Ecological roulette: The global transport and invasion of non-indigenous marine organisms. *Science* 261: 78–82. [19]

Carroll, S. P., and C. Boyd. 1992. Host race radiation in the soapberry bug: Natural history with the history. *Evolution* 46: 1052–1069. [6]

Carroll, S. P., H. Dingle, and S. P. Klassen. 1997. Genetic differentiation of fitness-associated traits among rapidly evolving populations of the soapberry bug. *Evolution* 51: 1182–1188. [6]

Carson, R. 1962. *Silent Spring*. Houghton Mifflin: Boston, MA. [21]

Carson, W. P. and R. B. Root. 2000. Herbivory and plant species coexistence: Community regulation by an outbreaking phytophagous insect. *Ecological Monographs* 70: 73–99. [1]

Cary, J. R., and L. B. Keith. 1979. Reproductive change in the ten-year cycle of snowshoe hares. *Canadian Journal of Zoology* 57: 375–390. [12]

Castro, J., R. Zamora, J. A. Hódar, and J. M. Gómez. 2002. The use of shrubs as nurse plants: A successional-friendly technique for pine reforestation in Mediterranean mountains. *Restoration Ecology* 10: 297–305. [15]

Caswell, H. 2001. *Matrix Population Models: Construction, Analysis, and Interpretation*. 2nd ed. Oxford University Press/Sinauer: Sunderland, MA. [11]

Caughley, G. 1994. Directions in conservation biology. *Journal of Animal Ecology* 63: 215–244. [23]

Caughley, G., N. Shepherd, and J. Short. 1987. *Kangaroos: Their Ecology and Management in the Sheep Rangelands of Australia*. Cambridge University Press: Cambridge. [9]

Ceballos, G., and P. R. Ehrlich. 2002. Mammal population losses and the extinction crisis. *Science* 296: 904–907. [23]

Cebrian, J., and J. Lartigue. 2004. Patterns of herbivory and decomposition in aquatic and terrestrial ecosystems. *Ecological Monographs* 74: 237–259. [21]

Cerling, T. E., J. M. Harris, B. J. MacFadden, M. G. Leakey, J. Quade, V. Eisenmann, and J. R. Ehleringer. 1997. Global vegetation change through the Miocene/Pliocene boundary. *Nature* 389: 153–158. [5]

Channell, R., and M. V. Lomolino. 2000. Trajectories to extinction: Spatial dynamics of the contraction of geographical ranges. *Journal of Biogeography* 27: 169–179. [23]

Chapin, F. S. III, L. Moilanen, and K. Kielland. 1993. Preferential use of organic nitrogen for growth by a non-mycorrhizal Arctic sedge. *Nature* 361: 150–153. [22]

Chapin, F. S. III, L. R. Walker, C. L. Fastie, and L. C. Sharman. 1994. Mechanisms of primary succession following deglaciation at Glacier Bay, Alaska. *Ecological Monographs* 64: 149–175. [17]

Chapin, F. S. III, P. A. Matson, and H. A. Mooney. 2002. *Principles of Terrestrial Ecosystem Ecology*. Springer: New York. [21, 25]

Chappelka, A., and L. Samuelson. 1998. Ambient ozone effects on forest trees of the eastern

Charnov, E. L. 1976. Optimal foraging, the marginal value theorem. *Theoretical Population Biology* 9: 129–136. [8]

Charnov, E. L. 1993. *Life History Invariants*. Oxford University Press: Oxford. [7]

Charnov, E. L., and D. Berrigan. 1990. Dimensionless numbers and life history evolution: Age of maturity versus the adult lifespan. *Evolutionary Ecology* 4: 273–275. [7]

Chase, T. N., R. Pielke Sr., T. G. F. Kittel, J. S. Baron and T. J. Stohlgren. 1999. Impacts on Colorado Rocky Mountain weather due to land use changes on the adjacent Great Plains. *Journal of Geophysical Research* 104: 16673–16690. [2]

Chen, J. Q., J. F. Franklin, and T. A. Spies. 1995. Growing-season microclimatic gradients from clear-cut edges into old-growth Douglas-fir forests. *Ecological Applications* 5: 74–86. [24]

Chen, J.-Y., and 10 others. 2009. Complex embryos displaying bilaterian characters from Precambrian Doushantuo phosphate deposits, Weng'an, Guizhou, China. *Proceedings of the National Academy of Sciences USA* 106: 19056–19060. [6]

Chen, M., M. Schliep, R. D. Willows, Z. Cai, B. A. Neilan, and H. Scheer. 2010. A red-shifted chlorophyll. *Science* 329: 1318–1319. [5]

Chesson, P. L., and R. R. Warner. 1981. Environmental variability promotes coexistence in lottery competitive systems. *American Naturalist* 117: 923–943. [19]

Christensen, N. L., and 12 others. 1996. The report of the Ecological Society of America Committee on the Scientific Basis for Ecosystem Management. *Ecological Applications* 6: 665–691. [24]

Ciais, P., and 15 others. 2013. Carbon and other biogeochemical cycles. In *Climate Change 2013: The Physical Science Basis*, T. F. Stocker et al. (Eds.), pp. 466–570. Cambridge University Press: Cambridge. [25]

Clay, C. A., E. M. Lehmer, S. S. Jeor, and M. D. Dearing. 2009. Sin Nombre virus and rodent species diversity: A test of the dilution and amplification hypotheses. *PLoS ONE* 4: e6467. [19]

Clements, F. E. 1916. *Plant Succession: An Analysis of the Development of Vegetation*. Publication 42. Carnegie Institution of Washington: Washington, DC. [17]

Cleveland, C. C., and 10 others. 1999. Global patterns of terrestrial biological nitrogen (N_2) fixation in natural ecosystems. *Global Biogeochemical Cycles* 13: 623–645. [25]

Clutton-Brock, T. H., F. E. Guinness, and S. D. Albon. 1983. The costs of reproduction to red deer hinds. *Journal of Animal Ecology* 52: 367–383. [6]

Cohen, J. E. 1995. *How Many People Can the Earth Support?* W. W. Norton: New York. [11]

Coltman, D. W., P. O'Donoghue, J. T. Jorgenson, J. T. Hogg, C. Strobeck, and M. Festa-Bianchet. 2003. Undesirable evolutionary consequences of trophy hunting. *Nature* 426: 655–658. [6, 8]

Connell, J. H. 1961a. The effects of competition, predation by *Thais lapillus*, and other factors on natural populations of the barnacle, *Balanus balanoides*. *Ecological Monographs* 31: 61–104. [14, 19]

Connell, J. H. 1961b. The influence of interspecific competition and other factors on the distribution of the barnacle *Chthamalus stellatus*. *Ecology* 42: 710–723. [14, 19]

Connell, J. H. 1970. A predator–prey system in the marine intertidal region. I. *Balanus glandula* and several predatory species of *Thais*. *Ecological Monographs* 40: 49–78. [11]

Connell, J. H. 1978. Diversity in tropical rain forests and coral reefs. *Science* 199: 1302–1310. [19]

Connell, J. H. 1983. On the prevalence and relative importance of interspecific competition: Evidence from field experiments. *American Naturalist* 122: 661–696. [14]

Connell, J. H., and R. O. Slatyer. 1977. Mechanisms of succession in natural communities and their role in community stability and organization. *American Naturalist* 111: 1119–1144. [17]

Connell, J. H., and W. P. Sousa. 1983. On the evidence needed to judge ecological stability or persistence. *American Naturalist* 121: 789–824. [17]

Cook, B. I., R. L. Miller, and R. Seager. 2009. Amplification of the North American "Dust Bowl" drought through human-induced land degradation. *Proceedings of the National Academy of Sciences USA* 106: 4997–5001. [25]

Cook, L. M., B. S. Grant, I. J. Saccheri, and J. Mallet. 2012. Selective bird predation on the peppered moth: The last experiment of Michael Majerus. *Biology Letters* 8: 609–612. [6]

Cooper, O. R., and 10 others. 2014. Global distribution and trends of tropospheric ozone: An observation-based review. *Elementa* 2: 29. [25]

Cooper, W. S. 1923a. The recent ecological history of Glacier Bay, Alaska: The interglacial forests of Glacier Bay. *Ecology* 4: 93–128. [17]

Cooper, W. S. 1923b. The recent ecological history of Glacier Bay, Alaska: Permanent quadrats at Glacier Bay: An initial report upon a long-period study. *Ecology* 4: 355–365. [17]

Copeyon, C. K., J. R. Walters, and J. H. Carter III. 1991. Induction of red-cockaded woodpecker group formation by artificial cavity construction. *Journal of Wildlife Management* 55: 549–556. [23]

Cornell, H. V., and J. H. Lawton. 1992. Species interactions, local and regional processes, and limits to the richness of ecological communities: A theoretical perspective. *Journal of Animal Ecology* 61: 1–12. [18]

Cornulier, T., and 20 others. 2013. Europe-wide dampening of population cycles in keystone herbivores. *Science* 340: 63–66. [10]

Cornwall, C. E., and 21 others. 2021. Global declines in coral reef calcium carbonate production under ocean acidification and warming. *Proceedings of the National Academy of Sciences USA* 118: 2015265118.

Costanza, R., and 7 others. 2014. Changes in the global value of ecosystem services. *Global Environmental Change* 26: 152–158. [23]

Costanzo, J. P., R. E. Lee Jr., A. L. DeVries, T. Wang, and J. R. Layne Jr. 1995. Survival mechanisms of vertebrate ectotherms at subfreezing temperatures: Applications in cryomedicine. *FASEB Journal* 9: 351–358. [4]

Costello, A., and 28 others. 2009. Managing the health effects of climate change. *The Lancet* 373: 1693–1733. [13]

Cotton, P. A., S. D. Rundle, and K. E. Smith. 2004. Trait compensation in marine gastropods: Shell shape, avoidance behavior, and susceptibility to predation. *Ecology* 85: 1581–1584. [12]

Courchamp, F., T. Clutton-Brock, and B. Grenfell. 1999. Inverse density dependence and the Allee effect. *Trends in Ecology and Evolution* 14: 405–410. [10]

Cowie, R. H. 2001. Decline and homogenization of Pacific faunas: The land snails of American Samoa. *Biological Conservation* 99: 207–222. [23]

Cowie, R. J. 1977. Optimal foraging in great tits (*Parus major*). *Nature* 268: 137–139. [8]

Cowles, H. C. 1899. The ecological relations of the vegetation on the sand dunes of Lake Michigan. *Botanical Gazette* 27: 95–117, 167–202, 281–308, 361–391. [17]

Creel, S., and D. Christianson. 2009. Wolf presence and increased willow consumption by Yellowstone elk: Implications for trophic cascades. *Ecology* 90: 2454–2466. [24]

Creel, S., and N. M. Creel. 1995. Communal hunting and pack size in African wild dogs, *Lycaon pictus*. *Animal Behaviour* 50: 1325–1339. [8]

Creel, S., J. Winnie Jr., B. Maxwell, K. Hamlin, and M. Creel. 2005. Elk alter habitat selection as an antipredator response to wolves. *Ecology* 86: 3387–3397. [8]

Crews, T. E., K. Kitayama, J. H. Fownes, D. A. Herbert, R. H. Riley, D. Mueller-Dombois, and P. M. Vitousek. 1995. Changes in soil phosphorus fractions and ecosystem dynamics across a long chronosequence in Hawaii. *Ecology* 76: 1407–1424. [22]

Crisafulli, C. M., J. A. MacMahon, and R. R. Parmenter. 2005. Small-mammal survival and colonization on the Mount St. Helens volcano: 1980–2005. In *Ecological Responses to the 1980 Eruption of Mount St. Helens*, V. H. Dale, F. J. Swanson, and C. M. Crisafulli (Eds.), pp. 183–197. Springer: New York. [17]

Croll, D. A., J. L. Maron, J. A. Estes, E. M. Danner, and G. V. Byrd. 2005. Introduced predators transform subarctic islands from grassland to tundra. *Science* 307: 1959–1961. [12]

Crouse, D. T., L. B. Crowder, and H. Caswell. 1987. A stage-based population model for loggerhead sea turtles and implications for conservation. *Ecology* 68: 1412–1423. [11]

Crouzeilles, R. and 8 others. 2017. Ecological restoration success is higher for natural regeneration than for active restoration in tropical forests. *Science Advances* 3: e1701345. [3]

Crowder, L. B., D. T. Crouse, S. S. Heppell, and T. H. Martin. 1994. Predicting the impact of turtle excluder devices on loggerhead sea turtle populations. *Ecological Applications* 4: 437–445. [11]

Crutzen, P. J., and E. F. Stoermer. 2000. The "Anthropocene." *Global Change Newsletter* 41: 12–13. [25]

Currie, C. R., and A. E. Stuart. 2001. Weeding and grooming of pathogens in agriculture by ants. *Proceedings of the Royal Society of London* 268B: 1033–1039. [15]

Currie, C. R., J. A. Scott, R. C. Summerbell, and D. Malloch. 1999b. Fungus-growing ants use antibiotic-producing bacteria to control garden parasites. *Nature* 398: 701–704. [15]

Currie, C. R., M. Poulsen, J. Mendenhall, J. J. Boomsma, and J. Billen. 2006. Coevolved crypts and exocrine glands support mutualistic bacteria in fungus-growing ants. *Science* 311: 81–83. [15]

Currie, C. R., U. G. Mueller, and D. Malloch. 1999a. The agricultural pathology of ant fungus gardens. *Proceedings of the National Academy of Sciences USA* 96: 7998–8002. [15]

D

Dahlgren, C. P., and D. B. Eggleston. 2000. Ecological processes underlying ontogenetic habitat shifts in a coral reef. *Ecology* 81: 2227–2240. [7]

Daily, G. C., and K. Ellison. 2002. *The New Economy of Nature: The Quest to Make Conservation Profitable.* Island Press: Washington, DC. [24]

Dale, V. H., F. J. Swanson, and C. M. Crisafulli, Eds. 2005. *Ecological Responses to the 1980 Eruption of Mount St. Helens.* Springer: New York. [17]

Dambrine, E., J. L. Dupouey, L. Laut, L. Humbert, M. Thinon, M. T. Beaufils, and H. Richard. 2007. Present forest biodiversity patterns in France related to former Roman agriculture. *Ecology* 88: 1430–1439. [24]

Damman, H., and M. L. Cain. 1998. Population growth and viability analyses of the clonal woodland herb, *Asarum canadense*. *Journal of Ecology* 86: 13–26. [11]

Dammerman, K. W. 1948. *The Fauna of Krakatau: 1883–1933.* Noord-Hollandsche Uitd.-Mij. [18]

Darimont, C. T., S. M. Carlson, M. T. Kinnison, P. C. Paquet, T. E. Reimchen, and C. C. Wilmers. 2009. Human predators outpace other agents of trait change in the wild. *Proceedings of the National Academy of Sciences USA* 106: 952–954. [6]

Darwin, C. 1859. *On the Origin of Species.* J. Murray: London. [6, 8, 12, 14, 16, 18, 19, 21]

Darwin, C. 1871. The Descent of Man and Selection in Relation to Sex. John Murray: London. [8]

Darwin, C. 1875. *Insectivorous Plants.* Appleton and Company: New York. [14]

Daskalov, G. M., A. N. Grishin, S. Rodionov, and V. Mihneva. 2007. Trophic cascades triggered by overfishing reveal possible mechanisms of ecosystem regime shifts. *Proceedings of the National Academy of Sciences USA* 104: 10518–10523. [10]

Davidson, J. 1938. On the growth of the sheep population in Tasmania. *Transactions of the Royal Society of South Australia* 62: 342–346. [10]

Davidson, J., and H. G. Andrewartha. 1948. The influence of rainfall, evaporation and atmospheric temperature of fluctuations in the size of a natural population of *Thrips imaginis* (Thysanoptera). *Journal of Animal Ecology* 17: 200–222. [11]

Davies, N. B., J. R. Krebs, and S. A. West. 2012. *An Introduction to Behavioural Ecology*. 4th ed. Wiley-Blackwell: Oxford. [8]

Davis, C. C. 1964. Evidence for the eutrophication of Lake Erie from phytoplankton records. *Limnology and Oceanography* 9: 275–283. [10]

Davis, M. B., J. Ford, and R. E. Moeller. 1985. Paleolimnology. In *An Ecosystem Approach to Aquatic Ecology: Mirror Lake and Its Environment*, G. E. Likens (Ed.), pp. 345–366. Springer: New York. [22]

Dawson, J., and R. Lucas. 2000. *Nature Guide to the New Zealand Forest*. A Godwit Book. Random House New Zealand: Auckland. [18]

Dayton, P. K. 1971. Competition, disturbance and community organization: The provision and subsequent utilization of space in a rocky intertidal community. *Ecological Monographs* 41: 351–389. [16]

De'ath, G., J. M. Lough, and K. E. Fabricus. 2009. Declining coral calcification in the Great Barrier Reef. *Science* 323: 116–119. [25]

Deevey, E. S. 1947. Life tables for natural populations. *Quarterly Review of Biology* 22: 283–314. [11]

Del Moral, R., D. M. Wood, and J. H. Titus. 2005. Proximity, microsites, and biotic interactions during early succession. In *Ecological Responses to the 1980 Eruption of Mount St. Helens*, V. H. Dale, F. J. Swanson, and C. M. Crisafulli (Eds.), pp. 93–107. Springer: New York. [17]

Delcourt, H. R. 2002. Creating landscape pattern. In *Learning Landscape Ecology*, S. E. Gergel and M. G. Turner (Eds.), pp. 62–82. Springer: New York. [24]

Delcourt, P. A., H. R. Delcourt, C. R. Ison, W. E. Sharp and K. J. Gremillion. 1998. Prehistoric human use of fire, the eastern agricultural complex, and Appalachian oak–chestnut forests: Paleoecology of Cliff Palace Pond, Kentucky. *American Antiquity* 63: 263–278. [3]

DellaSala, D. A., and J. E. Williams. 2006. The northwest forest plan: A global model of forest management in contentious times. *Conservation Biology* 20: 274–276. [24]

DeLucia, E. H., and 10 others. 1999. Net primary production of a forest ecosystem under experimental CO_2 enrichment. *Science* 284: 1177–1179. [25]

DeMaster, D. P., A. W. Trites, P. Clapham, S. Mizroch, P. Wade, R. J. Small, and J. Ver Hoef. 2006. The sequential megafaunal collapse hypothesis: Testing with existing data. *Progress in Oceanography* 68: 329–342. [9]

Dennison, P. E., S. C. Brewer, J. D. Arnold, and M. A. Moritz. 2014. Large wildfire trends in the western United States, 1984–2011. *Geophysical Research Letters* 41: 2928–2933. [25]

Denslow, J. S. 1987. Tropical rainforest gaps and tree species diversity. *Annual Review of Ecology and Systematics* 18: 431–451. [7]

Develey, P. F., and B. T. Phalan. 2021. Bird extinctions in Brazil's Atlantic rainforest and how they can be prevented. *Frontiers in Ecology and Evolution* 9:624587. doi: 10.3389/fevo.2021.624587 [23]

Dewailly, E., P. Ayotte, S. Bruneau, C. Laliberte, D. C. G. Muir, and R. J. Norstrom. 1993. Inuit exposure to organochlorines through the aquatic food-chain in Arctic Quebec. *Environmental Health Perspectives* 101: 618–620. [21]

Dewberry, E. P., J. N. Phillips, G. E. Dewberry, M. J. Blum, and D. Luther. 2020. Singing in a silent spring: Birds respond to half-century soundscape reversion during the COVID-19 shutdown. *Science* 370: 575–579. [8]

Diamond, J. M. 1997. *Guns, Germs, and Steel: The Fates of Human Societies*. W. W. Norton: New York. [13]

Dirzo, R. and P. H. Raven. 2003. Global state of biodiversity and loss. *Annual Review of Environment and Resources* 28: 137–167. [3]

Dixon, A. F. G. 1971. The role of aphids in wood formation. II. The effect of the lime aphid, *Eucallipterus tiliae* L. (Aphididae), on the growth of lime *Tilia vulgaris* Hayne. *Journal of Applied Ecology* 8: 393–399. [12]

Dixon, R. K., A. M. Solomon, S. Brown, R. A. Houghton, M. C. Trexier and J. Wisniewski. 1994. Carbon pools and flux of global forest ecosystems. *Science* 263: 185–190. [3]

Dizney, L. J., and L. A. Ruedas. 2009. Increased host species diversity and decreased prevalence of Sin Nombre virus. *Emerging Infectious Diseases* 15: 1012–1018. [19]

Dobson, A., and M. Meagher. 1996. The population dynamics of brucellosis in the Yellowstone National Park. *Ecology* 77: 1026–1036. [13]

Dodd, D. M. B. 1989. Reproductive isolation as a consequence of adaptive divergence in *Drosophila pseudoobscura*. *Evolution* 43: 1308–1311. [6]

Dodson, C. H., and A. H. Gentry. 1991. Biological extinction in western Ecuador. *Annals of the Missouri Botanical Garden* 78: 273–295. [23]

Dodson, S. 2004. *Introduction to Limnology*. McGraw Hill: New York. [2]

Donihue, C. M., and 7 others. 2018. Hurricane induced selection on the morphology of an island lizard. *Nature* 560: 88–91. [6]

Du, M., and J. D. Kessler. 2012. Assessment of the spatial and temporal variability of bulk hydrocarbon respiration following the Deepwater Horizon oil spill. *Environmental Science & Technology* 46: 10499–10507. [5]

Dugger, K. M., R. G. Anthony and L. S. Andrews. 2011. Transient dynamics of invasive competition: Barred owls, spotted owls, habitat, and the demons of competition present. *Ecological Applications* 21: 2459–2468. [9]

Duggins, D. O. 1980. Kelp beds and sea otters: An experimental approach. *Ecology* 61: 447–453. [9]

Duggins, D. O., C. A. Simenstad, and J. A. Estes. 1989. Magnification of secondary production by kelp detritus in coastal marine ecosystems. *Science* 245: 170–173. [9]

Dukes, J. S., and H. A. Mooney. 1999. Does global change increase the success of biological invaders? *Trends in Ecology and Evolution* 14: 135–139. [25]

Dybdahl, M. F., and C. M. Lively. 1998. Host–parasite coevolution: Evidence for rare advantage and time-lagged selection in a natural population. *Evolution* 52: 1057–1066. [13]

Dyer, L. A., and K. K. Letourneau. 1999a. Trophic cascades in a complex terrestrial community. *Proceedings of the National Academy of Sciences USA* 96: 5072–5076. [21]

Dyer, L. A., and K. K. Letourneau. 1999b. Relative strengths of top-down and bottom-up forces in a tropical forest community. *Oecologia* 119: 265–274. [21]

Eberhard, W. G. 2001. Under the influence: Webs and building behavior of *Plesiometa argyra* (Araneae, Tetragnathidae) when parasitized by *Hymenoepimecis argyraphaga* (Hymenoptera, Ichneumonidae). *Journal of Arachnology* 29: 354–366. [13]

E

Edmondson, W. T., and A. H. Litt. 1982. *Daphnia* in Lake Washington. *Limnology and Oceanography* 27: 272–293. [22]

Egan, T. 2006. The Worst Hard Time: The Untold Story of Those Who Survived the Great American Dust Bowl. Houghton Mifflin: Boston. [25]

Ehleringer, J. R., and C. S. Cook. 1990. Characteristics of *Encelia* species differing in leaf reflectance and transpiration rate under common garden conditions. *Oecologia* 82: 484–489. [4]

Ehleringer, J. R., T. E. Cerling, and B. R. Helliker. 1997. C_4 photosynthesis, atmospheric CO_2, and climate. *Oecologia* 112: 285–299. [5]

Ehrlich, P. R., and E. O. Wilson. 1991. Biodiversity studies: Science and policy. *Science* 253: 758–762. [23]

Eis, S., E. H. Garman, and L. F. Ebell. 1965. Relation between cone production and diameter increment of Douglas fir (*Pseudotsuga menziesii* (Mirb.) Franco), grand fir (*Abies grandis* (Dougl.) Lindl.), and western white pine (*Pinus monticola* Dougl.). *Canadian Journal of Botany* 43: 1553–1559. [7]

El-Sabaawi, R. W., M. C. Marshall, R. D. Bassar, A. López-Sepulcre, E. P. Palkovacs, and C. Dalton. 2015. Assessing the effects of guppy life history evolution on nutrient recycling: From experiments to the field. *Freshwater Biology* 60: 590–601. [6]

Elser, J. J., and 11 others. 2000. Nutritional constraints in terrestrial and freshwater food webs. *Nature* 408: 578–580. [21]

Elser, J. J., and 9 others. 2007. Global analysis of nitrogen and phosphorus limitation of primary producers in freshwater, marine and terrestrial ecosystems. *Ecology Letters* 10: 1135–1142. [20]

Elton, C. S. 1927. *Animal Ecology*. Sidgwick and Jackson: London. [17]

Elton, C. S. 1942. *Voles, Mice and Lemmings: Problems in Population Dynamics*. Oxford University Press: Oxford. [6]

Elton, C. S. 1958. *The Ecology of Invasions by Animals and Plants*. Chapman & Hall: London. [19, 21]

Emlen, S. T., and L. W. Oring. 1977. Ecology, sexual selection, and the evolution of mating systems. *Science* 197: 215–223. [8]

Endler, J. A. 1986. *Natural Selection in the Wild*. Princeton University Press: Princeton, NJ. [6]

Ensign, S. H., and M. W. Doyle. 2006. Nutrient spiraling in streams and river networks. *Journal of Geophysical Research* 111: G04009. [22]

Epstein, E., and A. J. Bloom. 2005. *Mineral Nutrition of Plants: Principles and Perspectives*. 2nd ed. Oxford University Press/Sinauer: Sunderland, MA. [22]

Epstein, P. R. 2000. Is global warming harmful to health? *Scientific American* 283: 50–57. [13]

Erickson, R. O. 1945. The *Clematis fremontii* var. *riehlii* population in the Ozarks. *Annals of the Missouri Botanical Garden* 32: 413–460. [9]

Eskew, W. A., S. J. Worth, J. E. Foley and B. D. Todd. 2015. American bullfrogs (*Lithobates catesbeianus*) resist infection by multiple isolates of *Batrachochytrium dendrobatidis*, including one implicated in wild mass mortality. *EcoHealth* 12: 513–518. [1]

Estes, J. A., and D. O. Duggins. 1995. Sea otters and kelp forests in Alaska: Generality and variation in a community ecology paradigm. *Ecological Monographs* 65: 75–100. [9]

Estes, J. A., M. T. Tinker, T. M. Williams, and D. F. Doak. 1998. Killer whale predation on sea otters linking oceanic and nearshore ecosystems. *Science* 282: 473–476. [9]

Evans, R. D., R. Rimer, L. Sperry, and J. Belnap. 2001. Exotic plant invasion alters nitrogen dynamics in an arid grassland. *Ecological Applications* 11: 1301–1310. [22]

Ewing, B., D. Moore, S. Goldfinger, A. Oursler, A. Reed, and M. Wackernagel. 2010. *The Ecological Footprint Atlas 2009*. Global Footprint Network: Oakland, CA. [11]

F

Fagan, W. F., R. S. Cantrell, and C. Cosner. 1999. How habitat edges change species interactions. *American Naturalist* 153: 165–182. [24]

Fahrig, L. 2019. Why do several small patches hold more species than few large patches? *Global Ecology and Biogeography* 29: 615–628. [24]

Faithfull, I. 1997. *Etiella behrii* (Zeller) (Lepidoptera: Pyralidae) bred from pods of gorse, *Ulex europaeus* L. (Fabaceae). *Victorian Entomologist* 27: 34. [19]

FAO and UNEP. 2020. The State of the World's Forests 2020. In brief. Forests, biodiversity and people. Rome.

FAO. 2020. Global Forest Resources Assessment 2020: Main report. Rome. https://doi.org/10.4060/ca9825en

FAO. *Global Forest Resources Assessment 2015. How Are the World's Forests Changing?* 2nd ed. Food and Agriculture Organization of the United Nations. www.fao.org/forestry/site/24690/en. [2]

Farrell, B. D. 1998. "Inordinate fondness" explained: Why are there so many beetles? *Science* 281: 555–559. [6]

Farrell, T. M. 1991. Models and mechanisms of succession: An example from a rocky intertidal community. *Ecological Monographs* 61: 95–113. [17]

Favreau, J. M., C. A. Drew, G. R. Hess, M. J. Rubino, F. H. Koch, and K. A. Eschelbach. 2006. Recommendations for assessing the effectiveness of surrogate species approaches. *Biodiversity and Conservation* 15: 3949–3969. [23]

Feder, J. L. 1998. The apple maggot fly, *Rhagoletis pomonella*: Flies in the face of conventional wisdom about speciation? In *Endless Forms: Species and Speciation*, D. J. Howard and S. H. Berlocher (Eds.), pp. 130–144. Oxford University Press: Oxford. [6]

Fedonkin, M. A., A. Simonetta, and A. Y. Ivantsov. 2007. New data on *Kimberella*, the Vendian mollusc-like organism (White Sea region, Russia): Palaeoecological and evolutionary implications. *Geological Society, London, Special Publications* 286: 157–179. [6]

Feely, R. A., C. L. Sabine, K. Lee, W. Berrelson, J. Kleypas, V. J. Fabry, and F. J. Millero. 2004. Impact of anthropogenic CO_2 on the $CaCO_3$ system in the oceans. *Science* 305: 362–366. [25]

Fenner, F., and F. N. Ratcliffe. 1965. *Myxomatosis*. Cambridge University Press: Cambridge. [13]

Fernández-Marín, H., J. K. Zimmerman, D. R. Nash, J. J. Boomsma, and W. T. Wcislo. 2009. Reduced biological control and enhanced chemical pest management in the evolution of fungus farming in ants. *Proceedings of the Royal Society of London* 276B: 2263–2269. [15]

Ferraz, G., G. J. Russell, P. C. Stouffer, R. O. Bierregaard Jr., S. L. Pimm, and T. E. Lovejoy. 2003. Rates of species loss from Amazonian forest fragments. *Proceedings of the National Academy of Sciences USA* 100: 14069–14073. [18]

Field, C. B., M. J. Behrenfeld, J. T. Randerson, and P. Falkowski. 1998. Primary productivity of the biosphere: Integrating terrestrial and oceanic components. *Science* 281: 237–240. [20]

Field, D. B., T. R. Baumgartner, C. D. Charles, V. Ferreira-Bartrina, and M. D. Ohman. 2006. Planktonic foraminifera of the California current reflect 20th-century warming. *Science* 311: 63–66. [25]

Field, J. P., and 9 others. 2010. The ecology of dust. *Frontiers in Ecology and the Environment* 8: 423–430. [25]

Findley, R., and J. P. Blair. 1990. Will we save our own? *National Geographic* 178: 106–136. [24]

Finkbeiner, E. M., B. P. Wallace, J. E. Moore, R. L. Lewison, L. B. Crowder, and A. J. Read. 2011. Cumulative estimates of sea turtle bycatch and mortality in USA fisheries between 1990 and 2007. *Biological Conservation* 144: 2719–2727. [11]

Finkelstein, M. E., and 7 others. 2012. Lead poisoning and the deceptive recovery of the critically endangered California condor. *Proceedings of the National Academy of Sciences USA* 109: 11449–11454. [23]

Fisher, S. G., and G. E. Likens. 1973. Energy budget in Bear Brook, New Hampshire: An integrative approach to stream ecosystem metabolism. *Ecological Monographs* 43: 421–439. [21]

Fitter, A. H. 2005. Darkness visible: Reflections on underground ecology. *Journal of Ecology* 93: 231–243. [15]

Fitter, A. H., and R. K. Hay. 1989. *Environmental Physiology of Plants*. Academic Press: London. [6]

Flecker, A. S., and C. R. Townsend. 1994. Community-wide consequences of trout introduction in New Zealand streams. *Ecological Applications* 4: 798–807. [21]

Fleishman, E., D. D. Murphy, and P. E. Brussard. 2000. A new method for selection of umbrella species for conservation planning. *Ecological Applications* 10: 569–579. [23]

Flessa, K. W. 1975. Area, continental drift and mammalian diversity. *Paleobiology* 1: 189–194. [18]

Foley, J. A., M. H. Costa, C. Delire, N. Ramankutty and P. Snyder. 2003. Green surprise? How terrestrial ecosystems could affect Earth's climate. *Frontiers in Ecology and Environment* 1: 38–44. [2]

Foo, D., J. M. Semmens, J. P. Y. Arnould, N. Dorville, A. J. Hoskins, K. Abernathy, G. J. Marshall, and M. A. Hindell. 2016. Testing optimal foraging theory models on benthic divers. *Animal Behaviour* 112: 127–138. [8]

Forbes, A. A., T. H. Q. Powell, L. L. Stelinski, J. J. Smith, and J. L. Feder. 2009. Sequential sympatric speciation across trophic levels. *Science* 323: 776–779. [6]

Forman, R. T. T. 1995. Some general principles of landscape and regional ecology. *Landscape Ecology* 10: 133–142. [24]

Forrester, M. L., G. Krotkov, and C. D. Nelson. 1966. Effect of oxygen on photosynthesis, photorespiration and respiration in detached leaves. II. Corn and other monocotyledons. *Plant Physiology* 41: 428–431. [5]

Foster, W. A., and J. E. Treherne. 1981. Evidence for the dilution effect in the selfish herd from fish predation on a marine insect. *Nature* 293: 466–467. [8]

Fowler, D., and 20 others. 2015. Effects of global change during the 21st century on the nitrogen cycle. *Atmospheric Chemistry and Physics* 15: 13849–13893. [25]

Frank, P. W., C. D. Boll, and R. W. Kelly. 1957. Vital statistics of laboratory cultures of *Daphnia pulex* De Geer as related to density. *Physiological Zoology* 30: 287–305. [11]

Franklin, J. F. 1996. Ecosystem management: An overview. In *Ecosystem Management: Applications for Sustainable Forest and Wildlife Resources*, M. S. Boyce and A. Haney (Eds.), pp. 21–53. Yale University Press: New Haven, CT. [24]

Franklin, J. F., and C. T. Dyrness. 1988. *Natural Vegetation of Oregon and Washington*. Oregon State University Press: Corvallis, OR. [18]

Franks, S. J., S. Sim, and A. E. Weis. 2007. Rapid evolution of flowering time by an annual plant in response to a climate fluctuation. *Proceedings of the National Academy of Sciences USA* 104: 1278–1282. [6]

Freeman, B. G., M. N. Scholer, V. Ruiz-Gutierrez, and J. W. Fitzpatrick. 2018. Climate change causes upslope shifts and mountaintop extirpations in a tropical bird community. *Proceedings of the National Academy of Sciences USA* 115:11982–11987.

Frölicher, T. L., M. Winton, and J. L. Sarmiento. 2014. Continued global warming after CO_2 emissions stoppage. *Nature Climate Change* 4: 40–44. [25]

Frost, C. C. 1993. Four centuries of changing landscape patterns in the longleaf pine ecosystem. *Proceedings of the Tall Timbers Fire Ecology Conference* 18: 17–44. [23]

Fry, B. 2007. *Stable Isotope Ecology*. Springer: New York. [5]

Funk, D. J., P. Nosil, and W. J. Etges. 2006. Ecological divergence exhibits consistently positive associations with reproductive isolation across disparate taxa. *Proceedings of the National Academy of Sciences USA* 103: 3209–3213. [6]

G

Galetti, M., and 14 others. 2013. Functional extinction of birds drives rapid evolutionary changes in seed size. *Science* 340: 1086–1090. [23]

Galloway, J. N., and 14 others. 2004. Nitrogen cycles: Past, present, and future. *Biogeochemistry* 70: 153–226. [25]

Galloway, J. N., and 8 others. 2008. Transformation of the nitrogen cycle: Recent trends, questions, and potential solutions. *Science* 320: 889–892. [25]

Garrison, V. H., and 9 others. 2003. African and Asian dust: From desert soils to coral reefs. *BioScience* 53: 469–480. [25]

Gascon, C., G. B. Williamson, and G. A. B. da Fonseca. 2000. Receding forest edges and vanishing reserves. *Science* 288: 1356–1358. [18]

Gasol, J. M., P. A. del Giorio, and C. M. Duarte. 1997. Biomass distribution in marine planktonic communities. *Limnology and Oceanography* 42: 1353–1363. [21]

Gaston, K. J. 2003. *The Structure and Dynamics of Geographic Ranges*. Oxford University Press: Oxford. [9]

Gaston, K. J., P. H. Williams, P. Eggleton, and C. J. Humphries. 1995. Large scale patterns of biodiversity: Spatial variation in family richness. *Proceedings of the Royal Society of London* 260B: 149–154. [18]

Gause, G. F. 1934. *The Struggle for Existence*. Williams & Wilkins: Baltimore, MD. [14]

Gause, G. F. 1934a. Experimental analysis of Vito Volterra's mathematical theory of the struggle for existence. *Science* 79: 16–17. [19]

Gause, G. F. 1935. Vérifications Expérimentales de la Théorie Mathématique de la Lutte pour la Vie. Hermann et Cie: Paris.

Geffeney, S. L., E. Fujimoto, E. D. Brodie III, E. D. Brodie Jr., and P. C. Ruben. 2005. Evolutionary diversification of TTX-resistant sodium channels in a predator–prey interaction. *Nature* 434: 759–763. [12]

Gilg, O., B. Sittler, and I. Hanski. 2009. Climate change and cyclic predator–prey population dynamics in the high Arctic. *Global Change Biology* 15: 2634–2652. [10]

Gilg, O., I. Hanski, and B. Sittler. 2003. Cyclic dynamics in a simple vertebrate predator-prey community. *Science* 302: 866-868. [10, 12]

Gilman, S. E., M. C. Urban, J. Tewksbury, G. W. Gilchrist, and R. D. Holt. 2010. A framework for community interactions under climate change. *Trends in Ecology and Evolution* 25: 325–331. [13]

Gleason, H. A. 1917. The structure and development of the plant association. *Bulletin of the Torrey Botanical Club* 44: 463–481. [17]

Glück, E. 1987. Benefits and costs of social foraging and optimal flock size in goldfinches (*Carduelis carduelis*). *Ethology* 74: 65–79. [8]

Godin, J. G. J., and S. E. Briggs. 1996. Female mate choice under predation risk in the guppy. *Animal Behaviour* 51: 117–130. [8]

González, C., O. Wang, S. E. Strutz, C. González-Salazar, V. Sánchez-Cordero, and S. Sarkar. 2010. Climate change and risk of leishmaniasis in North America: Predictions from ecological niche models of vector and reservoir species. *PLoS Neglected Tropical Diseases* 4: 1–16. [13]

Gotelli, N. J. and A. M. Ellison. 2013. *A Primer of Ecological Statistics*. 2nd ed. Oxford University Press/Sinauer: Sunderland, MA. [1]

Gough, L., G. R. Shaver, J. Carroll, D. L. Royer, and J. A. Laundre. 2000. Vascular plant species richness in Alaskan arctic tundra: The importance of soil pH. *Journal of Ecology* 88: 54–66. [22]

Goverde, M., K. Schweizer, B. Baur, and A. Erhardt. 2002. Small-scale habitat fragmentation effects on pollinator behaviour: Experimental evidence from the bumblebee *Bombus veteranus* on calcareous grasslands. *Biological Conservation* 104: 293–299. [24]

Grabherr, G., M. Gottfried, and H. Pauli. 1994. Climate effects on mountain plants. *Nature* 369: 448. [25]

Graham, I. M., and X. Lambin. 2002. The impact of weasel predation on cyclic field-vole survival: The specialist predator hypothesis contradicted. *Journal of Animal Ecology* 71: 946–956. [10]

Grant, B. R., and P. R. Grant. 2003. What Darwin's finches can teach us about the evolutionary origin and regulation of biodiversity. *BioScience* 53: 965–975. [6]

Grant, P. R., and B. R. Grant. 2006. Evolution of character displacement in Darwin's finches. *Science* 313: 224–226. [14]

Graven, H. D., and 13 others. 2013. Enhanced seasonal exchange of CO_2 by northern ecosystems since 1960. *Science* 341: 1085–1089. [25]

Grime, J. P. 1977. Evidence for the existence of three primary strategies in plants and its relevance to ecological and evolutionary theory. *American Naturalist* 111: 1169–1194. [7]

Groom, M. J., and 7 others. 1999. Buffer zones: Benefits and dangers of compatible stewardship. In *Continental Conservation: Scientific Foundations of Regional Reserve Networks*, M. E. Soulé and J. Terborgh (Eds.), pp. 171–197. Island Press: Washington, DC. [24]

Grousset, F. E., P. Ginoux, A. Bory, and P. E. Biscaye. 2003. Case study of a Chinese dust plume reaching the French Alps. *Geophysical Research Letters* 30: 1277. [25]

Grutter, A. S., J. M. Murphy, and J. H. Choat. 2003. Cleaner fish drives local fish diversity on coral reefs. *Current Biology* 13: 64–67. [15]

Guo, L. B. and R. M. Gifford. 2002. Soil carbon stocks and land use change: A meta analysis. *Global Change Biology* 8: 345–360. [3]

Gurevitch, J., L. L. Morrow, A. Wallace, and J. S. Walsh. 1992. A meta-analysis of competition in field experiments. *American Naturalist* 140: 539–572. [14]

Gustafsson, L., and T. Pärt. 1990. Acceleration of senescence in the collared flycatcher *Ficedula albicollis* by reproductive costs. *Nature* 347: 279–281. [7]

H

Haberl, H., K. Erb, and F. Krausmann. 2014. Human appropriation of net primary production: Patterns, trends, and planetary boundaries. *Annual Review of Environment and Resources* 39: 363–391. [23]

Hacker, S. D., and M. D. Bertness. 1996. Trophic consequences of a positive plant interaction. *American Naturalist* 148: 559–575. [16]

Hacker, S. D. and S. D. Gaines. 1997. Some implications of direct positive interactions for community species diversity. *Ecology* 78: 1990–2003. [19]

Hacker, S. D., and M. D. Bertness. 1999. Experimental evidence for the factors maintaining plant species diversity in a New England salt marsh. *Ecology* 80: 2064–2073. [19]

Haddad, N. M., and 11 others. 2014. A review of potential negative ecological effects of corridors. *Conservation Biology* 28: 1178–1187. [24]

Haddad, N. M., G. M. Crutsinger, K. Gross, J. Haarstad, and D. Tilman. 2011. Plant diversity and the stability of foodwebs. *Ecology Letters* 14: 42–46. [21]

Haines, B. L. 1978. Element and energy flows through colonies of the leaf-cutting ants, *Atta colombica*, in Panama. *Biotropica* 10: 270–277. [15]

Hairston Jr, N. G., and N. G. Hairston Sr. 1993. Cause-effect relationships in energy flow, trophic structure, and interspecific interactions. *American Naturalist* 142: 379–411. [21]

Halpern, B. S. and 18 others. 2008. A global map of human impact on marine ecosystems. *Science* 319: 948–952. [3, 23]

Hamilton, S. L., and J. E. Caselle. 2015. Exploitation and recovery of a sea urchin predator has implications for the resilience of southern California kelp forests. *Proceedings of the Royal Society of London* 282B: 20141817. [7]

Hansen, T. A. 1978. Larval dispersal and species longevity in lower Tertiary gastropods. *Science* 199: 885–887. [7]

Hanski, I. 1999. *Metapopulation Ecology*. Oxford University Press: New York. [9]

Harding, S. R., and J. R. Walters. 2002. Processes regulating the population dynamics of red-cockaded woodpecker cavities. *Journal of Wildlife Management* 66: 1083–1095. [23]

Hare, S. R. and R. C. Francis. 1994. Climate change and salmon production in the Northeast Pacific Ocean. In *Climate Change and Northern Fish Populations 121*, R. J. Beamish (Ed.), pp. 357–372. National Research Council of Canada: Ottawa. [2]

Harley, C. D. G. 2011. Climate change, keystone predation, and biodiversity loss. *Science* 334: 1124–1127. [23]

Harrison, G. W. 1995. Comparing predator-prey models to Luckinbill's experiment with *Didinium* and *Paramecium*. *Ecology* 76: 357–374. [12]

Harrison, P. 1987. *A Field Guide to Seabirds of the World*. Stephen Greene Press: Lexington, MA. [18]

Harrison, P. and F. Pearce. 2001. *AAAS Atlas of Population and the Environment*. University of California Press: Berkeley, CA. [3]

Hart, M. W., and R. R. Strathmann. 1994. Functional consequences of phenotypic plasticity in echinoid larvae. *Biological Bulletin* 186: 291–299. [7]

Hartl, D. L., and A. G. Clark. 2007. *Principles of Population Genetics*. 4th ed. Oxford University Press/Sinauer: Sunderland, MA. [6]

Hartmann, D. L., and 13 others. 2013. Observations: Atmosphere and surface. In *Climate Change 2013: The Physical Science Basis*, T. F. Stocker et al. (Eds.), pp. 159–254. Cambridge University Press: Cambridge. [25]

Hartnett, D. C., and G. W. T. Wilson. 1999. Mycorrhizae influence plant community structure and diversity in tallgrass prairie. *Ecology* 80: 1187–1195. [15]

Harvell, C. D., C. E. Mitchell, J. R. Ward, S. Altizer, A. P. Dobson, R. S. Ostfeld, and M. D. Samuel. 2002. Climate warming and disease risks for terrestrial and marine biota. *Science* 296: 2158–2162. [13]

Harvell, C. D., S. Altizer, I. M. Cattadori, L. Harrington, and E. Weil. 2009. Climate change and wildlife diseases: When does the host matter the most? *Ecology* 90: 912–920. [13]

Hattersley, P. W. 1983. The distribution of C_3 and C_4 grasses in Australia in relation to climate. *Oecologia* 57: 113–128. [5]

Havrilla, C.A., and 12 others. 2019. Towards a predictive framework for biocrust mediation of plant performance: A meta-analysis. *Journal of Ecology* 107: 2789–2807. [22]

Hawkins, C. E., and 11 others. 2006. Emerging disease and population decline of an island endemic, the Tasmanian devil *Sarcophilus harrisii*. *Biological Conservation* 131: 307–324. [23]

Hawlena, D., M. S. Strickland, M. A. Bradford, and O. J. Schmitz. 2012. Fear of predation slows plant-litter decomposition. *Science* 336: 1434–1438. [8]

Hays, J. D., J. Imbrie and N. J. Shackleton. 1976. Variations in the Earth's orbit: Pacemaker of the ice ages. *Science* 194: 1121–1132. [2]

Hearne, S. 1911. A Journey from Prince of Wales's Fort in Hudson's Bay to the Northern Ocean in the Years 1769, 1770, 1771 and 1772. The Champlain Society: Toronto. [4]

Hebblewhite, M., and 7 others. 2005. Human activity mediates a trophic cascade caused by wolves. *Ecology* 86: 2135–2144. [24]

Hejda, M., P. Pysek, and V. Jorosik. 2009. Impact of invasive plants on the species richness, diversity, and composition of invaded communities. *Journal of Ecology* 97: 393–403. [16]

Hellmann, J. J., J. E. Byers, B. G. Bierwagen, and J. S. Dukes. 2008. Five potential consequences of climate change for invasive species. *Conservation Biology* 22: 534–543. [19]

Henderson, S., P. Hattersley, S. von Caemmerer, and C. B. Osmond. 1995. Are C_4 pathway plants threatened by global climatic change? In *Ecophysiology of Photosynthesis*, E.-D. Schulze and M. M. Caldwell (Eds.), pp. 529–549. Springer: New York. [5]

Henry, M., J. M. Pons, and J. F. Cosson. 2007. Foraging behaviour of a frugivorous bat helps bridge landscape connectivity and ecological processes in a fragmented rainforest. *Journal of Animal Ecology* 76: 801–813. [24]

Herbert, D.A., J.H. Fownes, and P.M. Vitousek. 1999. Hurrican damage to a Hawaiian forest: nutrient supply affects resistance and resilience. *Ecology* 80: 908–920. [22]

Hersperger, A. M. 2006. Spatial adjacencies and interactions: Neighborhood mosaics for landscape ecological planning. *Landscape and Urban Planning* 77: 227–239. [24]

Hess, G. R. 1994. Conservation corridors and contagious disease: A cautionary note. *Conservation Biology* 8: 256–262. [24]

Hetrick, B. A. D., G. W. T. Wilson, and D. C. Hartnett. 1989. Relationship between mycorrhizal dependence and competitive ability of two tallgrass prairie grasses. *Canadian Journal of Botany* 67: 2608–2615. [15]

Hickman, J. E., S. Wu, L. J. Mickley, and M. T. Lerdau. 2010. Kudzu (*Pueraria montana*) invasion doubles emissions of nitric oxide and increases ozone pollution. *Proceedings of the National Academy of Sciences USA* 107: 10115–10119. [23]

Higuera, P. E., B. N. Schuman, and K. D. Wolf. 2021. Rocky Mountain subalpine forests now burning more than any time in recent millennia. *Proceedings of the National Academy of Science USA* 118: e2103135118.

Hik, D. S., and R. L. Jefferies. 1990. Increases in the net above-ground primary production of a salt-marsh forage grass: A test of the predictions of the herbivore-optimization model. *Journal of Ecology* 78: 180–195. [12]

Hilderbrand, G. V., S. D. Farley, C. T. Robbins, T. A. Hanley, K. Titus, and C. Servheen. 1996. Use of stable isotopes to determine diets of living and extinct bears. *Canadian Journal of Zoology* 74: 2080–2088. [20]

Hill, J., and R. Hill. 2001. Why are tropical rain forests so species rich? Classifying, reviewing, and evaluating theories. *Progress in Physical Geography* 25: 326–354. [19]

Hiller, N. L., S. Bhattacharjee, C. van Ooij, K. Liolios, T. Harrison, C. Lopez-Estraño, and K. Haldar. 2004. A host-targeting signal in virulence proteins reveals a secretome in malarial infection. *Science* 306: 1934–1937. [13]

Hobbs, R. J., E. Higgs, and J. A. Harris. 2009. Novel ecosystems: Implications for conservation and restoration. *Trends in Ecology and Evolution* 24: 599–605. [24]

Hochochka, P. W., and G. N. Somero. 2002. *Biochemical Adaptation, Mechanism and Process in Physiological Evolution.* Oxford University Press: Oxford. [4]

Hof, C., I. Levinsky, M. B. Araujo and C. Rahbek. 2011. Rethinking species' ability to cope with rapid climate change. *Global Change Biology* 17: 2987–2990. [1]

Hof, C., M. B. Araújo, W. Jetz and C. Rahbek. 2011. Additive threats from pathogens, climate and land-use change for global amphibian diversity. *Nature* 480: 516–519. [1]

Hoffmann, M., and 173 others. 2010. The impact of conservation on the status of the world's vertebrates. *Science* 330: 1503–1509. [23]

Hölldobler, B., and E. O. Wilson. 1990. *The Ants.* Belknap Press of Harvard University Press: Cambridge, MA. [15]

Hollinger, D. Y., and 14 others. 1998. Forest–atmosphere carbon dioxide exchange in eastern Siberia. *Agricultural and Forest Meteorology* 90: 291–306. [20]

Holt, B. G. and 14 others. 2013. An update of Wallace's zoogeographic regions of the world. *Science* 339: 74–78. [1, 18]

Hooper, R. G., W. E. Taylor, and S. C. Loeb. 2004. Long-term efficacy of artificial cavities for red-cockaded woodpeckers: Lessons learned from Hurricane Hugo. In *Red-Cockaded Woodpecker: Road to Recovery*, R. Costa and S. J. Daniels (Eds.), pp. 430–438. Hancock House: Blaine, WA. [23]

Hoover, K., M. Grove, M. Gardner, D. P. Hughes, J. McNeil, and J. Slavicek. 2011. A gene for an extended phenotype. *Science* 333: 1401. [13]

Houserová, P., V. Kubáň, S. Kráčmar, and J. Sitko. 2007. Total mercury and mercury species in birds and fish in an aquatic ecosystem in the Czech Republic. *Environmental Pollution* 145: 185–194. [21]

Howarth, R. W., and 14 others. 1996. Regional nitrogen budgets and riverine N & P fluxes for the drainages to the North Atlantic Ocean: Natural and human influences. *Biogeochemistry* 35: 75–139. [22, 25]

Hu, J., D. J. P. Moore, S. P. Burns, and R. K. Monson. 2010. Longer growing seasons lead to less carbon sequestration by a subalpine forest. *Global Change Biology* 16: 771–783. [25]

Hubbell, S. 2001. *The Unified Neutral Theory of Biodiversity and Biogeography.* Princeton University Press: Princeton, NJ. [19]

Hudson, P. J., A. P. Dobson, and D. Newborn. 1998. Prevention of population cycles by parasite removal. *Science* 282: 2256–2258. [13]

Huete, A. R. 2004. Remote Sensing for Environmental Monitoring, In *Environmental Monitoring and Characterization*, J. F. Artiola et al. [Eds.], pp. 183–206. Academic Press: Amsterdam

Huffaker, C. B. 1958. Experimental studies on predation: Dispersion factors and predator–prey oscillations. *Hilgardia* 27: 343–383. [12]

Huffaker, C. B., and C. E. Kennett. 1959. A ten-year study of vegetational changes associated with biological control of Klamath weed. *Journal of Range Management* 12: 69–82. [12]

Huffman, M. A. 1997. Current evidence for self-medication in primates: A multidisciplinary perspective. *Yearbook of Physical Anthropology* 40: 171–200. [13]

Hughes, F., P. M. Vitousek, and T. Tunison. 1991. Alien grass invasion and fire in the seasonal submontane zone of Hawai'i. *Ecology* 72: 743–746. [9]

Hughes, J. B., J. J. Hellman, T. H. Ricketts, and B. J. M. Bohannan. 2001. Counting the uncountable: Statistical approaches to estimating microbial diversity. *Applied and Environmental Microbiology* 67: 4399–4406. [16]

Humphreys, W. F. 1979. Production and respiration in animal populations. *Journal of Animal Ecology* 48: 427–454. [21]

Hunt, C. E., and S. Behrens Yamada. 2003. Biotic resistance experienced by an invasive crustacean in a temperate estuary. *Biological Invasions* 5: 33–43. [19]

Hunt, G. R. 1996. Manufacture and use of hook-tools by New Caledonian crows. *Nature* 379: 249–251. [5]

Hunt, G. R., and R. D. Gray. 2003. Diversification and cumulative evolution in New Caledonian crow tool manufacture. *Proceedings of the Royal Society of London* 270B: 867–874. [5]

Huston, M. 1979. A general hypothesis of species diversity. *American Naturalist* 113: 81–101. [19]

Hutchinson, G. E. 1959. Homage to Santa Rosalia or why are there so many kinds of animals? *American Naturalist* 93: 145–159. [18]

Hutchinson, G. E. 1961. The paradox of the plankton. *American Naturalist* 95: 137–145. [19]

I

Inouye, D. W. 2008. Effects of climate change on phenology, frost damage, and floral abundance of montane wildflowers. *Ecology* 89: 353–362. [25]

IPCC. 2013. *Climate Change 2013: The Physical Science Basis.* Cambridge University Press: Cambridge. Available at www.ipcc.ch/report/ar5/wg1/. [25]

IPCC. 2021: *Climate Change 2021: The Physical Science Basis.* Contribution of Working Group I to the Sixth Assessment Report of the Intergovernmental Panel on Climate Change [Masson-Delmotte, V., P. Zhai, A. Pirani, S.L.Connors, C. Péan, S. Berger, N. Caud, Y. Chen, L. Goldfarb, M.I. Gomis, M. Huang, K. Leitzell, E. Lonnoy, J.B.R.Matthews, T.K. Maycock, T. Waterfield, O. Yelekçi, R. Yu, and B. Zhou (eds.)]. Cambridge University Press, Cambridge, United Kingdom and New York, NY, USA, 2391 pp. doi:10.1017/9781009157896.

IUCN. 2022. Summary statistics table 1a. In: The IUCN Red List of Threatened Species, version 2022-1. https://www.iucnredlist.org/resources/summary-statistics. [23]

J

Jablonski, D. 1995. Extinctions in the fossil record. In *Extinction Rates*, J. H. Lawton and R. M. May (Eds.), pp. 25–44. Oxford University Press: Oxford. [6]

Jablonski, D. 2005. Mass extinctions and macroevolution. *Paleobiology* 31: 192–210. [23]

Jablonski, D., K. Roy, and J. W. Valentine. 2006. Out of the tropics: Evolutionary dynamics of the latitudinal diversity gradient. *Science* 314: 102–106. [18]

Jablonski, N. G. 2004. The evolution of human skin and skin color. *Annual Review of Anthropology* 33: 585–623. [25]

Jablonski, N. G. 2006. *Skin: A Natural History*. University of California Press: Berkeley, CA. [4]

Jacobson, M. Z. 2002. *Atmospheric Pollution, History, Science, and Regulation*. Cambridge University Press: Cambridge. [25]

Jaenike, J., and T. D. Brekke. 2011. Defensive endosymbionts: A cryptic trophic level in community ecology. *Ecology Letters* 14: 150–155. [13]

Jaenike, J., R. Unckless, S. N. Cockburn, L. M. Boelio, and S. J. Perlman. 2010. Adaptation via symbiosis: Recent spread of a *Drosophila* defensive symbiont. *Science* 329: 212–215. [15]

Jaffe, D., I. McKendry, T. Anderson, and H. Price. 2003. Six "new" episodes of trans-Pacific transport of air pollutants. *Atmospheric Environment* 37: 391–404. [25]

Janzen, D. H. 1966. Coevolution of mutualism between ants and acacias in Central America. *Evolution* 20: 249–275. [15]

Janzen, D. H. 1986. The future of tropical ecology. *Annual Review of Ecology and Systematics* 17: 305–324. [23]

Jefferies, R. L., R. F. Rockwell, and K. F. Abraham. 2003. The embarrassment of riches: Agricultural food subsidies, high goose numbers, and loss of Arctic wetlands—a continuing saga. *Environmental Reviews* 11: 193–232. [12]

Jentsch, A., J. Kreyling and C. Beierkuhnlein. 2007. A new generation of climate change experiments: Events, not trends. *Frontiers in Ecology and the Environment* 5: 315–324. [2]

Jeschke, J. M. 2007. When carnivores are "full and lazy." *Oecologia* 152: 357–364. [8]

Johnson, D. M., and 7 others. 2010. Climate change disrupts recurrent Alpine insect outbreaks. *Proceedings of the National Academy of Sciences USA* 107: 20576–20581. [10]

Johnson, P. T. J., K. B. Lunde, E. G. Ritchie and A. E. Launer. 1999. The effect of trematode infection on amphibian limb development and survivorship. *Science* 284: 802–804. [1]

Johnson, P. T. J. and 7 others. 2007. Aquatic eutrophication promotes pathogenic infection in amphibians. *Proceedings of the National Academy of Sciences USA* 104: 15781–15786. [1]

Johnson, P. T. J., P. Lund, R. B. Hartson, and T. Yoshino. 2009. Community diversity reduces *Schistosoma mansoni* transmission and human infection risk. *Proceedings of the Royal Society of London* 276B: 1657–1663. [19]

Johnson, R. G., and S. A. Temple. 1990. Nest predation and brood parasitism of tallgrass prairie birds. *Journal of Wildlife Management* 54: 106–111. [24]

Johnson, W. E., and 15 others. 2010. Genetic restoration of the Florida panther. *Science* 329: 1641–1645. [23]

Jones, C. G., J. H. Lawton, and M. Shachak. 1994. Organisms as ecosystem engineers. *Oikos* 69: 373–386. [16]

Jones, C. G., J. H. Lawton, and M. Shachak. 1997. Positive and negative effects of organisms as physical ecosystem engineers. *Ecology* 78: 1946–1957. [16]

Jones, H. L., and J. M. Diamond. 1976. Short-time-base studies of turnover in breeding bird populations on the California Channel Islands. *Condor* 78: 526–549. [10]

Jones, K. E., N. G. Patel, M. A. Levy, A. Storeygard, D. Balk, J. L. Gittleman, and P. Daszak. 2008. Global trends in emerging infectious diseases. *Nature* 451: 990–993. [19]

Jonkman, J. C. M. 1980. The external and internal structure and growth of nests of the leaf-cutting ant *Atta vollenweideri* Forel, 1893 (Hym.: Formicidae), part I. *Zeitschrift fur Angewandte Entomologie* 89:158-173. [15]

Jordan, M. J., and G. E. Likens. 1975. An organic carbon budget for an oligotrophic lake in New Hampshire. *Verhandlungen Internationale Vereinigung fur Theoretische and Angewandte Limnologie* 19: 994–1003. [21]

Jousson, O., and 8 others. 2000. Invasive alga reaches California. *Nature* 408: 157–158. [16]

K

Kacsoh, B. Z., Z. R. Lynch, N. T. Mortimer, and T. A. Schlenke. 2013. Fruit flies medicate offspring after seeing parasites. *Science* 339: 947–950. [8]

Karasov, W. H., and C. Martínez del Rio. 2007. *Physiological Ecology: How Animals Process Energy, Nutrients, and Toxins*. Princeton University Press: Princeton, NJ. [5]

Karban, R., and G. English-Loeb. 1997. Tachinid parasitoids affect host plant choice by caterpillars to increase caterpillar survival. *Ecology* 78: 603–611. [13]

Karell, P., K. Ahola, T. Karstinen, J. Valkama, and J. E. Brommer. 2011. Climate change drives microevolution in a wild bird. *Nature Communications* 2: 208. [6]

Kauffman, M. J., J. F. Brodie, and E. S. Jules. 2010. Are wolves saving Yellowstone's aspen? A landscape-level test of a behaviorally mediated trophic cascade. *Ecology* 91: 2742–2755. [24]

Keeling, C. D., S. C. Piper, R. B. Bacastow, M. Wahlen, T. P. Whorf, M. Heimann, and H. A. Meijer. 2001. Exchanges of atmospheric CO_2 and $13CO_2$ with the terrestrial biosphere and oceans from 1978 to 2000. *I. Global aspects*, SIO Reference Series, No 01-06. Scripps Institution of Oceanography: San Diego, CA. Data last updated August 2019. [2]

Keenan, R.J., G. A. Reams, F. Achard, J. V. de Freitas, A. Grainger, E. Lindquist. 2015. Dynamics of global forest area: Results from the FAO Global Forest Resources Assessment. Forest Ecology and Management 352: 9-20.

Keesing, F., and 10 others. 2010. Impacts of biodiversity on the emergence and transmission of infectious diseases. *Nature* 468: 647–652. [19]

Keever, K. 1953. Present composition of some stands of the former oak–chestnut forest in the southern Blue Ridge Mountains. *Ecology* 34: 44–54. [13]

Keller, I., and C. R. Largiadèr. 2003. Recent habitat fragmentation caused by major roads leads to reduction of gene flow and loss of genetic variability in ground beetles. *Proceedings of the Royal Society of London* 270B: 417–423. [24]

Kenward, B., A. A. S. Weir, C. Rutz, and A. Kacelnik. 2005. Tool manufacture by naive juvenile crows. *Nature* 433: 121. [5]

Kenward, R. E. 1978. Hawks and doves: Factors affecting success and selection in goshawk attacks on woodpigeons. *Journal of Animal Ecology* 47: 449–460. [8]

Kerr, J. T., and 13 others. 2015. Climate change impacts on bumblebees converge across continents. *Science* 349: 177–180. [9]

Kerr, P. J., and S. M. Best. 1998. Myxoma virus in rabbits. *Revue scientifique et technique – Office international des epizooties* 17: 256–268. [13]

Kessler, A., and I. T. Baldwin. 2001. Defensive function of herbivore-induced plant volatile emissions in nature. *Science* 291: 2141–2144. [12]

Kettlewell, H. B. D. 1955. Selection experiments on industrial melanism in the Lepidoptera. *Heredity* 9: 323–342. [6]

Kettlewell, H. B. D. 1956. Further selection experiments on industrial melanism in the Lepidoptera. *Heredity* 10: 287–301. [6]

Kideys, A. E. 2002. Fall and rise of the Black Sea ecosystem. *Science* 297: 1482–1484. [10]

Kiehl, J. T. and K. E. Trenberth. 1997. Earth's annual global mean energy budget. *Bulletin of the American Meteorological Society* 78: 197–208. [2]

Kiers, E. T., and 14 others. 2011. Reciprocal rewards stabilize cooperation in the mycorrhizal symbiosis. *Science* 333: 880–882. [15]

Kiesecker, J. M. 2002. Synergism between trematode infection and pesticide exposure: A link to amphibian limb deformities in nature? *Proceedings of the National Academy of Sciences USA* 99: 9900–9904. [1]

Kinchy, A. J. 2006. On the borders of post-war ecology: Struggles over the Ecological Society of America's preservation committee, 1917–1946. *Science as Culture* 15: 23–44. [23]

King, A. A., and W. M. Schaffer. 2001. The geometry of a population cycle: A mechanistic model of snowshoe hare demography. *Ecology* 82: 814–830. [12]

Kingsland, S. E. 1991. Defining ecology as a science. In *Foundations of Ecology: Classic Papers with Commentaries*, L. A. Real and J. H. Brown (Eds.), pp. 1–13. University of Chicago Press: Chicago. [17]

Kinnison, M. T., and A. P. Hendry. 2001. The pace of modern life II: From rates of contemporary evolution to pattern and process. *Genetica* 112–113: 145–164. [6]

Kinzig, A. P., and J. Harte. 2000. Implications of endemics–area relationships for estimates of species extinctions. *Ecology* 81: 3305–3311. [23]

Kitzes, J., and M. Wackernagel. 2009. Answers to common questions in Ecological Footprint accounting. *Ecological Indicators* 9: 812–817. [11]

Knapp, A. K. and 9 others. 2002. Rainfall variability, carbon cycling and plant species diversity in a mesic grassland. *Science* 298: 2202–2205. [3]

Kodric-Brown, A., and J. M. Brown. 1993. Highly structured fish communities in Australian desert springs. *Ecology* 74: 1847–1855. [18]

Kohler, S. L., and M. J. Wiley. 1997. Pathogen outbreaks reveal large-scale effects of competition in stream communities. *Ecology* 78: 2164–2176. [13]

Köhler, W. 1927. *The Mentality of Apes*. Routledge & Kegan Paul: London. [5]

Komdeur, J. 1992. Importance of habitat saturation and territory quality for evolution of cooperative breeding in the Seychelles warbler. *Nature* 358: 493–495. [9]

Kondoh, M. 2003. Foraging adaptation and the relationship between food-web complexity and stability. *Science* 299: 1388–1391. [21]

Kopp, R. E., and 9 others. 2016. Temperature-driven global sea-level variability in the Common Era. *Proceedings of the National Academy of Sciences USA* 113: E1434–E1441. [25]

Korpimäki, E., and K. Norrdahl. 1998. Experimental reduction of predators reverses the crash phase of small-rodent cycles. *Ecology* 79: 2448–2455. [10]

Koskela, T., S. Puustinen, V. Salonen, and P. Mutikainen. 2002. Resistance and tolerance in a host plant–holoparasitic plant interaction: Genetic variation and costs. *Evolution* 56: 899–908. [5]

Kozaki, A., and G. Takeba. 1996. Photorespiration protects C_3 plants from photooxidation. *Nature* 384: 557–580. [5]

Kraaijeveld, A. R., K. A. Hutcheson, E. C. Limentani, and H. C. J. Godfray. 2001. Costs of counterdefenses to host resistance in a parasitoid of *Drosophila*. *Evolution* 55: 1815–1821. [13]

Krebs, C. J., and 7 others. 1995. Impact of food and predation on the snowshoe hare cycle. *Science* 269: 1112–1115. [12]

Krebs, C. J. 1999. *Ecological Methodology*. Addison Wesley: Menlo Park, CA. [9]

Krebs, J. B., J. T. Eriksen, M. T. Webber, and E. L. Charnov. 1977. Optimal prey selection in the great tit (*Parus major*). *Animal Behaviour* 25: 30–38. [8]

Kremen, C. 2014. Giving back: Nature conservation in Madagascar. *Journal of Research Practice* 10: 8. [24]

Kremen, C., V. Razafimahatratra, R. P. Guillery, J. Rakotomalala, A. Weiss, and J. S. Ratsisompatrarivo. 1999. Designing the Masoala National Park in Madagascar based on biological and socioeconomic data. *Conservation Biology* 13: 1055–1068. [24]

Kricher, J. 1998. *A Field Guide to Eastern Forests*. Illustrated by G. Morrison. Houghton Mifflin: New York. [18]

Krümmel, E. M., and 7 others. 2003. Delivery of pollutants by spawning salmon. *Nature* 425: 255. [21]

Krützen, M., J. Mann, M. R. Heithaus, R. C. Connor, L. Bejder, and W. B. Sherwin. 2005. Cultural transmission of tool use in bottlenose dolphins. *Proceedings of the National Academy of Sciences USA* 102: 8939–8943. [5]

Kuhnlein, H. V., O. Receveur, D. C. G. Muir, H. M. Chan, and R. Soueida. 1995. Arctic indigenous women consume greater than acceptable levels of organochlorines. *Journal of Nutrition* 125: 2501–2510. [21]

Kurz, W. A., and 7 others. 2008. Mountain pine beetle and forest carbon feedback to climate change. *Nature* 452: 987–990. [20]

L

Lack, D. 1947. *Darwin's Finches*. Cambridge University Press: Cambridge. [14]

Lack, D. 1947. The significance of clutch size. *Ibis* 89: 302–352. [7]

Lambers, H., T. L. Pons, and F. S. Chapin III. 2008. *Plant Physiological Ecology*. 2nd ed. Springer: New York. [5]

Lande, R. 1988. Demographic models of the northern spotted owl (*Strix occidentalis caurina*). *Oecologia* 75: 601–607. [9]

Landesman, W. J., B. F. Allan, R. B. Langerhans, T. M. Knight and J. M. Chase. 2007. Inter-annual associations between precipitation and human incidence of West Nile virus in the United States. *Vector-Borne and Zoonotic Diseases* 7: 337–343. [1]

Langerhans, R. B., M. E. Gifford, and E. O. Joseph. 2007. Ecological speciation in *Gambusia* fishes. *Evolution* 61: 2056–2074. [6]

Larcher, W. 1980. *Physiological Plant Ecology*. Springer: Berlin. [20]

Lauenroth, W. K., and O. E. Sala. 1992. Long-term forage production of North American shortgrass steppe. *Ecological Applications* 2: 397–403. [20]

Laurance, W. F. 2001. The hyper-diverse flora of the Central Amazon. In *Lessons from Amazonia: The Ecology and Conservation of a Fragmented Forest*, R. O. Bierregaard Jr. et al. (Eds.), pp. 47–53. Yale University Press: New Haven, CT. [18]

Laurance, W. F., and 10 others. 2002. Ecosystem decay of Amazonian forest fragments: A 22-year investigation. *Conservation Biology* 16: 605–618. [18, 24]

Law, K.L. 2017. Plastics in the marine environment. *Annual Review of Marine Science* 9:205–229. [3]

Law, R. 1975. Colonization and the evolution of life histories in *Poa annua*. Unpublished PhD thesis, University of Liverpool. [11]

Law, R., A. D. Bradshaw, and P. D. Putwain. 1977. Life-history variation in *Poa annua*. *Evolution* 31: 233–246. [7]

Lawler, S. P. 1993. Species richness, species composition and population dynamics of protists in experimental microcosms. *Journal of Animal Ecology* 62: 711–719. [21]

Lawton, J. H. 2000. *Community Ecology in a Changing World*. Excellence in Ecology, O. Kinne (Ed.), no. 11. Ecology Institute: Luhe, Germany. [19]

Layne, J. R., and R. E. Lee. 1995. Adaptations of frogs to survive freezing. *Climate Research* 5: 53–59. [4]

Lennartsson, T., P. Nilsson, and J. Tuomi. 1998. Induction of overcompensation in the field gentian, *Gentianella campestris*. *Ecology* 79: 1061–1072. [12]

Lenoir, J. and J. C. Svenning. 2015. Climate-related range shifts–a global multidimensional synthesis and new research directions. *Ecography* 38: 15–28. [1]

Levin, S. A. 1992. The problem of pattern and scale in ecology. *Ecology* 73: 1943–1967. [24]

Levins, R. 1969. Some demographic and genetic consequences of environmental heterogeneity for biological control. *Bulletin of the Entomological Society of America* 15: 237–240. [9]

Levins, R. 1970. Extinction. In *Some Mathematical Problems in Biology*, M. Gerstenhaber (Ed.), pp. 75–107. American Mathematical Society: Providence, RI. [9]

Lewison, R. L., L. B. Crowder, A. J. Read, and S. A. Freeman. 2004. Understanding impacts of fisheries bycatch on marine megafauna. *Trends in Ecology and Evolution* 19: 598–604. [23]

Lewontin, R. C. 1969. The meaning of stability. In *Diversity and Stability in Ecological Systems*, pp. 13–24. Brookhaven Symposia in Biology, no. 22. Brookhaven National Laboratories: Brookhaven, NY. [17]

Likens, G. E., and F. H. Bormann. 1995. *Biogeochemistry of a Forested Ecosystem*. Springer: New York. [22]

Lima, S. L. 1998. Nonlethal effects in the ecology of predator–prey interactions. *BioScience* 48: 25–34. [8]

Lindeman, R. L. 1942. The trophic–dynamic aspect of ecology. *Ecology* 23: 399–418. [20]

Lindenmayer, D. B., and M. A. McCarthy. 2006. Evaluation of PVA models of arboreal marsupials: Coupling models with long-term monitoring data. *Biodiversity and Conservation* 15: 4079–4096. [23]

Lindqvist, C., and 13 others. 2010. Complete mitochondrial genome of a Pleistocene jawbone unveils the origin of polar bear. *Proceedings of the National Academy of Sciences USA* 107: 5053–5057. [9]

Ling, S. D. 2008. Range expansion of a habitat-modifying species leads to loss of taxonomic diversity: A new and impoverished reef state. *Oecologia* 156: 883–894. [9]

Ling, S. D., C. R. Johnson, S. D. Frusher, and K. R. Ridgway. 2009. Over-fishing reduces resilience of kelp beds to climate-driven catastrophic phase shift. *Proceedings of the National Academy of Sciences USA* 106: 22341–22345. [9]

Lippincott, C. L. 2000. Effects of *Imperata cylindrica* (L.) Beauv. (cogongrass) invasion on fire regime in Florida sandhill. *Natural Areas Journal* 20: 140–149. [23]

Little, D. P., and D. W. Stevenson. 2007. A comparison of algorithms for the identification of specimens using DNA barcodes: Examples from gymnosperms. *Cladistics* 23: 1–21. [23]

Little, E. L., Jr. 1970. *Atlas of United States Trees*. U.S. Department of Agriculture, Forest Service, and other publications. [13]

Liu, H., E. S. Menges, and P. F. Quintana-Ascencio. 2005. Population viability analyses of *Chamaecrista keyensis*: Effects of fire season and frequency. *Ecological Applications* 15: 210–221. [23]

Livdahl, T. P., and M. S. Willey. 1991. Prospects for an invasion: Competition between *Aedes albopictus* and native *Aedes triseriatus*. *Science* 253: 189–191. [14]

Lively, C. M. 1989. Adaptation by a parasitic trematode to local populations of its snail host. *Evolution* 43: 1663–1671. [13]

Locey, K. J., and J. T. Lennon. 2016. Scaling laws predict global microbial diversity. *Proceedings of the National Academy of Sciences USA* 113: 5970–5975. [23]

Logan, J. A., and J. A. Powell. 2001. Ghost forest, global warming, and the mountain pine beetle (Coleoptera: Scolytidae). *American Entomologist* 47: 160–172. [24]

Lomolino, M. V., B. R. Riddle, and J. H. Brown. 2006. *Biogeography*, 3rd ed. Oxford University Press/Sinauer: Sunderland, MA. [18]

Lomolino, M. V., J. H. Brown, and R. Davis. 1989. Island biogeography of montane forest mammals in the American Southwest. *Ecology* 70: 180–194. [18]

Lotka, A. J. 1932. The growth of mixed populations: Two species competing for a common food supply. *Journal of the Washington Academy of Sciences* 22: 461–469. [14]

Louda, S. M., D. Kendall, J. Conner, and D. Simberloff. 1997. Ecological effects of an insect introduced for the biological control of weeds. *Science* 277: 1088–1090. [9]

Lukas, D., and T. H. Clutton-Brock. 2013. The evolution of social monogamy in mammals. *Science* 341: 526–530. [8]

Lüthi, D., and 10 others. 2008. High resolution carbon dioxide concentrations record 650,000–800,000 years before present. *Nature* 453: 379–382. [25]

M

MacArthur, R. H. 1955. Fluctuations in animal populations and a measure of community stability. *Ecology* 36: 533–536. [19]

MacArthur, R. H. 1958. Population ecology of some warblers of Northeastern coniferous forests. *Ecology* 39: 599–619. [19]

MacArthur, R. H., and E. O. Wilson. 1963. An equilibrium theory of insular zoogeography. *Evolution* 17: 373–387. [18]

MacArthur, R. H., and E. O. Wilson. 1967. *The Theory of Island Biogeography*. Princeton University Press: Princeton, NJ. [7, 18]

MacArthur, R. H., and J. W. MacArthur. 1961. On bird species diversity. *Ecology* 42: 594–598. [19]

MacDonald, I. A. W., L. L. Loope, M. B. Usher, and O. Hamann. 1989. Wildlife conservation and the invasion of nature reserves by introduced species: A global perspective. In *Biological Invasions: A Global Perspective*, J. A. Drake et al. (Eds.), pp. 215–256. SCOPE (Scientific Committee on Problems of the Environment), 37. John Wiley & Sons: Chichester, UK. [23]

MacLulich, D. A. 1936. Sunspots and the abundance of animals. *Journal of the Royal Astronomical Society of Canada* 30: 233–246. [12]

Macnair, M. R., and P. Christie. 1983. Reproductive isolation as a pleiotropic effect of copper tolerance in *Mimulus guttatus*. *Heredity* 50: 295–302. [6]

MacPhee, R. D. E. and P. A. Marx. 1997. Humans, hyperdisease, and first-contact extinctions. In *Natural Change and Human Impact in Madagascar*, S. M. Goodman and B. D. Patterson (Eds.), pp. 169–217. Smithsonian Institution Press: Washington, DC. [3]

Madigan, M. T., and J. M. Martinko. 2005. *Brock Biology of Microorganisms*. Prentice Hall: Upper Saddle River, NJ. [5]

Madronich, S., R. L. McKenzie, L. O. Björn, and M. M. Caldwell. 1998. Changes in biologically active ultraviolet radiation reaching the Earth's surface. *Journal of Photochemistry and Photobiology: Biology* 46B: 5–19. [25]

Mahowald, N., and 8 others. 2005. The atmospheric global dust cycle and iron inputs to the ocean. *Global Biogeochemical Cycles* 19: GB4025. [25]

Maitland, D. P. 1994. A parasitic fungus infecting yellow dungflies manipulates host perching behaviour. *Proceedings of the Royal Society of London* 258B: 187–193. [13]

Mannion, P. D., P. Upchurch, R. B. J. Benson, and A. Goswami. 2014. The latitudinal biodiversity gradient through deep time. *Trends in Ecology and Evolution* 29: 42–50. [18]

Mantua, N. J. 2001. The Pacific Decadal Oscillation. In *The Encyclopedia of Global Environmental Change*, vol. 1, *The Earth System: Physical and Chemical Dimensions of Global Environmental Change*, M. C. McCracken and J. S. Perry (Eds.), pp. 592–594. Wiley: New York. [2]

Mantua, N. J., S. R. Hare, Y. Zhang, J. M. Wallace and R. C. Francis. 1997. A Pacific interdecadal climate oscillation with impacts on salmon production. *Bulletin of the American Meteorological Society* 78: 1069–1079. [2]

Mantua, N. J. and S. R. Hare. 2002. The Pacific Decadal Oscillation. *Journal of Oceanography* 58: 35–44. [2]

Margoluis, R., and N. Salafsky. 1998. Measures of Success: Designing, Managing, and Monitoring Conservation and Development Projects. Island Press: Washington, DC. [24]

Maron, J. L., and M. Vila. 2001. When do herbivores affect plant invasion? Evidence for the natural enemies and biotic resistance hypotheses. *Oikos* 95: 361–373. [19]

Marr, J. W. 1967. *Ecosystems of the East Slope of the Front Range of Colorado*. University of Colorado Studies, Series in Biology, no. 8. University of Colorado Press: Boulder, CO. [3]

Marschner, H. 1995. *Mineral Nutrition of Higher Plants*. Academic Press: San Diego, CA. [22]

Marti, M., R. T. Good, M. Rug, E. Knuepfer, and A. F. Cowman. 2004. Targeting malaria virulence and remodeling proteins to the host erythrocyte. *Science* 306: 1930–1933. [13]

Martin, J. H., and 43 others. 1994. Testing the iron hypothesis in ecosystems of the equatorial Pacific Ocean. *Nature* 371: 123–129. [20]

Martin, J. K., and A. A. Martin. 2007. Resource distribution influences mating system in the bobuck (*Trichosurus cunninghami*: Marsupialia). *Oecologia* 154: 227–236. [8]

Martin, P. A., A. C. Newton and J. M. Bullock. 2013. Carbon pools recover more quickly than plant diversity in tropical secondary forests. *Proceedings of the Royal Society of London* 280B: 20132236. [3]

Martin, P. S. 1984. Prehistoric overkill: The global model. In *Quaternary Extinctions: A Prehistoric Revolution*, P. S. Martin and R. G. Klein (Eds.), pp. 354–403. University of Arizona Press: Tucson, AZ. [3]

Martin, P. S. 2005. *Twilight of the Mammoths: Ice Age Extinctions and the Rewilding of America*. University of California Press: Berkeley, CA. [3]

Maslin, M. A., and E. Thomas. 2003. Balancing the deglacial global carbon budget: The hydrate factor. *Quaternary Science Reviews* 22: 1729–1736. [5]

Mass, C., and A. Robock. 1982. The influence of the Mount St. Helens volcanic eruption on surface temperature in the northwest United States. *Monthly Weather Review* 110: 614–622. [17]

Matsui, H., and 8 others. 2016. Adaptive bill morphology for enhanced tool manipulation in New Caledonian crows. *Scientific Reports* 6: 22776. [5]

Mattson, W. J. Jr. 1980. Herbivory in relation to plant nitrogen content. *Annual Review of Ecology and Systematics* 11: 119–161. [12]

May, R. M. 1973. *Stability and Complexity in Model Ecosystems*. Princeton University Press: Princeton, NJ. [21]

May, R. M. 2011. Why worry about how many species and their loss? *PLOS Biology* 9: e1001130. [23]

May, R. M., and R. M. Anderson. 1983. Epidemiology and genetics in the coevolution of parasites and hosts. *Proceedings of the Royal Society of London* 219B: 281–313. [13]

May, R. M., J. H. Lawton, and N. E. Stork. 1995. Assessing extinction rates. In *Extinction Rates*, J. H. Lawton and R. M. May (Eds.), pp. 1–24. Oxford University Press: Oxford. [23]

McCann, K., A. Hastings, and G. R. Huxel. 1998. Weak trophic interactions and the balance of nature. *Nature* 395: 794–798. [21]

McClain, M. E., and 11 others. 2003. Biogeochemical hot spots and hot moments at the interface of terrestrial and aquatic ecosystems. *Ecosystems* 6: 301–312. [24]

McClenachan, L. 2009. Documenting the loss of large trophy fish from the Florida Keys with historical photographs. *Conservation Biology* 23: 636–643. [23]

McCracken, K. G., and 9 others. 2009. Parallel evolution in the major haemoglobin genes of eight species of Andean waterfowl. *Molecular Ecology* 18: 3992–4005. [6]

McCreless, E., and 7 others. 2016. Past and estimated future impact of invasive alien mammals on insular threatened vertebrate populations. *Nature Communications* 7: 12488. https://doi.org/10.1038/ncomms12488

McDonald, M. J., D. P. Rice, and M. M. Desai. 2016. Sex speeds adaptation by altering the dynamics of molecular evolution. *Nature* 531: 233–236. [7]

McEvoy, P., C. Cox, and E. Coombs. 1991. Successful biological control of ragwort, *Senecio jacobaea*, by introduced insects in Oregon. *Ecological Applications* 1: 430–442. [14]

McGraw, W. S. 2005. Update on the search for Miss Waldron's red colobus monkey. *International Journal of Primatology* 26: 605–619. [23]

McIntyre, S., and R. Hobbs. 1999. A framework for conceptualizing human effects on landscapes and its relevance to management and research models. *Conservation Biology* 13: 1282. [24]

McKane, R. B., and 10 others. 2002. Resource-based niches provide a basis for plant species diversity and dominance in Arctic tundra. *Nature* 415: 68–71. [22]

McKinney, S. T., C. E. Fiedler, and D. F. Tomback. 2009. Invasive pathogen threatens bird-pine mutualism: Implications for sustaining a high-elevation ecosystem. *Ecological Applications* 19: 597–607. [24]

McMahon, T. A. and 16 others. 2014. Amphibians acquire resistance to live and dead fungus overcoming fungal immunosuppression. *Nature* 511: 224–227. [1]

McNeill, W. H. 1976. *Plagues and Peoples*. Anchor Press/Doubleday: Garden City, NY. [13]

McNeilly, T. 1968. Evolution in closely adjacent plant populations. III. *Agrostis tenuis* on a small copper mine. *Heredity* 23: 99–108. [6]

Meine, C., M. Soulé, and R. F. Noss. 2006. "A mission-driven discipline": The growth of conservation biology. *Conservation Biology* 20: 631–651. [23]

Meinesz, A. 1997. *Le roman noir de l'algue "tueuse"*: Caulerpa taxifolia *contre la Méditerranée*. Belin Éditeur: Paris. [16]

Meinesz, A. 2001. *Killer Algae*. University of Chicago Press: Chicago. [16]

Meire, P. M., and A. Ervynck. 1986. Are oystercatchers (*Haematopus ostralegus*) selecting the most profitable mussels (*Mytilus edulis*)? *Animal Behaviour* 34: 1427–1435. [8]

Melillo, J. M., J. D. Aber, and J. F. Muratore. 1982. Nitrogen and lignin control of hardwood leaf litter decomposition dynamics. *Ecology* 63: 621–626. [22]

Mellars, P. 1989. Major issues in the emergence of modern humans. *Current Anthropology* 30: 349–385. [5]

Menge, B. A. 1995. Indirect effects in marine rocky intertidal interaction webs: Patterns and importance. *Ecological Monographs* 65: 21–74. [16]

Menge, B. A., and J. P. Sutherland. 1987. Community regulation: Variation in disturbance, competition, and predation in relation to environmental stress and recruitment. *American Naturalist* 130: 730–757. [19]

Menge, B. A., B. Daley, and P. A. Wheeler. 1996. Control of interaction strength in marine benthic communities. In *Food Webs: Integration of Patterns and Dynamics*, G. A. Polis and K. O. Winemiller (Eds.), pp. 258–274. Chapman & Hall: New York. [16]

Merrick, R.L., M.K. Chumbley, and G.V. Byrd. 1997. Diet diversity of stellar sea lions (*Eumetopias jubatus*) and their population decline in Alaska: a potential relationship. *Canadian Journal of Fisheries and Aquatic Sciences* 54: 1342–1348. [21]

Millennium Ecosystem Assessment. 2005. *Ecosystems and Human Well-Being: Biodiversity Synthesis*. World Resources Institute: Washington, DC. [18]

Mills, L. S., M. Zimova, J. Oyler, S. Running, J. T. Abatzoglou, and P. M. Lukacs. 2013. Camouflage mismatch in seasonal coat color due to decreased snow duration. *Proceedings of the National Academy of Sciences USA* 110: 7360–7365. [7]

Minorsky, P. V. 2002. Allelopathy and grain crop production. *Plant Physiology* 130: 1745–1746. [14]

Mittelbach, G. G., and 21 others. 2007. Evolution and the latitudinal diversity gradient: Speciation, extinction and biogeography. *Ecology Letters* 10: 315–331. [18]

Mitton, J. B., and S. M. Ferrenberg. 2012. Mountain pine beetle develops an unprecedented summer generation in response to climate warming. *American Naturalist* 179: E163–E171. [25]

Mokany, K., R. J. Raison, and A. S. Prokushkin. 2006. Critical analysis of root:shoot ratios in terrestrial biomes. *Global Change Biology* 12: 84–96. [4]

Molina, M. J., and F. S. Rowland. 1974. Stratospheric sink for chlorofluoromethanes: Chlorine atom-catalysed destruction of ozone. *Nature* 249: 810–812. [25]

Molnár, O., Bajer, K., Mészáros, B., Török, J., and Herczeg, G. 2013. Negative correlation between nuptial throat colour and blood parasite load in male European green lizards supports the Hamilton-Zuk hypothesis. *Naturwissenschaften* 100: 551–558. [8]

Mooney, H. A. 1986. In *Plant Ecology*, M. J. Crawley (Ed.). Blackwell Science Ltd: Oxford. [5]

Mooney, H. A., C. Field, W. E. Williams, J. A. Berry, and O. Björkman. 1983. Photosynthetic characteristics of plants of a Californian cool coastal environment. *Oecologia* 57: 38–42. [5]

Mooney, H. A., O. Björkman, J. Ehrlinger, and J. Berry. 1976. Photosynthetic capacity of in situ Death Valley plants. *Carnegie Institution Year Book* 75: 410–413. [5]

Moore, J. A. 1957. An embryologist's view of the species concept. In *The Species Problem*, E. Mayr (Ed.), pp. 325–388. American Association for the Advancement of Science: Washington, DC. [6]

Moore, S. E., T. J. Cole, E. M. E. Poskitt, B. J. Sonko, R. G. Whitehead, I. A. McGregor, and A. M. Prentice. 1997. Season of birth predicts mortality in rural Gambia. *Nature* 388: 434. [11]

Morford, S. L., B. Z. Houlton, and R. A. Dahlgren. 2016. Direct quantification of long-term rock nitrogen inputs to temperate forest ecosystems. *Ecology* 97: 54–64. [25]

Morin, X., C. Augspurger, and I. Chuine. 2007. Process-based modeling of species' distributions: What limits temperate tree species' range boundaries? *Ecology* 88: 2280–2291. [4]

Morran, L. T., O. G. Schmidt, I. A. Gelarden, R. C. Parrish II, and C. M. Lively. 2011. Running with the Red Queen: Host–parasite coevolution selects for biparental sex. *Science* 333: 216–218. [7]

Morris, W. F., and D. F. Doak. 2002. *Quantitative Conservation Biology: Theory and Practice of Population Viability*. Oxford University Press/Sinauer: Sunderland, MA. [23]

Mouritsen, K. N., and R. Poulin. 2002. Parasitism, community structure and biodiversity in intertidal ecosystems. *Parasitology* 124: S101–S117. [13]

Mouritsen, K. N., L. T. Mouritsen, and K. T. Jensen. 1998. Changes of topography and sediment characteristics on an intertidal mud-flat following mass-mortality of the amphipod *Corophium volutator*. *Journal of the Marine Biological Association of the United Kingdom* 78: 1167–1180. [13]

Moutinho, P., C. D. Napstad, and E. A. Davidson. 2003. Influence of leaf-cutting ant nests on secondary forest growth and soil properties in Amazonia. *Ecology* 84: 1265–1276. [15]

Muir, J. 1915. *Travels in Alaska*. Houghton Mifflin: Boston. [17]

Munger, J. 1984. Optimal foraging? Patch use by horned lizards (Iguanidae: *Phrynosoma*). *American Naturalist* 123: 654–680. [8]

Murdoch, W. W., S. Avery, and M. E. B Smith. 1975. Switching in a predatory fish. *Ecology* 56: 1094–1105. [12]

Murphy, D. D., and C. M. Knopp. 2000. *Lake Tahoe Watershed Assessment*. Vol. 1. General Technical Report PSW-GTR-175. Pacific Southwest Research Station, Forest Service, U.S. Department of Agriculture: Albany, CA. [22]

Mussmann, S. M., and 6 others. 2017. Genetic rescue, the greater prairie chicken and the problem of conservation reliance in the Anthropocene. *Royal Society Open Science* 4: 160736. [6]

Myers, J. H., and D. Bazely. 1991. Thorns, spines, prickles, and hairs: Are they stimulated by herbivory and do they deter herbivores? In *Phytochemical Induction by Herbivores*, D. W. Tallamy and M. J. Raupp (Eds.), pp. 325–344. Wiley: New York. [12]

Myers, R. A., and B. Worm. 2003. Rapid worldwide depletion of predatory fish communities. *Nature* 423: 280–283. [23]

N

Nabhan, G. P. and A. R. Holdsworth. 1998. State of the Sonoran Desert Biome: Uniqueness, Biodiversity, Threats and the Adequacy of Protection in the Sonoran Bioregion. The Wildlands Project: Tucson, AZ. [3]

Naeem, S., L. J. Thompson, S. P. Lawler, J. H. Lawton, and R. M. Woodfin. 1995. Empirical evidence that declining species diversity may alter the performance of terrestrial systems. *Philosophical Transactions of the Royal Society of London* 347B: 249–262. [19]

Nager, R. G., P. Monaghan, and D. Houston. 2000. Within-clutch trade-offs between the number and quality of eggs: Experimental manipulations in gulls. *Ecology* 81: 1339–1350. [7]

Naiman, R. J., C. A. Johnston, and J. C. Kelley. 1988. Alterations of North American streams by beaver. *BioScience* 38: 753–762. [16]

National Academies of Sciences, Engineering, and Medicine. 2016. *Attribution of Extreme Weather Events in the Context of Climate Change*. National Academies Press: Washington, DC. [2]

National Research Council. 2002. *Effects of Trawling and Dredging on Seafloor Habitat*. Committee on Ecosystem Effects of Fishing: Phase I—Effects of Bottom Trawling on Seafloor Habitats, Ocean Studies Board, Division on Earth and Life Studies, National Research Council. National Academies Press: Washington, DC. [23]

NatureServe. 2013. www.natureserve.org/conservation-tools/natureserve-hotspots-map. [23]

Naughton-Treves, L., J. L. Mena, A. Treves, N. Alvarez, and V. C. Radeloff. 2003. Wildlife survival beyond park boundaries: The impact of slash-and-burn agriculture and hunting on mammals in Tambopata, Peru. *Conservation Biology* 17: 1106–1117. [24]

Nee, S., N. Colegrave, S. A. West, and A. Grafen. 2005. The illusion of invariant quantities in life histories. *Nature* 309: 1236–1239. [7]

Neff, J. C., R. L. Reynolds, J. Belnap, and P. Lamothe. 2005. Multi-decadal impacts of grazing on soil physical and biogeochemical properties in southeastern Utah. *Ecological Applications* 15: 87–95. [22]

Nemani, R. R., and 7 others. 2003. Climate-driven increases in global terrestrial net primary production from 1982 to 1999. *Science* 300: 1560–1563. [25]

Neutel, A.-M., J. A. P. Heesterbeek, and P. C. de Ruiter. 2002. Stability in real food webs: Weak links in long loops. *Science* 296: 1120–1123. [21]

New, M. G., M. Hulme, and P. D. Jones. 2000. Representing twentieth-century space–time climate variability. Part II: Development of 1901–06 monthly grids of terrestrial surface climate. *Journal of Climate* 13: 2217–2238. [2]

Newbold, J. D., J. W. Elwood, R. V. O'Neill, and A. L. Sheldon. 1983. Phosphorus dynamics in a woodland stream ecosystem: A study of nutrient spiralling. *Ecology* 64: 1249–1265. [22]

Nicholson, A. J. 1957. The self-adjustment of populations to change. *Cold Spring Harbor Symposia on Quantitative Biology* 22: 153–173. [10]

Norby, R. J., and 13 others. 2005. Forest response to elevated CO_2 is conserved across a broad range of productivity. *Proceedings of the National Academy of Sciences USA* 102: 18052–18056. [25]

Noss, R., E. Dinerstein, B. Gilbert, M. Gilpin, B. J. Miller, J. Terborgh, and S. Trombulak. 1999. Core areas: Where nature reigns. In *Continental Conservation: Scientific Foundations of Regional Reserve Networks*, M. E. Soulé and J. Terborgh (Eds.), pp. 99–128. Island Press: Washington, DC. [24]

Nürnberg, D. J., and 11 others. Photochemistry beyond the red limit in chlorophyll *f*–containing photosystems. *Science* 360: 1210–1213. [5]

O

Oates, J. F., M. Abedi-Lartey, W. S. McGraw, T. T. Struhsaker, and G. H. Whitesides. 2000. Extinction of a West African red colobus monkey. *Conservation Biology* 14: 1526–1532. [23]

Odum, H. T., and C. F. Jordan. 1970. Metabolism and evapotranspiration of the lower forest in a giant plastic cylinder. In *A Tropical Rain Forest*, H. T. Odum and R. F. Pigeon (Eds.), pp. 165–189. Division of Technical Information, U.S. Atomic Energy Commission: Washington, DC. [20]

Oechel, W. C., G. L. Vourlitis, S. J. Hastings, R. C. Zulueta, L. Hinzman, and D. Kane. 2000. Acclimation of ecosystem CO_2 exchange in the Alaskan Arctic in response to decadal climate warming. *Nature* 406: 978–981. [25]

Oechel, W. C., S. J. Hastings, G. Vourlitis, M. Jenkins, G. Riechers, and N. Grulke. 1993. Recent change of Arctic tundra ecosystems from a net carbon dioxide sink to a source. *Nature* 361: 520–523. [25]

Ogren, W. L. 1984. Photorespiration: Pathways, regulation, and modification. *Annual Review of Plant Physiology* 35: 415–442. [5]

Okin, G. S., N. Mahowald, O. A. Chadwick, and P. Artaxo. 2004. Impact of desert dust on the biogeochemistry of phosphorus in terrestrial ecosystems. *Global Biogeochemical Cycles* 18: GB2005. [25]

Olden, J. D., N. L. Poff, M. R. Douglas, M. E. Douglas, and K. D. Fausch. 2004. Ecological and evolutionary consequences of biotic homogenization. *Trends in Ecology and Evolution* 19: 18–24. [23]

Orr, J. C. and 26 others. 2005. Anthropogenic ocean acidification over the twenty-first century and its impact on calcifying organisms. *Nature* 437: 681–686. [2, 3, 25]

Osborn, S. 2007. Distribution of the cattle egret (*Bubulcus ibis*) in North America. In *The Birds of North America Online*, A. Poole (Ed.). Cornell Lab of Ornithology: Ithaca, NY. [10]

Osmond, C. B., K. Winter, and H. Ziegler. 1982. Functional significance of different pathways of photosynthesis. In *Physiological Plant Ecology*, Vol. 2, O. L. Lange et al. (Eds.), pp. 479–547. *Encyclopedia of Plant Physiology*, new series, vol. 12B. Springer: New York. [5]

Ostfeld, R. S. 2009. Climate change and the distribution and intensity of infectious diseases. *Ecology* 90: 903–905. [13]

Overpeck, J. T., R. S. Webb, and T. Webb. 1992. Mapping eastern North American vegetation change of the past 18,000 years: No-analogs and the future. *Geology* 20: 1071–1074. [25]

Owen-Smith, N. 1987. Pleistocene extinctions: The pivotal role of megaherbivores. *Paleobiology* 13: 351–362. [3]

P

Pace, M. L., and 8 others. 2004. Whole-lake carbon-13 additions reveal terrestrial support of aquatic food webs. *Nature* 427: 240–243. [21]

Pacific Fisherman. 1915. *September 1915*. M. Freeman Publications: Portland, OR. [2]

Pacific Fisherman. 1939. *1939 Yearbook*. M. Freeman Publications: Portland, OR. [2]

Pacific Fishery Management Council. 2008. Review of 2007 Ocean Salmon Fisheries. (Document prepared for the Council and its advisory entities.) Pacific Fishery Management Council: Portland, OR. [2]

Packer, C., A. E. Pusey, H. Rowley, D. A. Gilbert, J. Martenson, and S. J. O'Brien. 1991. Case study of a population bottleneck: Lions of the Ngorongoro Crater. *Conservation Biology* 5: 219–230. [10]

Paine, R. T. 1966. Food web complexity and species diversity. *American Naturalist* 100: 65–75. [19, 21]

Paine, R. T. 1974. Intertidal community structure: Experimental studies on the relations between a dominant competitor and its principal predator. *Oecologia* 15: 93–120. [12]

Paine, R. T. 1979. Disaster, catastrophe, and local persistence of the sea palm, *Postelsia palmaeformis*. *Science* 205: 685–687. [14]

Paine, R. T., J. C. Castillo, and J. Cancino. 1985. Perturbation and recovery patterns of starfish-dominated intertidal assemblages in Chile, New Zealand, and Washington State. *American Naturalist* 125: 679–691. [21]

Palumbi, S. R. 2001. Humans as the world's greatest evolutionary force. *Science* 293: 1786–1790. [6]

Park, T. 1948. Experimental studies of interspecies competition. I. Competition between populations of the flour beetles, *Tribolium confusum* Duvall and *Tribolium castaneum* Herbst. *Ecological Monographs* 18: 267–307. [13]

Parmenter, A. W., and 8 others. 2003. Land use and land cover change in the Greater Yellowstone Ecosystem: 1975–1995. *Ecological Applications* 13: 687–703. [24]

Parmesan, C. 2006. Ecological and evolutionary responses to recent climate change. *Annual Review of Ecology, Evolution, and Systematics* 37: 637–669. [1, 6, 23, 25]

Parmesan, C., and 12 others. 1999. Poleward shifts in geographical ranges of butterfly species associated with regional warming. *Nature* 399: 579–583. [25]

Parmesan, C., and G. Yohe. 2003. A globally coherent fingerprint of climate change impacts across natural systems. *Nature* 421: 37–42. [9, 25]

Partridge, L., and M. Farquhar. 1981. Sexual activity reduces lifespan of male fruitflies. *Nature* 294: 580–582. [7]

Pastor, J., R. J. Naiman, B. Dewey, and P. Mcinnes. 1988. Moose, microbes, and the boreal forest. *BioScience* 38: 770–777. [24]

Pastore, M. A., T. D. Lee, S. E. Hobbie, and P. B. Reich. 2019. Strong photosynthetic acclimation and enhanced water-use efficiency in grassland functional groups persist over 21 years of CO_2 enrichment, independent of nitrogen supply. *Global Change Biology* 25: 3031–3044. [25]

Paton, P. W. C. 1994. The effect of edge on avian nest success: How strong is the evidence? *Conservation Biology* 8: 17–26. [24]

Patterson, B. D. 1980. Montane mammalian biogeography in New Mexico. *Southwestern Naturalist* 25: 33–40. [14]

Patterson, B. D. 1981. Morphological shifts of some isolated populations of *Eutamias* (Rodentia: Sciuridae) in different congeneric assemblages. *Evolution* 35: 53–66. [14]

Paul, E. A., and F. E. Clark. 1996. *Soil Microbiology and Biochemistry*. Academic Press: San Diego, CA. [22]

Paul, N. D., and D. Gwynn-Jones. 2003. Ecological roles of solar UV radiation: Towards an integrated approach. *Trends in Ecology and Evolution* 18: 48–55. [25]

Paulay, G. 1997. Diversity and distribution of reef organisms. In *Life and Death of Coral Reefs*, C. Birkeland (Ed.), pp. 298–353. Chapman & Hall: New York. [3]

Paullin, C. O. 1932. *Atlas of Historical Geography of the United States*. Greenwood Press: Westport, CT. [24]

Pearce, F. 1997. Why is the apparently pristine Arctic full of toxic chemicals that started off thousands of kilometres away? *New Scientist* 154: 24–27. [21]

Pearcy, R. W. 1977. Acclimation of photosynthetic and respiratory carbon dioxide exchange to growth temperature in *Atriplex lentiformis* (Torr) Wats. *Plant Physiology* 59: 795–799. [5]

Pearl, R., and L. J. Reed. 1920. On the rate of growth of the population of the United States since 1790 and its mathematical representation. *Proceedings of the National Academy of Sciences USA* 6: 275–288. [11]

Peek, A. S., R. A. Feldman, R. A. Lutz, and R. C. Vrijenhoek. 1998. Cospeciation of chemoautotrophic bacteria and deep sea clams. *Proceedings of the National Academy of Sciences USA* 95: 9962–9966. [20]

Pellmyr, O. 2003. Yuccas, yucca moths and coevolution: A review. *Annals of the Missouri Botanical Garden* 90: 35–55. [15]

Pellmyr, O., and C. J. Huth. 1994. Evolutionary stability of mutualism between yuccas and yucca moths. *Nature* 372: 257–260. [15]

Pellmyr, O., C. C. Labandeira, and C. M. Herrera, Eds. 2002. *Plant–Animal Interactions: An Evolutionary Approach.* Blackwell Science: Malden, MA. [12]

Perez, J. M., J. E. Granados, R. C. Soriguer, P. Fandos, F. J. Marquez, and J. P. Crampe. 2002. Distribution, status and conservation problems of the Spanish Ibex, *Capra pyrenaica* (Mammalia: Artiodactyla). *Mammal Review* 32: 26–39. [23]

Peterson, B. J., and B. Fry. 1987. Stable isotopes in ecological studies. *Annual Review of Ecology and Systematics* 18: 293–320. [20]

Peterson, T. C., and R. S. Vose. 1997. An overview of the Global Historical Climatology Network temperature database. *Bulletin of the American Meteorological Society* 78: 2837–2849. [2, 3]

Peterson, W. T., C. A. Morgan, J. O. Peterson, J. L. Fisher, B. J. Burke and K. Fresh. 2013. *Ocean Ecosystem Indicators of Salmon Marine Survival in the Northern California Current.* Fish Ecology Division, National Oceanic and Atmospheric Administration, National Marine Fisheries Service, Northwest Fisheries Science Center: Seattle, WA. [2]

Pfennig, D. W. 1992. Proximate and functional causes of polyphenism in an anuran tadpole. *Functional Ecology* 6: 167–174. [7]

Pfennig, D. W., A. M. Rice, and R. A. Martin. 2007. Field and experimental evidence for competition's role in phenotypic divergence. *Evolution* 61: 257–271. [14]

Pickett, S. T. A. 1989. Space-for-time substitution as an alternative to long-term studies. In *Long-term Studies in Ecology: Approaches and Alternatives*, G. E. Likens (Ed.), pp. 110–135. Springer: New York. [17]

Pielou, E. C. 1979. *Biogeography.* Wiley: Hoboken, NJ. [18]

Pigeon, G., M. Festa-Bianchet, D. W. Coltman, and F. Pelletier. 2016. Intense selective hunting leads to artificial evolution in horn size. *Evolutionary Applications* 9: 521–530. [6]

Pimm, S. L. 2002. *Food Webs.* University of Chicago Press: Chicago. [21]

Pimm, S. L. and C. Jenkins. 2005. Sustaining the variety of life. *Scientific American* 293: 66–73. [3]

Pimm, S. L., and J. H. Lawton. 1977. The number of trophic levels in ecological communities. *Nature* 275: 542–544. [21]

Pimm, S. L., H. L. Jones, and J. Diamond. 1988. On the risk of extinction. *American Naturalist* 132: 757–785. [10]

Pinder, A. W., K. B. Storey, and G. R. Ultsch. 1992. Estivation and hibernation. In *Environmental Physiology of the Amphibia*, M. E. Feder and W. W. Burggren (Eds.), pp. 250–274. University of Chicago Press: Chicago. [4]

Pinto-Tomás, A. A., and 7 others. 2009. Symbiotic nitrogen fixation in the fungus gardens of leaf-cutter ants. *Science* 326: 1120–1123. [15]

Polis, G. A. 1991. Complex trophic interactions in deserts: An empirical critique of food web theory. *American Naturalist* 138: 123–155. [21]

Polis, G. A., M. E. Power, and G. R. Huxel. 2004. *Food Webs at the Landscape Level.* University of Chicago Press: Chicago. [21]

Pollock, M. M., R. J. Naiman, and T. A. Hanley. 1998. Plant species richness in riparian wetlands: A test of biodiversity theory. *Ecology* 79: 94–105. [19]

Post, D. M. 2002a. Using stable isotopes to estimate trophic position: Models, methods, and assumptions. *Ecology* 83: 703–718. [21]

Post, D. M. 2002b. The long and short of food-chain length. *Trends in Ecology and Evolution* 17: 269–277. [21]

Post, D. M., and G. Takimoto. 2007. Proximate structural mechanisms for variation in food-chain length. *Oikos* 116: 775–782. [21]

Powell, K. I., J. M. Chase, and T. M. Knight. 2013. Invasive plants have scale-dependent effects on diversity by altering species-area relationships. *Science* 339: 316–318. [18]

Power, M. E., and 9 others. 1996b. Challenges in the quest for keystones. *BioScience* 46: 609–620. [16]

Power, M. E., M. S. Parker, and J. T. Wootton. 1996a. Disturbance and food chain length in rivers. In *Food Webs: Integration of Patterns and Dynamics*, G. A. Polis and K. O. Winemiller (Eds.), pp. 286–297. Chapman & Hall: New York. [16]

Power, M. E., M. S. Parker, and W. E. Dietrich. 2008. Seasonal reassembly of a river food web: Floods, droughts, and impacts on fish. *Ecological Monographs* 78: 263–282. [16]

Pride, E. 2005. Optimal group size and seasonal stress in ring-tailed lemurs (*Lemur catta*). *Behavioral Ecology* 16: 550–560. [8]

Primack, R. B., and A. A. Sher. 2016. *An Introduction to Conservation Biology.* Sinauer Associates: Sunderland, MA.

Prugh, L. R., K. E. Hodges, A. R. E. Sinclair, and J. S. Brashares. 2008. Effect of habitat area and isolation on fragmented animal populations. *Proceedings of the National Academy of Sciences USA* 105: 20770–20775. [24]

Pyke, G. H. 2019. Optimal foraging theory—Introduction. In *Encyclopedia of Animal Behavior*, M. D. Breed and J. Moore (Eds.), pp. 111–117. Academic Press, Oxford. [8]

Pyke, G. H., H. R. Pulliam, and E. L. Charnov. 1977. Optimal foraging: A selective review of theory and tests. *Quarterly Review of Biology* 52: 137–154. [8]

R

Raab, T. K., D. A. Lipson, and R. K. Monson. 1996. Non-mycorrhizal uptake of amino acids by roots of the alpine sedge *Kobresia myosuroides*: Implications for the alpine nitrogen cycle. *Oecologia* 108: 488–494. [22]

Rabinowitz, D., S. Cairns, and T. Dillon. 1986. Seven forms of rarity and their frequency in the flora of the British Isles. In *Conservation Biology: The Science of Scarcity and Diversity*, M. E. Soulé (Ed.), pp. 182–204. Oxford University Press/Sinauer: Sunderland, MA. [23]

Raffel, T. R., J. M. Romansic, N. T. Halstead, T. A. McMahon, M. D. Venesky and J. R. Rohr. 2012. Disease and thermal acclimation in a more variable and unpredictable climate. *Nature Climate Change* 3: 146–151. [1]

Rahel, F. J. 2000. Homogenization of fish faunas across the United States. *Science* 288: 854–856. [23]

Ralls, K., and J. D. Ballou. 2004. Genetic status and management of California Condors. *Condor* 106: 215–228. [23]

Ranta, P., T. Blom, J. Niemela, E. Joensuu, and M. Siitonen. 1998. The fragmented Atlantic rain forest of Brazil: Size, shape and distribution of forest fragments. *Biodiversity and Conservation* 7: 385–403. [23]

Rantanen, M., and 7 others. 2022. The Arctic has warmed nearly four times faster than the globe since 1979. *Communications Earth & Environment* 3: 168.

Raxworthy, C. J., E. Martinez-Meyer, N. Horning, R. A. Nussbaum, G. E. Schneider, M. A. Ortega-Huerta, and A. T. Peterson. 2003. Predicting distributions of known and unknown reptile species in Madagascar. *Nature* 426: 837–841. [9]

Raymond, M., C. Chevillon, T. Guillemaud, T. Lenormand, and N. Pasteur. 1998. An overview of the evolution of overproduced esterases in the mosquito *Culex pipiens*. *Philosophical Transactions of the Royal Society of London* 353B: 1707–1711. [6]

Rayne, S., M. G. Ikonomou, and B. Antcliffe. 2003. Rapidly increasing polybrominated diphenyl ether concentrations in the Columbia River system from 1992 to 2000. *Environmental Science and Technology* 37: 2847–2854. [23]

Read, D. J. 1991. Mycorrhiza in ecosystems. *Experientia* 47: 376–391. [15]

Read, D. J., J. R. Leake, and J. Perez-Moreno. 2004. Mycorrhizal fungi as drivers of ecosystem processes in heathland and boreal forest biomes. *Canadian Journal of Botany* 82: 1243–1263. [15]

Réale, D., A. G. McAdam, S. Boutin, and D. Berteaux. 2003. Genetic and plastic responses of a northern mammal to climate change. *Proceedings of the Royal Society of London* 270B: 591–596. [6]

Redford, K. H. 1992. The empty forest. *BioScience* 42: 412–422. [23]

Redman, R. S., K. B. Sheehan, R. G. Stout, R. J. Rodriguez, and J. M. Henson. 2002. Thermotolerance generated by plant/fungal symbiosis. *Science* 298: 1581. [15]

Rees, W. E. 1992. Ecological footprints and appropriated carrying capacity: What urban economics leaves out. *Environment and Urbanization* 4: 121–130. [11]

Reiners, W. A., I. A. Worley, and D. B. Lawrence. 1971. Plant diversity in a chronosequence at Glacier Bay, Alaska. *Ecology* 52: 55–69. [17]

Reisen, W. K., Y. Fang and V. M. Martinez. 2006. Effects of temperature on the transmission of West Nile virus by *Culex tarsalis* (Diptera; Culicidae). *Journal of Medical Entomology* 43: 309–317. [1]

Relyea, R. A. 2003. Predator cues and pesticides: A double dose of danger for amphibians. *Ecological Applications* 13: 1515–1521. [1]

Resasco, J. 2019. Meta-analysis on a decade of testing corridor efficacy: What new have we learned? *Current Landscape Ecology Reports* 4: 61–69. [24]

Ricklefs, R. E. 1977. On the evolution of reproductive strategies in birds: Reproductive effort. *The American Naturalist* 111: 453–478. [7]

Rieseberg, L. H., and 9 others. 2003. Major ecological transitions in wild sunflowers facilitated by hybridization. *Science* 301: 1211–1216. [6]

Ripple, W. J., and R. L. Beschta. 2004. Wolves and the ecology of fear: Can predation risk structure ecosystems? *BioScience* 54: 755–766. [24]

Ripple, W. J., and R. L. Beschta. 2007. Restoring Yellowstone's aspen with wolves. *Biological Conservation* 138: 514–519. [24]

Rizki, R. M., and T. M. Rizki. 1990. Parasitoid virus-like particles destroy *Drosophila* cellular immunity. *Proceedings of the National Academy of Sciences USA* 87: 8388–8392. [13]

Roberts, L. 2011. 9 billion? *Science* 333: 540–543. [11]

Robertson, G. P., M. A. Huston, F. C. Evans, and J. M. Tiedje. 1988. Spatial variability in a successional plant community: Patterns of nitrogen availability. *Ecology* 69: 1517–1524. [19]

Rochman C.M., M.A. Browne, A.J. Underwood, J.A. van Franeker, R.C. Thompson, and L.A. Amaral-Zettler. 2016. The ecological impacts of marine debris: Unraveling the demonstrated evidence from what is perceived. *Ecology* 97: 302–312. [3]

Rodriguez, R. J., J. F. White Jr., A. E. Arnold, and R. S. Redman. 2009. Fungal endophytes: Diversity and functional roles. *New Phytologist* 182: 314–330. [15]

Rodriguez-Caballero, E., J. Belnap, B. Büdel, P. J. Crutzen, M. O. Andreae, U. Pöschl, and B. Weber. 2018. Dryland photoautotrophic soil surface communities endangered by global change. *Nature Geoscience* 11: 185–189. [22]

Rohr, J. R. and 11 others. 2008. Agrochemicals increase trematode infections in a declining amphibian species. *Nature* 455: 1235–1239. [1]

Rohr, J. R. and T. R. Raffel. 2010. Linking global climate and temperature variability to widespread amphibian declines putatively caused by disease. *Proceedings of the National Academy of Sciences USA* 107: 8269–8274. [1]

Root, T. L., J. T. Price, K. R. Hall, S. H. Schneider, C. Rosenzweig, and J. A. Pounds. 2003. Fingerprints of global warming on wild animals and plants. *Nature* 421: 57–60. [25]

Rosenzweig, M. L. 1992. Species diversity gradients: We know more and less than we thought. *Journal of Mammalogy* 73: 715–730. [18]

Rosenzweig, M. L. 1995. *Species Diversity in Space and Time.* Cambridge University Press: Cambridge. [18]

Ross, P. S. 2006. Fireproof killer whales (*Orcinus orca*): Flame-retardant chemicals and the conservation imperative in the charismatic icon of British Columbia, Canada. *Canadian Journal of Fisheries and Aquatic Sciences* 63: 224–234. [23]

Ross, P. S., G. M. Ellis, M. G. Ikonomou, L. Barrett-Lennard, and R. F. Addison. 2000. High PCB concentrations in free-ranging Pacific killer whales. *Marine Pollution Bulletin* 40: 504–515. [23]

Rovira, A. D., C. D. Bowen, and R. C. Foster. 1983. The significance of rhizosphere microflora and mycorrhizas in plant nutrition. In *Inorganic Plant Nutrition (Encyclopedia of Plant Physiology*, new series, Vol. 15B), A. Läuchli and R. L. Bieleski (Eds.), pp. 61–93. Springer: Berlin. [15]

Rutter, C. 1902. Natural history of the quinnat salmon. A report of investigations in the Sacramento River, 1896–1901. *Bulletin of the United States Fish Commission* 22: 65–141. [6]

Rutz, C., and 7 others. 2010. The ecological significance of tool use in New Caledonian crows. *Science* 329: 1523–1526. [5]

Rybczynski, N., M. R. Dawson, and R. H. Tedford. 2009. A semi-aquatic Arctic mammalian carnivore from the Miocene epoch and origin of Pinnipedia. *Nature* 458: 1021–1024. [6]

S

Sale, P. F. 1977. Maintenance of high diversity in coral reef fish communities. *American Naturalist* 111: 337–359. [19]

Sale, P. F. 1979. Recruitment, loss and coexistence in a guild of territorial coral reef fishes. *Oecologia* 42: 159–177. [19]

Salisbury, F. B., and C. Ross. 1992. *Plant Physiology*. 4th ed. Wadsworth: Belmont, CA. [22]

Sanderson, E. W., M. Jaiteh, M. A. Levy, K. H. Redford, A. V. Wannebo and G. Woolmer. 2002. The human footprint and the last of the wild. *BioScience* 52: 891–904. [3]

Sandquist, D. R., and J. R. Ehleringer. 2003. Population- and family-level variation of brittlebush (*Encelia farinosa*, Asteraceae) pubescence: Its relation to drought and implications for selection in variable environments. *American Journal of Botany* 90: 1481–1486. [4]

Saugier, B., J. Roy, and H. A. Mooney. 2001. Estimations of global terrestrial productivity: Converging toward a single number? In *Terrestrial Global Productivity*, J. Roy, B. Saugier, and H. A. Mooney (Eds.), pp. 543–557. Academic Press: San Diego, CA. [20]

Sax, D. F., and S. D. Gaines. 2003. Species diversity: From global decreases to local increases. *Trends in Ecology and Evolution* 18: 561–566. [23]

Scarborough, C. L., J. Ferrari, and H. C. J. Godfray. 2005. Aphid protected from pathogen by endosymbiont. *Science* 310: 1781. [13]

Schall, J. J. 1992. Parasite-mediated competition in *Anolis* lizards. *Oecologia* 92: 58–64. [13]

Scheffers, B. R., L. N. Joppa, S. L. Pimm, and W. F. Laurance. 2012. What we know and don't know about Earth's missing biodiversity. *Trends in Ecology and Evolution* 27: 501–510. [23]

Schimel, J. P., and J. Bennett. 2004. Nitrogen mineralization: Challenges of a changing paradigm. *Ecology* 85: 591–602. [22]

Schindler, D. W. 1974. Eutrophication and recovery in experimental lakes: Implications for lake management. *Science* 184: 897–899. [20]

Schipper, J., and 132 others. 2008. The status of the world's land and marine mammals: Diversity, threats, and knowledge. *Science* 322: 225–230. [23]

Schlaepfer, M. A., D. F. Sax, and J. D. Olden. 2011. The potential conservation value of non-native species. *Conservation Biology* 25: 428–437. [23]

Schlesinger, W. H. 1997. *Biogeochemistry: An Analysis of Global Change.* 2nd ed. Academic Press: San Diego, CA. [22]

Schlesinger, W. H., and E. S. Bernhardt. 2013. *Biogeochemistry: An Analysis of Global Change.* 3rd ed. Academic Press: Cambridge, MA. [25]

Schlesinger, W. H., and E. S. Bernhardt. 2020. *Biogeochemistry: An Analysis of Global Change.* 4th ed. Academic Press: Waltham, MA. [25]

Schluter, D. 1994. Experimental evidence that competition promotes divergence in adaptive radiation. *Science* 266: 798–801. [14]

Schluter, D. 1998. Ecological causes of speciation. In *Endless Forms: Species and Speciation*, D. J. Howard and S. H. Berlocher (Eds.), pp. 114–129. Oxford University Press: Oxford. [6]

Schluter, D., T. D. Price, and P. R. Grant. 1985. Ecological character displacement in Darwin's finches. *Science* 227: 1056–1059. [14]

Schmidt-Nielsen, B., and K. Schmidt-Nielsen. 1951. A complete account of the water metabolism in kangaroo rats and an experimental verification. *Journal of Cellular and Comparative Physiology* 38: 165–182. [4]

Schmidt-Nielsen, K. 1964. *Desert Animals: Physiological Problems of Heat and Water*. Clarendon Press: Oxford. [4]

Schmidt-Nielsen, K. 1997. *Animal Physiology, Adaptation and Environment*. 5th ed. Cambridge University Press: Cambridge, UK. [4]

Schnee, C., T. G. Kollner, M. Held, T. C. J. Turlings, J. Gershenzon, and J. Degenhardt. 2006. The products of a single maize sesquiterpene synthase form a volatile defense signal that attracts natural enemies of maize herbivores. *Proceedings of the National Academy of Sciences USA* 103: 1129–1134. [12]

Schoener, T. W. 1971. Theory of feeding strategies. *Annual Review of Ecology and Systematics* 2: 370–404. [8]

Schoener, T. W. 1974. Resource partitioning in ecological communities. *Science* 185: 27–39. [14]

Schoener, T. W. 1983. Field experiments on interspecific competition. *American Naturalist* 122: 240–285. [14]

Schoener, T. W., and D. A. Spiller. 1996. Devastation of prey diversity by experimentally introduced predators in the field. *Nature* 381: 691–694. [12]

Scholander, P. F., V. Walters, R. Hock, and L. Irving. 1950. Heat regulation in some arctic and tropical mammals and birds. *Biological Bulletin* 99: 237–258. [4]

Schtickzelle, N., and M. Baguette. 2003. Behavioural responses to habitat patch boundaries restrict dispersal and generate emigration–patch area relationships in fragmented landscapes. *Journal of Animal Ecology* 72: 533–545. [24]

Schulte, D. M., R. P. Burke, and R. N. Lipcius. 2009. Unprecedented restoration of a native oyster metapopulation. *Science* 325: 1124–1128. [24]

Schuur, E. A. G. 2003. Productivity and global climate revisited: The sensitivity of tropical forest growth to precipitation. *Ecology* 84: 1165–1170. [20]

Schwarz, C. J., and G. A. F. Seber. 1999. Estimating animal abundance: Review III. *Statistical Science* 14: 427–456. [9]

Seehausen, O., J. J. M. van Alphen, and F. Witte. 1997. Cichlid fish diversity threatened by eutrophication that curbs sexual selection. *Science* 277: 1808–1811. [23]

Seifert, K. A., R. A. Samson, and I. H. Chapela. 1995. *Escovopsis aspergilloides*, a rediscovered Hyphomycete from leaf-cutting ant nests. *Mycologia* 87: 407–413. [15]

Selosse, M.-A., and F. Le Tacon. 1998. The land flora: A phototroph–fungus partnership? *Trends in Ecology and Evolution* 13: 15–20. [15]

Seneviratne, S. I., X. Zhang, M. Adnan, W. Badi, C. Dereczynski, A. Di Luca, S. Ghosh, I. Iskandar, J. Kossin, S. Lewis, F. Otto, I. Pinto, M. Satoh, S. M. Vicente-Serrano, M. Wehner, B. Zhou, 2021, Weather 46 and Climate Extreme Events in a Changing Climate. In: Climate Change 2021: *The Physical Science Basis. Contribution of Working Group I to the Sixth Assessment Report of the Intergovernmental Panel on Climate 48 Change* [Masson-Delmotte, V., P. Zhai, A. Pirani, S. L. Connors, C. Péan, S. Berger, N. Caud, Y. Chen, L., Goldfarb, M. I. Gomis, M. Huang, K. Leitzell, E. Lonnoy, J. B. R. Matthews, T. K. Maycock, T. Waterfield, 50 O. Yelekçi, R. Yu and B. Zhou (eds.)]. Cambridge University Press. [2]

Sessions, S. K. and S. B. Ruth. 1990. Explanation for naturally occurring supernumerary limbs in amphibians. *Journal of Experimental Zoology* 254: 38–47. [1]

Shafer, C. L. 1997. Terrestrial nature reserve design at the urban/ rural interface. In M. W. Schwartz (Ed.), *Conservation in Highly Fractured Landscapes*, pp. 345–347. Chapman and Hall: New York.

Sheriff, M. J., C. J. Krebs, and R. Boonstra. 2009. The sensitive hare: Sublethal effects of predator stress on reproduction in snowshoe hares. *Journal of Animal Ecology* 78: 1249–1258. [12]

Sheriff, M. J., C. J. Krebs, and R. Boonstra. 2010. The ghosts of predators past: Population cycles and the role of maternal programming under fluctuating predation risk. *Ecology* 91: 2983–2994. [12]

Shmida, A., and M. V. Wilson. 1985. Biological determinants of species diversity. *Journal of Biogeography* 12: 1–20. [18]

Shurin, J. B. 2001. Interactive effects of predation and dispersal of zooplankton communities. *Ecology* 82: 3404–3416. [19]

Shurin, J. B., and 7 others. 2002. A cross-ecosystem comparison of the strength of trophic cascades. *Ecology Letters* 5: 785–791. [21]

Shurin, J. B., D. S. Gruner, and H. Hillebrand. 2006. All wet or dried up? Real differences between aquatic and terrestrial food webs. *Proceedings of the Royal Society of London* 273B: 1–9. [21]

Sidor, C. A. 2003. Evolutionary trends and the origin of the mammalian lower jaw. *Paleobiology* 29: 605–640. [6]

Sih, A., and B. Christensen. 2001. Optimal diet theory: When does it work, and when and why does it fail? *Animal Behaviour* 61: 379–390. [8]

Silverman, J., and D. N. Bieman. 1993. Glucose aversion in the German cockroach, *Blattella germanica*. *Journal of Insect Physiology* 39: 925–933. [8]

Simberloff, D. 2003. Eradication: Preventing invasions at the outset. *Weed Science* 51: 247–253. [23]

Simberloff, D. S., and E. O. Wilson. 1969. Experimental zoogeography of islands: The colonization of empty islands. *Ecology* 50: 278–296. [18]

Simberloff, D., and J. Cox. 1987. Consequences and costs of conservation corridors. *Conservation Biology* 1: 63–71. [24]

Simenstad, C. A., J. A. Estes, and K. W. Kenyon. 1978. Aleuts, sea otters, and alternative communities. *Science* 200: 403–411. [16]

Simkin, S. M., and 16 others. 2016. Conditional vulnerability of plant diversity to atmospheric nitrogen deposition across the United States. *Proceedings of the National Academy of Sciences USA* 113: 4086–4091. [25]

Sinervo, B. 1990. The evolution of maternal investment in lizards: An experimental and comparative analysis of egg size and its effects on offspring performance. *Evolution* 44: 279–294. [7]

Sinervo, B., and 25 others. 2010. Erosion of lizard diversity by climate change and altered thermal niches. *Science* 328: 894–899. [23, 25]

Singer, M. S., I. H. Lichter-Marck, T. E. Farkas, E. Aaron, K. D. Whitney, and K. A. Mooney. 2014. Herbivore diet breadth mediates the cascading effects of carnivores in food webs. *Proceedings of the National Academy of Sciences USA* 111: 9521–9526. [21]

Skerratt, L. F. and 7 others. 2007. Spread of chytridiomycosis has caused the rapid global decline and extinction of frogs. *EcoHealth* 4: 125–134. [1]

Skerratt, L. F., and 7 others. 2007. Spread of chytridiomycosis has caused the rapid global decline and extinction of frogs. *EcoHealth* 4: 125–134. [23]

Slatyer, R. D. 1967. *Plant-Water Relationships*. Academic Press: Cambridge, MA. [6]

Smith, F. D. M., R. M. May, R. Pellew, T. H. Johnson, and K. R. Walter. 1993. How much do we know about the current extinction rate? *Trends in Ecology and Evolution* 8: 375–378. [23]

Smith, F. E. 1961. Density-dependence in the Australian thrips. *Ecology* 42: 403–407. [11]

Smith, J. W., and C. W. Benkman. 2007. A coevolutionary arms race causes ecological speciation in crossbills. *American Naturalist* 169: 455–465.

Smith, K. F., M. Behrens, L. M. Schloegel, N. Marano, S. Burgiel, and P. Daszak. 2009. Reducing the risks of the wildlife trade. *Science* 324: 594–595. [23]

Smith, S. E., and D. J. Read. 2008. *Mycorrhizal Symbiosis*. 3rd ed. Academic Press: San Diego, CA. [15]

Smith, T. B. 1993. Disruptive selection and the genetic basis of bill size polymorphism in the African finch *Pyrenestes*. *Nature* 363: 618–620. [6]

Snyder, N. F. R., S. R. Derrickson, S. R. Beissinger, J. W. Wiley, T. B. Smith, W. D. Toone, and B. Miller. 1996. Limitations of captive breeding in endangered species recovery. *Conservation Biology* 10: 338–348. [23]

Sokal, R. R. and F. J. Rohlf. 2011. *Biometry*. 4th ed. W. H. Freeman: New York. [1]

Solé, R. V., and J. Montoya. 2001. Complexity and fragility in ecological networks. *Proceedings of the Royal Society of London* 268B: 2039–2045. [23]

Soulé, M. E. 1985. What is conservation biology? *BioScience* 35: 727–734. [23]

Soulé, M. E., and B. A. Wilcox, Eds. 1980. *Conservation Biology: An Evolutionary–Ecological Perspective*. Oxford University Press/Sinauer: Sunderland, MA. [23]

Soulé, M. E., and M. A. Sanjayan. 1998. Conservation targets: Do they help? *Science* 279: 2060–2061. [24]

Sousa, W. P. 1979. Experimental investigations of disturbance and ecological succession in a rocky intertidal algal community. *Ecological Monographs* 49: 227–254. [17]

Sousa, W. P. 1979a. Disturbance in marine intertidal boulder fields: The non-equilibrium maintenance of species diversity. *Ecology* 60: 1225–1239. [19]

Sousa, W. P. 2001. Natural disturbance and the dynamics of marine benthic communities. In *Marine Community Ecology*, M. D. Bertness, S. D. Gaines, and M. E. Hay (Eds.), pp. 85–130. Oxford University Press/Sinauer: Sunderland, MA. [17]

Spencer, K. A. 1972. *Handbooks for the Identification of British Insects, Diptera, Agromyzidae* (Vol. X, part 5g). Royal Entomological Society: London. [12]

Springer, A. M., and 7 others. 2003. Sequential megafaunal collapse in the North Pacific Ocean: An ongoing legacy of industrial whaling? *Proceedings of the National Academy of Sciences USA* 100: 12223–12228. [9, 21]

Stachowicz, J. J., and M. E. Hay. 1999. Reducing predation through chemically mediated camouflage: Indirect effects of plant defenses on herbivores. *Ecology* 80: 495–509. [8]

Steadman, D. W. 1995. Prehistoric extinctions of Pacific island birds: Biodiversity meets zooarchaeology. *Science* 267: 1123–1131. [23]

Stebbins, G. L. 1974. *Flowering Plants: Evolution above the Species Level*. Belknap: Cambridge, MA. [18]

Steltzer, H., C. Landry, T. H. Painter, J. Anderson, and E. Ayers. 2009. Biological consequences of earlier snowmelt from desert dust deposition in alpine landscapes. *Proceedings of the National Academy of Sciences USA* 106: 11629–11634. [25]

Steneck, R. S., S. D. Hacker, and M. N. Dethier. 1991. Mechanisms of competitive dominance between crustose coralline algae: An herbivore-mediated competitive reversal. *Ecology* 72: 938–950. [14]

Stevens, C. J., N. B. Dise, J. O. Mountford, and D. J. Gowling. 2004. Impact of nitrogen deposition on the species richness of grasslands. *Science* 303: 1876–1879. [25]

Stevens, O. A. 1932. The number and weight of seeds produced by weeds. *American Journal of Botany* 19: 784–794. [7]

Stiling, P. D. 2002. *Ecology: Theories and Applications*. 4th ed. Prentice-Hall: Upper Saddle River, NJ. [13]

Stinner, J. N., and V. H. Shoemaker. 1987. Cutaneous gas exchange and low evaporative water loss in the frogs *Phyllomedusa sauvagei* and *Chiromantis xerampelina*. *Journal of Comparative Physiology* 157: 423–427. [4]

Stomp, M., and 8 others. 2004. Adaptive divergence in pigment composition promotes phytoplankton biodiversity. *Nature* 432: 104–107. [14]

Stomp, M., J. Huisman, L. Vörös, F. R. Pick, M. Laamanen, T. Haverkamp, and L. J. Stal. 2007. Colourful coexistence of red and green picocyanobacteria in lakes and seas. *Ecology Letters* 10: 290–298. [14]

Storey, K. B. 1990. Life in a frozen state: Adaptive strategies for natural freeze tolerance in amphibians and reptiles. *American Journal of Physiology* 258: R559–R568. [4]

Strahler, A. H. and A. N. Strahler. 2005. *Physical Geography*, 3rd ed. John Wiley and Sons: Hoboken, NJ. Compiled by John E. Oliver. [2]

Strayer, D. L., and H. M. Malcom. 2007. Effects of zebra mussels (*Dreissena polymorpha*) on native bivalves: The beginning of the end or the end of the beginning? *Journal of the North American Benthological Society* 26: 111–122. [23]

Strebel, P. M., and S. L. Cochi. 2001. Waving goodbye to measles. *Nature* 414: 695–696. [13]

Stuart, S. N., J. S. Chanson, N. A. Cox, B. E. Young, A. S. L. Rodrigues, D. L. Fischman and R. W. Waller. 2004. Status and trends of amphibian declines and extinctions worldwide. *Science* 306: 1783–1786. [1]

Stuart, Y. E., T. S. Campbell, P. A. Hohenlohe, R. G. Reynolds, L. J. Revell, and J. B. Losos. 2014. Rapid invasion of a native species following invasion by a congener. *Science* 346: 463–466. [6]

Suding, K., and 7 others. 2005. Functional and abundance-based mechanisms explain diversity loss due to nitrogen fertilization. *Proceedings of the National Academy of Sciences USA* 102: 4387–4392. [25]

Sultan, S. E., T. Horgan-Kobelski, L. M. Nichols, C. E. Riggs, and R. K. Waples. 2013. A resurrection study reveals rapid adaptive evolution within populations of an invasive plant. *Evolutionary Applications* 6: 266–278. [6]

Summerhayes, V. S., and P. H. Williams. 1926. Studies on the ecology of English heaths. *Journal of Ecology* 14: 203–243. [17]

Sunday, J. M., A. E. Bates, and N. K. Dulvy. 2012. Thermal tolerance and the global redistribution of animals. *Nature Climate Change* 2: 686–690. [9]

Surovell, T.A., S.R. Pelton, R.Anderson-Sprecher, and A. D. Myers. 2016. Test of Martin's overkill hypothesis using radiocarbon dates on extinct megafauna. *Proceedings of the National Academy of Sciences*. 113: 886-891 [23]

Sutherland, J. P. 1974. Multiple stable points in natural communities. *American Naturalist* 108: 859–873. [17]

Suzán, G., and 8 others. 2009. Experimental evidence for reduced rodent diversity causing increased hantavirus prevalence. *PLoS ONE* 4: e5461. [19]

Svane, I. B. 1984. Observations on the long-term population dynamics of the perennial ascidian, *Ascidia mentula* O. F. Müller, on the Swedish west coast. *Biological Bulletin* 167: 630–646. [10]

T

Takimoto, G., D. A. Spiller, and D. M. Post. 2008. Ecosystem size, but not disturbance, determines food chain length on islands of the Bahamas. *Ecology* 89: 3001–3007. [21]

Tanabe, S. 2002. Contamination and toxic effects of persistent endocrine disrupters in marine mammals and birds. *Marine Pollution Bulletin* 45: 69–77. [23]

Tansley, A. G. 1917. On competition between *Galium saxatile* L. (*G. hercynicum* Weig.) and *Galium sylvestre* Poll. (*G. asperum* Schreb.) on different types of soil. *Journal of Ecology* 5: 173–179. [14]

Tansley, A. G. 1935. The use and abuse of vegetational concepts and terms. *Ecology* 16: 284–307. [20]

Tebbich, S., M. Taborsky, B. Fessl, and D. Blomqvist. 2001. Do woodpecker finches acquire tool-use by social learning? *Proceedings of the Royal Society of London* 268B: 2189–2193. [5]

Tegen, I., and I. Fung. 1995. Contribution to the atmospheric mineral aerosol load from land surface modification. *Journal of Geophysical Research* 100: 18707–18726. [25]

Templeton A.R. 1986. Coadaptation and outbreeding depression. In: Conservation Biology: The Science of Scarcity and Diversity (ed.Soule M), pp. 105–116. Sinauer Assoc., Sunderland, Massachusetts. [23]

Terborgh, J. 1973. On the notion of favorableness in plant ecology. *American Naturalist* 107: 481–501. [18]

Terborgh, J., K. Feeley, M. Silman, P. Nunez, and B. Balukjian. 2006. Vegetation dynamics of predator-free land-bridge islands. *Journal of Ecology* 94: 253–263. [24]

Terhaar, J., R. Lauerwald, P. Regnier, N. Gruber, and L. Bopp. 2021. Around one third of current Arctic Ocean primary production sustained by rivers and coastal erosion. *Nature Communications* 12: Article 169.

Tewksbury, J. J., and 9 others. 2002. Corridors affect plants, animals, and their interactions in fragmented landscapes. *Proceedings of the National Academy of Sciences USA* 99: 12923–12926. [24]

Thomas, C. D. and T. M. Jones. 1993. Partial recovery of a skipper butterfly (*Hesperia comma*) from population refuges: Lessons for conservation in a fragmented landscape. *Journal of Animal Ecology* 62: 472–481. [9]

Thomas, C. D., and 10 others. 2004. Extinction risk from climate change. *Nature* 427: 145–148. [23]

Thomas, C. D., J. A. Thomas, M. S. Warren. 1992. Distributions of occupied and vacant butterfly habitats in fragmented landscapes. *Oecologia* 92: 563–567. [9]

Thomas, F., A. Schmidt-Rhaesa, G. Martin, C. Manu, P. Durand, and F. Renaud. 2002. Do hairworms (Nematomorpha) manipulate the water seeking behaviour of their terrestrial hosts? *Journal of Evolutionary Biology* 15: 356–361. [13]

Thomas, F., and 7 others. 2003. Biochemical and histological changes in the brain of the cricket *Nemobius sylvestris* infected by the manipulative parasite *Paragordius tricuspidatus* (Nematomorpha). *International Journal for Parasitology* 33: 435–443. [13]

Thomas, R. Q., C. D. Canham, K. C. Weathers, and C. L. Goodale. 2010. Increased tree carbon storage in response to nitrogen deposition in the U.S. *Nature Geoscience* 3: 13–17. [25]

Thompson, J. N. 1998. Rapid evolution as an ecological process. *Trends in Ecology and Evolution* 13: 329–332. [6]

Thorpe, K. E., P. Joski, and K. Johnson. 2018. Antibiotic-resistant infection treatment costs have doubled since 2002, now exceeding $2 billion annually. *Health Affairs* 37(4): 662–669. [6]

Tilman, D. 1977. Resource competition between plankton algae: An experimental and theoretical approach. *Ecology* 58: 338–348. [19]

Tilman, D., and J. A. Downing. 1994. Biodiversity and stability in grasslands. *Nature* 367: 363–365. [19]

Tilman, D., D. Wedin, and J. Knops. 1996. Productivity and sustainability influenced by biodiversity in grassland ecosystems. *Nature* 379: 718–720. [19]

Tilman, D., M. Mattson, and S. Langer. 1981. Competition and nutrient kinetics along a temperature gradient: An experimental test of a mechanistic approach to niche theory. *Limnology and Oceanography* 26: 1020–1033. [14, 19]

Tinker, D. B., W. H. Romme, and D. G. Despain. 2003. Historic range of variability in landscape structure in subalpine forests of the Greater Yellowstone Area, USA. *Landscape Ecology* 18: 427–439. [24]

Tomback, D. F. 1982. Dispersal of whitebark pine seeds by Clark's nutcracker: A mutualism hypothesis. *Journal of Animal Ecology* 51: 451–467. [24]

Tomback, D. F., and P. Achuff. 2010. Blister rust and western forest biodiversity: Ecology, values and outlook for white pines. *Forest Pathology* 40: 186–225. [24]

Tracy, R. L., and G. E. Walsberg. 2002. Kangaroo rats revisited: Re-evaluating a classic case of desert survival. *Oecologia* 133: 449–457. [4]

Trites, A. W., and C. P. Donnelly. 2003. The decline of Steller sea lions *Eumetopias jubatus* in Alaska: A review of the nutritional stress hypothesis. *Mammal Review* 33: 3–28. [21]

Tsukaya, H., K. Fujikawa, and S. Wu. 2002. Thermal insulation and accumulation of heat in the downy inflorescences of *Saussurea medusa* (Asteraceae) at high elevation in Yunnan, China. *Journal of Plant Research* 115: 263–268. [4]

Turchin, P. 2003. *Complex Population Dynamics: A Theoretical/Empirical Synthesis*. Princeton University Press: Princeton, NJ. [12]

Turner, M. G., R. H. Gardner, and R. V. O'Neill. 2001. *Landscape Ecology in Theory and Practice: Pattern and Process*. Springer: New York. [24]

Turner, M. G., W. H. Romme, and D. B. Tinker. 2003. Surprises and lessons from the 1988 Yellowstone fires. *Frontiers in Ecology and the Environment* 1: 351–358. [24]

Tyerman, J. G., M. Bertrand, C. C. Spencer, and M. Doebeli. 2008. Experimental demonstration of ecological character displacement. *BMC Evolutionary Biology* 8: 34. [14]

Tylianakis, J. M., R. K. Didham, J. Bascompte, and D. A. Wardle. 2008. Global change and species interactions in terrestrial ecosystems. *Ecology Letters* 11: 1351–1363. [13]

U

U.S. Congress, Office of Technology Assessment. 1993. *Harmful Non-Indigenous Species in the United States*. OTA-F-565. U.S. Government Printing Office: Washington, DC. [23]

U.S. Environmental Protection Agency. 2005. *Evaluating Ozone Control Programs in the Eastern United States: Focus on the NO_x Budget Trading Program, 2004*. EPA454-K-05–001. www.epa.gov/airtrends/2005/ozonenbp.pdf. [25]

U.S. Fish and Wildlife Service. 2003. *Recovery Plan for the Red-Cockaded Woodpecker* (*Picoides borealis*): Second revision. U.S. Fish and Wildlife Service: Atlanta, GA. [23]

U.S. Fish and Wildlife Service. 2008. *Birds of Conservation Concern 2008*. U.S. Fish and Wildlife Service, Division of Migratory Bird Management: Arlington, VA. [24]

Uddenberg, N., T. Fagerström and A. Jeffner. 1995. Kontakten med urkällan: Naturen, människan och religionen. In *Det Stora Sammanhanget*, N. Uddenberg (Ed.), pp. 141–186. Doxa: Nora, Sweden. [1]

Umina, P. A., A. R. Weeks, M. R. Kearney, S. W. McKechnie, and A. A. Hoffmann. 2005. A rapid shift in a classic clinal pattern in *Drosophila* reflecting climate change. *Science* 308: 691–693. [6]

United Nations, Department of Economic and Social Affairs, Population Division. 2019. *World Population Prospects 2019: Data Booklet*. [11]

Urban, M. C. 2015. Accelerating extinction risk from climate change. *Science* 348: 571–573. [25]

Uthicke, S., P. Momigliano, and K. E. Fabricius. 2013. High risk of extinction of benthic foraminifera in this century due to ocean acidification. *Scientific Reports* 3: 1–5. [25]

V

Valentine, D. L., and 12 others. 2010. Propane respiration jump-starts microbial response to a deep oil spill. *Science* 330: 208–211. [5]

Van Breemen, N., and A. C. Finzi. 1998. Plant-soil interactions: Ecological aspects and evolutionary implications. *Biogeochemistry* 42: 1–19. [22]

Van der Heijden, M. G. A., and 7 others. 1998. Mycorrhizal fungal diversity determines plant biodiversity, ecosystem variability and productivity. *Nature* 396: 69–72. [15]

Van der Veken, S., J. Rogister, K. Verheyen, M. Hermy, and R. Nathan. 2007. Over the (range) edge: A 45-year transplant experiment with the perennial forest herb *Hyacinthoides non-scripta*. *Journal of Ecology* 95: 343–351. [9]

Van Dover, C. L. 2000. *The Ecology of Deep-Sea Hydrothermal Vents*. Princeton University Press: Princeton, NJ. [20]

Van Lear, D. H., W. D. Carroll, P. R. Kapeluck, and R. Johnson. 2005. History and restoration of the longleaf pine–grassland ecosystem: Implications for species at risk. *Forest Ecology and Management* 211: 150–165. [23]

Van Mantgem, P. J., and 10 others. 2009. Widespread increase of tree mortality rates in the western United States. *Science* 323: 521–524. [11]

Van Oers, K., and D. L. Sinn. 2013. Quantitative and molecular genetics of animal personality. In *Animal Personalities: Behavior, Physiology, and Evolution*, C. Carere and D. Maestripieri (Eds.), pp. 149–200. University of Chicago Press: Chicago. [8]

van Wijk, S. J., and 7 others. 2013. Experimental harvesting of fish populations drives genetically based shifts in body size and maturation. *Frontiers in Ecology and the Environment* 11: 181–187. [6]

Vannote, R. L., G. W. Minshall, K. W. Cummins, K. R. Sedell and C. E. Cushing. 1980. The River Continuum Concept. *Canadian Journal of Fisheries and Aquatic Sciences* 37: 130–137. [3]

Varley, N., and M. S. Boyce. 2006. Adaptive management for reintroductions: Updating a wolf recovery model for Yellowstone National Park. *Ecological Modelling* 193: 315–339. [24]

Veblen, T. T., T. Kitzberger and J. Donnegan. 2000. Climatic and human influences on fire regimes in ponderosa pine forests in the Colorado Front Range. *Ecological Applications* 10: 1178–1195. [2]

Vellend, M., J. A. Myers, S. Gardescu, and P. L. Marks. 2003. Dispersal of *Trillium* seeds by deer: Implications for long-distance migration of forest herbs. *Ecology* 84: 1067–1072. [15]

Venter, O., Sanderson, E.W., Magrach, A., Allan, J. R., Beher, J., Jones, K. R., Possingham, H. P., Laurance, W. F., Wood, P., Fekete, B. M., Levy, M. A., and Watson, J. E. M. 2016. Sixteen years of change in the global terrestrial human footprint and implications for biodiversity conservation. *Nature Communications*. 7:12558. https://doi.org/10.1038/ncomms1255

Verdolin, J. L. 2006. Meta-analysis of foraging and predation risk trade-offs in terrestrial systems. *Behavioral Ecology and Sociobiology* 60: 457–464. [8]

Vitousek, P. M. 1977. The regulation of element concentrations in mountain streams in the northeastern United States. *Ecological Monographs* 47: 65–87. [22]

Vitousek, P. M., and H. Farrington. 1997. Nutrient limitation and soil development: Experimental test of a biogeochemical theory. *Biogeochemistry* 37: 63–75. [22]

Vitousek, P.M., L.R. Walker, L.D. Whiteaker, and P.A. Matson. 1993. Nutrient limitations to plant growth during primary succession in Hawaii Volcanoes National Park. *Biogeochemistry* 23: 197–215. [22]

Volterra, V. 1926. Variazioni e fluttuazioni del numero d'individui in specie animali conviventi. *Memorie della R. Accademia dei Lincei* 2: 31–113. Reprinted 1931 as "Variations and fluctuations of the number of individuals in animal species living together," in R. N. Chapman, *Animal Ecology*. McGraw-Hill: New York. [14]

Vonlanthen, P., and 9 others. 2012. Eutrophication causes speciation reversal in whitefish adaptive radiations. *Nature* 482: 357–362. [6]

Voyles, J. and 10 others. 2009. Pathogenesis of chytridiomycosis, a cause of catastrophic amphibian declines. *Science* 326: 582–585. [1]

Voyles, J., D.C. Woodhams and 16 others. 2018. Shifts in disease dynamics in a tropical amphibian assemblage are not due to pathogen attenuation. *Science* 359: 1517–1519. [1]

Vredenburg, V. T. 2004. Reversing introduced species effects: Experimental removal of introduced fish leads to rapid recovery of a declining frog. *Proceedings of the National Academy of Sciences USA* 101: 7646–7650. [1]

W

Wackernagel, M., and 7 others. 1999. National natural capital accounting with the ecological footprint concept. *Ecological Economics* 29: 375–390. [11]

Wada-Katsumata, A., J. Silverman, and C. Schal. 2013. Changes in taste neurons support the emergence of an adaptive behavior in cockroaches. *Science* 340: 972–975. [8]

Wagner, W. L., M. M. Bruegmann, D. R. Herbst, and J. Q. C. Lau. 1999. Hawaiian vascular plants at risk: 1999. *Bishop Museum Occasional Papers* 60: 1–58. [23]

Wahlenberg, W. G. 1946. Longleaf pine: Its use, ecology, regeneration, protection, growth, and management. C. L. Pack Forestry Foundation & USDA Forest Service; USDA Natural Resource Conservation Service, Longleaf Pine Initiative: Washington, DC. [23]

Walker, T. W., and J. K. Syers. 1976. The fate of phosphorus during pedogenesis. *Geoderma* 15: 1–19. [22]

Wallace, A. R. 1855. On the law which has regulated the introduction of new species. *Annals and Magazine of Natural History* 16. [18]

Wallace, A. R. 1858. On the tendency of varieties to depart indefinitely from the original type. *Journal of the Proceedings of the Linnean Society, Zoology* 3: 53–62. [18]

Wallace, A. R. 1860. On the zoological geography of the Malay Archipelago. *Journal of the Linnean Society of London* 4: 172–184. [9]

Wallace, A. R. 1869. *The Malay Archipelago, the Land of the Orangutan and the Bird of Paradise: A Narrative of Travel with Studies of Man and Nature.* Macmillan: London. [18, 23]

Wallace, A. R. 1876. *The Geographic Distribution of Animals.* Harper and Brothers: New York. [18]

Wallace, A. R. 1878. *Tropical Nature and Other Essays.* Macmillan: London. [18]

Waller, N. L., I. C. Gynther, A. B. Freeman, T. H. Lavery, and L. K.-P. Leung. 2017. The Bramble Cay melomys *Melomys rubicola* (Rodentia: Muridae): A first mammalian extinction caused by human-induced climate change? *Wildlife Research* 44, 9–21. [23]

Walter, H. and H. Lieth. 1967. *Klimadiagram-Weltatlas.* VEB Gustav Fischer Verlag: Jena. [3]

Walters, J. R., C. K. Copeyon, and J. H. Carter. 1992. Test of the ecological basis of cooperative breeding in red-cockaded woodpeckers. *Auk* 109: 90–97. [23]

Walther, G. 2010. Community and ecosystem responses to recent climate change. *Philosophical Transactions of the Royal Society of London* 365B: 2019–2025. [25]

Wand, S. J. E., G. F. Midgley, M. H. Jones, and P. S. Curtis. 1999. Responses of wild C_4 and C_3 grass (Poaceae) species to elevated atmospheric CO_2 concentration: A meta-analytic test of current theories and perceptions. *Global Change Biology* 5: 723–741. [5]

Wang, X., Z. Dong, J. Zhang, and L. Liu. 2004. Modern dust storms in China: An overview. *Journal of Arid Environments* 58: 559–574. [25]

Ward, M. P., J. S. Milledge, and J. B. West. 1995. *High Altitude Medicine and Physiology.* Chapman & Hall: London. [4]

Wasser, S. K., C. Mailand, R. Booth, B. Mutayoba, E. Kisamo, B. Clark, and M. Stephens. 2007. Using DNA to track the origin of the largest ivory seizure since the 1989 trade ban. *Proceedings of the National Academy of Sciences USA* 104: 4228–4233. [23]

Watanabe, Y. Y., M. Ito, and A. Takahashi. 2014. Testing optimal foraging theory in a penguin–krill system. *Proceedings of the Royal Society of London* 281B: 20132376. [8]

Watson, H. C. 1859. *Cybele Brittanica: or British Plants and Their Geographical Relations* 4: 379. Longman and Company: United Kingdom. [18]

Weathers, K. C., M. L. Cadenasso, and S. T. A. Pickett. 2001. Forest edges as nutrient and pollutant concentrators. *Conservation Biology* 15: 1506–1514. [24]

Webb, E. E., and 7 others. 2016. Increased wintertime CO_2 loss as a result of sustained tundra warming. *Journal of Geophysical Research: Biogeosciences* 121: 249–265. [25]

Webb, N. R., and L. E. Haskins. 1980. An ecological survey of heathlands in the Poole Basin, Dorset, England, in 1978. *Biological Conservation* 17: 281–296. [9]

Webberley, K. M., and 8 others. 2004. Host reproduction and a sexually transmitted disease: Causes and consequences of *Coccipolipus hippodamiae* distribution on coccinellid beetles. *Journal of Animal Ecology* 73: 1–10. [13]

Weber, J. N., B. K. Peterson, and H. E. Hoekstra. 2013. Discrete genetic modules are responsible for complex burrow evolution in *Peromyscus* mice. *Nature* 493: 402–405. [8]

Weber, N. A. 1966. Fungus-growing ants. *Science* 153: 587–604. [15]

Weber, N. A. 1979. Fungus-culturing by ants. In *Insect-fungus symbiosis: mutualism and commensalism,* L. R. Batra (Ed.), pp. 77–116. Allanheld and Osmum: Montclair, N. J. [15]

Weber, R. E. 2007. High-altitude adaptations in vertebrate hemoglobins. *Respiratory Physiology and Neurobiology* 158: 132–142. [6]

Wedell, N., M. J. G. Gage, and G. A. Parker. 2002. Sperm competition, male prudence and sperm-limited females. *Trends in Ecology and Evolution* 17: 313–320. [8]

Weir, A. A. S., J. Chappell, and A. Kacelnik. 2002. Shaping of hooks in New Caledonian crows. *Science* 297: 981. [5]

Weis, A. E., and W. G. Abrahamson. 1986. Evolution of host-plant manipulation by gall makers: Ecology and genetic factors in the *Solidago–Eurosta* system. *American Naturalist* 127: 681–695. [6]

Weiss, R. A. and A. J. McMichael. 2004. Social and environmental risk factors in the emergence of infectious disease. *Nature Medicine* 10: S70–S76. [1]

Weldon, A. J. 2006. How corridors reduce Indigo Bunting nest success. *Conservation Biology* 20: 1300–1305. [24]

Wernberg, T., and 21 others. 2016. Climate-driven regime shift of a temperate marine ecosystem. *Science* 353: 169–172. [9]

Werner, E. E. 1988. Size scaling and the evolution of complex life cycles. In *Size-Structured Populations*, B. Ebenman and L. Persson (Eds.), pp. 61–81. Springer: Berlin. [7]

Werner, E. E., J. F. Gilliam, D. J. Hall, and G. G. Mittelbach. 1983. An experimental test of the effects of predation risk on habitat use in fish. *Ecology* 64: 1540–1548. [8]

Westemeier, R. L., and 8 others. 1998. Tracking the long-term decline and recovery of an isolated population. *Science* 282: 1695–1698. [6]

Westoby, M. 1997. What does "ecology" mean? *Trends in Ecology and Evolution* 12: 166. [1]

White, J., J. Wilson, and K. Horskins. 1997. The role of adjacent habitats in rodent damage levels in Australian macadamia orchard systems. *Crop Protection* 16: 727–732. [24]

Whittaker, R. H., K. J. Willis, and R. Field. 2001. Towards a general, hierarchical theory of species diversity. *Journal of Biogeography* 28: 453–470. [18]

Wiens, J. J. 2016. Climate-related extinctions are already widespread among plant and animal species. *PLOS Biology* 14: e2001104. [23]

Wilbur, H. M. 1997. Experimental ecology of food webs: Complex systems in temporary ponds. The Robert H. MacArthur Award Lecture. *Ecology* 78: 2279–2302. [21]

Wilkie, D. S., and J. F. Carpenter. 1999. Bushmeat hunting in the Congo Basin: An assessment of impacts and options for mitigation. *Biodiversity and Conservation* 8: 927–955. [23]

Williams, A. P., and 14 others. 2013. Temperature as a potent driver of regional forest drought stress and tree mortality. *Nature Climate Change* 3: 292–297. [11]

Williams, B. K., J. D. Nichols, and M. J. Conroy. 2002. *Analysis and Management of Animal Populations.* Academic Press: San Diego, CA. [9]

Williams, C. B. 1943. Area and number of species. *Nature* 152: 264–267. [18]

Williams, C. B. 1964. *Patterns in the Balance of Nature.* Academic Press: London. [18]

Williams, J. C., C. S. ReVelle, and S. A. Levin. 2005. Spatial attributes and reserve design models: A review. *Environmental Modeling and Assessment* 10: 163–181. [24]

Willig, M. R., D. M. Kaufman, and R. D. Stevens. 2003. Latitudinal gradients of biodiversity: Pattern, process, scale, and synthesis. *Annual Review of Ecology and Systematics* 34: 273–309. [18]

Willmer, P., G. Stone, and I. Johnston. 2005. *Environmental Physiology of Animals.* 2nd ed. Blackwell: Malden, MA. [4]

Wilson, E. O. 1994. *Naturalist.* Island Press: Washington, DC. [18]

Wilson, E. O., and D. S. Simberloff. 1969. Experimental zoogeography of islands: Defaunation and monitoring techniques. *Ecology* 50: 267–278. [18]

Wilson, S. D., and D. Tilman. 1993. Plant competition and resource availability in response to disturbance and fertilization. *Ecology* 74: 599–611. [14]

Windsor, D. A. 1998. Most of the species on Earth are parasites. *International Journal for Parasitology* 28: 1939–1941. [13]

Winkler, K., R. Fuchs, M. Rounsevell, and M. Herold. 2021. Global land use changes are four times greater than previously estimated. *Nature Communications* 12: 2501. doi.org/10.1038/s41467-021-22702-2 [23]

Winterhalder, B. P. 1980. Canadian fur bearer cycles and Cree-Ojibwa hunting and trapping practices. *American Naturalist* 115: 870–879. [12]

With, K. A. 2019. *Essentials of Landscape Ecology.* Oxford University Press: New York.

Witman, J. D., R. J. Etter, and F. Smith. 2004. The relationship between regional and local species diversity in marine benthic communities: A global perspective. *Proceedings of the National Academy of Sciences USA* 101: 15664–15669. [18]

Wittemyer, G., J. M. Northrup, J. Blanc, I. Douglas-Hamilton, P. Omondi, and K. P. Burnham. 2014. Illegal killing for ivory drives global decline in African elephants. *Proceedings of the National Academy of Sciences USA* 111: 13117–13121. [11]

Wolf, S., W. Eugster, C. Potvin, B. L. Turner, and N. Buchmann. 2011. Carbon sequestration potential of tropical pasture compared with afforestation in Panama. *Global Change Biology* 17: 2763–2780. [20]

Woodroffe, R. 1999. Managing disease threats to wild mammals. *Animal Conservation* 2: 185–193. [23]

Woodward, G., D. M. Perkins, and L. E. Brown. 2010. Climate change and freshwater ecosystems: Impacts across multiple levels of organization. *Philosophical Transactions of the Royal Society* 365B: 2093–2106. [23]

Wright, S. J. 1981. Intra-archipelago vertebrate distributions: The slope of the species–area relation. *American Naturalist* 118: 726–748. [18]

Wright, S. J. 2005. Tropical forests in a changing environment. *Trends in Ecology and Evolution* 10: 553–560. [3]

Wrong, O. M., C. J. Edmonds, and V. S. Chadwick. 1981. *The Large Intestine: Its Role in Mammalian Nutrition and Homeostasis.* Halsted: New York. [5]

Y

Yoda, J. A., T. Kira, H. Ogawa, and K. Hozumi. 1963. Self-thinning in overcrowded pure stands under cultivated and natural conditions. *Journal of Biology, Osaka City University* 14: 107–120. [11]

Z

Zanette, L. Y., A. F. White, M. C. Allen, and M. Clinchy. 2011. Perceived predation risk reduces the number of offspring songbirds produce per year. *Science* 334: 1398–1401. [8]

Zar, J. H. 2009. *Biostatistical Analysis.* 5th ed. Pearson: New York. [1]

Zhang, Z., and 11 others. 2017. Effect of climate warming on the annual terrestrial net ecosystem CO_2 exchange globally in the boreal and temperate regions. *Scientific Reports* 7: Article 3108. [25]

Zhao, M., and S. W. Running. 2010. Drought-induced reduction in global terrestrial net primary production from 2000 through 2009. *Science* 329: 940–943. [25]

Zimova, M., L. S. Mills, and J. J. Nowak. 2016. High fitness costs of climate-change induced camouflage mismatch. *Ecology Letters* 19: 299–307. [7]

Züst, T., C. Heichinger, U. Grossniklaus, R. Harrington, D. J. Kliebenstein, and L. A. Turnbull. 2012. Natural enemies drive geographical variation in plant defenses. *Science* 338: 116–119. [12]

Index

Page numbers in *italic* type indicate the information will be found in an illustration.